**Student's Solutions Manual
to accompany Jon Rogawski's**

Single Variable
CALCULUS

BRIAN BRADIE

Christopher Newport University

With Chapter 12 contributed by
GREGORY P. DRESDEN

Washington and Lee University

and additional contributions by
Art Belmonte
Cindy Chang-Fricke
Benjamin G. Jones
Kerry Marsack
Katherine Socha
Jill Zarestky
Kenneth Zimmerman

W. H. Freeman and Company
New York

© 2008 by W. H. Freeman and Company

ISBN-13: 978-0-7167-9866-8
ISBN-10: 0-7167-9866-2

Printed in the United States of America

Second printing

W. H. Freeman and Company, 41 Madison Avenue, New York, NY 10010
Houndmills, Basingstoke RG21 6XS, England
www.whfreeman.com

CONTENTS

1 | PRECALCULUS REVIEW

1.1 Real Numbers, Functions, and Graphs

Preliminary Questions

1. Give an example of numbers a and b such that $a < b$ and $|a| > |b|$.

SOLUTION Take $a = -3$ and $b = 1$. Then $a < b$ but $|a| = 3 > 1 = |b|$.

2. Which numbers satisfy $|a| = a$? Which satisfy $|a| = -a$? What about $|-a| = a$?

SOLUTION The numbers $a \geq 0$ satisfy $|a| = a$ and $|-a| = a$. The numbers $a \leq 0$ satisfy $|a| = -a$.

3. Give an example of numbers a and b such that $|a + b| < |a| + |b|$.

SOLUTION Take $a = -3$ and $b = 1$. Then

$$|a + b| = |-3 + 1| = |-2| = 2, \qquad \text{but} \qquad |a| + |b| = |-3| + |1| = 3 + 1 = 4.$$

Thus, $|a + b| < |a| + |b|$.

4. What are the coordinates of the point lying at the intersection of the lines $x = 9$ and $y = -4$?

SOLUTION The point $(9, -4)$ lies at the intersection of the lines $x = 9$ and $y = -4$.

5. In which quadrant do the following points lie?

(a) $(1, 4)$

(b) $(-3, 2)$

(c) $(4, -3)$

(d) $(-4, -1)$

SOLUTION

(a) Because both the x- and y-coordinates of the point $(1, 4)$ are positive, the point $(1, 4)$ lies in the first quadrant.

(b) Because the x-coordinate of the point $(-3, 2)$ is negative but the y-coordinate is positive, the point $(-3, 2)$ lies in the second quadrant.

(c) Because the x-coordinate of the point $(4, -3)$ is positive but the y-coordinate is negative, the point $(4, -3)$ lies in the fourth quadrant.

(d) Because both the x- and y-coordinates of the point $(-4, -1)$ are negative, the point $(-4, -1)$ lies in the third quadrant.

6. What is the radius of the circle with equation $(x - 9)^2 + (y - 9)^2 = 9$?

SOLUTION The circle with equation $(x - 9)^2 + (y - 9)^2 = 9$ has radius 3.

7. The equation $f(x) = 5$ has a solution if (choose one):

(a) 5 belongs to the domain of f.

(b) 5 belongs to the range of f.

SOLUTION The correct response is **(b)**: the equation $f(x) = 5$ has a solution if 5 belongs to the range of f.

8. What kind of symmetry does the graph have if $f(-x) = -f(x)$?

SOLUTION If $f(-x) = -f(x)$, then the graph of f is symmetric with respect to the origin.

Exercises

1. Use a calculator to find a rational number r such that $|r - \pi^2| < 10^{-4}$.

SOLUTION r must satisfy $\pi^2 - 10^{-4} < r < \pi^2 + 10^{-4}$, or $9.869504 < r < 9.869705$. $r = 9.8696 = \frac{12337}{1250}$ would be one such number.

In Exercises 3–8, express the interval in terms of an inequality involving absolute value.

3. $[-2, 2]$

SOLUTION $|x| \leq 2$

5. $(0, 4)$

SOLUTION The midpoint of the interval is $c = (0+4)/2 = 2$, and the radius is $r = (4-0)/2 = 2$; therefore, $(0, 4)$ can be expressed as $|x - 2| < 2$.

7. $[1, 5]$

SOLUTION The midpoint of the interval is $c = (1+5)/2 = 3$, and the radius is $r = (5-1)/2 = 2$; therefore, the interval $[1, 5]$ can be expressed as $|x - 3| \le 2$.

In Exercises 9–12, write the inequality in the form $a < x < b$ for some numbers a, b.

9. $|x| < 8$

SOLUTION $-8 < x < 8$

11. $|2x + 1| < 5$

SOLUTION $-5 < 2x + 1 < 5$ so $-6 < 2x < 4$ and $-3 < x < 2$

In Exercises 13–18, express the set of numbers x satisfying the given condition as an interval.

13. $|x| < 4$

SOLUTION $(-4, 4)$

15. $|x - 4| < 2$

SOLUTION The expression $|x - 4| < 2$ is equivalent to $-2 < x - 4 < 2$. Therefore, $2 < x < 6$, which represents the interval $(2, 6)$.

17. $|4x - 1| \le 8$

SOLUTION The expression $|4x - 1| \le 8$ is equivalent to $-8 \le 4x - 1 \le 8$ or $-7 \le 4x \le 9$. Therefore, $-\frac{7}{4} \le x \le \frac{9}{4}$, which represents the interval $[-\frac{7}{4}, \frac{9}{4}]$.

In Exercises 19–22, describe the set as a union of finite or infinite intervals.

19. $\{x : |x - 4| > 2\}$

SOLUTION $x - 4 > 2$ or $x - 4 < -2 \Rightarrow x > 6$ or $x < 2 \Rightarrow (-\infty, 2) \cup (6, \infty)$

21. $\{x : |x^2 - 1| > 2\}$

SOLUTION $x^2 - 1 > 2$ or $x^2 - 1 < -2 \Rightarrow x^2 > 3$ or $x^2 < -1$ (this will never happen) $\Rightarrow x > \sqrt{3}$ or $x < -\sqrt{3} \Rightarrow$
$(-\infty, -\sqrt{3}) \cup (\sqrt{3}, \infty)$.

23. Match the inequalities (a)–(f) with the corresponding statements (i)–(vi).

(a) $a > 3$

(b) $|a - 5| < \dfrac{1}{3}$

(c) $\left|a - \dfrac{1}{3}\right| < 5$

(d) $|a| > 5$

(e) $|a - 4| < 3$

(f) $1 < a < 5$

  (i) a lies to the right of 3.

 (ii) a lies between 1 and 7.

(iii) The distance from a to 5 is less than $\frac{1}{3}$.

(iv) The distance from a to 3 is at most 2.

 (v) a is less than 5 units from $\frac{1}{3}$.

(vi) a lies either to the left of -5 or to the right of 5.

SOLUTION
(a) On the number line, numbers greater than 3 appear to the right; hence, $a > 3$ is equivalent to the numbers to the right of 3: **(i)**.

(b) $|a - 5|$ measures the distance from a to 5; hence, $|a - 5| < \frac{1}{3}$ is satisfied by those numbers less than $\frac{1}{3}$ of a unit from 5: **(iii)**.

(c) $|a - \frac{1}{3}|$ measures the distance from a to $\frac{1}{3}$; hence, $|a - \frac{1}{3}| < 5$ is satisfied by those numbers less than 5 units from $\frac{1}{3}$: **(v)**.

(d) The inequality $|a| > 5$ is equivalent to $a > 5$ or $a < -5$; that is, either a lies to the right of 5 or to the left of -5: **(vi)**.

(e) The interval described by the inequality $|a - 4| < 3$ has a center at 4 and a radius of 3; that is, the interval consists of those numbers between 1 and 7: **(ii)**.

(f) The interval described by the inequality $1 < x < 5$ has a center at 3 and a radius of 2; that is, the interval consists of those numbers less than 2 units from 3: **(iv)**.

25. Show that if $a > b$, then $b^{-1} > a^{-1}$, provided that a and b have the same sign. What happens if $a > 0$ and $b < 0$?

SOLUTION Case 1a: If a and b are both positive, then $a > b \Rightarrow 1 > \frac{b}{a} \Rightarrow \frac{1}{b} > \frac{1}{a}$.

Case 1b: If a and b are both negative, then $a > b \Rightarrow 1 < \frac{b}{a}$ (since a is negative) $\Rightarrow \frac{1}{b} > \frac{1}{a}$ (again, since b is negative).

Case 2: If $a > 0$ and $b < 0$, then $\frac{1}{a} > 0$ and $\frac{1}{b} < 0$ so $\frac{1}{b} < \frac{1}{a}$. (See Exercise 2f for an example of this).

27. Show that if $|a - 5| < \frac{1}{2}$ and $|b - 8| < \frac{1}{2}$, then $|(a + b) - 13| < 1$. *Hint:* Use the triangle inequality.

SOLUTION

$$|a + b - 13| = |(a - 5) + (b - 8)|$$
$$\leq |a - 5| + |b - 8| \quad \text{(by the triangle inequality)}$$
$$< \frac{1}{2} + \frac{1}{2} = 1.$$

29. Suppose that $|x - 4| \leq 1$.
(a) What is the maximum possible value of $|x + 4|$?
(b) Show that $|x^2 - 16| \leq 9$.

SOLUTION

(a) $|x - 4| \leq 1$ guarantees $3 \leq x \leq 5$. Thus, $7 \leq x + 4 \leq 9$, so $|x + 4| \leq 9$.
(b) $|x^2 - 16| = |x - 4| \cdot |x + 4| \leq 1 \cdot 9 = 9$.

31. Express $r_1 = 0.\overline{27}$ as a fraction. *Hint:* $100r_1 - r_1$ is an integer. Then express $r_2 = 0.2666\ldots$ as a fraction.

SOLUTION Let $r_1 = .\overline{27}$. We observe that $100r_1 = 27.\overline{27}$. Therefore, $100r_1 - r_1 = 27.\overline{27} - .\overline{27} = 27$ and

$$r_1 = \frac{27}{99} = \frac{3}{11}.$$

Now, let $r_2 = .2\overline{666}$. Then $10r_2 = 2.\overline{666}$ and $100r_2 = 26.\overline{666}$. Therefore, $100r_2 - 10r_2 = 26.\overline{666} - 2.\overline{666} = 24$ and

$$r_2 = \frac{24}{90} = \frac{4}{15}.$$

33. The text states the following: *If the decimal expansions of two real numbers a and b agree to k places, then the distance $|a - b| \leq 10^{-k}$.* Show that the converse is not true, that is, for any k we can find real numbers a and b whose decimal expansions *do not agree at all* but $|a - b| \leq 10^{-k}$.

SOLUTION Let $a = 1$ and $b = .\overline{9}$ (see the discussion before Example 1). The decimal expansions of a and b do not agree, but $|1 - .\overline{9}| < 10^{-k}$ for all k.

35. Determine the equation of the circle with center $(2, 4)$ and radius 3.

SOLUTION The equation of the indicated circle is $(x - 2)^2 + (y - 4)^2 = 3^2 = 9$.

37. Find all points with integer coordinates located at a distance 5 from the origin. Then find all points with integer coordinates located at a distance 5 from $(2, 3)$.

SOLUTION

- To be located a distance 5 from the origin, the points must lie on the circle $x^2 + y^2 = 25$. This leads to 12 points with integer coordinates:

$$\begin{array}{cccc}
(5, 0) & (-5, 0) & (0, 5) & (0, -5) \\
(3, 4) & (-3, 4) & (3, -4) & (-3, -4) \\
(4, 3) & (-4, 3) & (4, -3) & (-4, -3)
\end{array}$$

- To be located a distance 5 from the point $(2, 3)$, the points must lie on the circle $(x - 2)^2 + (y - 3)^2 = 25$, which implies that we must shift the points listed above two units to the right and three units up. This gives the 12 points:

$$\begin{array}{cccc}
(7, 3) & (-3, 3) & (2, 8) & (2, -2) \\
(5, 7) & (-1, 7) & (5, -1) & (-1, -1) \\
(6, 6) & (-2, 6) & (6, 0) & (-2, 0)
\end{array}$$

39. Give an example of a function whose domain D has three elements and range R has two elements. Does a function exist whose domain D has two elements and range has three elements?

SOLUTION Define f by $f : \{a, b, c\} \rightarrow \{1, 2\}$ where $f(a) = 1$, $f(b) = 1$, $f(c) = 2$.

There is no function whose domain has two elements and range has three elements. If that happened, one of the domain elements would get assigned to more than one element of the range, which would contradict the definition of a function.

In Exercises 40–48, find the domain and range of the function.

41. $g(t) = t^4$

SOLUTION D : all reals; R : $\{y : y \geq 0\}$

43. $g(t) = \sqrt{2 - t}$

SOLUTION D : $\{t : t \leq 2\}$; R : $\{y : y \geq 0\}$

45. $h(s) = \dfrac{1}{s}$

SOLUTION D : $\{s : s \neq 0\}$; R : $\{y : y \neq 0\}$

47. $g(t) = \sqrt{2 + t^2}$

SOLUTION D : all reals; R : $\{y : y \geq \sqrt{2}\}$

In Exercises 49–52, find the interval on which the function is increasing.

49. $f(x) = |x + 1|$

SOLUTION A graph of the function $y = |x + 1|$ is shown below. From the graph, we see that the function is increasing on the interval $(-1, \infty)$.

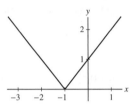

51. $f(x) = x^4$

SOLUTION A graph of the function $y = x^4$ is shown below. From the graph, we see that the function is increasing on the interval $(0, \infty)$.

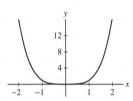

In Exercises 53–58, find the zeros of the function and sketch its graph by plotting points. Use symmetry and increase/decrease information where appropriate.

53. $f(x) = x^2 - 4$

SOLUTION Zeros: ± 2
Increasing: $x > 0$
Decreasing: $x < 0$
Symmetry: $f(-x) = f(x)$ (even function). So, y-axis symmetry.

55. $f(x) = x^3 - 4x$

SOLUTION Zeros: $0, \pm 2$; Symmetry: $f(-x) = -f(x)$ (odd function). So origin symmetry.

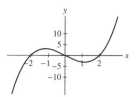

57. $f(x) = 2 - x^3$

SOLUTION This is an x-axis reflection of x^3 translated up 2 units. There is one zero at $x = \sqrt[3]{2}$.

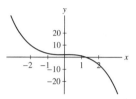

59. Which of the curves in Figure 26 is the graph of a function?

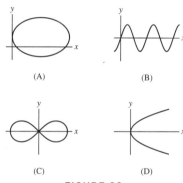

(A)　　　　　　　　(B)

(C)　　　　　　　　(D)

FIGURE 26

SOLUTION (B) is the graph of a function. (A), (C), and (D) all fail the vertical line test.

In Exercises 61–66, let $f(x)$ be the function whose graph is shown in Figure 27.

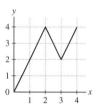

FIGURE 27

61. What are the domain and range of $f(x)$?

SOLUTION $D : [0, 4]; R : [0, 4]$

63. Sketch the graphs of $f(2x)$, $f\left(\frac{1}{2}x\right)$, and $2f(x)$.

SOLUTION The graph of $y = f(2x)$ is obtained by compressing the graph of $y = f(x)$ horizontally by a factor of 2 (see the graph below on the left). The graph of $y = f\left(\frac{1}{2}x\right)$ is obtained by stretching the graph of $y = f(x)$ horizontally by a factor of 2 (see the graph below in the middle). The graph of $y = 2f(x)$ is obtained by stretching the graph of $y = f(x)$ vertically by a factor of 2 (see the graph below on the right).

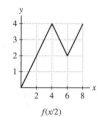

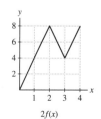

$f(2x)$　　　　　　$f(x/2)$　　　　　　$2f(x)$

65. Extend the graph of $f(x)$ to $[-4, 4]$ so that it is an even function.

SOLUTION   To continue the graph of $f(x)$ to the interval $[-4, 4]$ as an even function, reflect the graph of $f(x)$ across the y-axis (see the graph below).

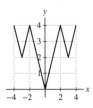

67. Suppose that $f(x)$ has domain $[4, 8]$ and range $[2, 6]$. What are the domain and range of:

(a) $f(x) + 3$ **(b)** $f(x + 3)$

(c) $f(3x)$ **(d)** $3f(x)$

SOLUTION

(a) $f(x) + 3$ is obtained by shifting $f(x)$ upward three units. Therefore, the domain remains $[4, 8]$, while the range becomes $[5, 9]$.

(b) $f(x + 3)$ is obtained by shifting $f(x)$ left three units. Therefore, the domain becomes $[1, 5]$, while the range remains $[2, 6]$.

(c) $f(3x)$ is obtained by compressing $f(x)$ horizontally by a factor of three. Therefore, the domain becomes $[\frac{4}{3}, \frac{8}{3}]$, while the range remains $[2, 6]$.

(d) $3f(x)$ is obtained by stretching $f(x)$ vertically by a factor of three. Therefore, the domain remains $[4, 8]$, while the range becomes $[6, 18]$.

69. Suppose that the graph of $f(x) = \sin x$ is compressed horizontally by a factor of 2 and then shifted 5 units to the right.

(a) What is the equation for the new graph?

(b) What is the equation if you first shift by 5 and then compress by 2?

(c) $\boxed{\text{GU}}$ Verify your answers by plotting your equations.

SOLUTION

(a) Let $f(x) = \sin x$. After compressing the graph of f horizontally by a factor of 2, we obtain the function $g(x) = f(2x) = \sin 2x$. Shifting the graph 5 units to the right then yields

$$h(x) = g(x - 5) = \sin 2(x - 5) = \sin(2x - 10).$$

(b) Let $f(x) = \sin x$. After shifting the graph 5 units to the right, we obtain the function $g(x) = f(x - 5) = \sin(x - 5)$. Compressing the graph horizontally by a factor of 2 then yields

$$h(x) = g(2x) = \sin(2x - 5).$$

(c) The figure below at the top left shows the graphs of $y = \sin x$ (the dashed curve), the sine graph compressed horizontally by a factor of 2 (the dash, double dot curve) and then shifted right 5 units (the solid curve). Compare this last graph with the graph of $y = \sin(2x - 10)$ shown at the bottom left.

    The figure below at the top right shows the graphs of $y = \sin x$ (the dashed curve), the sine graph shifted to the right 5 units (the dash, double dot curve) and then compressed horizontally by a factor of 2 (the solid curve). Compare this last graph with the graph of $y = \sin(2x - 5)$ shown at the bottom right.

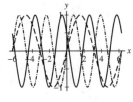
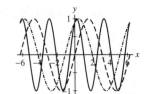
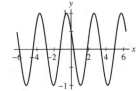
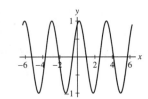

71. Sketch the graph of $f(2x)$ and $f\left(\frac{1}{2}x\right)$, where $f(x) = |x| + 1$ (Figure 28).

SOLUTION The graph of $y = f(2x)$ is obtained by compressing the graph of $y = f(x)$ horizontally by a factor of 2 (see the graph below on the left). The graph of $y = f\left(\frac{1}{2}x\right)$ is obtained by stretching the graph of $y = f(x)$ horizontally by a factor of 2 (see the graph below on the right).

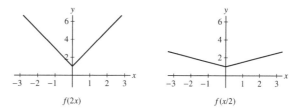

$f(2x)$ $\qquad$ $f(x/2)$

73. Define $f(x)$ to be the larger of x and $2 - x$. Sketch the graph of $f(x)$. What are its domain and range? Express $f(x)$ in terms of the absolute value function.

SOLUTION

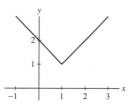

The graph of $y = f(x)$ is shown above. Clearly, the domain of f is the set of all real numbers while the range is $\{y \mid y \geq 1\}$. Notice the graph has the standard V-shape associated with the absolute value function, but the base of the V has been translated to the point $(1, 1)$. Thus, $f(x) = |x - 1| + 1$.

75. Show that the sum of two even functions is even and the sum of two odd functions is odd.

SOLUTION Even: $(f + g)(-x) = f(-x) + g(-x) \overset{\text{even}}{=} f(x) + g(x) = (f + g)(x)$

Odd: $(f + g)(-x) = f(-x) + g(-x) \overset{\text{odd}}{=} -f(x) + -g(x) = -(f + g)(x)$

77. Give an example of a curve that is symmetrical with respect to both the y-axis and the origin. Can the graph of a function have both symmetries? *Hint:* Prove algebraically that $f(x) = 0$ is the only such function.

SOLUTION A circle of radius 1 with its center at the origin is symmetrical both with respect to the y-axis and the origin.

The only function having both symmetries is $f(x) = 0$. For if f is symmetric with respect to the y-axis, then $f(-x) = f(x)$. If f is also symmetric with respect to the origin, then $f(-x) = -f(x)$. Thus $f(x) = -f(x)$ or $2f(x) = 0$. Finally, $f(x) = 0$.

Further Insights and Challenges

79. Show that if $r = a/b$ is a fraction in lowest terms, then r has a *finite* decimal expansion if and only if $b = 2^n 5^m$ for some $n, m \geq 0$. *Hint:* Observe that r has a finite decimal expansion when $10^N r$ is an integer for some $N \geq 0$ (and hence b divides 10^N).

SOLUTION Suppose r has a finite decimal expansion. Then there exists an integer $N \geq 0$ such that $10^N r$ is an integer, call it k. Thus, $r = k/10^N$. Because the only prime factors of 10 are 2 and 5, it follows that when r is written in lowest terms, its denominator must be of the form $2^n 5^m$ for some integers $n, m \geq 0$.

Conversely, suppose $r = \frac{a}{b}$ in lowest with $b = 2^n 5^m$ for some integers $n, m \geq 0$. Then $r = \frac{a}{b} = \frac{a}{2^n 5^m}$ or $2^n 5^m r = a$. If $m \geq n$, then $2^m 5^m r = a2^{m-n}$ or $r = \frac{a2^{m-n}}{10^m}$ and thus r has a finite decimal expansion (less than or equal to m terms, to be precise). On the other hand, if $n > m$, then $2^n 5^n r = a5^{n-m}$ or $r = \frac{a5^{n-m}}{10^n}$ and once again r has a finite decimal expansion.

81. A function $f(x)$ is symmetrical with respect to the vertical line $x = a$ if $f(a - x) = f(a + x)$.

(a) Draw the graph of a function that is symmetrical with respect to $x = 2$.

(b) Show that if $f(x)$ is symmetrical with respect to $x = a$, then $g(x) = f(x + a)$ is even.

SOLUTION

(a) There are many possibilities, two of which may be

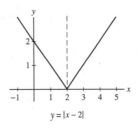

$y = |x - 2|$

(b) Let $g(x) = f(x + a)$. Then

$$g(-x) = f(-x + a) = f(a - x)$$
$$= f(a + x) \qquad \text{symmetry with respect to } x = a$$
$$= g(x)$$

Thus, $g(x)$ is even.

1.2 Linear and Quadratic Functions

Preliminary Questions

1. What is the slope of the line $y = -4x - 9$?

SOLUTION   The slope of the line $y = -4x - 9$ is -4, given by the coefficient of x.

2. Are the lines $y = 2x + 1$ and $y = -2x - 4$ perpendicular?

SOLUTION   The slopes of perpendicular lines are negative reciprocals of one another. Because the slope of $y = 2x + 1$ is 2 and the slope of $y = -2x - 4$ is -2, these two lines are *not* perpendicular.

3. When is the line $ax + by = c$ parallel to the y-axis? To the x-axis?

SOLUTION   The line $ax + by = c$ will be parallel to the y-axis when $b = 0$ and parallel to the x-axis when $a = 0$.

4. Suppose $y = 3x + 2$. What is Δy if x increases by 3?

SOLUTION   Because $y = 3x + 2$ is a linear function with slope 3, increasing x by 3 will lead to $\Delta y = 3(3) = 9$.

5. What is the minimum of $f(x) = (x + 3)^2 - 4$?

SOLUTION   Because $(x + 3)^2 \geq 0$, it follows that $(x + 3)^2 - 4 \geq -4$. Thus, the minimum value of $(x + 3)^2 - 4$ is -4.

6. What is the result of completing the square for $f(x) = x^2 + 1$?

SOLUTION   Because there is no x term in $x^2 + 1$, completing the square on this expression leads to $(x - 0)^2 + 1$.

Exercises

In Exercises 1–4, find the slope, the y-intercept, and the x-intercept of the line with the given equation.

1. $y = 3x + 12$

SOLUTION   Because the equation of the line is given in slope-intercept form, the slope is the coefficient of x and the y-intercept is the constant term: that is, $m = 3$ and the y-intercept is 12. To determine the x-intercept, substitute $y = 0$ and then solve for x: $0 = 3x + 12$ or $x = -4$.

3. $4x + 9y = 3$

SOLUTION   To determine the slope and y-intercept, we first solve the equation for y to obtain the slope-intercept form. This yields $y = -\frac{4}{9}x + \frac{1}{3}$. From here, we see that the slope is $m = -\frac{4}{9}$ and the y-intercept is $\frac{1}{3}$. To determine the x-intercept, substitute $y = 0$ and solve for x: $4x = 3$ or $x = \frac{3}{4}$.

In Exercises 5–8, find the slope of the line.

5. $y = 3x + 2$

SOLUTION $m = 3$

7. $3x + 4y = 12$

SOLUTION   First solve the equation for y to obtain the slope-intercept form. This yields $y = -\frac{3}{4}x + 3$. The slope of the line is therefore $m = -\frac{3}{4}$.

In Exercises 9–20, find the equation of the line with the given description.

9. Slope 3, y-intercept 8

SOLUTION   Using the slope-intercept form for the equation of a line, we have $y = 3x + 8$.

11. Slope 3, passes through $(7, 9)$

SOLUTION   Using the point-slope form for the equation of a line, we have $y - 9 = 3(x - 7)$ or $y = 3x - 12$.

13. Horizontal, passes through $(0, -2)$

SOLUTION   A horizontal line has a slope of 0. Using the point-slope form for the equation of a line, we have $y - (-2) = 0(x - 0)$ or $y = -2$.

15. Parallel to $y = 3x - 4$, passes through $(1, 1)$

SOLUTION   Because the equation $y = 3x - 4$ is in slope-intercept form, we can readily identify that it has a slope of 3. Parallel lines have the same slope, so the slope of the requested line is also 3. Using the point-slope form for the equation of a line, we have $y - 1 = 3(x - 1)$ or $y = 3x - 2$.

17. Perpendicular to $3x + 5y = 9$, passes through $(2, 3)$

SOLUTION   We start by solving the equation $3x + 5y = 9$ for y to obtain the slope-intercept form for the equation of a line. This yields

$$y = -\frac{3}{5}x + \frac{9}{5},$$

from which we identify the slope as $-\frac{3}{5}$. Perpendicular lines have slopes that are negative reciprocals of one another, so the slope of the desired line is $m_\perp = \frac{5}{3}$. Using the point-slope form for the equation of a line, we have $y - 3 = \frac{5}{3}(x - 2)$ or $y = \frac{5}{3}x - \frac{1}{3}$.

19. Horizontal, passes through $(8, 4)$

SOLUTION   A horizontal line has slope 0. Using the point slope form for the equation of a line, we have $y - 4 = 0(x - 8)$ or $y = 4$.

21. Find the equation of the perpendicular bisector of the segment joining $(1, 2)$ and $(5, 4)$ (Figure 11). *Hint:* The midpoint Q of the segment joining (a, b) and (c, d) is $\left(\dfrac{a + c}{2}, \dfrac{b + d}{2}\right)$.

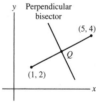

FIGURE 11

SOLUTION   The slope of the segment joining $(1, 2)$ and $(5, 4)$ is

$$m = \frac{4 - 2}{5 - 1} = \frac{1}{2}$$

and the midpoint of the segment (Figure 11) is

$$\text{midpoint} = \left(\frac{1 + 5}{2}, \frac{2 + 4}{2}\right) = (3, 3)$$

The perpendicular bisector has slope $-1/m = -2$ and passes through $(3, 3)$, so its equation is: $y - 3 = -2(x - 3)$ or $y = -2x + 9$.

23. Find the equation of the line with x-intercept $x = 4$ and y-intercept $y = 3$.

SOLUTION   From Exercise 22, $\frac{x}{4} + \frac{y}{3} = 1$ or $3x + 4y = 12$.

25. Determine whether there exists a constant c such that the line $x + cy = 1$

(a) Has slope 4   **(b)** Passes through $(3, 1)$

(c) Is horizontal   **(d)** Is vertical

SOLUTION

(a) Rewriting the equation of the line in slope-intercept form gives $y = -\frac{x}{c} + \frac{1}{c}$. To have slope 4 requires $-\frac{1}{c} = 4$ or $c = -\frac{1}{4}$.

(b) Substituting $x = 3$ and $y = 1$ into the equation of the line gives $3 + c = 1$ or $c = -2$.

(c) From (a), we know the slope of the line is $-\frac{1}{c}$. There is no value for c that will make this slope equal to 0.

(d) With $c = 0$, the equation becomes $x = 1$. This is the equation of a vertical line.

27. Materials expand when heated. Consider a metal rod of length L_0 at temperature T_0. If the temperature is changed by an amount ΔT, then the rod's length changes by $\Delta L = \alpha L_0 \Delta T$, where α is the thermal expansion coefficient. For steel, $\alpha = 1.24 \times 10^{-5}\,°C^{-1}$.

(a) A steel rod has length $L_0 = 40$ cm at $T_0 = 40°C$. What is its length at $T = 90°C$?

(b) Find its length at $T = 50°C$ if its length at $T_0 = 100°C$ is 65 in.

(c) Express length L as a function of T if $L_0 = 65$ in. at $T_0 = 100°C$.

SOLUTION

(a) With $T = 90°C$ and $T_0 = 40°C$, $\Delta T = 50°C$. Therefore,

$$\Delta L = \alpha L_0 \Delta T = (1.24 \times 10^{-5})(40)(50) = .0248 \quad \text{and} \quad L = L_0 + \Delta L = 40.0248 \text{ cm.}$$

(b) With $T = 50°C$ and $T_0 = 100°C$, $\Delta T = -50°C$. Therefore,

$$\Delta L = \alpha L_0 \Delta T = (1.24 \times 10^{-5})(65)(-50) = -.0403 \quad \text{and} \quad L = L_0 + \Delta L = 64.9597 \text{ in.}$$

(c) $L = L_0 + \Delta L = L_0 + \alpha L_0 \Delta T = L_0(1 + \alpha \Delta T) = 65(1 + \alpha(T - 100))$

29. Find b such that $(2, -1)$, $(3, 2)$, and $(b, 5)$ lie on a line.

SOLUTION   The slope of the line determined by the points $(2, -1)$ and $(3, 2)$ is

$$\frac{2 - (-1)}{3 - 2} = 3.$$

To lie on the same line, the slope between $(3, 2)$ and $(b, 5)$ must also be 3. Thus, we require

$$\frac{5 - 2}{b - 3} = \frac{3}{b - 3} = 3,$$

or $b = 4$.

31. The period T of a pendulum is measured for pendulums of several different lengths L. Based on the following data, does T appear to be a linear function of L?

L (ft)	2	3	4	5
T (s)	1.57	1.92	2.22	2.48

SOLUTION   Examine the slope between consecutive data points. The first pair of data points yields a slope of

$$\frac{1.92 - 1.57}{3 - 2} = .35,$$

while the second pair of data points yields a slope of

$$\frac{2.22 - 1.92}{4 - 3} = .3,$$

and the last pair of data points yields a slope of

$$\frac{2.48 - 2.22}{5 - 4} = .26$$

Because the three slopes are not equal, T does not appear to be a linear function of L.

33. Find the roots of the quadratic polynomials:

(a) $4x^2 - 3x - 1$

(b) $x^2 - 2x - 1$

SOLUTION

(a) $x = \dfrac{3 \pm \sqrt{9 - 4(4)(-1)}}{2(4)} = \dfrac{3 \pm \sqrt{25}}{8} = 1 \text{ or } -\dfrac{1}{4}$

(b) $x = \dfrac{2 \pm \sqrt{4 - (4)(1)(-1)}}{2} = \dfrac{2 \pm \sqrt{8}}{2} = 1 \pm \sqrt{2}$

In Exercises 34–41, complete the square and find the minimum or maximum value of the quadratic function.

35. $y = x^2 - 6x + 9$

SOLUTION $y = (x - 3)^2$; therefore, the minimum value of the quadratic polynomial is 0, and this occurs at $x = 3$.

37. $y = x^2 + 6x + 2$

SOLUTION $y = x^2 + 6x + 9 - 9 + 2 = (x + 3)^2 - 7$; therefore, the minimum value of the quadratic polynomial is -7, and this occurs at $x = -3$.

39. $y = -4x^2 + 3x + 8$

SOLUTION $y = -4x^2 + 3x + 8 = -4(x^2 - \frac{3}{4}x + \frac{9}{64}) + 8 + \frac{9}{16} = -4(x - \frac{3}{8})^2 + \frac{137}{16}$; therefore, the maximum value of the quadratic polynomial is $\frac{137}{16}$, and this occurs at $x = \frac{3}{8}$.

41. $y = 4x - 12x^2$

SOLUTION $y = -12(x^2 - \frac{x}{3}) = -12(x^2 - \frac{x}{3} + \frac{1}{36}) + \frac{1}{3} = -12(x - \frac{1}{6})^2 + \frac{1}{3}$; therefore, the maximum value of the quadratic polynomial is $\frac{1}{3}$, and this occurs at $x = \frac{1}{6}$.

43. Sketch the graph of $y = x^2 + 4x + 6$ by plotting the minimum point, the y-intercept, and one other point.

SOLUTION $y = x^2 + 4x + 4 - 4 + 6 = (x + 2)^2 + 2$ so the minimum occurs at $(-2, 2)$. If $x = 0$, then $y = 6$ and if $x = -4$, $y = 6$. This is the graph of x^2 moved left 2 units and up 2 units.

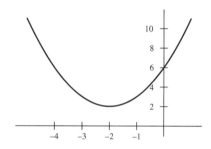

45. For which values of c does $f(x) = x^2 + cx + 1$ have a double root? No real roots?

SOLUTION   A double root occurs when $c^2 - 4(1)(1) = 0$ or $c^2 = 4$. Thus, $c = \pm 2$.
There are no real roots when $c^2 - 4(1)(1) < 0$ or $c^2 < 4$. Thus, $-2 < c < 2$.

47. Prove that $x + \dfrac{1}{x} \geq 2$ for all $x > 0$. *Hint:* Consider $(x^{1/2} - x^{-1/2})^2$.

SOLUTION   Let $x > 0$. Then

$$\left(x^{1/2} - x^{-1/2}\right)^2 = x - 2 + \frac{1}{x}.$$

Because $(x^{1/2} - x^{-1/2})^2 \geq 0$, it follows that

$$x - 2 + \frac{1}{x} \geq 0 \qquad \text{or} \qquad x + \frac{1}{x} \geq 2.$$

49. If objects of weights x and w_1 are suspended from the balance in Figure 13(A), the cross-beam is horizontal if $bx = aw_1$. If the lengths a and b are known, we may use this equation to determine an unknown weight x by selecting w_1 so that the cross-beam is horizontal. If a and b are not known precisely, we might proceed as follows. First balance x by w_1 on the left as in (A). Then switch places and balance x by w_2 on the right as in (B). The average $\bar{x} = \frac{1}{2}(w_1 + w_2)$ gives an estimate for x. Show that $\bar{x}$ is greater than or equal to the true weight x.

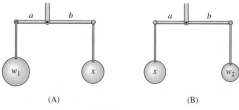

(A)   (B)

FIGURE 13

SOLUTION First note $bx = aw_1$ and $ax = bw_2$. Thus,

$$\bar{x} = \frac{1}{2}(w_1 + w_2)$$

$$= \frac{1}{2}\left(\frac{bx}{a} + \frac{ax}{b}\right)$$

$$= \frac{x}{2}\left(\frac{b}{a} + \frac{a}{b}\right)$$

$$\geq \frac{x}{2}(2) \qquad \text{by Exercise 47}$$

$$= x$$

51. Find a pair of numbers whose sum and product are both equal to 8.

SOLUTION Let x and y be numbers whose sum and product are both equal to 8. Then $x + y = 8$ and $xy = 8$. From the second equation, $y = \frac{8}{x}$. Substituting this expression for y in the first equation gives $x + \frac{8}{x} = 8$ or $x^2 - 8x + 8 = 0$. By the quadratic formula,

$$x = \frac{8 \pm \sqrt{64 - 32}}{2} = 4 \pm 2\sqrt{2}.$$

If $x = 4 + 2\sqrt{2}$, then

$$y = \frac{8}{4 + 2\sqrt{2}} = \frac{8}{4 + 2\sqrt{2}} \cdot \frac{4 - 2\sqrt{2}}{4 - 2\sqrt{2}} = 4 - 2\sqrt{2}.$$

On the other hand, if $x = 4 - 2\sqrt{2}$, then

$$y = \frac{8}{4 - 2\sqrt{2}} = \frac{8}{4 - 2\sqrt{2}} \cdot \frac{4 + 2\sqrt{2}}{4 + 2\sqrt{2}} = 4 + 2\sqrt{2}.$$

Thus, the two numbers are $4 + 2\sqrt{2}$ and $4 - 2\sqrt{2}$.

Further Insights and Challenges

53. Show that if $f(x)$ and $g(x)$ are linear, then so is $f(x) + g(x)$. Is the same true of $f(x)g(x)$?

SOLUTION If $f(x) = mx + b$ and $g(x) = nx + d$, then

$$f(x) + g(x) = mx + b + nx + d = (m + n)x + (b + d),$$

which is linear. $f(x)g(x)$ is not generally linear. Take, for example, $f(x) = g(x) = x$. Then $f(x)g(x) = x^2$.

55. Show that the ratio $\Delta y/\Delta x$ for the function $f(x) = x^2$ over the interval $[x_1, x_2]$ is not a constant, but depends on the interval. Determine the exact dependence of $\Delta y/\Delta x$ on x_1 and x_2.

SOLUTION For x^2, $\dfrac{\Delta y}{\Delta x} = \dfrac{x_2^2 - x_1^2}{x_2 - x_1} = x_2 + x_1.$

57. Let $a, c \neq 0$. Show that the roots of $ax^2 + bx + c = 0$ and $cx^2 + bx + a = 0$ are reciprocals of each other.

SOLUTION Let r_1 and r_2 be the roots of $ax^2 + bx + c$ and r_3 and r_4 be the roots of $cx^2 + bx + a$. Without loss of generality, let

$$r_1 = \frac{-b + \sqrt{b^2 - 4ac}}{2a} \Rightarrow \frac{1}{r_1} = \frac{2a}{-b + \sqrt{b^2 - 4ac}} \cdot \frac{-b - \sqrt{b^2 - 4ac}}{-b - \sqrt{b^2 - 4ac}}$$

$$= \frac{2a(-b - \sqrt{b^2 - 4ac})}{b^2 - b^2 + 4ac} = \frac{-b - \sqrt{b^2 - 4ac}}{2c} = r_4.$$

Similarly, you can show $\dfrac{1}{r_2} = r_3$.

59. Prove **Viète's Formulas**, which state that the quadratic polynomial with given numbers α and β as roots is $x^2 + bx + c$, where $b = -\alpha - \beta$ and $c = \alpha\beta$.

SOLUTION If a quadratic polynomial has roots α and β, then the polynomial is

$$(x - \alpha)(x - \beta) = x^2 - \alpha x - \beta x + \alpha\beta = x^2 + (-\alpha - \beta)x + \alpha\beta.$$

Thus, $b = -\alpha - \beta$ and $c = \alpha\beta$.

1.3 The Basic Classes of Functions

Preliminary Questions

1. Give an example of a rational function.

SOLUTION   One example is $\dfrac{3x^2 - 2}{7x^3 + x - 1}$.

2. Is $|x|$ a polynomial function? What about $|x^2 + 1|$?

SOLUTION $|x|$ is not a polynomial; however, because $x^2 + 1 > 0$ for all x, it follows that $|x^2 + 1| = x^2 + 1$, which is a polynomial.

3. What is unusual about the domain of $f \circ g$ for $f(x) = x^{1/2}$ and $g(x) = -1 - |x|$?

SOLUTION   Recall that $(f \circ g)(x) = f(g(x))$. Now, for any real number x, $g(x) = -1 - |x| \le -1 < 0$. Because we cannot take the square root of a negative number, it follows that $f(g(x))$ is not defined for any real number. In other words, the domain of $f(g(x))$ is the empty set.

4. Is $f(x) = \left(\frac{1}{2}\right)^x$ increasing or decreasing?

SOLUTION   The function $f(x) = \left(\frac{1}{2}\right)^x$ is an exponential function with base $b = \frac{1}{2} < 1$. Therefore, f is a decreasing function.

5. Give an example of a transcendental function.

SOLUTION   One possibility is $f(x) = e^x - \sin x$.

Exercises

In Exercises 1–12, determine the domain of the function.

1. $f(x) = x^{1/4}$

SOLUTION $x \ge 0$

3. $f(x) = x^3 + 3x - 4$

SOLUTION   All reals

5. $g(t) = \dfrac{1}{t + 2}$

SOLUTION $t \ne -2$

7. $G(u) = \dfrac{1}{u^2 - 4}$

SOLUTION $u \ne \pm 2$

9. $f(x) = x^{-4} + (x - 1)^{-3}$

SOLUTION $x \ne 0, 1$

11. $g(y) = 10^{\sqrt{y} + y^{-1}}$

SOLUTION $y > 0$

In Exercises 13–24, identify each of the following functions as polynomial, rational, algebraic, or transcendental.

13. $f(x) = 4x^3 + 9x^2 - 8$

SOLUTION   Polynomial

15. $f(x) = \sqrt{x}$

SOLUTION   Algebraic

17. $f(x) = \dfrac{x^2}{x + \sin x}$

SOLUTION   Transcendental

19. $f(x) = \dfrac{2x^3 + 3x}{9 - 7x^2}$

SOLUTION   Rational

21. $f(x) = \sin(x^2)$

SOLUTION Transcendental

23. $f(x) = x^2 + 3x^{-1}$

SOLUTION Rational

25. Is $f(x) = 2^{x^2}$ a transcendental function?

SOLUTION Yes.

In Exercises 27–34, calculate the composite functions $f \circ g$ and $g \circ f$, and determine their domains.

27. $f(x) = \sqrt{x}$, $g(x) = x + 1$

SOLUTION $f(g(x)) = \sqrt{x+1}$; D: $x \geq -1$, $g(f(x)) = \sqrt{x} + 1$; D: $x \geq 0$

29. $f(x) = 2^x$, $g(x) = x^2$

SOLUTION $f(g(x)) = 2^{x^2}$; D: **R**, $g(f(x)) = (2^x)^2 = 2^{2x}$; D: **R**

31. $f(\theta) = \cos \theta$, $g(x) = x^3 + x^2$

SOLUTION $f(g(x)) = \cos(x^3 + x^2)$; D: **R**, $g(f(\theta)) = \cos^3 \theta + \cos^2 \theta$; D: **R**

33. $f(t) = \dfrac{1}{\sqrt{t}}$, $g(t) = -t^2$

SOLUTION $f(g(t)) = \dfrac{1}{\sqrt{-t^2}}$; D: Not valid for any t, $g(f(t)) = -\left(\dfrac{1}{\sqrt{t}}\right)^2 = -\dfrac{1}{t}$; D: $t > 0$

35. The population (in millions) of a country as a function of time t (years) is $P(t) = 30 \cdot 2^{kt}$, with $k = 0.1$. Show that the population doubles every 10 years. Show more generally that for any nonzero constants a and k, the function $g(t) = a2^{kt}$ doubles after $1/k$ years.

SOLUTION Let $P(t) = 30 \cdot 2^{0.1t}$. Then

$$P(t + 10) = 30 \cdot 2^{0.1(t+10)} = 30 \cdot 2^{0.1t+1} = 2(30 \cdot 2^{0.1t}) = 2P(t).$$

Hence, the population doubles in size every 10 years. In the more general case, let $g(t) = a2^{kt}$. Then

$$g\left(t + \frac{1}{k}\right) = a2^{k(t+1/k)} = a2^{kt+1} = 2a2^{kt} = 2g(t).$$

Hence, the function g doubles after $1/k$ years.

Further Insights and Challenges

In Exercises 37–43, we define the first difference δf of a function $f(x)$ by $\delta f(x) = f(x + 1) - f(x)$.

37. Show that if $f(x) = x^2$, then $\delta f(x) = 2x + 1$. Calculate δf for $f(x) = x$ and $f(x) = x^3$.

SOLUTION $f(x) = x^2$: $\delta f(x) = f(x + 1) - f(x) = (x + 1)^2 - x^2 = 2x + 1$
$f(x) = x$: $\delta f(x) = x + 1 - x = 1$
$f(x) = x^3$: $\delta f(x) = (x + 1)^3 - x^3 = 3x^2 + 3x + 1$

39. Show that for any two functions f and g, $\delta(f + g) = \delta f + \delta g$ and $\delta(cf) = c\delta(f)$, where c is any constant.

SOLUTION $\delta(f + g) = (f(x + 1) + g(x + 1)) - (f(x) - g(x))$

$$= (f(x + 1) - f(x)) + (g(x + 1) - g(x)) = \delta f(x) + \delta g(x)$$

$$\delta(cf) = cf(x + 1) - cf(x) = c(f(x + 1) - f(x)) = c\delta f(x).$$

41. First show that $P(x) = \dfrac{x(x + 1)}{2}$ satisfies $\delta P = (x + 1)$. Then apply Exercise 40 to conclude that

$$1 + 2 + 3 + \cdots + n = \frac{n(n + 1)}{2}$$

SOLUTION Let $P(x) = x(x + 1)/2$. Then

$$\delta P(x) = P(x + 1) - P(x) = \frac{(x + 1)(x + 2)}{2} - \frac{x(x + 1)}{2} = \frac{(x + 1)(x + 2 - x)}{2} = x + 1.$$

Also, note that $P(0) = 0$. Thus, by Exercise 40, with $k = 1$, it follows that

$$P(n) = \frac{n(n + 1)}{2} = 1 + 2 + 3 + \cdots + n.$$

43. This exercise combined with Exercise 40 shows that for all k, there exists a polynomial $P(x)$ satisfying Eq. (1). The solution requires proof by induction and the Binomial Theorem (see Appendix C).

(a) Show that

$$\delta(x^{k+1}) = (k+1)x^k + \cdots$$

where the dots indicate terms involving smaller powers of x.

(b) Show by induction that for all whole numbers k, there exists a polynomial of degree $k+1$ with leading coefficient $1/(k+1)$:

$$P(x) = \frac{1}{k+1}x^{k+1} + \cdots$$

such that $\delta P = (x+1)^k$ and $P(0) = 0$.

SOLUTION

(a) By the Binomial Theorem:

$$\delta(x^{n+1}) = (x+1)^{n+1} - x^{n+1} = \left(x^{n+1} + \binom{n+1}{1}x^n + \binom{n+1}{2}x^{n-1} + \cdots + 1\right) - x^{n+1}$$

$$= \binom{n+1}{1}x^n + \binom{n+1}{2}x^{n-1} + \cdots + 1$$

Thus,

$$\delta(x^{n+1}) = (n+1)x^n + \cdots$$

where the dots indicate terms involving smaller powers of x.

(b) For $k = 0$, note that $P(x) = x$ satisfies $\delta P = (x+1)^0 = 1$ and $P(0) = 0$.

Now suppose the polynomial

$$P(x) = \frac{1}{k}x^k + p_{k-1}x^{k-1} + \cdots + p_1 x$$

which clearly satisfies $P(0) = 0$ also satisfies $\delta P = (x+1)^{k-1}$. We try to prove the existence of

$$Q(x) = \frac{1}{k+1}x^{k+1} + q_k x^k + \cdots + q_1 x$$

such that $\delta Q = (x+1)^k$. Observe that $Q(0) = 0$.

If $\delta Q = (x+1)^k$ and $\delta P = (x+1)^{k-1}$, then

$$\delta Q = (x+1)^k = (x+1)\delta P = x\delta P(x) + \delta P$$

By the linearity of δ (Exercise 39), we find $\delta Q - \delta P = x\delta P$ or $\delta(Q - P) = x\delta P$. By definition,

$$Q - P = \frac{1}{k+1}x^{k+1} + \left(q_k - \frac{1}{k}\right)x^k + \cdots + (q_1 - p_1)x,$$

so, by the linearity of δ,

$$\delta(Q - P) = \frac{1}{k+1}\delta(x^{k+1}) + \left(q_k - \frac{1}{k}\right)\delta(x^k) + \cdots + (q_1 - p_1) = x(x+1)^{k-1} \tag{1}$$

By part (a),

$$\delta(x^{k+1}) = (k+1)x^k + L_{k-1,k-1}x^{k-1} + \ldots + L_{k-1,1}x + 1$$

$$\delta(x^k) = kx^{k-1} + L_{k-2,k-2}x^{k-2} + \ldots + L_{k-2,1}x + 1$$

$$\vdots$$

$$\delta(x^2) = 2x + 1$$

where the $L_{i,j}$ are real numbers for each i, j.

To construct Q, we have to group like powers of x on both sides of (1). This yields the system of equations

$$\frac{1}{k+1}\left((k+1)x^k\right) = x^k$$

$$\frac{1}{k+1}L_{k-1,k-1}x^{k-1} + \left(q_k - \frac{1}{k}\right)kx^{k-1} = (k-1)x^{k-1}$$

$$\vdots$$

$$\frac{1}{k+1} + \left(q_k - \frac{1}{k}\right) + (q_{k-1} - p_{k-1}) + \cdots + (q_1 - p_1) = 0.$$

The first equation is identically true, and the second equation can be solved immediately for q_k. Substituting the value of q_k into the third equation of the system, we can then solve for q_{k-1}. We continue this process until we substitute the values of $q_k, q_{k-1}, \ldots q_2$ into the last equation, and then solve for q_1.

1.4 Trigonometric Functions

Preliminary Questions

1. How is it possible for two different rotations to define the same angle?

SOLUTION   Working from the same initial radius, two rotations that differ by a whole number of full revolutions will have the same ending radius; consequently, the two rotations will define the same angle even though the measures of the rotations will be different.

2. Give two different positive rotations that define the angle $\frac{\pi}{4}$.

SOLUTION   The angle $\pi/4$ is defined by any rotation of the form $\frac{\pi}{4} + 2\pi k$ where k is an integer. Thus, two different positive rotations that define the angle $\pi/4$ are

$$\frac{\pi}{4} + 2\pi(1) = \frac{9\pi}{4} \quad \text{and} \quad \frac{\pi}{4} + 2\pi(5) = \frac{41\pi}{4}.$$

3. Give a negative rotation that defines the angle $\frac{\pi}{3}$.

SOLUTION   The angle $\pi/3$ is defined by any rotation of the form $\frac{\pi}{3} + 2\pi k$ where k is an integer. Thus, a negative rotation that defines the angle $\pi/3$ is

$$\frac{\pi}{3} + 2\pi(-1) = -\frac{5\pi}{3}.$$

4. The definition of $\cos\theta$ using right triangles applies when (choose the correct answer):

(a) $0 < \theta < \dfrac{\pi}{2}$ $\qquad\qquad$ **(b)** $0 < \theta < \pi$ $\qquad\qquad$ **(c)** $0 < \theta < 2\pi$

SOLUTION   The correct response is **(a)**: $0 < \theta < \frac{\pi}{2}$.

5. What is the unit circle definition of $\sin\theta$?

SOLUTION   Let O denote the center of the unit circle, and let P be a point on the unit circle such that the radius $\overline{OP}$ makes an angle θ with the positive x-axis. Then, $\sin\theta$ is the y-coordinate of the point P.

6. How does the periodicity of $\sin\theta$ and $\cos\theta$ follow from the unit circle definition?

SOLUTION   Let O denote the center of the unit circle, and let P be a point on the unit circle such that the radius $\overline{OP}$ makes an angle θ with the positive x-axis. Then, $\cos\theta$ and $\sin\theta$ are the x- and y-coordinates, respectively, of the point P. The angle $\theta + 2\pi$ is obtained from the angle θ by making one full revolution around the circle. The angle $\theta + 2\pi$ will therefore have the radius $\overline{OP}$ as its terminal side. Thus

$$\cos(\theta + 2\pi) = \cos\theta \quad \text{and} \quad \sin(\theta + 2\pi) = \sin\theta.$$

In other words, $\sin\theta$ and $\cos\theta$ are periodic functions.

Exercises

1. Find the angle between 0 and 2π that is equivalent to $13\pi/4$.

SOLUTION   Because $13\pi/4 > 2\pi$, we repeatedly subtract 2π until we arrive at a radian measure that is between 0 and 2π. After one subtraction, we have $13\pi/4 - 2\pi = 5\pi/4$. Because $0 < 5\pi/4 < 2\pi$, $5\pi/4$ is the angle measure between 0 and 2π that is equivalent to $13\pi/4$.

3. Convert from radians to degrees:

(a) 1 $\qquad$ **(b)** $\dfrac{\pi}{3}$ $\qquad$ **(c)** $\dfrac{5}{12}$ $\qquad$ **(d)** $-\dfrac{3\pi}{4}$

SOLUTION

(a) $1\left(\dfrac{180°}{\pi}\right) = \dfrac{180°}{\pi} \approx 57.1°$

(b) $\dfrac{\pi}{3}\left(\dfrac{180°}{\pi}\right) = 60°$

(c) $\dfrac{5}{12}\left(\dfrac{180°}{\pi}\right) = \dfrac{75°}{\pi} \approx 23.87°$

(d) $-\dfrac{3\pi}{4}\left(\dfrac{180°}{\pi}\right) = -135°$

5. Find the lengths of the arcs subtended by the angles θ and ϕ radians in Figure 20.

FIGURE 20 Circle of radius 4.

SOLUTION $s = r\theta = 4(.9) = 3.6;\ s = r\phi = 4(2) = 8$

7. Fill in the remaining values of $(\cos\theta, \sin\theta)$ for the points in Figure 22.

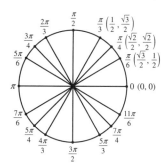

FIGURE 22

SOLUTION

θ	$\frac{\pi}{2}$	$\frac{2\pi}{3}$	$\frac{3\pi}{4}$	$\frac{5\pi}{6}$	π	$\frac{7\pi}{6}$
$(\cos\theta, \sin\theta)$	$(0,1)$	$\left(\frac{-1}{2}, \frac{\sqrt{3}}{2}\right)$	$\left(\frac{-\sqrt{2}}{2}, \frac{\sqrt{2}}{2}\right)$	$\left(\frac{-\sqrt{3}}{2}, \frac{1}{2}\right)$	$(-1,0)$	$\left(\frac{-\sqrt{3}}{2}, \frac{-1}{2}\right)$

θ	$\frac{5\pi}{4}$	$\frac{4\pi}{3}$	$\frac{3\pi}{2}$	$\frac{5\pi}{3}$	$\frac{7\pi}{4}$	$\frac{11\pi}{6}$
$(\cos\theta, \sin\theta)$	$\left(\frac{-\sqrt{2}}{2}, \frac{-\sqrt{2}}{2}\right)$	$\left(\frac{-1}{2}, \frac{-\sqrt{3}}{2}\right)$	$(0,-1)$	$\left(\frac{1}{2}, \frac{-\sqrt{3}}{2}\right)$	$\left(\frac{\sqrt{2}}{2}, \frac{-\sqrt{2}}{2}\right)$	$\left(\frac{\sqrt{3}}{2}, \frac{-1}{2}\right)$

In Exercises 9–14, use Figure 22 to find all angles between 0 and 2π satisfying the given condition.

9. $\cos\theta = \frac{1}{2}$

SOLUTION $\theta = \frac{\pi}{3}, \frac{5\pi}{3}$

11. $\tan\theta = -1$

SOLUTION $\theta = \frac{3\pi}{4}, \frac{7\pi}{4}$

13. $\sin x = \dfrac{\sqrt{3}}{2}$

SOLUTION $x = \frac{\pi}{3}, \frac{2\pi}{3}$

15. Fill in the following table of values:

θ	$\frac{\pi}{6}$	$\frac{\pi}{4}$	$\frac{\pi}{3}$	$\frac{\pi}{2}$	$\frac{2\pi}{3}$	$\frac{3\pi}{4}$	$\frac{5\pi}{6}$
$\tan\theta$							
$\sec\theta$							

SOLUTION

θ	$\dfrac{\pi}{6}$	$\dfrac{\pi}{4}$	$\dfrac{\pi}{3}$	$\dfrac{\pi}{2}$	$\dfrac{2\pi}{3}$	$\dfrac{3\pi}{4}$	$\dfrac{5\pi}{6}$
$\tan\theta$	$\dfrac{1}{\sqrt{3}}$	1	$\sqrt{3}$	und	$-\sqrt{3}$	-1	$-\dfrac{1}{\sqrt{3}}$
$\sec\theta$	$\dfrac{2}{\sqrt{3}}$	$\sqrt{2}$	2	und	-2	$-\sqrt{2}$	$-\dfrac{2}{\sqrt{3}}$

17. Show that if $\tan\theta = c$ and $0 \le \theta < \pi/2$, then $\cos\theta = 1/\sqrt{1+c^2}$. *Hint:* Draw a right triangle whose opposite and adjacent sides have lengths c and 1.

SOLUTION Because $0 \le \theta < \pi/2$, we can use the definition of the trigonometric functions in terms of right triangles. $\tan\theta$ is the ratio of the length of the side opposite the angle θ to the length of the adjacent side. With $c = \frac{c}{1}$, we label the length of the opposite side as c and the length of the adjacent side as 1 (see the diagram below). By the Pythagorean theorem, the length of the hypotenuse is $\sqrt{1+c^2}$. Finally, we use the fact that $\cos\theta$ is the ratio of the length of the adjacent side to the length of the hypotenuse to obtain

$$\cos\theta = \frac{1}{\sqrt{1+c^2}}.$$

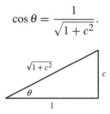

In Exercises 19–24, assume that $0 \le \theta < \pi/2$.

19. Find $\sin\theta$ and $\tan\theta$ if $\cos\theta = \frac{5}{13}$.

SOLUTION Consider the triangle below. The lengths of the side adjacent to the angle θ and the hypotenuse have been labeled so that $\cos\theta = \frac{5}{13}$. The length of the side opposite the angle θ has been calculated using the Pythagorean theorem: $\sqrt{13^2 - 5^2} = 12$. From the triangle, we see that

$$\sin\theta = \frac{12}{13} \qquad \text{and} \qquad \tan\theta = \frac{12}{5}.$$

21. Find $\sin\theta$, $\sec\theta$, and $\cot\theta$ if $\tan\theta = \frac{2}{7}$.

SOLUTION If $\tan\theta = \frac{2}{7}$, then $\cot\theta = \frac{7}{2}$. For the remaining trigonometric functions, consider the triangle below. The lengths of the sides opposite and adjacent to the angle θ have been labeled so that $\tan\theta = \frac{2}{7}$. The length of the hypotenuse has been calculated using the Pythagorean theorem: $\sqrt{2^2 + 7^2} = \sqrt{53}$. From the triangle, we see that

$$\sin\theta = \frac{2}{\sqrt{53}} = \frac{2\sqrt{53}}{53} \qquad \text{and} \qquad \sec\theta = \frac{\sqrt{53}}{7}.$$

23. Find $\cos 2\theta$ if $\sin\theta = \frac{1}{5}$.

SOLUTION Using the double angle formula $\cos 2\theta = \cos^2\theta - \sin^2\theta$ and the fundamental identity $\sin^2\theta + \cos^2\theta = 1$, we find that $\cos 2\theta = 1 - 2\sin^2\theta$. Thus, $\cos 2\theta = 1 - 2(1/25) = 23/25$.

25. Find $\cos\theta$ and $\tan\theta$ if $\sin\theta = 0.4$ and $\pi/2 \le \theta < \pi$.

SOLUTION   We can determine the "magnitude" of $\cos\theta$ and $\tan\theta$ using the triangle shown below. The lengths of the side opposite the angle θ and the hypotenuse have been labeled so that $\sin\theta = 0.4 = \frac{2}{5}$. The length of the side adjacent to the angle θ was calculated using the Pythagorean theorem: $\sqrt{5^2 - 2^2} = \sqrt{21}$. From the triangle, we see that

$$|\cos\theta| = \frac{\sqrt{21}}{5} \quad \text{and} \quad |\tan\theta| = \frac{2}{\sqrt{21}} = \frac{2\sqrt{21}}{21}.$$

Because $\pi/2 \le \theta < \pi$, both $\cos\theta$ and $\tan\theta$ are negative; consequently,

$$\cos\theta = -\frac{\sqrt{21}}{5} \quad \text{and} \quad \tan\theta = -\frac{2\sqrt{21}}{21}.$$

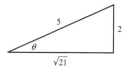

27. Find the values of $\sin\theta$, $\cos\theta$, and $\tan\theta$ at the eight points in Figure 23.

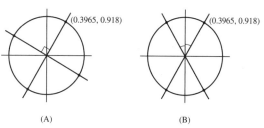

FIGURE 23

SOLUTION   Let's start with the four points in Figure 23(A).

- The point in the first quadrant has coordinates $(0.3965, 0.918)$. Therefore,

$$\sin\theta = 0.918, \quad \cos\theta = 0.3965, \quad \text{and} \quad \tan\theta = \frac{0.918}{0.3965} = 2.3153.$$

- The coordinates of the point in the second quadrant are $(-0.918, 0.3965)$. Therefore,

$$\sin\theta = 0.3965, \quad \cos\theta = -0.918, \quad \text{and} \quad \tan\theta = \frac{0.3965}{-0.918} = -0.4319.$$

- Because the point in the third quadrant is symmetric to the point in the first quadrant with respect to the origin, its coordinates are $(-0.3965, -0.918)$. Therefore,

$$\sin\theta = -0.918, \quad \cos\theta = -0.3965, \quad \text{and} \quad \tan\theta = \frac{-0.918}{-0.3965} = 2.3153.$$

- Because the point in the fourth quadrant is symmetric to the point in the second quadrant with respect to the origin, its coordinates are $(0.918, -0.3965)$. Therefore,

$$\sin\theta = -0.3965, \quad \cos\theta = 0.918, \quad \text{and} \quad \tan\theta = \frac{-0.3965}{0.918} = -0.4319.$$

Now consider the four points in Figure 23(B).

- The point in the first quadrant has coordinates $(0.3965, 0.918)$. Therefore,

$$\sin\theta = 0.918, \quad \cos\theta = 0.3965, \quad \text{and} \quad \tan\theta = \frac{0.918}{0.3965} = 2.3153.$$

- The point in the second quadrant is a reflection through the y-axis of the point in the first quadrant. Its coordinates are therefore $(-0.3965, 0.918)$ and

$$\sin\theta = 0.918, \quad \cos\theta = -0.3965, \quad \text{and} \quad \tan\theta = \frac{0.918}{0.3965} = -2.3153.$$

- Because the point in the third quadrant is symmetric to the point in the first quadrant with respect to the origin, its coordinates are $(-0.3965, -0.918)$. Therefore,

$$\sin \theta = -0.918, \quad \cos \theta = -0.3965, \quad \text{and} \quad \tan \theta = \frac{-0.918}{-0.3965} = 2.3153.$$

- Because the point in the fourth quadrant is symmetric to the point in the second quadrant with respect to the origin, its coordinates are $(0.3965, -0.918)$. Therefore,

$$\sin \theta = -0.918, \quad \cos \theta = 0.3965, \quad \text{and} \quad \tan \theta = \frac{-0.918}{0.3965} = -2.3153.$$

29. Refer to Figure 24(B). Compute $\cos \psi$, $\sin \psi$, $\cot \psi$, and $\csc \psi$.

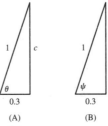

(A)   (B)

FIGURE 24

SOLUTION By the Pythagorean theorem, the length of the side opposite the angle ψ in Figure 24 (B) is $\sqrt{1 - 0.3^2} = \sqrt{0.91}$. Consequently,

$$\cos \psi = \frac{0.3}{1} = 0.3, \quad \sin \psi = \frac{\sqrt{0.91}}{1} = \sqrt{0.91}, \quad \cot \psi = \frac{0.3}{\sqrt{0.91}} \quad \text{and} \quad \csc \psi = \frac{1}{\sqrt{0.91}}.$$

In Exercises 31–34, sketch the graph over $[0, 2\pi]$.

31. $2 \sin 4\theta$

SOLUTION

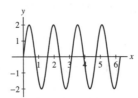

33. $\cos\left(2\theta - \frac{\pi}{2}\right)$

SOLUTION

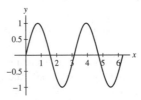

35. How many points lie on the intersection of the horizontal line $y = c$ and the graph of $y = \sin x$ for $0 \le x < 2\pi$? *Hint:* The answer depends on c.

SOLUTION Recall that for any x, $-1 \le \sin x \le 1$. Thus, if $|c| > 1$, the horizontal line $y = c$ and the graph of $y = \sin x$ never intersect. If $c = +1$, then $y = c$ and $y = \sin x$ intersect at the peak of the sine curve; that is, they intersect at $x = \frac{\pi}{2}$. On the other hand, if $c = -1$, then $y = c$ and $y = \sin x$ intersect at the bottom of the sine curve; that is, they intersect at $x = \frac{3\pi}{2}$. Finally, if $|c| < 1$, the graphs of $y = c$ and $y = \sin x$ intersect twice.

In Exercises 37–40, solve for $0 \le \theta < 2\pi$ (see Example 6).

37. $\sin 2\theta + \sin 3\theta = 0$

SOLUTION $\sin \alpha = -\sin \beta$ when $\alpha = -\beta + 2\pi k$ or $\alpha = \pi + \beta + 2\pi k$. Substituting $\alpha = 2\theta$ and $\beta = 3\theta$, we have either $2\theta = -3\theta + 2\pi k$ or $2\theta = \pi + 3\theta + 2\pi k$. Solving each of these equations for θ yields $\theta = \frac{2}{5}\pi k$ or $\theta = -\pi - 2\pi k$. The solutions on the interval $0 \leq \theta < 2\pi$ are then

$$\theta = 0, \frac{2\pi}{5}, \frac{4\pi}{5}, \pi, \frac{6\pi}{5}, \frac{8\pi}{5}.$$

39. $\cos 4\theta + \cos 2\theta = 0$

SOLUTION $\cos \alpha = -\cos \beta$ when $\alpha + \beta = \pi + 2\pi k$ or $\alpha = \beta + \pi + 2\pi k$. Substituting $\alpha = 4\theta$ and $\beta = 2\theta$, we have either $6\theta = \pi + 2\pi k$ or $4\theta = 2\theta + \pi + 2\pi k$. Solving each of these equations for θ yields $\theta = \frac{\pi}{6} + \frac{\pi}{3}k$ or $\theta = \frac{\pi}{2} + \pi k$. The solutions on the interval $0 \leq \theta < 2\pi$ are then

$$\theta = \frac{\pi}{6}, \frac{\pi}{2}, \frac{5\pi}{6}, \frac{7\pi}{6}, \frac{3\pi}{2}, \frac{11\pi}{6}.$$

In Exercises 41–50, derive the identities using the identities listed in this section.

41. $\cos 2\theta = 2\cos^2 \theta - 1$

SOLUTION Starting from the double angle formula for cosine, $\cos^2 \theta = \frac{1}{2}(1 + \cos 2\theta)$, we solve for $\cos 2\theta$. This gives $2\cos^2 \theta = 1 + \cos 2\theta$ and then $\cos 2\theta = 2\cos^2 \theta - 1$.

43. $\sin \frac{\theta}{2} = \sqrt{\dfrac{1 - \cos \theta}{2}}$

SOLUTION Substitute $x = \theta/2$ into the double angle formula for sine, $\sin^2 x = \frac{1}{2}(1 - \cos 2x)$ to obtain $\sin^2 \left(\frac{\theta}{2} \right) = \dfrac{1 - \cos \theta}{2}$. Taking the square root of both sides yields $\sin \left(\frac{\theta}{2} \right) = \sqrt{\dfrac{1 - \cos \theta}{2}}$.

45. $\cos(\theta + \pi) = -\cos \theta$

SOLUTION From the addition formula for the cosine function, we have

$$\cos(\theta + \pi) = \cos \theta \cos \pi - \sin \theta \sin \pi = \cos \theta(-1) = -\cos \theta$$

47. $\tan(\pi - \theta) = -\tan \theta$

SOLUTION Using Exercises 44 and 45,

$$\tan(\pi - \theta) = \frac{\sin(\pi - \theta)}{\cos(\pi - \theta)} = \frac{\sin(\pi + (-\theta))}{\cos(\pi + (-\theta))} = \frac{-\sin(-\theta)}{-\cos(-\theta)} = \frac{\sin \theta}{-\cos \theta} = -\tan \theta.$$

The second to last equality occurs because $\sin x$ is an odd function and $\cos x$ is an even function.

49. $\tan x = \dfrac{\sin 2x}{1 + \cos 2x}$

SOLUTION Using the addition formula for the sine function, we find

$$\sin 2x = \sin(x + x) = \sin x \cos x + \cos x \sin x = 2\sin x \cos x.$$

By Exercise 41, we know that $\cos 2x = 2\cos^2 x - 1$. Therefore,

$$\frac{\sin 2x}{1 + \cos 2x} = \frac{2\sin x \cos x}{1 + 2\cos^2 x - 1} = \frac{2\sin x \cos x}{2\cos^2 x} = \frac{\sin x}{\cos x} = \tan x.$$

51. Use Exercises 44 and 45 to show that $\tan \theta$ and $\cot \theta$ are periodic with period π.

SOLUTION By Exercises 44 and 45,

$$\tan(\theta + \pi) = \frac{\sin(\theta + \pi)}{\cos(\theta + \pi)} = \frac{-\sin \theta}{-\cos \theta} = \tan \theta,$$

and

$$\cot(\theta + \pi) = \frac{\cos(\theta + \pi)}{\sin(\theta + \pi)} = \frac{-\cos \theta}{-\sin \theta} = \cot \theta.$$

Thus, both $\tan \theta$ and $\cot \theta$ are periodic with period π.

53. Use the Law of Cosines to find the distance from P to Q in Figure 26.

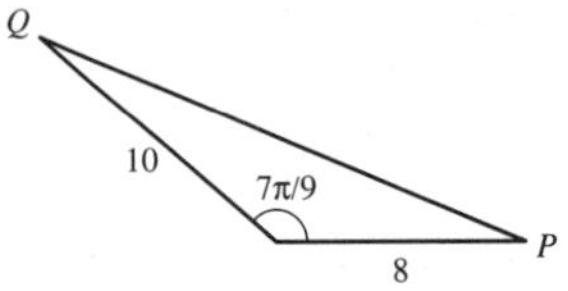

FIGURE 26

SOLUTION By the Law of Cosines, the distance from P to Q is

$$\sqrt{10^2 + 8^2 - 2(10)(8)\cos\frac{7\pi}{9}} = 16.928.$$

Further Insights and Challenges

55. Use the addition formulas for sine and cosine to prove

$$\tan(a+b) = \frac{\tan a + \tan b}{1 - \tan a \tan b}$$

$$\cot(a-b) = \frac{\cot a \cot b + 1}{\cot b - \cot a}$$

SOLUTION

$$\tan(a+b) = \frac{\sin(a+b)}{\cos(a+b)} = \frac{\sin a \cos b + \cos a \sin b}{\cos a \cos b - \sin a \sin b} = \frac{\frac{\sin a \cos b}{\cos a \cos b} + \frac{\cos a \sin b}{\cos a \cos b}}{\frac{\cos a \cos b}{\cos a \cos b} - \frac{\sin a \sin b}{\cos a \cos b}} = \frac{\tan a + \tan b}{1 - \tan a \tan b}$$

$$\cot(a-b) = \frac{\cos(a-b)}{\sin(a-b)} = \frac{\cos a \cos b + \sin a \sin b}{\sin a \cos b - \cos a \sin b} = \frac{\frac{\cos a \cos b}{\sin a \sin b} + \frac{\sin a \sin b}{\sin a \sin b}}{\frac{\sin a \cos b}{\sin a \sin b} - \frac{\cos a \sin b}{\sin a \sin b}} = \frac{\cot a \cot b + 1}{\cot b - \cot a}$$

57. Let L_1 and L_2 be the lines of slope m_1 and m_2 [Figure 27(B)]. Show that the angle θ between L_1 and L_2 satisfies $\cot\theta = \dfrac{m_2 m_1 + 1}{m_2 - m_1}$.

SOLUTION Measured from the positive x-axis, let α and β satisfy $\tan\alpha = m_1$ and $\tan\beta = m_2$. Without loss of generality, let $\beta \geq \alpha$. Then the angle between the two lines will be $\theta = \beta - \alpha$. Then from Exercise 55,

$$\cot\theta = \cot(\beta - \alpha) = \frac{\cot\beta \cot\alpha + 1}{\cot\alpha - \cot\beta} = \frac{(\frac{1}{m_1})(\frac{1}{m_2}) + 1}{\frac{1}{m_1} - \frac{1}{m_2}} = \frac{1 + m_1 m_2}{m_2 - m_1}$$

59. Apply the double-angle formula to prove:

(a) $\cos\dfrac{\pi}{8} = \dfrac{1}{2}\sqrt{2 + \sqrt{2}}$

(b) $\cos\dfrac{\pi}{16} = \dfrac{1}{2}\sqrt{2 + \sqrt{2 + \sqrt{2}}}$

Guess the values of $\cos\dfrac{\pi}{32}$ and of $\cos\dfrac{\pi}{2^n}$ for all n.

SOLUTION

(a) $\cos\dfrac{\pi}{8} = \cos\dfrac{\pi/4}{2} = \sqrt{\dfrac{1 + \cos\frac{\pi}{4}}{2}} = \sqrt{\dfrac{1 + \frac{\sqrt{2}}{2}}{2}} = \dfrac{1}{2}\sqrt{2 + \sqrt{2}}.$

(b) $\cos\dfrac{\pi}{16} = \sqrt{\dfrac{1 + \cos\frac{\pi}{8}}{2}} = \sqrt{\dfrac{1 + \frac{1}{2}\sqrt{2 + \sqrt{2}}}{2}} = \dfrac{1}{2}\sqrt{2 + \sqrt{2 + \sqrt{2}}}.$

(c) Observe that $8 = 2^3$ and $\cos\frac{\pi}{8}$ involves two nested square roots of 2; further, $16 = 2^4$ and $\cos\frac{\pi}{16}$ involves three nested square roots of 2. Since $32 = 2^5$, it seems plausible that

$$\cos\frac{\pi}{32} = \frac{1}{2}\sqrt{2 + \sqrt{2 + \sqrt{2 + \sqrt{2}}}},$$

and that $\cos\frac{\pi}{2^n}$ involves $n - 1$ nested square roots of 2. Note that the general case can be proven by induction.

1.5 Technology: Calculators and Computers

Preliminary Questions

1. Is there a definite way of choosing the optimal viewing rectangle, or is it best to experiment until you find a viewing rectangle appropriate to the problem at hand?

SOLUTION It is best to experiment with the window size until one is found that is appropriate for the problem at hand.

2. Describe the calculator screen produced when the function $y = 3 + x^2$ is plotted with viewing rectangle:

(a) $[-1, 1] \times [0, 2]$ **(b)** $[0, 1] \times [0, 4]$

SOLUTION

(a) Using the viewing rectangle $[-1, 1]$ by $[0, 2]$, the screen will display nothing as the minimum value of $y = 3 + x^2$ is $y = 3$.

(b) Using the viewing rectangle $[0, 1]$ by $[0, 4]$, the screen will display the portion of the parabola between the points $(0, 3)$ and $(1, 4)$.

3. According to the evidence in Example 4, it appears that $f(n) = (1 + 1/n)^n$ never takes on a value greater than 3 for $n \geq 0$. Does this evidence *prove* that $f(n) \leq 3$ for $n \geq 0$?

SOLUTION No, this evidence does not constitute a proof that $f(n) \leq 3$ for $n \geq 0$.

4. How can a graphing calculator be used to find the minimum value of a function?

SOLUTION Experiment with the viewing window to zoom in on the lowest point on the graph of the function. The y-coordinate of the lowest point on the graph is the minimum value of the function.

Exercises

The exercises in this section should be done using a graphing calculator or computer algebra system.

1. Plot $f(x) = 2x^4 + 3x^3 - 14x^2 - 9x + 18$ in the appropriate viewing rectangles and determine its roots.

SOLUTION Using a viewing rectangle of $[-4, 3]$ by $[-20, 20]$, we obtain the plot below.

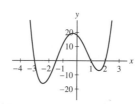

Now, the roots of $f(x)$ are the x-intercepts of the graph of $y = f(x)$. From the plot, we can identify the x-intercepts as $-3, -1.5, 1$, and 2. The roots of $f(x)$ are therefore $x = -3$, $x = -1.5$, $x = 1$, and $x = 2$.

3. How many *positive* solutions does $x^3 - 12x + 8 = 0$ have?

SOLUTION The graph of $y = x^3 - 12x + 8$ shown below has two x-intercepts to the right of the origin; therefore the equation $x^3 - 12x + 8 = 0$ has two positive solutions.

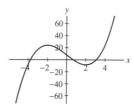

5. Find all the solutions of $\sin x = \sqrt{x}$ for $x > 0$.

SOLUTION Solutions to the equation $\sin x = \sqrt{x}$ correspond to points of intersection between the graphs of $y = \sin x$ and $y = \sqrt{x}$. The two graphs are shown below; the only point of intersection is at $x = 0$. Therefore, there are no solutions of $\sin x = \sqrt{x}$ for $x > 0$.

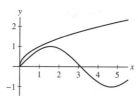

7. Let $f(x) = (x - 100)^2 + 1,000$. What will the display show if you graph $f(x)$ in the viewing rectangle $[-10, 10]$ by $[-10, -10]$? What would be an appropriate viewing rectangle?

SOLUTION Because $(x - 100)^2 \geq 0$ for all x, it follows that $f(x) = (x - 100)^2 + 1000 \geq 1000$ for all x. Thus, using a viewing rectangle of $[-10, 10]$ by $[-10, 10]$ will display nothing. The minimum value of the function occurs when $x = 100$, so an appropriate viewing rectangle would be $[50, 150]$ by $[1000, 2000]$.

9. Plot the graph of $f(x) = \dfrac{x}{4 - x}$ in a viewing rectangle that clearly displays the vertical and horizontal asymptotes.

SOLUTION From the graph of $y = \dfrac{x}{4 - x}$ shown below, we see that the vertical asymptote is $x = 4$ and the horizontal asymptote is $y = -1$.

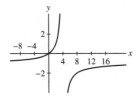

11. Plot $f(x) = \cos(x^2) \sin x$ for $0 \leq x \leq 2\pi$. Then illustrate local linearity at $x = 3.8$ by choosing appropriate viewing rectangles.

SOLUTION The following three graphs display $f(x) = \cos(x^2) \sin x$ over the intervals $[0, 2\pi]$, $[3.5, 4.1]$ and $[3.75, 3.85]$. The final graph looks like a straight line.

 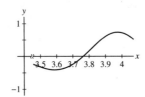 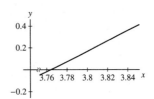

In Exercises 13–18, investigate the behavior of the function as n or x grows large by making a table of function values and plotting a graph (see Example 4). Describe the behavior in words.

13. $f(n) = n^{1/n}$

SOLUTION The table and graphs below suggest that as n gets large, $n^{1/n}$ approaches 1.

n	$n^{1/n}$
10	1.258925412
10^2	1.047128548
10^3	1.006931669
10^4	1.000921458
10^5	1.000115136
10^6	1.000013816

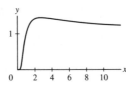

15. $f(n) = \left(1 + \dfrac{1}{n}\right)^{n^2}$

SOLUTION The table and graphs below suggest that as n gets large, $f(n)$ tends toward ∞.

n	$\left(1 + \dfrac{1}{n}\right)^{n^2}$
10	13780.61234
10^2	$1.635828711 \times 10^{43}$
10^3	$1.195306603 \times 10^{434}$
10^4	$5.341783312 \times 10^{4342}$
10^5	$1.702333054 \times 10^{43429}$
10^6	$1.839738749 \times 10^{434294}$

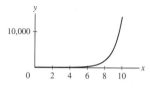 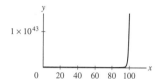

17. $f(x) = \left(x \tan \dfrac{1}{x}\right)^x$

SOLUTION The table and graphs below suggest that as x gets large, $f(x)$ approaches 1.

x	$\left(x \tan \dfrac{1}{x}\right)^x$
10	1.033975759
10^2	1.003338973
10^3	1.000333389
10^4	1.000033334
10^5	1.000003333
10^6	1.000000333

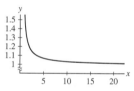 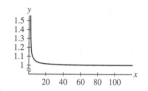

19. The graph of $f(\theta) = A \cos \theta + B \sin \theta$ is a sinusoidal wave for any constants A and B. Confirm this for $(A, B) = (1, 1)$, $(1, 2)$, and $(3, 4)$ by plotting $f(\theta)$.

SOLUTION The graphs of $f(\theta) = \cos \theta + \sin \theta$, $f(\theta) = \cos \theta + 2 \sin \theta$ and $f(\theta) = 3 \cos \theta + 4 \sin \theta$ are shown below.

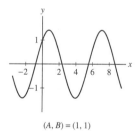 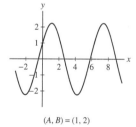 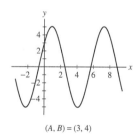

$(A, B) = (1, 1)$ $(A, B) = (1, 2)$ $(A, B) = (3, 4)$

21. Find the intervals on which $f(x) = x(x + 2)(x - 3)$ is positive by plotting a graph.

SOLUTION The function $f(x) = x(x + 2)(x - 3)$ is positive when the graph of $y = x(x + 2)(x - 3)$ lies above the x-axis. The graph of $y = x(x + 2)(x - 3)$ is shown below. Clearly, the graph lies above the x-axis and the function is positive for $x \in (-2, 0) \cup (3, \infty)$.

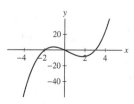

Further Insights and Challenges

23. $\boxed{\text{C R S}}$ Let $f_1(x) = x$ and define a sequence of functions by $f_{n+1}(x) = \frac{1}{2}(f_n(x) + x/f_n(x))$. For example, $f_2(x) = \frac{1}{2}(x + 1)$. Use a computer algebra system to compute $f_n(x)$ for $n = 3, 4, 5$ and plot $f_n(x)$ together with $\sqrt{x}$ for $x \geq 0$. What do you notice?

SOLUTION With $f_1(x) = x$ and $f_2(x) = \frac{1}{2}(x + 1)$, we calculate

$$f_3(x) = \frac{1}{2}\left(\frac{1}{2}(x+1) + \frac{x}{\frac{1}{2}(x+1)}\right) = \frac{x^2 + 6x + 1}{4(x+1)}$$

$$f_4(x) = \frac{1}{2}\left(\frac{x^2 + 6x + 1}{4(x+1)} + \frac{x}{\frac{x^2+6x+1}{4(x+1)}}\right) = \frac{x^4 + 28x^3 + 70x^2 + 28x + 1}{8(1+x)(1 + 6x + x^2)}$$

and

$$f_5(x) = \frac{1 + 120x + 1820x^2 + 8008x^3 + 12870x^4 + 8008x^5 + 1820x^6 + 120x^7 + x^8}{16(1+x)(1 + 6x + x^2)(1 + 28x + 70x^2 + 28x^3 + x^4)}.$$

A plot of $f_1(x)$, $f_2(x)$, $f_3(x)$, $f_4(x)$, $f_5(x)$ and $\sqrt{x}$ is shown below, with the graph of $\sqrt{x}$ shown as a dashed curve. It seems as if the f_n are asymptotic to $\sqrt{x}$.

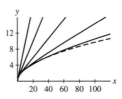

CHAPTER REVIEW EXERCISES

1. Express $(4, 10)$ as a set $\{x : |x - a| < c\}$ for suitable a and c.

SOLUTION The center of the interval $(4, 10)$ is $\frac{4+10}{2} = 7$ and the radius is $\frac{10-4}{2} = 3$. Therefore, the interval $(4, 10)$ is equivalent to the set $\{x : |x - 7| < 3\}$.

3. Express $\{x : 2 \leq |x - 1| \leq 6\}$ as a union of two intervals.

SOLUTION The set $\{x : 2 \leq |x - 1| \leq 6\}$ consists of those numbers that are at least 2 but at most 6 units from 1. The numbers larger than 1 that satisfy these conditions are $3 \leq x \leq 7$, while the numbers smaller than 1 that satisfy these conditions are $-5 \leq x \leq -1$. Therefore $\{x : 2 \leq |x - 1| \leq 6\} = [-5, -1] \cup [3, 7]$.

5. Describe the pairs of numbers x, y such that $|x + y| = x - y$.

SOLUTION First consider the case when $x + y \geq 0$. Then $|x + y| = x + y$ and we obtain the equation $x + y = x - y$. The solution of this equation is $y = 0$. Thus, the pairs $(x, 0)$ with $x \geq 0$ satisfy $|x + y| = x - y$. Next, consider the case when $x + y < 0$. Then $|x + y| = -(x + y) = -x - y$ and we obtain the equation $-x - y = x - y$. The solution of this equation is $x = 0$. Thus, the pairs $(0, y)$ with $y < 0$ also satisfy $|x + y| = x - y$.

In Exercises 7–10, let $f(x)$ be the function whose graph is shown in Figure 1.

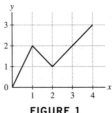

FIGURE 1

7. Sketch the graphs of $y = f(x) + 2$ and $y = f(x + 2)$.

SOLUTION The graph of $y = f(x) + 2$ is obtained by shifting the graph of $y = f(x)$ up 2 units (see the graph below at the left). The graph of $y = f(x + 2)$ is obtained by shifting the graph of $y = f(x)$ to the left 2 units (see the graph below at the right).

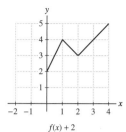

$f(x) + 2$

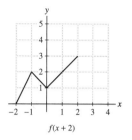

$f(x + 2)$

9. Continue the graph of $f(x)$ to the interval $[-4, 4]$ as an even function.

SOLUTION   To continue the graph of $f(x)$ to the interval $[-4, 4]$ as an even function, reflect the graph of $f(x)$ across the y-axis (see the graph below).

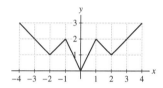

In Exercises 11–14, find the domain and range of the function.

11. $f(x) = \sqrt{x + 1}$

SOLUTION   The domain of the function $f(x) = \sqrt{x + 1}$ is $\{x : x \geq -1\}$ and the range is $\{y : y \geq 0\}$.

13. $f(x) = \dfrac{2}{3 - x}$

SOLUTION   The domain of the function $f(x) = \dfrac{2}{3 - x}$ is $\{x : x \neq 3\}$ and the range is $\{y : y \neq 0\}$.

15. Determine whether the function is increasing, decreasing, or neither:

(a) $f(x) = 3^{-x}$

(b) $f(x) = \dfrac{1}{x^2 + 1}$

(c) $g(t) = t^2 + t$

(d) $g(t) = t^3 + t$

SOLUTION

(a) The function $f(x) = 3^{-x}$ can be rewritten as $f(x) = (\frac{1}{3})^x$. This is an exponential function with a base less than 1; therefore, this is a decreasing function.

(b) From the graph of $y = 1/(x^2 + 1)$ shown below, we see that this function is neither increasing nor decreasing for all x (though it is increasing for $x < 0$ and decreasing for $x > 0$).

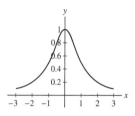

(c) The graph of $y = t^2 + t$ is an upward opening parabola; therefore, this function is neither increasing nor decreasing for all t. By completing the square we find $y = (t + \frac{1}{2})^2 - \frac{1}{4}$. The vertex of this parabola is then at $t = -\frac{1}{2}$, so the function is decreasing for $t < -\frac{1}{2}$ and increasing for $t > -\frac{1}{2}$.

(d) From the graph of $y = t^3 + t$ shown below, we see that this is an increasing function.

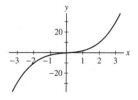

In Exercises 17–22, find the equation of the line.

17. Line passing through $(-1, 4)$ and $(2, 6)$

SOLUTION   The slope of the line passing through $(-1, 4)$ and $(2, 6)$ is

$$m = \frac{6-4}{2-(-1)} = \frac{2}{3}.$$

The equation of the line passing through $(-1, 4)$ and $(2, 6)$ is therefore $y - 4 = \frac{2}{3}(x + 1)$ or $2x - 3y = -14$.

19. Line of slope 6 through $(9, 1)$

SOLUTION   Using the point-slope form for the equation of a line, the equation of the line of slope 6 and passing through $(9, 1)$ is $y - 1 = 6(x - 9)$ or $6x - y = 53$.

21. Line through $(2, 3)$ parallel to $y = 4 - x$

SOLUTION   The equation $y = 4 - x$ is in slope-intercept form; it follows that the slope of this line is -1. Any line parallel to $y = 4 - x$ will have the same slope, so we are looking for the equation of the line of slope -1 and passing through $(2, 3)$. The equation of this line is $y - 3 = -(x - 2)$ or $x + y = 5$.

23. Does the following table of market data suggest a linear relationship between price and number of homes sold during a one-year period? Explain.

Price (thousands of $)	180	195	220	240
No. of homes sold	127	118	103	91

SOLUTION   Examine the slope between consecutive data points. The first pair of data points yields a slope of

$$\frac{118 - 127}{195 - 180} = -\frac{9}{15} = -\frac{3}{5},$$

while the second pair of data points yields a slope of

$$\frac{103 - 118}{220 - 195} = -\frac{15}{25} = -\frac{3}{5}$$

and the last pair of data points yields a slope of

$$\frac{91 - 103}{240 - 220} = -\frac{12}{20} = -\frac{3}{5}.$$

Because all three slopes are equal, the data does suggest a linear relationship between price and the number of homes sold.

25. Find the roots of $f(x) = x^4 - 4x^2$ and sketch its graph. On which intervals is $f(x)$ decreasing?

SOLUTION   The roots of $f(x) = x^4 - 4x^2$ are obtained by solving the equation $x^4 - 4x^2 = x^2(x - 2)(x + 2) = 0$, which yields $x = -2$, $x = 0$ and $x = 2$. The graph of $y = f(x)$ is shown below. From this graph we see that $f(x)$ is decreasing for x less than approximately -1.4 and for x between 0 and approximately 1.4.

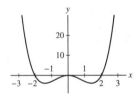

27. Let $f(x)$ be the square of the distance from the point $(2, 1)$ to a point $(x, 3x + 2)$ on the line $y = 3x + 2$. Show that $f(x)$ is a quadratic function and find its minimum value by completing the square.

SOLUTION   Let $f(x)$ denote the square of the distance from the point $(2, 1)$ to a point $(x, 3x + 2)$ on the line $y = 3x + 2$. Then

$$f(x) = (x - 2)^2 + (3x + 2 - 1)^2 = x^2 - 4x + 4 + 9x^2 + 6x + 1 = 10x^2 + 2x + 5,$$

which is a quadratic function. Completing the square, we find

$$f(x) = 10\left(x^2 + \frac{1}{5}x + \frac{1}{100}\right) + 5 - \frac{1}{10} = 10\left(x + \frac{1}{10}\right)^2 + \frac{49}{10}.$$

Because $(x + \frac{1}{10})^2 \geq 0$ for all x, it follows that $f(x) \geq \frac{49}{10}$ for all x. Hence, the minimum value of $f(x)$ is $\frac{49}{10}$.

In Exercises 29–34, sketch the graph by hand.

29. $y = t^4$

SOLUTION

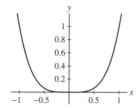

31. $y = \sin \dfrac{\theta}{2}$

SOLUTION

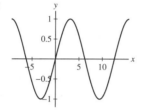

33. $y = x^{1/3}$

SOLUTION

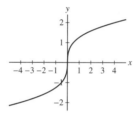

35. Show that the graph of $y = f\left(\frac{1}{3}x - b\right)$ is obtained by shifting the graph of $y = f\left(\frac{1}{3}x\right)$ to the right $3b$ units. Use this observation to sketch the graph of $y = \left|\frac{1}{3}x - 4\right|$.

SOLUTION   Let $g(x) = f\left(\frac{1}{3}x\right)$. Then

$$g(x - 3b) = f\left(\frac{1}{3}(x - 3b)\right) = f\left(\frac{1}{3}x - b\right).$$

Thus, the graph of $y = f\left(\frac{1}{3}x - b\right)$ is obtained by shifting the graph of $y = f\left(\frac{1}{3}x\right)$ to the right $3b$ units.

The graph of $y = \left|\frac{1}{3}x - 4\right|$ is the graph of $y = \left|\frac{1}{3}x\right|$ shifted right 12 units (see the graph below).

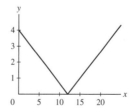

37. Find functions f and g such that the function

$$f(g(t)) = (12t + 9)^4$$

SOLUTION   One possible choice is $f(t) = t^4$ and $g(t) = 12t + 9$. Then

$$f(g(t)) = f(12t + 9) = (12t + 9)^4$$

as desired.

39. What is the period of the function $g(\theta) = \sin 2\theta + \sin \dfrac{\theta}{2}$?

SOLUTION   The function $\sin 2\theta$ has a period of π, and the function $\sin(\theta/2)$ has a period of 4π. Because 4π is a multiple of π, the period of the function $g(\theta) = \sin 2\theta + \sin \theta/2$ is 4π.

41. Give an example of values a, b such that

(a) $\cos(a + b) \neq \cos a + \cos b$ **(b)** $\cos \dfrac{a}{2} \neq \dfrac{\cos a}{2}$

SOLUTION

(a) Take $a = b = \pi/2$. Then $\cos(a + b) = \cos \pi = -1$ but

$$\cos a + \cos b = \cos \frac{\pi}{2} + \cos \frac{\pi}{2} = 0 + 0 = 0.$$

(b) Take $a = \pi$. Then

$$\cos\left(\frac{a}{2}\right) = \cos\left(\frac{\pi}{2}\right) = 0$$

but

$$\frac{\cos a}{2} = \frac{\cos \pi}{2} = \frac{-1}{2} = -\frac{1}{2}.$$

43. Solve $\sin 2x + \cos x = 0$ for $0 \leq x < 2\pi$.

SOLUTION   Using the double angle formula for the sine function, we rewrite the equation as $2 \sin x \cos x + \cos x = \cos x (2 \sin x + 1) = 0$. Thus, either $\cos x = 0$ or $\sin x = -1/2$. From here we see that the solutions are $x = \pi/2$, $x = 7\pi/6$, $x = 3\pi/2$ and $x = 11\pi/6$.

45. $\boxed{\text{GU}}$ Use a graphing calculator to determine whether the equation $\cos x = 5x^2 - 8x^4$ has any solutions.

SOLUTION   The graphs of $y = \cos x$ and $y = 5x^2 - 8x^4$ are shown below. Because the graphs do not intersect, there are no solutions to the equation $\cos x = 5x^2 - 8x^4$.

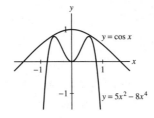

2 | LIMITS

2.1 Limits, Rates of Change, and Tangent Lines

Preliminary Questions

1. Average velocity is defined as a ratio of which two quantities?

SOLUTION Average velocity is defined as the ratio of distance traveled to time elapsed.

2. Average velocity is equal to the slope of a secant line through two points on a graph. Which graph?

SOLUTION Average velocity is the slope of a secant line through two points on the graph of position as a function of time.

3. Can instantaneous velocity be defined as a ratio? If not, how is instantaneous velocity computed?

SOLUTION Instantaneous velocity cannot be defined as a ratio. It is defined as the limit of average velocity as time elapsed shrinks to zero.

4. What is the graphical interpretation of instantaneous velocity at a moment $t = t_0$?

SOLUTION Instantaneous velocity at time $t = t_0$ is the slope of the line tangent to the graph of position as a function of time at $t = t_0$.

5. What is the graphical interpretation of the following statement: The average ROC approaches the instantaneous ROC as the interval $[x_0, x_1]$ shrinks to x_0?

SOLUTION The slope of the secant line over the interval $[x_0, x_1]$ approaches the slope of the tangent line at $x = x_0$.

6. The ROC of atmospheric temperature with respect to altitude is equal to the slope of the tangent line to a graph. Which graph? What are possible units for this rate?

SOLUTION The rate of change of atmospheric temperature with respect to altitude is the slope of the line tangent to the graph of atmospheric temperature as a function of altitude. Possible units for this rate of change are °F/ft or °C/m.

Exercises

1. A ball is dropped from a state of rest at time $t = 0$. The distance traveled after t seconds is $s(t) = 16t^2$ ft.

(a) How far does the ball travel during the time interval $[2, 2.5]$?

(b) Compute the average velocity over $[2, 2.5]$.

(c) Compute the average velocity over time intervals $[2, 2.01]$, $[2, 2.005]$, $[2, 2.001]$, $[2, 2.00001]$. Use this to estimate the object's instantaneous velocity at $t = 2$.

SOLUTION

(a) Galileo's formula is $s(t) = 16t^2$. The ball thus travels $\Delta s = s(2.5) - s(2) = 16(2.5)^2 - 16(2)^2 = 36$ ft.

(b) The average velocity over $[2, 2.5]$ is

$$\frac{\Delta s}{\Delta t} = \frac{s(2.5) - s(2)}{2.5 - 2} = \frac{36}{0.5} = 72 \text{ ft/s}.$$

(c)

time interval	[2, 2.01]	[2, 2.005]	[2, 2.001]	[2, 2.00001]
average velocity	64.16	64.08	64.016	64.00016

The instantaneous velocity at $t = 2$ is 64 ft/s.

3. Let $v = 20\sqrt{T}$ as in Example 2. Estimate the instantaneous ROC of v with respect to T when $T = 300$ K.

SOLUTION

T interval	[300, 300.01]	[300, 300.005]
average rate of change	0.577345	0.577348
T interval	[300, 300.001]	[300, 300.00001]
average ROC	0.57735	0.57735

The instantaneous rate of change is approximately 0.57735 m/(s · K).

In Exercises 5–6, a stone is tossed in the air from ground level with an initial velocity of 15 m/s. Its height at time t is $h(t) = 15t - 4.9t^2$ *m.*

5. Compute the stone's average velocity over the time interval [0.5, 2.5] and indicate the corresponding secant line on a sketch of the graph of $h(t)$.

SOLUTION The average velocity is equal to

$$\frac{h(2.5) - h(0.5)}{2} = 0.3.$$

The secant line is plotted with $h(t)$ below.

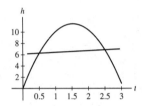

7. With an initial deposit of $100, the balance in a bank account after t years is $f(t) = 100(1.08)^t$ dollars.
(a) What are the units of the ROC of $f(t)$?
(b) Find the average ROC over [0, 0.5] and [0, 1].
(c) Estimate the instantaneous rate of change at $t = 0.5$ by computing the average ROC over intervals to the left and right of $t = 0.5$.

SOLUTION
(a) The units of the rate of change of $f(t)$ are dollars/year or $/yr.
(b) The average rate of change of $f(t) = 100(1.08)^t$ over the time interval $[t_1, t_2]$ is given by

$$\frac{\Delta f}{\Delta t} = \frac{f(t_2) - f(t_1)}{t_2 - t_1}.$$

time interval	[0, .5]	[0, 1]
average rate of change	7.8461	8

(c)

time interval	[.5, .51]	[.5, .501]	[.5, .5001]
average rate of change	8.0011	7.9983	7.9981
time interval	[.49, .5]	[.499, .5]	[.4999, .5]
average ROC	7.9949	7.9977	7.998

The rate of change at $t = 0.5$ is approximately $8/yr.

In Exercises 9–12, estimate the instantaneous rate of change at the point indicated.

9. $P(x) = 4x^2 - 3; \ x = 2$

SOLUTION

x interval	[2, 2.01]	[2, 2.001]	[2, 2.0001]	[1.99, 2]	[1.999, 2]	[1.9999, 2]
average rate of change	16.04	16.004	16.0004	15.96	15.996	15.9996

The rate of change at $x = 2$ is approximately 16.

11. $y(x) = \dfrac{1}{x + 2}; \ x = 2$

SOLUTION

x interval	[2, 2.01]	[2, 2.001]	[2, 2.0001]	[1.99, 2]	[1.999, 2]	[1.9999, 2]
average ROC	−.0623	−.0625	−.0625	−.0627	−.0625	−.0625

The rate of change at $x = 2$ is approximately -0.06.

13. The atmospheric temperature T (in °F) above a certain point on earth is $T = 59 - 0.00356h$, where h is the altitude in feet (valid for $h \le 37,000$). What are the average and instantaneous rates of change of T with respect to h? Why are they the same? Sketch the graph of T for $h \le 37,000$.

SOLUTION The average and instantaneous rates of change of T with respect to h are both -0.00356°F/ft. The rates of change are the same because T is a linear function of h with slope -0.00356.

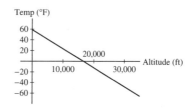

15. The number $P(t)$ of E. coli cells at time t (hours) in a petri dish is plotted in Figure 9.

(a) Calculate the average ROC of $P(t)$ over the time interval $[1, 3]$ and draw the corresponding secant line.

(b) Estimate the slope m of the line in Figure 9. What does m represent?

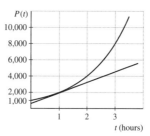

FIGURE 9 Number of E. coli cells at time t.

SOLUTION

(a) Looking at the graph, we can estimate $P(1) = 2000$ and $P(3) = 8000$. Assuming these values of $P(t)$, the average rate of change is

$$\frac{P(3) - P(1)}{3 - 1} = \frac{6000}{2} = 3000 \text{ cells/hour.}$$

The secant line is here:

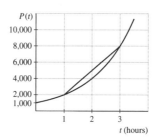

(b) The line in Figure 9 goes through two points with approximate coordinates $(1, 2000)$ and $(2.5, 4000)$. This line has approximate slope

$$m = \frac{4000 - 2000}{2.5 - 1} = \frac{4000}{3} \text{ cells/hour.}$$

m is close to the slope of the line tangent to the graph of $P(t)$ at $t = 1$, and so m represents the instantaneous rate of change of $P(t)$ at $t = 1$ hour.

17. Assume that the period T (in seconds) of a pendulum (the time required for a complete back-and-forth cycle) is $T = \frac{3}{2}\sqrt{L}$, where L is the pendulum's length (in meters).

(a) What are the units for the ROC of T with respect to L? Explain what this rate measures.

(b) Which quantities are represented by the slopes of lines A and B in Figure 10?

(c) Estimate the instantaneous ROC of T with respect to L when $L = 3$ m.

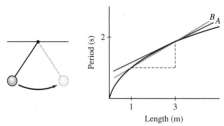

FIGURE 10 The period T is the time required for a pendulum to swing back and forth.

SOLUTION

(a) The units for the rate of change of T with respect to L are seconds per meter. This rate measures the sensitivity of the period of the pendulum to a change in the length of the pendulum.

(b) The slope of the line B represents the average rate of change in T from $L = 1$ m to $L = 3$ m. The slope of the line A represents the instantaneous rate of change of T at $L = 3$ m.

(c)

time interval	[3, 3.01]	[3, 3.001]	[3, 3.0001]	[2.99, 3]	[2.999, 3]	[2.9999, 3]
average velocity	0.4327	0.4330	0.4330	0.4334	0.4330	0.4330

The instantaneous rate of change at $L = 1$ m is approximately 0.4330 s/m.

19. The graphs in Figure 12 represent the positions s of moving particles as functions of time t. Match each graph with one of the following statements:

(a) Speeding up
(b) Speeding up and then slowing down
(c) Slowing down
(d) Slowing down and then speeding up

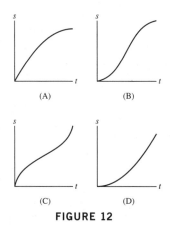

FIGURE 12

SOLUTION When a particle is speeding up over a time interval, its graph is bent upward over that interval. When a particle is slowing down, its graph is bent downward over that interval. Accordingly,

- In graph (A), the particle is (c) slowing down.
- In graph (B), the particle is (b) speeding up and then slowing down.
- In graph (C), the particle is (d) slowing down and then speeding up.
- In graph (D), the particle is (a) speeding up.

21. The fraction of a city's population infected by a flu virus is plotted as a function of time (in weeks) in Figure 14.

(a) Which quantities are represented by the slopes of lines A and B? Estimate these slopes.
(b) Is the flu spreading more rapidly at $t = 1, 2,$ or 3?
(c) Is the flu spreading more rapidly at $t = 4, 5,$ or 6?

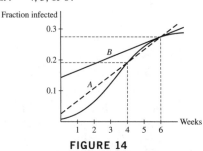

FIGURE 14

SOLUTION

(a) The slope of line A is the average rate of change over the interval $[4, 6]$, whereas the slope of the line B is the instantaneous rate of change at $t = 6$. Thus, the slope of the line $A \approx (0.28 - 0.19)/2 = 0.045/\text{week}$, whereas the slope of the line $B \approx (0.28 - 0.15)/6 = 0.0217/\text{week}$.

(b) Among times $t = 1, 2, 3$, the flu is spreading most rapidly at $t = 3$ since the slope is greatest at that instant; hence, the rate of change is greatest at that instant.

(c) Among times $t = 4, 5, 6$, the flu is spreading most rapidly at $t = 4$ since the slope is greatest at that instant; hence, the rate of change is greatest at that instant.

23. 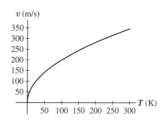 Let $v = 20\sqrt{T}$ as in Example 2. Is the ROC of v with respect to T greater at low temperatures or high temperatures? Explain in terms of the graph.

SOLUTION

As the graph progresses to the right, the graph bends progressively downward, meaning that the slope of the tangent lines becomes smaller. This means that the ROC of v with respect to T is lower at high temperatures.

25. 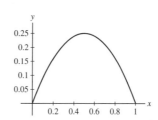 Sketch the graph of $f(x) = x(1 - x)$ over $[0, 1]$. Refer to the graph and, without making any computations, find:

(a) The average ROC over $[0, 1]$

(b) The (instantaneous) ROC at $x = \frac{1}{2}$

(c) The values of x at which the ROC is positive

SOLUTION

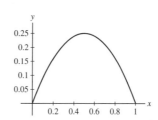

(a) $f(0) = f(1)$, so there is no change between $x = 0$ and $x = 1$. The average ROC is zero.

(b) The tangent line to the graph of $f(x)$ is horizontal at $x = \frac{1}{2}$; the instantaneous ROC is zero at this point.

(c) The ROC is positive at all points where the graph is rising, because the slope of the tangent line is positive at these points. This is so for all x between $x = 0$ and $x = 0.5$.

Further Insights and Challenges

27. The height of a projectile fired in the air vertically with initial velocity 64 ft/s is $h(t) = 64t - 16t^2$ ft.

(a) Compute $h(1)$. Show that $h(t) - h(1)$ can be factored with $(t - 1)$ as a factor.

(b) Using part (a), show that the average velocity over the interval $[1, t]$ is $-16(t - 3)$.

(c) Use this formula to find the average velocity over several intervals $[1, t]$ with t close to 1. Then estimate the instantaneous velocity at time $t = 1$.

SOLUTION

(a) With $h(t) = 64t - 16t^2$, we have $h(1) = 48$ ft, so

$$h(t) - h(1) = -16t^2 + 64t - 48.$$

Taking out the common factor of -16 and factoring the remaining quadratic, we get

$$h(t) - h(1) = -16(t^2 - 4t + 3) = -16(t - 1)(t - 3).$$

(b) The average velocity over the interval $[1, t]$ is

$$\frac{h(t) - h(1)}{t - 1} = \frac{-16(t - 1)(t - 3)}{t - 1} = -16(t - 3).$$

(c)

t	1.1	1.05	1.01	1.001
average velocity over $[1, t]$	30.4	31.2	31.84	31.984

The instantaneous velocity is approximately 32 ft/s. Plugging $t = 1$ second into the formula in (b) yields $-16(1 - 3) = 32$ ft/s exactly.

29. Show that the average ROC of $f(x) = x^3$ over $[1, x]$ is equal to $x^2 + x + 1$. Use this to estimate the instantaneous ROC of $f(x)$ at $x = 1$.

SOLUTION The average ROC is

$$\frac{f(x) - f(1)}{x - 1} = \frac{x^3 - 1}{x - 1}.$$

Factoring the numerator as the difference of cubes means the average rate of change is

$$\frac{(x - 1)(x^2 + x + 1)}{x - 1} = x^2 + x + 1$$

(for all $x \neq 1$). The closer x gets to 1, the closer the average ROC gets to $1^2 + 1 + 1 = 3$. The instantaneous ROC is 3.

31. Let $T = \frac{3}{2}\sqrt{L}$ as in Exercise 17. The numbers in the second column of Table 4 are increasing and those in the last column are decreasing. Explain why in terms of the graph of T as a function of L. Also, explain graphically why the instantaneous ROC at $L = 3$ lies between 0.4329 and 0.4331.

TABLE 4   Average Rates of Change of T with Respect to L

Interval	Average ROC	Interval	Average ROC
[3, 3.2]	0.42603	[2.8, 3]	0.44048
[3, 3.1]	0.42946	[2.9, 3]	0.43668
[3, 3.001]	0.43298	[2.999, 3]	0.43305
[3, 3.0005]	0.43299	[2.9995, 3]	0.43303

SOLUTION Since the average ROC is increasing on the intervals $[3, L]$ as L get close to 3, we know that the slopes of the secant lines between points on the graph over these intervals are increasing. The more rows we add with smaller intervals, the greater the average ROC. This means that the instantaneous ROC is probably greater than all of the numbers in this column.

Likewise, since the average ROC is *decreasing* on the intervals $[L, 3]$ as L gets closer to 3, we know that the slopes of the secant lines between points over these intervals are decreasing. This means that the instantaneous ROC is probably less than all of the numbers in this column.

The tangent slope is somewhere between the greatest value in the first column and the least value in the second column. Hence, it is between .43299 and .43303. The first column underestimates the instantaneous ROC by secant slopes; this estimate improves as L decreases toward $L = 3$. The second column overestimates the instantaneous ROC by secant slopes; this estimate improves as L increases toward $L = 3$.

2.2 Limits: A Numerical and Graphical Approach

Preliminary Questions

1. What is the limit of $f(x) = 1$ as $x \to \pi$?

SOLUTION $\lim_{x \to \pi} 1 = 1$.

2. What is the limit of $g(t) = t$ as $t \to \pi$?

SOLUTION $\lim_{t \to \pi} t = \pi$.

3. Can $f(x)$ approach a limit as $x \to c$ if $f(c)$ is undefined? If so, give an example.

SOLUTION Yes. The limit of a function f as $x \to c$ does not depend on what happens *at $x = c$*, only on the behavior of f as $x \to c$. As an example, consider the function

$$f(x) = \frac{x^2 - 1}{x - 1}.$$

The function is clearly not defined at $x = 1$ but

$$\lim_{x \to 1} f(x) = \lim_{x \to 1} \frac{x^2 - 1}{x - 1} = \lim_{x \to 1} (x + 1) = 2.$$

4. Is $\lim_{x \to 10} 20$ equal to 10 or 20?

SOLUTION $\lim_{x \to 10} 20 = 20$.

5. What does the following table suggest about $\lim_{x \to 1-} f(x)$ and $\lim_{x \to 1+} f(x)$?

x	0.9	0.99	0.999	1.1	1.01	1.001
$f(x)$	7	25	4317	3.0126	3.0047	3.00011

SOLUTION The values in the table suggest that $\lim_{x \to 1-} f(x) = \infty$ and $\lim_{x \to 1+} f(x) = 3$.

6. Is it possible to tell if $\lim_{x \to 5} f(x)$ exists by only examining values $f(x)$ for x close to but *greater* than 5? Explain.

SOLUTION No. By examining values of $f(x)$ for x close to but greater than 5, we can determine whether the one-sided limit $\lim_{x \to 5+} f(x)$ exists. To determine whether $\lim_{x \to 5} f(x)$ exists, we must examine value of $f(x)$ on both sides of $x = 5$.

7. If you know in advance that $\lim_{x \to 5} f(x)$ exists, can you determine its value just knowing the values of $f(x)$ for all $x > 5$?

SOLUTION Yes. If $\lim_{x \to 5} f(x)$ exists, then both one-sided limits must exist and be equal.

8. Which of the following pieces of information is sufficient to determine whether $\lim_{x \to 5} f(x)$ exists? Explain.

(a) The values of $f(x)$ for all x
(b) The values of $f(x)$ for x in $[4.5, 5.5]$
(c) The values of $f(x)$ for all x in $[4.5, 5.5]$ other than $x = 5$
(d) The values of $f(x)$ for all $x \geq 5$
(e) $f(5)$

SOLUTION To determine whether $\lim_{x \to 5} f(x)$ exists, we must know the values of $f(x)$ for values of x near 5, both smaller than and larger than 5. Thus, the information in (a), (b) or (c) would be sufficient. The information in (d) does not include values of f for $x < 5$, so this information is not sufficient to determine whether the limit exists. The limit does not depend at all on the value $f(5)$, so the information in (e) is also not sufficient to determine whether the limit exists.

Exercises

In Exercises 1–4, fill in the tables and guess the value of the limit.

1. $\lim_{x \to 1} f(x)$, where $f(x) = \dfrac{x^3 - 1}{x^2 - 1}$.

x	$f(x)$	x	$f(x)$
1.002		0.998	
1.001		0.999	
1.0005		0.9995	
1.00001		0.99999	

SOLUTION

x	0.998	0.999	0.9995	0.99999	1.00001	1.0005	1.001	1.002
$f(x)$	1.498501	1.499250	1.499625	1.499993	1.500008	1.500375	1.500750	1.501500

The limit as $x \to 1$ is $\frac{3}{2}$.

3. $\lim_{y \to 2} f(y)$, where $f(y) = \dfrac{y^2 - y - 2}{y^2 + y - 6}$.

y	$f(y)$	y	$f(y)$
2.002		1.998	
2.001		1.999	
2.0001		1.9999	

SOLUTION

y	1.998	1.999	1.9999	2.0001	2.001	2.02
$f(y)$	0.59984	0.59992	0.599992	0.600008	0.60008	0.601594

The limit as $y \to 2$ is $\frac{3}{5}$.

5. Determine $\lim_{x \to 0.5} f(x)$ for the function $f(x)$ shown in Figure 8.

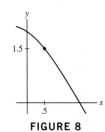

FIGURE 8

SOLUTION The graph suggests that $f(x) \to 1.5$ as $x \to .5$.

In Exercises 7–8, evaluate the limit.

7. $\lim_{x \to 21} x$

SOLUTION As $x \to 21$, $f(x) = x \to 21$. You can see this, for example, on the graph of $f(x) = x$.

In Exercises 9–18, verify each limit using the limit definition. For example, in Exercise 9, show that $|2x - 6|$ can be made as small as desired by taking x close to 3.

9. $\lim_{x \to 3} 2x = 6$

SOLUTION $|2x - 6| = 2|x - 3|$. $|2x - 6|$ can be made arbitrarily small by making x close enough to 3, thus making $|x - 3|$ small.

11. $\lim_{x \to 2} (4x + 3) = 11$

SOLUTION $|(4x + 3) - 11| = |4x - 8| = 4|x - 2|$. Therefore, if you make $|x - 2|$ small enough, you can make $|(4x + 3) - 11|$ as small as desired.

13. $\lim_{x \to 9} (-2x) = -18$

SOLUTION We have $|-2x - (-18)| = |(-2)(x - 9)| = 2|x - 9|$. If you make $|x - 9|$ small enough, you can make $2|x - 9| = |-2x - (-18)|$ as small as desired.

15. $\lim_{x \to 0} x^2 = 0$

SOLUTION As $x \to 0$, we have $|x^2 - 0| = |x + 0||x - 0|$. To simplify things, suppose that $|x| < 1$, so that $|x + 0||x - 0| = |x||x| < |x|$. By making $|x|$ sufficiently small, so that $|x + 0||x - 0| = x^2$ is even smaller, you can make $|x^2 - 0|$ as small as desired.

17. $\lim_{x \to 0} (x^2 + 2x + 3) = 3$

SOLUTION As $x \to 0$, we have $|x^2 + 2x + 3 - 3| = |x^2 + 2x| = |x||x + 2|$. If $|x| < 1$, $|x + 2|$ can be no bigger than 3, so $|x||x + 2| < 3|x|$. Therefore, by making $|x - 0| = |x|$ sufficiently small, you can make $|x^2 + 2x + 3 - 3| = |x||x + 2|$ as small as desired.

In Exercises 19–32, estimate the limit numerically or state that the limit does not exist.

19. $\lim\limits_{x \to 1} \dfrac{\sqrt{x} - 1}{x - 1}$

SOLUTION

x	.9995	.99999	1.00001	1.0005
$f(x)$	.500063	.500001	.49999	.499938

The limit as $x \to 1$ is $\frac{1}{2}$.

21. $\lim\limits_{x \to 2} \dfrac{x^2 + x - 6}{x^2 - x - 2}$

SOLUTION

x	1.999	1.99999	2.00001	2.001
$f(x)$	1.666889	1.666669	1.666664	1.666445

The limit as $x \to 2$ is $\frac{5}{3}$.

23. $\lim\limits_{x \to 0} \dfrac{\sin 2x}{x}$

SOLUTION

x	-0.01	-0.005	0.005	0.01
$f(x)$	1.999867	1.999967	1.999967	1.999867

The limit as $x \to 0$ is 2.

25. $\lim\limits_{x \to 0} \dfrac{\sin x}{x^2}$

SOLUTION

x	$-.01$	$-.001$	$-.0001$	.0001	.001	.01
$f(x)$	-99.9983	-999.9998	-10000.0	10000.0	999.9998	99.9983

The limit does not exist. As $x \to 0-$, $f(x) \to -\infty$; similarly, as $x \to 0+$, $f(x) \to \infty$.

27. $\lim\limits_{h \to 0} \cos \dfrac{1}{h}$

SOLUTION

h	± 0.1	± 0.01	± 0.001	± 0.0001
$f(h)$	-0.839072	0.862319	0.562379	-0.952155

The limit does not exist since $\cos(1/h)$ oscillates infinitely often as $h \to 0$.

29. $\lim\limits_{h \to 0} \dfrac{2^h - 1}{h}$

SOLUTION

h	$-.05$	$-.001$	.001	.05
$f(h)$	.681273	.692907	.693387	.705298

The limit as $x \to 0$ is approximately 0.693. (The exact answer is $\ln 2$.)

31. $\lim\limits_{h \to 2} \dfrac{5^h - 25}{h - 2}$

SOLUTION

x	1.95	1.999	2.001	2.05
$f(x)$	38.6596	40.2036	40.2683	41.8992

The limit as $h \to 2$ is approximately 40.2. (The exact answer is $25 \ln 5$.)

33. Determine $\lim\limits_{x \to 2+} f(x)$ and $\lim\limits_{x \to 2-} f(x)$ for the function shown in Figure 10.

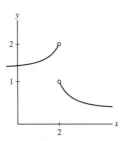

FIGURE 10

SOLUTION The left-hand limit is $\lim\limits_{x \to 2-} f(x) = 2$, whereas the right-hand limit is $\lim\limits_{x \to 2+} f(x) = 1$. Accordingly, the two-sided limit does not exist.

35. The greatest integer function is defined by $[x] = n$, where n is the unique integer such that $n \leq x < n + 1$. See Figure 12.

(a) For which values of c does $\lim\limits_{x \to c-} [x]$ exist? What about $\lim\limits_{x \to c+} [x]$?

(b) For which values of c does $\lim\limits_{x \to c} [x]$ exist?

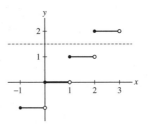

FIGURE 12 Graph of $y = [x]$.

SOLUTION

(a) The one-sided limits exist for all real values of c.

(b) For each integer value of c, the one-sided limits differ. In particular, $\lim\limits_{x \to c-} [x] = c - 1$, whereas $\lim\limits_{x \to c+} [x] = c$. (For noninteger values of c, the one-sided limits both equal $[c]$.) The limit $\lim\limits_{x \to c} [x]$ exists when

$$\lim_{x \to c-} [x] = \lim_{x \to c+} [x],$$

namely for noninteger values of c: $n < c < n + 1$, where n is an integer.

In Exercises 37–39, determine the one-sided limits numerically.

37. $\lim\limits_{x \to 0\pm} \dfrac{\sin x}{|x|}$

SOLUTION

x	$-.2$	$-.02$	.02	.2
$f(x)$	$-.993347$	$-.999933$	.999933	.993347

The left-hand limit is $\lim\limits_{x \to 0-} f(x) = -1$, whereas the right-hand limit is $\lim\limits_{x \to 0+} f(x) = 1$.

39. $\lim\limits_{x \to 0\pm} \dfrac{x - \sin(|x|)}{x^3}$

SOLUTION

x	$-.1$	$-.01$	.01	.1
$f(x)$	199.853	19999.8	.166666	.166583

The left-hand limit is $\lim\limits_{x \to 0-} f(x) = \infty$, whereas the right-hand limit is $\lim\limits_{x \to 0+} f(x) = \dfrac{1}{6}$.

41. Determine the one-sided limits of $f(x)$ at $c = 2$ and $c = 4$, for the function shown in Figure 14.

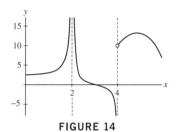

FIGURE 14

SOLUTION

- For $c = 2$, we have $\lim\limits_{x \to 2-} f(x) = \infty$ and $\lim\limits_{x \to 2+} f(x) = \infty$.
- For $c = 4$, we have $\lim\limits_{x \to 4-} f(x) = -\infty$ and $\lim\limits_{x \to 4+} f(x) = 10$.

In Exercises 43–46, draw the graph of a function with the given limits.

43. $\lim\limits_{x \to 1} f(x) = 2, \ \lim\limits_{x \to 3-} f(x) = 0, \ \lim\limits_{x \to 3+} f(x) = 4$

SOLUTION

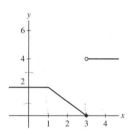

45. $\lim\limits_{x \to 2+} f(x) = f(2) = 3, \ \lim\limits_{x \to 2-} f(x) = -1, \ \lim\limits_{x \to 4} f(x) = 2 \neq f(4)$

SOLUTION

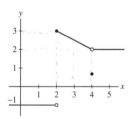

[GU] *In Exercises 47–52, graph the function and use the graph to estimate the value of the limit.*

47. $\lim\limits_{\theta \to 0} \dfrac{\sin 3\theta}{\sin 2\theta}$

SOLUTION

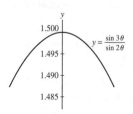

$y = \dfrac{\sin 3\theta}{\sin 2\theta}$

The limit as $\theta \to 0$ is $\dfrac{3}{2}$.

49. $\lim\limits_{x\to 0} \dfrac{2^x - \cos x}{x}$

SOLUTION

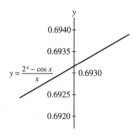

The limit as $x \to 0$ is approximately 0.693. (The exact answer is $\ln 2$.)

51. $\lim\limits_{\theta\to 0} \dfrac{\cos 3\theta - \cos 4\theta}{\theta^2}$

SOLUTION

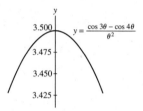

The limit as $\theta \to 0$ is 3.5.

Further Insights and Challenges

53. Light waves of frequency λ passing through a slit of width a produce a **Fraunhofer diffraction pattern** of light and dark fringes (Figure 16). The intensity as a function of the angle θ is given by

$$I(\theta) = I_m \left(\frac{\sin(R \sin \theta)}{R \sin \theta} \right)^2$$

where $R = \pi a / \lambda$ and I_m is a constant. Show that the intensity function is not defined at $\theta = 0$. Then check numerically that $I(\theta)$ approaches I_m as $\theta \to 0$ for any two values of R (e.g., choose two integer values).

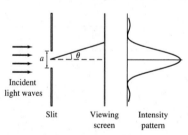

FIGURE 16 Fraunhofer diffraction pattern.

SOLUTION If you plug in $\theta = 0$, you get a division by zero in the expression

$$\frac{\sin\left(R \sin \theta\right)}{R \sin \theta};$$

thus, $I(0)$ is undefined. If $R = 2$, a table of values as $\theta \to 0$ follows:

θ	-0.01	-0.005	0.005	0.01
$I(\theta)$	$0.998667\ I_m$	$0.9999667\ I_m$	$0.9999667\ I_m$	$0.9998667\ I_m$

The limit as $\theta \to 0$ is $1 \cdot I_m = I_m$.
    If $R = 3$, the table becomes:

θ	-0.01	-0.005	0.005	0.01
$I(\theta)$	$0.999700\ I_m$	$0.999925\ I_m$	$0.999925\ I_m$	$0.999700\ I_m$

Again, the limit as $\theta \to 0$ is $1 I_m = I_m$.

55. Show numerically that $\lim\limits_{x\to 0} \dfrac{b^x - 1}{x}$ for $b = 3, 5$ appears to equal $\ln 3, \ln 5$, where $\ln x$ is the natural logarithm. Then make a conjecture (guess) for the value in general and test your conjecture for two additional values of b (these results are explained in Chapter 7).

SOLUTION

-

x	$-.1$	$-.01$	$-.001$	$.001$	$.01$	$.1$
$\dfrac{5^x - 1}{x}$	1.486601	1.596556	1.608144	1.610734	1.622459	1.746189

We have $\ln 5 \approx 1.6094$.

-

x	$-.1$	$-.01$	$-.001$	$.001$	$.01$	$.1$
$\dfrac{3^x - 1}{x}$	1.040415	1.092600	1.098009	1.099216	1.104669	1.161232

We have $\ln 3 \approx 1.0986$.

- We conjecture that $\lim\limits_{x\to 0} \dfrac{b^x - 1}{x} = \ln b$ for any positive number b. Here are two additional test cases.

x	$-.1$	$-.01$	$-.001$	$.001$	$.01$	$.1$
$\dfrac{\left(\frac{1}{2}\right)^x - 1}{x}$	$-.717735$	$-.695555$	$-.693387$	$-.692907$	$-.690750$	$-.669670$

We have $\ln \frac{1}{2} \approx -0.69315$.

x	$-.1$	$-.01$	$-.001$	$.001$	$.01$	$.1$
$\dfrac{7^x - 1}{x}$	1.768287	1.927100	1.944018	1.947805	1.964966	2.148140

We have $\ln 7 \approx 1.9459$.

57. Investigate $\lim\limits_{x\to 1} \dfrac{x^n - 1}{x^m - 1}$ for (m, n) equal to $(2, 1)$, $(1, 2)$, $(2, 3)$, and $(3, 2)$. Then guess the value of the limit in general and check your guess for at least three additional pairs.

SOLUTION

-

x	$.99$	$.9999$	1.0001	1.01
$\dfrac{x - 1}{x^2 - 1}$	.502513	.500025	.499975	.497512

The limit as $x \to 1$ is $\frac{1}{2}$.

x	$.99$	$.9999$	1.0001	1.01
$\dfrac{x^2 - 1}{x - 1}$	1.99	1.9999	2.0001	2.01

The limit as $x \to 1$ is 2.

x	$.99$	$.9999$	1.0001	1.01
$\dfrac{x^2 - 1}{x^3 - 1}$	0.670011	0.666700	0.666633	0.663344

The limit as $x \to 1$ is $\frac{2}{3}$.

x	.99	.9999	1.0001	1.01
$\dfrac{x^3 - 1}{x^2 - 1}$	1.492513	1.499925	1.500075	1.507512

The limit as $x \to 1$ is $\frac{3}{2}$.

- For general m and n, we have $\lim\limits_{x \to 1} \dfrac{x^n - 1}{x^m - 1} = \dfrac{n}{m}$.

-

x	.99	.9999	1.0001	1.01
$\dfrac{x - 1}{x^3 - 1}$	.336689	.333367	.333300	.330022

The limit as $x \to 1$ is $\frac{1}{3}$.

x	.99	.9999	1.0001	1.01
$\dfrac{x^3 - 1}{x - 1}$	2.9701	2.9997	3.0003	3.0301

The limit as $x \to 1$ is 3.

x	.99	.9999	1.0001	1.01
$\dfrac{x^3 - 1}{x^7 - 1}$	.437200	.428657	.428486	.420058

The limit as $x \to 1$ is $\frac{3}{7} \approx 0.428571$.

59. Sketch a graph of $f(x) = \dfrac{2^x - 8}{x - 3}$ with a graphing calculator. Observe that $f(3)$ is not defined.

(a) Zoom in on the graph to estimate $L = \lim\limits_{x \to 3} f(x)$.

(b) Observe that the graph of $f(x)$ is increasing. Explain how this implies that

$$f(2.99999) \le L \le f(3.00001)$$

Use this to determine L to three decimal places.

SOLUTION

(a)

(b) It is clear that the graph of f rises as we move to the right. Mathematically, we may express this observation as: whenever $u < v$, $f(u) < f(v)$. Because

$$2.99999 < 3 = \lim\limits_{x \to 3} f(x) < 3.00001,$$

it follows that

$$f(2.99999) < L = \lim\limits_{x \to 3} f(x) < f(3.00001).$$

With $f(2.99999) \approx 5.54516$ and $f(3.00001) \approx 5.545195$, the above inequality becomes $5.54516 < L < 5.545195$; hence, to three decimal places, $L = 5.545$.

61. 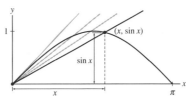 Show that $f(x) = \dfrac{\sin x}{x}$ is equal to the slope of a secant line through the origin and the point $(x, \sin x)$ on the graph of $y = \sin x$ (Figure 17). Use this to give a geometric interpretation of $\lim_{x \to 0} f(x)$.

FIGURE 17 Graph of $y = \sin x$.

SOLUTION The slope of a secant line through the points $(0, 0)$ and $(x, \sin x)$ is given by $\dfrac{\sin x - 0}{x - 0} = \dfrac{\sin x}{x}$. Since $f(x)$ is the slope of the secant line, the limit $\lim_{x \to 0} f(x)$ is equal to the slope of the tangent line to $y = \sin x$ at $x = 0$.

2.3 Basic Limit Laws

Preliminary Questions

1. State the Sum Law and Quotient Law.

SOLUTION Suppose $\lim_{x \to c} f(x)$ and $\lim_{x \to c} g(x)$ both exist. The Sum Law states that

$$\lim_{x \to c} (f(x) + g(x)) = \lim_{x \to c} f(x) + \lim_{x \to c} g(x).$$

Provided $\lim_{x \to c} g(x) \neq 0$, the Quotient Law states that

$$\lim_{x \to c} \frac{f(x)}{g(x)} = \frac{\lim_{x \to c} f(x)}{\lim_{x \to c} g(x)}.$$

2. Which of the following is a verbal version of the Product Law?
(a) The product of two functions has a limit.
(b) The limit of the product is the product of the limits.
(c) The product of a limit is a product of functions.
(d) A limit produces a product of functions.

SOLUTION The verbal version of the Product Law is **(b)**: The limit of the product is the product of the limits.

3. Which of the following statements are incorrect (k and c are constants)?
(a) $\lim_{x \to c} k = c$
(b) $\lim_{x \to c} k = k$
(c) $\lim_{x \to c^2} x = c^2$
(d) $\lim_{x \to c} x = x$

SOLUTION Statements **(a)** and **(d)** are incorrect. Because k is constant, statement **(a)** should read $\lim_{x \to c} k = k$. Statement **(d)** should be $\lim_{x \to c} x = c$.

4. Which of the following statements are incorrect?
(a) The Product Law does not hold if the limit of one of the functions is zero.
(b) The Quotient Law does not hold if the limit of the denominator is zero.
(c) The Quotient Law does not hold if the limit of the numerator is zero.

SOLUTION Statements **(a)** and **(c)** are incorrect. The Product Law remains valid when the limit of one or both of the functions is zero, and the Quotient Law remains valid when the limit of the numerator is zero.

Exercises

In Exercises 1–22, evaluate the limits using the Limit Laws and the following two facts, where c and k are constants:

$$\lim_{x \to c} x = c, \qquad \lim_{x \to c} k = k$$

1. $\lim_{x \to 9} x$

SOLUTION $\lim_{x \to 9} x = 9.$

3. $\lim_{x \to 9} 14$

SOLUTION $\lim_{x \to 9} 14 = 14.$

5. $\lim_{x \to -3} (3x + 4)$

SOLUTION We apply the Laws for Sums, Products, and Constants:

$$\lim_{x \to -3} (3x + 4) = \lim_{x \to -3} 3x + \lim_{x \to -3} 4$$
$$= 3 \lim_{x \to -3} x + \lim_{x \to -3} 4 = 3(-3) + 4 = -5.$$

7. $\lim_{y \to -3} (y + 14)$

SOLUTION $\lim_{y \to -3} (y + 14) = \lim_{y \to -3} y + \lim_{y \to -3} 14 = -3 + 14 = 11.$

9. $\lim_{t \to 4} (3t - 14)$

SOLUTION $\lim_{t \to 4} (3t - 14) = 3 \lim_{t \to 4} t - \lim_{t \to 4} 14 = 3 \cdot 4 - 14 = -2.$

11. $\lim_{x \to \frac{1}{2}} (4x + 1)(2x - 1)$

SOLUTION

$$\lim_{x \to 1/2} (4x + 1)(2x - 1) = \left(4 \lim_{x \to 1/2} x + \lim_{x \to 1/2} 1 \right) \left(2 \lim_{x \to 1/2} x - \lim_{x \to 1/2} 1 \right)$$
$$= \left(4 \left(\frac{1}{2} \right) + 1 \right) \left(2 \left(\frac{1}{2} \right) - 1 \right) = 3 \cdot 0 = 0.$$

13. $\lim_{x \to 2} x(x + 1)(x + 2)$

SOLUTION We apply the Product Law and Sum Law:

$$\lim_{x \to 2} x(x + 1)(x + 2) = \left(\lim_{x \to 2} x \right) \left(\lim_{x \to 2} (x + 1) \right) \left(\lim_{x \to 2} (x + 2) \right)$$
$$= 2 \left(\lim_{x \to 2} x + \lim_{x \to 2} 1 \right) \left(\lim_{x \to 2} x + \lim_{x \to 2} 2 \right)$$
$$= 2(2 + 1)(2 + 2) = 24$$

15. $\lim_{t \to 9} \dfrac{t}{t + 1}$

SOLUTION $\lim_{t \to 9} \dfrac{t}{t + 1} = \dfrac{\lim_{t \to 9} t}{\lim_{t \to 9} t + \lim_{t \to 9} 1} = \dfrac{9}{9 + 1} = \dfrac{9}{10}.$

17. $\lim_{x \to 3} \dfrac{1 - x}{1 + x}$

SOLUTION $\lim_{x \to 3} \dfrac{1 - x}{1 + x} = \dfrac{\lim_{x \to 3} 1 - \lim_{x \to 3} x}{\lim_{x \to 3} 1 + \lim_{x \to 3} x} = \dfrac{1 - 3}{1 + 3} = \dfrac{-2}{4} = -\dfrac{1}{2}.$

19. $\lim_{t \to 2} t^{-1}$

SOLUTION We apply the definition of t^{-1}, and then the Quotient Law.

$$\lim_{t \to 2} t^{-1} = \lim_{t \to 2} \frac{1}{t} = \frac{\lim_{t \to 2} 1}{\lim_{t \to 2} t} = \frac{1}{2}.$$

21. $\lim_{x \to 3} (x^2 + 9x^{-3})$

SOLUTION We apply the Sum, Product, and Quotient Laws. The Product Law is applied to the exponentiations $x^2 = x \cdot x$ and $x^3 = x \cdot x \cdot x$.

$$\lim_{x \to 3} (x^2 + 9x^{-3}) = \lim_{x \to 3} x^2 + \lim_{x \to 3} 9x^{-3} = \left(\lim_{x \to 3} x \right)^2 + 9 \left(\lim_{x \to 3} \frac{1}{x^3} \right)$$

$$= 9 + 9 \left(\frac{\lim_{x \to 3} 1}{(\lim_{x \to 3} x)^3} \right) = 9 + 9 \left(\frac{1}{27} \right) = \frac{28}{3}.$$

23. Use the Quotient Law to prove that if $\lim_{x \to c} f(x)$ exists and is nonzero, then

$$\lim_{x \to c} \frac{1}{f(x)} = \frac{1}{\lim_{x \to c} f(x)}$$

SOLUTION Since $\lim_{x \to c} f(x)$ is nonzero, we can apply the Quotient Law:

$$\lim_{x \to c} \left(\frac{1}{f(x)} \right) = \frac{\left(\lim_{x \to c} 1 \right)}{\left(\lim_{x \to c} f(x) \right)} = \frac{1}{\lim_{x \to c} f(x)}.$$

In Exercises 25–28, evaluate the limit assuming that $\lim_{x \to -4} f(x) = 3$ *and* $\lim_{x \to -4} g(x) = 1$.

25. $\lim_{x \to -4} f(x)g(x)$

SOLUTION $\lim_{x \to -4} f(x)g(x) = \lim_{x \to -4} f(x) \lim_{x \to -4} g(x) = 3 \cdot 1 = 3.$

27. $\lim_{x \to -4} \frac{g(x)}{x^2}$

SOLUTION Since $\lim_{x \to -4} x^2 \neq 0$, we may apply the Quotient Law, then applying the Product Law (from $x^2 = x \cdot x$):

$$\lim_{x \to -4} \frac{g(x)}{x^2} = \frac{\lim_{x \to -4} g(x)}{\lim_{x \to -4} x^2} = \frac{1}{\left(\lim_{x \to -4} x \right)^2} = \frac{1}{16}.$$

29. Can the Quotient Law be applied to evaluate $\lim_{x \to 0} \frac{\sin x}{x}$? Explain.

SOLUTION The limit Quotient Law *cannot* be applied to evaluate $\lim_{x \to 0} \frac{\sin x}{x}$ since $\lim_{x \to 0} x = 0$. This violates a condition of the Quotient Law. Accordingly, the rule *cannot* be employed.

31. Give an example where $\lim_{x \to 0} (f(x) + g(x))$ exists but neither $\lim_{x \to 0} f(x)$ nor $\lim_{x \to 0} g(x)$ exists.

SOLUTION Let $f(x) = 1/x$ and $g(x) = -1/x$. Then $\lim_{x \to 0} (f(x) + g(x)) = \lim_{x \to 0} 0 = 0$ However, $\lim_{x \to 0} f(x) = \lim_{x \to 0} 1/x$ and $\lim_{x \to 0} g(x) = \lim_{x \to 0} -1/x$ do not exist.

33. 📖 Use the Limit Laws and the result $\lim_{x \to c} x = c$ to show that $\lim_{x \to c} x^n = c^n$ for all whole numbers n. If you are familiar with induction, give a formal proof by induction.

SOLUTION Correct answers can vary. An example is given:

Let $P[n]$ be the proposition : $\lim_{x \to c} x^n = c^n$, and proceed by induction.

$P[1]$ is true, as $\lim_{x \to c} x = c$. Suppose that $P[n]$ is true, so that $\lim_{x \to c} x^n = c^n$. We must prove $P[n + 1]$, that is, that $\lim_{x \to c} x^{n+1} = c^{n+1}$.

Applying the Product Law and $P[n]$, we see:

$$\lim_{x \to c} x^{n+1} = \lim_{x \to c} x^n \cdot x = \left(\lim_{x \to c} x^n \right) \left(\lim_{x \to c} x \right) = c^n c = c^{n+1}.$$

Therefore, $P[n]$ true implies that $P[n + 1]$ true. By induction, the limit is equal to c^n for all n.

Further Insights and Challenges

35. Show that if both $\lim\limits_{x \to c} f(x)\,g(x)$ and $\lim\limits_{x \to c} g(x)$ exist and $\lim\limits_{x \to c} g(x)$ is nonzero, then $\lim\limits_{x \to c} f(x)$ exists. *Hint:* Write $f(x) = (f(x)\,g(x))/g(x)$ and apply the Quotient Law.

SOLUTION   Given that $\lim\limits_{x \to c} f(x)g(x) = L$ and $\lim\limits_{x \to c} g(x) = M \neq 0$ both exist, observe that

$$\lim_{x \to c} f(x) = \lim_{x \to c} \frac{f(x)g(x)}{g(x)} = \frac{\lim\limits_{x \to c} f(x)g(x)}{\lim\limits_{x \to c} g(x)} = \frac{L}{M}$$

also exists.

37. Prove that if $\lim\limits_{t \to 3} \dfrac{h(t)}{t} = 5$, then $\lim\limits_{t \to 3} h(t) = 15$.

SOLUTION   Given that $\lim\limits_{t \to 3} \dfrac{h(t)}{t} = 5$, observe that $\lim\limits_{t \to 3} t = 3$. Now use the Product Law:

$$\lim_{t \to 3} h(t) = \lim_{t \to 3} t \,\frac{h(t)}{t} = \left(\lim_{t \to 3} t \right) \left(\lim_{t \to 3} \frac{h(t)}{t} \right) = 3 \cdot 5 = 15.$$

39. Prove that if $\lim\limits_{x \to c} f(x) = L \neq 0$ and $\lim\limits_{x \to c} g(x) = 0$, then $\lim\limits_{x \to c} \dfrac{f(x)}{g(x)}$ does not exist.

SOLUTION   Suppose that $\lim\limits_{x \to c} \dfrac{f(x)}{g(x)}$ exists. Then

$$L = \lim_{x \to c} f(x) = \lim_{x \to c} g(x) \cdot \frac{f(x)}{g(x)} = \lim_{x \to c} g(x) \cdot \lim_{x \to c} \frac{f(x)}{g(x)} = 0 \cdot \lim_{x \to c} \frac{f(x)}{g(x)} = 0.$$

But, we were given that $L \neq 0$, so we have arrived at a contradiction. Thus, $\lim\limits_{x \to c} \dfrac{f(x)}{g(x)}$ does not exist.

2.4 Limits and Continuity

Preliminary Questions

1. Which property of $f(x) = x^3$ allows us to conclude that $\lim\limits_{x \to 2} x^3 = 8$?

SOLUTION   We can conclude that $\lim_{x \to 2} x^3 = 8$ because the function x^3 is continuous at $x = 2$.

2. What can be said about $f(3)$ if f is continuous and $\lim\limits_{x \to 3} f(x) = \frac{1}{2}$?

SOLUTION   If f is continuous and $\lim_{x \to 3} f(x) = \frac{1}{2}$, then $f(3) = \frac{1}{2}$.

3. Suppose that $f(x) < 0$ if x is positive and $f(x) > 1$ if x is negative. Can f be continuous at $x = 0$?

SOLUTION   Since $f(x) < 0$ when x is positive and $f(x) > 1$ when x is negative, it follows that

$$\lim_{x \to 0+} f(x) \leq 0 \quad \text{and} \quad \lim_{x \to 0-} f(x) \geq 1.$$

Thus, $\lim_{x \to 0} f(x)$ does not exist, so f cannot be continuous at $x = 0$.

4. Is it possible to determine $f(7)$ if $f(x) = 3$ for all $x < 7$ and f is right-continuous at $x = 7$?

SOLUTION   No. To determine $f(7)$, we need to combine either knowledge of the values of $f(x)$ for $x < 7$ with *left*-continuity or knowledge of the values of $f(x)$ for $x > 7$ with right-continuity.

5. Are the following true or false? If false, state a correct version.

(a) $f(x)$ is continuous at $x = a$ if the left- and right-hand limits of $f(x)$ as $x \to a$ exist and are equal.

(b) $f(x)$ is continuous at $x = a$ if the left- and right-hand limits of $f(x)$ as $x \to a$ exist and equal $f(a)$.

(c) If the left- and right-hand limits of $f(x)$ as $x \to a$ exist, then f has a removable discontinuity at $x = a$.

(d) If $f(x)$ and $g(x)$ are continuous at $x = a$, then $f(x) + g(x)$ is continuous at $x = a$.

(e) If $f(x)$ and $g(x)$ are continuous at $x = a$, then $f(x)/g(x)$ is continuous at $x = a$.

SOLUTION

(a) False. The correct statement is "$f(x)$ is continuous at $x = a$ if the left- and right-hand limits of $f(x)$ as $x \to a$ exist and equal $f(a)$."

(b) True.

(c) False. The correct statement is "If the left- and right-hand limits of $f(x)$ as $x \to a$ are equal but not equal to $f(a)$, then f has a removable discontinuity at $x = a$."

(d) True.

(e) False. The correct statement is "If $f(x)$ and $g(x)$ are continuous at $x = a$ and $g(a) \neq 0$, then $f(x)/g(x)$ is continuous at $x = a$."

Exercises

1. Find the points of discontinuity of the function shown in Figure 14 and state whether it is left- or right-continuous (or neither) at these points.

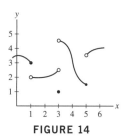

FIGURE 14

SOLUTION

- The function f is discontinuous at $x = 1$; it is left-continuous there.
- The function f is discontinuous at $x = 3$; it is neither left-continuous nor right-continuous there.
- The function f is discontinuous at $x = 5$; it is left-continuous there.

In Exercises 2–4, refer to the function $f(x)$ in Figure 15.

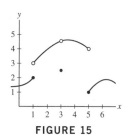

FIGURE 15

3. At which point c does $f(x)$ have a removable discontinuity? What value should be assigned to $f(c)$ to make f continuous at $x = c$?

SOLUTION Because $\lim_{x \to 3} f(x)$ exists, the function f has a removable discontinuity at $x = 3$. Assigning $f(3) = 4.5$ makes f continuous at $x = 3$.

5. (a) For the function shown in Figure 16, determine the one-sided limits at the points of discontinuity.

(b) Which of these discontinuities is removable and how should f be redefined to make it continuous at this point?

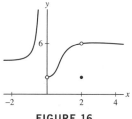

FIGURE 16

SOLUTION

(a) The function f is discontinuous at $x = 0$, at which $\lim_{x \to 0-} f(x) = \infty$ and $\lim_{x \to 0+} f(x) = 2$. The function f is also discontinuous at $x = 2$, at which $\lim_{x \to 2-} f(x) = 6$ and $\lim_{x \to 2+} f(x) = 6$.

(b) The discontinuity at $x = 2$ is removable. Assigning $f(2) = 6$ makes f continuous at $x = 2$.

In Exercises 7–14, use the Laws of Continuity and Theorems 2–3 to show that the function is continuous.

7. $f(x) = x + \sin x$

SOLUTION Since x and $\sin x$ are continuous, so is $x + \sin x$ by Continuity Law (i).

9. $f(x) = 3x + 4\sin x$

SOLUTION Since x and $\sin x$ are continuous, so are $3x$ and $4\sin x$ by Continuity Law (iii). Thus $3x + 4\sin x$ is continuous by Continuity Law (i).

11. $f(x) = \dfrac{1}{x^2 + 1}$

SOLUTION

- Since x is continuous, so is x^2 by Continuity Law (ii).
- Recall that constant functions, such as 1, are continuous. Thus $x^2 + 1$ is continuous.
- Finally, $\dfrac{1}{x^2 + 1}$ is continuous by Continuity Law (iv) because $x^2 + 1$ is never 0.

13. $f(x) = \dfrac{3^x}{1 + 4^x}$

SOLUTION The functions 3^x, 1 and 4^x are each continuous. Therefore, $1 + 4^x$ is continuous by Continuity Law (i). Because $1 + 4^x$ is never zero, it follows that $\dfrac{3^x}{1 + 4^x}$ is continuous by Continuity Law (iv).

In Exercises 15–32, determine the points at which the function is discontinuous and state the type of discontinuity: removable, jump, infinite, or none of these.

15. $f(x) = \dfrac{1}{x}$

SOLUTION The function $1/x$ is discontinuous at $x = 0$, at which there is an infinite discontinuity.

17. $f(x) = \dfrac{x - 2}{|x - 1|}$

SOLUTION The function $\dfrac{x - 2}{|x - 1|}$ is discontinuous at $x = 1$, at which there is an infinite discontinuity.

19. $f(x) = [x]$

SOLUTION This function has a jump discontinuity at $x = n$ for every integer n. It is continuous at all other values of x. For every integer n,

$$\lim_{x \to n+} [x] = n$$

since $[x] = n$ for all x between n and $n + 1$. This shows that $[x]$ is *right-continuous* at $x = n$. On the other hand,

$$\lim_{x \to n-} [x] = n - 1$$

since $[x] = n - 1$ for all x between $n - 1$ and n. Thus $[x]$ is not left-continuous.

21. $g(t) = \dfrac{1}{t^2 - 1}$

SOLUTION The function $f(t) = \dfrac{1}{t^2 - 1} = \dfrac{1}{(t - 1)(t + 1)}$ is discontinuous at $t = -1$ and $t = 1$, at which there are infinite discontinuities.

23. $f(x) = 3x^{3/2} - 9x^3$

SOLUTION The function $f(x) = 3x^{3/2} - 9x^3$ is continuous for $x > 0$. At $x = 0$ it is right-continuous. (It is not defined for $x < 0$.)

25. $h(z) = \dfrac{1 - 2z}{z^2 - z - 6}$

SOLUTION The function $f(z) = \dfrac{1 - 2z}{z^2 - z - 6} = \dfrac{1 - 2z}{(z + 2)(z - 3)}$ is discontinuous at $z = -2$ and $z = 3$, at which there are infinite discontinuities.

27. $f(x) = \dfrac{x^2 - 3x + 2}{|x - 2|}$

SOLUTION $\dfrac{x^2 - 3x + 2}{|x - 2|} = \dfrac{(x - 2)(x - 1)}{|x - 2|}$. For $x > 2$, the function

$$f(x) = \frac{(x - 2)(x - 1)}{|x - 2|} = \frac{(x - 2)(x - 1)}{(x - 2)} = x - 1.$$

For $x < 2$, $f(x) = \dfrac{(x - 2)(x - 1)}{2 - x} = -(x - 1)$. This function has a jump discontinuity at $x = 2$.

29. $f(x) = \csc x^2$

SOLUTION The function $f(x) = \csc(x^2) = \dfrac{1}{\sin(x^2)}$ is discontinuous whenever $\sin(x^2) = 0$; i.e., whenever $x^2 = n\pi$ or $x = \pm\sqrt{n\pi}$, where n is a positive integer. At every such value of x there is an infinite discontinuity.

31. $f(x) = \tan(\sin x)$

SOLUTION The function $f(x) = \tan(\sin x)$ is continuous everywhere. Reason: $\sin x$ is continuous everywhere and $\tan u$ is continuous on $\left(-\frac{\pi}{2}, \frac{\pi}{2}\right)$—and in particular on $-1 \le u = \sin x \le 1$. Continuity of $\tan(\sin x)$ follows by the continuity of composite functions.

In Exercises 33–46, determine the domain of the function and prove that it is continuous on its domain using the Laws of Continuity and the facts quoted in this section.

33. $f(x) = \sqrt{9 - x^2}$

SOLUTION The domain of $\sqrt{9 - x^2}$ is all x such that $9 - x^2 \ge 0$, or $|x| \le 3$. Since $\sqrt{x}$ and the polynomial $9 - x^2$ are both continuous on this domain, so is the composite function $\sqrt{9 - x^2}$.

35. $f(x) = \sqrt{x}\sin x$

SOLUTION This function is defined as long as $x \ge 0$. Since $\sqrt{x}$ and $\sin x$ are continuous, so is $\sqrt{x}\sin x$ by Continuity Law (ii).

37. $f(x) = x^{2/3}2^x$

SOLUTION The domain of $x^{2/3}2^x$ is all real numbers as the denominator of the rational exponent is odd. Both $x^{2/3}$ and 2^x are continuous on this domain, so $x^{2/3}2^x$ is continuous by Continuity Law (ii).

39. $f(x) = x^{-4/3}$

SOLUTION This function is defined for all $x \ne 0$. Because the function $x^{4/3}$ is continuous and not equal to zero for $x \ne 0$, it follows that

$$x^{-4/3} = \frac{1}{x^{4/3}}$$

is continuous for $x \ne 0$ by Continuity Law (iv).

41. $f(x) = \tan^2 x$

SOLUTION The domain of $\tan^2 x$ is all $x \ne \pm(2n - 1)\pi/2$ where n is a positive integer. Because $\tan x$ is continuous on this domain, it follows from Continuity Law (ii) that $\tan^2 x$ is also continuous on this domain.

43. $f(x) = (x^4 + 1)^{3/2}$

SOLUTION The domain of $(x^4 + 1)^{3/2}$ is all real numbers as $x^4 + 1 > 0$ for all x. Because $x^{3/2}$ and the polynomial $x^4 + 1$ are both continuous, so is the composite function $(x^4 + 1)^{3/2}$.

45. $f(x) = \dfrac{\cos(x^2)}{x^2 - 1}$

SOLUTION The domain for this function is all $x \ne \pm1$. Because the functions $\cos x$ and x^2 are continuous on this domain, so is the composite function $\cos(x^2)$. Finally, because the polynomial $x^2 - 1$ is continuous and not equal to zero for $x \ne \pm1$, the function $\dfrac{\cos(x^2)}{x^2 - 1}$ is continuous by Continuity Law (iv).

47. Suppose that $f(x) = 2$ for $x > 0$ and $f(x) = -4$ for $x < 0$. What is $f(0)$ if f is left-continuous at $x = 0$? What is $f(0)$ if f is right-continuous at $x = 0$?

SOLUTION Let $f(x) = 2$ for positive x and $f(x) = -4$ for negative x.

- If f is left-continuous at $x = 0$, then $f(0) = \lim_{x \to 0-} f(x) = -4$.
- If f is right-continuous at $x = 0$, then $f(0) = \lim_{x \to 0+} f(x) = 2$.

In Exercises 49–52, draw the graph of a function on [0, 5] *with the given properties.*

49. $f(x)$ is not continuous at $x = 1$, but $\lim\limits_{x \to 1+} f(x)$ and $\lim\limits_{x \to 1-} f(x)$ exist and are equal.

SOLUTION

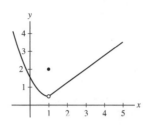

51. $f(x)$ has a removable discontinuity at $x = 1$, a jump discontinuity at $x = 2$, and

$$\lim_{x \to 3-} f(x) = -\infty, \qquad \lim_{x \to 3+} f(x) = 2$$

SOLUTION

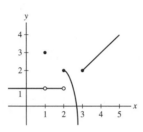

53. Each of the following statements is *false*. For each statement, sketch the graph of a function that provides a counterexample.

(a) If $\lim\limits_{x \to a} f(x)$ exists, then $f(x)$ is continuous at $x = a$.

(b) If $f(x)$ has a jump discontinuity at $x = a$, then $f(a)$ is equal to either $\lim\limits_{x \to a-} f(x)$ or $\lim\limits_{x \to a+} f(x)$.

(c) If $f(x)$ has a discontinuity at $x = a$, then $\lim\limits_{x \to a-} f(x)$ and $\lim\limits_{x \to a+} f(x)$ exist but are not equal.

(d) The one-sided limits $\lim\limits_{x \to a-} f(x)$ and $\lim\limits_{x \to a+} f(x)$ always exist, even if $\lim\limits_{x \to a} f(x)$ does not exist.

SOLUTION   Refer to the four figures shown below.

(a) The figure at the top left shows a function for which $\lim\limits_{x \to a} f(x)$ exists, but the function is not continuous at $x = a$ because the function is not defined at $x = a$.

(b) The figure at the top right shows a function that has a jump discontinuity at $x = a$ but $f(a)$ is not equal to either $\lim\limits_{x \to a-} f(x)$ or $\lim\limits_{x \to a-} f(x)$.

(c) This statement can be false either when the two one-sided limits exist and are equal or when one or both of the one-sided limits does not exist. The figure at the top left shows a function that has a discontinuity at $x = a$ with both one-sided limits being equal; the figure at the bottom left shows a function that has a discontinuity at $x = a$ with a one-sided limit that does not exist.

(d) The figure at the bottom left shows a function for which $\lim\limits_{x \to a} f(x)$ does not exist and one of the one-sided limits also does not exist; the figure at the bottom right shows a function for which $\lim\limits_{x \to a} f(x)$ does not exist and neither of the one-sided limits exists.

In Exercises 55–70, evaluate the limit using the substitution method.

55. $\lim\limits_{x \to 5} x^2$

SOLUTION $\lim\limits_{x \to 5} x^2 = 5^2 = 25$.

57. $\lim\limits_{x \to -1} (2x^3 - 4)$

SOLUTION $\lim\limits_{x \to -1} (2x^3 - 4) = 2(-1)^3 - 4 = -6$.

59. $\lim\limits_{x \to 0} \dfrac{x + 9}{x - 9}$

SOLUTION $\lim\limits_{x \to 0} \dfrac{x + 9}{x - 9} = \dfrac{0 + 9}{0 - 9} = -1$.

61. $\lim\limits_{x \to \pi} \sin\left(\dfrac{x}{2} - \pi\right)$

SOLUTION $\lim\limits_{x \to \pi} \sin(\frac{x}{2} - \pi) = \sin(-\frac{\pi}{2}) = -1$.

63. $\lim\limits_{x \to \frac{\pi}{4}} \tan(3x)$

SOLUTION $\lim\limits_{x \to \frac{\pi}{4}} \tan(3x) = \tan(3 \cdot \frac{\pi}{4}) = \tan(\frac{3\pi}{4}) = -1$

65. $\lim\limits_{x \to 4} x^{-5/2}$

SOLUTION $\lim\limits_{x \to 4} x^{-5/2} = 4^{-5/2} = \dfrac{1}{32}$.

67. $\lim\limits_{x \to -1} (1 - 8x^3)^{3/2}$

SOLUTION $\lim\limits_{x \to -1} (1 - 8x^3)^{3/2} = (1 - 8(-1)^3)^{3/2} = 27$.

69. $\lim\limits_{x \to 3} 10^{x^2 - 2x}$

SOLUTION $\lim\limits_{x \to 3} 10^{x^2 - 2x} = 10^{3^2 - 2(3)} = 1000$.

In Exercises 71–74, sketch the graph of the given function. At each point of discontinuity, state whether f is left- or right-continuous.

71. $f(x) = \begin{cases} x^2 & \text{for } x \leq 1 \\ 2 - x & \text{for } x > 1 \end{cases}$

SOLUTION

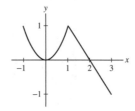

The function f is continuous everywhere.

73. $f(x) = \begin{cases} |x - 3| & \text{for } x \leq 3 \\ x - 3 & \text{for } x > 3 \end{cases}$

SOLUTION

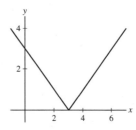

The function f is continuous everywhere.

In Exercises 75–76, find the value of c that makes the function continuous.

75. $f(x) = \begin{cases} x^2 - c & \text{for } x < 5 \\ 4x + 2c & \text{for } x \geq 5 \end{cases}$

SOLUTION As $x \to 5-$, we have $x^2 - c \to 25 - c = L$. As $x \to 5+$, we have $4x + 2c \to 20 + 2c = R$. Match the limits: $L = R$ or $25 - c = 20 + 2c$ implies $c = \frac{5}{3}$.

77. Find all constants a, b such that the following function has no discontinuities:

$$f(x) = \begin{cases} ax + \cos x & \text{for } x \leq \dfrac{\pi}{4} \\[2mm] bx + 2 & \text{for } x > \dfrac{\pi}{4} \end{cases}$$

SOLUTION As $x \to \frac{\pi}{4}-$, we have $ax + \cos x \to \frac{a\pi}{4} + \frac{1}{\sqrt{2}} = L$. As $x \to \frac{\pi}{4}+$, we have $bx + 2 \to \frac{b\pi}{4} + 2 = R$. Match the limits: $L = R$ or $\frac{a\pi}{4} + \frac{1}{\sqrt{2}} = \frac{b\pi}{4} + 2$. This implies $\frac{\pi}{4}(a - b) = 2 - \frac{1}{\sqrt{2}}$ or $a - b = \frac{8 - 2\sqrt{2}}{\pi}$.

79. In 1993, the amount $T(x)$ of federal income tax owed on an income of x dollars was determined by the formula

$$T(x) = \begin{cases} 0.15x & \text{for } 0 \leq x < 21{,}450 \\ 3{,}217.50 + 0.28(x - 21{,}450) & \text{for } 21{,}450 \leq x < 51{,}900 \\ 11{,}743.50 + 0.31(x - 51{,}900) & \text{for } x \geq 51{,}900 \end{cases}$$

Sketch the graph of $T(x)$ and determine if it has any discontinuities. Explain why, if $T(x)$ had a jump discontinuity, it might be advantageous in some situations to earn *less* money.

SOLUTION $T(x)$, the amount of federal income tax owed on an income of x dollars in 1993, might be a discontinuous function depending upon how the tax tables are constructed (as determined by that year's regulations). Here is a graph of $T(x)$ for that particular year.

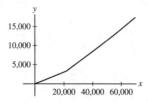

If $T(x)$ had a jump discontinuity (say at $x = c$), it might be advantageous to earn slightly less income than c (say $c - \epsilon$) and be taxed at a lower rate than to earn c or more and be taxed at a higher rate. Your net earnings may actually be more in the former case than in the latter one.

Further Insights and Challenges

81. Give an example of functions $f(x)$ and $g(x)$ such that $f(g(x))$ is continuous but $g(x)$ has at least one discontinuity.

SOLUTION   Answers may vary. The simplest examples are the functions $f(g(x))$ where $f(x) = C$ is a constant function, and $g(x)$ is defined for all x. In these cases, $f(g(x)) = C$. For example, if $f(x) = 3$ and $g(x) = [x]$, g is discontinuous at all integer values $x = n$, but $f(g(x)) = 3$ is continuous.

83. Let $f(x) = 1$ if x is rational and $f(x) = -1$ if x is irrational. Show that $f(x)$ is discontinuous at all points, whereas $f(x)^2$ is continuous at all points.

SOLUTION $\lim\limits_{x \to c} f(x)$ does not exist for any c. If c is irrational, then there is always a rational number r arbitrarily close to c so that $|f(c) - f(r)| = 2$. If, on the other hand, c is rational, there is always an *irrational* number z arbitrarily close to c so that $|f(c) - f(z)| = 2$.

$f(x)^2$, on the other hand, is a constant function that always has value 1, which is obviously continuous.

2.5 Evaluating Limits Algebraically

Preliminary Questions

1. Which of the following is indeterminate at $x = 1$?

$$\frac{x^2 + 1}{x - 1}, \qquad \frac{x^2 - 1}{x + 2}, \qquad \frac{x^2 - 1}{\sqrt{x + 3} - 2}, \qquad \frac{x^2 + 1}{\sqrt{x + 3} - 2}$$

SOLUTION   At $x = 1$, $\frac{x^2 - 1}{\sqrt{x + 3} - 2}$ is of the form $\frac{0}{0}$; hence, this function is indeterminate. None of the remaining functions is indeterminate at $x = 1$: $\frac{x^2 + 1}{x - 1}$ and $\frac{x^2 + 1}{\sqrt{x + 3} - 2}$ are undefined because the denominator is zero but the numerator is not, while $\frac{x^2 - 1}{x + 2}$ is equal to 0.

2. Give counterexamples to show that each of the following statements is false:

(a) If $f(c)$ is indeterminate, then the right- and left-hand limits as $x \to c$ are not equal.

(b) If $\lim\limits_{x \to c} f(x)$ exists, then $f(c)$ is not indeterminate.

(c) If $f(x)$ is undefined at $x = c$, then $f(x)$ has an indeterminate form at $x = c$.

SOLUTION

(a) Let $f(x) = \frac{x^2 - 1}{x - 1}$. At $x = 1$, f is indeterminate of the form $\frac{0}{0}$ but

$$\lim_{x \to 1-} \frac{x^2 - 1}{x - 1} = \lim_{x \to 1-} (x + 1) = 2 = \lim_{x \to 1+} (x + 1) = \lim_{x \to 1+} \frac{x^2 - 1}{x - 1}.$$

(b) Again, let $f(x) = \frac{x^2 - 1}{x - 1}$. Then

$$\lim_{x \to 1} f(x) = \lim_{x \to 1} \frac{x^2 - 1}{x - 1} = \lim_{x \to 1} (x + 1) = 2$$

but $f(1)$ is indeterminate of the form $\frac{0}{0}$.

(c) Let $f(x) = \frac{1}{x}$. Then f is undefined at $x = 0$ but does not have an indeterminate form at $x = 0$.

3. Although the method for evaluating limits discussed in this section is sometimes called "simplify and plug in," explain how it actually relies on the property of continuity.

SOLUTION   If f is continuous at $x = c$, then, by definition, $\lim_{x \to c} f(x) = f(c)$; in other words, the limit of a continuous function at $x = c$ is the value of the function at $x = c$. The "simplify and plug-in" strategy is based on simplifying a function which is indeterminate to a continuous function. Once the simplification has been made, the limit of the remaining continuous function is obtained by evaluation.

Exercises

In Exercises 1–4, show that the limit leads to an indeterminate form. Then carry out the two-step procedure: Transform the function algebraically and evaluate using continuity.

1. $\displaystyle\lim_{x \to 5} \frac{x^2 - 25}{x - 5}$

SOLUTION When we substitute $x = 5$ into $\frac{x^2-25}{x-5}$, we obtain the indeterminate form $\frac{0}{0}$. Upon factoring the numerator and simplifying, we find

$$\lim_{x \to 5} \frac{x^2 - 25}{x - 5} = \lim_{x \to 5} \frac{(x - 5)(x + 5)}{x - 5} = \lim_{x \to 5} (x + 5) = 10.$$

3. $\displaystyle\lim_{t \to 7} \frac{2t - 14}{5t - 35}$

SOLUTION When we substitute $t = 7$ into $\frac{2t-14}{5t-35}$, we obtain the indeterminate form $\frac{0}{0}$. Upon dividing out the common factor of $t - 7$ from both the numerator and denominator, we find

$$\lim_{t \to 7} \frac{2t - 14}{5t - 35} = \lim_{t \to 7} \frac{2(t - 7)}{5(t - 7)} = \lim_{t \to 7} \frac{2}{5} = \frac{2}{5}.$$

In Exercises 5–32, evaluate the limit or state that it does not exist.

5. $\displaystyle\lim_{x \to 8} \frac{x^2 - 64}{x - 8}$

SOLUTION $\displaystyle\lim_{x \to 8} \frac{x^2 - 64}{x - 8} = \lim_{x \to 8} \frac{(x + 8)(x - 8)}{x - 8} = \lim_{x \to 8} (x + 8) = 16.$

7. $\displaystyle\lim_{x \to 2} \frac{x^2 - 3x + 2}{x - 2}$

SOLUTION $\displaystyle\lim_{x \to 2} \frac{x^2 - 3x + 2}{x - 2} = \lim_{x \to 2} \frac{(x - 1)(x - 2)}{x - 2} = \lim_{x \to 2} (x - 1) = 1.$

9. $\displaystyle\lim_{x \to 2} \frac{x - 2}{x^3 - 4x}$

SOLUTION $\displaystyle\lim_{x \to 2} \frac{x - 2}{x^3 - 4x} = \lim_{x \to 2} \frac{x - 2}{x(x - 2)(x + 2)} = \lim_{x \to 2} \frac{1}{x(x + 2)} = \frac{1}{8}.$

11. $\displaystyle\lim_{h \to 0} \frac{(1 + h)^3 - 1}{h}$

SOLUTION

$$\lim_{h \to 0} \frac{(1 + h)^3 - 1}{h} = \lim_{h \to 0} \frac{1 + 3h + 3h^2 + h^3 - 1}{h} = \lim_{h \to 0} \frac{3h + 3h^2 + h^3}{h}$$

$$= \lim_{h \to 0} 3 + 3h + h^2 = 3 + 3(0) + 0^2 = 3.$$

13. $\displaystyle\lim_{x \to 2} \frac{3x^2 - 4x - 4}{2x^2 - 8}$

SOLUTION $\displaystyle\lim_{x \to 2} \frac{3x^2 - 4x - 4}{2x^2 - 8} = \lim_{x \to 2} \frac{(3x + 2)(x - 2)}{2(x - 2)(x + 2)} = \lim_{x \to 2} \frac{3x + 2}{2(x + 2)} = \frac{8}{8} = 1.$

15. $\displaystyle\lim_{y \to 2} \frac{(y - 2)^3}{y^3 - 5y + 2}$

SOLUTION $\displaystyle\lim_{y \to 2} \frac{(y - 2)^3}{y^3 - 5y + 2} = \lim_{y \to 2} \frac{(y - 2)^3}{(y - 2)\left(y^2 + 2y - 1\right)} = \lim_{y \to 2} \frac{(y - 2)^2}{y^2 + 2y - 1} = 0.$

17. $\displaystyle\lim_{h \to 0} \frac{\dfrac{1}{3 + h} - \dfrac{1}{3}}{h}$

SOLUTION $\lim\limits_{h \to 0} \dfrac{\frac{1}{3+h} - \frac{1}{3}}{h} = \lim\limits_{h \to 0} \dfrac{1}{h} \dfrac{3 - (3+h)}{3(3+h)} = \lim\limits_{h \to 0} \dfrac{-1}{3(3+h)} = -\dfrac{1}{9}.$

19. $\lim\limits_{h \to 0} \dfrac{\sqrt{2+h} - 2}{h}$

SOLUTION $\lim\limits_{h \to 0} \dfrac{\sqrt{h+2} - 2}{h}$ does not exist.

- As $h \to 0+$, we have $\dfrac{\sqrt{h+2} - 2}{h} = \dfrac{\left(\sqrt{h+2} - 2\right)\left(\sqrt{h+2} + 2\right)}{h(\sqrt{h+2} + 2)} = \dfrac{h - 2}{h(\sqrt{h+2} + 2)} \to -\infty.$

- As $h \to 0-$, we have $\dfrac{\sqrt{h+2} - 2}{h} = \dfrac{\left(\sqrt{h+2} - 2\right)\left(\sqrt{h+2} + 2\right)}{h(\sqrt{h+2} + 2)} = \dfrac{h - 2}{h(\sqrt{h+2} + 2)} \to \infty.$

21. $\lim\limits_{x \to 2} \dfrac{x - 2}{\sqrt{x} - \sqrt{4 - x}}$

SOLUTION

$$\lim_{x \to 2} \frac{x - 2}{\sqrt{x} - \sqrt{4 - x}} = \lim_{x \to 2} \frac{(x-2)(\sqrt{x} + \sqrt{4-x})}{(\sqrt{x} - \sqrt{4-x})(\sqrt{x} + \sqrt{4-x})} = \lim_{x \to 2} \frac{(x-2)(\sqrt{x} + \sqrt{4-x})}{x - (4-x)}$$

$$= \lim_{x \to 2} \frac{(x-2)(\sqrt{x} + \sqrt{4-x})}{2x - 4} = \lim_{x \to 2} \frac{(x-2)(\sqrt{x} + \sqrt{4-x})}{2(x-2)}$$

$$= \lim_{x \to 2} \frac{(\sqrt{x} + \sqrt{4-x})}{2} = \frac{\sqrt{2} + \sqrt{2}}{2} = \sqrt{2}.$$

23. $\lim\limits_{x \to 2} \dfrac{\sqrt{x^2 - 1} - \sqrt{x + 1}}{x - 3}$

SOLUTION $\lim\limits_{x \to 2} \dfrac{\sqrt{x^2 - 1} - \sqrt{x + 1}}{x - 3} = \dfrac{\sqrt{3} - \sqrt{3}}{-1} = 0.$

25. $\lim\limits_{x \to 4} \left(\dfrac{1}{\sqrt{x} - 2} - \dfrac{4}{x - 4} \right)$

SOLUTION $\lim\limits_{x \to 4} \left(\dfrac{1}{\sqrt{x} - 2} - \dfrac{4}{x - 4} \right) = \lim\limits_{x \to 4} \dfrac{\sqrt{x} + 2 - 4}{\left(\sqrt{x} - 2\right)\left(\sqrt{x} + 2\right)} = \lim\limits_{x \to 4} \dfrac{\sqrt{x} - 2}{\left(\sqrt{x} - 2\right)\left(\sqrt{x} + 2\right)} = \dfrac{1}{4}.$

27. $\lim\limits_{x \to 0} \dfrac{\cot x}{\csc x}$

SOLUTION $\lim\limits_{x \to 0} \dfrac{\cot x}{\csc x} = \lim\limits_{x \to 0} \dfrac{\cos x}{\sin x} \cdot \sin x = \cos 0 = 1.$

29. $\lim\limits_{x \to \frac{\pi}{4}} \dfrac{\sin x - \cos x}{\tan x - 1}$

SOLUTION $\lim\limits_{x \to \frac{\pi}{4}} \dfrac{\sin x - \cos x}{\tan x - 1} \cdot \dfrac{\cos x}{\cos x} = \lim\limits_{x \to \frac{\pi}{4}} \dfrac{(\sin x - \cos x)\cos x}{\sin x - \cos x} = \cos \dfrac{\pi}{4} = \dfrac{\sqrt{2}}{2}.$

31. $\lim\limits_{x \to \frac{\pi}{3}} \dfrac{2\cos^2 x + 3\cos x - 2}{2\cos x - 1}$

SOLUTION

$$\lim_{x \to \frac{\pi}{3}} \frac{2\cos^2 x + 3\cos x - 2}{2\cos x - 1} = \lim_{x \to \frac{\pi}{3}} \frac{(2\cos x - 1)(\cos x + 2)}{2\cos x - 1} = \lim_{x \to \frac{\pi}{3}} \cos x + 2 = \cos \frac{\pi}{3} + 2 = \frac{5}{2}.$$

33. ⬜ **GU**   Use a plot of $f(x) = \dfrac{x - 2}{\sqrt{x} - \sqrt{4 - x}}$ to estimate $\lim\limits_{x \to 2} f(x)$ to two decimal places. Compare with the answer obtained algebraically in Exercise 21.

SOLUTION   Let $f(x) = \dfrac{x-2}{\sqrt{x} - \sqrt{4-x}}$. From the plot of $f(x)$ shown below, we estimate $\lim\limits_{x \to 2} f(x) \approx 1.41$; to two decimal places, this matches the value of $\sqrt{2}$ obtained in Exercise 21.

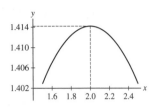

In Exercises 35–40, use the identity $a^3 - b^3 = (a - b)(a^2 + ab + b^2)$.

35. $\lim\limits_{x \to 1} \dfrac{x^3 - 1}{x - 1}$

SOLUTION $\lim\limits_{x \to 1} \dfrac{x^3 - 1}{x - 1} = \lim\limits_{x \to 1} \dfrac{(x - 1)\left(x^2 + x + 1\right)}{x - 1} = \lim\limits_{x \to 1} \left(x^2 + x + 1\right) = 3.$

37. $\lim\limits_{x \to 1} \dfrac{x^2 - 3x + 2}{x^3 - 1}$

SOLUTION $\lim\limits_{x \to 1} \dfrac{x^2 - 3x + 2}{x^3 - 1} = \lim\limits_{x \to 1} \dfrac{(x - 1)(x - 2)}{(x - 1)\left(x^2 + x + 1\right)} = \lim\limits_{x \to 1} \dfrac{x - 2}{x^2 + x + 1} = -\dfrac{1}{3}.$

39. $\lim\limits_{x \to 1} \dfrac{x^4 - 1}{x^3 - 1}$

SOLUTION

$$\lim\limits_{x \to 1} \dfrac{x^4 - 1}{x^3 - 1} = \lim\limits_{x \to 1} \dfrac{(x^2 - 1)(x^2 + 1)}{(x - 1)(x^2 + x + 1)} = \lim\limits_{x \to 1} \dfrac{(x - 1)(x + 1)(x^2 + 1)}{(x - 1)(x^2 + x + 1)} = \lim\limits_{x \to 1} \dfrac{(x + 1)(x^2 + 1)}{(x^2 + x + 1)} = \dfrac{4}{3}.$$

In Exercises 41–50, evaluate the limits in terms of the constants involved.

41. $\lim\limits_{x \to 0} (2a + x)$

SOLUTION $\lim\limits_{x \to 0} (2a + x) = 2a.$

43. $\lim\limits_{t \to -1} (4t - 2at + 3a)$

SOLUTION $\lim\limits_{t \to -1} (4t - 2at + 3a) = -4 + 5a.$

45. $\lim\limits_{x \to 0} \dfrac{2(x + h)^2 - 2x^2}{h}$

SOLUTION $\lim\limits_{x \to 0} \dfrac{2(x + h)^2 - 2x^2}{h} = \lim\limits_{x \to 0} \dfrac{4hx + 2h^2}{h} = \lim\limits_{x \to 0} (4x + 2h) = 2h.$

47. $\lim\limits_{x \to a} \dfrac{\sqrt{x} - \sqrt{a}}{x - a}$

SOLUTION $\lim\limits_{x \to a} \dfrac{\sqrt{x} - \sqrt{a}}{x - a} = \lim\limits_{x \to a} \dfrac{\sqrt{x} - \sqrt{a}}{\left(\sqrt{x} - \sqrt{a}\right)\left(\sqrt{x} + \sqrt{a}\right)} = \lim\limits_{x \to a} \dfrac{1}{\sqrt{x} + \sqrt{a}} = \dfrac{1}{2\sqrt{a}}.$

49. $\lim\limits_{x \to 0} \dfrac{(x + a)^3 - a^3}{x}$

SOLUTION $\lim\limits_{x \to 0} \dfrac{(x + a)^3 - a^3}{x} = \lim\limits_{x \to 0} \dfrac{x^3 + 3x^2 a + 3xa^2 + a^3 - a^3}{x} = \lim\limits_{x \to 0} (x^2 + 3xa + 3a^2) = 3a^2.$

Further Insights and Challenges

In Exercises 51–52, find all values of c such that the limit exists.

51. $\lim\limits_{x \to c} \dfrac{x^2 - 5x - 6}{x - c}$

SOLUTION $\lim\limits_{x \to c} \dfrac{x^2 - 5x - 6}{x - c}$ will exist provided that $x - c$ is a factor of the numerator. (Otherwise there will be an infinite discontinuity at $x = c$.) Since $x^2 - 5x - 6 = (x + 1)(x - 6)$, this occurs for $c = -1$ and $c = 6$.

53. For which sign $\pm$ does the following limit exist?

$$\lim_{x \to 0} \left(\frac{1}{x} \pm \frac{1}{x(x-1)} \right)$$

SOLUTION

- The limit $\lim\limits_{x \to 0} \left(\dfrac{1}{x} + \dfrac{1}{x(x-1)} \right) = \lim\limits_{x \to 0} \dfrac{(x-1)+1}{x(x-1)} = \lim\limits_{x \to 0} \dfrac{1}{x-1} = -1.$

- The limit $\lim\limits_{x \to 0} \left(\dfrac{1}{x} - \dfrac{1}{x(x-1)} \right)$ does not exist.

  - As $x \to 0+$, we have $\dfrac{1}{x} - \dfrac{1}{x(x-1)} = \dfrac{(x-1)-1}{x(x-1)} = \dfrac{x-2}{x(x-1)} \to \infty.$
  - As $x \to 0-$, we have $\dfrac{1}{x} - \dfrac{1}{x(x-1)} = \dfrac{(x-1)-1}{x(x-1)} = \dfrac{x-2}{x(x-1)} \to -\infty.$

2.6 Trigonometric Limits

Preliminary Questions

1. Assume that $-x^4 \le f(x) \le x^2$. What is $\lim\limits_{x \to 0} f(x)$? Is there enough information to evaluate $\lim\limits_{x \to \frac{1}{2}} f(x)$? Explain.

SOLUTION Since $\lim_{x \to 0} -x^4 = \lim_{x \to 0} x^2 = 0$, the squeeze theorem guarantees that $\lim_{x \to 0} f(x) = 0$. Since $\lim_{x \to \frac{1}{2}} -x^4 = -\frac{1}{16} \ne \frac{1}{4} = \lim_{x \to \frac{1}{2}} x^2$, we do not have enough information to determine $\lim_{x \to \frac{1}{2}} f(x)$.

2. State the Squeeze Theorem carefully.

SOLUTION Assume that for $x \ne c$ (in some open interval containing c),

$$l(x) \le f(x) \le u(x)$$

and that $\lim\limits_{x \to c} l(x) = \lim\limits_{x \to c} u(x) = L$. Then $\lim\limits_{x \to c} f(x)$ exists and

$$\lim_{x \to c} f(x) = L.$$

3. Suppose that $f(x)$ is squeezed at $x = c$ on an open interval I by two *constant* functions $u(x)$ and $l(x)$. Is $f(x)$ necessarily constant on I?

SOLUTION Yes. Because $f(x)$ is squeezed at $x = c$ on an open interval I by the two functions $u(x)$ and $l(x)$, we know that

$$\lim_{x \to c} u(x) = \lim_{x \to c} l(x).$$

Combining this relation with the fact that $u(x)$ and $l(x)$ are constant functions, it follows that $u(x) = l(x)$ for all x on I. Finally, to have

$$l(x) \le f(x) \le u(x) = l(x),$$

it must be the case that $f(x) = l(x)$ for all x on I. In other words, $f(x)$ is constant on I.

4. If you want to evaluate $\lim\limits_{h \to 0} \dfrac{\sin 5h}{3h}$, it is a good idea to rewrite the limit in terms of the variable (choose one):

(a) $\theta = 5h$ **(b)** $\theta = 3h$ **(c)** $\theta = \dfrac{5h}{3}$

SOLUTION To match the given limit to the pattern of

$$\lim_{\theta \to 0} \frac{\sin \theta}{\theta},$$

it is best to substitute for the argument of the sine function; thus, rewrite the limit in terms of **(a):** $\theta = 5h.$

Exercises

1. In Figure 6, is $f(x)$ squeezed by $u(x)$ and $l(x)$ at $x = 3$? At $x = 2$?

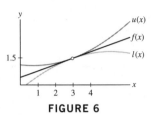

FIGURE 6

SOLUTION Because there is an open interval containing $x = 3$ on which $l(x) \le f(x) \le u(x)$ and $\lim_{x \to 3} l(x) = \lim_{x \to 3} u(x)$, $f(x)$ is *squeezed* by $u(x)$ and $l(x)$ at $x = 3$. Because there is an open interval containing $x = 2$ on which $l(x) \le f(x) \le u(x)$ but $\lim_{x \to 2} l(x) \ne \lim_{x \to 2} u(x)$, $f(x)$ is *trapped* by $u(x)$ and $l(x)$ at $x = 2$ but not *squeezed*.

3. What information about $f(x)$ does the Squeeze Theorem provide if we assume that the graphs of $f(x)$, $u(x)$, and $l(x)$ are related as in Figure 7 and that $\lim_{x \to 7} u(x) = \lim_{x \to 7} l(x) = 6$? Note that the inequality $f(x) \le u(x)$ is not satisfied for all x. Does this affect the validity of your conclusion?

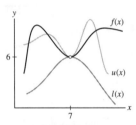

FIGURE 7

SOLUTION The Squeeze Theorem does not require that the inequalities $l(x) \le f(x) \le u(x)$ hold for all x, only that the inequalities hold on some open interval containing $x = c$. In Figure 7, it is clear that $l(x) \le f(x) \le u(x)$ on some open interval containing $x = 7$. Because $\lim_{x \to 7} u(x) = \lim_{x \to 7} l(x) = 6$, the Squeeze Theorem guarantees that $\lim_{x \to 7} f(x) = 6$.

In Exercises 5–8, use the Squeeze Theorem to evaluate the limit.

5. $\lim_{x \to 0} x \cos \dfrac{1}{x}$

SOLUTION We prove the limit to the right first, and then the limit to the left. Suppose $x > 0$. Since $-1 \le \cos \frac{1}{x} \le 1$, multiplication by x (positive) yields $-x \le x \cos \frac{1}{x} \le x$. The squeeze theorem implies that $\lim_{x \to 0+} -x \le \lim_{x \to 0+} x \cos \frac{1}{x} \le \lim_{x \to 0+} x$, so $0 \le \lim_{x \to 0+} x \cos \frac{1}{x} \le 0$. This gives us $\lim_{x \to 0+} x \cos \frac{1}{x} = 0$. On the other hand, suppose $x < 0$. $-1 \le \cos \frac{1}{x} \le 1$ multiplied by x gives us $x \le x \cos \frac{1}{x} \le -x$, and we can apply the squeeze theorem again to get $\lim_{x \to 0-} x \cos \frac{1}{x} = 0$. Therefore, $\lim_{x \to 0} x \cos \frac{1}{x} = 0$.

7. $\lim_{x \to 0} \cos \left(\dfrac{1}{x} \right) \sin x$

SOLUTION Since $-1 \le \cos \left(\frac{1}{x} \right) \le 1$, observe that $-\sin x \le \cos \left(\frac{1}{x} \right) \sin x \le \sin x$. Since $\lim_{x \to 0} \sin x = 0 = \lim_{x \to 0} -\sin x$, the Squeeze Theorem shows that

$$\lim_{x \to 0} \cos \left(\frac{1}{x} \right) \sin x = 0.$$

In Exercises 9–16, evaluate the limit using Theorem 2 as necessary.

9. $\lim_{x \to 0} \dfrac{\sin x \cos x}{x}$

SOLUTION $\lim_{x \to 0} \dfrac{\sin x \cos x}{x} = \lim_{x \to 0} \dfrac{\sin x}{x} \cdot \lim_{x \to 0} \cos x = 1 \cdot 1 = 1.$

11. $\lim_{t \to 0} \dfrac{\sin^2 t}{t}$

SOLUTION $\displaystyle\lim_{t\to 0}\frac{\sin^2 t}{t} = \lim_{t\to 0}\frac{\sin t}{t}\sin t = \lim_{t\to 0}\frac{\sin t}{t}\cdot\lim_{t\to 0}\sin t = 1\cdot 0 = 0.$

13. $\displaystyle\lim_{x\to 0}\frac{x^2}{\sin^2 x}$

SOLUTION $\displaystyle\lim_{x\to 0}\frac{x^2}{\sin^2 x} = \lim_{x\to 0}\frac{1}{\frac{\sin x}{x}\frac{\sin x}{x}} = \lim_{x\to 0}\frac{1}{\frac{\sin x}{x}}\cdot\lim_{x\to 0}\frac{1}{\frac{\sin x}{x}} = \frac{1}{1}\cdot\frac{1}{1} = 1.$

15. $\displaystyle\lim_{t\to\frac{\pi}{4}}\frac{\sin t}{t}$

SOLUTION $\dfrac{\sin t}{t}$ is continuous at $t = \dfrac{\pi}{4}$. Hence, by substitution

$$\lim_{t\to\frac{\pi}{4}}\frac{\sin t}{t} = \frac{\frac{\sqrt{2}}{2}}{\frac{\pi}{4}} = \frac{2\sqrt{2}}{\pi}.$$

17. Let $L = \displaystyle\lim_{x\to 0}\frac{\sin 10x}{x}$.

(a) Show, by letting $\theta = 10x$, that $L = \displaystyle\lim_{\theta\to 0}10\frac{\sin\theta}{\theta}$.

(b) Compute L.

SOLUTION Since $\theta = 10x$, $x = \frac{\theta}{10}$. $\theta\to 0$ as $x\to 0$, so we get:

$$\lim_{x\to 0}\frac{\sin 10x}{x} = \lim_{\theta\to 0}\frac{\sin\theta}{(\theta/10)} = \lim_{\theta\to 0}10\frac{\sin\theta}{\theta} = 10\lim_{\theta\to 0}\frac{\sin\theta}{\theta} = 10.$$

In Exercises 19–40, evaluate the limit.

19. $\displaystyle\lim_{h\to 0}\frac{\sin 6h}{h}$

SOLUTION $\displaystyle\lim_{h\to 0}\frac{\sin 6h}{h} = \lim_{h\to 0}6\frac{\sin 6h}{6h} = 6.$

21. $\displaystyle\lim_{h\to 0}\frac{\sin 6h}{6h}$

SOLUTION Substitute $x = 6h$. As $h\to 0$, $x\to 0$, so: $\displaystyle\lim_{h\to 0}\frac{\sin 6h}{6h} = \lim_{x\to 0}\frac{\sin x}{x} = 1.$

23. $\displaystyle\lim_{x\to 0}\frac{\sin 7x}{3x}$

SOLUTION $\displaystyle\lim_{x\to 0}\frac{\sin 7x}{3x} = \lim_{x\to 0}\frac{7}{3}\frac{\sin 7x}{7x} = \frac{7}{3}.$

25. $\displaystyle\lim_{x\to 0}\frac{\tan 4x}{9x}$

SOLUTION $\displaystyle\lim_{x\to 0}\frac{\tan 4x}{9x} = \lim_{x\to 0}\frac{1}{9}\cdot\frac{\sin 4x}{4x}\cdot\frac{4}{\cos 4x} = \frac{4}{9}.$

27. $\displaystyle\lim_{t\to 0}\frac{\tan 4t}{t\sec t}$

SOLUTION $\displaystyle\lim_{t\to 0}\frac{\tan 4t}{t\sec t} = \lim_{t\to 0}\frac{4\sin 4t}{4t\cos(4t)\sec(t)} = \lim_{t\to 0}\frac{4\cos t}{\cos 4t}\cdot\frac{\sin 4t}{4t} = 4.$

29. $\displaystyle\lim_{z\to 0}\frac{\sin(z/3)}{\sin z}$

SOLUTION $\displaystyle\lim_{z\to 0}\frac{\sin(z/3)}{\sin z}\cdot\frac{z/3}{z/3} = \lim_{z\to 0}\frac{1}{3}\cdot\frac{z}{\sin z}\cdot\frac{\sin(z/3)}{z/3} = \frac{1}{3}.$

31. $\displaystyle\lim_{x\to 0}\frac{\tan 4x}{\tan 9x}$

SOLUTION $\displaystyle\lim_{x\to 0}\frac{\tan 4x}{\tan 9x} = \lim_{x\to 0}\frac{\cos 9x}{\cos 4x}\cdot\frac{\sin 4x}{4x}\cdot\frac{4}{9}\cdot\frac{9x}{\sin 9x} = \frac{4}{9}.$

33. $\lim\limits_{x \to 0} \dfrac{\sin 5x \sin 2x}{\sin 3x \sin 5x}$

SOLUTION $\lim\limits_{x \to 0} \dfrac{\sin 5x \sin 2x}{\sin 3x \sin 5x} = \lim\limits_{x \to 0} \dfrac{\sin 2x}{2x} \cdot \dfrac{2}{3} \cdot \dfrac{3x}{\sin 3x} = \dfrac{2}{3}$.

35. $\lim\limits_{h \to 0} \dfrac{1 - \cos 2h}{h}$

SOLUTION $\lim\limits_{h \to 0} \dfrac{1 - \cos 2h}{h} = \lim\limits_{h \to 0} 2\dfrac{1 - \cos 2h}{2h} = 2 \lim\limits_{h \to 0} \dfrac{1 - \cos 2h}{2h} = 2 \cdot 0 = 0$.

37. $\lim\limits_{t \to 0} \dfrac{1 - \cos t}{\sin t}$

SOLUTION A single multiplication by $\frac{t}{t}$ turns this limit into the quotient of two familiar limits.

$$\lim\limits_{t \to 0} \dfrac{1 - \cos t}{\sin t} = \lim\limits_{t \to 0} \dfrac{t(1 - \cos t)}{t \sin t} = \lim\limits_{t \to 0} \dfrac{t}{\sin t} \dfrac{(1 - \cos t)}{t} = \lim\limits_{t \to 0} \dfrac{t}{\sin t} \lim\limits_{t \to 0} \dfrac{(1 - \cos t)}{t} = 1 \cdot 0 = 0.$$

39. $\lim\limits_{h \to \frac{\pi}{2}} \dfrac{1 - \cos 3h}{h}$

SOLUTION The function is continuous at $\frac{\pi}{2}$, so we may use substitution:

$$\lim\limits_{h \to \frac{\pi}{2}} \dfrac{1 - \cos 3h}{h} = \dfrac{1 - \cos 3\frac{\pi}{2}}{\frac{\pi}{2}} = \dfrac{1 - 0}{\frac{\pi}{2}} = \dfrac{2}{\pi}.$$

41. Calculate (a) $\lim\limits_{x \to 0+} \dfrac{\sin x}{|x|}$ and (b) $\lim\limits_{x \to 0-} \dfrac{\sin x}{|x|}$.

SOLUTION

(a) $\lim\limits_{x \to 0+} \dfrac{\sin x}{|x|} = \lim\limits_{x \to 0+} \dfrac{\sin x}{x} = 1$.

(b) $\lim\limits_{x \to 0-} \dfrac{\sin x}{|x|} = \lim\limits_{x \to 0-} \dfrac{\sin x}{-x} = -1$.

43. Show that $-|\tan x| \le \tan x \cos \frac{1}{x} \le |\tan x|$. Then evaluate $\lim\limits_{x \to 0} \tan x \cos \frac{1}{x}$ using the Squeeze Theorem.

SOLUTION Since $\left|\cos\left(\frac{1}{x}\right)\right| \le 1$, it follows that $\left|\tan x \cos\left(\frac{1}{x}\right)\right| \le |\tan x|$, which is equivalent to $-|\tan x| \le \tan x \cos\left(\frac{1}{x}\right) \le |\tan x|$. Because $\lim_{x \to 0} -|\tan x| = \lim_{x \to 0} |\tan x| = 0$, the Squeeze Theorem gives

$$\lim\limits_{x \to 0} \tan x \cos\left(\dfrac{1}{x}\right) = 0.$$

45. [GU] Plot the graphs of $u(x) = 1 + |x - \frac{\pi}{2}|$ and $l(x) = \sin x$ on the same set of axes. What can you say about $\lim\limits_{x \to \frac{\pi}{2}} f(x)$ if $f(x)$ is squeezed by $l(x)$ and $u(x)$ at $x = \frac{\pi}{2}$?

SOLUTION

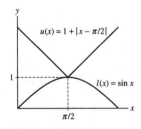

$\lim\limits_{x \to \pi/2} u(x) = 1$ and $\lim\limits_{x \to \pi/2} l(x) = 1$, so any function $f(x)$ satisfying $l(x) \le f(x) \le u(x)$ for all x near $\pi/2$ will satisfy $\lim\limits_{x \to \pi/2} f(x) = 1$.

Further Insights and Challenges

47. GU Investigate $\lim\limits_{h\to 0} \dfrac{1 - \cos h}{h^2}$ numerically (and graphically if you have a graphing utility). Then prove that the limit is equal to $\frac{1}{2}$. Hint: see the hint for Exercise 42.

SOLUTION

-

h	$-.1$	$-.01$	$.01$	$.1$
$\dfrac{1 - \cos h}{h^2}$	.499583	.499996	.499996	.499583

The limit is $\frac{1}{2}$.

-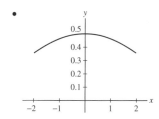

- $\lim\limits_{h\to 0} \dfrac{1 - \cos h}{h^2} = \lim\limits_{h\to 0} \dfrac{1 - \cos^2 h}{h^2(1 + \cos h)} = \lim\limits_{h\to 0} \left(\dfrac{\sin h}{h}\right)^2 \dfrac{1}{1 + \cos h} = \dfrac{1}{2}.$

49. Evaluate $\lim\limits_{h\to 0} \dfrac{\cos 3h - 1}{\cos 2h - 1}$.

SOLUTION

$$\lim_{h\to 0} \frac{\cos 3h - 1}{\cos 2h - 1} = \lim_{h\to 0} \frac{\cos 3h - 1}{\cos 2h - 1} \cdot \frac{\cos 3h + 1}{\cos 3h + 1} \cdot \frac{\cos 2h + 1}{\cos 2h + 1} = \lim_{h\to 0} \frac{\cos^2 3h - 1}{\cos^2 2h - 1} \cdot \frac{\cos 2h + 1}{\cos 3h + 1}$$

$$= \lim_{h\to 0} \frac{-\sin^2 3h}{-\sin^2 2h} \cdot \frac{\cos 2h + 1}{\cos 3h + 1} = \frac{9}{4} \lim_{h\to 0} \frac{\sin 3h}{3h} \cdot \frac{\sin 3h}{3h} \cdot \frac{2h}{\sin 2h} \cdot \frac{2h}{\sin 2h} \frac{\cos 2h + 1}{\cos 3h + 1}$$

$$= \frac{9}{4} \cdot 1 \cdot 1 \cdot 1 \cdot 1 \cdot \frac{2}{2} = \frac{9}{4}.$$

51. Using a diagram of the unit circle and the Pythagorean Theorem, show that

$$\sin^2 \theta \le (1 - \cos \theta)^2 + \sin^2 \theta \le \theta^2$$

Conclude that $\sin^2 \theta \le 2(1 - \cos \theta) \le \theta^2$ and use this to give an alternate proof of Eq. (7) in Exercise 42. Then give an alternate proof of the result in Exercise 47.

SOLUTION

- Consider the unit circle shown below. The triangle BDA is a right triangle. It has base $1 - \cos \theta$, altitude $\sin \theta$, and hypotenuse h. Observe that the hypotenuse h is less than the arc length $AB = $ radius $\cdot$ angle $= 1 \cdot \theta = \theta$. Apply the Pythagorean Theorem to obtain $(1 - \cos \theta)^2 + \sin^2 \theta = h^2 \le \theta^2$. The inequality $\sin^2 \theta \le (1 - \cos \theta)^2 + \sin^2 \theta$ follows from the fact that $(1 - \cos \theta)^2 \ge 0$.

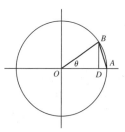

- Note that

$$(1 - \cos \theta)^2 + \sin^2 \theta = 1 - 2\cos \theta + \cos^2 \theta + \sin^2 \theta = 2 - 2\cos \theta = 2(1 - \cos \theta).$$

Therefore,

$$\sin^2 \theta \leq 2(1 - \cos \theta) \leq \theta^2.$$

- Divide the previous inequality by 2θ to obtain

$$\frac{\sin^2 \theta}{2\theta} \leq \frac{1 - \cos \theta}{\theta} \leq \frac{\theta}{2}.$$

Because

$$\lim_{\theta \to 0} \frac{\sin^2 \theta}{2\theta} = \frac{1}{2} \lim_{\theta \to 0} \frac{\sin \theta}{\theta} \cdot \lim_{\theta \to 0} \sin \theta = \frac{1}{2}(1)(0) = 0,$$

and $\lim_{h \to 0} \dfrac{\theta}{2} = 0$, it follows by the Squeeze Theorem that

$$\lim_{\theta \to 0} \frac{1 - \cos \theta}{\theta} = 0.$$

- Divide the inequality

$$\sin^2 \theta \leq 2(1 - \cos \theta) \leq \theta^2$$

by $2\theta^2$ to obtain

$$\frac{\sin^2 \theta}{2\theta^2} \leq \frac{1 - \cos \theta}{\theta^2} \leq \frac{1}{2}.$$

Because

$$\lim_{\theta \to 0} \frac{\sin^2 \theta}{2\theta^2} = \frac{1}{2} \lim_{\theta \to 0} \left(\frac{\sin \theta}{\theta} \right)^2 = \frac{1}{2}(1^2) = \frac{1}{2},$$

and $\lim_{h \to 0} \dfrac{1}{2} = \dfrac{1}{2}$, it follows by the Squeeze Theorem that

$$\lim_{\theta \to 0} \frac{1 - \cos \theta}{\theta^2} = \frac{1}{2}.$$

53. Let $A(n)$ be the area of a regular n-gon inscribed in a unit circle (Figure 8).

(a) Prove that $A(n) = \dfrac{1}{2} n \sin\left(\dfrac{2\pi}{n} \right)$.

(b) Intuitively, why might we expect $A(n)$ to converge to the area of the unit circle as $n \to \infty$?

(c) Use Theorem 2 to evaluate $\lim_{n \to \infty} A(n)$.

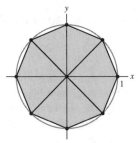

FIGURE 8 Regular n-gon inscribed in a unit circle.

SOLUTION

(a) $A(n) = n \cdot a(n)$, where $a(n)$ is the area of one section of the n-gon as shown. The sections into which the n-gon is divided are identical. By looking at the section lying just atop the x-axis, we can see that each section has a base length of 1 and a height of $\sin \theta$, where θ is the interior angle of the triangle.

$$\theta = \frac{2\pi}{n},$$

so

$$a(n) = \frac{1}{2} \sin \frac{2\pi}{n}.$$

This means

$$A(n) = n\frac{1}{2}\sin\left(\frac{2\pi}{n}\right) = \frac{1}{2}n\sin\left(\frac{2\pi}{n}\right).$$

(b) As n increases, the difference between the n-gon and the unit circle shrinks toward zero; hence, as n increases, $A(n)$ approaches the area of the unit circle.

(c) Substitute $x = \frac{1}{n}$, $n = \frac{1}{x}$, so $A(n) = \frac{1}{2}\frac{1}{x}\sin(2\pi x)$. $x \to 0$ as $n \to \infty$, so

$$\lim_{n\to\infty} A(n) = \lim_{x\to 0} \frac{1}{2}\frac{\sin(2\pi x)}{x} = \frac{1}{2}(2\pi) = \pi.$$

2.7 Intermediate Value Theorem

Preliminary Questions

1. Explain why $f(x) = x^2$ takes on the value 0.5 in the interval $[0, 1]$.

SOLUTION Observe that $f(x) = x^2$ is continuous on $[0, 1]$ with $f(0) = 0$ and $f(1) = 1$. Because $f(0) < 0.5 < f(1)$, the Intermediate Value Theorem guarantees there is a $c \in [0, 1]$ such that $f(c) = 0.5$.

2. The temperature in Vancouver was $46°$ at 6 AM and rose to $68°$ at noon. What must we assume about temperature to conclude that the temperature was $60°$ at some moment of time between 6 AM and noon?

SOLUTION We must assume that temperature is a continuous function of time.

3. What is the graphical interpretation of the IVT?

SOLUTION If f is continuous on $[a, b]$, then the horizontal line $y = k$ for every k between $f(a)$ and $f(b)$ intersects the graph of $y = f(x)$ at least once.

4. Show that the following statement is false by drawing a graph that provides a counterexample: *If $f(x)$ is continuous and has a root in $[a, b]$, then $f(a)$ and $f(b)$ have opposite signs.*

SOLUTION

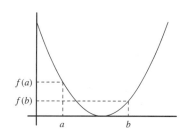

Exercises

1. Use the IVT to show that $f(x) = x^3 + x$ takes on the value 9 for some x in $[1, 2]$.

SOLUTION Observe that $f(1) = 2$ and $f(2) = 10$. Since f is a polynomial, it is continuous everywhere; in particular on $[1, 2]$. Therefore, by the IVT there is a $c \in [1, 2]$ such that $f(c) = 9$.

3. Show that $g(t) = t^2 \tan t$ takes on the value $\frac{1}{2}$ for some t in $\left[0, \frac{\pi}{4}\right]$.

SOLUTION $g(0) = 0$ and $g(\frac{\pi}{4}) = \frac{\pi^2}{16}$. $g(t)$ is continuous for all t between 0 and $\frac{\pi}{4}$, and $0 < \frac{1}{2} < \frac{\pi^2}{16}$; therefore, by the IVT, there is a $c \in [0, \frac{\pi}{4}]$ such that $g(c) = \frac{1}{2}$.

5. Show that $\cos x = x$ has a solution in the interval $[0, 1]$. *Hint:* Show that $f(x) = x - \cos x$ has a zero in $[0, 1]$.

SOLUTION Let $f(x) = x - \cos x$. Observe that f is continuous with $f(0) = -1$ and $f(1) = 1 - \cos 1 \approx .46$. Therefore, by the IVT there is a $c \in [0, 1]$ such that $f(c) = c - \cos c = 0$. Thus $c = \cos c$ and hence the equation $\cos x = x$ has a solution c in $[0, 1]$.

In Exercises 7–14, use the IVT to prove each of the following statements.

7. $\sqrt{c} + \sqrt{c + 1} = 2$ for some number c.

SOLUTION Let $f(x) = \sqrt{x} + \sqrt{x+1} - 2$. Note that f is continuous on $\left[\frac{1}{4}, 1\right]$ with $f(\frac{1}{4}) = \sqrt{\frac{1}{4}} + \sqrt{\frac{5}{4}} - 2 \approx -.38$ and $f(1) = \sqrt{2} - 1 \approx .41$. Therefore, by the IVT there is a $c \in \left[\frac{1}{4}, 1\right]$ such that $f(c) = \sqrt{c} + \sqrt{c+1} - 2 = 0$. Thus $\sqrt{c} + \sqrt{c+1} = 2$ and hence the equation $\sqrt{x} + \sqrt{x+1} = 2$ has a solution c in $\left[\frac{1}{4}, 1\right]$.

9. $\sqrt{2}$ exists. *Hint: Consider* $f(x) = x^2$.

SOLUTION Let $f(x) = x^2$. Observe that f is continuous with $f(1) = 1$ and $f(2) = 4$. Therefore, by the IVT there is a $c \in [1, 2]$ such that $f(c) = c^2 = 2$. This proves the existence of $\sqrt{2}$, a number whose square is 2.

11. For all positive integers k, there exists x such that $\cos x = x^k$.

SOLUTION For each positive integer k, let $f(x) = x^k - \cos x$. Observe that f is continuous on $\left[0, \frac{\pi}{2}\right]$ with $f(0) = -1$ and $f(\frac{\pi}{2}) = \left(\frac{\pi}{2}\right)^k > 0$. Therefore, by the IVT there is a $c \in \left[0, \frac{\pi}{2}\right]$ such that $f(c) = c^k - \cos(c) = 0$. Thus $\cos c = c^k$ and hence the equation $\cos x = x^k$ has a solution c in the interval $\left[0, \frac{\pi}{2}\right]$.

13. $2^x = b$ has a solution for all $b > 0$ (treat $b \geq 1$ first).

SOLUTION Let $b \geq 1$. Let $f(x) = 2^x$. f is continuous for all x. $f(0) = 1$, and $f(x) > b$ for some $x > 0$ (since $f(x)$ increases without bound). Hence, by the IVT, $f(x) = b$ for some $x > 0$.

On the other hand, suppose $b < 1$. Since $\lim_{x \to -\infty} 2^x = 0$, there is some $x < 0$ such that $f(x) = 2^x < b$. $f(0) = 1 > b$, so the IVT states that $f(x) = b$ for some $x < 0$.

15. Carry out three steps of the Bisection Method for $f(x) = 2^x - x^3$ as follows:
(a) Show that $f(x)$ has a zero in $[1, 1.5]$.
(b) Show that $f(x)$ has a zero in $[1.25, 1.5]$.
(c) Determine whether $[1.25, 1.375]$ or $[1.375, 1.5]$ contains a zero.

SOLUTION Note that $f(x)$ is continuous for all x.

(a) $f(1) = 1$, $f(1.5) = 2^{1.5} - (1.5)^3 < 3 - 3.375 < 0$. Hence, $f(x) = 0$ for some x between 0 and 1.5.
(b) $f(1.25) \approx 0.4253 > 0$ and $f(1.5) < 0$. Hence, $f(x) = 0$ for some x between 1.25 and 1.5.
(c) $f(1.375) \approx -0.0059$. Hence, $f(x) = 0$ for some x between 1.25 and 1.375.

17. Find an interval of length $\frac{1}{4}$ in $[0, 1]$ containing a root of $x^5 - 5x + 1 = 0$.

SOLUTION Let $f(x) = x^5 - 5x + 1$. Observe that f is continuous with $f(0) = 1$ and $f(1) = -3$. Therefore, by the IVT there is a $c \in [0, 1]$ such that $f(c) = 0$. $f(.5) = -1.46875 < 0$, so $f(c) = 0$ for some $c \in [0, .5]$. $f(.25) = -.0.249023 < 0$, and so $f(c) = 0$ for some $c \in [0, .25]$. This means that $[0, .25]$ is an interval of length 0.25 containing a root of $f(x)$.

In Exercises 19–22, draw the graph of a function $f(x)$ on $[0, 4]$ with the given property.

19. Jump discontinuity at $x = 2$ and does not satisfy the conclusion of the IVT.

SOLUTION The function graphed below has a jump discontinuity at $x = 2$. Note that while $f(0) = 2$ and $f(4) = 4$, there is no point c in the interval $[0, 4]$ such that $f(c) = 3$. Accordingly, the conclusion of the IVT is *not* satisfied.

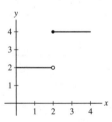

21. Infinite one-sided limits at $x = 2$ and does not satisfy the conclusion of the IVT.

SOLUTION The function graphed below has infinite one-sided limits at $x = 2$. Note that while $f(0) = 2$ and $f(4) = 4$, there is no point c in the interval $[0, 4]$ such that $f(c) = 3$. Accordingly, the conclusion of the IVT is *not* satisfied.

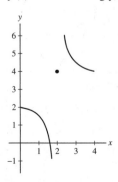

23. ✏️ Corollary 2 is not foolproof. Let $f(x) = x^2 - 1$ and explain why the corollary fails to detect the roots $x = \pm 1$ if $[a, b]$ contains $[-1, 1]$.

SOLUTION If $[a, b]$ contains $[-1, 1]$, $f(a) > 0$ and $f(b) > 0$. If we were using the corollary, we would guess that $f(x)$ has no roots in $[a, b]$. However, we can easily see that $f(1) = 0$ and $f(-1) = 0$, so that any interval $[a, b]$ containing $[-1, 1]$ contains two roots.

Further Insights and Challenges

25. ✏️ Assume that $f(x)$ is continuous and that $0 \le f(x) \le 1$ for $0 \le x \le 1$ (see Figure 5). Show that $f(c) = c$ for some c in $[0, 1]$.

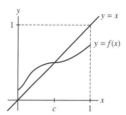

FIGURE 5 A function satisfying $0 \le f(x) \le 1$ for $0 \le x \le 1$.

SOLUTION If $f(0) = 0$, the proof is done with $c = 0$. We may assume that $f(0) > 0$. Let $g(x) = f(x) - x$. $g(0) = f(0) - 0 = f(0) > 0$. Since $f(x)$ is continuous, the Rule of Differences dictates that $g(x)$ is continuous. We need to prove that $g(c) = 0$ for some $c \in [0, 1]$. Since $f(1) \le 1$, $g(1) = f(1) - 1 \le 0$. If $g(1) = 0$, the proof is done with $c = 1$, so let's assume that $g(1) < 0$.

We now have a continuous function $g(x)$ on the interval $[0, 1]$ such that $g(0) > 0$ and $g(1) < 0$. From the IVT, there must be some $c \in [0, 1]$ so that $g(c) = 0$, so $f(c) - c = 0$ and so $f(c) = c$.

This is a simple case of a very general, useful, and beautiful theorem called the **Brouwer fixed point theorem**.

27. ✏️ Figure 6(B) shows a slice of ham on a piece of bread. Prove that it is possible to slice this open-faced sandwich so that each part has equal amounts of ham and bread. *Hint:* By Exercise 26, for all $0 \le \theta \le \pi$ there is a line $L(\theta)$ of incline θ (which we assume is unique) that divides the ham into two equal pieces. For $\theta \ne 0$ or π, let $B(\theta)$ denote the amount of bread to the left of $L(\theta)$ minus the amount to the right. Notice that $L(\pi) = L(0)$ since $\theta = 0$ and $\theta = \pi$ give the same line, but $B(\pi) = -B(0)$ since left and right get interchanged as the angle moves from 0 to π. Assume that $B(\theta)$ is continuous and apply the IVT. (By a further extension of this argument, one can prove the full "Ham Sandwich Theorem," which states that if you allow the knife to cut at a slant, then it is possible to cut a sandwich consisting of a slice of ham and two slices of bread so that all three layers are divided in half.)

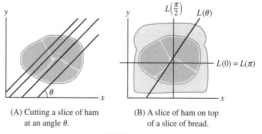

(A) Cutting a slice of ham at an angle θ.

(B) A slice of ham on top of a slice of bread.

FIGURE 6

SOLUTION For each angle θ, $0 \le \theta < \pi$, let $L(\theta)$ be the line at angle θ to the x-axis that slices the ham exactly in half, as shown in figure 6. Let $L(0) = L(\pi)$ be the horizontal line cutting the ham in half, also as shown. For θ and $L(\theta)$ thus defined, let $B(\theta) = $ the amount of bread to the left of $L(\theta)$ minus that to the right of $L(\theta)$.

To understand this argument, one must understand what we mean by "to the left" or "to the right". Here, we mean to the left or right of the line as viewed in the direction θ. Imagine you are walking along the line in direction θ (directly right if $\theta = 0$, directly left if $\theta = \pi$, etc).

We will further accept the fact that B is continuous as a function of θ, which seems intuitively obvious. We need to prove that $B(c) = 0$ for some angle c.

Since $L(0)$ and $L(\pi)$ are drawn in opposite direction, $B(0) = -B(\pi)$. If $B(0) > 0$, we apply the IVT on $[0, \pi]$ with $B(0) > 0$, $B(\pi) < 0$, and B continuous on $[0, \pi]$; by IVT, $B(c) = 0$ for some $c \in [0, \pi]$. On the other hand, if $B(0) < 0$, then we apply the IVT with $B(0) < 0$ and $B(\pi) > 0$. If $B(0) = 0$, we are also done; $L(0)$ is the appropriate line.

2.8 The Formal Definition of a Limit

Preliminary Questions

1. Given that $\lim\limits_{x \to 0} \cos x = 1$, which of the following statements is true?

(a) If $|\cos x - 1|$ is very small, then x is close to 0.

(b) There is an $\epsilon > 0$ such that $|x| < 10^{-5}$ if $0 < |\cos x - 1| < \epsilon$.

(c) There is a $\delta > 0$ such that $|\cos x - 1| < 10^{-5}$ if $0 < |x| < \delta$.

(d) There is a $\delta > 0$ such that $|\cos x| < 10^{-5}$ if $0 < |x - 1| < \delta$.

SOLUTION   The true statement is **(c)**: There is a $\delta > 0$ such that $|\cos x - 1| < 10^{-5}$ if $0 < |x| < \delta$.

2. Suppose that for a given ϵ and δ, it is known that $|f(x) - 2| < \epsilon$ if $0 < |x - 3| < \delta$. Which of the following statements must also be true?

(a) $|f(x) - 2| < \epsilon$ if $0 < |x - 3| < 2\delta$

(b) $|f(x) - 2| < 2\epsilon$ if $0 < |x - 3| < \delta$

(c) $|f(x) - 2| < \dfrac{\epsilon}{2}$ if $0 < |x - 3| < \dfrac{\delta}{2}$

(d) $|f(x) - 2| < \epsilon$ if $0 < |x - 3| < \dfrac{\delta}{2}$

SOLUTION   Statements **(b)** and **(d)** are true.

Exercises

1. Consider $\lim\limits_{x \to 4} f(x)$, where $f(x) = 8x + 3$.

(a) Show that $|f(x) - 35| = 8|x - 4|$.

(b) Show that for any $\epsilon > 0$, $|f(x) - 35| < \epsilon$ if $|x - 4| < \delta$, where $\delta = \frac{\epsilon}{8}$. Explain how this proves rigorously that $\lim\limits_{x \to 4} f(x) = 35$.

SOLUTION

(a) $|f(x) - 35| = |8x + 3 - 35| = |8x - 32| = |8(x - 4)| = 8\,|x - 4|$. (Remember that the last step is justified because $8 > 0$).

(b) Let $\epsilon > 0$. Let $\delta = \epsilon/8$ and suppose $|x - 4| < \delta$. By part **(a)**, $|f(x) - 35| = 8|x - 4| < 8\delta$. Substituting $\delta = \epsilon/8$, we see $|f(x) - 35| < 8\epsilon/8 = \epsilon$. We see that, for any $\epsilon > 0$, we found an appropriate δ so that $|x - 4| < \delta$ implies $|f(x) - 35| < \epsilon$. Hence $\lim\limits_{x \to 4} f(x) = 35$.

3. Consider $\lim\limits_{x \to 2} x^2 = 4$ (refer to Example 2).

(a) Show that $|x^2 - 4| < 0.05$ if $0 < |x - 2| < 0.01$.

(b) Show that $|x^2 - 4| < 0.0009$ if $0 < |x - 2| < 0.0002$.

(c) Find a value of δ such that $|x^2 - 4|$ is less than 10^{-4} if $0 < |x - 2| < \delta$.

SOLUTION

(a) If $0 < |x - 2| < \delta = .01$, then $|x| < 3$ and $\left|x^2 - 4\right| = |x - 2||x + 2| \leq |x - 2|\,(|x| + 2) < 5|x - 2| < .05$.

(b) If $0 < |x - 2| < \delta = .0002$, then $|x| < 2.0002$ and

$$\left|x^2 - 4\right| = |x - 2||x + 2| \leq |x - 2|\,(|x| + 2) < 4.0002|x - 2| < .00080004 < .0009.$$

(c) Note that $\left|x^2 - 4\right| = |(x + 2)(x - 2)| \leq |x + 2|\,|x - 2|$. Since $|x - 2|$ can get arbitrarily small, we can require $|x - 2| < 1$ so that $1 < x < 3$. This ensures that $|x + 2|$ is at most 5. Now we know that $\left|x^2 - 4\right| \leq 5|x - 2|$. Let $\delta = 10^{-5}$. Then, if $|x - 2| < \delta$, we get $\left|x^2 - 4\right| \leq 5|x - 2| < 5 \times 10^{-5} < 10^{-4}$ as desired.

5. Refer to Example 3 to find a value of $\delta > 0$ such that

$$\left|\frac{1}{x} - \frac{1}{3}\right| < 10^{-4} \qquad \text{if} \qquad 0 < |x - 3| < \delta$$

SOLUTION   The Example shows that for any $\epsilon > 0$ we have

$$\left|\frac{1}{x} - \frac{1}{3}\right| \leq \epsilon \quad \text{if } |x - 3| < \delta$$

where δ is the smaller of the numbers 6ϵ and 1. In our case, we may take $\delta = 6 \times 10^{-4}$.

7. GU   Plot $f(x) = \tan x$ together with the horizontal lines $y = 0.99$ and $y = 1.01$. Use this plot to find a value of $\delta > 0$ such that $|\tan x - 1| < 0.01$ if $|x - \frac{\pi}{4}| < \delta$.

SOLUTION   From the plot below, we see that $\delta = 0.005$ will guarantee that $|\tan x - 1| < 0.01$ whenever $|x - \frac{\pi}{4}| \leq \delta$.

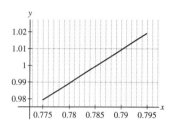

9. Consider $\lim\limits_{x \to 2} \frac{1}{x}$.

(a) Show that if $|x - 2| < 1$, then

$$\left| \frac{1}{x} - \frac{1}{2} \right| < \frac{1}{2}|x - 2|$$

(b) Let δ be the smaller of 1 and 2ϵ. Prove the following statement:

$$\left| \frac{1}{x} - \frac{1}{2} \right| < \epsilon \quad \text{if} \quad 0 < |x - 2| < \delta$$

(c) Find a $\delta > 0$ such that $\left| \frac{1}{x} - \frac{1}{2} \right| < 0.01$ if $|x - 2| < \delta$.

(d) Prove rigorously that $\lim\limits_{x \to 2} \frac{1}{x} = \frac{1}{2}$.

SOLUTION

(a) Since $|x - 2| < 1$, it follows that $1 < x < 3$, in particular that $x > 1$. Because $x > 1$, then $\frac{1}{x} < 1$ and

$$\left| \frac{1}{x} - \frac{1}{2} \right| = \left| \frac{2 - x}{2x} \right| = \frac{|x - 2|}{2x} < \frac{1}{2}|x - 2|.$$

(b) Let $\delta = \min\{1, 2\epsilon\}$ and suppose that $|x - 2| < \delta$. Then by part (a) we have

$$\left| \frac{1}{x} - \frac{1}{2} \right| < \frac{1}{2}|x - 2| < \frac{1}{2}\delta < \frac{1}{2} \cdot 2\epsilon = \epsilon.$$

(c) Choose $\delta = .02$. Then $\left| \frac{1}{x} - \frac{1}{2} \right| < \frac{1}{2}\delta = .01$ by part (b).

(d) Let $\epsilon > 0$ be given. Then whenever $0 < |x - 2| < \delta = \min\{1, 2\epsilon\}$, we have

$$\left| \frac{1}{x} - \frac{1}{2} \right| < \frac{1}{2}\delta \leq \epsilon.$$

Since ϵ was arbitrary, we conclude that $\lim\limits_{x \to 2} \frac{1}{x} = \frac{1}{2}$.

11. Based on the information conveyed in Figure 5(A), find values of L, ϵ, and $\delta > 0$ such that the following statement holds: $|f(x) - L| < \epsilon$ if $|x| < \delta$.

SOLUTION   We see $-.1 < x < .1$ forces $3.3 < f(x) < 4.7$. Rewritten, this means that $|x - 0| < .1$ implies that $|f(x) - 4| < .7$. Looking at the limit definition $|x - c| < \delta$ implies $|f(x) - L| < \epsilon$, we can replace so that $L = 4$, $\epsilon = .7$, $c = 0$, and $\delta = .1$.

13. ✏️   A calculator gives the following values for $f(x) = \sin x$:

$$f\left(\frac{\pi}{4} - 0.1\right) \approx 0.633, \quad f\left(\frac{\pi}{4}\right) \approx 0.707, \quad f\left(\frac{\pi}{4} + 0.1\right) \approx 0.774$$

Use these values and the fact that $f(x)$ is increasing on $[0, \frac{\pi}{2}]$ to justify the statement

$$\left| f(x) - f\left(\frac{\pi}{4}\right) \right| < 0.08 \quad \text{if} \quad \left| x - \frac{\pi}{4} \right| < 0.1$$

Then draw a figure like Figure 3 to illustrate this statement.

SOLUTION Since $f(x)$ is increasing on the interval, the three $f(x)$ values tell us that $.633 \leq f(x) \leq .774$ for all x between $\frac{\pi}{4} - .1$ and $\frac{\pi}{4} + .1$. We may subtract $f(\frac{\pi}{4})$ from the inequality for $f(x)$. This show that, for $\frac{\pi}{4} - .1 < x < \frac{\pi}{4} + .1$, $.633 - f(\frac{\pi}{4}) \leq f(x) - f(\frac{\pi}{4}) \leq .774 - f(\frac{\pi}{4})$. This means that, if $|x - \frac{\pi}{4}| < .1$, then $.633 - .707 \leq f(x) - f(\frac{\pi}{4}) \leq .774 - .707$, so $-0.074 \leq f(x) - f(\frac{\pi}{4}) \leq 0.067$. Then $-0.08 < f(x) - f(\frac{\pi}{4}) < 0.08$ follows from this, so $|x - \frac{\pi}{4}| < 0.1$ implies $|f(x) - f(\frac{\pi}{4})| < .08$. The figure below illustrates this.

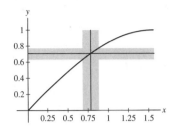

15. Adapt the argument in Example 2 to prove rigorously that $\lim\limits_{x \to c} x^2 = c^2$ for all c.

SOLUTION To relate the gap to $|x - c|$, we take

$$\left|x^2 - c^2\right| = |(x + c)(x - c)| = |x + c|\,|x - c|.$$

We choose δ in two steps. First, since we are requiring $|x - c|$ to be small, we require $\delta < |c|$, so that x lies between 0 and $2c$. This means that $|x + c| < 3|c|$, so $|x - c||x + c| < 3|c|\delta$. Next, we require that $\delta < \dfrac{\epsilon}{3|c|}$, so

$$|x - c||x + c| < \frac{\epsilon}{3|c|}3|c| = \epsilon,$$

and we are done.

Therefore, given $\epsilon > 0$, we let

$$\delta = \min\left\{|c|, \frac{\epsilon}{3|c|}\right\}.$$

Then, for $|x - c| < \delta$, we have

$$|x^2 - c^2| = |x - c|\,|x + c| < 3|c|\delta < 3|c|\frac{\epsilon}{3|c|} = \epsilon.$$

In Exercises 17–22, use the formal definition of the limit to prove the statement rigorously.

17. $\lim\limits_{x \to 4} \sqrt{x} = 2$

SOLUTION Let $\epsilon > 0$ be given. We bound $|\sqrt{x} - 2|$ by multiplying $\dfrac{\sqrt{x} + 2}{\sqrt{x} + 2}$.

$$|\sqrt{x} - 2| = \left|\sqrt{x} - 2\left(\frac{\sqrt{x} + 2}{\sqrt{x} + 2}\right)\right| = \left|\frac{x - 4}{\sqrt{x} + 2}\right| = |x - 4|\left|\frac{1}{\sqrt{x} + 2}\right|.$$

We can assume $\delta < 1$, so that $|x - 4| < 1$, and hence $\sqrt{x} + 2 > \sqrt{3} + 2 > 3$. This gives us

$$|\sqrt{x} - 2| = |x - 4|\left|\frac{1}{\sqrt{x} + 2}\right| < |x - 4|\frac{1}{3}.$$

Let $\delta = \min(1, 3\epsilon)$. If $|x - 4| < \delta$,

$$|\sqrt{x} - 2| = |x - 4|\left|\frac{1}{\sqrt{x} + 2}\right| < |x - 4|\frac{1}{3} < \delta\frac{1}{3} < 3\epsilon\frac{1}{3} = \epsilon,$$

thus proving the limit rigorously.

19. $\lim\limits_{x \to 1} x^3 = 1$

SOLUTION Let $\epsilon > 0$ be given. We bound $\left|x^3 - 1\right|$ by factoring the difference of cubes:

$$\left|x^3 - 1\right| = \left|(x^2 + x + 1)(x - 1)\right| = |x - 1|\left|x^2 + x + 1\right|.$$

Let $\delta = \min(1, \frac{\epsilon}{7})$, and assume $|x - 1| < \delta$. Since $\delta < 1$, $0 < x < 2$. Since $x^2 + x + 1$ increases as x increases for $x > 0$, $x^2 + x + 1 < 7$ for $0 < x < 2$, and so

$$\left| x^3 - 1 \right| = |x - 1| \left| x^2 + x + 1 \right| < 7|x - 1| < 7\frac{\epsilon}{7} = \epsilon$$

and the limit is rigorously proven.

21. $\lim_{x \to c} |x| = |c|$

SOLUTION Let $\epsilon > 0$ be given. First we bound $|x| - |c|$ by multiplying by $\dfrac{|x| + |c|}{|x| + |c|}$.

$$\big| |x| - |c| \big| < \left| \left(|x| - |c| \right) \frac{|x| + |c|}{|x| + |c|} \right| = \frac{\big| |x|^2 - |c|^2 \big|}{|x| + |c|} = \left| \frac{x^2 - c^2}{|x| + |c|} \right| = \left| \frac{(x + c)(x - c)}{|x| + |c|} \right| \leq |x - c| \frac{|x + c|}{|x| + |c|}.$$

Let $\delta = \epsilon$. We apply the *Triangle Inequality*—for all a, b:

$$|a + b| \leq |a| + |b|.$$

This yields:

$$\big| |x| - |c| \big| \leq |x - c| \frac{|x + c|}{|x| + |c|} \leq |x - c| = \epsilon,$$

and the limit is rigorously proven.

23. Let $f(x) = \dfrac{x}{|x|}$. Prove rigorously that $\lim_{x \to 0} f(x)$ does not exist. *Hint:* Show that no number L qualifies as the limit because there always exists some x such that $|x| < \delta$ but $|f(x) - L| \geq \frac{1}{2}$, no matter how small δ is taken.

SOLUTION Let L be any real number. Let $\delta > 0$ be any small positive number. Let $x = \frac{\delta}{2}$, which satisfies $|x| < \delta$, and $f(x) = 1$. We consider two cases:

- $(|f(x) - L| \geq \frac{1}{2})$: we are done.
- $(|f(x) - L| < \frac{1}{2})$: This means $\frac{1}{2} < L < \frac{3}{2}$. In this case, let $x = -\frac{\delta}{2}$. $f(x) = -1$, and so $\frac{3}{2} < L - f(x)$.

In either case, there exists an x such that $|x| < \frac{\delta}{2}$, but $|f(x) - L| \geq \frac{1}{2}$.

25. Use the identity

$$\sin x + \sin y = 2 \sin \left(\frac{x + y}{2} \right) \cos \left(\frac{x - y}{2} \right)$$

to verify the relation

$$\sin(a + h) - \sin a = h \frac{\sin(h/2)}{h/2} \cos \left(a + \frac{h}{2} \right) \qquad \boxed{6}$$

Conclude that $|\sin(a + h) - \sin a| < |h|$ for all a and prove rigorously that $\lim_{x \to a} \sin x = \sin a$. You may use the inequality $\left| \dfrac{\sin x}{x} \right| \leq 1$ for $x \neq 0$.

SOLUTION We first write

$$\sin(a + h) - \sin a = \sin(a + h) + \sin(-a).$$

Applying the identity with $x = a + h$, $y = -a$, yields:

$$\sin(a + h) - \sin a = \sin(a + h) + \sin(-a) = 2 \sin \left(\frac{a + h - a}{2} \right) \cos \left(\frac{2a + h}{2} \right)$$

$$= 2 \sin \left(\frac{h}{2} \right) \cos \left(a + \frac{h}{2} \right) = 2 \left(\frac{h}{h} \right) \sin \left(\frac{h}{2} \right) \cos \left(a + \frac{h}{2} \right) = h \frac{\sin(h/2)}{h/2} \cos \left(a + \frac{h}{2} \right).$$

Therefore,

$$|\sin(a + h) - \sin a| = |h| \left| \frac{\sin(h/2)}{h/2} \right| \left| \cos \left(a + \frac{h}{2} \right) \right|.$$

Using the fact that $\left| \dfrac{\sin \theta}{\theta} \right| < 1$ and that $|\cos \theta| \leq 1$, we have Making the substitution $h = x - a$, we see that this last relation is equivalent to

$$|\sin x - \sin a| < |x - a|.$$

Now, to prove the desired limit, let $\epsilon > 0$, and take $\delta = \epsilon$. If $|x - a| < \delta$, then

$$|\sin x - \sin a| < |x - a| < \delta = \epsilon,$$

Therefore, a δ was found for arbitrary ϵ, and the proof is complete.

Further Insights and Challenges

27. Uniqueness of the Limit   Show that a function converges to at most one limiting value. In other words, use the limit definition to show that if $\lim_{x \to c} f(x) = L_1$ and $\lim_{x \to c} f(x) = L_2$, then $L_1 = L_2$.

SOLUTION   Let $\epsilon > 0$ be given. Since $\lim_{x \to c} f(x) = L_1$, there exists δ_1 such that if $|x - c| < \delta_1$ then $|f(x) - L_1| < \epsilon$. Similarly, since $\lim_{x \to c} f(x) = L_2$, there exists δ_2 such that if $|x - c| < \delta_2$ then $|f(x) - L_2| < \epsilon$. Now let $|x - c| < \min(\delta_1, \delta_2)$ and observe that

$$\begin{aligned}
|L_1 - L_2| &= |L_1 - f(x) + f(x) - L_2| \\
&\leq |L_1 - f(x)| + |f(x) - L_2| \\
&= |f(x) - L_1| + |f(x) - L_2| < 2\epsilon.
\end{aligned}$$

So, $|L_1 - L_2| < 2\epsilon$ for any $\epsilon > 0$. We have $|L_1 - L_2| = \lim_{\epsilon \to 0} |L_1 - L_2| < \lim_{\epsilon \to 0} 2\epsilon = 0$. Therefore, $|L_1 - L_2| = 0$ and, hence, $L_1 = L_2$.

In Exercises 28–30, prove the statement using the formal limit definition.

29.   The Squeeze Theorem. (Theorem 1 in Section 2.6, p. 89)

SOLUTION   *Proof of the Squeeze Theorem.* Suppose that (i) the inequalities $h(x) \leq f(x) \leq g(x)$ hold for all x near (but not equal to) a and (ii) $\lim_{x \to a} h(x) = \lim_{x \to a} g(x) = L$. Let $\epsilon > 0$ be given.

- By (i), there exists a $\delta_1 > 0$ such that $h(x) \leq f(x) \leq g(x)$ whenever $0 < |x - a| < \delta_1$.
- By (ii), there exist $\delta_2 > 0$ and $\delta_3 > 0$ such that $|h(x) - L| < \epsilon$ whenever $0 < |x - a| < \delta_2$ and $|g(x) - L| < \epsilon$ whenever $0 < |x - a| < \delta_3$.
- Choose $\delta = \min\{\delta_1, \delta_2, \delta_3\}$. Then whenever $0 < |x - a| < \delta$ we have $L - \epsilon < h(x) \leq f(x) \leq g(x) < L + \epsilon$; i.e., $|f(x) - L| < \epsilon$. Since ϵ was arbitrary, we conclude that $\lim_{x \to a} f(x) = L$.

31.   Let $f(x) = 1$ if x is rational and $f(x) = 0$ if x is irrational. Prove that $\lim_{x \to c} f(x)$ does not exist for any c.

SOLUTION   Let c be any number, and let $\delta > 0$ be an arbitrary small number. We will prove that there is an x such that $|x - c| < \delta$, but $|f(x) - f(c)| > \frac{1}{2}$. c must be either irrational or rational. If c is rational, then $f(c) = 1$. Since the irrational numbers are dense, there is at least one irrational number z such that $|z - c| < \delta$. $|f(z) - f(c)| = 1 > \frac{1}{2}$, so the function is discontinuous at $x = c$. On the other hand, if c is irrational, then there is a *rational* number q such that $|q - c| < \delta$. $|f(q) - f(c)| = |1 - 0| = 1 > \frac{1}{2}$, so the function is discontinuous at $x = c$.

CHAPTER REVIEW EXERCISES

1.   A particle's position at time t (s) is $s(t) = \sqrt{t^2 + 1}$ m. Compute its average velocity over $[2, 5]$ and estimate its instantaneous velocity at $t = 2$.

SOLUTION   Let $s(t) = \sqrt{t^2 + 1}$. The average velocity over $[2, 5]$ is

$$\frac{s(5) - s(2)}{5 - 2} = \frac{\sqrt{26} - \sqrt{5}}{3} \approx 0.954 \text{ m/s.}$$

From the data in the table below, we estimate that the instantaneous velocity at $t = 2$ is approximately 0.894 m/s.

interval	[1.9, 2]	[1.99, 2]	[1.999, 2]	[2, 2.001]	[2, 2.01]	[2, 2.1]
average ROC	0.889769	0.893978	0.894382	0.894472	0.894873	0.898727

3.   For a whole number n, let $P(n)$ be the number of *partitions* of n, that is, the number of ways of writing n as a sum of one or more whole numbers. For example, $P(4) = 5$ since the number 4 can be partitioned in five different ways: 4, $3 + 1$, $2 + 2$, $2 + 1 + 1$, $1 + 1 + 1 + 1$. Treating $P(n)$ as a continuous function, use Figure 1 to estimate the rate of change of $P(n)$ at $n = 12$.

SOLUTION   The tangent line drawn in the figure appears to pass through the points (15, 140) and (10.5, 40). We therefore estimate that the rate of change of $P(n)$ at $n = 12$ is

$$\frac{140 - 40}{15 - 10.5} = \frac{100}{4.5} = \frac{200}{9}.$$

In Exercises 5–8, estimate the limit numerically to two decimal places or state that the limit does not exist.

5. $\displaystyle\lim_{x \to 0} \frac{1 - \cos^3(x)}{x^2}$

SOLUTION   Let $f(x) = \frac{1 - \cos^3 x}{x^2}$. The data in the table below suggests that

$$\lim_{x \to 0} \frac{1 - \cos^3 x}{x^2} \approx 1.50.$$

In constructing the table, we take advantage of the fact that f is an even function.

x	± 0.001	± 0.01	± 0.1
$f(x)$	1.500000	1.499912	1.491275

(The exact value is $\frac{3}{2}$.)

7. $\displaystyle\lim_{x \to 2} \frac{x^x - 4}{x^2 - 4}$

SOLUTION   Let $f(x) = \frac{x^x - 4}{x^2 - 4}$. The data in the table below suggests that

$$\lim_{x \to 2} \frac{x^x - 4}{x^2 - 4} \approx 1.69.$$

x	1.9	1.99	1.999	2.001	2.01	2.1
$f(x)$	1.575461	1.680633	1.691888	1.694408	1.705836	1.828386

(The exact value is $1 + \ln 2$.)

In Exercises 9–42, evaluate the limit if possible or state that it does not exist.

9. $\displaystyle\lim_{x \to 4} (3 + x^{1/2})$

SOLUTION $\displaystyle\lim_{x \to 4} (3 + x^{1/2}) = 3 + \sqrt{4} = 5.$

11. $\displaystyle\lim_{x \to -2} \frac{4}{x^3}$

SOLUTION $\displaystyle\lim_{x \to -2} \frac{4}{x^3} = \frac{4}{(-2)^3} = -\frac{1}{2}.$

13. $\displaystyle\lim_{x \to 1} \frac{x^3 - x}{x - 1}$

SOLUTION $\displaystyle\lim_{x \to 1} \frac{x^3 - x}{x - 1} = \lim_{x \to 1} \frac{x(x - 1)(x + 1)}{x - 1} = \lim_{x \to 1} x(x + 1) = 1(1 + 1) = 2.$

15. $\displaystyle\lim_{t \to 9} \frac{\sqrt{t} - 3}{t - 9}$

SOLUTION $\displaystyle\lim_{t \to 9} \frac{\sqrt{t} - 3}{t - 9} = \lim_{t \to 9} \frac{\sqrt{t} - 3}{(\sqrt{t} - 3)(\sqrt{t} + 3)} = \lim_{t \to 9} \frac{1}{\sqrt{t} + 3} = \frac{1}{\sqrt{9} + 3} = \frac{1}{6}.$

17. $\displaystyle\lim_{x \to 3} \frac{\sqrt{x + 1} - 2}{x - 3}$

SOLUTION

$$\lim_{x \to 3} \frac{\sqrt{x + 1} - 2}{x - 3} = \lim_{x \to 3} \frac{\sqrt{x + 1} - 2}{x - 3} \cdot \frac{\sqrt{x + 1} + 2}{\sqrt{x + 1} + 2} = \lim_{x \to 3} \frac{(x + 1) - 4}{(x - 3)(\sqrt{x + 1} + 2)}$$

$$= \lim_{x \to 3} \frac{1}{\sqrt{x + 1} + 2} = \frac{1}{\sqrt{3 + 1} + 2} = \frac{1}{4}.$$

19. $\lim\limits_{h \to 0} \dfrac{2(a+h)^2 - 2a^2}{h}$

SOLUTION

$$\lim_{h \to 0} \frac{2(a+h)^2 - 2a^2}{h} = \lim_{h \to 0} \frac{2a^2 + 4ah + 2h^2 - 2a^2}{h} = \lim_{h \to 0} \frac{h(4a + 2h)}{h} = \lim_{h \to 0} (4a + 2h) = 4a + 2(0) = 4a.$$

21. $\lim\limits_{t \to 3} \dfrac{1}{t^2 - 9}$

SOLUTION Because the one-sided limits

$$\lim_{t \to 3-} \frac{1}{t^2 - 9} = -\infty \qquad \text{and} \qquad \lim_{t \to 3+} \frac{1}{t^2 - 9} = \infty,$$

are not equal, the two-sided limit

$$\lim_{t \to 3} \frac{1}{t^2 - 9} \qquad \text{does not exist.}$$

23. $\lim\limits_{a \to b} \dfrac{a^2 - 3ab + 2b^2}{a - b}$

SOLUTION $\quad \lim\limits_{a \to b} \dfrac{a^2 - 3ab + 2b^2}{a - b} = \lim\limits_{a \to b} \dfrac{(a - b)(a - 2b)}{a - b} = \lim\limits_{a \to b} (a - 2b) = b - 2b = -b.$

25. $\lim\limits_{x \to 0} \left(\dfrac{1}{3x} - \dfrac{1}{x(x+3)} \right)$

SOLUTION $\quad \lim\limits_{x \to 0} \left(\dfrac{1}{3x} - \dfrac{1}{x(x+3)} \right) = \lim\limits_{x \to 0} \dfrac{(x+3) - 3}{3x(x+3)} = \lim\limits_{x \to 0} \dfrac{1}{3(x+3)} = \dfrac{1}{3(0+3)} = \dfrac{1}{9}.$

27. $\lim\limits_{x \to 1.5} \dfrac{[x]}{x}$

SOLUTION $\quad \lim\limits_{x \to 1.5} \dfrac{[x]}{x} = \dfrac{[1.5]}{1.5} = \dfrac{1}{1.5} = \dfrac{2}{3}.$

29. $\lim\limits_{t \to 0-} \dfrac{[x]}{x}$

SOLUTION For x sufficiently close to zero but negative, $[x] = -1$. Therefore,

$$\lim_{x \to 0-} \frac{[x]}{x} = \lim_{x \to 0-} \frac{-1}{x} = \infty.$$

31. $\lim\limits_{\theta \to \frac{\pi}{4}} \sec \theta$

SOLUTION

$$\lim_{\theta \to \frac{\pi}{4}} \sec \theta = \sec \frac{\pi}{4} = \sqrt{2}.$$

33. $\lim\limits_{\theta \to 0} \dfrac{\cos \theta - 2}{\theta}$

SOLUTION Because the one-sided limits

$$\lim_{\theta \to 0-} \frac{\cos \theta - 2}{\theta} = \infty \qquad \text{and} \qquad \lim_{\theta \to 0+} \frac{\cos \theta - 2}{\theta} = -\infty$$

are not equal, the two-sided limit

$$\lim_{\theta \to 0} \frac{\cos \theta - 2}{\theta} \qquad \text{does not exist.}$$

35. $\lim\limits_{\theta \to 0} \dfrac{\sin 4x}{\sin 3x}$

SOLUTION

$$\lim_{\theta\to 0}\frac{\sin 4\theta}{\sin 3\theta}=\frac{4}{3}\lim_{\theta\to 0}\frac{\sin 4\theta}{4\theta}\cdot\frac{3\theta}{\sin 3\theta}=\frac{4}{3}\lim_{\theta\to 0}\frac{\sin 4\theta}{4\theta}\cdot\lim_{\theta\to 0}\frac{3\theta}{\sin 3\theta}=\frac{4}{3}(1)(1)=\frac{4}{3}.$$

37. $\displaystyle\lim_{x\to\frac{\pi}{2}}\tan x$

SOLUTION   Because the one-sided limits

$$\lim_{x\to\frac{\pi}{2}-}\tan x=\infty\qquad\text{and}\qquad\lim_{x\to\frac{\pi}{2}+}\tan x=-\infty$$

are not equal, the two-sided limit

$$\lim_{x\to\frac{\pi}{2}}\tan x\qquad\text{does not exist.}$$

39. $\displaystyle\lim_{t\to 0+}\sqrt{t}\cos\frac{1}{t}$

SOLUTION   For $t>0$,

$$-1\le\cos\left(\frac{1}{t}\right)\le 1,$$

so

$$-\sqrt{t}\le\sqrt{t}\cos\left(\frac{1}{t}\right)\le\sqrt{t}.$$

Because

$$\lim_{t\to 0+}-\sqrt{t}=\lim_{t\to 0+}\sqrt{t}=0,$$

it follows from the Squeeze Theorem that

$$\lim_{t\to 0+}\sqrt{t}\cos\left(\frac{1}{t}\right)=0.$$

41. $\displaystyle\lim_{x\to 0}\frac{\cos x-1}{\sin x}$

SOLUTION

$$\lim_{x\to 0}\frac{\cos x-1}{\sin x}=\lim_{x\to 0}\frac{\cos x-1}{\sin x}\cdot\frac{\cos x+1}{\cos x+1}=\lim_{x\to 0}\frac{-\sin^2 x}{\sin x(\cos x+1)}=-\lim_{x\to 0}\frac{\sin x}{\cos x+1}=-\frac{0}{1+1}=0.$$

43. Find the left- and right-hand limits of the function $f(x)$ in Figure 2 at $x=0,2,4$. State whether $f(x)$ is left- or right-continuous (or both) at these points.

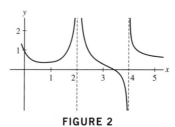

FIGURE 2

SOLUTION   According to the graph of $f(x)$,

$$\lim_{x\to 0-}f(x)=\lim_{x\to 0+}f(x)=1$$

$$\lim_{x\to 2-}f(x)=\lim_{x\to 2+}f(x)=\infty$$

$$\lim_{x\to 4-}f(x)=-\infty$$

$$\lim_{x\to 4+}f(x)=\infty.$$

The function is both left- and right-continuous at $x=0$ and neither left- nor right-continuous at $x=2$ and $x=4$.

45. Sketch the graph of a function $g(x)$ such that

$$\lim_{x \to -3-} g(x) = \infty, \qquad \lim_{x \to -3+} g(x) = -\infty, \qquad \lim_{x \to 4} g(x) = \infty$$

SOLUTION

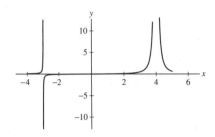

47. Find a constant b such that $h(x)$ is continuous at $x = 2$, where

$$h(x) = \begin{cases} x + 1 & \text{for } |x| < 2 \\ b - x^2 & \text{for } |x| \geq 2 \end{cases}$$

With this choice of b, find all points of discontinuity.

SOLUTION   To make $h(x)$ continuous at $x = 2$, we must have the two one-sided limits as x approaches 2 be equal. With

$$\lim_{x \to 2-} h(x) = \lim_{x \to 2-} (x + 1) = 2 + 1 = 3$$

and

$$\lim_{x \to 2+} h(x) = \lim_{x \to 2+} (b - x^2) = b - 4,$$

it follows that we must choose $b = 7$. Because $x + 1$ is continuous for $-2 < x < 2$ and $7 - x^2$ is continuous for $x \leq -2$ and for $x \geq 2$, the only possible point of discontinuity is $x = -2$. At $x = -2$,

$$\lim_{x \to -2+} h(x) = \lim_{x \to -2+} (x + 1) = -2 + 1 = -1$$

and

$$\lim_{x \to -2-} h(x) = \lim_{x \to -2-} (7 - x^2) = 7 - (-2)^2 = 3,$$

so $h(x)$ has a jump discontinuity at $x = -2$.

49. Let $f(x)$ and $g(x)$ be functions such that $g(x) \neq 0$ for $x \neq a$, and let

$$A = \lim_{x \to a} f(x), \qquad B = \lim_{x \to a} g(x), \qquad L = \lim_{x \to a} \frac{f(x)}{g(x)}$$

Prove that if the limits A, B, and L exist and $L = 1$, then $A = B$. *Hint:* You cannot use the Quotient Law if $B = 0$, so apply the Product Law to L and B instead.

SOLUTION   Suppose the limits A, B, and L all exist and $L = 1$. Then

$$B = B \cdot 1 = B \cdot L = \lim_{x \to a} g(x) \cdot \lim_{x \to a} \frac{f(x)}{g(x)} = \lim_{x \to a} g(x) \frac{f(x)}{g(x)} = \lim_{x \to a} f(x) = A.$$

51. Let $f(x)$ be a function defined for all real numbers. Which of the following statements *must be true*, *might be true*, or *are never true*.

(a) $\lim_{x \to 3} f(x) = f(3)$.

(b) If $\lim_{x \to 0} \frac{f(x)}{x} = 1$, then $f(0) = 0$.

(c) If $\lim_{x \to -7} f(x) = 8$, then $\lim_{x \to -7} \frac{1}{f(x)} = \frac{1}{8}$.

(d) If $\lim_{x \to 5+} f(x) = 4$ and $\lim_{x \to 5-} f(x) = 8$, then $\lim_{x \to 5} f(x) = 6$.

(e) If $\lim_{x \to 0} \frac{f(x)}{x} = 1$, then $\lim_{x \to 0} f(x) = 0$

(f) If $\lim_{x \to 5} f(x) = 2$, then $\lim_{x \to 5} f(x)^3 = 8$.

SOLUTION

(a) This statement might be true. If $f(x)$ is continuous at $x = 3$, then the statement will be true; if $f(x)$ is not continuous at $x = 3$, then the statement will not be true.

(b) This statement might be true. If $f(x)$ is continuous at $x = 0$, then the statement will be true; if $f(x)$ is not continuous at $x = 0$, then the statement will not be true.

(c) This statement is always true.

(d) This statement is never true. If the two one-sided limits are not equal, then the two-sided limit does not exist.

(e) This statement is always true.

(f) This statement is always true.

53. Let $f(x) = x \left[\frac{1}{x}\right]$, where $[x]$ is the greatest integer function.

(a) Sketch the graph of $f(x)$ on the interval $[\frac{1}{4}, 2]$.

(b) Show that for $x \neq 0$,

$$\frac{1}{x} - 1 < \left[\frac{1}{x}\right] \le \frac{1}{x}$$

Then use the Squeeze Theorem to prove that

$$\lim_{x \to 0} x \left[\frac{1}{x}\right] = 1$$

SOLUTION

(a) The graph of $f(x) = x \left[\frac{1}{x}\right]$ over $[\frac{1}{4}, 2]$ is shown below.

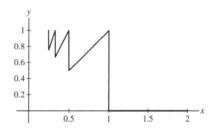

(b) Let y be any real number. From the definition of the greatest integer function, it follows that $y - 1 < [y] \le y$, with equality holding if and only if y is an integer. If $x \neq 0$, then $\frac{1}{x}$ is a real number, so

$$\frac{1}{x} - 1 < \left[\frac{1}{x}\right] \le \frac{1}{x}.$$

Upon multiplying this inequality through by x, we find

$$1 - x < x \left[\frac{1}{x}\right] \le 1.$$

Because

$$\lim_{x \to 0} (1 - x) = \lim_{x \to 0} 1 = 1,$$

it follows from the Squeeze Theorem that

$$\lim_{x \to 0} x \left[\frac{1}{x}\right] = 1.$$

55. Let $f(x) = \frac{1}{x+2}$.

(a) Show that $\left| f(x) - \frac{1}{4} \right| < \frac{|x-2|}{12}$ if $|x - 2| < 1$. *Hint:* Observe that $|4(x + 2)| > 12$ if $|x - 2| < 1$.

(b) Find $\delta > 0$ such that $\left| f(x) - \frac{1}{4} \right| < 0.01$ for $|x - 2| < \delta$.

(c) Prove rigorously that $\lim_{x \to 2} f(x) = \frac{1}{4}$.

FIGURE 3

SOLUTION

(a) Let $f(x) = \frac{1}{x+2}$. Then

$$\left| f(x) - \frac{1}{4} \right| = \left| \frac{1}{x+2} - \frac{1}{4} \right| = \left| \frac{4 - (x+2)}{4(x+2)} \right| = \frac{|x-2|}{|4(x+2)|}.$$

If $|x - 2| < 1$, then $1 < x < 3$, so $3 < x + 2 < 5$ and $12 < 4(x + 2) < 20$. Hence,

$$\frac{1}{|4(x+2)|} < \frac{1}{12} \qquad \text{and} \qquad \left| f(x) - \frac{1}{4} \right| < \frac{|x-2|}{12}.$$

(b) If $|x - 2| < \delta$, then by part (a),

$$\left| f(x) - \frac{1}{4} \right| < \frac{\delta}{12}.$$

Choosing $\delta = 0.12$ will then guarantee that $|f(x) - \frac{1}{4}| < 0.01$.

(c) Let $\epsilon > 0$ and take $\delta = \min\{1, 12\epsilon\}$. Then, whenever $|x - 2| < \delta$,

$$\left| f(x) - \frac{1}{4} \right| = \left| \frac{1}{x+2} - \frac{1}{4} \right| = \frac{|2-x|}{4|x+2|} \le \frac{|x-2|}{12} < \frac{\delta}{12} = \epsilon.$$

57. Prove rigorously that $\lim_{x \to -1} (4 + 8x) = -4$.

SOLUTION Let $\epsilon > 0$ and take $\delta = \epsilon/8$. Then, whenever $|x - (-1)| = |x + 1| < \delta$,

$$|f(x) - (-4)| = |4 + 8x + 4| = 8|x + 1| < 8\delta = \epsilon.$$

59. Use the IVT to prove that the curves $y = x^2$ and $y = \cos x$ intersect.

SOLUTION Let $f(x) = x^2 - \cos x$. Note that any root of $f(x)$ corresponds to a point of intersection between the curves $y = x^2$ and $y = \cos x$. Now, $f(x)$ is continuous over the interval $[0, \frac{\pi}{2}]$, $f(0) = -1 < 0$ and $f(\frac{\pi}{2}) = \frac{\pi^2}{4} > 0$. Therefore, by the Intermediate Value Theorem, there exists a $c \in (0, \frac{\pi}{2})$ such that $f(c) = 0$; consequently, the curves $y = x^2$ and $y = \cos x$ intersect.

61. Use the Bisection Method to locate a root of $x^2 - 7 = 0$ to two decimal places.

SOLUTION Let $f(x) = x^2 - 7$. By trial and error, we find that $f(2.6) = -0.24 < 0$ and $f(2.7) = 0.29 > 0$. Because $f(x)$ is continuous on $[2.6, 2.7]$, it follows from the Intermediate Value Theorem that $f(x)$ has a root on $(2.6, 2.7)$. We approximate the root by the midpoint of the interval: $x = 2.65$. Now, $f(2.65) = 0.0225 > 0$. Because $f(2.6)$ and $f(2.65)$ are of opposite sign, the root must lie on $(2.6, 2.65)$. The midpoint of this interval is $x = 2.625$ and $f(2.625) < 0$; hence, the root must be on the interval $(2.625, 2.65)$. Continuing in this fashion, we construct the following sequence of intervals and midpoints.

interval	midpoint
(2.625, 2.65)	2.6375
(2.6375, 2.65)	2.64375
(2.64375, 2.65)	2.646875
(2.64375, 2.646875)	2.6453125
(2.6453125, 2.646875)	2.64609375

At this point, we note that, to two decimal places, one root of $x^2 - 7 = 0$ is 2.65.

3 | DIFFERENTIATION

3.1 Definition of the Derivative

Preliminary Questions

1. Which of the lines in Figure 10 are tangent to the curve?

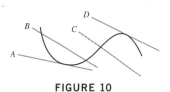

FIGURE 10

SOLUTION Lines A and D are tangent to the curve.

2. What are the two ways of writing the difference quotient?

SOLUTION The difference quotient may be written either as

$$\frac{f(x) - f(a)}{x - a}$$

or as

$$\frac{f(a + h) - f(a)}{h}.$$

3. For which value of x is

$$\frac{f(x) - f(3)}{x - 3} = \frac{f(7) - f(3)}{4}?$$

SOLUTION With $x = 7$,

$$\frac{f(x) - f(3)}{x - 3} = \frac{f(7) - f(3)}{4}.$$

4. What do the following quantities represent in terms of the graph of $f(x) = \sin x$?

(a) $\sin 1.3 - \sin 0.9$

(b) $\dfrac{\sin 1.3 - \sin 0.9}{0.4}$

(c) $f'(1.3)$

SOLUTION Consider the graph of $y = \sin x$.

(a) The quantity $\sin 1.3 - \sin .9$ represents the difference in height between the points $(.9, \sin .9)$ and $(1.3, \sin 1.3)$.

(b) The quantity $\dfrac{\sin 1.3 - \sin .9}{.4}$ represents the slope of the secant line between the points $(.9, \sin .9)$ and $(1.3, \sin 1.3)$ on the graph.

(c) The quantity $f'(1.3)$ represents the slope of the tangent line to the graph at $x = 1.3$.

5. For which values of a and h is $\dfrac{f(a + h) - f(a)}{h}$ equal to the slope of the secant line between the points $(3, f(3))$ and $(5, f(5))$ on the graph of $f(x)$?

SOLUTION With $a = 3$ and $h = 2$, $\dfrac{f(a + h) - f(a)}{h}$ is equal to the slope of the secant line between the points $(3, f(3))$ and $(5, f(5))$ on the graph of $f(x)$.

6. To which derivative is the quantity

$$\frac{\tan\left(\frac{\pi}{4} + 0.00001\right) - 1}{0.00001}$$

a good approximation?

SOLUTION $\dfrac{\tan\left(\frac{\pi}{4} + 0.00001\right) - 1}{0.00001}$ is a good approximation to the derivative of the function $f(x) = \tan x$ at $x = \frac{\pi}{4}$.

Exercises

1. Let $f(x) = 3x^2$. Show that

$$f(2 + h) = 3h^2 + 12h + 12$$

Then show that

$$\frac{f(2 + h) - f(2)}{h} = 3h + 12$$

and compute $f'(2)$ by taking the limit as $h \to 0$.

SOLUTION   With $f(x) = 3x^2$, it follows that

$$f(2 + h) = 3(2 + h)^2 = 3(4 + 4h + h^2) = 12 + 12h + 3h^2.$$

Using this result, we find

$$\frac{f(2 + h) - f(2)}{h} = \frac{12 + 12h + 3h^2 - 3 \cdot 4}{h} = \frac{12 + 12h + 3h^2 - 12}{h} = \frac{12h + 3h^2}{h} = 12 + 3h.$$

As $h \to 0$, $12 + 3h \to 12$, so $f'(2) = 12$.

In Exercises 3–6, compute $f'(a)$ in two ways, using Eq. (1) and Eq. (2).

3. $f(x) = x^2 + 9x, \quad a = 0$

SOLUTION   Let $f(x) = x^2 + 9x$. Then

$$f'(0) = \lim_{h \to 0} \frac{f(0 + h) - f(0)}{h} = \lim_{h \to 0} \frac{(0 + h)^2 + 9(0 + h) - 0}{h} = \lim_{h \to 0} \frac{9h + h^2}{h} = \lim_{h \to 0} (9 + h) = 9.$$

Alternately,

$$f'(0) = \lim_{x \to 0} \frac{f(x) - f(0)}{x - 0} = \lim_{x \to 0} \frac{x^2 + 9x - 0}{x} = \lim_{x \to 0} (x + 9) = 9.$$

5. $f(x) = 3x^2 + 4x + 2, \quad a = -1$

SOLUTION   Let $f(x) = 3x^2 + 4x + 2$. Then

$$f'(-1) = \lim_{h \to 0} \frac{f(-1 + h) - f(-1)}{h} = \lim_{h \to 0} \frac{3(-1 + h)^2 + 4(-1 + h) + 2 - 1}{h}$$

$$= \lim_{h \to 0} \frac{3h^2 - 2h}{h} = \lim_{h \to 0} (3h - 2) = -2.$$

Alternately,

$$f'(-1) = \lim_{x \to -1} \frac{f(x) - f(-1)}{x - (-1)} = \lim_{x \to -1} \frac{3x^2 + 4x + 2 - 1}{x + 1}$$

$$= \lim_{x \to -1} \frac{(3x + 1)(x + 1)}{x + 1} = \lim_{x \to -1} (3x + 1) = -2.$$

In Exercises 7–10, refer to the function whose graph is shown in Figure 11.

7. Calculate the slope of the secant line through the points on the graph where $x = 0$ and $x = 2.5$.

SOLUTION $f(0) \approx 6$ and $f(2.5) \approx 2$, so the slope of the secant line connecting $(0, f(0))$ and $(2.5, f(2.5))$ is

$$\frac{f(2.5) - f(0)}{2.5 - 0} \approx \frac{2 - 6}{2.5} = \frac{-4}{2.5} = -1.6.$$

9. Estimate $\dfrac{f(2 + h) - f(2)}{h}$ for $h = 0.5$ and $h = -0.5$. Are these numbers larger or smaller than $f'(2)$?

Explain.

SOLUTION $f(2) \approx 1$. If $h = .5$, $f(2 + h) = f(2.5) \approx 2$, so

$$\frac{f(2+h) - f(2)}{h} \approx \frac{2-1}{.5} = 2.$$

This number is probably greater than $f'(2)$. Since the curve is bending upward slightly at $(2, f(2))$, the secant line lies inside and over the curve, showing it has higher slope than the tangent line.

If $h = -.5$, $f(2 + h) = f(1.5) \approx .5$, so

$$\frac{f(2+h) - f(2)}{h} \approx \frac{.5-1}{-.5} = 1.$$

This number is smaller than $f'(2)$, since the curve is bending upwards near $(2, f(2))$.

In Exercises 11–14, let $f(x)$ be the function whose graph is shown in Figure 12.

FIGURE 12 Graph of $f(x)$.

11. Determine $f'(a)$ for $a = 1, 2, 4, 7$.

SOLUTION   Remember that the value of the derivative of f at $x = a$ can be interpreted as the slope of the line tangent to the graph of $y = f(x)$ at $x = a$. From Figure 12, we see that the graph of $y = f(x)$ is a horizontal line (that is, a line with zero slope) on the interval $0 \leq x \leq 3$. Accordingly, $f'(1) = f'(2) = 0$. On the interval $3 \leq x \leq 5$, the graph of $y = f(x)$ is a line of slope $\frac{1}{2}$; thus, $f'(4) = \frac{1}{2}$. Finally, the line tangent to the graph of $y = f(x)$ at $x = 7$ is horizontal, so $f'(7) = 0$.

13. Which is larger: $f'(5.5)$ or $f'(6.5)$?

SOLUTION   The line tangent to the graph of $y = f(x)$ at $x = 5.5$ has a larger slope than the line tangent to the graph of $y = f(x)$ at $x = 6.5$. Therefore, $f'(5.5)$ is larger than $f'(6.5)$.

In Exercises 15–18, use the limit definition to find the derivative of the linear function. (Note: The derivative does not depend on a.)

15. $f(x) = 3x - 2$

SOLUTION

$$\lim_{h \to 0} \frac{f(a+h) - f(a)}{h} = \lim_{h \to 0} \frac{3(a+h) - 2 - (3a - 2)}{h} = \lim_{h \to 0} 3 = 3.$$

17. $g(t) = 9 - t$

SOLUTION

$$\lim_{h \to 0} \frac{g(a+h) - g(a)}{h} = \lim_{h \to 0} \frac{9 - (a+h) - (9 - a)}{h} = \lim_{h \to 0} \frac{-h}{h} = \lim_{h \to 0} (-1) = -1.$$

19. Let $f(x) = \frac{1}{x}$. Does $f(-2 + h)$ equal $\frac{1}{-2 + h}$ or $\frac{1}{-2} + \frac{1}{h}$? Compute the difference quotient for $f(x)$ at $a = -2$ with $h = 0.5$.

SOLUTION   Let $f(x) = \frac{1}{x}$. Then

$$f(-2 + h) = \frac{1}{-2 + h}.$$

With $a = -2$ and $h = .5$, the difference quotient is

$$\frac{f(a+h) - f(a)}{h} = \frac{f(-1.5) - f(-2)}{.5} = \frac{\frac{1}{-1.5} - \frac{1}{-2}}{.5} = -\frac{1}{3}.$$

21. Let $f(x) = \sqrt{x}$. Show that

$$\frac{f(9+h) - f(9)}{h} = \frac{1}{\sqrt{9+h} + 3}$$

Then use this formula to compute $f'(9)$ (by taking the limit).

SOLUTION   We multiply numerator and denominator by $\sqrt{9+h} + 3$ to make a difference of squares. Thus,

$$\frac{f(9+h) - f(9)}{h} = \frac{\sqrt{9+h} - 3}{h} = \frac{(\sqrt{9+h} - 3)(\sqrt{9+h} + 3)}{h(\sqrt{9+h} + 3)}$$

$$= \frac{9 + h - 9}{h(\sqrt{9+h} + 3)} = \frac{h}{h(\sqrt{9+h} + 3)} = \frac{1}{\sqrt{9+h} + 3}.$$

Applying this formula we find:

$$f'(9) = \lim_{h \to 0} \frac{f(9+h) - f(9)}{h} = \lim_{h \to 0} \frac{1}{\sqrt{9+h} + 3} = \frac{1}{\sqrt{9} + 3} = \frac{1}{6}.$$

In Exercises 23–40, compute the derivative at $x = a$ using the limit definition and find an equation of the tangent line.

23. $f(x) = 3x^2 + 2x, \quad a = 2$

SOLUTION   Let $f(x) = 3x^2 + 2x$. Then

$$f'(2) = \lim_{h \to 0} \frac{f(2+h) - f(2)}{h} = \lim_{h \to 0} \frac{3(2+h)^2 + 2(2+h) - 16}{h}$$

$$= \lim_{h \to 0} \frac{12 + 12h + 3h^2 + 4 + 2h - 16}{h} = \lim_{h \to 0} (14 + 3h) = 14.$$

At $a = 2$, the tangent line is

$$y = f'(2)(x - 2) + f(2) = 14(x - 2) + 16 = 14x - 12.$$

25. $f(x) = x^3, \quad a = 2$

SOLUTION   Let $f(x) = x^3$. Then

$$f'(2) = \lim_{h \to 0} \frac{f(2+h) - f(2)}{h} = \lim_{h \to 0} \frac{(2+h)^3 - 2^3}{h} = \lim_{h \to 0} \frac{8 + 12h + 6h^2 + h^3 - 8}{h} = \lim_{h \to 0} (12 + 6h + h^2) = 12$$

At $a = 2$, the tangent line is

$$y = f'(2)(x - 2) + f(2) = 12(x - 2) + 8 = 12x - 16.$$

27. $f(x) = x^3 + x, \quad a = 0$

SOLUTION   Let $f(x) = x^3 + x$. Then

$$f'(0) = \lim_{h \to 0} \frac{f(0+h) - f(0)}{h} = \lim_{h \to 0} \frac{(0+h)^3 + (0+h) - 0}{h} = \lim_{h \to 0} \frac{h^3 + h}{h} = \lim_{h \to 0} \left(h^2 + 1 \right) = 1.$$

The tangent line at $a = 0$ is $y = f(0) + f'(0)(x - 0) = 0 + 1(x - 0)$ or $y = x$.

29. $f(x) = x^{-1}, \quad a = 3$

SOLUTION   Let $f(x) = x^{-1}$. Then

$$f'(3) = \lim_{h \to 0} \frac{f(3+h) - f(3)}{h} = \lim_{h \to 0} \frac{\frac{1}{3+h} - \left(\frac{1}{3}\right)}{h} = \lim_{h \to 0} \frac{\frac{3 - 3 - h}{3(3+h)}}{h} = \lim_{h \to 0} \frac{-h}{(9 + 3h)h} = -\frac{1}{9}$$

The tangent at $a = 3$ is

$$y = f'(3)(x - 3) + f(3) = -\frac{1}{9}(x - 3) + \frac{1}{3} = -\frac{1}{9}x + \frac{2}{3}.$$

31. $f(x) = 9x - 4, \quad a = -7$

SOLUTION Let $f(x) = 9x - 4$. Then

$$f'(-7) = \lim_{h \to 0} \frac{f(-7+h) - f(-7)}{h} = \lim_{h \to 0} \frac{(9(-7+h) - 4) - (-67)}{h}$$

$$= \lim_{h \to 0} \frac{(-63 + 9h - 4) + 67}{h} = \lim_{h \to 0} \frac{9h}{h} = \lim_{h \to 0} 9 = 9.$$

The tangent line at $a = -7$ is

$$y = f'(-7)(x + 7) + f(-7) = 9(x + 7) - 67 = 9x - 4.$$

In general, we can take a shorter path. The tangent line to a line is the line itself.

33. $f(x) = \dfrac{1}{x + 3}$, $a = -2$

SOLUTION Let $f(x) = \frac{1}{x+3}$. Then

$$f'(-2) = \lim_{h \to 0} \frac{f(-2+h) - f(-2)}{h} = \lim_{h \to 0} \frac{\frac{1}{-2+h+3} - 1}{h} = \lim_{h \to 0} \frac{\frac{1}{1+h} - 1}{h} = \lim_{h \to 0} \frac{-h}{h(1+h)} = \lim_{h \to 0} \frac{-1}{1+h} = -1.$$

The tangent line at $a = -2$ is

$$y = f'(-2)(x + 2) + f(-2) = -1(x + 2) + 1 = -x - 1.$$

35. $f(t) = \dfrac{2}{1 - t}$, $a = -1$

SOLUTION Let $f(t) = \frac{2}{1-t}$. Then

$$f'(-1) = \lim_{h \to 0} \frac{f(-1+h) - f(-1)}{h} = \lim_{h \to 0} \frac{\frac{2}{1-(-1+h)} - 1}{h} = \lim_{h \to 0} \frac{2 - (2 - h)}{h(2 - h)} = \lim_{h \to 0} \frac{1}{2 - h} = \frac{1}{2}.$$

At $a = -1$, the tangent line is

$$y = f'(-1)(x + 1) + f(-1) = \frac{1}{2}(x + 1) + 1 = \frac{1}{2}x + \frac{3}{2}.$$

37. $f(t) = t^{-3}$, $a = 1$

SOLUTION Let $f(t) = \dfrac{1}{t^3}$. Then

$$f'(1) = \lim_{h \to 0} \frac{f(1+h) - f(h)}{h} = \lim_{h \to 0} \frac{\frac{1}{(1+h)^3} - 1}{h} = \lim_{h \to 0} \frac{\frac{-h(3+3h+h^2)}{(1+h)^3}}{h} = \lim_{h \to 0} \frac{-(3 + 3h + h^2)}{(1 + h)^3} = -3.$$

The tangent line at $a = 1$ is

$$y = f'(1)(t - 1) + f(1) = -3(t - 1) + 1 = -3t + 4.$$

39. $f(x) = \dfrac{1}{\sqrt{x}}$, $a = 9$

SOLUTION Let $f(x) = \dfrac{1}{\sqrt{x}}$. Then

$$f'(9) = \lim_{h \to 0} \frac{f(9+h) - f(9)}{h} = \lim_{h \to 0} \frac{\frac{1}{\sqrt{9+h}} - \frac{1}{3}}{h} = \lim_{h \to 0} \frac{\frac{3 - \sqrt{9+h}}{3\sqrt{9+h}} \cdot \frac{3 + \sqrt{9+h}}{3 + \sqrt{9+h}}}{h} = \lim_{h \to 0} \frac{\frac{9 - 9 - h}{9\sqrt{9+h} + 3(9+h)}}{h}$$

$$= \lim_{h \to 0} \frac{-1}{9\sqrt{9 + h} + 3(9 + h)} = -\frac{1}{54}.$$

At $a = 9$ the tangent line is

$$y = f'(9)(x - 9) + f(9) = -\frac{1}{54}(x - 9) + \frac{1}{3} = -\frac{1}{54}x + \frac{1}{2}.$$

41. What is an equation of the tangent line at $x = 3$, assuming that $f(3) = 5$ and $f'(3) = 2$?

SOLUTION By definition, the equation of the tangent line to the graph of $f(x)$ at $x = 3$ is $y = f(3) + f'(3)(x - 3) = 5 + 2(x - 3) = 2x - 1$.

43. Consider the "curve" $y = 2x + 8$. What is the tangent line at the point $(1, 10)$? Describe the tangent line at an arbitrary point.

SOLUTION Since $y = 2x + 8$ represents a straight line, the tangent line at any point is the line itself, $y = 2x + 8$.

45. **GU** Verify that $P = \left(1, \frac{1}{2}\right)$ lies on the graphs of both $f(x) = \dfrac{1}{1+x^2}$ and $L(x) = \frac{1}{2} + m(x - 1)$ for every slope m. Plot $f(x)$ and $L(x)$ on the same axes for several values of m. Experiment until you find a value of m for which $y = L(x)$ appears tangent to the graph of $f(x)$. What is your estimate for $f'(1)$?

SOLUTION Let $f(x) = \dfrac{1}{1+x^2}$ and $L(x) = \frac{1}{2} + m(x - 1)$. Because

$$f(1) = \frac{1}{1+1^2} = \frac{1}{2} \quad \text{and} \quad L(1) = \frac{1}{2} + m(1 - 1) = \frac{1}{2},$$

it follows that $P = \left(1, \frac{1}{2}\right)$ lies on the graphs of both functions. A plot of $f(x)$ and $L(x)$ on the same axes for several values of m is shown below. The graph of $L(x)$ with $m = -\frac{1}{2}$ appears to be tangent to the graph of $f(x)$ at $x = 1$. We therefore estimate $f'(1) = -\frac{1}{2}$.

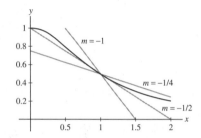

47. **GU** Plot $f(x) = x^x$ and the line $y = x + c$ on the same set of axes for several values of c. Experiment until you find a value c_0 such that the line is tangent to the graph. Then:

(a) Use your plot to estimate the value x_0 such that $f'(x_0) = 1$.

(b) Verify that $\dfrac{f(x_0 + h) - f(x_0)}{h}$ is close to 1 for $h = 0.01, 0.001, 0.0001$.

SOLUTION The figure below shows the graphs of the function $f(x) = x^x$ together with the lines $y = x - 0.5$, $y = x$, and $y = x + 0.5$.

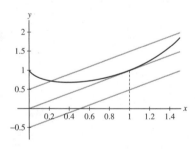

(a) The graph of $y = x$ appears to be tangent to the graph of $f(x)$ at $x = 1$. We therefore estimate that $f'(1) = 1$.

(b) With $x_0 = 1$, we generate the table

h	0.01	0.001	0.0001
$(f(x_0 + h) - f(x_0))/h$	1.010050	1.001001	1.000100

49. The vapor pressure of water is defined as the atmospheric pressure P at which no net evaporation takes place. The following table and Figure 13 give P (in atmospheres) as a function of temperature T in kelvins.

(a) Which is larger: $P'(300)$ or $P'(350)$? Answer by referring to the graph.

(b) Estimate $P'(T)$ for $T = 303, 313, 323, 333, 343$ using the table and the average of the difference quotients for $h = 10$ and -10:

$$P'(T) \approx \frac{P(T + 10) - P(T - 10)}{20} \qquad \boxed{4}$$

T (K)	293	303	313	323	333	343	353
P (atm)	0.0278	0.0482	0.0808	0.1311	0.2067	0.3173	0.4754

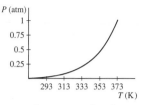

FIGURE 13 Vapor pressure of water as a function temperature T in kelvins.

SOLUTION

(a) Consider the graph of vapor pressure as a function of temperature. If we draw the tangent line at $T = 300$ and another at $T = 350$, it is clear that the latter has a steeper slope. Therefore, $P'(350)$ is larger than $P'(300)$.

(b) Using equation (4),

$$P'(303) \approx \frac{P(313) - P(293)}{20} = \frac{.0808 - .0278}{20} = .00265 \text{ atm/K};$$

$$P'(313) \approx \frac{P(323) - P(303)}{20} = \frac{.1311 - .0482}{20} = .004145 \text{ atm/K};$$

$$P'(323) \approx \frac{P(333) - P(313)}{20} = \frac{.2067 - .0808}{20} = .006295 \text{ atm/K};$$

$$P'(333) \approx \frac{P(343) - P(323)}{20} = \frac{.3173 - .1311}{20} = .00931 \text{ atm/K};$$

$$P'(343) \approx \frac{P(353) - P(333)}{20} = \frac{.4754 - .2067}{20} = .013435 \text{ atm/K}$$

In Exercises 50–51, traffic speed S along a certain road (in mph) varies as a function of traffic density q (number of cars per mile on the road). Use the following data to answer the questions:

q (density)	100	110	120	130	140
S (speed)	45	42	39.5	37	35

51. The quantity $V = qS$ is called *traffic volume*. Explain why V is equal to the number of cars passing a particular point per hour. Use the data to compute values of V as a function of q and estimate $V'(q)$ when $q = 120$.

SOLUTION The traffic speed S has units of miles/hour, and the traffic density has units of cars/mile. Therefore, the traffic volume $V = Sq$ has units of cars/hour. A table giving the values of V follows.

q	100	110	120	130	140
V	4500	4620	4740	4810	4900

To estimate dV/dq, we take the difference quotient.

h	−20	−10	10	20
$\frac{V(120 + h) - V(120)}{h}$	12	12	7	8

The mean of the difference quotients is 9.75. Hence $dV/dq \approx 9.75$ mph when $q = 120$.

In Exercises 53–58, each of the limits represents a derivative $f'(a)$. Find $f(x)$ and a.

53. $\displaystyle \lim_{h \to 0} \frac{(5 + h)^3 - 125}{h}$

SOLUTION The difference quotient $\dfrac{(5 + h)^3 - 125}{h}$ has the form $\dfrac{f(a + h) - f(a)}{h}$ where $f(x) = x^3$ and $a = 5$.

55. $\displaystyle \lim_{h \to 0} \frac{\sin(\frac{\pi}{6} + h) - .5}{h}$

SOLUTION The difference quotient $\dfrac{\sin(\frac{\pi}{6}+h)-.5}{h}$ has the form $\dfrac{f(a+h)-f(a)}{h}$ where $f(x)=\sin x$ and $a=\frac{\pi}{6}$.

57. $\displaystyle\lim_{h\to 0}\dfrac{5^{2+h}-25}{h}$

SOLUTION The difference quotient $\dfrac{5^{(2+h)}-25}{h}$ has the form $\dfrac{f(a+h)-f(a)}{h}$ where $f(x)=5^x$ and $a=2$.

59. Sketch the graph of $f(x)=\sin x$ on $[0,\pi]$ and guess the value of $f'(\frac{\pi}{2})$. Then calculate the slope of the secant line between $x=\frac{\pi}{2}$ and $x=\frac{\pi}{2}+h$ for at least three small positive and negative values of h. Are these calculations consistent with your guess?

SOLUTION Here is the graph of $y=\sin x$ on $[0,\pi]$.

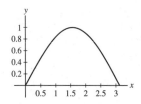

At $x=\frac{\pi}{2}$, we're at the peak of the sine graph. The tangent line appears to be horizontal, so the slope is 0; hence, $f'(\frac{\pi}{2})$ appears to be 0.

h	$-.01$	$-.001$	$-.0001$	$.0001$	$.001$	$.01$
$\dfrac{\sin(\frac{\pi}{2}+h)-1}{h}$	$.005$	$.0005$	$.00005$	$-.00005$	$-.0005$	$-.005$

These numerical calculations are consistent with our guess.

61. [GU] Let $f(x)=\dfrac{4}{1+2^x}$.

(a) Plot $f(x)$ over $[-2,2]$. Then zoom in near $x=0$ until the graph appears straight and estimate the slope $f'(0)$.

(b) Use your estimate to find an approximate equation to the tangent line at $x=0$. Plot this line and the graph on the same set of axes.

SOLUTION

(a) The figure below at the left shows the graph of $f(x)=\frac{4}{1+2^x}$ over $[-2,2]$. The figure below at the right is a close-up near $x=0$. From the close-up, we see that the graph is nearly straight and passes through the points $(-0.22,2.15)$ and $(0.22,1.85)$. We therefore estimate

$$f'(0)\approx\frac{1.85-2.15}{0.22-(-0.22)}=\frac{-0.3}{0.44}=-0.68$$

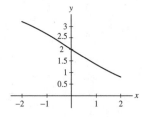
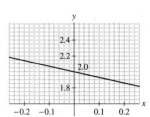

(b) Using the estimate for $f'(0)$ obtained in part (a), the approximate equation of the tangent line is

$$y=f'(0)(x-0)+f(0)=-0.68x+2.$$

The figure below shows the graph of $f(x)$ and the approximate tangent line.

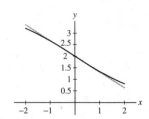

63. GU Let $f(x) = \cot x$. Estimate $f'(\frac{\pi}{2})$ graphically by zooming in on a plot of $f(x)$ near $x = \frac{\pi}{2}$.

SOLUTION The figure below shows a close-up of the graph of $f(x) = \cot x$ near $x = \frac{\pi}{2} \approx 1.5708$. From the close-up, we see that the graph is nearly straight and passes through the points $(1.53, 0.04)$ and $(1.61, -0.04)$. We therefore estimate

$$f'\left(\frac{\pi}{2}\right) \approx \frac{-0.04 - 0.04}{1.61 - 1.53} = \frac{-0.08}{0.08} = -1$$

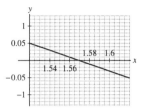

65. Apply the method of Example 6 to $f(x) = \sin x$ to determine $f'\left(\frac{\pi}{4}\right)$ accurately to four decimal places.

SOLUTION We know that

$$f'(\pi/4) = \lim_{h \to 0} \frac{f(\pi/4 + h) - f(\pi/4)}{h} = \lim_{h \to 0} \frac{\sin(\pi/4 + h) - \sqrt{2}/2}{h}.$$

Creating a table of values of h close to zero:

h	$-.001$	$-.0001$	$-.00001$	$.00001$	$.0001$	$.001$
$\dfrac{\sin(\frac{\pi}{4} + h) - (\sqrt{2}/2)}{h}$	.7074602	.7071421	.7071103	.7071033	.7070714	.7067531

Accurate up to four decimal places, $f'(\frac{\pi}{4}) \approx .7071$.

67. GU Sketch the graph of $f(x) = x^{5/2}$ on $[0, 6]$.
(a) Use the sketch to justify the inequalities for $h > 0$:

$$\frac{f(4) - f(4 - h)}{h} \leq f'(4) \leq \frac{f(4 + h) - f(4)}{h}$$

(b) Use part (a) to compute $f'(4)$ to four decimal places.
(c) Use a graphing utility to plot $f(x)$ and the tangent line at $x = 4$ using your estimate for $f'(4)$.

SOLUTION
(a) The slope of the secant line between points $(4, f(4))$ and $(4 + h, f(4 + h))$ is

$$\frac{f(4 + h) - f(4)}{h}.$$

$x^{5/2}$ is a smooth curve increasing at a faster rate as $x \to \infty$. Therefore, if $h > 0$, then the slope of the secant line is greater than the slope of the tangent line at $f(4)$, which happens to be $f'(4)$. Likewise, if $h < 0$, the slope of the secant line is less than the slope of the tangent line at $f(4)$, which happens to be $f'(4)$.
(b) We know that

$$f'(4) = \lim_{h \to 0} \frac{f(4 + h) - f(4)}{h} = \lim_{h \to 0} \frac{(4 + h)^{5/2} - 32}{h}.$$

Creating a table with values of h close to zero:

h	$-.0001$	$-.00001$	$.00001$	$.0001$
$\dfrac{(4 + h)^{5/2} - 32}{h}$	19.999625	19.99999	20.0000	20.0000375

Thus, $f'(4) \approx 20.0000$.
(c) Using the estimate for $f'(4)$ obtained in part (b), the equation of the line tangent to $f(x) = x^{5/2}$ at $x = 4$ is

$$y = f'(4)(x - 4) + f(4) = 20(x - 4) + 32 = 20x - 48.$$

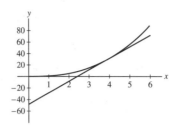

69. The graph in Figure 18 (based on data collected by the biologist Julian Huxley, 1887–1975) gives the average antler weight W of a red deer as a function of age t. Estimate the slope of the tangent line to the graph at $t = 4$. For which values of t is the slope of the tangent line equal to zero? For which values is it negative?

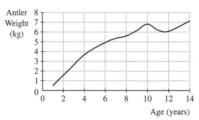

FIGURE 18

SOLUTION Let $W(t)$ denote the antler weight as a function of age. The "tangent line" sketched in the figure below passes through the points $(1, 1)$ and $(6, 5.5)$. Therefore

$$W'(4) \approx \frac{5.5 - 1}{6 - 1} = 0.9 \text{ kg/year.}$$

If the slope of the tangent is zero, the tangent line is horizontal. This appears to happen at roughly $t = 10$ and at $t = 11.6$. The slope of the tangent line is negative when the height of the graph decreases as we move to the right. For the graph in Figure 18, this occurs for $10 < t < 11.6$.

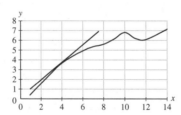

In Exercises 70–72, $i(t)$ is the current (in amperes) at time t (seconds) flowing in the circuit shown in Figure 19. According-ing to Kirchhoff's law, $i(t) = Cv'(t) + R^{-1}v(t)$, where $v(t)$ is the voltage (in volts) at time t, C the capacitance (in farads), and R the resistance (in ohms, Ω).

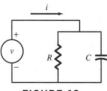

FIGURE 19

71. Use the following table to estimate $v'(10)$. For a better estimate, take the *average* of the difference quotients for h and $-h$ as described in Exercise 48. Then estimate $i(10)$, assuming $C = 0.03$ and $R = 1,000$.

t	9.8	9.9	10	10.1	10.2
$v(t)$	256.52	257.32	258.11	258.9	259.69

SOLUTION We generate a table of difference quotients at 10 using the following table:

h	−0.2	−0.1	0.1	0.2
$\dfrac{v(10 + h) - v(10)}{h}$	7.95	7.9	7.9	7.9

The mean of the difference quotients is 7.9125, so $v'(10) \approx 7.9125$ volts/s. Thus,

$$i(10) = 0.03(7.9125) + \frac{1}{1000}(258.11) = 0.495485 \text{ amperes.}$$

Further Insights and Challenges

In Exercises 73–76, we define the symmetric difference quotient (SDQ) at $x = a$ for $h \neq 0$ by

$$\frac{f(a+h) - f(a-h)}{2h}$$

$\boxed{5}$

73. Explain how SDQ can be interpreted as the slope of a secant line.

SOLUTION The symmetric difference quotient

$$\frac{f(a+h) - f(a-h)}{2h}$$

is the slope of the secant line connecting the points $(a - h, f(a - h))$ and $(a + h, f(a + h))$ on the graph of f; the difference in the function values is divided by the difference in the x-values.

75. Show that if $f(x)$ is a quadratic polynomial, then the SDQ at $x = a$ (for any $h \neq 0$) is *equal* to $f'(a)$. Explain the graphical meaning of this result.

SOLUTION Let $f(x) = px^2 + qx + r$ be a quadratic polynomial. We compute the SDQ at $x = a$.

$$\frac{f(a+h) - f(a-h)}{2h} = \frac{p(a+h)^2 + q(a+h) + r - (p(a-h)^2 + q(a-h) + r)}{2h}$$

$$= \frac{pa^2 + 2pah + ph^2 + qa + qh + r - pa^2 + 2pah - ph^2 - qa + qh - r}{2h}$$

$$= \frac{4pah + 2qh}{2h} = \frac{2h(2pa + q)}{2h} = 2pa + q$$

Since this doesn't depend on h, the limit, which is equal to $f'(a)$, is also $2pa + q$. Graphically, this result tells us that the secant line to a parabola passing through points chosen symmetrically about $x = a$ is always parallel to the tangent line at $x = a$.

77. Which of the two functions in Figure 20 satisfies the inequality

$$\frac{f(a+h) - f(a-h)}{2h} \le \frac{f(a+h) - f(a)}{h}$$

for $h > 0$? Explain in terms of secant lines.

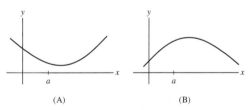

(A)   (B)

FIGURE 20

SOLUTION Figure (A) satisfies the inequality

$$\frac{f(a+h) - f(a-h)}{2h} \le \frac{f(a+h) - f(a)}{h}$$

since in this graph the symmetric difference quotient has a larger negative slope than the ordinary right difference quotient. [In figure (B), the symmetric difference quotient has a larger positive slope than the ordinary right difference quotient and therefore does *not* satisfy the stated inequality.]

3.2 The Derivative as a Function

Preliminary Questions

1. What is the slope of the tangent line through the point $(2, f(2))$ if f is a function such that $f'(x) = x^3$?

SOLUTION The slope of the tangent line through the point $(2, f(2))$ is given by $f'(2)$. Since $f'(x) = x^3$, it follows that $f'(2) = 2^3 = 8$.

2. Evaluate $(f - g)'(1)$ and $(3f + 2g)'(1)$ assuming that $f'(1) = 3$ and $g'(1) = 5$. Can we evaluate $(fg)'(1)$ using the information given and the rules presented in this section?

SOLUTION $(f - g)'(1) = f'(1) - g'(1) = 3 - 5 = -2$ and $(3f + 2g)'(1) = 3f'(1) + 2g'(1) = 3(3) + 2(5) = 19$. Using the rules of this section, we cannot evaluate $(fg)'(1)$.

3. Which of the following functions can be differentiated using the rules covered in this section? Explain.

(a) $f(x) = x^2$

(b) $f(x) = 2^x$

(c) $f(x) = \dfrac{1}{\sqrt{x}}$

(d) $f(x) = x^{-4/5}$

(e) $f(x) = \sin x$

(f) $f(x) = (x + 5)^3$

SOLUTION

(a) Yes. x^2 is a power function, so the Power Rule can be applied.

(b) No. 2^x is an exponential function (the base is constant while the exponent is a variable), so the Power Rule does not apply.

(c) Yes. $\dfrac{1}{\sqrt{x}}$ can be rewritten as $x^{-1/2}$. This is a power function, so the Power Rule can be applied.

(d) Yes. $x^{-4/5}$ is a power function, so the Power Rule can be applied.

(e) No. The Power Rule does not apply to the trigonometric function $\sin x$.

(f) Yes. $(x + 5)^3$ can be expanded to $x^3 + 15x^2 + 75x + 125$. The Power Rule, Rule for Sums, and Rule for Constant Multiples can be applied to this function.

4. Which algebraic identity is used to prove the Power Rule for positive integer exponents? Explain how it is used.

SOLUTION The algebraic identity

$$x^n - a^n = (x - a)(x^{n-1} + x^{n-2}a + x^{n-3}a^2 + \cdots + xa^{n-2} + a^{n-1})$$

is used to prove the Power Rule for positive integer exponents. With this identity, the difference quotient

$$\frac{x^n - a^n}{x - a}$$

can be simplified to

$$x^{n-1} + x^{n-2}a + x^{n-3}a^2 + \cdots + xa^{n-2} + a^{n-1}.$$

This expression is continuous at $x = a$, so the limit as $x \to a$ can be evaluated by substitution.

5. Does the Power Rule apply to $f(x) = \sqrt[5]{x}$? Explain.

SOLUTION Yes. $\sqrt[5]{x} = x^{1/5}$, which is a power function.

6. In which of the following two cases does the derivative not exist?

(a) Horizontal tangent

(b) Vertical tangent

SOLUTION The derivative does not exist when there is a vertical tangent. At a horizontal tangent, the derivative is zero.

Exercises

In Exercises 1–8, compute $f'(x)$ using the limit definition.

1. $f(x) = 4x - 3$

SOLUTION Let $f(x) = 4x - 3$. Then,

$$f'(x) = \lim_{h \to 0} \frac{f(x + h) - f(x)}{h} = \lim_{h \to 0} \frac{4(x + h) - 3 - (4x - 3)}{h} = \lim_{h \to 0} \frac{4h}{h} = 4.$$

3. $f(x) = 1 - 2x^2$

SOLUTION Let $f(x) = 1 - 2x^2$. Then,

$$f'(x) = \lim_{h \to 0} \frac{f(x+h) - f(x)}{h} = \lim_{h \to 0} \frac{1 - 2(x+h)^2 - (1 - 2x^2)}{h}$$

$$= \lim_{h \to 0} \frac{-4xh - 2h^2}{h} = \lim_{h \to 0} (-4x - 2h) = -4x.$$

5. $f(x) = x^{-1}$

SOLUTION Let $f(x) = x^{-1}$. Then,

$$f'(x) = \lim_{h \to 0} \frac{f(x+h) - f(x)}{h} = \lim_{h \to 0} \frac{\frac{1}{x+h} - \frac{1}{x}}{h} = \lim_{h \to 0} \frac{\frac{x-x-h}{x(x+h)}}{h} = \lim_{h \to 0} \frac{-1}{x(x+h)} = -\frac{1}{x^2}.$$

7. $f(x) = \sqrt{x}$

SOLUTION Let $f(x) = \sqrt{x}$. Then,

$$f'(x) = \lim_{h \to 0} \frac{f(x+h) - f(x)}{h} = \lim_{h \to 0} \frac{\sqrt{x+h} - \sqrt{x}}{h} = \lim_{h \to 0} \frac{\sqrt{x+h} - \sqrt{x}}{h} \cdot \left(\frac{\sqrt{x+h} + \sqrt{x}}{\sqrt{x+h} + \sqrt{x}} \right)$$

$$= \lim_{h \to 0} \frac{(x+h) - x}{h(\sqrt{x+h} + \sqrt{x})} = \lim_{h \to 0} \frac{1}{\sqrt{x+h} + \sqrt{x}} = \frac{1}{2\sqrt{x}}.$$

In Exercises 9–16, use the Power Rule to compute the derivative.

9. $\dfrac{d}{dx} x^4 \Big|_{x=-2}$

SOLUTION $\dfrac{d}{dx}\left(x^4 \right) = 4x^3$ so $\dfrac{d}{dx} x^4 \Big|_{x=-2} = 4(-2)^3 = -32.$

11. $\dfrac{d}{dt} t^{2/3} \Big|_{t=8}$

SOLUTION $\dfrac{d}{dt}\left(t^{2/3} \right) = \dfrac{2}{3} t^{-1/3}$ so $\dfrac{d}{dt} t^{2/3} \Big|_{t=8} = \dfrac{2}{3}(8)^{-1/3} = \dfrac{1}{3}.$

13. $\dfrac{d}{dx} x^{0.35}$

SOLUTION $\dfrac{d}{dx}\left(x^{0.35} \right) = 0.35(x^{0.35-1}) = 0.35x^{-0.65}.$

15. $\dfrac{d}{dt} t^{\sqrt{17}}$

SOLUTION $\dfrac{d}{dt}\left(t^{\sqrt{17}} \right) = \sqrt{17} t^{\sqrt{17}-1}$

In Exercises 17–20, compute $f'(a)$ and find an equation of the tangent line to the graph at $x = a$.

17. $f(x) = x^5, \quad a = 1$

SOLUTION Let $f(x) = x^5$. Then, by the Power Rule, $f'(x) = 5x^4$. The equation to the tangent line to the graph of $f(x)$ at $x = 1$ is

$$y = f'(1)(x - 1) + f(1) = 5(x - 1) + 1 = 5x - 4.$$

19. $f(x) = 3\sqrt{x} + 8x, \quad a = 9$

SOLUTION Let $f(x) = 3x^{1/2} + 8x$. Then $f'(x) = \frac{3}{2}x^{-1/2} + 8$. In particular, $f'(9) = \frac{17}{2}$. The tangent line at $x = 9$ is

$$y = f'(9)(x - 9) + f(9) = \frac{17}{2}(x - 9) + 81 = \frac{17}{2}x + \frac{9}{2}.$$

In Exercises 21–32, calculate the derivative of the function.

21. $f(x) = x^3 + x^2 - 12$

SOLUTION $\dfrac{d}{dx}\left(x^3+x^2-12\right) = 3x^2 + 2x.$

23. $f(x) = 2x^3 - 10x^{-1}$

SOLUTION $\dfrac{d}{dx}\left(2x^3 - 10x^{-1}\right) = 6x^2 + 10x^{-2}.$

25. $g(z) = 7z^{-3} + z^2 + 5$

SOLUTION $\dfrac{d}{dz}\left(7z^{-3} + z^2 + 5\right) = -21z^{-4} + 2z.$

27. $f(s) = \sqrt[4]{s} + \sqrt[3]{s}$

SOLUTION $f(s) = \sqrt[4]{s} + \sqrt[3]{s} = s^{1/4} + s^{1/3}.$ In this form, we can apply the Sum and Power Rules.

$$\frac{d}{ds}\left(s^{1/4} + s^{1/3}\right) = \frac{1}{4}(s^{(1/4)-1}) + \frac{1}{3}(s^{(1/3)-1}) = \frac{1}{4}s^{-3/4} + \frac{1}{3}s^{-2/3}.$$

29. $f(x) = (x+1)^3$ (*Hint:* Expand)

SOLUTION $\dfrac{d}{dx}\left((x+1)^3\right) = \dfrac{d}{dx}\left(x^3 + 3x^2 + 3x + 1\right) = 3x^2 + 6x + 3.$

31. $P(z) = (3z-1)(2z+1)$

SOLUTION $\dfrac{d}{dz}((3z-1)(2z+1)) = \dfrac{d}{dz}\left(6z^2 + z - 1\right) = 12z + 1.$

In Exercises 33–38, calculate the derivative indicated.

33. $f'(2), \quad f(x) = \dfrac{3}{x^4}$

SOLUTION With $f(x) = 3x^{-4}$, we have $f'(x) = -12x^{-5}$, so

$$f'(2) = -12(2)^{-5} = -\frac{3}{8}.$$

35. $\dfrac{dT}{dC}\Big|_{C=8}, \quad T = 3C^{2/3}$

SOLUTION With $T(C) = 3C^{2/3}$, we have $\dfrac{dT}{dC} = 2C^{-1/3}.$ Therefore,

$$\frac{dT}{dC}\Big|_{C=8} = 2(8)^{-1/3} = 1.$$

37. $\dfrac{ds}{dz}\Big|_{z=2}, \quad s = 4z - 16z^2$

SOLUTION With $s = 4z - 16z^2$, we have $\dfrac{ds}{dz} = 4 - 32z.$ Therefore,

$$\frac{ds}{dz}\Big|_{z=2} = 4 - 32(2) = -60.$$

39. Match the functions in graphs (A)–(D) with their derivatives (I)–(III) in Figure 11. Note that two of the functions have the same derivative. Explain why.

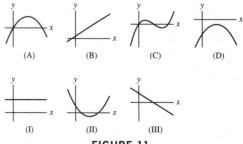

FIGURE 11

SOLUTION

- Consider the graph in (A). On the left side of the graph, the slope of the tangent line is positive but on the right side the slope of the tangent line is negative. Thus the derivative should transition from positive to negative with increasing x. This matches the graph in (III).
- Consider the graph in (B). This is a linear function, so its slope is constant. Thus the derivative is constant, which matches the graph in (I).
- Consider the graph in (C). Moving from left to right, the slope of the tangent line transitions from positive to negative then back to positive. The derivative should therefore be negative in the middle and positive to either side. This matches the graph in (II).
- Consider the graph in (D). On the left side of the graph, the slope of the tangent line is positive but on the right side the slope of the tangent line is negative. Thus the derivative should transition from positive to negative with increasing x. This matches the graph in (III).

Note that the functions whose graphs are shown in (A) and (D) have the same derivative. This happens because the graph in (D) is just a vertical translation of the graph in (A), which means the two functions differ by a constant. The derivative of a constant is zero, so the two functions end up with the same derivative.

41. Sketch the graph of $f'(x)$ for $f(x)$ as in Figure 13, omitting points where $f(x)$ is not differentiable.

FIGURE 13 Graph of $f(x)$.

SOLUTION   On the interval $0 \le x \le 3$, the function is constant, so $f'(x) = 0$. On the interval $3 \le x \le 5$, the function is linear with slope $\frac{1}{2}$, so $f'(x) = \frac{1}{2}$. From $x = 5$ to $x = 7$, the derivative should be positive but decreasing to zero at $x = 7$. Finally, for $7 < x \le 9$, the derivative is negative and decreasing. A sketch of $f'(x)$ is given below.

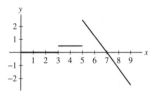

43. Let R be a variable and r a constant. Compute the derivatives

(a) $\dfrac{d}{dR} R$

(b) $\dfrac{d}{dR} r$

(c) $\dfrac{d}{dR} r^2 R^3$

SOLUTION

(a) $\dfrac{d}{dR} R = 1$, since R is a linear function of R with slope 1.

(b) $\dfrac{d}{dR} r = 0$, since r is a constant.

(c) We apply the Linearity and Power Rules:

$$\frac{d}{dR} r^2 R^3 = r^2 \frac{d}{dR} R^3 = r^2 \left(3(R^2)\right) = 3r^2 R^2.$$

45. Find the points on the curve $y = x^2 + 3x - 7$ at which the slope of the tangent line is equal to 4.

SOLUTION   Let $y = x^2 + 3x - 7$. Solving $dy/dx = 2x + 3 = 4$ yields $x = \frac{1}{2}$.

47. Find all values of x where the tangent lines to $y = x^3$ and $y = x^4$ are parallel.

SOLUTION   Let $f(x) = x^3$ and let $g(x) = x^4$. The two graphs have parallel tangent lines at all x where $f'(x) = g'(x)$.

$$f'(x) = g'(x)$$
$$3x^2 = 4x^3$$
$$3x^2 - 4x^3 = 0$$
$$x^2(3 - 4x) = 0$$

hence, $x = 0$ or $x = \frac{3}{4}$.

49. Determine coefficients a and b such that $p(x) = x^2 + ax + b$ satisfies $p(1) = 0$ and $p'(1) = 4$.

SOLUTION   Let $p(x) = x^2 + ax + b$ satisfy $p(1) = 0$ and $p'(1) = 4$. Now, $p'(x) = 2x + a$. Therefore $0 = p(1) = 1 + a + b$ and $4 = p'(1) = 2 + a$; i.e., $a = 2$ and $b = -3$.

51. Let $f(x) = x^3 - 3x + 1$. Show that $f'(x) \geq -3$ for all x, and that for every $m > -3$, there are precisely two points where $f'(x) = m$. Indicate the position of these points and the corresponding tangent lines for one value of m in a sketch of the graph of $f(x)$.

SOLUTION   Let $P = (a, b)$ be a point on the graph of $f(x) = x^3 - 3x + 1$.

- The derivative satisfies $f'(x) = 3x^2 - 3 \geq -3$ since $3x^2$ is nonnegative.
- Suppose the slope m of the tangent line is greater than -3. Then $f'(a) = 3a^2 - 3 = m$, whence

$$a^2 = \frac{m+3}{3} > 0 \quad \text{and thus} \quad a = \pm\sqrt{\frac{m+3}{3}}.$$

- The two parallel tangent lines with slope 2 are shown with the graph of $f(x)$ here.

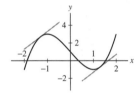

53. Compute the derivative of $f(x) = x^{-2}$ using the limit definition. *Hint:* Show that

$$\frac{f(x+h) - f(x)}{h} = -\frac{1}{x^2(x+h)^2} \frac{(x+h)^2 - x^2}{h}$$

SOLUTION

$$\frac{f(x+h) - f(x)}{h} = \frac{\frac{1}{(x+h)^2} - \frac{1}{x^2}}{h} = \frac{\frac{x^2 - (x+h)^2}{x^2(x+h)^2}}{h} = \frac{1}{h}\left(\frac{x^2 - (x+h)^2}{1}\right)\left(\frac{1}{x^2(x+h)^2}\right)$$

$$= -\frac{1}{x^2(x+h)^2} \frac{(x+h)^2 - x^2}{h}.$$

We now compute the limits as $h \to 0$ of each factor. By continuity, we see that, if $x \neq 0$,

$$\lim_{h \to 0} \frac{-1}{x^2(x+h)^2} = -\frac{1}{x^4}.$$

The second factor can be recognized easily as the limit definition of the derivative of x^2, which is $2x$. Applying the Product Rule, we get that:

$$\frac{d}{dx}x^{-2} = -\frac{1}{x^4}(2x) = -\frac{2}{x^3}.$$

55. The average speed (in meters per second) of a gas molecule is $v_{avg} = \sqrt{8RT/(\pi M)}$, where T is the temperature (in kelvin), M is the molar mass (kg/mol) and $R = 8.31$. Calculate dv_{avg}/dT at $T = 300$ K for oxygen, which has a molar mass of 0.032 kg/mol.

SOLUTION   Using the form $v_{av} = (8RT/(\pi M))^{1/2} = \sqrt{8R/(\pi M)}T^{1/2}$, where M and R are constant, we use the Power Rule to compute the derivative dv_{av}/dT.

$$\frac{d}{dT}\sqrt{8R/(\pi M)}T^{1/2} = \sqrt{8R/(\pi M)}\frac{d}{dT}T^{1/2} = \sqrt{8R/(\pi M)}\frac{1}{2}(T^{(1/2)-1}).$$

In particular, if $T = 300°$K,

$$\frac{d}{dT}v_{av} = \sqrt{8(8.31)/(\pi(0.032))}\frac{1}{2}(300)^{-1/2} = 0.74234 \text{ m/(s} \cdot \text{K).}$$

57. Some studies suggest that kidney mass K in mammals (in kilograms) is related to body mass m (in kilograms) by the approximate formula $K = 0.007m^{0.85}$. Calculate dK/dm at $m = 68$. Then calculate the derivative with respect to m of the relative kidney-to-mass ratio K/m at $m = 68$.

SOLUTION

$$\frac{dK}{dm} = 0.007(0.85)m^{-0.15} = 0.00595m^{-0.15};$$

hence,

$$\frac{dK}{dm}\bigg|_{m=68} = 0.00595(68)^{-0.15} = 0.00315966.$$

Because

$$\frac{K}{m} = 0.007\frac{m^{0.85}}{m} = 0.007m^{-0.15},$$

we find

$$\frac{d}{dm}\left(\frac{K}{m}\right) = 0.007\frac{d}{dm}m^{-0.15} = -0.00105m^{-1.15},$$

and

$$\frac{d}{dm}\left(\frac{K}{m}\right)\bigg|_{m=68} = -8.19981 \times 10^{-6} \text{ kg}^{-1}.$$

59. Let L be a tangent line to the hyperbola $xy = 1$ at $x = a$, where $a > 0$. Show that the area of the triangle bounded by L and the coordinate axes does not depend on a.

SOLUTION Let $f(x) = x^{-1}$. The tangent line to f at $x = a$ is $y = f'(a)(x - a) + f(a) = -\frac{1}{a^2}(x - a) + \frac{1}{a}$. The y-intercept of this line (where $x = 0$) is $\frac{2}{a}$. Its x-intercept (where $y = 0$) is $2a$. Hence the area of the triangle bounded by the tangent line and the coordinate axes is $A = \frac{1}{2}bh = \frac{1}{2}(2a)\left(\frac{2}{a}\right) = 2$, which is independent of a.

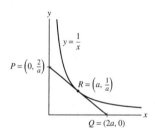

61. Match the functions (A)–(C) with their derivatives (I)–(III) in Figure 16.

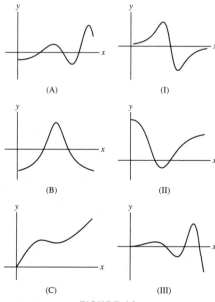

FIGURE 16

SOLUTION   Note that the graph in (A) has three locations with a horizontal tangent line. The derivative must therefore cross the *x*-axis in three locations, which matches (III).

The graph in (B) has only one location with a horizontal tangent line, so its derivative should cross the *x*-axis only once. Thus, (I) is the graph corresponding to the derivative of (B).

Finally, the graph in (B) has two locations with a horizontal tangent line, so its derivative should cross the *x*-axis twice. Thus, (II) is the graph corresponding to the derivative of (C).

63. Make a rough sketch of the graph of the derivative of the function shown in Figure 17(A).

SOLUTION   The graph has a tangent line with negative slope approximately on the interval (1, 3.6), and has a tangent line with a positive slope elsewhere. This implies that the derivative must be negative on the interval (1, 3.6) and positive elsewhere. The graph may therefore look like this:

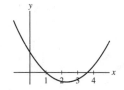

65. [icon] Of the two functions f and g in Figure 18, which is the derivative of the other? Justify your answer.

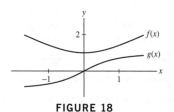

FIGURE 18

SOLUTION $g(x)$ is the derivative of $f(x)$. For $f(x)$ the slope is negative for negative values of x until $x = 0$, where there is a horizontal tangent, and then the slope is positive for positive values of x. Notice that $g(x)$ is negative for negative values of x, goes through the origin at $x = 0$, and then is positive for positive values of x.

In Exercises 67–72, find the points c (if any) such that $f'(c)$ does not exist.

67. $f(x) = |x - 1|$

SOLUTION

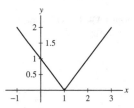

Here is the graph of $f(x) = |x - 1|$. Its derivative does not exist at $x = 1$. At that value of x there is a sharp corner.

69. $f(x) = x^{2/3}$

SOLUTION   Here is the graph of $f(x) = x^{2/3}$. Its derivative does not exist at $x = 0$. At that value of x, there is a sharp corner or "cusp".

71. $f(x) = |x^2 - 1|$

SOLUTION   Here is the graph of $f(x) = \left|x^2 - 1\right|$. Its derivative does not exist at $x = -1$ or at $x = 1$. At these values of x, the graph has sharp corners.

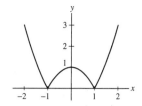

GU   *In Exercises 73–78, zoom in on a plot of* $f(x)$ *at the point* $(a, f(a))$ *and state whether or not* $f(x)$ *appears to be differentiable at* $x = a$. *If nondifferentiable, state whether the tangent line appears to be vertical or does not exist.*

73. $f(x) = (x - 1)|x|$, $a = 0$

SOLUTION   The graph of $f(x) = (x - 1)|x|$ for x near 0 is shown below. Because the graph has a sharp corner at $x = 0$, it appears that f is not differentiable at $x = 0$. Moreover, the tangent line does not exist at this point.

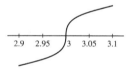

75. $f(x) = (x - 3)^{1/3}$, $a = 3$

SOLUTION   The graph of $f(x) = (x - 3)^{1/3}$ for x near 3 is shown below. From this graph, it appears that f is not differentiable at $x = 3$. Moreover, the tangent line appears to be vertical.

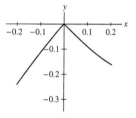

77. $f(x) = |\sin x|$, $a = 0$

SOLUTION   The graph of $f(x) = |\sin x|$ for x near 0 is shown below. Because the graph has a sharp corner at $x = 0$, it appears that f is not differentiable at $x = 0$. Moreover, the tangent line does not exist at this point.

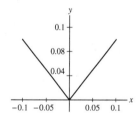

79.   Is it true that a nondifferentiable function is not continuous? If not, give a counterexample.

SOLUTION   This statement is false. Indeed, the function $|x|$ is not differentiable at $x = 0$ but it is continuous.

Further Insights and Challenges

81. Prove the following theorem of Apollonius of Perga (the Greek mathematician born in 262 BCE who gave the parabola, ellipse, and hyperbola their names): The tangent to the parabola $y = x^2$ at $x = a$ intersects the x-axis at the midpoint between the origin and $(a, 0)$. Draw a diagram.

SOLUTION   Let $f(x) = x^2$. The tangent line to f at $x = a$ is

$$y = f'(a)(x - a) + f(a) = 2a(x - a) + a^2 = 2ax - a^2.$$

The x-intercept of this line (where $y = 0$) is $\frac{a}{2}$, which is halfway between the origin and the point $(a, 0)$.

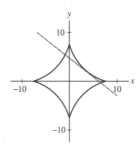

83. CAS   Plot the graph of $f(x) = (4 - x^{2/3})^{3/2}$ (the "astroid"). Let L be a tangent line to a point on the graph in the first quadrant. Show that the portion of L in the first quadrant has a constant length 8.

SOLUTION

- Here is a graph of the astroid.

- Let $f(x) = (4 - x^{2/3})^{3/2}$. Since we have not yet encountered the Chain Rule, we use Maple throughout this exercise. The tangent line to f at $x = a$ is

$$ y = -\frac{\sqrt{4 - a^{2/3}}}{a^{1/3}}(x - a) + \left(4 - a^{2/3}\right)^{3/2}. $$

The y-intercept of this line is the point $P = \left(0, 4\sqrt{4 - a^{2/3}}\right)$, its x-intercept is the point $Q = \left(4a^{1/3}, 0\right)$, and the distance between P and Q is 8.

85.   A vase is formed by rotating $y = x^2$ around the y-axis. If we drop in a marble, it will either touch the bottom point of the vase or be suspended above the bottom by touching the sides (Figure 21). How small must the marble be to touch the bottom?

FIGURE 21

SOLUTION   Suppose a circle is tangent to the parabola $y = x^2$ at the point (t, t^2). The slope of the parabola at this point is $2t$, so the slope of the radius of the circle at this point is $-\frac{1}{2t}$ (since it is perpendicular to the tangent line of the circle). Thus the center of the circle must be where the line given by $y = -\frac{1}{2t}(x - t) + t^2$ crosses the y-axis. We can find the y-coordinate by setting $x = 0$: we get $y = \frac{1}{2} + t^2$. Thus, the radius extends from $(0, \frac{1}{2} + t^2)$ to (t, t^2) and

$$ r = \sqrt{\left(\frac{1}{2} + t^2 - t^2\right)^2 + t^2} = \sqrt{\frac{1}{4} + t^2}. $$

This radius is greater than $\frac{1}{2}$ whenever $t > 0$; so, if a marble has radius $> 1/2$ it sits on the edge of the vase, but if it has radius $\leq 1/2$ it rolls all the way to the bottom.

87. Negative Exponents   Let n be a whole number. Use the Power Rule for x^n to calculate the derivative of $f(x) = x^{-n}$ by showing that

$$ \frac{f(x + h) - f(x)}{h} = \frac{-1}{x^n(x + h)^n} \frac{(x + h)^n - x^n}{h} $$

SOLUTION   Let $f(x) = x^{-n}$ where n is a positive integer.

- The difference quotient for f is

$$\frac{f(x+h) - f(x)}{h} = \frac{(x+h)^{-n} - x^{-n}}{h} = \frac{\frac{1}{(x+h)^n} - \frac{1}{x^n}}{h} = \frac{\frac{x^n - (x+h)^n}{x^n(x+h)^n}}{h}$$

$$= \frac{-1}{x^n(x+h)^n} \frac{(x+h)^n - x^n}{h}.$$

- Therefore,

$$f'(x) = \lim_{h \to 0} \frac{f(x+h) - f(x)}{h} = \lim_{h \to 0} \frac{-1}{x^n(x+h)^n} \frac{(x+h)^n - x^n}{h}$$

$$= \lim_{h \to 0} \frac{-1}{x^n(x+h)^n} \lim_{h \to 0} \frac{(x+h)^n - x^n}{h} = -x^{-2n} \frac{d}{dx}\left(x^n\right).$$

- From above, we continue: $f'(x) = -x^{-2n} \dfrac{d}{dx}\left(x^n\right) = -x^{-2n} \cdot nx^{n-1} = -nx^{-n-1}$. Since n is a positive integer, $k = -n$ is a negative integer and we have $\dfrac{d}{dx}\left(x^k\right) = \dfrac{d}{dx}\left(x^{-n}\right) = -nx^{-n-1} = kx^{k-1}$; i.e. $\dfrac{d}{dx}\left(x^k\right) = kx^{k-1}$ for negative integers k.

89. Infinitely Rapid Oscillations   Define

$$f(x) = \begin{cases} x \sin \dfrac{1}{x} & \text{if } x \neq 0 \\ 0 & \text{if } x = 0 \end{cases}$$

Show that $f(x)$ is continuous at $x = 0$ but $f'(0)$ does not exist (see Figure 10).

SOLUTION   Let $f(x) = \begin{cases} x \sin\left(\frac{1}{x}\right) & \text{if } x \neq 0 \\ 0 & \text{if } x = 0 \end{cases}$. As $x \to 0$,

$$|f(x) - f(0)| = \left| x \sin\left(\frac{1}{x}\right) - 0 \right| = |x| \left| \sin\left(\frac{1}{x}\right) \right| \to 0$$

since the values of the sine lie between -1 and 1. Hence, by the Squeeze Theorem, $\lim_{x \to 0} f(x) = f(0)$ and thus f is continuous at $x = 0$.

As $x \to 0$, the difference quotient at $x = 0$,

$$\frac{f(x) - f(0)}{x - 0} = \frac{x \sin\left(\frac{1}{x}\right) - 0}{x - 0} = \sin\left(\frac{1}{x}\right)$$

does *not* converge to a limit since it oscillates infinitely through every value between -1 and 1. Accordingly, $f'(0)$ does not exist.

3.3 Product and Quotient Rules

Preliminary Questions

1. Are the following statements true or false? If false, state the correct version.

(a) The notation fg denotes the function whose value at x is $f(g(x))$.

(b) The notation f/g denotes the function whose value at x is $f(x)/g(x)$.

(c) The derivative of the product is the product of the derivatives.

SOLUTION

(a) False. The notation fg denotes the function whose value at x is $f(x)g(x)$.

(b) True.

(c) False. The derivative of the product fg is $f'(x)g(x) + f(x)g'(x)$.

2. Are the following equations true or false? If false, state the correct version.

(a) $\dfrac{d}{dx}(fg)\bigg|_{x=4} = f(4)g'(4) - g(4)f'(4)$

(b) $\dfrac{d}{dx}\dfrac{f}{g}\bigg|_{x=4} = \dfrac{f(4)g'(4) + g(4)f'(4)}{(g(4))^2}$

(c) $\dfrac{d}{dx}(fg)\bigg|_{x=0} = f(0)g'(0) + g(0)f'(0)$

SOLUTION

(a) False. $\dfrac{d}{dx}(fg)\bigg|_{x=4} = f(4)g'(4) + g(4)f'(4)$.

(b) False. $\dfrac{d}{dx}(f/g)\bigg|_{x=4} = [g(4)f'(4) - f(4)g'(4)]/g(4)^2$.

(c) True.

3. What is the derivative of f/g at $x = 1$ if $f(1) = f'(1) = g(1) = 2$, and $g'(1) = 4$?

SOLUTION $\dfrac{d}{dx}(f/g)\big|_{x=1} = [g(1)f'(1) - f(1)g'(1)]/g(1)^2 = [2(2) - 2(4)]/2^2 = -1$.

4. Suppose that $f(1) = 0$ and $f'(1) = 2$. Find $g(1)$, assuming that $(fg)'(1) = 10$.

SOLUTION $(fg)'(1) = f(1)g'(1) + f'(1)g(1)$, so $10 = 0 \cdot g'(1) + 2g(1)$ and $g(1) = 5$.

Exercises

In Exercises 1–4, use the Product Rule to calculate the derivative.

1. $f(x) = x(x^2 + 1)$

SOLUTION Let $f(x) = x(x^2 + 1)$. Then

$$f'(x) = x\frac{d}{dx}(x^2 + 1) + (x^2 + 1)\frac{d}{dx}x = x(2x) + (x^2 + 1) = 3x^2 + 1.$$

3. $\dfrac{dy}{dt}\bigg|_{t=3},\quad y = (t^2 + 1)(t + 9)$

SOLUTION Let $y = (t^2 + 1)(t + 9)$. Then

$$\frac{dy}{dt} = (t^2 + 1)\frac{d}{dt}(t + 9) + (t + 9)\frac{d}{dt}(t^2 + 1) = (t^2 + 1) + (t + 9)(2t) = 3t^2 + 18t + 1.$$

Therefore,

$$\frac{dy}{dt}\bigg|_{t=3} = 3(3)^2 + 18(3) + 1 = 82.$$

In Exercises 5–8, use the Quotient Rule to calculate the derivative.

5. $f(x) = \dfrac{x}{x - 2}$

SOLUTION Let $f(x) = \frac{x}{x-2}$. Then

$$f'(x) = \frac{(x-2)\frac{d}{dx}x - x\frac{d}{dx}(x-2)}{(x-2)^2} = \frac{(x-2) - x}{(x-2)^2} = \frac{-2}{(x-2)^2}.$$

7. $\dfrac{dg}{dt}\bigg|_{t=-2},\quad g(t) = \dfrac{t^2 + 1}{t^2 - 1}$

SOLUTION Let $g(t) = \dfrac{t^2 + 1}{t^2 - 1}$. Then

$$\frac{dg}{dt} = \frac{(t^2 - 1)\frac{d}{dt}(t^2 + 1) - (t^2 + 1)\frac{d}{dt}(t^2 - 1)}{(t^2 - 1)^2} = \frac{(t^2 - 1)(2t) - (t^2 + 1)(2t)}{(t^2 - 1)^2} = -\frac{4t}{(t^2 - 1)^2}.$$

Therefore,

$$\frac{dg}{dt}\bigg|_{t=-2} = -\frac{4(-2)}{((-2)^2 - 1)^2} = \frac{8}{9}.$$

In Exercises 9–12, calculate the derivative in two ways: First use the Product or Quotient Rule, then rewrite the function algebraically and apply the Power Rule directly.

9. $f(t) = (2t + 1)(t^2 - 2)$

SOLUTION Let $f(t) = (2t + 1)(t^2 - 2)$. Then, using the Product Rule,

$$f'(t) = (2t + 1)(2t) + (t^2 - 2)(2) = 6t^2 + 2t - 4.$$

Multiplying out first, we find $f(t) = 2t^3 + t^2 - 4t - 2$. Therefore, $f'(t) = 6t^2 + 2t - 4$.

11. $g(x) = \dfrac{x^3 + 2x^2 + 3x^{-1}}{x}$

SOLUTION Let $g(x) = \dfrac{x^3 + 2x^2 + 3x^{-1}}{x}$. Using the quotient rule and the sum and power rules, and simplifying

$$g'(x) = \frac{x(3x^2 + 4x - 3x^{-2}) - (x^3 + 2x^2 + 3x^{-1})1}{x^2} = \frac{1}{x^2}\left(2x^3 + 2x^2 - 6x^{-1}\right) = 2x + 2 - 6x^{-3}.$$

Simplifying first yields $g(x) = x^2 + 2x + 3x^{-2}$, from which we calculate $g'(x) = 2x + 2 - 6x^{-3}$.

In Exercises 13–32, calculate the derivative using the appropriate rule or combination of rules.

13. $f(x) = (x^4 - 4)(x^2 + x + 1)$

SOLUTION Let $f(x) = \left(x^4 - 4\right)\left(x^2 + x + 1\right)$. Then

$$f'(x) = (x^4 - 4)(2x + 1) + (x^2 + x + 1)(4x^3) = 6x^5 + 5x^4 + 4x^3 - 8x - 4.$$

15. $\dfrac{dy}{dx}\Big|_{x=2}, \quad y = \dfrac{1}{x+4}$

SOLUTION Let $y = \frac{1}{x+4}$. Using the quotient rule:

$$\frac{dy}{dx} = \frac{(x + 4)(0) - 1(1)}{(x + 4)^2} = -\frac{1}{(x + 4)^2}.$$

Therefore,

$$\frac{dy}{dx}\Big|_{x=2} = -\frac{1}{(2 + 4)^2} = -\frac{1}{36}.$$

17. $f(x) = (\sqrt{x} + 1)(\sqrt{x} - 1)$

SOLUTION Let $f(x) = (\sqrt{x} + 1)(\sqrt{x} - 1)$. Multiplying through first yields $f(x) = x - 1$ for $x \geq 0$. Therefore, $f'(x) = 1$ for $x \geq 0$. If we carry out the product rule on $f(x) = (x^{1/2} + 1)(x^{1/2} - 1)$, we get

$$f'(x) = (x^{1/2} + 1)\left(\frac{1}{2}(x^{-1/2})\right) + (x^{1/2} - 1)\left(\frac{1}{2}x^{-1/2}\right) = \frac{1}{2} + \frac{1}{2}x^{-1/2} + \frac{1}{2} - \frac{1}{2}x^{-1/2} = 1.$$

19. $\dfrac{dy}{dx}\Big|_{x=2}, \quad y = \dfrac{x^4 - 4}{x^2 - 5}$

SOLUTION Let $y = \dfrac{x^4 - 4}{x^2 - 5}$. Then

$$\frac{dy}{dx} = \frac{\left(x^2 - 5\right)\left(4x^3\right) - \left(x^4 - 4\right)(2x)}{\left(x^2 - 5\right)^2} = \frac{2x^5 - 20x^3 + 8x}{\left(x^2 - 5\right)^2}.$$

Therefore,

$$\frac{dy}{dx}\Big|_{x=2} = \frac{2(2)^5 - 20(2)^3 + 8(2)}{(2^2 - 5)^2} = -80.$$

21. $\dfrac{dz}{dx}\Big|_{x=1}, \quad z = \dfrac{1}{x^3 + 1}$

SOLUTION   Let $z = \frac{1}{x^3+1}$. Using the quotient rule:

$$\frac{dz}{dx} = \frac{(x^3+1)(0) - 1(3x^2)}{(x^3+1)^2} = -\frac{3x^2}{(x^3+1)^2}.$$

Therefore,

$$\frac{dz}{dx}\bigg|_{x=1} = -\frac{3(1)^2}{(1^3+1)^2} = -\frac{3}{4}.$$

23. $h(t) = \dfrac{t}{(t^4+t^2)(t^7+1)}$

SOLUTION   Let $h(t) = \dfrac{t}{(t^4+t^2)(t^7+1)} = \dfrac{t}{t^{11}+t^9+t^4+t^2}$. Then

$$h'(t) = \frac{\left(t^{11}+t^9+t^4+t^2\right)(1) - t\left(11t^{10}+9t^8+4t^3+2t\right)}{\left(t^{11}+t^9+t^4+t^2\right)^2} = -\frac{10t^{11}+8t^9+3t^4+t^2}{\left(t^{11}+t^9+t^4+t^2\right)^2}.$$

25. $f(t) = 3^{1/2} \cdot 5^{1/2}$

SOLUTION   Let $f(t) = \sqrt{3}\sqrt{5}$. Then $f'(t) = 0$, since $f(t)$ is a *constant* function!

27. $f(x) = (x+3)(x-1)(x-5)$

SOLUTION   Let $f(x) = (x+3)(x-1)(x-5)$. Using the Product Rule inside the Product Rule with a first factor of $(x+3)$ and a second factor of $(x-1)(x-5)$, we find

$$f'(x) = (x+3)\left((x-1)(1) + (x-5)(1)\right) + (x-1)(x-5)(1) = 3x^2 - 6x - 13.$$

Alternatively,

$$f(x) = (x+3)\left(x^2 - 6x + 5\right) = x^3 - 3x^2 - 13x + 15.$$

Therefore, $f'(x) = 3x^2 - 6x - 13$.

29. $g(z) = \left(\dfrac{z^2-4}{z-1}\right)\left(\dfrac{z^2-1}{z+2}\right)$ *Hint:* Simplify first.

SOLUTION   Let

$$g(z) = \left(\frac{z^2-4}{z-1}\right)\left(\frac{z^2-1}{z+2}\right) = \left(\frac{(z+2)(z-2)}{z-1}\frac{(z+1)(z-1)}{z+2}\right) = (z-2)(z+1)$$

for $z \neq -2$ and $z \neq 1$. Then,

$$g'(z) = (z+1)(1) + (z-2)(1) = 2z - 1.$$

31. $\dfrac{d}{dt}\left(\dfrac{xt-4}{t^2-x}\right)$ (x constant)

SOLUTION   Let $f(t) = \frac{xt-4}{t^2-x}$. Using the quotient rule:

$$f'(t) = \frac{(t^2-x)(x) - (xt-4)(2t)}{(t^2-x)^2} = \frac{xt^2 - x^2 - 2xt^2 + 8t}{(t^2-x)^2} = \frac{-xt^2 + 8t - x^2}{(t^2-x)^2}.$$

33. $\boxed{\text{GU}}$ Plot the derivative of $f(x) = \dfrac{x}{x^2+1}$ over $[-4, 4]$. Use the graph to determine the intervals on which $f'(x) > 0$ and $f'(x) < 0$. Then plot $f(x)$ and describe how the sign of $f'(x)$ is reflected in the graph of $f(x)$.

SOLUTION   Let $f(x) = \dfrac{x}{x^2+1}$. Then

$$f'(x) = \frac{(x^2+1)(1) - x(2x)}{(x^2+1)^2} = \frac{1-x^2}{(x^2+1)^2}.$$

The derivative is shown in the figure below at the left. From this plot we see that $f'(x) > 0$ for $-1 < x < 1$ and $f'(x) < 0$ for $|x| > 1$. The original function is plotted in the figure below at the right. Observe that the graph of $f(x)$ is increasing whenever $f'(x) > 0$ and that $f(x)$ is decreasing whenever $f'(x) < 0$.

35. Let $P = \dfrac{V^2 R}{(R + r)^2}$ as in Example 5. Calculate dP/dr, assuming that r is variable and R is constant.

SOLUTION   Note that V is also constant. Let

$$f(r) = \frac{V^2 R}{(R + r)^2} = \frac{V^2 R}{R^2 + 2Rr + r^2}.$$

Using the quotient rule:

$$f'(r) = \frac{(R^2 + 2Rr + r^2)(0) - (V^2 R)(2R + 2r)}{(R + r)^4} = -\frac{2V^2 R(R + r)}{(R + r)^4} = -\frac{2V^2 R}{(R + r)^3}.$$

37. Find an equation of the tangent line to the graph $y = \dfrac{x}{x + x^{-1}}$ at $x = 2$.

SOLUTION   Let $f(x) = \frac{x}{x + x^{-1}}$. The equation of the tangent line to the graph of $f(x)$ at $x = 2$ is $y = f'(2)(x - 2) + f(2)$. Using the quotient rule, we compute:

$$f'(x) = \frac{(x + x^{-1})(1) - (x)(1 - x^{-2})}{(x + x^{-1})^2} = \frac{1}{(x + x^{-1})^2}\left(x + x^{-1} - x + x^{-1}\right) = \frac{2}{x(x + x^{-1})^2}.$$

Therefore, $f'(2) = \frac{2}{\frac{25}{2}} = \frac{4}{25}$. This makes the equation of the tangent line

$$y = \frac{4}{25}(x - 2) + \frac{2}{\frac{5}{2}} = \frac{4}{25}x + \frac{12}{5}.$$

39. Let $f(x) = g(x) = x$. Show that $(f/g)' \neq f'/g'$.

SOLUTION $(f/g) = (x/x) = 1$, so $(f/g)' = 0$. On the other hand, $(f'/g') = (x'/x') = (1/1) = 1$. We see that $0 \neq 1$.

41. Show that $(f^3)' = 3f^2 f'$.

SOLUTION   Let $g = f^3 = fff$. Then

$$g' = \left(f^3\right)' = [f(ff)]' = f\left(ff' + ff'\right) + ff(f') = 3f^2 f'.$$

In Exercises 42–45, use the following function values:

$f(4)$	$f'(4)$	$g(4)$	$g'(4)$
2	-3	5	-1

43. Calculate $F'(4)$, where $F(x) = xf(x)$.

SOLUTION   Let $F(x) = xf(x)$. Then $F'(x) = xf'(x) + f(x)$, and

$$F'(4) = 4f'(4) + f(4) = 4(-3) + 2 = -10.$$

45. Calculate $H'(4)$, where $H(x) = \dfrac{x}{g(x)f(x)}$.

SOLUTION   Let $H(x) = \dfrac{x}{g(x)f(x)}$. Then

$$H'(x) = \frac{g(x)f(x) \cdot 1 - x\left(g(x)f'(x) + f(x)g'(x)\right)}{(g(x)f(x))^2}.$$

Therefore,

$$H'(4) = \frac{(5)(2) - 4 \cdot ((5)(-3) + (2)(-1))}{((5)(2))^2} = \frac{78}{100} = \frac{39}{50}.$$

47. Proceed as in Exercise 46 to calculate $F'(0)$, where

$$F(x) = \left(1 + x + x^{4/3} + x^{5/3}\right) \frac{3x^5 + 5x^4 + 5x + 1}{8x^9 - 7x^4 + 1}$$

SOLUTION   Write $F(x) = f(x)(g(x)/h(x))$, where

$$f(x) = (1 + x + x^{4/3} + x^{5/3})$$
$$g(x) = 3x^5 + 5x^4 + 5x + 1$$

and

$$h(x) = 8x^9 - 7x^4 + 1.$$

Now, $f'(x) = 1 + \frac{4}{3}x^{\frac{1}{3}} + \frac{5}{3}x^{\frac{2}{3}}$, $g'(x) = 15x^4 + 20x^3 + 5$, and $h'(x) = 72x^8 - 28x^3$. Moreover, $f(0) = 1$, $f'(0) = 1$, $g(0) = 1$, $g'(0) = 5$, $h(0) = 1$, and $h'(0) = 0$. From the product and quotient rules,

$$F'(0) = f(0)\frac{h(0)g'(0) - g(0)h'(0)}{h(0)^2} + f'(0)(g(0)/h(0)) = 1\frac{1(5) - 1(0)}{1} + 1(1/1) = 6.$$

Further Insights and Challenges

49. Prove the Quotient Rule using the limit definition of the derivative.

SOLUTION   Let $p = \dfrac{f}{g}$. Suppose that f and g are differentiable at $x = a$ and that $g(a) \neq 0$. Then

$$p'(a) = \lim_{h \to 0} \frac{p(a + h) - p(a)}{h} = \lim_{h \to 0} \frac{\dfrac{f(a+h)}{g(a+h)} - \dfrac{f(a)}{g(a)}}{h} = \lim_{h \to 0} \frac{\dfrac{f(a+h)g(a) - f(a)g(a+h)}{g(a+h)g(a)}}{h}$$

$$= \lim_{h \to 0} \frac{f(a+h)g(a) - f(a)g(a) + f(a)g(a) - f(a)g(a+h)}{hg(a+h)g(a)}$$

$$= \lim_{h \to 0} \left(\frac{1}{g(a+h)g(a)}\left(g(a)\frac{f(a+h) - f(a)}{h} - f(a)\frac{g(a+h) - g(a)}{h}\right)\right)$$

$$= \left(\lim_{h \to 0}\frac{1}{g(a+h)g(a)}\right)\left(\left(g(a)\lim_{h \to 0}\frac{f(a+h) - f(a)}{h}\right) - \left(f(a)\lim_{h \to 0}\frac{g(a+h) - g(a)}{h}\right)\right)$$

$$= \frac{1}{(g(a))^2}\left(g(a)f'(a) - f(a)g'(a)\right) = \frac{g(a)f'(a) - f(a)g'(a)}{(g(a))^2}$$

In other words, $p' = \left(\dfrac{f}{g}\right)' = \dfrac{gf' - fg'}{g^2}$.

51. Derive the Quotient Rule using Eq. (6) and the Product Rule.

SOLUTION   Let $h(x) = \frac{f(x)}{g(x)}$. We can write $h(x) = f(x)\frac{1}{g(x)}$. Applying Eq. (6),

$$h'(x) = f(x)\left(\left(\frac{1}{g(x)}\right)'\right) + f'(x)\left(\frac{1}{g(x)}\right) = -f(x)\left(\frac{g'(x)}{(g(x))^2}\right) + \frac{f'(x)}{g(x)} = \frac{-f(x)g'(x) + f'(x)g(x)}{(g(x))^2}.$$

53. Carry out the details of Agnesi's proof of the Quotient Rule from her book on calculus, published in 1748: Assume that f, g, and $h = f/g$ are differentiable. Compute the derivative of $hg = f$ using the Product Rule and solve for h'.

SOLUTION   Suppose that f, g, and h are differentiable functions with $h = f/g$.

- Then $hg = f$ and via the product rule $hg' + gh' = f'$.

- Solving for h' yields $h' = \dfrac{f' - hg'}{g} = \dfrac{f' - \dfrac{f}{g}g'}{g} = \dfrac{gf' - fg'}{g^2}$.

In Exercises 55–56, let $f(x)$ be a polynomial. A basic fact of algebra states that c is a root of $f(x)$ if and only if $f(x) = (x - c)g(x)$ for some polynomial $g(x)$. We say that c is a multiple root if $f(x) = (x - c)^2 h(x)$, where $h(x)$ is a polynomial.

55. Show that c is a multiple root of $f(x)$ if and only if c is a root of both $f(x)$ and $f'(x)$.

SOLUTION Assume first that $f(c) = f'(c) = 0$ and let us show that c is a multiple root of $f(x)$. We have $f(x) = (x - c)g(x)$ for some polynomial $g(x)$ and so $f'(x) = (x - c)g'(x) + g(x)$. However, $f'(c) = 0 + g(c) = 0$, so c is also a root of $g(x)$ and hence $g(x) = (x - c)h(x)$ for some polynomial $h(x)$. We conclude that $f(x) = (x - c)^2 h(x)$, which shows that c is a multiple root of $f(x)$.

Conversely, assume that c is a multiple root. Then $f(c) = 0$ and $f(x) = (x - c)^2 g(x)$ for some polynomial $g(x)$. Then $f'(x) = (x - c)^2 g'(x) + 2g(x)(x - c)$. Therefore, $f'(c) = (c - c)^2 g'(c) + 2g(c)(c - c) = 0$.

57. 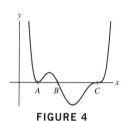 Figure 4 is the graph of a polynomial with roots at A, B, and C. Which of these is a multiple root? Explain your reasoning using the result of Exercise 55.

FIGURE 4

SOLUTION A on the figure is a multiple root. It is a multiple root because $f(x) = 0$ at A and because the tangent line to the graph at A is horizontal, so that $f'(x) = 0$ at A. For the same reasons, f also has a multiple root at C.

3.4 Rates of Change

Preliminary Questions

1. What units might be used to measure the ROC of:

(a) Pressure (in atmospheres) in a water tank with respect to depth?

(b) The reaction rate of a chemical reaction (the ROC of concentration with respect to time), where concentration is measured in moles per liter?

SOLUTION

(a) The rate of change of pressure with respect to depth might be measured in atmospheres/meter.

(b) The reaction rate of a chemical reaction might be measured in moles/(liter·hour).

2. Suppose that $f(2) = 4$ and the average ROC of f between 2 and 5 is 3. What is $f(5)$?

SOLUTION The average rate of change of f between 2 and 5 is given by

$$\frac{f(5) - f(2)}{5 - 2} = \frac{f(5) - 4}{3}.$$

Setting this equal to 3 and solving for $f(5)$ yields $f(5) = 13$.

3. Two trains travel from New Orleans to Memphis in 4 hours. The first train travels at a constant velocity of 90 mph, but the velocity of the second train varies. What was the second train's average velocity during the trip?

SOLUTION Since both trains travel the same distance in the same amount of time, they have the same average velocity: 90 mph.

4. Estimate $f(26)$, assuming that $f(25) = 43$ and $f'(25) = 0.75$.

SOLUTION $f(x) \approx f(25) + f'(25)(x - 25)$, so $f(26) \approx 43 + 0.75(26 - 25) = 43.75$.

5. The population $P(t)$ of Freedonia in 1933 was $P(1933) = 5$ million.

(a) What is the meaning of the derivative $P'(1933)$?

(b) Estimate $P(1934)$ if $P'(1933) = 0.2$. What if $P'(1933) = 0$?

SOLUTION

(a) Because $P(t)$ measures the population of Freedonia as a function of time, the derivative $P'(1933)$ measures the rate of change of the population of Freedonia in the year 1933.

(b) $P(1934) \approx P(1933) + P'(1933)$. Thus, if $P'(1933) = 0.2$, then $P(1934) \approx 5.2$ million. On the other hand, if $P'(1933) = 0$, then $P(1934) \approx 5$ million.

Exercises

1. Find the ROC of the area of a square with respect to the length of its side s when $s = 3$ and $s = 5$.

SOLUTION Let the area be $A = f(s) = s^2$. Then the rate of change of A with respect to s is $d/ds(s^2) = 2s$. When $s = 3$, the area changes at a rate of 6 square units per unit increase. When $s = 5$, the area changes at a rate of 10 square units per unit increase. (Draw a 5×5 square on graph paper and trace the area added by increasing each side length by 1, excluding the corner, to see what this means.)

3. Find the ROC of $y = x^{-1}$ with respect to x for $x = 1, 10$.

SOLUTION The ROC of change is $dy/dx = -x^{-2}$. If $x = 1$, the ROC is -1. If $x = 10$, the ROC is $-\frac{1}{100}$.

In Exercises 5–8, calculate the ROC.

5. $\dfrac{dV}{dr}$, where V is the volume of a cylinder whose height is equal to its radius (the volume of a cylinder of height h and radius r is $\pi r^2 h$)

SOLUTION The volume of the cylinder is $V = \pi r^2 h = \pi r^3$. Thus $dV/dr = 3\pi r^2$.

7. ROC of the volume V of a sphere with respect to its radius (the volume of a sphere is $V = \frac{4}{3}\pi r^3$)

SOLUTION The volume of a sphere of radius r is $V = \frac{4}{3}\pi r^3$. Thus $dV/dr = 4\pi r^2$.

In Exercises 9–10, refer to Figure 10, which shows the graph of distance (in kilometers) versus time (in hours) for a car trip.

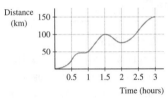

FIGURE 10 Graph of distance versus time for a car trip.

9. (a) Estimate the average velocity over [0.5, 1].

(b) Is average velocity greater over [1, 2] or [2, 3]?

(c) At what time is velocity at a maximum?

SOLUTION

(a) The average velocity over the interval [.5, 1] is

$$\frac{50 - 25}{1 - .5} = 50 \text{ km/h.}$$

(b) The average velocity over the interval [1, 2],

$$\frac{75 - 50}{2 - 1} = 25 \text{ km/h,}$$

which is less than that over the interval [2, 3],

$$\frac{150 - 75}{3 - 2} = 75 \text{ km/h.}$$

(c) The car's velocity is maximum when the slope of the distance versus time curve is most positive. This appears to happen when $t = 0.5$ h, $t = 1.25$ h, or $t = 2.5$ h.

11. Figure 11 displays the voltage across a capacitor as a function of time while the capacitor is being charged. Estimate the ROC of voltage at $t = 20$ s. Indicate the values in your calculation and include proper units. Does voltage change more quickly or more slowly as time goes on? Explain in terms of tangent lines.

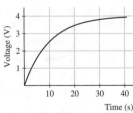

FIGURE 11

SOLUTION The tangent line sketched in the figure below appears to pass through the points (10, 3) and (30, 4). Thus, the ROC of voltage at $t = 20$ seconds is approximately

$$\frac{4-3}{30-10} = 0.05 \text{ V/s}.$$

As we move to the right of the graph, the tangent lines to it grow shallower, indicating that the voltage changes more slowly as time goes on.

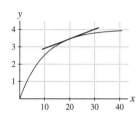

13. A stone is tossed vertically upward with an initial velocity of 25 ft/s from the top of a 30-ft building.

(a) What is the height of the stone after 0.25 s?

(b) Find the velocity of the stone after 1 s.

(c) When does the stone hit the ground?

SOLUTION We employ Galileo's formula, $s(t) = s_0 + v_0 t - \frac{1}{2}gt^2 = 30 + 25t - 16t^2$, where the time t is in seconds (s) and the height s is in feet (ft).

(a) The height of the stone after .25 seconds is $s(.25) = 35.25$ ft.

(b) The velocity at time t is $s'(t) = 25 - 32t$. When $t = 1$, this is -7 ft/s.

(c) When the stone hits the ground, its height is zero. Solve $30 + 25t - 16t^2 = 0$ to obtain $t = \dfrac{-25 \pm \sqrt{2545}}{-32}$ or $t \approx 2.36$ s. (The other solution, $t \approx -0.79$, we discard since it represents a time before the stone was thrown.)

15. The temperature of an object (in degrees Fahrenheit) as a function of time (in minutes) is $T(t) = \frac{3}{4}t^2 - 30t + 340$ for $0 \le t \le 20$. At what rate does the object cool after 10 min (give correct units)?

SOLUTION Let $T(t) = \frac{3}{4}t^2 - 30t + 340$, $0 \le t \le 20$. Then $T'(t) = \frac{3}{2}t - 30$, so $T'(10) = -15°$F/min.

17. The earth exerts a gravitational force of $F(r) = \dfrac{2.99 \times 10^{16}}{r^2}$ (in Newtons) on an object with a mass of 75 kg, where r is the distance (in meters) from the center of the earth. Find the ROC of force with respect to distance at the surface of the earth, assuming the radius of the earth is 6.77×10^6 m.

SOLUTION The rate of change of force is $F'(r) = -5.98 \times 10^{16}/r^3$. Therefore,

$$F'(6.77 \times 10^6) = -5.98 \times 10^{16}/(6.77 \times 10^6)^3 = -1.93 \times 10^{-4} \text{ N/m}.$$

19. The power delivered by a battery to an apparatus of resistance R (in ohms) is $P = \dfrac{2.25R}{(R+0.5)^2}$ W. Find the rate of change of power with respect to resistance for $R = 3$ and $R = 5$ Ω.

SOLUTION

$$P'(R) = \frac{(R+.5)^2 2.25 - 2.25R(2R+1)}{(R+.5)^4}.$$

Therefore, $P'(3) = -0.1312$ W/Ω and $P'(5) = -0.0609$ W/Ω.

21. By Faraday's Law, if a conducting wire of length ℓ meters moves at velocity v m/s perpendicular to a magnetic field of strength B (in teslas), a voltage of size $V = -B\ell v$ is induced in the wire. Assume that $B = 2$ and $\ell = 0.5$.

(a) Find the rate of change dV/dv.

(b) Find the rate of change of V with respect to time t if $v = 4t + 9$.

SOLUTION

(a) Assuming that $B = 2$ and $l = 0.5$, $V = -2(.5)v = -v$. Therefore,

$$\frac{dV}{dv} = -1.$$

(b) If $v = 4t + 9$, then $V = -2(.5)(4t + 9) = -(4t + 9)$. Therefore, $\frac{dV}{dt} = -4$.

23. The population $P(t)$ of a city (in millions) is given by the formula $P(t) = 0.00005t^2 + 0.01t + 1$, where t denotes the number of years since 1990.

(a) How large is the population in 1996 and how fast is it growing?

(b) When does the population grow at a rate of 12,000 people per year?

SOLUTION   Let $P(t) = (0.00005)t^2 + (0.01)t + 1$ be the population of a city in millions. Here t is the number of years past 1990.

(a) In 1996 ($t = 6$ years after 1990), the population is $P(6) = 1.0618$ million. The rate of growth of population is $P'(t) = 0.0001t + 0.01$. In 1996, this corresponds to a growth rate of $P'(6) = 0.0106$ million per year or 10,600 people per year.

(b) When the growth rate is 12,000 people per year (or 0.012 million per year), we have $P'(t) = 0.0001t + 0.01 = 0.012$. Solving for t gives $t = 20$. This corresponds to the year 2010; i.e., 20 years past 1990.

25. ⬛ Ethan finds that with h hours of tutoring, he is able to answer correctly $S(h)$ percent of the problems on a math exam. What is the meaning of the derivative $S'(h)$? Which would you expect to be larger: $S'(3)$ or $S'(30)$? Explain.

SOLUTION   The derivative $S'(h)$ measures the rate at which the percent of problems Ethan answers correctly changes with respect to the number of hours of tutoring he receives.

One possible graph of $S(h)$ is shown in the figure below on the left. This graph indicates that in the early hours of working with the tutor, Ethan makes rapid progress in learning the material but eventually approaches either the limit of his ability to learn the material or the maximum possible score on the exam. In this scenario, $S'(3)$ would be larger than $S'(30)$.

An alternative graph of $S(h)$ is shown below on the right. Here, in the early hours of working with the tutor little progress is made (perhaps the tutor is assessing how much Ethan already knows, his learning style, his personality, etc.). This is followed by a period of rapid improvement and finally a leveling off as Ethan reaches his maximum score. In this scenario, $S'(3)$ and $S'(30)$ might be roughly equal.

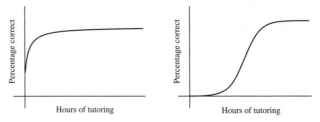

27. ⬛ Table 2 gives the total U.S. population during each month of 1999 as determined by the U.S. Department of Commerce.

(a) Estimate $P'(t)$ for each of the months January–November.

(b) Plot these data points for $P'(t)$ and connect the points by a smooth curve.

(c) Write a newspaper headline describing the information contained in this plot.

TABLE 2   Total U.S. Population in 1999

t	$P(t)$ in Thousands
January	271,841
February	271,987
March	272,142
April	272,317
May	272,508
June	272,718
July	272,945
August	273,197
September	273,439
October	273,672
November	273,891
December	274,076

SOLUTION   The table in the text gives the growing population $P(t)$ of the United States.

(a) Here are estimates of $P'(t)$ in thousands/month for January–November. The estimates are computed using the estimate $f(t + 1) - f(t)$.

t	Jan	Feb	Mar	Apr	May	Jun	Jul	Aug	Sep	Oct	Nov
$P'(t)$	146	155	175	191	210	227	252	242	233	219	185

(b) Here is a plot of these estimates (1 = Jan, 2 = Feb, etc.)

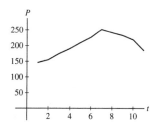

(c) "U.S. Growth Rate Declines After Midsummer Peak"

29. According to a formula widely used by doctors to determine drug dosages, a person's body surface area (BSA) (in meters squared) is given by the formula BSA $= \sqrt{hw}/60$, where h is the height in centimeters and w the weight in kilograms. Calculate the ROC of BSA with respect to weight for a person of constant height $h = 180$. What is this ROC for $w = 70$ and $w = 80$? Express your result in the correct units. Does BSA increase more rapidly with respect to weight at lower or higher body weights?

SOLUTION Assuming constant height $h = 180$ cm, let $f(w) = \sqrt{hw}/60 = \frac{\sqrt{5}}{10}\sqrt{w}$ be the formula for body surface area in terms of weight. The ROC of BSA with respect to weight is

$$f'(w) = \frac{\sqrt{5}}{10}\left(\frac{1}{2}w^{-1/2}\right) = \frac{\sqrt{5}}{20\sqrt{w}}.$$

If $w = 70$ kg, this is

$$f'(70) = \frac{\sqrt{5}}{20\sqrt{70}} = \frac{\sqrt{14}}{280} \approx 0.0133631 \ \frac{\text{m}^2}{\text{kg}}.$$

If $w = 80$ kg,

$$f'(80) = \frac{\sqrt{5}}{20\sqrt{80}} = \frac{1}{20\sqrt{16}} = \frac{1}{80} \ \frac{\text{m}^2}{\text{kg}}.$$

Because the rate of change of BSA depends on $1/\sqrt{w}$, it is clear that BSA increases more rapidly at lower body weights.

31. What is the velocity of an object dropped from a height of 300 m when it hits the ground?

SOLUTION We employ Galileo's formula, $s(t) = s_0 + v_0 t - \frac{1}{2}gt^2 = 300 - 4.9t^2$, where the time t is in seconds (s) and the height s is in meters (m). When the ball hits the ground its height is 0. Solve $s(t) = 300 - 4.9t^2 = 0$ to obtain $t \approx 7.8246$ s. (We discard the negative time, which took place before the ball was dropped.) The velocity at impact is $v(7.8246) = -9.8(7.8246) \approx -76.68$ m/s. This signifies that the ball is *falling* at 76.68 m/s.

33. A ball is tossed up vertically from ground level and returns to earth 4 s later. What was the initial velocity of the stone and how high did it go?

SOLUTION Galileo's formula gives $s(t) = s_0 + v_0 t - \frac{1}{2}gt^2 = v_0 t - 4.9t^2$, where the time t is in seconds (s) and the height s is in meters (m). When the ball hits the ground after 4 seconds its height is 0. Solve $0 = s(4) = 4v_0 - 4.9(4)^2$ to obtain $v_0 = 19.6$ m/s. The stone reaches its maximum height when $s'(t) = 0$, that is, when $19.6 - 9.8t = 0$, or $t = 2$ s. At this time, $t = 2$ s,

$$s(2) = 0 + 19.6(2) - \frac{1}{2}(9.8)(4) = 19.6 \text{ m}.$$

35. A man on the tenth floor of a building sees a bucket (dropped by a window washer) pass his window and notes that it hits the ground 1.5 s later. Assuming a floor is 16 ft high (and neglecting air friction), from which floor was the bucket dropped?

SOLUTION A falling object moves $16t^2$ feet in t seconds. Suppose H is the unknown height from which the bucket fell starting at time zero. The man saw it at some time t (also unknown to us) and it hit the ground, 160 feet down at time $t + 1.5$. Thus

$$(H - 16t^2) - (H - 16(t + 1.5)^2) = 160.$$

The H's cancel as do the t^2 terms and solving for t gives $t = 31/12$ s. Thus the bucket fell $16(31/12)^2 \approx 16 \cdot 6.67$ feet before the man saw it. Since there are 16 feet in a floor the bucket was dropped from 6.67 floors above the 10th: either the 16th or 17th floor.

37. Show that for an object rising and falling according to Galileo's formula in Eq. (3), the average velocity over any time interval $[t_1, t_2]$ is equal to the average of the instantaneous velocities at t_1 and t_2.

SOLUTION   The simplest way to proceed is to compute both values and show that they are equal. The average velocity over $[t_1, t_2]$ is

$$\frac{s(t_2) - s(t_1)}{t_2 - t_1} = \frac{(s_0 + v_0 t_2 - \frac{1}{2} g t_2^2) - (s_0 + v_0 t_1 - \frac{1}{2} g t_1^2)}{t_2 - t_1} = \frac{v_0(t_2 - t_1) + \frac{g}{2}(t_2^2 - t_1^2)}{t_2 - t_1}$$

$$= \frac{v_0(t_2 - t_1)}{t_2 - t_1} - \frac{g}{2}(t_2 + t_1) = v_0 - \frac{g}{2}(t_2 + t_1)$$

Whereas the average of the instantaneous velocities at the beginning and end of $[t_1, t_2]$ is

$$\frac{s'(t_1) + s'(t_2)}{2} = \frac{1}{2}\Big((v_0 - g t_1) + (v_0 - g t_2)\Big) = \frac{1}{2}(2v_0) - \frac{g}{2}(t_2 + t_1) = v_0 - \frac{g}{2}(t_2 + t_1).$$

The two quantities are the same.

In Exercises 39–46, use Eq. (2) to estimate the unit change.

39. Estimate $\sqrt{2} - \sqrt{1}$ and $\sqrt{101} - \sqrt{100}$. Compare your estimates with the actual values.

SOLUTION   Let $f(x) = \sqrt{x} = x^{1/2}$. Then $f'(x) = \frac{1}{2}(x^{-1/2})$. We are using the derivative to estimate the average ROC. That is,

$$\frac{\sqrt{x+h} - \sqrt{x}}{h} \approx f'(x),$$

so that

$$\sqrt{x+h} - \sqrt{x} \approx h f'(x).$$

Thus, $\sqrt{2} - \sqrt{1} \approx 1 f'(1) = \frac{1}{2}(1) = \frac{1}{2}$. The actual value, to six decimal places, is 0.414214. Also, $\sqrt{101} - \sqrt{100} \approx 1 f'(100) = \frac{1}{2}\left(\frac{1}{10}\right) = .05$. The actual value, to six decimal places, is 0.0498756.

41. Let $F(s) = 1.1s + 0.03s^2$ be the stopping distance as in Example 3. Calculate $F(65)$ and estimate the increase in stopping distance if speed is increased from 65 to 66 mph. Compare your estimate with the actual increase.

SOLUTION   Let $F(s) = 1.1s + .03s^2$ be as in Example 3. $F'(s) = 1.1 + 0.06s$.

- Then $F(65) = 198.25$ ft and $F'(65) = 5.00$ ft/mph.
- $F'(65) \approx F(66) - F(65)$ is approximately equal to the change in stopping distance per 1 mph increase in speed when traveling at 65 mph. Increasing speed from 65 to 66 therefore increases stopping distance by approximately 5 ft.
- The actual increase in stopping distance when speed increases from 65 mph to 66 mph is $F(66) - F(65) = 203.28 - 198.25 = 5.03$ feet, which differs by less than one percent from the estimate found using the derivative.

43. The dollar cost of producing x bagels is $C(x) = 300 + 0.25x - 0.5(x/1{,}000)^3$. Determine the cost of producing 2,000 bagels and estimate the cost of the 2001st bagel. Compare your estimate with the actual cost of the 2001st bagel.

SOLUTION   Expanding the power of 3 yields

$$C(x) = 300 + .25x - 5 \times 10^{-10} x^3.$$

This allows us to get the derivative $C'(x) = .25 - 1.5 \times 10^{-9} x^2$. The cost of producing 2000 bagels is

$$C(2000) = 300 + 0.25(2000) - 0.5(2000/1000)^3 = 796$$

dollars. The cost of the 2001st bagel is, by definition, $C(2001) - C(2000)$. By the derivative estimate, $C(2001) - C(2000) \approx C'(2000)(1)$, so the cost of the 2001st bagel is approximately

$$C'(2000) = .25 - 1.5 \times 10^{-9}(2000^2) = \$.244.$$

$C(2001) = 796.244$, so the *exact* cost of the 2001st bagel is indistinguishable from the estimated cost. The function is very nearly linear at this point.

45. The demand for a commodity generally decreases as the price is raised. Suppose that the demand for oil (per capita per year) is $D(p) = 900/p$ barrels, where p is the price per barrel in dollars. Find the demand when $p = \$40$. Estimate the decrease in demand if p rises to $41 and the increase if p is decreased to $39.

SOLUTION $D(p) = 900p^{-1}$, so $D'(p) = -900p^{-2}$. When the price is $40 a barrel, the per capita demand is $D(40) = 22.5$ barrels per year. With an increase in price from $40 to $41 a barrel, the change in demand $D(41) - D(40)$ is approximately $D'(40) = -900(40^{-2}) = -.5625$ barrels a year. With a decrease in price from $40 to $39 a barrel, the change in demand $D(39) - D(40)$ is approximately $-D'(40) = +.5625$. An increase in oil prices of a dollar leads to a decrease in demand of .5625 barrels a year, and a decrease of a dollar leads to an *increase* in demand of .5625 barrels a year.

47. Let $A = s^2$. Show that the estimate of $A(s + 1) - A(s)$ provided by Eq. (2) has error exactly equal to 1. Explain this result using Figure 14.

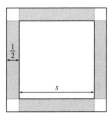

FIGURE 14

SOLUTION Let $A = s^2$. Then

$$A(s + 1) - A(s) = (s + 1)^2 - s^2 = s^2 + 2s + 1 - s^2 = 2s + 1,$$

while $A'(s) = 2s$. Therefore, regardless of the value of s,

$$A(s + 1) - A(s) - A'(s) = (2s + 1) - 2s = 1.$$

To understand this result, consider Figure 14, which illustrates a square of side length s expanded to a square of side length $s + 1$ by increasing the side length by $1/2$ unit in each direction. The difference $A(s + 1) - A(s)$ is the total area between the two squares. On the other hand, $A'(s) = 2s$ is the total area of the four rectangles of dimension s by $1/2$ bordering the sides of the original square. The error in approximating $A(s + 1) - A(s)$ by $A'(s)$ is given by the four $1/2$ by $1/2$ squares in the corners of the figure. The total area of these four small squares is always exactly 1.

49. Let $M(t)$ be the mass (in kilograms) of a plant as a function of time (in years). Recent studies by Niklas and Enquist have suggested that for a remarkably wide range of plants (from algae and grass to palm trees), the growth rate during the life span of the organism satisfies a *three-quarter power law*, that is, $dM/dt = CM^{3/4}$ for some constant C.
(a) If a tree has a growth rate of 6 kg/year when $M = 100$ kg, what is its growth rate when $M = 125$ kg?
(b) If $M = 0.5$ kg, how much more mass must the plant acquire to double its growth rate?

SOLUTION

(a) Suppose a tree has a growth rate dM/dt of 6 kg/yr when $M = 100$, then $6 = C(100^{3/4}) = 10C\sqrt{10}$, so that $C = \frac{3\sqrt{10}}{50}$. When $M = 125$,

$$\frac{dM}{dt} = C(125^{3/4}) = \frac{3\sqrt{10}}{50} 25(5^{1/4}) = 7.09306.$$

(b) The growth rate when $M = .5$ kg is $dM/dt = C(.5^{3/4})$. To double the rate, we must find M so that $dM/dt = CM^{3/4} = 2C(.5^{3/4})$. We solve for M.

$$CM^{3/4} = 2C(.5^{3/4})$$
$$M^{3/4} = 2(.5^{3/4})$$
$$M = (2(.5^{3/4}))^{4/3} = 1.25992.$$

The plant must acquire the difference $1.25992 - .5 = .75992$ kg in order to double its growth rate.
Note that a doubling of growth rate requires *more* than a doubling of mass.

Further Insights and Challenges

51. The size of a certain animal population $P(t)$ at time t (in months) satisfies $\dfrac{dP}{dt} = 0.2(300 - P)$.

(a) Is P growing or shrinking when $P = 250$? when $P = 350$?
(b) Sketch the graph of dP/dt as a function of P for $0 \le P \le 300$.
(c) Which of the graphs in Figure 15 is the graph of $P(t)$ if $P(0) = 200$?

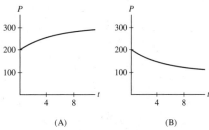

(A)   (B)

FIGURE 15

SOLUTION Let $P'(t) = dP/dt = 0.2(300 - P)$.

(a) Since $P'(250) = 10$, the population is growing when $P = 250$.
  Since $P'(350) = -10$, the population is shrinking when $P = 350$.

(b) Here is a graph of dP/dt for $0 \leq t \leq 300$.

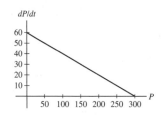

(c) If $P(0) = 200$, as in graph (A), then $P'(0) = 20 > 0$ and the population is growing as depicted. Accordingly, graph (A) has the correct shape for $P(t)$.
  If $P(0) = 200$, as in graph (B), then $P'(0) = 20 > 0$ and the population is growing, contradicting what is depicted. Thus graph (B) cannot be the correct shape for $P(t)$.

In Exercises 53–54, the average cost per unit at production level x is defined as $C_{avg}(x) = C(x)/x$, where $C(x)$ is the cost function. Average cost is a measure of the efficiency of the production process.

53. Show that $C_{avg}(x)$ is equal to the slope of the line through the origin and the point $(x, C(x))$ on the graph of $C(x)$. Using this interpretation, determine whether average cost or marginal cost is greater at points A, B, C, D in Figure 16.

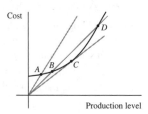

FIGURE 16 Graph of $C(x)$.

SOLUTION By definition, the slope of the line through the origin and $(x, C(x))$, that is, between $(0, 0)$ and $(x, C(x))$ is

$$\frac{C(x) - 0}{x - 0} = \frac{C(x)}{x} = C_{av}.$$

At point A, average cost is greater than marginal cost, as the line from the origin to A is steeper than the curve at this point (we see this because the line, tracing from the origin, crosses the curve from below). At point B, the average cost is still greater than the marginal cost. At the point C, the average cost and the marginal cost are nearly the same, since the tangent line and the line from the origin are nearly the same. The line from the origin to D crosses the cost curve from above, and so is less steep than the tangent line to the curve at D; the average cost at this point is less than the marginal cost.

3.5 Higher Derivatives

Preliminary Questions

1. An economist who announces that "America's economic growth is slowing" is making a statement about the gross national product (GNP) as a function of time. Is the second derivative of the GNP positive? What about the first derivative?

SOLUTION If America's economic growth is slowing, then the GNP is increasing, but the rate of increase is decreasing. Thus, the second derivative of GNP is negative, while the first derivative is positive.

2. On September 4, 2003, the *Wall Street Journal* printed the headline "Stocks Go Higher, Though the Pace of Their Gains Slows." Rephrase as a statement about the first and second time derivatives of stock prices and sketch a possible graph.

SOLUTION Because stocks are going higher, stock prices are increasing and the first derivative of stock prices must therefore be positive. On the other hand, because the pace of gains is slowing, the second derivative of stock prices must be negative.

3. Is the following statement true or false? The third derivative of position with respect to time is zero for an object falling to earth under the influence of gravity. Explain.

SOLUTION   This statement is true. The acceleration of an object falling to earth under the influence of gravity is constant; hence, the second derivative of position with respect to time is constant. Because the third derivative is just the derivative of the second derivative and the derivative of a constant is zero, it follows that the third derivative is zero.

4. Which type of polynomial satisfies $f''(x) = 0$ for all x?

SOLUTION   The second derivative of all linear polynomials (polynomials of the form $ax + b$ for some constants a and b) is equal to 0 for all x.

Exercises

In Exercises 1–12, calculate the second and third derivatives.

1. $y = 14x^2$

SOLUTION   Let $y = 14x^2$. Then $y' = 28x$, $y'' = 28$, and $y''' = 0$.

3. $y = x^4 - 25x^2 + 2x$

SOLUTION   Let $y = x^4 - 25x^2 + 2x$. Then $y' = 4x^3 - 50x + 2$, $y'' = 12x^2 - 50$, and $y''' = 24x$.

5. $y = \dfrac{4}{3}\pi r^3$

SOLUTION   Let $y = \frac{4}{3}\pi r^3$. Then $y' = 4\pi r^2$, $y'' = 8\pi r$, and $y''' = 8\pi$.

7. $y = 20t^{4/5} - 6t^{2/3}$

SOLUTION   Let $y = 20t^{4/5} - 6t^{2/3}$. Then $y' = 16t^{-1/5} - 4t^{-1/3}$, $y'' = -\frac{16}{5}t^{-6/5} + \frac{4}{3}t^{-4/3}$, and $y''' = \frac{96}{25}t^{-11/15} - \frac{16}{9}t^{-7/3}$.

9. $y = z - \dfrac{1}{z}$

SOLUTION   Let $y = z - z^{-1}$. Then $y' = 1 + z^{-2}$, $y'' = -2z^{-3}$, and $y''' = 6z^{-4}$.

11. $y = (x^2 + x)(x^3 + 1)$

SOLUTION   Since we don't want to apply the product rule to an ever growing list of products, we multiply through first. Let $y = (x^2 + x)(x^3 + 1) = x^5 + x^4 + x^2 + x$. Then $y' = 5x^4 + 4x^3 + 2x + 1$, $y'' = 20x^3 + 12x^2 + 2$, and $y''' = 60x^2 + 24x$.

In Exercises 13–24, calculate the derivative indicated.

13. $f^{(4)}(1)$, $f(x) = x^4$

SOLUTION   Let $f(x) = x^4$. Then $f'(x) = 4x^3$, $f''(x) = 12x^2$, $f'''(x) = 24x$, and $f^{(4)}(x) = 24$. Thus $f^{(4)}(1) = 24$.

15. $\left.\dfrac{d^2 y}{dt^2}\right|_{t=1}$, $y = 4t^{-3} + 3t^2$

SOLUTION   Let $y = 4t^{-3} + 3t^2$. Then $\frac{dy}{dt} = -12t^{-4} + 6t$ and $\frac{d^2 y}{dt^2} = 48t^{-5} + 6$. Hence

$$\left.\frac{d^2 y}{dt^2}\right|_{t=1} = 48(1)^{-5} + 6 = 54.$$

17. $h'''(9)$, $h(x) = \sqrt{x}$

SOLUTION   Let $h(x) = \sqrt{x} = x^{1/2}$. Then $h'(x) = \frac{1}{2}x^{-1/2}$, $h''(x) = -\frac{1}{4}x^{-3/2}$, and $h'''(x) = \frac{3}{8}x^{-5/2}$. Thus $h'''(9) = \frac{1}{648}$.

19. $\left.\dfrac{d^4x}{dt^4}\right|_{t=16}$, $x = t^{-3/4}$

SOLUTION   Let $x(t) = t^{-3/4}$. Then $\frac{dx}{dt} = -\frac{3}{4}t^{-7/4}$, $\frac{d^2x}{dt^2} = \frac{21}{16}t^{-11/4}$, $\frac{d^3x}{dt^3} = -\frac{231}{64}t^{-15/4}$, and $\frac{d^4x}{dt^4} = \frac{3465}{256}t^{-19/4}$. Thus

$$\left.\frac{d^4x}{dt^4}\right|_{t=16} = \frac{3465}{256}16^{-19/4} = \frac{3465}{134217728}.$$

21. $g''(1)$, $g(x) = \dfrac{x}{x+1}$

SOLUTION   Let $g(x) = \dfrac{x}{x+1}$. Then

$$g'(x) = \frac{(x+1)(1) - (x)(1)}{(x+1)^2} = \frac{1}{(x+1)^2} = \frac{1}{x^2 + 2x + 1}$$

and

$$g''(x) = \frac{(x^2+2x+1)(0) - 1(2x+2)}{(x^2+2x+1)^2} = -\frac{2x+2}{(x+1)^4} = -\frac{2}{(x+1)^3}.$$

Thus, $g''(1) = -\dfrac{1}{4}$.

23. $h''(1)$, $h(x) = \dfrac{1}{\sqrt{x}+1}$

SOLUTION   Let $h(x) = \dfrac{1}{\sqrt{x}+1}$. Then

$$h'(x) = \frac{-\frac{1}{2}x^{-1/2}}{\left(\sqrt{x}+1\right)^2} = \frac{-\frac{1}{2}x^{-1/2}}{x + 2\sqrt{x} + 1},$$

and

$$h''(x) = \frac{(x + 2\sqrt{x} + 1)\frac{1}{4}x^{-3/2} + \frac{1}{2}x^{-1/2}(1 + x^{-1/2})}{(x + 2\sqrt{x} + 1)^2} = \frac{\frac{3}{4}x^{-1/2} + x^{-1} + \frac{1}{4}x^{-3/2}}{(\sqrt{x}+1)^4}$$

$$= \frac{(\sqrt{x}+1)\left(\frac{3}{4}x^{-1} + \frac{1}{4}x^{-3/2}\right)}{(\sqrt{x}+1)^4} = \frac{\frac{3}{4}x^{-1} + \frac{1}{4}x^{-3/2}}{(\sqrt{x}+1)^3}.$$

Accordingly, $h''(1) = \frac{(3/4)+(1/4)}{2^3} = \frac{1}{8}$.

25. Calculate $y^{(k)}(0)$ for $0 \le k \le 5$, where $y = x^4 + ax^3 + bx^2 + cx + d$ (with a, b, c, d the constants).

SOLUTION   Applying the power, constant multiple, and sum rules at each stage, we get (note $y^{(0)}$ is y by convention):

k	$y^{(k)}$
0	$x^4 + ax^3 + bx^2 + cx + d$
1	$4x^3 + 3ax^2 + 2bx + c$
2	$12x^2 + 6ax + 2b$
3	$24x + 6a$
4	24
5	0

from which we get $y^{(0)}(0) = d$, $y^{(1)}(0) = c$, $y^{(2)}(0) = 2b$, $y^{(3)}(0) = 6a$, $y^{(4)}(0) = 24$, and $y^{(5)}(0) = 0$.

27. Use the result in Example 2 to find $\dfrac{d^6}{dx^6}x^{-1}$.

SOLUTION   The equation in Example 2 indicates that

$$\frac{d^6}{dx^6}x^{-1} = (-1)^6 6! x^{-6-1}.$$

$(-1)^6 = 1$ and $6! = 6 \times 5 \times 4 \times 3 \times 2 \times 1 = 720$, so

$$\frac{d^6}{dx^6} x^{-1} = 720x^{-7}.$$

In Exercises 29–32, find a general formula for $f^{(n)}(x)$.

29. $f(x) = (x+1)^{-1}$

SOLUTION Let $f(x) = (x+1)^{-1} = \frac{1}{x+1}$. By Exercise 12, $f'(x) = -1(x+1)^{-2}$, $f''(x) = 2(x+1)^{-3}$, $f'''(x) = -6(x+1)^{-4}$, $f^{(4)}(x) = 24(x+1)^{-5}, \dots$ From this we conclude that the nth derivative can be written as $f^{(n)}(x) = (-1)^n n!(x+1)^{-(n+1)}$.

31. $f(x) = x^{-1/2}$

SOLUTION $f'(x) = \frac{-1}{2}x^{-3/2}$. We will avoid simplifying numerators and denominators to find the pattern:

$$f''(x) = \frac{-3}{2}\frac{-1}{2}x^{-5/2} = (-1)^2 \frac{3 \times 1}{2^2}x^{-5/2}$$

$$f'''(x) = -\frac{5}{2}\frac{3 \times 1}{2^2}x^{-7/2} = (-1)^3 \frac{5 \times 3 \times 1}{2^3}x^{-7/2}$$

$$\vdots$$

$$f^{(n)}(x) = (-1)^n \frac{(2n-1) \times (2n-3) \times \dots \times 1}{2^n}x^{-(2n+1)/2}.$$

33. (a) Find the acceleration at time $t = 5$ min of a helicopter whose height (in feet) is $h(t) = -3t^3 + 400t$.
(b) [GU] Plot the acceleration $h''(t)$ for $0 \le t \le 6$. How does this graph show that the helicopter is slowing down during this time interval?

SOLUTION

(a) Let $h(t) = -3t^3 + 400t$, with t in minutes and h in feet. The velocity is $v(t) = h'(t) = -9t^2 + 400$ and acceleration is $a(t) = h''(t) = -18t$. Thus $a(5) = -90$ ft/min^2.
(b) The acceleration of the helicopter for $0 \le t \le 6$ is shown in the figure below. As the acceleration of the helicopter is negative, the velocity of the helicopter must be decreasing. Because the velocity is positive for $0 \le t \le 6$, the helicopter is slowing down.

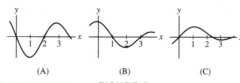

35. Figure 5 shows f, f', and f''. Determine which is which.

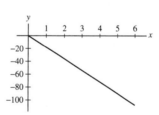

FIGURE 5

SOLUTION (a) f'' (b) f' (c) f.

The tangent line to (c) is horizontal at $x = 1$ and $x = 3$, where (b) has roots. The tangent line to (b) is horizontal at $x = 2$ and $x = 0$, where (a) has roots.

37. Figure 7 shows the graph of the position of an object as a function of time. Determine the intervals on which the acceleration is positive.

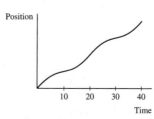

FIGURE 7

SOLUTION   Roughly from time 10 to time 20 and from time 30 to time 40. The acceleration is positive over the same intervals over which the graph is bending upward.

39. Find a polynomial $f(x)$ satisfying the equation $xf''(x) + f(x) = x^2$.

SOLUTION   Since $xf''(x) + f(x) = x^2$, and x^2 is a polynomial, it seems reasonable to assume that $f(x)$ is a polynomial of some degree, call it n. The degree of $f''(x)$ is $n - 2$, so the degree of $xf''(x)$ is $n - 1$, and the degree of $xf''(x) + f(x)$ is n. Hence, $n = 2$, since the degree of x^2 is 2. Therefore, let $f(x) = ax^2 + bx + c$. Then $f'(x) = 2ax + b$ and $f''(x) = 2a$. Substituting into the equation $xf''(x) + f(x) = x^2$ yields $ax^2 + (2a + b)x + c = x^2$, an identity in x. Equating coefficients, we have $a = 1$, $2a + b = 0$, $c = 0$. Therefore, $b = -2$ and $f(x) = x^2 - 2x$.

41. A servomotor controls the vertical movement of a drill bit that will drill a pattern of holes in sheet metal. The maximum vertical speed of the drill bit is 4 in./s, and while drilling the hole, it must move no more than 2.6 in./s to avoid warping the metal. During a cycle, the bit begins and ends at rest, quickly approaches the sheet metal, and quickly returns to its initial position after the hole is drilled. Sketch possible graphs of the drill bit's vertical velocity and acceleration. Label the point where the bit enters the sheet metal.

SOLUTION   There will be multiple cycles, each of which will be more or less identical. Let $v(t)$ be the *downward* vertical velocity of the drill bit, and let $a(t)$ be the vertical acceleration. From the narrative, we see that $v(t)$ can be no greater than 4 and no greater than 2.6 while drilling is taking place. During each cycle, $v(t) = 0$ initially, $v(t)$ goes to 4 quickly. When the bit hits the sheet metal, $v(t)$ goes down to 2.6 quickly, at which it stays until the sheet metal is drilled through. As the drill pulls out, it reaches maximum non-drilling upward speed ($v(t) = -4$) quickly, and maintains this speed until it returns to rest. A possible plot follows:

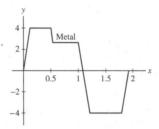

A graph of the acceleration is extracted from this graph:

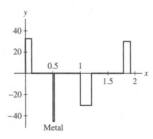

In Exercises 42–43, refer to the following. In their 1997 study, Boardman and Lave related the traffic speed S on a two-lane road to traffic density Q (number of cars per mile of road) by the formula $S = 2{,}882Q^{-1} - 0.052Q + 31.73$ for $60 \le Q \le 400$ (Figure 9).

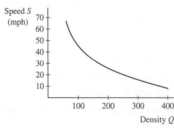

FIGURE 9 Speed as a function of traffic density.

43. (a) 📖   Explain intuitively why we should expect that $dS/dQ < 0$.

(b) Show that $d^2S/dQ^2 > 0$. Then use the fact that $dS/dQ < 0$ and $d^2S/dQ^2 > 0$ to justify the following statement: *A one-unit increase in traffic density slows down traffic more when Q is small than when Q is large.*

(c) GU Plot dS/dQ. Which property of this graph shows that $d^2S/dQ^2 > 0$?

SOLUTION

(a) Traffic speed must be reduced when the road gets more crowded so we expect dS/dQ to be negative. This is indeed the case since $dS/dQ = -.052 - 2882/Q^2 < 0$.

(b) The decrease in speed due to a one-unit increase in density is approximately dS/dQ (a negative number). Since $d^2S/dQ^2 = 5764Q^{-3} > 0$ is positive, this tells us that dS/dQ gets larger as Q increases—and a negative number which gets larger is getting closer to zero. So the decrease in speed is smaller when Q is larger, that is, a one-unit increase in traffic density has a smaller effect when Q is large.

(c) dS/dQ is plotted below. The fact that this graph is increasing shows that $d^2S/dQ^2 > 0$.

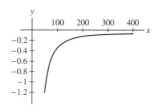

45. ⌐*R5* Use a computer algebra system to compute $f^{(k)}(x)$ for $k = 1, 2, 3$ for the functions:

(a) $f(x) = (1 + x^3)^{5/3}$

(b) $f(x) = \dfrac{1 - x^4}{1 - 5x - 6x^2}$

SOLUTION

(a) Let $f(x) = (1 + x^3)^{5/3}$. Using a computer algebra system,

$$f'(x) = 5x^2(1 + x^3)^{2/3};$$

$$f''(x) = 10x(1 + x^3)^{2/3} + 10x^4(1 + x^3)^{-1/3}; \text{ and}$$

$$f'''(x) = 10(1 + x^3)^{2/3} + 60x^3(1 + x^3)^{-1/3} - 10x^6(1 + x^3)^{-4/3}.$$

(b) Let $f(x) = \dfrac{1 - x^4}{1 - 5x - 6x^2}$. Using a computer algebra system,

$$f'(x) = \frac{12x^3 - 9x^2 + 2x + 5}{(6x - 1)^2};$$

$$f''(x) = \frac{2(36x^3 - 18x^2 + 3x - 31)}{(6x - 1)^3}; \text{ and}$$

$$f'''(x) = \frac{1110}{(6x - 1)^4}.$$

Further Insights and Challenges

47. Find the 100th derivative of

$$p(x) = (x + x^5 + x^7)^{10}(1 + x^2)^{11}(x^3 + x^5 + x^7)$$

SOLUTION This is a polynomial of degree $70 + 22 + 7 = 99$, so its 100th derivative is zero.

49. Use the Product Rule twice to find a formula for $(fg)''$ in terms of the first and second derivative of f and g.

SOLUTION Let $h = fg$. Then $h' = fg' + gf' = f'g + fg'$ and

$$h'' = f'g' + gf'' + fg'' + g'f' = f''g + 2f'g' + fg''.$$

51. Compute

$$\Delta f(x) = \lim_{h \to 0} \frac{f(x + h) + f(x - h) - 2f(x)}{h^2}$$

for the following functions:

(a) $f(x) = x$ **(b)** $f(x) = x^2$ **(c)** $f(x) = x^3$

Based on these examples, can you formulate a conjecture about what Δf is?

SOLUTION   For $f(x) = x$, we have

$$f(x + h) + f(x - h) - 2f(x) = (x + h) + (x - h) - 2x = 0.$$

Hence, $\Delta(x) = 0$. For $f(x) = x^2$,

$$f(x + h) + f(x - h) - 2f(x) = (x + h)^2 + (x - h)^2 - 2x^2$$
$$= x^2 + 2xh + h^2 + x^2 - 2xh + h^2 - 2x^2 = 2h^2,$$

so $\Delta(x^2) = 2$. Working in a similar fashion, we find $\Delta(x^3) = 6x$. One can prove that for twice differentiable functions, $\Delta f = f''$. It is an interesting fact of more advanced mathematics that there are functions f for which Δf exists at all points, but the function is not differentiable.

3.6 Trigonometric Functions

Preliminary Questions

1. Determine the sign $\pm$ that yields the correct formula for the following:

(a) $\dfrac{d}{dx}(\sin x + \cos x) = \pm \sin x \pm \cos x$

(b) $\dfrac{d}{dx}\tan x = \pm \sec^2 x$

(c) $\dfrac{d}{dx}\sec x = \pm \sec x \tan x$

(d) $\dfrac{d}{dx}\cot x = \pm \csc^2 x$

SOLUTION   The correct formulas are

(a) $\dfrac{d}{dx}(\sin x + \cos x) = -\sin x + \cos x$

(b) $\dfrac{d}{dx}\tan x = \sec^2 x$

(c) $\dfrac{d}{dx}\sec x = \sec x \tan x$

(d) $\dfrac{d}{dx}\cot x = -\csc^2 x$

2. Which of the following functions can be differentiated using the rules we have covered so far?

(a) $y = 3\cos x \cot x$ **(b)** $y = \cos(x^2)$ **(c)** $y = x^2 \cos x$

SOLUTION

(a) $3\cos x \cot x$ is a product of functions whose derivatives are known. This function can therefore be differentiated using the Product Rule.

(b) $\cos(x^2)$ is a composition of the functions $\cos x$ and x^2. We have not yet discussed how to differentiate composite functions.

(c) $x^2 \cos x$ is a product of functions whose derivatives are known. This function can therefore be differentiated using the Product Rule.

3. Compute $\dfrac{d}{dx}(\sin^2 x + \cos^2 x)$ without using the derivative formulas for $\sin x$ and $\cos x$.

SOLUTION   Recall that $\sin^2 x + \cos^2 x = 1$ for all x. Thus,

$$\frac{d}{dx}(\sin^2 x + \cos^2 x) = \frac{d}{dx}1 = 0.$$

4. How is the addition formula used in deriving the formula $(\sin x)' = \cos x$?

SOLUTION   The difference quotient for the function $\sin x$ involves the expression $\sin(x + h)$. The addition formula for the sine function is used to expand this expression as $\sin(x + h) = \sin x \cos h + \sin h \cos x$.

Exercises

In Exercises 1–4, find an equation of the tangent line at the point indicated.

1. $y = \sin x$, $x = \dfrac{\pi}{4}$

SOLUTION   Let $f(x) = \sin x$. Then $f'(x) = \cos x$ and the equation of the tangent line is

$$y = f'\left(\frac{\pi}{4}\right)\left(x - \frac{\pi}{4}\right) + f\left(\frac{\pi}{4}\right) = \frac{\sqrt{2}}{2}\left(x - \frac{\pi}{4}\right) + \frac{\sqrt{2}}{2} = \frac{\sqrt{2}}{2}x + \frac{\sqrt{2}}{2}\left(1 - \frac{\pi}{4}\right).$$

3. $y = \tan x$, $x = \dfrac{\pi}{4}$

SOLUTION   Let $f(x) = \tan x$. Then $f'(x) = \sec^2 x$ and the equation of the tangent line is

$$y = f'\left(\frac{\pi}{4}\right)\left(x - \frac{\pi}{4}\right) + f\left(\frac{\pi}{4}\right) = 2\left(x - \frac{\pi}{4}\right) + 1 = 2x + 1 - \frac{\pi}{2}.$$

In Exercises 5–26, use the Product and Quotient Rules as necessary to find the derivative of each function.

5. $f(x) = \sin x \cos x$

SOLUTION   Let $f(x) = \sin x \cos x$. Then

$$f'(x) = \sin x(-\sin x) + \cos x(\cos x) = -\sin^2 x + \cos^2 x.$$

7. $f(x) = \sin^2 x$

SOLUTION   Let $f(x) = \sin^2 x = \sin x \sin x$. Then

$$f'(x) = \sin x(\cos x) + \sin x(\cos x) = 2\sin x \cos x.$$

9. $f(x) = x^3 \sin x$

SOLUTION   Let $f(x) = x^3 \sin x$. Then $f'(x) = x^3 \cos x + 3x^2 \sin x$.

11. $f(\theta) = \tan \theta \sec \theta$

SOLUTION   Let $f(\theta) = \tan \theta \sec \theta$. Then

$$f'(\theta) = \tan \theta \sec \theta \tan \theta + \sec \theta \sec^2 \theta = \sec \theta \tan^2 \theta + \sec^3 \theta = \left(\tan^2 \theta + \sec^2 \theta\right)\sec \theta.$$

13. $h(\theta) = \cos^2 \theta$

SOLUTION   Let $h(\theta) = \cos^2 \theta = \cos \theta \cos \theta$. Then

$$h'(\theta) = \cos \theta(-\sin \theta) + \cos \theta(-\sin \theta) = -2\cos \theta \sin \theta.$$

15. $f(x) = (x - x^2)\cot x$

SOLUTION   Let $f(x) = (x - x^2)\cot x$. Then

$$f'(x) = (x - x^2)(-\csc^2 x) + \cot x(1 - 2x).$$

17. $f(x) = \dfrac{\sec x}{x^2}$

SOLUTION   Let $f(x) = \dfrac{\sec x}{x^2}$. Then $f'(x) = \dfrac{(x^2)\sec x \tan x - 2x \sec x}{x^4} = \dfrac{x \sec x \tan x - 2\sec x}{x^3}$.

19. $g(t) = \sin t - \dfrac{2}{\cos t}$

SOLUTION   Let $g(t) = \sin t - \dfrac{2}{\cos t} = \sin t - 2\sec t$. Then $g'(t) = \cos t - 2\sec t \tan t$.

21. $f(x) = \dfrac{x}{\sin x + 2}$

SOLUTION   Let $f(x) = \dfrac{x}{2 + \sin x}$. Then

$$f'(x) = \frac{(2 + \sin x)(1) - x \cos x}{(2 + \sin x)^2} = \frac{2 + \sin x - x \cos x}{(2 + \sin x)^2}.$$

23. $f(x) = \dfrac{1 + \sin x}{1 - \sin x}$

SOLUTION   Let $f(x) = \dfrac{1 + \sin x}{1 - \sin x}$. Then

$$f'(x) = \frac{(1 - \sin x)(\cos x) - (1 + \sin x)(-\cos x)}{(1 - \sin x)^2} = \frac{2 \cos x}{(1 - \sin x)^2}.$$

25. $g(x) = \dfrac{\sec x}{x}$

SOLUTION   Let $g(x) = \dfrac{\sec x}{x}$. Then

$$g'(x) = \frac{x \sec x \tan x - (\sec x)(1)}{x^2} = \frac{(x \tan x - 1) \sec x}{x^2}.$$

In Exercises 27–30, calculate the second derivative.

27. $f(x) = 3 \sin x + 4 \cos x$

SOLUTION   Let $f(x) = 3 \sin x + 4 \cos x$. Then $f'(x) = 3 \cos x - 4 \sin x$ and $f''(x) = -3 \sin x - 4 \cos x$.

29. $g(\theta) = \theta \sin \theta$

SOLUTION   Let $g(\theta) = \theta \sin \theta$. Then, by the product rule, $g'(\theta) = \theta \cos \theta + \sin \theta$ and

$$g''(\theta) = -\theta \sin \theta + \cos \theta + \cos \theta = 2 \cos \theta - \theta \sin \theta.$$

In Exercises 31–36, find an equation of the tangent line at the point specified.

31. $y = x^2 + \sin x, \quad x = 0$

SOLUTION   Let $f(x) = x^2 + \sin x$. Then $f'(x) = 2x + \cos x$ and $f'(0) = 1$. The tangent line at $x = 0$ is

$$y = f'(0)(x - 0) + f(0) = 1(x - 0) + 0 = x.$$

33. $y = 2 \sin x + 3 \cos x, \quad x = \dfrac{\pi}{3}$

SOLUTION   Let $f(x) = 2 \sin x + 3 \cos x$. Then $f'(x) = 2 \cos x - 3 \sin x$ and $f'\left(\frac{\pi}{3}\right) = 1 - \frac{3\sqrt{3}}{2}$. The tangent line at $x = \frac{\pi}{3}$ is

$$y = f'\left(\frac{\pi}{3}\right)\left(x - \frac{\pi}{3}\right) + f\left(\frac{\pi}{3}\right) = \left(1 - \frac{3\sqrt{3}}{2}\right)\left(x - \frac{\pi}{3}\right) + \sqrt{3} + \frac{3}{2}$$

$$= \left(1 - \frac{3\sqrt{3}}{2}\right)x + \sqrt{3} + \frac{3}{2} + \frac{\sqrt{3}}{2}\pi - \frac{\pi}{3}.$$

35. $y = \csc x - \cot x, \quad x = \dfrac{\pi}{4}$

SOLUTION   Let $f(x) = \csc x - \cot x$. Then

$$f'(x) = \csc^2 x - \csc x \cot x$$

and

$$f'\left(\frac{\pi}{4}\right) = 2 - \sqrt{2} \cdot 1 = 2 - \sqrt{2}.$$

Hence the tangent line is

$$y = f'\left(\frac{\pi}{4}\right)\left(x - \frac{\pi}{4}\right) + f\left(\frac{\pi}{4}\right) = \left(2 - \sqrt{2}\right)\left(x - \frac{\pi}{4}\right) + \left(\sqrt{2} - 1\right)$$

$$= \left(2 - \sqrt{2}\right)x + \sqrt{2} - 1 + \frac{\pi}{4}\left(\sqrt{2} - 2\right).$$

In Exercises 37–39, verify the formula using $(\sin x)' = \cos x$ *and* $(\cos x)' = -\sin x$.

37. $\dfrac{d}{dx}\cot x = -\csc^2 x$

SOLUTION $\cot x = \dfrac{\cos x}{\sin x}$. Using the quotient rule and the derivative formulas, we compute:

$$\frac{d}{dx}\cot x = \frac{d}{dx}\frac{\cos x}{\sin x} = \frac{\sin x(-\sin x) - \cos x(\cos x)}{\sin^2 x} = \frac{-(\sin^2 x + \cos^2 x)}{\sin^2 x} = \frac{-1}{\sin^2 x} = -\csc^2 x.$$

39. $\dfrac{d}{dx}\csc x = -\csc x \cot x$

SOLUTION   Since $\csc x = \dfrac{1}{\sin x}$, we can apply the quotient rule and the two known derivatives to get:

$$\frac{d}{dx}\csc x = \frac{d}{dx}\frac{1}{\sin x} = \frac{\sin x(0) - 1(\cos x)}{\sin^2 x} = \frac{-\cos x}{\sin^2 x} = -\frac{\cos x}{\sin x}\frac{1}{\sin x} = -\cot x \csc x.$$

41. Calculate the first five derivatives of $f(x) = \cos x$. Then determine $f^{(8)}$ and $f^{(37)}$.

SOLUTION   Let $f(x) = \cos x$.

- Then $f'(x) = -\sin x$, $f''(x) = -\cos x$, $f'''(x) = \sin x$, $f^{(4)}(x) = \cos x$, and $f^{(5)}(x) = -\sin x$.
- Accordingly, the successive derivatives of f cycle among

$$\{-\sin x, -\cos x, \sin x, \cos x\}$$

    in that order. Since 8 is a multiple of 4, we have $f^{(8)}(x) = \cos x$.
- Since 36 is a multiple of 4, we have $f^{(36)}(x) = \cos x$. Therefore, $f^{(37)}(x) = -\sin x$.

43. Calculate $f''(x)$ and $f'''(x)$, where $f(x) = \tan x$.

SOLUTION   Let $f(x) = \tan x$. Then

$$f'(x) = \sec^2 x = \sec x \sec x$$
$$f''(x) = 2\sec x \sec x \tan x = 2\sec^2 x \tan x,$$

and

$$f'''(x) = 2\left(\sec^2 x \sec^2 x + (\tan x)(2\sec^2 x \tan x)\right) = 2\sec^4 x + 4\sec^2 x \tan^2 x$$

45. CAS Let $f(x) = \dfrac{\sin x}{x}$ for $x \neq 0$ and $f(0) = 1$.

(a) Plot $f(x)$ on $[-3\pi, 3\pi]$.

(b) Show that $f'(c) = 0$ if $c = \tan c$. Use the numerical root finder on a computer algebra system to find a good approximation to the smallest *positive* value c_0 such that $f'(c_0) = 0$.

(c) Verify that the horizontal line $y = f(c_0)$ is tangent to the graph of $y = f(x)$ at $x = c_0$ by plotting them on the same set of axes.

SOLUTION

(a) Here is the graph of $f(x)$ over $[-3\pi, 3\pi]$.

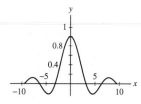

(b) Let $f(x) = \dfrac{\sin x}{x}$. Then

$$f'(x) = \frac{x\cos x - \sin x}{x^2}.$$

To have $f'(c) = 0$, it follows that $c\cos c - \sin c = 0$, or

$$\tan c = c.$$

Using a computer algebra system, we find that the smallest positive value c_0 such that $f'(c_0) = 0$ is $c_0 = 4.493409$.

(c) The horizontal line $y = f(c_0) = -0.217234$ and the function $y = f(x)$ are both plotted below. The horizontal line is clearly tangent to the graph of $f(x)$.

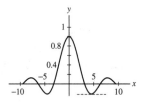

47. The height at time t (s) of a weight, oscillating up and down at the end of a spring, is $s(t) = 300 + 40 \sin t$ cm. Find the velocity and acceleration at $t = \frac{\pi}{3}$ s.

SOLUTION   Let $s(t) = 300 + 40 \sin t$ be the height. Then the velocity is

$$v(t) = s'(t) = 40 \cos t$$

and the acceleration is

$$a(t) = v'(t) = -40 \sin t.$$

At $t = \frac{\pi}{3}$, the velocity is $v\left(\frac{\pi}{3}\right) = 20$ cm/sec and the acceleration is $a\left(\frac{\pi}{3}\right) = -20\sqrt{3}$ cm/sec^2.

49. If you stand 1 m from a wall and mark off points on the wall at equal increments δ of angular elevation (Figure 4), then these points grow increasingly far apart. Explain how this illustrates the fact that the derivative of $\tan \theta$ is increasing.

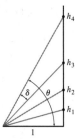

FIGURE 4

SOLUTION   Let $D(\theta) = \tan \theta$ represent the distance from the horizontal base to the point on the wall at angular elevation θ. Based on the derivative estimate of the rate of change, for a small change δ in the angle of elevation, the change in the distance can be approximated by:

$$\frac{D(\theta + \delta) - D(\theta)}{\delta} \approx D'(\theta) = \frac{d}{d\theta} \tan \theta.$$

The change in distance resulting from a δ increase in angle is then

$$D(\theta + \delta) - D(\theta) \approx \delta \frac{d}{d\theta} \tan \theta.$$

Knowing that $D(\theta + \delta) - D(\theta)$ increases as θ increases, it follows that $D'(\theta) = \frac{d}{d\theta} \tan \theta$ must increase as θ increases.

Further Insights and Challenges

51. Show that a nonzero polynomial function $y = f(x)$ *cannot* satisfy the equation $y'' = -y$. Use this to prove that neither $\sin x$ nor $\cos x$ is a polynomial.

SOLUTION

- Let p be a nonzero polynomial of degree n and *assume* that p satisfies the differential equation $y'' + y = 0$. Then $p'' + p = 0$ for all x. There are exactly three cases.

   (a) If $n = 0$, then p is a constant polynomial and thus $p'' = 0$. Hence $0 = p'' + p = p$ or $p \equiv 0$ (i.e., p is equal to 0 for all x or p is identically 0). This is a contradiction, since p is a *nonzero* polynomial.

   (b) If $n = 1$, then p is a linear polynomial and thus $p'' = 0$. Once again, we have $0 = p'' + p = p$ or $p \equiv 0$, a contradiction since p is a nonzero polynomial.

   (c) If $n \geq 2$, then p is at least a quadratic polynomial and thus p'' is a polynomial of degree $n - 2 \geq 0$. Thus $q = p'' + p$ is a polynomial of degree $n \geq 2$. By assumption, however, $p'' + p = 0$. Thus $q \equiv 0$, a polynomial of degree 0. This is a contradiction, since the degree of q is $n \geq 2$.

CONCLUSION: In all cases, we have reached a contradiction. Therefore the *assumption* that p satisfies the differential equation $y'' + y = 0$ is *false*. Accordingly, a nonzero polynomial *cannot* satisfy the stated differential equation.

- Let $y = \sin x$. Then $y' = \cos x$ and $y'' = -\sin x$. Therefore, $y'' = -y$. Now, let $y = \cos x$. Then $y' = -\sin x$ and $y'' = -\cos x$. Therefore, $y'' = -y$. Because $\sin x$ and $\cos x$ are nonzero functions that satisfy $y'' = -y$, it follows that neither $\sin x$ nor $\cos x$ is a polynomial.

53. Let $f(x) = x \sin x$ and $g(x) = x \cos x$.

(a) Show that $f'(x) = g(x) + \sin x$ and $g'(x) = -f(x) + \cos x$.

(b) Verify that $f''(x) = -f(x) + 2\cos x$ and $g''(x) = -g(x) - 2\sin x$.

(c) By further experimentation, try to find formulas for all higher derivatives of f and g. *Hint:* The kth derivative depends on whether $k = 4n$, $4n+1$, $4n+2$, or $4n+3$.

SOLUTION Let $f(x) = x \sin x$ and $g(x) = x \cos x$.

(a) We examine first derivatives: $f'(x) = x \cos x + (\sin x) \cdot 1 = g(x) + \sin x$ and $g'(x) = (x)(-\sin x) + (\cos x) \cdot 1 = -f(x) + \cos x$; i.e., $f'(x) = g(x) + \sin x$ and $g'(x) = -f(x) + \cos x$.

(b) Now look at second derivatives: $f''(x) = g'(x) + \cos x = -f(x) + 2\cos x$ and $g''(x) = -f'(x) - \sin x = -g(x) - 2\sin x$; i.e., $f''(x) = -f(x) + 2\cos x$ and $g''(x) = -g(x) - 2\sin x$.

(c)
- The third derivatives are $f'''(x) = -f'(x) - 2\sin x = -g(x) - 3\sin x$ and $g'''(x) = -g'(x) - 2\cos x = f(x) - 3\cos x$; i.e., $f'''(x) = -g(x) - 3\sin x$ and $g'''(x) = f(x) - 3\cos x$.
- The fourth derivatives are $f^{(4)}(x) = -g'(x) - 3\cos x = f(x) - 4\cos x$ and $g^{(4)}(x) = f'(x) + 3\sin x = g(x) + 4\sin x$; i.e., $f^{(4)} = f(x) - 4\cos x$ and $g^{(4)}(x) = g(x) + 4\sin x$.
- We can now see the pattern for the derivatives, which are summarized in the following table. Here $n = 0, 1, 2, \ldots$

k	$4n$	$4n+1$	$4n+2$	$4n+3$
$f^{(k)}(x)$	$f(x) - k\cos x$	$g(x) + k\sin x$	$-f(x) + k\cos x$	$-g(x) - k\sin x$
$g^{(k)}(x)$	$g(x) + k\sin x$	$-f(x) + k\cos x$	$-g(x) - k\sin x$	$f(x) - k\cos x$

3.7 The Chain Rule

Preliminary Questions

1. Identify the outside and inside functions for each of these composite functions.

(a) $y = \sqrt{4x + 9x^2}$ **(b)** $y = \tan(x^2 + 1)$ **(c)** $y = \sec^5 x$

SOLUTION

(a) The outer function is $\sqrt{x}$, and the inner function is $4x + 9x^2$.

(b) The outer function is $\tan x$, and the inner function is $x^2 + 1$.

(c) The outer function is x^5, and the inner function is $\sec x$.

2. Which of the following can be differentiated easily *without* using the Chain Rule?

$$y = \tan(7x^2 + 2), \qquad y = \frac{x}{x+1},$$
$$y = \sqrt{x} \cdot \sec x, \qquad y = \sqrt{x \cos x}$$

SOLUTION The function $\frac{x}{x+1}$ can be differentiated using the Quotient Rule, and the function $\sqrt{x} \cdot \sec x$ can be differentiated using the Product Rule. The functions $\tan(7x^2 + 2)$ and $\sqrt{x \cos x}$ require the Chain Rule.

3. Which is the derivative of $f(5x)$?

(a) $5f'(x)$ **(b)** $5f'(5x)$ **(c)** $f'(5x)$

SOLUTION The correct answer is **(b)**: $5f'(5x)$.

4. How many times must the Chain Rule be used to differentiate each function?

(a) $y = \cos(x^2 + 1)$ **(b)** $y = \cos((x^2+1)^4)$

(c) $y = \sqrt{\cos((x^2+1)^4)}$

SOLUTION

(a) To differentiate $\cos(x^2 + 1)$, the Chain Rule must be used once.

(b) To differentiate $\cos((x^2+1)^4)$, the Chain Rule must be used twice.

(c) To differentiate $\sqrt{\cos((x^2+1)^4)}$, the Chain Rule must be used three times.

5. Suppose that $f'(4) = g(4) = g'(4) = 1$. Do we have enough information to compute $F'(4)$, where $F(x) = f(g(x))$? If not, what is missing?

SOLUTION If $F(x) = f(g(x))$, then $F'(x) = f'(g(x))g'(x)$ and $F'(4) = f'(g(4))g'(4)$. Thus, we do not have enough information to compute $F'(4)$. We are missing the value of $f'(1)$.

Exercises

In Exercises 1–4, fill in a table of the following type:

$f(g(x))$	$f'(u)$	$f'(g(x))$	$g'(x)$	$(f \circ g)'$

1. $f(u) = u^{3/2}, \quad g(x) = x^4 + 1$

SOLUTION

$f(g(x))$	$f'(u)$	$f'(g(x))$	$g'(x)$	$(f \circ g)'$
$(x^4+1)^{3/2}$	$\frac{3}{2}u^{1/2}$	$\frac{3}{2}(x^4+1)^{1/2}$	$4x^3$	$6x^3(x^4+1)^{1/2}$

3. $f(u) = \tan u, \quad g(x) = x^4$

SOLUTION

$f(g(x))$	$f'(u)$	$f'(g(x))$	$g'(x)$	$(f \circ g)'$
$\tan(x^4)$	$\sec^2 u$	$\sec^2(x^4)$	$4x^3$	$4x^3\sec^2(x^4)$

In Exercises 5–6, write the function as a composite $f(g(x))$ and compute the derivative using the Chain Rule.

5. $y = (x + \sin x)^4$

SOLUTION Let $f(x) = x^4$, $g(x) = x + \sin x$, and $y = f(g(x)) = (x + \sin x)^4$. Then

$$\frac{dy}{dx} = f'(g(x))g'(x) = 4(x + \sin x)^3(1 + \cos x).$$

7. Calculate $\dfrac{d}{dx}\cos u$ for the following choices of $u(x)$:

(a) $u = 9 - x^2$ **(b)** $u = x^{-1}$ **(c)** $u = \tan x$

SOLUTION

(a) $\cos(u(x)) = \cos(9 - x^2)$.

$$\frac{d}{dx}\cos(u(x)) = -\sin(u(x))u'(x) = -\sin(9 - x^2)(-2x) = 2x\sin(9 - x^2).$$

(b) $\cos(u(x)) = \cos(x^{-1})$.

$$\frac{d}{dx}\cos(u(x)) = -\sin(u(x))u'(x) = -\sin(x^{-1})\left(-\frac{1}{x^2}\right) = \frac{\sin(x^{-1})}{x^2}.$$

(c) $\cos(u(x)) = \cos(\tan x)$.

$$\frac{d}{dx}\cos(u(x)) = -\sin(u(x))u'(x) = -\sin(\tan x)(\sec^2 x) = -\sec^2 x\sin(\tan x).$$

In Exercises 9–14, use the General Power Rule or the Shifting and Scaling Rule to find the derivative.

9. $y = (x^2 + 9)^4$

SOLUTION Let $g(x) = x^2 + 9$. We apply the general power rule.

$$\frac{d}{dx}g(x)^4 = \frac{d}{dx}(x^2 + 9)^4 = 4(x^2 + 9)^3\frac{d}{dx}(x^2 + 9) = 4(x^2 + 9)^3(2x) = 8x(x^2 + 9)^3.$$

11. $y = \sqrt{11x + 4}$

SOLUTION Let $g(x) = 11x + 4$. We apply the general power rule.

$$\frac{d}{dx}g(x)^{1/2} = \frac{d}{dx}\sqrt{11x + 4} = \frac{1}{2}(11x + 4)^{-1/2}\frac{d}{dx}(11x + 4) = \frac{1}{2}(11x + 4)^{-1/2}(11) = \frac{11}{2\sqrt{11x + 4}}.$$

Alternately, let $f(x) = \sqrt{x}$ and apply the shifting and scaling rule. Then

$$\frac{d}{dx}f(11x + 4) = \frac{d}{dx}\sqrt{11x + 4} = (11)\left(\frac{1}{2}(11x + 4)^{-1/2}\right) = \frac{11}{2\sqrt{11x + 4}}.$$

13. $y = \sin(1 - 4x)$

SOLUTION Let $f(x) = \sin x$. We apply the shifting and scaling rule.

$$\frac{d}{dx}f(1 - 4x) = \frac{d}{dx}\sin(1 - 4x) = -4\cos(1 - 4x).$$

In Exercises 15–18, find the derivative of $f \circ g$.

15. $f(u) = \sin u, \quad g(x) = 2x + 1$

SOLUTION Let $h(x) = f(g(x)) = \sin(2x + 1)$. Then, applying the shifting and scaling rule, $h'(x) = 2\cos(2x + 1)$. Alternately,

$$\frac{d}{dx}f(g(x)) = f'(g(x))g'(x) = \cos(2x + 1) \cdot 2 = 2\cos(2x + 1).$$

17. $f(u) = u^9, \quad g(x) = x + x^{-1}$

SOLUTION Let $h(x) = f(g(x)) = (x + x^{-1})^9$. Then, applying the general power rule, $h'(x) = 9(x + x^{-1})^8(1 - \frac{1}{x^2})$. Alternately,

$$\frac{d}{dx}f(g(x)) = f'(g(x))g'(x) = 9(x + x^{-1})^8(1 - x^{-2}).$$

In Exercises 19–20, find the derivatives of $f(g(x))$ and $g(f(x))$.

19. $f(u) = \cos u, \quad g(x) = x^2 + 1$

SOLUTION

$$\frac{d}{dx}f(g(x)) = f'(g(x))g'(x) = -\sin(x^2 + 1)(2x) = -2x\sin(x^2 + 1).$$

$$\frac{d}{dx}g(f(x)) = g'(f(x))f'(x) = 2(\cos x)(-\sin x) = -2\sin x\cos x.$$

In Exercises 21–32, use the Chain Rule to find the derivative.

21. $y = \sin(x^2)$

SOLUTION Let $y = \sin\left(x^2\right)$. Then $y' = \cos\left(x^2\right) \cdot 2x = 2x\cos\left(x^2\right)$.

23. $y = \cot(4t^2 + 9)$

SOLUTION Let $y = \cot\left(4t^2 + 9\right)$. Then

$$y' = -\csc^2\left(4t^2 + 9\right) \cdot 8t = -8t\csc^2\left(4t^2 + 9\right).$$

25. $y = (t^2 + 3t + 1)^{-5/2}$

SOLUTION Let $y = \left(t^2 + 3t + 1\right)^{-5/2}$. Then

$$y' = -\frac{5}{2}\left(t^2 + 3t + 1\right)^{-7/2}(2t + 3) = -\frac{5(2t + 3)}{2(t^2 + 3t + 1)^{7/2}}.$$

27. $y = \left(\dfrac{x+1}{x-1}\right)^4$

SOLUTION   Let $y = \left(\dfrac{x+1}{x-1}\right)^4$. Then

$$y' = 4\left(\frac{x+1}{x-1}\right)^3 \cdot \frac{(x-1)\cdot 1 - (x+1)\cdot 1}{(x-1)^2} = -\frac{8\,(x+1)^3}{(x-1)^5} = \frac{8(1+x)^3}{(1-x)^5}.$$

29. $y = \cos^3(\theta^2)$

SOLUTION   Let $y = \cos^3\left(\theta^2\right) = \left(\cos\left(\theta^2\right)\right)^3$. Here, we note that calculating the derivative of the inside function, $\cos(\theta^2)$, requires the chain rule. After two applications of the chain rule, we have

$$y' = 3\left(\cos\left(\theta^2\right)\right)^2\left(-\sin\left(\theta^2\right)\cdot 2\theta\right) = -6\theta\sin\left(\theta^2\right)\cos^2\left(\theta^2\right).$$

31. $y = \dfrac{1}{\sqrt{\cos(x^2)+1}}$

SOLUTION   Let $y = \left(1 + \cos\left(x^2\right)\right)^{-1/2}$. Here, we note that calculating the derivative of the inside function, $1 + \cos(x^2)$, requires the chain rule. After two applications of the chain rule, we have

$$y' = -\frac{1}{2}\left(1 + \cos\left(x^2\right)\right)^{-3/2}\cdot\left(-\sin\left(x^2\right)\cdot(2x)\right) = \frac{x\sin\left(x^2\right)}{\left(1 + \cos\left(x^2\right)\right)^{3/2}}.$$

In Exercises 33–62, find the derivative using the appropriate rule or combination of rules.

33. $y = \tan 5x$

SOLUTION   Let $y = \tan 5x$. By the scaling and shifting rule,

$$\frac{dy}{dx} = 5\sec^2(5x).$$

35. $y = x\cos(1 - 3x)$

SOLUTION   Let $y = x\cos(1 - 3x)$. Applying the product rule and then the scaling and shifting rule,

$$y' = x\left(-\sin(1 - 3x)\right)\cdot(-3) + \cos(1 - 3x)\cdot 1 = 3x\sin(1 - 3x) + \cos(1 - 3x).$$

37. $y = (4t + 9)^{1/2}$

SOLUTION   Let $y = (4t + 9)^{1/2}$. By the shifting and scaling rule,

$$\frac{dy}{dt} = 4\left(\frac{1}{2}\right)(4t + 9)^{-1/2} = 2(4t + 9)^{-1/2}.$$

39. $y = (x^3 + \cos x)^{-4}$

SOLUTION   Let $y = (x^3 + \cos x)^{-4}$. By the general power rule,

$$y' = -4(x^3 + \cos x)^{-5}(3x^2 - \sin x) = 4(\sin x - 3x^2)(x^3 + \cos x)^{-5}.$$

41. $y = \sqrt{\sin x \cos x}$

SOLUTION   We start by using a trig identity to rewrite

$$y = \sqrt{\sin x \cos x} = \sqrt{\frac{1}{2}\sin 2x} = \frac{1}{\sqrt{2}}(\sin 2x)^{1/2}.$$

Then, after two applications of the chain rule,

$$y' = \frac{1}{\sqrt{2}}\cdot\frac{1}{2}(\sin 2x)^{-1/2}\cdot\cos 2x \cdot 2 = \frac{\cos 2x}{\sqrt{2}\sin 2x}.$$

43. $y = (z + 1)^4 (2z - 1)^3$

SOLUTION Let $y = (z + 1)^4 (2z - 1)^3$. Applying the product rule and the general power rule,

$$\frac{dy}{dz} = (z + 1)^4 (3(2z - 1)^2)(2) + (2z - 1)^3 (4(z + 1)^3)(1) = (z + 1)^3 (2z - 1)^2 (6(z + 1) + 4(2z - 1))$$

$$= (z + 1)^3 (2z - 1)^2 (14z + 2).$$

45. $y = (x + x^{-1})\sqrt{x + 1}$

SOLUTION Let $y = (x + x^{-1})\sqrt{x + 1}$. Applying the product rule and the shifting and scaling rule, and then factoring, we get:

$$y' = (x + x^{-1})\frac{1}{2}(x + 1)^{-1/2} + \sqrt{x + 1}(1 - x^{-2}) = \frac{1}{2\sqrt{x + 1}}(x + x^{-1} + 2(x + 1)(1 - x^{-2}))$$

$$= \frac{1}{2x^2\sqrt{x + 1}}(x^3 + x + 2x^3 + 2x^2 - 2x - 2) = \frac{1}{2x^2\sqrt{x + 1}}(3x^3 + 2x^2 - x - 2).$$

47. $y = (\cos 6x + \sin x^2)^{1/2}$

SOLUTION Let $y = (\cos 6x + \sin(x^2))^{1/2}$. Applying the general power rule followed by both the scaling and shifting rule and the chain rule,

$$y' = \frac{1}{2}(\cos 6x + \sin(x^2))^{-1/2}(-\sin 6x \cdot 6 + \cos(x^2) \cdot 2x) = \frac{x\cos(x^2) - 3\sin 6x}{\sqrt{\cos 6x + \sin(x^2)}}.$$

49. $y = \tan^3 x + \tan(x^3)$

SOLUTION Let $y = \tan^3 x + \tan(x^3) = (\tan x)^3 + \tan(x^3)$. Applying the general power rule to the first term and the chain rule to the second term,

$$y' = 3(\tan x)^2 \sec^2 x + \sec^2(x^3) \cdot 3x^2 = 3(x^2 \sec^2(x^3) + \sec^2 x \tan^2 x).$$

51. $y = \sqrt{\dfrac{z + 1}{z - 1}}$

SOLUTION Let $y = \left(\dfrac{z + 1}{z - 1}\right)^{1/2}$. Applying the general power rule followed by the quotient rule,

$$\frac{dy}{dz} = \frac{1}{2}\left(\frac{z + 1}{z - 1}\right)^{-1/2} \cdot \frac{(z - 1) \cdot 1 - (z + 1) \cdot 1}{(z - 1)^2} = \frac{-1}{\sqrt{z + 1}(z - 1)^{3/2}}.$$

53. $y = \dfrac{\cos(1 + x)}{1 + \cos x}$

SOLUTION Let

$$y = \frac{\cos(1 + x)}{1 + \cos x}.$$

Then, applying the quotient rule and the shifting and scaling rule,

$$\frac{dy}{dx} = \frac{-(1 + \cos x)\sin(1 + x) + \cos(1 + x)\sin x}{(1 + \cos x)^2} = \frac{\cos(1 + x)\sin x - \cos x \sin(1 + x) - \sin(1 + x)}{(1 + \cos x)^2}$$

$$= \frac{\sin(-1) - \sin(1 + x)}{(1 + \cos x)^2}.$$

The last line follows from the identity

$$\sin(A - B) = \sin A \cos B - \cos A \sin B$$

with $A = x$ and $B = 1 + x$.

55. $y = \cot^7(x^5)$

SOLUTION Let $y = \cot^7\left(x^5\right)$. Applying the general power rule followed by the chain rule,

$$\frac{dy}{dx} = 7\cot^6\left(x^5\right) \cdot \left(-\csc^2\left(x^5\right)\right) \cdot 5x^4 = -35x^4 \cot^6\left(x^5\right)\csc^2\left(x^5\right).$$

57. $y = (1 + (x^2 + 2)^5)^3$

SOLUTION Let $y = (1 + (x^2 + 2)^5)^3$. Then, applying the general power rule twice, we obtain:

$$\frac{dy}{dx} = 3(1 + (x^2 + 2)^5)^2(5(x^2 + 2)^4(2x)) = 30x(1 + (x^2 + 2)^5)^2(x^2 + 2)^4.$$

59. $y = \sqrt{1 + \sqrt{1 + \sqrt{x}}}$

SOLUTION Let $y = \left(1 + \left(1 + x^{1/2}\right)^{1/2}\right)^{1/2}$. Applying the general power rule twice,

$$\frac{dy}{dx} = \frac{1}{2}\left(1 + \left(1 + x^{1/2}\right)^{1/2}\right)^{-1/2} \cdot \frac{1}{2}\left(1 + x^{1/2}\right)^{-1/2} \cdot \frac{1}{2}x^{-1/2} = \frac{1}{8\sqrt{x}\sqrt{1 + \sqrt{x}}\sqrt{1 + \sqrt{1 + \sqrt{x}}}}.$$

61. $y = \sqrt{kx + b}$; k and b any constants

SOLUTION Let $y = (kx + b)^{1/2}$, where b and k are constants. By the scaling and shifting rule,

$$y' = \frac{1}{2}(kx + b)^{-1/2} \cdot k = \frac{k}{2\sqrt{kx + b}}.$$

63. Compute $\dfrac{df}{dx}$ if $\dfrac{df}{du} = 2$ and $\dfrac{du}{dx} = 6$.

SOLUTION Assuming f is a function of u, which is in turn a function of x,

$$\frac{df}{dx} = \frac{df}{du} \cdot \frac{du}{dx} = 2(6) = 12.$$

65. With notation as in Example 5, calculate

(a) $\dfrac{d}{d\theta}\sin\theta\Big|_{\theta=60°}$

(b) $\dfrac{d}{d\theta}(\theta + \tan\theta)\Big|_{\theta=45°}$

SOLUTION

(a) $\dfrac{d}{d\theta}\sin\theta\Big|_{\theta=60°} = \dfrac{d}{d\theta}\sin\left(\dfrac{\pi}{180}\theta\right)\Big|_{\theta=60°} = \left(\dfrac{\pi}{180}\right)\cos\left(\dfrac{\pi}{180}(60)\right) = \dfrac{\pi}{180}\dfrac{1}{2} = \dfrac{\pi}{360}.$

(b) $\dfrac{d}{d\theta}(\theta + \tan\theta)\Big|_{\theta=45°} = \dfrac{d}{d\theta}\left(\theta + \tan\left(\dfrac{\pi}{180}\theta\right)\right)\Big|_{\theta=45°} = 1 + \dfrac{\pi}{180}\sec^2\left(\dfrac{\pi}{4}\right) = 1 + \dfrac{\pi}{90}.$

67. Compute the derivative of $h(\sin x)$ at $x = \frac{\pi}{6}$, assuming that $h'(0.5) = 10$.

SOLUTION Let $u = \sin x$ and suppose that $h'(0.5) = 10$. Then

$$\frac{d}{dx}(h(u)) = \frac{dh}{du}\frac{du}{dx} = \frac{dh}{du}\cos x.$$

When $x = \frac{\pi}{6}$, we have $u = .5$. Accordingly, the derivative of $h(\sin x)$ at $x = \frac{\pi}{6}$ is $10\cos\left(\frac{\pi}{6}\right) = 5\sqrt{3}$.

In Exercises 69–72, use the table of values to calculate the derivative of the function at the given point.

x	1	4	6
$f(x)$	4	0	6
$f'(x)$	5	7	4
$g(x)$	4	1	6
$g'(x)$	5	$\frac{1}{2}$	3

69. $f(g(x))$, $x = 6$

SOLUTION $\dfrac{d}{dx} f(g(x))\Big|_{x=6} = f'(g(6))g'(6) = f'(6)g'(6) = 4 \times 3 = 12.$

71. $g(\sqrt{x})$, $x = 16$

SOLUTION $\dfrac{d}{dx} g(\sqrt{x})\Big|_{x=16} = g'(4)\left(\dfrac{1}{2}\right)(1/\sqrt{16}) = \left(\dfrac{1}{2}\right)\left(\dfrac{1}{2}\right)\left(\dfrac{1}{4}\right) = \dfrac{1}{16}.$

In Exercises 73–76, compute the indicated higher derivatives.

73. $\dfrac{d^2}{dx^2} \sin(x^2)$

SOLUTION Let $f(x) = \sin\left(x^2\right)$. Then, by the chain rule, $f'(x) = 2x \cos\left(x^2\right)$ and, by the product rule and the chain rule,

$$f''(x) = 2x\left(-\sin\left(x^2\right)\cdot 2x\right) + 2\cos\left(x^2\right) = 2\cos\left(x^2\right) - 4x^2 \sin\left(x^2\right).$$

75. $\dfrac{d^3}{dx^3}(3x+9)^{11}$

SOLUTION Let $f(x) = (3x+9)^{11}$. Then, by repeated use of the scaling and shifting rule,

$$f'(x) = 11(3x+9)^{10}\cdot 3 = 33(3x+9)^{10}$$
$$f''(x) = 330(3x+9)^9 \cdot 3 = 990(3x+9)^9,$$
$$f'''(x) = 8910(3x+9)^8 \cdot 3 = 26730\,(3x+9)^8.$$

77. CAS Use a computer algebra system to compute $f^{(k)}(x)$ for $k = 1, 2, 3$ for the following functions:

(a) $f(x) = \cot(x^2)$ **(b)** $f(x) = \sqrt{x^3+1}$

SOLUTION

(a) Let $f(x) = \cot(x^2)$. Using a computer algebra system,

$$f'(x) = -2x \csc^2(x^2);$$
$$f''(x) = 2\csc^2(x^2)(4x^2 \cot(x^2) - 1); \text{ and}$$
$$f'''(x) = -8x \csc^2(x^2)\left(6x^2 \cot^2(x^2) - 3\cot(x^2) + 2x^2\right).$$

(b) Let $f(x) = \sqrt{x^3+1}$. Using a computer algebra system,

$$f'(x) = \dfrac{3x^2}{2\sqrt{x^3+1}};$$
$$f''(x) = \dfrac{3x(x^3+4)}{4(x^3+1)^{3/2}}; \text{ and}$$
$$f'''(x) = -\dfrac{3(x^6+20x^3-8)}{8(x^3+1)^{5/2}}.$$

79. Compute the second derivative of $\sin(g(x))$ at $x = 2$, assuming that $g(2) = \frac{\pi}{4}$, $g'(2) = 5$, and $g''(2) = 3$.

SOLUTION Let $f(x) = \sin(g(x))$. Then $f'(x) = \cos(g(x))g'(x)$ and

$$f''(x) = \cos(g(x))g''(x) + g'(x)(-\sin(g(x)))g'(x) = \cos(g(x))g''(x) - (g'(x))^2 \sin(g(x)).$$

Therefore,

$$f''(2) = g''(2)\cos(g(2)) - (g'(2))^2 \sin(g(2)) = 3\cos\left(\tfrac{\pi}{4}\right) - (5)^2 \sin\left(\tfrac{\pi}{4}\right) = -22\cdot\tfrac{\sqrt{2}}{2} = -11\sqrt{2}$$

81. The power P in a circuit is $P = Ri^2$, where R is resistance and i the current. Find dP/dt at $t = 2$ if $R = 1{,}000\ \Omega$ and i varies according to $i = \sin(4\pi t)$ (time in seconds).

SOLUTION $\dfrac{d}{dt}\left(Ri^2\right)\Big|_{t=2} = 2Ri\dfrac{di}{dt}\Big|_{t=2} = 2(1000)4\pi \sin(4\pi t)\cos(4\pi t)|_{t=2} = 0.$

83. The force F (in Newtons) between two charged objects is $F = 100/r^2$, where r is the distance (in meters) between them. Find dF/dt at $t = 10$ if the distance at time t (in seconds) is $r = 1 + 0.4t^2$.

SOLUTION Let $F(r) = 100/r^2$, and let $r(t) = 1 + .4t^2$. We compute $F'(r) = -200/r^3$, and $r'(t) = .8t$. This gives us $r(10) = 41$ and $r'(10) = 8$.

$$\frac{dF}{dt} = \frac{dF}{dr}\frac{dr}{dt},$$

so

$$\left.\frac{dF}{dt}\right|_{t=10} = -200(41)^{-3}(8) \approx -0.0232 \text{ N/s}.$$

85. Conservation of Energy The position at time t (in seconds) of a weight of mass m oscillating at the end of a spring is $x(t) = L\sin(2\pi\omega t)$. Here, L is the maximum length of the spring, and ω the frequency (number of oscillations per second). Let v and a be the velocity and acceleration of the weight.

(a) By Hooke's Law, the spring exerts a force of magnitude $F = -kx$ on the weight, where k is the *spring constant*. Use Newton's Second Law, $F = ma$, to show that $2\pi\omega = \sqrt{k/m}$.

(b) The weight has kinetic energy $K = \frac{1}{2}mv^2$ and potential energy $U = \frac{1}{2}kx^2$. Prove that the total energy $E = K + U$ is conserved, that is, $dE/dt = 0$.

SOLUTION

(a) Let $x(t) = L\sin(2\pi\omega t)$. Then $v(t) = x'(t) = 2\pi\omega L\cos(2\pi\omega t)$ and $a(t) = v'(t) = -(2\pi\omega)^2 L\sin(2\pi\omega t)$. Equating Hooke's Law, $F = -kx$, to Newton's second law, $F = ma$, yields

$$-kL\sin(2\pi\omega t) = -m(2\pi\omega)^2 L\sin(2\pi\omega t)$$

or $k/m = (2\pi\omega)^2$. The desired result follows upon taking square roots.

(b) We have

$$\frac{dU}{dt} = kx\frac{dx}{dt} = kxv$$

and

$$\frac{dK}{dt} = mv\frac{dv}{dt} = mva$$

But $F = ma = -kx$ by Hooke's Law, so $a = (-k/m)x$ and we get

$$\frac{dK}{dt} = mva = mv(-k/m)x = -kvx.$$

This shows that the derivative of $U + K$ is zero.

Further Insights and Challenges

87. 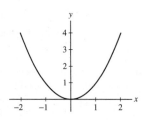 Recall that $f(x)$ is *even* if $f(-x) = f(x)$ and *odd* if $f(-x) = -f(x)$.

(a) Sketch a graph of any even function and explain graphically why its derivative is an odd function.

(b) Show that differentiation reverses parity: If f is even, then f' is odd, and if f is odd, then f' is even. *Hint:* Differentiate $f(-x)$.

(c) Suppose that f' is even. Is f necessarily odd? *Hint:* Check if this is true for linear functions.

SOLUTION A function is *even* if $f(-x) = f(x)$ and *odd* if $f(-x) = -f(x)$.

(a) The graph of an even function is symmetric with respect to the y-axis. Accordingly, its image in the left half-plane is a mirror reflection of that in the right half-plane through the y-axis. If at $x = a \geq 0$, the slope of f exists and is equal to m, then by reflection its slope at $x = -a \leq 0$ is $-m$. That is, $f'(-a) = -f'(a)$. *Note:* This means that if $f'(0)$ exists, then it equals 0.

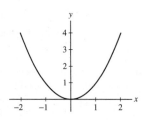

(b) By the chain rule, $\frac{d}{dx} f(-x) = -f'(-x)$. Now suppose that f is even. Then $f(-x) = f(x)$ and

$$\frac{d}{dx} f(-x) = \frac{d}{dx} f(x) = f'(x).$$

Hence, when f is even, $-f'(-x) = f'(x)$ or $f'(-x) = -f'(x)$ and f' is odd. On the other hand, suppose f is odd. Then $f(-x) = -f(x)$ and

$$\frac{d}{dx} f(-x) = -\frac{d}{dx} f(x) = -f'(x).$$

Hence, when f is odd, $-f'(-x) = -f'(x)$ or $f'(-x) = f'(x)$ and f' is even.

(c) Suppose that f' is even. Then f is not necessarily odd. Let $f(x) = 4x + 7$. Then $f'(x) = 4$, an even function. But f is not odd. For example, $f(2) = 15$, $f(-2) = -1$, but $f(-2) \neq -f(2)$.

In Exercises 89–91, use the following fact (proved in Chapter 4): If a differentiable function f satisfies $f'(x) = 0$ for all x, then f is a constant function.

89. Differential Equation of Sine and Cosine   Suppose that $f(x)$ satisfies the following equation (called a **differential equation**):

$$f''(x) = -f(x) \qquad \boxed{3}$$

(a) Show that $f(x)^2 + f'(x)^2 = f(0)^2 + f'(0)^2$. *Hint:* Show that the function on the left has zero derivative.
(b) Verify that $\sin x$ and $\cos x$ satisfy Eq. (3) and deduce that $\sin^2 x + \cos^2 x = 1$.

SOLUTION

(a) Let $g(x) = f(x)^2 + f'(x)^2$. Then

$$g'(x) = 2f(x)f'(x) + 2f'(x)f''(x) = 2f(x)f'(x) + 2f'(x)(-f(x)) = 0,$$

where we have used the fact that $f''(x) = -f(x)$. Because $g'(0) = 0$ for all x, $g(x) = f(x)^2 + f'(x)^2$ must be a constant function. In other words, $f(x)^2 + f'(x)^2 = C$ for some constant C. To determine the value of C, we can substitute any number for x. In particular, for this problem, we want to substitute $x = 0$ and find $C = f(0)^2 + f'(0)^2$. Hence,

$$f(x)^2 + f'(x)^2 = f(0)^2 + f'(0)^2.$$

(b) Let $f(x) = \sin x$. Then $f'(x) = \cos x$ and $f''(x) = -\sin x$, so $f''(x) = -f(x)$. Next, let $f(x) = \cos x$. Then $f'(x) = -\sin x$, $f''(x) = -\cos x$, and we again have $f''(x) = -f(x)$. Finally, if we take $f(x) = \sin x$, the result from part (a) guarantees that

$$\sin^2 x + \cos^2 x = \sin^2 0 + \cos^2 0 = 0 + 1 = 1.$$

91. Use the result of Exercise 90 to show the following: $f(x) = \sin x$ is the unique solution of Eq. (3) such that $f(0) = 0$ and $f'(0) = 1$; and $g(x) = \cos x$ is the unique solution such that $g(0) = 1$ and $g'(0) = 0$. This result provides a means of defining and developing all the properties of the trigonometric functions without reference to triangles or the unit circle.

SOLUTION   In part (b) of Exercise 89, it was shown that $f(x) = \sin x$ satisfies Eq. (3), and we can directly calculate that $f(0) = \sin 0 = 0$ and $f'(0) = \cos 0 = 1$. Suppose there is another function, call it $F(x)$, that satisfies Eq. (3) with the same initial conditions: $F(0) = 0$ and $F'(0) = 1$. By Exercise 90, it follows that $F(x) = \sin x$ for all x. Hence, $f(x) = \sin x$ is the unique solution of Eq. (3) satisfying $f(0) = 0$ and $f'(0) = 1$. The proof that $g(x) = \cos x$ is the unique solution of Eq. (3) satisfying $g(0) = 1$ and $g'(0) = 0$ is carried out in a similar manner.

93. Chain Rule   This exercise proves the Chain Rule without the special assumption made in the text. For any number b, define a new function

$$F(u) = \frac{f(u) - f(b)}{u - b} \qquad \text{for all } u \neq b$$

Observe that $F(u)$ is equal to the slope of the secant line through $(b, f(b))$ and $(u, f(u))$.
(a) Show that if we define $F(b) = f'(b)$, then $F(u)$ is continuous at $u = b$.
(b) Take $b = g(a)$. Show that if $x \neq a$, then for all u,

$$\frac{f(u) - f(g(a))}{x - a} = F(u)\frac{u - g(a)}{x - a} \qquad \boxed{4}$$

Note that both sides are zero if $u = g(a)$.

(c) Substitute $u = g(x)$ in Eq. (4) to obtain

$$\frac{f(g(x)) - f(g(a))}{x - a} = F(g(x))\,\frac{g(x) - g(a)}{x - a}$$

Derive the Chain Rule by computing the limit of both sides as $x \to a$.

SOLUTION   For any differentiable function f and any number b, define

$$F(u) = \frac{f(u) - f(b)}{u - b}$$

for all $u \ne b$.

(a) Define $F(b) = f'(b)$. Then

$$\lim_{u \to b} F(u) = \lim_{u \to b} \frac{f(u) - f(b)}{u - b} = f'(b) = F(b),$$

i.e., $\lim\limits_{u \to b} F(u) = F(b)$. Therefore, F is continuous at $u = b$.

(b) Let g be a differentiable function and take $b = g(a)$. Let x be a number distinct from a. If we substitute $u = g(a)$ into Eq. (4), both sides evaluate to 0, so equality is satisfied. On the other hand, if $u \ne g(a)$, then

$$\frac{f(u) - f(g(a))}{x - a} = \frac{f(u) - f(g(a))}{u - g(a)}\,\frac{u - g(a)}{x - a} = \frac{f(u) - f(b)}{u - b}\,\frac{u - g(a)}{x - a} = F(u)\,\frac{u - g(a)}{x - a}.$$

(c) Hence for all u, we have

$$\frac{f(u) - f(g(a))}{x - a} = F(u)\,\frac{u - g(a)}{x - a}.$$

(d) Substituting $u = g(x)$ in Eq. (4), we have

$$\frac{f(g(x)) - f(g(a))}{x - a} = F(g(x))\,\frac{g(x) - g(a)}{x - a}.$$

Letting $x \to a$ gives

$$\lim_{x \to a} \frac{f(g(x)) - f(g(a))}{x - a} = \lim_{x \to a} \left(F(g(x))\,\frac{g(x) - g(a)}{x - a} \right) = F(g(a))g'(a) = F(b)g'(a) = f'(b)g'(a)$$

$$= f'(g(a))g'(a).$$

Therefore $(f \circ g)'(a) = f'(g(a))g'(a)$, which is the Chain Rule.

3.8 Implicit Differentiation

Preliminary Questions

1. Which differentiation rule is used to show $\dfrac{d}{dx} \sin y = \cos y\, \dfrac{dy}{dx}$?

SOLUTION   The chain rule is used to show that $\frac{d}{dx} \sin y = \cos y \frac{dy}{dx}$.

2. One of (a)–(c) is incorrect. Find the mistake and correct it.

(a) $\dfrac{d}{dy} \sin(y^2) = 2y \cos(y^2)$

(b) $\dfrac{d}{dx} \sin(x^2) = 2x \cos(x^2)$

(c) $\dfrac{d}{dx} \sin(y^2) = 2y \cos(y^2)$

SOLUTION

(a) This is correct. Note that the differentiation is with respect to the variable y.

(b) This is correct. Note that the differentiation is with respect to the variable x.

(c) This is incorrect. Because the differentiation is with respect to the variable x, the chain rule is needed to obtain

$$\frac{d}{dx} \sin(y^2) = 2y \cos(y^2)\frac{dy}{dx}.$$

3. On an exam, Jason was asked to differentiate the equation

$$x^2 + 2xy + y^3 = 7$$

What are the errors in Jason's answer?

$$2x + 2xy' + 3y^2 = 0$$

SOLUTION There are two mistakes in Jason's answer. First, Jason should have applied the product rule to the second term to obtain

$$\frac{d}{dx}(2xy) = 2x\frac{dy}{dx} + 2y.$$

Second, he should have applied the general power rule to the third term to obtain

$$\frac{d}{dx}y^3 = 3y^2\frac{dy}{dx}.$$

4. Which of (a) or (b) is equal to $\frac{d}{dx}(x\sin t)$?

(a) $(x\cos t)\dfrac{dt}{dx}$

(b) $(x\cos t)\dfrac{dt}{dx} + \sin t$

SOLUTION Using the product rule and the chain rule we see that

$$\frac{d}{dx}(x\sin t) = x\cos t\frac{dt}{dx} + \sin t,$$

so the correct answer is **(b)**.

Exercises

1. Show that if you differentiate both sides of $x^2 + 2y^3 = 6$, the result is $2x + 6y^2y' = 0$. Then solve for y' and calculate dy/dx at the point $(2, 1)$.

SOLUTION

$$\frac{d}{dx}(x^2 + 2y^3) = \frac{d}{dx}6$$

$$2x + 6y^2y' = 0$$

$$2x + 6y^2y' = 0$$

$$6y^2y' = -2x$$

$$y' = \frac{-2x}{6y^2}.$$

At $(2, 1)$, $\frac{dy}{dx} = \frac{-4}{6} = -\frac{2}{3}$.

In Exercises 3–8, differentiate the expression with respect to x.

3. x^2y^3

SOLUTION Assuming that y depends on x, then

$$\frac{d}{dx}\left(x^2y^3\right) = x^2 \cdot 3y^2y' + y^3 \cdot 2x = 3x^2y^2y' + 2xy^3.$$

5. $\dfrac{y^3}{x}$

SOLUTION Assuming that y depends on x, then

$$\frac{d}{dx}\left(\frac{y^3}{x}\right) = \frac{x(3y^2y') - y^3}{x^2} = \frac{-y^3}{x^2} + \frac{3y^2y'}{x}.$$

7. $z + z^2$

SOLUTION Assuming that z depends on x, then $\dfrac{d}{dx}(z + z^2) = z' + 2z(z')$

In Exercises 9–24, calculate the derivative of y (or other variable) with respect to x.

9. $3y^3 + x^2 = 5$

SOLUTION   Let $3y^3 + x^2 = 5$. Then $9y^2y' + 2x = 0$, and $y' = -\dfrac{2x}{9y^2}$.

11. $x^2y + 2xy^2 = x + y$

SOLUTION   Let $x^2y + 2xy^2 = x + y$. Then

$$x^2y' + 2xy + 2x \cdot 2yy' + 2y^2 = 1 + y'$$
$$x^2y' + 4xyy' - y' = 1 - 2xy - 2y^2$$
$$y' = \frac{1 - 2xy - 2y^2}{x^2 + 4xy - 1}.$$

13. $x^2y + y^4 - 3x = 8$

SOLUTION   We apply the product rule and the chain rule together:

$$\frac{d}{dx}(x^2y + y^4 - 3x) = \frac{d}{dx}8$$
$$x^2y' + 2xy + 4y^3y' - 3 = 0$$
$$x^2y' + 4y^3y' = 3 - 2xy$$
$$y'(x^2 + 4y^3) = 3 - 2xy$$
$$y' = \frac{3 - 2xy}{x^2 + 4y^3}.$$

15. $x^4 + z^4 = 1$

SOLUTION   Let $x^4 + z^4 = 1$. Then $4x^3 + 4z^3z' = 0$, and $z' = -x^3/z^3$.

17. $y^{-3/2} + x^{3/2} = 1$

SOLUTION   Let $y^{-3/2} + x^{3/2} = 1$. Then $-\frac{3}{2}y^{-5/2}y' + \frac{3}{2}x^{1/2} = 0$, and $y' = x^{1/2}y^{5/2} = \sqrt{xy^5}$.

19. $\sqrt{x+s} = \dfrac{1}{x} + \dfrac{1}{s}$

SOLUTION   Let $(x+s)^{1/2} = x^{-1} + s^{-1}$. Then

$$\frac{1}{2}(x+s)^{-1/2}\left(1 + s'\right) = -x^{-2} - s^{-2}s'.$$

Multiplying by $2x^2s^2\sqrt{x+s}$ and then solving for s' gives

$$x^2s^2\left(1 + s'\right) = -2s^2\sqrt{x+s} - 2x^2s'\sqrt{x+s}$$
$$x^2s^2s' + 2x^2s'\sqrt{x+s} = -2s^2\sqrt{x+s} - x^2s^2$$
$$x^2\left(s^2 + 2\sqrt{x+s}\right)s' = -s^2\left(x^2 + 2\sqrt{x+s}\right)$$
$$s' = -\frac{s^2\left(x^2 + 2\sqrt{x+s}\right)}{x^2\left(s^2 + 2\sqrt{x+s}\right)}.$$

21. $y + \dfrac{x}{y} = 1$

SOLUTION   Let $y + xy^{-1} = 1$. Then $y' + x\left(-y^{-2}y'\right) + y^{-1} \cdot 1 = 0$, and

$$y' = \frac{y^{-1}}{xy^{-2} - 1} \cdot \frac{y^2}{y^2} = \frac{y}{x - y^2}.$$

23. $\sin(x + y) = x + \cos y$

SOLUTION Let $\sin(x+y) = x + \cos y$. Then

$$(1+y')\cos(x+y) = 1 - y'\sin y$$
$$\cos(x+y) + y'\cos(x+y) = 1 - y'\sin y$$
$$(\cos(x+y) + \sin y)y' = 1 - \cos(x+y)$$
$$y' = \frac{1 - \cos(x+y)}{\cos(x+y) + \sin y}.$$

In Exercises 25–26, find dy/dx at the given point.

25. $(x+2)^2 - 6(2y+3)^2 = 3$, $(1, -1)$

SOLUTION By the scaling and shifting rule,

$$2(x+2) - 24(2y+3)y' = 0.$$

If $x = 1$ and $y = -1$, then

$$2(3) - 24(1)y' = 0.$$

so that $24y' = 6$, or $y' = \frac{1}{4}$.

In Exercises 27–30, find an equation of the tangent line at the given point.

27. $xy - 2y = 1$, $(3, 1)$

SOLUTION Taking the derivative of both sides of $xy - 2y = 1$ yields $xy' + y - 2y' = 0$. Substituting $x = 3$, $y = 1$ yields $3y' + 1 - 2y' = 0$. Solving, we get:

$$3y' + 1 - 2y' = 0$$
$$y' + 1 = 0$$
$$y' = -1.$$

Hence, the equation of the tangent line at $(3, 1)$ is $y - 1 = -(x - 3)$, or $y = 4 - x$.

29. $x^{2/3} + y^{2/3} = 2$, $(1, 1)$

SOLUTION Taking the derivative of both sides of $x^{2/3} + y^{2/3} = 2$ yields

$$\frac{2}{3}x^{-1/3} + \frac{2}{3}y^{-1/3}y' = 0.$$

Substituting $x = 1$, $y = 1$ yields $\frac{2}{3} + \frac{2}{3}y' = 0$, so that $1 + y' = 0$, or $y' = -1$. Hence, the equation of the tangent line at $(1, 1)$ is $y - 1 = -(x - 1)$, or $y = 2 - x$.

31. Find the points on the graph of $y^2 = x^3 - 3x + 1$ (Figure 5) where the tangent line is horizontal.

(a) First show that $2yy' = 3x^2 - 3$, where $y' = dy/dx$.

(b) Do not solve for y'. Rather, set $y' = 0$ and solve for x. This gives two possible values of x where the slope may be zero.

(c) Show that the positive value of x does not correspond to a point on the graph.

(d) The negative value corresponds to the two points on the graph where the tangent line is horizontal. Find the coordinates of these two points.

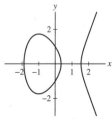

FIGURE 5 Graph of $y^2 = x^3 - 3x + 1$.

SOLUTION

(a) Applying implicit differentiation to $y^2 = x^3 - 3x + 1$, we have

$$2y\frac{dy}{dx} = 3x^2 - 3.$$

(b) Setting $y' = 0$ we have $0 = 3x^2 - 3$, so $x = 1$ or $x = -1$.

(c) If we return to the equation $y^2 = x^3 - 3x + 1$ and substitute $x = 1$, we obtain the equation $y^2 = -1$, which has no real solutions.

(d) Substituting $x = -1$ into $y^2 = x^3 - 3x + 1$ yields

$$y^2 = (-1)^3 - 3(-1) + 1 = -1 + 3 + 1 = 3,$$

so $y = \sqrt{3}$ or $-\sqrt{3}$. The tangent is horizontal at the points $(-1, \sqrt{3})$ and $(-1, -\sqrt{3})$.

33. Show that no point on the graph of $x^2 - 3xy + y^2 = 1$ has a horizontal tangent line.

SOLUTION   Let the implicit curve $x^2 - 3xy + y^2 = 1$ be given. Then

$$2x - 3xy' - 3y + 2yy' = 0,$$

so

$$y' = \frac{2x - 3y}{3x - 2y}.$$

Setting $y' = 0$ leads to $y = \frac{2}{3}x$. Substituting $y = \frac{2}{3}x$ into the equation of the implicit curve gives

$$x^2 - 3x\left(\frac{2}{3}x\right) + \left(\frac{2}{3}x\right)^2 = 1,$$

or $-\frac{5}{9}x^2 = 1$, which has *no* real solutions. Accordingly, there are *no* points on the implicit curve where the tangent line has slope zero.

35. If the derivative dx/dy exists at a point and $dx/dy = 0$, then the tangent line is vertical. Calculate dx/dy for the equation $y^4 + 1 = y^2 + x^2$ and find the points on the graph where the tangent line is vertical.

SOLUTION   Let $y^4 + 1 = y^2 + x^2$. Differentiating this equation with respect to y yields

$$4y^3 = 2y + 2x\frac{dx}{dy},$$

so

$$\frac{dx}{dy} = \frac{4y^3 - 2y}{2x} = \frac{y(2y^2 - 1)}{x}.$$

Thus, $\dfrac{dx}{dy} = 0$ when $y = 0$ and when $y = \pm\dfrac{\sqrt{2}}{2}$. Substituting $y = 0$ into the equation $y^4 + 1 = y^2 + x^2$ gives $1 = x^2$, so $x = \pm 1$. Substituting $y = \pm\dfrac{\sqrt{2}}{2}$, gives $x^2 = 3/4$, so $x = \pm\dfrac{\sqrt{3}}{2}$. Thus, there are six points on the graph of $y^4 + 1 = y^2 + x^2$ where the tangent line is vertical:

$$(1, 0), (-1, 0), \left(\frac{\sqrt{3}}{2}, \frac{\sqrt{2}}{2}\right), \left(-\frac{\sqrt{3}}{2}, \frac{\sqrt{2}}{2}\right), \left(\frac{\sqrt{3}}{2}, -\frac{\sqrt{2}}{2}\right), \left(-\frac{\sqrt{3}}{2}, -\frac{\sqrt{2}}{2}\right).$$

37. Differentiate the equation $x^3 + 3xy^2 = 1$ with respect to the variable t and express dy/dt in terms of dx/dt, as in Exercise 36.

SOLUTION   Let $x^3 + 3xy^2 = 1$. Then

$$3x^2\frac{dx}{dt} + 6xy\frac{dy}{dt} + 3y^2\frac{dx}{dt} = 0,$$

and

$$\frac{dy}{dt} = -\frac{x^2 + y^2}{2xy}\frac{dx}{dt}.$$

In Exercises 38–39, differentiate the equation with respect to t to calculate dy/dt in terms of dx/dt.

39. $y^4 + 2xy + x^2 = 0$

SOLUTION Let $y^4 + 2xy + x^2 = 0$. Then

$$4y^3 \frac{dy}{dt} + 2x \frac{dy}{dt} + 2y \frac{dx}{dt} + 2x \frac{dx}{dt} = 0,$$

and

$$\frac{dy}{dt} = -\frac{x + y}{x + 2y^3} \frac{dx}{dt}.$$

41. The **folium of Descartes** is the curve with equation $x^3 + y^3 = 3xy$ (Figure 7). It was first discussed in 1638 by the French philosopher-mathematician René Descartes. The name "folium" means leaf. Descartes's scientific colleague Gilles de Roberval called it the "jasmine flower." Both men believed incorrectly that the leaf shape in the first quadrant was repeated in each quadrant, giving the appearance of petals of a flower. Find an equation of the tangent line to this curve at the point $\left(\frac{2}{3}, \frac{4}{3}\right)$.

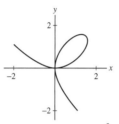

FIGURE 7 Folium of Descartes: $x^3 + y^3 = 3xy$.

SOLUTION Let $x^3 + y^3 = 3xy$. Then $3x^2 + 3y^2y' = 3xy' + 3y$, and $y' = \dfrac{x^2 - y}{x - y^2}$. At the point $\left(\frac{2}{3}, \frac{4}{3}\right)$, we have

$$y' = \frac{\frac{4}{9} - \frac{4}{3}}{\frac{2}{3} - \frac{16}{9}} = \frac{-\frac{8}{9}}{-\frac{10}{9}} = \frac{4}{5}.$$

The tangent line at P is thus $y - \frac{4}{3} = \frac{4}{5}\left(x - \frac{2}{3}\right)$ or $y = \frac{4}{5}x + \frac{4}{5}$.

43. ⬚GU⬚ Plot the equation $x^3 + y^3 = 3xy + b$ for several values of b.

(a) Describe how the graph changes as $b \to 0$.

(b) Compute dy/dx (in terms of b) at the point on the graph where $y = 0$. How does this value change as $b \to \infty$? Do your plots confirm this conclusion?

SOLUTION

(a) Consider the first row of figures below. When $b < 0$, the graph of $x^3 + y^3 = 3xy + b$ consists of two pieces. As $b \to 0-$, the two pieces move closer to intersecting at the origin. From the second row of figures, we see that the graph of $x^3 + y^3 = 3xy + b$ when $b > 0$ consists of a single piece that has a "loop" in the first quadrant. As $b \to 0+$, the loop comes closer to "pinching off" at the origin.

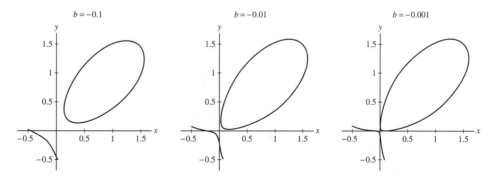

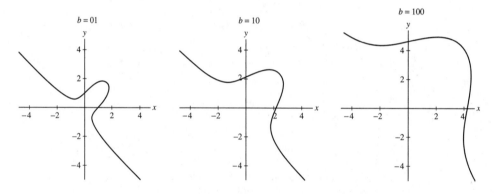

(b) Differentiating the equation $x^3 + y^3 = 3xy + b$ with respect to x yields $3x^2 + 3y^2 y' = 3xy' + 3y$, so

$$y' = \frac{y - x^2}{y^2 - x}.$$

Substituting $y = 0$ into $x^3 + y^3 = 3xy + b$ leads to $x^3 = b$, or $x = \sqrt[3]{b}$. Thus, at the point on the graph where $y = 0$,

$$y' = \frac{0 - x^2}{0^2 - x} = x = \sqrt[3]{b}.$$

Consequently, as $b \to \infty$, $y' \to \infty$ at the point on the graph where $y = 0$. This conclusion is supported by the figures shown below, which correspond to $b = 1$, $b = 10$, and $b = 100$.

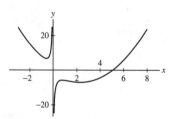

45. The equation $xy = x^3 - 5x^2 + 2x - 1$ defines a *trident curve* (Figure 9), named by Isaac Newton in his treatise on curves published in 1710. Find the points where the tangent to the trident is horizontal as follows.

(a) Show that $xy' + y = 3x^2 - 10x + 2$.

(b) Set $y' = 0$ in (a), replace y by $x^{-1}(x^3 - 5x^2 + 2x - 1)$, and solve the resulting equation for x.

FIGURE 9 Trident curve: $xy = x^3 - 5x^2 + 2x - 1$.

SOLUTION   Consider the equation of a trident curve:

$$xy = x^3 - 5x^2 + 2x - 1$$

or

$$y = x^{-1}(x^3 - 5x^2 + 2x - 1).$$

(a) Taking the derivative of $xy = x^3 - 5x^2 + 2x - 1$ yields

$$xy' + y = 3x^2 - 10x + 2.$$

(b) Setting $y' = 0$ in (a) gives $y = 3x^2 - 10x + 2$. Thus, we have

$$x^{-1}(x^3 - 5x^2 + 2x - 1) = 3x^2 - 10x + 2.$$

Collecting like terms and setting to zero, we have

$$0 = 2x^3 - 5x^2 + 1 = (2x - 1)(x^2 - 2x - 1).$$

Hence, $x = \frac{1}{2}, 1 \pm \sqrt{2}$.

47. Find equations of the tangent lines at the points where $x = 1$ on the so-called folium (Figure 11):

$$(x^2 + y^2)^2 = \frac{25}{4}xy^2$$

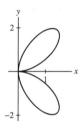

FIGURE 11

SOLUTION First, find the points $(1, y)$ on the curve. Setting $x = 1$ in the equation $(x^2 + y^2)^2 = \frac{25}{4}xy^2$ yields

$$(1 + y^2)^2 = \frac{25}{4}y^2$$

$$y^4 + 2y^2 + 1 = \frac{25}{4}y^2$$

$$4y^4 + 8y^2 + 4 = 25y^2$$

$$4y^4 - 17y^2 + 4 = 0$$

$$(4y^2 - 1)(y^2 - 4) = 0$$

$$y^2 = \frac{1}{4} \text{ or } y^2 = 4$$

Hence $y = \pm\frac{1}{2}$ or $y = \pm 2$. Taking $\frac{d}{dx}$ of both sides of the original equation yields

$$2(x^2 + y^2)(2x + 2yy') = \frac{25}{4}y^2 + \frac{25}{2}xyy'$$

$$4(x^2 + y^2)x + 4(x^2 + y^2)yy' = \frac{25}{4}y^2 + \frac{25}{2}xyy'$$

$$(4(x^2 + y^2) - \frac{25}{2}x)yy' = \frac{25}{4}y^2 - 4(x^2 + y^2)x$$

$$y' = \frac{\frac{25}{4}y^2 - 4(x^2 + y^2)x}{y(4(x^2 + y^2) - \frac{25}{2}x)}$$

- At $(1, 2)$, $x^2 + y^2 = 5$, and

$$y' = \frac{\frac{25}{4}2^2 - 4(5)(1)}{2(4(5) - \frac{25}{2}(1))} = \frac{1}{3}.$$

Hence, at $(1, 2)$, the equation of the tangent line is $y - 2 = \frac{1}{3}(x - 1)$ or $y = \frac{1}{3}x + \frac{5}{3}$.
- At $(1, -2)$, $x^2 + y^2 = 5$ as well, and

$$y' = \frac{\frac{25}{4}(-2)^2 - 4(5)(1)}{-2(4(5) - \frac{25}{2}(1))} = -\frac{1}{3}.$$

Hence, at $(1, -2)$, the equation of the tangent line is $y + 2 = -\frac{1}{3}(x - 1)$ or $y = -\frac{1}{3}x - \frac{5}{3}$.

• At $(1, \frac{1}{2})$, $x^2 + y^2 = \frac{5}{4}$, and

$$y' = \frac{\frac{25}{4}\left(\frac{1}{2}\right)^2 - 4\left(\frac{5}{4}\right)(1)}{\frac{1}{2}\left(4\left(\frac{5}{4}\right) - \frac{25}{2}(1)\right)} = \frac{11}{12}.$$

Hence, at $(1, \frac{1}{2})$, the equation of the tangent line is $y - \frac{1}{2} = \frac{11}{12}(x - 1)$ or $y = \frac{11}{12}x - \frac{5}{12}$.

• At $(1, -\frac{1}{2})$, $x^2 + y^2 = \frac{5}{4}$, and

$$y' = \frac{\frac{25}{4}\left(-\frac{1}{2}\right)^2 - 4\left(\frac{5}{4}\right)(1)}{-\frac{1}{2}\left(4\left(\frac{5}{4}\right) - \frac{25}{2}(1)\right)} = -\frac{11}{12}.$$

Hence, at $(1, -\frac{1}{2})$, the equation of the tangent line is $y + \frac{1}{2} = -\frac{11}{12}(x - 1)$ or $y = -\frac{11}{12}x + \frac{5}{12}$.

The folium and its tangent lines are plotted below:

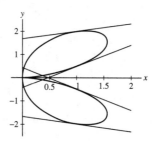

49. CAS Use a computer algebra system to plot $y^2 = x^3 - 4x$ for $-4 \le x, y \le 4$. Show that if $dx/dy = 0$, then $y = 0$. Conclude that the tangent line is vertical at the points where the curve intersects the x-axis. Does your plot confirm this conclusion?

SOLUTION   A plot of the curve $y^2 = x^3 - 4x$ is shown below.

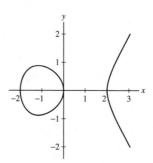

Differentiating the equation $y^2 = x^3 - 4x$ with respect to y yields

$$2y = 3x^2\frac{dx}{dy} - 4\frac{dx}{dy},$$

or

$$\frac{dx}{dy} = \frac{2y}{3x^2 - 4}.$$

From here, it follows that $\frac{dx}{dy} = 0$ when $y = 0$, so the tangent line to this curve is vertical at the points where the curve intersects the x-axis. This conclusion is confirmed by the plot of the curve shown above.

In Exercises 50–53, use implicit differentiation to calculate higher derivatives.

51. Use the method of the previous exercise to show that $y'' = -y^{-3}$ if $x^2 + y^2 = 1$.

SOLUTION   Let $x^2 + y^2 = 1$. Then $2x + 2yy' = 0$, and $y' = -\frac{x}{y}$. Thus

$$y'' = -\frac{y \cdot 1 - xy'}{y^2} = -\frac{y - x\left(-\frac{x}{y}\right)}{y^2} = -\frac{y^2 + x^2}{y^3} = -\frac{1}{y^3} = -y^{-3}.$$

53. Use the method of the previous exercise to compute y'' at the point $\left(\frac{2}{3}, \frac{4}{3}\right)$ on the folium of Descartes with the equation $x^3 + y^3 = 3xy$.

SOLUTION Let $x^3 + y^3 = 3xy$. Then $3x^2 + 3y^2y' = 3xy' + 3y$, and $y' = \dfrac{x^2 - y}{x - y^2}$. At $(x, y) = \left(\frac{2}{3}, \frac{4}{3}\right)$, we find

$$y' = \frac{\frac{4}{9} - \frac{4}{3}}{\frac{2}{3} - \frac{16}{9}} = \frac{4 - 12}{6 - 16} = \frac{-8}{-10} = \frac{4}{5}.$$

Similarly,

$$y'' = \frac{\left(x - y^2\right)\left(2x - y'\right) - \left(x^2 - y\right)\left(1 - 2yy'\right)}{\left(x - y^2\right)^2} = -\frac{162}{125}$$

when $(x, y) = \left(\frac{2}{3}, \frac{4}{3}\right)$ and $y' = \frac{4}{5}$.

Further Insights and Challenges

55. The *lemniscate curve* $(x^2 + y^2)^2 = 4(x^2 - y^2)$ was discovered by Jacob Bernoulli in 1694, who noted that it is "shaped like a figure 8, or a knot, or the bow of a ribbon." Find the coordinates of the four points at which the tangent line is horizontal (Figure 12).

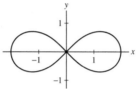

FIGURE 12 Lemniscate curve: $(x^2 + y^2)^2 = 4(x^2 - y^2)$.

SOLUTION Consider the equation of a lemniscate curve: $\left(x^2 + y^2\right)^2 = 4\left(x^2 - y^2\right)$. Taking the derivative of both sides of this equation, we have

$$2\left(x^2 + y^2\right)\left(2x + 2yy'\right) = 4\left(2x - 2yy'\right).$$

Therefore,

$$y' = \frac{8x - 4x\left(x^2 + y^2\right)}{8y + 4y\left(x^2 + y^2\right)} = -\frac{\left(x^2 + y^2 - 2\right)x}{\left(x^2 + y^2 + 2\right)y}.$$

If $y' = 0$, then either $x = 0$ or $x^2 + y^2 = 2$.

- If $x = 0$ in the lemniscate curve, then $y^4 = -4y^2$ or $y^2\left(y^2 + 4\right) = 0$. If y is real, then $y = 0$. The formula for y' in (a) is not defined at the origin (0/0). An alternative parametric analysis shows that the slopes of the tangent lines to the curve at the origin are ± 1.

- If $x^2 + y^2 = 2$ or $y^2 = 2 - x^2$, then plugging this into the lemniscate equation gives $4 = 4\left(2x^2 - 2\right)$ which yields $x = \pm\sqrt{\frac{3}{2}} = \pm\frac{\sqrt{6}}{2}$. Thus $y = \pm\sqrt{\frac{1}{2}} = \pm\frac{\sqrt{2}}{2}$. Accordingly, the four points at which the tangent lines to the lemniscate curve are horizontal are $\left(-\frac{\sqrt{6}}{2}, -\frac{\sqrt{2}}{2}\right), \left(-\frac{\sqrt{6}}{2}, \frac{\sqrt{2}}{2}\right), \left(\frac{\sqrt{6}}{2}, -\frac{\sqrt{2}}{2}\right)$, and $\left(\frac{\sqrt{6}}{2}, \frac{\sqrt{2}}{2}\right)$.

3.9 Related Rates

Preliminary Questions

1. Assign variables and restate the following problem in terms of known and unknown derivatives (but do not solve it): How fast is the volume of a cube increasing if its side increases at a rate of 0.5 cm/s?

SOLUTION Let s and V denote the length of the side and the corresponding volume of a cube, respectively. Determine $\frac{dV}{dt}$ if $\frac{ds}{dt} = 0.5$ cm/s.

2. What is the relation between dV/dt and dr/dt if

$$V = \left(\frac{4}{3}\right)\pi r^3$$

SOLUTION   Applying the general power rule, we find $\frac{dV}{dt} = 4\pi r^2 \frac{dr}{dt}$.

In Questions 3–4, suppose that water pours into a cylindrical glass of radius 4 cm. The variables V and h denote the volume and water level at time t, respectively.

3. Restate in terms of the derivatives dV/dt and dh/dt: How fast is the water level rising if water pours in at a rate of $2 \text{ cm}^3/\text{min}$?

SOLUTION   Determine $\frac{dh}{dt}$ if $\frac{dV}{dt} = 2 \text{ cm}^3/\text{min}$.

4. Repeat the same for this problem: At what rate is water pouring in if the water level rises at a rate of 1 cm/min?

SOLUTION   Determine $\frac{dV}{dt}$ if $\frac{dh}{dt} = 1$ cm/min.

Exercises

In Exercises 1–2, consider a rectangular bathtub whose base is 18 ft².

1. How fast is the water level rising if water is filling the tub at a rate of $0.7 \text{ ft}^3/\text{min}$?

SOLUTION   Let h be the height of the water in the tub and V be the volume of the water. Then $V = 18h$ and $\frac{dV}{dt} = 18\frac{dh}{dt}$. Thus

$$\frac{dh}{dt} = \frac{1}{18}\frac{dV}{dt} = \frac{1}{18}(.7) \approx .039 \text{ ft/min}.$$

3. The radius of a circular oil slick expands at a rate of 2 m/min.

(a) How fast is the area of the oil slick increasing when the radius is 25 m?

(b) If the radius is 0 at time $t = 0$, how fast is the area increasing after 3 min?

SOLUTION   Let r be the radius of the oil slick and A its area.

(a) Then $A = \pi r^2$ and $\frac{dA}{dt} = 2\pi r \frac{dr}{dt}$. Substituting $r = 25$ and $\frac{dr}{dt} = 2$, we find

$$\frac{dA}{dt} = 2\pi(25)(2) = 100\pi \approx 314.16 \text{ m}^2/\text{min}.$$

(b) Since $\frac{dr}{dt} = 2$ and $r(0) = 0$, it follows that $r(t) = 2t$. Thus, $r(3) = 6$ and

$$\frac{dA}{dt} = 2\pi(6)(2) = 24\pi \approx 75.40 \text{ m}^2/\text{min}.$$

In Exercises 5–8, assume that the radius r of a sphere is expanding at a rate of 14 in./min. The volume of a sphere is $V = \frac{4}{3}\pi r^3$ and its surface area is $4\pi r^2$.

5. Determine the rate at which the volume is changing with respect to time when $r = 8$ in.

SOLUTION   As the radius is expanding at 14 inches per minute, we know that $\frac{dr}{dt} = 14$ in./min. Taking $\frac{d}{dt}$ of the equation $V = \frac{4}{3}\pi r^3$ yields

$$\frac{dV}{dt} = \frac{4}{3}\pi\left(3r^2\frac{dr}{dt}\right) = 4\pi r^2\frac{dr}{dt}.$$

Substituting $r = 8$ and $\frac{dr}{dt} = 14$ yields

$$\frac{dV}{dt} = 4\pi(8)^2(14) = 3584\pi \text{ in.}^3/\text{min}.$$

7. Determine the rate at which the surface area is changing when the radius is $r = 8$ in.

SOLUTION   Taking the derivative of both sides of $A = 4\pi r^2$ with respect to t yields $\frac{dA}{dt} = 8\pi r\frac{dr}{dt}$. $\frac{dr}{dt} = 14$, so

$$\frac{dA}{dt} = 8\pi(8)(14) = 896\pi \text{ in.}^2/\text{min}.$$

9. A road perpendicular to a highway leads to a farmhouse located 1 mile away (Figure 9). An automobile travels past the farmhouse at a speed of 60 mph. How fast is the distance between the automobile and the farmhouse increasing when the automobile is 3 miles past the intersection of the highway and the road?

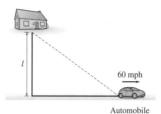

60 mph

Automobile

FIGURE 9

SOLUTION Let l denote the distance between the automobile and the farmhouse, and let s denote the distance past the intersection of the highway and the road. Then $l^2 = 1 + s^2$. Taking the derivative of both sides of this equation yields $2l\frac{dl}{dt} = 2s\frac{ds}{dt}$, so

$$\frac{dl}{dt} = \frac{s}{l}\frac{ds}{dt}.$$

When the auto is 3 miles past the intersection, we have

$$\frac{dl}{dt} = \frac{3 \cdot 60}{\sqrt{1^2 + 3^2}} = \frac{180}{\sqrt{10}} = 18\sqrt{10} \approx 56.92 \text{ mph}.$$

11. Follow the same set-up as Exercise 10, but assume that the water level is rising at a rate of 0.3 m/min when it is 2 m. At what rate is water flowing in?

SOLUTION Consider the cone of water in the tank at a certain instant. Let r be the radius of its (inverted) base, h its height, and V its volume. By similar triangles, $\frac{r}{h} = \frac{2}{3}$ or $r = \frac{2}{3}h$ and thus $V = \frac{1}{3}\pi r^2 h = \frac{4}{27}\pi h^3$. Accordingly,

$$\frac{dV}{dt} = \frac{4}{9}\pi h^2 \frac{dh}{dt}.$$

Substituting $h = 2$ and $\frac{dh}{dt} = 0.3$ yields

$$\frac{dV}{dt} = \frac{4}{9}\pi (2)^2 (.3) \approx 1.68 \text{ m}^3/\text{min}.$$

13. Answer (a) and (b) in Exercise 12 assuming that Sonya begins moving 1 minute after Isaac takes off.

SOLUTION With Isaac x miles east of the center of the lake and Sonya y miles south of its center, let h be the distance between them.

(a) After 12 minutes or $\frac{12}{60} = \frac{1}{5}$ hour, Isaac has traveled $\frac{1}{5} \times 27 = \frac{27}{5}$ miles. After 11 minutes or $\frac{11}{60}$ hour, Sonya has traveled $\frac{11}{60} \times 32 = \frac{88}{15}$ miles.

(b) We have $h^2 = x^2 + y^2$ and $2h\frac{dh}{dt} = 2x\frac{dx}{dt} + 2y\frac{dy}{dt}$. Thus,

$$\frac{dh}{dt} = \frac{x\frac{dx}{dt} + y\frac{dy}{dt}}{h} = \frac{x\frac{dx}{dt} + y\frac{dy}{dt}}{\sqrt{x^2 + y^2}}.$$

Substituting $x = \frac{27}{5}$, $\frac{dx}{dt} = 27$, $y = \frac{88}{15}$, and $\frac{dy}{dt} = 32$ yields

$$\frac{dh}{dt} = \frac{\left(\frac{27}{5}\right)(27) + \left(\frac{88}{15}\right)(32)}{\sqrt{\left(\frac{27}{5}\right)^2 + \left(\frac{88}{15}\right)^2}} = \frac{5003}{\sqrt{14305}} \approx 41.83 \text{ mph}.$$

15. At a given moment, a plane passes directly above a radar station at an altitude of 6 miles.

(a) If the plane's speed is 500 mph, how fast is the distance between the plane and the station changing half an hour later?

(b) How fast is the distance between the plane and the station changing when the plane passes directly above the station?

SOLUTION Let x be the distance of the plane from the station along the ground and h the distance through the air.

(a) By the Pythagorean Theorem, we have

$$h^2 = x^2 + 6^2 = x^2 + 36.$$

Thus $2h\dfrac{dh}{dt} = 2x\dfrac{dx}{dt}$, and $\dfrac{dh}{dt} = \dfrac{x}{h}\dfrac{dx}{dt}$. After an half hour, $x = \frac{1}{2} \times 500 = 250$ miles. With $x = 250$, $h = \sqrt{250^2 + 36}$, and $\frac{dx}{dt} = 500$,

$$\frac{dh}{dt} = \frac{250}{\sqrt{250^2 + 36}} \times 500 \approx 499.86 \text{ mph.}$$

(b) When the plane is directly above the station, $x = 0$, so the distance between the plane and the station is not changing, for at this instant we have

$$\frac{dh}{dt} = \frac{0}{6} \times 500 = 0 \text{ mph.}$$

17. A hot air balloon rising vertically is tracked by an observer located 2 miles from the lift-off point. At a certain moment, the angle between the observer's line-of-sight and the horizontal is $\frac{\pi}{5}$, and it is changing at a rate of 0.2 rad/min. How fast is the balloon rising at this moment?

SOLUTION Let y be the height of the balloon (in miles) and θ the angle between the line-of-sight and the horizontal. Via trigonometry, we have $\tan\theta = \dfrac{y}{2}$. Therefore,

$$\sec^2\theta \cdot \frac{d\theta}{dt} = \frac{1}{2}\frac{dy}{dt},$$

and

$$\frac{dy}{dt} = 2\frac{d\theta}{dt}\sec^2\theta.$$

Using $\frac{d\theta}{dt} = 0.2$ and $\theta = \frac{\pi}{5}$ yields

$$\frac{dy}{dt} = 2\,(.2)\frac{1}{\cos^2(\pi/5)} \approx .61 \text{ mi/min.}$$

In Exercises 19–23, refer to a 16-ft ladder sliding down a wall, as in Figures 1 and 2. The variable h is the height of the ladder's top at time t, and x is the distance from the wall to the ladder's bottom.

19. Assume the bottom slides away from the wall at a rate of 3 ft/s. Find the velocity of the top of the ladder at $t = 2$ if the bottom is 5 ft from the wall at $t = 0$.

SOLUTION Let x denote the distance from the base of the ladder to the wall, and h denote the height of the top of the ladder from the floor. The ladder is 16 ft long, so $h^2 + x^2 = 16^2$. At any time t, $x = 5 + 3t$. Therefore, at time $t = 2$, the base is $x = 5 + 3(2) = 11$ ft from the wall. Furthermore, we have

$$2h\frac{dh}{dt} + 2x\frac{dx}{dt} = 0 \quad \text{so} \quad \frac{dh}{dt} = -\frac{x}{h}\frac{dx}{dt}.$$

Substituting $x = 11$, $h = \sqrt{16^2 - 11^2}$ and $\frac{dx}{dt} = 3$, we obtain

$$\frac{dh}{dt} = -\frac{11}{\sqrt{16^2 - 11^2}}(3) = -\frac{11}{\sqrt{15}} \text{ ft/s} \approx -2.84 \text{ ft/s.}$$

21. Suppose that $h(0) = 12$ and the top slides down the wall at a rate of 4 ft/s. Calculate x and dx/dt at $t = 2$ s.

SOLUTION Let h and x be the height of the ladder's top and the distance from the wall of the ladder's bottom, respectively. After 2 seconds, $h = 12 + 2(-4) = 4$ ft. Since $h^2 + x^2 = 16^2$,

$$x = \sqrt{16^2 - 4^2} = 4\sqrt{15} \text{ ft.}$$

Furthermore, we have $2h\dfrac{dh}{dt} + 2x\dfrac{dx}{dt} = 0$, so that $\dfrac{dx}{dt} = -\dfrac{h}{x}\dfrac{dh}{dt}$. Substituting $h = 4$, $x = 4\sqrt{15}$, and $\frac{dh}{dt} = -4$, we find

$$\frac{dx}{dt} = -\frac{4}{4\sqrt{15}}(-4) = \frac{4}{\sqrt{15}} \approx 1.03 \text{ ft/s.}$$

23. Show that the velocity dh/dt approaches infinity as the ladder slides down to the ground (assuming dx/dt is constant). This suggests that our mathematical description is unrealistic, at least for small values of h. What would, in fact, happen as the top of the ladder approaches the ground?

SOLUTION Let L be the (constant) length of the ladder, let x be the distance from the base of the ladder to the point of contact of the ladder with the ground, and let h be the height of the point of contact of the ladder with the wall. By Pythagoras' theorem,

$$x^2 + h^2 = L^2.$$

Taking derivatives with respect to t yields:

$$2x\frac{dx}{dt} + 2h\frac{dh}{dt} = 0.$$

Therefore,

$$\frac{dh}{dt} = -\frac{x}{h}\frac{dx}{dt}.$$

As $\frac{dx}{dt}$ is constant and positive and x approaches the length of the ladder as it falls, $-\frac{x}{h}\frac{dx}{dt}$ gets arbitrarily large as $h \to 0$.

In a real situation, the top of the ladder would slide down the wall only part of the way. At some point it would lose contact with the wall and fall down freely with acceleration g.

25. Suppose that both the radius r and height h of a circular cone change at a rate of 2 cm/s. How fast is the volume of the cone increasing when $r = 10$ and $h = 20$?

SOLUTION Let r be the radius, h be the height, and V be the volume of a right circular cone. Then $V = \frac{1}{3}\pi r^2 h$, and

$$\frac{dV}{dt} = \frac{1}{3}\pi\left(r^2\frac{dh}{dt} + 2hr\frac{dr}{dt}\right).$$

When $r = 10$, $h = 20$, and $\frac{dr}{dt} = \frac{dh}{dt} = 2$, we find

$$\frac{dV}{dt} = \frac{\pi}{3}\left(10^2 \cdot 2 + 2 \cdot 20 \cdot 10 \cdot 2\right) = \frac{1000\pi}{3} \approx 1047.20 \text{ cm}^3/\text{s}.$$

27. A searchlight rotates at a rate of 3 revolutions per minute. The beam hits a wall located 10 miles away and produces a dot of light that moves horizontally along the wall. How fast is this dot moving when the angle θ between the beam and the line through the searchlight perpendicular to the wall is $\frac{\pi}{6}$? Note that $d\theta/dt = 3(2\pi) = 6\pi$.

SOLUTION Let y be the distance between the dot of light and the point of intersection of the wall and the line through the searchlight perpendicular to the wall. Let θ be the angle between the beam of light and the line. Using trigonometry, we have $\tan\theta = \frac{y}{10}$. Therefore,

$$\sec^2\theta \cdot \frac{d\theta}{dt} = \frac{1}{10}\frac{dy}{dt},$$

and

$$\frac{dy}{dt} = 10\frac{d\theta}{dt}\sec^2\theta.$$

With $\theta = \frac{\pi}{6}$ and $\frac{d\theta}{dt} = 6\pi$, we find

$$\frac{dy}{dt} = 10\,(6\pi)\,\frac{1}{\cos^2(\pi/6)} = 80\pi \approx 251.33 \text{ mi/min} \approx 15{,}079.64 \text{ mph}.$$

29. A plane traveling at an altitude of 20,000 ft passes directly overhead at time $t = 0$. One minute later you observe that the angle between the vertical and your line of sight to the plane is 1.14 rad and that this angle is changing at a rate of 0.38 rad/min. Calculate the velocity of the airplane.

SOLUTION Let x be the distance of the plane from you along the ground and θ the angle between the vertical and your line of sight to the plane. Then $\tan\theta = \frac{x}{20000}$ and

$$\frac{dx}{dt} = 20000\sec^2\theta \cdot \frac{d\theta}{dt}.$$

Substituting $\theta = 1.14$ and $\frac{d\theta}{dt} = 0.38$, we find

$$\frac{dx}{dt} = \frac{20000}{\cos^2(1.14)}\,(.38) \approx 43581.69 \text{ ft/min}$$

or roughly 495.25 mph.

31. A jogger runs around a circular track of radius 60 ft. Let (x, y) be her coordinates, where the origin is at the center of the track. When the jogger's coordinates are $(36, 48)$, her x-coordinate is changing at a rate of 14 ft/s. Find dy/dt.

SOLUTION We have $x^2 + y^2 = 60^2$. Thus $2x\dfrac{dx}{dt} + 2y\dfrac{dy}{dt} = 0$, and $\dfrac{dy}{dt} = -\dfrac{x}{y}\dfrac{dx}{dt}$. With $x = 36$, $y = 48$, and $\dfrac{dx}{dt} = 14$,

$$\frac{dy}{dt} = -\frac{36}{48}(14) = -\frac{21}{2} = -10.5 \text{ ft/s}.$$

In Exercises 33–34, assume that the pressure P (in kilopascals) and volume V (in cm^3) of an expanding gas are related by $PV^b = C$, where b and C are constants (this holds in adiabatic *expansion, without heat gain or loss).*

33. Find dP/dt if $b = 1.2$, $P = 8$ kPa, $V = 100$ cm^2, and $dV/dt = 20$ cm^3/min.

SOLUTION Let $PV^b = C$. Then

$$PbV^{b-1}\frac{dV}{dt} + V^b\frac{dP}{dt} = 0,$$

and

$$\frac{dP}{dt} = -\frac{Pb}{V}\frac{dV}{dt}.$$

Substituting $b = 1.2$, $P = 8$, $V = 100$, and $\frac{dV}{dt} = 20$, we find

$$\frac{dP}{dt} = -\frac{(8)(1.2)}{100}(20) = -1.92 \text{ kPa/min}.$$

35. A point moves along the parabola $y = x^2 + 1$. Let $\ell(t)$ be the distance between the point and the origin. Calculate $\ell'(t)$, assuming that the x-coordinate of the point is increasing at a rate of 9 ft/s.

SOLUTION A point moves along the parabola $y = x^2 + 1$. Let ℓ be the distance between the point and the origin. By the distance formula, we have $\ell = (x^2 + (x^2 + 1)^2)^{1/2}$. Hence

$$\frac{d\ell}{dt} = \frac{1}{2}\left(x^2 + (x^2 + 1)^2\right)^{-1/2}\left(2x\frac{dx}{dt} + 2\left(x^2 + 1\right) \cdot 2x\frac{dx}{dt}\right) = \frac{(2x^2 + 3)x\frac{dx}{dt}}{\sqrt{x^4 + 3x^2 + 1}}.$$

With $\frac{dx}{dt} = 9$, we have

$$\ell'(t) = \frac{9x(2x^2 + 3)}{\sqrt{x^4 + 3x^2 + 1}} \text{ ft/s}.$$

37. A water tank in the shape of a right circular cone of radius 300 cm and height 500 cm leaks water from the vertex at a rate of 10 cm^3/min. Find the rate at which the water level is decreasing when it is 200 cm.

SOLUTION Consider the cone of water in the tank at a certain instant. Let r be the radius of its (inverted) base, h its height, and V its volume. By similar triangles, $\frac{r}{h} = \frac{300}{500}$ or $r = \frac{3}{5}h$ and thus $V = \frac{1}{3}\pi r^2 h = \frac{3}{25}\pi h^3$. Therefore,

$$\frac{dV}{dt} = \frac{9}{25}\pi h^2\frac{dh}{dt},$$

and

$$\frac{dh}{dt} = \frac{25}{9\pi h^2}\frac{dV}{dt}.$$

We are given $\frac{dV}{dt} = -10$. When $h = 200$, it follows that

$$\frac{dh}{dt} = \frac{25}{9\pi(200)^2}(-10) = -\frac{1}{1440\pi} \approx -2.21 \times 10^{-4} \text{ cm/min}.$$

Thus, the water level is decreasing at the rate of 2.21×10^{-4} cm/min.

Further Insights and Challenges

39. Henry is pulling on a rope that passes through a pulley on a 10-ft pole and is attached to a wagon (Figure 13). Assume that the rope is attached to a loop on the wagon 2 ft off the ground. Let x be the distance between the loop and the pole.

(a) Find a formula for the speed of the wagon in terms of x and the rate at which Henry pulls the rope.

(b) Find the speed of the wagon when it is 12 ft from the pole, assuming that Henry pulls the rope at a rate of 1.5 ft/sec.

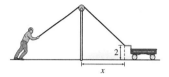

FIGURE 13

SOLUTION Let h be the distance from the pulley to the loop on the wagon. (Note that the rate at which Henry pulls the rope is the rate at which h is *decreasing*.) Using the Pythagorean Theorem, we have $h^2 = x^2 + (10 - 2)^2 = x^2 + 8^2$.

(a) Thus $2h\dfrac{dh}{dt} = 2x\dfrac{dx}{dt}$, and $\dfrac{dx}{dt} = \dfrac{h}{x}\dfrac{dh}{dt}$.

(b) As Henry pulls the rope at the rate of $1.5 = \frac{3}{2}$ ft/s, the distance h is decreasing at that rate; i.e., $dh/dt = -\frac{3}{2}$. When the wagon is 12 feet from the pole, we thus have

$$\frac{dx}{dt} = \frac{\sqrt{12^2 + 8^2}}{12}\left(-\frac{3}{2}\right) = -\sqrt{13}/2 \text{ ft/s}.$$

Thus, the speed of the wagon is $\dfrac{\sqrt{13}}{2}$ ft/s.

41. Using a telescope, you track a rocket that was launched 2 miles away, recording the angle θ between the telescope and the ground at half-second intervals. Estimate the velocity of the rocket if $\theta(10) = 0.205$ and $\theta(10.5) = 0.225$.

SOLUTION Let h be the height of the vertically ascending rocket. Using trigonometry, $\tan\theta = \dfrac{h}{2}$, so

$$\frac{dh}{dt} = 2\sec^2\theta \cdot \frac{d\theta}{dt}.$$

We are given $\theta(10) = 0.205$, and we can estimate

$$\left.\frac{d\theta}{dt}\right|_{t=10} \approx \frac{\theta(10.5) - \theta(10)}{0.5} = 0.04.$$

Thus,

$$\frac{dh}{dt} = 2\sec^2(0.205) \cdot (0.04) \approx 0.083 \text{ mi/s},$$

or roughly 300 mph.

43. A baseball player runs from home plate toward first base at 20 ft/s. How fast is the player's distance from second base changing when the player is halfway to first base? See Figure 15.

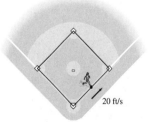

FIGURE 15 Baseball diamond.

SOLUTION In baseball, the distance between bases is 90 feet. Let x be the distance of the player from home plate and h the player's distance from second base. Using the Pythagorean theorem, we have $h^2 = 90^2 + (90 - x)^2$. Therefore,

$$2h\frac{dh}{dt} = 2(90 - x)\left(-\frac{dx}{dt}\right),$$

and

$$\frac{dh}{dt} = -\frac{90 - x}{h} \frac{dx}{dt}.$$

We are given $\frac{dx}{dt} = 20$. When the player is halfway to first base, $x = 45$ and $h = \sqrt{90^2 + 45^2}$, so

$$\frac{dh}{dt} = -\frac{45}{\sqrt{90^2 + 45^2}} (20) = -4\sqrt{5} \approx -8.94 \text{ ft/s}.$$

45. A spectator seated 300 m away from the center of a circular track of radius 100 m watches an athlete run laps at a speed of 5 m/s. How fast is the distance between the spectator and athlete changing when the runner is approaching the spectator and the distance between them is 250 m? *Hint:* The diagram for this problem is similar to Figure 16, with $r = 100$ and $x = 300$.

SOLUTION   From the diagram, the coordinates of P are $(r \cos \theta, r \sin \theta)$ and those of Q are $(x, 0)$.

- The distance formula gives

$$L = \sqrt{(x - r \cos \theta)^2 + (-r \sin \theta)^2}.$$

Thus,

$$L^2 = (x - r \cos \theta)^2 + r^2 \sin^2 \theta.$$

Note that x (the distance of the spectator from the center of the track) and r (the radius of the track) are constants.
- Differentiating with respect to t gives

$$2L \frac{dL}{dt} = 2 (x - r \cos \theta) r \sin \theta \frac{d\theta}{dt} + 2r^2 \sin \theta \cos \theta \frac{d\theta}{dt}.$$

Thus,

$$\frac{dL}{dt} = \frac{rx}{L} \sin \theta \frac{d\theta}{dt}.$$

- Recall the relation between arc length s and angle θ, namely $s = r\theta$. Thus $\frac{d\theta}{dt} = \frac{1}{r} \frac{ds}{dt}$. Given $r = 100$ and $\frac{ds}{dt} = -5$, we have

$$\frac{d\theta}{dt} = \frac{1}{100} (-5) = -\frac{1}{20} \text{ rad/s}.$$

(*Note:* In this scenario, the runner traverses the track in a *clockwise* fashion and approaches the spectator from Quadrant 1.)
- Next, the Law of Cosines gives $L^2 = r^2 + x^2 - 2rx \cos \theta$, so

$$\cos \theta = \frac{r^2 + x^2 - L^2}{2rx} = \frac{100^2 + 300^2 - 250^2}{2 (100) (300)} = \frac{5}{8}.$$

Accordingly,

$$\sin \theta = \sqrt{1 - \left(\frac{5}{8}\right)^2} = \frac{\sqrt{39}}{8}.$$

- Finally

$$\frac{dL}{dt} = \frac{(300) (100)}{250} \left(\frac{\sqrt{39}}{8}\right) \left(-\frac{1}{20}\right) = -\frac{3\sqrt{39}}{4} \approx -4.68 \text{ m/s}.$$

CHAPTER REVIEW EXERCISES

In Exercises 1–4, refer to the function $f(x)$ whose graph is shown in Figure 1.

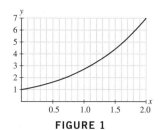

FIGURE 1

1. Compute the average ROC of $f(x)$ over $[0, 2]$. What is the graphical interpretation of this average ROC?

SOLUTION   The average rate of change of $f(x)$ over $[0, 2]$ is

$$\frac{f(2) - f(0)}{2 - 0} = \frac{7 - 1}{2 - 0} = 3.$$

Graphically, this average rate of change represents the slope of the secant line through the points $(2, 7)$ and $(0, 1)$ on the graph of $f(x)$.

3. Estimate $\dfrac{f(0.7 + h) - f(0.7)}{h}$ for $h = 0.3$. Is this number larger or smaller than $f'(0.7)$?

SOLUTION   For $h = 0.3$,

$$\frac{f(0.7 + h) - f(0.7)}{h} = \frac{f(1) - f(0.7)}{0.3} \approx \frac{2.8 - 2}{0.3} = \frac{8}{3}.$$

Because the curve is concave up, the slope of the secant line is larger than the slope of the tangent line, so the value of the difference quotient should be larger than the value of the derivative.

In Exercises 5–8, compute $f'(a)$ using the limit definition and find an equation of the tangent line to the graph of $f(x)$ at $x = a$.

5. $f(x) = x^2 - x, \quad a = 1$

SOLUTION   Let $f(x) = x^2 - x$ and $a = 1$. Then

$$f'(a) = \lim_{h \to 0} \frac{f(a + h) - f(a)}{h} = \lim_{h \to 0} \frac{(1 + h)^2 - (1 + h) - (1^2 - 1)}{h}$$

$$= \lim_{h \to 0} \frac{1 + 2h + h^2 - 1 - h}{h} = \lim_{h \to 0} (1 + h) = 1$$

and the equation of the tangent line to the graph of $f(x)$ at $x = a$ is

$$y = f'(a)(x - a) + f(a) = 1(x - 1) + 0 = x - 1.$$

7. $f(x) = x^{-1}, \quad a = 4$

SOLUTION   Let $f(x) = x^{-1}$ and $a = 4$. Then

$$f'(a) = \lim_{h \to 0} \frac{f(a + h) - f(a)}{h} = \lim_{h \to 0} \frac{\frac{1}{4+h} - \frac{1}{4}}{h} = \lim_{h \to 0} \frac{4 - (4 + h)}{4h(4 + h)}$$

$$= \lim_{h \to 0} \frac{-1}{4(4 + h)} = -\frac{1}{4(4 + 0)} = -\frac{1}{16}$$

and the equation of the tangent line to the graph of $f(x)$ at $x = a$ is

$$y = f'(a)(x - a) + f(a) = -\frac{1}{16}(x - 4) + \frac{1}{4} = -\frac{1}{16}x + \frac{1}{2}.$$

In Exercises 9–12, compute dy/dx using the limit definition.

9. $y = 4 - x^2$

SOLUTION   Let $y = 4 - x^2$. Then

$$\frac{dy}{dx} = \lim_{h \to 0} \frac{4 - (x + h)^2 - (4 - x^2)}{h} = \lim_{h \to 0} \frac{4 - x^2 - 2xh - h^2 - 4 + x^2}{h} = \lim_{h \to 0} (-2x - h) = -2x - 0 = -2x.$$

11. $y = \dfrac{1}{2-x}$

SOLUTION Let $y = \dfrac{1}{2-x}$. Then

$$\frac{dy}{dx} = \lim_{h \to 0} \frac{\frac{1}{2-(x+h)} - \frac{1}{2-x}}{h} = \lim_{h \to 0} \frac{(2-x) - (2-x-h)}{h(2-x-h)(2-x)} = \lim_{h \to 0} \frac{1}{(2-x-h)(2-x)} = \frac{1}{(2-x)^2}.$$

In Exercises 13–16, express the limit as a derivative.

13. $\displaystyle\lim_{h \to 0} \dfrac{\sqrt{1+h} - 1}{h}$

SOLUTION Let $f(x) = \sqrt{x}$. Then

$$\lim_{h \to 0} \frac{\sqrt{1+h} - 1}{h} = \lim_{h \to 0} \frac{f(1+h) - f(1)}{h} = f'(1).$$

15. $\displaystyle\lim_{t \to \pi} \dfrac{\sin t \cos t}{t - \pi}$

SOLUTION Let $f(t) = \sin t \cos t$ and note that $f(\pi) = \sin \pi \cos \pi = 0$. Then

$$\lim_{t \to \pi} \frac{\sin t \cos t}{t - \pi} = \lim_{t \to \pi} \frac{f(t) - f(\pi)}{t - \pi} = f'(\pi).$$

17. Find $f(4)$ and $f'(4)$ if the tangent line to the graph of $f(x)$ at $x = 4$ has equation $y = 3x - 14$.

SOLUTION The equation of the tangent line to the graph of $f(x)$ at $x = 4$ is $y = f'(4)(x - 4) + f(4) = f'(4)x + (f(4) - 4f'(4))$. Matching this to $y = 3x - 14$, we see that $f'(4) = 3$ and $f(4) - 4(3) = -14$, so $f(4) = -2$.

19. Which of (A), (B), or (C) is the graph of the derivative of the function $f(x)$ shown in Figure 3?

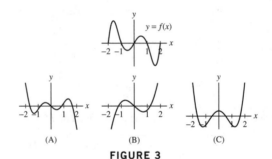

FIGURE 3

SOLUTION The graph of $f(x)$ has four horizontal tangent lines on $[-2, 2]$, so the graph of its derivative must have four x-intercepts on $[-2, 2]$. This eliminates (B). Moreover, $f(x)$ is increasing at both ends of the interval, so its derivative must be positive at both ends. This eliminates (A) and identifies (C) as the graph of $f'(x)$.

21. A girl's height $h(t)$ (in centimeters) is measured at time t (years) for $0 \le t \le 14$:

$$52, 75.1, 87.5, 96.7, 104.5, 111.8, 118.7, 125.2,$$

$$131.5, 137.5, 143.3, 149.2, 155.3, 160.8, 164.7$$

(a) What is the girl's average growth rate over the 14-year period?

(b) Is the average growth rate larger over the first half or the second half of this period?

(c) Estimate $h'(t)$ (in centimeters per year) for $t = 3, 8$.

SOLUTION

(a) The average growth rate over the 14-year period is

$$\frac{164.7 - 52}{14} = 8.05 \text{ cm/year.}$$

(b) Over the first half of the 14-year period, the average growth rate is

$$\frac{125.2 - 52}{7} \approx 10.46 \text{ cm/year,}$$

which is larger than the average growth rate over the second half of the 14-year period:

$$\frac{164.7 - 125.2}{7} \approx 5.64 \text{ cm/year.}$$

(c) For $t = 3$,

$$h'(3) \approx \frac{h(4) - h(3)}{4 - 3} = \frac{104.5 - 96.7}{1} = 7.8 \text{ cm/year;}$$

for $t = 8$,

$$h'(8) \approx \frac{h(9) - h(8)}{9 - 8} = \frac{137.5 - 131.5}{1} = 6.0 \text{ cm/year.}$$

In Exercises 23–24, use the following table of values for the number $A(t)$ of automobiles (in millions) manufactured in the United States in year t.

t	1970	1971	1972	1973	1974	1975	1976
$A(t)$	6.55	8.58	8.83	9.67	7.32	6.72	8.50

23. What is the interpretation of $A'(t)$? Estimate $A'(1971)$. Does $A'(1974)$ appear to be positive or negative?

SOLUTION   Because $A(t)$ measures the number of automobiles manufactured in the United States in year t, $A'(t)$ measures the rate of change in automobile production in the United States. For $t = 1971$,

$$A'(1971) \approx \frac{A(1972) - A(1971)}{1972 - 1971} = \frac{8.83 - 8.58}{1} = 0.25 \text{ million automobiles/year.}$$

Because $A(t)$ decreases from 1973 to 1974 and from 1974 to 1975, it appears that $A'(1974)$ would be negative.

In Exercises 25–50, compute the derivative.

25. $y = 3x^5 - 7x^2 + 4$

SOLUTION   Let $y = 3x^5 - 7x^2 + 4$. Then

$$\frac{dy}{dx} = 15x^4 - 14x.$$

27. $y = t^{-7.3}$

SOLUTION   Let $y = t^{-7.3}$. Then

$$\frac{dy}{dt} = -7.3t^{-8.3}.$$

29. $y = \dfrac{x + 1}{x^2 + 1}$

SOLUTION   Let $y = \dfrac{x + 1}{x^2 + 1}$. Then

$$\frac{dy}{dx} = \frac{(x^2 + 1)(1) - (x + 1)(2x)}{(x^2 + 1)^2} = \frac{1 - 2x - x^2}{(x^2 + 1)^2}.$$

31. $y = (x^4 - 9x)^6$

SOLUTION   Let $y = (x^4 - 9x)^6$. Then

$$\frac{dy}{dx} = 6(x^4 - 9x)^5 \frac{d}{dx}(x^4 - 9x) = 6(4x^3 - 9)(x^4 - 9x)^5.$$

33. $y = (2 + 9x^2)^{3/2}$

SOLUTION   Let $y = (2 + 9x^2)^{3/2}$. Then

$$\frac{dy}{dx} = \frac{3}{2}(2 + 9x^2)^{1/2}\frac{d}{dx}(2 + 9x^2) = 27x(2 + 9x^2)^{1/2}.$$

35. $y = \dfrac{z}{\sqrt{1-z}}$

SOLUTION   Let $y = \dfrac{z}{\sqrt{1-z}}$. Then

$$\frac{dy}{dz} = \frac{\sqrt{1-z} - (-\frac{z}{2})\frac{1}{\sqrt{1-z}}}{1-z} = \frac{1-z+\frac{z}{2}}{(1-z)^{3/2}} = \frac{2-z}{2(1-z)^{3/2}}.$$

37. $y = \dfrac{x^4 + \sqrt{x}}{x^2}$

SOLUTION   Let

$$y = \frac{x^4 + \sqrt{x}}{x^2} = x^2 + x^{-3/2}.$$

Then

$$\frac{dy}{dx} = 2x - \frac{3}{2}x^{-5/2}.$$

39. $y = \tan(t^{-3})$

SOLUTION   Let $y = \tan(t^{-3})$. Then

$$\frac{dy}{dt} = \sec^2(t^{-3})\frac{d}{dt}t^{-3} = -3t^{-4}\sec^2(t^{-3}).$$

41. $y = \sin(2x)\cos^2 x$

SOLUTION   Let $y = \sin(2x)\cos^2 x = 2\sin x\cos^3 x$. Then

$$\frac{dy}{dx} = -6\sin^2 x\cos^2 x + 2\cos^4 x.$$

43. $y = \tan^3 \theta$

SOLUTION   Let $y = \tan^3 \theta$. Then

$$\frac{dy}{d\theta} = 3\tan^2 \theta\frac{d}{d\theta}\tan \theta = 3\tan^2 \theta\sec^2 \theta.$$

45. $y = \dfrac{t}{1 + \sec t}$

SOLUTION   Let $y = \dfrac{t}{1 + \sec t}$. Then

$$\frac{dy}{dt} = \frac{1 + \sec t - t\sec t\tan t}{(1 + \sec t)^2}.$$

47. $y = \dfrac{8}{1 + \cot \theta}$

SOLUTION   Let $y = \dfrac{8}{1 + \cot \theta} = 8(1 + \cot \theta)^{-1}$. Then

$$\frac{dy}{d\theta} = -8(1 + \cot \theta)^{-2}\frac{d}{d\theta}(1 + \cot \theta) = \frac{8\csc^2 \theta}{(1 + \cot \theta)^2}.$$

49. $y = \sqrt{x + \sqrt{x + \sqrt{x}}}$

SOLUTION   Let $y = \sqrt{x + \sqrt{x + \sqrt{x}}}$. Then

$$\frac{dy}{dx} = \frac{1}{2}\left(x + \sqrt{x + \sqrt{x}}\right)^{-1/2} \frac{d}{dx}\left(x + \sqrt{x + \sqrt{x}}\right)$$

$$= \frac{1}{2}\left(x + \sqrt{x + \sqrt{x}}\right)^{-1/2}\left(1 + \frac{1}{2}\left(x + \sqrt{x}\right)^{-1/2}\frac{d}{dx}\left(x + \sqrt{x}\right)\right)$$

$$= \frac{1}{2}\left(x + \sqrt{x + \sqrt{x}}\right)^{-1/2}\left(1 + \frac{1}{2}\left(x + \sqrt{x}\right)^{-1/2}\left(1 + \frac{1}{2}x^{-1/2}\right)\right).$$

In Exercises 51–56, use the table of values to calculate the derivative of the given function at $x = 2$.

x	$f(x)$	$g(x)$	$f'(x)$	$g'(x)$
2	5	4	−3	9
4	3	2	−2	3

51. $S(x) = 3f(x) - 2g(x)$

SOLUTION   Let $S(x) = 3f(x) - 2g(x)$. Then $S'(x) = 3f'(x) - 2g'(x)$ and

$$S'(2) = 3f'(2) - 2g'(2) = 3(-3) - 2(9) = -27.$$

53. $R(x) = \dfrac{f(x)}{g(x)}$

SOLUTION   Let $R(x) = f(x)/g(x)$. Then

$$R'(x) = \frac{g(x)f'(x) - f(x)g'(x)}{g(x)^2}$$

and

$$R'(2) = \frac{g(2)f'(2) - f(2)g'(2)}{g(2)^2} = \frac{4(-3) - 5(9)}{4^2} = -\frac{57}{16}.$$

55. $F(x) = f(g(2x))$

SOLUTION   Let $F(x) = f(g(2x))$. Then $F'(x) = 2f'(g(2x))g'(2x)$ and

$$F'(2) = 2f'(g(4))g'(4) = 2f'(2)g'(4) = 2(-3)(3) = -18.$$

In Exercise 57–60, let $f(x) = x^3 - 3x^2 + x + 4$.

57. Find the points on the graph of $f(x)$ where the tangent line has slope 10.

SOLUTION   Let $f(x) = x^3 - 3x^2 + x + 4$. Then $f'(x) = 3x^2 - 6x + 1$. The tangent line to the graph of $f(x)$ will have slope 10 when $f'(x) = 10$. Solving the quadratic equation $3x^2 - 6x + 1 = 10$ yields $x = -1$ and $x = 3$. Thus, the points on the graph of $f(x)$ where the tangent line has slope 10 are $(-1, -1)$ and $(3, 7)$.

59. Find all values of b such that $y = 25x + b$ is tangent to the graph of $f(x)$.

SOLUTION   Let $f(x) = x^3 - 3x^2 + x + 4$. The equation $y = 25x + b$ represents a line with slope 25; in order for this line to be tangent to the graph of $f(x)$, we must therefore have $f'(x) = 3x^2 - 6x + 1 = 25$. Solving for x yields $x = 4$ and $x = -2$. The equation of the line tangent to the graph of $f(x)$ at $x = 4$ is

$$y = f'(4)(x - 4) + f(4) = 25(x - 4) + 24 = 25x - 76,$$

while the equation of the tangent line at $x = -2$ is

$$y = f'(-2)(x + 2) + f(-2) = 25(x + 2) - 18 = 25x + 32.$$

Thus, when $b = -76$ and when $b = 32$, the line $y = 25x + b$ is tangent to the graph of $f(x)$.

61. **(a)** Show that there is a unique value of a such that $f(x) = x^3 - 2x^2 + x + 1$ has the same slope at both a and $a + 1$.

(b) $\boxed{\text{GU}}$ Plot $f(x)$ together with the tangent lines at $x = a$ and $x = a + 1$ and confirm your answer to part (a).

SOLUTION

(a) Let $f(x) = x^3 - 2x^2 + x + 1$. Then $f'(x) = 3x^2 - 4x + 1$ and the slope of the tangent line at $x = a$ is $f'(a) = 3a^2 - 4a + 1$, while the slope of the tangent line at $x = a + 1$ is

$$f'(a+1) = 3(a+1)^2 - 4(a+1) + 1 = 3(a^2 + 2a + 1) - 4a - 4 + 1 = 3a^2 + 2a.$$

In order for the tangent lines at $x = a$ and $x = a + 1$ to have the same slope, we must have $f'(a) = f'(a+1)$, or

$$3a^2 - 4a + 1 = 3a^2 + 2a.$$

The only solution to this equation is $a = \frac{1}{6}$.

(b) The equation of the tangent line at $x = \frac{1}{6}$ is

$$y = f'\left(\frac{1}{6}\right)\left(x - \frac{1}{6}\right) + f\left(\frac{1}{6}\right) = \frac{5}{12}\left(x - \frac{1}{6}\right) + \frac{241}{216} = \frac{5}{12}x + \frac{113}{108},$$

and the equation of the tangent line at $x = \frac{7}{6}$ is

$$y = f'\left(\frac{7}{6}\right)\left(x - \frac{7}{6}\right) + f\left(\frac{7}{6}\right) = \frac{5}{12}\left(x - \frac{7}{6}\right) + \frac{223}{216} = \frac{5}{12}x + \frac{59}{108}.$$

The graphs of $f(x)$ and the two tangent lines appear below.

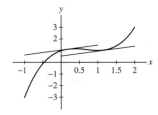

In Exercises 63–68, calculate y''.

63. $y = 12x^3 - 5x^2 + 3x$

SOLUTION   Let $y = 12x^3 - 5x^2 + 3x$. Then

$$y' = 36x^2 - 10x + 3 \quad \text{and} \quad y'' = 72x - 10.$$

65. $y = \sqrt{2x + 3}$

SOLUTION   Let $y = \sqrt{2x + 3} = (2x + 3)^{1/2}$. Then

$$y' = \frac{1}{2}(2x + 3)^{-1/2}\frac{d}{dx}(2x + 3) = (2x + 3)^{-1/2} \quad \text{and} \quad y'' = -\frac{1}{2}(2x + 3)^{-3/2}\frac{d}{dx}(2x + 3) = -(2x + 3)^{-3/2}.$$

67. $y = \tan(x^2)$

SOLUTION   Let $y = \tan(x^2)$. Then

$$y' = 2x \sec^2(x^2) \quad \text{and}$$

$$y'' = 2x\left(2\sec(x^2)\frac{d}{dx}\sec(x^2)\right) + 2\sec^2(x^2) = 8x^2 \sec^2(x^2)\tan(x^2) + 2\sec^2(x^2).$$

69. In Figure 5, label the graphs f, f', and f''.

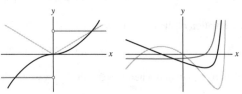

FIGURE 5

SOLUTION   First consider the plot on the left. Observe that the green curve is nonnegative whereas the red curve is increasing, suggesting that the green curve is the derivative of the red curve. Moreover, the green curve is linear with negative slope for $x < 0$ and linear with positive slope for $x > 0$ while the blue curve is a negative constant for $x < 0$ and a positive constant for $x > 0$, suggesting the blue curve is the derivative of the green curve. Thus, the red, green and blue curves, respectively, are the graphs of f, f' and f''.

Now consider the plot on the right. Because the red curve is decreasing when the blue curve is negative and increasing when the blue curve is positive and the green curve is decreasing when the red curve is negative and increasing when the red curve is positive, it follows that the green, red and blue curves, respectively, are the graphs of f, f' and f''.

71. 📖   The number q of frozen chocolate cakes that a commercial bakery can sell per week depends on the price p. In this setting, the percentage ROC of q with respect to p is called the *price elasticity of demand $E(p)$*. Assume that $q = 50p(10 - p)$ for $5 < p < 10$.

(a) Show that $E(p) = \dfrac{2p - 10}{p - 10}$.

(b) Compute the value of $E(8)$ and explain how it justifies the following statement: If the price is set at \$8, then a 1% increase in price reduces demand by 3%.

SOLUTION

(a) Let $q = 50p(10 - p) = 500p - 50p^2$. Then $q'(p) = 500 - 100p$ and

$$E(p) = \left(\frac{p}{q}\right)\frac{dq}{dp} = \frac{p}{50p(10 - p)}(500 - 100p) = \frac{10 - 2p}{10 - p} = \frac{2p - 10}{p - 10}.$$

(b) From part (a),

$$E(8) = \frac{2(8) - 10}{8 - 10} = -3.$$

Thus, with the price set at \$8, a 1% increase in price results in a 3% decrease in demand.

In Exercises 73–80, compute $\dfrac{dy}{dx}$.

73. $x^3 - y^3 = 4$

SOLUTION   Consider the equation $x^3 - y^3 = 4$. Differentiating with respect to x yields

$$3x^2 - 3y^2\frac{dy}{dx} = 0.$$

Therefore,

$$\frac{dy}{dx} = \frac{x^2}{y^2}.$$

75. $y = xy^2 + 2x^2$

SOLUTION   Consider the equation $y = xy^2 + 2x^2$. Differentiating with respect to x yields

$$\frac{dy}{dx} = 2xy\frac{dy}{dx} + y^2 + 4x.$$

Therefore,

$$\frac{dy}{dx} = \frac{y^2 + 4x}{1 - 2xy}.$$

77. $x + y = \sqrt{3x^2 + 2y^2}$

SOLUTION   Squaring the equation $x + y = \sqrt{3x^2 + 2y^2}$ and combining like terms leaves $2xy = 2x^2 + y^2$. Differentiating with respect to x yields

$$2x\frac{dy}{dx} + 2y = 4x + 2y\frac{dy}{dx}.$$

Therefore,

$$\frac{dy}{dx} = \frac{y - 2x}{y - x}.$$

79. $\tan(x + y) = xy$

SOLUTION   Consider the equation $\tan(x + y) = xy$. Differentiating with respect to x yields

$$\sec^2(x + y)\left(1 + \frac{dy}{dx}\right) = x\frac{dy}{dx} + y.$$

Therefore,

$$\frac{dy}{dx} = \frac{y - \sec^2(x + y)}{\sec^2(x + y) - x}.$$

81. Find the points on the graph of $x^3 - y^3 = 3xy - 3$ where the tangent line is horizontal.

SOLUTION   Suppose $x^3 - y^3 = 3xy - 3$. Differentiating with respect to x leads to

$$3x^2 - 3y^2\frac{dy}{dx} = 3x\frac{dy}{dx} + 3y,$$

or

$$\frac{dy}{dx} = \frac{x^2 - y}{x + y^2}.$$

Tangents to the curve are therefore horizontal when $y = x^2$. Substituting $y = x^2$ into the equation for the curve yields $x^3 - x^6 = 3x^3 - 3$ or

$$x^6 + 2x^3 - 3 = (x^3 - 1)(x^3 + 3) = 0.$$

Thus, $x = 1$ or $x = -\sqrt[3]{3}$. The corresponding y-coordinates are $y = 1$ and $y = \sqrt[3]{9}$. Thus, the points along the curve $x^3 - y^3 = 3xy - 3$ where the tangent line is horizontal are:

$$(1, 1) \quad \text{and} \quad (-\sqrt[3]{3}, \sqrt[3]{9}).$$

83. For which values of α is $f(x) = |x|^\alpha$ differentiable at $x = 0$?

SOLUTION   Let $f(x) = |x|^\alpha$. If $\alpha < 0$, then $f(x)$ is not continuous at $x = 0$ and therefore cannot be differentiable at $x = 0$. If $\alpha = 0$, then the function reduces to $f(x) = 1$, which is differentiable at $x = 0$. Now, suppose $\alpha > 0$ and consider the limit

$$\lim_{x \to 0} \frac{f(x) - f(0)}{x - 0} = \lim_{x \to 0} \frac{|x|^\alpha}{x}.$$

If $0 < \alpha < 1$, then

$$\lim_{x \to 0-} \frac{|x|^\alpha}{x} = -\infty \quad \text{while} \quad \lim_{x \to 0+} \frac{|x|^\alpha}{x} = \infty$$

and $f'(0)$ does not exist. If $\alpha = 1$, then

$$\lim_{x \to 0-} \frac{|x|}{x} = -1 \quad \text{while} \quad \lim_{x \to 0+} \frac{|x|}{x} = 1$$

and $f'(0)$ again does not exist. Finally, if $\alpha > 1$, then

$$\lim_{x \to 0} \frac{|x|^\alpha}{x} = 0,$$

so $f'(0)$ does exist.
   In summary, $f(x) = |x|^\alpha$ is differentiable at $x = 0$ when $\alpha = 0$ and when $\alpha > 1$.

85. Water pours into the tank in Figure 7 at a rate of 20 m^3/min. How fast is the water level rising when the water level is $h = 4$ m?

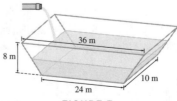

FIGURE 7

SOLUTION   When the water level is at height h, the length of the upper surface of the water is $24 + \frac{3}{2}h$ and the volume of water in the trough is

$$V = \frac{1}{2}h\left(24 + 24 + \frac{3}{2}h\right)(10) = 240h + \frac{15}{2}h^2.$$

Therefore,

$$\frac{dV}{dt} = (240 + 15h)\frac{dh}{dt} = 20 \text{ m}^3/\text{min}.$$

When $h = 4$, we have

$$\frac{dh}{dt} = \frac{20}{240 + 15(4)} = \frac{1}{15} \text{ m/min}.$$

87. A light moving at 3 ft/s approaches a 6-ft man standing 12 ft from a wall (Figure 8). The light is 3 ft above the ground. How fast is the tip P of the man's shadow moving when the light is 24 ft from the wall?

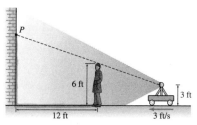

FIGURE 8

SOLUTION   Let x denote the distance between the man and the light. Using similar triangles, we find

$$\frac{3}{x} = \frac{P - 3}{12 + x} \quad \text{or} \quad P = \frac{36}{x} + 6.$$

Therefore,

$$\frac{dP}{dt} = -\frac{36}{x^2}\frac{dx}{dt}.$$

When the light is 24 feet from the wall, $x = 12$. With $\frac{dx}{dt} = -3$, we have

$$\frac{dP}{dt} = -\frac{36}{12^2}(-3) = 0.75 \text{ ft/s}.$$

89. (a) Side x of the triangle in Figure 9 is increasing at 2 cm/s and side y is increasing at 3 cm/s. Assume that θ decreases in such a way that the area of the triangle has the constant value 4 cm². How fast is θ decreasing when $x = 4$, $y = 4$?
(b) How fast is the distance between P and Q changing when $x = 4$, $y = 4$?

FIGURE 9

SOLUTION
(a) The area of the triangle is

$$A = \frac{1}{2}xy\sin\theta = 4.$$

Differentiating with respect to t, we obtain

$$\frac{dA}{dt} = \frac{1}{2}xy\cos\theta\frac{d\theta}{dt} + \frac{1}{2}y\sin\theta\frac{dx}{dt} + x\sin\theta\frac{dy}{dt} = 0.$$

When $x = y = 4$, we have $\frac{1}{2}(4)(4)\sin\theta = 4$, so $\sin\theta = \frac{1}{2}$. Thus, $\theta = \frac{\pi}{6}$ and

$$\frac{1}{2}(4)(4)\frac{\sqrt{3}}{2}\frac{d\theta}{dt} + \frac{1}{2}(4)\left(\frac{1}{2}\right)(2) + \frac{1}{2}(4)\left(\frac{1}{2}\right)(3) = 0.$$

Solving for $d\theta/dt$, we find

$$\frac{d\theta}{dt} = -\frac{5}{4\sqrt{3}} \approx -0.72 \text{ rad/s}.$$

(b) By the Law of Cosines, the distance D between P and Q satisfies

$$D^2 = x^2 + y^2 - 2xy \cos \theta,$$

so

$$2D\frac{dD}{dt} = 2x\frac{dx}{dt} + 2y\frac{dy}{dt} + 2xy \sin \theta \frac{d\theta}{dt} - 2x \cos \theta \frac{dy}{dt} - 2y \cos \theta \frac{dx}{dt}.$$

With $x = y = 4$ and $\theta = \frac{\pi}{6}$,

$$D = \sqrt{4^2 + 4^2 - 2(4)(4)\frac{\sqrt{3}}{2}} = 4\sqrt{2 - \sqrt{3}}.$$

Therefore,

$$\frac{dD}{dt} = \frac{16 + 24 - \frac{20}{\sqrt{3}} - 12\sqrt{3} - 8\sqrt{3}}{8\sqrt{2 - \sqrt{3}}} \approx -1.50 \text{ cm/s}.$$

4 | APPLICATIONS OF THE DERIVATIVE

4.1 Linear Approximation and Applications

Preliminary Questions

1. Estimate $g(1.2) - g(1)$ if $g'(1) = 4$.

SOLUTION Using the Linear Approximation,

$$g(1.2) - g(1) \approx g'(1)(1.2 - 1) = 4(0.2) = 0.8.$$

2. Estimate $f(2.1)$, assuming that $f(2) = 1$ and $f'(2) = 3$.

SOLUTION Using the Linear Approximation,

$$f(2.1) \approx f(2) + f'(2)(2.1 - 2) = 1 + 3(0.1) = 1.3$$

3. The velocity of a train at a given instant is 110 ft/s. How far does the train travel during the next half-second (use the Linear Approximation)?

SOLUTION Using the Linear Approximation, we estimate that the train travels at the constant velocity of 110 ft/sec over the next half-second; hence, we estimate that the train will travel 55 ft over the next half-second.

4. Discuss how the Linear Approximation makes the following statement more precise: The sensitivity of the output to a small change in input depends on the derivative.

SOLUTION The Linear Approximation tells us that the change in the output value of a function is approximately equal to the value of the derivative times the change in the input value.

5. Suppose that the linearization of $f(x)$ at $a = 2$ is $L(x) = 2x + 4$. What are $f(2)$ and $f'(2)$?

SOLUTION The linearization of $f(x)$ at $a = 2$ is

$$L(x) = f(2) + f'(2)(x - 2) = f'(2) \cdot x + (f(2) - 2f'(2))$$

Matching this to the given expression, $L(x) = 2x + 4$, it follows that $f'(2) = 2$. Moreover, $f(2) - 2f'(2) = 4$, so $f(2) = 8$.

Exercises

In Exercises 1–4, use the Linear Approximation to estimate $\Delta f = f(3.02) - f(3)$ for the given function.

1. $f(x) = x^2$

SOLUTION Let $f(x) = x^2$. Then $f'(x) = 2x$ and $\Delta f \approx f'(3)\Delta x = 6(.02) = 0.12$.

3. $f(x) = x^{-1}$

SOLUTION Let $f(x) = x^{-1}$. Then $f'(x) = -x^{-2}$ and $\Delta f \approx f'(3)\Delta x = -\frac{1}{9}(.02) = -.00222$.

In Exercises 5–10, estimate Δf using the Linear Approximation and use a calculator to compute both the error and the percentage error.

5. $f(x) = x - 2x^2$, $a = 5$, $\Delta x = -0.4$

SOLUTION Let $f(x) = x - 2x^2$, $a = 5$, and $\Delta x = -0.4$. Then $f'(x) = 1 - 4x$, $f'(a) = f'(5) = -19$ and $\Delta f \approx f'(a)\Delta x = (-19)(-0.4) = 7.6$. The actual change is

$$\Delta f = f(a + \Delta x) - f(a) = f(4.6) - f(5) = -37.72 - (-45) = 7.28.$$

The error in the Linear Approximation is therefore $|7.28 - 7.6| = 0.32$; in percentage terms, the error is

$$\frac{0.32}{7.28} \times 100\% \approx 4.40\%$$

7. $f(x) = \dfrac{1}{1+x^2}$, $\quad a = 3$, $\quad \Delta x = 0.5$

SOLUTION Let $f(x) = \frac{1}{1+x^2}$, $a = 3$, and $\Delta x = .5$. Then $f'(x) = -\frac{2x}{(1+x^2)^2}$, $f'(a) = f'(3) = -.06$ and $\Delta f \approx f'(a)\Delta x = -.06(.5) = -0.03$. The actual change is

$$\Delta f = f(a + \Delta x) - f(a) = f(3.5) - f(3) \approx -.0245283.$$

The error in the Linear Approximation is therefore $|-.0245283 - (-.03)| = .0054717$; in percentage terms, the error is

$$\left| \frac{.0054717}{-.0245283} \right| \times 100\% \approx 22.31\%$$

9. $f(x) = \tan x$, $\quad a = \frac{\pi}{4}$, $\quad \Delta x = 0.013$

SOLUTION Let $f(x) = \tan x$, $a = \frac{\pi}{4}$, and $\Delta x = .013$. Then $f'(x) = \sec^2 x$, $f'(a) = f'(\frac{\pi}{4}) = 2$ and $\Delta f \approx f'(a)\Delta x = 2(.013) = .026$. The actual change is

$$\Delta f = f(a + \Delta x) - f(a) = f\left(\frac{\pi}{4} + .013\right) - f\left(\frac{\pi}{4}\right) \approx 1.026344 - 1 = .026344.$$

The error in the Linear Approximation is therefore $|.026344 - .026| = .000344$; in percentage terms, the error is

$$\frac{.000344}{.026344} \times 100\% \approx 1.31\%$$

In Exercises 11–16, estimate the quantity using the Linear Approximation and find the error using a calculator.

11. $\sqrt{26} - \sqrt{25}$

SOLUTION Let $f(x) = \sqrt{x}$, $a = 25$, and $\Delta x = 1$. Then $f'(x) = \frac{1}{2}x^{-1/2}$ and $f'(a) = f'(25) = \frac{1}{10}$.

- The Linear Approximation is $\Delta f \approx f'(a)\Delta x = \frac{1}{10}(1) = .1$.
- The actual change is $\Delta f = f(a + \Delta x) - f(a) = f(26) - f(25) \approx .0990195$.
- The error in this estimate is $|.0990195 - .1| = .000980486$.

13. $\dfrac{1}{\sqrt{101}} - \dfrac{1}{10}$

SOLUTION Let $f(x) = \frac{1}{\sqrt{x}}$, $a = 100$, and $\Delta x = 1$. Then $f'(x) = \frac{d}{dx}(x^{-1/2}) = -\frac{1}{2}x^{-3/2}$ and $f'(a) = -\frac{1}{2}(\frac{1}{1000}) = -0.0005$.

- The Linear Approximation is $\Delta f \approx f'(a)\Delta x = -0.0005(1) = -0.0005$.
- The actual change is

$$\Delta f = f(a + \Delta x) - f(a) = \frac{1}{\sqrt{101}} - \frac{1}{10} = -0.000496281.$$

- The error in this estimate is $|-0.0005 - (-0.000496281)| = 3.71902 \times 10^{-6}$.

15. $\sin(0.023)$ *Hint:* Estimate $\sin(0.023) - \sin(0)$.

SOLUTION Let $f(x) = \sin x$, $a = 0$, and $\Delta x = .023$. Then $f'(x) = \cos x$ and $f'(a) = f'(0) = 1$.

- The Linear Approximation is $\Delta f \approx f'(a)\Delta x = 1(.023) = .023$.
- The actual change is $\Delta f = f(a + \Delta x) - f(a) = f(0.023) - f(0) = 0.02299797$.
- The error in this estimate is $|.023 - .02299797| \approx 2.03 \times 10^{-6}$.

17. The cube root of 27 is 3. How much larger is the cube root of 27.2? Estimate using the Linear Approximation.

SOLUTION Let $f(x) = x^{1/3}$, $a = 27$, and $\Delta x = .2$. Then $f'(x) = \frac{1}{3}x^{-2/3}$ and $f'(a) = f'(27) = \frac{1}{27}$. The Linear Approximation is

$$\Delta f \approx f'(a)\Delta x = \frac{1}{27}(.2) = .0074074$$

19. Estimate $\sin 61° - \sin 60°$ using the Linear Approximation. *Note:* You must express $\Delta\theta$ in radians.

SOLUTION   Let $f(x) = \sin x$, $a = \frac{\pi}{3}$, and $\Delta x = \frac{\pi}{180}$. Then $f'(x) = \cos x$ and $f'(a) = f'(\frac{\pi}{3}) = \frac{1}{2}$. Finally, the Linear Approximation is

$$\Delta f \approx f'(a) \Delta x = \frac{1}{2} \left(\frac{\pi}{180} \right) = \frac{\pi}{360} \approx .008727$$

21. Atmospheric pressure P at altitude $h = 40{,}000$ ft is 390 lb/ft^2. Estimate P at $h = 41{,}000$ if $dP/dh|_{h=40{,}000} = -0.0188$ lb/ft^3.

SOLUTION   Let $P(h)$ by the pressure P at altitude h. By the Linear Approximation,

$$P(41000) - P(40000) \approx \frac{dP}{dh}\bigg|_{h=40000} \Delta h = -.0188 \text{ lbs/ft}^3 (1000 \text{ ft}) = -18.8 \text{ lbs/ft}^2.$$

Hence, $P(41000) \approx 390 \text{ lbs/ft}^2 - 18.8 \text{ lbs/ft}^2 = 371.2 \text{ lbs/ft}^2$.

23. The side s of a square carpet is measured at 6 ft. Estimate the maximum error in the area A of the carpet if s is accurate to within half an inch.

SOLUTION   Let s be the length in feet of the side of the square carpet. Then $A(s) = s^2$ is the area of the carpet. With $a = 6$ and $\Delta s = \frac{1}{24}$ (note that 1 inch equals $\frac{1}{12}$ foot), an estimate of the size of the error in the area is given by the Linear Approximation:

$$\Delta A \approx A'(6)\Delta s = 12 \left(\frac{1}{24} \right) = .5 \text{ ft}^2$$

25. A stone tossed vertically in the air with initial velocity v ft/s reaches a maximum height of $h = v^2/64$ ft.
(a) Estimate Δh if v is increased from 25 to 26 ft/s.
(b) Estimate Δh if v is increased from 30 to 31 ft/s.
(c) In general, does a 1 ft/s increase in initial velocity cause a greater change in maximum height at low or high initial velocities? Explain.

SOLUTION   A stone tossed vertically with initial velocity v ft/s attains a maximum height of $h = v^2/64$ ft.
(a) If $v = 25$ and $\Delta v = 1$, then $\Delta h \approx h'(v)\Delta v = \frac{1}{32}(25)(1) = .78125$ ft.
(b) If $v = 30$ and $\Delta v = 1$, then $\Delta h \approx h'(v)\Delta v = \frac{1}{32}(30)(1) = .9375$ ft.
(c) A one foot per second increase in initial velocity v increases the maximum height by approximately $v/32$ ft. Accordingly, there is a bigger effect at higher velocities.

27. The *stopping distance* for an automobile (after applying the brakes) is approximately $F(s) = 1.1s + 0.054s^2$ ft, where s is the speed in mph. Use the Linear Approximation to estimate the change in stopping distance per additional mph when $s = 35$ and when $s = 55$.

SOLUTION   Let $F(s) = 1.1s + .054s^2$.

- The Linear Approximation at $s = 35$ mph is

$$\Delta F \approx F'(35)\Delta s = (1.1 + .108 \times 35)\Delta s = 4.88\Delta s \text{ ft}$$

  The change in stopping distance per additional mph for $s = 35$ mph is approximately 4.88 ft.
- The Linear Approximation at $s = 55$ mph is

$$\Delta F \approx F'(55)\Delta s = (1.1 + .108 \times 55)\Delta s = 7.04\Delta s \text{ ft}$$

  The change in stopping distance per additional mph for $s = 55$ mph is approximately 7.04 ft.

29. Estimate the weight loss per mile of altitude gained for a 130-lb pilot. At which altitude would she weigh 129.5 lb? See Example 4.

SOLUTION   From the discussion in the text, the weight loss ΔW at altitude h (in miles) for a person weighing W_0 at the surface of the earth is approximately

$$\Delta W \approx -\frac{1}{1980} W_0 h$$

If $W_0 = 130$ pounds, then $\Delta W \approx -0.066h$. Accordingly, the pilot loses approximately 0.066 pounds per mile of altitude gained. She will weigh 129.5 pounds at the altitude h such that $-0.066h = -0.5$, or $h = 0.5/0.066 \approx 7.6$ miles.

31. The volume of a certain gas (in liters) is related to pressure P (in atmospheres) by the formula $PV = 24$. Suppose that $V = 5$ with a possible error of ± 0.5 L.

(a) Compute P and estimate the possible error.

(b) Estimate the maximum allowable error in V if P must have an error of at most 0.5 atm.

SOLUTION

(a) We have $P = 24V^{-1}$. For $V = 5$ L, $P = 24/5 = 4.8$ atm. If the error in computing the volume is ± 0.5 liters, then the error in computing the pressure is approximately

$$|\Delta P| \approx |P'(V)\Delta V| = |-24V^{-2}\Delta V| = \left|-\frac{24}{25}(\pm .5)\right| = \frac{12}{25} = .48 \text{ atm}$$

(b) When the volume is $V = 5$ L, the error in computing the pressure is approximately

$$|\Delta P| \approx |P'(V)\Delta V| = |-24V^{-2}\Delta V| = \left|-\frac{24}{25}\Delta V\right| = \frac{24}{25}|\Delta V| = .5 \text{ atm}$$

Thus, the maximum allowable error in V is approximately $\Delta V = \pm .520833$ L.

In Exercises 33–38, find the linearization at $x = a$.

33. $y = \cos x \sin x, \quad a = 0$

SOLUTION Let $f(x) = \sin x \cos x = \frac{1}{2}\sin 2x$. Then $f'(x) = \cos 2x$. The linearization at $a = 0$ is

$$L(x) = f'(a)(x - a) + f(a) = 1(x - 0) + 0 = x.$$

35. $y = (1 + x)^{-1/2}, \quad a = 0$

SOLUTION Let $f(x) = (1 + x)^{-1/2}$. Then $f'(x) = -\frac{1}{2}(1 + x)^{-3/2}$. The linearization at $a = 0$ is

$$L(x) = f'(a)(x - a) + f(a) = -\frac{1}{2}x + 1.$$

37. $y = (1 + x^2)^{-1/2}, \quad a = 0$

SOLUTION Let $f(x) = (1 + x^2)^{-1/2}$. Then $f'(x) = -x(1 + x^2)^{-3/2}$, $f(a) = 1$ and $f'(a) = 0$, so the linearization at a is

$$L(x) = f'(a)(x - a) + f(a) = 1.$$

39. GU Estimate $\sqrt{16.2}$ using the linearization $L(x)$ of $f(x) = \sqrt{x}$ at $a = 16$. Plot $f(x)$ and $L(x)$ on the same set of axes and determine if the estimate is too large or too small.

SOLUTION Let $f(x) = x^{1/2}$, $a = 16$, and $\Delta x = .2$. Then $f'(x) = \frac{1}{2}x^{-1/2}$ and $f'(a) = f'(16) = \frac{1}{8}$. The linearization to $f(x)$ is

$$L(x) = f'(a)(x - a) + f(a) = \frac{1}{8}(x - 16) + 4 = \frac{1}{8}x + 2.$$

Thus, we have $\sqrt{16.2} \approx L(16.2) = 4.025$. Graphs of $f(x)$ and $L(x)$ are shown below. Because the graph of $L(x)$ lies above the graph of $f(x)$, we expect that the estimate from the Linear Approximation is too large.

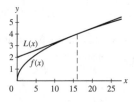

In Exercises 41–46, approximate using linearization and use a calculator to compute the percentage error.

41. $\sqrt{17}$

SOLUTION Let $f(x) = x^{1/2}$, $a = 16$, and $\Delta x = 1$. Then $f'(x) = \frac{1}{2}x^{-1/2}$, $f'(a) = f'(16) = \frac{1}{8}$ and the linearization to $f(x)$ is

$$L(x) = f'(a)(x - a) + f(a) = \frac{1}{8}(x - 16) + 4 = \frac{1}{8}x + 2.$$

Thus, we have $\sqrt{17} \approx L(17) = 4.125$. The percentage error in this estimate is

$$\left|\frac{\sqrt{17} - 4.125}{\sqrt{17}}\right| \times 100\% \approx .046\%$$

43. $(17)^{1/4}$

SOLUTION Let $f(x) = x^{1/4}$, $a = 16$, and $\Delta x = 1$. Then $f'(x) = \frac{1}{4}x^{-3/4}$, $f'(a) = f'(16) = \frac{1}{32}$ and the linearization to $f(x)$ is

$$L(x) = f'(a)(x - a) + f(a) = \frac{1}{32}(x - 16) + 2 = \frac{1}{32}x + \frac{3}{2}.$$

Thus, we have $(17)^{1/4} \approx L(17) = 2.03125$. The percentage error in this estimate is

$$\left| \frac{(17)^{1/4} - 2.03125}{(17)^{1/4}} \right| \times 100\% \approx .035\%$$

45. $(27.001)^{1/3}$

SOLUTION Let $f(x) = x^{1/3}$, $a = 27$, and $\Delta x = .001$. Then $f'(x) = \frac{1}{3}x^{-2/3}$, $f'(a) = f'(27) = \frac{1}{27}$ and the linearization to $f(x)$ is

$$L(x) = f'(a)(x - a) + f(a) = \frac{1}{27}(x - 27) + 3 = \frac{1}{27}x + 2.$$

Thus, we have $(27.001)^{1/3} \approx L(27.001) \approx 3.0000370370370$. The percentage error in this estimate is

$$\left| \frac{(27.001)^{1/3} - 3.0000370370370}{(27.001)^{1/3}} \right| \times 100\% \approx .000000015\%$$

47. GU Plot $f(x) = \tan x$ and its linearization $L(x)$ at $a = \dfrac{\pi}{4}$ on the same set of axes.

(a) Does the linearization overestimate or underestimate $f(x)$?

(b) Show, by graphing $y = f(x) - L(x)$ and $y = 0.1$ on the same set of axes, that the error $|f(x) - L(x)|$ is at most 0.1 for $0.55 \le x \le 0.95$.

(c) Find an interval of x-values on which the error is at most 0.05.

SOLUTION Let $f(x) = \tan x$ and $a = \frac{\pi}{4}$. Then $f'(x) = \sec^2 x$, $f'(a) = 2$ and

$$L(x) = f(a) + f'(a)(x - a) = 1 + 2\left(x - \frac{\pi}{4}\right).$$

Graphs of $f(x)$ and $L(x)$ are shown below.

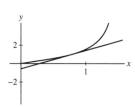

(a) From the figure above, we see that the graph of the linearization $L(x)$ lies below the graph of $f(x)$; thus, the linearization underestimates $f(x)$.

(b) The graph of $y = f(x) - L(x)$ is shown below at the left. Below at the right, portions of the graphs of $y = f(x) - L(x)$ and $y = 0.1$ are shown. From the graph on the right, we see that $|f(x) - L(x)| \le 0.1$ for $0.55 \le x \le 0.95$.

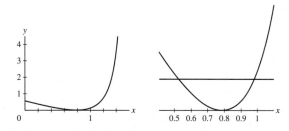

(c) Portions of the graphs of $y = f(x) - L(x)$ and $y = 0.05$ are shown below. From the graph, we see that $|f(x) - L(x)| \le 0.05$ roughly for $0.62 < x < 0.93$.

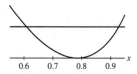

In Exercises 49–50, use the following fact derived from Newton's Laws: An object released at an angle θ with initial velocity v ft/s travels a total distance $s = \frac{1}{32}v^2 \sin 2\theta$ ft (Figure 8).

49. A player located 18.1 ft from a basket launches a successful jump shot from a height of 10 ft (level with the rim of the basket), at an angle $\theta = 34°$ and initial velocity of $v = 25$ ft/s.

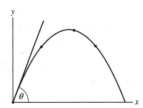

FIGURE 8 Trajectory of an object released at an angle θ.

(a) Show that the distance s of the shot changes by approximately $0.255\Delta\theta$ ft if the angle changes by an amount $\Delta\theta$. Remember to convert the angles to radians in the Linear Approximation.

(b) Is it likely that the shot would have been successful if the angle were off by $2°$?

SOLUTION Using Newton's laws and the given initial velocity of $v = 25$ ft/s, the shot travels $s = \frac{1}{32}v^2 \sin 2t = \frac{625}{32}\sin 2t$ ft, where t is in radians.

(a) If $\theta = 34°$ (i.e., $t = \frac{17}{90}\pi$), then

$$\Delta s \approx s'(t)\Delta t = \frac{625}{16}\cos\left(\frac{17}{45}\pi\right)\Delta t = \frac{625}{16}\cos\left(\frac{17}{45}\pi\right)\Delta\theta \cdot \frac{\pi}{180} \approx 0.255\Delta\theta.$$

(b) If $\Delta\theta = 2°$, this gives $\Delta s \approx 0.51$ ft, in which case the shot would not have been successful, having been off half a foot.

51. Compute the linearization of $f(x) = 3x - 4$ at $a = 0$ and $a = 2$. Prove more generally that a linear function coincides with its linearization at $x = a$ for all a.

SOLUTION Let $f(x) = 3x - 4$. Then $f'(x) = 3$. With $a = 0$, $f(a) = -4$ and $f'(a) = 3$, so the linearization of $f(x)$ at $a = 0$ is

$$L(x) = -4 + 3(x - 0) = 3x - 4 = f(x).$$

With $a = 2$, $f(a) = 2$ and $f'(a) = 3$, so the linearization of $f(x)$ at $a = 2$ is

$$L(x) = 2 + 3(x - 2) = 2 + 3x - 6 = 3x - 4 = f(x).$$

More generally, let $g(x) = bx + c$ be any linear function. The linearization $L(x)$ of $g(x)$ at $x = a$ is

$$L(x) = g'(a)(x - a) + g(a) = b(x - a) + ba + c = bx + c = g(x);$$

i.e., $L(x) = g(x)$.

53. Show that the Linear Approximation to $f(x) = \tan x$ at $x = \frac{\pi}{4}$ yields the estimate $\tan(\frac{\pi}{4} + h) - 1 \approx 2h$. Compute the error E for $h = 10^{-n}$, $1 \le n \le 4$, and verify that E satisfies the Error Bound (3) with $K = 6.2$.

SOLUTION Let $f(x) = \tan x$. Then $f(\frac{\pi}{4}) = 1$, $f'(x) = \sec^2 x$ and $f'(\frac{\pi}{4}) = 2$. Therefore, by the Linear Approximation,

$$f\left(\frac{\pi}{4} + h\right) - f\left(\frac{\pi}{4}\right) = \tan\left(\frac{\pi}{4} + h\right) - 1 \approx 2h.$$

From the following table, we see that for $h = 10^{-n}$, $1 \le n \le 4$, $E \le 6.2h^2$.

h	$E = \lvert\tan(\frac{\pi}{4} + h) - 1 - 2h\rvert$	$6.2h^2$
10^{-1}	2.305×10^{-2}	6.20×10^{-2}
10^{-2}	2.027×10^{-4}	6.20×10^{-4}
10^{-3}	2.003×10^{-6}	6.20×10^{-6}
10^{-4}	2.000×10^{-8}	6.20×10^{-8}

Further Insights and Challenges

55. GU (a) Show that $f(x) = \sin x$ and $g(x) = \tan x$ have the same linearization at $a = 0$.

(b) Which function is approximated more accurately? Explain using a graph over $[0, \frac{\pi}{6}]$.

(c) Calculate the error in these linearizations at $x = \frac{\pi}{6}$. Does the answer confirm your conclusion in (b)?

SOLUTION Let $f(x) = \sin x$ and $g(x) = \tan x$.

(a) The Linear Approximation of $f(x)$ at $a = 0$ is $L(x) = f'(0)(x - 0) + f(0) = \cos 0 \cdot (x - 0) + \sin 0 = x$, whereas the Linear Approximation of $g(x)$ at $a = 0$ is $L(x) = f'(0)(x - 0) + f(0) = \sec^2 0 \cdot (x - 0) + \tan 0 = x$.

(b) The linearization $L(x) = x$ more closely approximates the function $f(x) = \sin x$ than $g(x) = \tan x$, since the graph of x lies closer to that of $\sin x$ than to $\tan x$.

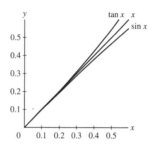

(c) The error in the Linear Approximation for f at $x = \frac{\pi}{6}$ is $\sin(\frac{\pi}{6}) - \frac{\pi}{6} \approx -.02360$; the error in the Linear Approximation for g at $x = \frac{\pi}{6}$ is $\tan(\frac{\pi}{6}) - \frac{\pi}{6} \approx .05375$. These results confirm what the graphs of the curves told us in (b).

57. Let $f(x) = x^{-1}$ and let $E = |\Delta f - f'(1)h|$ be the error in the Linear Approximation at $a = 1$. Show directly that $E = h^2/(1 + h)$. Then prove that $E \le 2h^2$ if $-\frac{1}{2} \le h \le \frac{1}{2}$. *Hint:* In this case, $\frac{1}{2} \le 1 + h \le \frac{3}{2}$.

SOLUTION Let $f(x) = x^{-1}$. Then

$$\Delta f = f(1 + h) - f(1) = \frac{1}{1 + h} - 1 = -\frac{h}{1 + h}$$

and

$$E = |\Delta f - f'(1)h| = \left| -\frac{h}{1 + h} + h \right| = \frac{h^2}{1 + h}.$$

If $-\frac{1}{2} \le h \le \frac{1}{2}$, then $\frac{1}{2} \le 1 + h \le \frac{3}{2}$ and $\frac{2}{3} \le \frac{1}{1+h} \le 2$. Thus, $E \le 2h^2$ for $-\frac{1}{2} \le h \le \frac{1}{2}$.

4.2 Extreme Values

Preliminary Questions

1. If $f(x)$ is continuous on $(0, 1)$, then (choose the correct statement):

(a) $f(x)$ has a minimum value on $(0, 1)$.

(b) $f(x)$ has no minimum value on $(0, 1)$.

(c) $f(x)$ might not have a minimum value on $(0, 1)$.

SOLUTION The correct response is **(c)**: $f(x)$ might not have a minimum value on $(0, 1)$. The key is that the interval $(0, 1)$ is not closed. The function might have a minimum value but it does not have to have a minimum value.

2. If $f(x)$ is not continuous on $[0, 1]$, then (choose the correct statement):

(a) $f(x)$ has no extreme values on $[0, 1]$.

(b) $f(x)$ might not have any extreme values on $[0, 1]$.

SOLUTION The correct response is **(b)**: $f(x)$ might not have any extreme values on $[0, 1]$. Although $[0, 1]$ is closed, because f is not continuous, the function is not guaranteed to have any extreme values on $[0, 1]$.

3. What is the definition of a critical point?

SOLUTION A critical point is a value of the independent variable x in the domain of a function f at which either $f'(x) = 0$ or $f'(x)$ does not exist.

4. True or false: If $f(x)$ is differentiable and $f'(x) = 0$ has no solutions, then $f(x)$ has no local minima or maxima.

SOLUTION True. If $f(x)$ is differentiable but $f'(x) = 0$ has no solutions, then $f(x)$ has no critical points. As critical points are the only candidates for local minima and local maxima, it follows that $f(x)$ has no local extrema.

5. Fermat's Theorem *does not* claim that if $f'(c) = 0$, then $f(c)$ is a local extreme value (this is false). What *does* Fermat's Theorem assert?

SOLUTION   Fermat's Theorem claims: If $f(c)$ is a local extreme value, then either $f'(c) = 0$ or $f'(c)$ does not exist.

6. If $f(x)$ is continuous but has no critical points in $[0, 1]$, then (choose the correct statement):

(a) $f(x)$ has no min or max on $[0, 1]$.

(b) Either $f(0)$ or $f(1)$ is the minimum value on $[0, 1]$.

SOLUTION   The correct response is **(b)**: either $f(0)$ or $f(1)$ is the minimum value on $[0, 1]$. Remember that extreme values occur either at critical points or endpoints. If a continuous function on a closed interval has no critical points, the extreme values must occur at the endpoints.

Exercises

1. The following questions refer to Figure 15.

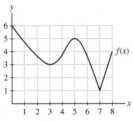

FIGURE 15

(a) How many critical points does $f(x)$ have?

(b) What is the maximum value of $f(x)$ on $[0, 8]$?

(c) What are the local maximum values of $f(x)$?

(d) Find a closed interval on which both the minimum and maximum values of $f(x)$ occur at critical points.

(e) Find an interval on which the minimum value occurs at an endpoint.

SOLUTION

(a) $f(x)$ has three critical points on the interval $[0, 8]$: at $x = 3$, $x = 5$ and $x = 7$. Two of these, $x = 3$ and $x = 5$, are where the derivative is zero and one, $x = 7$, is where the derivative does not exist.

(b) The maximum value of $f(x)$ on $[0, 8]$ is 6; the function takes this value at $x = 0$.

(c) $f(x)$ achieves a local maximum of 5 at $x = 5$.

(d) Answers may vary. One example is the interval $[4, 8]$. Another is $[2, 6]$.

(e) Answers may vary. The easiest way to ensure this is to choose an interval on which the graph takes no local minimum. One example is $[0, 2]$.

In Exercises 3–10, find all critical points of the function.

3. $f(x) = x^2 - 2x + 4$

SOLUTION   Let $f(x) = x^2 - 2x + 4$. Then $f'(x) = 2x - 2 = 0$ implies that $x = 1$ is the lone critical point of f.

5. $f(x) = x^3 - \dfrac{9}{2}x^2 - 54x + 2$

SOLUTION   Let $f(x) = x^3 - \frac{9}{2}x^2 - 54x + 2$. Then $f'(x) = 3x^2 - 9x - 54 = 3(x + 3)(x - 6) = 0$ implies that $x = -3$ and $x = 6$ are the critical points of f.

7. $f(x) = \dfrac{x}{x^2 + 1}$

SOLUTION   Let $f(x) = \dfrac{x}{x^2 + 1}$. Then $f'(x) = \dfrac{1 - x^2}{(x^2 + 1)^2} = 0$ implies that $x = \pm 1$ are the critical points of f.

9. $f(x) = x^{1/3}$

SOLUTION   Let $f(x) = x^{1/3}$. Then $f'(x) = \frac{1}{3}x^{-2/3}$. The derivative is never zero but does not exist at $x = 0$. Thus, $x = 0$ is the only critical point of f.

11. Let $f(x) = x^2 - 4x + 1$.

(a) Find the critical point c of $f(x)$ and compute $f(c)$.

(b) Compute the value of $f(x)$ at the endpoints of the interval $[0, 4]$.

(c) Determine the min and max of $f(x)$ on $[0, 4]$.

(d) Find the extreme values of $f(x)$ on $[0, 1]$.

SOLUTION Let $f(x) = x^2 - 4x + 1$.

(a) Then $f'(c) = 2c - 4 = 0$ implies that $c = 2$ is the sole critical point of f. We have $f(2) = -3$.

(b) $f(0) = f(4) = 1$.

(c) Using the results from (a) and (b), we find the maximum value of f on $[0, 4]$ is 1 and the minimum value is -3.

(d) We have $f(1) = -2$. Hence the maximum value of f on $[0, 1]$ is 1 and the minimum value is -2.

13. Find the critical points of $f(x) = \sin x + \cos x$ and determine the extreme values on $[0, \frac{\pi}{2}]$.

SOLUTION

- Let $f(x) = \sin x + \cos x$. Then on the interval $[0, \frac{\pi}{2}]$, we have $f'(x) = \cos x - \sin x = 0$ at $x = \frac{\pi}{4}$, the only critical point of f in this interval.
- Since $f(\frac{\pi}{4}) = \sqrt{2}$ and $f(0) = f(\frac{\pi}{2}) = 0$, the maximum value of f on $[0, \frac{\pi}{2}]$ is $\sqrt{2}$, while the minimum value is 0.

15. [GU] Plot $f(x) = 2\sqrt{x} - x$ on $[0, 4]$ and determine the maximum value graphically. Then verify your answer using calculus.

SOLUTION The graph of $y = 2\sqrt{x} - x$ over the interval $[0, 4]$ is shown below. From the graph, we see that at $x = 1$, the function achieves its maximum value of 1.

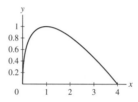

To verify the information obtained from the plot, let $f(x) = 2\sqrt{x} - x$. Then $f'(x) = x^{-1/2} - 1$. Solving $f'(x) = 0$ yields the critical points $x = 0$ and $x = 1$. Because $f(0) = f(4) = 0$ and $f(1) = 1$, we see that the maximum value of f on $[0, 4]$ is 1.

In Exercises 17–45, find the maximum and minimum values of the function on the given interval.

17. $y = 2x^2 - 4x + 2$, $[0, 3]$

SOLUTION Let $f(x) = 2x^2 - 4x + 2$. Then $f'(x) = 4x - 4 = 0$ implies that $x = 1$ is a critical point of f. On the interval $[0, 3]$, the minimum value of f is $f(1) = 0$, whereas the maximum value of f is $f(3) = 8$. (*Note:* $f(0) = 2$.)

19. $y = x^2 - 6x - 1$, $[-2, 2]$

SOLUTION Let $f(x) = x^2 - 6x - 1$. Then $f'(x) = 2x - 6 = 0$ implies that $x = 3$ is a critical point of f. The minimum of f on the interval $[-2, 2]$ is $f(2) = -9$, whereas its maximum is $f(-2) = 15$. (*Note:* The critical point $x = 3$ is outside the interval $[-2, 2]$.)

21. $y = -4x^2 + 3x + 4$, $[-1, 1]$

SOLUTION Let $f(x) = -4x^2 + 3x + 4$. Then $f'(x) = -8x + 3 = 0$ implies that $x = \frac{3}{8}$ is a critical point of f. The minimum of f on the interval $[-1, 1]$ is $f(-1) = -3$, whereas its maximum is $f(\frac{3}{8}) = 4.5625$. (*Note:* $f(1) = 3$.)

23. $y = x^3 - 6x + 1$, $[-1, 1]$

SOLUTION Let $f(x) = x^3 - 6x + 1$. Then $f'(x) = 3x^2 - 6 = 0$ implies that $x = \pm\sqrt{2}$ are the critical points of f. The minimum of f on the interval $[-1, 1]$ is $f(1) = -4$, whereas its maximum is $f(-1) = 6$. (*Note:* The critical points $x = \pm\sqrt{2}$ are not in the interval $[-1, 1]$.)

25. $y = x^3 + 3x^2 - 9x + 2$, $[-1, 1]$

SOLUTION Let $f(x) = x^3 + 3x^2 - 9x + 2$. Then $f'(x) = 3x^2 + 6x - 9 = 3(x + 3)(x - 1) = 0$ implies that $x = 1$ and $x = -3$ are the critical points of f. The minimum of f on the interval $[-1, 1]$ is $f(1) = -3$, whereas its maximum is $f(-1) = 13$. (*Note:* The critical point $x = -3$ is not in the interval $[-1, 1]$.)

27. $y = x^3 + 3x^2 - 9x + 2$, $[-4, 4]$

SOLUTION Let $f(x) = x^3 + 3x^2 - 9x + 2$. Then $f'(x) = 3x^2 + 6x - 9 = 3(x + 3)(x - 1) = 0$ implies that $x = 1$ and $x = -3$ are the critical points of f. The minimum of f on the interval $[-4, 4]$ is $f(1) = -3$, whereas its maximum is $f(4) = 78$. (*Note:* $f(-4) = 22$ and $f(-3) = 29$.)

29. $y = x^5 - 3x^2$, $[-1, 5]$

SOLUTION   Let $f(x) = x^5 - 3x^2$. Then $f'(x) = 5x^4 - 6x = 0$ implies that $x = 0$ and $x = \frac{1}{5}(150)^{1/3} \approx 1.06$ are critical points of f. Over the interval $[-1, 5]$, the minimum value of f is $f(-1) = -4$, whereas its maximum value is $f(5) = 3050$. (*Note:* $f(0) = 0$ and $f(\frac{1}{5}(150)^{1/3}) = -\frac{9}{125}(150)^{2/3} \approx -2.03$).

31. $y = \dfrac{x^2 + 1}{x - 4}$, $[5, 6]$

SOLUTION   Let $f(x) = \dfrac{x^2 + 1}{x - 4}$. Then

$$f'(x) = \frac{(x - 4) \cdot 2x - (x^2 + 1) \cdot 1}{(x - 4)^2} = \frac{x^2 - 8x - 1}{(x - 4)^2} = 0$$

implies $x = 4 \pm \sqrt{17}$ are critical points of f. $x = 4$ is not a critical point because $x = 4$ is not in the domain of f. On the interval $[5, 6]$, the minimum of f is $f(6) = \frac{37}{2} = 18.5$, whereas the maximum of f is $f(5) = 26$. (*Note:* The critical points $x = 4 \pm \sqrt{17}$ are not in the interval $[5, 6]$.)

33. $y = x - \dfrac{4x}{x + 1}$, $[0, 3]$

SOLUTION   Let $f(x) = x - \dfrac{4x}{x + 1}$. Then

$$f'(x) = 1 - \frac{4}{(x + 1)^2} = \frac{(x - 1)(x + 3)}{(x + 1)^2} = 0$$

implies that $x = 1$ and $x = -3$ are critical points of f. $x = -1$ is not a critical point because $x = -1$ is not in the domain of f. The minimum of f on the interval $[0, 3]$ is $f(1) = -1$, whereas the maximum is $f(0) = f(3) = 0$. (*Note:* The critical point $x = -3$ is not in the interval $[0, 3]$.)

35. $y = (2 + x)\sqrt{2 + (2 - x)^2}$, $[0, 2]$

SOLUTION   Let $f(x) = (2 + x)\sqrt{2 + (2 - x)^2}$. Then

$$f'(x) = \sqrt{2 + (2 - x)^2} - (2 + x)(2 + (2 - x)^2)^{-1/2}(2 - x) = \frac{2(x - 1)^2}{\sqrt{2 + (2 - x)^2}} = 0$$

implies that $x = 1$ is the critical point of f. On the interval $[0, 2]$, the minimum is $f(0) = 2\sqrt{6} \approx 4.898979$ and the maximum is $f(2) = 4\sqrt{2} \approx 5.656854$. (*Note:* $f(1) = 3\sqrt{3} \approx 5.196152$.)

37. $y = \sqrt{x + x^2} - 2\sqrt{x}$, $[0, 4]$

SOLUTION   Let $f(x) = \sqrt{x + x^2} - 2\sqrt{x}$. Then

$$f'(x) = \frac{1}{2}(x + x^2)^{-1/2}(1 + 2x) - x^{-1/2} = \frac{1 + 2x - 2\sqrt{1 + x}}{2\sqrt{x}\sqrt{1 + x}} = 0$$

implies that $x = 0$ and $x = \frac{\sqrt{3}}{2}$ are the critical points of f. Neither $x = -1$ nor $x = -\frac{\sqrt{3}}{2}$ is a critical point because neither is in the domain of f. On the interval $[0, 4]$, the minimum of f is $f\left(\frac{\sqrt{3}}{2}\right) \approx -.589980$ and the maximum is $f(4) \approx .472136$. (*Note:* $f(0) = 0$.)

39. $y = \sin x \cos x$, $[0, \frac{\pi}{2}]$

SOLUTION   Let $f(x) = \sin x \cos x = \frac{1}{2}\sin 2x$. On the interval $[0, \frac{\pi}{2}]$, $f'(x) = \cos 2x = 0$ when $x = \frac{\pi}{4}$. The minimum of f on this interval is $f(0) = f(\frac{\pi}{2}) = 0$, whereas the maximum is $f(\frac{\pi}{4}) = \frac{1}{2}$.

41. $y = \sqrt{2}\,\theta - \sec\theta$, $[0, \frac{\pi}{3}]$

SOLUTION   Let $f(\theta) = \sqrt{2}\theta - \sec\theta$. On the interval $[0, \frac{\pi}{3}]$, $f'(\theta) = \sqrt{2} - \sec\theta\tan\theta = 0$ at $\theta = \frac{\pi}{4}$. The minimum value of f on this interval is $f(0) = -1$, whereas the maximum value over this interval is $f(\frac{\pi}{4}) = \sqrt{2}(\frac{\pi}{4} - 1) \approx -.303493$. (*Note:* $f(\frac{\pi}{3}) = \sqrt{2}\frac{\pi}{3} - 2 \approx -.519039$.)

43. $y = \theta - 2\sin\theta$, $[0, 2\pi]$

SOLUTION   Let $g(\theta) = \theta - 2\sin\theta$. On the interval $[0, 2\pi]$, $g'(\theta) = 1 - 2\cos\theta = 0$ at $\theta = \frac{\pi}{3}$ and $\theta = \frac{5}{3}\pi$. The minimum of g on this interval is $g(\frac{\pi}{3}) = \frac{\pi}{3} - \sqrt{3} \approx -.685$ and the maximum is $g(\frac{5}{3}\pi) = \frac{5}{3}\pi + \sqrt{3} \approx 6.968$. (*Note:* $g(0) = 0$ and $g(2\pi) = 2\pi \approx 6.283$.)

45. $y = \tan x - 2x$, $[0, 1]$

SOLUTION Let $f(x) = \tan x - 2x$. Then on the interval $[0, 1]$, $f'(x) = \sec^2 x - 2 = 0$ at $x = \frac{\pi}{4}$. The minimum of f is $f(\frac{\pi}{4}) = 1 - \frac{\pi}{2} \approx -.570796$ and the maximum is $f(0) = 0$. (*Note:* $f(1) = \tan 1 - 2 \approx -.442592$.)

47. GU   Find the critical points of $f(x) = 2\cos 3x + 3\cos 2x$. Check your answer against a graph of $f(x)$.

SOLUTION $f(x)$ is differentiable for all x, so we are looking for points where $f'(x) = 0$ only. Setting $f'(x) = -6\sin 3x - 6\sin 2x$, we get $\sin 3x = -\sin 2x$. Looking at a unit circle, we find the relationship between angles y and x such that $\sin y = -\sin x$. This technique is also used in Exercise 46.

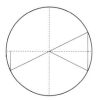

From the diagram, we see that $\sin y = -\sin x$ if y is either (i.) the point antipodal to x ($y = \pi + x + 2\pi k$) or (ii.) the point obtained by reflecting x through the horizontal axis ($y = -x + 2\pi k$).

Since $\sin 3x = -\sin 2x$, we get either $3x = \pi + 2x + 2\pi k$ or $3x = -2x + 2\pi k$. Solving each of these equations for x yields $x = \pi + 2\pi k$ and $x = \frac{2\pi}{5}k$, respectively. The values of x between 0 and 2π are 0, $\frac{2\pi}{5}$, $\frac{4\pi}{5}$, π, $\frac{6\pi}{5}$, $\frac{8\pi}{5}$, and 2π.

The graph is shown below. As predicted, it has horizontal tangent lines at $\frac{2\pi}{5}k$ and at $x = \frac{\pi}{2}$. Each of these points is a local extremum.

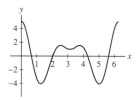

In Exercises 48–51, find the critical points and the extreme values on $[0, 3]$. In Exercises 50 and 51, refer to Figure 18.

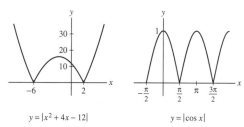

$$y = |x^2 + 4x - 12| \qquad\qquad y = |\cos x|$$

FIGURE 18

49. $y = |3x - 9|$

SOLUTION Let $f(x) = |3x - 9| = 3|x - 3|$. For $x < 3$, we have $f'(x) = -3$. For $x > 3$, we have $f'(x) = 3$. Now as $x \to 3-$, we have $\dfrac{f(x) - f(3)}{x - 3} = \dfrac{3(3 - x) - 0}{x - 3} \to -3$; whereas as $x \to 3+$, we have $\dfrac{f(x) - f(3)}{x - 3} = \dfrac{3(x - 3) - 0}{x - 3} \to 3$. Therefore, $f'(3) = \lim\limits_{x \to 3} \dfrac{f(x) - f(3)}{x - 3}$ does not exist and the lone critical point of f is $x = 3$. Alternately, we examine the graph of $f(x) = |3x - 9|$ shown below.

To find the extrema of $f(x)$ on $[0, 3]$, we test the values of $f(x)$ at the critical point and the endpoints. $f(0) = 9$ and $f(3) = 0$, so $f(x)$ takes its minimum value of 0 at $x = 3$, and its maximum value of 9 at $x = 0$.

51. $y = |\cos x|$

SOLUTION Let $f(x) = |\cos x|$. There are two types of critical points: points of the form πn where the derivative is zero and points of the form $n\pi + \pi/2$ where the derivative does not exist.

To find points where $f'(x)$ does not exist, we look at points where the inside of the absolute value function is equal to zero, and take derivatives in either direction, to find out whether they are the same.

$\cos x = 0$ at the points $x = -\frac{\pi}{2}$, $x = \frac{\pi}{2}$, and $x = \frac{3\pi}{2}$, etc. Only one of these, $x = \frac{\pi}{2}$ is in the interval $[0, 3]$, so we check the one critical point and the endpoints. $f(0) = 1$, $f(3) = |\cos 3| \approx .98992$, and $f(\frac{\pi}{2}) = 0$, so $f(x)$ takes its maximum value of 1 at $x = 0$ and its minimum of 0 at $x = \frac{\pi}{2}$.

In Exercises 53–56, verify Rolle's Theorem for the given interval.

53. $f(x) = x + x^{-1}$, $[\frac{1}{2}, 2]$

SOLUTION   Because f is continuous on $[\frac{1}{2}, 2]$, differentiable on $(\frac{1}{2}, 2)$ and

$$f\left(\frac{1}{2}\right) = \frac{1}{2} + \frac{1}{\frac{1}{2}} = \frac{5}{2} = 2 + \frac{1}{2} = f(2),$$

we may conclude from Rolle's Theorem that there exists a $c \in (\frac{1}{2}, 2)$ at which $f'(c) = 0$. Here, $f'(x) = 1 - x^{-2} = \frac{x^2-1}{x^2}$, so we may take $c = 1$.

55. $f(x) = \dfrac{x^2}{8x - 15}$, $[3, 5]$

SOLUTION   Because f is continuous on $[3, 5]$, differentiable on $(3, 5)$ and $f(3) = f(5) = 1$, we may conclude from Rolle's Theorem that there exists a $c \in (3, 5)$ at which $f'(c) = 0$. Here,

$$f'(x) = \frac{(8x - 15)(2x) - 8x^2}{(8x - 15)^2} = \frac{2x(4x - 15)}{(8x - 15)^2},$$

so we may take $c = \frac{15}{4}$.

57.   Use Rolle's Theorem to prove that $f(x) = x^5 + 2x^3 + 4x - 12$ has at most one real root.

SOLUTION   We use proof by contradiction. Suppose $f(x) = x^5 + 2x^3 + 4x - 12$ has two real roots, $x = a$ and $x = b$. Then $f(a) = f(b) = 0$ and Rolle's Theorem guarantees that there exists a $c \in (a, b)$ at which $f'(c) = 0$. However, $f'(x) = 5x^4 + 6x^2 + 4 \geq 4$ for all x, so there is no $c \in (a, b)$ at which $f'(c) = 0$. Based on this contradiction, we conclude that $f(x) = x^5 + 2x^3 + 4x - 12$ cannot have more than one real root.

59.   The concentration $C(t)$ (in mg/cm^3) of a drug in a patient's bloodstream after t hours is

$$C(t) = \frac{0.016t}{t^2 + 4t + 4}$$

Find the maximum concentration and the time at which it occurs.

SOLUTION

$$C'(t) = \frac{.016(t^2 + 4t + 4) - (.016t(2t + 4))}{(t^2 + 4t + 4)^2} = .016\frac{-t^2 + 4}{(t^2 + 4t + 4)^2} = .016\frac{2 - t}{(t + 2)^3}.$$

$C'(t)$ exists for all $t \geq 0$, so we are looking for points where $C'(t) = 0$. $C'(t) = 0$ when $t = 2$, so we check $C(2) = .002\frac{\text{mg}}{\text{cm}^3}$. Since $C(0) = 0$ and $C(t) \to 0$ as $t \to \infty$, the point $(2, .002)$ is a local maximum of $C(t)$.

Hence, the maximum concentration occurs 2 hours after being administered, and that concentration is $.002\frac{\text{mg}}{\text{cm}^3}$.

61.   Bees build honeycomb structures out of cells with a hexagonal base and three rhombus-shaped faces on top as in Figure 20. Using geometry, we can show that the surface area of this cell is

$$A(\theta) = 6hs + \frac{3}{2}s^2(\sqrt{3}\csc\theta - \cot\theta)$$

where h, s, and θ are as indicated in the figure. It is a remarkable fact that bees "know" which angle θ minimizes the surface area (and therefore requires the least amount of wax).

(a) Show that this angle is approximately 54.7° by finding the critical point of $A(\theta)$ for $0 < \theta < \pi/2$ (assume h and s are constant).

(b) [GU]   Confirm, by graphing $f(\theta) = \sqrt{3}\csc\theta - \cot\theta$, that the critical point indeed minimizes the surface area.

FIGURE 20 A cell in a honeycomb constructed by bees.

SOLUTION

(a) Because h and s are constant relative to θ, we have $A'(\theta) = \frac{3}{2}s^2(-\sqrt{3}\csc\theta\cot\theta + \csc^2\theta) = 0$. From this, we get $\sqrt{3}\csc\theta\cot\theta = \csc^2\theta$, or $\cos\theta = \frac{1}{\sqrt{3}}$, whence $\theta = \cos^{-1}\left(\frac{1}{\sqrt{3}}\right) = .955317$ radians $= 54.736°$.

(b) The plot of $\sqrt{3}\csc\theta - \cot\theta$, where θ is given in degrees, is given below. We can see that the minimum occurs just below $55°$.

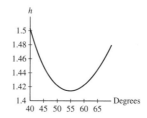

63. Find the maximum of $y = x^a - x^b$ on $[0, 1]$ where $0 < a < b$. In particular, find the maximum of $y = x^5 - x^{10}$ on $[0, 1]$.

SOLUTION

- Let $f(x) = x^a - x^b$. Then $f'(x) = ax^{a-1} - bx^{b-1}$. Since $a < b$, $f'(x) = x^{a-1}(a - bx^{b-a}) = 0$ implies critical points $x = 0$ and $x = \left(\frac{a}{b}\right)^{1/(b-a)}$, which is in the interval $[0, 1]$ as $a < b$ implies $\frac{a}{b} < 1$ and consequently $x = \left(\frac{a}{b}\right)^{1/(b-a)} < 1$. Also, $f(0) = f(1) = 0$ and $a < b$ implies $x^a > x^b$ on the interval $[0, 1]$, which gives $f(x) > 0$ and thus the maximum value of f on $[0, 1]$ is

$$f\left(\left(\frac{a}{b}\right)^{1/(b-a)}\right) = \left(\frac{a}{b}\right)^{a/(b-a)} - \left(\frac{a}{b}\right)^{b/(b-a)}.$$

- Let $f(x) = x^5 - x^{10}$. Then by part (a), the maximum value of f on $[0, 1]$ is

$$f\left(\left(\frac{1}{2}\right)^{1/5}\right) = \left(\frac{1}{2}\right) - \left(\frac{1}{2}\right)^2 = \frac{1}{2} - \frac{1}{4} = \frac{1}{4}.$$

In Exercises 64–66, plot the function using a graphing utility and find its critical points and extreme values on $[-5, 5]$.

65. GU $\quad y = \dfrac{1}{1 + |x - 1|} + \dfrac{1}{1 + |x - 4|}$

SOLUTION Let

$$f(x) = \frac{1}{1 + |x - 1|} + \frac{1}{1 + |x - 4|}.$$

The plot follows:

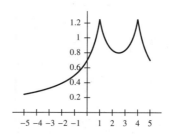

We can see on the plot that the critical points of $f(x)$ lie at the cusps at $x = 1$ and $x = 4$ and at the location of the horizontal tangent line at $x = \frac{5}{2}$. With $f(-5) = \frac{17}{70}$, $f(1) = f(4) = \frac{5}{4}$, $f(\frac{5}{2}) = \frac{4}{5}$ and $f(5) = \frac{7}{10}$, it follows that the maximum value of $f(x)$ on $[-5, 5]$ is $f(1) = f(4) = \frac{5}{4}$ and the minimum value is $f(-5) = \frac{17}{70}$.

67. (a) Use implicit differentiation to find the critical points on the curve $27x^2 = (x^2 + y^2)^3$.

(b) $\boxed{\text{GU}}$ Plot the curve and the horizontal tangent lines on the same set of axes.

SOLUTION

(a) Differentiating both sides of the equation $27x^2 = (x^2 + y^2)^3$ with respect to x yields

$$54x = 3(x^2 + y^2)^2 \left(2x + 2y\frac{dy}{dx}\right).$$

Solving for dy/dx we obtain

$$\frac{dy}{dx} = \frac{27x - 3x(x^2 + y^2)^2}{3y(x^2 + y^2)^2} = \frac{x(9 - (x^2 + y^2)^2)}{y(x^2 + y^2)^2}.$$

Thus, the derivative is zero when $x^2 + y^2 = 3$. Substituting into the equation for the curve, this yields $x^2 = 1$, or $x = \pm 1$. There are therefore four points at which the derivative is zero:

$$(-1, -\sqrt{2}),\ (-1, \sqrt{2}),\ (1, -\sqrt{2}),\ (1, \sqrt{2}).$$

There are also critical points where the derivative does not exist. This occurs when $y = 0$ and gives the following points with vertical tangents:

$$(0, 0),\ (\pm\sqrt[4]{27}, 0).$$

(b) The curve $27x^2 = (x^2 + y^2)^3$ and its horizontal tangents are plotted below.

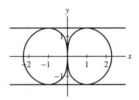

69. Sketch the graph of a continuous function on $(0, 4)$ having a local minimum but no absolute minimum.

SOLUTION   Here is the graph of a function f on $(0, 4)$ with a local minimum value [between $x = 2$ and $x = 4$] but no absolute minimum [since $f(x) \to -\infty$ as $x \to 0+$].

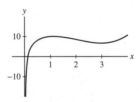

71. Sketch the graph of a function $f(x)$ on $[0, 4]$ with a discontinuity such that $f(x)$ has an absolute minimum but no absolute maximum.

SOLUTION   Here is the graph of a function f on $[0, 4]$ that (a) has a discontinuity [at $x = 4$] and (b) has an absolute minimum [at $x = 0$] but no absolute maximum [since $f(x) \to \infty$ as $x \to 4-$].

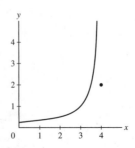

Further Insights and Challenges

73. Show, by considering its minimum, that $f(x) = x^2 - 2x + 3$ takes on only positive values. More generally, find the conditions on r and s under which the quadratic function $f(x) = x^2 + rx + s$ takes on only positive values. Give examples of r and s for which f takes on both positive and negative values.

SOLUTION

- Observe that $f(x) = x^2 - 2x + 3 = (x-1)^2 + 2 > 0$ for all x. Let $f(x) = x^2 + rx + s$. Completing the square, we note that $f(x) = (x + \frac{1}{2}r)^2 + s - \frac{1}{4}r^2 > 0$ for all x provided that $s > \frac{1}{4}r^2$.

- Let $f(x) = x^2 - 4x + 3 = (x-1)(x-3)$. Then f takes on both positive and negative values. Here, $r = -4$ and $s = 3$.

75. Generalize Exercise 74: Show that if the horizontal line $y = c$ intersects the graph of $f(x) = x^2 + rx + s$ at two points $(x_1, f(x_1))$ and $(x_2, f(x_2))$, then $f(x)$ takes its minimum value at the midpoint $M = \dfrac{x_1 + x_2}{2}$ (Figure 21).

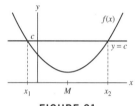

FIGURE 21

SOLUTION Suppose that a horizontal line $y = c$ intersects the graph of a quadratic function $f(x) = x^2 + rx + s$ in two points $(x_1, f(x_1))$ and $(x_2, f(x_2))$. Then of course $f(x_1) = f(x_2) = c$. Let $g(x) = f(x) - c$. Then $g(x_1) = g(x_2) = 0$. By Exercise 74, g takes on its minimum value at $x = \frac{1}{2}(x_1 + x_2)$. Hence so does $f(x) = g(x) + c$.

77. Find the minimum and maximum values of $f(x) = x^p(1-x)^q$ on $[0, 1]$, where p and q are positive numbers.

SOLUTION Let $f(x) = x^p(1-x)^q$, $0 \le x \le 1$, where p and q are positive numbers. Then

$$f'(x) = x^p q(1-x)^{q-1}(-1) + (1-x)^q p x^{p-1}$$

$$= x^{p-1}(1-x)^{q-1}(p(1-x) - qx) = 0 \quad \text{at} \quad x = 0, 1, \frac{p}{p+q}$$

The minimum value of f on $[0, 1]$ is $f(0) = f(1) = 0$, whereas its maximum value is

$$f\left(\frac{p}{p+q}\right) = \frac{p^p q^q}{(p+q)^{p+q}}.$$

4.3 The Mean Value Theorem and Monotonicity

Preliminary Questions

1. Which value of m makes the following statement correct? If $f(2) = 3$ and $f(4) = 9$, where $f(x)$ is differentiable, then the graph of f has a tangent line of slope m.

SOLUTION The Mean Value Theorem guarantees that the function must have a tangent line with slope equal to

$$\frac{f(4) - f(2)}{4 - 2} = \frac{9 - 3}{4 - 2} = 3.$$

Hence, $m = 3$ makes the statement correct.

2. Which of the following conclusions does *not* follow from the MVT (assume that f is differentiable)?

(a) If f has a secant line of slope 0, then $f'(c) = 0$ for some value of c.

(b) If $f(5) < f(9)$, then $f'(c) > 0$ for some $c \in (5, 9)$.

(c) If $f'(c) = 0$ for some value of c, then there is a secant line whose slope is 0.

(d) If $f'(x) > 0$ for all x, then every secant line has positive slope.

SOLUTION Conclusion **(c)** does not follow from the Mean Value Theorem. As a counterexample, consider the function $f(x) = x^3$. Note that $f'(0) = 0$, but no secant line has zero slope.

3. Can a function that takes on only negative values have a positive derivative? Sketch an example or explain why no such functions exist.

SOLUTION   Yes. The figure below displays a function that takes on only negative values but has a positive derivative.

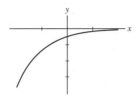

4. (a) Use the graph of $f'(x)$ in Figure 10 to determine whether $f(c)$ is a local minimum or maximum.
(b) Can you conclude from Figure 10 that $f(x)$ is a decreasing function?

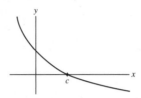

FIGURE 10 Graph of derivative $f'(x)$.

SOLUTION

(a) To the left of $x = c$, the derivative is positive, so f is increasing; to the right of $x = c$, the derivative is negative, so f is decreasing. Consequently, $f(c)$ must be a local maximum.

(b) No. The derivative is a decreasing function, but as noted in part (a), $f(x)$ is increasing for $x < c$ and decreasing for $x > c$.

Exercises

In Exercises 1–6, find a point c satisfying the conclusion of the MVT for the given function and interval.

1. $y = x^{-1}$, $[1, 4]$

SOLUTION   Let $f(x) = x^{-1}$, $a = 1$, $b = 4$. Then $f'(x) = -x^{-2}$, and by the MVT, there exists a $c \in (1, 4)$ such that

$$-\frac{1}{c^2} = f'(c) = \frac{f(b) - f(a)}{b - a} = \frac{\frac{1}{4} - \frac{1}{1}}{4 - 1} = -\frac{1}{4}.$$

Thus $c^2 = 4$ and $c = \pm 2$. Choose $c = 2 \in (1, 4)$.

3. $y = (x - 1)(x - 3)$, $[1, 3]$

SOLUTION   Let $f(x) = (x - 1)(x - 3)$, $a = 1$, $b = 3$. Then $f'(x) = 2x - 4$, and by the MVT, there exists a $c \in (1, 3)$ such that

$$2c - 4 = f'(c) = \frac{f(b) - f(a)}{b - a} = \frac{0 - 0}{3 - 1} = 0.$$

Thus $2c - 4 = 0$ and $c = 2 \in (1, 3)$.

5. $y = \dfrac{x}{x + 1}$, $[3, 6]$

SOLUTION   Let $f(x) = x/(x + 1)$, $a = 3$, $b = 6$. Then $f'(x) = \frac{1}{(x+1)^2}$, and by the MVT, there exists a $c \in (3, 6)$ such that

$$\frac{1}{(c + 1)^2} = f'(c) = \frac{f(b) - f(a)}{b - a} = \frac{\frac{6}{7} - \frac{3}{4}}{6 - 3} = \frac{1}{28}.$$

Thus $(c + 1)^2 = 28$ and $c = -1 \pm 2\sqrt{7}$. Choose $c = 2\sqrt{7} - 1 \approx 4.29 \in (3, 6)$.

7. ⬛GU   Let $f(x) = x^5 + x^2$. Check that the secant line between $x = 0$ and $x = 1$ has slope 2. By the MVT, $f'(c) = 2$ for some $c \in (0, 1)$. Estimate c graphically as follows. Plot $f(x)$ and the secant line on the same axes. Then plot the lines $y = 2x + b$ for different values of b until you find a value of b for which it is tangent to $y = f(x)$. Zoom in on the point of tangency to find its x-coordinate.

SOLUTION Let $f(x) = x^5 + x^2$. The slope of the secant line between $x = 0$ and $x = 1$ is

$$\frac{f(1) - f(0)}{1 - 0} = \frac{2 - 0}{1} = 2.$$

A plot of $f(x)$, the secant line between $x = 0$ and $x = 1$, and the line $y = 2x - 0.764$ is shown below at the left. The line $y = 2x - 0.764$ appears to be tangent to the graph of $y = f(x)$. Zooming in on the point of tangency (see below at the right), it appears that the x-coordinate of the point of tangency is approximately 0.62.

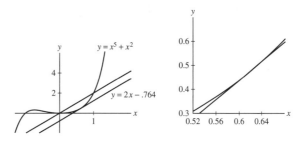

9. Determine the intervals on which $f(x)$ is increasing or decreasing, assuming that Figure 11 is the graph of the derivative $f'(x)$.

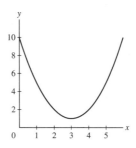

FIGURE 11

SOLUTION $f(x)$ is increasing on every interval (a, b) over which $f'(x) > 0$, and is decreasing on every interval over which $f'(x) < 0$. If the graph of $f'(x)$ is given in Figure 11, then $f(x)$ is increasing on the intervals $(0, 2)$ and $(4, 6)$, and is decreasing on the interval $(2, 4)$.

In Exercises 11–14, sketch the graph of a function $f(x)$ whose derivative $f'(x)$ has the given description.

11. $f'(x) > 0$ for $x > 3$ and $f'(x) < 0$ for $x < 3$.

SOLUTION Here is the graph of a function f for which $f'(x) > 0$ for $x > 3$ and $f'(x) < 0$ for $x < 3$.

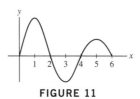

13. $f'(x)$ is negative on $(1, 3)$ and positive everywhere else.

SOLUTION Here is the graph of a function f for which $f'(x)$ is negative on $(1, 3)$ and positive elsewhere.

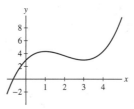

In Exercises 15–18, use the First Derivative Test to determine whether the function attains a local minimum or local maximum (or neither) at the given critical point.

15. $y = 7 + 4x - x^2$, $c = 2$

SOLUTION   Let $f(x) = 7 + 4x - x^2$. Then $f'(c) = 4 - 2c = 0$ implies $c = 2$ is a critical point of f. Since f' makes the sign transition $+, -$ as x increases through $c = 2$, we conclude that $f(2) = 11$ is a local maximum of f.

17. $y = \dfrac{x^2}{x+1}, \quad c = 0$

SOLUTION   Let $f(x) = \frac{x^2}{x+1}$. Then

$$f'(c) = \frac{2c(c+1) - c^2}{(c+1)^2} = \frac{c(c+2)}{(c+1)^2} = 0 \quad \text{at } c = 0.$$

Since f' makes the sign transition $-, +$ as x increases through $c = 0$, we conclude that $f(0) = 0$ is a local minimum of f.

19. Assuming that Figure 11 is the graph of the derivative $f'(x)$, state whether $f(2)$ and $f(4)$ are local minima or maxima.

SOLUTION

- $f'(x)$ makes a transition from positive to negative at $x = 2$, so $f(2)$ is a local maximum.
- $f'(x)$ makes a transition from negative to positive at $x = 4$, so $f(4)$ is a local minimum.

In Exercises 21–40, find the critical points and the intervals on which the function is increasing or decreasing, and apply the First Derivative Test to each critical point.

SOLUTION   *Here is a table legend for Exercises 21–40.*

SYMBOL	MEANING
$-$	The entity is negative on the given interval.
0	The entity is zero at the specified point.
$+$	The entity is positive on the given interval.
U	The entity is undefined at the specified point.
$\nearrow$	f is increasing on the given interval.
$\searrow$	f is decreasing on the given interval.
M	f has a local maximum at the specified point.
m	f has a local minimum at the specified point.
$\neg$	There is no local extremum here.

21. $y = -x^2 + 7x - 17$

SOLUTION   Let $f(x) = -x^2 + 7x - 17$. Then $f'(x) = 7 - 2x = 0$ yields the critical point $c = \frac{7}{2}$.

x	$\left(-\infty, \frac{7}{2}\right)$	$7/2$	$\left(\frac{7}{2}, \infty\right)$
f'	$+$	0	$-$
f	$\nearrow$	M	$\searrow$

23. $y = x^3 - 6x^2$

SOLUTION   Let $f(x) = x^3 - 6x^2$. Then $f'(x) = 3x^2 - 12x = 3x(x - 4) = 0$ yields critical points $c = 0, 4$.

x	$(-\infty, 0)$	0	$(0, 4)$	4	$(4, \infty)$
f'	$+$	0	$-$	0	$+$
f	$\nearrow$	M	$\searrow$	m	$\nearrow$

25. $y = 3x^4 + 8x^3 - 6x^2 - 24x$

SOLUTION Let $f(x) = 3x^4 + 8x^3 - 6x^2 - 24x$. Then

$$f'(x) = 12x^3 + 24x^2 - 12x - 24$$
$$= 12x^2(x + 2) - 12(x + 2) = 12(x + 2)(x^2 - 1)$$
$$= 12(x - 1)(x + 1)(x + 2) = 0$$

yields critical points $c = -2, -1, 1$.

x	$(-\infty, -2)$	-2	$(-2, -1)$	-1	$(-1, 1)$	1	$(1, \infty)$
f'	$-$	0	$+$	0	$-$	0	$+$
f	$\searrow$	m	$\nearrow$	M	$\searrow$	m	$\nearrow$

27. $y = \frac{1}{3}x^3 + \frac{3}{2}x^2 + 2x + 4$

SOLUTION Let $f(x) = \frac{1}{3}x^3 + \frac{3}{2}x^2 + 2x + 4$. Then $f'(x) = x^2 + 3x + 2 = (x + 1)(x + 2) = 0$ yields critical points $c = -2, -1$.

x	$(-\infty, -2)$	-2	$(-2, -1)$	-1	$(-1, \infty)$
f'	$+$	0	$-$	0	$+$
f	$\nearrow$	M	$\searrow$	m	$\nearrow$

29. $y = x^4 + x^3$

SOLUTION Let $f(x) = x^4 + x^3$. Then $f'(x) = 4x^3 + 3x^2 = x^2(4x + 3)$ yields critical points $c = 0, -\frac{3}{4}$.

x	$\left(-\infty, -\frac{3}{4}\right)$	$-\frac{3}{4}$	$\left(-\frac{3}{4}, 0\right)$	0	$(0, \infty)$
f'	$-$	0	$+$	0	$+$
f	$\searrow$	m	$\nearrow$	$\neg$	$\nearrow$

31. $y = \dfrac{1}{x^2 + 1}$

SOLUTION Let $f(x) = \left(x^2 + 1\right)^{-1}$. Then $f'(x) = -2x\left(x^2 + 1\right)^{-2} = 0$ yields critical point $c = 0$.

x	$(-\infty, 0)$	0	$(0, \infty)$
f'	$+$	0	$-$
f	$\nearrow$	M	$\searrow$

33. $y = x + x^{-1}$ $(x > 0)$

SOLUTION Let $f(x) = x + x^{-1}$ for $x > 0$. Then $f'(x) = 1 - x^{-2} = 0$ yields the critical point $c = 1$. (*Note: c = -1* is not in the interval under consideration.)

x	$(0, 1)$	1	$(1, \infty)$
f'	$-$	0	$+$
f	$\searrow$	m	$\nearrow$

35. $y = x^{5/2} - x^2$ $(x > 0)$

SOLUTION Let $f(x) = x^{5/2} - x^2$. Then $f'(x) = \frac{5}{2}x^{3/2} - 2x = x(\frac{5}{2}x^{1/2} - 2) = 0$, so the critical point is $c = \frac{16}{25}$. (*Note: c = 0* is not in the interval under consideration.)

x	$(0, \frac{16}{25})$	$\frac{16}{25}$	$(\frac{16}{25}, \infty)$
f'	$-$	0	$+$
f	$\searrow$	m	$\nearrow$

37. $y = \sin \theta \cos \theta$, $[0, 2\pi]$

SOLUTION   Let $f(\theta) = \sin \theta \cos \theta$. Then $f'(\theta) = -\sin^2 \theta + \cos^2 \theta = 0$ implies that $\tan^2 \theta = 1$. On the interval $[0, 2\pi]$, this yields $c = \frac{\pi}{4}, \frac{3\pi}{4}, \frac{5\pi}{4}, \frac{7\pi}{4}$.

x	$(0, \frac{\pi}{4})$	$\frac{\pi}{4}$	$(\frac{\pi}{4}, \frac{3\pi}{4})$	$\frac{3\pi}{4}$	$(\frac{3\pi}{4}, \frac{5\pi}{4})$	$\frac{5\pi}{4}$	$(\frac{5\pi}{4}, \frac{7\pi}{4})$	$\frac{7\pi}{4}$	$(\frac{7\pi}{4}, 2\pi)$
f'	+	0	−	0	+	0	−	0	+
f	↗	M	↘	m	↗	M	↘	m	↗

39. $y = \theta + \cos \theta$, $[0, 2\pi]$

SOLUTION   Let $f(\theta) = \theta + \cos \theta$. Then $f'(\theta) = 1 - \sin \theta = 0$, which yields $c = \frac{\pi}{2}$ on the interval $[0, 2\pi]$.

x	$(0, \frac{\pi}{2})$	$\frac{\pi}{2}$	$(\frac{\pi}{2}, 2\pi)$
f'	+	0	+
f	↗	¬	↗

41. Show that $f(x) = x^2 + bx + c$ is decreasing on $(-\infty, -\frac{b}{2})$ and increasing on $(-\frac{b}{2}, \infty)$.

SOLUTION   Let $f(x) = x^2 + bx + c$. Then $f'(x) = 2x + b = 0$ yields the critical point $c = -\frac{b}{2}$.

- For $x < -\frac{b}{2}$, we have $f'(x) < 0$, so f is decreasing on $\left(-\infty, -\frac{b}{2}\right)$.
- For $x > -\frac{b}{2}$, we have $f'(x) > 0$, so f is increasing on $\left(-\frac{b}{2}, \infty\right)$.

43. Find conditions on a and b that ensure that $f(x) = x^3 + ax + b$ is increasing on $(-\infty, \infty)$.

SOLUTION   Let $f(x) = x^3 + ax + b$.

- If $a > 0$, then $f'(x) = 3x^2 + a > 0$ and f is increasing for all x.
- If $a = 0$, then

$$f(x_2) - f(x_1) = (3x_2^3 + b) - (3x_1^3 + b) = 3(x_2 - x_1)(x_2^2 + x_2 x_1 + x_1^2) > 0$$

  whenever $x_2 > x_1$. Thus, f is increasing for all x.
- If $a < 0$, then $f'(x) = 3x^2 + a < 0$ and f is decreasing for $|x| < \sqrt{-\frac{a}{3}}$.

In summary, $f(x) = x^3 + ax + b$ is increasing on $(-\infty, \infty)$ whenever $a \geq 0$.

45. Sam made two statements that Deborah found dubious.

(a) "Although the average velocity for my trip was 70 mph, at no point in time did my speedometer read 70 mph."

(b) "Although a policeman clocked me going 70 mph, my speedometer never read 65 mph."

In each case, which theorem did Deborah apply to prove Sam's statement false: the Intermediate Value Theorem or the Mean Value Theorem? Explain.

SOLUTION

(a) Deborah is applying the Mean Value Theorem here. Let $s(t)$ be Sam's distance, in miles, from his starting point, let a be the start time for Sam's trip, and let b be the end time of the same trip. Sam is claiming that at no point was

$$s'(t) = \frac{s(b) - s(a)}{b - a}.$$

This violates the MVT.

(b) Deborah is applying the Intermediate Value Theorem here. Let $v(t)$ be Sam's velocity in miles per hour. Sam started out at rest, and reached a velocity of 70 mph. By the IVT, he should have reached a velocity of 65 mph at some point.

47. Show that $f(x) = 1 - |x|$ satisfies the conclusion of the MVT on $[a, b]$ if both a and b are positive or negative, but not if $a < 0$ and $b > 0$.

SOLUTION   Let $f(x) = 1 - |x|$.

- If a and b (where $a < b$) are both positive (or both negative), then f is continuous on $[a, b]$ and differentiable on (a, b). Accordingly, the hypotheses of the MVT are met and the theorem does apply. Indeed, in these cases, any point $c \in (a, b)$ satisfies the conclusion of the MVT (since f' is constant on $[a, b]$ in these instances).

- For $a = -2$ and $b = 1$, we have $\dfrac{f(b) - f(a)}{b - a} = \dfrac{0 - (-1)}{1 - (-2)} = \dfrac{1}{3}$. Yet there is no point $c \in (-2, 1)$ such that $f'(c) = \frac{1}{3}$. Indeed, $f'(x) = 1$ for $x < 0$, $f'(x) = -1$ for $x > 0$, and $f'(0)$ is undefined. The MVT does not apply in this case, since f is not differentiable on the open interval $(-2, 1)$.

49. Show that if f is a quadratic polynomial, then the midpoint $c = \dfrac{a + b}{2}$ satisfies the conclusion of the MVT on $[a, b]$ for any a and b.

SOLUTION   Let $f(x) = px^2 + qx + r$ with $p \neq 0$ and consider the interval $[a, b]$. Then $f'(x) = 2px + q$, and by the MVT we have

$$2pc + q = f'(c) = \frac{f(b) - f(a)}{b - a} = \frac{\left(pb^2 + qb + r\right) - \left(pa^2 + qa + r\right)}{b - a}$$

$$= \frac{(b - a)\,(p\,(b + a) + q)}{b - a} = p\,(b + a) + q$$

Thus $2pc + q = p(a + b) + q$, and $c = \dfrac{a + b}{2}$.

51. Suppose that $f(2) = -2$ and $f'(x) \geq 5$. Show that $f(4) \geq 8$.

SOLUTION   The MVT, applied to the interval $[2, 4]$, guarantees there exists a $c \in (2, 4)$ such that

$$f'(c) = \frac{f(4) - f(2)}{4 - 2} \qquad \text{or} \qquad f(4) - f(2) = 2f'(c).$$

Because $f'(x) \geq 5$, it follows that $f(4) - f(2) \geq 10$, or $f(4) \geq f(2) + 10 = 8$.

Further Insights and Challenges

53. Prove that if $f(0) = g(0)$ and $f'(x) \leq g'(x)$ for $x \geq 0$, then $f(x) \leq g(x)$ for all $x \geq 0$. *Hint:* Show that $f(x) - g(x)$ is nonincreasing.

SOLUTION   Let $h(x) = f(x) - g(x)$. By the sum rule, $h'(x) = f'(x) - g'(x)$. Since $f'(x) \leq g'(x)$ for all $x \geq 0$, $h'(x) \leq 0$ for all $x \geq 0$. This implies that h is nonincreasing. Since $h(0) = f(0) - g(0) = 0$, $h(x) \leq 0$ for all $x \geq 0$ (as h is nonincreasing, it cannot climb above zero). Hence $f(x) - g(x) \leq 0$ for all $x \geq 0$, and so $f(x) \leq g(x)$ for $x \geq 0$.

55. Use Exercises 53 and 54 to establish the following assertions for all $x \geq 0$ (each assertion follows from the previous one):

(a) $\cos x \geq 1 - \frac{1}{2}x^2$

(b) $\sin x \geq x - \frac{1}{6}x^3$

(c) $\cos x \leq 1 - \frac{1}{2}x^2 + \frac{1}{24}x^4$

(d) Can you guess the next inequality in the series?

SOLUTION

(a) We prove this using Exercise 53: Let $g(x) = \cos x$ and $f(x) = 1 - \frac{1}{2}x^2$. Then $f(0) = g(0) = 1$ and $g'(x) = -\sin x \geq -x = f'(x)$ for $x \geq 0$ by Exercise 54. Now apply Exercise 53 to conclude that $\cos x \geq 1 - \frac{1}{2}x^2$ for $x \geq 0$.

(b) Let $g(x) = \sin x$ and $f(x) = x - \frac{1}{6}x^3$. Then $f(0) = g(0) = 0$ and $g'(x) = \cos x \geq 1 - \frac{1}{2}x^2 = f'(x)$ for $x \geq 0$ by part (a). Now apply Exercise 53 to conclude that $\sin x \geq x - \frac{1}{6}x^3$ for $x \geq 0$.

(c) Let $g(x) = 1 - \frac{1}{2}x^2 + \frac{1}{24}x^4$ and $f(x) = \cos x$. Then $f(0) = g(0) = 1$ and $g'(x) = -x + \frac{1}{6}x^3 \geq -\sin x = f'(x)$ for $x \geq 0$ by part (b). Now apply Exercise 53 to conclude that $\cos x \leq 1 - \frac{1}{2}x^2 + \frac{1}{24}x^4$ for $x \geq 0$.

(d) The next inequality in the series is $\sin x \leq x - \frac{1}{6}x^3 + \frac{1}{120}x^5$, valid for $x \geq 0$. To construct (d) from (c), we note that the derivative of $\sin x$ is $\cos x$, and look for a polynomial (which we currently must do by educated guess) whose derivative is $1 - \frac{1}{2}x^2 + \frac{1}{24}x^4$. We know the derivative of x is 1, and that a term whose derivative is $-\frac{1}{2}x^2$ should be of the form Cx^3. $\frac{d}{dx}Cx^3 = 3Cx^2 = -\frac{1}{2}x^2$, so $C = -\frac{1}{6}$. A term whose derivative is $\frac{1}{24}x^4$ should be of the form Dx^5. From this, $\frac{d}{dx}Dx^5 = 5Dx^4 = \frac{1}{24}x^4$, so that $5D = \frac{1}{24}$, or $D = \frac{1}{120}$.

57. Define $f(x) = x^3 \sin(\frac{1}{x})$ for $x \neq 0$ and $f(0) = 0$.

(a) Show that $f'(x)$ is continuous at $x = 0$ and $x = 0$ is a critical point of f.

(b) GU   Examine the graphs of $f(x)$ and $f'(x)$. Can the First Derivative Test be applied?

(c) Show that $f(0)$ is neither a local min nor max.

SOLUTION

(a) Let $f(x) = x^3 \sin(\frac{1}{x})$. Then

$$f'(x) = 3x^2 \sin\left(\frac{1}{x}\right) + x^3 \cos\left(\frac{1}{x}\right)(-x^{-2}) = x\left(3x \sin\left(\frac{1}{x}\right) - \cos\left(\frac{1}{x}\right)\right).$$

This formula is not defined at $x = 0$, but its limit is. Since $-1 \le \sin x \le 1$ and $-1 \le \cos x \le 1$ for all x,

$$|f'(x)| = |x|\left|3x \sin\left(\frac{1}{x}\right) - \cos\left(\frac{1}{x}\right)\right| \le |x|\left(\left|3x \sin\left(\frac{1}{x}\right)\right| + \left|\cos\left(\frac{1}{x}\right)\right|\right) \le |x|(3|x| + 1)$$

so, by the Squeeze Theorem, $\lim_{x \to 0} |f'(x)| = 0$. But does $f'(0) = 0$? We check using the limit definition of the derivative:

$$f'(0) = \lim_{x \to 0} \frac{f(x) - f(0)}{x - 0} = \lim_{x \to 0} x^2 \sin\left(\frac{1}{x}\right) = 0.$$

Thus $f'(x)$ is continuous at $x = 0$, and $x = 0$ is a critical point of f.

(b) The figure below at the left shows $f(x)$, and the figure below at the right shows $f'(x)$. Note how the two functions oscillate near $x = 0$, which implies that the First Derivative Test cannot be applied.

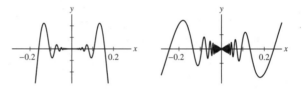

(c) As x approaches 0 from either direction, $f(x)$ alternates between positive and negative arbitrarily close to $x = 0$. This means that $f(0)$ cannot be a local minimum (since $f(x)$ gets lower than $f(0)$ arbitrarily close to 0), nor can $f(0)$ be a local maximum (since $f(x)$ takes values higher than $f(0)$ arbitrarily close to $x = 0$). Therefore $f(0)$ is neither a local minimum nor a local maximum of f.

4.4 The Shape of a Graph

Preliminary Questions

1. Choose the correct answer: If f is concave up, then f' is:

(a) increasing   **(b)** decreasing

SOLUTION The correct response is **(a)**: increasing. If the function is concave up, then f'' is positive. Since f'' is the derivative of f', it follows that the derivative of f' is positive and f' must therefore be increasing.

2. If x_0 is a critical point and f is concave down, then $f(x_0)$ is a local:

(a) min   **(b)** max   **(c)** undetermined

SOLUTION By the Second Derivative Test, the correct response is **(b)**: maximum.

In Questions 3–8, state whether true or false and explain. Assume that $f''(x)$ exists for all x.

3. If $f'(c) = 0$ and $f''(c) < 0$, then $f(c)$ is a local minimum.

SOLUTION False. By the Second Derivative Test, the correct conclusion would be that $f(c)$ is a local maximum.

4. A function that is concave down on $(-\infty, \infty)$ can have no minimum value.

SOLUTION True.

5. If $f''(c) = 0$, then f must have a point of inflection at $x = c$.

SOLUTION False. f has an inflection point at $x = c$ provided concavity changes at $x = c$. Points where $f''(c) = 0$ are simply candidates for inflection points. The function $f(x) = x^4$ provides a counterexample. Here, $f''(x) = 12x^2$. Thus, $f''(0) = 0$ but $f''(x)$ does not change sign at $x = 0$. Therefore, $f(x) = x^4$ does not have an inflection point at $x = 0$.

6. If f has a point of inflection at $x = c$, then $f''(c) = 0$.

SOLUTION True. In general, if f has an inflection point at $x = c$, then either $f''(c) = 0$ or $f''(c)$ does not exist. Because we are assuming that $f''(x)$ exists for all x, we must have $f''(c) = 0$.

7. If f is concave up and f' changes sign at $x = c$, then f' changes sign from negative to positive at $x = c$.

SOLUTION True. If f is concave up, then f'' is positive and f' is increasing; therefore, f' must go from negative to positive at $x = c$.

8. If $f(c)$ is a local maximum, then $f''(c)$ must be negative.

SOLUTION True. In general, it could be that either $f''(c)$ does not exist or $f''(c)$ is negative. Because we are assuming that $f''(x)$ exists for all x, we must have $f''(c) < 0$.

9. Suppose that $f''(c) = 0$ and $f''(x)$ changes sign from $+$ to $-$ at $x = c$. Which of the following statements are correct?

(a) $f(x)$ has a local maximum at $x = a$.

(b) $f'(x)$ has a local minimum at $x = a$.

(c) $f'(x)$ has a local maximum at $x = a$.

(d) $f(x)$ has a point of inflection at $x = a$.

SOLUTION Statements **(c)** and **(d)** are correct. Because $f''(x)$ goes from positive to negative, it follows that $f'(x)$ goes from increasing to decreasing and so must have a local maximum at $x = c$. Also, because $f''(c) = 0$ and $f''(x)$ changes sign at $x = c$, it follows that f has a point of inflection at $x = c$.

Exercises

1. Match the graphs in Figure 12 with the description:

(a) $f''(x) < 0$ for all x.

(b) $f''(x)$ goes from $+$ to $-$.

(c) $f''(x) > 0$ for all x.

(d) $f''(x)$ goes from $-$ to $+$.

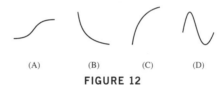

(A)   (B)   (C)   (D)

FIGURE 12

SOLUTION

(a) In C, we have $f''(x) < 0$ for all x.

(b) In A, $f''(x)$ goes from $+$ to $-$.

(c) In B, we have $f''(x) > 0$ for all x.

(d) In D, $f''(x)$ goes from $-$ to $+$.

3. Sketch the graph of an increasing function such that $f''(x)$ changes from $+$ to $-$ at $x = 2$ and from $-$ to $+$ at $x = 4$. Do the same for a decreasing function.

SOLUTION The graph shown below at the left is an increasing function which changes from concave up to concave down at $x = 2$ and from concave down to concave up at $x = 4$. The graph shown below at the right is a decreasing function which changes from concave up to concave down at $x = 2$ and from concave down to concave up at $x = 4$.

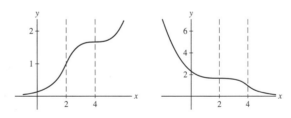

5. If Figure 14 is the graph of the *derivative* $f'(x)$, where do the points of inflection of $f(x)$ occur, and on which interval is $f(x)$ concave down?

SOLUTION Points of inflection occur when $f''(x)$ changes sign. Consequently, points of inflection occur when $f'(x)$ changes from increasing to decreasing or from decreasing to increasing. In Figure 14, this occurs at $x = b$ and at $x = e$; therefore, $f(x)$ has an inflection point at $x = b$ and another at $x = e$. The function $f(x)$ will be concave down when $f''(x) < 0$ or when $f'(x)$ is decreasing. Thus, $f(x)$ is concave down for $b < x < e$.

In Exercises 7–14, determine the intervals on which the function is concave up or down and find the points of inflection.

7. $y = x^2 + 7x + 10$

SOLUTION Let $f(x) = x^2 + 7x + 10$. Then $f'(x) = 2x + 7$ and $f''(x) = 2 > 0$ for all x. Therefore, f is concave up everywhere, and there are no points of inflection.

9. $y = x - 2\cos x$

SOLUTION Let $f(x) = x - 2\cos x$. Then $f'(x) = 1 + 2\sin x$ and $f''(x) = 2\cos x = 0$ at $x = \frac{1}{2}(2n+1)\pi$, where n is an integer. Now, f is concave up on the intervals

$$\left(-\frac{\pi}{2} + 2n\pi, \frac{\pi}{2} + 2n\pi \right)$$

where n is any integer since $f''(x) > 0$ there. Moreover, f is concave down on the intervals

$$\left(\frac{\pi}{2} + 2n\pi, \frac{3\pi}{2} + 2n\pi \right)$$

where n is any integer since $f''(x) < 0$ there. Finally, because $f''(x)$ changes sign at each $x = \frac{1}{2}(2n+1)\pi$, there is a point of inflection at each of these locations.

11. $y = x(x - 8\sqrt{x})$

SOLUTION Let $f(x) = x(x - 8\sqrt{x}) = x^2 - 8x^{3/2}$. Then $f'(x) = 2x - 12x^{1/2}$ and $f''(x) = 2 - 6x^{-1/2}$. Now, f is concave down for $0 < x < 9$ since $f''(x) < 0$ there. Moreover, f is concave up for $x > 9$ since $f''(x) > 0$ there. Finally, because $f''(x)$ changes sign at $x = 9$, $f(x)$ has a point of inflection at $x = 9$.

13. $y = \dfrac{1}{x^2 + 3}$

SOLUTION Let $f(x) = \dfrac{1}{x^2 + 3}$. Then $f'(x) = -\dfrac{2x}{(x^2+3)^2}$ and

$$f''(x) = -\frac{2(x^2+3)^2 - 8x^2(x^2+3)}{(x^2+3)^4} = \frac{6x^2 - 6}{(x^2+3)^3}.$$

Now, f is concave up for $|x| > 1$ since $f''(x) > 0$ there. Moreover, f is concave down for $|x| < 1$ since $f''(x) < 0$ there. Finally, because $f''(x)$ changes sign at both $x = -1$ and $x = 1$, $f(x)$ has a point of inflection at both $x = -1$ and $x = 1$.

15. Sketch the graph of $f(x) = x^4$ and state whether f has any points of inflection. Verify your conclusion by showing that $f''(x)$ does not change sign.

SOLUTION From the plot of $f(x) = x^4$ below, it appears that f has no points of inflection. Indeed, $f'(x) = 4x^3$ and $f''(x) = 12x^2 = 0$ at $x = 0$, but $f''(x)$ does *not* change sign as x increases through 0 because $12x^2 \geq 0$ for all x.

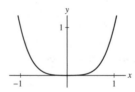

17. 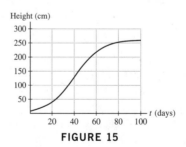 The growth of a sunflower during its first 100 days is modeled well by the *logistic curve* $y = h(t)$ shown in Figure 15. Estimate the growth rate at the point of inflection and explain its significance. Then make a rough sketch of the first and second derivatives of $h(t)$.

Height (cm)

300
250
200
150
100
50

20 40 60 80 100 t (days)

FIGURE 15

SOLUTION The point of inflection in Figure 15 appears to occur at $t = 40$ days. The graph below shows the logistic curve with an approximate tangent line drawn at $t = 40$. The approximate tangent line passes roughly through the points $(20, 20)$ and $(60, 240)$. The growth rate at the point of inflection is thus

$$\frac{240 - 20}{60 - 20} = \frac{220}{40} = 5.5 \text{ cm/day}.$$

Because the logistic curve changes from concave up to concave down at $t = 40$, the growth rate at this point is the maximum growth rate for the sunflower plant.

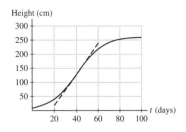

Sketches of the first and second derivative of $h(t)$ are shown below at the left and at the right, respectively.

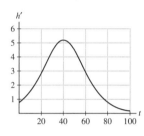
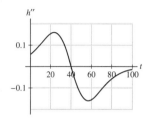

In Exercises 19–28, find the critical points of $f(x)$ and use the Second Derivative Test (if possible) to determine whether each corresponds to a local minimum or maximum.

19. $f(x) = x^3 - 12x^2 + 45x$

SOLUTION Let $f(x) = x^3 - 12x^2 + 45x$. Then $f'(x) = 3x^2 - 24x + 45 = 3(x-3)(x-5)$, and the critical points are $x = 3$ and $x = 5$. Moreover, $f''(x) = 6x - 24$, so $f''(3) = -6 < 0$ and $f''(5) = 6 > 0$. Therefore, by the Second Derivative Test, $f(3) = 54$ is a local maximum, and $f(5) = 50$ is a local minimum.

21. $f(x) = 3x^4 - 8x^3 + 6x^2$

SOLUTION Let $f(x) = 3x^4 - 8x^3 + 6x^2$. Then $f'(x) = 12x^3 - 24x^2 + 12x = 12x(x-1)^2 = 0$ at $x = 0, 1$ and $f''(x) = 36x^2 - 48x + 12$. Thus, $f''(0) > 0$, which implies $f(0)$ is a local minimum; however, $f''(1) = 0$, which is inconclusive.

23. $f(x) = x^5 - x^3$

SOLUTION Let $f(x) = x^5 - x^3$. Then $f'(x) = 5x^4 - 3x^2 = x^2(5x^2 - 3) = 0$ at $x = 0$, $x = \pm\sqrt{\frac{3}{5}}$ and $f''(x) = 20x^3 - 6x = x(20x^2 - 6)$. Thus, $f''\left(\sqrt{\frac{3}{5}}\right) > 0$, which implies $f\left(\sqrt{\frac{3}{5}}\right)$ is a local minimum, and $f''\left(-\sqrt{\frac{3}{5}}\right) < 0$, which implies that $f\left(-\sqrt{\frac{3}{5}}\right)$ is a local maximum; however, $f''(0) = 0$, which is inconclusive.

25. $f(x) = \dfrac{1}{\cos x + 2}$

SOLUTION Let $f(x) = \dfrac{1}{\cos x + 2}$. Then $f'(x) = \dfrac{\sin x}{(\cos x + 2)^2} = 0$ at $x = 0, \pm\pi, \pm2\pi, \pm3\pi, \ldots$, and

$$f''(x) = \frac{\cos^2 x + 2\cos x + 2\sin^2 x}{(\cos x + 2)^3}.$$

Thus, $f''(0) > 0$, $f''(\pm2\pi) > 0$ and $f''(\pm4\pi) > 0$, which implies that $f(x)$ has local minima when x is an even multiple of π. Similarly, $f''(\pm\pi) < 0$, $f''(\pm3\pi) < 0$, and $f''(\pm5\pi) < 0$, which implies that $f(x)$ has local maxima when x is an odd multiple of π.

27. $f(x) = 3x^{3/2} - x^{1/2}$

SOLUTION Let $f(x) = 3x^{3/2} - x^{1/2}$. Then $f'(x) = \frac{9}{2}x^{1/2} - \frac{1}{2}x^{-1/2} = \frac{1}{2}x^{-1/2}(9x - 1)$, so there are two critical points: $x = 0$ and $x = \frac{1}{9}$. Now,

$$f''(x) = \frac{9}{4}x^{-1/2} + \frac{1}{4}x^{-3/2} = \frac{1}{4}x^{-3/2}(9x + 1).$$

Thus, $f''\left(\frac{1}{9}\right) > 0$, which implies $f\left(\frac{1}{9}\right)$ is a local minimum. $f''(x)$ is undefined at $x = 0$, so the Second Derivative Test cannot be applied there.

In Exercises 29–40, find the intervals on which f is concave up or down, the points of inflection, and the critical points, and determine whether each critical point corresponds to a local minimum or maximum (or neither).

SOLUTION Here is a table legend for Exercises 29–40.

SYMBOL	MEANING
−	The entity is negative on the given interval.
0	The entity is zero at the specified point.
+	The entity is positive on the given interval.
U	The entity is undefined at the specified point.
↗	The function (f, g, etc.) is increasing on the given interval.
↘	The function (f, g, etc.) is decreasing on the given interval.
⌣	The function (f, g, etc.) is concave up on the given interval.
⌢	The function (f, g, etc.) is concave down on the given interval.
M	The function (f, g, etc.) has a local maximum at the specified point.
m	The function (f, g, etc.) has a local minimum at the specified point.
I	The function (f, g, etc.) has an inflection point here.
¬	There is no local extremum or inflection point here.

29. $f(x) = x^3 - 2x^2 + x$

SOLUTION Let $f(x) = x^3 - 2x^2 + x$.

- Then $f'(x) = 3x^2 - 4x + 1 = (x - 1)(3x - 1) = 0$ yields $x = 1$ and $x = \frac{1}{3}$ as candidates for extrema.
- Moreover, $f''(x) = 6x - 4 = 0$ gives a candidate for a point of inflection at $x = \frac{2}{3}$.

x	$(-\infty, \frac{1}{3})$	$\frac{1}{3}$	$(\frac{1}{3}, 1)$	1	$(1, \infty)$
f'	+	0	−	0	+
f	↗	M	↘	m	↗

x	$(-\infty, \frac{2}{3})$	$\frac{2}{3}$	$(\frac{2}{3}, \infty)$
f''	−	0	+
f	⌢	I	⌣

31. $f(t) = t^2 - t^3$

SOLUTION Let $f(t) = t^2 - t^3$.

- Then $f'(t) = 2t - 3t^2 = t(2 - 3t) = 0$ yields $t = 0$ and $t = \frac{2}{3}$ as candidates for extrema.
- Moreover, $f''(t) = 2 - 6t = 0$ gives a candidate for a point of inflection at $t = \frac{1}{3}$.

t	$(-\infty, 0)$	0	$(0, \frac{2}{3})$	$\frac{2}{3}$	$(\frac{2}{3}, \infty)$
f'	−	0	+	0	−
f	↘	m	↗	M	↘

t	$(-\infty, \frac{1}{3})$	$\frac{1}{3}$	$(\frac{1}{3}, \infty)$
f''	+	0	−
f	⌣	I	⌢

33. $f(x) = x^2 - x^{1/2}$

SOLUTION Let $f(x) = x^2 - x^{1/2}$. Note that the domain of f is $x \geq 0$.

- Then $f'(x) = 2x - \frac{1}{2}x^{-1/2} = \frac{1}{2}x^{-1/2}\left(4x^{3/2} - 1\right) = 0$ yields $x = 0$ and $x = \left(\frac{1}{4}\right)^{2/3}$ as candidates for extrema.
- Moreover, $f''(x) = 2 + \frac{1}{4}x^{-3/2} > 0$ for all $x \geq 0$, which means there are no inflection points.

x	0	$\left(0, \left(\frac{1}{4}\right)^{2/3}\right)$	$\left(\frac{1}{4}\right)^{2/3}$	$\left(\left(\frac{1}{4}\right)^{2/3}, \infty\right)$
f'	U	−	0	+
f	M	↘	m	↗

35. $f(t) = \dfrac{1}{t^2 + 1}$

SOLUTION Let $f(t) = \dfrac{1}{t^2 + 1}$.

- Then $f'(t) = -\dfrac{2t}{\left(t^2 + 1\right)^2} = 0$ yields $t = 0$ as a candidate for an extremum.

- Moreover, $f''(t) = -\dfrac{\left(t^2 + 1\right)^2 (2) - 2t \cdot 2\left(t^2 + 1\right)(2t)}{\left(t^2 + 1\right)^4} = \dfrac{8t^2 - 2t^2 - 2}{\left(t^2 + 1\right)^3} = \dfrac{2\left(3t^2 - 1\right)}{\left(t^2 + 1\right)^3} = 0$ gives candidates

  for a point of inflection at $t = \pm\sqrt{\frac{1}{3}}$.

t	$(-\infty, 0)$	0	$(0, \infty)$
f'	$+$	0	$-$
f	↗	M	↘

t	$\left(-\infty, -\sqrt{\frac{1}{3}}\right)$	$-\sqrt{\frac{1}{3}}$	$\left(-\sqrt{\frac{1}{3}}, \sqrt{\frac{1}{3}}\right)$	$\sqrt{\frac{1}{3}}$	$\left(\sqrt{\frac{1}{3}}, \infty\right)$
f''	$+$	0	$-$	0	$+$
f	⌣	I	⌢	I	⌣

37. $f(\theta) = \theta + \sin\theta$ for $0 \le \theta \le 2\pi$

SOLUTION Let $f(\theta) = \theta + \sin\theta$ on $[0, 2\pi]$.

- Then $f'(\theta) = 1 + \cos\theta = 0$ yields $\theta = \pi$ as a candidate for an extremum.
- Moreover, $f''(\theta) = -\sin\theta = 0$ gives candidates for a point of inflection at $\theta = 0$, at $\theta = \pi$, and at $\theta = 2\pi$.

θ	$(0, \pi)$	π	$(\pi, 2\pi)$
f'	$+$	0	$+$
f	↗	¬	↗

θ	0	$(0, \pi)$	π	$(\pi, 2\pi)$	2π
f''	0	$-$	0	$+$	0
f	¬	⌢	I	⌣	¬

39. $f(x) = x - \sin x$ for $0 \le x \le 2\pi$

SOLUTION Let $f(x) = x - \sin x$ on $[0, 2\pi]$.

- Then $f'(x) = 1 - \cos x > 0$ on $(0, 2\pi)$.
- Moreover, $f''(x) = \sin x = 0$ gives a candidate for a point of inflection at $x = \pi$.

x	$(0, 2\pi)$
f'	$+$
f	↗

x	$(0, \pi)$	π	$(\pi, 2\pi)$
f''	$+$	0	$-$
f	⌣	I	⌢

41. An infectious flu spreads slowly at the beginning of an epidemic. The infection process accelerates until a majority of the susceptible individuals are infected, at which point the process slows down.

(a) If $R(t)$ is the number of individuals infected at time t, describe the concavity of the graph of R near the beginning and end of the epidemic.

(b) Write a one-sentence news bulletin describing the status of the epidemic on the day that $R(t)$ has a point of inflection.

SOLUTION

(a) Near the beginning of the epidemic, the graph of R is concave up. Near the epidemic's end, R is concave down.

(b) "Epidemic subsiding: number of new cases declining."

43. Water is pumped into a sphere at a variable rate in such a way that the water level rises at a constant rate c (Figure 17). Let $V(t)$ be the volume of water at time t. Sketch the graph of $V(t)$ (approximately, but with the correct concavity). Where does the point of inflection occur?

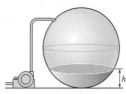

FIGURE 17

SOLUTION Because water is entering the sphere in such a way that the water level rises at a constant rate, we expect the volume to increase more slowly near the bottom and top of the sphere where the sphere is not as "wide" and to increase more rapidly near the middle of the sphere. The graph of $V(t)$ should therefore start concave up and change to concave down when the sphere is half full; that is, the point of inflection should occur when the water level is equal to the radius of the sphere. A possible graph of $V(t)$ is shown below.

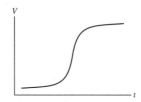

In Exercises 45–47, sketch the graph of a function $f(x)$ satisfying all of the given conditions.

45. $f'(x) > 0$ and $f''(x) < 0$ for all x.

SOLUTION Here is the graph of a function $f(x)$ satisfying $f'(x) > 0$ for all x and $f''(x) < 0$ for all x.

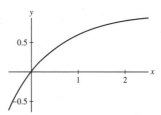

47. **(i)** $f'(x) < 0$ for $x < 0$ and $f'(x) > 0$ for $x > 0$, and
(ii) $f''(x) < 0$ for $|x| > 2$, and $f''(x) > 0$ for $|x| < 2$.

SOLUTION

Interval	$(-\infty, -2)$	$(-2, 0)$	$(0, 2)$	$(2, \infty)$
Direction	↘	↘	↗	↗
Concavity	⌢	⌣	⌣	⌢

One potential graph with this shape is the following:

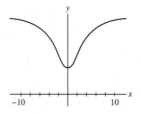

Further Insights and Challenges

In Exercises 48–50, assume that $f(x)$ is differentiable.

49. 📖 Assume that $f''(x)$ exists and $f''(x) > 0$. Prove that the graph of $f(x)$ "sits above" its tangent lines as follows.

(a) For any c, set $G(x) = f(x) - f'(c)(x - c) - f(c)$. It is sufficient to prove that $G(x) \geq 0$ for all c. Explain why with a sketch.

(b) Show that $G(c) = G'(c) = 0$ and $G''(x) > 0$ for all x. Use this to conclude that $G'(x) < 0$ for $x < c$ and $G'(x) > 0$ for $x > c$. Then deduce, using the MVT, that $G(x) > G(c)$ for $x \neq c$.

SOLUTION

(a) Let c be any number. Then $y = f'(c)(x - c) + f(c)$ is the equation of the line tangent to the graph of $f(x)$ at $x = c$ and $G(x) = f(x) - f'(c)(x - c) - f(c)$ measures the amount by which the value of the function exceeds the value of the tangent line (see the figure below). Thus, to prove that the graph of $f(x)$ "sits above" its tangent lines, it is sufficient to prove that $G(x) \geq 0$ for all c.

(b) Note that $G(c) = f(c) - f'(c)(c - c) - f(c) = 0$, $G'(x) = f'(x) - f'(c)$ and $G'(c) = f'(c) - f'(c) = 0$. Moreover, $G''(x) = f''(x) > 0$ for all x. Now, because $G'(c) = 0$ and $G'(x)$ is increasing, it must be true that $G'(x) < 0$ for $x < c$ and that $G'(x) > 0$ for $x > c$. Therefore, $G(x)$ is decreasing for $x < c$ and increasing for $x > c$. This implies that $G(c) = 0$ is a minimum; consequently $G(x) > G(c) = 0$ for $x \neq c$.

51. Let $C(x)$ be the cost of producing x units of a certain good. Assume that the graph of $C(x)$ is concave up.

(a) Show that the average cost $A(x) = C(x)/x$ is minimized at that production level x_0 for which average cost equals marginal cost.

(b) Show that the line through $(0, 0)$ and $(x_0, C(x_0))$ is tangent to the graph of $C(x)$.

SOLUTION Let $C(x)$ be the cost of producing x units of a commodity. Assume the graph of C is concave up.

(a) Let $A(x) = C(x)/x$ be the average cost and let x_0 be the production level at which average cost is minimized. Then $A'(x_0) = \dfrac{x_0 C'(x_0) - C(x_0)}{x_0^2} = 0$ implies $x_0 C'(x_0) - C(x_0) = 0$, whence $C'(x_0) = C(x_0)/x_0 = A(x_0)$. In other words, $A(x_0) = C'(x_0)$ or average cost equals marginal cost at production level x_0. To confirm that x_0 corresponds to a local minimum of A, we use the Second Derivative Test. We find

$$A''(x_0) = \frac{x_0^2 C''(x_0) - 2(x_0 C'(x_0) - C(x_0))}{x_0^3} = \frac{C''(x_0)}{x_0} > 0$$

because C is concave up. Hence, x_0 corresponds to a local minimum.

(b) The line between $(0, 0)$ and $(x_0, C(x_0))$ is

$$\frac{C(x_0) - 0}{x_0 - 0}(x - x_0) + C(x_0) = \frac{C(x_0)}{x_0}(x - x_0) + C(x_0) = A(x_0)(x - x_0) + C(x_0)$$

$$= C'(x_0)(x - x_0) + C(x_0)$$

which is the tangent line to C at x_0.

53. Critical Points and Inflection Points If $f'(c) = 0$ and $f(c)$ is neither a local min or max, must $x = c$ be a point of inflection? This is true of most "reasonable" examples (including the examples in this text), but it is not true in general. Let

$$f(x) = \begin{cases} x^2 \sin \frac{1}{x} & \text{for } x \neq 0 \\ 0 & \text{for } x = 0 \end{cases}$$

(a) Use the limit definition of the derivative to show that $f'(0)$ exists and $f'(0) = 0$.

(b) Show that $f(0)$ is neither a local min nor max.

(c) Show that $f'(x)$ changes sign infinitely often near $x = 0$ and conclude that $f(x)$ does not have a point of inflection at $x = 0$.

SOLUTION Let $f(x) = \begin{cases} x^2 \sin(1/x) & \text{for } x \neq 0 \\ 0 & \text{for } x = 0 \end{cases}$.

(a) Now $f'(0) = \lim_{x \to 0} \dfrac{f(x) - f(0)}{x - 0} = \lim_{x \to 0} \dfrac{x^2 \sin(1/x)}{x} = \lim_{x \to 0} x \sin\left(\dfrac{1}{x}\right) = 0$ by the Squeeze Theorem: as $x \to 0$ we have

$$\left| x \sin\left(\frac{1}{x}\right) - 0 \right| = |x| \left| \sin\left(\frac{1}{x}\right) \right| \to 0,$$

since $|\sin u| \leq 1$.

(b) Since $\sin(\frac{1}{x})$ oscillates through every value between -1 and 1 with increasing frequency as $x \to 0$, in any open interval $(-\delta, \delta)$ there are points a and b such that $f(a) = a^2 \sin(\frac{1}{a}) < 0$ and $f(b) = b^2 \sin(\frac{1}{b}) > 0$. Accordingly, $f(0) = 0$ can neither be a local minimum value nor a local maximum value of f.

(c) In part (a) it was shown that $f'(0) = 0$. For $x \neq 0$, we have

$$f'(x) = x^2 \cos\left(\frac{1}{x}\right)\left(-\frac{1}{x^2}\right) + 2x \sin\left(\frac{1}{x}\right) = 2x \sin\left(\frac{1}{x}\right) - \cos\left(\frac{1}{x}\right).$$

As $x \to 0$, $f'(x)$ oscillates increasingly rapidly; consequently, $f'(x)$ changes sign infinitely often near $x = 0$. From this we conclude that $f(x)$ does not have a point of inflection at $x = 0$.

4.5 Graph Sketching and Asymptotes

Preliminary Questions

1. Sketch an arc where f' and f'' have the sign combination $++$. Do the same for $-+$.

SOLUTION An arc with the sign combination $++$ (increasing, concave up) is shown below at the left. An arc with the sign combination $-+$ (decreasing, concave up) is shown below at the right.

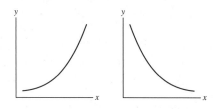

2. If the sign combination of f' and f'' changes from $++$ to $+-$ at $x = c$, then (choose the correct answer):

(a) $f(c)$ is a local min   **(b)** $f(c)$ is a local max

(c) c is a point of inflection

SOLUTION Because the sign of the second derivative changes at $x = c$, the correct response is **(c)**: c is a point of inflection.

3. What are the following limits?

(a) $\lim\limits_{x \to \infty} x^3$ **(b)** $\lim\limits_{x \to -\infty} x^3$ **(c)** $\lim\limits_{x \to -\infty} x^4$

SOLUTION

(a) $\lim_{x \to \infty} x^3 = \infty$

(b) $\lim_{x \to -\infty} x^3 = -\infty$

(c) $\lim_{x \to -\infty} x^4 = \infty$

4. What is the sign of a_2 if $f(x) = a_2 x^2 + a_1 x + a_0$ satisfies $\lim\limits_{x \to -\infty} f(x) = -\infty$?

SOLUTION The behavior of $f(x) = a_2 x^2 + a_1 x + a_0$ as $x \to -\infty$ is controlled by the leading term; that is, $\lim_{x \to -\infty} f(x) = \lim_{x \to -\infty} a_2 x^2$. Because $x^2 \to +\infty$ as $x \to -\infty$, a_2 must be negative to have $\lim_{x \to -\infty} f(x) = -\infty$.

5. What is the sign of the leading coefficient a_7 if $f(x)$ is a polynomial of degree 7 such that $\lim\limits_{x \to -\infty} f(x) = \infty$?

SOLUTION The behavior of $f(x)$ as $x \to -\infty$ is controlled by the leading term; that is, $\lim_{x \to -\infty} f(x) = \lim_{x \to -\infty} a_7 x^7$. Because $x^7 \to -\infty$ as $x \to -\infty$, a_7 must be negative to have $\lim_{x \to -\infty} f(x) = \infty$.

6. The second derivative of the function $f(x) = (x - 4)^{-1}$ is $f''(x) = 2(x - 4)^{-3}$. Although $f''(x)$ changes sign at $x = 4$, $f(x)$ does not have a point of inflection at $x = 4$. Why not?

SOLUTION The function f does not have a point of inflection at $x = 4$ because $x = 4$ is not in the domain of f.

Exercises

1. Determine the sign combinations of f' and f'' for each interval A–G in Figure 18.

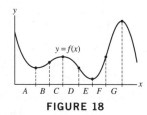

FIGURE 18

SOLUTION

- In A, f is decreasing and concave up, so $f' < 0$ and $f'' > 0$.
- In B, f is increasing and concave up, so $f' > 0$ and $f'' > 0$.
- In C, f is increasing and concave down, so $f' > 0$ and $f'' < 0$.
- In D, f is decreasing and concave down, so $f' < 0$ and $f'' < 0$.
- In E, f is decreasing and concave up, so $f' < 0$ and $f'' > 0$.

- In F, f is increasing and concave up, so $f' > 0$ and $f'' > 0$.
- In G, f is increasing and concave down, so $f' > 0$ and $f'' < 0$.

In Exercises 3–6, draw the graph of a function for which f' and f'' take on the given sign combinations.

3. ++,   +−,   −−

SOLUTION   This function changes from concave up to concave down at $x = -1$ and from increasing to decreasing at $x = 0$.

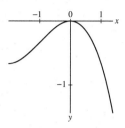

5. −+,   −−,   −+

SOLUTION   The function is decreasing everywhere and changes from concave up to concave down at $x = -1$ and from concave down to concave up at $x = -\frac{1}{2}$.

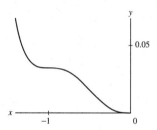

7. Sketch the graph of $y = x^2 - 2x + 3$.

SOLUTION   Let $f(x) = x^2 - 2x + 3$. Then $f'(x) = 2x - 2$ and $f''(x) = 2$. Hence f is decreasing for $x < 1$, is increasing for $x > 1$, has a local minimum at $x = 1$ and is concave up everywhere.

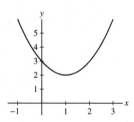

9. Sketch the graph of the cubic $f(x) = x^3 - 3x^2 + 2$. For extra accuracy, plot the zeros of $f(x)$, which are $x = 1$ and $x = 1 \pm \sqrt{3}$ or $x \approx -0.73, 2.73$.

SOLUTION   Let $f(x) = x^3 - 3x^2 + 2$. Then $f'(x) = 3x^2 - 6x = 3x(x - 2) = 0$ yields $x = 0, 2$ and $f''(x) = 6x - 6$. Thus f is concave down for $x < 1$, is concave up for $x > 1$, has an inflection point at $x = 1$, is increasing for $x < 0$ and for $x > 2$, is decreasing for $0 < x < 2$, has a local maximum at $x = 0$, and has a local minimum at $x = 2$.

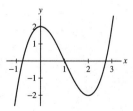

11. Extend the sketch of the graph of $f(x) = \cos x + \frac{1}{2}x$ over $[0, \pi]$ in Example 4 to the interval $[0, 5\pi]$.

SOLUTION   Let $f(x) = \cos x + \frac{1}{2}x$. Then $f'(x) = -\sin x + \frac{1}{2} = 0$ yields critical points at $x = \frac{\pi}{6}, \frac{5\pi}{6}, \frac{13\pi}{6}, \frac{17\pi}{6}, \frac{25\pi}{6}$, and $\frac{29\pi}{6}$. Moreover, $f''(x) = -\cos x$ so there are points of inflection at $x = \frac{\pi}{2}, \frac{3\pi}{2}, \frac{5\pi}{2}, \frac{7\pi}{2}$, and $\frac{9\pi}{2}$.

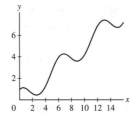

In Exercises 13–30, sketch the graph of the function. Indicate the transition points (local extrema and points of inflection).

13. $y = x^3 + \frac{3}{2}x^2$

SOLUTION Let $f(x) = x^3 + \frac{3}{2}x^2$. Then $f'(x) = 3x^2 + 3x = 3x(x+1)$ and $f''(x) = 6x + 3$. This shows that f has critical points at $x = 0$ and $x = -1$ and a candidate for an inflection point at $x = -\frac{1}{2}$.

Interval	$(-\infty, -1)$	$(-1, -\frac{1}{2})$	$(-\frac{1}{2}, 0)$	$(0, \infty)$
Signs of f' and f''	$+-$	$--$	$-+$	$++$

Thus, there is a local maximum at $x = -1$, a local minimum at $x = 0$, and an inflection point at $x = -\frac{1}{2}$. Here is a graph of f with these transition points highlighted as in the graphs in the textbook.

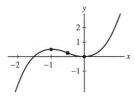

15. $y = x^2 - 4x^3$

SOLUTION Let $f(x) = x^2 - 4x^3$. Then $f'(x) = 2x - 12x^2 = 2x(1 - 6x)$ and $f''(x) = 2 - 24x$. Critical points are at $x = 0$ and $x = \frac{1}{6}$, and the sole candidate point of inflection is at $x = \frac{1}{12}$.

Interval	$(-\infty, 0)$	$(0, \frac{1}{12})$	$(\frac{1}{12}, \frac{1}{6})$	$(\frac{1}{6}, \infty)$
Signs of f' and f''	$-+$	$++$	$+-$	$--$

Thus, $f(0)$ is a local minimum, $f(\frac{1}{6})$ is a local maximum, and there is a point of inflection at $x = \frac{1}{12}$. Here is the graph of f with transition points highlighted as in the textbook:

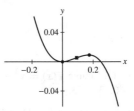

17. $y = 4 - 2x^2 + \frac{1}{6}x^4$

SOLUTION Let $f(x) = \frac{1}{6}x^4 - 2x^2 + 4$. Then $f'(x) = \frac{2}{3}x^3 - 4x = \frac{2}{3}x\left(x^2 - 6\right)$ and $f''(x) = 2x^2 - 4$. This shows that f has critical points at $x = 0$ and $x = \pm\sqrt{6}$ and has candidates for points of inflection at $x = \pm\sqrt{2}$.

Interval	$(-\infty, -\sqrt{6})$	$(-\sqrt{6}, -\sqrt{2})$	$(-\sqrt{2}, 0)$	$(0, \sqrt{2})$	$(\sqrt{2}, \sqrt{6})$	$(\sqrt{6}, \infty)$
Signs of f' and f''	$-+$	$++$	$+-$	$--$	$-+$	$++$

Thus, f has local minima at $x = \pm\sqrt{6}$, a local maximum at $x = 0$, and inflection points at $x = \pm\sqrt{2}$. Here is a graph of f with transition points highlighted.

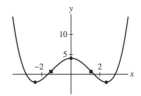

19. $y = x^5 + 5x$

SOLUTION Let $f(x) = x^5 + 5x$. Then $f'(x) = 5x^4 + 5 = 5(x^4 + 1)$ and $f''(x) = 20x^3$. $f'(x) > 0$ for all x, so the graph has no critical points and is always increasing. $f''(x) = 0$ at $x = 0$. Sign analyses reveal that $f''(x)$ changes from negative to positive at $x = 0$, so that the graph of $f(x)$ has an inflection point at $(0, 0)$. Here is a graph of f with transition points highlighted.

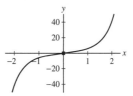

21. $y = x^4 - 3x^3 + 4x$

SOLUTION Let $f(x) = x^4 - 3x^3 + 4x$. Then $f'(x) = 4x^3 - 9x^2 + 4 = (4x^2 - x - 2)(x - 2)$ and $f''(x) = 12x^2 - 18x = 6x(2x - 3)$. This shows that f has critical points at $x = 2$ and $x = \dfrac{1 \pm \sqrt{33}}{8}$ and candidate points of inflection at $x = 0$ and $x = \frac{3}{2}$. Sign analyses reveal that $f'(x)$ changes from negative to positive at $x = \frac{1-\sqrt{33}}{8}$, from positive to negative at $x = \frac{1+\sqrt{33}}{8}$, and again from negative to positive at $x = 2$. Therefore, $f(\frac{1-\sqrt{33}}{8})$ and $f(2)$ are local minima of $f(x)$, and $f(\frac{1+\sqrt{33}}{8})$ is a local maximum. Further sign analyses reveal that $f''(x)$ changes from positive to negative at $x = 0$ and from negative to positive at $x = \frac{3}{2}$, so that there are points of inflection both at $x = 0$ and $x = \frac{3}{2}$. Here is a graph of $f(x)$ with transition points highlighted.

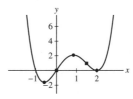

23. $y = 6x^7 - 7x^6$

SOLUTION Let $f(x) = 6x^7 - 7x^6$. Then $f'(x) = 42x^6 - 42x^5 = 42x^5 (x - 1)$ and $f''(x) = 252x^5 - 210x^4 = 42x^4 (6x - 5)$. Critical points are at $x = 0$ and $x = 1$, and candidate inflection points are at $x = 0$ and $x = \frac{5}{6}$. Sign analyses reveal that $f'(x)$ changes from positive to negative at $x = 0$ and from negative to positive at $x = 1$. Therefore $f(0)$ is a local maximum and $f(1)$ is a local minimum. Also, $f''(x)$ changes from negative to positive at $x = \frac{5}{6}$. Therefore, there is a point of inflection at $x = \frac{5}{6}$. Here is a graph of f with transition points highlighted.

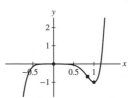

25. $y = x - \sqrt{x}$

SOLUTION Let $f(x) = x - \sqrt{x} = x - x^{1/2}$. Then $f'(x) = 1 - \frac{1}{2}x^{-1/2}$. This shows that f has critical points at $x = 0$ (where the derivative does not exist) and at $x = \frac{1}{4}$ (where the derivative is zero). Because $f'(x) < 0$ for $0 < x < \frac{1}{4}$ and $f'(x) > 0$ for $x > \frac{1}{4}$, $f\left(\frac{1}{4}\right)$ is a local minimum. Now $f''(x) = \frac{1}{4}x^{-3/2} > 0$ for all $x > 0$, so the graph is always concave up. Here is a graph of f with transition points highlighted.

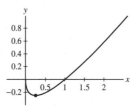

27. $y = \sqrt{x} + \sqrt{9 - x}$

SOLUTION Let $f(x) = \sqrt{x} + \sqrt{9 - x} = x^{1/2} + (9 - x)^{1/2}$. Note that the domain of f is $[0, 9]$. Now, $f'(x) = \frac{1}{2}x^{-1/2} - \frac{1}{2}(9 - x)^{-1/2}$ and $f''(x) = -\frac{1}{4}x^{-3/2} - \frac{1}{4}(9 - x)^{-3/2}$. Thus, the critical points are $x = 0$, $x = \frac{9}{2}$ and $x = 9$. Sign analysis reveals that $f'(x) > 0$ for $0 < x < \frac{9}{2}$ and $f'(x) < 0$ for $\frac{9}{2} < x < 9$, so f has a local maximum at $x = \frac{9}{2}$. Further, $f''(x) < 0$ on $(0, 9)$, so the graph is always concave down. Here is a graph of f with the transition point highlighted.

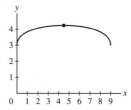

29. $y = (x^2 - x)^{1/3}$

SOLUTION Let $f(x) = (x^2 - x)^{1/3}$. Then

$$f'(x) = \frac{1}{3}(2x - 1)(x^2 - x)^{-2/3}$$

and

$$f''(x) = \frac{1}{3}\left(2(x^2 - x)^{-2/3} - \frac{2}{3}(2x - 1)(2x - 1)(x^2 - x)^{-5/3}\right)$$

$$= \frac{2}{9}(x^2 - x)^{-5/3}\left(3(x^2 - x) - (2x - 1)^2\right) = -\frac{2}{9}(x^2 - x)^{-5/3}(x^2 - x + 1).$$

Critical points of $f(x)$ are points where the numerator $2x - 1 = 0$ or where $f'(x)$ doesn't exist, that is, at $x = \frac{1}{2}$ and points where $x^2 - x = 0$ so that $x = 1$ or $x = 0$.

Candidate points of inflection lie at points where $f''(x) = 0$ (of which there are none), and points where $f''(x)$ does not exist (at $x = 0$ and $x = 1$). Sign analyses reveal that $f''(x) < 0$ for $x < 0$ and for $x > 1$, while $f''(x) > 0$ for $0 < x < 1$. Therefore, the graph of $f(x)$ has points of inflection at $x = 0$ and $x = 1$.

Since $(x^2 - x)^{-2/3}$ is positive wherever it is defined, the sign of $f'(x)$ depends solely on the sign of $2x - 1$. Hence, $f'(x)$ does not change sign at $x = 0$ or $x = 1$, and goes from negative to positive at $x = \frac{1}{2}$. $f(\frac{1}{2})$ is, in that case, a local minimum.

Here is a graph of $f(x)$ with the transition points indicated.

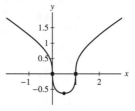

31. Sketch the graph of $f(x) = 18(x - 3)(x - 1)^{2/3}$ using the following formulas:

$$f'(x) = \frac{30(x - \frac{9}{5})}{(x - 1)^{1/3}}, \qquad f''(x) = \frac{20(x - \frac{3}{5})}{(x - 1)^{4/3}}$$

SOLUTION

$$f'(x) = \frac{30(x - \frac{9}{5})}{(x - 1)^{1/3}}$$

yields critical points at $x = \frac{9}{5}$, $x = 1$.

$$f''(x) = \frac{20(x - \frac{3}{5})}{(x - 1)^{4/3}}$$

yields potential inflection points at $x = \frac{3}{5}$, $x = 1$.

Interval	signs of f' and f''
$(-\infty, \frac{3}{5})$	$+-$
$(\frac{3}{5}, 1)$	$++$
$(1, \frac{9}{5})$	$-+$
$(\frac{9}{5}, \infty)$	$++$

The graph has an inflection point at $x = \frac{3}{5}$, a local maximum at $x = 1$ (at which the graph has a cusp), and a local minimum at $x = \frac{9}{5}$. The sketch looks something like this.

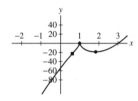

In Exercises 32–37, sketch the graph over the given interval. Indicate the transition points.

33. $y = \sin x + \cos x$, $[0, 2\pi]$

SOLUTION   Let $f(x) = \sin x + \cos x$. Setting $f'(x) = \cos x - \sin x = 0$ yields $\sin x = \cos x$, so that $\tan x = 1$, and $x = \frac{\pi}{4}, \frac{5\pi}{4}$. Setting $f''(x) = -\sin x - \cos x = 0$ yields $\sin x = -\cos x$, so that $-\tan x = 1$, and $x = \frac{3\pi}{4}$, $x = \frac{7\pi}{4}$.

Interval	signs of f' and f''
$(0, \frac{\pi}{4})$	$+-$
$(\frac{\pi}{4}, \frac{3\pi}{4})$	$--$
$(\frac{3\pi}{4}, \frac{5\pi}{4})$	$-+$
$(\frac{5\pi}{4}, \frac{7\pi}{4})$	$++$
$(\frac{7\pi}{4}, 2\pi)$	$+-$

The graph has a local maximum at $x = \frac{\pi}{4}$, a local minimum at $x = \frac{5\pi}{4}$, and inflection points at $x = \frac{3\pi}{4}$ and $x = \frac{7\pi}{4}$. Here is a sketch of the graph of $f(x)$:

35. $y = \sin x + \frac{1}{2}x$, $[0, 2\pi]$

SOLUTION   Let $f(x) = \sin x + \frac{1}{2}x$. Setting $f'(x) = \cos x + \frac{1}{2} = 0$ yields $x = \frac{2\pi}{3}$ or $\frac{4\pi}{3}$. Setting $f''(x) = -\sin x = 0$ yields potential points of inflection at $x = 0, \pi, 2\pi$.

Interval	signs of f' and f''
$(0, \frac{2\pi}{3})$	+−
$(\frac{2\pi}{3}, \pi)$	−−
$(\pi, \frac{4\pi}{3})$	−+
$(\frac{4\pi}{3}, 2\pi)$	++

The graph has a local maximum at $x = \frac{2\pi}{3}$, a local minimum at $x = \frac{4\pi}{3}$, and an inflection point at $x = \pi$. Here is a graph of f *without* transition points highlighted.

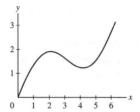

37. $y = \sin x - \dfrac{1}{2}\sin 2x, \quad [0, \pi]$

Hint for Exercise 37: We find numerically that there is one point of inflection in the interval, occurring at $x \approx 1.3182$.

SOLUTION   Let $f(x) = \sin x - \frac{1}{2}\sin 2x$. Setting $f'(x) = \cos x - \cos 2x = 0$ yields $\cos 2x = \cos x$. Using the double angle formula for cosine, this gives $2\cos^2 x - 1 = \cos x$ or $(2\cos x + 1)(\cos x - 1) = 0$. Solving for $x \in [0, \pi]$, we find $x = 0$ or $\frac{2\pi}{3}$.

Setting $f''(x) = -\sin x + 2\sin 2x = 0$ yields $4\sin x \cos x = \sin x$, so $\sin x = 0$ or $\cos x = \frac{1}{4}$. Hence, there are potential points of inflection at $x = 0$, $x = \pi$ and $x = \cos^{-1}\frac{1}{4} \approx 1.31812$.

Interval	Sign of f' and f''
$(0, \cos^{-1}\frac{1}{4})$	++
$(\cos^{-1}\frac{1}{4}, \frac{2\pi}{3})$	+−
$(\frac{2\pi}{3}, \pi)$	−−

The graph of $f(x)$ has a local maximum at $x = \frac{2\pi}{3}$ and a point of inflection at $x = \cos^{-1}\frac{1}{4}$.

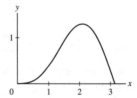

39. Suppose that f is twice differentiable satisfying (i) $f(0) = 1$, (ii) $f'(x) > 0$ for all $x \neq 0$, and (iii) $f''(x) < 0$ for $x < 0$ and $f''(x) > 0$ for $x > 0$. Let $g(x) = f(x^2)$.

(a) Sketch a possible graph of $f(x)$.

(b) Prove that $g(x)$ has no points of inflection and a unique local extreme value at $x = 0$. Sketch a possible graph of $g(x)$.

SOLUTION

(a) To produce a possible sketch, we give the direction and concavity of the graph over every interval.

Interval	$(-\infty, 0)$	$(0, \infty)$
Direction	↗	↗
Concavity	⌢	⌣

A sketch of one possible such function appears here:

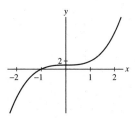

(b) Let $g(x) = f(x^2)$. Then $g'(x) = 2xf'(x^2)$. If $g'(x) = 0$, either $x = 0$ or $f'(x^2) = 0$, which implies that $x = 0$ as well. Since $f'(x^2) > 0$ for all $x \neq 0$, $g'(x) < 0$ for $x < 0$ and $g'(x) > 0$ for $x > 0$. This gives $g(x)$ a unique local extreme value at $x = 0$, a minimum. $g''(x) = 2f'(x^2) + 4x^2 f''(x^2)$. For all $x \neq 0$, $x^2 > 0$, and so $f''(x^2) > 0$ and $f'(x^2) > 0$. Thus $g''(x) > 0$, and so $g''(x)$ does not change sign, and can have no inflection points. A sketch of $g(x)$ based on the sketch we made for $f(x)$ follows: indeed, this sketch shows a unique local minimum at $x = 0$.

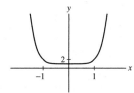

In Exercises 41–50, calculate the following limits (divide the numerator and denominator by the highest power of x appearing in the denominator).

41. $\displaystyle\lim_{x \to \infty} \frac{x}{x+9}$

SOLUTION

$$\lim_{x \to \infty} \frac{x}{x+9} = \lim_{x \to \infty} \frac{x^{-1}(x)}{x^{-1}(x+9)} = \lim_{x \to \infty} \frac{1}{1 + \frac{9}{x}} = \frac{1}{1+0} = 1.$$

43. $\displaystyle\lim_{x \to \infty} \frac{3x^2 + 20x}{2x^4 + 3x^3 - 29}$

SOLUTION

$$\lim_{x \to \infty} \frac{3x^2 + 20x}{2x^4 + 3x^3 - 29} = \lim_{x \to \infty} \frac{x^{-4}(3x^2 + 20x)}{x^{-4}(2x^4 + 3x^3 - 29)} = \lim_{x \to \infty} \frac{\frac{3}{x^2} + \frac{20}{x^3}}{2 + \frac{3}{x} - \frac{29}{x^4}} = \frac{0}{2} = 0.$$

45. $\displaystyle\lim_{x \to \infty} \frac{7x - 9}{4x + 3}$

SOLUTION

$$\lim_{x \to \infty} \frac{7x - 9}{4x + 3} = \lim_{x \to \infty} \frac{x^{-1}(7x - 9)}{x^{-1}(4x + 3)} = \lim_{x \to \infty} \frac{7 - \frac{9}{x}}{4 + \frac{3}{x}} = \frac{7}{4}.$$

47. $\displaystyle\lim_{x \to -\infty} \frac{7x^2 - 9}{4x + 3}$

SOLUTION

$$\lim_{x \to -\infty} \frac{7x^2 - 9}{4x + 3} = \lim_{x \to -\infty} \frac{x^{-1}(7x^2 - 9)}{x^{-1}(4x + 3)} = \lim_{x \to -\infty} \frac{7x - \frac{9}{x}}{4 + \frac{3}{x}} = -\infty.$$

49. $\displaystyle\lim_{x \to -\infty} \frac{x^2 - 1}{x + 4}$

SOLUTION

$$\lim_{x \to -\infty} \frac{x^2 - 1}{x + 4} = \lim_{x \to -\infty} \frac{x^{-1}(x^2 - 1)}{x^{-1}(x + 4)} = \lim_{x \to -\infty} \frac{x - \frac{1}{x}}{1 + \frac{4}{x}} = -\infty.$$

In Exercises 51–56, calculate the limit.

51. $\lim\limits_{x\to\infty} \dfrac{\sqrt{x^2+1}}{x+1}$

SOLUTION Looking at only the highest powers in the top and bottom of $\frac{\sqrt{x^2+1}}{x+1}$, it seems this quotient behaves like $\frac{\sqrt{x^2}}{x} = \frac{x}{x}$. Multiplying top and bottom by x^{-1} yields

$$\frac{\sqrt{x^2+1}}{x+1} = \frac{\sqrt{1+\frac{1}{x^2}}}{1+\frac{1}{x}}.$$

From this we get:

$$\lim_{x\to\infty} \frac{\sqrt{x^2+1}}{x+1} = \frac{\lim\limits_{x\to\infty}\sqrt{1+1/x^2}}{\lim\limits_{x\to\infty}1+1/x} = \frac{1}{1} = 1.$$

53. $\lim\limits_{x\to\infty} \dfrac{x+1}{\sqrt[3]{x^2+1}}$

SOLUTION We can see that the denominator behaves like $x^{2/3}$. Dividing top and bottom by $x^{2/3}$ yields:

$$\lim_{x\to\infty} \frac{x+1}{\sqrt[3]{x^2+1}} = \frac{\lim\limits_{x\to\infty} x^{1/3}+1/x^{2/3}}{\lim\limits_{x\to\infty}\sqrt[3]{1+1/x^2}} = \frac{\infty}{1} = \infty.$$

55. $\lim\limits_{x\to-\infty} \dfrac{x}{(x^6+1)^{1/3}}$

SOLUTION Looking at highest exponents of x, we see the denominator behaves like x^2. Dividing top and bottom by x^2 yields:

$$\lim_{x\to-\infty} \frac{x}{(x^6+1)^{1/3}} = \frac{\lim\limits_{x\to-\infty} 1/x}{\lim\limits_{x\to-\infty}(1+1/x^6)^{1/3}} = \frac{0}{1} = 0.$$

57. Which curve in Figure 21 is the graph of $f(x) = \dfrac{2x^4-1}{1+x^4}$? Explain on the basis of horizontal asymptotes.

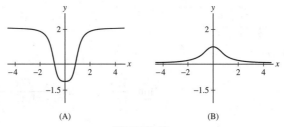

(A)   (B)

FIGURE 21

SOLUTION Since

$$\lim_{x\to\pm\infty} \frac{2x^4-1}{1+x^4} = \frac{2}{1}\cdot\lim_{x\to\pm\infty}1 = 2$$

the graph has left and right horizontal asymptotes at $y=2$, so the left curve is the graph of $f(x) = \dfrac{2x^4-1}{1+x^4}$.

59. Match the functions with their graphs in Figure 23.

(a) $y = \dfrac{1}{x^2-1}$ **(b)** $y = \dfrac{x^2}{x^2+1}$

(c) $y = \dfrac{1}{x^2+1}$ **(d)** $y = \dfrac{x}{x^2-1}$

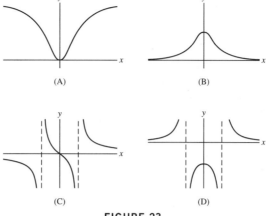

FIGURE 23

SOLUTION

(a) The graph of $\frac{1}{x^2-1}$ should have a horizontal asymptote at $y = 0$ and vertical asymptotes at $x = \pm 1$. Further, the graph should consist of positive values for $|x| > 1$ and negative values for $|x| < 1$. Hence, the graph of $\frac{1}{x^2-1}$ is (D).

(b) The graph of $\frac{x^2}{x^2+1}$ should have a horizontal asymptote at $y = 1$ and no vertical asymptotes. Hence, the graph of $\frac{x^2}{x^2+1}$ is (A).

(c) The graph of $\frac{1}{x^2+1}$ should have a horizontal asymptote at $y = 0$ and no vertical asymptotes. Hence, the graph of $\frac{1}{x^2+1}$ is (B).

(d) The graph of $\frac{x}{x^2-1}$ should have a horizontal asymptote at $y = 0$ and vertical asymptotes at $x = \pm 1$. Further, the graph should consist of positive values for $-1 < x < 0$ and $x > 1$ and negative values for $x < 1$ and $0 < x < 1$. Hence, the graph of $\frac{x}{x^2-1}$ is (C).

61. Sketch the graph of $f(x) = \dfrac{x}{x^2 + 1}$ using the formulas

$$f'(x) = \frac{1 - x^2}{(1 + x^2)^2}, \qquad f''(x) = \frac{2x(x^2 - 3)}{(x^2 + 1)^3}.$$

SOLUTION   Let $f(x) = \dfrac{x}{x^2 + 1}$.

- Because $\displaystyle\lim_{x \to \pm\infty} f(x) = \frac{1}{1} \cdot \lim_{x \to \pm\infty} x^{-1} = 0$, $y = 0$ is a horizontal asymptote for f.

- Now $f'(x) = \dfrac{1 - x^2}{\left(x^2 + 1\right)^2}$ is negative for $x < -1$ and $x > 1$, positive for $-1 < x < 1$, and 0 at $x = \pm 1$.

  Accordingly, f is decreasing for $x < -1$ and $x > 1$, is increasing for $-1 < x < 1$, has a local minimum value at $x = -1$ and a local maximum value at $x = 1$.

- Moreover,

$$f''(x) = \frac{2x\left(x^2 - 3\right)}{\left(x^2 + 1\right)^3}.$$

Here is a sign chart for the second derivative, similar to those constructed in various exercises in Section 4.4. (The legend is on page 184.)

x	$\left(-\infty, -\sqrt{3}\right)$	$-\sqrt{3}$	$\left(-\sqrt{3}, 0\right)$	0	$\left(0, \sqrt{3}\right)$	$\sqrt{3}$	$\left(\sqrt{3}, \infty\right)$
f''	$-$	0	$+$	0	$-$	0	$+$
f	$\frown$	I	$\smile$	I	$\frown$	I	$\smile$

- Here is a graph of $f(x) = \dfrac{x}{x^2 + 1}$.

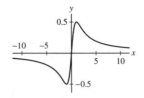

In Exercises 62–77, sketch the graph of the function. Indicate the asymptotes, local extrema, and points of inflection.

63. $y = \dfrac{x}{2x - 1}$

SOLUTION   Let $f(x) = \dfrac{x}{2x - 1}$. Then $f'(x) = \dfrac{-1}{(2x - 1)^2}$, so that f is decreasing for all $x \neq \frac{1}{2}$. Moreover, $f''(x) = \dfrac{4}{(2x - 1)^3}$, so that f is concave up for $x > \frac{1}{2}$ and concave down for $x < \frac{1}{2}$. Because $\displaystyle\lim_{x \to \pm\infty} \dfrac{x}{2x - 1} = \dfrac{1}{2}$, f has a horizontal asymptote at $y = \frac{1}{2}$. Finally, f has a vertical asymptote at $x = \frac{1}{2}$ with

$$\lim_{x \to \frac{1}{2}-} \frac{x}{2x - 1} = -\infty \qquad \text{and} \qquad \lim_{x \to \frac{1}{2}+} \frac{x}{2x - 1} = \infty.$$

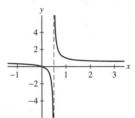

65. $y = \dfrac{x + 3}{x - 2}$

SOLUTION   Let $f(x) = \dfrac{x + 3}{x - 2}$. Then $f'(x) = \dfrac{-5}{(x - 2)^2}$, so that f is decreasing for all $x \neq 2$. Moreover, $f''(x) = \dfrac{10}{(x - 2)^3}$, so that f is concave up for $x > 2$ and concave down for $x < 2$. Because $\displaystyle\lim_{x \to \pm\infty} \dfrac{x + 3}{x - 2} = 1$, f has a horizontal asymptote at $y = 1$. Finally, f has a vertical asymptote at $x = 2$ with

$$\lim_{x \to 2-} \frac{x + 3}{x - 2} = -\infty \qquad \text{and} \qquad \lim_{x \to 2+} \frac{x + 3}{x - 2} = \infty.$$

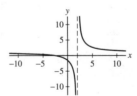

67. $y = \dfrac{1}{x} + \dfrac{1}{x - 1}$

SOLUTION   Let $f(x) = \dfrac{1}{x} + \dfrac{1}{x - 1}$. Then $f'(x) = -\dfrac{2x^2 - 2x + 1}{x^2 (x - 1)^2}$, so that f is decreasing for all $x \neq 0, 1$. Moreover, $f''(x) = \dfrac{2\left(2x^3 - 3x^2 + 3x - 1\right)}{x^3 (x - 1)^3}$, so that f is concave up for $0 < x < \frac{1}{2}$ and $x > 1$ and concave down for $x < 0$ and $\frac{1}{2} < x < 1$. Because $\displaystyle\lim_{x \to \pm\infty} \left(\dfrac{1}{x} + \dfrac{1}{x - 1}\right) = 0$, f has a horizontal asymptote at $y = 0$. Finally, f has vertical asymptotes at $x = 0$ and $x = 1$ with

$$\lim_{x \to 0-} \left(\frac{1}{x} + \frac{1}{x - 1}\right) = -\infty \qquad \text{and} \qquad \lim_{x \to 0+} \left(\frac{1}{x} + \frac{1}{x - 1}\right) = \infty$$

and

$$\lim_{x \to 1-} \left(\frac{1}{x} + \frac{1}{x - 1}\right) = -\infty \qquad \text{and} \qquad \lim_{x \to 1+} \left(\frac{1}{x} + \frac{1}{x - 1}\right) = \infty.$$

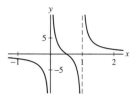

69. $y = 1 - \dfrac{3}{x} + \dfrac{4}{x^3}$

SOLUTION   Let $f(x) = 1 - \dfrac{3}{x} + \dfrac{4}{x^3}$. Then

$$f'(x) = \frac{3}{x^2} - \frac{12}{x^4} = \frac{3(x-2)(x+2)}{x^4},$$

so that f is increasing for $|x| > 2$ and decreasing for $-2 < x < 0$ and for $0 < x < 2$. Moreover,

$$f''(x) = -\frac{6}{x^3} + \frac{48}{x^5} = \frac{6(8 - x^2)}{x^5},$$

so that f is concave down for $-2\sqrt{2} < x < 0$ and for $x > 2\sqrt{2}$, while f is concave up for $x < -2\sqrt{2}$ and for $0 < x < 2\sqrt{2}$. Because

$$\lim_{x \to \pm\infty} \left(1 - \frac{3}{x} + \frac{4}{x^3} \right) = 1,$$

f has a horizontal asymptote at $y = 1$. Finally, f has a vertical asymptote at $x = 0$ with

$$\lim_{x \to 0-} \left(1 - \frac{3}{x} + \frac{4}{x^3} \right) = -\infty \quad \text{and} \quad \lim_{x \to 0+} \left(1 - \frac{3}{x} + \frac{4}{x^3} \right) = \infty.$$

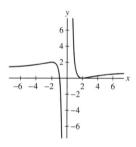

71. $y = \dfrac{1}{x^2} + \dfrac{1}{(x-2)^2}$

SOLUTION   Let $f(x) = \dfrac{1}{x^2} + \dfrac{1}{(x-2)^2}$. Then

$$f'(x) = -2x^{-3} - 2(x-2)^{-3} = -\frac{4(x-1)(x^2 - 2x + 4)}{x^3(x-2)^3},$$

so that f is increasing for $x < 0$ and for $1 < x < 2$, is decreasing for $0 < x < 1$ and for $x > 2$, and has a local minimum at $x = 1$. Moreover, $f''(x) = 6x^{-4} + 6(x-2)^{-4}$, so that f is concave up for all $x \neq 0, 2$. Because $\lim\limits_{x \to \pm\infty} \left(\dfrac{1}{x^2} + \dfrac{1}{(x-2)^2} \right) = 0$, f has a horizontal asymptote at $y = 0$. Finally, f has vertical asymptotes at $x = 0$ and $x = 2$ with

$$\lim_{x \to 0-} \left(\frac{1}{x^2} + \frac{1}{(x-2)^2} \right) = \infty \quad \text{and} \quad \lim_{x \to 0+} \left(\frac{1}{x^2} + \frac{1}{(x-2)^2} \right) = \infty$$

and

$$\lim_{x \to 2-} \left(\frac{1}{x^2} + \frac{1}{(x-2)^2} \right) = \infty \quad \text{and} \quad \lim_{x \to 2+} \left(\frac{1}{x^2} + \frac{1}{(x-2)^2} \right) = \infty.$$

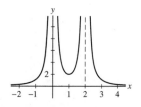

73. $y = \dfrac{4}{x^2 - 9}$

SOLUTION   Let $f(x) = \dfrac{4}{x^2 - 9}$. Then $f'(x) = -\dfrac{8x}{\left(x^2 - 9\right)^2}$, so that f is increasing for $x < -3$ and for $-3 < x < 0$,

is decreasing for $0 < x < 3$ and for $x > 3$, and has a local maximum at $x = 0$. Moreover, $f''(x) = \dfrac{24\left(x^2 + 3\right)}{\left(x^2 - 9\right)^3}$, so

that f is concave up for $x < -3$ and for $x > 3$ and is concave down for $-3 < x < 3$. Because $\displaystyle\lim_{x \to \pm\infty} \dfrac{4}{x^2 - 9} = 0$, f

has a horizontal asymptote at $y = 0$. Finally, f has vertical asymptotes at $x = -3$ and $x = 3$, with

$$\lim_{x \to -3-} \left(\frac{4}{x^2 - 9}\right) = \infty \quad \text{and} \quad \lim_{x \to -3+} \left(\frac{4}{x^2 - 9}\right) = -\infty$$

and

$$\lim_{x \to 3-} \left(\frac{4}{x^2 - 9}\right) = -\infty \quad \text{and} \quad \lim_{x \to 3+} \left(\frac{4}{x^2 - 9}\right) = \infty.$$

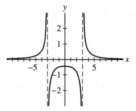

75. $y = \dfrac{x^2}{(x^2 - 1)(x^2 + 1)}$

SOLUTION   Let

$$f(x) = \frac{x^2}{(x^2 - 1)(x^2 + 1)}.$$

Then

$$f'(x) = -\frac{2x(1 + x^4)}{(x - 1)^2(x + 1)^2(x^2 + 1)^2},$$

so that f is increasing for $x < -1$ and for $-1 < x < 0$, is decreasing for $0 < x < 1$ and for $x > 1$, and has a local
maximum at $x = 0$. Moreover,

$$f''(x) = \frac{2 + 24x^4 + 6x^8}{(x - 1)^3(x + 1)^3(x^2 + 1)^3},$$

so that f is concave up for $|x| > 1$ and concave down for $|x| < 1$. Because $\displaystyle\lim_{x \to \pm\infty} \dfrac{x^2}{(x^2 - 1)(x^2 + 1)} = 0$, f has a

horizontal asymptote at $y = 0$. Finally, f has vertical asymptotes at $x = -1$ and $x = 1$, with

$$\lim_{x \to -1-} \frac{x^2}{(x^2 - 1)(x^2 + 1)} = \infty \quad \text{and} \quad \lim_{x \to -1+} \frac{x^2}{(x^2 - 1)(x^2 + 1)} = -\infty$$

and

$$\lim_{x \to 1-} \frac{x^2}{(x^2 - 1)(x^2 + 1)} = -\infty \quad \text{and} \quad \lim_{x \to 1+} \frac{x^2}{(x^2 - 1)(x^2 + 1)} = \infty.$$

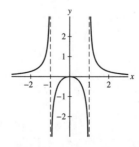

77. $y = \dfrac{x}{\sqrt{x^2 + 1}}$

SOLUTION Let

$$f(x) = \frac{x}{\sqrt{x^2 + 1}}.$$

Then

$$f'(x) = (x^2 + 1)^{-3/2} \quad \text{and} \quad f''(x) = \frac{-3x}{(x^2 + 1)^{5/2}}.$$

Thus, f is increasing for all x, is concave up for $x < 0$, is concave down for $x > 0$, and has a point of inflection at $x = 0$. Because

$$\lim_{x \to \infty} \frac{x}{\sqrt{x^2 + 1}} = 1 \quad \text{and} \quad \lim_{x \to -\infty} \frac{x}{\sqrt{x^2 + 1}} = -1,$$

f has horizontal asymptotes of $y = -1$ and $y = 1$. There are no vertical asymptotes.

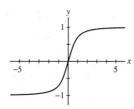

Further Insights and Challenges

*In Exercises 78–82, we explore functions whose graphs approach a nonhorizontal line as $x \to \infty$. A line $y = ax + b$ is called a **slant asymptote** if*

$$\lim_{x \to \infty} (f(x) - (ax + b)) = 0$$

or

$$\lim_{x \to -\infty} (f(x) - (ax + b)) = 0.$$

79. If $f(x) = P(x)/Q(x)$, where P and Q are polynomials of degrees $m + 1$ and m, then by long division, we can write

$$f(x) = (ax + b) + P_1(x)/Q(x)$$

where P_1 is a polynomial of degree $< m$. Show that $y = ax + b$ is the slant asymptote of $f(x)$. Use this procedure to find the slant asymptotes of the functions:

(a) $y = \dfrac{x^2}{x + 2}$

(b) $y = \dfrac{x^3 + x}{x^2 + x + 1}$

SOLUTION Since $\deg(P_1) < \deg(Q)$,

$$\lim_{x \to \pm\infty} \frac{P_1(x)}{Q(x)} = 0.$$

Thus

$$\lim_{x \to \pm\infty} (f(x) - (ax + b)) = 0$$

and $y = ax + b$ is a slant asymptote of f.

(a) $\dfrac{x^2}{x + 2} = x - 2 + \dfrac{4}{x + 2}$; hence $y = x - 2$ is a slant asymptote of $\dfrac{x^2}{x + 2}$.

(b) $\dfrac{x^3 + x}{x^2 + x + 1} = (x - 1) + \dfrac{x + 1}{x^2 - 1}$; hence, $y = x - 1$ is a slant asymptote of $\dfrac{x^3 + x}{x^2 + x + 1}$.

81. Show that $y = 3x$ is a slant asymptote for $f(x) = 3x + x^{-2}$. Determine whether $f(x)$ approaches the slant asymptote from above or below and make a sketch of the graph.

SOLUTION   Let $f(x) = 3x + x^{-2}$. Then

$$\lim_{x \to \pm\infty} (f(x) - 3x) = \lim_{x \to \pm\infty} (3x + x^{-2} - 3x) = \lim_{x \to \pm\infty} x^{-2} = 0$$

which implies that $3x$ is the slant asymptote of $f(x)$. Since $f(x) - 3x = x^{-2} > 0$ as $x \to \pm\infty$, $f(x)$ approaches the slant asymptote from above in both directions. Moreover, $f'(x) = 3 - 2x^{-3}$ and $f''(x) = 6x^{-4}$. Sign analyses reveal a local minimum at $x = \left(\frac{3}{2}\right)^{-1/3} \approx .87358$ and that f is concave up for all $x \neq 0$. Limit analyses give a vertical asymptote at $x = 0$.

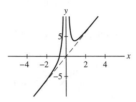

83. Assume that $f'(x)$ and $f''(x)$ exist for all x and let c be a critical point of $f(x)$. Show that $f(x)$ cannot make a transition from $++$ to $-+$ at $x = c$. *Hint:* Apply the MVT to $f'(x)$.

SOLUTION   Let $f(x)$ be a function such that $f''(x) > 0$ for all x and such that it transitions from $++$ to $-+$ at a critical point c where $f'(c)$ is defined. That is, $f'(c) = 0$, $f'(x) > 0$ for $x < c$ and $f'(x) < 0$ for $x > c$. Let $g(x) = f'(x)$. The previous statements indicate that $g(c) = 0$, $g(x_0) > 0$ for some $x_0 < c$, and $g(x_1) < 0$ for some $x_1 > c$. By the Mean Value Theorem,

$$\frac{g(x_1) - g(x_0)}{x_1 - x_0} = g'(c_0),$$

for some c_0 between x_0 and x_1. Because $x_1 > c > x_0$ and $g(x_1) < 0 < g(x_0)$,

$$\frac{g(x_1) - g(x_0)}{x_1 - x_0} < 0.$$

But, on the other hand $g'(c_0) = f''(c_0) > 0$, so there is a contradiction. This means that our assumption of the existence of such a function $f(x)$ must be in error, so no function can transition from $++$ to $-+$.

If we drop the requirement that $f'(c)$ exist, such a function can be found. The following is a graph of $f(x) = -x^{2/3}$. $f''(x) > 0$ wherever $f''(x)$ is defined, and $f'(x)$ transitions from positive to negative at $x = 0$.

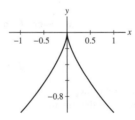

4.6 Applied Optimization

Preliminary Questions

1. The problem is to find the right triangle of perimeter 10 whose area is as large as possible. What is the constraint equation relating the base b and height h of the triangle?

SOLUTION   The perimeter of a right triangle is the sum of the lengths of the base, the height and the hypotenuse. If the base has length b and the height is h, then the length of the hypotenuse is $\sqrt{b^2 + h^2}$ and the perimeter of the triangle is $P = b + h + \sqrt{b^2 + h^2}$. The requirement that the perimeter be 10 translates to the constraint equation

$$b + h + \sqrt{b^2 + h^2} = 10.$$

2. What are the relevant variables if the problem is to find a right circular cone of surface area 20 and maximum volume?

SOLUTION   The relevant variables are the radius r and the height h of the cone. We are asked to maximize the volume

$$V = \frac{1}{3}\pi r^2 h$$

subject to the constraint

$$S = \pi r \sqrt{r^2 + h^2} + \pi r^2 = 20.$$

3. Does a continuous function on an open interval always have a maximum value?

SOLUTION   No, it is possible for a continuous function on an open interval to have no maximum value. As an example, consider the function $f(x) = x^{-1}$ on the open interval $(0, 1)$. Because f is a decreasing function on $(0, 1)$ and

$$\lim_{x \to 0+} f(x) = \lim_{x \to 0+} \frac{1}{x} = \infty,$$

it follows that f does not have a maximum value on $(0, 1)$.

4. Describe a way of showing that a continuous function on an open interval (a, b) has a minimum value.

SOLUTION   If the function tends to infinity at the endpoints of the interval, then the function must take on a minimum value at a critical point.

Exercises

1. Find the dimensions of the rectangle of maximum area that can be formed from a 50-in. piece of wire.
(a) What is the constraint equation relating the lengths x and y of the sides?
(b) Find a formula for the area in terms of x alone.
(c) Does this problem require optimization over an open interval or a closed interval?
(d) Solve the optimization problem.

SOLUTION
(a) The perimeter of the rectangle is 50 inches, so $50 = 2x + 2y$, which is equivalent to $y = 25 - x$.
(b) Using part (a), $A = xy = x(25 - x) = 25x - x^2$.
(c) This problem requires optimization over the closed interval $[0, 25]$, since both x and y must be non-negative.
(d) $A'(x) = 25 - 2x = 0$, which yields $x = \frac{25}{2}$ and consequently, $y = \frac{25}{2}$. Because $A(0) = A(25) = 0$ and $A(\frac{25}{2}) = 156.25$, the maximum area 156.25 in^2 is achieved with $x = y = \frac{25}{2}$ inches.

3. Find the positive number x such that the sum of x and its reciprocal is as small as possible. Does this problem require optimization over an open interval or a closed interval?

SOLUTION   Let $x > 0$ and $f(x) = x + x^{-1}$. Here we require optimization over the open interval $(0, \infty)$. Solve $f'(x) = 1 - x^{-2} = 0$ for $x > 0$ to obtain $x = 1$. Since $f(x) \to \infty$ as $x \to 0+$ and as $x \to \infty$, we conclude that f has an absolute minimum of $f(1) = 2$ at $x = 1$.

5. Find positive numbers x, y such that $xy = 16$ and $x + y$ is as small as possible.

SOLUTION   Let $x, y > 0$. Now $xy = 16$ implies $y = \frac{16}{x}$. Let $f(x) = x + y = x + 16x^{-1}$. Solve $f'(x) = 1 - 16x^{-2} = 0$ for $x > 0$ to obtain $x = 4$ and, consequently, $y = 4$. Since $f(x) \to \infty$ as $x \to 0+$ and as $x \to \infty$, we conclude that f has an absolute minimum of $f(4) = 8$ at $x = y = 4$.

7. Let S be the set of all rectangles with area 100.
(a) What are the dimensions of the rectangle in S with the least perimeter?
(b) Is there a rectangle in S with the greatest perimeter? Explain.

SOLUTION   Consider the set of all rectangles with area 10.

(a) Let $x, y > 0$ be the lengths of the sides. Now $xy = 100$, so that $y = 100/x$. Let $p(x) = 2x + 2y = 2x + 200x^{-1}$ be the perimeter. Solve $p'(x) = 2 - 200x^{-2} = 0$ for $x > 0$ to obtain $x = 10$. Since $p(x) \to \infty$ as $x \to 0+$ and as $x \to \infty$, the least perimeter is $p(10) = 40$ when $x = 10$ and $y = 10$.
(b) There is no rectangle in this set with greatest perimeter. For as $x \to 0+$ or as $x \to \infty$, we have $p(x) = 2x + 200x^{-1} \to \infty$.

9. Suppose that 600 ft of fencing are used to enclose a corral in the shape of a rectangle with a semicircle whose diameter is a side of the rectangle as in Figure 10. Find the dimensions of the corral with maximum area.

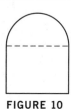

FIGURE 10

SOLUTION Let x be the width of the corral and therefore the diameter of the semicircle, and let y be the height of the rectangular section. Then the perimeter of the corral can be expressed by the equation $2y + x + \frac{\pi}{2}x = 2y + (1 + \frac{\pi}{2})x = 600$ ft or equivalently, $y = \frac{1}{2}\left(600 - (1 + \frac{\pi}{2})x\right)$. Since x and y must both be nonnegative, it follows that x must be restricted to the interval $[0, \frac{600}{1+\pi/2}]$. The area of the corral is the sum of the area of the rectangle and semicircle, $A = xy + \frac{\pi}{8}x^2$. Making the substitution for y from the constraint equation,

$$A(x) = \frac{1}{2}x\left(600 - (1 + \frac{\pi}{2})x\right) + \frac{\pi}{8}x^2 = 300x - \frac{1}{2}\left(1 + \frac{\pi}{2}\right)x^2 + \frac{\pi}{8}x^2.$$

Now, $A'(x) = 300 - \left(1 + \frac{\pi}{2}\right)x + \frac{\pi}{4}x = 0$ implies $x = \frac{300}{(1+\frac{\pi}{4})} \approx 168.029746$ feet. With $A(0) = 0$ ft^2,

$$A\left(\frac{300}{1 + \pi/4}\right) \approx 25204.5 \text{ ft}^2 \quad \text{and} \quad A\left(\frac{600}{1 + \pi/2}\right) \approx 21390.8 \text{ ft}^2,$$

it follows that the corral of maximum area has dimensions

$$x = \frac{300}{1 + \pi/4} \text{ ft} \quad \text{and} \quad y = \frac{150}{1 + \pi/4} \text{ ft}.$$

11. A landscape architect wishes to enclose a rectangular garden on one side by a brick wall costing \$30/ft and on the other three sides by a metal fence costing \$10/ft. If the area of the garden is 1,000 ft^2, find the dimensions of the garden that minimize the cost.

SOLUTION Let x be the length of the brick wall and y the length of an adjacent side with $x, y > 0$. With $xy = 1000$ or $y = \frac{1000}{x}$, the total cost is

$$C(x) = 30x + 10(x + 2y) = 40x + 20000x^{-1}.$$

Solve $C'(x) = 40 - 20000x^{-2} = 0$ for $x > 0$ to obtain $x = 10\sqrt{5}$. Since $C(x) \to \infty$ as $x \to 0+$ and as $x \to \infty$, the minimum cost is $C(10\sqrt{5}) = 800\sqrt{5} \approx \$1788.85$ when $x = 10\sqrt{5} \approx 22.36$ ft and $y = 20\sqrt{5} \approx 44.72$ ft.

13. Find the point P on the parabola $y = x^2$ closest to the point $(3, 0)$ (Figure 12).

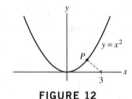

FIGURE 12

SOLUTION With $y = x^2$, let's equivalently minimize the square of the distance,

$$f(x) = (x - 3)^2 + y^2 = x^4 + x^2 - 6x + 9.$$

Then

$$f'(x) = 4x^3 + 2x - 6 = 2(x - 1)(2x^2 + 2x + 3),$$

so that $f'(x) = 0$ when $x = 1$ (plus two complex solutions, which we discard). Since $f(x) \to \infty$ as $x \to \pm\infty$, $P = (1, 1)$ is the point on $y = x^2$ closest to $(3, 0)$.

15. Find the dimensions of the rectangle of maximum area that can be inscribed in a circle of radius r (Figure 13).

FIGURE 13

SOLUTION Place the center of the circle at the origin with the sides of the rectangle (of lengths $2x > 0$ and $2y > 0$) parallel to the coordinate axes. By the Pythagorean Theorem, $x^2 + y^2 = r^2$, so that $y = \sqrt{r^2 - x^2}$. Thus the area of the rectangle is $A(x) = 2x \cdot 2y = 4x\sqrt{r^2 - x^2}$. To guarantee both x and y are real and nonnegative, we must restrict x to the interval $[0, r]$. Solve

$$A'(x) = 4\sqrt{r^2 - x^2} - \frac{4x^2}{\sqrt{r^2 - x^2}} = 0$$

for $x > 0$ to obtain $x = \frac{r}{\sqrt{2}}$. Since $A(0) = A(r) = 0$ and $A(r/\sqrt{2}) = 2r^2$, the rectangle of maximum area has dimensions $x = y = \frac{r}{\sqrt{2}}$.

17. Find the angle θ that maximizes the area of the isosceles triangle whose legs have length ℓ (Figure 14).

FIGURE 14

SOLUTION The area of the triangle is

$$A(\theta) = \frac{1}{2}\ell^2 \sin\theta,$$

where $0 \le \theta \le \pi$. Setting

$$A'(\theta) = \frac{1}{2}\ell^2 \cos\theta = 0$$

yields $\theta = \frac{\pi}{2}$. Since $A(0) = A(\pi) = 0$ and $A(\frac{\pi}{2}) = \frac{1}{2}\ell^2$, the angle that maximizes the area of the isosceles triangle is $\theta = \frac{\pi}{2}$.

19. Rice production requires both labor and capital investment in equipment and land. Suppose that if x dollars per acre are invested in labor and y dollars per acre are invested in equipment and land, then the yield P of rice per acre is given by the formula $P = 100\sqrt{x} + 150\sqrt{y}$. If a farmer invests \$40/acre, how should he divide the \$40 between labor and capital investment in order to maximize the amount of rice produced?

SOLUTION Since $x + y = 40$, we have $P(x) = 100\sqrt{x} + 150\sqrt{y} = 100\sqrt{x} + 150\sqrt{40 - x}$. Solve $P'(x) = \frac{50}{\sqrt{x}} - \frac{75}{\sqrt{40 - x}} = 0$ to obtain $x = \frac{160}{13} \approx \$12.31$. Since $P(0) = 300\sqrt{10} \approx 948.68$ and $P(40) = 200\sqrt{10} \approx 632.46$, we have that the maximum yield is $P\left(\frac{160}{13}\right) = 100\sqrt{13}\sqrt{10} \approx 1140.18$ when $x = \frac{160}{13} \approx \$12.31$ and $y = \frac{360}{13} \approx \$27.69$.

21. Find the dimensions x and y of the rectangle inscribed in a circle of radius r that maximizes the quantity xy^2.

SOLUTION Place the center of the circle of radius r at the origin with the sides of the rectangle (of lengths $x > 0$ and $y > 0$) parallel to the coordinate axes. By the Pythagorean Theorem, we have $(\frac{x}{2})^2 + (\frac{y}{2})^2 = r^2$, whence $y^2 = 4r^2 - x^2$. Let $f(x) = xy^2 = 4xr^2 - x^3$. Allowing for degenerate rectangles, we have $0 \le x \le 2r$. Solve $f'(x) = 4r^2 - 3x^2$ for $x \ge 0$ to obtain $x = \frac{2r}{\sqrt{3}}$. Since $f(0) = f(2r) = 0$, the maximal value of f is $f(\frac{2r}{\sqrt{3}}) = \frac{16}{9}\sqrt{3}r^3$ when $x = \frac{2r}{\sqrt{3}}$ and $y = 2\sqrt{\frac{2}{3}}r$.

23. Consider a rectangular industrial warehouse consisting of three separate spaces of equal size as in Figure 17. Assume that the wall materials cost \$200 per linear ft and the company allocates \$2,400,000 for the project.

(a) Which dimensions maximize the total area of the warehouse?

(b) What is the area of each compartment in this case?

FIGURE 17

SOLUTION Let the dimensions of one compartment be height x and width y. Then the perimeter of the warehouse is given by $P = 4x + 6y$ and the constraint equation is cost $= 2,400,000 = 200(4x + 6y)$, which gives $y = 2000 - \frac{2}{3}x$.

(a) Area is given by $A = 3xy = 3x\left(2000 - \frac{2}{3}x\right) = 6000x - 2x^2$, where $0 \le x \le 3000$. Then $A'(x) = 6000 - 4x = 0$ yields $x = 1500$ and consequently $y = 1000$. Since $A(0) = A(3000) = 0$ and $A(1500) = 4,500,000$, the area of the warehouse is maximized with a height of 1500 feet and a total width of 3000 feet.

(b) The area of one compartment is $1500 \cdot 1000 = 1,500,000$ square feet.

25. The amount of light reaching a point at a distance r from a light source A of intensity I_A is I_A/r^2. Suppose that a second light source B of intensity $I_B = 4I_A$ is located 10 ft from A. Find the point on the segment joining A and B where the total amount of light is at a minimum.

SOLUTION   Place the segment in the xy-plane with A at the origin and B at $(10, 0)$. Let x be the distance from A. Then $10 - x$ is the distance from B. The total amount of light is

$$f(x) = \frac{I_A}{x^2} + \frac{I_B}{(10-x)^2} = I_A\left(\frac{1}{x^2} + \frac{4}{(10-x)^2}\right).$$

Solve

$$f'(x) = I_A\left(\frac{8}{(10-x)^3} - \frac{2}{x^3}\right) = 0$$

for $0 \leq x \leq 10$ to obtain

$$4 = \frac{(10-x)^3}{x^3} = \left(\frac{10}{x} - 1\right)^3 \quad \text{or} \quad x = \frac{10}{1 + \sqrt[3]{4}} \approx 3.86 \text{ ft.}$$

Since $f(x) \to \infty$ as $x \to 0+$ and $x \to 10-$ we conclude that the minimal amount of light occurs 3.86 feet from A.

27. According to postal regulations, a carton is classified as "oversized" if the sum of its height and girth (the perimeter of its base) exceeds 108 in. Find the dimensions of a carton with square base that is not oversized and has maximum volume.

SOLUTION   Let h denote the height of the carton and s denote the side length of the square base. Clearly the volume will be maximized when the sum of the height and girth equals 108; i.e., $4s + h = 108$, whence $h = 108 - 4s$. Allowing for degenerate cartons, the carton's volume is $V(s) = s^2 h = s^2(108 - 4s)$, where $0 \leq s \leq 27$. Solve $V'(s) = 216s - 12s^3 = 0$ for s to obtain $s = 0$ or $s = 18$. Since $V(0) = V(27) = 0$, the maximum volume is $V(18) = 11664 \text{ in}^3$ when $s = 18$ in and $h = 36$ in.

29. What is the area of the largest rectangle that can be circumscribed around a rectangle of sides L and H? *Hint:* Express the area of the circumscribed rectangle in terms of the angle θ (Figure 19).

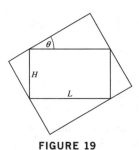

FIGURE 19

SOLUTION   Position the $L \times H$ rectangle in the first quadrant of the xy-plane with its "northwest" corner at the origin. Let θ be the angle the base of the circumscribed rectangle makes with the positive x-axis, where $0 \leq \theta \leq \frac{\pi}{2}$. Then the area of the circumscribed rectangle is $A = LH + 2 \cdot \frac{1}{2}(H \sin \theta)(H \cos \theta) + 2 \cdot \frac{1}{2}(L \sin \theta)(L \cos \theta) = LH + \frac{1}{2}(L^2 + H^2) \sin 2\theta$, which has a maximum value of $LH + \frac{1}{2}(L^2 + H^2)$ when $\theta = \frac{\pi}{4}$ because $\sin 2\theta$ achieves its maximum when $\theta = \frac{\pi}{4}$.

31. An 8-billion-bushel corn crop brings a price of \$2.40/bushel. A commodity broker uses the following rule of thumb: If the crop is reduced by x percent, then the price increases by $10x$ cents. Which crop size results in maximum revenue and what is the price per bushel? *Hint:* Revenue is equal to price times crop size.

SOLUTION   Let x denote the percentage reduction in crop size. Then the price for corn is $2.40 + 0.10x$, the crop size is $8(1 - 0.01x)$ and the revenue (in billions of dollars) is

$$R(x) = (2.4 + .1x)8(1 - .01x) = 8(-.001x^2 + .076x + 2.4),$$

where $0 \leq x \leq 100$. Solve

$$R'(x) = -.002x + .076 = 0$$

to obtain $x = 38$ percent. Since $R(0) = 19.2$, $R(38) = 30.752$, and $R(100) = 0$, revenue is maximized when $x = 38$. So we reduce the crop size to

$$8(1 - .38) = 4.96 \text{ billion bushels.}$$

The price would be $\$2.40 + .10(38) = 2.40 + 3.80 = \$6.20$.

33. Let $P = (a, b)$ be a point in the first quadrant.

(a) Find the slope of the line through P such that the triangle bounded by this line and the axes in the first quadrant has minimal area.

(b) Show that P is the midpoint of the hypotenuse of this triangle.

SOLUTION   Let $P(a, b)$ be a point in the first quadrant (thus $a, b > 0$) and $y - b = m(x - a)$, $-\infty < m < 0$, be a line through P that cuts the positive x- and y-axes. Then $y = L(x) = m(x - a) + b$.

(a) The line $L(x)$ intersects the y-axis at $H(0, b - am)$ and the x-axis at $W\left(a - \frac{b}{m}, 0\right)$. Hence the area of the triangle is

$$A(m) = \frac{1}{2}(b - am)\left(a - \frac{b}{m}\right) = ab - \frac{1}{2}a^2 m - \frac{1}{2}b^2 m^{-1}.$$

Solve $A'(m) = \frac{1}{2}b^2 m^{-2} - \frac{1}{2}a^2 = 0$ for $m < 0$ to obtain $m = -\frac{b}{a}$. Since $A \to \infty$ as $m \to -\infty$ or $m \to 0-$, we conclude that the minimal triangular area is obtained when $m = -\frac{b}{a}$.

(b) For $m = -b/a$, we have $H(0, 2b)$ and $W(2a, 0)$. The midpoint of the line segment connecting H and W is thus $P(a, b)$.

35. Figure 20 shows a rectangular plot of size 100×200 feet. Pipe is to be laid from A to a point P on side BC and from there to C. The cost of laying pipe through the lot is \$30/ft (since it must be underground) and the cost along the side of the plot is \$15/ft.

(a) Let $f(x)$ be the total cost, where x is the distance from P to B. Determine $f(x)$, but note that f is discontinuous at $x = 0$ (when $x = 0$, the cost of the entire pipe is \$15/ft).

(b) What is the most economical way to lay the pipe? What if the cost along the sides is \$24/ft?

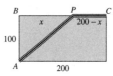

FIGURE 20

SOLUTION

(a) Let x be the distance from P to B. If $x > 0$, then the length of the underground pipe is $\sqrt{100^2 + x^2}$ and the length of the pipe along the side of the plot is $200 - x$. The total cost is

$$f(x) = 30\sqrt{100^2 + x^2} + 15(200 - x).$$

If $x = 0$, all of the pipe is along the side of the plot and $f(0) = 15(200 + 100) = \$4500$.

(b) To locate the critical points of f, solve

$$f'(x) = \frac{30x}{\sqrt{100^2 + x^2}} - 15 = 0.$$

We find $x = \pm 100/\sqrt{3}$. Note that only the positive value is in the domain of the problem. Because $f(0) = \$4500$, $f(100/\sqrt{3}) = \$5598.08$ and $f(200) = \$6708.20$, the most economical way to lay the pipe is to place the pipe along the side of the plot.

If the cost of laying the pipe along the side of the plot is \$24 per foot, then

$$f(x) = 30\sqrt{100^2 + x^2} + 24(200 - x)$$

and

$$f'(x) = \frac{30x}{\sqrt{100^2 + x^2}} - 24.$$

The only critical point in the domain of the problem is $x = 400/3$. Because $f(0) = \$7200$, $f(400/3) = \$6600$ and $f(200) = \$6708.20$, the most economical way to lay the pipe is place the underground pipe from A to a point 133.3 feet to the right of B and continuing to C along the side of the plot.

37. In Example 6 in this section, find the x-coordinate of the point P where the light beam strikes the mirror if $h_1 = 10$, $h_2 = 5$, and $L = 20$.

SOLUTION   Substitute $h_1 = 10$ feet, $h_2 = 5$ feet, and $L = 20$ feet into

$$\frac{x}{\sqrt{x^2 + h_1^2}} = \frac{L - x}{\sqrt{(L - x)^2 + h_2^2}} \quad \text{to obtain} \quad \frac{x}{\sqrt{x^2 + 10^2}} = \frac{20 - x}{\sqrt{(20 - x)^2 + 5^2}}.$$

Solving for x gives $x = \frac{40}{3}$ feet.

In Exercises 38–40, a box (with no top) is to be constructed from a piece of cardboard of sides A and B by cutting out squares of length h from the corners and folding up the sides (Figure 21).

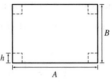

FIGURE 21

39. Which value of h maximizes the volume if $A = B$?

SOLUTION   When $A = B$, the volume of the box is

$$V(h) = hxy = h(A - 2h)^2 = 4h^3 - 4Ah^2 + A^2h,$$

where $0 \le h \le \frac{A}{2}$ (allowing for degenerate boxes). Solve $V'(h) = 12h^2 - 8Ah + A^2 = 0$ for h to obtain $h = \frac{A}{2}$ or $h = \frac{A}{6}$. Because $V(0) = V(\frac{A}{2}) = 0$ and $V(\frac{A}{6}) = \frac{2}{27}A^3$, volume is maximized when $h = \frac{A}{6}$.

41. The monthly output P of a light bulb factory is given by the formula $P = 350LK$, where L is the amount invested in labor and K the amount invested in equipment (in thousands of dollars). If the company needs to produce 10,000 units per month, how should the investment be divided among labor and equipment to minimize the cost of production? The cost of production is $L + K$.

SOLUTION   Since $P = 10000$ and $P = 350LK$, we have $L = \dfrac{200}{7K}$. Accordingly, the cost of production is

$$C(K) = L + K = K + \frac{200}{7K}.$$

Solve $C'(K) = 1 - \dfrac{200}{7K^2}$ for $K \ge 0$ to obtain $K = \frac{10}{7}\sqrt{14}$. Since $C(K) \to \infty$ as $K \to 0+$ and as $K \to \infty$, the minimum cost of production is achieved for $K = L = \frac{10}{7}\sqrt{14} \approx 5.345$ or \$5435 invested in both labor and equipment.

43. Janice can swim 3 mph and run 8 mph. She is standing at one bank of a river that is 300 ft wide and wants to reach a point located 200 ft downstream on the other side as quickly as possible. She will swim diagonally across the river and then jog along the river bank. Find the best route for Janice to take.

SOLUTION   Let lengths be in feet, times in seconds, and speeds in ft/s. Let x be the distance from the point directly opposite Janice on the other shore to the point where she *starts* to run after having swum. Then $\sqrt{x^2 + 300^2}$ is the distance she swims and $200 - x$ is the distance she runs. Janice swims at $\frac{22}{5}$ ft/s (i.e., 3 mph) and runs at $\frac{176}{15}$ ft/s (i.e., 8 mph). The total time she travels is

$$f(x) = \frac{\sqrt{x^2 + 300^2}}{22/5} + \frac{200 - x}{176/15} = \frac{5}{22}\sqrt{x^2 + 90000} + \frac{375}{22} - \frac{15}{176}x, \quad 0 \le x \le 200.$$

Solve

$$f'(x) = \frac{5}{22}\frac{x}{\sqrt{x^2 + 90000}} - \frac{15}{176} = 0$$

to obtain $x = \frac{180}{11}\sqrt{55}$. Since $f(0) = \frac{4875}{22} \approx 221.59$ and $f(200) = \frac{750}{11}\sqrt{13} \approx 245.83$, we conclude that the minimum amount of time $f\left(\frac{180}{11}\sqrt{55}\right) = \frac{375}{44}\sqrt{55} + \frac{375}{22} \approx 80.35$ s occurs when Janice swims $\sqrt{x^2 + 300^2} = \frac{480}{11}\sqrt{55} \approx 323.62$ ft and runs $200 - x = 200 - \frac{180}{11}\sqrt{55} \approx 78.64$ ft.

45. **(a)** Find the radius and height of a cylindrical can of total surface area A whose volume is as large as possible.
(b) Can you design a cylinder with total surface area A and minimal total volume?

SOLUTION   Let a closed cylindrical can be of radius r and height h.

(a) Its total surface area is $S = 2\pi r^2 + 2\pi rh = A$, whence $h = \dfrac{A}{2\pi r} - r$. Its volume is thus $V(r) = \pi r^2 h = \frac{1}{2}Ar - \pi r^3$, where $0 < r \le \sqrt{\frac{A}{2\pi}}$. Solve $V'(r) = \frac{1}{2}A - 3\pi r^2$ for $r > 0$ to obtain $r = \sqrt{\dfrac{A}{6\pi}}$. Since $V(0) = V(\sqrt{\frac{A}{2\pi}}) = 0$ and

$$V\left(\sqrt{\frac{A}{6\pi}}\right) = \frac{\sqrt{6}A^{3/2}}{18\sqrt{\pi}},$$

the maximum volume is achieved when

$$r = \sqrt{\frac{A}{6\pi}} \quad \text{and} \quad h = \frac{1}{3}\sqrt{\frac{6A}{\pi}}.$$

(b) For a can of total surface area A, there are cans of arbitrarily small volume since $\lim\limits_{r \to 0+} V(r) = 0$.

47. A billboard of height b is mounted on the side of a building with its bottom edge at a distance h from the street. At what distance x should an observer stand from the wall to maximize the angle of observation θ (Figure 23)?

(a) Find x using calculus. You may wish to use the addition formula

$$\cot(a - b) = \frac{1 + \cot a \cot b}{\cot b - \cot a}.$$

(b) Solve the problem again using geometry (without any calculation!). There is a unique circle passing through points B and C which is tangent to the street. Let R be the point of tangency. Show that θ is maximized at the point R. *Hint:* The two angles labeled ψ are, in fact, equal because they subtend equal arcs on the circle. Let A be the intersection of the circle with PC and show that $\psi = \theta + \angle PBA > \theta$.

(c) Prove that the two answers in (a) and (b) agree.

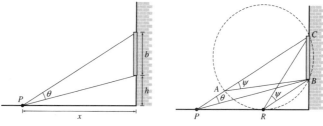

FIGURE 23

SOLUTION

(a) From the leftmost diagram in Figure 23 and the addition formula for the cotangent function, we see that

$$\cot \theta = \frac{1 + \frac{x}{b+h}\frac{x}{h}}{\frac{x}{h} - \frac{x}{b+h}} = \frac{x^2 + h(b+h)}{bx},$$

where b and h are constant. Now, differentiate with respect to x and solve

$$-\csc^2 \theta \frac{d\theta}{dx} = \frac{x^2 - h(b+h)}{bx^2} = 0$$

to obtain $x = \sqrt{bh + h^2}$. Since this is the only critical point, and since $\theta \to 0$ as $x \to 0+$ and $\theta \to 0$ as $x \to \infty$, $\theta(x)$ reaches its maximum at $x = \sqrt{bh + h^2}$.

(b) Following the directions, and mindful of the diagram in Figure 23, let C be the point at the top of the painting along the wall. For every radius r, there is a unique circle of radius r through B and C with center to the front of the billboard. Only one of these is tangent to the street (all the others are either too big or too small). Draw such a circle (call it D), as shown in the right half of Figure 23, and let R be the point where the circle touches the street.

Let P be any other such point along the street. We will prove that the angle of observation θ at P is less than the angle of observation ψ at R, thus proving that R is the point with maximum angle of observation along the floor. P, C, and B form a triangle. Let A be the topmost point of intersection of the triangle and circle D as shown. Because the two angles both subtend the arc BC, the measure of angle $\angle CAB$ is equal to the measure of angle $\angle CRB$. By supplementarity, $\psi + \angle PAB = 180°$. Because they form a triangle, $\theta + \angle PAB + \angle PBA = 180°$. Thus, $\psi = \theta + \angle PBA > \theta$.

(c) To show that the two answers agree, let O be the center of the circle. One observes that if d is the distance from R to the wall, then O has coordinates $(-d, \frac{b}{2} + h)$. This is because the height of the center is equidistant from points B and C and because the center must lie directly above R if the circle is tangent to the floor.

Now we can solve for d. The radius of the circle is clearly $\frac{b}{2} + h$, by the distance formula:

$$\overline{OB}^2 = d^2 + \left(\frac{b}{2} + h - h\right)^2 = \left(\frac{b}{2} + h\right)^2$$

This gives

$$d^2 = \left(\frac{b}{2} + h\right)^2 - \left(\frac{b}{2}\right)^2 = bh + h^2$$

or $d = \sqrt{bh + h^2}$ as claimed.

49. Snell's Law, derived in Exercise 48, explains why it is impossible to see above the water if the angle of vision θ_2 is such that $\sin \theta_2 > v_2/v_1$.

(a) Show that if this inequality holds, then there is no angle θ_1 for which Snell's Law holds.

(b) What will you see if you look through the water with an angle of vision θ_2 such that $\sin \theta_2 > v_2/v_1$?

SOLUTION

(a) Suppose $\sin \theta_2 > \dfrac{v_2}{v_1}$. By Snell's Law, $\dfrac{\sin \theta_1}{v_1} = \dfrac{\sin \theta_2}{v_2}$, whence $\sin \theta_1 = \dfrac{v_1}{v_2} \sin \theta_2 > \dfrac{v_1}{v_2} \cdot \dfrac{v_2}{v_1} = 1$, a contradiction, since $\sin \theta \leq 1$ for all real values of θ. Accordingly, there is no real value θ_1 for which Snell's Law holds.

(b) The underwater observer sees what appears to be a silvery mirror due to total internal reflection.

51. Vascular Branching A small blood vessel of radius r branches off at an angle θ from a larger vessel of radius R to supply blood along a path from A to B. According to Poiseuille's Law, the total resistance to blood flow is proportional to

$$T = \left(\frac{a - b \cot \theta}{R^4} + \frac{b \csc \theta}{r^4} \right)$$

where a and b are as in Figure 25. Show that the total resistance is minimized when $\cos \theta = (r/R)^4$.

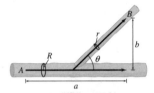

FIGURE 25

SOLUTION With $a, b, r, R > 0$ and $R > r$, let $T(\theta) = \left(\dfrac{a - b \cot \theta}{R^4} + \dfrac{b \csc \theta}{r^4} \right)$. Set

$$T'(\theta) = \left(\frac{b \csc^2 \theta}{R^4} - \frac{b \csc \theta \cot \theta}{r^4} \right) = 0.$$

Then

$$\frac{b \left(r^4 - R^4 \cos \theta \right)}{R^4 r^4 \sin^2 \theta} = 0,$$

so that $\cos \theta = \left(\dfrac{r}{R} \right)^4$. Since $\lim\limits_{\theta \to 0+} T(\theta) = \infty$ and $\lim\limits_{\theta \to \pi-} T(\theta) = \infty$, the minimum value of $T(\theta)$ occurs when $\cos \theta = \left(\dfrac{r}{R} \right)^4$.

53. Let (a, b) be a fixed point in the first quadrant and let $S(d)$ be the sum of the distances from $(d, 0)$ to the points $(0, 0)$, (a, b), and $(a, -b)$.

(a) Find the value of d for which $S(d)$ is minimal. The answer depends on whether $b < \sqrt{3}a$ or $b \geq \sqrt{3}a$. *Hint:* Show that $d = 0$ when $b \geq \sqrt{3}a$.

(b) [GU] Let $a = 1$. Plot $S(d)$ for $b = 0.5, \sqrt{3}, 3$ and describe the position of the minimum.

SOLUTION

(a) If $d < 0$, then the distance from $(d, 0)$ to the other three points can all be reduced by increasing the value of d. Similarly, if $d > a$, then the distance from $(d, 0)$ to the other three points can all be reduced by decreasing the value of d. It follows that the minimum of $S(d)$ must occur for $0 \leq d \leq a$. Restricting attention to this interval, we find

$$S(d) = d + 2\sqrt{(d - a)^2 + b^2}.$$

Solving

$$S'(d) = 1 + \frac{2(d - a)}{\sqrt{(d - a)^2 + b^2}} = 0$$

yields the critical point $d = a - b/\sqrt{3}$. If $b < \sqrt{3}a$, then $d = a - b/\sqrt{3} > 0$ and the minimum occurs at this value of d. On the other hand, if $b \geq \sqrt{3}a$, then the minimum occurs at the endpoint $d = 0$.

(b) Let $a = 1$. Plots of $S(d)$ for $b = 0.5, b = \sqrt{3}$ and $b = 3$ are shown below. For $b = 0.5$, the results of (a) indicate the minimum should occur for $d = 1 - 0.5/\sqrt{3} \approx 0.711$, and this is confirmed in the plot. For both $b = \sqrt{3}$ and $b = 3$, the results of (a) indicate that the minimum should occur at $d = 0$, and both of these conclusions are confirmed in the plots.

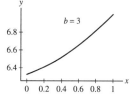

55. In the setting of Exercise 54, show that for any f the minimal force required is proportional to $1/\sqrt{1+f^2}$.

SOLUTION   We minimize $F(\theta)$ by finding the maximum value $g(\theta) = \cos\theta + f\sin\theta$. The angle θ is restricted to the interval $[0, \frac{\pi}{2}]$. We solve for the critical points:

$$g'(\theta) = -\sin\theta + f\cos\theta = 0$$

We obtain

$$f\cos\theta = \sin\theta \Rightarrow \tan\theta = f$$

From the figure below we find that $\cos\theta = 1/\sqrt{1+f^2}$ and $\sin\theta = f/\sqrt{1+f^2}$. Hence

$$g(\theta) = \frac{1}{f} + \frac{f^2}{\sqrt{1+f^2}} = \frac{1+f^2}{\sqrt{1+f^2}} = \sqrt{1+f^2}$$

The values at the endpoints are

$$g(0) = 1, \qquad g\left(\frac{\pi}{2}\right) = f$$

Both of these values are less than $\sqrt{1+f^2}$. Therefore the maximum value of $g(\theta)$ is $\sqrt{1+f^2}$ and the minimum value of $F(\theta)$ is

$$F = \frac{fmg}{g(\theta)} = \frac{fmg}{\sqrt{1+f^2}}$$

57. Find the maximum length of a pole that can be carried horizontally around a corner joining corridors of widths 8 ft and 4 ft (Figure 29).

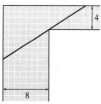

FIGURE 29

SOLUTION   In order to find the length of the *longest* pole that can be carried around the corridor, we have to find the *shortest* length from the left wall to the top wall touching the corner of the inside wall. Any pole that does not fit in this shortest space cannot be carried around the corner, so an exact fit represents the longest possible pole.

Let θ be the angle between the pole and a horizontal line to the right. Let c_1 be the length of pole in the 8 ft corridor and let c_2 be the length of pole in the 4 foot corridor. By the definitions of sine and cosine,

$$\frac{4}{c_2} = \sin\theta \quad \text{and} \quad \frac{8}{c_1} = \cos\theta,$$

so that $c_1 = \frac{8}{\cos \theta}$, $c_2 = \frac{4}{\sin \theta}$. What must be minimized is the total length, given by

$$f(\theta) = \frac{8}{\cos \theta} + \frac{4}{\sin \theta}.$$

Setting $f'(\theta) = 0$ yields

$$\frac{8 \sin \theta}{\cos^2 \theta} - \frac{4 \cos \theta}{\sin^2 \theta} = 0$$

$$\frac{8 \sin \theta}{\cos^2 \theta} = \frac{4 \cos \theta}{\sin^2 \theta}$$

$$8 \sin^3 \theta = 4 \cos^3 \theta$$

As $\theta < \frac{\pi}{2}$ (the pole is being turned around a corner, after all), we can divide both sides by $\cos^3 \theta$, getting $\tan^3 \theta = \frac{1}{2}$. This implies that $\tan \theta = \left(\frac{1}{2}\right)^{1/3}$ ($\tan \theta > 0$ as the angle is acute).

Since $f(\theta) \to \infty$ as $\theta \to 0+$ and as $\theta \to \frac{\pi}{2}-$, we can tell that the *minimum* is attained at θ_0 where $\tan \theta_0 = \left(\frac{1}{2}\right)^{1/3}$. Because

$$\tan \theta_0 = \frac{\text{opposite}}{\text{adjacent}} = \frac{1}{2^{1/3}},$$

we draw a triangle with opposite side 1 and adjacent side $2^{1/3}$.

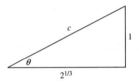

By Pythagoras, $c = \sqrt{1 + 2^{2/3}}$, so

$$\sin \theta_0 = \frac{1}{\sqrt{1 + 2^{2/3}}} \quad \text{and} \quad \cos \theta_0 = \frac{2^{1/3}}{\sqrt{1 + 2^{2/3}}}.$$

From this, we get

$$f(\theta_0) = \frac{8}{\cos \theta_0} + \frac{4}{\sin \theta_0} = \frac{8}{2^{1/3}}\sqrt{1 + 2^{2/3}} + 4\sqrt{1 + 2^{2/3}} = 4\sqrt{1 + 2^{2/3}}\left(2^{2/3} + 1\right).$$

59. Find the isosceles triangle of smallest area that circumscribes a circle of radius 1 (from Thomas Simpson's *The Doctrine and Application of Fluxions*, a calculus text that appeared in 1750). See Figure 30.

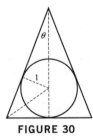

FIGURE 30

SOLUTION   From the diagram, we see that the height h and base b of the triangle are $h = 1 + \csc \theta$ and $b = 2h \tan \theta = 2(1 + \csc \theta) \tan \theta$. Thus, the area of the triangle is

$$A(\theta) = \frac{1}{2}hb = (1 + \csc \theta)^2 \tan \theta,$$

where $0 < \theta < \pi$. We now set the derivative equal to zero:

$$A'(\theta) = (1 + \csc \theta)(-2 \csc \theta + \sec^2 \theta(1 + \csc \theta)) = 0.$$

The first factor gives $\theta = 3\pi/2$ which is not in the domain of the problem. To find the roots of the second factor, multiply through by $\cos^2 \theta \sin \theta$ to obtain

$$-2 \cos^2 \theta + \sin \theta + 1 = 0,$$

or

$$2 \sin^2 \theta + \sin \theta - 1 = 0.$$

This is a quadratic equation in $\sin \theta$ with roots $\sin \theta = -1$ and $\sin \theta = 1/2$. Only the second solution is relevant and gives us $\theta = \pi/6$. Since $A(\theta) \to \infty$ as $\theta \to 0+$ and as $\theta \to \pi-$, we see that the minimum area occurs when the triangle is an equilateral triangle.

Further Insights and Challenges

61. Bird Migration Power P is the rate at which energy E is consumed per unit time. Ornithologists have found that the power consumed by a certain pigeon flying at velocity v m/s is described well by the function $P(v) = 17v^{-1} + 10^{-3}v^3$ J/s. Assume that the pigeon can store 5×10^4 J of usable energy as body fat.

(a) Find the velocity v_{pmin} that *minimizes* power consumption.

(b) Show that a pigeon flying at velocity v and using all of its stored energy can fly a total distance of $D(v) = (5 \times 10^4)v/P(v)$.

(c) Migrating birds are smart enough to fly at the velocity that maximizes distance traveled rather than minimizes power consumption. Show that the velocity v_{dmax} which maximizes total distance $D(v)$ satisfies $P'(v) = P(v)/v$. Show that v_{dmax} is obtained graphically as the velocity coordinate of the point where a line through the origin is tangent to the graph of $P(v)$ (Figure 32).

(d) Find v_{dmax} and the maximum total distance that the bird can fly.

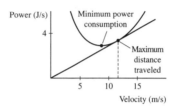

FIGURE 32

SOLUTION

(a) Let $P(v) = 17v^{-1} + 10^{-3}v^3$. Then $P'(v) = -17v^{-2} + 0.003v^2 = 0$ implies $v_{\text{pmin}} = \left(\frac{17}{.003}\right)^{1/4} \approx 8.676247$. This critical point is a minimum, because it is the only critical point and $P(v) \to \infty$ both as $v \to 0+$ and as $v \to \infty$.

(b) Flying at a velocity v, the birds will exhaust their energy store after $T = \dfrac{5 \cdot 10^4 \text{ joules}}{P(v) \text{ joules/sec}} = \dfrac{5 \cdot 10^4 \text{ sec}}{P(v)}$. The total distance traveled is then $D(v) = vT = \dfrac{5 \cdot 10^4 v}{P(v)}$.

(c) $D'(v) = \dfrac{P(v) \cdot 5 \cdot 10^4 - 5 \cdot 10^4 v \cdot P'(v)}{(P(v))^2} = 5 \cdot 10^4 \dfrac{P(v) - vP'(v)}{(P(v))^2} = 0$ implies $P(v) - vP'(v) = 0$, or $P'(v) = \dfrac{P(v)}{v}$. Since $D(v) \to 0$ as $v \to 0$ and as $v \to \infty$, the critical point determined by $P'(v) = P(v)/v$ corresponds to a maximum.

Graphically, the expression

$$\frac{P(v)}{v} = \frac{P(v) - 0}{v - 0}$$

is the slope of the line passing through the origin and $(v, P(v))$. The condition $P'(v) = P(v)/v$ which defines v_{dmax} therefore indicates that v_{dmax} is the velocity component of the point where a line through the origin is tangent to the graph of $P(v)$.

(d) Using $P'(v) = \dfrac{P(v)}{v}$ gives

$$-17v^{-2} + .003v^2 = \frac{17v^{-1} + .001v^3}{v} = 17v^{-2} + .001v^2,$$

which simplifies to $.002v^4 = 34$ and thus $v_{\text{dmax}} \approx 11.418583$. The maximum total distance is given by $D(v_{\text{dmax}}) = \dfrac{5 \cdot 10^4 \cdot v_{\text{dmax}}}{P(v_{\text{dmax}})} = 191.741$ kilometers.

Seismic Prospecting *Exercises 62–64 are concerned with determining the thickness d of a layer of soil that lies on top of a rock formation. Geologists send two sound pulses from point A to point D separated by a distance s. The first pulse travels directly from A to D along the surface of the earth. The second pulse travels down to the rock formation, then along its surface, and then back up to D (path ABCD), as in Figure 33. The pulse travels with velocity v_1 in the soil and v_2 in the rock.*

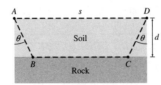

FIGURE 33

63. In this exercise, assume that $v_2/v_1 \geq \sqrt{1 + 4(d/s)^2}$.

(a) Show that inequality (1) holds if $\sin\theta = v_1/v_2$.

(b) Show that the minimal time for the second pulse is

$$t_2 = \frac{2d}{v_1}(1 - k^2)^{1/2} + \frac{s}{v_2}$$

where $k = v_1/v_2$.

(c) Conclude that $\dfrac{t_2}{t_1} = \dfrac{2d(1 - k^2)^{1/2}}{s} + k$.

SOLUTION

(a) If $\sin\theta = \frac{v_1}{v_2}$, then

$$\tan\theta = \frac{v_1}{\sqrt{v_2^2 - v_1^2}} = \frac{1}{\sqrt{\left(\frac{v_2}{v_1}\right)^2 - 1}}.$$

Because $\frac{v_2}{v_1} \geq \sqrt{1 + 4(\frac{d}{s})^2}$, it follows that

$$\sqrt{\left(\frac{v_2}{v_1}\right)^2 - 1} \geq \sqrt{1 + 4\left(\frac{d}{s}\right)^2 - 1} = \frac{2d}{s}.$$

Hence, $\tan\theta \leq \frac{s}{2d}$ as required.

(b) For the time-minimizing choice of θ, we have $\sin\theta = \frac{v_1}{v_2}$ from which $\sec\theta = \frac{v_2}{\sqrt{v_2^2 - v_1^2}}$ and $\tan\theta = \frac{v_1}{\sqrt{v_2^2 - v_1^2}}$.

Thus

$$t_2 = \frac{2d}{v_1}\sec\theta + \frac{s - 2d\tan\theta}{v_2} = \frac{2d}{v_1}\frac{v_2}{\sqrt{v_2^2 - v_1^2}} + \frac{s - 2d\frac{v_1}{\sqrt{v_2^2 - v_1^2}}}{v_2}$$

$$= \frac{2d}{v_1}\left(\frac{v_2}{\sqrt{v_2^2 - v_1^2}} - \frac{v_1^2}{v_2\sqrt{v_2^2 - v_1^2}}\right) + \frac{s}{v_2}$$

$$= \frac{2d}{v_1}\left(\frac{v_2^2 - v_1^2}{v_2\sqrt{v_2^2 - v_1^2}}\right) + \frac{s}{v_2} = \frac{2d}{v_1}\left(\frac{\sqrt{v_2^2 - v_1^2}}{\sqrt{v_2^2}}\right) + \frac{s}{v_2}$$

$$= \frac{2d}{v_1}\sqrt{1 - \left(\frac{v_1}{v_2}\right)^2} + \frac{s}{v_2} = \frac{2d\left(1 - k^2\right)^{1/2}}{v_1} + \frac{s}{v_2}.$$

(c) Recall that $t_1 = \frac{s}{v_1}$. We therefore have

$$\frac{t_2}{t_1} = \frac{\frac{2d\left(1 - k^2\right)^{1/2}}{v_1} + \frac{s}{v_2}}{\frac{s}{v_1}}$$

$$= \frac{2d\left(1 - k^2\right)^{1/2}}{s} + \frac{v_1}{v_2} = \frac{2d\left(1 - k^2\right)^{1/2}}{s} + k.$$

65. Three towns A, B, and C are to be joined by an underground fiber cable as illustrated in Figure 34(A). Assume that C is located directly below the midpoint of $\overline{AB}$. Find the junction point P that minimizes the total amount of cable used.

(a) First show that P must lie directly above C. Show that if the junction is placed at a point Q as in Figure 34(B), then we can reduce the cable length by moving Q horizontally over to the point P lying above C. You may want to use the result of Example 6.

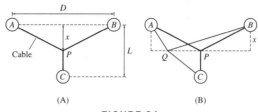

FIGURE 34

(b) With x as in Figure 34(A), let $f(x)$ be the total length of cable used. Show that $f(x)$ has a unique critical point c. Compute c and show that $0 \le c \le L$ if and only if $D \le 2\sqrt{3}\,L$.

(c) Find the minimum of $f(x)$ on $[0, L]$ in two cases: $D = 2$, $L = 4$ and $D = 8$, $L = 2$.

SOLUTION

(a) Look at diagram 34(B). Let T be the point directly above Q on $\overline{AB}$. Let $s = AT$ and $D = AB$ so that $TB = D - s$. Let ℓ be the total length of cable from A to Q and B to Q. By the Pythagorean Theorem applied to $\triangle AQT$ and $\triangle BQT$, we get:

$$\ell^2 = s^2 + x^2 + (D - s)^2 + x^2$$
$$= 2x^2 + D^2 - 2Ds + 2s^2 = 2x^2 + D^2 + 2s^2 - 2Ds$$
$$= 2x^2 + D^2 + 2(s - D/2)^2 - D^2/2.$$

We wish to find the value of s producing the minimum value of ℓ^2. Since D and x are not being changed, this clearly occurs where $s = D/2$, that is, where $Q = P$. Since it is obvious that $PC \le QC$ (QC is the hypotenuse of the triangle $\triangle PQC$), it follows that total cable length is minimized at $Q = P$.

(b) Let $f(x)$ be the total cable length. From diagram 34(A), we get:

$$f(x) = (L - x) + 2\sqrt{x^2 + D^2/4}.$$

Then

$$f'(x) = -1 + \frac{2x}{\sqrt{x^2 + D^2/4}} = 0$$

gives

$$2x = \sqrt{x^2 + D^2/4}$$

or

$$4x^2 = x^2 + D^2/4$$

and the critical point is

$$c = D/2\sqrt{3}.$$

This is the only critical point of f. It lies in the interval $[0, L]$ if and only if $c \le L$, or

$$D \le 2\sqrt{3}L.$$

(c) The minimum of f will depend on whether $D \le 2\sqrt{3}L$.

- $D = 2$, $L = 4$; $2\sqrt{3}L = 8\sqrt{3} > D$, so $c = D/(2\sqrt{3}) = \sqrt{3}/3 \in [0, L]$. $f(0) = L + D = 6$, $f(L) = 2\sqrt{L^2 + D^2/4} = 2\sqrt{17} \approx 8.24621$, and $f(c) = 4 - (\sqrt{3}/3) + 2\sqrt{\frac{1}{3} + 1} = 4 + \sqrt{3} \approx 5.73204$. Therefore, the total length is minimized where $x = c = \sqrt{3}/3$.
- $D = 8$, $L = 2$; $2\sqrt{3}L = 4\sqrt{3} < D$, so c does not lie in the interval $[0, L]$. $f(0) = 2 + 2\sqrt{64/4} = 10$, and $f(L) = 0 + 2\sqrt{4 + 64/4} = 2\sqrt{20} = 4\sqrt{5} \approx 8.94427$. Therefore, the total length is minimized were $x = L$, or where $P = C$.

4.7 Newton's Method

Preliminary Questions

1. How many iterations of Newton's Method are required to compute a root if $f(x)$ is a linear function?

SOLUTION   Remember that Newton's Method uses the linear approximation of a function to estimate the location of a root. If the original function is linear, then only one iteration of Newton's Method will be required to compute the root.

2. What happens in Newton's Method if your initial guess happens to be a zero of f?

SOLUTION   If x_0 happens to be a zero of f, then

$$x_1 = x_0 - \frac{f(x_0)}{f'(x_0)} = x_0 - 0 = x_0;$$

in other words, every term in the Newton's Method sequence will remain x_0.

3. What happens in Newton's Method if your initial guess happens to be a local min or max of f?

SOLUTION   Assuming that the function is differentiable, then the derivative is zero at a local maximum or a local minimum. If Newton's Method is started with an initial guess such that $f'(x_0) = 0$, then Newton's Method will fail in the sense that x_1 will not be defined. That is, the tangent line will be parallel to the x-axis and will never intersect it.

4. Is the following a reasonable description of Newton's Method: "A root of the equation of the tangent line to $f(x)$ is used as an approximation to a root of $f(x)$ itself"? Explain.

SOLUTION   Yes, that is a reasonable description. The iteration formula for Newton's Method was derived by solving the equation of the tangent line to $y = f(x)$ at x_0 for its x-intercept.

Exercises

In Exercises 1–4, use Newton's Method with the given function and initial value x_0 to calculate x_1, x_2, x_3.

1. $f(x) = x^2 - 2$, $x_0 = 1$

SOLUTION   Let $f(x) = x^2 - 2$ and define

$$x_{n+1} = x_n - \frac{f(x_n)}{f'(x_n)} = x_n - \frac{x_n^2 - 2}{2x_n}.$$

With $x_0 = 1$, we compute

n	1	2	3
x_n	1.5	1.416666667	1.414215686

3. $f(x) = x^3 - 5$, $x_0 = 1.6$

SOLUTION   Let $f(x) = x^3 - 5$ and define

$$x_{n+1} = x_n - \frac{f(x_n)}{f'(x_n)} = x_n - \frac{x_n^3 - 5}{3x_n^2}.$$

With $x_0 = 1.6$ we compute

n	1	2	3
x_n	1.717708333	1.710010702	1.709975947

5. Use Figure 6 to choose an initial guess x_0 to the unique real root of $x^3 + 2x + 5 = 0$. Then compute the first three iterates of Newton's Method.

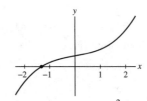

FIGURE 6 Graph of $y = x^3 + 2x + 5$.

SOLUTION   Let $f(x) = x^3 + 2x + 5$ and define

$$x_{n+1} = x_n - \frac{f(x_n)}{f'(x_n)} = x_n - \frac{x_n^3 + 2x_n + 5}{3x_n^2 + 2}.$$

We take $x_0 = -1.4$, based on the figure, and then calculate

n	1	2	3
x_n	-1.330964467	-1.328272820	-1.328268856

In Exercises 7–10, use Newton's Method to approximate the root to three decimal places and compare with the value obtained from a calculator.

7. $\sqrt{10}$

SOLUTION   Let $f(x) = x^2 - 10$, and let $x_0 = 3$. Newton's Method yields:

n	1	2	3
x_n	3.16667	3.162280702	3.16227766

A calculator yields 3.16227766.

9. $5^{1/3}$

SOLUTION   Let $f(x) = x^3 - 5$, and let $x_0 = 2$. Here are approximations to the root of $f(x)$, which is $5^{1/3}$.

n	1	2	3	4
x_n	1.75	1.710884354	1.709976429	1.709975947

A calculator yields 1.709975947.

11. Use Newton's Method to approximate the largest positive root of $f(x) = x^4 - 6x^2 + x + 5$ to within an error of at most 10^{-4}. Refer to Figure 5.

SOLUTION   Figure 5 from the text suggests the largest positive root of $f(x) = x^4 - 6x^2 + x + 5$ is near 2. So let $f(x) = x^4 - 6x^2 + x + 5$ and take $x_0 = 2$.

n	1	2	3	4
x_n	2.111111111	2.093568458	2.093064768	2.093064358

The largest positive root of $x^4 - 6x^2 + x + 5$ is approximately 2.093064358.

13. GU   Use a graphing calculator to choose an initial guess for the unique positive root of $x^4 + x^2 - 2x - 1 = 0$. Calculate the first three iterates of Newton's Method.

SOLUTION   Let $f(x) = x^4 + x^2 - 2x - 1$. The graph of $f(x)$ shown below suggests taking $x_0 = 1$. Starting from $x_0 = 1$, the first three iterates of Newton's Method are:

n	1	2	3
x_n	1.25	1.189379699	1.184171279

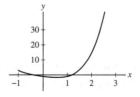

15. GU   Estimate the smallest positive solution of $\dfrac{\sin \theta}{\theta} = 0.9$ to three decimal places. Use a graphing calculator to choose the initial guess.

SOLUTION Let

$$f(\theta) = \frac{\sin \theta}{\theta} - .9.$$

We use the following plot to obtain an initial guess:

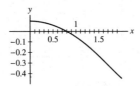

$\theta = 0.8$ seems a good first guess for the location of a zero of the function. A series of approximations based on $\theta_0 = 0.8$ follows:

n	1	2	3
θ_n	0.7867796879	0.7866830771	0.7866830718

A three digit approximation for the point where $\frac{\sin \theta}{\theta} = .9$ is 0.787. Plugging $\sin(0.787)/0.787$ into a calculator yields 0.900 to three decimal places.

17. Let x_1, x_2 be the estimates to a root obtained by applying Newton's Method with $x_0 = 1$ to the function graphed in Figure 7. Estimate the numerical values of x_1 and x_2, and draw the tangent lines used to obtain them.

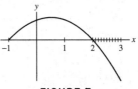

FIGURE 7

SOLUTION The graph with tangent lines drawn on it appears below. The tangent line to the curve at $(x_0, f(x_0))$ has an x-intercept at approximately $x_1 = 3.0$. The tangent line to the curve at $(x_1, f(x_1))$ has an x-intercept at approximately $x_2 = 2.2$.

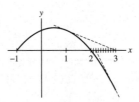

19. Find the x-coordinate to two decimal places of the first point in the region $x > 0$ where $y = x$ intersects $y = \tan x$ (draw a graph).

SOLUTION Here is a plot of $\tan x$ and x on the same axes:

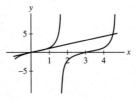

The first intersection with $x > 0$ lies on the second "branch" of $y = \tan x$, between $x = \frac{5\pi}{4}$ and $x = \frac{3\pi}{2}$. Let $f(x) = \tan x - x$. The graph suggests an initial guess $x_0 = \frac{5\pi}{4}$, from which we get the following table:

n	1	2	3	4
x_n	6.85398	21.921	4480.8	7456.27

This is clearly leading nowhere, so we need to try a better initial guess. *Note: This happens with Newton's Method—it is sometimes difficult to choose an initial guess.* We try the point directly between $\frac{5\pi}{4}$ and $\frac{3\pi}{2}$, $x_0 = \frac{11\pi}{8}$:

n	1	2	3	4	5	6	7
x_n	4.64662	4.60091	4.54662	4.50658	4.49422	4.49341	4.49341

The first point where $y = x$ and $y = \tan x$ cross is at approximately $x = 4.49341$, which is approximately 1.4303π.

Newton's Method is often used to determine interest rates in financial calculations. In Exercises 20–22, r denotes a yearly interest rate expressed as a decimal (rather than as a percent).

21. If you borrow L dollars for N years at a yearly interest rate r, your monthly payment of P dollars is calculated using the equation

$$L = P\left(\frac{1 - b^{-12N}}{b - 1}\right) \qquad \text{where } b = 1 + \frac{r}{12}.$$

(a) What is the monthly payment if $L = \$5,000$, $N = 3$, and $r = 0.08$ (8%)?
(b) You are offered a loan of $L = \$5,000$ to be paid back over 3 years with monthly payments of $P = \$200$. Use Newton's Method to compute b and find the implied interest rate r of this loan. *Hint:* Show that $(L/P)b^{12N+1} - (1 + L/P)b^{12N} + 1 = 0$.

SOLUTION

(a) $b = (1 + .08/12) = 1.00667$

$$P = L\left(\frac{b - 1}{1 - b^{-12N}}\right) = 5000\left(\frac{1.00667 - 1}{1 - 1.00667^{-36}}\right) \approx \$156.69$$

(b) Starting from

$$L = P\left(\frac{1 - b^{-12N}}{b - 1}\right),$$

divide by P, multiply by $b - 1$, multiply by b^{12N} and collect like terms to arrive at

$$(L/P)b^{12N+1} - (1 + L/P)b^{12N} + 1 = 0.$$

Since $L/P = 5000/200 = 25$, we must solve

$$25b^{37} - 26b^{36} + 1 = 0.$$

Newton's Method gives $b \approx 1.02121$ and

$$r = 12(b - 1) = 12(.02121) \approx .25452$$

So the interest rate is around 25.45%.

23. Kepler's Problem   Although planetary motion is beautifully described by Kepler's three laws (see Section 14.6), there is no simple formula for the position of a planet P along its elliptical orbit as a function of time. Kepler developed a method for locating P at time t by drawing the auxiliary dashed circle in Figure 9 and introducing the angle θ (note that P determines θ, which is the central angle of the point B on the circle). Let $a = OA$ and $e = OS/OA$ (the eccentricity of the orbit).

(a) Show that sector BSA has area $(a^2/2)(\theta - e \sin \theta)$.
(b) It follows from Kepler's Second Law that the area of sector BSA is proportional to the time t elapsed since the planet passed point A. More precisely, since the circle has area πa^2, BSA has area $(\pi a^2)(t/T)$, where T is the period of the orbit. Deduce that

$$\frac{2\pi t}{T} = \theta - e \sin \theta.$$

(c) The eccentricity of Mercury's orbit is approximately $e = 0.2$. Use Newton's Method to find θ after a quarter of Mercury's year has elapsed ($t = T/4$). Convert θ to degrees. Has Mercury covered more than a quarter of its orbit at $t = T/4$?

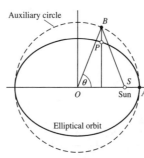

FIGURE 9

SOLUTION

(a) The sector SAB is the slice OAB with the triangle OPS removed. OAB is a central sector with arc θ and radius $\overline{OA} = a$, and therefore has area $\frac{a^2\theta}{2}$. OPS is a triangle with height $a\sin\theta$ and base length $\overline{OS} = ea$. Hence, the area of the sector is

$$\frac{a^2}{2}\theta - \frac{1}{2}ea^2\sin\theta = \frac{a^2}{2}(\theta - e\sin\theta).$$

(b) Since Kepler's second law indicates that the area of the sector is proportional to the time t since the planet passed point A, we get

$$\pi a^2(t/T) = a^2/2(\theta - e\sin\theta)$$

$$2\pi\frac{t}{T} = \theta - e\sin\theta.$$

(c) If $t = T/4$, the last equation in (b) gives:

$$\frac{\pi}{2} = \theta - e\sin\theta = \theta - .2\sin\theta.$$

Let $f(\theta) = \theta - .2\sin\theta - \frac{\pi}{2}$. We will use Newton's Method to find the point where $f(\theta) = 0$. Since a quarter of the year on Mercury has passed, a good first estimate θ_0 would be $\frac{\pi}{2}$.

n	1	2	3	4
x_n	1.7708	1.76696	1.76696	1.76696

From the point of view of the Sun, Mercury has traversed an angle of approximately 1.76696 radians $= 101.24°$. Mercury has therefore traveled more than one fourth of the way around (from the point of view of central angle) during this time.

25. What happens when you apply Newton's Method to the equation $x^3 - 20x = 0$ with the unlucky initial guess $x_0 = 2$?

SOLUTION Let $f(x) = x^3 - 20x$. Define

$$x_{n+1} = x_n - \frac{f(x_n)}{f'(x_n)} = x_n - \frac{x_n^3 - 20x_n}{3x_n^2 - 20}.$$

Take $x_0 = 2$. Then the sequence of iterates is $-2, 2, -2, 2, \dots$, which diverges by oscillation.

Further Insights and Challenges

27. The roots of $f(x) = \frac{1}{3}x^3 - 4x + 1$ to three decimal places are -3.583, 0.251, and 3.332 (Figure 10). Determine the root to which Newton's Method converges for the initial choices $x_0 = 1.85$, 1.7, and 1.55. The answer shows that a small change in x_0 can have a significant effect on the outcome of Newton's Method.

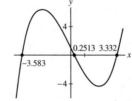

FIGURE 10 Graph of $f(x) = \frac{1}{3}x^3 - 4x + 1$.

SOLUTION Let $f(x) = \frac{1}{3}x^3 - 4x + 1$, and define

$$x_{n+1} = x_n - \frac{f(x_n)}{f'(x_n)} = x_n - \frac{\frac{1}{3}x_n^3 - 4x_n + 1}{x_n^2 - 4}.$$

- Taking $x_0 = 1.85$, we have

n	1	2	3	4	5	6	7
x_n	-5.58	-4.31	-3.73	-3.59	-3.58294362	-3.582918671	-3.58291867

- Taking $x_0 = 1.7$, we have

n	1	2	3	4	5	6	7	8	9
x_n	-2.05	-33.40	-22.35	-15.02	-10.20	-7.08	-5.15	-4.09	-3.66

n	10	11	12	13
x_n	-3.585312288	-3.582920989	-3.58291867	-3.58291867

- Taking $x_0 = 1.55$, we have

n	1	2	3	4	5	6
x_n	-0.928	0.488	0.245	0.251320515	0.251322863	0.251322863

In Exercises 28–29, consider a metal rod of length L inches that is fastened at both ends. If you cut the rod and weld on an additional m inches of rod, leaving the ends fixed, the rod will bow up into a circular arc of radius R (unknown), as indicated in Figure 11.

29. Let $L = 3$ and $m = 1$. Apply Newton's Method to Eq. (2) to estimate θ and use this to estimate h.

SOLUTION We let $L = 3$ and $m = 1$. We want the solution of:

$$\frac{\sin \theta}{\theta} = \frac{L}{L+m}$$

$$\frac{\sin \theta}{\theta} - \frac{L}{L+m} = 0$$

$$\frac{\sin \theta}{\theta} - \frac{3}{4} = 0.$$

Let $f(\theta) = \frac{\sin \theta}{\theta} - \frac{3}{4}$.

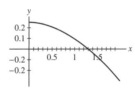

The figure above suggests that $\theta_0 = 1.5$ would be a good initial guess. The Newton's Method approximations for the solution follow:

n	1	2	3	4
θ_n	1.2854388	1.2757223	1.2756981	1.2756981

The angle where $\frac{\sin \theta}{\theta} = \frac{L}{L+m}$ is approximately 1.2757. Hence

$$h = L \frac{1 - \cos \theta}{2 \sin \theta} \approx 1.11181.$$

4.8 Antiderivatives

Preliminary Questions

1. Find an antiderivative of the function $f(x) = 0$.

SOLUTION Since the derivative of any constant is zero, any constant function is an antiderivative for the function $f(x) = 0$.

2. What is the difference, if any, between finding the general antiderivative of a function $f(x)$ and evaluating $\int f(x)\,dx$?

SOLUTION No difference. The indefinite integral is the symbol for denoting the general antiderivative.

3. Jacques happens to know that $f(x)$ and $g(x)$ have the same derivative, and he would like to know if $f(x) = g(x)$. Does Jacques have sufficient information to answer his question?

SOLUTION   No. Knowing that the two functions have the same derivative is only good enough to tell Jacques that the functions may differ by at most an additive constant. To determine whether the functions are equal for all x, Jacques needs to know the value of each function for a single value of x. If the two functions produce the same output value for a single input value, they must take the same value for all input values.

4. Write any two antiderivatives of $\cos x$. Which initial conditions do they satisfy at $x = 0$?

SOLUTION   Antiderivatives for $\cos x$ all have the form $\sin x + C$ for some constant C. At $x = 0$, $\sin x + C$ has the value C.

5. Suppose that $F'(x) = f(x)$ and $G'(x) = g(x)$. Are the following statements true or false? Explain.
(a) If $f = g$, then $F = G$.
(b) If F and G differ by a constant, then $f = g$.
(c) If f and g differ by a constant, then $F = G$.

SOLUTION
(a) False. Even if $f(x) = g(x)$, the antiderivatives F and G may differ by an additive constant.
(b) True. This follows from the fact that the derivative of any constant is 0.
(c) False. If the functions f and g are different, then the antiderivatives F and G differ by a linear function: $F(x) - G(x) = ax + b$ for some constants a and b.

6. Determine if $y = x^2$ is a solution to the differential equation with initial condition

$$\frac{dy}{dx} = 2x, \qquad y(0) = 1$$

SOLUTION   Although $\frac{d}{dx}x^2 = 2x$, x^2 takes the value 0 when $x = 0$, so $y = x^2$ is *not* a solution of the indicated initial value problem.

Exercises

In Exercises 1–6, find the general antiderivative of $f(x)$ and check your answer by differentiating.

1. $f(x) = 12x$

SOLUTION

$$\int 12x \, dx = 12 \int x \, dx = 12 \cdot \frac{1}{2}x^2 + C = 6x^2 + C.$$

As a check, we have

$$\frac{d}{dx}(6x^2 + C) = 12x$$

as needed.

3. $f(x) = x^2 + 3x + 2$

SOLUTION

$$\int (x^2 + 3x + 2) \, dx = \int x^2 dx + 3 \int x^1 \, dx + 2 \int x^0 \, dx = \frac{1}{3}x^3 + 3 \cdot \frac{1}{2}x^2 + 2 \cdot \frac{1}{1}x^1 + C$$

$$= \frac{1}{3}x^3 + \frac{3}{2}x^2 + 2x + C.$$

As a check, we have

$$\frac{d}{dx}\left(\frac{1}{3}x^3 + \frac{3}{2}x^2 + 2x + C\right) = x^2 + 3x + 2$$

as needed.

5. $f(x) = 8x^{-4}$

SOLUTION

$$\int 8x^{-4} \, dx = 8 \int x^{-4} \, dx = 8 \cdot \frac{1}{-3}x^{-3} + C = -\frac{8}{3}x^{-3} + C.$$

As a check, we have

$$\frac{d}{dx}\left(-\frac{8}{3}x^{-3} + C\right) = 8x^{-4}$$

as needed.

In Exercises 7–10, match the function with its antiderivative (a)–(d).

(a) $F(x) = \cos(1 - x)$

(b) $F(x) = -\cos x$

(c) $F(x) = -\frac{1}{2}\cos(x^2)$

(d) $F(x) = \sin x - x\cos x$

7. $f(x) = \sin x$

SOLUTION   An antiderivative of $\sin x$ is $-\cos x$, which is **(b)**. As a check, we have $\frac{d}{dx}(-\cos x) = -(-\sin x) = \sin x$.

9. $f(x) = \sin(1 - x)$

SOLUTION   An antiderivative of $\sin(1 - x)$ is $\cos(1 - x)$ or **(a)**. As a check, we have $\frac{d}{dx}\cos(1 - x) = -\sin(1 - x) \cdot (-1) = \sin(1 - x)$.

In Exercises 11–36, evaluate the indefinite integral.

11. $\displaystyle\int (x + 1)\, dx$

SOLUTION $\displaystyle\int (x + 1)\, dx = \frac{1}{2}x^2 + x + C.$

13. $\displaystyle\int (t^5 + 3t + 2)\, dt$

SOLUTION $\displaystyle\int (t^5 + 3t + 2)\, dt = \frac{1}{6}t^6 + \frac{3}{2}t^2 + 2t + C.$

15. $\displaystyle\int t^{-9/5}\, dt$

SOLUTION $\displaystyle\int t^{-9/5}\, dt = -\frac{5}{4}t^{-4/5} + C.$

17. $\displaystyle\int 2\, dx$

SOLUTION $\displaystyle\int 2\, dx = 2x + C.$

19. $\displaystyle\int (5t - 9)\, dt$

SOLUTION $\displaystyle\int (5t - 9)\, dt = \frac{5}{2}t^2 - 9t + C.$

21. $\displaystyle\int x^{-2}\, dx$

SOLUTION $\displaystyle\int x^{-2}\, dx = \frac{1}{-1}x^{-1} + C = -\frac{1}{x} + C$

23. $\displaystyle\int (x + 3)^{-2}\, dx$

SOLUTION $\displaystyle\int (x + 3)^{-2}\, dx = \frac{1}{-1}(x + 3)^{-1} + C = -\frac{1}{x + 3} + C.$

25. $\displaystyle\int \frac{3}{z^5}\, dz$

SOLUTION

$$\int \frac{3}{z^5}\, dz = \int 3z^{-5}\, dz = 3\left(-\frac{1}{4}z^{-4}\right) + C = -\frac{3}{4}z^{-4} + C.$$

27. $\displaystyle\int \sqrt{x}(x - 1)\, dx$

SOLUTION   To perform this indefinite integral, we need to distribute $\sqrt{x} = x^{1/2}$ over $(x - 1)$:

$$\sqrt{x}(x - 1) = x^{1/2}(x - 1) = x^{3/2} - x^{1/2}.$$

Written in this form, we can take the indefinite integral:

$$\int \sqrt{x}(x - 1)\, dx = \int \left(x^{3/2} - x^{1/2}\right) dx = \left(\frac{1}{5/2}\right)x^{5/2} - \left(\frac{1}{3/2}\right)x^{3/2} + C = \frac{2}{5}x^{5/2} - \frac{2}{3}x^{3/2} + C.$$

29. $\int \dfrac{t-7}{\sqrt{t}}\, dt$

SOLUTION To proceed, we first distribute $\dfrac{1}{\sqrt{t}} = t^{-1/2}$ over $(t-7)$:

$$\dfrac{t-7}{\sqrt{t}} = (t-7)t^{-1/2} = t^{1/2} - 7t^{-1/2}.$$

From this, we get:

$$\int \dfrac{t-7}{\sqrt{t}}\, dt = \int (t^{1/2} - 7t^{-1/2})\, dt = \dfrac{1}{3/2}t^{3/2} - 7\left(\dfrac{1}{1/2}t^{1/2}\right) + C = \dfrac{2}{3}t^{3/2} - 14t^{1/2} + C.$$

31. $\int (4\sin x - 3\cos x)\, dx$

SOLUTION $\int (4\sin x - 3\cos x)\, dx = -4\cos x - 3\sin x + C.$

33. $\int \cos(6t+4)\, dt$

SOLUTION Using the integral formula for $\cos(kt+b)$, we get:

$$\int \cos(6t+4)\, dt = \dfrac{1}{6}\sin(6t+4) + C.$$

35. $\int \cos(3-4t)\, dt$

SOLUTION From the integral formula for $\cos(kt+b)$ with $k = -4$, $b = 3$, we get:

$$\int \cos(3-4t)\, dt = \dfrac{1}{-4}\sin(3-4t) + C = -\dfrac{1}{4}\sin(3-4t) + C.$$

37. In Figure 2, which of (A) or (B) is the graph of an antiderivative of $f(x)$?

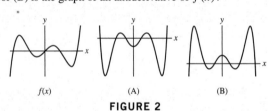

$f(x)$ (A) (B)

FIGURE 2

SOLUTION Let $F(x)$ be an antiderivative of $f(x)$. By definition, this means $F'(x) = f(x)$. In other words, $f(x)$ provides information as to the increasing/decreasing behavior of $F(x)$. Since, moving left to right, $f(x)$ transitions from $-$ to $+$ to $-$ to $+$ to $-$ to $+$, it follows that $F(x)$ must transition from decreasing to increasing to decreasing to increasing to decreasing to increasing. This describes the graph in (A)!

39. Use the formulas for the derivatives of $f(x) = \tan x$ and $f(x) = \sec x$ to evaluate the integrals.

(a) $\int \sec^2(3x)\, dx$ **(b)** $\int \sec(x+3)\tan(x+3)\, dx$

SOLUTION Recall that $\dfrac{d}{dx}(\tan x) = \sec^2 x$ and $\dfrac{d}{dx}(\sec x) = \sec x \tan x$.

(a) Accordingly, we have $\int \sec^2(3x)\, dx = \dfrac{1}{3}\tan(3x) + C.$

(b) Moreover, we see that $\int \sec(x+3)\tan(x+3)\, dx = \sec(x+3) + C.$

In Exercises 41–52, solve the differential equation with initial condition.

41. $\dfrac{dy}{dx} = \cos 2x,\quad y(0) = 3$

SOLUTION Since $\dfrac{dy}{dx} = \cos 2x$, we have

$$y = \int \cos 2x\, dx = \dfrac{1}{2}\sin 2x + C.$$

Thus

$$3 = y(0) = \dfrac{1}{2}\sin(2\cdot 0) + C = C,$$

so that $C = 3$. Therefore, $y = \dfrac{1}{2}\sin 2x + 3.$

43. $\dfrac{dy}{dx} = x$, $y(0) = 5$

SOLUTION Since $\frac{dy}{dx} = x$, we have

$$y = \int x \, dx = \frac{1}{2}x^2 + C.$$

Thus $5 = y(0) = 0 + C$, so that $C = 5$. Therefore, $y = \frac{1}{2}x^2 + 5$.

45. $\dfrac{dy}{dt} = 5 - 2t^2$, $y(1) = 2$

SOLUTION Since $\frac{dy}{dt} = 5 - 2t^2$, we have

$$y = \int (5 - 2t^2) \, dt = 5t - \frac{2}{3}t^3 + C.$$

Thus,

$$2 = y(1) = 5(1) - \frac{2}{3}(1^3) + C,$$

so that $C = -3 + \frac{2}{3} = -\frac{7}{3}$. Therefore $y = 5t - \frac{2}{3}t^3 - \frac{7}{3}$.

47. $\dfrac{dy}{dt} = 4t + 9$, $y(0) = 1$

SOLUTION Since $\frac{dy}{dt} = 4t + 9$, we have

$$y = \int (4t + 9) \, dt = 2t^2 + 9t + C.$$

Thus $1 = y(0) = 0 + C$, so that $C = 1$. Therefore, $y = 2t^2 + 9t + 1$.

49. $\dfrac{dy}{dx} = \sin x$, $y\left(\dfrac{\pi}{2}\right) = 1$

SOLUTION Since $\frac{dy}{dx} = \sin x$, we have

$$y = \int \sin x \, dx = -\cos x + C.$$

Thus

$$1 = y\left(\frac{\pi}{2}\right) = 0 + C,$$

so that $C = 1$. Therefore, $y = 1 - \cos x$.

51. $\dfrac{dy}{dx} = \cos 5x$, $y(\pi) = 3$

SOLUTION Since $\frac{dy}{dx} = \cos 5x$, we have

$$y = \int \cos 5x \, dx = \frac{1}{5}\sin 5x + C.$$

Thus $3 = y(\pi) = 0 + C$, so that $C = 3$. Therefore, $y = 3 + \frac{1}{5}\sin 5x$.

In Exercises 53–58, first find f' and then find f.

53. $f''(x) = x$, $f'(0) = 1$, $f(0) = 0$

SOLUTION Let $g(x) = f'(x)$. $g'(x) = x$ and $g(0) = 1$, so $g(x) = \frac{1}{2}x^2 + C$ with $g(0) = 1 = C$. Hence $f'(x) = g(x) = \frac{1}{2}x^2 + 1$. $f'(x) = \frac{1}{2}x^2 + 1$ and $f(0) = 0$, so $f(x) = \frac{1}{2}(\frac{1}{3}x^3) + x + C = \frac{1}{6}x^3 + x + C$, and $f(0) = 0 = C$. Therefore $f(x) = \frac{1}{6}x^3 + x$.

55. $f''(x) = x^3 - 2x + 1$, $f'(1) = 0$, $f(1) = 4$

SOLUTION Let $g(x) = f'(x)$. The problem statement gives us $g'(x) = x^3 - 2x + 1$, $g(0) = 0$. From $g'(x)$, we get $g(x) = \frac{1}{4}x^4 - x^2 + x + C$, and from $g(1) = 0$, we get $0 = \frac{1}{4} - 1 + 1 + C$, so that $C = -\frac{1}{4}$. This gives $f'(x) = g(x) = \frac{1}{4}x^4 - x^2 + x - \frac{1}{4}$. From $f'(x)$, we get $f(x) = \frac{1}{4}(\frac{1}{5}x^5) - \frac{1}{3}x^3 + \frac{1}{2}x^2 - \frac{1}{4}x + C = \frac{1}{20}x^5 - \frac{1}{3}x^3 + \frac{1}{2}x^2 - \frac{1}{4}x + C$. From $f(1) = 4$, we get

$$\frac{1}{20} - \frac{1}{3} + \frac{1}{2} - \frac{1}{4} + C = 4,$$

so that $C = \frac{121}{30}$. Hence,

$$f(x) = \frac{1}{20}x^5 - \frac{1}{3}x^3 + \frac{1}{2}x^2 - \frac{1}{4}x + \frac{121}{30}.$$

57. $f''(\theta) = \cos \theta$, $f'\left(\frac{\pi}{2}\right) = 1$, $f\left(\frac{\pi}{2}\right) = 6$

SOLUTION Let $g(\theta) = f'(\theta)$. The problem statement gives

$$g'(\theta) = \cos \theta, \qquad g\left(\frac{\pi}{2}\right) = 1.$$

From $g'(\theta)$ we get $g(\theta) = \sin \theta + C$. From $g(\frac{\pi}{2}) = 1$ we get $1 + C = 1$, so $C = 0$. Hence $f'(\theta) = g(\theta) = \sin \theta$. From $f'(\theta)$ we get $f(\theta) = -\cos \theta + C$. From $f(\frac{\pi}{2}) = 6$ we get $C = 6$, so

$$f(\theta) = -\cos \theta + 6.$$

59. Show that $f(x) = \tan^2 x$ and $g(x) = \sec^2 x$ have the same derivative. What can you conclude about the relation between f and g? Verify this conclusion directly.

SOLUTION Let $f(x) = \tan^2 x$ and $g(x) = \sec^2 x$. Then $f'(x) = 2 \tan x \sec^2 x$ and $g'(x) = 2 \sec x \cdot \sec x \tan x = 2 \tan x \sec^2 x$; hence $f'(x) = g'(x)$. Accordingly, $f(x)$ and $g(x)$ must differ by a constant; i.e., $f(x) - g(x) = \tan^2 x - \sec^2 x = C$ for some constant C. To see that this is true directly, divide the identity $\sin^2 x + \cos^2 x = 1$ by $\cos^2 x$. This yields $\tan^2 x + 1 = \sec^2 x$, so that $\tan^2 x - \sec^2 x = -1$.

61. A particle located at the origin at $t = 0$ begins moving along the x-axis with velocity $v(t) = \frac{1}{2}t^2 - t$ ft/s. Let $s(t)$ be its position at time t. State the differential equation with initial condition satisfied by $s(t)$ and find $s(t)$.

SOLUTION Given that $v(t) = \frac{ds}{dt}$ and that the particle starts at the origin, the differential equation is

$$\frac{ds}{dt} = \frac{1}{2}t^2 - t, \qquad s(0) = 0.$$

Accordingly, we have

$$s(t) = \int \left(\frac{1}{2}t^2 - t\right) dt = \frac{1}{2} \int t^2 \, dt - \int t \, dt = \frac{1}{2} \cdot \frac{1}{3}t^3 - \frac{1}{2}t^2 + C = \frac{1}{6}t^3 - \frac{1}{2}t^2 + C.$$

Thus $0 = s(0) = \frac{1}{6}(0)^3 - \frac{1}{2}(0)^2 + C$, so that $C = 0$. Therefore, the position is $s(t) = \frac{1}{6}t^3 - \frac{1}{2}t^2$.

63. A particle moves along the x-axis with velocity $v(t) = 25t - t^2$ ft/s. Let $s(t)$ be the position at time t.
(a) Find $s(t)$, assuming that the particle is located at $x = 5$ at time $t = 0$.
(b) Find $s(t)$, assuming that the particle is located at $x = 5$ at time $t = 2$.

SOLUTION The differential equation is $v(t) = \frac{ds}{dt} = 25t - t^2$.
(a) Suppose $s(0) = 5$. Then

$$s(t) = \int (25t - t^2) \, dt = 25 \int t \, dt - \int t^2 \, dt$$

$$= 25 \cdot \frac{1}{2}t^2 - \frac{1}{3}t^3 + C = \frac{25}{2}t^2 - \frac{1}{3}t^3 + C.$$

Thus $5 = s(0) = \frac{25}{2}(0)^2 - \frac{1}{3}(0)^3 + C$, so that $C = 5$. Therefore, the position is

$$s(t) = \frac{25}{2}t^2 - \frac{1}{3}t^3 + 5.$$

(b) From (a), we originally ascertained that $s(t) = \frac{25}{2}t^2 - \frac{1}{3}t^3 + C$. Thus $5 = s(2) = \frac{25}{2}(2)^2 - \frac{1}{3}(2)^3 + C = \frac{142}{3} + C$, so that $C = 5 - \frac{142}{3} = -\frac{127}{3}$. Therefore, the position is

$$s(t) = \frac{25}{2}t^2 - \frac{1}{3}t^3 - \frac{127}{3}.$$

65. A car traveling 84 ft/s begins to decelerate at a constant rate of 14 ft/s². After how many seconds does the car come to a stop and how far will the car have traveled before stopping?

SOLUTION Since the acceleration of the car is a constant -14ft/s², v is given by the differential equation:

$$\frac{dv}{dt} = -14, \qquad v(0) = 84.$$

From $\frac{dv}{dt}$, we get $v(t) = \int -14\,dt = -14t + C$. Since $v(0) = 84$, $C = 84$. From this, $v(t) = -14t + 84\ \frac{\text{ft}}{\text{s}}$. To find the time until the car stops, we must solve $v(t) = 0$:

$$-14t + 84 = 0$$

$$14t = 84$$

$$t = 84/14 = 6 \text{ s}.$$

Now we have a differential equation for $s(t)$. Since we want to know how far the car has traveled from the beginning of its deceleration at time $t = 0$, we have $s(0) = 0$ by definition, so:

$$\frac{ds}{dt} = v(t) = -14t + 84, \qquad s(0) = 0.$$

From this, $s(t) = \int(-14t + 84)\,dt = -7t^2 + 84t + C$. Since $s(0) = 0$, we have $C = 0$, and

$$s(t) = -7t^2 + 84t.$$

At stopping time $t = 6$ s, the car has traveled

$$s(6) = -7(36) + 84(6) = 252 \text{ ft}.$$

67. A 900-kg rocket is released from a spacecraft. As the rocket burns fuel, its mass decreases and its velocity increases. Let $v(m)$ be the velocity (in meters per second) as a function of mass m. Find the velocity when $m = 500$ if $dv/dm = -50m^{-1/2}$. Assume that $v(900) = 0$.

SOLUTION Since $\frac{dv}{dm} = -50m^{-1/2}$, we have $v(m) = \int -50m^{-1/2}\,dm = -100m^{1/2} + C$. Thus $0 = v(900) = -100\sqrt{900} + C = -3000 + C$, and $C = 3000$. Therefore, $v(m) = 3000 - 100\sqrt{m}$. Accordingly,

$$v(500) = 3000 - 100\sqrt{500} = 3000 - 1000\sqrt{5} \approx 764 \text{ meters/sec}.$$

69. Find constants c_1 and c_2 such that $F(x) = c_1x \sin x + c_2 \cos x$ is an antiderivative of $f(x) = x \cos x$.

SOLUTION Let $F(x) = c_1x \sin x + c_2 \cos x$. If $F(x)$ is to be an antiderivative of $f(x) = x \cos x$, we must have $F'(x) = f(x)$ for all x. Hence $c_1 (x \cos x + \sin x) - c_2 \sin x = x \cos x$ for all x. Equating coefficients on the left- and right-hand sides, we have $c_1 = 1$ (i.e., the coefficients of $x \cos x$ are equal) and $c_1 - c_2 = 0$ (i.e., the coefficients of $\sin x$ are equal). Thus $c_1 = c_2 = 1$ and hence $F(x) = x \sin x + \cos x$. As a check, we have $F'(x) = x \cos x + \sin x - \sin x = x \cos x = f(x)$, as required.

71. Verify the linearity properties of the indefinite integral stated in Theorem 3.

SOLUTION To verify the Sum Rule, let $F(x)$ and $G(x)$ be any antiderivatives of $f(x)$ and $g(x)$, respectively. Because

$$\frac{d}{dx}(F(x) + G(x)) = \frac{d}{dx}F(x) + \frac{d}{dx}G(x) = f(x) + g(x),$$

it follows that $F(x) + G(x)$ is an antiderivative of $f(x) + g(x)$; i.e.,

$$\int (f(x) + g(x))\,dx = \int f(x)\,dx + \int g(x)\,dx.$$

To verify the Multiples Rule, again let $F(x)$ be any antiderivative of $f(x)$ and let c be a constant. Because

$$\frac{d}{dx}(cF(x)) = c\frac{d}{dx}F(x) = cf(x),$$

it follows that $cF(x)$ is and antiderivative of $cf(x)$; i.e.,

$$\int (cf(x))\,dx = c\int f(x)\,dx.$$

Further Insights and Challenges

73. Suppose that $F'(x) = f(x)$.

(a) Show that $\frac{1}{2}F(2x)$ is an antiderivative of $f(2x)$.

(b) Find the general antiderivative of $f(kx)$ for any constant k.

SOLUTION   Let $F'(x) = f(x)$.

(a) By the Chain Rule, we have

$$\frac{d}{dx}\left(\frac{1}{2}F(2x)\right) = \frac{1}{2}F'(2x) \cdot 2 = F'(2x) = f(2x).$$

Thus $\frac{1}{2}F(2x)$ is an antiderivative of $f(2x)$.

(b) For nonzero constant k, the Chain Rules gives

$$\frac{d}{dx}\left(\frac{1}{k}F(kx)\right) = \frac{1}{k}F'(kx) \cdot k = F'(kx) = f(kx).$$

Thus $\frac{1}{k}F(kx)$ is an antiderivative of $f(kx)$. Hence the general antiderivative of $f(kx)$ is $\frac{1}{k}F(kx) + C$, where C is a constant.

75. The Power Rule for antiderivatives does not apply to $f(x) = x^{-1}$. Which of the graphs in Figure 4 could plausibly represent an antiderivative of $f(x) = x^{-1}$?

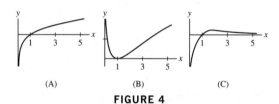

(A)   (B)   (C)

FIGURE 4

SOLUTION   Let $F(x)$ be an antiderivative of $f(x) = x^{-1}$. Then for $x > 0$, we have $F'(x) = f(x) = x^{-1} > 0$. In other words, the graph of $F(x)$ is increasing for $x > 0$. The only graph for which this is true is (A). Accordingly, neither graph (B) nor graph (C) is the graph of $F(x)$.

CHAPTER REVIEW EXERCISES

In Exercises 1–6, estimate using the Linear Approximation or linearization and use a calculator to compute the error.

1. $8.1^{1/3} - 2$

SOLUTION   Let $f(x) = x^{1/3}$, $a = 8$ and $\Delta x = 0.1$. Then $f'(x) = \frac{1}{3}x^{-2/3}$, $f'(a) = \frac{1}{12}$ and, by the Linear Approximation,

$$\Delta f = 8.1^{1/3} - 2 \approx f'(a)\Delta x = \frac{1}{12}(0.1) = 0.00833333.$$

Using a calculator, $8.1^{1/3} - 2 = 0.00829885$. The error in the Linear Approximation is therefore

$$|0.00829885 - 0.00833333| = 3.445 \times 10^{-5}.$$

3. $625^{1/4} - 624^{1/4}$

SOLUTION   Let $f(x) = x^{1/4}$, $a = 625$ and $\Delta x = -1$. Then $f'(x) = \frac{1}{4}x^{-3/4}$, $f'(a) = \frac{1}{500}$ and, by the Linear Approximation,

$$\Delta f = 624^{1/4} - 625^{1/4} \approx f'(a)\Delta x = \frac{1}{500}(-1) = -0.002.$$

Thus $625^{1/4} - 624^{1/4} \approx 0.002$. Using a calculator,

$$625^{1/4} - 624^{1/4} = 0.00200120.$$

The error in the Linear Approximation is therefore

$$|0.00200120 - (0.002)| = 1.201 \times 10^{-6}.$$

5. $\dfrac{1}{1.02}$

SOLUTION Let $f(x) = x^{-1}$ and $a = 1$. Then $f(a) = 1$, $f'(x) = -x^{-2}$ and $f'(a) = -1$. The linearization of $f(x)$ at $a = 1$ is therefore

$$L(x) = f(a) + f'(a)(x - a) = 1 - (x - 1) = 2 - x,$$

and $\dfrac{1}{1.02} \approx L(1.02) = 0.98$. Using a calculator, $\dfrac{1}{1.02} = 0.980392$, so the error in the Linear Approximation is

$$|0.980392 - 0.98| = 3.922 \times 10^{-4}.$$

In Exercises 7–10, find the linearization at the point indicated.

7. $y = \sqrt{x}, \quad a = 25$

SOLUTION Let $y = \sqrt{x}$ and $a = 25$. Then $y(a) = 5$, $y' = \frac{1}{2}x^{-1/2}$ and $y'(a) = \frac{1}{10}$. The linearization of y at $a = 25$ is therefore

$$L(x) = y(a) + y'(a)(x - 25) = 5 + \frac{1}{10}(x - 25).$$

9. $A(r) = \frac{4}{3}\pi r^3, \quad a = 3$

SOLUTION Let $A(r) = \frac{4}{3}\pi r^3$ and $a = 3$. Then $A(a) = 36\pi$, $A'(r) = 4\pi r^2$ and $A'(a) = 36\pi$. The linearization of $A(r)$ at $a = 3$ is therefore

$$L(r) = A(a) + A'(a)(r - a) = 36\pi + 36\pi(r - 3) = 36\pi(r - 2).$$

In Exercises 11–15, use the Linear Approximation.

11. The position of an object in linear motion at time t is $s(t) = 0.4t^2 + (t + 1)^{-1}$. Estimate the distance traveled over the time interval $[4, 4.2]$.

SOLUTION Let $s(t) = 0.4t^2 + (t + 1)^{-1}$, $a = 4$ and $\Delta t = 0.2$. Then $s'(t) = 0.8t - (t + 1)^{-2}$ and $s'(a) = 3.16$. Using the Linear Approximation, the distance traveled over the time interval $[4, 4.2]$ is approximately

$$\Delta s = s(4.2) - s(4) \approx s'(a)\Delta t = 3.16(0.2) = 0.632.$$

13. A store sells 80 MP3 players per week when the players are priced at $P = \$75$. Estimate the number N sold if P is raised to $\$80$, assuming that $dN/dP = -4$. Estimate N if the price is lowered to $\$69$.

SOLUTION If P is raised to $\$80$, then $\Delta P = 5$. With the assumption that $dN/dP = -4$, we estimate, using the Linear Approximation, that

$$\Delta N \approx \frac{dN}{dP}\Delta P = (-4)(5) = -20;$$

therefore, we estimate that only 60 MP3 players will be sold per week when the price is $\$80$. On the other hand, if the price is lowered to $\$69$, then $\Delta P = -6$ and $\Delta N \approx (-4)(-6) = 24$. We therefore estimate that 104 MP3 players will be sold per week when the price is $\$69$.

15. Show that $\sqrt{a^2 + b} \approx a + \frac{b}{2a}$ if b is small. Use this to estimate $\sqrt{26}$ and find the error using a calculator.

SOLUTION Let $a > 0$ and let $f(b) = \sqrt{a^2 + b}$. Then

$$f'(b) = \frac{1}{2\sqrt{a^2 + b}}.$$

By the Linear Approximation, $f(b) \approx f(0) + f'(0)b$, so

$$\sqrt{a^2 + b} \approx a + \frac{b}{2a}.$$

To estimate $\sqrt{26}$, let $a = 5$ and $b = 1$. Then

$$\sqrt{26} = \sqrt{5^2 + 1} \approx 5 + \frac{1}{10} = 5.1.$$

The error in this estimate is $|\sqrt{26} - 5.1| = 9.80 \times 10^{-4}$.

17. Show that $f(x) = 2x^3 + 2x + \sin x + 1$ has precisely one real root.

SOLUTION We have $f(0) = 1$ and $f(-1) = -3 + \sin(-1) = -3.84 < 0$. Therefore $f(x)$ has a root in the interval $[-1, 0]$. Now, suppose that $f(x)$ has two real roots, say a and b. Because $f(x)$ is continuous on $[a, b]$ and differentiable on (a, b) and $f(a) = f(b) = 0$, Rolle's Theorem guarantees that there exists $c \in (a, b)$ such that $f'(c) = 0$. However

$$f'(x) = 6x^2 + 2 + \cos x > 0$$

for all x (since $2 + \cos x \geq 0$). We have reached a contradiction. Consequently, $f(x)$ must have precisely one real root.

19. Suppose that $f(1) = 5$ and $f'(x) \geq 2$ for $x \geq 1$. Use the MVT to show that $f(8) \geq 19$.

SOLUTION Because f is continuous on $[1, 8]$ and differentiable on $(1, 8)$, the Mean Value Theorem guarantees there exists a $c \in (1, 8)$ such that

$$f'(c) = \frac{f(8) - f(1)}{8 - 1} \quad \text{or} \quad f(8) = f(1) + 7f'(c).$$

Now, we are given that $f(1) = 5$ and that $f'(x) \geq 2$ for $x \geq 1$. Therefore,

$$f(8) \geq 5 + 7(2) = 19.$$

In Exercises 21–24, find the local extrema and determine whether they are minima, maxima, or neither.

21. $f(x) = x^3 - 4x^2 + 4x$

SOLUTION Let $f(x) = x^3 - 4x^2 + 4x$. Then $f'(x) = 3x^2 - 8x + 4 = (3x - 2)(x - 2)$, so that $x = \frac{2}{3}$ and $x = 2$ are critical points. Next, $f''(x) = 6x - 8$, so $f''(\frac{2}{3}) = -4 < 0$ and $f''(2) = 4 > 0$. Therefore, by the Second Derivative Test, $f(\frac{2}{3})$ is a local maximum while $f(2)$ is a local minimum.

23. $f(x) = x^2(x + 2)^3$

SOLUTION Let $f(x) = x^2(x + 2)^3$. Then

$$f'(x) = 3x^2(x + 2)^2 + 2x(x + 2)^3 = x(x + 2)^2(3x + 2x + 4) = x(x + 2)^2(5x + 4),$$

so that $x = 0$, $x = -2$ and $x = -\frac{4}{5}$ are critical points. The sign of the first derivative on the intervals surrounding the critical points is indicated in the table below. Based on this information, $f(-2)$ is neither a local maximum nor a local minimum, $f(-\frac{4}{5})$ is a local maximum and $f(0)$ is a local minimum.

Interval	$(-\infty, -2)$	$(-2, -\frac{4}{5})$	$(-\frac{4}{5}, 0)$	$(0, \infty)$
Sign of f'	+	+	−	+

In Exercises 25–30, find the extreme values on the interval.

25. $f(x) = x(10 - x)$, $[-1, 3]$

SOLUTION Let $f(x) = x(10 - x) = 10x - x^2$. Then $f'(x) = 10 - 2x$, so that $x = 5$ is the only critical point. As this critical point is not in the interval $[-1, 3]$, we only need to check the value of f at the endpoints to determine the extreme values. Because $f(-1) = -11$ and $f(3) = 21$, the maximum value of $f(x) = x(10 - x)$ on the interval $[-1, 3]$ is 21 while the minimum value is -11.

27. $g(\theta) = \sin^2 \theta - \cos \theta$, $[0, 2\pi]$

SOLUTION Let $g(\theta) = \sin^2 \theta - \cos \theta$. Then

$$g'(\theta) = 2 \sin \theta \cos \theta + \sin \theta = \sin \theta(2 \cos \theta + 1) = 0$$

when $\theta = 0, \frac{2\pi}{3}, \pi, \frac{4\pi}{3}, 2\pi$. The table below lists the value of g at each of the critical points and the endpoints of the interval $[0, 2\pi]$. Based on this information, the minimum value of $g(\theta)$ on the interval $[0, 2\pi]$ is -1 and the maximum value is $\frac{5}{4}$.

θ	0	$2\pi/3$	π	$4\pi/3$	2π
$g(\theta)$	−1	5/4	1	5/4	−1

29. $f(x) = x^{2/3} - 2x^{1/3}$, $[-1, 3]$

SOLUTION   Let $f(x) = x^{2/3} - 2x^{1/3}$. Then $f'(x) = \frac{2}{3}x^{-1/3} - \frac{2}{3}x^{-2/3} = \frac{2}{3}x^{-2/3}(x^{1/3} - 1)$, so that the critical points are $x = 0$ and $x = 1$. With $f(-1) = 3$, $f(0) = 0$, $f(1) = -1$ and $f(3) = \sqrt[3]{9} - 2\sqrt[3]{3} \approx -0.804$, it follows that the minimum value of $f(x)$ on the interval $[-1, 3]$ is -1 and the maximum value is 3.

31. Find the critical points and extreme values of $f(x) = |x - 1| + |2x - 6|$ in $[0, 8]$.

SOLUTION   Let

$$f(x) = |x - 1| + |2x - 6| = \begin{cases} 7 - 3x, & x < 1 \\ 5 - x, & 1 \le x < 3 \\ 3x - 7, & x \ge 3 \end{cases}.$$

The derivative of $f(x)$ is never zero but does not exist at the transition points $x = 1$ and $x = 3$. Thus, the critical points of f are $x = 1$ and $x = 3$. With $f(0) = 7$, $f(1) = 4$, $f(3) = 2$ and $f(8) = 17$, it follows that the minimum value of $f(x)$ on the interval $[0, 8]$ is 2 and the maximum value is 17.

In Exercises 33–36, find the points of inflection.

33. $y = x^3 - 4x^2 + 4x$

SOLUTION   Let $y = x^3 - 4x^2 + 4x$. Then $y' = 3x^2 - 8x + 4$ and $y'' = 6x - 8$. Thus, $y'' > 0$ and y is concave up for $x > \frac{4}{3}$, while $y'' < 0$ and y is concave down for $x < \frac{4}{3}$. Hence, there is a point of inflection at $x = \frac{4}{3}$.

35. $y = \dfrac{x^2}{x^2 + 4}$

SOLUTION   Let $y = \dfrac{x^2}{x^2 + 4} = 1 - \dfrac{4}{x^2 + 4}$. Then $y' = \dfrac{8x}{(x^2 + 4)^2}$ and

$$y'' = \frac{(x^2 + 4)^2(8) - 8x(2)(2x)(x^2 + 4)}{(x^2 + 4)^4} = \frac{8(4 - 3x^2)}{(x^2 + 4)^3}.$$

Thus, $y'' > 0$ and y is concave up for

$$-\frac{2}{\sqrt{3}} < x < \frac{2}{\sqrt{3}},$$

while $y'' < 0$ and y is concave down for

$$|x| \ge \frac{2}{\sqrt{3}}.$$

Hence, there are points of inflection at

$$x = \pm\frac{2}{\sqrt{3}}.$$

37. Match the description of $f(x)$ with the graph of its derivative $f'(x)$ in Figure 1.

(a) $f(x)$ is increasing and concave up.

(b) $f(x)$ is decreasing and concave up.

(c) $f(x)$ is increasing and concave down.

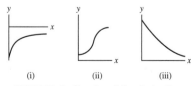

(i)   (ii)   (iii)

FIGURE 1 Graphs of the derivative.

SOLUTION

(a) If $f(x)$ is increasing and concave up, then $f'(x)$ is positive and increasing. This matches the graph in (ii).

(b) If $f(x)$ is decreasing and concave up, then $f'(x)$ is negative and increasing. This matches the graph in (i).

(c) If $f(x)$ is increasing and concave down, then $f'(x)$ is positive and decreasing. This matches the graph in (iii).

In Exercises 39–52, evaluate the limit.

39. $\displaystyle\lim_{x \to \infty} (9x^4 - 12x^3)$

SOLUTION   Because the leading term in the polynomial has a positive coefficient,

$$\lim_{x\to\infty} (9x^4 - 12x^3) = \infty.$$

41. $\lim_{x\to\infty} \dfrac{x^3 + 2x}{4x^2 - 9}$

SOLUTION

$$\lim_{x\to\infty} \frac{x^3 + 2x}{4x^2 - 9} = \lim_{x\to\infty} \frac{(x^3 + 2x)x^{-2}}{(4x^2 - 9)x^{-2}} = \lim_{x\to\infty} \frac{x + 2x^{-1}}{4 - 9x^{-2}} = \infty.$$

43. $\lim_{x\to\infty} \dfrac{x^3 + 2x}{4x^3 - 9}$

SOLUTION

$$\lim_{x\to\infty} \frac{x^3 + 2x}{4x^3 - 9} = \lim_{x\to\infty} \frac{(x^3 + 2x)x^{-3}}{(4x^3 - 9)x^{-3}} = \lim_{x\to\infty} \frac{1 + 2x^{-2}}{4 - 9x^{-3}} = \frac{1}{4}.$$

45. $\lim_{x\to\infty} \dfrac{x^2 - 9x}{\sqrt{x^2 + 4x}}$

SOLUTION   For $x > 0$, $\sqrt{x^{-2}} = |x|^{-1} = x^{-1}$. Then,

$$\lim_{x\to\infty} \frac{x^2 - 9x}{\sqrt{x^2 + 4x}} = \lim_{x\to\infty} \frac{(x^2 - 9x)x^{-1}}{\sqrt{x^2 + 4x}\sqrt{x^{-2}}} = \lim_{x\to\infty} \frac{x - 9}{\sqrt{1 + 4x^{-1}}} = \infty.$$

47. $\lim_{x\to\infty} \dfrac{x^{1/2}}{\sqrt{4x - 9}}$

SOLUTION

$$\lim_{x\to\infty} \frac{x^{1/2}}{\sqrt{4x - 9}} = \lim_{x\to\infty} \frac{x^{1/2}x^{-1/2}}{\sqrt{4x - 9}\sqrt{x^{-1}}} = \lim_{x\to\infty} \frac{1}{\sqrt{4 - 9x^{-1}}} = \frac{1}{2}.$$

49. $\lim_{x\to3+} \dfrac{1 - 2x}{x - 3}$

SOLUTION   As $x \to 3+$, $1 - 2x \to -5$ and $x - 3 \to 0+$, so

$$\lim_{x\to3+} \frac{1 - 2x}{x - 3} = -\infty.$$

51. $\lim_{x\to2+} \dfrac{x - 5}{x - 2}$

SOLUTION   As $x \to 2+$, $x - 5 \to -3$ and $x - 2 \to 0+$, so

$$\lim_{x\to2+} \frac{x - 5}{x - 2} = -\infty.$$

In Exercises 53–62, sketch the graph, noting the transition points and asymptotic behavior.

53. $y = 12x - 3x^2$

SOLUTION   Let $y = 12x - 3x^2$. Then $y' = 12 - 6x$ and $y'' = -6$. It follows that the graph of $y = 12x - 3x^2$ is increasing for $x < 2$, decreasing for $x > 2$, has a local maximum at $x = 2$ and is concave down for all x. Because

$$\lim_{x\to\pm\infty} (12x - 3x^2) = -\infty,$$

the graph has no horizontal asymptotes. There are also no vertical asymptotes. The graph is shown below.

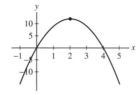

55. $y = x^3 - 2x^2 + 3$

SOLUTION   Let $y = x^3 - 2x^2 + 3$. Then $y' = 3x^2 - 4x$ and $y'' = 6x - 4$. It follows that the graph of $y = x^3 - 2x^2 + 3$ is increasing for $x < 0$ and $x > \frac{4}{3}$, is decreasing for $0 < x < \frac{4}{3}$, has a local maximum at $x = 0$, has a local minimum at $x = \frac{4}{3}$, is concave up for $x > \frac{2}{3}$, is concave down for $x < \frac{2}{3}$ and has a point of inflection at $x = \frac{2}{3}$. Because

$$\lim_{x \to -\infty} (x^3 - 2x^2 + 3) = -\infty \quad \text{and} \quad \lim_{x \to \infty} (x^3 - 2x^2 + 3) = \infty,$$

the graph has no horizontal asymptotes. There are also no vertical asymptotes. The graph is shown below.

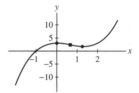

57. $y = \dfrac{x}{x^3 + 1}$

SOLUTION   Let $y = \dfrac{x}{x^3 + 1}$. Then

$$y' = \frac{x^3 + 1 - x(3x^2)}{(x^3 + 1)^2} = \frac{1 - 2x^3}{(x^3 + 1)^2}$$

and

$$y'' = \frac{(x^3 + 1)^2(-6x^2) - (1 - 2x^3)(2)(x^3 + 1)(3x^2)}{(x^3 + 1)^4} = -\frac{6x^2(2 - x^3)}{(x^3 + 1)^3}.$$

It follows that the graph of $y = \dfrac{x}{x^3 + 1}$ is increasing for $x < -1$ and $-1 < x < \sqrt[3]{\frac{1}{2}}$, is decreasing for $x > \sqrt[3]{\frac{1}{2}}$, has a local maximum at $x = \sqrt[3]{\frac{1}{2}}$, is concave up for $x < -1$ and $x > \sqrt[3]{2}$, is concave down for $-1 < x < 0$ and $0 < x < \sqrt[3]{2}$ and has a point of inflection at $x = \sqrt[3]{2}$. Note that $x = -1$ is not an inflection point because $x = -1$ is not in the domain of the function. Now,

$$\lim_{x \to \pm\infty} \frac{x}{x^3 + 1} = 0,$$

so $y = 0$ is a horizontal asymptote. Moreover,

$$\lim_{x \to -1-} \frac{x}{x^3 + 1} = \infty \quad \text{and} \quad \lim_{x \to -1+} \frac{x}{x^3 + 1} = -\infty,$$

so $x = -1$ is a vertical asymptote. The graph is shown below.

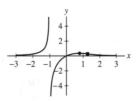

59. $y = \dfrac{1}{|x + 2| + 1}$

SOLUTION Let $y = \dfrac{1}{|x + 2| + 1}$. Because

$$\lim_{x \to \pm\infty} \frac{1}{|x + 2| + 1} = 0,$$

the graph of this function has a horizontal asymptote of $y = 0$. The graph has no vertical asymptotes as $|x + 2| + 1 \geq 1$ for all x. The graph is shown below. From this graph we see there is a local maximum at $x = -2$.

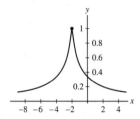

61. $y = 2 \sin x - \cos x$ on $[0, 2\pi]$

SOLUTION Let $y = \sqrt{3} \sin x - \cos x$. Then $y' = \sqrt{3} \cos x + \sin x$ and $y'' = -\sqrt{3} \sin x + \cos x$. It follows that the graph of $y = \sqrt{3} \sin x - \cos x$ is increasing for $0 < x < 5\pi/6$ and $11\pi/6 < x < 2\pi$, is decreasing for $5\pi/6 < x < 11\pi/6$, has a local maximum at $x = 5\pi/6$, has a local minimum at $x = 11\pi/6$, is concave up for $0 < x < \pi/3$ and $4\pi/3 < x < 2\pi$, is concave down for $\pi/3 < x < 4\pi/3$ and has points of inflection at $x = \pi/3$ and $x = 4\pi/3$. The graph is shown below.

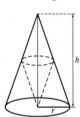

63. Find the maximum volume of a right-circular cone placed upside-down in a right-circular cone of radius R and height H (Figure 3). The volume of a cone of radius r and height h is $\frac{4}{3} \pi r^2 h$.

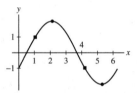

FIGURE 3

SOLUTION Let r denote the radius and h the height of the upside down cone. By similar triangles, we obtain the relation

$$\frac{H - h}{r} = \frac{H}{R} \qquad \text{so} \qquad h = H \left(1 - \frac{r}{R} \right)$$

and the volume of the upside down cone is

$$V(r) = \frac{4}{3} \pi r^2 h = \frac{4}{3} \pi H \left(r^2 - \frac{r^3}{R} \right)$$

for $0 \leq r \leq R$. Thus,

$$\frac{dV}{dr} = \frac{4}{3} \pi H \left(2r - \frac{3r^2}{R} \right),$$

and the critical points are $r = 0$ and $r = 2R/3$. Because $V(0) = V(R) = 0$ and

$$V \left(\frac{2R}{3} \right) = \frac{4}{3} \pi H \left(\frac{4R^2}{9} - \frac{8R^2}{27} \right) = \frac{16}{81} \pi R^2 H,$$

the maximum volume of a right-circular cone placed upside down in a right-circular cone of radius R and height H is

$$\frac{16}{81} \pi R^2 H.$$

65. Show that the maximum area of a parallelogram ADEF inscribed in a triangle $\triangle ABC$, as in Figure 4, is equal to one-half the area of $\triangle ABC$.

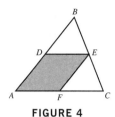

FIGURE 4

SOLUTION Let θ denote the measure of angle BAC. Then the area of the parallelogram is given by $\overline{AD} \cdot \overline{AF} \sin \theta$. Now, suppose that

$$\overline{BE}/\overline{BC} = x.$$

Then, by similar triangles, $\overline{AD} = (1 - x)\overline{AB}$, $\overline{AF} = \overline{DE} = x\overline{AC}$, and the area of the parallelogram becomes $\overline{AB} \cdot \overline{AC}x(1 - x) \sin \theta$. The function $x(1 - x)$ achieves its maximum value of $\frac{1}{4}$ when $x = \frac{1}{2}$. Thus, the maximum area of a parallelogram inscribed in a triangle $\triangle ABC$ is

$$\frac{1}{4}\overline{AB} \cdot \overline{AC} \sin \theta = \frac{1}{2}\left(\frac{1}{2}\overline{AB} \cdot \overline{AC} \sin \theta\right) = \frac{1}{2}\,(\text{area of } \triangle ABC).$$

67. Let $f(x)$ be a function whose graph does not pass through the x-axis and let $Q = (a, 0)$. Let $P = (x_0, f(x_0))$ be the point on the graph closest to Q (Figure 6). Prove that $\overline{PQ}$ is perpendicular to the tangent line to the graph of x_0. *Hint:* Let $q(x)$ be the distance from $(x, f(x))$ to $(a, 0)$ and observe that x_0 is a critical point of $q(x)$.

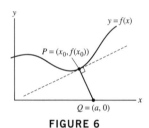

FIGURE 6

SOLUTION Let $P = (a, 0)$ and let $Q = (x_0, f(x_0))$ be the point on the graph of $y = f(x)$ closest to P. The slope of the segment joining P and Q is then

$$\frac{f(x_0)}{x_0 - a}.$$

Now, let

$$q(x) = \sqrt{(x - a)^2 + (f(x))^2},$$

the distance from the arbitrary point $(x, f(x))$ on the graph of $y = f(x)$ to the point P. As $(x_0, f(x_0))$ is the point closest to P, we must have

$$q'(x_0) = \frac{2(x_0 - a) + 2f(x_0)f'(x_0)}{\sqrt{(x_0 - a)^2 + (f(x_0))^2}} = 0.$$

Thus,

$$f'(x_0) = -\frac{x_0 - a}{f(x_0)} = -\left(\frac{f(x_0)}{x_0 - a}\right)^{-1}.$$

In other words, the slope of the segment joining P and Q is the negative reciprocal of the slope of the line tangent to the graph of $y = f(x)$ at $x = x_0$; hence; the two lines are perpendicular.

69. Use Newton's Method to estimate $\sqrt[3]{25}$ to four decimal places.

SOLUTION Let $f(x) = x^3 - 25$ and define

$$x_{n+1} = x_n - \frac{f(x_n)}{f'(x_n)} = x_n - \frac{x_n^3 - 25}{3x_n^2}.$$

With $x_0 = 3$, we find

n	1	2	3
x_n	2.925925926	2.924018982	2.924017738

Thus, to four decimal places $\sqrt[3]{25} = 2.9240$.

In Exercises 71–80, calculate the indefinite integral.

71. $\displaystyle\int \left(4x^3 - 2x^2\right) dx$

SOLUTION $\displaystyle\int (4x^3 - 2x^2)\,dx = x^4 - \frac{2}{3}x^3 + C.$

73. $\displaystyle\int \sin(\theta - 8)\,d\theta$

SOLUTION $\displaystyle\int \sin(\theta - 8)\,d\theta = -\cos(\theta - 8) + C.$

75. $\displaystyle\int (4t^{-3} - 12t^{-4})\,dt$

SOLUTION $\displaystyle\int (4t^{-3} - 12t^{-4})\,dt = -2t^{-2} + 4t^{-3} + C.$

77. $\displaystyle\int \sec^2 x\,dx$

SOLUTION $\displaystyle\int \sec^2 x\,dx = \tan x + C.$

79. $\displaystyle\int (y + 2)^4\,dy$

SOLUTION $\displaystyle\int (y + 2)^4\,dy = \frac{1}{5}(y + 2)^5 + C.$

In Exercises 81–84, solve the differential equation with initial condition.

81. $\dfrac{dy}{dx} = 4x^3, \quad y(1) = 4$

SOLUTION Let $\frac{dy}{dx} = 4x^3$. Then

$$y(x) = \int 4x^3\,dx = x^4 + C.$$

Using the initial condition $y(1) = 4$, we find $y(1) = 1^4 + C = 4$, so $C = 3$. Thus, $y(x) = x^4 + 3$.

83. $\dfrac{dy}{dx} = x^{-1/2}, \quad y(1) = 1$

SOLUTION Let $\frac{dy}{dx} = x^{-1/2}$. Then

$$y(x) = \int x^{-1/2}\,dx = 2x^{1/2} + C.$$

Using the initial condition $y(1) = 1$, we find $y(1) = 2\sqrt{1} + C = 1$, so $C = -1$. Thus, $y(x) = 2x^{1/2} - 1$.

85. Find $f(t)$, assuming that $f''(t) = 1 - 2t$, $f(0) = 2$, and $f'(0) = -1$.

SOLUTION Suppose $f''(t) = 1 - 2t$. Then

$$f'(t) = \int f''(t)\,dt = \int (1 - 2t)\,dt = t - t^2 + C.$$

Using the initial condition $f'(0) = -1$, we find $f'(0) = 0 - 0^2 + C = -1$, so $C = -1$. Thus, $f'(t) = t - t^2 - 1$. Now,

$$f(t) = \int f'(t)\,dt = \int (t - t^2 - 1)\,dt = \frac{1}{2}t^2 - \frac{1}{3}t^3 - t + C.$$

Using the initial condition $f(0) = 2$, we find $f(0) = \frac{1}{2}0^2 - \frac{1}{3}0^3 - 0 + C = 2$, so $C = 2$. Thus,

$$f(t) = \frac{1}{2}t^2 - \frac{1}{3}t^3 - t + 2.$$

5 | THE INTEGRAL

5.1 Approximating and Computing Area

Preliminary Questions

1. Suppose that [2, 5] is divided into six subintervals. What are the right and left endpoints of the subintervals?

SOLUTION If the interval [2, 5] is divided into six subintervals, the length of each subinterval is $\frac{5-2}{6} = \frac{1}{2}$. The right endpoints of the subintervals are then $\frac{5}{2}, 3, \frac{7}{2}, 4, \frac{9}{2}, 5$, while the left endpoints are $2, \frac{5}{2}, 3, \frac{7}{2}, 4, \frac{9}{2}$.

2. If $f(x) = x^{-2}$ on [3, 7], which is larger: R_2 or L_2?

SOLUTION On [3, 7], the function $f(x) = x^{-2}$ is a decreasing function; hence, for any subinterval of [3, 7], the function value at the left endpoint is larger than the function value at the right endpoint. Consequently, L_2 must be larger than R_2.

3. Which of the following pairs of sums are *not* equal?

(a) $\sum_{i=1}^{4} i, \quad \sum_{\ell=1}^{4} \ell$

(b) $\sum_{j=1}^{4} j^2, \quad \sum_{k=2}^{5} k^2$

(c) $\sum_{j=1}^{4} j, \quad \sum_{i=2}^{5} (i-1)$

(d) $\sum_{i=1}^{4} i(i+1), \quad \sum_{j=2}^{5} (j-1)j$

SOLUTION

(a) Only the name of the index variable has been changed, so these two sums *are* the same.

(b) These two sums are *not* the same; the second squares the numbers two through five while the first squares the numbers one through four.

(c) These two sums *are* the same. Note that when i ranges from two through five, the expression $i - 1$ ranges from one through four.

(d) These two sums *are* the same. Both sums are $1 \cdot 2 + 2 \cdot 3 + 3 \cdot 4 + 4 \cdot 5$.

4. Explain why $\sum_{j=1}^{100} j$ is equal to $\sum_{j=0}^{100} j$ but $\sum_{j=1}^{100} 1$ is not equal to $\sum_{j=0}^{100} 1$.

SOLUTION The first term in the sum $\sum_{j=0}^{100} j$ is equal to zero, so it may be dropped. More specifically,

$$\sum_{j=0}^{100} j = 0 + \sum_{j=1}^{100} j = \sum_{j=1}^{100} j.$$

On the other hand, the first term in $\sum_{j=0}^{100} 1$ is not zero, so this term cannot be dropped. In particular,

$$\sum_{j=0}^{100} 1 = 1 + \sum_{j=1}^{100} 1 \neq \sum_{j=1}^{100} 1.$$

5. We divide the interval [1, 5] into 16 subintervals.

(a) What are the left endpoints of the first and last subintervals?

(b) What are the right endpoints of the first two subintervals?

SOLUTION Note that each of the 16 subintervals has length $\frac{5-1}{16} = \frac{1}{4}$.

(a) The left endpoint of the first subinterval is 1, and the left endpoint of the last subinterval is $5 - \frac{1}{4} = \frac{19}{4}$.

(b) The right endpoints of the first two subintervals are $1 + \frac{1}{4} = \frac{5}{4}$ and $1 + 2\left(\frac{1}{4}\right) = \frac{3}{2}$.

6. Are the following statements true or false?

(a) The right-endpoint rectangles lie below the graph if $f(x)$ is increasing.

(b) If $f(x)$ is monotonic, then the area under the graph lies between R_N and L_N.

(c) If $f(x)$ is constant, then the right-endpoint rectangles all have the same height.

SOLUTION

(a) False. If f is increasing, then the right-endpoint rectangles lie above the graph.

(b) True. If $f(x)$ is increasing, then the area under the graph is larger than L_N but smaller than R_N; on the other hand, if $f(x)$ is decreasing, then the area under the graph is larger than R_N but smaller than L_N.

(c) True. The height of the right-endpoint rectangles is given by the value of the function, which, for a constant function, is always the same.

Exercises

1. An athlete runs with velocity 4 mph for half an hour, 6 mph for the next hour, and 5 mph for another half-hour. Compute the total distance traveled and indicate on a graph how this quantity can be interpreted as an area.

SOLUTION The figure below displays the velocity of the runner as a function of time. The area of the shaded region equals the total distance traveled. Thus, the total distance traveled is $(4)(0.5) + (6)(1) + (5)(0.5) = 10.5$ miles.

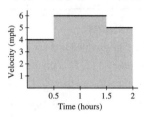

3. A rainstorm hit Portland, Maine, in October 1996, resulting in record rainfall. The rainfall rate $R(t)$ on October 21 is recorded, in inches per hour, in the following table, where t is the number of hours since midnight. Compute the total rainfall during this 24-hour period and indicate on a graph how this quantity can be interpreted as an area.

t	0–2	2–4	4–9	9–12	12–20	20–24
$R(t)$	0.2	0.1	0.4	1.0	0.6	0.25

SOLUTION Over each interval, the total rainfall is the time interval in hours times the rainfall in inches per hour. Thus

$$R = 2(.2) + 2(.1) + 5(.4) + 3(1.0) + 8(.6) + 4(.25) = 11.4 \text{ inches.}$$

The figure below is a graph of the rainfall as a function of time. The area of the shaded region represents the total rainfall.

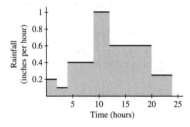

5. Compute R_6, L_6, and M_3 to estimate the distance traveled over $[0, 3]$ if the velocity at half-second intervals is as follows:

t (s)	0	0.5	1	1.5	2	2.5	3
v (ft/s)	0	12	18	25	20	14	20

SOLUTION For R_6 and L_6, $\Delta t = \frac{3-0}{6} = .5$. For M_3, $\Delta t = \frac{3-0}{3} = 1$. Then

$$R_6 = 0.5 \sec (12 + 18 + 25 + 20 + 14 + 20) \text{ ft/sec} = .5(109) \text{ ft} = 54.5 \text{ ft,}$$

$$L_6 = 0.5 \sec (0 + 12 + 18 + 25 + 20 + 14) \text{ ft/sec} = .5(89) \text{ ft} = 44.5 \text{ ft,}$$

and

$$M_3 = 1 \sec (12 + 25 + 14) \text{ ft/sec} = 51 \text{ ft.}$$

7. Consider $f(x) = 2x + 3$ on $[0, 3]$.

(a) Compute R_6 and L_6 over $[0, 3]$.

(b) Find the error in these approximations by computing the area exactly using geometry.

SOLUTION Let $f(x) = 2x + 3$ on $[0, 3]$.

(a) We partition $[0, 3]$ into 6 equally-spaced subintervals. The left endpoints of the subintervals are $\left\{0, \frac{1}{2}, 1, \frac{3}{2}, 2, \frac{5}{2}\right\}$ whereas the right endpoints are $\left\{\frac{1}{2}, 1, \frac{3}{2}, 2, \frac{5}{2}, 3\right\}$.

- Let $a = 0$, $b = 3$, $n = 6$, $\Delta x = (b - a)/n = \frac{1}{2}$, and $x_k = a + k\Delta x$, $k = 0, 1, \ldots, 5$ (left endpoints). Then

$$L_6 = \sum_{k=0}^{5} f(x_k)\Delta x = \Delta x \sum_{k=0}^{5} f(x_k) = \frac{1}{2}(3 + 4 + 5 + 6 + 7 + 8) = 16.5.$$

- With $x_k = a + k\Delta x$, $k = 1, 2, \ldots, 6$ (right endpoints), we have

$$R_6 = \sum_{k=1}^{6} f(x_k)\Delta x = \Delta x \sum_{k=1}^{6} f(x_k) = \frac{1}{2}(4 + 5 + 6 + 7 + 8 + 9) = 19.5.$$

(b) Via geometry (see figure below), the exact area is $A = \frac{1}{2}(3)(6) + 3^2 = 18$. Thus, L_6 underestimates the true area ($L_6 - A = -1.5$), while R_6 overestimates the true area ($R_6 - A = +1.5$).

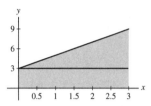

9. Estimate R_6, L_6, and M_6 over $[0, 1.5]$ for the function in Figure 15.

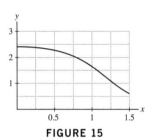

FIGURE 15

SOLUTION   Let $f(x)$ on $[0, \frac{3}{2}]$ be given by Figure 15. For $n = 6$, $\Delta x = (\frac{3}{2} - 0)/6 = \frac{1}{4}$, $\{x_k\}_{k=0}^{6} = \left\{0, \frac{1}{4}, \frac{1}{2}, \frac{3}{4}, 1, \frac{5}{4}, \frac{3}{2}\right\}$. Therefore

$$L_6 = \frac{1}{4}\sum_{k=0}^{5} f(x_k) = \frac{1}{4}(2.4 + 2.35 + 2.25 + 2 + 1.65 + 1.05) = 2.925,$$

$$R_6 = \frac{1}{4}\sum_{k=1}^{6} f(x_k) = \frac{1}{4}(2.35 + 2.25 + 2 + 1.65 + 1.05 + 0.65) = 2.4875,$$

$$M_6 = \frac{1}{4}\sum_{k=1}^{6} f\left(x_k - \frac{1}{2}\Delta x\right) = \frac{1}{4}(2.4 + 2.3 + 2.2 + 1.85 + 1.45 + 0.8) = 2.75.$$

11. Let $f(x) = \sqrt{x^2 + 1}$ and $\Delta x = \frac{1}{3}$. Sketch the graph of $f(x)$ and draw the rectangles whose area is represented by the sum $\sum_{i=1}^{6} f(1 + i\Delta x)\Delta x$.

SOLUTION   Because the summation index runs from $i = 1$ through $i = 6$, we will treat this as a right-endpoint approximation to the area under the graph of $y = \sqrt{x^2 + 1}$. With $\Delta x = \frac{1}{3}$, it follows that the right endpoints of the subintervals are $x_1 = \frac{4}{3}$, $x_2 = \frac{5}{3}$, $x_3 = 2$, $x_4 = \frac{7}{3}$, $x_5 = \frac{8}{3}$ and $x_6 = 3$. The sketch of the graph with the rectangles represented by the sum $\sum_{i=1}^{6} f(1 + i\Delta x)\Delta x$ is given below.

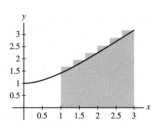

In Exercises 13–24, calculate the approximation for the given function and interval.

13. R_8, $f(x) = 7 - x$, $[3, 5]$

SOLUTION   Let $f(x) = 7 - x$ on $[3, 5]$. For $n = 8$, $\Delta x = (5 - 3)/8 = \frac{1}{4}$, and $\{x_k\}_{k=0}^{8} = \left\{3, 3\frac{1}{4}, 3\frac{1}{2}, 3\frac{3}{4}, 4, 4\frac{1}{4}, 4\frac{1}{2}, 4\frac{3}{4}, 5\right\}$. Therefore

$$R_8 = \frac{1}{4}\sum_{k=1}^{8}(7 - x_k) = \frac{1}{4}(3.75 + 3.5 + 3.25 + 3 + 2.75 + 2.5 + 2.25 + 2) = \frac{1}{4}(23) = 5.75.$$

15. M_4, $f(x) = x^2$, $[0, 1]$

SOLUTION   Let $f(x) = x^2$ on $[0, 1]$. For $n = 4$, $\Delta x = (1 - 0)/4 = \frac{1}{4}$ and $\{x_k^*\}_{k=0}^{3} = \{.125, .375, .625, .875\}$. Therefore

$$M_4 = \frac{1}{4}\sum_{k=0}^{3}(x_k^*)^2 = \frac{1}{4}(0.125^2 + 0.375^2 + 0.625^2 + 0.875^2) = .328125.$$

17. R_6, $f(x) = 2x^2 - x + 2$, $[1, 4]$

SOLUTION   Let $f(x) = 2x^2 - x + 2$ on $[1, 4]$. For $n = 6$, $\Delta x = (4 - 1)/6 = \frac{1}{2}$, $\{x_k\}_{k=0}^{6} = \left\{1, 1\frac{1}{2}, 2, 2\frac{1}{2}, 3, 3\frac{1}{2}, 4\right\}$. Therefore

$$R_6 = \frac{1}{2}\sum_{k=1}^{6}(2x_k^2 - x_k + 2) = \frac{1}{2}(5 + 8 + 12 + 17 + 23 + 30) = 47.5.$$

19. L_5, $f(x) = x^{-1}$, $[1, 2]$

SOLUTION   Let $f(x) = x^{-1}$ on $[1, 2]$. For $n = 5$, $\Delta x = \frac{(2-1)}{5} = \frac{1}{5}$, $\{x_k\}_{k=0}^{5} = \left\{1, \frac{6}{5}, \frac{7}{5}, \frac{8}{5}, \frac{9}{5}, 2\right\}$. Therefore

$$L_5 = \frac{1}{5}\sum_{k=0}^{4}(x_k)^{-1} = \frac{1}{5}\left(1 + \frac{5}{6} + \frac{5}{7} + \frac{5}{8} + \frac{5}{9}\right) \approx .745635.$$

21. L_4, $f(x) = \cos x$, $[\frac{\pi}{4}, \frac{\pi}{2}]$

SOLUTION   Let $f(x) = \cos x$ on $[\frac{\pi}{4}, \frac{\pi}{2}]$. For $n = 4$,

$$\Delta x = \frac{(\pi/2 - \pi/4)}{4} = \frac{\pi}{16} \quad \text{and} \quad \{x_k\}_{k=0}^{4} = \left\{\frac{\pi}{4}, \frac{5\pi}{16}, \frac{3\pi}{8}, \frac{7\pi}{16}, \frac{\pi}{2}\right\}.$$

Therefore

$$L_4 = \frac{\pi}{16}\sum_{k=0}^{3}\cos x_k \approx .361372.$$

23. M_4, $f(x) = \dfrac{1}{x^2 + 1}$, $[1, 5]$

SOLUTION   Let $f(x) = \frac{1}{x^2+1}$. For $n = 4$, $\Delta x = \frac{5-1}{4} = 1$, and $\{x_k^*\}_{k=0}^{3} = \{1.5, 2.5, 3.5, 4.5\}$. Therefore

$$M_4 = 1\sum_{k=0}^{3}\frac{1}{(x_k^*)^2 + 1} = \left(\frac{4}{13} + \frac{4}{29} + \frac{4}{53} + \frac{4}{85}\right) \approx .568154.$$

In Exercises 25–28, use the Graphical Insight on page 254 to obtain bounds on the area.

25. Let A be the area under the graph of $f(x) = \sqrt{x}$ over $[0, 1]$. Prove that $0.51 \leq A \leq 0.77$ by computing R_4 and L_4. Explain your reasoning.

SOLUTION   For $n = 4$, $\Delta x = \frac{1-0}{4} = \frac{1}{4}$ and $\{x_i\}_{i=0}^{4} = \{0 + i\Delta x\} = \{0, \frac{1}{4}, \frac{1}{2}, \frac{3}{4}, 1\}$. Therefore,

$$R_4 = \Delta x\sum_{i=1}^{4}f(x_i) = \frac{1}{4}\left(\frac{1}{2} + \frac{\sqrt{2}}{2} + \frac{\sqrt{3}}{2} + 1\right) \approx .768$$

$$L_4 = \Delta x\sum_{i=0}^{3}f(x_i) = \frac{1}{4}\left(0 + \frac{1}{2} + \frac{\sqrt{2}}{2} + \frac{\sqrt{3}}{2}\right) \approx .518.$$

In the plot below, you can see the rectangles whose area is represented by L_4 under the graph and the top of those whose area is represented by R_4 above the graph. The area A under the curve is somewhere between L_4 and R_4, so

$$.518 \le A \le .768.$$

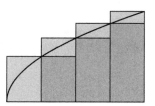

L_4, R_4 and the graph of $f(x)$.

27. Use R_4 and L_4 to show that the area A under the graph of $y = \sin x$ over $[0, \pi/2]$ satisfies $0.79 \le A \le 1.19$.

SOLUTION Let $f(x) = \sin x$. $f(x)$ is increasing over the interval $[0, \pi/2]$, so the Insight on page 254 applies, which indicates that $L_4 \le A \le R_4$. For $n = 4$, $\Delta x = \frac{\pi/2 - 0}{4} = \frac{\pi}{8}$ and $\{x_i\}_{i=0}^4 = \{0 + i\Delta x\}_{i=0}^4 = \{0, \frac{\pi}{8}, \frac{\pi}{4}, \frac{3\pi}{8}, \frac{\pi}{2}\}$. From this,

$$L_4 = \frac{\pi}{8}\sum_{i=0}^3 f(x_i) \approx .79, \qquad R_4 = \frac{\pi}{8}\sum_{i=1}^4 f(x_i) \approx 1.18.$$

Hence A is between .79 and 1.19.

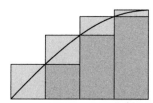

Left and Right endpoint approximations to A.

29. **CAS** Show that the area A in Exercise 25 satisfies $L_N \le A \le R_N$ for all N. Then use a computer algebra system to calculate L_N and R_N for $N = 100$ and 150. Which of these calculations allows you to conclude that $A \approx 0.66$ to two decimal places?

SOLUTION On $[0, 1]$, $f(x) = \sqrt{x}$ is an increasing function; therefore, $L_N \le A \le R_N$ for all N. Now,

$$L_{100} = .6614629 \qquad R_{100} = .6714629$$
$$L_{150} = .6632220 \qquad R_{150} = .6698887$$

Using the values obtained with $N = 150$, it follows that $.6632220 \le A \le .6698887$. Thus, to two decimal places, $A \approx .66$.

31. Calculate the following sums:

(a) $\displaystyle\sum_{i=1}^5 3$

(b) $\displaystyle\sum_{i=0}^5 3$

(c) $\displaystyle\sum_{k=2}^4 k^3$

(d) $\displaystyle\sum_{j=3}^4 \sin\left(j\frac{\pi}{2}\right)$

(e) $\displaystyle\sum_{k=2}^4 \frac{1}{k-1}$

(f) $\displaystyle\sum_{j=0}^3 3^j$

SOLUTION

(a) $\displaystyle\sum_{i=1}^5 3 = 3 + 3 + 3 + 3 + 3 = 15$. Alternatively, $\displaystyle\sum_{i=1}^5 3 = 3\sum_{i=1}^5 1 = (3)(5) = 15$.

(b) $\displaystyle\sum_{i=0}^5 3 = 3 + 3 + 3 + 3 + 3 + 3 = 18$. Alternatively, $\displaystyle\sum_{i=0}^5 3 = 3\sum_{i=0}^5 = (3)(6) = 18$.

(c) $\displaystyle\sum_{k=2}^4 k^3 = 2^3 + 3^3 + 4^3 = 99$. Alternatively,

$$\sum_{k=2}^4 k^3 = \left(\sum_{k=1}^4 k^3\right) - \left(\sum_{k=1}^1 k^3\right) = \left(\frac{4^4}{4} + \frac{4^3}{2} + \frac{4^2}{4}\right) - \left(\frac{1^4}{4} + \frac{1^3}{2} + \frac{1^2}{4}\right) = 99.$$

(d) $\displaystyle\sum_{j=3}^{4} \sin\left(\frac{j\pi}{2}\right) = \sin\left(\frac{3\pi}{2}\right) + \sin\left(\frac{4\pi}{2}\right) = -1 + 0 = -1.$

(e) $\displaystyle\sum_{k=2}^{4} \frac{1}{k-1} = 1 + \frac{1}{2} + \frac{1}{3} = \frac{11}{6}.$

(f) $\displaystyle\sum_{j=0}^{3} 3^j = 1 + 3 + 3^2 + 3^3 = 40.$

33. Calculate $\displaystyle\sum_{j=101}^{200} j$ by writing it as a difference of two sums and using formula (3).

SOLUTION

$$\sum_{j=101}^{200} j = \sum_{j=1}^{200} j - \sum_{j=1}^{100} j = \left(\frac{200^2}{2} + \frac{200}{2}\right) - \left(\frac{100^2}{2} + \frac{100}{2}\right) = 20100 - 5050 = 15050.$$

In Exercises 34–39, write the sum in summation notation.

35. $(2^2 + 2) + (3^2 + 3) + (4^2 + 4) + (5^2 + 5)$

SOLUTION   The first term is $2^2 + 2$, and the last term is $5^2 + 5$, so it seems that the sum limits are 2 and 5, and the kth term is $k^2 + k$. Therefore, the sum is:

$$\sum_{k=2}^{5} (k^2 + k).$$

37. $\sqrt{1 + 1^3} + \sqrt{2 + 2^3} + \cdots + \sqrt{n + n^3}$

SOLUTION   The first term is $\sqrt{1 + 1^3}$ and the last term is $\sqrt{n + n^3}$, so it seems the summation limits are 1 through n, and the k-th term is $\sqrt{k + k^3}$. Therefore, the sum is

$$\sum_{k=1}^{n} \sqrt{k + k^3}.$$

39. $\sin(\pi) + \sin\left(\frac{\pi}{2}\right) + \sin\left(\frac{\pi}{3}\right) + \cdots + \sin\left(\frac{\pi}{n+1}\right)$

SOLUTION   The first summand is $\sin\left(\frac{\pi}{1}\right) = \sin\left(\frac{\pi}{0+1}\right)$ hence,

$$\sin(\pi) + \sin\left(\frac{\pi}{2}\right) + \sin\left(\frac{\pi}{3}\right) + \cdots + \sin\left(\frac{\pi}{n+1}\right) = \sum_{i=0}^{n} \sin\left(\frac{\pi}{i+1}\right).$$

In Exercises 40–47, use linearity and formulas (3)–(5) to rewrite and evaluate the sums.

41. $\displaystyle\sum_{k=1}^{20} (2k + 1)$

SOLUTION $\displaystyle\sum_{k=1}^{20} (2k + 1) = 2 \sum_{k=1}^{20} k + \sum_{k=1}^{20} 1 = 2\left(\frac{20^2}{2} + \frac{20}{2}\right) + 20 = 440.$

43. $\displaystyle\sum_{k=100}^{200} k^3$

SOLUTION   By rewriting the sum as a difference of two power sums,

$$\sum_{k=100}^{200} k^3 = \sum_{k=1}^{200} k^3 - \sum_{k=1}^{99} k^3 = \left(\frac{200^4}{4} + \frac{200^3}{2} + \frac{200^2}{4}\right) - \left(\frac{99^4}{4} + \frac{99^3}{2} + \frac{99^2}{4}\right) = 379507500.$$

45. $\displaystyle\sum_{j=2}^{30} \left(6j + \frac{4j^2}{3}\right)$

SOLUTION

$$\sum_{j=2}^{30}\left(6j+\frac{4j^2}{3}\right) = 6\sum_{j=2}^{30}j + \frac{4}{3}\sum_{j=2}^{30}j^2 = 6\left(\sum_{j=1}^{30}j - \sum_{j=1}^{1}j\right) + \frac{4}{3}\left(\sum_{j=1}^{30}j^2 - \sum_{j=1}^{1}j^2\right)$$

$$= 6\left(\frac{30^2}{2}+\frac{30}{2}-1\right) + \frac{4}{3}\left(\frac{30^3}{3}+\frac{30^2}{2}+\frac{30}{6}-1\right)$$

$$= 6\,(464) + \frac{4}{3}\,(9454) = 2784 + \frac{37816}{3} = \frac{46168}{3}.$$

47. $\displaystyle\sum_{s=1}^{30}(3s^2 - 4s - 1)$

SOLUTION

$$\sum_{s=1}^{30}(3s^2 - 4s - 1) = 3\sum_{s=1}^{30}s^2 - 4\sum_{s=1}^{30}s - \sum_{s=1}^{30}1 = 3\left(\frac{30^3}{3}+\frac{30^2}{2}+\frac{30}{6}\right) - 4\left(\frac{30^2}{2}+\frac{30}{2}\right) - 30 = 26475.$$

In Exercises 48–51, calculate the sum, assuming that $a_1 = -1$, $\displaystyle\sum_{i=1}^{10}a_i = 10$, and $\displaystyle\sum_{i=1}^{10}b_i = 7$.

49. $\displaystyle\sum_{i=1}^{10}(a_i - b_i)$

SOLUTION $\displaystyle\sum_{i=1}^{10}(a_i - b_i) = \sum_{i=1}^{10}a_i - \sum_{i=1}^{10}b_i = 10 - 7 = 3.$

51. $\displaystyle\sum_{i=2}^{10}a_i$

SOLUTION $\displaystyle\sum_{i=2}^{10}a_i = \left(\sum_{i=1}^{10}a_i\right) - a_1 = 10 - (-1) = 11.$

In Exercises 52–55, use formulas (3)–(5) to evaluate the limit.

53. $\displaystyle\lim_{N\to\infty}\sum_{j=1}^{N}\frac{j^3}{N^4}$

SOLUTION Let $s_N = \displaystyle\sum_{j=1}^{N}\frac{j^3}{N^4}$. Then

$$s_N = \frac{1}{N^4}\sum_{j=1}^{N}j^3 = \frac{1}{N^4}\left(\frac{N^4}{4}+\frac{N^3}{2}+\frac{N^2}{4}\right) = \frac{1}{4}+\frac{1}{2N}+\frac{1}{4N^2}.$$

Therefore, $\displaystyle\lim_{N\to\infty}s_N = \frac{1}{4}.$

55. $\displaystyle\lim_{N\to\infty}\sum_{i=1}^{N}\left(\frac{i^3}{N^4}-\frac{20}{N}\right)$

SOLUTION Let $s_N = \displaystyle\sum_{i=1}^{N}\left(\frac{i^3}{N^4}-\frac{20}{N}\right)$. Then

$$s_N = \frac{1}{N^4}\sum_{i=1}^{N}i^3 - \frac{20}{N}\sum_{i=1}^{N}1 = \frac{1}{N^4}\left(\frac{N^4}{4}+\frac{N^3}{2}+\frac{N^2}{4}\right) - 20 = \frac{1}{4}+\frac{1}{2N}+\frac{1}{4N^2} - 20.$$

Therefore, $\displaystyle\lim_{N\to\infty}s_N = \frac{1}{4} - 20 = -\frac{79}{4}.$

In Exercises 56–59, calculate the limit for the given function and interval. Verify your answer by using geometry.

57. $\lim\limits_{N\to\infty} L_N$, $f(x) = 5x$, $[1, 3]$

SOLUTION Let $f(x) = 5x$ on $[1, 3]$. Let $N > 0$ be an integer, and set $a = 1$, $b = 3$, and $\Delta x = (b-a)/N = 2/N$. Also, let $x_k = a + k\Delta x = 1 + \frac{2k}{N}$, $k = 0, 1, \ldots N - 1$ be the left endpoints of the N subintervals of $[1, 3]$. Then

$$L_N = \Delta x \sum_{k=0}^{N-1} f(x_k) = \frac{2}{N} \sum_{k=0}^{N-1} 5\left(1 + \frac{2k}{N}\right) = \frac{10}{N} \sum_{k=0}^{N-1} 1 + \frac{20}{N} \sum_{k=0}^{N-1} k$$

$$= \frac{10}{N} N + \frac{20}{N^2}\left(\frac{(N-1)^2}{2} + \frac{N-1}{2}\right) = 20 - \frac{30}{N} + \frac{20}{N^2}.$$

The area under the graph is

$$\lim_{N\to\infty} L_N = 20.$$

The region under the curve is a trapezoid with base width 2 and heights 5 and 15. Therefore the area is $\frac{1}{2}(2)(5 + 15) = 20$, which agrees with the value obtained from the limit of the left-endpoint approximations.

59. $\lim\limits_{N\to\infty} M_N$, $f(x) = x$, $[0, 1]$

SOLUTION Let $f(x) = x$ on $[0, 1]$. Let $N > 0$ be an integer and set $a = 0$, $b = 1$, and $\Delta x = (b-a)/N = \frac{1}{N}$. Also, let $x_k^* = 0 + (k - \frac{1}{2})\Delta x = \frac{2k-1}{2N}$, $k = 1, 2, \ldots N$ be the midpoints of the N subintervals of $[0, 1]$. Then

$$M_N = \Delta x \sum_{k=1}^{N} f(x_k^*) = \frac{1}{N} \sum_{k=1}^{N} \frac{2k-1}{2N} = \frac{1}{2N^2} \sum_{k=1}^{N} (2k - 1)$$

$$= \frac{1}{2N^2}\left(2 \sum_{k=1}^{N} k - N\right) = \frac{1}{N^2}\left(\frac{N^2}{2} + \frac{N}{2}\right) - \frac{1}{2N} = \frac{1}{2}.$$

The area under the curve over $[0, 1]$ is

$$\lim_{N\to\infty} M_N = \frac{1}{2}.$$

The region under the curve over $[0, 1]$ is a triangle with base and height 1, and thus area $\frac{1}{2}$, which agrees with the answer obtained from the limit of the midpoint approximations.

In Exercises 60–69, find a formula for R_N for the given function and interval. Then compute the area under the graph as a limit.

61. $f(x) = x^3$, $[0, 1]$

SOLUTION Let $f(x) = x^3$ on the interval $[0, 1]$. Then $\Delta x = \dfrac{1-0}{N} = \dfrac{1}{N}$ and $a = 0$. Hence,

$$R_N = \Delta x \sum_{j=1}^{N} f(0 + j\Delta x) = \frac{1}{N} \sum_{j=1}^{N} \left(j^3 \frac{1}{N^3}\right) = \frac{1}{N^4} \sum_{j=1}^{N} j^3$$

$$= \frac{1}{N^4}\left(\frac{N^4}{4} + \frac{N^3}{2} + \frac{N^2}{4}\right) = \frac{1}{4} + \frac{1}{2N} + \frac{1}{4N^2}$$

and

$$\lim_{N\to\infty} R_N = \lim_{N\to\infty}\left(\frac{1}{4} + \frac{1}{2N} + \frac{1}{4N^2}\right) = \frac{1}{4}.$$

63. $f(x) = 1 - x^3$, $[0, 1]$

SOLUTION Let $f(x) = 1 - x^3$ on the interval $[0, 1]$. Then $\Delta x = \dfrac{1-0}{N} = \dfrac{1}{N}$ and $a = 0$. Hence,

$$R_N = \Delta x \sum_{j=1}^{N} f(0 + j\Delta x) = \frac{1}{N} \sum_{j=1}^{N} \left(1 - j^3 \frac{1}{N^3}\right)$$

$$= \frac{1}{N} \sum_{j=1}^{N} 1 - \frac{1}{N^4} \sum_{j=1}^{N} j^3 = \frac{1}{N} N - \frac{1}{N^4} \left(\frac{N^4}{4} + \frac{N^3}{2} + \frac{N^2}{4} \right) = 1 - \frac{1}{4} - \frac{1}{2N} - \frac{1}{4N^2}$$

and

$$\lim_{N \to \infty} R_N = \lim_{N \to \infty} \left(\frac{3}{4} - \frac{1}{2N} - \frac{1}{4N^2} \right) = \frac{3}{4}.$$

65. $f(x) = 3x^2 - x + 4$, $[1, 5]$

SOLUTION Let $f(x) = 3x^2 - x + 4$ on the interval $[1, 5]$. Then $\Delta x = \frac{5-1}{N} = \frac{4}{N}$ and $a = 1$. Hence,

$$R_N = \Delta x \sum_{j=1}^{N} f(1 + j\Delta x) = \frac{4}{N} \sum_{j=1}^{N} \left(j^2 \frac{48}{N^2} + j \frac{20}{N} + 6 \right) = \frac{192}{N^3} \sum_{j=1}^{N} j^2 + \frac{80}{N^2} \sum_{j=1}^{N} j + \frac{24}{N} \sum_{j=1}^{N} 1$$

$$= \frac{192}{N^3} \left(\frac{N^3}{3} + \frac{N^2}{2} + \frac{N}{6} \right) + \frac{80}{N^2} \left(\frac{N^2}{2} + \frac{N}{2} \right) + \frac{24}{N} N = 64 + \frac{96}{N} + \frac{32}{N^2} + 40 + \frac{40}{N} + 24$$

and

$$\lim_{N \to \infty} R_N = \lim_{N \to \infty} \left(128 + \frac{136}{N} + \frac{32}{N^2} \right) = 128.$$

67. $f(x) = x^2$, $[2, 4]$

SOLUTION Let $f(x) = x^2$ on the interval $[2, 4]$. Then $\Delta x = \frac{4-2}{N} = \frac{2}{N}$ and $a = 2$. Hence,

$$R_N = \Delta x \sum_{j=1}^{N} f(2 + j\Delta x) = \frac{2}{N} \sum_{j=1}^{N} \left(4 + j \frac{8}{N} + j^2 \frac{4}{N^2} \right) = \frac{8}{N} \sum_{j=1}^{N} 1 + \frac{16}{N^2} \sum_{j=1}^{N} j + \frac{8}{N^3} \sum_{j=1}^{N} j^2$$

$$= \frac{8}{N} N + \frac{16}{N^2} \left(\frac{N^2}{2} + \frac{N}{2} \right) + \frac{8}{N^3} \left(\frac{N^3}{3} + \frac{N^2}{2} + \frac{N}{6} \right) = 8 + 8 + \frac{8}{N} + \frac{8}{3} + \frac{4}{N} + \frac{4}{3N^2}$$

and

$$\lim_{N \to \infty} R_N = \lim_{N \to \infty} \left(\frac{56}{3} + \frac{12}{N} + \frac{4}{3N^2} \right) = \frac{56}{3}.$$

69. $f(x) = x^2$, $[a, b]$ (a, b constants with $a < b$)

SOLUTION Let $f(x) = x^2$ on the interval $[a, b]$. Then $\Delta x = \frac{b-a}{N}$. Hence,

$$R_N = \Delta x \sum_{j=1}^{N} f(a + j\Delta x) = \frac{(b-a)}{N} \sum_{j=1}^{N} \left(a^2 + 2aj \frac{(b-a)}{N} + j^2 \frac{(b-a)^2}{N^2} \right)$$

$$= \frac{a^2(b-a)}{N} \sum_{j=1}^{N} 1 + \frac{2a(b-a)^2}{N^2} \sum_{j=1}^{N} j + \frac{(b-a)^3}{N^3} \sum_{j=1}^{N} j^2$$

$$= \frac{a^2(b-a)}{N} N + \frac{2a(b-a)^2}{N^2} \left(\frac{N^2}{2} + \frac{N}{2} \right) + \frac{(b-a)^3}{N^3} \left(\frac{N^3}{3} + \frac{N^2}{2} + \frac{N}{6} \right)$$

$$= a^2(b-a) + a(b-a)^2 + \frac{a(b-a)^2}{N} + \frac{(b-a)^3}{3} + \frac{(b-a)^3}{2N} + \frac{(b-a)^3}{6N^2}$$

and

$$\lim_{N \to \infty} R_N = \lim_{N \to \infty} \left(a^2(b-a) + a(b-a)^2 + \frac{a(b-a)^2}{N} + \frac{(b-a)^3}{3} + \frac{(b-a)^3}{2N} + \frac{(b-a)^3}{6N^2} \right)$$

$$= a^2(b-a) + a(b-a)^2 + \frac{(b-a)^3}{3} = \frac{1}{3}b^3 - \frac{1}{3}a^3.$$

In Exercises 70–73, describe the area represented by the limits.

71. $\displaystyle\lim_{N\to\infty}\frac{3}{N}\sum_{j=1}^{N}\left(2+\frac{3j}{N}\right)^{4}$

SOLUTION   The limit

$$\lim_{N\to\infty}R_N=\lim_{N\to\infty}\frac{3}{N}\sum_{j=1}^{N}\left(2+j\cdot\frac{3}{N}\right)^{4}$$

represents the area between the graph of $f(x)=x^4$ and the x-axis over the interval $[2,5]$.

73. $\displaystyle\lim_{N\to\infty}\frac{\pi}{2N}\sum_{j=1}^{N}\sin\left(\frac{\pi}{3}+\frac{j\pi}{2N}\right)$

SOLUTION   The limit

$$\lim_{N\to\infty}\frac{\pi}{2N}\sum_{j=1}^{N}\sin\left(\frac{\pi}{3}+\frac{j\pi}{2N}\right)$$

represents the area between the graph of $f(x)=\sin x$ and the x axis over the interval $[\frac{\pi}{3},\frac{5\pi}{6}]$.

In Exercises 75–80, use the approximation indicated (in summation notation) to express the area under the graph as a limit but do not evaluate.

75. R_N, $f(x)=\sin x$ over $[0,\pi]$

SOLUTION   Let $f(x)=\sin x$ over $[0,\pi]$ and set $a=0$, $b=\pi$, and $\Delta x=(b-a)/N=\pi/N$. Then

$$R_N=\Delta x\sum_{k=1}^{N}f(x_k)=\frac{\pi}{N}\sum_{k=1}^{N}\sin\left(\frac{k\pi}{N}\right).$$

Hence

$$\lim_{N\to\infty}R_N=\lim_{N\to\infty}\frac{\pi}{N}\sum_{k=1}^{N}\sin\left(\frac{k\pi}{N}\right)$$

is the area between the graph of $f(x)=\sin x$ and the x-axis over $[0,\pi]$.

77. M_N, $f(x)=\tan x$ over $[\frac{1}{2},1]$

SOLUTION   Let $f(x)=\tan x$ over the interval $[\frac{1}{2},1]$. Then $\Delta x=\frac{1-\frac{1}{2}}{N}=\frac{1}{2N}$ and $a=\frac{1}{2}$. Hence

$$M_N=\Delta x\sum_{j=1}^{N}f\left(\frac{1}{2}+\left(j-\frac{1}{2}\right)\Delta x\right)=\frac{1}{2N}\sum_{j=1}^{N}\tan\left(\frac{1}{2}+\frac{1}{2N}\left(j-\frac{1}{2}\right)\right)$$

and so

$$\lim_{N\to\infty}M_N=\lim_{N\to\infty}\frac{1}{2N}\sum_{j=1}^{N}\tan\left(\frac{1}{2}+\frac{1}{2N}\left(j-\frac{1}{2}\right)\right)$$

is the area between the graph of $f(x)=\tan x$ and the x-axis over $[\frac{1}{2},1]$.

79. L_N, $f(x)=\cos x$ over $[\frac{\pi}{8},\pi]$

SOLUTION   Let $f(x)=\cos x$ over the interval $[\frac{\pi}{8},\pi]$. Then $\Delta x=\dfrac{\pi-\frac{\pi}{8}}{N}=\dfrac{7\pi}{8N}$ and $a=\frac{\pi}{8}$. Hence,

$$L_N=\Delta x\sum_{j=0}^{N-1}f\left(\frac{\pi}{8}+j\Delta x\right)=\frac{7\pi}{8N}\sum_{j=0}^{N-1}\cos\left(\frac{\pi}{8}+j\frac{7\pi}{8N}\right)$$

and

$$\lim_{N\to\infty}L_N=\lim_{N\to\infty}\frac{7\pi}{8N}\sum_{j=0}^{N-1}\cos\left(\frac{\pi}{8}+j\frac{7\pi}{8N}\right)$$

is the area between the graph of $f(x)=\cos x$ and the x-axis over $[\frac{\pi}{8},\pi]$.

In Exercises 81–83, let $f(x) = x^2$ and let R_N, L_N, and M_N be the approximations for the interval [0, 1].

81. 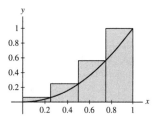 Show that $R_N = \frac{1}{3} + \frac{1}{2N} + \frac{1}{6N^2}$. Interpret the quantity $\frac{1}{2N} + \frac{1}{6N^2}$ as the area of a region.

SOLUTION Let $f(x) = x^2$ on [0, 1]. Let $N > 0$ be an integer and set $a = 0$, $b = 1$ and $\Delta x = \frac{1-0}{N} = \frac{1}{N}$. Then

$$R_N = \Delta x \sum_{j=1}^{N} f(0 + j\Delta x) = \frac{1}{N} \sum_{j=1}^{N} j^2 \frac{1}{N^2} = \frac{1}{N^3}\left(\frac{N^3}{3} + \frac{N^2}{2} + \frac{N}{6}\right) = \frac{1}{3} + \frac{1}{2N} + \frac{1}{6N^2}.$$

The quantity

$$\frac{1}{2N} + \frac{6}{N^2} \quad \text{in} \quad R_N = \frac{1}{3} + \frac{1}{2N} + \frac{1}{6N^2}$$

represents the collective area of the parts of the rectangles that lie above the graph of $f(x)$. It is the error between R_N and the true area $A = \frac{1}{3}$.

83. For each of R_N, L_N, and M_N, find the smallest integer N for which the error is less than 0.001.

SOLUTION

- For R_N, the error is less than .001 when:

$$\frac{1}{2N} + \frac{1}{6N^2} < .001.$$

We find an adequate solution in N:

$$\frac{1}{2N} + \frac{1}{6N^2} < .001$$

$$3N + 1 < .006(N^2)$$

$$0 < .006N^2 - 3N - 1,$$

in particular, if $N > \frac{3+\sqrt{9.024}}{.012} = 500.333$. Hence R_{501} is within .001 of A.

- For L_N, the error is less than .001 if

$$\left|-\frac{1}{2N} + \frac{1}{6N^2}\right| < .001.$$

We solve this equation for N:

$$\left|\frac{1}{2N} - \frac{1}{6N^2}\right| < .001$$

$$\left|\frac{3N - 1}{6N^2}\right| < .001$$

$$3N - 1 < .006N^2$$

$$0 < .006N^2 - 3N + 1,$$

which is satisfied if $N > \frac{3+\sqrt{9-.024}}{.012} = 499.666$. Therefore, L_{500} is within .001 units of A.

- For M_N, the error is given by $-\frac{1}{12N^2}$, so the error is less than .001 if

$$\frac{1}{12N^2} < .001$$

$$1000 < 12N^2$$

$$9.13 < N$$

Therefore, M_{10} is within .001 units of the correct answer.

Further Insights and Challenges

85. Draw the graph of a positive continuous function on an interval such that R_2 and L_2 are both smaller than the exact area under the graph. Can such a function be monotonic?

SOLUTION In the plot below, the area under the saw-tooth function $f(x)$ is 3, whereas $L_2 = R_2 = 2$. Thus L_2 and R_2 are both smaller than the exact area. Such a function cannot be monotonic; if $f(x)$ is increasing, then L_N underestimates and R_N overestimates the area for all N, and, if $f(x)$ is decreasing, then L_N overestimates and R_N underestimates the area for all N.

Left/right-endpoint approximation, n = 2

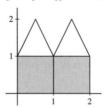

87. 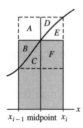 Assume that $f(x)$ is monotonic. Prove that M_N lies between R_N and L_N and that M_N is closer to the actual area under the graph than both R_N and L_N. *Hint:* Argue from Figure 18; the part of the error in R_N due to the ith rectangle is the sum of the areas $A + B + D$, and for M_N it is $|B - E|$.

x_{i-1} midpoint x_i

FIGURE 18

SOLUTION Suppose $f(x)$ is monotonic increasing on the interval $[a, b]$, $\Delta x = \dfrac{b - a}{N}$,

$$\{x_k\}_{k=0}^{N} = \{a, a + \Delta x, a + 2\Delta x, \ldots, a + (N - 1)\Delta x, b\}$$

and

$$\{x_k^*\}_{k=0}^{N-1} = \left\{ \frac{a + (a + \Delta x)}{2}, \frac{(a + \Delta x) + (a + 2\Delta x)}{2}, \ldots, \frac{(a + (N - 1)\Delta x) + b}{2} \right\}.$$

Note that $x_i < x_i^* < x_{i+1}$ implies $f(x_i) < f(x_i^*) < f(x_{i+1})$ for all $0 \le i < N$ because $f(x)$ is monotone increasing. Then

$$\left(L_N = \frac{b - a}{N} \sum_{k=0}^{N-1} f(x_k) \right) < \left(M_N = \frac{b - a}{N} \sum_{k=0}^{N-1} f(x_k^*) \right) < \left(R_N = \frac{b - a}{N} \sum_{k=1}^{N} f(x_k) \right)$$

Similarly, if $f(x)$ is monotone decreasing,

$$\left(L_N = \frac{b - a}{N} \sum_{k=0}^{N-1} f(x_k) \right) > \left(M_N = \frac{b - a}{N} \sum_{k=0}^{N-1} f(x_k^*) \right) > \left(R_N = \frac{b - a}{N} \sum_{k=1}^{N} f(x_k) \right)$$

Thus, if $f(x)$ is monotonic, then M_N always lies in between R_N and L_N.

Now, as in Figure 18, consider the typical subinterval $[x_{i-1}, x_i]$ and its midpoint x_i^*. We let A, B, C, D, E, and F be the areas as shown in Figure 18. Note that, by the fact that x_i^* is the midpoint of the interval, $A = D + E$ and $F = B + C$. Let E_R represent the right endpoint approximation error ($= A + B + D$), let E_L represent the left endpoint approximation error ($= C + F + E$) and let E_M represent the midpoint approximation error ($= |B - E|$).

- If $B > E$, then $E_M = B - E$. In this case,

$$E_R - E_M = A + B + D - (B - E) = A + D + E > 0,$$

so $E_R > E_M$, while

$$E_L - E_M = C + F + E - (B - E) = C + (B + C) + E - (B - E) = 2C + 2E > 0,$$

so $E_L > E_M$. Therefore, the midpoint approximation is more accurate than either the left or the right endpoint approximation.

- If $B < E$, then $E_M = E - B$. In this case,

$$E_R - E_M = A + B + D - (E - B) = D + E + D - (E - B) = 2D + B > 0,$$

so that $E_R > E_M$ while

$$E_L - E_M = C + F + E - (E - B) = C + F + B > 0,$$

so $E_L > E_M$. Therefore, the midpoint approximation is more accurate than either the right or the left endpoint approximation.

- If $B = E$, the midpoint approximation is exactly equal to the area.

Hence, for $B < E$, $B > E$, or $B = E$, the midpoint approximation is more accurate than either the left endpoint or the right endpoint approximation.

89. In this exercise, we prove that the limits $\lim_{N \to \infty} R_N$ and $\lim_{N \to \infty} L_N$ exist and are equal if $f(x)$ is positive and increasing [the case of $f(x)$ decreasing is similar]. We use the concept of a least upper bound discussed in Appendix B.

(a) Explain with a graph why $L_N \leq R_M$ for all $N, M \geq 1$.

(b) By part (a), the sequence $\{L_N\}$ is bounded by R_M for any M, so it has a least upper bound L. By definition, L is the smallest number such that $L_N \leq L$ for all N. Show that $L \leq R_M$ for all M.

(c) According to part (b), $L_N \leq L \leq R_N$ for all N. Use Eq. (8) to show that $\lim_{N \to \infty} L_N = L$ and $\lim_{N \to \infty} R_N = L$.

SOLUTION

(a) Let $f(x)$ be positive and increasing, and let N and M be positive integers. From the figure below at the left, we see that L_N underestimates the area under the graph of $y = f(x)$, while from the figure below at the right, we see that R_M overestimates the area under the graph. Thus, for all $N, M \geq 1$, $L_N \leq R_M$.

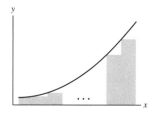 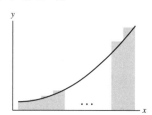

(b) Because the sequence $\{L_N\}$ is bounded above by R_M for any M, each R_M is an upper bound for the sequence. Furthermore, the sequence $\{L_N\}$ must have a least upper bound, call it L. By definition, the least upper bound must be no greater than any other upper bound; consequently, $L \leq R_M$ for all M.

(c) Since $L_N \leq L \leq R_N$, $R_N - L \leq R_N - L_N$, so $|R_N - L| \leq |R_N - L_N|$. From this,

$$\lim_{N \to \infty} |R_N - L| \leq \lim_{N \to \infty} |R_N - L_N|.$$

By Eq. (8),

$$\lim_{N \to \infty} |R_N - L_N| = \lim_{N \to \infty} \frac{1}{N} |(b - a)(f(b) - f(a))| = 0,$$

so $\lim_{N \to \infty} |R_N - L| \leq |R_N - L_N| = 0$, hence $\lim_{N \to \infty} R_N = L$.

Similarly, $|L_N - L| = L - L_N \leq R_N - L_N$, so

$$|L_N - L| \leq |R_N - L_N| = \frac{(b - a)}{N} (f(b) - f(a)).$$

This gives us that

$$\lim_{N \to \infty} |L_N - L| \leq \lim_{N \to \infty} \frac{1}{N} |(b - a)(f(b) - f(a))| = 0,$$

so $\lim_{N \to \infty} L_N = L$.

This proves $\lim_{N \to \infty} L_N = \lim_{N \to \infty} R_N = L$.

In Exercises 91–92, use Eq. (9) to find a value of N such that $|R_N - A| < 10^{-4}$ for the given function and interval.

91. $f(x) = \sqrt{x}$, $[1, 4]$

SOLUTION Let $f(x) = \sqrt{x}$ on $[1, 4]$. Then $b = 4$, $a = 1$, and

$$|R_N - A| \leq \frac{4 - 1}{N} (f(4) - f(1)) = \frac{3}{N} (2 - 1) = \frac{3}{N}.$$

We need $\frac{3}{N} < 10^{-4}$, which gives $N > 30000$. Thus $|R_{30001} - A| < 10^{-4}$ for $f(x) = \sqrt{x}$ on $[1, 4]$.

5.2 The Definite Integral

Preliminary Questions

1. What is $\int_a^b dx$ [here the function is $f(x) = 1$]?

SOLUTION $\int_a^b dx = \int_a^b 1 \cdot dx = 1(b-a) = b - a$.

2. Are the following statements true or false [assume that $f(x)$ is continuous]?

(a) $\int_a^b f(x)\, dx$ is the area between the graph and the x-axis over $[a, b]$.

(b) $\int_a^b f(x)\, dx$ is the area between the graph and the x-axis over $[a, b]$ if $f(x) \geq 0$.

(c) If $f(x) \leq 0$, then $-\int_a^b f(x)\, dx$ is the area between the graph of $f(x)$ and the x-axis over $[a, b]$.

SOLUTION

(a) False. $\int_a^b f(x)\, dx$ is the *signed* area between the graph and the x-axis.

(b) True.

(c) True.

3. Explain graphically why $\int_0^\pi \cos x \, dx = 0$.

SOLUTION Because $\cos(\pi - x) = -\cos x$, the "negative" area between the graph of $y = \cos x$ and the x-axis over $[\frac{\pi}{2}, \pi]$ exactly cancels the "positive" area between the graph and the x-axis over $[0, \frac{\pi}{2}]$.

4. Is $\int_{-5}^{-1} 8\, dx$ negative?

SOLUTION No, the integrand is the positive constant 8, so the value of the integral is 8 times the length of the integration interval $(-1 - (-5) = 4)$, or 32.

5. What is the largest possible value of $\int_0^6 f(x)\, dx$ if $f(x) \leq \frac{1}{3}$?

SOLUTION Because $f(x) \leq \frac{1}{3}$, $\int_0^6 f(x)\, dx \leq \frac{1}{3}(6 - 0) = 2$.

Exercises

In Exercises 1–10, draw a graph of the signed area represented by the integral and compute it using geometry.

1. $\int_{-3}^3 2x \, dx$

SOLUTION The region bounded by the graph of $y = 2x$ and the x-axis over the interval $[-3, 3]$ consists of two right triangles. One has area $\frac{1}{2}(3)(6) = 9$ below the axis, and the other has area $\frac{1}{2}(3)(6) = 9$ above the axis. Hence,

$$\int_{-3}^3 2x \, dx = 9 - 9 = 0.$$

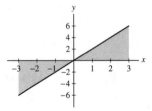

3. $\int_{-2}^1 (3x + 4) \, dx$

SOLUTION   The region bounded by the graph of $y = 3x + 4$ and the x-axis over the interval $[-2, 1]$ consists of two right triangles. One has area $\frac{1}{2}(\frac{2}{3})(2) = \frac{2}{3}$ below the axis, and the other has area $\frac{1}{2}(\frac{7}{3})(7) = \frac{49}{6}$ above the axis. Hence,

$$\int_{-2}^{1} (3x + 4)\,dx = \frac{49}{6} - \frac{2}{3} = \frac{15}{2}.$$

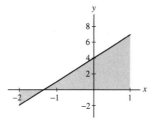

5. $\displaystyle\int_{6}^{8} (7 - x)\,dx$

SOLUTION   The region bounded by the graph of $y = 7 - x$ and the x-axis over the interval $[6, 8]$ consists of two right triangles. One triangle has area $\frac{1}{2}(1)(1) = \frac{1}{2}$ above the axis, and the other has area $\frac{1}{2}(1)(1) = \frac{1}{2}$ below the axis. Hence,

$$\int_{6}^{8} (7 - x)\,dx = \frac{1}{2} - \frac{1}{2} = 0.$$

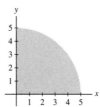

7. $\displaystyle\int_{0}^{5} \sqrt{25 - x^2}\,dx$

SOLUTION   The region bounded by the graph of $y = \sqrt{25 - x^2}$ and the x-axis over the interval $[0, 5]$ is one-quarter of a circle of radius 5. Hence,

$$\int_{0}^{5} \sqrt{25 - x^2}\,dx = \frac{1}{4}\pi(5)^2 = \frac{25\pi}{4}.$$

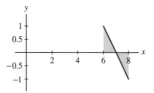

9. $\displaystyle\int_{-2}^{2} (2 - |x|)\,dx$

SOLUTION   The region bounded by the graph of $y = 2 - |x|$ and the x-axis over the interval $[-2, 2]$ is a triangle above the axis with base 4 and height 2. Consequently,

$$\int_{-2}^{2} (2 - |x|)\,dx = \frac{1}{2}(2)(4) = 4.$$

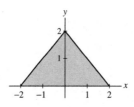

11. Calculate $\int_0^6 (4 - x)\, dx$ in two ways:

(a) As the limit $\lim\limits_{N \to \infty} R_N$

(b) By sketching the relevant signed area and using geometry

SOLUTION Let $f(x) = 4 - x$ over $[0, 6]$. Consider the integral $\int_0^6 f(x)\, dx = \int_0^6 (4 - x)\, dx$.

(a) Let N be a positive integer and set $a = 0$, $b = 6$, $\Delta x = (b - a)/N = 6/N$. Also, let $x_k = a + k\Delta x = 6k/N$, $k = 1, 2, \ldots, N$ be the right endpoints of the N subintervals of $[0, 6]$. Then

$$R_N = \Delta x \sum_{k=1}^{N} f(x_k) = \frac{6}{N} \sum_{k=1}^{N} \left(4 - \frac{6k}{N}\right) = \frac{6}{N} \left(4\left(\sum_{k=1}^{N} 1\right) - \frac{6}{N}\left(\sum_{k=1}^{N} k\right)\right)$$

$$= \frac{6}{N}\left(4N - \frac{6}{N}\left(\frac{N^2}{2} + \frac{N}{2}\right)\right) = 6 - \frac{18}{N}.$$

Hence $\lim\limits_{N \to \infty} R_N = \lim\limits_{N \to \infty} \left(6 - \frac{18}{N}\right) = 6.$

(b) The region bounded by the graph of $y = 4 - x$ and the x-axis over the interval $[0, 6]$ consists of two right triangles. One triangle has area $\frac{1}{2}(4)(4) = 8$ above the axis, and the other has area $\frac{1}{2}(2)(2) = 2$ below the axis. Hence,

$$\int_0^6 (4 - x)\, dx = 8 - 2 = 6.$$

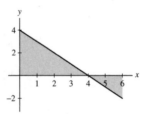

13. Evaluate the integrals for $f(x)$ shown in Figure 13.

(a) $\displaystyle\int_0^2 f(x)\, dx$ 　　　　　　　　　　　**(b)** $\displaystyle\int_0^6 f(x)\, dx$

(c) $\displaystyle\int_1^4 f(x)\, dx$ 　　　　　　　　　　　**(d)** $\displaystyle\int_1^6 |f(x)|\, dx$

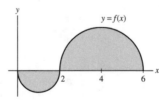

FIGURE 13 The two parts of the graph are semicircles.

SOLUTION Let $f(x)$ be given by Figure 13.

(a) The definite integral $\int_0^2 f(x)\, dx$ is the signed area of a semicircle of radius 1 which lies below the x-axis. Therefore,

$$\int_0^2 f(x)\, dx = -\frac{1}{2}\pi (1)^2 = -\frac{\pi}{2}.$$

(b) The definite integral $\int_0^6 f(x)\, dx$ is the signed area of a semicircle of radius 1 which lies below the x-axis and a semicircle of radius 2 which lies above the x-axis. Therefore,

$$\int_0^6 f(x)\, dx = \frac{1}{2}\pi (2)^2 - \frac{1}{2}\pi (1)^2 = \frac{3\pi}{2}.$$

(c) The definite integral $\int_1^4 f(x)\, dx$ is the signed area of one-quarter of a circle of radius 1 which lies below the x-axis and one-quarter of a circle of radius 2 which lies above the x-axis. Therefore,

$$\int_1^4 f(x)\, dx = \frac{1}{4}\pi (2)^2 - \frac{1}{4}\pi (1)^2 = \frac{3}{4}\pi.$$

(d) The definite integral $\int_1^6 |f(x)|\, dx$ is the signed area of one-quarter of a circle of radius 1 and a semicircle of radius 2, both of which lie above the x-axis. Therefore,

$$\int_1^6 |f(x)|\, dx = \frac{1}{2}\pi (2)^2 + \frac{1}{4}\pi (1)^2 = \frac{9\pi}{4}.$$

In Exercises 14–15, refer to Figure 14.

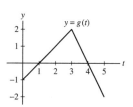

FIGURE 14

15. Find a, b, and c such that $\int_0^a g(t)\, dt$ and $\int_b^c g(t)\, dt$ are as large as possible.

SOLUTION   To make the value of $\int_0^a g(t)\, dt$ as large as possible, we want to include as much positive area as possible. This happens when we take $a = 4$. Now, to make the value of $\int_b^c g(t)\, dt$ as large as possibe, we want to make sure to include all of the positive area and only the positive area. This happens when we take $b = 1$ and $c = 4$.

In Exercises 17–20, sketch the signed area represented by the integral. Indicate the regions of positive and negative area.

17. $\int_0^2 (x - x^2)\, dx$

SOLUTION   Here is a sketch of the signed area represented by the integral $\int_0^2 (x - x^2)\, dx$.

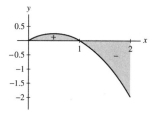

19. $\int_\pi^{2\pi} \sin x\, dx$

SOLUTION   Here is a sketch of the signed area represented by the integral $\int_\pi^{2\pi} \sin x\, dx$.

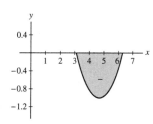

In Exercises 21–24, determine the sign of the integral without calculating it. Draw a graph if necessary.

21. $\int_{-2}^1 x^4\, dx$

SOLUTION   The integrand is always positive. The integral must therefore be positive, since the signed area has only positive part.

23. $\boxed{\text{GU}}$ $\int_0^{2\pi} x \sin x\, dx$

SOLUTION   As you can see from the graph below, the area below the axis is greater than the area above the axis. Thus, the definite integral is negative.

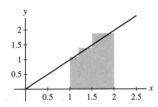

In Exercises 25–28, calculate the Riemann sum $R(f, P, C)$ for the given function, partition, and choice of intermediate points. Also, sketch the graph of f and the rectangles corresponding to $R(f, P, C)$.

25. $f(x) = x$, $P = \{1, 1.2, 1.5, 2\}$, $C = \{1.1, 1.4, 1.9\}$

SOLUTION   Let $f(x) = x$. With

$$P = \{x_0 = 1, x_1 = 1.2, x_3 = 1.5, x_4 = 2\} \quad \text{and} \quad C = \{c_1 = 1.1, c_2 = 1.4, c_3 = 1.9\},$$

we get

$$R(f, P, C) = \Delta x_1 f(c_1) + \Delta x_2 f(c_2) + \Delta x_3 f(c_3)$$
$$= (1.2 - 1)(1.1) + (1.5 - 1.2)(1.4) + (2 - 1.5)(1.9) = 1.59.$$

Here is a sketch of the graph of f and the rectangles.

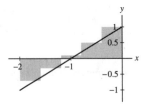

27. $f(x) = x + 1$, $P = \{-2, -1.6, -1.2, -0.8, -0.4, 0\}$,
   $C = \{-1.7, -1.3, -0.9, -0.5, 0\}$

SOLUTION   Let $f(x) = x + 1$. With

$$P = \{x_0 = -2, x_1 = -1.6, x_3 = -1.2, x_4 = -.8, x_5 = -.4, x_6 = 0\}$$

and

$$C = \{c_1 = -1.7, c_2 = -1.3, c_3 = -.9, c_4 = -.5, c_5 = 0\},$$

we get

$$R(f, P, C) = \Delta x_1 f(c_1) + \Delta x_2 f(c_2) + \Delta x_3 f(c_3) + \Delta x_4 f(c_4) + \Delta x_5 f(c_5)$$
$$= (-1.6 - (-2))(-.7) + (-1.2 - (-1.6))(-.3) + (-0.8 - (-1.2))(.1)$$
$$+ (-0.4 - (-0.8))(.5) + (0 - (-0.4))(1) = .24.$$

Here is a sketch of the graph of f and the rectangles.

In Exercises 29–36, use the basic properties of the integral and the formulas in the summary to calculate the integrals.

29. $\displaystyle\int_0^4 x^2 \, dx$

SOLUTION   By formula (4), $\displaystyle\int_0^4 x^2 \, dx = \frac{1}{3}(4)^3 = \frac{64}{3}$.

31. $\displaystyle\int_0^3 (3t+4)\,dt$

SOLUTION $\displaystyle\int_0^3 (3t+4)\,dt = 3\int_0^3 t\,dt + 4\int_0^3 1\,dt = 3\cdot\frac{1}{2}(3)^2 + 4(3-0) = \frac{51}{2}.$

33. $\displaystyle\int_0^1 (u^2 - 2u)\,du$

SOLUTION

$$\int_0^1 (u^2 - 2u)\,du = \int_0^1 u^2\,du - 2\int_0^1 u\,du = \frac{1}{3}(1)^3 - 2\left(\frac{1}{2}\right)(1)^2 = \frac{1}{3} - 1 = -\frac{2}{3}.$$

35. $\displaystyle\int_{-a}^1 (x^2+x)\,dx$

SOLUTION First, $\int_0^b (x^2+x)\,dx = \int_0^b x^2\,dx + \int_0^b x\,dx = \frac{1}{3}b^3 + \frac{1}{2}b^2$. Therefore

$$\int_{-a}^1 (x^2+x)\,dx = \int_{-a}^0 (x^2+x)\,dx + \int_0^1 (x^2+x)\,dx = \int_0^1 (x^2+x)\,dx - \int_0^{-a} (x^2+x)\,dx$$

$$= \left(\frac{1}{3}\cdot 1^3 + \frac{1}{2}\cdot 1^2\right) - \left(\frac{1}{3}(-a)^3 + \frac{1}{2}(-a)^2\right) = \frac{1}{3}a^3 - \frac{1}{2}a^2 + \frac{5}{6}.$$

37. Prove by computing the limit of right-endpoint approximations:

$$\int_0^b x^3\,dx = \frac{b^4}{4}. \qquad \boxed{7}$$

SOLUTION Let $f(x) = x^3$, $a = 0$ and $\Delta x = (b-a)/N = b/N$. Then

$$R_N = \Delta x \sum_{k=1}^N f(x_k) = \frac{b}{N}\sum_{k=1}^N \left(k^3\cdot\frac{b^3}{N^3}\right) = \frac{b^4}{N^4}\left(\sum_{k=1}^N k^3\right) = \frac{b^4}{N^4}\left(\frac{N^4}{4} + \frac{N^3}{2} + \frac{N^2}{4}\right) = \frac{b^4}{4} + \frac{b^4}{2N} + \frac{b^4}{4N^2}.$$

Hence $\displaystyle\int_0^b x^3\,dx = \lim_{N\to\infty} R_N = \lim_{N\to\infty}\left(\frac{b^4}{4} + \frac{b^4}{2N} + \frac{b^4}{4N^2}\right) = \frac{b^4}{4}.$

In Exercises 38–45, use the formulas in the summary and Eq. (7) to evaluate the integral.

39. $\displaystyle\int_0^2 (x^2+2x)\,dx$

SOLUTION Applying the linearity of the definite integral and the formulas from Examples 4 and 5,

$$\int_0^2 (x^2+2x)\,dx = \int_0^2 x^2\,dx + 2\int_0^2 x\,dx = \frac{1}{3}(2)^3 + 2\cdot\frac{1}{2}(2)^2 = \frac{20}{3}.$$

41. $\displaystyle\int_0^2 (x - x^3)\,dx$

SOLUTION Applying the linearity of the definite integral, the formula from Example 5 and Eq. (7):

$$\int_0^2 (x - x^3)\,dx = \int_0^2 x\,dx - \int_0^2 x^3\,dx = \frac{1}{2}(2)^2 - \frac{1}{4}(2)^4 = -2.$$

43. $\displaystyle\int_{-3}^0 (2x - 5)\,dx$

SOLUTION Applying the linearity of the definite integral, reversing the limits of integration, and using the formulas for the integral of x and of a constant:

$$\int_{-3}^0 (2x-5)\,dx = 2\int_{-3}^0 x\,dx - \int_{-3}^0 5\,dx = -2\int_0^{-3} x\,dx - \int_{-3}^0 5\,dx = -2\cdot\frac{1}{2}(-3)^2 - 15 = -24.$$

45. $\int_1^2 (x - x^3) \, dx$

SOLUTION   Applying the linearity and the additivity of the definite integral:

$$\int_1^2 (x - x^3) \, dx = \int_1^2 x \, dx - \int_1^2 x^3 \, dx = \int_0^2 x \, dx - \int_0^1 x \, dx - \left(\int_0^2 x^3 \, dx - \int_0^1 x^3 \, dx \right)$$

$$= \frac{1}{2}(2^2) - \frac{1}{2}(1^2) - \left(\frac{1}{4}(2)^4 - \frac{1}{4}(1)^4 \right) = \frac{3}{2} - \frac{15}{4} = -\frac{9}{4}.$$

In Exercises 46–50, calculate the integral, assuming that

$$\int_0^5 f(x) \, dx = 5, \qquad \int_0^5 g(x) \, dx = 12$$

47. $\int_0^5 (f(x) + 4g(x)) \, dx$

SOLUTION $\int_0^5 (f(x) + 4g(x)) \, dx = \int_0^5 f(x) \, dx + 4 \int_0^5 g(x) \, dx = 5 + 4(12) = 53.$

49. $\int_0^5 (3f(x) - 5g(x)) \, dx$

SOLUTION $\int_0^5 (3f(x) - 5g(x)) \, dx = 3 \int_0^5 f(x) \, dx - 5 \int_0^5 g(x) \, dx = 3(5) - 5(12) = -45.$

In Exercises 51–54, calculate the integral, assuming that

$$\int_0^1 f(x) \, dx = 1, \qquad \int_0^2 f(x) \, dx = 4, \qquad \int_1^4 f(x) \, dx = 7$$

51. $\int_0^4 f(x) \, dx$

SOLUTION $\int_0^4 f(x) \, dx = \int_0^1 f(x) \, dx + \int_1^4 f(x) \, dx = 1 + 7 = 8.$

53. $\int_4^1 f(x) \, dx$

SOLUTION $\int_4^1 f(x) \, dx = - \int_1^4 f(x) \, dx = -7.$

In Exercises 55–58, express each integral as a single integral.

55. $\int_0^3 f(x) \, dx + \int_3^7 f(x) \, dx$

SOLUTION $\int_0^3 f(x) \, dx + \int_3^7 f(x) \, dx = \int_0^7 f(x) \, dx.$

57. $\int_2^9 f(x) \, dx - \int_2^5 f(x) \, dx$

SOLUTION $\int_2^9 f(x) \, dx - \int_2^5 f(x) \, dx = \left(\int_2^5 f(x) \, dx + \int_5^9 f(x) \, dx \right) - \int_2^5 f(x) \, dx = \int_5^9 f(x) \, dx.$

In Exercises 59–62, calculate the integral, assuming that f is an integrable function such that $\int_1^b f(x) \, dx = 1 - b^{-1}$ for all $b > 0$.

59. $\int_1^3 f(x) \, dx$

SOLUTION $\int_1^3 f(x) \, dx = 1 - 3^{-1} = \frac{2}{3}.$

61. $\int_1^4 (4f(x) - 2) \, dx$

SOLUTION $\int_1^4 (4f(x) - 2)\,dx = 4\int_1^4 f(x)\,dx - 2\int_1^4 1\,dx = 4(1 - 4^{-1}) - 2(4 - 1) = -3.$

63. ✏️ Explain the difference in graphical interpretation between $\int_a^b f(x)\,dx$ and $\int_a^b |f(x)|\,dx$.

SOLUTION When $f(x)$ takes on both positive and negative values on $[a, b]$, $\int_a^b f(x)\,dx$ represents the signed area between $f(x)$ and the x-axis, whereas $\int_a^b |f(x)|\,dx$ represents the total (unsigned) area between $f(x)$ and the x-axis. Any negatively signed areas that were part of $\int_a^b f(x)\,dx$ are regarded as positive areas in $\int_a^b |f(x)|\,dx$. Here is a graphical example of this phenomenon.

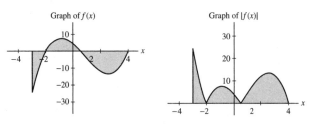

In Exercises 65–68, calculate the integral.

65. $\int_0^6 |3 - x|\,dx$

SOLUTION Over the interval, the region between the curve and the interval $[0, 6]$ consists of two triangles above the x axis, each of which has height 3 and width 3, and so area $\frac{9}{2}$. The total area, hence the definite integral, is 9.

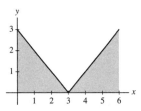

Alternately,

$$\int_0^6 |3 - x|\,dx = \int_0^3 (3 - x)\,dx + \int_3^6 (x - 3)\,dx$$

$$= 3\int_0^3 dx - \int_0^3 x\,dx + \left(\int_0^6 x\,dx - \int_0^3 x\,dx\right) - 3\int_3^6 dx$$

$$= 9 - \frac{1}{2}3^2 + \frac{1}{2}6^2 - \frac{1}{2}3^2 - 9 = 9.$$

67. $\int_{-1}^1 |x^3|\,dx$

SOLUTION

$$|x^3| = \begin{cases} x^3 & x \geq 0 \\ -x^3 & x < 0. \end{cases}$$

Therefore,

$$\int_{-1}^1 |x^3|\,dx = \int_{-1}^0 -x^3\,dx + \int_0^1 x^3\,dx = \int_0^{-1} x^3\,dx + \int_0^1 x^3\,dx = \frac{1}{4}(-1)^4 + \frac{1}{4}(1)^4 = \frac{1}{2}.$$

69. Use the Comparison Theorem to show that

$$\int_0^1 x^5\,dx \leq \int_0^1 x^4\,dx, \qquad \int_1^2 x^4\,dx \leq \int_1^2 x^5\,dx$$

SOLUTION   On the interval $[0, 1]$, $x^5 \le x^4$, so, by Theorem 5,

$$\int_0^1 x^5 \, dx \le \int_0^1 x^4 \, dx.$$

On the other hand, $x^4 \le x^5$ for $x \in [1, 2]$, so, by the same Theorem,

$$\int_1^2 x^4 \, dx \le \int_1^2 x^5 \, dx.$$

71. Prove that $0.0198 \le \int_{0.2}^{0.3} \sin x \, dx \le 0.0296$. *Hint:* Show that $0.198 \le \sin x \le 0.296$ for x in $[0.2, 0.3]$.

SOLUTION   For $0 \le x \le \frac{\pi}{6} \approx 0.52$, we have $\frac{d}{dx}(\sin x) = \cos x > 0$. Hence $\sin x$ is increasing on $[0.2, 0.3]$. Accordingly, for $0.2 \le x \le 0.3$, we have

$$m = 0.198 \le 0.19867 \approx \sin 0.2 \le \sin x \le \sin 0.3 \approx 0.29552 \le 0.296 = M$$

Therefore, by the Comparison Theorem, we have

$$0.0198 = m(0.3 - 0.2) = \int_{0.2}^{0.3} m \, dx \le \int_{0.2}^{0.3} \sin x \, dx \le \int_{0.2}^{0.3} M \, dx = M(0.3 - 0.2) = 0.0296.$$

73. $\boxed{\text{GU}}$ Prove that

$$\int_{\pi/4}^{\pi/2} \frac{\sin x}{x} \, dx \le \frac{\sqrt{2}}{2}$$

Hint: Graph $y = \dfrac{\sin x}{x}$ and observe that it is decreasing on $[\frac{\pi}{4}, \frac{\pi}{2}]$.

SOLUTION   Let

$$f(x) = \frac{\sin x}{x}.$$

As we can see in the sketch below, $f(x)$ is decreasing on the interval $[\pi/4, \pi/2]$. Therefore $f(x) \le f(\pi/4)$ for all x in $[\pi/4, \pi/2]$. $f(\pi/4) = \frac{2\sqrt{2}}{\pi}$, so:

$$\int_{\pi/4}^{\pi/2} \frac{\sin x}{x} \, dx \le \int_{\pi/4}^{\pi/2} \frac{2\sqrt{2}}{\pi} \, dx = \frac{\pi}{4} \frac{2\sqrt{2}}{\pi} = \frac{\sqrt{2}}{2}.$$

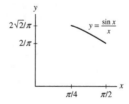

75. 📝   Suppose that $f(x) \le g(x)$ on $[a, b]$. By the Comparison Theorem, $\int_a^b f(x) \, dx \le \int_a^b g(x) \, dx$. Is it also true that $f'(x) \le g'(x)$ for $x \in [a, b]$? If not, give a counterexample.

SOLUTION   The assertion $f'(x) \le g'(x)$ is false. Consider $a = 0$, $b = 1$, $f(x) = x$, $g(x) = 2$. $f(x) \le g(x)$ for all x in the interval $[0, 1]$, but $f'(x) = 1$ while $g'(x) = 0$ for all x.

Further Insights and Challenges

77. Explain graphically: $\int_{-a}^{a} f(x) \, dx = 0$ if $f(x)$ is an odd function.

SOLUTION   If f is an odd function, then $f(-x) = -f(x)$ for all x. Accordingly, for every positively signed area in the right half-plane where f is above the x-axis, there is a corresponding negatively signed area in the left half-plane where f is below the x-axis. Similarly, for every negatively signed area in the right half-plane where f is below the x-axis, there is a corresponding positively signed area in the left half-plane where f is above the x-axis. We conclude that the net area between the graph of f and the x-axis over $[-a, a]$ is 0, since the positively signed areas and negatively signed areas cancel each other out exactly.

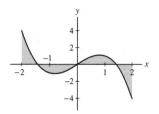

79. Let k and b be positive. Show, by comparing the right-endpoint approximations, that

$$\int_0^b x^k \, dx = b^{k+1} \int_0^1 x^k \, dx$$

SOLUTION Let k and b be any positive numbers. Let $f(x) = x^k$ on $[0, b]$. Since f is continuous, both $\int_0^b f(x)\,dx$ and $\int_0^1 f(x)\,dx$ exist. Let N be a positive integer and set $\Delta x = (b - 0)/N = b/N$. Let $x_j = a + j\Delta x = bj/N$, $j = 1, 2, \dots, N$ be the right endpoints of the N subintervals of $[0, b]$. Then the right-endpoint approximation to $\int_0^b f(x)\,dx = \int_0^b x^k\,dx$ is

$$R_N = \Delta x \sum_{j=1}^N f(x_j) = \frac{b}{N} \sum_{j=1}^N \left(\frac{bj}{N}\right)^k = b^{k+1}\left(\frac{1}{N^{k+1}} \sum_{j=1}^N j^k\right).$$

In particular, if $b = 1$ above, then the right-endpoint approximation to $\int_0^1 f(x)\,dx = \int_0^1 x^k\,dx$ is

$$S_N = \Delta x \sum_{j=1}^N f(x_j) = \frac{1}{N} \sum_{j=1}^N \left(\frac{j}{N}\right)^k = \frac{1}{N^{k+1}} \sum_{j=1}^N j^k = \frac{1}{b^{k+1}} R_N$$

In other words, $R_N = b^{k+1} S_N$. Therefore,

$$\int_0^b x^k \, dx = \lim_{N \to \infty} R_N = \lim_{N \to \infty} b^{k+1} S_N = b^{k+1} \lim_{N \to \infty} S_N = b^{k+1} \int_0^1 x^k \, dx.$$

81. Show that Eq. (4) holds for $b \le 0$.

SOLUTION Let $c = -b$. Since $b < 0$, $c > 0$, so by Eq. (4),

$$\int_0^c x^2 \, dx = \frac{1}{3} c^3.$$

Furthermore, x^2 is an even function, so symmetry of the areas gives

$$\int_{-c}^0 x^2 \, dx = \int_0^c x^2 \, dx.$$

Finally,

$$\int_0^b x^2 \, dx = \int_0^{-c} x^2 \, dx = -\int_{-c}^0 x^2 \, dx = -\int_0^c x^2 \, dx = -\frac{1}{3}c^3 = \frac{1}{3}b^3.$$

5.3 The Fundamental Theorem of Calculus, Part I

Preliminary Questions

1. Assume that $f(x) \ge 0$. What is the area under the graph of $f(x)$ over $[0, 2]$ if $f(x)$ has an antiderivative $F(x)$ such that $F(0) = 3$ and $F(2) = 7$?

SOLUTION Because $f(x) \ge 0$, the area under the graph of $y = f(x)$ over the interval $[0, 2]$ is

$$\int_0^2 f(x)\,dx = F(2) - F(0) = 7 - 3 = 4.$$

2. Suppose that $F(x)$ is an antiderivative of $f(x)$. What is the graphical interpretation of $F(4) - F(1)$ if $f(x)$ takes on both positive and negative values?

SOLUTION   Because $F(x)$ is an antiderivative of $f(x)$, it follows that $F(4) - F(1) = \int_1^4 f(x)\,dx$. Hence, $F(4) - F(1)$ represents the signed area between the graph of $y = f(x)$ and the x-axis over the interval $[1, 4]$.

3. Evaluate $\int_0^7 f(x)\,dx$ and $\int_2^7 f(x)\,dx$, assuming that $f(x)$ has an antiderivative $F(x)$ with values from the following table:

x	0	2	7
$F(x)$	3	7	9

SOLUTION   Because $F(x)$ is an antiderivative of $f(x)$,

$$\int_0^7 f(x)\,dx = F(7) - F(0) = 9 - 3 = 6$$

and

$$\int_2^7 f(x)\,dx = F(7) - F(2) = 9 - 7 = 2.$$

4. Are the following statements true or false? Explain.

(a) The FTC I is only valid for positive functions.

(b) To use the FTC I, you have to choose the right antiderivative.

(c) If you cannot find an antiderivative of $f(x)$, then the definite integral does not exist.

SOLUTION

(a) False. The FTC I is valid for continuous functions.

(b) False. The FTC I works for any antiderivative of the integrand.

(c) False. If you cannot find an antiderivative of the integrand, you cannot use the FTC I to evaluate the definite integral, but the definite integral may still exist.

5. What is the value of $\int_2^9 f'(x)\,dx$ if $f(x)$ is differentiable and $f(2) = f(9) = 4$?

SOLUTION   Because f is differentiable, $\int_2^9 f'(x)\,dx = f(9) - f(2) = 4 - 4 = 0$.

Exercises

In Exercises 1–4, sketch the region under the graph of the function and find its area using the FTC I.

1. $f(x) = x^2$, $[0, 1]$

SOLUTION

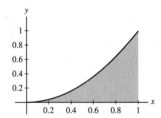

We have the area

$$A = \int_0^1 x^2\,dx = \frac{1}{3}x^3\Big|_0^1 = \frac{1}{3}.$$

3. $f(x) = \sin x$, $[0, \pi/2]$

SOLUTION

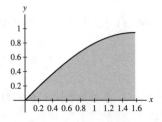

Let A be the area indicated. Then

$$A = \int_0^{\pi/2} \sin x \, dx = -\cos x \Big|_0^{\pi/2} = 0 - (-1) = 1.$$

In Exercises 5–34, evaluate the integral using the FTC I.

5. $\displaystyle\int_3^6 x \, dx$

SOLUTION $\displaystyle\int_3^6 x \, dx = \frac{1}{2}x^2 \Big|_3^6 = \frac{1}{2}(6)^2 - \frac{1}{2}(3)^2 = \frac{27}{2}.$

7. $\displaystyle\int_{-3}^2 u^2 \, du$

SOLUTION $\displaystyle\int_{-3}^2 u^2 \, du = \frac{1}{3}u^3 \Big|_{-3}^2 = \frac{1}{3}(2)^3 - \frac{1}{3}(-3)^3 = \frac{35}{3}.$

9. $\displaystyle\int_1^3 (t^3 - t^2) \, dt$

SOLUTION $\displaystyle\int_1^3 (t^3 - t^2) \, dt = \left(\frac{1}{4}t^4 - \frac{1}{3}t^3\right) \Big|_1^3 = \left(\frac{1}{4}(3)^4 - \frac{1}{3}(3)^3\right) - \left(\frac{1}{4} - \frac{1}{3}\right) = \frac{34}{3}.$

11. $\displaystyle\int_{-3}^4 (x^2 + 2) \, dx$

SOLUTION

$$\int_{-3}^4 (x^2 + 2) \, dx = \left(\frac{1}{3}x^3 + 2x\right) \Big|_{-3}^4 = \left(\frac{1}{3}(4)^3 + 2(4)\right) - \left(\frac{1}{3}(-3)^3 + 2(-3)\right) = \frac{133}{3}.$$

13. $\displaystyle\int_{-2}^2 (10x^9 + 3x^5) \, dx$

SOLUTION $\displaystyle\int_{-2}^2 (10x^9 + 3x^5) \, dx = \left(x^{10} + \frac{1}{2}x^6\right) \Big|_{-2}^2 = \left(2^{10} + \frac{1}{2}2^6\right) - \left(2^{10} + \frac{1}{2}2^6\right) = 0.$

15. $\displaystyle\int_1^3 (4t^{3/2} + t^{7/2}) \, dt$

SOLUTION

$$\int_1^3 (4t^{3/2} + t^{7/2}) \, dt = \left(\frac{8}{5}t^{5/2} + \frac{2}{9}t^{9/2}\right) \Big|_1^3 = \left(\frac{72\sqrt{3}}{5} + 18\sqrt{3}\right) - \left(\frac{8}{5} + \frac{2}{9}\right) = \frac{162\sqrt{3}}{5} - \frac{82}{45}.$$

17. $\displaystyle\int_1^4 \frac{1}{t^2} \, dt$

SOLUTION $\displaystyle\int_1^4 \frac{1}{t^2} \, dt = \int_1^4 t^{-2} \, dt = \left(-t^{-1}\right) \Big|_1^4 = \left(-(4)^{-1}\right) - \left(-(1)^{-1}\right) = \frac{3}{4}.$

19. $\displaystyle\int_1^{27} x^{1/3} \, dx$

SOLUTION $\displaystyle\int_1^{27} x^{1/3} \, dx = \frac{3}{4}x^{4/3} \Big|_1^{27} = \frac{3}{4}(81) - \frac{3}{4} = 60.$

21. $\displaystyle\int_1^9 t^{-1/2} \, dt$

SOLUTION $\displaystyle\int_1^9 t^{-1/2} \, dt = 2t^{1/2} \Big|_1^9 = 2(9)^{1/2} - 2(1)^{1/2} = 4.$

23. $\displaystyle\int_{-2}^{-1} \frac{1}{x^3} \, dx$

SOLUTION $\int_{-2}^{-1} \frac{1}{x^3} \, dx = -\frac{1}{2}x^{-2}\Big|_{-2}^{-1} = -\frac{1}{2}(-1)^{-2} + \frac{1}{2}(-2)^{-2} = -\frac{3}{8}.$

25. $\int_{1}^{27} \frac{t+1}{\sqrt{t}} \, dt$

SOLUTION

$$\int_{1}^{27} \frac{t+1}{\sqrt{t}} \, dt = \int_{1}^{27} (t^{1/2} + t^{-1/2}) \, dt = \left(\frac{2}{3}t^{3/2} + 2t^{1/2}\right)\Big|_{1}^{27}$$
$$= \left(\frac{2}{3}(81\sqrt{3}) + 6\sqrt{3}\right) - \left(\frac{2}{3} + 2\right) = 60\sqrt{3} - \frac{8}{3}.$$

27. $\int_{-\pi/2}^{\pi/2} \cos x \, dx$

SOLUTION $\int_{-\pi/2}^{\pi/2} \cos x \, dx = \sin x \Big|_{-\pi/2}^{\pi/2} = 1 - (-1) = 2.$

29. $\int_{\pi/4}^{3\pi/4} \sin \theta \, d\theta$

SOLUTION $\int_{\pi/4}^{3\pi/4} \sin \theta \, d\theta = -\cos \theta \Big|_{\pi/4}^{3\pi/4} = \frac{\sqrt{2}}{2} + \frac{\sqrt{2}}{2} = \sqrt{2}.$

31. $\int_{0}^{\pi/4} \sec^2 t \, dt$

SOLUTION $\int_{0}^{\pi/4} \sec^2 t \, dt = \tan t \Big|_{0}^{\pi/4} = \tan \frac{\pi}{4} - \tan 0 = 1.$

33. $\int_{\pi/6}^{\pi/3} \csc x \cot x \, dx$

SOLUTION $\int_{\pi/6}^{\pi/3} \csc x \cot x \, dx = (-\csc x)\Big|_{\pi/6}^{\pi/3} = \left(-\csc \frac{\pi}{3}\right) - \left(-\csc \frac{\pi}{6}\right) = 2 - \frac{2}{3}\sqrt{3}.$

In Exercises 35–40, write the integral as a sum of integrals without absolute values and evaluate.

35. $\int_{-2}^{1} |x| \, dx$

SOLUTION

$$\int_{-2}^{1} |x| \, dx = \int_{-2}^{0} (-x) \, dx + \int_{0}^{1} x \, dx = -\frac{1}{2}x^2\Big|_{-2}^{0} + \frac{1}{2}x^2\Big|_{0}^{1} = 0 - \left(-\frac{1}{2}(4)\right) + \frac{1}{2} = \frac{5}{2}.$$

37. $\int_{-2}^{3} |x^3| \, dx$

SOLUTION

$$\int_{-2}^{3} |x^3| \, dx = \int_{-2}^{0} (-x^3) \, dx + \int_{0}^{3} x^3 \, dx = -\frac{1}{4}x^4\Big|_{-2}^{0} + \frac{1}{4}x^4\Big|_{0}^{3}$$
$$= 0 + \frac{1}{4}(-2)^4 + \frac{1}{4}3^4 - 0 = \frac{97}{4}.$$

39. $\int_{0}^{\pi} |\cos x| \, dx$

SOLUTION

$$\int_{0}^{\pi} |\cos x| \, dx = \int_{0}^{\pi/2} \cos x \, dx + \int_{\pi/2}^{\pi} (-\cos x) \, dx = \sin x \Big|_{0}^{\pi/2} - \sin x \Big|_{\pi/2}^{\pi} = 1 - 0 - (-1 - 0) = 2.$$

In Exercises 41–44, evaluate the integral in terms of the constants.

41. $\displaystyle\int_1^b x^3\,dx$

SOLUTION $\displaystyle\int_1^b x^3\,dx = \frac{1}{4}x^4\Big|_1^b = \frac{1}{4}b^4 - \frac{1}{4}(1)^4 = \frac{1}{4}\left(b^4 - 1\right)$ for any number b.

43. $\displaystyle\int_1^b x^5\,dx$

SOLUTION $\displaystyle\int_1^b x^5\,dx = \frac{1}{6}x^6\Big|_1^b = \frac{1}{6}b^6 - \frac{1}{6}(1)^6 = \frac{1}{6}(b^6 - 1)$ for any number b.

45. Use the FTC I to show that $\displaystyle\int_{-1}^1 x^n\,dx = 0$ if n is an odd whole number. Explain graphically.

SOLUTION We have

$$\int_{-1}^1 x^n\,dx = \frac{x^{n+1}}{n+1}\Big|_{-1}^1 = \frac{(1)^{n+1}}{n+1} - \frac{(-1)^{n+1}}{n+1}.$$

Because n is odd, $n + 1$ is even, which means that $(-1)^{n+1} = (1)^{n+1} = 1$. Hence

$$\frac{(1)^{n+1}}{n+1} - \frac{(-1)^{n+1}}{n+1} = \frac{1}{n+1} - \frac{1}{n+1} = 0.$$

Graphically speaking, for an odd function such as x^3 shown here, the positively signed area from $x = 0$ to $x = 1$ cancels the negatively signed area from $x = -1$ to $x = 0$.

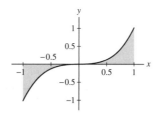

47. Show that the area of a parabolic arch (the shaded region in Figure 5) is equal to four-thirds the area of the triangle shown.

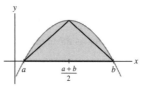

FIGURE 5 Graph of $y = (x - a)(b - x)$.

SOLUTION We first calculate the area of the parabolic arch:

$$\int_a^b (x - a)(b - x)\,dx = -\int_a^b (x - a)(x - b)\,dx = -\int_a^b (x^2 - ax - bx + ab)\,dx$$

$$= -\left(\frac{1}{3}x^3 - \frac{a}{2}x^2 - \frac{b}{2}x^2 + abx\right)\Big|_a^b$$

$$= -\frac{1}{6}\left(2x^3 - 3ax^2 - 3bx^2 + 6abx\right)\Big|_a^b$$

$$= -\frac{1}{6}\left((2b^3 - 3ab^2 - 3b^3 + 6ab^2) - (2a^3 - 3a^3 - 3ba^2 + 6a^2b)\right)$$

$$= -\frac{1}{6}\left((-b^3 + 3ab^2) - (-a^3 + 3a^2b)\right)$$

$$= -\frac{1}{6}\left(a^3 + 3ab^2 - 3a^2b - b^3\right) = \frac{1}{6}(b - a)^3.$$

The indicated triangle has a base of length $b - a$ and a height of

$$\left(\frac{a+b}{2} - a\right)\left(b - \frac{a+b}{2}\right) = \left(\frac{b-a}{2}\right)^2.$$

Thus, the area of the triangle is

$$\frac{1}{2}(b-a)\left(\frac{b-a}{2}\right)^2 = \frac{1}{8}(b-a)^3.$$

Finally, we note that

$$\frac{1}{6}(b-a)^3 = \frac{4}{3} \cdot \frac{1}{8}(b-a)^3,$$

as required.

49. Calculate $\displaystyle\int_{-2}^{3} f(x)\,dx$, where

$$f(x) = \begin{cases} 12 - x^2 & \text{for } x \le 2 \\ x^3 & \text{for } x > 2 \end{cases}$$

SOLUTION

$$\int_{-2}^{3} f(x)\,dx = \int_{-2}^{2} f(x)\,dx + \int_{2}^{3} f(x)\,dx = \int_{-2}^{2} (12 - x^2)\,dx + \int_{2}^{3} x^3\,dx$$

$$= \left(12x - \frac{1}{3}x^3\right)\Big|_{-2}^{2} + \frac{1}{4}x^4\Big|_{2}^{3}$$

$$= \left(12(2) - \frac{1}{3}(2)^3\right) - \left(12(-2) - \frac{1}{3}(-2)^3\right) + \frac{1}{4}3^4 - \frac{1}{4}2^4$$

$$= \frac{128}{3} + \frac{65}{4} = \frac{707}{12}.$$

Further Insights and Challenges

51. In this exercise, we generalize the result of Exercise 47 by proving the famous result of Archimedes: For $r < s$, the area of the shaded region in Figure 6 is equal to four-thirds the area of triangle $\triangle ACE$, where C is the point on the parabola at which the tangent line is parallel to secant line $\overline{AE}$.

(a) Show that C has x-coordinate $(r + s)/2$.

(b) Show that $ABDE$ has area $(s - r)^3/4$ by viewing it as a parallelogram of height $s - r$ and base of length $\overline{CF}$.

(c) Show that $\triangle ACE$ has area $(s - r)^3/8$ by observing that it has the same base and height as the parallelogram.

(d) Compute the shaded area as the area under the graph minus the area of a trapezoid and prove Archimedes's result.

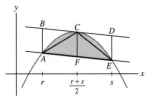

FIGURE 6 Graph of $f(x) = (x - a)(b - x)$.

SOLUTION

(a) The slope of the secant line $\overline{AE}$ is

$$\frac{f(s) - f(r)}{s - r} = \frac{(s - a)(b - s) - (r - a)(b - r)}{s - r} = a + b - (r + s)$$

and the slope of the tangent line along the parabola is

$$f'(x) = a + b - 2x.$$

If C is the point on the parabola at which the tangent line is parallel to the secant line $\overline{AE}$, then its x-coordinate must satisfy

$$a + b - 2x = a + b - (r + s) \qquad \text{or} \qquad x = \frac{r + s}{2}.$$

(b) Parallelogram $ABDE$ has height $s - r$ and base of length $\overline{CF}$. Since the equation of the secant line $\overline{AE}$ is

$$y = [a + b - (r + s)](x - r) + (r - a)(b - r),$$

the length of the segment $\overline{CF}$ is

$$\left(\frac{r+s}{2} - a\right)\left(b - \frac{r+s}{2}\right) - [a + b - (r+s)]\left(\frac{r+s}{2} - r\right) - (r-a)(b-r) = \frac{(s-r)^2}{4}.$$

Thus, the area of $ABDE$ is $\frac{(s-r)^3}{4}$.

(c) Triangle ACE is comprised of $\triangle ACF$ and $\triangle CEF$. Each of these smaller triangles has height $\frac{s-r}{2}$ and base of length $\frac{(s-r)^2}{4}$. Thus, the area of $\triangle ACE$ is

$$\frac{1}{2}\frac{s-r}{2}\cdot\frac{(s-r)^2}{4} + \frac{1}{2}\frac{s-r}{2}\cdot\frac{(s-r)^2}{4} = \frac{(s-r)^3}{8}.$$

(d) The area under the graph of the parabola between $x = r$ and $x = s$ is

$$\int_r^s (x-a)(b-x)\,dx = \left(-abx + \frac{1}{2}(a+b)x^2 - \frac{1}{3}x^3\right)\Big|_r^s$$

$$= -abs + \frac{1}{2}(a+b)s^2 - \frac{1}{3}s^3 + abr - \frac{1}{2}(a+b)r^2 + \frac{1}{3}r^3$$

$$= ab(r-s) + \frac{1}{2}(a+b)(s-r)(s+r) + \frac{1}{3}(r-s)(r^2+rs+s^2),$$

while the area of the trapezoid under the shaded region is

$$\frac{1}{2}(s-r)\left[(s-a)(b-s) + (r-a)(b-r)\right]$$

$$= \frac{1}{2}(s-r)\left[-2ab + (a+b)(r+s) - r^2 - s^2\right]$$

$$= ab(r-s) + \frac{1}{2}(a+b)(s-r)(r+s) + \frac{1}{2}(r-s)(r^2+s^2).$$

Thus, the area of the shaded region is

$$(r-s)\left(\frac{1}{3}r^2 + \frac{1}{3}rs + \frac{1}{3}s^2 - \frac{1}{2}r^2 - \frac{1}{2}s^2\right) = (s-r)\left(\frac{1}{6}r^2 - \frac{1}{3}rs + \frac{1}{6}s^2\right) = \frac{1}{6}(s-r)^3,$$

which is four-thirds the area of the triangle ACE.

53. Use the method of Exercise 52 to prove that

$$1 - \frac{x^2}{2} \le \cos x \le 1 - \frac{x^2}{2} + \frac{x^4}{24}$$

$$x - \frac{x^3}{6} \le \sin x \le x - \frac{x^3}{6} + \frac{x^5}{120} \quad \text{(for } x \ge 0\text{)}$$

Verify these inequalities for $x = 0.1$. Why have we specified $x \ge 0$ for $\sin x$ but not $\cos x$?

SOLUTION By Exercise 52, $t - \frac{1}{6}t^3 \le \sin t \le t$ for $t > 0$. Integrating this inequality over the interval $[0, x]$, and then solving for $\cos x$, yields:

$$\frac{1}{2}x^2 - \frac{1}{24}x^4 \le 1 - \cos x \le \frac{1}{2}x^2$$

$$1 - \frac{1}{2}x^2 \le \cos x \le 1 - \frac{1}{2}x^2 + \frac{1}{24}x^4.$$

These inequalities apply for $x \ge 0$. Since $\cos x$, $1 - \frac{x^2}{2}$, and $1 - \frac{x^2}{2} + \frac{x^4}{24}$ are all even functions, they also apply for $x \le 0$.

Having established that

$$1 - \frac{t^2}{2} \le \cos t \le 1 - \frac{t^2}{2} + \frac{t^4}{24},$$

for all $t \ge 0$, we integrate over the interval $[0, x]$, to obtain:

$$x - \frac{x^3}{6} \le \sin x \le x - \frac{x^3}{6} + \frac{x^5}{120}.$$

The functions $\sin x$, $x - \frac{1}{6}x^3$ and $x - \frac{1}{6}x^3 + \frac{1}{120}x^5$ are all odd functions, so the inequalities are reversed for $x < 0$. Evaluating these inequalities at $x = .1$ yields

$$0.995000000 \le 0.995004165 \le 0.995004167$$

$$0.0998333333 \le 0.0998334166 \le 0.0998334167,$$

both of which are true.

55. Assume that $|f'(x)| \le K$ for $x \in [a, b]$. Use FTC I to prove that $|f(x) - f(a)| \le K|x - a|$ for $x \in [a, b]$.

SOLUTION Let $a > b$ be real numbers, and let $f(x)$ be such that $|f'(x)| \le K$ for $x \in [a, b]$. By FTC,

$$\int_a^x f'(t)\, dt = f(x) - f(a).$$

Since $f'(x) \ge -K$ for all $x \in [a, b]$, we get:

$$f(x) - f(a) = \int_a^x f'(t)\, dt \ge -K(x - a).$$

Since $f'(x) \le K$ for all $x \in [a, b]$, we get:

$$f(x) - f(a) = \int_a^x f'(t)\, dt \le K(x - a).$$

Combining these two inequalities yields

$$-K(x - a) \le f(x) - f(a) \le K(x - a),$$

so that, by definition,

$$|f(x) - f(a)| \le K|x - a|.$$

5.4 The Fundamental Theorem of Calculus, Part II

Preliminary Questions

1. What is $A(-2)$, where $A(x) = \int_{-2}^{x} f(t)\, dt$?

SOLUTION By definition, $A(-2) = \int_{-2}^{-2} f(t)\, dt = 0$.

2. Let $G(x) = \int_{4}^{x} \sqrt{t^3 + 1}\, dt$.
(a) Is the FTC needed to calculate $G(4)$?
(b) Is the FTC needed to calculate $G'(4)$?

SOLUTION

(a) No. $G(4) = \int_{4}^{4} \sqrt{t^3 + 1}\, dt = 0$.
(b) Yes. By the FTC II, $G'(x) = \sqrt{x^3 + 1}$, so $G'(4) = \sqrt{65}$.

3. Which of the following defines an antiderivative $F(x)$ of $f(x) = x^2$ satisfying $F(2) = 0$?

(a) $\int_{2}^{x} 2t\, dt$ **(b)** $\int_{0}^{2} t^2\, dt$ **(c)** $\int_{2}^{x} t^2\, dt$

SOLUTION The correct answer is **(c)**: $\int_{2}^{x} t^2\, dt$.

4. True or false? Some continuous functions do not have antiderivatives. Explain.

SOLUTION False. All continuous functions have an antiderivative, namely $\int_{a}^{x} f(t)\, dt$.

5. Let $G(x) = \int_{4}^{x^3} \sin t\, dt$. Which of the following statements are correct?

(a) $G(x)$ is the composite function $\sin(x^3)$.

(b) $G(x)$ is the composite function $A(x^3)$, where

$$A(x) = \int_4^x \sin(t)\,dt.$$

(c) $G(x)$ is too complicated to differentiate.

(d) The Product Rule is used to differentiate $G(x)$.

(e) The Chain Rule is used to differentiate $G(x)$.

(f) $G'(x) = 3x^2 \sin(x^3)$.

SOLUTION Statements **(b)**, **(e)**, and **(f)** are correct.

6. Trick question: Find the derivative of $\int_1^3 t^3\,dt$ at $x = 2$.

SOLUTION Note that the definite integral $\int_1^3 t^3\,dt$ does not depend on x; hence the derivative with respect to x is 0 for any value of x.

Exercises

1. Write the area function of $f(x) = 2x + 4$ with lower limit $a = -2$ as an integral and find a formula for it.

SOLUTION Let $f(x) = 2x + 4$. The area function with lower limit $a = -2$ is

$$A(x) = \int_a^x f(t)\,dt = \int_{-2}^x (2t + 4)\,dt.$$

Carrying out the integration, we find

$$\int_{-2}^x (2t + 4)\,dt = (t^2 + 4t)\Big|_{-2}^x = (x^2 + 4x) - ((-2)^2 + 4(-2)) = x^2 + 4x + 4$$

or $(x + 2)^2$. Therefore, $A(x) = (x + 2)^2$.

3. Let $G(x) = \int_1^x (t^2 - 2)\,dt$.

(a) What is $G(1)$?

(b) Use FTC II to find $G'(1)$ and $G'(2)$.

(c) Find a formula for $G(x)$ and use it to verify your answers to (a) and (b).

SOLUTION Let $G(x) = \int_1^x (t^2 - 2)\,dt$.

(a) Then $G(1) = \int_1^1 (t^2 - 2)\,dt = 0$.

(b) Now $G'(x) = x^2 - 2$, so that $G'(1) = -1$ and $G'(2) = 2$.

(c) We have

$$\int_1^x (t^2 - 2)\,dt = \left(\frac{1}{3}t^3 - 2t\right)\Big|_1^x = \left(\frac{1}{3}x^3 - 2x\right) - \left(\frac{1}{3}(1)^3 - 2(1)\right) = \frac{1}{3}x^3 - 2x + \frac{5}{3}.$$

Thus $G(x) = \frac{1}{3}x^3 - 2x + \frac{5}{3}$ and $G'(x) = x^2 - 2$. Moreover, $G(1) = \frac{1}{3}(1)^3 - 2(1) + \frac{5}{3} = 0$, as in (a), and $G'(1) = -1$ and $G'(2) = 2$, as in (b).

5. Find $G(1)$, $G'(0)$, and $G'(\pi/4)$, where $G(x) = \int_1^x \tan t\,dt$.

SOLUTION By definition, $G(1) = \int_1^1 \tan t\,dt = 0$. By FTC, $G'(x) = \tan x$, so that $G'(0) = \tan 0 = 0$ and $G'(\frac{\pi}{4}) = \tan \frac{\pi}{4} = 1$.

In Exercises 7–14, find formulas for the functions represented by the integrals.

7. $\int_2^x u^3\,du$

SOLUTION $F(x) = \int_2^x u^3\,du = \frac{1}{4}u^4\Big|_2^x = \frac{1}{4}x^4 - 4$.

9. $\int_1^{x^2} t\,dt$

SOLUTION $F(x) = \int_1^{x^2} t\,dt = \frac{1}{2}t^2\Big|_1^{x^2} = \frac{1}{2}x^4 - \frac{1}{2}$.

11. $\int_x^5 (4t - 1)\, dt$

SOLUTION $F(x) = \int_x^5 (4t - 1)\, dt = (2t^2 - t)\Big|_x^5 = 45 + x - 2x^2.$

13. $\int_{-\pi/4}^x \sec^2 \theta\, d\theta$

SOLUTION $F(x) = \int_{-\pi/4}^x \sec^2 \theta\, d\theta = \tan \theta\Big|_{-\pi/4}^x = \tan x - \tan(-\pi/4) = \tan x + 1.$

In Exercises 15–18, express the antiderivative $F(x)$ of $f(x)$ satisfying the given initial condition as an integral.

15. $f(x) = \sqrt{x^4 + 1}, \quad F(3) = 0$

SOLUTION The antiderivative $F(x)$ of $f(x) = \sqrt{x^4 + 1}$ satisfying $F(3) = 0$ is $F(x) = \int_3^x \sqrt{t^4 + 1}\, dt.$

17. $f(x) = \sec x, \quad F(0) = 0$

SOLUTION The antiderivative $F(x)$ of $f(x) = \sec x$ satisfying $F(0) = 0$ is $F(x) = \int_0^x \sec t\, dt.$

In Exercises 19–22, calculate the derivative.

19. $\dfrac{d}{dx} \int_0^x (t^3 - t)\, dt$

SOLUTION By FTC II, $\dfrac{d}{dx} \int_0^x (t^3 - t)\, dt = x^3 - x.$

21. $\dfrac{d}{dt} \int_{100}^t \cos 5x\, dx$

SOLUTION By FTC II, $\dfrac{d}{dt} \int_{100}^t \cos(5x)\, dx = \cos 5t.$

23. Sketch the graph of $A(x) = \int_0^x f(t)\, dt$ for each of the functions shown in Figure 10.

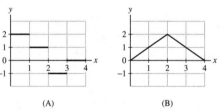

(A) (B)

FIGURE 10

SOLUTION

- Remember that $A'(x) = f(x)$. It follows from Figure 10(A) that $A'(x)$ is constant and consequently $A(x)$ is linear on the intervals $[0, 1]$, $[1, 2]$, $[2, 3]$ and $[3, 4]$. With $A(0) = 0$, $A(1) = 2$, $A(2) = 3$, $A(3) = 2$ and $A(4) = 2$, we obtain the graph shown below at the left.
- Since the graph of $y = f(x)$ in Figure 10(B) lies above the x-axis for $x \in [0, 4]$, it follows that $A(x)$ is increasing over $[0, 4]$. For $x \in [0, 2]$, area accumulates more rapidly with increasing x, while for $x \in [2, 4]$, area accumulates more slowly. This suggests $A(x)$ should be concave up over $[0, 2]$ and concave down over $[2, 4]$. A sketch of $A(x)$ is shown below at the right.

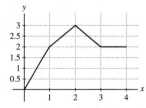 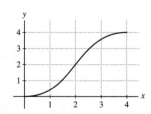

25. Make a rough sketch of the graph of the area function of $g(x)$ shown in Figure 12.

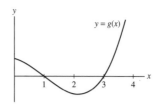

FIGURE 12

SOLUTION   The graph of $y = g(x)$ lies above the x-axis over the interval $[0, 1]$, below the x-axis over $[1, 3]$, and above the x-axis over $[3, 4]$. The corresponding area function should therefore be increasing on $(0, 1)$, decreasing on $(1, 3)$ and increasing on $(3, 4)$. Further, it appears from Figure 12 that the local minimum of the area function at $x = 3$ should be negative. One possible graph of the area function is the following.

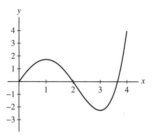

27. Find $G'(x)$, where $G(x) = \int_3^{x^3} \tan t \, dt$.

SOLUTION   By combining the FTC and the chain rule, we have

$$G'(x) = \tan(x^3) \cdot 3x^2 = 3x^2 \tan(x^3).$$

In Exercises 29–36, calculate the derivative.

29. $\dfrac{d}{dx} \displaystyle\int_0^{x^2} \sin^2 t \, dt$

SOLUTION   Let $G(x) = \displaystyle\int_0^{x^2} \sin^2 t \, dt$. By applying the Chain Rule and FTC, we have

$$G'(x) = \sin^2(x^2) \cdot 2x = 2x \sin^2(x^2).$$

31. $\dfrac{d}{ds} \displaystyle\int_{-6}^{\cos s} (u^4 - 3u) \, du$

SOLUTION   Let $G(s) = \displaystyle\int_{-6}^{s} (u^4 - 3u) \, du$. Then, by the chain rule,

$$\frac{d}{ds} \int_{-6}^{\cos s} (u^4 - 3u) \, du = \frac{d}{ds} G(\cos s) = -\sin s (\cos^4 s - 3 \cos s).$$

33. $\dfrac{d}{dx} \displaystyle\int_{x^3}^{0} \sin^2 t \, dt$

SOLUTION   Let $F(x) = \displaystyle\int_0^{x} \sin^2 t \, dt$. Then $\displaystyle\int_{x^3}^{0} \sin^2 t \, dt = -\int_0^{x^3} \sin^2 t \, dt = -F(x^3)$. From this,

$$\frac{d}{dx} \int_{x^3}^{0} \sin^2 t \, dt = \frac{d}{dx}(-F(x^3)) = -3x^2 F'(x^3) = -3x^2 \sin^2 x^3.$$

35. $\dfrac{d}{dx} \displaystyle\int_{\sqrt{x}}^{x^2} \tan t \, dt$

SOLUTION Let

$$G(x) = \int_{\sqrt{x}}^{x^2} \tan t \, dt = \int_0^{x^2} \tan t \, dt - \int_0^{\sqrt{x}} \tan t \, dt.$$

Applying the Chain Rule combined with FTC twice, we have

$$G'(x) = \tan(x^2) \cdot 2x - \tan(\sqrt{x}) \cdot \frac{1}{2}x^{-1/2} = 2x \tan(x^2) - \frac{\tan(\sqrt{x})}{2\sqrt{x}}.$$

In Exercises 37–38, let $A(x) = \displaystyle\int_0^x f(t)\,dt$ and $B(x) = \displaystyle\int_2^x f(t)\,dt$, with $f(x)$ as in Figure 13.

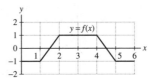

FIGURE 13

37. Find the min and max of $A(x)$ on $[0, 6]$.

SOLUTION The minimum values of $A(x)$ on $[0, 6]$ occur where $A'(x) = f(x)$ goes from negative to positive. This occurs at one place, where $x = 1.5$. The minimum value of $A(x)$ is therefore $A(1.5) = -1.25$. The maximum values of $A(x)$ on $[0, 6]$ occur where $A'(x) = f(x)$ goes from positive to negative. This occurs at one place, where $x = 4.5$. The maximum value of $A(x)$ is therefore $A(4.5) = 1.25$.

39. Let $A(x) = \displaystyle\int_0^x f(t)\,dt$, with $f(x)$ as in Figure 14.

(a) Does $A(x)$ have a local maximum at P?

(b) Where does $A(x)$ have a local minimum?

(c) Where does $A(x)$ have a local maximum?

(d) True or false? $A(x) < 0$ for all x in the interval shown.

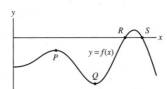

FIGURE 14 Graph of $f(x)$.

SOLUTION

(a) In order for $A(x)$ to have a local maximum, $A'(x) = f(x)$ must transition from positive to negative. As this does not happen at P, $A(x)$ does not have a local maximum at P.

(b) $A(x)$ will have a local minimum when $A'(x) = f(x)$ transitions from negative to positive. This happens at R, so $A(x)$ has a local minimum at R.

(c) $A(x)$ will have a local maximum when $A'(x) = f(x)$ transitions from positive to negative. This happens at S, so $A(x)$ has a local maximum at S.

(d) It is true that $A(x) < 0$ on I since the signed area from 0 to x is clearly always negative from the figure.

In Exercises 41–42, let $A(x) = \displaystyle\int_a^x f(t)\,dt$, where $f(x)$ is continuous.

41. **Area Functions and Concavity** Explain why the following statements are true. Assume $f(x)$ is differentiable.

(a) If c is an inflection point of $A(x)$, then $f'(c) = 0$.

(b) $A(x)$ is concave up if $f(x)$ is increasing.

(c) $A(x)$ is concave down if $f(x)$ is decreasing.

SOLUTION

(a) If $x = c$ is an inflection point of $A(x)$, then $A''(c) = f'(c) = 0$.

(b) If $A(x)$ is concave up, then $A''(x) > 0$. Since $A(x)$ is the area function associated with $f(x)$, $A'(x) = f(x)$ by FTC II, so $A''(x) = f'(x)$. Therefore $f'(x) > 0$, so $f(x)$ is increasing.

(c) If $A(x)$ is concave down, then $A''(x) < 0$. Since $A(x)$ is the area function associated with $f(x)$, $A'(x) = f(x)$ by FTC II, so $A''(x) = f'(x)$. Therefore, $f'(x) < 0$ and so $f(x)$ is decreasing.

43. Let $A(x) = \displaystyle\int_0^x f(t)\,dt$, with $f(x)$ as in Figure 15. Determine:

(a) The intervals on which $A(x)$ is increasing and decreasing

(b) The values x where $A(x)$ has a local min or max

(c) The inflection points of $A(x)$

(d) The intervals where $A(x)$ is concave up or concave down

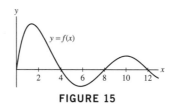

FIGURE 15

SOLUTION

(a) $A(x)$ is increasing when $A'(x) = f(x) > 0$, which corresponds to the intervals $(0, 4)$ and $(8, 12)$. $A(x)$ is decreasing when $A'(x) = f(x) < 0$, which corresponds to the intervals $(4, 8)$ and $(12, \infty)$.

(b) $A(x)$ has a local minimum when $A'(x) = f(x)$ changes from $-$ to $+$, corresponding to $x = 8$. $A(x)$ has a local maximum when $A'(x) = f(x)$ changes from $+$ to $-$, corresponding to $x = 4$ and $x = 12$.

(c) Inflection points of $A(x)$ occur where $A''(x) = f'(x)$ changes sign, or where f changes from increasing to decreasing or vice versa. Consequently, $A(x)$ has inflection points at $x = 2$, $x = 6$, and $x = 10$.

(d) $A(x)$ is concave up when $A''(x) = f'(x)$ is positive or $f(x)$ is increasing, which corresponds to the intervals $(0, 2)$ and $(6, 10)$. Similarly, $A(x)$ is concave down when $f(x)$ is decreasing, which corresponds to the intervals $(2, 6)$ and $(10, \infty)$.

45. Sketch the graph of an increasing function $f(x)$ such that both $f'(x)$ and $A(x) = \displaystyle\int_0^x f(t)\,dt$ are decreasing.

SOLUTION If $f'(x)$ is decreasing, then $f''(x)$ must be negative. Furthermore, if $A(x) = \displaystyle\int_0^x f(t)\,dt$ is decreasing, then $A'(x) = f(x)$ must also be negative. Thus, we need a function which is negative but increasing and concave down. The graph of one such function is shown below.

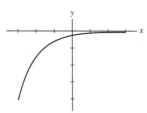

47. [GU] Find the smallest positive inflection point of

$$F(x) = \int_0^x \cos(t^{3/2})\,dt$$

Use a graph of $y = \cos(x^{3/2})$ to determine whether the concavity changes from up to down or vice versa at this point of inflection.

SOLUTION Candidate inflection points of $F(x)$ occur where $F''(x) = 0$. By FTC, $F'(x) = \cos(x^{3/2})$, so $F''(x) = -(3/2)x^{1/2}\sin(x^{3/2})$. Finding the smallest positive solution of $F''(x) = 0$, we get:

$$-(3/2)x^{1/2}\sin(x^{3/2}) = 0$$
$$\sin(x^{3/2}) = 0 \quad (\text{since } x > 0)$$
$$x^{3/2} = \pi$$
$$x = \pi^{2/3} \approx 2.14503.$$

From the plot below, we see that $F'(x) = \cos(x^{3/2})$ changes from decreasing to increasing at $\pi^{2/3}$, so $F(x)$ changes from concave down to concave up at that point.

49. Determine the function $g(x)$ and all values of c such that

$$\int_c^x g(t)\,dt = x^2 + x - 6$$

SOLUTION By the FTC II we have

$$g(x) = \frac{d}{dx}(x^2 + x - 6) = 2x + 1$$

and therefore,

$$\int_c^x g(t)\,dt = x^2 + x - (c^2 + c)$$

We must choose c so that $c^2 + c = 6$. We can take $c = 2$ or $c = -3$.

Further Insights and Challenges

51. Proof of FTC I FTC I asserts that $\int_a^b f(t)\,dt = F(b) - F(a)$ if $F'(x) = f(x)$. Assume FTC II and give a new proof of FTC I as follows. Set $A(x) = \int_a^x f(t)\,dt$.

(a) Show that $F(x) = A(x) + C$ for some constant.

(b) Show that $F(b) - F(a) = A(b) - A(a) = \int_a^b f(t)\,dt$.

SOLUTION Let $F'(x) = f(x)$ and $A(x) = \int_a^x f(t)\,dt$.

(a) Then by the FTC, Part II, $A'(x) = f(x)$ and thus $A(x)$ and $F(x)$ are both antiderivatives of $f(x)$. Hence $F(x) = A(x) + C$ for some constant C.

(b)

$$F(b) - F(a) = (A(b) + C) - (A(a) + C) = A(b) - A(a)$$

$$= \int_a^b f(t)\,dt - \int_a^a f(t)\,dt = \int_a^b f(t)\,dt - 0 = \int_a^b f(t)\,dt$$

which proves the FTC, Part I.

53. Find the values $a \le b$ such that $\int_a^b (x^2 - 9)\,dx$ has minimal value.

SOLUTION Let a be given, and let $F_a(x) = \int_a^x (t^2 - 9)\,dt$. Then $F_a'(x) = x^2 - 9$, and the critical points are $x = \pm 3$. Because $F_a''(-3) = -6$ and $F_a''(3) = 6$, we see that $F_a(x)$ has a minimum at $x = 3$. Now, we find a minimizing $\int_a^3 (x^2 - 9)\,dx$. Let $G(x) = \int_x^3 (x^2 - 9)\,dx$. Then $G'(x) = -(x^2 - 9)$, yielding critical points $x = 3$ or $x = -3$. With $x = -3$,

$$G(-3) = \int_{-3}^3 (x^2 - 9)\,dx = \left(\frac{1}{3}x^3 - 9x\right)\Big|_{-3}^3 = -36.$$

With $x = 3$,

$$G(3) = \int_3^3 (x^2 - 9)\,dx = 0.$$

Hence $a = -3$ and $b = 3$ are the values minimizing $\int_a^b (x^2 - 9)\,dx$.

5.5 Net or Total Change as the Integral of a Rate

Preliminary Questions

1. An airplane makes the 350-mile trip from Los Angeles to San Francisco in 1 hour. Assuming that the plane's velocity at time t is $v(t)$ mph, what is the value of the integral $\int_0^1 v(t)\,dt$?

SOLUTION The definite integral $\int_0^1 v(t)\,dt$ represents the total distance traveled by the airplane during the one hour flight from Los Angeles to San Francisco. Therefore the value of $\int_0^1 v(t)\,dt$ is 350 miles.

2. A hot metal object is submerged in cold water. The rate at which the object cools (in degrees per minute) is a function $f(t)$ of time. Which quantity is represented by the integral $\int_0^T f(t)\,dt$?

SOLUTION The definite integral $\int_0^T f(t)\,dt$ represents the total drop in temperature of the metal object in the first T minutes after being submerged in the cold water.

3. Which of the following quantities would be naturally represented as derivatives and which as integrals?
(a) Velocity of a train
(b) Rainfall during a 6-month period
(c) Mileage per gallon of an automobile
(d) Increase in the population of Los Angeles from 1970 to 1990

SOLUTION Quantities **(a)** and **(c)** involve rates of change, so these would naturally be represented as derivatives. Quantities **(b)** and **(d)** involve an accumulation, so these would naturally be represented as integrals.

4. Two airplanes take off at $t = 0$ from the same place and in the same direction. Their velocities are $v_1(t)$ and $v_2(t)$, respectively. What is the physical interpretation of the area between the graphs of $v_1(t)$ and $v_2(t)$ over an interval $[0, T]$?

SOLUTION The area between the graphs of $v_1(t)$ and $v_2(t)$ over an interval $[0, T]$ represents the difference in distance traveled by the two airplanes in the first T hours after take off.

Exercises

1. Water flows into an empty reservoir at a rate of $3{,}000 + 5t$ gal/hour. What is the quantity of water in the reservoir after 5 hours?

SOLUTION The quantity of water in the reservoir after five hours is

$$\int_0^5 (3000 + 5t)\,dt = \left(3000t + \frac{5}{2}t^2\right)\Big|_0^5 = \frac{30125}{2} = 15{,}062.5 \text{ gallons.}$$

3. A population of insects increases at a rate of $200 + 10t + 0.25t^2$ insects per day. Find the insect population after 3 days, assuming that there are 35 insects at $t = 0$.

SOLUTION The increase in the insect population over three days is

$$\int_0^3 200 + 10t + \frac{1}{4}t^2\,dt = \left(200t + 5t^2 + \frac{1}{12}t^3\right)\Big|_0^3 = \frac{2589}{4} = 647.25.$$

Accordingly, the population after 3 days is $35 + 647.25 = 682.25$ or 682 insects.

5. A factory produces bicycles at a rate of $95 + 0.1t^2 - t$ bicycles per week (t in weeks). How many bicycles were produced from day 8 to 21?

SOLUTION The rate of production is $r(t) = 95 + \frac{1}{10}t^2 - t$ bicycles per week and the period between days 8 and 21 corresponds to the second and third weeks of production. Accordingly, the number of bikes produced between days 8 and 21 is

$$\int_1^3 r(t)\,dt = \int_1^3 \left(95 + \frac{1}{10}t^2 - t\right)dt = \left(95t + \frac{1}{30}t^3 - \frac{1}{2}t^2\right)\Big|_1^3 = \frac{2803}{15} \approx 186.87$$

or 187 bicycles.

7. A cat falls from a tree (with zero initial velocity) at time $t = 0$. How far does the cat fall between $t = 0.5$ and $t = 1$ s? Use Galileo's formula $v(t) = -32t$ ft/s.

SOLUTION Given $v(t) = -32$ ft/s, the total distance the cat falls during the interval $[\frac{1}{2}, 1]$ is

$$\int_{1/2}^{1} |v(t)| \, dt = \int_{1/2}^{1} 32t \, dt = 16t^2 \Big|_{1/2}^{1} = 16 - 4 = 12 \text{ ft.}$$

In Exercises 9–12, assume that a particle moves in a straight line with given velocity. Find the total displacement and total distance traveled over the time interval, and draw a motion diagram like Figure 3 (with distance and time labels).

9. $12 - 4t$ ft/s, $[0, 5]$

SOLUTION Total displacement is given by $\int_0^5 (12 - 4t) \, dt = (12t - 2t^2) \Big|_0^5 = 10$ ft, while total distance is given by

$$\int_0^5 |12 - 4t| \, dt = \int_0^3 (12 - 4t) \, dt + \int_3^5 (4t - 12) \, dt = (12t - 2t^2) \Big|_0^3 + (2t^2 - 12t) \Big|_3^5 = 26 \text{ ft.}$$

The displacement diagram is given here.

11. $t^{-2} - 1$ m/s, $[0.5, 2]$

SOLUTION Total displacement is given by $\int_{.5}^2 (t^{-2} - 1) \, dt = (-t^{-1} - t) \Big|_{.5}^2 = 0$ m, while total distance is given by

$$\int_{.5}^2 \left| t^{-2} - 1 \right| dt = \int_{.5}^1 (t^{-2} - 1) \, dt + \int_1^2 (1 - t^{-2}) \, dt = (-t^{-1} - t) \Big|_{.5}^1 + (t + t^{-1}) \Big|_1^2 = 1 \text{ m.}$$

The displacement diagram is given here.

13. The rate (in liters per minute) at which water drains from a tank is recorded at half-minute intervals. Use the average of the left- and right-endpoint approximations to estimate the total amount of water drained during the first 3 min.

t (min)	0	0.5	1	1.5	2	2.5	3
l/min	50	48	46	44	42	40	38

SOLUTION Let $\Delta t = .5$. Then

$$R_N = .5(48 + 46 + 44 + 42 + 40 + 38) = 129.0 \text{ liters}$$

$$L_N = .5(50 + 48 + 46 + 44 + 42 + 40) = 135.0 \text{ liters}$$

The average of R_N and L_N is $\frac{1}{2}(129 + 135) = 132$ liters.

15. Let $a(t)$ be the acceleration of an object in linear motion at time t. Explain why $\int_{t_1}^{t_2} a(t) \, dt$ is the net change in velocity over $[t_1, t_2]$. Find the net change in velocity over $[1, 6]$ if $a(t) = 24t - 3t^2$ ft/s^2.

SOLUTION Let $a(t)$ be the acceleration of an object in linear motion at time t. Let $v(t)$ be the velocity of the object. We know that $v'(t) = a(t)$. By FTC,

$$\int_{t_1}^{t_2} a(t) \, dt = (v(t) + C) \Big|_{t_1}^{t_2} = v(t_2) + C - (v_{t_1} + C) = v(t_2) - v(t_1),$$

which is the net change in velocity over $[t_1, t_2]$. Let $a(t) = 24t - 3t^2$. The net change in velocity over $[1, 6]$ is

$$\int_1^6 (24t - 3t^2) \, dt = (12t^2 - t^3) \Big|_1^6 = 205 \text{ ft/s.}$$

17. The traffic flow rate past a certain point on a highway is $q(t) = 3,000 + 2,000t - 300t^2$, where t is in hours and $t = 0$ is 8 AM. How many cars pass by during the time interval from 8 to 10 AM?

SOLUTION The number of cars is given by

$$\int_0^2 q(t)\,dt = \int_0^2 (3000 + 2000t - 300t^2)\,dt = \left(3000t + 1000t^2 - 100t^3\right)\Big|_0^2$$
$$= 3000(2) + 1000(4) - 100(8) = 9200 \text{ cars.}$$

19. Carbon Tax To encourage manufacturers to reduce pollution, a carbon tax on each ton of CO_2 released into the atmosphere has been proposed. To model the effects of such a tax, policymakers study the *marginal cost of abatement* $B(x)$, defined as the cost of increasing CO_2 reduction from x to $x + 1$ tons (in units of ten thousand tons—Figure 4). Which quantity is represented by $\displaystyle\int_0^3 B(t)\,dt$?

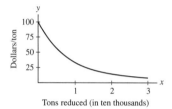

FIGURE 4 Marginal cost of abatement $B(x)$.

SOLUTION The quantity $\displaystyle\int_0^3 B(t)\,dt$ represents the total cost of reducing the amount of CO_2 released into the atmosphere by 3 tons.

21. Figure 6 shows the migration rate $M(t)$ of Ireland during the period 1988–1998. This is the rate at which people (in thousands per year) move in or out of the country.

(a) What does $\displaystyle\int_{1988}^{1991} M(t)\,dt$ represent?

(b) Did migration over the 11-year period 1988–1998 result in a net influx or outflow of people from Ireland? Base your answer on a rough estimate of the positive and negative areas involved.

(c) During which year could the Irish prime minister announce, "We are still losing population but we've hit an inflection point—the trend is now improving."

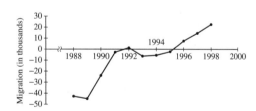

FIGURE 6 Irish migration rate (in thousands per year).

SOLUTION

(a) The amount $\displaystyle\int_{1988}^{1991} M(t)\,dt$ represents the net migration in thousands of people during the period from 1988–1991.

(b) Via linear interpolation and using the midpoint approximation with $n = 10$, the migration (in thousands of people) over the period 1988 – 1998 is estimated to be

$$1 \cdot (-43 - 33.5 - 12 + 0.5 - 2.5 - 6 - 3.5 + 3 + 11.5 + 19) = -66.5$$

That is, there was a net outflow of 66,500 people from Ireland during this period.

(c) "The trend is now improving" implies that the population is decreasing, but that the rate of decrease is approaching zero. The population is decreasing with an improving trend in part of the years 1989, 1990, 1991, 1993, and 1994. "We've hit an inflection point" implies that the rate of population has changed from decreasing to increasing. There are two years in which the trend improves after it was getting worse: 1989 and 1993. During only one of these, 1989, was the population declining for the entire previous year.

23. Heat Capacity The heat capacity $C(T)$ of a substance is the amount of energy (in joules) required to raise the temperature of 1 g by 1°C at temperature T.

(a) Explain why the energy required to raise the temperature from T_1 to T_2 is the area under the graph of $C(T)$ over $[T_1, T_2]$.

(b) How much energy is required to raise the temperature from 50 to 100°C if $C(T) = 6 + 0.2\sqrt{T}$?

SOLUTION

(a) Since $C(T)$ is the energy required to raise the temperature of one gram of a substance by one degree when its temperature is T, the total energy required to raise the temperature from T_1 to T_2 is given by the definite integral $\int_{T_1}^{T_2} C(T)\,dT$. As $C(T) > 0$, the definite integral also represents the area under the graph of $C(T)$.

(b) If $C(T) = 6 + .2\sqrt{T} = 6 + \frac{1}{5}T^{1/2}$, then the energy required to raise the temperature from 50°C to 100°C is $\int_{50}^{100} C(T)\,dT$ or

$$\int_{50}^{100}\left(6 + \frac{1}{5}T^{1/2}\right)dT = \left(6T + \frac{2}{15}T^{3/2}\right)\Bigg|_{50}^{100} = \left(6(100) + \frac{2}{15}(100)^{3/2}\right) - \left(6(50) + \frac{2}{15}(50)^{3/2}\right)$$

$$= \frac{1300 - 100\sqrt{2}}{3} \approx 386.19 \text{ Joules}$$

In Exercises 24 and 25, consider the following. Paleobiologists have studied the extinction of marine animal families during the phanerozoic period, which began 544 million years ago. A recent study suggests that the extinction rate $r(t)$ may be modeled by the function $r(t) = 3{,}130/(t + 262)$ for $0 \le t \le 544$. Here, t is time elapsed (in millions of years) since the beginning of the phanerozoic period. Thus, $t = 544$ refers to the present time, $t = 540$ is 4 million years ago, etc.

25. **CAS**   Estimate the total number of extinct families from $t = 0$ to the present, using M_N with $N = 544$.

SOLUTION   We are estimating

$$\int_0^{544} \frac{3130}{(t + 262)}\,dt$$

using M_N with $N = 544$. If $N = 544$, $\Delta t = \dfrac{544 - 0}{544} = 1$ and $\{t_i^*\}_{i=1,\ldots N} = i\Delta t - (\Delta t/2) = i - \frac{1}{2}$.

$$M_N = \Delta t \sum_{i=1}^{N} r(t_i^*) = 1 \cdot \sum_{i=1}^{544} \frac{3130}{261.5 + i} = 3517.3021.$$

Thus, we estimate that 3517 families have become extinct over the past 544 million years.

Further Insights and Challenges

27. A particle located at the origin at $t = 0$ moves along the x-axis with velocity $v(t) = (t + 1)^{-2}$. Show that the particle will never pass the point $x = 1$.

SOLUTION   The particle's velocity is $v(t) = s'(t) = (t + 1)^{-2}$, an antiderivative for which is $F(t) = -(t + 1)^{-1}$. Hence its position at time t is

$$s(t) = \int_0^t s'(u)\,du = F(u)\Bigg|_0^t = F(t) - F(0) = 1 - \frac{1}{t + 1} < 1$$

for all $t \ge 0$. Thus the particle will never pass the point $x = 1$.

5.6 Substitution Method

Preliminary Questions

1. Which of the following integrals is a candidate for the Substitution Method?

(a) $\displaystyle\int 5x^4 \sin(x^5)\,dx$ (b) $\displaystyle\int \sin^5 x \, \cos x \, dx$ (c) $\displaystyle\int x^5 \sin x \, dx$

SOLUTION   The function in (c): $x^5 \sin x$ is not of the form $g(u(x))u'(x)$. The function in (a) meets the prescribed pattern with $g(u) = \sin u$ and $u(x) = x^5$. Similarly, the function in (b) meets the prescribed pattern with $g(u) = u^5$ and $u(x) = \sin x$.

2. Write each of the following functions in the form $cg(u(x))u'(x)$, where c is a constant.

(a) $x(x^2 + 9)^4$ (b) $x^2 \sin(x^3)$ (c) $\sin x \, \cos^2 x$

SOLUTION

(a) $x(x^2 + 9)^4 = \frac{1}{2}(2x)(x^2 + 9)^4$; hence, $c = \frac{1}{2}$, $g(u) = u^4$, and $u(x) = x^2 + 9$.

(b) $x^2 \sin(x^3) = \frac{1}{3}(3x^2)\sin(x^3)$; hence, $c = \frac{1}{3}$, $g(u) = \sin u$, and $u(x) = x^3$.

(c) $\sin x \cos^2 x = -(-\sin x)\cos^2 x$; hence, $c = -1$, $g(u) = u^2$, and $u(x) = \cos x$.

3. Which of the following is equal to $\int_0^2 x^2(x^3 + 1)\,dx$ for a suitable substitution?

(a) $\dfrac{1}{3}\displaystyle\int_0^2 u\,du$
(b) $\displaystyle\int_0^9 u\,du$
(c) $\dfrac{1}{3}\displaystyle\int_1^9 u\,du$

SOLUTION With the substitution $u = x^3 + 1$, the definite integral $\int_0^2 x^2(x^3 + 1)\,dx$ becomes $\frac{1}{3}\int_1^9 u\,du$. The correct answer is (c).

Exercises

In Exercises 1–6, calculate du for the given function.

1. $u = 1 - x^2$

SOLUTION Let $u = 1 - x^2$. Then $du = -2x\,dx$.

3. $u = x^3 - 2$

SOLUTION Let $u = x^3 - 2$. Then $du = 3x^2\,dx$.

5. $u = \cos(x^2)$

SOLUTION Let $u = \cos(x^2)$. Then $du = -\sin(x^2) \cdot 2x\,dx = -2x\sin(x^2)\,dx$.

In Exercises 7–20, write the integral in terms of u and du. Then evaluate.

7. $\displaystyle\int (x - 7)^3\,dx, \quad u = x - 7$

SOLUTION Let $u = x - 7$. Then $du = dx$. Hence

$$\int (x - 7)^3\,dx = \int u^3\,du = \frac{1}{4}u^4 + C = \frac{1}{4}(x - 7)^4 + C.$$

9. $\displaystyle\int (x + 1)^{-2}\,dx, \quad u = x + 1$

SOLUTION Let $u = x + 1$. Then $du = dx$. Hence

$$\int (x + 1)^{-2}\,dx = \int u^{-2}\,du = -u^{-1} + C = -(x + 1)^{-1} + C = -\frac{1}{x + 1} + C.$$

11. $\displaystyle\int \sin(2x - 4)\,dx, \quad u = 2x - 4$

SOLUTION Let $u = 2x - 4$. Then $du = 2\,dx$ or $\frac{1}{2}\,du = dx$. Hence

$$\int \sin(2x - 4)\,dx = \frac{1}{2}\int \sin u\,du = -\frac{1}{2}\cos u + C = -\frac{1}{2}\cos(2x - 4) + C.$$

13. $\displaystyle\int \frac{x + 1}{(x^2 + 2x)^3}\,dx, \quad u = x^2 + 2x$

SOLUTION Let $u = x^2 + 2x$. Then $du = (2x + 2)\,dx$ or $\frac{1}{2}du = (x + 1)\,dx$. Hence

$$\int \frac{x + 1}{(x^2 + 2x)^3}\,dx = \frac{1}{2}\int \frac{1}{u^3}\,du = \frac{1}{2}\left(-\frac{1}{2}u^{-2}\right) + C = -\frac{1}{4}(x^2 + 2x)^{-2} + C = \frac{-1}{4(x^2 + 2x)^2} + C.$$

15. $\displaystyle\int \sqrt{4x - 1}\,dx, \quad u = 4x - 1$

SOLUTION Let $u = 4x - 1$. Then $du = 4\,dx$ or $\frac{1}{4}du = dx$. Hence

$$\int \sqrt{4u - 1}\,dx = \frac{1}{4}\int u^{1/2}\,du = \frac{1}{4}\left(\frac{2}{3}u^{3/2}\right) + C = \frac{1}{6}(4x - 1)^{3/2} + C.$$

17. $\int x^2\sqrt{4x-1}\,dx$, $\quad u = 4x - 1$

SOLUTION Let $u = 4x - 1$. Then $x = \frac{1}{4}(u+1)$ and $du = 4\,dx$ or $\frac{1}{4}du = dx$. Hence

$$\int x^2\sqrt{4x-1}\,dx = \frac{1}{4}\int \left(\frac{1}{4}(u+1)\right)^2 u^{1/2}\,du = \frac{1}{64}\int (u^{5/2} + 2u^{3/2} + u^{1/2})\,du$$

$$= \frac{1}{64}\left(\frac{2}{7}u^{7/2}\right) + \frac{1}{64}\left(\frac{2}{5}u^{5/2}\right) + \frac{1}{64}\left(\frac{2}{3}u^{3/2}\right) + C$$

$$= \frac{1}{224}(4x-1)^{7/2} + \frac{1}{160}(4x-1)^{5/2} + \frac{1}{96}(4x-1)^{3/2} + C.$$

19. $\int \sin^2 x \cos x \, dx$, $\quad u = \sin x$

SOLUTION Let $u = \sin x$. Then $du = \cos x \, dx$. Hence

$$\int \sin^2 x \cos x \, dx = \int u^2 \, du = \frac{1}{3}u^3 + C = \frac{1}{3}\sin^3 x + C.$$

In Exercises 21–24, show that each of the following integrals is equal to a multiple of $\sin(u(x)) + C$ *for an appropriate choice of* $u(x)$.

21. $\int x^3 \cos(x^4)\,dx$

SOLUTION Let $u = x^4$. Then $du = 4x^3\,dx$ or $\frac{1}{4}\,du = x^3 dx$. Hence

$$\int x^3 \cos(x^4)\,dx = \frac{1}{4}\int \cos u \, du = \frac{1}{4}\sin u + C,$$

which is a multiple of $\sin(u(x))$.

23. $\int x^{1/2}\cos(x^{3/2})\,dx$

SOLUTION Let $u = x^{3/2}$. Then $du = \frac{3}{2}x^{1/2}\,dx$ or $\frac{2}{3}\,du = x^{1/2}\,dx$. Hence

$$\int x^{1/2}\cos(x^{3/2})\,dx = \frac{2}{3}\int \cos u \, du = \frac{2}{3}\sin u + C,$$

which is a multiple of $\sin(u(x))$.

In Exercises 25–51, evaluate the indefinite integral.

25. $\int (4x+3)^4\,dx$

SOLUTION Let $u = 4x + 3$. Then $du = 4\,dx$ or $\frac{1}{4}\,du = dx$. Hence

$$\int (4x+3)^4\,dx = \frac{1}{4}\int u^4\,du = \frac{1}{4}\left(\frac{1}{5}u^5\right) + C = \frac{1}{20}(4x+3)^5 + C.$$

27. $\int \frac{1}{\sqrt{x-7}}\,dx$

SOLUTION Let $u = x - 7$. Then $du = dx$. Hence

$$\int (x-7)^{-1/2}\,dx = \int u^{-1/2}\,du = 2u^{1/2} + C = 2\sqrt{x-7} + C.$$

29. $\int x\sqrt{x^2-4}\,dx$

SOLUTION Let $u = x^2 - 4$. Then $du = 2x\,dx$ or $\frac{1}{2}\,du = x\,dx$. Hence

$$\int x\sqrt{x^2-4}\,dx = \frac{1}{2}\int \sqrt{u}\,du = \frac{1}{2}\left(\frac{2}{3}u^{3/2}\right) + C = \frac{1}{3}(x^2-4)^{3/2} + C.$$

31. $\displaystyle\int \frac{dx}{(x+9)^2}$

SOLUTION Let $u = x + 9$, then $du = dx$. Hence

$$\int \frac{dx}{(x+9)^2} = \int \frac{du}{u^2} = -\frac{1}{u} + C = -\frac{1}{x+9} + C.$$

33. $\displaystyle\int \frac{2x^2 + x}{(4x^3 + 3x^2)^2}\, dx$

SOLUTION Let $u = 4x^3 + 3x^2$. Then $du = (12x^2 + 6x)\, dx$ or $\frac{1}{6} du = (2x^2 + x)\, dx$. Hence

$$\int (4x^3 + 3x^2)^{-2}(2x^2 + x)\, dx = \frac{1}{6}\int u^{-2}\, du = -\frac{1}{6}u^{-1} + C = -\frac{1}{6}(4x^3 + 3x^2)^{-1} + C.$$

35. $\displaystyle\int \frac{5x^4 + 2x}{(x^5 + x^2)^3}\, dx$

SOLUTION Let $u = x^5 + x^2$. Then $du = (5x^4 + 2x)\, dx$. Hence

$$\int \frac{5x^4 + 2x}{(x^5 + x^2)^3}\, dx = \int \frac{1}{u^3}\, du = -\frac{1}{2}\frac{1}{u^2} + C = -\frac{1}{2}\frac{1}{(x^5 + x^2)^2} + C.$$

37. $\displaystyle\int (3x + 9)^{10}\, dx$

SOLUTION Let $u = 3x + 9$. Then $du = 3\, dx$ or $\frac{1}{3} du = dx$. Hence

$$\int (3x + 9)^{10}\, dx = \frac{1}{3}\int u^{10}\, du = \frac{1}{3}\left(\frac{1}{11}u^{11}\right) + C = \frac{1}{33}(3x + 9)^{11} + C.$$

39. $\displaystyle\int x(x + 1)^{1/4}\, dx$

SOLUTION Let $u = x + 1$. Then $u - 1 = x$ and $du = dx$. Hence

$$\int x(x + 1)^{1/4}\, dx = \int (u - 1)u^{1/4}\, du$$

$$= \int (u^{5/4} - u^{1/4})\, du = \frac{4}{9}u^{9/4} - \frac{4}{5}u^{5/4} + C$$

$$= \frac{4}{9}(x + 1)^{9/4} - \frac{4}{5}(x + 1)^{5/4} + C.$$

41. $\displaystyle\int x^3(x^2 - 1)^{3/2}\, dx$

SOLUTION Let $u = x^2 - 1$. Then $u + 1 = x^2$ and $du = 2x\, dx$ or $\frac{1}{2} du = x\, dx$. Hence

$$\int x^3(x^2 - 1)^{3/2}\, dx = \int x^2 \cdot x(x^2 - 1)^{3/2}\, dx$$

$$= \frac{1}{2}\int (u + 1)u^{3/2}\, du = \frac{1}{2}\int (u^{5/2} + u^{3/2})\, du$$

$$= \frac{1}{2}\left(\frac{2}{7}u^{7/2}\right) + \frac{1}{2}\left(\frac{2}{5}u^{5/2}\right) + C = \frac{1}{7}(x^2 - 1)^{7/2} + \frac{1}{5}(x^2 - 1)^{5/2} + C.$$

43. $\displaystyle\int \sin^5 x \cos x\, dx$

SOLUTION Let $u = \sin x$. Then $du = \cos x\, dx$. Hence

$$\int \sin^5 x \cos x\, dx = \int u^5\, du = \frac{1}{6}u^6 + C = \frac{1}{6}\sin^6 x + C.$$

45. $\displaystyle\int \sec^2(4x + 9)\, dx$

SOLUTION Let $u = 4x + 9$. Then $du = 4\,dx$ or $\frac{1}{4}\,du = dx$. Hence

$$\int \sec^2(4x + 9)\,dx = \frac{1}{4}\int \sec^2 u\,du = \frac{1}{4}\tan u + C = \frac{1}{4}\tan(4x + 9) + C.$$

47. $\displaystyle\int \frac{\cos 2x}{(1 + \sin 2x)^2}\,dx$

SOLUTION Let $u = 1 + \sin 2x$. Then $du = 2\cos 2x$ or $\frac{1}{2}du = \cos 2x\,dx$. Hence

$$\int (1 + \sin 2x)^{-2}\cos 2x\,dx = \frac{1}{2}\int u^{-2}\,du = -\frac{1}{2}u^{-1} + C = -\frac{1}{2}(1 + \sin 2x)^{-1} + C.$$

49. $\displaystyle\int \cos x(3\sin x - 1)\,dx$

SOLUTION Let $u = 3\sin x - 1$. Then $du = 3\cos x\,dx$ or $\frac{1}{3}du = \cos x\,dx$. Hence

$$\int (3\sin x - 1)\cos x\,dx = \frac{1}{3}\int u\,du = \frac{1}{3}\left(\frac{1}{2}u^2\right) + C = \frac{1}{6}(3\sin x - 1)^2 + C.$$

51. $\displaystyle\int \sec^2 x(4\tan^3 x - 3\tan^2 x)\,dx$

SOLUTION Let $u = \tan x$. Then $du = \sec^2 x\,dx$. Hence

$$\int \sec^2 x(4\tan^3 x - 3\tan^2 x)\,dx = \int (4u^3 - 3u^2)\,du = u^4 - u^3 + C = \tan^4 x - \tan^3 x + C.$$

53. Evaluate $\displaystyle\int (x^3 + 1)^{1/4}x^5\,dx$.

SOLUTION Let $u = x^3 + 1$. Then $x^3 = u - 1$ and $du = 3x^2\,dx$ or $\frac{1}{3}du = x^2\,dx$. Hence

$$\int x^5(x^3 + 1)^{1/4}\,dx = \frac{1}{3}\int u^{1/4}(u - 1)\,du = \frac{1}{3}\int (u^{5/4} - u^{1/4})\,du$$

$$= \frac{1}{3}\left(\frac{4}{9}u^{9/4} - \frac{4}{5}u^{5/4}\right) + C = \frac{4}{27}(x^3 + 1)^{9/4} - \frac{4}{15}(x^3 + 1)^{5/4} + C.$$

55. Evaluate $\displaystyle\int \sin x\cos x\,dx$ using substitution in two different ways: first using $u = \sin x$ and then $u = \cos x$. Reconcile the two different answers.

SOLUTION First, let $u = \sin x$. Then $du = \cos x\,dx$ and

$$\int \sin x\cos x\,dx = \int u\,du = \frac{1}{2}u^2 + C_1 = \frac{1}{2}\sin^2 x + C_1.$$

Next, let $u = \cos x$. Then $du = -\sin x\,dx$ or $-du = \sin x\,dx$. Hence,

$$\int \sin x\cos x\,dx = -\int u\,du = -\frac{1}{2}u^2 + C_2 = -\frac{1}{2}\cos^2 x + C_2.$$

To reconcile these two seemingly different answers, recall that any two antiderivatives of a specified function differ by a constant. To show that this is true here, note that $(\frac{1}{2}\sin^2 x + C_1) - (-\frac{1}{2}\cos^2 x + C_2) = \frac{1}{2} + C_1 - C_2$, a constant. Here we used the trigonometric identity $\sin^2 x + \cos^2 x = 1$.

57. What are the new limits of integration if we apply the substitution $u = 3x + \pi$ to the integral $\displaystyle\int_0^\pi \sin(3x + \pi)\,dx$?

SOLUTION The new limits of integration are $u(0) = 3 \cdot 0 + \pi = \pi$ and $u(\pi) = 3\pi + \pi = 4\pi$.

In Exercises 59–70, use the Change of Variables Formula to evaluate the definite integral.

59. $\displaystyle\int_1^3 (x + 2)^3\,dx$

SOLUTION   Let $u = x + 2$. Then $du = dx$. Hence

$$\int_1^3 (x+2)^3\,dx = \int_3^5 u^3\,du = \frac{1}{4}u^4\Big|_3^5 = \frac{5^4}{4} - \frac{3^4}{4} = 136.$$

61. $\displaystyle \int_0^1 \frac{x}{(x^2+1)^3}\,dx$

SOLUTION   Let $u = x^2 + 1$. Then $du = 2x\,dx$ or $\frac{1}{2}\,du = x\,dx$. Hence

$$\int_0^1 \frac{x}{(x^2+1)^3}\,dx = \frac{1}{2}\int_1^2 \frac{1}{u^3}\,du = \frac{1}{2}\left(-\frac{1}{2}u^{-2}\right)\Big|_1^2 = -\frac{1}{16} + \frac{1}{4} = \frac{3}{16} = 0.1875.$$

63. $\displaystyle \int_0^4 x\sqrt{x^2+9}\,dx$

SOLUTION   Let $u = x^2 + 9$. Then $du = 2x\,dx$ or $\frac{1}{2}\,du = x\,dx$. Hence

$$\int_0^4 \sqrt{x^2+9}\,dx = \frac{1}{2}\int_9^{25} \sqrt{u}\,du = \frac{1}{2}\left(\frac{2}{3}u^{3/2}\right)\Big|_9^{25} = \frac{1}{3}(125 - 27) = \frac{98}{3}.$$

65. $\displaystyle \int_1^2 (x+1)(x^2+2x)^3\,dx$

SOLUTION   Let $u = x^2 + 2x$. Then $du = (2x+2)dx$ and so $\frac{1}{2}du = (x+1)dx$. Hence

$$\int_1^2 (x+1)(x^2+2x)^3\,dx = \frac{1}{2}\int_3^8 u^3\,du = \frac{1}{2}\left(\frac{1}{4}u^4\right)\Big|_3^8 = \frac{1}{8}(8^4 - 3^4) = \frac{4015}{8}.$$

67. $\displaystyle \int_0^{\pi/2} \cos 3x\,dx$

SOLUTION   Let $u = 3x$. Then $du = 3\,dx$ or $\frac{1}{3}du = dx$. Hence

$$\int_0^{\pi/2} \cos 3x\,dx = \frac{1}{3}\int_0^{3\pi/2} \cos u\,du = \frac{1}{3}\sin u\Big|_0^{3\pi/2} = -\frac{1}{3} - 0 = -\frac{1}{3}.$$

69. $\displaystyle \int_0^{\pi/2} \cos^3 x \sin x\,dx$

SOLUTION   Let $u = \cos x$. Then $du = -\sin x\,dx$. Hence

$$\int_0^{\pi/2} \cos^3 x \sin x\,dx = -\int_1^0 u^3\,du = \int_0^1 u^3\,du = \frac{1}{4}u^4\Big|_0^1 = \frac{1}{4} - 0 = \frac{1}{4}.$$

71. Evaluate $\displaystyle \int \frac{dx}{(2+\sqrt{x})^3}$ using $u = 2 + \sqrt{x}$.

SOLUTION   Let $u = 2 + \sqrt{x}$. Then $du = \frac{1}{2\sqrt{x}}dx$, so that

$$2\sqrt{x}\,du = dx$$
$$2(u-2)\,du = dx.$$

From this, we get:

$$\int \frac{dx}{(2+\sqrt{x})^3} = \int 2\frac{u-2}{u^3}\,du = 2\int \left(u^{-2} - 2u^{-3}\right)\,du = 2\left(-u^{-1} + u^{-2}\right) + C$$

$$= 2\left(-\frac{1}{2+\sqrt{x}} + \frac{1}{(2+\sqrt{x})^2}\right) + C = 2\left(\frac{-2-\sqrt{x}+1}{(2+\sqrt{x})^2}\right) + C = -2\frac{1+\sqrt{x}}{(2+\sqrt{x})^2} + C.$$

In Exercises 73–74, use substitution to evaluate the integral in terms of $f(x)$.

73. $\int f(x)^3 f'(x) \, dx$

SOLUTION Let $u = f(x)$. Then $du = f'(x) \, dx$. Hence

$$\int f(x)^3 f'(x) \, dx = \int u^3 \, du = \frac{1}{4}u^4 + C = \frac{1}{4}f(x)^4 + C.$$

75. Show that $\displaystyle\int_0^{\pi/6} f(\sin \theta) \, d\theta = \int_0^{1/2} f(u) \frac{1}{\sqrt{1 - u^2}} \, du.$

SOLUTION Let $u = \sin \theta$. Then $u(\pi/6) = 1/2$ and $u(0) = 0$, as required. Furthermore, $du = \cos \theta \, d\theta$, so that

$$d\theta = \frac{du}{\cos \theta}.$$

If $\sin \theta = u$, then $u^2 + \cos^2 \theta = 1$, so that $\cos \theta = \sqrt{1 - u^2}$. Therefore $d\theta = du/\sqrt{1 - u^2}$. This gives

$$\int_0^{\pi/6} f(\sin \theta) \, d\theta = \int_0^{1/2} f(u) \frac{1}{\sqrt{1 - u^2}} \, du.$$

Further Insights and Challenges

77. Use the substitution $u = 1 + x^{1/n}$ to show that

$$\int \sqrt{1 + x^{1/n}} \, dx = n \int u^{1/2}(u - 1)^{n-1} \, du$$

Evaluate for $n = 2, 3$.

SOLUTION Let $u = 1 + x^{1/n}$. Then $x = (u - 1)^n$ and $dx = n(u - 1)^{n-1} \, du$. Accordingly, $\int \sqrt{1 + x^{1/n}} \, dx = n \int u^{1/2}(u - 1)^{n-1} \, du.$

For $n = 2$, we have

$$\int \sqrt{1 + x^{1/2}} \, dx = 2 \int u^{1/2}(u - 1)^1 \, du = 2 \int (u^{3/2} - u^{1/2}) \, du$$

$$= 2 \left(\frac{2}{5}u^{5/2} - \frac{2}{3}u^{3/2} \right) + C = \frac{4}{5}(1 + x^{1/2})^{5/2} - \frac{4}{3}(1 + x^{1/2})^{3/2} + C.$$

For $n = 3$, we have

$$\int \sqrt{1 + x^{1/3}} \, dx = 3 \int u^{1/2}(u - 1)^2 \, du = 3 \int (u^{5/2} - 2u^{3/2} + u^{1/2}) \, du$$

$$= 3 \left(\frac{2}{7}u^{7/2} - (2) \left(\frac{2}{5} \right) u^{5/2} + \frac{2}{3}u^{3/2} \right) + C$$

$$= \frac{6}{7}(1 + x^{1/3})^{7/2} - \frac{12}{5}(1 + x^{1/3})^{5/2} + 2(1 + x^{1/3})^{3/2} + C.$$

79. Show that $\displaystyle\int_{-a}^{a} f(x) \, dx = 0$ if f is an odd function.

SOLUTION We assume that f is continuous. If $f(x)$ is an odd function, then $f(-x) = -f(x)$. Let $u = -x$. Then $x = -u$ and $du = -dx$ or $-du = dx$. Accordingly,

$$\int_{-a}^{a} f(x) \, dx = \int_{-a}^{0} f(x) \, dx + \int_0^{a} f(x) \, dx = -\int_a^{0} f(-u) \, du + \int_0^{a} f(x) \, dx$$

$$= \int_0^{a} f(x) \, dx - \int_0^{a} f(u) \, du = 0.$$

81. Show that the two regions in Figure 4 have the same area. Then use the identity $\cos^2 u = \frac{1}{2}(1 + \cos 2u)$ to compute the second area.

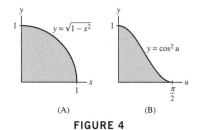

FIGURE 4

SOLUTION   The area of the region in Figure 4(A) is given by $\int_0^1 \sqrt{1-x^2}\,dx$. Let $x = \sin u$. Then $dx = \cos u\,du$ and $\sqrt{1-x^2} = \sqrt{1 - \sin^2 u} = \cos u$. Hence,

$$\int_0^1 \sqrt{1-x^2}\,dx = \int_0^{\pi/2} \cos u \cdot \cos u\,du = \int_0^{\pi/2} \cos^2 u\,du.$$

This last integral represents the area of the region in Figure 4(B). The two regions in Figure 4 therefore have the same area.

Let's now focus on the definite integral $\int_0^{\pi/2} \cos^2 u\,du$. Using the trigonometric identity $\cos^2 u = \frac{1}{2}(1 + \cos 2u)$, we have

$$\int_0^{\pi/2} \cos^2 u\,du = \frac{1}{2}\int_0^{\pi/2} 1 + \cos 2u\,du = \frac{1}{2}\left(u + \frac{1}{2}\sin 2u\right)\Big|_0^{\pi/2} = \frac{1}{2}\cdot\frac{\pi}{2} - 0 = \frac{\pi}{4}.$$

83. Area of an Ellipse   Prove the formula $A = \pi ab$ for the area of the ellipse with equation

$$\frac{x^2}{a^2} + \frac{y^2}{b^2} = 1$$

Hint: Show that $A = 2b\int_{-a}^{a} \sqrt{1 - (x/a)^2}\,dx$, change variables, and use the formula for the area of a circle (Figure 5).

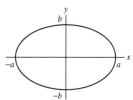

FIGURE 5 Graph of $\dfrac{x^2}{a^2} + \dfrac{y^2}{b^2} = 1$.

SOLUTION   Consider the ellipse with equation $\frac{x^2}{a^2} + \frac{y^2}{b^2} = 1$; here $a, b > 0$. The area between the part of the ellipse in the upper half-plane, $y = f(x) = \sqrt{b^2\left(1 - \frac{x^2}{a^2}\right)}$, and the x-axis is $\int_{-a}^{a} f(x)\,dx$. By symmetry, the part of the elliptical region in the lower half-plane has the same area. Accordingly, the area enclosed by the ellipse is

$$2\int_{-a}^{a} f(x)\,dx = 2\int_{-a}^{a} \sqrt{b^2\left(1 - \frac{x^2}{a^2}\right)}\,dx = 2b\int_{-a}^{a} \sqrt{1 - (x/a)^2}\,dx$$

Now, let $u = x/a$. Then $x = au$ and $a\,du = dx$. Accordingly,

$$2b\int_{-a}^{a} \sqrt{1 - \left(\frac{x}{a}\right)^2}\,dx = 2ab\int_{-1}^{1} \sqrt{1 - u^2}\,du = 2ab\left(\frac{\pi}{2}\right) = \pi ab$$

Here we recognized that $\int_{-1}^{1} \sqrt{1 - u^2}\,du$ represents the area of the upper unit semicircular disk, which by Exercise 81 is $2(\frac{\pi}{4}) = \frac{\pi}{2}$.

CHAPTER REVIEW EXERCISES

In Exercises 1–4, refer to the function $f(x)$ whose graph is shown in Figure 1.

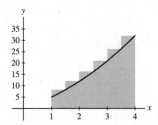

FIGURE 1

1. Estimate L_4 and M_4 on $[0, 4]$.

SOLUTION With $n = 4$ and an interval of $[0, 4]$, $\Delta x = \frac{4-0}{4} = 1$. Then,

$$L_4 = \Delta x(f(0) + f(1) + f(2) + f(3)) = 1\left(\frac{1}{4} + 1 + \frac{5}{2} + 2\right) = \frac{23}{4}$$

and

$$M_4 = \Delta x\left(f\left(\frac{1}{2}\right) + f\left(\frac{3}{2}\right) + f\left(\frac{5}{2}\right) + f\left(\frac{7}{2}\right)\right) = 1\left(\frac{1}{2} + 2 + \frac{9}{4} + \frac{9}{4}\right) = 7.$$

3. Find an interval $[a, b]$ on which R_4 is larger than $\int_a^b f(x)\,dx$. Do the same for L_4.

SOLUTION In general, R_N is larger than $\int_a^b f(x)\,dx$ on any interval $[a, b]$ over which $f(x)$ is increasing. Given the graph of $f(x)$, we may take $[a, b] = [0, 2]$. In order for L_4 to be larger than $\int_a^b f(x)\,dx$, $f(x)$ must be decreasing over the interval $[a, b]$. We may therefore take $[a, b] = [2, 3]$.

In Exercises 5–8, let $f(x) = x^2 + 4x$.

5. Calculate R_6, M_6, and L_6 for $f(x)$ on the interval $[1, 4]$. Sketch the graph of $f(x)$ and the corresponding rectangles for each approximation.

SOLUTION Let $f(x) = x^2 + 4x$. A uniform partition of $[1, 4]$ with $N = 6$ subintervals has

$$\Delta x = \frac{4-1}{6} = \frac{1}{2}, \qquad x_j = a + j\Delta x = 1 + \frac{j}{2},$$

and

$$x_j^* = a + \left(j - \frac{1}{2}\right)\Delta x = \frac{3}{4} + \frac{j}{2}.$$

Now,

$$R_6 = \Delta x \sum_{j=1}^{6} f(x_j) = \frac{1}{2}\left(f\left(\frac{3}{2}\right) + f(2) + f\left(\frac{5}{2}\right) + f(3) + f\left(\frac{7}{2}\right) + f(4)\right)$$

$$= \frac{1}{2}\left(\frac{33}{4} + 12 + \frac{65}{4} + 21 + \frac{105}{4} + 32\right) = \frac{463}{8}.$$

The rectangles corresponding to this approximation are shown below.

Next,

$$M_6 = \Delta x \sum_{j=1}^{6} f(x_j^*) = \frac{1}{2}\left(f\left(\frac{5}{4}\right) + f\left(\frac{7}{4}\right) + f\left(\frac{9}{4}\right) + f\left(\frac{11}{4}\right) + f\left(\frac{13}{4}\right) + f\left(\frac{15}{4}\right)\right)$$

$$= \frac{1}{2}\left(\frac{105}{16} + \frac{161}{16} + \frac{225}{16} + \frac{297}{16} + \frac{377}{16} + \frac{465}{16}\right) = \frac{1630}{32} = \frac{815}{16}.$$

The rectangles corresponding to this approximation are shown below.

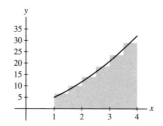

Finally,

$$L_6 = \Delta x \sum_{j=0}^{5} f(x_j) = \frac{1}{2}\left(f(1) + f\left(\frac{3}{2}\right) + f(2) + f\left(\frac{5}{2}\right) + f(3) + f\left(\frac{7}{2}\right) \right)$$

$$= \frac{1}{2}\left(5 + \frac{33}{4} + 12 + \frac{65}{4} + 21 + \frac{105}{4} \right) = \frac{355}{8}.$$

The rectangles corresponding to this approximation are shown below.

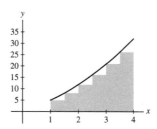

7. Find a formula for L_N for $f(x)$ on $[0, 2]$ and compute $\int_0^2 f(x)\,dx$ by taking the limit.

SOLUTION Let $f(x) = x^2 + 4x$ and N be a positive integer. Then

$$\Delta x = \frac{2-0}{N} = \frac{2}{N}$$

and

$$x_j = a + j\Delta x = 0 + \frac{2j}{N} = \frac{2j}{N}$$

for $0 \le j \le N$. Thus,

$$L_N = \Delta x \sum_{j=0}^{N-1} f(x_j) = \frac{2}{N} \sum_{j=0}^{N-1} \left(\frac{4j^2}{N^2} + \frac{8j}{N} \right) = \frac{8}{N^3} \sum_{j=0}^{N-1} j^2 + \frac{16}{N^2} \sum_{j=0}^{N-1} j$$

$$= \frac{4(N-1)(2N-1)}{3N^2} + \frac{8(N-1)}{N} = \frac{32}{3} + \frac{12}{N} + \frac{4}{3N^2}.$$

Finally,

$$\int_0^2 f(x)\,dx = \lim_{N \to \infty} \left(\frac{32}{3} + \frac{12}{N} + \frac{4}{3N^2} \right) = \frac{32}{3}.$$

9. Calculate R_6, M_6, and L_6 for $f(x) = (x^2 + 1)^{-1}$ on the interval $[0, 1]$.

SOLUTION Let $f(x) = (x^2 + 1)^{-1}$. A uniform partition of $[0, 1]$ with $N = 6$ subintervals has

$$\Delta x = \frac{1-0}{6} = \frac{1}{6}, \qquad x_j = a + j\Delta x = \frac{j}{6},$$

and

$$x_j^* = a + \left(j - \frac{1}{2} \right)\Delta x = \frac{2j-1}{12}.$$

Now,

$$R_6 = \Delta x \sum_{j=1}^{6} f(x_j) = \frac{1}{6}\left(f\left(\frac{1}{6}\right) + f\left(\frac{1}{3}\right) + f\left(\frac{1}{2}\right) + f\left(\frac{2}{3}\right) + f\left(\frac{5}{6}\right) + f(1) \right)$$

$$= \frac{1}{6}\left(\frac{36}{37} + \frac{9}{10} + \frac{4}{5} + \frac{9}{13} + \frac{36}{61} + \frac{1}{2}\right) \approx 0.742574.$$

Next,

$$M_6 = \Delta x \sum_{j=1}^{6} f(x_j^*) = \frac{1}{6}\left(f\left(\frac{1}{12}\right) + f\left(\frac{1}{4}\right) + f\left(\frac{5}{12}\right) + f\left(\frac{7}{12}\right) + f\left(\frac{3}{4}\right) + f\left(\frac{11}{12}\right)\right)$$

$$= \frac{1}{6}\left(\frac{144}{145} + \frac{16}{17} + \frac{144}{169} + \frac{144}{193} + \frac{16}{25} + \frac{144}{265}\right) \approx 0.785977.$$

Finally,

$$L_6 = \Delta x \sum_{j=0}^{5} f(x_j) = \frac{1}{6}\left(f(0) + f\left(\frac{1}{6}\right) + f\left(\frac{1}{3}\right) + f\left(\frac{1}{2}\right) + f\left(\frac{2}{3}\right) + f\left(\frac{5}{6}\right)\right)$$

$$= \frac{1}{6}\left(1 + \frac{36}{37} + \frac{9}{10} + \frac{4}{5} + \frac{9}{13} + \frac{36}{61}\right) \approx 0.825907.$$

11. Which approximation to the area is represented by the shaded rectangles in Figure 3? Compute R_5 and L_5.

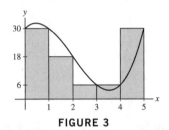

FIGURE 3

SOLUTION   There are five rectangles and the height of each is given by the function value at the right endpoint of the subinterval. Thus, the area represented by the shaded rectangles is R_5.

From the figure, we see that $\Delta x = 1$. Then

$$R_5 = 1(30 + 18 + 6 + 6 + 30) = 90 \quad \text{and} \quad L_5 = 1(30 + 30 + 18 + 6 + 6) = 90.$$

In Exercises 13–34, evaluate the integral.

13. $\int (6x^3 - 9x^2 + 4x)\,dx$

SOLUTION $\int (6x^3 - 9x^2 + 4x)\,dx = \frac{3}{2}x^4 - 3x^3 + 2x^2 + C.$

15. $\int (2x^3 - 1)^2\,dx$

SOLUTION $\int (2x^3 - 1)^2\,dx = \int (4x^6 - 4x^3 + 1)\,dx = \frac{4}{7}x^7 - x^4 + x + C.$

17. $\int \frac{x^4 + 1}{x^2}\,dx$

SOLUTION $\int \frac{x^4 + 1}{x^2}\,dx = \int (x^2 + x^{-2})\,dx = \frac{1}{3}x^3 - x^{-1} + C.$

19. $\int_{-1}^{4} |x^2 - 9|\,dx$

SOLUTION

$$\int_{-1}^{4} |x^2 - 9|\,dx = \int_{-1}^{3} (9 - x^2)\,dx + \int_{3}^{4} (x^2 - 9)\,dx = \left(9x - \frac{1}{3}x^3\right)\Big|_{-1}^{3} + \left(\frac{1}{3}x^3 - 9x\right)\Big|_{3}^{4}$$

$$= (27 - 9) - \left(-9 + \frac{1}{3}\right) + \left(\frac{64}{3} - 36\right) - (9 - 27) = 30.$$

21. $\int \csc^2 \theta\,d\theta$

SOLUTION $\int \csc^2 \theta \, d\theta = -\cot \theta + C.$

23. $\int \sec^2(9t - 4) \, dt$

SOLUTION Let $u = 9t - 4$. Then $du = 9dt$ and

$$\int \sec^2(9t - 4) \, dt = \frac{1}{9} \int \sec^2 u \, du = \frac{1}{9} \tan u + C = \frac{1}{9} \tan(9t - 4) + C.$$

25. $\int (9t - 4)^{11} \, dt$

SOLUTION Let $u = 9t - 4$. Then $du = 9dt$ and

$$\int (9t - 4)^{11} \, dt = \frac{1}{9} \int u^{11} \, du = \frac{1}{108} u^{12} + C = \frac{1}{108}(9t - 4)^{12} + C.$$

27. $\int \sin^2(3\theta) \cos(3\theta) \, d\theta$

SOLUTION Let $u = \sin(3\theta)$. Then $du = 3\cos(3\theta)d\theta$ and

$$\int \sin^2(3\theta) \cos(3\theta) \, d\theta = \frac{1}{3} \int u^2 \, du = \frac{1}{9} u^3 + C = \frac{1}{9} \sin^3(3\theta) + C.$$

29. $\int \frac{(2x^3 + 3x) \, dx}{(3x^4 + 9x^2)^5}$

SOLUTION Let $u = 3x^4 + 9x^2$. Then $du = (12x^3 + 18x) \, dx = 6(2x^3 + 3x) \, dx$ and

$$\int \frac{(2x^3 + 3x) \, dx}{(3x^4 + 9x^2)^5} = \frac{1}{6} \int u^{-5} \, du = -\frac{1}{24} u^{-4} + C = -\frac{1}{24}(3x^4 + 9x^2)^{-4} + C.$$

31. $\int \sin \theta \sqrt{4 - \cos \theta} \, d\theta$

SOLUTION Let $u = 4 - \cos \theta$. Then $du = \sin \theta \, d\theta$ and

$$\int \sin \theta \sqrt{4 - \cos \theta} \, d\theta = \int u^{1/2} \, du = \frac{2}{3} u^{3/2} + C = \frac{2}{3}(4 - \cos \theta)^{3/2} + C.$$

33. $\int y\sqrt{2y + 3} \, dy$

SOLUTION Let $u = 2y + 3$. Then $du = 2dy$, $y = \frac{1}{2}(u - 3)$ and

$$\int y\sqrt{2y + 3} \, dy = \frac{1}{4} \int (u - 3)\sqrt{u} \, du = \frac{1}{4} \int (u^{3/2} - 3u^{1/2}) \, du$$

$$= \frac{1}{4}\left(\frac{2}{5}u^{5/2} - 2u^{3/2}\right) + C = \frac{1}{10}(2y + 3)^{5/2} - \frac{1}{2}(2y + 3)^{3/2} + C.$$

35. Combine to write as a single integral

$$\int_0^8 f(x) \, dx + \int_{-2}^0 f(x) \, dx + \int_8^6 f(x) \, dx$$

SOLUTION First, rewrite

$$\int_0^8 f(x) \, dx = \int_0^6 f(x) \, dx + \int_6^8 f(x) \, dx$$

and observe that

$$\int_8^6 f(x) \, dx = -\int_6^8 f(x) \, dx.$$

Thus,

$$\int_0^8 f(x)\,dx + \int_8^6 f(x)\,dx = \int_0^6 f(x)\,dx.$$

Finally,

$$\int_0^8 f(x)\,dx + \int_{-2}^0 f(x)\,dx + \int_8^6 f(x)\,dx = \int_0^6 f(x)\,dx + \int_{-2}^0 f(x)\,dx = \int_{-2}^6 f(x)\,dx.$$

37. Find inflection points of $A(x) = \displaystyle\int_3^x \frac{t\,dt}{t^2+1}$.

SOLUTION   Let

$$A(x) = \int_3^x \frac{t\,dt}{t^2+1}.$$

Then

$$A'(x) = \frac{x}{x^2+1}$$

and

$$A''(x) = \frac{(x^2+1)(1) - x(2x)}{(x^2+1)^2} = \frac{1-x^2}{(x^2+1)^2}.$$

Thus, $A(x)$ is concave down for $|x| > 1$ and concave up for $|x| < 1$. $A(x)$ therefore has inflection points at $x = \pm 1$.

39. On a typical day, a city consumes water at the rate of $r(t) = 100 + 72t - 3t^2$ (in thousands of gallons per hour), where t is the number of hours past midnight. What is the daily water consumption? How much water is consumed between 6 PM and midnight?

SOLUTION   With a consumption rate of $r(t) = 100 + 72t - 3t^2$ thousand gallons per hour, the daily consumption of water is

$$\int_0^{24} (100 + 72t - 3t^2)\,dt = \left(100t + 36t^2 - t^3\right)\Big|_0^{24} = 100(24) + 36(24)^2 - (24)^3 = 9312,$$

or 9.312 million gallons. From 6 PM to midnight, the water consumption is

$$\int_{18}^{24} (100 + 72t - 3t^2)\,dt = \left(100t + 36t^2 - t^3\right)\Big|_{18}^{24}$$

$$= 100(24) + 36(24)^2 - (24)^3 - \left(100(18) + 36(18)^2 - (18)^3\right)$$

$$= 9312 - 7632 = 1680,$$

or 1.68 million gallons.

41. Cost engineers at NASA have the task of projecting the cost P of major space projects. It has been found that the cost C of developing a projection increases with P at the rate $dC/dP \approx 21P^{-0.65}$, where C is in thousands of dollars and P in millions of dollars. What is the cost of developing a projection for a project whose cost turns out to be $P = \$35$ million?

SOLUTION   Assuming it costs nothing to develop a projection for a project with a cost of $0, the cost of developing a projection for a project whose cost turns out to be $35 million is

$$\int_0^{35} 21P^{-0.65}\,dP = 60P^{0.35}\Big|_0^{35} = 60(35)^{0.35} \approx 208.245,$$

or $208,245.

43. Let $f(x)$ be a positive increasing continuous function on $[a, b]$, where $0 \le a < b$ as in Figure 7. Show that the shaded region has area

$$I = bf(b) - af(a) - \int_a^b f(x)\,dx \qquad \boxed{1}$$

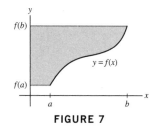

FIGURE 7

SOLUTION   We can construct the shaded region in Figure 7 by taking a rectangle of length b and height $f(b)$ and removing a rectangle of length a and height $f(a)$ as well as the region between the graph of $y = f(x)$ and the x-axis over the interval $[a, b]$. The area of the resulting region is then the area of the large rectangle minus the area of the small rectangle and minus the area under the curve $y = f(x)$; that is,

$$I = bf(b) - af(a) - \int_a^b f(x)\,dx.$$

In Exercises 45–49, express the limit as an integral (or multiple of an integral) and evaluate.

45. $\displaystyle \lim_{N\to\infty} \frac{2}{N} \sum_{j=1}^{N} \sin\left(\frac{2j}{N}\right)$

SOLUTION   Let $f(x) = \sin x$ and N be a positive integer. A uniform partition of the interval $[0, 2]$ with N subintervals has

$$\Delta x = \frac{2}{N} \qquad \text{and} \qquad x_j = \frac{2j}{N}$$

for $0 \le j \le N$. Then

$$\frac{2}{N} \sum_{j=1}^{N} \sin\left(\frac{2j}{N}\right) = \Delta x \sum_{j=1}^{N} f(x_j) = R_N;$$

consequently,

$$\lim_{N\to\infty} \frac{2}{N} \sum_{j=1}^{N} \sin\left(\frac{2j}{N}\right) = \int_0^2 \sin x\,dx = -\cos x \Big|_0^2 = 1 - \cos 2.$$

47. $\displaystyle \lim_{N\to\infty} \frac{\pi}{N} \sum_{j=0}^{N-1} \sin\left(\frac{\pi}{2} + \frac{\pi j}{N}\right)$

SOLUTION   Let $f(x) = \sin x$ and N be a positive integer. A uniform partition of the interval $[\pi/2, 3\pi/2]$ with N subintervals has

$$\Delta x = \frac{\pi}{N} \qquad \text{and} \qquad x_j = \frac{\pi}{2} + \frac{\pi j}{N}$$

for $0 \le j \le N$. Then

$$\frac{\pi}{N} \sum_{j=0}^{N-1} \sin\left(\frac{\pi}{2} + \frac{\pi j}{N}\right) = \Delta x \sum_{j=0}^{N-1} f(x_j) = L_N;$$

consequently,

$$\lim_{N\to\infty} \frac{\pi}{N} \sum_{j=0}^{N-1} \sin\left(\frac{\pi}{2} + \frac{\pi j}{N}\right) = \int_{\pi/2}^{3\pi/2} \sin x\,dx = -\cos x \Big|_{\pi/2}^{3\pi/2} = 0.$$

49. $\displaystyle \lim_{N\to\infty} \frac{1^k + 2^k + \cdots N^k}{N^{k+1}} \quad (k > 0)$

SOLUTION   Observe that

$$\frac{1^k + 2^k + 3^k + \cdots + N^k}{N^{k+1}} = \frac{1}{N}\left[\left(\frac{1}{N}\right)^k + \left(\frac{2}{N}\right)^k + \left(\frac{3}{N}\right)^k + \cdots \left(\frac{N}{N}\right)^k\right] = \frac{1}{N} \sum_{j=1}^{N} \left(\frac{j}{N}\right)^k.$$

Now, let $f(x) = x^k$ and N be a positive integer. A uniform partition of the interval $[0, 1]$ with N subintervals has

$$\Delta x = \frac{1}{N} \quad \text{and} \quad x_j = \frac{j}{N}$$

for $0 \le j \le N$. Then

$$\frac{1}{N} \sum_{j=1}^{N} \left(\frac{j}{N}\right)^k = \Delta x \sum_{j=1}^{N} f(x_j) = R_N;$$

consequently,

$$\lim_{N \to \infty} \frac{1}{N} \sum_{j=1}^{N} \left(\frac{j}{N}\right)^k = \int_0^1 x^k \, dx = \frac{1}{k+1} x^{k+1} \Big|_0^1 = \frac{1}{k+1}.$$

51. Evaluate $\displaystyle\int_0^1 f(x) \, dx$, assuming that $f(x)$ is an even continuous function such that

$$\int_1^2 f(x) \, dx = 5, \qquad \int_{-2}^1 f(x) \, dx = 8$$

SOLUTION   Using the given information

$$\int_{-2}^2 f(x) \, dx = \int_{-2}^1 f(x) \, dx + \int_1^2 f(x) \, dx = 13.$$

Because $f(x)$ is an even function, it follows that

$$\int_{-2}^0 f(x) \, dx = \int_0^2 f(x) \, dx,$$

so

$$\int_0^2 f(x) \, dx = \frac{13}{2}.$$

Finally,

$$\int_0^1 f(x) \, dx = \int_0^2 f(x) \, dx - \int_1^2 f(x) \, dx = \frac{13}{2} - 5 = \frac{3}{2}.$$

53. Show that

$$\int x \, f(x) \, dx = x F(x) - G(x)$$

where $F'(x) = f(x)$ and $G'(x) = F(x)$. Use this to evaluate $\displaystyle\int x \cos x \, dx$.

SOLUTION   Suppose $F'(x) = f(x)$ and $G'(x) = F(x)$. Then

$$\frac{d}{dx} (x F(x) - G(x)) = x F'(x) + F(x) - G'(x) = x f(x) + F(x) - F(x) = x f(x).$$

Therefore, $x F(x) - G(x)$ is an antiderivative of $x f(x)$ and

$$\int x f(x) \, dx = x F(x) - G(x) + C.$$

To evaluate $\int x \cos x \, dx$, note that $f(x) = \cos x$. Thus, we may take $F(x) = \sin x$ and $G(x) = -\cos x$. Finally,

$$\int x \cos x \, dx = x \sin x + \cos x + C.$$

55. [GU]   Plot the graph of $f(x) = x^{-2} \sin x$ and show that $0.2 \le \displaystyle\int_1^2 f(x) \, dx \le 0.9$.

SOLUTION   Let $f(x) = x^{-2} \sin x$. From the figure below, we see that

$$0.2 \le f(x) \le 0.9$$

for $1 \le x \le 2$. Therefore,

$$0.2 = \int_0^1 0.2 \, dx \le \int_0^1 f(x) \, dx \le \int_0^1 0.9 \, dx = 0.9.$$

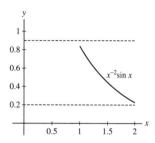

In Exercises 57–62, find the derivative.

57. $A'(x)$, where $A(x) = \displaystyle\int_3^x \sin(t^3) \, dt$

SOLUTION   Let $A(x) = \displaystyle\int_3^x \sin(t^3) \, dt$. Then $A'(x) = \sin(x^3)$.

59. $\dfrac{d}{dy} \displaystyle\int_{-2}^y 3^x \, dx$

SOLUTION $\dfrac{d}{dy} \displaystyle\int_{-2}^y 3^x \, dx = 3^y$.

61. $G'(2)$, where $G(x) = \displaystyle\int_0^{x^3} \sqrt{t+1} \, dt$

SOLUTION   Let $G(x) = \displaystyle\int_0^{x^3} \sqrt{t+1} \, dt$. Then

$$G'(x) = \sqrt{x^3 + 1}\,\frac{d}{dx}x^3 = 3x^2\sqrt{x^3 + 1}$$

and $G'(2) = 3(2)^2\sqrt{8+1} = 36$.

63. Explain with a graph: If $f(x)$ is increasing and concave up on $[a, b]$, then L_N is more accurate than R_N. Which is more accurate if $f(x)$ is increasing and concave down?

SOLUTION   Consider the figure below, which displays a portion of the graph of an increasing, concave up function.

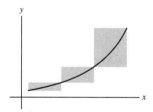

The shaded rectangles represent the differences between the right-endpoint approximation R_N and the left-endpoint approximation L_N. In particular, the portion of each rectangle that lies below the graph of $y = f(x)$ is the amount by which L_N underestimates the area under the graph, whereas the portion of each rectangle that lies above the graph of $y = f(x)$ is the amount by which R_N overestimates the area. Because the graph of $y = f(x)$ is increasing and concave up, the lower portion of each shaded rectangle is smaller than the upper portion. Therefore, L_N is more accurate (introduces less error) than R_N. By similar reasoning, if $f(x)$ is increasing and concave down, then R_N is more accurate than L_N.

6 APPLICATIONS OF THE INTEGRAL

6.1 Area Between Two Curves

Preliminary Questions

1. What is the area interpretation of $\int_a^b \left(f(x) - g(x) \right) dx$ if $f(x) \geq g(x)$?

SOLUTION Because $f(x) \geq g(x)$, $\int_a^b (f(x) - g(x))\, dx$ represents the area of the region bounded between the graphs of $y = f(x)$ and $y = g(x)$, bounded on the left by the vertical line $x = a$ and on the right by the vertical line $x = b$.

2. Is $\int_a^b \left(f(x) - g(x) \right) dx$ still equal to the area between the graphs of f and g if $f(x) \geq 0$ but $g(x) \leq 0$?

SOLUTION Yes. Since $f(x) \geq 0$ and $g(x) \leq 0$, it follows that $f(x) - g(x) \geq 0$.

3. Suppose that $f(x) \geq g(x)$ on $[0, 3]$ and $g(x) \geq f(x)$ on $[3, 5]$. Express the area between the graphs over $[0, 5]$ as a sum of integrals.

SOLUTION Remember that to calculate an area between two curves, one must subtract the equation for the lower curve from the equation for the upper curve. Over the interval $[0, 3]$, $y = f(x)$ is the upper curve. On the other hand, over the interval $[3, 5]$, $y = g(x)$ is the upper curve. The area between the graphs over the interval $[0, 5]$ is therefore given by

$$\int_0^3 (f(x) - g(x))\, dx + \int_3^5 (g(x) - f(x))\, dx.$$

4. Suppose that the graph of $x = f(y)$ lies to the left of the y-axis. Is $\int_a^b f(y)\, dy$ positive or negative?

SOLUTION If the graph of $x = f(y)$ lies to the left of the y-axis, then for each value of y, the corresponding value of x is less than zero. Hence, the value of $\int_a^b f(y)\, dy$ is negative.

Exercises

1. Find the area of the region between $y = 3x^2 + 12$ and $y = 4x + 4$ over $[-3, 3]$ (Figure 8).

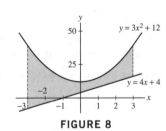

FIGURE 8

SOLUTION As the graph of $y = 3x^2 + 12$ lies above the graph of $y = 4x + 4$ over the interval $[-3, 3]$, the area between the graphs is

$$\int_{-3}^3 \left((3x^2 + 12) - (4x + 4) \right) dx = \int_{-3}^3 (3x^2 - 4x + 8)\, dx = \left(x^3 - 2x^2 + 8x \right)\Big|_{-3}^3 = 102.$$

3. Let $f(x) = x$ and $g(x) = 2 - x^2$ [Figure 9(B)].
(a) Find the points of intersection of the graphs.
(b) Find the area enclosed by the graphs of f and g.

SOLUTION

(a) Setting $f(x) = g(x)$ yields $2 - x^2 = x$, which simplifies to

$$0 = x^2 + x - 2 = (x + 2)(x - 1).$$

Thus, the graphs of $y = f(x)$ and $y = g(x)$ intersect at $x = -2$ and $x = 1$.

(b) As the graph of $y = x$ lies below the graph of $y = 2 - x^2$ over the interval $[-2, 1]$, the area between the graphs is

$$\int_{-2}^{1} \left((2 - x^2) - x \right) dx = \left(2x - \frac{1}{3}x^3 - \frac{1}{2}x^2 \right)\Big|_{-2}^{1} = \frac{9}{2}.$$

In Exercises 5–7, find the area between $y = \sin x$ and $y = \cos x$ over the interval. Sketch the curves if necessary.

5. $\left[0, \frac{\pi}{4} \right]$

SOLUTION   Over the interval $[0, \frac{\pi}{4}]$, the graph of $y = \sin x$ lies below that of $y = \cos x$. Hence, the area between the two curves is

$$\int_{0}^{\pi/4} (\cos x - \sin x) \, dx = (\sin x + \cos x)\Big|_{0}^{\pi/4} = \frac{\sqrt{2}}{2} + \frac{\sqrt{2}}{2} - (0 + 1) = \sqrt{2} - 1.$$

7. $[0, \pi]$

SOLUTION   Over the interval $[0, \frac{\pi}{4}]$, the graph of $y = \sin x$ lies below that of $y = \cos x$, while over the interval $[\frac{\pi}{4}, \pi]$, the orientation of the graphs is reversed. The area between the graphs over $[0, \pi]$ is then

$$\int_{0}^{\pi/4} (\cos x - \sin x) \, dx + \int_{\pi/4}^{\pi} (\sin x - \cos x) \, dx$$

$$= (\sin x + \cos x)\Big|_{0}^{\pi/4} + (-\cos x - \sin x)\Big|_{\pi/4}^{\pi}$$

$$= \frac{\sqrt{2}}{2} + \frac{\sqrt{2}}{2} - (0 + 1) + (1 - 0) - \left(-\frac{\sqrt{2}}{2} - \frac{\sqrt{2}}{2} \right) = 2\sqrt{2}.$$

In Exercises 8–10, let $f(x) = 20 + x - x^2$ and $g(x) = x^2 - 5x$.

9. Find the area of the region enclosed by the two graphs.

SOLUTION   Setting $f(x) = g(x)$ gives $20 + x - x^2 = x^2 - 5x$, which simplifies to

$$0 = 2x^2 - 6x - 20 = 2(x - 5)(x + 2).$$

Thus, the curves intersect at $x = -2$ and $x = 5$. With $y = 20 + x - x^2$ being the upper curve, the area between the two curves is

$$\int_{-2}^{5} \left((20 + x - x^2) - (x^2 - 5x) \right) dx = \int_{-2}^{5} \left(20 + 6x - 2x^2 \right) dx = \left(20x + 3x^2 - \frac{2}{3}x^3 \right)\Big|_{-2}^{5} = \frac{343}{3}.$$

In Exercises 11–14, find the area of the shaded region in the figure.

11.

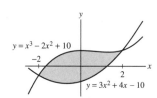

$y = x^3 - 2x^2 + 10$

$y = 3x^2 + 4x - 10$

FIGURE 10

SOLUTION   As the graph of $y = x^3 - 2x^2 + 10$ lies above the graph of $y = 3x^2 + 4x - 10$, the area of the shaded region is

$$\int_{-2}^{2} \left((x^3 - 2x^2 + 10) - (3x^2 + 4x - 10) \right) dx = \int_{-2}^{2} \left(x^3 - 5x^2 - 4x + 20 \right) dx$$

$$= \left(\frac{1}{4}x^4 - \frac{5}{3}x^3 - 2x^2 + 20x \right)\Big|_{-2}^{2} = \frac{160}{3}.$$

13.

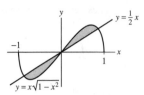

FIGURE 12

SOLUTION Setting $\frac{1}{2}x = x\sqrt{1-x^2}$ yields $x = 0$ or $\frac{1}{2} = \sqrt{1-x^2}$, so that $x = \pm\frac{\sqrt{3}}{2}$. Over the interval $[-\frac{\sqrt{3}}{2}, 0]$, $y = \frac{1}{2}x$ is the upper curve but over the interval $[0, \frac{\sqrt{3}}{2}]$, $y = x\sqrt{1-x^2}$ is the upper curve. The area of the shaded region is then

$$\int_{-\sqrt{3}/2}^{0} \left(\frac{1}{2}x - x\sqrt{1-x^2}\right) dx + \int_{0}^{\sqrt{3}/2} \left(x\sqrt{1-x^2} - \frac{1}{2}x\right) dx$$

$$= \left(\frac{1}{4}x^2 + \frac{1}{3}(1-x^2)^{3/2}\right)\Big|_{-\sqrt{3}/2}^{0} + \left(-\frac{1}{3}(1-x^2)^{3/2} - \frac{1}{4}x^2\right)\Big|_{0}^{\sqrt{3}/2} = \frac{5}{48} + \frac{5}{48} = \frac{5}{24}.$$

15. Find the area of the region enclosed by the curves $y = x^3 - 6x$ and $y = 8 - 3x^2$.

SOLUTION Setting $x^3 - 6x = 8 - 3x^2$ yields $(x + 1)(x + 4)(x - 2) = 0$, so the two curves intersect at $x = -4$, $x = -1$ and $x = 2$. Over the interval $[-4, -1]$, $y = x^3 - 6x$ is the upper curve, while $y = 8 - 3x^2$ is the upper curve over the interval $[-1, 2]$. The area of the region enclosed by the two curves is then

$$\int_{-4}^{-1} \left((x^3 - 6x) - (8 - 3x^2)\right) dx + \int_{-1}^{2} \left((8 - 3x^2) - (x^3 - 6x)\right) dx$$

$$= \left(\frac{1}{4}x^4 - 3x^2 - 8x + x^3\right)\Big|_{-4}^{-1} + \left(8x - x^3 - \frac{1}{4}x^4 + 3x^2\right)\Big|_{-1}^{2} = \frac{81}{4} + \frac{81}{4} = \frac{81}{2}.$$

In Exercises 17–18, find the area between the graphs of $x = \sin y$ and $x = 1 - \cos y$ over the given interval (Figure 14).

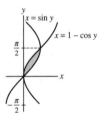

FIGURE 14

17. $0 \leq y \leq \frac{\pi}{2}$

SOLUTION As shown in the figure, the graph on the right is $x = \sin y$ and the graph on the left is $x = 1 - \cos y$. Therefore, the area between the two curves is given by

$$\int_{0}^{\pi/2} (\sin y - (1 - \cos y)) \, dy = (-\cos y - y + \sin y)\Big|_{0}^{\pi/2} = \left(-\frac{\pi}{2} + 1\right) - (-1) = 2 - \frac{\pi}{2}.$$

19. Find the area of the region lying to the right of $x = y^2 + 4y - 22$ and the left of $x = 3y + 8$.

SOLUTION Setting $y^2 + 4y - 22 = 3y + 8$ yields

$$0 = y^2 + y - 30 = (y + 6)(y - 5),$$

so the two curves intersect at $y = -6$ and $y = 5$. The area in question is then given by

$$\int_{-6}^{5} \left((3y + 8) - (y^2 + 4y - 22)\right) dy = \int_{-6}^{5} \left(-y^2 - y + 30\right) dy = \left(-\frac{y^3}{3} - \frac{y^2}{2} + 30y\right)\Big|_{-6}^{5} = \frac{1331}{6}.$$

21. Calculate the area enclosed by $x = 9 - y^2$ and $x = 5$ in two ways: as an integral along the y-axis and as an integral along the x-axis.

SOLUTION Along the y-axis, we have points of intersection at $y = \pm 2$. Therefore, the area enclosed by the two curves is

$$\int_{-2}^{2} \left(9 - y^2 - 5\right) dy = \int_{-2}^{2} \left(4 - y^2\right) dy = \left(4y - \frac{1}{3}y^3\right)\Big|_{-2}^{2} = \frac{32}{3}.$$

Along the x-axis, we have integration limits of $x = 5$ and $x = 9$. Therefore, the area enclosed by the two curves is

$$\int_{5}^{9} 2\sqrt{9 - x}\, dx = -\frac{4}{3}(9 - x)^{3/2}\Big|_{5}^{9} = 0 - \left(-\frac{32}{3}\right) = \frac{32}{3}.$$

In Exercises 23–24, find the area of the region using the method (integration along either the x- or y-axis) that requires you to evaluate just one integral.

23. Region between $y^2 = x + 5$ and $y^2 = 3 - x$

SOLUTION From the figure below, we see that integration along the x-axis would require two integrals, but integration along the y-axis requires only one integral. Setting $y^2 - 5 = 3 - y^2$ yields points of intersection at $y = \pm 2$. Thus, the area is given by

$$\int_{-2}^{2} \left((3 - y^2) - (y^2 + 5)\right) dy = \int_{-2}^{2} \left(8 - 2y^2\right) dy = \left(8y - \frac{2}{3}y^3\right)\Big|_{-2}^{2} = \frac{64}{3}.$$

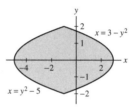

In Exercises 25–41, sketch the region enclosed by the curves and compute its area as an integral along the x- or y-axis.

25. $y = 4 - x^2$, $\quad y = x^2 - 4$

SOLUTION Setting $4 - x^2 = x^2 - 4$ yields $2x^2 = 8$ or $x^2 = 4$. Thus, the curves $y = 4 - x^2$ and $y = x^2 - 4$ intersect at $x = \pm 2$. From the figure below, we see that $y = 4 - x^2$ lies above $y = x^2 - 4$ over the interval $[-2, 2]$; hence, the area of the region enclosed by the curves is

$$\int_{-2}^{2} \left((4 - x^2) - (x^2 - 4)\right) dx = \int_{-2}^{2} (8 - 2x^2)\, dx = \left(8x - \frac{2}{3}x^3\right)\Big|_{-2}^{2} = \frac{64}{3}.$$

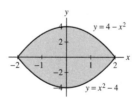

27. $x = \sin y$, $\quad x = \dfrac{2}{\pi}y$

SOLUTION Here, integration along the y-axis will require less work than integration along the x-axis. The curves intersect when $\frac{2y}{\pi} = \sin y$ or when $y = 0, \pm\frac{\pi}{2}$. From the graph below, we see that both curves are symmetric with respect to the origin. It follows that the portion of the region enclosed by the curves in the first quadrant is identical to the region enclosed in the third quadrant. We can therefore determine the total area enclosed by the two curves by doubling the area enclosed in the first quadrant. In the first quadrant, $x = \sin y$ lies to the right of $x = \frac{2y}{\pi}$, so the total area enclosed by the two curves is

$$2\int_{0}^{\pi/2} \left(\sin y - \frac{2}{\pi}y\right) dy = 2\left(-\cos y - \frac{1}{\pi}y^2\right)\Big|_{0}^{\pi/2} = 2\left[\left(0 - \frac{\pi}{4}\right) - (-1 - 0)\right] = 2 - \frac{\pi}{2}.$$

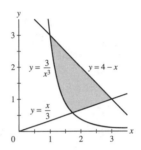

29. $y = 3x^{-3}$, $y = 4 - x$, $y = \dfrac{x}{3}$

SOLUTION The curves $y = 3x^{-3}$ and $y = 4 - x$ intersect at $x = 1$, the curves $y = 3x^{-3}$ and $y = x/3$ intersect at $x = \sqrt{3}$ and the curves $y = 4 - x$ and $y = x/3$ intersect at $x = 3$. From the graph below, we see that the top of the region enclosed by the three curves is always bounded by $y = 4 - x$. The bottom of the region is bounded by $y = 3x^{-3}$ for $1 \le x \le \sqrt{3}$ and by $y = x/3$ for $\sqrt{3} \le x \le 3$. The total area of the region is then

$$\int_1^{\sqrt{3}} \left((4 - x) - 3x^{-3} \right) dx + \int_{\sqrt{3}}^3 \left((4 - x) - \frac{1}{3}x \right) dx = \left(4x - \frac{1}{2}x^2 + \frac{3}{2}x^{-2} \right)\Big|_1^{\sqrt{3}} + \left(4x - \frac{2}{3}x^2 \right)\Big|_{\sqrt{3}}^3 = 2.$$

31. $y = x\sqrt{x - 2}$, $y = -x\sqrt{x - 2}$, $x = 4$

SOLUTION Note that $y = x\sqrt{x - 2}$ and $y = -x\sqrt{x - 2}$ are the upper and lower branches, respectively, of the curve $y^2 = x^2(x - 2)$. The area enclosed by this curve and the vertical line $x = 4$ is

$$\int_2^4 \left(x\sqrt{x - 2} - (-x\sqrt{x - 2}) \right) dx = \int_2^4 2x\sqrt{x - 2}\, dx.$$

Substitute $u = x - 2$. Then $du = dx$, $x = u + 2$ and

$$\int_2^4 2x\sqrt{x - 2}\, dx = \int_0^2 2(u + 2)\sqrt{u}\, du = \int_0^2 \left(2u^{3/2} + 4u^{1/2} \right) du = \left(\frac{4}{5}u^{5/2} + \frac{8}{3}u^{3/2} \right)\Big|_0^2 = \frac{128\sqrt{2}}{15}.$$

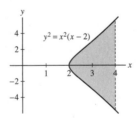

33. $x = |y|$, $x = 6 - y^2$

SOLUTION From the graph below, we see that integration along the y-axis will require less work than integration along the x-axis. Moreover, the region is symmetric with respect to the x-axis, so the total area can be determined by doubling the area of the upper portion of the region. For $y > 0$, setting $y = 6 - y^2$ yields

$$0 = y^2 + y - 6 = (y - 2)(y + 3),$$

so the curves intersect at $y = 2$. Because $x = 6 - y^2$ lies to the right of $x = |y| = y$ in the first quadrant, we find the total area of the region is

$$2\int_0^2 (6 - y^2 - y)\, dy = 2\left(6y - \frac{1}{3}y^3 - \frac{1}{2}y^2 \right)\Big|_0^2 = 2\left(12 - \frac{8}{3} - 2 \right) = \frac{44}{3}.$$

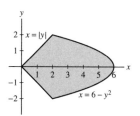

35. $x = 12 - y$, $x = y$, $x = 2y$

SOLUTION   From the graph below, we see that the bottom of the region enclosed by the three curves is always bounded by $y = \frac{x}{2}$. On the other hand, the top of the region is bounded by $y = x$ for $0 \le x \le 6$ and by $y = 12 - x$ for $6 \le x \le 8$. The area of the region is then

$$\int_0^6 \left(x - \frac{x}{2} \right) dx + \int_6^8 \left((12 - x) - \frac{x}{2} \right) dx = \frac{1}{4} x^2 \Big|_0^6 + \left(12x - \frac{3}{4} x^2 \right) \Big|_6^8 = 9 + (96 - 48) - (72 - 27) = 12.$$

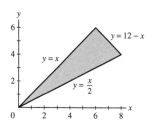

37. $x = 2y$, $x + 1 = (y - 1)^2$

SOLUTION   Setting $2y = (y - 1)^2 - 1$ yields

$$0 = y^2 - 4y = y(y - 4),$$

so the two curves intersect at $y = 0$ and at $y = 4$. From the graph below, we see that $x = 2y$ lies to the right of $x + 1 = (y - 1)^2$ over the interval $[0, 4]$ along the y-axis. Thus, the area of the region enclosed by the two curves is

$$\int_0^4 \left(2y - ((y - 1)^2 - 1) \right) dy = \int_0^4 \left(4y - y^2 \right) dy = \left(2y^2 - \frac{1}{3} y^3 \right) \Big|_0^4 = \frac{32}{3}.$$

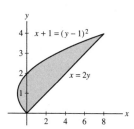

39. $y = 6$, $y = x^{-2} + x^2$ (in the region $x > 0$)

SOLUTION   Setting $6 = x^{-2} + x^2$ yields

$$0 = x^4 - 6x^2 + 1,$$

which is a quadratic equation in the variable x^2. By the quadratic formula,

$$x^2 = \frac{6 \pm \sqrt{36 - 4}}{2} = 3 \pm 2\sqrt{2}.$$

Now,

$$3 + 2\sqrt{2} = (\sqrt{2} + 1)^2 \quad \text{and} \quad 3 - 2\sqrt{2} = (\sqrt{2} - 1)^2,$$

so the two curves intersect at $x = \sqrt{2} - 1$ and $x = \sqrt{2} + 1$. Note there are also two points of intersection with $x < 0$, but as the problem specifies the region is for $x > 0$, we neglect these other two values. From the graph below, we see that $y = 6$ lies above $y = x^{-2} + x^2$, so the area of the region enclosed by the two curves for $x > 0$ is

$$\int_{\sqrt{2}-1}^{\sqrt{2}+1} \left(6 - (x^{-2} + x^2) \right) dx = \left(6x + x^{-1} - \frac{1}{3} x^3 \right) \Big|_{\sqrt{2}-1}^{\sqrt{2}+1} = \frac{16}{3}.$$

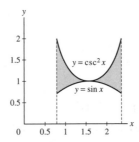

41. $y = \sin x$, $y = \csc^2 x$, $x = \dfrac{\pi}{4}$, $x = \dfrac{3\pi}{4}$

SOLUTION Over the interval $[\frac{\pi}{4}, \frac{3\pi}{4}]$, $y = \csc^2 x$ lies above $y = \sin x$. The area of the region enclosed by the two curves is then

$$\int_{\pi/4}^{3\pi/4} \left(\csc^2 x - \sin x\right) dx = \left(-\cot x + \cos x\right)\Big|_{\pi/4}^{3\pi/4} = \left(1 - \frac{\sqrt{2}}{2}\right) - \left(-1 + \frac{\sqrt{2}}{2}\right) = 2 - \sqrt{2}.$$

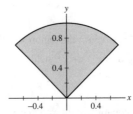

43. Sketch a region whose area is represented by

$$\int_{-\sqrt{2}/2}^{\sqrt{2}/2} \left(\sqrt{1 - x^2} - |x|\right) dx$$

and evaluate using geometry.

SOLUTION Matching the integrand $\sqrt{1 - x^2} - |x|$ with the $y_{\text{TOP}} - y_{\text{BOT}}$ template for calculating area, we see that the region in question is bounded along the top by the curve $y = \sqrt{1 - x^2}$ (the upper half of the unit circle) and is bounded along the bottom by the curve $y = |x|$. Hence, the region is $\frac{1}{4}$ of the unit circle (see the figure below). The area of the region must then be

$$\frac{1}{4} \pi (1)^2 = \frac{\pi}{4}.$$

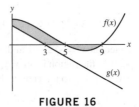

45. Express the area (not signed) of the shaded region in Figure 16 as a sum of three integrals involving the functions f and g.

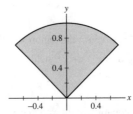

FIGURE 16

SOLUTION Because either the curve bounding the top of the region or the curve bounding the bottom of the region or both change at $x = 3$ and at $x = 5$, the area is calculated using three integrals. Specifically, the area is

$$\int_0^3 (f(x) - g(x)) \, dx + \int_3^5 (f(x) - 0) \, dx + \int_5^9 (0 - f(x)) \, dx$$

$$= \int_0^3 (f(x) - g(x)) \, dx + \int_3^5 f(x) \, dx - \int_5^9 f(x) \, dx.$$

47. Set up (but do not evaluate) an integral that expresses the area between the circles $x^2 + y^2 = 2$ and $x^2 + (y - 1)^2 = 1$.

SOLUTION Setting $2 - y^2 = 1 - (y - 1)^2$ yields $y = 1$. The two circles therefore intersect at the points $(1, 1)$ and $(-1, 1)$. From the graph below, we see that over the interval $[-1, 1]$, the upper half of the circle $x^2 + y^2 = 2$ lies above the lower half of the circle $x^2 + (y - 1)^2 = 1$. The area enclosed by the two circles is therefore given by the integral

$$\int_{-1}^1 \left(\sqrt{2 - x^2} - (1 - \sqrt{1 - x^2}) \right) \, dx.$$

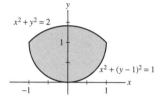

49. **CAS** Find a numerical approximation to the area above $y = 1 - (x/\pi)$ and below $y = \sin x$ (find the points of intersection numerically).

SOLUTION The region in question is shown in the figure below. Using a computer algebra system, we find that $y = 1 - x/\pi$ and $y = \sin x$ intersect on the left at $x = 0.8278585215$. Analytically, we determine the two curves intersect on the right at $x = \pi$. The area above $y = 1 - x/\pi$ and below $y = \sin x$ is then

$$\int_{0.8278585215}^{\pi} \left(\sin x - \left(1 - \frac{x}{\pi} \right) \right) \, dx = 0.8244398727,$$

where the definite integral was evaluated using a computer algebra system.

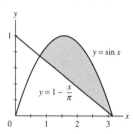

51. **CAS** Use a computer algebra system to find a numerical approximation to the number c (besides zero) in $[0, \frac{\pi}{2}]$, where the curves $y = \sin x$ and $y = \tan^2 x$ intersect. Then find the area enclosed by the graphs over $[0, c]$.

SOLUTION The region in question is shown in the figure below. Using a computer algebra system, we find that $y = \sin x$ and $y = \tan^2 x$ intersect at $x = 0.6662394325$. The area of the region enclosed by the two curves is then

$$\int_0^{0.6662394325} \left(\sin x - \tan^2 x \right) \, dx = 0.09393667698,$$

where the definite integral was evaluated using a computer algebra system.

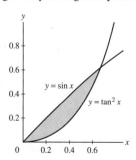

Further Insights and Challenges

53. Find the line $y = mx$ that divides the area under the curve $y = x(1 - x)$ over [0, 1] into two regions of equal area.

SOLUTION   First note that

$$\int_0^1 x(1 - x)\, dx = \int_0^1 \left(x - x^2\right) dx = \left(\frac{1}{2}x^2 - \frac{1}{3}x^3\right)\Big|_0^1 = \frac{1}{6}.$$

Now, the line $y = mx$ and the curve $y = x(1 - x)$ intersect when $mx = x(1 - x)$, or at $x = 0$ and at $x = 1 - m$. The area of the region enclosed by the two curves is then

$$\int_0^{1-m} (x(1 - x) - mx)\, dx = \int_0^{1-m} \left((1 - m)x - x^2\right) dx = \left((1 - m)\frac{x^2}{2} - \frac{1}{3}x^3\right)\Big|_0^{1-m} = \frac{1}{6}(1 - m)^3.$$

To have $\frac{1}{6}(1 - m)^3 = \frac{1}{2} \cdot \frac{1}{6}$ requires

$$m = 1 - \left(\frac{1}{2}\right)^{1/3} \approx 0.206299.$$

55. ✏️   Explain geometrically (without calculation) why the following holds for any $n > 0$:

$$\int_0^1 x^n\, dx + \int_0^1 x^{1/n}\, dx = 1$$

SOLUTION   Let A_1 denote the area of region 1 in the figure below. Define A_2 and A_3 similarly. It is clear from the figure that

$$A_1 + A_2 + A_3 = 1.$$

Now, note that x^n and $x^{1/n}$ are inverses of each other. Therefore, the graphs of $y = x^n$ and $y = x^{1/n}$ are symmetric about the line $y = x$, so regions 1 and 3 are also symmetric about $y = x$. This guarantees that $A_1 = A_3$. Finally,

$$\int_0^1 x^n\, dx + \int_0^1 x^{1/n}\, dx = A_3 + (A_2 + A_3) = A_1 + A_2 + A_3 = 1.$$

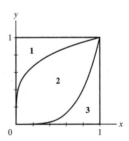

6.2 Setting Up Integrals: Volume, Density, Average Value

Preliminary Questions

1. What is the average value of $f(x)$ on [1, 4] if the area between the graph of $f(x)$ and the x-axis is equal to 9?

SOLUTION   Assuming that $f(x) \geq 0$ over the interval [1, 4], the fact that the area between the graph of f and the x-axis is equal to 9 indicates that $\int_1^4 f(x)\, dx = 9$. The average value of f over the interval [1, 4] is then

$$\frac{\int_1^4 f(x)\, dx}{4 - 1} = \frac{9}{3} = 3.$$

2. Find the volume of a solid extending from $y = 2$ to $y = 5$ if the cross section at y has area $A(y) = 5$ for all y.

SOLUTION   Because the cross-sectional area of the solid is constant, the volume is simply the cross-sectional area times the length, or $5 \times 3 = 15$.

3. Describe the horizontal cross sections of an ice cream cone and the vertical cross sections of a football (when it is held horizontally).

SOLUTION The horizontal cross sections of an ice cream cone, as well as the vertical cross sections of a football (when held horizontally), are circles.

4. What is the formula for the total population within a circle of radius R around a city center if the population has a radial function?

SOLUTION Because the population density is a radial function, the total population within a circle of radius R is

$$2\pi \int_0^R r\rho(r)\, dr,$$

where $\rho(r)$ is the radial population density function.

5. What is the definition of flow rate?

SOLUTION The flow rate of a fluid is the volume of fluid that passes through a cross-sectional area at a given point per unit time.

6. Which assumption about fluid velocity did we use to compute the flow rate as an integral?

SOLUTION To express flow rate as an integral, we assumed that the fluid velocity depended only on the radial distance from the center of the tube.

Exercises

1. Let V be the volume of a pyramid of height 20 whose base is a square of side 8.

(a) Use similar triangles as in Example 1 to find the area of the horizontal cross section at a height y.

(b) Calculate V by integrating the cross-sectional area.

SOLUTION

(a) We can use similar triangles to determine the side length, s, of the square cross section at height y. Using the diagram below, we find

$$\frac{8}{20} = \frac{s}{20 - y} \qquad \text{or} \qquad s = \frac{2}{5}(20 - y).$$

The area of the cross section at height y is then given by $\frac{4}{25}(20 - y)^2$.

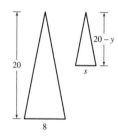

(b) The volume of the pyramid is

$$\int_0^{20} \frac{4}{25}(20 - y)^2\, dy = -\frac{4}{75}(20 - y)^3 \bigg|_0^{20} = \frac{1280}{3}.$$

3. Use the method of Exercise 2 to find the formula for the volume of a right circular cone of height h whose base is a circle of radius r (Figure 16).

SOLUTION

(a) From similar triangles (see Figure 16),

$$\frac{h}{h - y} = \frac{r}{r_0},$$

where r_0 is the radius of the cone at a height of y. Thus, $r_0 = r - \frac{ry}{h}$.

(b) The volume of the cone is

$$\pi \int_0^h \left(r - \frac{ry}{h}\right)^2 dy = \frac{-h\pi}{r} \frac{\left(r - \frac{ry}{h}\right)^3}{3} \bigg|_0^h = \frac{h\pi}{r} \frac{r^3}{3} = \frac{\pi r^2 h}{3}.$$

5. Find the volume of liquid needed to fill a sphere of radius R to height h (Figure 18).

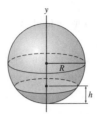

FIGURE 18 Sphere filled with liquid to height h.

SOLUTION The radius r at any height y is given by $r = \sqrt{R^2 - (R - y)^2}$. Thus, the volume of the filled portion of the sphere is

$$\pi \int_0^h r^2 \, dy = \pi \int_0^h \left(R^2 - (R - y)^2 \right) dy = \pi \int_0^h (2Ry - y^2) \, dy = \pi \left(Ry^2 - \frac{y^3}{3} \right) \Big|_0^h = \pi \left(Rh^2 - \frac{h^3}{3} \right).$$

7. Derive a formula for the volume of the wedge in Figure 19(B) in terms of the constants a, b, and c.

SOLUTION The line from c to a is given by the equation $(z/c) + (x/a) = 1$ and the line from b to a is given by $(y/b) + (x/a) = 1$. The cross sections perpendicular to the x-axis are right triangles with height $c(1 - x/a)$ and base $b(1 - x/a)$. Thus we have

$$\int_0^a \frac{1}{2} bc \, (1 - x/a)^2 \, dx = -\frac{1}{6} abc \left(1 - \frac{x}{a} \right)^3 \Big|_0^a = \frac{1}{6} abc.$$

In Exercises 9–14, find the volume of the solid with given base and cross sections.

9. The base is the unit circle $x^2 + y^2 = 1$ and the cross sections perpendicular to the x-axis are triangles whose height and base are equal.

SOLUTION At each location x, the side of the triangular cross section that lies in the base of the solid extends from the top half of the unit circle (with $y = \sqrt{1 - x^2}$) to the bottom half (with $y = -\sqrt{1 - x^2}$). The triangle therefore has base and height equal to $2\sqrt{1 - x^2}$ and area $2(1 - x^2)$. The volume of the solid is then

$$\int_{-1}^1 2(1 - x^2) \, dx = 2 \left(x - \frac{1}{3} x^3 \right) \Big|_{-1}^1 = \frac{8}{3}.$$

11. The base is the semicircle $y = \sqrt{9 - x^2}$, where $-3 \le x \le 3$. The cross sections perpendicular to the x-axis are squares.

SOLUTION For each x, the base of the square cross section extends from the semicircle $y = \sqrt{9 - x^2}$ to the x-axis. The square therefore has a base with length $\sqrt{9 - x^2}$ and an area of $\left(\sqrt{9 - x^2} \right)^2 = 9 - x^2$. The volume of the solid is then

$$\int_{-3}^3 \left(9 - x^2 \right) dx = \left(9x - \frac{1}{3} x^3 \right) \Big|_{-3}^3 = 36.$$

13. The base is the region enclosed by $y = x^2$ and $y = 3$. The cross sections perpendicular to the y-axis are squares.

SOLUTION At any location y, the distance to the parabola from the y-axis is $\sqrt{y}$. Thus the base of the square will have length $2\sqrt{y}$. Therefore the volume is

$$\int_0^3 \left(2\sqrt{y} \right) \left(2\sqrt{y} \right) dy = \int_0^3 4y \, dy = 2y^2 \Big|_0^3 = 18.$$

15. Find the volume of the solid whose base is the region $|x| + |y| \le 1$ and whose vertical cross sections perpendicular to the y-axis are semicircles (with diameter along the base).

SOLUTION The region R in question is a diamond shape connecting the points $(1, 0)$, $(0, -1)$, $(-1, 0)$, and $(0, 1)$. Thus, in the lower half of the xy-plane, the radius of the circles is $y + 1$ and in the upper half, the radius is $1 - y$. Therefore, the volume is

$$\frac{\pi}{2} \int_{-1}^0 (y + 1)^2 \, dy + \frac{\pi}{2} \int_0^1 (1 - y)^2 \, dy = \frac{\pi}{2} \left(\frac{1}{3} + \frac{1}{3} \right) = \frac{\pi}{3}.$$

17. Find the volume V of a *regular* tetrahedron whose face is an equilateral triangle of side s (Figure 20).

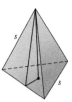

FIGURE 20 Regular tetrahedron.

SOLUTION   Our first task is to determine the relationship between the height of the tetrahedron, h, and the side length of the equilateral triangles, s. Let B be the orthocenter of the tetrahedron (the point directly below the apex), and let b denote the distance from B to each corner of the base triangle. By the Law of Cosines, we have

$$s^2 = b^2 + b^2 - 2b^2 \cos 120° = 3b^2,$$

so $b^2 = \frac{1}{3}s^2$. Thus

$$h^2 = s^2 - b^2 = \frac{2}{3}s^2 \quad \text{or} \quad h = s\sqrt{\frac{2}{3}}.$$

Therefore, using similar triangles, the side length of the equilateral triangle at height z above the base is

$$s\left(\frac{h-z}{h}\right) = s - \frac{z}{\sqrt{2/3}}.$$

The volume of the tetrahedron is then given by

$$\int_0^{s\sqrt{2/3}} \frac{\sqrt{3}}{4}\left(s - \frac{z}{\sqrt{2/3}}\right)^2 dz = -\frac{\sqrt{2}}{12}\left(s - \frac{z}{\sqrt{2/3}}\right)^3 \Bigg|_0^{s\sqrt{2/3}} = \frac{s^3\sqrt{2}}{12}.$$

19. A frustum of a pyramid is a pyramid with its top cut off [Figure 22(A)]. Let V be the volume of a frustum of height h whose base is a square of side a and top is a square of side b with $a > b \geq 0$.

(a) Show that if the frustum were continued to a full pyramid, it would have height $\dfrac{ha}{a-b}$ [Figure 22(B)].

(b) Show that the cross section at height x is a square of side $(1/h)(a(h-x)+bx)$.

(c) Show that $V = \frac{1}{3}h(a^2 + ab + b^2)$. A papyrus dating to the year 1850 BCE indicates that Egyptian mathematicians had discovered this formula almost 4,000 years ago.

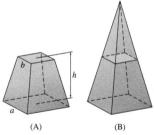

FIGURE 22

SOLUTION

(a) Let H be the height of the full pyramid. Using similar triangles, we have the proportion

$$\frac{H}{a} = \frac{H-h}{b}$$

which gives

$$H = \frac{ha}{a-b}.$$

(b) Let w denote the side length of the square cross section at height x. By similar triangles, we have

$$\frac{a}{H} = \frac{w}{H-x}.$$

Substituting the value for H from part (a) gives

$$w = \frac{a(h-x)+bx}{h}.$$

(c) The volume of the frustrum is

$$\int_0^h \left(\frac{1}{h}(a(h-x)+bx)\right)^2 dx = \frac{1}{h^2}\int_0^h \left(a^2(h-x)^2 + 2ab(h-x)x + b^2x^2\right) dx$$

$$= \frac{1}{h^2}\left(-\frac{a^2}{3}(h-x)^3 + abhx^2 - \frac{2}{3}abx^3 + \frac{1}{3}b^2x^3\right)\Big|_0^h = \frac{h}{3}\left(a^2 + ab + b^2\right).$$

21. Figure 24 shows the solid S obtained by intersecting two cylinders of radius r whose axes are perpendicular.

(a) The horizontal cross section of each cylinder at distance y from the central axis is a rectangular strip. Find the strip's width.

(b) Find the area of the horizontal cross section of S at distance y.

(c) Find the volume of S as a function of r.

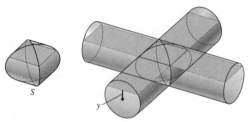

FIGURE 24 Intersection of two cylinders intersecting at right angles.

SOLUTION

(a) The horizontal cross section at distance y from the central axis (for $-r \le y \le r$) is a square of width $w = 2\sqrt{r^2 - y^2}$.

(b) The area of the horizontal cross section of S at distance y from the central axis is $w^2 = 4(r^2 - y^2)$.

(c) The volume of the solid S is then

$$4\int_{-r}^r \left(r^2 - y^2\right) dy = 4\left(r^2 y - \frac{1}{3}y^3\right)\Big|_{-r}^r = \frac{16}{3}r^3.$$

23. Calculate the volume of a cylinder inclined at an angle $\theta = 30°$ whose height is 10 and whose base is a circle of radius 4 (Figure 25).

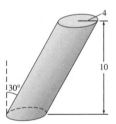

FIGURE 25 Cylinder inclined at an angle $\theta = 30°$.

SOLUTION The area of each circular cross section is $\pi(4)^2 = 16\pi$, hence the volume of the cylinder is

$$\int_0^{10} 16\pi\, dx = (16\pi x)\Big|_0^{10} = 160\pi$$

25. Find the total mass of a 2-m rod whose linear density function is $\rho(x) = 1 + 0.5\sin(\pi x)$ kg/m for $0 \le x \le 2$.

SOLUTION The total mass of the rod is

$$\int_0^2 \rho(x)\,dx = \int_0^2 (1 + .5\sin \pi x)\,dx = \left(x - .5\frac{\cos \pi x}{\pi}\right)\Big|_0^2 = 2 \text{ kg,}$$

27. Calculate the population within a 10-mile radius of the city center if the radial population density is $\rho(r) = 4(1 + r^2)^{1/3}$ (in thousands per square mile).

SOLUTION   The total population is

$$2\pi \int_0^{10} r \cdot \rho(r)\, dr = 2\pi \int_0^{10} 4r(1 + r^2)^{1/3}\, dr = 3\pi(1 + r^2)^{4/3}\Big|_0^{10}$$

$$\approx 4423.59 \text{ thousand} \approx 4.4 \text{ million}.$$

29. Table 1 lists the population density (in people per squared kilometer) as a function of distance r (in kilometers) from the center of a rural town. Estimate the total population within a 2-km radius of the center by taking the average of the left- and right-endpoint approximations.

TABLE 1   Population Density

r	$\rho(r)$	r	$\rho(r)$
0.0	125.0	1.2	37.6
0.2	102.3	1.4	30.8
0.4	83.8	1.6	25.2
0.6	68.6	1.8	20.7
0.8	56.2	2.0	16.9
1.0	46.0		

SOLUTION   The total population is given by

$$2\pi \int_0^2 r \cdot \rho(r)\, dr.$$

With $\Delta r = 0.2$, the left- and right-endpoint approximations to the required definite integral are

$$L_{10} = .2(2\pi)[0(125) + (.2)(102.3) + (.4)(83.8) + (.6)(68.6) + (.8)(56.2) + (1)(46)$$
$$+ (1.2)(37.6) + (1.4)(30.8) + (1.6)(25.2) + (1.8)(20.7)] = 442.24;$$

$$R_{10} = .2(2\pi)[(.2)(102.3) + (.4)(83.8) + (.6)(68.6) + (0.8)(56.2) + (1)(46)$$
$$+ (1.2)(37.6) + (1.4)(30.8) + (1.6)(25.2) + (1.8)(20.7) + (2)(16.9)] = 484.71.$$

This gives an average of 463.475. Thus, there are roughly 463 people within a 2-km radius of the town center.

31. The density of deer in a forest is the radial function $\rho(r) = 150(r^2 + 2)^{-2}$ deer per km^2, where r is the distance (in kilometers) to a small meadow. Calculate the number of deer in the region $2 \leq r \leq 5$ km.

SOLUTION   The number of deer in the region $2 \leq r \leq 5$ km is

$$2\pi \int_2^5 r\,(150)\left(r^2 + 2\right)^{-2}\, dr = -150\pi \left(\frac{1}{r^2 + 2}\right)\Big|_2^5 = -150\pi \left(\frac{1}{27} - \frac{1}{6}\right) \approx 61 \text{ deer.}$$

33. Find the flow rate through a tube of radius 4 cm, assuming that the velocity of fluid particles at a distance r cm from the center is $v(r) = 16 - r^2$ cm/s.

SOLUTION   The flow rate is

$$2\pi \int_0^R r v(r)\, dr = 2\pi \int_0^4 r \left(16 - r^2\right) dr = 2\pi \left(8r^2 - \frac{1}{4}r^4\right)\Big|_0^4 = 128\pi \ \frac{\text{cm}^3}{\text{s}}.$$

35. A solid rod of radius 1 cm is placed in a pipe of radius 3 cm so that their axes are aligned. Water flows through the pipe and around the rod. Find the flow rate if the velocity of the water is given by the radial function $v(r) = 0.5(r - 1)(3 - r)$ cm/s.

SOLUTION   The flow rate is

$$2\pi \int_1^3 r(.5)(r - 1)(3 - r)\, dr = \pi \int_1^3 \left(-r^3 + 4r^2 - 3r\right) dr = \pi \left(-\frac{1}{4}r^4 + \frac{4}{3}r^3 - \frac{3}{2}r^2\right)\Big|_1^3 = \frac{8\pi}{3} \ \frac{\text{cm}^3}{\text{s}}.$$

In Exercises 37–44, calculate the average over the given interval.

37. $f(x) = x^3$, $[0, 1]$

SOLUTION The average is

$$\frac{1}{1-0} \int_0^1 x^3 \, dx = \int_0^1 x^3 \, dx = \frac{1}{4} x^4 \Big|_0^1 = \frac{1}{4}.$$

39. $f(x) = \cos x$, $[0, \frac{\pi}{2}]$

SOLUTION The average is

$$\frac{1}{\pi/2 - 0} \int_0^{\pi/2} \cos x \, dx = \frac{2}{\pi} \left(\sin t \Big|_0^{\pi/2} \right) = \frac{2}{\pi}.$$

41. $f(s) = s^{-2}$, $[2, 5]$

SOLUTION The average is

$$\frac{1}{5-2} \int_2^5 s^{-2} \, ds = -\frac{1}{3} s^{-1} \Big|_2^5 = \frac{1}{10}.$$

43. $f(x) = 2x^3 - 3x^2$, $[-1, 3]$

SOLUTION The average is

$$\frac{1}{3-(-1)} \int_{-1}^3 \left(2x^3 - 3x^2 \right) dx = \frac{1}{4} \left(\frac{1}{2} x^4 - x^3 \right) \Big|_{-1}^3 = 3.$$

45. Let M be the average value of $f(x) = x^3$ on $[0, A]$, where $A > 0$. Which theorem guarantees that $f(c) = M$ has a solution c in $[0, A]$? Find c.

SOLUTION The Mean Value Theorem for Integrals guarantees that $f(c) = M$ has a solution c in $[0, A]$. With $f(x) = x^3$ on $[0, A]$,

$$M = \frac{1}{A - 0} \int_0^A x^3 \, dx = \frac{1}{A} \frac{1}{4} x^4 \Big|_0^A = \frac{A^3}{4}.$$

Solving $f(c) = c^3 = \frac{A^3}{4}$ for c yields

$$c = \frac{A}{\sqrt[3]{4}}.$$

47. Which of $f(x) = x \sin^2 x$ and $g(x) = x^2 \sin^2 x$ has a larger average value over $[0, 1]$? Over $[1, 2]$?

SOLUTION The functions f and g differ only in the power of x multiplying $\sin^2 x$. It is also important to note that $\sin^2 x \geq 0$ for all x. Now, for each $x \in (0, 1)$, $x > x^2$ so

$$f(x) = x \sin^2 x > x^2 \sin^2 x = g(x).$$

Thus, over $[0, 1]$, $f(x)$ will have a larger average value than $g(x)$. On the other hand, for each $x \in (1, 2)$, $x^2 > x$, so

$$g(x) = x^2 \sin^2 x > x \sin^2 x = f(x).$$

Thus, over $[1, 2]$, $g(x)$ will have the larger average value.

49. Sketch the graph of a function $f(x)$ such that $f(x) \geq 0$ on $[0, 1]$ and $f(x) \leq 0$ on $[1, 2]$, whose average on $[0, 2]$ is negative.

SOLUTION Many solutions will exist. One could be

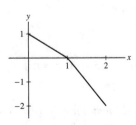

51. The temperature $T(t)$ at time t (in hours) in an art museum varies according to $T(t) = 70 + 5\cos\left(\frac{\pi}{12}t\right)$. Find the average over the time periods $[0, 24]$ and $[2, 6]$.

SOLUTION

- The average temperature over the 24-hour period is

$$\frac{1}{24-0}\int_0^{24}\left(70 + 5\cos\left(\frac{\pi}{12}t\right)\right)dt = \frac{1}{24}\left(70t + \frac{60}{\pi}\sin\left(\frac{\pi}{12}t\right)\right)\Big|_0^{24} = 70^\circ\text{F}.$$

- The average temperature over the 4-hour period is

$$\frac{1}{6-2}\int_2^6\left(70 + 5\cos\left(\frac{\pi}{12}t\right)\right)dt = \frac{1}{4}\left(70t + \frac{60}{\pi}\sin\left(\frac{\pi}{12}t\right)\right)\Big|_2^6 = 72.4^\circ\text{F}.$$

53. What is the average area of the circles whose radii vary from 0 to 1?

SOLUTION The average area is

$$\frac{1}{1-0}\int_0^1 \pi r^2\, dr = \frac{\pi}{3}r^3\Big|_0^1 = \frac{\pi}{3}.$$

55. The acceleration of a particle is $a(t) = t - t^3$ m/s^2 for $0 \le t \le 1$. Compute the average acceleration and average velocity over the time interval $[0, 1]$, assuming that the particle's initial velocity is zero.

SOLUTION The average acceleration is

$$\frac{1}{1-0}\int_0^1\left(t - t^3\right)dt = \left(\frac{1}{2}t^2 - \frac{1}{4}t^4\right)\Big|_0^1 = \frac{1}{4}\text{ m/s}^2.$$

An acceleration $a(t) = t - t^3$ with zero initial velocity gives

$$v(t) = \frac{1}{2}t^2 - \frac{1}{4}t^4.$$

Thus the average velocity is given by

$$\frac{1}{1-0}\int_0^1\left(\frac{1}{2}t^2 - \frac{1}{4}t^4\right)dt = \left(\frac{1}{6}t^3 - \frac{1}{20}t^5\right)\Big|_0^1 = \frac{7}{60}\text{ m/s}.$$

57. Let $f(x) = \sqrt{x}$. Find a value of c in $[4, 9]$ such that $f(c)$ is equal to the average of f on $[4, 9]$.

SOLUTION The average value is

$$\frac{1}{9-4}\int_4^9 \sqrt{x}\, dx = \frac{1}{5}\int_4^9 \sqrt{x}\, dx = \frac{2}{15}x^{3/2}\Big|_4^9 = \frac{38}{15}.$$

Then $f(c) = \sqrt{c} = \frac{38}{15}$ implies

$$c = \left(\frac{38}{15}\right)^2 = \frac{1444}{225} \approx 6.417778.$$

Further Insights and Challenges

59. An object is tossed in the air vertically from ground level with initial velocity v_0 ft/s at time $t = 0$. Find the average speed of the object over the time interval $[0, T]$, where T is the time the object returns to earth.

SOLUTION The height is given by $h(t) = v_0 t - 16t^2$. The ball is at ground level at time $t = 0$ and $T = v_0/16$. The velocity is given by $v(t) = v_0 - 32t$ and thus the speed is given by $s(t) = |v_0 - 32t|$. The average speed is

$$\frac{1}{v_0/16 - 0}\int_0^{v_0/16}|v_0 - 32t|\, dt = \frac{16}{v_0}\int_0^{v_0/32}(v_0 - 32t)\, dt + \frac{16}{v_0}\int_{v_0/32}^{v_0/16}(32t - v_0)\, dt$$

$$= \frac{16}{v_0}\left(v_0 t - 16t^2\right)\Big|_0^{v_0/32} + \frac{16}{v_0}\left(16t^2 - v_0 t\right)\Big|_{v_0/32}^{v_0/16} = v_0/2.$$

6.3 Volumes of Revolution

Preliminary Questions

1. Which of the following is a solid of revolution?

(a) Sphere   **(b)** Pyramid   **(c)** Cylinder   **(d)** Cube

SOLUTION The sphere and the cylinder have circular cross sections; hence, these are solids of revolution. The pyramid and cube do not have circular cross sections, so these are not solids of revolution.

2. True or false? When a solid is formed by rotating the region under a graph about the x-axis, the cross sections perpendicular to the x-axis are circular disks.

SOLUTION True. The cross sections will be disks with radius equal to the value of the function.

3. True or false? When a solid is formed by rotating the region between two graphs about the x-axis, the cross sections perpendicular to the x-axis are circular disks.

SOLUTION False. The cross sections may be washers.

4. Which of the following integrals expresses the volume of the solid obtained by rotating the area between $y = f(x)$ and $y = g(x)$ over $[a, b]$ around the x-axis [assume $f(x) \geq g(x) \geq 0$]?

(a) $\pi \int_a^b \left(f(x) - g(x)\right)^2 dx$

(b) $\pi \int_a^b \left(f(x)^2 - g(x)^2\right) dx$

SOLUTION The correct answer is **(b)**. Cross sections of the solid will be washers with outer radius $f(x)$ and inner radius $g(x)$. The area of the washer is then $\pi f(x)^2 - \pi g(x)^2 = \pi(f(x)^2 - g(x)^2)$.

Exercises

In Exercises 1–4, (a) sketch the solid obtained by revolving the region under the graph of $f(x)$ about the x-axis over the given interval, (b) describe the cross section perpendicular to the x-axis located at x, and (c) calculate the volume of the solid.

1. $f(x) = x + 1$, $[0, 3]$

SOLUTION

(a) A sketch of the solid of revolution is shown below:

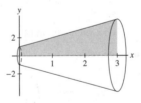

(b) Each cross section is a disk with radius $x + 1$.

(c) The volume of the solid of revolution is

$$\pi \int_0^3 (x + 1)^2 dx = \pi \int_0^3 (x^2 + 2x + 1) dx = \pi \left(\frac{1}{3}x^3 + x^2 + x\right)\Big|_0^3 = 21\pi.$$

3. $f(x) = \sqrt{x + 1}$, $[1, 4]$

SOLUTION

(a) A sketch of the solid of revolution is shown below:

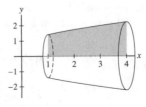

(b) Each cross section is a disk with radius $\sqrt{x + 1}$.

(c) The volume of the solid of revolution is

$$\pi \int_1^4 (\sqrt{x+1})^2 \, dx = \pi \int_1^4 (x+1) \, dx = \pi \left(\frac{1}{2}x^2 + x \right) \Big|_1^4 = \frac{21\pi}{2}.$$

In Exercises 5–12, find the volume of the solid obtained by rotating the region under the graph of the function about the x-axis over the given interval.

5. $f(x) = x^2 - 3x$, $[0, 3]$

SOLUTION The volume of the solid of revolution is

$$\pi \int_0^3 (x^2 - 3x)^2 \, dx = \pi \int_0^3 (x^4 - 6x^3 + 9x^2) \, dx = \pi \left(\frac{1}{5}x^5 - \frac{3}{2}x^4 + 3x^3 \right) \Big|_0^3 = \frac{81\pi}{10}.$$

7. $f(x) = x^{5/3}$, $[1, 8]$

SOLUTION The volume of the solid of revolution is

$$\pi \int_1^8 (x^{5/3})^2 \, dx = \pi \int_1^8 x^{10/3} \, dx = \frac{3\pi}{13}x^{13/3} \Big|_1^8 = \frac{3\pi}{13}(2^{13} - 1) = \frac{24573\pi}{13}.$$

9. $f(x) = \dfrac{2}{x+1}$, $[1, 3]$

SOLUTION The volume of the solid of revolution is

$$\pi \int_1^3 \left(\frac{2}{x+1} \right)^2 \, dx = 4\pi \int_1^3 (x+1)^{-2} \, dx = -4\pi (x+1)^{-1} \Big|_1^3 = \pi.$$

11. $f(x) = \sqrt{\cos x + 1}$, $[0, \pi]$

SOLUTION The volume of the solid of revolution is

$$\pi \int_0^\pi (\sqrt{\cos x + 1})^2 \, dx = \pi \int_0^\pi (\cos x + 1) \, dx = \pi (\sin x + x) \Big|_0^\pi = \pi^2.$$

In Exercises 13–18, (a) sketch the region enclosed by the curves, (b) describe the cross section perpendicular to the x-axis located at x, and (c) find the volume of the solid obtained by rotating the region about the x-axis.

13. $y = x^2 + 2$, $y = 10 - x^2$

SOLUTION

(a) Setting $x^2 + 2 = 10 - x^2$ yields $2x^2 = 8$, or $x^2 = 4$. The two curves therefore intersect at $x = \pm 2$. The region enclosed by the two curves is shown in the figure below.

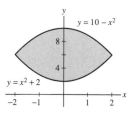

(b) When the region is rotated about the x-axis, each cross section is a washer with outer radius $R = 10 - x^2$ and inner radius $r = x^2 + 2$.

(c) The volume of the solid of revolution is

$$\pi \int_{-2}^2 \left((10 - x^2)^2 - (x^2 + 2)^2 \right) dx = \pi \int_{-2}^2 (96 - 24x^2) \, dx = \pi \left(96x - 8x^3 \right) \Big|_{-2}^2 = 256\pi.$$

15. $y = 16 - x$, $y = 3x + 12$, $x = -1$

SOLUTION

(a) Setting $16 - x = 3x + 12$, we find that the two lines intersect at $x = 1$. The region enclosed by the two curves is shown in the figure below.

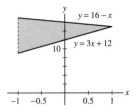

(b) When the region is rotated about the x-axis, each cross section is a washer with outer radius $R = 16 - x$ and inner radius $r = 3x + 12$.

(c) The volume of the solid of revolution is

$$\pi \int_{-1}^{1} \left((16 - x)^2 - (3x + 12)^2 \right) dx = \pi \int_{-1}^{1} (112 - 104x - 8x^2) \, dx = \pi \left(112x - 52x^2 - \frac{8}{3}x^3 \right) \Big|_{-1}^{1} = \frac{656\pi}{3}.$$

17. $y = \sec x, \quad y = 0, \quad x = -\dfrac{\pi}{4}, \quad x = \dfrac{\pi}{4}$

SOLUTION

(a) The region in question is shown in the figure below.

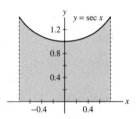

(b) When the region is rotated about the x-axis, each cross section is a circular disk with radius $R = \sec x$.

(c) The volume of the solid of revolution is

$$\pi \int_{-\pi/4}^{\pi/4} (\sec x)^2 \, dx = \pi (\tan x) \Big|_{-\pi/4}^{\pi/4} = 2\pi.$$

In Exercises 19–22, find the volume of the solid obtained by rotating the region enclosed by the graphs about the y-axis over the given interval.

19. $x = \sqrt{y}, \quad x = 0; \quad 1 \le y \le 4$

SOLUTION When the region in question (shown in the figure below) is rotated about the y-axis, each cross section is a disk with radius $\sqrt{y}$. The volume of the solid of revolution is

$$\pi \int_{1}^{4} \left(\sqrt{y} \right)^2 dy = \frac{\pi y^2}{2} \Big|_{1}^{4} = \frac{15\pi}{2}.$$

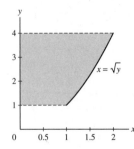

21. $x = y^2, \quad x = \sqrt{y}; \quad 0 \le y \le 1$

SOLUTION When the region in question (shown in the figure below) is rotated about the y-axis, each cross section is a washer with outer radius $R = \sqrt{y}$ and inner radius $r = y^2$. The volume of the solid of revolution is

$$\pi \int_{0}^{1} \left((\sqrt{y})^2 - (y^2)^2 \right) dy = \pi \left(\frac{y^2}{2} - \frac{y^5}{5} \right) \Big|_{0}^{1} = \frac{3\pi}{10}.$$

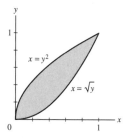

In Exercises 23–28, find the volume of the solid obtained by rotating region A in Figure 10 about the given axis.

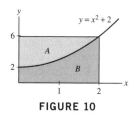

FIGURE 10

23. x-axis

SOLUTION Rotating region A about the x-axis produces a solid whose cross sections are washers with outer radius $R = 6$ and inner radius $r = x^2 + 2$. The volume of the solid of revolution is

$$\pi \int_0^2 \left((6)^2 - (x^2 + 2)^2 \right) dx = \pi \int_0^2 (32 - 4x^2 - x^4) \, dx = \pi \left(32x - \frac{4}{3}x^3 - \frac{1}{5}x^5 \right) \Big|_0^2 = \frac{704\pi}{15}.$$

25. $y = 2$

SOLUTION Rotating the region A about $y = 2$ produces a solid whose cross sections are washers with outer radius $R = 6 - 2 = 4$ and inner radius $r = x^2 + 2 - 2 = x^2$. The volume of the solid of revolution is

$$\pi \int_0^2 \left(4^2 - (x^2)^2 \right) dx = \pi \left(16x - \frac{1}{5}x^5 \right) \Big|_0^2 = \frac{128\pi}{5}.$$

27. $x = -3$

SOLUTION Rotating region A about $x = -3$ produces a solid whose cross sections are washers with outer radius $R = \sqrt{y - 2} - (-3) = \sqrt{y - 2} + 3$ and inner radius $r = 0 - (-3) = 3$. The volume of the solid of revolution is

$$\pi \int_2^6 \left((3 + \sqrt{y - 2})^2 - (3)^2 \right) dy = \pi \int_2^6 (6\sqrt{y - 2} + y - 2) \, dy = \pi \left(4(y - 2)^{3/2} + \frac{1}{2}y^2 - 2y \right) \Big|_2^6 = 40\pi.$$

In Exercises 29–34, find the volume of the solid obtained by rotating region B in Figure 10 about the given axis.

29. x-axis

SOLUTION Rotating region B about the x-axis produces a solid whose cross sections are disks with radius $R = x^2 + 2$. The volume of the solid of revolution is

$$\pi \int_0^2 (x^2 + 2)^2 \, dx = \pi \int_0^2 (x^4 + 4x^2 + 4) \, dx = \pi \left(\frac{1}{5}x^5 + \frac{4}{3}x^3 + 4x \right) \Big|_0^2 = \frac{376\pi}{15}.$$

31. $y = 6$

SOLUTION Rotating region B about $y = 6$ produces a solid whose cross sections are washers with outer radius $R = 6 - 0 = 6$ and inner radius $r = 6 - (x^2 + 2) = 4 - x^2$. The volume of the solid of revolution is

$$\pi \int_0^2 \left(6^2 - (4 - x^2)^2 \right) dy = \pi \int_0^2 \left(20 + 8x^2 - x^4 \right) dy = \pi \left(20x + \frac{8}{3}x^3 - \frac{1}{5}x^5 \right) \Big|_0^2 = \frac{824\pi}{15}.$$

33. $x = 2$

SOLUTION Rotating region B about $x = 2$ produces a solid with two different cross sections. For each $y \in [0, 2]$, the cross section is a disk with radius $R = 2$; for each $y \in [2, 6]$, the cross section is a disk with radius $R = 2 - \sqrt{y - 2}$. The volume of the solid of revolution is

$$\pi \int_0^2 (2)^2 \, dy + \pi \int_2^6 (2 - \sqrt{y - 2})^2 \, dy = \pi \int_0^2 4 \, dy + \pi \int_2^6 (2 + y - 4\sqrt{y - 2}) \, dy$$

$$= \pi (4y) \Big|_0^2 + \pi \left(2y + \frac{1}{2}y^2 - \frac{8}{3}(y - 2)^{3/2} \right) \Big|_2^6 = \frac{32\pi}{3}.$$

In Exercises 35–46, find the volume of the solid obtained by rotating the region enclosed by the graphs about the given axis.

35. $y = x^2$, $y = 12 - x$, $x = 0$, about $y = -2$

SOLUTION Rotating the region enclosed by $y = x^2$, $y = 12 - x$ and the y-axis (shown in the figure below) about $y = -2$ produces a solid whose cross sections are washers with outer radius $R = 12 - x - (-2) = 14 - x$ and inner radius $r = x^2 - (-2) = x^2 + 2$. The volume of the solid of revolution is

$$\pi \int_0^3 \left((14 - x)^2 - (x^2 + 2)^2 \right) dx = \pi \int_0^3 (192 - 28x - 3x^2 - x^4) \, dx$$

$$= \pi \left(192x - 14x^2 - x^3 - \frac{1}{5}x^5 \right) \Big|_0^3 = \frac{1872\pi}{5}.$$

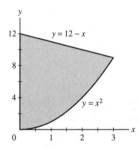

37. $y = 16 - x$, $y = 3x + 12$, $x = 0$, about y-axis

SOLUTION Rotating the region enclosed by $y = 16 - x$, $y = 3x + 12$ and the y-axis (shown in the figure below) about the y-axis produces a solid with two different cross sections. For each $y \in [12, 15]$, the cross section is a disk with radius $R = \frac{1}{3}(y - 12)$; for each $y \in [15, 16]$, the cross section is a disk with radius $R = 16 - y$. The volume of the solid of revolution is

$$\pi \int_{12}^{15} \left(\frac{1}{3}(y - 12) \right)^2 dy + \pi \int_{15}^{16} (16 - y)^2 \, dy$$

$$= \pi \int_{12}^{15} \frac{1}{9}(y^2 - 24y + 144) \, dy + \pi \int_{15}^{16} (y^2 - 32y + 256) \, dy$$

$$= \frac{\pi}{9} \left(\frac{1}{3}y^3 - 12y^2 + 144y \right) \Big|_{12}^{15} + \pi \left(\frac{1}{3}y^3 - 16y^2 + 256y \right) \Big|_{15}^{16} = \frac{4}{3}\pi.$$

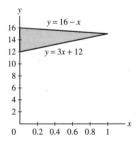

39. $y = \dfrac{9}{x^2}$, $y = 10 - x^2$, about x-axis

SOLUTION The region enclosed by the two curves is shown in the figure below. Note that the region consists of two pieces that are symmetric with respect to the y-axis. We may therefore compute the volume of the solid of revolution by considering one of the pieces and doubling the result. Rotating the portion of the region in the first quadrant about the x-axis produces a solid whose cross sections are washers with outer radius $R = 10 - x^2$ and inner radius $r = 9x^{-2}$. The volume of the solid of revolution is

$$2\pi \int_1^3 \left((10 - x^2)^2 - (9x^{-2})^2\right) dx = 2\pi \int_1^3 \left(100 - 20x^2 + x^4 - 81x^{-4}\right) dx$$

$$= 2\pi \left(100x - \frac{20}{3}x^3 + \frac{1}{5}x^5 + 27x^{-3}\right)\Big|_1^3 = \frac{1472\pi}{15}.$$

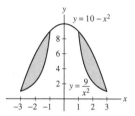

41. $y = \dfrac{1}{x}$, $y = \dfrac{5}{2} - x$, about y-axis

SOLUTION Rotating the region enclosed by $y = x^{-1}$ and $y = \frac{5}{2} - x$ (shown in the figure below) about the y-axis produces a solid whose cross sections are washers with outer radius $R = \frac{5}{2} - y$ and inner radius $r = y^{-1}$. The volume of the solid of revolution is

$$\pi \int_{1/2}^2 \left(\left(\frac{5}{2} - y\right)^2 - (y^{-1})^2\right) dy = \pi \int_{1/2}^2 \left(\frac{25}{4} - 5y + y^2 - y^{-2}\right) dy$$

$$= \pi \left(\frac{25}{4}y - \frac{5}{2}y^2 + \frac{1}{3}y^3 + y^{-1}\right)\Big|_{1/2}^2 = \frac{9\pi}{8}.$$

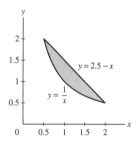

43. $y = x^3$, $y = x^{1/3}$, about y-axis

SOLUTION Rotating the region enclosed by $y = x^3$ and $y = x^{1/3}$ (shown in the figure below) about the y-axis produces a solid whose cross sections are washers with outer radius $R = y^{1/3}$ and inner radius $r = y^3$. The volume of the solid of revolution is

$$\pi \int_{-1}^1 \left((y^{1/3})^2 - (y^3)^2\right) dy = \pi \int_{-1}^1 (y^{2/3} - y^6) dy = \pi \left(\frac{3}{5}y^{5/3} - \frac{1}{7}y^7\right)\Big|_{-1}^1 = \frac{32\pi}{35}.$$

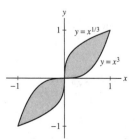

45. $y^2 = 4x$, $y = x$, about x-axis

SOLUTION Rotating the region enclosed by $y^2 = 4x$ and $y = x$ (shown in the figure below) about the x-axis produces a solid whose cross sections are washers with outer radius $R = 2\sqrt{x}$ and inner radius $r = x$. The volume of the solid of revolution is

$$\pi \int_0^4 \left((2\sqrt{x})^2 - x^2 \right) dx = \pi \left(2x^2 - \frac{1}{3}x^3 \right)\Big|_0^4 = \frac{32\pi}{3}.$$

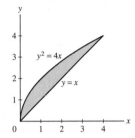

47. GU   Sketch the hypocycloid $x^{2/3} + y^{2/3} = 1$ and find the volume of the solid obtained by revolving it about the x-axis.

SOLUTION   A sketch of the hypocycloid is shown below.

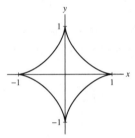

For the hypocycloid, $y = \pm\left(1 - x^{2/3}\right)^{3/2}$. Rotating this region about the x-axis will produce a solid whose cross sections are disks with radius $R = \left(1 - x^{2/3}\right)^{3/2}$. Thus the volume of the solid of revolution will be

$$\pi \int_{-1}^1 \left((1 - x^{2/3})^{3/2} \right)^2 dx = \pi \left(\frac{-x^3}{3} + \frac{9}{7}x^{7/3} - \frac{9}{5}x^{5/3} + x \right)\Big|_{-1}^1 = \frac{32\pi}{105}.$$

49. A "bead" is formed by removing a cylinder of radius r from the center of a sphere of radius R (Figure 12). Find the volume of the bead with $r = 1$ and $R = 2$.

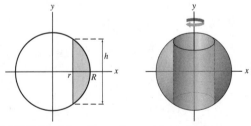

FIGURE 12 A bead is a sphere with a cylinder removed.

SOLUTION   The equation of the outer circle is $x^2 + y^2 = 2^2$, and the inner cylinder intersects the sphere when $y = \pm\sqrt{3}$. Each cross section of the bead is a washer with outer radius $\sqrt{4 - y^2}$ and inner radius 1, so the volume is given by

$$\pi \int_{-\sqrt{3}}^{\sqrt{3}} \left(\left(\sqrt{4 - y^2} \right)^2 - 1^2 \right) dy = \pi \int_{-\sqrt{3}}^{\sqrt{3}} \left(3 - y^2 \right) dy = 4\pi\sqrt{3}.$$

Further Insights and Challenges

51. The solid generated by rotating the region inside the ellipse with equation $\left(\frac{x}{a}\right)^2 + \left(\frac{y}{b}\right)^2 = 1$ around the x-axis is called an **ellipsoid**. Show that the ellipsoid has volume $\frac{4}{3}\pi ab^2$. What is the volume if the ellipse is rotated around the y-axis?

SOLUTION

- Rotating the ellipse about the x-axis produces an ellipsoid whose cross sections are disks with radius $R = b\sqrt{1-(x/a)^2}$. The volume of the ellipsoid is then

$$\pi \int_{-a}^{a} \left(b\sqrt{1-(x/a)^2} \right)^2 \, dx = b^2 \pi \int_{-a}^{a} \left(1 - \frac{1}{a^2}x^2 \right) \, dx = b^2 \pi \left(x - \frac{1}{3a^2}x^3 \right)\Big|_{-a}^{a} = \frac{4}{3}\pi a b^2.$$

- Rotating the ellipse about the y-axis produces an ellipsoid whose cross sections are disks with radius $R = a\sqrt{1-(y/b)^2}$. The volume of the ellipsoid is then

$$\int_{-b}^{b} \left(a\sqrt{1-(y/b)^2} \right)^2 \, dy = a^2 \pi \int_{-b}^{b} \left(1 - \frac{1}{b^2}y^2 \right) \, dy = a^2 \pi \left(y - \frac{1}{3b^2}y^3 \right)\Big|_{-b}^{b} = \frac{4}{3}\pi a^2 b.$$

53. The curve $y = f(x)$ in Figure 14, called a **tractrix**, has the following property: the tangent line at each point (x, y) on the curve has slope

$$\frac{dy}{dx} = \frac{-y}{\sqrt{1-y^2}}.$$

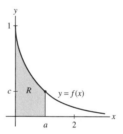

FIGURE 14 The tractrix.

Let R be the shaded region under the graph of $0 \le x \le a$ in Figure 14. Compute the volume V of the solid obtained by revolving R around the x-axis in terms of the constant $c = f(a)$. *Hint:* Use the disk method and the substitution $u = f(x)$ to show that

$$V = \pi \int_{c}^{1} u\sqrt{1-u^2} \, du$$

SOLUTION Let $y = f(x)$ be the tractrix depicted in Figure 14. Rotating the region R about the x-axis produces a solid whose cross sections are disks with radius $f(x)$. The volume of the resulting solid is then

$$V = \pi \int_{0}^{a} [f(x)]^2 \, dx.$$

Now, let $u = f(x)$. Then

$$du = f'(x) \, dx = \frac{-f(x)}{\sqrt{1-[f(x)]^2}} \, dx = \frac{-u}{\sqrt{1-u^2}} \, dx;$$

hence,

$$dx = -\frac{\sqrt{1-u^2}}{u} \, du,$$

and

$$V = \pi \int_{1}^{c} u^2 \left(-\frac{\sqrt{1-u^2}}{u} \, du \right) = \pi \int_{c}^{1} u\sqrt{1-u^2} \, du.$$

Carrying out the integration, we find

$$V = -\frac{\pi}{3}(1-u^2)^{3/2}\Big|_{c}^{1} = \frac{\pi}{3}(1-c^2)^{3/2}.$$

55. Let R be the region in the unit circle lying above the cut with the line $y = mx + b$ (Figure 16). Assume the points where the line intersects the circle lie above the x-axis. Use the method of Exercise 54 to show that the solid obtained by rotating R about the x-axis has volume $V = \dfrac{\pi}{6} hd^2$, with h and d as in the figure.

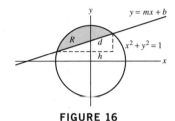

FIGURE 16

SOLUTION Let x_1 and x_2 denote the x-coordinates of the points of intersection between the circle $x^2 + y^2 = 1$ and the line $y = mx + b$ with $x_1 < x_2$. Rotating the region enclosed by the two curves about the x-axis produces a solid whose cross sections are washers with outer radius $R = \sqrt{1 - x^2}$ and inner radius $r = mx + b$. The volume of the resulting solid is then

$$V = \pi \int_{x_1}^{x_2} \left((1 - x^2) - (mx + b)^2 \right) dx$$

Because x_1 and x_2 are roots of the equation $(1 - x^2) - (mx + b)^2 = 0$ and $(1 - x^2) - (mx + b)^2$ is a quadratic polynomial in x with leading coefficient $-(1 + m^2)$, it follows that $(1 - x^2) - (mx + b)^2 = -(1 + m^2)(x - x_1)(x - x_2)$. Therefore,

$$V = -\pi(1 + m^2) \int_{x_1}^{x_2} (x - x_1)(x - x_2) \, dx = \frac{\pi}{6} (1 + m^2)(x_2 - x_1)^3.$$

From the diagram, we see that $h = x_2 - x_1$. Moreover, by the Pythagorean theorem, $d^2 = h^2 + (mh)^2 = (1 + m^2)h^2$. Thus,

$$V = \frac{\pi}{6}(1 + m^2)h^3 = \frac{\pi}{6} h \left[(1 + m^2)h^2 \right] = \frac{\pi}{6} hd^2.$$

6.4 The Method of Cylindrical Shells

Preliminary Questions

1. Consider the region $\mathcal{R}$ under the graph of the constant function $f(x) = h$ over the interval $[0, r]$. What are the height and radius of the cylinder generated when $\mathcal{R}$ is rotated about:

(a) the x-axis   **(b)** the y-axis

SOLUTION

(a) When the region is rotated about the x-axis, each shell will have radius h and height r.

(b) When the region is rotated about the y-axis, each shell will have radius r and height h.

2. Let V be the volume of a solid of revolution about the y-axis.

(a) Does the Shell Method for computing V lead to an integral with respect to x or y?

(b) Does the Disk or Washer Method for computing V lead to an integral with respect to x or y?

SOLUTION

(a) The Shell method requires slicing the solid parallel to the axis of rotation. In this case, that will mean slicing the solid in the vertical direction, so integration will be with respect to x.

(b) The Disk or Washer method requires slicing the solid perpendicular to the axis of rotation. In this case, that means slicing the solid in the horizontal direction, so integration will be with respect to y.

Exercises

In Exercises 1–10, sketch the solid obtained by rotating the region underneath the graph of the function over the given interval about the y-axis and find its volume.

1. $f(x) = x^3$, $[0, 1]$

SOLUTION   A sketch of the solid is shown below. Each shell has radius x and height x^3, so the volume of the solid is

$$2\pi \int_0^1 x \cdot x^3 \, dx = 2\pi \int_0^1 x^4 \, dx = 2\pi \left(\frac{1}{5}x^5\right)\Big|_0^1 = \frac{2}{5}\pi.$$

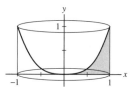

3. $f(x) = 3x + 2$, $[2, 4]$

SOLUTION   A sketch of the solid is shown below. Each shell has radius x and height $3x + 2$, so the volume of the solid is

$$2\pi \int_2^4 x(3x + 2) \, dx = 2\pi \int_2^4 (3x^2 + 2x) \, dx = 2\pi \left(x^3 + x^2\right)\Big|_2^4 = 136\pi.$$

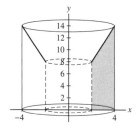

5. $f(x) = 4 - x^2$, $[0, 2]$

SOLUTION   A sketch of the solid is shown below. Each shell has radius x and height $4 - x^2$, so the volume of the solid is

$$2\pi \int_0^2 x(4 - x^2) \, dx = 2\pi \int_0^2 (4x - x^3) \, dx = 2\pi \left(2x^2 - \frac{1}{4}x^4\right)\Big|_0^2 = 8\pi.$$

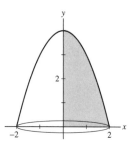

7. $f(x) = \sin(x^2)$, $[0, \sqrt{\pi}]$

SOLUTION   A sketch of the solid is shown below. Each shell has radius x and height $\sin x^2$, so the volume of the solid is

$$2\pi \int_0^{\sqrt{\pi}} x \sin(x^2) \, dx.$$

Let $u = x^2$. Then $du = 2x \, dx$ and

$$2\pi \int_0^{\sqrt{\pi}} x \sin(x^2) \, dx = \pi \int_0^{\pi} \sin u \, du = -\pi \left(\cos u\right)\Big|_0^{\pi} = 2\pi.$$

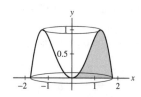

9. $f(x) = x + 1 - 2x^2$, [0, 1]

SOLUTION A sketch of the solid is shown below. Each shell has radius x and height $x + 1 - 2x^2$, so the volume of the solid is

$$2\pi \int_0^1 x(x + 1 - 2x^2)\, dx = 2\pi \int_0^1 (x^2 + x - 2x^3)\, dx = 2\pi \left(\frac{1}{3}x^3 + \frac{1}{2}x^2 - \frac{1}{2}x^4\right)\Big|_0^1 = \frac{2}{3}\pi.$$

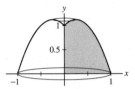

In Exercises 11–14, use the Shell Method to compute the volume of the solids obtained by rotating the region enclosed by the graphs of the functions about the y-axis.

11. $y = x^2$, $y = 8 - x^2$, $x = 0$

SOLUTION The region enclosed by $y = x^2$, $y = 8 - x^2$ and the y-axis is shown below. When rotating this region about the y-axis, each shell has radius x and height $8 - x^2 - x^2 = 8 - 2x^2$. The volume of the resulting solid is

$$2\pi \int_0^2 x(8 - 2x^2)\, dx = 2\pi \int_0^2 (8x - 2x^3)\, dx = 2\pi \left(4x^2 - \frac{1}{2}x^4\right)\Big|_0^2 = 16\pi.$$

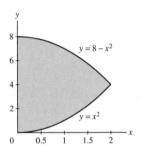

13. $y = \sqrt{x}$, $y = x^2$

SOLUTION The region enclosed by $y = \sqrt{x}$ and $y = x^2$ is shown below. When rotating this region about the y-axis, each shell has radius x and height $\sqrt{x} - x^2$. The volume of the resulting solid is

$$2\pi \int_0^1 x(\sqrt{x} - x^2)\, dx = 2\pi \int_0^1 (x^{3/2} - x^3)\, dx = 2\pi \left(\frac{2}{5}x^{5/2} - \frac{1}{4}x^4\right)\Big|_0^1 = \frac{3}{10}\pi.$$

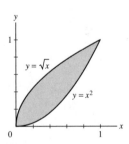

$\boxed{GU}$ *In Exercises 15–16, use the Shell Method to compute the volume of rotation of the region enclosed by the curves about the y-axis. Use a computer algebra system or graphing utility to find the points of intersection numerically.*

15. $y = \frac{1}{2}x^2$, $y = \sin(x^2)$

SOLUTION The region enclosed by $y = \frac{1}{2}x^2$ and $y = \sin x^2$ is shown below. When rotating this region about the y-axis, each shell has radius x and height $\sin x^2 - \frac{1}{2}x^2$. Using a computer algebra system, we find that the x-coordinate of the point of intersection on the right is $x = 1.376769504$. Thus, the volume of the resulting solid of revolution is

$$2\pi \int_0^{1.376769504} x\left(\sin x^2 - \frac{1}{2}x^2\right)\, dx = 1.321975576.$$

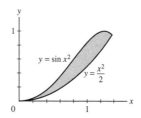

In Exercises 17–22, sketch the solid obtained by rotating the region underneath the graph of the function over the interval about the given axis and calculate its volume using the Shell Method.

17. $f(x) = x^3$, $[0, 1]$, $x = 2$

SOLUTION A sketch of the solid is shown below. Each shell has radius $2 - x$ and height x^3, so the volume of the solid is

$$2\pi \int_0^1 (2 - x)\left(x^3\right) dx = 2\pi \int_0^1 (2x^3 - x^4)\, dx = 2\pi \left(\frac{x^4}{2} - \frac{x^5}{5}\right)\bigg|_0^1 = \frac{3\pi}{5}.$$

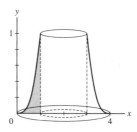

19. $f(x) = x^{-4}$, $[-3, -1]$, $x = 4$

SOLUTION A sketch of the solid is shown below. Each shell has radius $4 - x$ and height x^{-4}, so the volume of the solid is

$$2\pi \int_{-3}^{-1} (4 - x)\left(x^{-4}\right) dx = 2\pi \int_{-3}^{-1} (4x^{-4} - x^{-3})\, dx = 2\pi \left(\frac{1}{2}x^{-2} - \frac{4}{3}x^{-3}\right)\bigg|_{-3}^{-1} = \frac{280\pi}{81}.$$

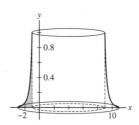

21. $f(x) = a - bx$, $[0, a/b]$, $x = -1$, $a, b > 0$

SOLUTION A sketch of the solid is shown below. Each shell has radius $x - (-1) = x + 1$ and height $a - bx$, so the volume of the solid is

$$2\pi \int_0^{a/b} (x + 1)(a - bx)\, dx = 2\pi \int_0^{a/b} \left(a + (a - b)x - bx^2\right) dx$$

$$= 2\pi \left(ax + \frac{a - b}{2}x^2 - \frac{b}{3}x^3\right)\bigg|_0^{a/b}$$

$$= 2\pi \left(\frac{a^2}{b} + \frac{a^2(a - b)}{2b^2} - \frac{a^3}{3b^2}\right) = \frac{a^2(a + 3b)}{3b^2}\pi.$$

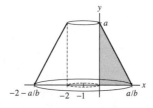

In Exercises 23–28, use the Shell Method to calculate the volume of rotation about the x-axis for the region underneath the graph.

23. $y = x, \quad 0 \le x \le 1$

SOLUTION When the region shown below is rotated about the x-axis, each shell has radius y and height $1 - y$. The volume of the resulting solid is

$$2\pi \int_0^1 y(1 - y)\, dy = 2\pi \int_0^1 (y - y^2)\, dy = 2\pi \left(\frac{1}{2} y^2 - \frac{1}{3} y^3 \right) \Big|_0^1 = \frac{\pi}{3}.$$

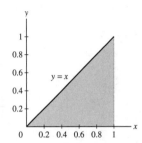

25. $y = x^{1/3} - 2, \quad 8 \le x \le 27$

SOLUTION When the region shown below is rotated about the x-axis, each shell has radius y and height $27 - (y + 2)^3$. The volume of the resulting solid is

$$2\pi \int_0^1 y \cdot \left(27 - (y + 2)^3 \right) dy = 2\pi \int_0^1 \left(19y - 12y^2 - 6y^3 - y^4 \right) dy$$

$$= 2\pi \left(\frac{19}{2} y^2 - 4y^3 - \frac{3}{2} y^4 - \frac{1}{5} y^5 \right) \Big|_0^1 = \frac{38\pi}{5}.$$

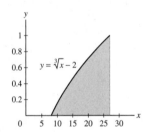

27. $y = x^{-2}, \quad 2 \le x \le 4$

SOLUTION When the region shown below is rotated about the x-axis, two different shells are generated. For each $y \in [0, \frac{1}{16}]$, the shell has radius y and height $4 - 2 = 2$; for each $y \in [\frac{1}{16}, \frac{1}{4}]$, the shell has radius y and height $\frac{1}{\sqrt{y}} - 2$. The volume of the resulting solid is

$$2\pi \int_0^{1/16} 2y\, dy + 2\pi \int_{1/16}^{1/4} y(y^{-1/2} - 2)\, dy = 2\pi \int_0^{1/16} 2y\, dy + 2\pi \int_{1/16}^{1/4} (y^{1/2} - 2y)\, dy$$

$$= 2\pi \left(y^2 \right) \Big|_0^{1/16} + 2\pi \left(\frac{2}{3} y^{3/2} - y^2 \right) \Big|_{1/16}^{1/4}$$

$$= \frac{\pi}{128} + \frac{11\pi}{384} = \frac{7\pi}{192}.$$

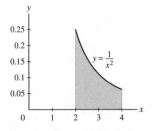

29. Use both the Shell and Disk Methods to calculate the volume of the solid obtained by rotating the region under the graph of $f(x) = 8 - x^3$ for $0 \le x \le 2$ about:

(a) the x-axis

(b) the y-axis

SOLUTION

(a) x-axis: Using the disk method, the cross sections are disks with radius $R = 8 - x^3$; hence the volume of the solid is

$$\pi \int_0^2 (8 - x^3)^2 \, dx = \pi \left(64x - 4x^4 + \frac{1}{7}x^7 \right) \Big|_0^2 = \frac{576\pi}{7}.$$

With the shell method, each shell has radius y and height $(8 - y)^{1/3}$. The volume of the solid is

$$2\pi \int_0^8 y \, (8 - y)^{1/3} \, dy$$

Let $u = 8 - y$. Then $dy = -du$, $y = 8 - u$ and

$$2\pi \int_0^8 y \, (8 - y)^{1/3} \, dy = 2\pi \int_0^8 (8 - u) \cdot u^{1/3} \, du = 2\pi \int_0^8 (8u^{1/3} - u^{4/3}) \, du$$

$$= 2\pi \left(6u^{4/3} - \frac{3}{7}u^{7/3} \right) \Big|_0^8 = \frac{576\pi}{7}.$$

(b) y-axis: With the shell method, each shell has radius x and height $8 - x^3$. The volume of the solid is

$$2\pi \int_0^2 x(8 - x^3) \, dx = 2\pi \left(4x^2 - \frac{1}{5}x^5 \right) \Big|_0^2 = \frac{96\pi}{5}.$$

Using the disk method, the cross sections are disks with radius $R = (8 - y)^{1/3}$. The volume is then given by

$$\pi \int_0^8 (8 - y)^{2/3} \, dy = -\frac{3\pi}{5}(8 - y)^{5/3} \Big|_0^8 = \frac{96\pi}{5}.$$

31. Assume that the graph in Figure 9(A) can be described by both $y = f(x)$ and $x = h(y)$. Let V be the volume of the solid obtained by rotating the region under the curve about the y-axis.

(a) Describe the figures generated by rotating segments $\overline{AB}$ and $\overline{CB}$ about the y-axis.

(b) Set up integrals that compute V by the Shell and Disk Methods.

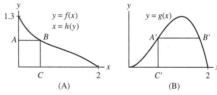

FIGURE 9

SOLUTION

(a) When rotated about the y-axis, the segment $\overline{AB}$ generates a disk with radius $R = h(y)$ and the segment $\overline{CB}$ generates a shell with radius x and height $f(x)$.

(b) Based on Figure 9(A) and the information from part (a), when using the Shell Method,

$$V = 2\pi \int_0^2 x f(x) \, dx;$$

when using the Disk Method,

$$V = \pi \int_0^{1.3} (h(y))^2 \, dy.$$

In Exercises 33–38, use the Shell Method to find the volume of the solid obtained by rotating region A in Figure 10 about the given axis.

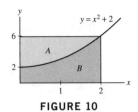

FIGURE 10

33. *y*-axis

SOLUTION   When rotating region *A* about the *y*-axis, each shell has radius *x* and height $6 - (x^2 + 2) = 4 - x^2$. The volume of the resulting solid is

$$2\pi \int_0^2 x(4 - x^2)\,dx = 2\pi \int_0^2 (4x - x^3)\,dx = 2\pi \left(2x^2 - \frac{1}{4}x^4\right)\Big|_0^2 = 8\pi.$$

35. *x* = 2

SOLUTION   When rotating region *A* about *x* = 2, each shell has radius $2 - x$ and height $6 - (x^2 + 2) = 4 - x^2$. The volume of the resulting solid is

$$2\pi \int_0^2 (2 - x)\left(4 - x^2\right)\,dx = 2\pi \int_0^2 \left(8 - 2x^2 - 4x + x^3\right)\,dx = 2\pi \left(8x - \frac{2}{3}x^3 - 2x^2 + \frac{1}{4}x^4\right)\Big|_0^2 = \frac{40\pi}{3}.$$

37. *y* = −2

SOLUTION   When rotating region *A* about *y* = −2, each shell has radius $y - (-2) = y + 2$ and height $\sqrt{y - 2}$. The volume of the resulting solid is

$$2\pi \int_2^6 (y + 2)\sqrt{y - 2}\,dy$$

Let $u = y - 2$. Then $du = dy$, $y + 2 = u + 4$ and

$$2\pi \int_2^6 (y + 2)\sqrt{y - 2}\,dy = 2\pi \int_0^4 (u + 4)\sqrt{u}\,du = 2\pi \left(\frac{2}{5}u^{5/2} + \frac{8}{3}u^{3/2}\right)\Big|_0^4 = \frac{1024\pi}{15}.$$

In Exercises 39–44, use the Shell Method to find the volumes of the solids obtained by rotating region B in Figure 10 about the given axis.

39. *y*-axis

SOLUTION   When rotating region *B* about the *y*-axis, each shell has radius *x* and height $x^2 + 2$. The volume of the resulting solid is

$$2\pi \int_0^2 x(x^2 + 2)\,dx = 2\pi \int_0^2 (x^3 + 2x)\,dx = 2\pi \left(\frac{1}{4}x^4 + x^2\right)\Big|_0^2 = 16\pi.$$

41. *x* = 2

SOLUTION   When rotating region *B* about *x* = 2, each shell has radius $2 - x$ and height $x^2 + 2$. The volume of the resulting solid is

$$2\pi \int_0^2 (2 - x)\left(x^2 + 2\right)\,dx = 2\pi \int_0^2 \left(2x^2 - x^3 + 4 - 2x\right)\,dx = 2\pi \left(\frac{2}{3}x^3 - \frac{1}{4}x^4 + 4x - x^2\right)\Big|_0^2 = \frac{32\pi}{3}.$$

43. *y* = −2

SOLUTION   When rotating region *B* about *y* = −2, two different shells are generated. For each $y \in [0, 2]$, the resulting shell has radius $y - (-2) = y + 2$ and height 2; for each $y \in [2, 6]$, the resulting shell has radius $y - (-2) = y + 2$ and height $2 - \sqrt{y - 2}$. The volume of the solid is then

$$2\pi \int_0^2 2(y + 2)\,dy + 2\pi \int_2^6 (y + 2)(2 - \sqrt{y - 2})\,dy = 2\pi \int_0^6 2(y + 2)\,dy - 2\pi \int_2^6 (y + 2)\sqrt{y - 2}\,dy$$

$$= 120\pi - 2\pi \int_2^6 (y + 2)\sqrt{y - 2}\,dy.$$

In the remaining integral, let $u = y - 2$, so $du = dy$ and $y + 2 = u + 4$. Then

$$2\pi \int_2^6 (y+2)\sqrt{y-2}\, dy = 2\pi \int_0^4 (u+4)\sqrt{u}\, du = 2\pi \left(\frac{2}{5}u^{5/2} + \frac{8}{3}u^{3/2} \right)\Bigg|_0^4 = \frac{1024\pi}{15}.$$

Finally, the volume of the solid is

$$120\pi - \frac{1024\pi}{15} = \frac{776\pi}{15}.$$

45. Use the Shell Method to compute the volume of a sphere of radius r.

SOLUTION A sphere of radius r can be generated by rotating the region under the semicircle $y = \sqrt{r^2 - x^2}$ around the x-axis. Each shell has radius y and height

$$\sqrt{r^2 - y^2} - \left(-\sqrt{r^2 - y^2}\right) = 2\sqrt{r^2 - y^2}.$$

Thus, the volume of the sphere is

$$2\pi \int_0^r 2y\sqrt{r^2 - y^2}\, dy.$$

Let $u = r^2 - y^2$. Then $du = -2y\, dy$ and

$$2\pi \int_0^r 2y\sqrt{r^2 - y^2}\, dy = 2\pi \int_0^{r^2} \sqrt{u}\, du = 2\pi \left(\frac{2}{3}u^{3/2} \right)\Bigg|_0^{r^2} = \frac{4}{3}\pi r^3.$$

47. Use the Shell Method to compute the volume of the torus obtained by rotating the interior of the circle $(x-a)^2 + y^2 = r^2$ about the y-axis, where $a > r$. *Hint:* Evaluate the integral by interpreting part of it as the area of a circle.

SOLUTION When rotating the region enclosed by the circle $(x-a)^2 + y^2 = r^2$ about the y-axis each shell has radius x and height

$$\sqrt{r^2 - (x-a)^2} - \left(-\sqrt{r^2 - (x-a)^2}\right) = 2\sqrt{r^2 - (x-a)^2}.$$

The volume of the resulting torus is then

$$2\pi \int_{a-r}^{a+r} 2x\sqrt{r^2 - (x-a)^2}\, dx.$$

Let $u = x - a$. Then $du = dx$, $x = u + a$ and

$$2\pi \int_{a-r}^{a+r} 2x\sqrt{r^2 - (x-a)^2}\, dx = 2\pi \int_{-r}^r 2(u+a)\sqrt{r^2 - u^2}\, du$$

$$= 4\pi \int_{-r}^r u\sqrt{r^2 - u^2}\, du + 4a\pi \int_{-r}^r \sqrt{r^2 - u^2}\, du.$$

Now,

$$\int_{-r}^r u\sqrt{r^2 - u^2}\, du = 0$$

because the integrand is an odd function and the integration interval is symmetric with respect to zero. Moreover, the other integral is one-half the area of a circle of radius r; thus,

$$\int_{-r}^r \sqrt{r^2 - u^2}\, du = \frac{1}{2}\pi r^2.$$

Finally, the volume of the torus is

$$4\pi(0) + 4a\pi \left(\frac{1}{2}\pi r^2 \right) = 2\pi^2 ar^2.$$

49. Use the most convenient method to compute the volume of the solid obtained by rotating the region in Figure 12 about the axis:

(a) $x = 4$ **(b)** $y = -2$

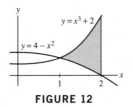

FIGURE 12

SOLUTION   Examine Figure 12. If the indicated region is sliced vertically, then the top of the slice lies along the curve $y = x^3 + 2$ and the bottom lies along the curve $y = 4 - x^2$. On the other hand, the left end of a horizontal slice switches from $y = 4 - x^2$ to $y = x^3 + 2$ at $y = 3$. Here, vertical slices will be more convenient.

(a) Now, suppose the region in Figure 12 is rotated about $x = 4$. Because a vertical slice is parallel to $x = 4$, we will calculate the volume of the resulting solid using the shell method. Each shell has radius $4 - x$ and height $x^3 + 2 - (4 - x^2) = x^3 + x^2 - 2$, so the volume is

$$2\pi \int_1^2 (4 - x)(x^3 + x^2 - 2)\, dx = 2\pi \left(-\frac{1}{5}x^5 + \frac{3}{4}x^4 + \frac{4}{3}x^3 + x^2 - 8x \right)\Big|_1^2 = \frac{563\pi}{30}.$$

(b) Now suppose the region is rotated about $y = -2$. Because a vertical slice is perpendicular to $y = -2$, we will calculate the volume of the resulting solid using the disk method. Each cross section is a washer with outer radius $R = x^3 + 2 - (-2) = x^3 + 4$ and inner radius $r = 4 - x^2 - (-2) = 6 - x^2$, so the volume is

$$\pi \int_1^2 \left((x^3 + 4)^2 - (6 - x^2)^2 \right) dx = \pi \left(\frac{1}{7}x^7 - \frac{1}{5}x^5 + 2x^4 + 4x^3 - 20x \right)\Big|_1^2 = \frac{1748\pi}{35}.$$

Further Insights and Challenges

51. Let R be the region bounded by the ellipse $\left(\frac{x}{a}\right)^2 + \left(\frac{y}{b}\right)^2 = 1$ (Figure 13). Show that the solid obtained by rotating R about the y-axis (called an **ellipsoid**) has volume $\frac{4}{3}\pi a^2 b$.

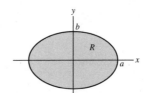

FIGURE 13 The ellipse $\left(\frac{x}{a}\right)^2 + \left(\frac{y}{b}\right)^2 = 1$.

SOLUTION   Let's slice the portion of the ellipse in the first and fourth quadrants horizontally and rotate the slices about the y-axis. The resulting ellipsoid has cross sections that are disks with radius

$$R = \sqrt{a^2 - \frac{a^2 y^2}{b^2}}.$$

Thus, the volume of the ellipsoid is

$$\pi \int_{-b}^{b} \left(a^2 - \frac{a^2 y^2}{b^2} \right) dy = \pi \left(a^2 y - \frac{a^2 y^3}{3b^2} \right)\Big|_{-b}^{b} = \pi \left[\left(a^2 b - \frac{a^2 b}{3} \right) - \left(-a^2 b + \frac{a^2 b}{3} \right) \right] = \frac{4}{3}\pi a^2 b.$$

6.5 Work and Energy

Preliminary Questions

1. Why is integration needed to compute the work performed in stretching a spring?

SOLUTION   Recall that the force needed to extend or compress a spring depends on the amount by which the spring has already been extended or compressed from its equilibrium position. In other words, the force needed to move a spring is variable. Whenever the force is variable, work needs to be computed with an integral.

2. Why is integration needed to compute the work performed in pumping water out of a tank but not to compute the work performed in lifting up the tank?

SOLUTION   To lift a tank through a vertical distance d, the force needed to move the tank remains constant; hence, no integral is needed to calculate the work done in lifting the tank. On the other hand, pumping water from a tank requires that different layers of the water be moved through different distances, and, depending on the shape of the tank, may require different forces. Thus, pumping water from a tank requires that an integral be evaluated.

3. Which of the following represents the work required to stretch a spring (with spring constant k) a distance x beyond its equilibrium position: kx, $-kx$, $\frac{1}{2}mk^2$, $\frac{1}{2}kx^2$, or $\frac{1}{2}mx^2$?

SOLUTION   The work required to stretch a spring with spring constant k a distance x beyond its equilibrium position is

$$\int_0^x ky \, dy = \frac{1}{2}ky^2 \Big|_0^x = \frac{1}{2}kx^2.$$

Exercises

1. How much work is done raising a 4-kg mass to a height of 16 m above ground?

SOLUTION   The force needed to lift a 4-kg object is a constant

$$(4 \text{ kg})(9.8 \text{ m/s}^2) = 39.2 \text{ N}.$$

The work done in lifting the object to a height of 16 m is then

$$(39.2 \text{ N})(16 \text{ m}) = 627.2 \text{ J}.$$

In Exercises 3–6, compute the work (in joules) required to stretch or compress a spring as indicated, assuming that the spring constant is $k = 150$ kg/s^2.

3. Stretching from equilibrium to 12 cm past equilibrium

SOLUTION   The work required to stretch the spring 12 cm past equilibrium is

$$\int_0^{.12} 150x \, dx = 75x^2 \Big|_0^{.12} = 1.08 \text{ J}.$$

5. Stretching from 5 to 15 cm past equilibrium

SOLUTION   The work required to stretch the spring from 5 cm to 15 cm past equilibrium is

$$\int_{.05}^{.15} 150x \, dx = 75x^2 \Big|_{.05}^{.15} = 1.5 \text{ J}.$$

7. If 5 J of work are needed to stretch a spring 10 cm beyond equilibrium, how much work is required to stretch it 15 cm beyond equilibrium?

SOLUTION   First, we determine the value of the spring constant as follows:

$$\int_0^{0.1} kx \, dx = \frac{1}{2}kx^2 \Big|_0^{0.1} = 0.005k = 5 \text{ J}.$$

Thus, $k = 1000$ kg/s^2. Next, we calculate the work required to stretch the spring 15 cm beyond equilibrium:

$$\int_0^{0.15} 1000x \, dx = 500x^2 \Big|_0^{0.15} = 11.25 \text{ J}.$$

9. If 10 ft-lb of work are needed to stretch a spring 1 ft beyond equilibrium, how far will the spring stretch if a 10-lb weight is attached to its end?

SOLUTION   First, we determine the value of the spring constant as follows:

$$\int_0^1 kx \, dx = \frac{1}{2}kx^2 \Big|_0^1 = \frac{1}{2}k = 10 \text{ ft-lb}.$$

Thus $k = 20$ lb/ft. Balancing the forces acting on the weight, we have $10 \text{ lb} = kd = 20d$, which implies $d = 0.5$ ft. A 10-lb weight will therefore stretch the spring 6 inches.

In Exercises 11–14, calculate the work against gravity required to build the structure out of brick using the method of Examples 2 and 3. Assume that brick has density 80 lb/ft^3.

11. A tower of height 20 ft and square base of side 10 ft

SOLUTION The volume of one layer is $100\Delta y$ ft^3 and so the weight of one layer is $8000\Delta y$ lb. Thus, the work done against gravity to build the tower is

$$W = \int_0^{20} 8000y\, dy = 4000y^2 \Big|_0^{20} = 1.6 \times 10^6 \text{ ft-lb.}$$

13. A 20-ft-high tower in the shape of a right circular cone with base of radius 4 ft

SOLUTION From similar triangles, the area of one layer is $\pi\left(4 - \frac{y}{5}\right)^2$ ft^2, so the volume of each small layer is $\pi\left(4 - \frac{y}{5}\right)^2 \Delta y$ ft^3. The weight of one layer is then $80\pi\left(4 - \frac{y}{5}\right)^2 \Delta y$ lb. Finally, the total work done against gravity to build the tower is

$$\int_0^{20} 80\pi\left(4 - \frac{y}{5}\right)^2 y\, dy = \frac{128,000\pi}{3} \text{ ft-lb.}$$

15. Built around 2600 BCE, the Great Pyramid of Giza in Egypt is 485 ft high (due to erosion, its current height is slightly less) and has a square base of side 755.5 ft (Figure 6). Find the work needed to build the pyramid if the density of the stone is estimated at 125 lb/ft^3.

FIGURE 6 The Great Pyramid in Giza, Egypt.

SOLUTION From similar triangles, the area of one layer is

$$\left(755.5 - \frac{755.5}{485}y\right)^2 \text{ ft}^2,$$

so the volume of each small layer is

$$\left(755.5 - \frac{755.5}{485}y\right)^2 \Delta y \text{ ft}^3.$$

The weight of one layer is then

$$125\left(755.5 - \frac{755.5}{485}y\right)^2 \Delta y \text{ lb.}$$

Finally, the total work needed to build the pyramid was

$$\int_0^{485} 125\left(755.5 - \frac{755.5}{485}y\right)^2 y\, dy = 1.399 \times 10^{12} \text{ ft-lb.}$$

In Exercises 16–20, calculate the work (in joules) required to pump all of the water out of the tank. Assume that the tank is full, distances are measured in meters, and the density of water is 1,000 kg/m^3.

17. The hemisphere in Figure 8; water exits from the spout as shown.

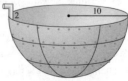

FIGURE 8

SOLUTION Place the origin at the center of the hemisphere, and let the positive y-axis point downward. The radius of a layer of water at depth y is $\sqrt{100 - y^2}$ m, so the volume of the layer is $\pi(100 - y^2)\Delta y$ m^3, and the force needed to lift the layer is $9800\pi(100 - y^2)\Delta y$ N. The layer must be lifted $y + 2$ meters, so the total work needed to empty the tank is

$$\int_0^{10} 9800\pi(100 - y^2)(y + 2)\,dy = \frac{112700000\pi}{3} \text{ J} \approx 1.18 \times 10^8 \text{ J}.$$

19. The horizontal cylinder in Figure 10; water exits from a small hole at the top. *Hint:* Evaluate the integral by interpreting part of it as the area of a circle.

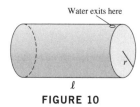

Water exits here

ℓ

FIGURE 10

SOLUTION Place the origin along the axis of the cylinder. At location y, the layer of water is a rectangular slab of length ℓ, width $2\sqrt{r^2 - y^2}$ and thickness Δy. Thus, the volume of the layer is $2\ell\sqrt{r^2 - y^2}\Delta y$, and the force needed to lift the layer is $19600\ell\sqrt{r^2 - y^2}\Delta y$. The layer must be lifted a distance $r - y$, so the total work needed to empty the tank is given by

$$\int_{-r}^{r} 19600\ell\sqrt{r^2 - y^2}(r - y)\,dy = 19600\ell r\int_{-r}^{r}\sqrt{r^2 - y^2}\,dy - 19600\ell\int_{-r}^{r} y\sqrt{r^2 - y^2}\,dy.$$

Now,

$$\int_{-r}^{r} y\sqrt{r^2 - y^2}\,du = 0$$

because the integrand is an odd function and the integration interval is symmetric with respect to zero. Moreover, the other integral is one-half the area of a circle of radius r; thus,

$$\int_{-r}^{r}\sqrt{r^2 - y^2}\,dy = \frac{1}{2}\pi r^2.$$

Finally, the total work needed to empty the tank is

$$19600\ell r\left(\frac{1}{2}\pi r^2\right) - 19600\ell(0) = 9800\ell\pi r^3 \text{ J}.$$

21. Find the work W required to empty the tank in Figure 7 if it is half full of water.

SOLUTION Place the origin on the top of the box, and let the positive y-axis point downward. Note that with this coordinate system, the bottom half of the box corresponds to y values from 2.5 to 5. The volume of one layer of water is $32\Delta y$ m^3, so the force needed to lift each layer is

$$(9.8)(1000)32\Delta y = 313600\Delta y \text{ N}.$$

Each layer must be lifted y meters, so the total work needed to empty the tank is

$$\int_{2.5}^{5} 313600y\,dy = 156800y^2\Big|_{2.5}^{5} = 2.94 \times 10^6 \text{ J}.$$

23. Find the work required to empty the tank in Figure 9 if it is half full of water.

SOLUTION Place the origin at the vertex of the inverted cone, and let the positive y-axis point upward. Consider a layer of water at a height of y meters. From similar triangles, the area of the layer is

$$\pi\left(\frac{y}{2}\right)^2 \text{ m}^2,$$

so the volume is

$$\pi\left(\frac{y}{2}\right)^2 \Delta y \text{ m}^3.$$

Thus the weight of one layer is

$$9800\pi \left(\frac{y}{2}\right)^2 \Delta y \text{ N.}$$

The layer must be lifted $12 - y$ meters, so the total work needed to empty the half-full tank is

$$\int_0^5 9800\pi \left(\frac{y}{2}\right)^2 (12 - y)\, dy = \frac{1684375\pi}{2} \text{ J} \approx 2.65 \times 10^6 \text{ J.}$$

25. Assume that the tank in Figure 9 is full.

(a) Calculate the work $F(y)$ required to pump out water until the water level has reached level y.

(b) *CAS* Plot $F(y)$.

(c) What is the significance of $F'(y)$ as a rate of change?

(d) *CAS* If your goal is to pump out all of the water, at which water level y_0 will half of the work be done?

SOLUTION

(a) Place the origin at the vertex of the inverted cone, and let the positive y-axis point upward. Consider a layer of water at a height of y meters. From similar triangles, the area of the layer is

$$\pi \left(\frac{y}{2}\right)^2 \text{ m}^2,$$

so the volume is

$$\pi \left(\frac{y}{2}\right)^2 \Delta y \text{ m}^3.$$

Thus the weight of one layer is

$$9800\pi \left(\frac{y}{2}\right)^2 \Delta y \text{ N.}$$

The layer must be lifted $12 - y$ meters, so the total work needed to pump out water until the water level has reached level y is

$$\int_y^{10} 9800\pi \left(\frac{y}{2}\right)^2 (12 - y)\, dy = 3675000\pi - 9800\pi y^3 + \frac{1225\pi}{2} y^4 \text{ J.}$$

(b) A plot of $F(y)$ is shown below.

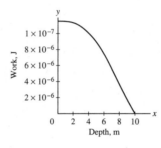

(c) First, note that $F'(y) < 0$; as y increases, less water is being pumped from the tank, so $F(y)$ decreases. Therefore, when the water level in the tank has reached level y, we can interpret $-F'(y)$ as the amount of work per meter needed to remove the next layer of water from the tank. In other words, $-F'(y)$ is a "marginal work" function.

(d) The amount of work needed to empty the tank is 3675000π J. Half of this work will be done when the water level reaches height y_0 satisfying

$$3675000\pi - 9800\pi y_0^3 + \frac{1225\pi}{2} y_0^4 = 1837500\pi.$$

Using a computer algebra system, we find $y_0 = 6.91$ m.

27. How much work is done lifting a 3-m chain over the side of a building if the chain has mass density 4 kg/m?

SOLUTION Consider a segment of the chain of length Δy located a distance y_j meters from the top of the building. The work needed to lift this segment of the chain to the top of the building is approximately

$$W_j \approx (4\Delta y)(9.8)y_j \text{ J.}$$

Summing over all segments of the chain and passing to the limit as $\Delta y \to 0$, it follows that the total work is

$$\int_0^3 4 \cdot 9.8y\, dy = 19.6y^2 \Big|_0^3 = 176.4 \text{ J}.$$

29. A 20-foot chain with mass density 3 lb/ft is initially coiled on the ground. How much work is performed in lifting the chain so that it is fully extended (and one end touches the ground)?

SOLUTION   Consider a segment of the chain of length Δy that must be lifted y_j feet off the ground. The work needed to lift this segment of the chain is approximately

$$W_j \approx (3\Delta y)y_j \text{ ft-lb}.$$

Summing over all segments of the chain and passing to the limit as $\Delta y \to 0$, it follows that the total work is

$$\int_0^{20} 3y\, dy = \frac{3}{2}y^2 \Big|_0^{20} = 600 \text{ ft-lb}.$$

31. A 1,000-lb wrecking ball hangs from a 30-ft cable of density 10 lb/ft attached to a crane. Calculate the work done if the crane lifts the ball from ground level to 30 ft in the air by drawing in the cable.

SOLUTION   We will treat the cable and the wrecking ball separately. Consider a segment of the cable of length Δy that must be lifted y_j feet. The work needed to lift the cable segment is approximately

$$W_j \approx (10\Delta y)y_j \text{ ft-lb}.$$

Summing over all of the segments of the cable and passing to the limit as $\Delta y \to 0$, it follows that lifting the cable requires

$$\int_0^{30} 10y\, dy = 5y^2 \Big|_0^{30} = 4500 \text{ ft-lb}.$$

Lifting the 1000 lb wrecking ball 30 feet requires an additional 30,000 ft-lb. Thus, the total work is 34,500 ft-lb.

In Exercises 32–34, use Newton's Universal Law of Gravity, according to which the gravitational force between two objects of mass m and M separated by a distance r has magnitude GMm/r^2, where $G = 6.67 \times 10^{-11} \text{ m}^3\text{kg}^{-1}\text{s}^{-1}$. Although the Universal Law refers to point masses, Newton proved that it also holds for uniform spherical objects, where r is the distance between their centers.

33. Use the result of Exercise 32 to calculate the work required to place a 2,000-kg satellite in an orbit 1,200 km above the surface of the earth. Assume that the earth is a sphere of mass $M_e = 5.98 \times 10^{24}$ kg and radius $r_e = 6.37 \times 10^6$ m. Treat the satellite as a point mass.

SOLUTION   The satellite will move from a distance $r_1 = r_e$ to a distance $r_2 = r_e + 1200000$. Thus, from Exercise 32,

$$W = (6.67 \times 10^{-11})(5.98 \times 10^{24})(2000)\left(\frac{1}{6.37 \times 10^6} - \frac{1}{6.37 \times 10^6 + 1200000}\right) \approx 1.99 \times 10^{10} \text{ J}.$$

35. Assume that the pressure P and volume V of the gas in a 30-in. cylinder of radius 3 in. with a movable piston are related by $PV^{1.4} = k$, where k is a constant (Figure 13). When the cylinder is full, the gas pressure is 200 lb/in.^{2}.
(a) Calculate k.
(b) Calculate the force on the piston as a function of the length x of the column of gas (the force is PA, where A is the piston's area).
(c) Calculate the work required to compress the gas column from 30 to 20 in.

FIGURE 13 Gas in a cylinder with a piston.

SOLUTION
(a) We have $P = 200$ and $V = 270\pi$. Thus

$$k = 200(270\pi)^{1.4} = 2.517382 \times 10^6 \text{ lb-in}^{2.2}.$$

(b) The area of the piston is $A = 9\pi$ and the volume of the cylinder as a function of x is $V = 9\pi x$, which gives $P = k/V^{1.4} = k/(9\pi x)^{1.4}$. Thus

$$F = PA = \frac{k}{(9\pi x)^{1.4}} 9\pi = k(9\pi)^{-0.4} x^{-1.4}.$$

(c) Since the force is pushing against the piston, in order to calculate work, we must calculate the integral of the opposite force, i.e., we have

$$W = -k(9\pi)^{-0.4} \int_{30}^{20} x^{-1.4}\, dx = -k(9\pi)^{-0.4} \frac{1}{-0.4} x^{-0.4} \Big|_{30}^{20} = 74677.8 \text{ in-lb}.$$

Further Insights and Challenges

37. Work-Kinetic Energy Theorem The **kinetic energy** of an object of mass m moving with velocity v is KE $= \frac{1}{2}mv^2$.

(a) Suppose that the object moves from x_1 to x_2 during the time interval $[t_1, t_2]$ due to a net force $F(x)$ acting along the interval $[x_1, x_2]$. Let $x(t)$ be the position of the object at time t. Use the Change of Variables formula to show that the work performed is equal to

$$W = \int_{x_1}^{x_2} F(x)\, dx = \int_{t_1}^{t_2} F(x(t))v(t)\, dt$$

(b) By Newton's Second Law, $F(x(t)) = ma(t)$, where $a(t)$ is the acceleration at time t. Show that

$$\frac{d}{dt}\left(\frac{1}{2}mv(t)^2\right) = F(x(t))v(t)$$

(c) Use the FTC to show that the change in kinetic energy during the time interval $[t_1, t_2]$ is equal to

$$\int_{t_1}^{t_2} F(x(t))v(t)\, dt$$

(d) Prove the Work-Kinetic Energy Theorem: The change in KE is equal to the work W performed.

SOLUTION

(a) Let $x_1 = x(t_1)$ and $x_2 = x(t_2)$, then $x = x(t)$ gives $dx = v(t)\, dt$. By substitution we have

$$W = \int_{x_1}^{x_2} F(x)\, dx = \int_{t_1}^{t_2} F(x(t))v(t)\, dt.$$

(b) Knowing $F(x(t)) = m \cdot a(t)$, we have

$$\frac{d}{dt}\left(\frac{1}{2}m \cdot v(t)^2\right) = m \cdot v(t)\, v'(t) \qquad \text{(Chain Rule)}$$

$$= m \cdot v(t)\, a(t)$$

$$= v(t) \cdot F(x(t)) \qquad \text{(Newton's 2nd law)}$$

(c) From the FTC,

$$\frac{1}{2}m \cdot v(t)^2 = \int F(x(t))\, v(t)\, dt.$$

Since $KE = \frac{1}{2}m v^2$,

$$\Delta KE = KE(t_2) - KE(t_1) = \frac{1}{2}m\, v(t_2)^2 - \frac{1}{2}m\, v(t_1)^2 = \int_{t_1}^{t_2} F(x(t))\, v(t)\, dt.$$

(d)

$$W = \int_{x_1}^{x_2} F(x)\, dx = \int_{t_1}^{t_2} F(x(t))\, v(t)\, dt \qquad \text{(Part (a))}$$

$$= KE(t_2) - KE(t_1) \qquad \text{(Part (c))}$$

$$= \Delta KE \qquad \text{(as required)}$$

39. With what initial velocity v_0 must we fire a rocket so it attains a maximum height r above the earth? *Hint:* Use the results of Exercises 32 and 37. As the rocket reaches its maximum height, its KE decreases from $\frac{1}{2}mv_0^2$ to zero.

SOLUTION   The work required to move the rocket a distance r from the surface of the earth is

$$W(r) = GM_e m \left(\frac{1}{r_e} - \frac{1}{r + r_e} \right).$$

As the rocket climbs to a height r, its kinetic energy is reduced by the amount $W(r)$. The rocket reaches its maximum height when its kinetic energy is reduced to zero, that is, when

$$\frac{1}{2}mv_0^2 = GM_e m \left(\frac{1}{r_e} - \frac{1}{r + r_e} \right).$$

Therefore, its initial velocity must be

$$v_0 = \sqrt{2GM_e \left(\frac{1}{r_e} - \frac{1}{r + r_e} \right)}.$$

41. Calculate **escape velocity,** the minimum initial velocity of an object to ensure that it will continue traveling into space and never fall back to earth (assuming that no force is applied after takeoff). *Hint:* Take the limit as $r \to \infty$ in Exercise 39.

SOLUTION   The result of the previous exercise leads to an interesting conclusion. The initial velocity v_0 required to reach a height r does not increase beyond all bounds as r tends to infinity; rather, it approaches a finite limit, called the escape velocity:

$$v_{esc} = \lim_{r \to \infty} \sqrt{2GM_e \left(\frac{1}{r_e} - \frac{1}{r + r_e} \right)} = \sqrt{\frac{2GM_e}{r_e}}$$

In other words, v_{esc} is large enough to insure that the rocket reaches a height r for every value of r! Therefore, a rocket fired with initial velocity v_{esc} never returns to earth. It continues traveling indefinitely into outer space.

Now, let's see how large escape velocity actually is:

$$v_{esc} = \left(\frac{2 \cdot 6.67 \times 10^{-11} \cdot 5.989 \times 10^{24}}{6.37 \times 10^6} \right)^{1/2} \approx 11{,}190 \text{ m/sec.}$$

Since one meter per second is equal to 2.236 miles per hour, escape velocity is approximately $11{,}190(2.236) = 25{,}020$ miles per hour.

CHAPTER REVIEW EXERCISES

In Exercises 1–4, find the area of the region bounded by the graphs of the functions.

1. $y = \sin x, \quad y = \cos x, \quad 0 \le x \le \dfrac{5\pi}{4}$

SOLUTION   The region bounded by the graphs of $y = \sin x$ and $y = \cos x$ over the interval $[0, \frac{5\pi}{4}]$ is shown below. For $x \in [0, \frac{\pi}{4}]$, the graph of $y = \cos x$ lies above the graph of $y = \sin x$, whereas, for $x \in [\frac{\pi}{4}, \frac{5\pi}{4}]$, the graph of $y = \sin x$ lies above the graph of $y = \cos x$. The area of the region is therefore given by

$$\int_0^{\pi/4} (\cos x - \sin x)\, dx + \int_{\pi/4}^{5\pi/4} (\sin x - \cos x)\, dx$$

$$= (\sin x + \cos x)\Big|_0^{\pi/4} + (-\cos x - \sin x)\Big|_{\pi/4}^{5\pi/4}$$

$$= \frac{\sqrt{2}}{2} + \frac{\sqrt{2}}{2} - (0 + 1) + \left(\frac{\sqrt{2}}{2} + \frac{\sqrt{2}}{2} \right) - \left(-\frac{\sqrt{2}}{2} - \frac{\sqrt{2}}{2} \right) = 3\sqrt{2} - 1.$$

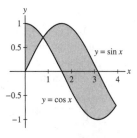

3. $f(x) = x^2 + 2x$, $\quad g(x) = x^2 - 1$, $\quad h(x) = x^2 + x - 2$

SOLUTION   The region bounded by the graphs of $y = x^2 + 2x$, $y = x^2 - 1$ and $y = x^2 + x - 2$ is shown below. For each $x \in [-2, -\frac{1}{2}]$, the graph of $y = x^2 + 2x$ lies above the graph of $y = x^2 + x - 2$, whereas, for each $x \in [-\frac{1}{2}, 1]$, the graph of $y = x^2 - 1$ lies above the graph of $y = x^2 + x - 2$. The area of the region is therefore given by

$$\int_{-2}^{-1/2} \left((x^2 + 2x) - (x^2 + x - 2) \right) dx + \int_{-1/2}^{1} \left((x^2 - 1) - (x^2 + x - 2) \right) dx$$

$$= \left(\frac{1}{2}x^2 + 2x \right)\Big|_{-2}^{-1/2} + \left(-\frac{1}{2}x^2 + x \right)\Big|_{-1/2}^{1}$$

$$= \left(\frac{1}{8} - 1 \right) - (2 - 4) + \left(-\frac{1}{2} + 1 \right) - \left(-\frac{1}{8} - \frac{1}{2} \right) = \frac{9}{4}.$$

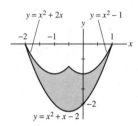

In Exercises 5–8, sketch the region bounded by the graphs of the functions and find its area.

5. $f(x) = x^3 - x^2 - x + 1$, $\quad g(x) = \sqrt{1 - x^2}$, $\quad 0 \le x \le 1$
   Hint: Use geometry to evaluate the integral.

SOLUTION   The region bounded by the graphs of $y = x^3 - x^2 - x + 1$ and $y = \sqrt{1 - x^2}$ is shown below. As the graph of $y = \sqrt{1 - x^2}$ lies above the graph of $y = x^3 - x^2 - x + 1$, the area of the region is given by

$$\int_0^1 \left(\sqrt{1 - x^2} - (x^3 - x^2 - x + 1) \right) dx.$$

Now, the region below the graph of $y = \sqrt{1 - x^2}$ but above the x-axis over the interval $[0, 1]$ is one-quarter of the unit circle; thus,

$$\int_0^1 \sqrt{1 - x^2}\, dx = \frac{1}{4}\pi.$$

Moreover,

$$\int_0^1 (x^3 - x^2 - x + 1)\, dx = \left(\frac{1}{4}x^4 - \frac{1}{3}x^3 - \frac{1}{2}x^2 + x \right)\Big|_0^1 = \frac{1}{4} - \frac{1}{3} - \frac{1}{2} + 1 = \frac{5}{12}.$$

Finally, the area of the region shown below is

$$\frac{1}{4}\pi - \frac{5}{12}.$$

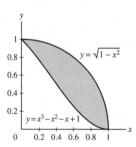

7. $y = 4 - x^2$, $\quad y = 3x$, $\quad y = 4$

SOLUTION   The region bounded by the graphs of $y = 4 - x^2$, $y = 3x$ and $y = 4$ is shown below. For $x \in [0, 1]$, the graph of $y = 4$ lies above the graph of $y = 4 - x^2$, whereas, for $x \in [1, \frac{4}{3}]$, the graph of $y = 4$ lies above the graph of $y = 3x$. The area of the region is therefore given by

$$\int_0^1 (4 - (4 - x^2))\, dx + \int_1^{4/3} (4 - 3x)\, dx = \frac{1}{3}x^3\Big|_0^1 + \left(4x - \frac{3}{2}x^2 \right)\Big|_1^{4/3} = \frac{1}{3} + \left(\frac{16}{3} - \frac{8}{3} \right) - \left(4 - \frac{3}{2} \right) = \frac{1}{2}.$$

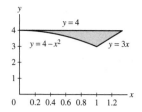

9. [GU] Use a graphing utility to locate the points of intersection of $y = e^{-x}$ and $y = 1 - x^2$ and find the area between the two curves (approximately).

SOLUTION The region bounded by the graphs of $y = e^{-x}$ and $y = 1 - x^2$ is shown below. One point of intersection clearly occurs at $x = 0$. Using a computer algebra system, we find that the other point of intersection occurs at $x = 0.7145563847$. As the graph of $y = 1 - x^2$ lies above the graph of $y = e^{-x}$, the area of the region is given by

$$\int_0^{0.7145563847} \left(1 - x^2 - e^{-x}\right) dx = 0.08235024596$$

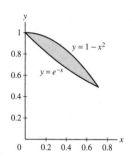

11. Find the total weight of a 3-ft metal rod of linear density

$$\rho(x) = 1 + 2x + \frac{2}{9}x^3 \text{ lb/ft}.$$

SOLUTION The total weight of the rod is

$$\int_0^3 \rho(x)\, dx = \left(x + x^2 + \frac{1}{18}x^4\right)\Big|_0^3 = 3 + 9 + \frac{9}{2} = \frac{33}{2} \text{ lb}.$$

In Exercises 13–16, find the average value of the function over the interval.

13. $f(x) = x^3 - 2x + 2$, $[-1, 2]$

SOLUTION The average value is

$$\frac{1}{2 - (-1)} \int_{-1}^2 \left(x^3 - 2x + 2\right) dx = \frac{1}{3}\left(\frac{1}{4}x^4 - x^2 + 2x\right)\Big|_{-1}^2 = \frac{1}{3}\left[(4 - 4 + 4) - \left(\frac{1}{4} - 1 - 2\right)\right] = \frac{9}{4}.$$

15. $f(x) = |x|$, $[-4, 4]$

SOLUTION The average value is

$$\frac{1}{4 - (-4)} \int_{-4}^4 |x|\, dx = \frac{1}{8}\left(\int_{-4}^0 (-x)\, dx + \int_0^4 x\, dx\right) = \frac{1}{8}\left(-\frac{1}{2}x^2\Big|_{-4}^0 + \frac{1}{2}x^2\Big|_0^4\right) = \frac{1}{8}[(0 + 8) + (8 - 0)] = 2.$$

17. The average value of $g(t)$ on $[2, 5]$ is 9. Find $\int_2^5 g(t)\, dt$.

SOLUTION The average value of the function $g(t)$ on $[2, 5]$ is given by

$$\frac{1}{5 - 2}\int_2^5 g(t)\, dt = \frac{1}{3}\int_2^5 g(t)\, dt.$$

Therefore,

$$\int_2^5 g(t)\, dt = 3(\text{average value}) = 3(9) = 27.$$

19. Use the Shell Method to find the volume of the solid obtained by revolving the region between $y = x^2$ and $y = mx$ about the x-axis (Figure 2).

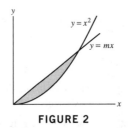

FIGURE 2

SOLUTION Setting $x^2 = mx$ yields $x(x - m) = 0$, so the two curves intersect at $(0, 0)$ and (m, m^2). To use the shell method, we must slice the solid parallel to the axis of rotation; as we are revolving about the x-axis, this implies a horizontal slice and integration in y. For each $y \in [0, m^2]$, the shell has radius y and height $\sqrt{y} - \frac{y}{m}$. The volume of the solid is therefore given by

$$2\pi \int_0^{m^2} y \left(\sqrt{y} - \frac{y}{m} \right) dy = 2\pi \left(\frac{2}{5} y^{5/2} - \frac{y^3}{3m} \right) \Bigg|_0^{m^2} = 2\pi \left(\frac{2m^5}{5} - \frac{m^5}{3} \right) = \frac{2\pi}{15} m^5.$$

21. Let R be the intersection of the circles of radius 1 centered at $(1, 0)$ and $(0, 1)$. Express as an integral (but do not evaluate): **(a)** the area of R and **(b)** the volume of revolution of R about the x-axis.

SOLUTION The region R is shown below.

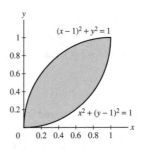

(a) A vertical slice of R has its top along the upper left arc of the circle $(x - 1)^2 + y^2 = 1$ and its bottom along the lower right arc of the circle $x^2 + (y - 1)^2 = 1$. The area of R is therefore given by

$$\int_0^1 \left(\sqrt{1 - (x - 1)^2} - (1 - \sqrt{1 - x^2}) \right) dx.$$

(b) If we revolve R about the x-axis and use the washer method, each cross section is a washer with outer radius $\sqrt{1 - (x - 1)^2}$ and inner radius $1 - \sqrt{1 - x^2}$. The volume of the solid is therefore given by

$$\pi \int_0^1 \left[(1 - (x - 1)^2) - (1 - \sqrt{1 - x^2})^2 \right] dx.$$

In Exercises 23–31, find the volume of the solid obtained by rotating the region enclosed by the curves about the given axis.

23. $y = 2x$, $y = 0$, $x = 8$; x-axis

SOLUTION The region bounded by the graphs of $y = 2x$, $y = 0$ and $x = 8$ is shown below. Let's choose to slice the region vertically. Because a vertical slice is perpendicular to the axis of rotation, we will use the washer method to calculate the volume of the solid of revolution. For each $x \in [0, 8]$, the cross section is a circular disk with radius $R = 2x$. The volume of the solid is therefore given by

$$\pi \int_0^8 (2x)^2 \, dx = \frac{4\pi}{3} x^3 \Bigg|_0^8 = \frac{2048\pi}{3}.$$

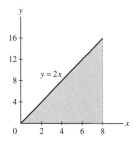

25. $y = x^2 - 1$, $y = 2x - 1$, axis $x = -2$

SOLUTION The region bounded by the graphs of $y = x^2 - 1$ and $y = 2x - 1$ is shown below. Let's choose to slice the region vertically. Because a vertical slice is parallel to the axis of rotation, we will use the shell method to calculate the volume of the solid of revolution. For each $x \in [0, 2]$, the shell has radius $x - (-2) = x + 2$ and height $(2x - 1) - (x^2 - 1) = 2x - x^2$. The volume of the solid is therefore given by

$$2\pi \int_0^2 (x + 2)(2x - x^2)\, dx = 2\pi \left(2x^2 - \frac{1}{4}x^4 \right)\Big|_0^2 = 2\pi(8 - 4) = 8\pi.$$

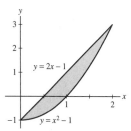

27. $y^2 = x^3$, $y = x$, $x = 8$; axis $x = -1$

SOLUTION The region bounded by the graphs of $y^2 = x^3$, $y = x$ and $x = 8$ is composed of two components, shown below. Let's choose to slice the region vertically. Because a vertical slice is parallel to the axis of rotation, we will use the shell method to calculate the volume of the solid of revolution. For each $x \in [0, 1]$, the shell as radius $x - (-1) = x + 1$ and height $x - x^{3/2}$; for each $x \in [1, 8]$, the shell also has radius $x + 1$, but the height is $x^{3/2} - x$. The volume of the solid is therefore given by

$$2\pi \int_0^1 (x + 1)(x - x^{3/2})\, dx + 2\pi \int_1^8 (x + 1)(x^{3/2} - x)\, dx = \frac{2\pi}{35}(12032\sqrt{2} - 7083).$$

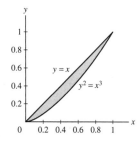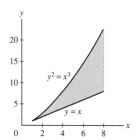

29. $y = -x^2 + 4x - 3$, $y = 0$; axis $y = -1$

SOLUTION The region bounded by the graph of $y = -x^2 + 4x - 3$ and the x-axis is shown below. Let's choose to slice the region vertically. Because a vertical slice is perpendicular to the axis of rotation, we will use the washer method to calculate the volume of the solid of revolution. For each $x \in [1, 3]$, the cross section is a washer with outer radius $R = -x^2 + 4x - 3 - (-1) = -x^2 + 4x - 2$ and inner radius $r = 0 - (-1) = 1$. The volume of the solid is therefore given by

$$\pi \int_1^3 \left((-x^2 + 4x - 2)^2 - 1 \right) dx = \pi \left(\frac{1}{5}x^5 - 2x^4 + \frac{20}{3}x^3 - 8x^2 + 3x \right)\Big|_1^3$$

$$= \pi \left[\left(\frac{243}{5} - 162 + 180 - 72 + 9 \right) - \left(\frac{1}{5} - 2 + \frac{20}{3} - 8 + 3 \right) \right] = \frac{56\pi}{15}.$$

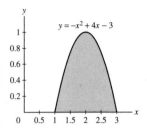

31. $y^2 = x^{-1}$, $x = 1$, $x = 3$; axis $x = -3$

SOLUTION The region bounded by the graphs of $y^2 = x^{-1}$, $x = 1$ and $x = 3$ is shown below. Let's choose to slice the region vertically. Because a vertical slice is parallel to the axis of rotation, we will use the shell method to calculate the volume of the solid of revolution. For each $x \in [1, 3]$, the shell has radius $x - (-3) = x + 3$ and height

$$\frac{1}{\sqrt{x}} - \left(-\frac{1}{\sqrt{x}}\right) = \frac{2}{\sqrt{x}}.$$

The volume of the solid is therefore given by

$$2\pi \int_1^3 (x + 3)\frac{2}{\sqrt{x}}\,dx = 2\pi\left(\frac{4}{3}x^{3/2} + 12x^{1/2}\right)\Big|_1^3 = 2\pi\left(16\sqrt{3} - \frac{40}{3}\right).$$

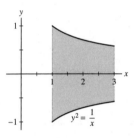

In Exercises 32–34, the regions refer to the graph of the hyperbola $y^2 - x^2 = 1$ in Figure 3. Calculate the volume of revolution about both the x- and y-axes.

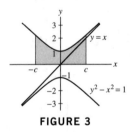

FIGURE 3

33. The region between the upper branch of the hyperbola and the line $y = x$ for $0 \le x \le c$.

SOLUTION

- *x*-axis: Let's choose to slice the region vertically. Because a vertical slice is perpendicular to the axis of rotation, we will use the washer method to calculate the volume of the solid of revolution. For each $x \in [0, c]$, cross sections are washers with outer radius $R = \sqrt{1 + x^2}$ and inner radius $r = x$. The volume of the solid is therefore given by

$$\pi \int_0^c \left((1 + x^2) - x^2\right) dx = \pi x\Big|_0^c = c\pi.$$

- *y*-axis: Let's choose to slice the region vertically. Because a vertical slice is parallel to the axis of rotation, we will use the shell method to calculate the volume of the solid of revolution. For each $x \in [0, c]$, the shell has radius x and height $\sqrt{1 + x^2} - x$. The volume of the solid is therefore given by

$$2\pi \int_0^c x\left(\sqrt{1 + x^2} - x\right) dx = \frac{2\pi}{3}\left((1 + x^2)^{3/2} - x^3\right)\Big|_0^c = \frac{2\pi}{3}\left((1 + c^2)^{3/2} - c^3 - 1\right).$$

35. Let $a > 0$. Show that when the region between $y = a\sqrt{x - ax^2}$ and the x-axis is rotated about the x-axis, the resulting volume is independent of the constant a.

SOLUTION   Setting $a\sqrt{x - ax^2} = 0$ yields $x = 0$ and $x = 1/a$. Using the washer method, cross sections are circular disks with radius $R = a\sqrt{x - ax^2}$. The volume of the solid is therefore given by

$$\pi \int_0^{1/a} a^2(x - ax^2)\,dx = \pi \left(\frac{1}{2}a^2x^2 - \frac{1}{3}a^3x^3\right)\Bigg|_0^{1/a} = \pi\left(\frac{1}{2} - \frac{1}{3}\right) = \frac{\pi}{6},$$

which is independent of the constant a.

In Exercises 37–38, water is pumped into a spherical tank of radius 5 ft from a source located 2 ft below a hole at the bottom (Figure 4). The density of water is 64.2 lb/ft³.

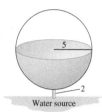

FIGURE 4

37. Calculate the work required to fill the tank.

SOLUTION   Place the origin at the base of the sphere with the positive y-axis pointing upward. The equation for the great circle of the sphere is then $x^2 + (y - 5)^2 = 25$. At location y, the horizontal cross section is a circle of radius $\sqrt{25 - (y - 5)^2} = \sqrt{10y - y^2}$; the volume of the layer is then $\pi(10y - y^2)\Delta y$ ft³, and the force needed to lift the layer is $64.2\pi(10y - y^2)\Delta y$ lb. The layer of water must be lifted $y + 2$ feet, so the work required to fill the tank is given by

$$64.2\pi \int_0^{10} (y + 2)(10y - y^2)\,dy = 64.2\pi \int_0^{10} (8y^2 + 20y - y^3)\,dy$$

$$= 64.2\pi \left(\frac{8}{3}y^3 + 10y^2 - \frac{1}{4}y^4\right)\Bigg|_0^{10}$$

$$= 74900\pi \approx 235{,}305 \text{ ft-lb.}$$

39. A container weighing 50 lb is filled with 20 ft³ of water. The container is raised vertically at a constant speed of 2 ft/s for 1 min, during which time it leaks water at a rate of $\frac{1}{3}$ ft³/s. Calculate the total work performed in raising the container. The density of water is 64.2 lb/ft³.

SOLUTION   Let t denote the elapsed time of the ascent of the container, and let y denote the height of the container. Given that the speed of ascent is 2 ft/s, $y = 2t$; moreover, the volume of water in the container is

$$20 - \frac{1}{3}t = 20 - \frac{1}{6}y \text{ ft}^3.$$

The force needed to lift the container and its contents is then

$$50 + 64.2\left(20 - \frac{1}{6}y\right) = 1334 - 10.7y \text{ lb,}$$

and the work required to lift the container and its contents is

$$\int_0^{120} (1334 - 10.7y)\,dy = (1334y - 5.35y^2)\Bigg|_0^{120} = 83040 \text{ ft-lb.}$$

7 | THE EXPONENTIAL FUNCTION

7.1 Derivative of b^x and the Number e

Preliminary Questions

1. Which of the following equations is incorrect?

(a) $3^2 \cdot 3^5 = 3^7$

(b) $(\sqrt{5})^{4/3} = 5^{2/3}$

(c) $3^2 \cdot 2^3 = 1$

(d) $(2^{-2})^{-2} = 16$

SOLUTION

(a) This equation is correct: $3^2 \cdot 3^5 = 3^{2+5} = 3^7$.

(b) This equation is correct: $(\sqrt{5})^{4/3} = (5^{1/2})^{4/3} = 5^{(1/2)\cdot(4/3)} = 5^{2/3}$.

(c) This equation is incorrect: $3^2 \cdot 2^3 = 9 \cdot 8 = 72 \neq 1$.

(d) this equation is correct: $(2^{-2})^{-2} = 2^{(-2)\cdot(-2)} = 2^4 = 16$.

2. Which of the following functions can be differentiated using the Power Rule?

(a) x^2 **(b)** 2^e **(c)** x^e **(d)** e^x

SOLUTION The Power Rule applies when the function has a variable base and a constant exponent. Therefore, the Power Rule applies to **(a)** x^2 and **(c)** x^e.

3. For which values of b does b^x have a negative derivative?

SOLUTION The function b^x has a negative derivative when $0 < b < 1$.

4. For which values of b is the graph of $y = b^x$ concave up?

SOLUTION The graph of $y = b^x$ is concave up for all $b > 0$ except $b = 1$.

5. Which point lies on the graph of $y = b^x$ for all b?

SOLUTION The point $(0, 1)$ lies on the graph of $y = b^x$ for all b.

6. Which of the following statements is not true?

(a) $(e^x)' = e^x$

(b) $\lim\limits_{h \to 0} \dfrac{e^h - 1}{h} = 1$

(c) The tangent line to $y = e^x$ at $x = 0$ has slope e.

(d) The tangent line to $y = e^x$ at $x = 0$ has slope 1.

SOLUTION

(a) This statement is true: $(e^x)' = e^x$.

(b) This statement is true:

$$\lim_{h \to 0} \frac{e^h - 1}{h} = \frac{d}{dx} e^x \bigg|_{x=0} = e^0 = 1.$$

(c) This statement is false: the tangent line to $y = e^x$ at $x = 0$ has slope $e^0 = 1$.

(d) This statement is true: the tangent line to $y = e^x$ at $x = 0$ has slope $e^0 = 1$.

Exercises

1. Rewrite as a whole number (without using a calculator):

(a) 7^0

(b) $10^2(2^{-2} + 5^{-2})$

(c) $\dfrac{(4^3)^5}{(4^5)^3}$

(d) $27^{4/3}$

(e) $8^{-1/3} \cdot 8^{5/3}$

(f) $3 \cdot 4^{1/4} - 12 \cdot 2^{-3/2}$

SOLUTION

(a) $7^0 = 1$.

(b) $10^2(2^{-2} + 5^{-2}) = 100(1/4 + 1/25) = 25 + 4 = 29.$

(c) $(4^3)^5/(4^5)^3 = 4^{15}/4^{15} = 1.$

(d) $(27)^{4/3} = (27^{1/3})^4 = 3^4 = 81.$

(e) $8^{-1/3} \cdot 8^{5/3} = (8^{1/3})^5/8^{1/3} = 2^5/2 = 2^4 = 16.$

(f) $3 \cdot 4^{1/4} - 12 \cdot 2^{-3/2} = 3 \cdot 2^{1/2} - 3 \cdot 2^2 \cdot 2^{-3/2} = 0.$

In Exercises 2–10, solve for the unknown variable.

3. $e^{2x} = e^{x+1}$

SOLUTION   If $e^{2x} = e^{x+1}$ then $2x = x + 1$, and $x = 1$.

5. $3^x = \left(\frac{1}{3}\right)^{x+1}$

SOLUTION   Rewrite $\left(\frac{1}{3}\right)^{x+1}$ as $(3^{-1})^{x+1} = 3^{-x-1}$. Then $3^x = 3^{-x-1}$, which requires $x = -x - 1$. Thus, $x = -1/2$.

7. $4^{-x} = 2^{x+1}$

SOLUTION   Rewrite 4^{-x} as $(2^2)^{-x} = 2^{-2x}$. Then $2^{-2x} = 2^{x+1}$, which requires $-2x = x + 1$. Solving for x gives $x = -1/3$.

9. $k^{3/2} = 27$

SOLUTION   Raise both sides of the equation to the two-thirds power. This gives $k = (27)^{2/3} = (27^{1/3})^2 = 3^2 = 9$.

In Exercises 11–14, determine the limit.

11. $\lim\limits_{x \to \infty} 4^x$

SOLUTION $\lim\limits_{x \to \infty} 4^x = \infty.$

13. $\lim\limits_{x \to \infty} \left(\frac{1}{4}\right)^{-x}$

SOLUTION $\lim\limits_{x \to \infty} \left(\frac{1}{4}\right)^{-x} = \lim\limits_{x \to \infty} 4^x = \infty.$

In Exercises 15–18, find the equation of the tangent line at the point indicated.

15. $y = 4e^x$, $x_0 = 0$

SOLUTION   Let $f(x) = 4e^x$. Then $f'(x) = 4e^x$ and $f'(0) = 4$. At $x_0 = 0$, $f(0) = 4$, so the equation of the tangent line is $y = 4(x - 0) + 4 = 4x + 4$.

17. $y = e^{x+2}$, $x_0 = -1$

SOLUTION   Let $f(x) = e^{x+2}$. Then $f'(x) = e^{x+2}$ and $f'(-1) = e^1$. At $x_0 = -1$, $f(-1) = e$, so the equation of the tangent line is $y = e(x + 1) + e = ex + 2e$.

In Exercises 19–41, find the derivative.

19. $f(x) = 7e^{2x} + 3e^{4x}$

SOLUTION $\dfrac{d}{dx}(7e^{2x} + 3e^{4x}) = 14e^{2x} + 12e^{4x}.$

21. $f(x) = e^{\pi x}$

SOLUTION $\dfrac{d}{dx}e^{\pi x} = \pi e^{\pi x}.$

23. $f(x) = 4e^{-x} + 7e^{-2x}$

SOLUTION $\dfrac{d}{dx}(4e^{-x} + 7e^{-2x}) = -4e^{-x} - 14e^{-2x}.$

25. $f(x) = x^2 e^{2x}$

SOLUTION $\dfrac{d}{dx}x^2 e^{2x} = 2xe^{2x} + 2x^2 e^{2x}.$

27. $f(x) = (2e^{3x} + 2e^{-2x})^4$

SOLUTION $\dfrac{d}{dx}(2e^{3x} + 2e^{-2x})^4 = 4(2e^{3x} + 2e^{-2x})^3(6e^{3x} - 4e^{-2x}).$

29. $f(x) = e^{1/x}$

SOLUTION $\dfrac{d}{dx} e^{1/x} = \dfrac{-e^{1/x}}{x^2}$.

31. $f(x) = e^{\sin x}$

SOLUTION $\dfrac{d}{dx} e^{\sin x} = \cos x e^{\sin x}$.

33. $f(x) = \sin(e^x)$

SOLUTION $\dfrac{d}{dx} \sin(e^x) = e^x \cos(e^x)$.

35. $f(t) = \dfrac{1}{1 - e^{-3t}}$

SOLUTION $\dfrac{d}{dt} \left(\dfrac{1}{1 - e^{-3t}} \right) = \dfrac{d}{dt} (1 - e^{-3t})^{-1} = -(1 - e^{-3t})^{-2} (3e^{-3t})$.

37. $f(t) = \cos(te^{-2t})$

SOLUTION $\dfrac{d}{dt} \cos(te^{-2t}) = -\sin(te^{-2t})(t(-2e^{-2t}) + e^{-2t}) = (2te^{-2t} - e^{-2t}) \sin(te^{-2t})$.

39. $f(x) = \tan(e^{5-6x})$

SOLUTION $\dfrac{d}{dx} \tan(e^{5-6x}) = -6e^{5-6x} \sec^2(e^{5-6x})$.

41. $f(x) = e^{e^x}$

SOLUTION $\dfrac{d}{dx} e^{e^x} = e^x e^{e^x}$.

In Exercises 42–47, find the critical points and determine whether they are local minima, maxima, or neither.

43. $f(x) = x + e^{-x}$

SOLUTION   Setting $f'(x) = 1 - e^{-x}$ equal to zero and solving for x gives $e^{-x} = 1$ which is true if and only if $x = 0$. $f''(x) = e^{-x}$, so $f''(0) = e^0 = 1 > 0$. Therefore, $x = 0$ corresponds to a local minimum.

45. $f(x) = x^2 e^x$

SOLUTION   Setting $f'(x) = (x^2 + 2x)e^x$ equal to zero and solving for x gives $(x^2 + 2x)e^x = 0$ which is true if and only if $x = 0$ or $x = -2$. Now, $f''(x) = (x^2 + 4x + 2)e^x$. Because $f''(0) = 2 > 0$, $x = 0$ corresponds to a local minimum. On the other hand, $f''(-2) = (4 - 8 + 2)e^{-2} = -2/e^2 < 0$, so $x = -2$ corresponds to a local maximum.

47. $g(t) = (t^3 - 2t)e^t$

SOLUTION

$$g'(t) = (t^3 - 2t)(e^t) + e^t(3t^2 - 2) = e^t(t^3 + 3t^2 - 2t - 2).$$

The critical points occur when $t^3 + 3t^2 - 2t - 2 = 0$. This occurs when $t = 1, -2 \pm \sqrt{2}$.

$$g''(t) = e^t(3t^2 + 6t - 2) + e^t(t^3 + 3t^2 - 2t - 2) = e^t(t^3 + 6t^2 + 4t - 4).$$

Thus,

- $g''(1) = 7e > 0$, so $t = 1$ corresponds to a local minimum;
- $g''(-2 + \sqrt{2}) \approx -2.497 < 0$, so $t = -2 + \sqrt{2}$ corresponds to a local maximum; and
- $g''(-2 - \sqrt{2}) \approx .4108 > 0$, so $t = -2 - \sqrt{2}$ corresponds to a local minimum.

In Exercises 48–53, find the critical points and points of inflection. Then sketch the graph.

49. $y = e^{-x} + e^x$

SOLUTION   Let $f(x) = e^{-x} + e^x$. Then

$$f'(x) = -e^{-x} + e^x = \dfrac{e^{2x} - 1}{e^x}.$$

Thus, $x = 0$ is a critical point, and $f(x)$ is increasing for $x > 0$ and decreasing for $x < 0$. Observe that $f''(x) = e^{-x} + e^x > 0$ for all x, so $f(x)$ is concave up for all x and there are no points of inflection. A graph of $y = f(x)$ is shown below.

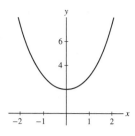

51. $y = e^{-x^2}$

SOLUTION   Let $f(x) = e^{-x^2}$. Then $f'(x) = -2xe^{-x^2}$ and $x = 0$ is the only critical point. Further, $f(x)$ is increasing for $x < 0$ and decreasing for $x > 0$. Now,

$$f''(x) = 4x^2 e^{-x^2} - 2e^{-x^2} = 2e^{-x^2}(2x^2 - 1),$$

so $f(x)$ is concave up for $|x| > \frac{\sqrt{2}}{2}$, is concave down for $|x| < \frac{\sqrt{2}}{2}$ and has points of inflection at $x = \pm\frac{\sqrt{2}}{2}$. A graph of $y = f(x)$ is shown below.

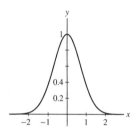

53. $y = x^2 e^{-x}$ on $[0, 10]$

SOLUTION   Let $f(x) = x^2 e^{-x}$. Then

$$f'(x) = -x^2 e^{-x} + 2xe^{-x} = x(2-x)e^{-x},$$

so $x = 0$ and $x = 2$ are critical points. Also, $f(x)$ is increasing for $0 < x < 2$ and decreasing for $2 < x < 10$. Since

$$f''(x) = -(2x - x^2)e^{-x} + (2 - 2x)e^{-x} = (x^2 - 4x + 2)e^{-x},$$

we find $f(x)$ is concave up for $0 < x < 2 - \sqrt{2}$ and for $2 + \sqrt{2} < x < 10$, is concave down for $2 - \sqrt{2} < x < 2 + \sqrt{2}$, and has inflection points at $x = 2 \pm \sqrt{2}$. A graph of $y = f(x)$ is shown below.

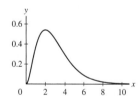

In Exercises 55–68, evaluate the indefinite integral.

55. $\displaystyle\int (e^x + 2)\, dx$

SOLUTION $\displaystyle\int (e^x + 2)\, dx = e^x + 2x + C.$

57. $\displaystyle\int y e^{y^2}\, dy$

SOLUTION   Use the substitution $u = y^2$, $du = 2y\, dy$. Then

$$\int y e^{y^2}\, dy = \frac{1}{2}\int e^u\, du = \frac{1}{2}e^u + C = \frac{1}{2}e^{y^2} + C.$$

59. $\displaystyle\int e^{-9t}\, dt$

SOLUTION Use the substitution $u = -9t$, $du = -9\,dt$. Then

$$\int e^{-9t}\,dt = -\frac{1}{9}\int e^u\,du = -\frac{1}{9}e^u + C = -\frac{1}{9}e^{-9t} + C.$$

61. $\displaystyle\int e^t\sqrt{e^t + 1}\,dt$

SOLUTION Use the substitution $u = e^t + 1$, $du = e^t\,dt$. Then

$$\int e^t\sqrt{e^t + 1}\,dt = \int \sqrt{u}\,du = \frac{2}{3}u^{3/2} + C = \frac{2}{3}(e^t + 1)^{3/2} + C.$$

63. $\displaystyle\int (7 - e^{10x})\,dx$

SOLUTION First, observe that

$$\int (7 - e^{10x})\,dx = \int 7\,dx - \int e^{10x}\,dx = 7x - \int e^{10x}\,dx.$$

In the remaining integral, use the substitution $u = 10x$, $du = 10\,dx$. Then

$$\int e^{10x}\,dx = \frac{1}{10}\int e^u\,du = \frac{1}{10}e^u + C = \frac{1}{10}e^{10x} + C.$$

Finally,

$$\int (7 - e^{10x})\,dx = 7x - \frac{1}{10}e^{10x} + C.$$

65. $\displaystyle\int xe^{-4x^2}\,dx$

SOLUTION Use the substitution $u = -4x^2$, $du = -8x\,dx$. Then

$$\int xe^{-4x^2}\,dx = -\frac{1}{8}\int e^u\,du = -\frac{1}{8}e^u + C = -\frac{1}{8}e^{-4x^2} + C.$$

67. $\displaystyle\int \frac{e^x}{\sqrt{e^x + 1}}\,dx$

SOLUTION Use the substitution $u = e^x + 1$, $du = e^x\,dx$. Then

$$\int \frac{e^x}{\sqrt{e^x + 1}}\,dx = \int \frac{du}{\sqrt{u}} = 2\sqrt{u} + C = 2\sqrt{e^x + 1} + C.$$

69. Find an approximation to m_4 using the limit definition and estimate the slope of the tangent line to $y = 4^x$ at $x = 0$ and $x = 2$.

SOLUTION Recall

$$m_4 = \lim_{h\to 0}\left(\frac{4^h - 1}{h}\right).$$

Using a table of values, we find

h	$\dfrac{4^h - 1}{h}$
.01	1.39595
.001	1.38726
.0001	1.38639
.00001	1.38630

Thus $m_4 \approx 1.386$. Knowing that $y'(x) = m_4 \cdot 4^x$, it follows that $y'(0) \approx 1.386$ and $y'(2) \approx 1.386 \cdot 16 = 22.176$.

71. Find the area between $y = e^x$ and $y = e^{-x}$ over $[0, 2]$.

SOLUTION Over $[0, 2]$, the graph of $y = e^x$ lies above the graph of $y = e^{-x}$. Hence, the area between the graphs is

$$\int_0^2 (e^x - e^{-x})\, dx = (e^x + e^{-x})\Big|_0^2 = e^2 + \frac{1}{e^2} - (1 + 1) = e^2 - 2 + \frac{1}{e^2}.$$

73. Find the volume obtained by revolving $y = e^x$ about the x-axis for $0 \le x \le 1$.

SOLUTION Each cross section of the solid is a disk with radius $R = e^x$. The volume is then

$$\pi \int_0^1 e^{2x}\, dx = \frac{\pi}{2} e^{2x}\Big|_0^1 = \frac{\pi}{2}(e^2 - 1).$$

Further Insights and Challenges

75. Prove that $f(x) = e^x$ is not a polynomial function. *Hint:* Differentiation lowers the degree of a polynomial by 1.

SOLUTION Assume $f(x) = e^x$ is a polynomial function of degree n. Then $f^{(n+1)}(x) = 0$. But we know that any derivative of e^x is e^x and $e^x \ne 0$. Hence, e^x cannot be a polynomial function.

77. Generalize Exercise 76; that is, use induction (if you are familiar with this method of proof) to prove that for all $n \ge 0$,

$$e^x \ge 1 + x + \frac{1}{2}x^2 + \frac{1}{6}x^3 + \cdots + \frac{1}{n!}x^n \quad (x \ge 0)$$

SOLUTION For $n = 1$, $e^x \ge 1 + x$ by Exercise 76. Assume the statement is true for $n = k$. We need to prove the statement is true for $n = k + 1$. By the Induction Hypothesis,

$$e^x \ge 1 + x + x^2/2 + \cdots + x^k/k!.$$

Integrating both sides of this inequality yields

$$\int_0^x e^t\, dt = e^x - 1 \ge x + x^2/2 + \cdots + x^{k+1}/(k+1)!$$

or

$$e^x \ge 1 + x + x^2/2 + \cdots + x^{k+1}/(k+1)!$$

as required.

79. Calculate the first three derivatives of $f(x) = xe^x$. Then guess the formula for $f^{(n)}(x)$ (use induction to prove it if you are familiar with this method of proof).

SOLUTION $f'(x) = e^x + xe^x$, $f''(x) = e^x + e^x + xe^x = 2e^x + xe^x$, $f'''(x) = 2e^x + e^x + xe^x = 3e^x + xe^x$. So one would guess that $f^{(n)}(x) = ne^x + xe^x$. Assuming this is true for $f^{(n)}(x)$, we verify that $f^{(n+1)}(x) = (f^{(n)}(x))' = ne^x + e^x + xe^x = (n+1)e^x + xe^x$.

81. Prove in two ways that the numbers m_a satisfy

$$m_{ab} = m_a + m_b$$

(a) First method: Use the limit definition of m_b and

$$\frac{(ab)^h - 1}{h} = b^h \left(\frac{a^h - 1}{h} \right) + \frac{b^h - 1}{h}$$

(b) Second method: Apply the Product Rule to $a^x b^x = (ab)^x$.

SOLUTION

(a) $m_{ab} = \lim\limits_{h \to 0} \dfrac{(ab)^h - 1}{h} = \lim\limits_{h \to 0} \dfrac{b^h(a^h - 1)}{h} + \dfrac{b^h - 1}{h} = \lim\limits_{h \to 0} b^h \lim\limits_{h \to 0} \dfrac{a^h - 1}{h} + \lim\limits_{h \to 0} \dfrac{b^h - 1}{h}$

$$= \lim\limits_{h \to 0} \frac{a^h - 1}{h} + \lim\limits_{h \to 0} \frac{b^h - 1}{h} = m_a + m_b.$$

So, $m_{ab} = m_a + m_b$.

(b) $m_{ab}(ab)^x = ((ab)^x)' = (a^x b^x)' = (a^x)' b^x + (b^x)' a^x = m_a a^x b^x + m_b a^x b^x$

$$= a^x b^x (m_a + m_b) = (ab)^x (m_a + m_b).$$

Therefore, we have $m_{ab}(ab)^x = (ab)^x (m_a + m_b)$. Dividing both sides by $(ab)^x$, we see that $m_{ab} = m_a + m_b$.

7.2 Inverse Functions

Preliminary Questions

1. Which of the following satisfy $f^{-1}(x) = f(x)$?

(a) $f(x) = x$ (b) $f(x) = 1 - x$

(c) $f(x) = 1$ (d) $f(x) = \sqrt{x}$

(e) $f(x) = |x|$ (f) $f(x) = x^{-1}$

SOLUTION The functions (a) $f(x) = x$, (b) $f(x) = 1 - x$ and (f) $f(x) = x^{-1}$ satisfy $f^{-1}(x) = f(x)$.

2. The graph of a function looks like the track of a roller coaster. Is the function one-to-one?

SOLUTION Because the graph looks like the track of a roller coaster, there will be several locations at which the graph has the same height. The graph will therefore fail the horizontal line test, meaning that the function is *not* one-to-one.

3. Consider the function f that maps teenagers in the United States to their last names. Explain why the inverse of f does not exist.

SOLUTION Many different teenagers will have the same last name, so this function will not be one-to-one. Consequently, the function does not have an inverse.

4. View the following fragment of a train schedule for the New Jersey Transit System as defining a function f from towns to times. Is f one-to-one? What is $f^{-1}(6{:}27)$?

Trenton	6:21
Hamilton Township	6:27
Princeton Junction	6:34
New Brunswick	6:38

SOLUTION This function is one-to-one, and $f^{-1}(6{:}27) =$ Hamilton Township.

5. A homework problem asks for a sketch of the graph of the *inverse* of $f(x) = x + \cos x$. Frank, after trying but failing to find a formula for $f^{-1}(x)$, says it's impossible to graph the inverse. Bianca hands in an accurate sketch without solving for f^{-1}. How did Bianca complete the problem?

SOLUTION The graph of the inverse function is the reflection of the graph of $y = f(x)$ through the line $y = x$.

Exercises

1. Show that $f(x) = 7x - 4$ is invertible by finding its inverse.

SOLUTION Solving $y = 7x - 4$ for x yields $x = \dfrac{y+4}{7}$. Thus, $f^{-1}(x) = \dfrac{x+4}{7}$.

3. What is the largest interval containing zero on which $f(x) = \sin x$ is one-to-one?

SOLUTION Looking at the graph of $\sin x$, the function is one-to-one on the interval $[-\pi/2, \pi/2]$.

5. Verify that $f(x) = x^3 + 3$ and $g(x) = (x - 3)^{1/3}$ are inverses by showing that $f(g(x)) = x$ and $g(f(x)) = x$.

SOLUTION

- $f(g(x)) = \left((x - 3)^{1/3}\right)^3 + 3 = x - 3 + 3 = x.$
- $g(f(x)) = \left(x^3 + 3 - 3\right)^{1/3} = \left(x^3\right)^{1/3} = x.$

7. The escape velocity from a planet of mass M and radius R is $v(R) = \sqrt{\dfrac{2GM}{R}}$, where G is the universal gravitational constant. Find the inverse of $v(R)$ as a function of R.

SOLUTION To find the inverse, we solve

$$y = \sqrt{\frac{2GM}{R}}$$

for R. This yields

$$R = \frac{2GM}{y^2}.$$

Therefore,

$$v^{-1}(R) = \frac{2GM}{R^2}.$$

In Exercises 9–16, find a domain on which f is one-to-one and a formula for the inverse of f restricted to this domain. Sketch the graphs of f and f^{-1}.

9. $f(x) = 3x - 2$

SOLUTION The linear function $f(x) = 3x - 2$ is one-to-one for all real numbers. Solving $y = 3x - 2$ for x gives $x = (y + 2)/3$. Thus,

$$f^{-1}(x) = \frac{x + 2}{3}.$$

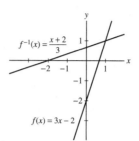

11. $f(x) = \dfrac{1}{x + 1}$

SOLUTION The graph of $f(x) = 1/(x + 1)$ given below shows that f passes the horizontal line test, and is therefore one-to-one, on its entire domain $\{x : x \neq -1\}$. Solving $y = \dfrac{1}{x + 1}$ for x gives $x = \dfrac{1}{y} - 1$. Thus, $f^{-1}(x) = \dfrac{1}{x} - 1$.

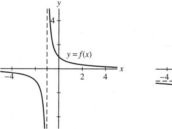

13. $f(s) = \dfrac{1}{s^2}$

SOLUTION To make $f(s) = s^{-2}$ one-to-one, we must restrict the domain to either $\{s : s > 0\}$ or $\{s : s < 0\}$. If we choose the domain $\{s : s > 0\}$, then solving $y = \dfrac{1}{s^2}$ for s yields $s = \dfrac{1}{\sqrt{y}}$. Hence, $f^{-1}(s) = \dfrac{1}{\sqrt{s}}$. Had we chosen the domain $\{s : s < 0\}$, the inverse would have been $f^{-1}(s) = -\dfrac{1}{\sqrt{s}}$.

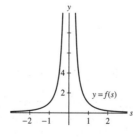
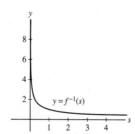

15. $f(z) = z^3$

SOLUTION The function $f(z) = z^3$ is one-to-one over its entire domain (see the graph below). Solving $y = z^3$ for z yields $y^{1/3} = z$. Thus, $f^{-1}(z) = z^{1/3}$.

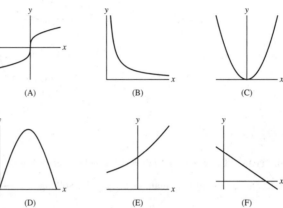

17. For each function shown in Figure 15, sketch the graph of the inverse (restrict the function's domain if necessary).

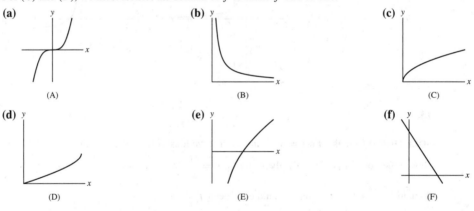

FIGURE 15

SOLUTION   Here, we apply the rule that the graph of f^{-1} is obtained by reflecting the graph of f across the line $y = x$. For (C) and (D), we must restrict the domain of f to make f one-to-one.

(a)   **(b)**   **(c)**

(A)   (B)   (C)

(d)   **(e)**   **(f)**

(D)   (E)   (F)

19. Let n be a non-zero integer. Find a domain on which $f(x) = (1 - x^n)^{1/n}$ coincides with its inverse. *Hint:* The answer depends on whether n is even or odd.

SOLUTION   First note

$$f(f(x)) = \left(1 - \left((1 - x^n)^{1/n}\right)^n\right)^{1/n} = \left(1 - (1 - x^n)\right)^{1/n} = (x^n)^{1/n} = x,$$

so $f(x)$ coincides with its inverse. For the domain and range of f, let's first consider the case when $n > 0$. If n is even, then $f(x)$ is defined only when $1 - x^n \geq 0$. Hence, the domain is $-1 \leq x \leq 1$. The range is $0 \leq y \leq 1$. If n is odd, then $f(x)$ is defined for all real numbers, and the range is also all real numbers. Now, suppose $n < 0$. Then $-n > 0$, and

$$f(x) = \left(1 - \frac{1}{x^{-n}}\right)^{-1/-n} = \left(\frac{x^{-n}}{x^{-n} - 1}\right)^{1/-n}.$$

If n is even, then $f(x)$ is defined only when $x^{-n} - 1 > 0$. Hence, the domain is $|x| > 1$. The range is $y > 1$. If n is odd, then $f(x)$ is defined for all real numbers except $x = 1$. The range is all real numbers except $y = 1$.

21. Show that the inverse of $f(x) = e^{-x}$ exists (without finding it explicitly). What is the domain of f^{-1}?

SOLUTION

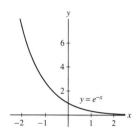

Notice that the graph of $f(x) = e^{-x}$ (shown above) passes the horizontal line test. That is, $f(x) = e^{-x}$ is one-to-one. By Theorem 3, $f^{-1}(x)$ exists and the domain of $f^{-1}(x)$ is the range of $f(x)$, namely $(0, \infty)$

23. Let $f(x) = x^2 - 2x$. Determine a domain on which f^{-1} exists and find a formula for f^{-1} on this domain.

SOLUTION   From the graph of $y = x^2 - 2x$ shown below, we see that if the domain of f is restricted to either $x \le 1$ or $x \ge 1$, then f is one-to-one and f^{-1} exists. To find a formula for f^{-1}, we solve $y = x^2 - 2x$ for x as follows:

$$y + 1 = x^2 - 2x + 1 = (x - 1)^2$$
$$x - 1 = \pm\sqrt{y + 1}$$
$$x = 1 \pm \sqrt{y + 1}$$

If the domain of f is restricted to $x \le 1$, then we choose the negative sign in front of the radical and $f^{-1}(x) = 1 - \sqrt{x + 1}$. If the domain of f is restricted to $x \ge 1$, we choose the positive sign in front of the radical and $f^{-1}(x) = 1 + \sqrt{x + 1}$.

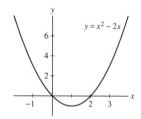

25. Find the inverse $g(x)$ of $f(x) = \sqrt{x^2 + 9}$ with domain $x \ge 0$ and calculate $g'(x)$ in two ways: using Theorem 2 and by direct calculation.

SOLUTION   To find a formula for $g(x) = f^{-1}(x)$, solve $y = \sqrt{x^2 + 9}$ for x. This yields $x = \pm\sqrt{y^2 - 9}$. Because the domain of f was restricted to $x \ge 0$, we must choose the positive sign in front of the radical. Thus

$$g(x) = f^{-1}(x) = \sqrt{x^2 - 9}.$$

Because $x^2 + 9 \ge 9$ for all x, it follows that $f(x) \ge 3$ for all x. Thus, the domain of $g(x) = f^{-1}(x)$ is $x \ge 3$. The range of g is the restricted domain of f: $y \ge 0$.

By Theorem 2,

$$g'(x) = \frac{1}{f'(g(x))}.$$

With

$$f'(x) = \frac{x}{\sqrt{x^2 + 9}},$$

it follows that

$$f'(g(x)) = \frac{\sqrt{x^2 - 9}}{\sqrt{(\sqrt{x^2 - 9})^2 + 9}} = \frac{\sqrt{x^2 - 9}}{\sqrt{x^2}} = \frac{\sqrt{x^2 - 9}}{x}$$

since the domain of g is $x \ge 3$. Thus,

$$g'(x) = \frac{1}{f'(g(x))} = \frac{x}{\sqrt{x^2 - 9}}.$$

This agrees with the answer we obtain by differentiating directly:

$$g'(x) = \frac{2x}{2\sqrt{x^2 - 9}} = \frac{x}{\sqrt{x^2 - 9}}.$$

In Exercises 27–32, use Theorem 2 to calculate $g'(x)$, where $g(x)$ is the inverse of $f(x)$.

27. $f(x) = 7x + 6$

SOLUTION Let $f(x) = 7x + 6$ then $f'(x) = 7$. Solving $y = 7x + 6$ for x and switching variables, we obtain the inverse $g(x) = (x - 6)/7$. Thus,

$$g'(x) = \frac{1}{f'(g(x))} = \frac{1}{7}.$$

29. $f(x) = x^{-5}$

SOLUTION Let $f(x) = x^{-5}$, then $f'(x) = -5x^{-6}$. Solving $y = x^{-5}$ for x and switching variables, we obtain the inverse $g(x) = x^{-1/5}$. Thus,

$$g'(x) = \frac{1}{-5(x^{-1/5})^{-6}} = -\frac{1}{5}x^{-6/5}.$$

31. $f(x) = \dfrac{x}{x + 1}$

SOLUTION Let $f(x) = \frac{x}{x+1}$, then

$$f'(x) = \frac{(x + 1) - x}{(x + 1)^2} = \frac{1}{(x + 1)^2}.$$

Solving $y = \frac{x}{x+1}$ for x and switching variables, we obtain the inverse $g(x) = \frac{x}{1-x}$. Thus

$$g'(x) = 1 \bigg/ \frac{1}{(x/(1 - x) + 1)^2} = \frac{1}{(1 - x)^2}.$$

33. Let $g(x)$ be the inverse of $f(x) = x^3 + 2x + 4$. Calculate $g(7)$ [without finding a formula for $g(x)$] and then calculate $g'(7)$.

SOLUTION Let $g(x)$ be the inverse of $f(x) = x^3 + 2x + 4$. Because

$$f(1) = 1^3 + 2(1) + 4 = 7,$$

it follows that $g(7) = 1$. Moreover, $f'(x) = 3x^2 + 2$, and

$$g'(7) = \frac{1}{f'(g(7))} = \frac{1}{f'(1)} = \frac{1}{5}.$$

In Exercises 35–40, calculate $g(b)$ and $g'(b)$, where g is the inverse of f (in the given domain, if indicated).

35. $f(x) = x + \cos x$, $b = 1$

SOLUTION $f(0) = 1$, so $g(1) = 0$. $f'(x) = 1 - \sin x$ so $f'(g(1)) = f'(0) = 1 - \sin 0 = 1$. Thus, $g'(1) = 1/1 = 1$.

37. $f(x) = \sqrt{x^2 + 6x}$ for $x \geq 0$, $b = 4$

SOLUTION To determine $g(4)$, we solve $f(x) = \sqrt{x^2 + 6x} = 4$ for x. This yields:

$$x^2 + 6x = 16$$
$$x^2 + 6x - 16 = 0$$
$$(x + 8)(x - 2) = 0$$

or $x = -8, 2$. Because the domain of f has been restricted to $x \geq 0$, we have $g(4) = 2$. With

$$f'(x) = \frac{x + 3}{\sqrt{x^2 + 6x}},$$

it then follows that

$$g'(4) = \frac{1}{f'(g(4))} = \frac{1}{f'(2)} = \frac{4}{5}.$$

39. $f(x) = \dfrac{1}{x + 1}$, $b = \dfrac{1}{4}$

SOLUTION $f(3) = 1/4$, so $g(1/4) = 3$. $f'(x) = \frac{-1}{(x+1)^2}$ so $f'(g(1/4)) = f'(3) = \frac{-1}{(3+1)^2} = -1/16$. Thus, $g'(1/4) = -16$.

41. Let $f(x) = x^n$ and $g(x) = x^{1/n}$. Compute $g'(x)$ using Theorem 2 and check your answer using the Power Rule.

SOLUTION Note that $g(x) = f^{-1}(x)$ (see problem 55 from 6.1). Therefore,

$$g'(x) = \frac{1}{f'(g(x))} = \frac{1}{n(g(x))^{n-1}} = \frac{1}{n(x^{1/n})^{n-1}} = \frac{1}{n(x^{1-1/n})} = \frac{x^{1/n-1}}{n} = \frac{1}{n}(x^{1/n-1})$$

which agrees with the Power Rule.

Further Insights and Challenges

43. Show that if $f(x)$ is odd and $f^{-1}(x)$ exists, then $f^{-1}(x)$ is odd. Show, on the other hand, that an even function does not have an inverse.

SOLUTION Suppose $f(x)$ is odd and $f^{-1}(x)$ exists. Because $f(x)$ is odd, $f(-x) = -f(x)$. Let $y = f^{-1}(x)$, then $f(y) = x$. Since $f(x)$ is odd, $f(-y) = -f(y) = -x$. Thus $f^{-1}(-x) = -y = -f^{-1}(x)$. Hence, f^{-1} is odd.

On the other hand, if $f(x)$ is even, then $f(-x) = f(x)$. Hence, f is not one-to-one and f^{-1} does not exist.

45. Let g be the inverse of a function f satisfying $f'(x) = f(x)$. Show that $g'(x) = x^{-1}$. We will apply this in the next section to show that the inverse of $f(x) = e^x$ (the natural logarithm) is an antiderivative of x^{-1}.

SOLUTION

$$g'(x) = \frac{1}{f'(g(x))} = \frac{1}{f'(f^{-1}(x))} = \frac{1}{f(f^{-1}(x))} = \frac{1}{x}.$$

7.3 Logarithms and Their Derivatives

Preliminary Questions

1. Compute $\log_{b^2}(b^4)$.

SOLUTION Because $b^4 = (b^2)^2$, $\log_{b^2}(b^4) = 2$.

2. When is $\ln x$ negative?

SOLUTION $\ln x$ is negative for $0 < x < 1$.

3. What is $\ln(-3)$? Explain.

SOLUTION $\ln(-3)$ is not defined.

4. Explain the phrase "the logarithm converts multiplication into addition."

SOLUTION This phrase is a verbal description of the general property of logarithms that states

$$\log(ab) = \log a + \log b.$$

5. What are the domain and range of $\ln x$?

SOLUTION The domain of $\ln x$ is $x > 0$ and the range is all real numbers.

6. Does x^{-1} have an antiderivative for $x < 0$? If so, describe one.

SOLUTION Yes, $\ln(-x)$ is an antiderivative of x^{-1} for $x < 0$.

Exercises

In Exercises 1–12, calculate directly (without using a calculator).

1. $\log_3 27$

SOLUTION $\log_3 27 = \log_3 3^3 = 3 \log_3 3 = 3$.

3. $\log_2(2^{5/3})$

SOLUTION $\log_2 2^{5/3} = \frac{5}{3} \log_2 2 = \frac{5}{3}$.

5. $\log_{64} 4$

SOLUTION $\log_{64} 4 = \log_{64} 64^{1/3} = \dfrac{1}{3} \log_{64} 64 = \dfrac{1}{3}$.

7. $\log_8 2 + \log_4 2$

SOLUTION $\log_8 2 + \log_4 2 = \log_8 8^{1/3} + \log_4 4^{1/2} = \dfrac{1}{3} + \dfrac{1}{2} = \dfrac{5}{6}$.

9. $\log_4 48 - \log_4 12$

SOLUTION $\log_4 48 - \log_4 12 = \log_4 \dfrac{48}{12} = \log_4 4 = 1$.

11. $\ln(e^3) + \ln(e^4)$

SOLUTION $\ln(e^3) + \ln(e^4) = 3 + 4 = 7$.

13. Write as the natural log of a single expression:

(a) $2 \ln 5 + 3 \ln 4$ **(b)** $5 \ln(x^{1/2}) + \ln(9x)$

SOLUTION

(a) $2 \ln 5 + 3 \ln 4 = \ln 5^2 + \ln 4^3 = \ln 25 + \ln 64 = \ln(25 \cdot 64) = \ln 1600$.

(b) $5 \ln x^{1/2} + \ln 9x = \ln x^{5/2} + \ln 9x = \ln(x^{5/2} \cdot 9x) = \ln(9x^{7/2})$.

In Exercises 15–20, solve for the unknown.

15. $7e^{5t} = 100$

SOLUTION Divide the equation by 7 and then take the natural logarithm of both sides. This gives

$$5t = \ln\left(\frac{100}{7}\right) \quad \text{or} \quad t = \frac{1}{5} \ln\left(\frac{100}{7}\right).$$

17. $2^{x^2 - 2x} = 8$

SOLUTION Since $8 = 2^3$, we have $x^2 - 2x - 3 = 0$ or $(x - 3)(x + 1) = 0$. Thus, $x = -1$ or $x = 3$.

19. $\ln(x^4) - \ln(x^2) = 2$

SOLUTION $\ln(x^4) - \ln(x^2) = \ln\left(\dfrac{x^4}{x^2}\right) = \ln(x^2) = 2 \ln x$. Thus, $2 \ln x = 2$ or $\ln x = 1$. Hence, $x = e$.

21. The population of a city (in millions) at time t (years) is $P(t) = 2.4e^{0.06t}$, where $t = 0$ is the year 2000. When will the population double from its size at $t = 0$?

SOLUTION Population doubles when $4.8 = 2.4e^{.06t}$. Thus, $0.06t = \ln 2$ or $t = \dfrac{\ln 2}{0.06} \approx 11.55$ years.

In Exercises 23–40, find the derivative.

23. $y = x \ln x$

SOLUTION $\dfrac{d}{dx} x \ln x = \ln x + \dfrac{x}{x} = \ln x + 1$.

25. $y = (\ln x)^2$

SOLUTION $\dfrac{d}{dx} (\ln x)^2 = (2 \ln x) \dfrac{1}{x} = \dfrac{2}{x} \ln x$.

27. $y = \ln(x^3 + 3x + 1)$

SOLUTION $\dfrac{d}{dx} \ln(x^3 + 3x + 1) = \dfrac{3x^2 + 3}{x^3 + 3x + 1}$.

29. $y = \ln(\sin t + 1)$

SOLUTION $\dfrac{d}{dt} \ln(\sin t + 1) = \dfrac{\cos t}{\sin t + 1}$.

31. $y = \dfrac{\ln x}{x}$

SOLUTION $\dfrac{d}{dx} \dfrac{\ln x}{x} = \dfrac{\frac{1}{x}(x) - \ln x}{x^2} = \dfrac{1 - \ln x}{x^2}$.

33. $y = \ln(\ln x)$

SOLUTION $\dfrac{d}{dx}\ln(\ln x) = \dfrac{1}{x\ln x}$.

35. $y = \ln((\ln x)^3)$

SOLUTION $\dfrac{d}{dx}\ln((\ln x)^3) = \dfrac{3(\ln x)^2}{x(\ln x)^3} = \dfrac{3}{x\ln x}$.

Alternately, because $\ln((\ln x)^3) = 3\ln(\ln x)$,

$$\frac{d}{dx}\ln((\ln x)^3) = 3\frac{d}{dx}\ln(\ln x) = 3\cdot\frac{1}{x\ln x}.$$

37. $y = \ln(\tan x)$

SOLUTION $\dfrac{d}{dx}\ln(\tan x) = \dfrac{1}{\tan x}\cdot\sec^2 x = \dfrac{1}{\sin x\cos x}$.

39. $y = 5^x$

SOLUTION $\dfrac{d}{dx}5^x = \ln 5\cdot 5^x$.

In Exercises 41–44, compute the derivative using Eq. (1).

41. $f'(x)$, $f(x) = \log_2 x$

SOLUTION $f(x) = \log_2 x = \dfrac{\ln x}{\ln 2}$. Thus, $f'(x) = \dfrac{1}{x}\cdot\dfrac{1}{\ln 2}$.

43. $f'(3)$, $f(x) = \log_5 x$

SOLUTION $f(x) = \dfrac{\ln x}{\ln 5}$, so $f'(x) = \dfrac{1}{x\ln 5}$. Thus, $f'(3) = \dfrac{1}{3\ln 5}$.

In Exercises 45–56, find an equation of the tangent line at the point indicated.

45. $f(x) = 4^x$, $x = 3$

SOLUTION Let $f(x) = 4^x$. Then $f(3) = 4^3 = 64$. $f'(x) = \ln 4\cdot 4^x$, so $f'(3) = \ln 4\cdot 64$. Therefore, the equation of the tangent line is $y = 64\ln 4(x - 3) + 64$.

47. $s(t) = 3^{7t}$, $t = 2$

SOLUTION Let $s(t) = 3^{7t}$. Then $s(2) = 3^{14}$. $s'(t) = 7\ln 3\cdot 3^{7t}$, so $s'(2) = 7\ln 3\cdot 3^{14}$. Therefore the equation of the tangent line is $y = 7\ln 3\cdot 3^{14}(t - 2) + 3^{14}$.

49. $f(x) = 5^{x^2-2x+9}$, $x = 1$

SOLUTION Let $f(x) = 5^{x^2-2x+9}$. Then $f(1) = 5^8$. $f'(x) = \ln 5\cdot 5^{x^2-2x+9}(2x - 2)$, so $f'(1) = \ln 5(0) = 0$. Therefore, the equation of the tangent line is $y = 5^8$.

51. $s(t) = \ln(8 - 4t)$, $t = 1$

SOLUTION Let $s(t) = \ln(8 - 4t)$. Then $s(1) = \ln(8 - 4) = \ln 4$. $s'(t) = \dfrac{-4}{8-4t}$, so $s'(1) = -4/4 = -1$. Therefore the equation of the tangent line is $y = -1(t - 1) + \ln 4$.

53. $f(x) = 4\ln(9x + 2)$, $x = 2$

SOLUTION Let $f(x) = 4\ln(9x + 2)$. Then $f(2) = 4\ln 20$. $f'(x) = \dfrac{36}{9x+2}$, so $f'(2) = 9/5$. Therefore the equation of the tangent line is $y = (9/5)(x - 2) + 4\ln 20$.

55. $f(x) = \log_5 x$, $x = 2$

SOLUTION Let $f(x) = \log_5 x$. Then $f(2) = \log_5 2 = \dfrac{\ln 2}{\ln 5}$. $f'(x) = \dfrac{1}{x\ln 5}$, so $f'(2) = \dfrac{1}{2\ln 5}$. Therefore the equation of the tangent line is

$$y = \frac{1}{2\ln 5}(x - 2) + \frac{\ln 2}{\ln 5}.$$

In Exercises 57–60, find the local extreme values in the domain $\{x : x > 0\}$ and use the Second Derivative Test to determine whether these values are local minima or maxima.

57. $g(x) = \dfrac{\ln x}{x}$

SOLUTION   Let $g(x) = \frac{\ln x}{x}$. Then

$$g'(x) = \frac{x(1/x) - \ln x}{x^2} = \frac{1 - \ln x}{x^2}.$$

We know $g'(x) = 0$ when $1 - \ln x = 0$, or when $x = e$.

$$g''(x) = \frac{x^2(-1/x) - (1 - \ln x)(2x)}{x^4} = \frac{-3 + 2\ln x}{x^3}$$

so $g''(e) = \frac{-1}{e^3} < 0$. Thus, $g(e)$ is a local maximum.

59. $g(x) = x - \ln x$

SOLUTION   Let $g(x) = x - \ln x$. Then $g'(x) = 1 - 1/x$, and $g'(x) = 0$ when $x = 1$. $g''(x) = 1/x^2$, so $g''(1) = 1 > 0$. Thus $g(1)$ is a local minimum.

In Exercises 61–66, find the derivative using the methods of Example 9.

61. $f(x) = x^{2x}$

SOLUTION   Method 1: $x^{2x} = e^{2x \ln x}$, so

$$\frac{d}{dx}x^{2x} = e^{2x \ln x}(2 + 2\ln x) = x^{2x}(2 + 2\ln x).$$

Method 2: Let $y = x^{2x}$. Then, $\ln y = 2x \ln x$. By logarithmic differentiation

$$\frac{y'}{y} = 2x \cdot \frac{1}{x} + 2\ln x,$$

so

$$y' = y(2 + 2\ln x) = x^{2x}(2 + 2\ln x).$$

63. $f(x) = x^{e^x}$

SOLUTION   Method 1: $x^{e^x} = e^{e^x \ln x}$, so

$$\frac{d}{dx}x^{e^x} = e^{e^x \ln x}\left(\frac{e^x}{x} + e^x \ln x\right) = x^{e^x}\left(\frac{e^x}{x} + e^x \ln x\right).$$

Method 2: Let $y = x^{e^x}$. Then $\ln y = e^x \ln x$. By logarithmic differentiation

$$\frac{y'}{y} = e^x \cdot \frac{1}{x} + e^x \ln x,$$

so

$$y' = y\left(\frac{e^x}{x} + e^x \ln x\right) = x^{e^x}\left(\frac{e^x}{x} + e^x \ln x\right).$$

65. $f(x) = x^{2^x}$

SOLUTION   Method 1: $x^{2^x} = e^{2^x \ln x}$, so

$$\frac{d}{dx}x^{2^x} = e^{2^x \ln x}\left(\frac{2^x}{x} + (\ln x)(\ln 2)2^x\right) = x^{2^x}\left(\frac{2^x}{x} + (\ln x)(\ln 2)2^x\right).$$

Method 2: Let $y = x^{2^x}$. Then $\ln y = 2^x \ln x$. By logarithmic differentiation

$$\frac{y'}{y} = 2^x \frac{1}{x} + (\ln x)(\ln 2)2^x,$$

so

$$y' = x^{2^x}\left(\frac{2^x}{x} + (\ln x)(\ln 2)2^x\right).$$

In Exercises 67–74, evaluate the derivative using logarithmic differentiation as in Example 8.

67. $y = (x + 2)(x + 4)$

SOLUTION Let $y = (x + 2)(x + 4)$. Then $\ln y = \ln((x + 2)(x + 4)) = \ln(x + 2) + \ln(x + 4)$. By logarithmic differentiation

$$\frac{y'}{y} = \frac{1}{x + 2} + \frac{1}{x + 4}$$

or

$$y' = (x + 2)(x + 4)\left(\frac{1}{x + 2} + \frac{1}{x + 4}\right) = (x + 4) + (x + 2) = 2x + 6.$$

69. $y = \dfrac{x(x + 1)^3}{(3x - 1)^2}$

SOLUTION Let $y = \dfrac{x(x+1)^3}{(3x-1)^2}$. Then $\ln y = \ln x + 3\ln(x + 1) - 2\ln(3x - 1)$. By logarithmic differentiation

$$\frac{y'}{y} = \frac{1}{x} + \frac{3}{x + 1} - \frac{6}{3x - 1},$$

so

$$y' = \frac{(x + 1)^3}{(3x - 1)^2} + \frac{3x(x + 1)^2}{(3x - 1)^2} - \frac{6x(x + 1)^3}{(3x - 1)^3}.$$

71. $y = (2x + 1)(4x^2)\sqrt{x - 9}$

SOLUTION Let $y = (2x + 1)(4x^2)\sqrt{x - 9}$. Then

$$\ln y = \ln(2x + 1) + \ln 4x^2 + \ln(x - 9)^{1/2} = \ln(2x + 1) + \ln 4 + 2\ln x + \frac{1}{2}\ln(x - 9).$$

By logarithmic differentiation

$$\frac{y'}{y} = \frac{2}{2x + 1} + \frac{2}{x} + \frac{1}{2(x - 9)},$$

so

$$y' = (2x + 1)(4x^2)\sqrt{x - 9}\left(\frac{2}{2x + 1} + \frac{2}{x} + \frac{1}{2(x - 9)}\right).$$

73. $y = (x^2 + 1)(x^2 + 2)(x^2 + 3)^2$

SOLUTION Let $y = (x^2 + 1)(x^2 + 2)(x^2 + 3)^2$. Then $\ln y = \ln(x^2 + 1) + \ln(x^2 + 2) + 2\ln(x^2 + 3)$. By logarithmic differentiation

$$\frac{y'}{y} = \frac{2x}{x^2 + 1} + \frac{2x}{x^2 + 2} + \frac{4x}{x^2 + 3},$$

so

$$y' = (x^2 + 1)(x^2 + 2)(x^2 + 3)^2\left(\frac{2x}{x^2 + 1} + \frac{2x}{x^2 + 2} + \frac{4x}{x^2 + 3}\right).$$

In Exercises 75–78, find the local extrema and points of inflection, and sketch the graph of $y = f(x)$ over the given interval.

75. $y = x^2 - \ln x$, $[\frac{1}{2}, 2]$

SOLUTION Let $y = x^2 - \ln x$. Then

$$y' = 2x - \frac{1}{x} = \frac{2x^2 - 1}{x}$$

and $y'' = 2 + \frac{1}{x^2}$. Thus, the function is decreasing on $(\frac{1}{2}, \frac{\sqrt{2}}{2})$, is increasing on $(\frac{\sqrt{2}}{2}, 2)$, has a local minimum at $x = \frac{\sqrt{2}}{2}$, and is concave up over the entire interval $(\frac{1}{2}, 2)$. A graph of $y = x^2 - \ln x$ is shown below.

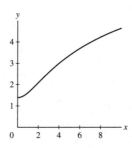

77. $y = \ln(x^2 + 4)$, $[0, \infty)$

SOLUTION Let $y = \ln(x^2 + 4)$. Then

$$y' = \frac{2x}{x^2 + 4} \quad \text{and} \quad y'' = \frac{(x^2 + 4)(2) - 2x(2x)}{(x^2 + 4)^2} = \frac{8 - 2x^2}{(x^2 + 4)^2}.$$

Thus, the function is increasing on $(0, \infty)$, is concave up on $(0, 2)$, is concave down on $(2, \infty)$ and has a point of inflection at $x = 2$. A graph of $y = \ln(x^2 + 4)$ is shown below.

In Exercises 79–96, evaluate the indefinite integral, using substitution if necessary.

79. $\displaystyle \int \frac{dx}{2x + 4}$

SOLUTION Let $u = 2x + 4$. Then $du = 2\,dx$, and

$$\int \frac{dx}{2x + 4} = \frac{1}{2} \int \frac{1}{u}\,du = \frac{1}{2} \ln|2x + 4| + C.$$

81. $\displaystyle \int \frac{x^2\,dx}{x^3 + 2}$

SOLUTION Let $u = x^3 + 2$. Then $du = 3x^2\,dx$, and

$$\int \frac{x^2\,dx}{x^3 + 2} = \frac{1}{3} \int \frac{du}{u} = \frac{1}{3} \ln|x^3 + 2| + C.$$

83. $\displaystyle \int \tan(4x + 1)\,dx$

SOLUTION First we rewrite $\int \tan(4x + 1)\,dx$ as $\int \frac{\sin(4x+1)}{\cos(4x+1)}\,dx$. Let $u = \cos(4x + 1)$. Then $du = -4 \sin(4x + 1)\,dx$, and

$$\int \frac{\sin(4x + 1)}{\cos(4x + 1)}\,dx = -\frac{1}{4} \int \frac{du}{u} = -\frac{1}{4} \ln|\cos(4x + 1)| + C.$$

85. $\displaystyle \int \frac{\cos x}{2 \sin x + 3}\,dx$

SOLUTION Let $u = 2 \sin x + 3$. Then $du = 2 \cos x\,dx$, and

$$\int \frac{\cos x}{2 \sin x + 3}\,dx = \frac{1}{2} \int \frac{du}{u} = \frac{1}{2} \ln(2 \sin x + 3) + C,$$

where we have used the fact that $2 \sin x + 3 \geq 1$ to drop the absolute value.

87. $\displaystyle \int \frac{4 \ln x + 5}{x}\,dx$

SOLUTION   Let $u = 4 \ln x + 5$. Then $du = (4/x)dx$, and

$$\int \frac{4 \ln x + 5}{x} dx = \frac{1}{4} \int u \, du = \frac{1}{8} u^2 + C = \frac{1}{8} (4 \ln x + 5)^2 + C.$$

89. $\int \dfrac{dx}{x \ln x}$

SOLUTION   Let $u = \ln x$. Then $du = (1/x)dx$, and

$$\int \frac{dx}{x \ln x} = \int \frac{1}{u} du = \ln |u| + C = \ln | \ln x| + C.$$

91. $\int \dfrac{\ln(\ln x)}{x \ln x} dx$

SOLUTION   Let $u = \ln(\ln x)$. Then $du = \dfrac{1}{\ln x} \cdot \dfrac{1}{x} dx$ and

$$\int \frac{\ln(\ln x)}{x \ln x} dx = \int u \, du = \frac{u^2}{2} + C = \frac{(\ln(\ln x))^2}{2} + C.$$

93. $\int 3^x \, dx$

SOLUTION $\int 3^x \, dx = \dfrac{3^x}{\ln 3} + C.$

95. $\int \cos x \, 3^{\sin x} \, dx$

SOLUTION   Let $u = \sin x$. Then $du = \cos x \, dx$, and

$$\int \cos x \, 3^{\sin x} \, dx = \int 3^u \, du = \frac{3^u}{\ln 3} + C = \frac{3^{\sin x}}{\ln 3} + C.$$

In Exercises 97–102, evaluate the definite integral.

97. $\displaystyle\int_1^2 \frac{1}{x} dx$

SOLUTION $\displaystyle\int_1^2 \frac{1}{x} dx = \ln |x| \Big|_1^2 = \ln 2 - \ln 1 = \ln 2.$

99. $\displaystyle\int_1^e \frac{1}{x} dx$

SOLUTION $\displaystyle\int_1^e \frac{1}{x} dx = \ln |x| \Big|_1^e = \ln e - \ln 1 = 1.$

101. $\displaystyle\int_{-e^2}^{-e} \frac{1}{t} dt$

SOLUTION $\displaystyle\int_{-e^2}^{-e} \frac{1}{t} dt = \ln |t| \Big|_{-e^2}^{-e} = \ln |-e| - \ln |-e^2| = \ln \frac{e}{e^2} = \ln(1/e) = -1.$

103. CAS   Find a good numerical approximation to the coordinates of the point on the graph of $y = \ln x - x$ closest to the origin (Figure 10).

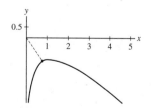

FIGURE 10 Graph of $y = \ln x - x$.

SOLUTION The distance from the origin to the point $(x, \ln x - x)$ on the graph of $y = \ln x - x$ is $d = \sqrt{x^2 + (\ln x - x)^2}$. As usual, we will minimize d^2. Let $d^2 = f(x) = x^2 + (\ln x - x)^2$. Then

$$f'(x) = 2x + 2(\ln x - x)\left(\frac{1}{x} - 1\right).$$

To determine x, we need to solve

$$4x + \frac{2\ln x}{x} - 2\ln x - 2 = 0.$$

This yields $x \approx .632784$. Thus, the point on the graph of $y = \ln x - x$ that is closest to the origin is approximately $(0.632784, -1.090410)$.

105. ✍️ Use the formula $(\ln f(x))' = f'(x)/f(x)$ to show that $\ln x$ and $\ln(2x)$ have the same derivative. Is there a simpler explanation of this result?

SOLUTION Observe

$$(\ln x)' = \frac{1}{x} \quad \text{and} \quad (\ln 2x)' = \frac{2}{2x} = \frac{1}{x}.$$

As an alternative explanation, note that $\ln(2x) = \ln 2 + \ln x$. Hence, $\ln x$ and $\ln(2x)$ differ by a constant, which implies the two functions have the same derivative.

107. What is b if the slope of the tangent line to the curve $y = b^x$ at $x = 0$ is 4?

SOLUTION $y' = \ln b \cdot b^x$ so $y'(0) = \ln b$. Setting this equal to 4 and solving for b yields $b = e^4$.

109. The energy E (in joules) radiated as seismic waves from an earthquake of Richter magnitude M is given by the formula $\log_{10} E = 4.8 + 1.5M$.
(a) Express E as a function of M.
(b) Show that when M increases by 1, the energy increases by a factor of approximately 31.
(c) Calculate dE/dM.

SOLUTION
(a) Solving $\log_{10} E = 4.8 + 1.5M$ for E yields

$$E = 10^{4.8+1.5M}.$$

(b) Using the formula from part (a), we find

$$\frac{E(M+1)}{E(M)} = \frac{10^{4.8+1.5(M+1)}}{10^{4.8+1.5M}} = \frac{10^{6.3+1.5M}}{10^{4.8+1.5M}} = 10^{1.5} \approx 31.6228.$$

(c) Again using the formula from part (a),

$$\frac{dE}{dM} = 1.5(\ln 10)10^{4.8+1.5M}.$$

Further Insights and Challenges

111. Show that $\log_a b \, \log_b a = 1$.

SOLUTION $\log_a b = \dfrac{\ln b}{\ln a}$ and $\log_b a = \dfrac{\ln a}{\ln b}$. Thus $\log_a b \cdot \log_b a = \dfrac{\ln b}{\ln a} \cdot \dfrac{\ln a}{\ln b} = 1$.

113. Use Exercise 112 to verify the formula

$$\frac{d}{dx} \log_b x = \frac{1}{(\ln b)x}$$

SOLUTION $\dfrac{d}{dx} \log_b x = \dfrac{d}{dx} \dfrac{\ln x}{\ln b} = \dfrac{1}{(\ln b)x}$.

115. Show that if $f(x)$ is strictly increasing and satisfies $f(xy) = f(x) + f(y)$, then its inverse $g(x)$ satisfies $g(x + y) = g(x)g(y)$.

SOLUTION Let $x = f(w)$ and $y = f(z)$. Then

$$g(x + y) = g(f(w) + f(z)) = g(f(wz)) = wz = g(x) \cdot g(y).$$

Exercises 114–116 provide a mathematically elegant approach to the exponential and logarithm functions, which avoids the problem of defining e^x for irrational x and of proving that e^x is differentiable.

7.4 Exponential Growth and Decay

Preliminary Questions

1. Two quantities increase exponentially with growth constants $k = 1.2$ and $k = 3.4$, respectively. Which quantity doubles more rapidly?

SOLUTION Doubling time is inversely proportional to the growth constant. Consequently, the quantity with $k = 3.4$ doubles more rapidly.

2. If you are given both the doubling time and the growth constant of a quantity that increases exponentially, can you determine the initial amount?

SOLUTION No. To determine the initial amount, we need to know the amount at one instant in time.

3. A cell population grows exponentially beginning with one cell. Does it take less time for the population to increase from one to two cells than from 10 million to 20 million cells?

SOLUTION Because growth from one cell to two cells and growth from 10 million to 20 million cells both involve a doubling of the population, both increases take exactly the same amount of time.

4. Referring to his popular book *A Brief History of Time*, the renowned physicist Stephen Hawking said, "Someone told me that each equation I included in the book would halve its sales." If this is so, write a differential equation satisfied by the sales function $S(n)$, where n is the number of equations in the book.

SOLUTION Let $S(0)$ denote the sales with no equations in the book. Translating Hawking's observation into an equation yields

$$S(n) = \frac{S(0)}{2^n}.$$

Differentiating with respect to n then yields

$$\frac{dS}{dn} = S(0)\frac{d}{dn}2^{-n} = -\ln 2 S(0)2^{-n} = -\ln 2 S(n).$$

5. Carbon dating is based on the assumption that the ratio R of C^{14} to C^{12} in the atmosphere has been constant over the past 50,000 years. If R were actually smaller in the past than it is today, would the age estimates produced by carbon dating be too ancient or too recent?

SOLUTION If R were actually smaller in the past than it is today, then we would be overestimating the amount of decay and therefore overestimating the age. Our estimates would be too ancient.

Exercises

1. A certain bacteria population P obeys the exponential growth law $P(t) = 2{,}000e^{1.3t}$ (t in hours).

(a) How many bacteria are present initially?

(b) At what time will there be 10,000 bacteria?

SOLUTION

(a) $P(0) = 2000e^0 = 2000$ bacteria initially.

(b) We solve $2000e^{1.3t} = 10{,}000$ for t. Thus, $e^{1.3t} = 5$ or

$$t = \frac{1}{1.3}\ln 5 \approx 1.24 \text{ hours.}$$

3. A certain RNA molecule has a doubling time of 3 minutes. Find the growth constant k and the differential equation for the number $N(t)$ of molecules present at time t (in minutes). Starting with one molecule, how many will be present after 10 min?

SOLUTION The doubling time is $\dfrac{\ln 2}{k}$ so $k = \dfrac{\ln 2}{\text{doubling time}}$. Thus, the differential equation is $N'(t) = kN(t) = \dfrac{\ln 2}{3}N(t)$. With one molecule initially,

$$N(t) = e^{(\ln 2/3)t} = 2^{t/3}.$$

Thus, after ten minutes, there are

$$N(10) = 2^{10/3} \approx 10.079,$$

or 10 molecules present.

5. The decay constant of Cobalt-60 is 0.13 years^{-1}. What is its half-life?

SOLUTION Half-life $= \dfrac{\ln 2}{0.13} \approx 5.33$ years.

7. Find all solutions to the differential equation $y' = -5y$. Which solution satisfies the initial condition $y(0) = 3.4$?

SOLUTION $y' = -5y$, so $y(t) = Ce^{-5t}$ for some constant C. The initial condition $y(0) = 3.4$ determines $C = 3.4$. Therefore, $y(t) = 3.4e^{-5t}$.

9. Find the solution to $y' = 3y$ satisfying $y(2) = 4$.

SOLUTION $y' = 3y$, so $y(t) = Ce^{3t}$ for some constant C. The initial condition $y(2) = 4$ determines $C = \dfrac{4}{e^6}$. Therefore, $y(t) = \dfrac{4}{e^6}e^{3t} = 4e^{3(t-2)}$.

11. The population of a city is $P(t) = 2 \cdot e^{0.06t}$ (in millions), where t is measured in years.
(a) Calculate the doubling time of the population.
(b) How long does it take for the population to triple in size?
(c) How long does it take for the population to quadruple in size?

SOLUTION
(a) Since $k = 0.06$, the doubling time is

$$\frac{\ln 2}{k} \approx 11.55 \text{ years.}$$

(b) The tripling time is calculated in the same way as the doubling time. Solve for Δ in the equation

$$P(t + \Delta) = 3P(t)$$

$$2 \cdot e^{0.06(t+\Delta)} = 3(2e^{0.06t})$$

$$2 \cdot e^{0.06t}e^{0.06\Delta} = 3(2e^{0.06t})$$

$$e^{0.06\Delta} = 3$$

$$0.06\Delta = \ln 3,$$

or $\Delta = \ln 3/0.06 \approx 18.31$ years.
(c) Since the population doubles every 11.55 years, it quadruples after

$$2 \times 11.55 = 23.10 \text{ years.}$$

13. Assuming that population growth is approximately exponential, which of the two sets of data is most likely to represent the population (in millions) of a city over a 5-year period?

Year	2000	2001	2002	2003	2004
Data I	3.14	3.36	3.60	3.85	4.11
Data II	3.14	3.24	3.54	4.04	4.74

SOLUTION If the population growth is approximately exponential, then the ratio between successive years' data needs to be approximately the same.

Year	2000	2001	2002	2003	2004
Data I	3.14	3.36	3.60	3.85	4.11
Ratios		1.07006	1.07143	1.06944	1.06753
Data II	3.14	3.24	3.54	4.04	4.74
Ratios		1.03185	1.09259	1.14124	1.17327

As you can see, the ratio of successive years in the data from "Data I" is very close to 1.07. Therefore, we would expect exponential growth of about $P(t) \approx (3.14)(1.07^t)$.

15. The **Beer–Lambert Law** is used in spectroscopy to determine the molar absorptivity α or the concentration c of a compound dissolved in a solution at low concentrations (Figure 10). The law states that the intensity I of light as it passes through the solution satisfies $\ln(I/I_0) = \alpha c x$, where I_0 is the initial intensity and x is the distance traveled by the light. Show that I satisfies a differential equation $dI/dx = -kI$ for some constant k.

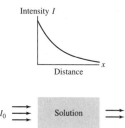

FIGURE 10 Light of intensity passing through a solution.

SOLUTION $\ln\left(\dfrac{I}{I_0}\right) = \alpha c x$ so $\dfrac{I}{I_0} = e^{\alpha c x}$ or $I = I_0 e^{\alpha c x}$. Therefore,

$$\frac{dI}{dx} = I_0 e^{\alpha c x}(\alpha c) = I(\alpha c) = -kI,$$

where $k = -\alpha c$ is a constant.

17. A 10-kg quantity of a radioactive isotope decays to 3 kg after 17 years. Find the decay constant of the isotope.

SOLUTION $P(t) = 10e^{-kt}$. Thus $P(17) = 3 = 10e^{-17k}$, so $k = \dfrac{\ln(3/10)}{-17} \approx 0.071$ years^{-1}.

19. Chauvet Caves In 1994, rock climbers in southern France stumbled on a cave containing prehistoric cave paintings. A C^{14}-analysis carried out by French archeologist Helene Valladas showed that the paintings are between 29,700 and 32,400 years old, much older than any previously known human art. Given that the C^{14} to C^{12} ratio of the atmosphere is $R = 10^{-12}$, what range of C^{14} to C^{12} ratios did Valladas find in the charcoal specimens?

SOLUTION The C^{14}-C^{12} ratio found in the specimens ranged from

$$10^{-12}e^{-0.000121(32400)} \approx 1.98 \times 10^{-14}$$

to

$$10^{-12}e^{-0.000121(29700)} \approx 2.75 \times 10^{-14}.$$

21. Atmospheric Pressure The atmospheric pressure $P(h)$ (in pounds per square inch) at a height h (in miles) above sea level on earth satisfies a differential equation $P' = -kP$ for some positive constant k.

(a) Measurements with a barometer show that $P(0) = 14.7$ and $P(10) = 2.13$. What is the decay constant k?

(b) Determine the atmospheric pressure 15 miles above sea level.

SOLUTION

(a) Because $P' = -kP$ for some positive constant k, $P(h) = Ce^{-kh}$ where $C = P(0) = 14.7$. Therefore, $P(h) = 14.7e^{-kh}$. We know that $P(10) = 14.7e^{-10k} = 2.13$. Solving for k yields

$$k = -\frac{1}{10}\ln\left(\frac{2.13}{14.7}\right) \approx 0.193 \text{ miles}^{-1}.$$

(b) $P(15) = 14.7e^{-0.193(15)} \approx 0.813$ pounds per square inch.

23. A quantity P increases exponentially with doubling time 6 hours. After how many hours has P increased by 50%?

SOLUTION The doubling time is $\dfrac{\ln 2}{k} = 6$ so $k \approx 0.1155$ hours^{-1}. P will have increased by 50% when $1.5P_0 = P_0 e^{0.1155t}$, or when $t = \dfrac{\ln 1.5}{0.1155} \approx 3.5$ hours.

25. Moore's Law In 1965, Gordon Moore predicted that the number N of transistors on a microchip would increase exponentially.

(a) Does the table of data below confirm Moore's prediction for the period from 1971 to 2000? If so, estimate the growth constant k.

(b) *CAS* Plot the data in the table.

(c) Let $N(t)$ be the number of transistors t years after 1971. Find an approximate formula $N(t) \approx Ce^{kt}$, where t is the number of years after 1971.

(d) Estimate the doubling time in Moore's Law for the period from 1971 to 2000.

(e) If Moore's Law continues to hold until the end of the decade, how many transistors will a chip contain in 2010?

(f) Can Moore have expected his prediction to hold indefinitely?

Transistors	Year	No. Transistors
4004	1971	2,250
8008	1972	2,500
8080	1974	5,000
8086	1978	29,000
286	1982	120,000
386 processor	1985	275,000
486 DX processor	1989	1,180,000
Pentium processor	1993	3,100,000
Pentium II processor	1997	7,500,000
Pentium III processor	1999	24,000,000
Pentium 4 processor	2000	42,000,000

SOLUTION

(a) Yes, the graph looks like an exponential graph especially towards the latter years. We estimate the growth constant by setting 1971 as our starting point, so $P_0 = 2250$. Therefore, $P(t) = 2250e^{kt}$. In 2000, $t = 29$. Therefore, $P(29) = 2250e^{29k} = 42000000$, so $k = \frac{\ln 18666.67}{29} \approx 0.339$. Note: A better estimate can be found by calculating k for each time period and then averaging the k values.

(b)

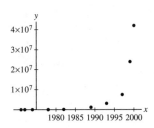

(c) $N(t) = 2250e^{0.339t}$

(d) The doubling time is $\ln 2/0.339 \approx 2.04$ years.

(e) In 2010, $t = 39$ years. Therefore, $N(39) = 2250e^{0.339(39)} \approx 1,241,623,327$.

(f) No, you can't make a microchip smaller than an atom.

*In Exercises 27–28, we consider the **Gompertz differential equation**:*

$$\frac{dy}{dt} = ky \ln\left(\frac{y}{M}\right)$$

(where M and k are constants), introduced in 1825 by the English mathematician Benjamin Gompertz and still used today to model aging and mortality.

27. Show that $y = Me^{ae^{kt}}$ is a solution for any constant a.

SOLUTION   Let $y = Me^{ae^{kt}}$. Then

$$\frac{dy}{dt} = M(kae^{kt})e^{ae^{kt}}$$

and, since

$$\ln(y/M) = ae^{kt},$$

we have

$$ky \ln(y/M) = Mkae^{kt}e^{ae^{kt}} = \frac{dy}{dt}.$$

29. A certain quantity increases quadratically: $P(t) = P_0 t^2$.

(a) Starting at time $t_0 = 1$, how long will it take for P to double in size? How long will it take starting at $t_0 = 2$ or 3?

(b) In general, starting at time t_0, how long will it take for P to double in size?

SOLUTION

(a) Starting from $t_0 = 1$, P doubles when $P(t) = 2P(1) = 2P_0$. Thus, $P_0 t^2 = 2P_0$ and $t = \sqrt{2}$. Starting from $t_0 = 2$, P doubles when

$$P(t) = P_0 t^2 = 2P(2) = 8P_0.$$

Thus, $t = 2\sqrt{2}$. Finally, starting from $t_0 = 3$, P doubles when

$$P(t) = P_0 t^2 = 2P(3) = 18P_0.$$

Thus, $t = 3\sqrt{2}$.

(b) Starting from $t = t_0$, P doubles when

$$P(t) = P_0 t^2 = 2P(t_0) = 2P_0 t_0^2.$$

Thus, $t = t_0 \sqrt{2}$.

Further Insights and Challenges

31.   **Isotopes for Dating**   Which of the following isotopes would be most suitable for dating extremely old rocks: Carbon-14 (half-life 5,570 years), Lead-210 (half-life 22.26 years), and Potassium-49 (half-life 1.3 billion years)? Explain why.

SOLUTION   For extremely old rocks, you need to have an isotope that decays very slowly. In other words, you want a very large half-life such as Potassium-49; otherwise, the amount of undecayed isotope in the rock sample would be too small to accurately measure.

33.   **Average Time of Decay**   Physicists use the radioactive decay law $R = R_0 e^{-kt}$ to compute the average or *mean time* M until an atom decays. Let $F(t) = R/R_0 = e^{-kt}$ be the fraction of atoms that have survived to time t without decaying.

(a) Find the inverse function $t(F)$.

(b) The error in the following approximation tends to zero as $N \to \infty$:

$$M = \text{mean time to decay} \approx \frac{1}{N} \sum_{j=1}^{N} t\left(\frac{j}{N}\right)$$

Argue that $M = \int_0^1 t(F) \, dF$.

(c) Verify the formula $\int \ln x \, dx = x \ln x - x$ by differentiation and use it to show that for $c > 0$,

$$\int_c^1 t(F) \, dF = \frac{1}{k} + \frac{1}{k}(c \ln c - c)$$

(d) Verify numerically that $\lim_{c \to 0} (c - c \ln c) = 0$ (we will prove this in Section 7.7).

(e) The integral defining M is "improper" because $t(0)$ is infinite. Show that $M = 1/k$ by computing the limit

$$M = \lim_{c \to 0} \int_c^1 t(F) \, dF$$

(f) What is the mean time to decay for Radon (with a half-life of 3.825 days)?

SOLUTION

(a) $F = e^{-kt}$ so $\ln F = -kt$ and $t(F) = \dfrac{\ln F}{-k}$

(b) $M \approx \dfrac{1}{N} \sum_{j=1}^{N} t(j/N)$. For the interval $[0, 1]$, from the approximation given, the subinterval length is $1/N$ and thus the right-hand endpoints have x-coordinate (j/N). Thus we have a Riemann sum and by definition,

$$\lim_{N \to \infty} \frac{1}{N} \sum_{j=1}^{N} t(j/N) = \int_0^1 t(F) \, dF.$$

(c) $\dfrac{d}{dx}(x \ln x - x) = x\left(\dfrac{1}{x}\right) + \ln x - 1 = \ln x$. Thus

$$\int_c^1 t(F) \, dF = -\frac{1}{k}(F \ln F - F)\Big|_c^1 = \frac{1}{k}(F - F \ln F)\Big|_c^1$$

$$= \frac{1}{k}(1 - 1 \ln 1 - (c - c \ln c)) = \frac{1}{k} + \frac{1}{k}(c \ln c - c).$$

(d) Let $g(c) = c \ln c - c$. Then,

c	0.01	0.001	0.0001	0.00001
$g(c)$	-0.056052	-0.007908	-0.001021	-0.000125

Thus, as $c \to 0+$, it appears that $g(c) \to 0$.

(e) $M = \lim\limits_{c \to 0} \int_c^1 t(F)\,dF = \lim\limits_{c \to 0} \left(\dfrac{1}{k} + \dfrac{1}{k}(c \ln c - c) \right) = \dfrac{1}{k}$.

(f) Since the half-life is 3.825 days, $k = \dfrac{\ln 2}{3.825}$ and $\dfrac{1}{k} = 5.52$. Thus, $M = 5.52$ days.

7.5 Compound Interest and Present Value

Preliminary Questions

1. Which is preferable: an interest rate of 12% compounded quarterly, or an interest rate of 11% compounded continuously?

SOLUTION   To answer this question, we need to determine the yearly multiplier associated with each interest rate. The multiplier associated with an interest rate of 12% compounded quarterly is

$$\left(1 + \frac{0.12}{4} \right)^4 \approx 1.1255,$$

while the multiplier associated with an interest rate of 11% compounded continuously is

$$e^{0.11} \approx 1.11627.$$

Thus, the compounded quarterly rate is preferable.

2. Find the yearly multiplier if $r = 9\%$ and interest is compounded (a) continuously and (b) quarterly.

SOLUTION   With $r = 9\%$, the yearly multiplier for continuously compounded interest is

$$e^{0.09} \approx 1.09417,$$

and the yearly multiplier for compounded quarterly interest is

$$\left(1 + \frac{0.09}{4} \right)^4 \approx 1.09308.$$

3. The PV of N dollars received at time T is (choose the correct answer):

(a) The value at time T of N dollars invested today

(b) The amount you would have to invest today in order to receive N dollars at time T

SOLUTION   The correct response is **(b)**: the PV of N dollars received at time T is the amount you would have to invest today in order to receive N dollars at time T.

4. A year from now, \$1 will be received. Will its PV increase or decrease if the interest rate goes up?

SOLUTION   If the interest rate goes up, the present value of \$1 a year from now will decrease.

5. Xavier expects to receive a check for \$1,000 1 year from today. Explain, using the concept of PV, whether he will be happy or sad to learn that the interest rate has just increased from 6% to 7%.

SOLUTION   If the interest rate goes up, the present value of \$1,000 one year from today decreases. Therefore, Xavier will be sad is the interest rate has just increased from 6 to 7%.

Exercises

1. Compute the balance after 10 years if \$2,000 is deposited in an account paying 9% interest and interest is compounded (a) quarterly, (b) monthly, and (c) continuously.

SOLUTION

(a) $P(10) = 2000(1 + .09/4)^{4(10)} = \$4870.38$

(b) $P(10) = 2000(1 + .09/12)^{12(10)} = \$4902.71$

(c) $P(10) = 2000e^{.09(10)} = \$4919.21$

3. A bank pays interest at a rate of 5%. What is the yearly multiplier if interest is compounded

(a) yearly? **(b)** three times a year?

(c) continuously?

SOLUTION

(a) $P(t) = P_0(1 + 0.05)^t$, so the yearly multiplier is 1.05.

(b) $P(t) = P_0\left(1 + \dfrac{0.05}{3}\right)^{3t}$, so the yearly multiplier is $\left(1 + \dfrac{0.05}{3}\right)^3 \approx 1.0508$.

(c) $P(t) = P_0 e^{0.05t}$, so the yearly multiplier is $e^{0.05} \approx 1.0513$.

5. Show that if interest is compounded continuously at a rate r, then an account doubles after $(\ln 2)/r$ years.

SOLUTION   The account doubles when $P(t) = 2P_0 = P_0 e^{rt}$, so $2 = e^{rt}$ and $t = \dfrac{\ln 2}{r}$.

7. An investment increases in value at a continuously compounded rate of 9%. How large must the initial investment be in order to build up a value of \$50,000 over a seven-year period?

SOLUTION   Solving $50{,}000 = P_0 e^{0.09(7)}$ for P_0 yields

$$P_0 = \frac{50000}{e^{0.63}} \approx \$26{,}629.59.$$

9. Is it better to receive \$1,000 today or \$1,300 in 4 years? Consider $r = 0.08$ and $r = 0.03$.

SOLUTION   Assuming continuous compounding, if $r = 0.08$, then the present value of \$1300 four years from now is $1300e^{-0.08(4)} = \$943.99$. It is better to get \$1,000 now. On the other hand, if $r = 0.03$, the present value of \$1300 four years from now is $1300e^{-0.03(4)} = \$1153.00$, so it is better to get the \$1,300 in four years.

11. If a company invests \$2 million to upgrade its factory, it will earn additional profits of \$500,000/year for 5 years. Is the investment worthwhile, assuming an interest rate of 6% (assume that the savings are received as a lump sum at the end of each year)?

SOLUTION   The present value of the stream of additional profits is

$$500{,}000(e^{-0.06} + e^{-0.12} + e^{-0.18} + e^{-0.24} + e^{-0.3}) = \$2{,}095{,}700.63.$$

This is more than the \$2 million cost of the upgrade, so the upgrade should be made.

13. After winning \$25 million in the state lottery, Jessica learns that she will receive five yearly payments of \$5 million beginning immediately.

(a) What is the PV of Jessica's prize if $r = 6\%$?

(b) How much more would the prize be worth if the entire amount were paid today?

SOLUTION

(a) The present value of the prize is

$$5{,}000{,}000(e^{-0.24} + e^{-0.18} + e^{-0.12} + e^{-0.06} + e^{-.06(0)}) = \$22{,}252{,}915.21.$$

(b) If the entire amount were paid today, the present value would be \$25 million, or \$2,747,084.79 more than the stream of payments made over five years.

15. Use Eq. (2) to compute the PV of an income stream paying out $R(t) = \$5,000/\text{year}$ continuously for 10 years and $r = 0.05$.

SOLUTION $PV = \displaystyle\int_0^{10} 5{,}000e^{-0.05t}\, dt = -100{,}000e^{-0.05t}\Big|_0^{10} = \$39{,}346.93.$

17. Find the PV of an investment that produces income continuously at a rate of \$800/year for 5 years, assuming an interest rate of $r = 0.08$.

SOLUTION $PV = \displaystyle\int_0^5 800e^{-0.08t}\, dt = -10{,}000e^{-0.08t}\Big|_0^5 = \$3296.80.$

19. Show that the PV of an investment that pays out R dollars/year continuously for T years is $R(1 - e^{-rT})/r$, where r is the interest rate.

SOLUTION   The present value of an investment that pays out R dollars/year continuously for T years is

$$PV = \int_0^T Re^{-rt}\, dt.$$

Let $u = -rt$, $du = -r\, dt$. Then

$$PV = -\frac{1}{r}\int_0^{-rT} Re^u\, du = -\frac{R}{r}e^u\Big|_0^{-rT} = -\frac{R}{r}(e^{-rT} - 1) = \frac{R}{r}(1 - e^{-rT}).$$

21. Suppose that $r = 0.06$. Use the result of Exercise 20 to estimate the payout rate R needed to produce an income stream whose PV is $20,000, assuming that the stream continues for a large number of years.

SOLUTION From Exercise 20, $PV = \dfrac{R}{r}$ so $20000 = \dfrac{R}{.06}$ or $R = \$1200$.

23. Use Eq. (5) to compute the PV of an investment that pays out income continuously at a rate $R(t) = (5,000 + 1,000t)e^{0.02t}$ dollars/year for 10 years and $r = 0.08$.

SOLUTION

$$PV - \int_0^{10} (5000 + 1000t)(e^{0.02t})e^{-0.08t}\, dt = \int_0^{10} 5000e^{-0.06t}\, dt + \int_0^{10} 1000te^{-0.06t}\, dt$$

$$= \frac{5000}{-0.06}(e^{-0.06(10)} - 1) - 1000\left(\frac{e^{-0.06(10)}(1 + 0.06(10))}{(0.06)^2}\right) + 1000\frac{1}{(0.06)^2}$$

$$= 37599.03 - 243916.28 + 277777.78 \approx \$71,460.53.$$

Further Insights and Challenges

25. The text proves that $e = \lim\limits_{n\to\infty} (1 + \frac{1}{n})^n$. Use a change of variables to show that for any x,

$$\lim_{n\to\infty} \left(1 + \frac{x}{n}\right)^n = \lim_{n\to\infty} \left(1 + \frac{1}{n}\right)^{nx}$$

Use this to conclude that $e^x = \lim\limits_{n\to\infty} (1 + \frac{x}{n})^n$.

SOLUTION Let $t = x/n$. Then

$$\lim_{n\to\infty} \left(1 + \frac{x}{n}\right)^n = \lim_{t\to\infty} \left(1 + \frac{1}{t}\right)^{tx} = \lim_{n\to\infty} \left(1 + \frac{1}{n}\right)^{nx}.$$

Since $e = \lim_{n\to\infty} \left(1 + \dfrac{1}{n}\right)^n$,

$$e^x = \lim_{n\to\infty} \left(1 + \frac{1}{n}\right)^{nx} = \lim_{n\to\infty} \left(1 + \frac{x}{n}\right)^n.$$

27. A bank pays interest at the rate r, compounded M times yearly. The **effective interest rate** r_e is the rate at which interest, if compounded annually, would have to be paid to produce the same yearly return.
(a) Find r_e if $r = 9\%$ compounded monthly.
(b) Show that $r_e = (1 + r/M)^M - 1$ and that $r_e = e^r - 1$ if interest is compounded continuously.
(c) Find r_e if $r = 11\%$ compounded continuously.
(d) Find the rate r, compounded weekly, that would yield an effective rate of 20%.

SOLUTION
(a) Compounded monthly, $P(t) = P_0(1 + r/12)^{12t}$. By the definition of r_e,

$$P_0(1 + 0.09/12)^{12t} = P_0(1 + r_e)^t$$

so

$$(1 + 0.09/12)^{12t} = (1 + r_e)^t \quad \text{or} \quad r_e = (1 + 0.09/12)^{12} - 1 = 0.0938,$$

or 9.38%
(b) In general,

$$P_0(1 + r/M)^{Mt} = P_0(1 + r_e)^t,$$

so $(1 + r/M)^{Mt} = (1 + r_e)^t$ or $r_e = (1 + r/M)^M - 1$. If interest is compounded continuously, then $P_0 e^{rt} = P_0(1 + r_e)^t$ so $e^{rt} = (1 + r_e)^t$ or $r_e = e^r - 1$.
(c) Using part (b), $r_e = e^{0.11} - 1 \approx 0.1163$ or 11.63%.
(d) Solving

$$0.20 = \left(1 + \frac{r}{52}\right)^{52} - 1$$

for r yields $r = 52(1.2^{1/52} - 1) = 0.1826$ or 18.26%.

7.6 Models Involving $y' = k(y - b)$

Preliminary Questions

1. What is the general solution to $y' = -k(y - b)$?

SOLUTION The general solution is $y(t) = b + Ce^{-kt}$ for any constant C.

2. Write down a solution to $y' = 4(y - 5)$ that tends to $-\infty$ as $t \to \infty$.

SOLUTION The general solution is $y(t) = 5 + Ce^{4t}$ for any constant C; thus the solution tends to $-\infty$ as $t \to \infty$ whenever $C < 0$. One specific example is $y(t) = 5 - e^{4t}$.

3. Does there exist a solution of $y' = -4(y - 5)$ that tends to ∞ as $t \to \infty$?

SOLUTION The general solution is $y(t) = 5 + Ce^{-4t}$ for any constant C. As $t \to \infty$, $y(t) \to 5$. Thus, there is no solution of $y' = -4(y - 5)$ that tends to ∞ as $t \to \infty$.

4. True or false? If $k > 0$, then all solutions of $y' = -k(y - b)$ approach the same limit as $t \to \infty$.

SOLUTION True. The general solution of $y' = -k(y - b)$ is $y(t) = b + Ce^{-kt}$ for any constant C. If $k > 0$, then $y(t) \to b$ as $t \to \infty$.

5. Suppose that material A cools more rapidly than material B. Which material has a larger cooling constant k in Newton's Law of Cooling, $y' = -k(y - T_0)$?

SOLUTION Because material A cools more rapidly, material A has the larger cooling constant.

Exercises

1. Find the general solution of $y' = 2(y - 10)$. Then find the two solutions satisfying $y(0) = 25$ and $y(0) = 5$, and sketch their graphs.

SOLUTION The general solution of $y' = 2(y - 10)$ is $y(t) = 10 + Ce^{2t}$ for any constant C. If $y(0) = 25$, then $10 + C = 25$, or $C = 15$; therefore, $y(t) = 10 + 15e^{2t}$. On the other hand, if $y(0) = 5$, then $10 + C = 5$, or $C = -5$; therefore, $y(t) = 10 - 5e^{2t}$. Graphs of these two functions are given below.

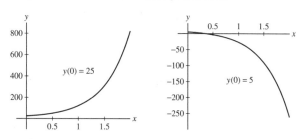

3. Verify directly that $y = b + Ce^{kt}$ satisfies $y' = k(y - b)$ for any constant C.

SOLUTION Let $y(t) = b + Ce^{kt}$ for any constant C. Then $y' = kCe^{kt}$ and

$$k(y - b) = k(b + Ce^{kt} - b) = kCe^{kt} = y'.$$

5. A hot metal bar is submerged in a large reservoir of water whose temperature is $60°$F. The temperature of the bar 20 s after submersion is $100°$F. After 1 min, the temperature has cooled to $80°$F.
(a) Determine the cooling constant k.
(b) What is the differential equation satisfied by the temperature $F(t)$ of the bar?
(c) What is the formula for $F(t)$?
(d) Determine the temperature of the bar at the moment it is submerged.

SOLUTION With $T_0 = 60°$F, the temperature of the bar is given by $F(t) = 60 + Ce^{-kt}$ for some constants C and k.

(a) Given $F(20) = 100°$F and $F(60) = 80°$F, it follows that $Ce^{-20k} = 40$ and $Ce^{-60k} = 20$. Dividing the first equation by the second leaves $e^{40k} = 2$, so

$$k = \frac{1}{40} \ln 2 \approx 0.017 \text{ seconds}^{-1}.$$

(b) $F'(t) = -0.017(F(t) - 60)$.

(c) The general solution of the equation in part (b) is $F(t) = 60 + Ce^{-0.017t}$ for some constant C. With $F(20) = 100 = 60 + Ce^{-0.017(20)}$ we have $40 = Ce^{-0.017(20)}$ so $C \approx 56.2$. Thus, $F(t) = 60 + 56.2e^{-0.017t}$.

(d) Using the formula found in part (c), $F(0) = 60 + 56.2 = 116.2°$F.

7. When a hot object is placed in a water bath whose temperature is 25°C, it cools from 100 to 50°C in 150 s. In another bath, the same cooling occurs in 120 s. Find the temperature of the second bath.

SOLUTION With $T_0 = 25°C$, the temperature of the object is given by $F(t) = 25 + Ce^{-kt}$ for some constants C and k. From the initial condition, $F(0) = 25 + C = 100$, so $C = 75$. After 150 seconds, $F(150) = 25 + 75e^{-150k} = 50$, so

$$k = -\frac{1}{150}\ln\left(\frac{25}{75}\right) \approx 0.0073 \text{ seconds}^{-1}.$$

If we place the same object with a temperature of 100°C into a second bath whose temperature is T_0, then the temperature of the object is given by

$$F(t) = T_0 + (100 - T_0)e^{-0.0073t}.$$

To cool from 100°C to 50°C in 120 seconds, T_0 must satisfy

$$T_0 + (100 - T_0)e^{-0.0073(120)} = 50.$$

Thus, $T_0 = 14.32°C$.

9. A cup of coffee, cooling off in a room at temperature 20°C, has cooling constant $k = 0.09 \text{ min}^{-1}$.

(a) How fast is the coffee cooling (in degrees per minute) when its temperature is $T = 80°C$?

(b) Use the Linear Approximation to estimate the change in temperature over the next 6 s when $T = 80°C$.

(c) The coffee is served at a temperature of 90°C. How long should you wait before drinking it if the optimal temperature is 65°C?

SOLUTION

(a) According to Newton's Law of Cooling, the coffee will cool at the rate $k(T - T_0)$, where k is the cooling constant of the coffee, T is the current temperature of the coffee and T_0 is the temperature of the surroundings. With $k = 0.09$ min^{-1}, $T = 80°C$ and $T_0 = 20°C$, the coffee is cooling at the rate

$$0.09(80 - 20) = 5.4°C/\text{min}.$$

(b) Using the result from part (a) and the Linear Approximation, we estimate that the coffee will cool

$$(5.4°C/\text{min})(0.1 \text{ min}) = 0.54°C$$

over the next 6 seconds.

(c) With $T_0 = 20°C$ and an initial temperature of 90°C, the temperature of the coffee at any time t is $T(t) = 20 + 70e^{-0.09t}$. Solving $20 + 70e^{-0.09t} = 65$ for t yields

$$t = -\frac{1}{0.09}\ln\left(\frac{45}{70}\right) \approx 4.91 \text{ minutes}.$$

In Exercises 11–14, use the model for free-fall with air resistance discussed in this section. If the weight w of an object (in pounds) is given rather than its mass (as in Example 2), then the differential equation for free fall with air resistance is $v' = -(kg/w)(v + w/k)$, where $g = 32 \text{ ft/s}^2$.

11. A 60-kg skydiver jumps out of an airplane. What is her terminal velocity in miles per hour, assuming that $k = 10 \text{ kg/s}$ for free-fall (no parachute)?

SOLUTION The free-fall terminal velocity is

$$\frac{-gm}{k} = \frac{-9.8(60)}{10} = -58.8 \text{ m/s} \approx -192.9 \text{ ft/s} \approx -131.5 \text{ mph}.$$

13. A 175-lb skydiver jumps out of an airplane (with zero initial velocity). Assume that $k = 0.7 \text{ lb-s/ft}$ with a closed parachute and $k = 5 \text{ lb-s/ft}$ with an open parachute. What is the skydiver's velocity at $t = 25$ s if the parachute opens after 20 seconds of free fall?

SOLUTION The skydiver's velocity $v(t)$ satisfies the differential equation $v' = -(kg/w)(v + w/k)$, which has general solution

$$v(t) = -\frac{w}{k} + Ce^{-(kg/w)t}.$$

While the parachute is closed, $k = 0.7$. Since $w = 175$ lbs, we have

$$\frac{w}{k} = \frac{175}{0.7} = 250 \text{ ft/sec}, \qquad \frac{kg}{w} = \frac{(0.7)(32)}{175} = 0.128 \text{ sec}^{-1}$$

and thus

$$v(t) = -250 + Ce^{-0.128t}.$$

From the initial condition $v(0) = -250 + C = 0$, we find that $C = 250$ and

$$v(t) = -250 + 250e^{-0.128t} = -250(1 - e^{-0.128t}).$$

At $t = 20$ sec, the skydiver's velocity is then

$$v(20) = -250(1 - e^{-0.128(20)}) \approx 231 \text{ ft/sec.}$$

After the parachute opens, we have $k = 5$, so

$$\frac{w}{k} = \frac{175}{5} = 35 \text{ ft/sec,} \qquad \frac{kg}{w} = \frac{(5)(32)}{175} \approx 0.91 \text{ sec}^{-1}$$

and

$$v(t) = -35 + Ce^{-0.91t},$$

where $t = 0$ now corresponds to the time when the parachute opens, Because the skydiver's velocity is 231 ft/sec when the parachute opens, we have

$$v(0) = -35 + C = 231 \quad \text{so that} \quad C = 266 \text{ ft/sec.}$$

Therefore

$$v(t) = -35 + 266e^{-0.91t}.$$

Finally, five seconds after the parachute opens, the skydiver's velocity is

$$v(t) = -35 + 266e^{-0.91(5)} = -35 + 266e^{-4.55} \approx -32 \text{ ft/sec.}$$

15. A continuous annuity with withdrawal rate $N = \$1{,}000$/year and interest rate $r = 5\%$ is funded by an initial deposit P_0.

(a) When will the annuity run out of funds if $P_0 = \$15{,}000$?

(b) Which initial deposit P_0 yields a constant balance?

SOLUTION

(a) Let $P(t)$ denote the balance of the annuity at time t measured in years. Then

$$P(t) = \frac{N}{r} + Ce^{rt} = \frac{1000}{0.05} + Ce^{0.05t} = 20000 + Ce^{0.05t}$$

for some constant C. If $P_0 = P(0) = 15000$, then $15000 = 20000 + C$ and $C = -5000$. To determine when the annuity runs out, we set $P(t) = 0$ and solve for t. This yields

$$t = \frac{1}{0.05} \ln 4 \approx 28 \text{ years.}$$

(b) From part (a), we know that $P(t) = 20000 + Ce^{0.05t}$. The balance of the annuity will remain constant provided $C = 0$. Then $P(t) = 20000$ for all t. Hence, $P_0 = \$20000$ leads to a constant balance.

17. Find the minimum initial deposit that will allow an annuity to pay out $\$500$/year indefinitely if it earns interest at a rate of 5%.

SOLUTION Let $P(t)$ denote the balance of the annuity at time t measured in years. Then

$$P(t) = \frac{N}{r} + Ce^{rt} = \frac{500}{0.05} + Ce^{0.05t} = 10000 + Ce^{0.05t}$$

for some constant C. To fund the annuity indefinitely, we must have $C \geq 0$. If the initial deposit is P_0, then $P_0 = 10000 + C$ and $C = P_0 - 10000$. Thus, to fund the annuity indefinitely, we must have $P_0 \geq \$10000$.

19. An initial deposit of $\$5{,}000$ is placed in a bank account. What is the minimum interest rate the bank must pay to allow continuous withdrawals at a rate of $\$500$/year to continue indefinitely?

SOLUTION Let $P(t)$ denote the balance of the annuity at time t measured in years. Then

$$P(t) = \frac{N}{r} + Ce^{rt} = \frac{500}{r} + Ce^{rt}$$

for some constant C. To fund the annuity indefinitely, we need $C \geq 0$. If the initial deposit is $\$5000$, then $5000 = \frac{500}{r} + C$ and $C = 5000 - \frac{500}{r}$. Thus, to fund the annuity indefinitely, we need $5000 - \frac{500}{r} \geq 0$, or $r \geq 0.1$. The bank must pay at least 10%.

21. Julie borrows $10,000 from a bank at an interest rate of 9% and pays back the loan continuously at a rate of N dollars/year. Let $P(t)$ denote the amount still owed at time t.

(a) Explain why $y = P(t)$ satisfies the differential equation

$$y' = 0.09y - N$$

(b) How long will it take Julie to pay back the loan if $N = \$1,200$?

(c) Will she ever be able to pay back the loan if $N = \$800$?

SOLUTION

(a)

$$\text{Rate of Change of Loan} = (\text{Amount still owed})(\text{Interest rate}) - (\text{Payback rate})$$

$$= P(t) \cdot r - N = r\left(P - \frac{N}{r}\right).$$

Therefore, if $y = P(t)$,

$$y' = r\left(y - \frac{N}{r}\right) = ry - N$$

(b) From the differential equation derived in part (a), we know that $P(t) = \frac{N}{r} + Ce^{rt} = 13333.33 + Ce^{0.09t}$. Since 10000 was initially borrowed, $P(0) = 13333.33 + C = 10000$, and $C = -3333.33$. The loan is paid off when $P(t) = 13333.33 - 3333.33e^{0.09t} = 0$. This yields

$$t = \frac{1}{0.09}\ln\left(\frac{13333.33}{3333.33}\right) \approx 15.4 \text{ years.}$$

(c) If the annual rate of payment is $800, then $P(t) = 800/0.09 + Ce^{0.09t} = 8888.89 + Ce^{.09t}$. With $P(0) = 8888.89 + C = 10000$, it follows that $C = 1111.11$. Since $C > 0$ and $e^{0.09t} \to \infty$ as $t \to \infty$, $P(t) \to \infty$, and the loan will never be paid back.

23. Current in a Circuit The electric current flowing in the circuit in Figure 6 (consisting of a battery of V volts, a resistor of R ohms, and an inductor) satisfies

$$\frac{dI}{dt} = -k(I - b)$$

for some constants k and b with $k > 0$. Initially, $I(0) = 0$ and $I(t)$ approaches a maximum level V/R as $t \to \infty$.

(a) What is the value of b?

(b) Find a formula for $I(t)$ in terms of k, V, and R.

(c) Show that $I(t)$ reaches approximately 63% of its maximum value at time $t = 1/k$.

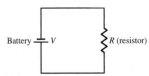

FIGURE 6 Current flow approaches the level $I_{\max} = V/R$.

SOLUTION

(a) Since the general solution has the form $I(t) = b + Ce^{-kt}$, we see that b is the final current level as $t \to \infty$. Therefore $b = V/R$.

(b) $I(t) = \dfrac{V}{R} - \dfrac{V}{R}e^{-kt} = \dfrac{V}{R}(1 - e^{-kt})$.

(c) As $1 - e^{-kt}$ is an increasing function of t, it follows that $I(t)$ achieves its maximum value as $t \to \infty$:

$$\text{maximum value} = \lim_{t\to\infty}\frac{V}{R}(1 - e^{-kt}) = \frac{V}{R}.$$

At time $t = 1/k$,

$$I\left(\frac{1}{k}\right) = \frac{V}{R}(1 - e^{-1}) \approx 0.6321\frac{V}{R},$$

or roughly 63% of its maximum value.

Further Insights and Challenges

25. Show that by Newton's Law of Cooling, the time required to cool an object from temperature A to temperature B is

$$t = \frac{1}{k} \ln\left(\frac{A - T_0}{B - T_0}\right)$$

where T_0 is the ambient temperature.

SOLUTION At any time t, the temperature of the object is $y(t) = T_0 + Ce^{-kt}$ for some constant C. Suppose the object is initially at temperature A and reaches temperature B at time t. Then $A = T_0 + C$, so $C = A - T_0$. Moreover,

$$B = T_0 + Ce^{-kt} = T_0 + (A - T_0)e^{-kt}.$$

Solving this last equation for t yields

$$t = \frac{1}{k} \ln\left(\frac{A - T_0}{B - T_0}\right).$$

7.7 L'Hôpital's Rule

Preliminary Questions

1. Which of the following two limits can be evaluated using L'Hôpital's Rule?

$$\lim_{x \to 4} \frac{3x - 12}{x^2 - 16}, \qquad \lim_{x \to 4} \frac{12x - 3}{x - 4}$$

SOLUTION As $x \to 4$,

$$\frac{3x - 12}{x^2 - 16}$$

is of the form $\frac{0}{0}$, so L'Hôpital's Rule can be used to evaluate

$$\lim_{x \to 4} \frac{3x - 12}{x^2 - 16}.$$

As $x \to 4$,

$$\frac{12x - 3}{x - 4}$$

is not of the form $\frac{0}{0}$ or $\frac{\infty}{\infty}$, so L'Hôpital's Rule cannot be used to evaluate

$$\lim_{x \to 4} \frac{12x - 3}{x - 4}.$$

2. What is wrong with evaluating

$$\lim_{x \to 0} \frac{x^2 - 2x}{3x - 2}$$

using L'Hôpital's Rule?

SOLUTION As $x \to 0$,

$$\frac{x^2 - 2x}{3x - 2}$$

is not of the form $\frac{0}{0}$ or $\frac{\infty}{\infty}$, so L'Hôpital's Rule cannot be used.

Exercises

In Exercises 1–10, show that L'Hôpital's Rule is applicable and use it to evaluate the limit.

1. $\displaystyle\lim_{x\to 1}\frac{2x^2+x-3}{x-1}$

SOLUTION The functions $2x^2+x-3$ and $x-1$ are differentiable, but the quotient is indeterminate at $x=1$,

$$\frac{2x^2+x-3}{x-1}\bigg|_{x=1}=\frac{2+1-3}{1-1}=\frac{0}{0},$$

so L'Hôpital's Rule applies. We find

$$\lim_{x\to 1}\frac{2x^2+x-3}{x-1}=\lim_{x\to 1}\frac{4x+1}{1}=4+1=5.$$

3. $\displaystyle\lim_{x\to -1}\frac{6x^3+13x^2+9x+2}{6x^3-x^2-5x+2}$

SOLUTION The functions $6x^3+13x^2+9x+2$ and $6x^3-x^2-5x+2$ are differentiable, but the quotient is indeterminate at $x=-1$,

$$\frac{6x^3+13x^2+9x+2}{6x^3-x^2-5x+2}\bigg|_{x=-1}=\frac{-6+13-9+2}{-6-1+5+2}=\frac{0}{0},$$

so L'Hôpital's Rule applies. We find

$$\lim_{x\to -1}\frac{6x^3+13x^2+9x+2}{6x^3-x^2-5x+2}=\lim_{x\to -1}\frac{18x^2+26x+9}{18x^2-2x-5}=\frac{18-26+9}{18+2-5}=\frac{1}{15}.$$

5. $\displaystyle\lim_{x\to 9}\frac{\sqrt{x}-3}{2x^2-17x-9}$

SOLUTION The functions $\sqrt{x}-3$ and $2x^2-17x-9$ are differentiable, but the quotient is indeterminate at $x=9$,

$$\frac{\sqrt{x}-3}{2x^2-17x-9}\bigg|_{x=9}=\frac{3-3}{162-153-9}=\frac{0}{0},$$

so L'Hôpital's Rule applies. We find

$$\lim_{x\to 9}\frac{\sqrt{x}-3}{2x^2-17x-9}=\lim_{x\to 9}\frac{1/(2\sqrt{x})}{4x-17}=\frac{1/6}{36-17}=\frac{1}{114}.$$

7. $\displaystyle\lim_{x\to 0}\frac{x^2}{1-\cos x}$

SOLUTION The functions x^2 and $1-\cos x$ are differentiable, but the quotient is indeterminate at $x=0$,

$$\frac{x^2}{1-\cos x}\bigg|_{x=0}=\frac{0}{1-1}=\frac{0}{0},$$

so L'Hôpital's Rule applies. Here, we use L'Hôpital's Rule twice to find

$$\lim_{x\to 0}\frac{x^2}{1-\cos x}=\lim_{x\to 0}\frac{2x}{\sin x}=\lim_{x\to 0}\frac{2}{\cos x}=2.$$

9. $\displaystyle\lim_{x\to 0}\frac{\cos x-\cos^2 x}{\sin x}$

SOLUTION The functions $\cos x-\cos^2 x$ and $\sin x$ are differentiable, but the quotient is indeterminate at $x=0$,

$$\frac{\cos x-\cos^2 x}{\sin x}\bigg|_{x=0}=\frac{1-1}{0}=\frac{0}{0},$$

so L'Hôpital's Rule applies. We find

$$\lim_{x\to 0}\frac{\cos x-\cos^2 x}{\sin x}=\lim_{x\to 0}\frac{-\sin x+2\cos x\sin x}{\cos x}=\frac{0}{1}=0.$$

In Exercises 11–16, show that L'Hôpital's Rule is applicable to the limit as $x \to \pm\infty$ and evaluate.

11. $\displaystyle\lim_{x\to\infty}\frac{3x-1}{7-12x}$

SOLUTION As $x \to \infty$, the quotient $\dfrac{3x-1}{7-12x}$ is of the form $\dfrac{\infty}{\infty}$, so L'Hôpital's Rule applies. We find

$$\lim_{x\to\infty}\frac{3x-1}{7-12x} = \lim_{x\to\infty}\frac{3}{-12} = -\frac{1}{4}.$$

13. $\displaystyle\lim_{x\to\infty}\frac{x}{e^x}$

SOLUTION As $x \to \infty$, the quotient $\dfrac{x}{e^x}$ is of the form $\dfrac{\infty}{\infty}$, so L'Hôpital's Rule applies. We find

$$\lim_{x\to\infty}\frac{x}{e^x} = \lim_{x\to\infty}\frac{1}{e^x} = 0.$$

15. $\displaystyle\lim_{x\to\infty}\frac{x^2}{e^x}$

SOLUTION As $x \to \infty$, the quotient $\dfrac{x^2}{e^x}$ is of the form $\dfrac{\infty}{\infty}$, so L'Hôpital's Rule applies. Here, we use L'Hôpital's Rule twice to find

$$\lim_{x\to\infty}\frac{x^2}{e^x} = \lim_{x\to\infty}\frac{2x}{e^x} = \lim_{x\to\infty}\frac{2}{e^x} = 0.$$

In Exercises 17–48, apply L'Hôpital's Rule to evaluate the limit. In some cases, it may be necessary to apply it more than once.

17. $\displaystyle\lim_{x\to0}\frac{\sin 4x}{\sin 3x}$

SOLUTION $\displaystyle\lim_{x\to0}\frac{\sin 4x}{\sin 3x} = \lim_{x\to0}\frac{4\cos 4x}{3\cos 3x} = \frac{4}{3}.$

19. $\displaystyle\lim_{x\to0}\frac{\tan x}{x}$

SOLUTION $\displaystyle\lim_{x\to0}\frac{\tan x}{x} = \lim_{x\to0}\frac{\sec^2 x}{1} = 1.$

21. $\displaystyle\lim_{x\to0}\left(\cot x - \frac{1}{x}\right)$

SOLUTION

$$\lim_{x\to0}\left(\cot x - \frac{1}{x}\right) = \lim_{x\to0}\frac{x\cos x - \sin x}{x\sin x} = \lim_{x\to0}\frac{-x\sin x + \cos x - \cos x}{x\cos x + \sin x} = \lim_{x\to0}\frac{-x\sin x}{x\cos x + \sin x}$$

$$= \lim_{x\to0}\frac{-x\cos x - x}{-x\sin x + \cos x + \cos x} = \frac{0}{2} = 0.$$

23. $\displaystyle\lim_{x\to-\infty}\frac{3x-2}{1-5x}$

SOLUTION $\displaystyle\lim_{x\to-\infty}\frac{3x-2}{1-5x} = \lim_{x\to-\infty}\frac{3}{-5} = -\frac{3}{5}.$

25. $\displaystyle\lim_{x\to-\infty}\frac{7x^2+4x}{9-3x^2}$

SOLUTION $\displaystyle\lim_{x\to-\infty}\frac{7x^2+4x}{9-3x^2} = \lim_{x\to-\infty}\frac{14x+4}{-6x} = \lim_{x\to-\infty}\frac{14}{-6} = -\frac{7}{3}.$

27. $\displaystyle\lim_{x\to\infty}\frac{3x^3+4x^2}{4x^3-7}$

SOLUTION $\displaystyle\lim_{x\to\infty}\frac{3x^3+4x^2}{4x^3-7} = \lim_{x\to\infty}\frac{9x^2+8x}{12x^2} = \lim_{x\to\infty}\frac{18x+8}{24x} = \frac{18}{24} = \frac{3}{4}.$

29. $\lim\limits_{x \to 1} \dfrac{x(\ln x - 1) + 1}{(x - 1)\ln x}$

SOLUTION

$$\lim\limits_{x \to 1} \frac{x(\ln x - 1) + 1}{(x - 1)\ln x} = \lim\limits_{x \to 1} \frac{x(\frac{1}{x}) + (\ln x - 1)}{(x - 1)(\frac{1}{x}) + \ln x} = \lim\limits_{x \to 1} \frac{\ln x}{1 - \frac{1}{x} + \ln x} = \lim\limits_{x \to 1} \frac{\frac{1}{x}}{\frac{1}{x^2} + \frac{1}{x}} = \frac{1}{1 + 1} = \frac{1}{2}.$$

31. $\lim\limits_{x \to 0} \dfrac{\cos(x + \frac{\pi}{2})}{\sin x}$

SOLUTION $\lim\limits_{x \to 0} \dfrac{\cos(x + \frac{\pi}{2})}{\sin x} = \lim\limits_{x \to 0} \dfrac{-\sin(x + \frac{\pi}{2})}{\cos x} = -1.$

33. $\lim\limits_{x \to 0} \dfrac{\sin x - x \cos x}{x - \sin x}$

SOLUTION

$$\lim\limits_{x \to 0} \frac{\sin x - x \cos x}{x - \sin x} = \lim\limits_{x \to 0} \frac{x \sin x}{1 - \cos x} = \lim\limits_{x \to 0} \frac{\sin x + x \cos x}{\sin x} = \lim\limits_{x \to 0} \frac{\cos x + \cos x - x \sin x}{\cos x} = 2.$$

35. $\lim\limits_{x \to 1} \dfrac{e^x - e}{\ln x}$

SOLUTION $\lim\limits_{x \to 1} \dfrac{e^x - e}{\ln x} = \lim\limits_{x \to 1} \dfrac{e^x}{x^{-1}} = \dfrac{e}{1} = e.$

37. $\lim\limits_{x \to \pi/2} \dfrac{\cos x}{\sin(2x)}$

SOLUTION $\lim\limits_{x \to \pi/2} \dfrac{\cos x}{\sin(2x)} = \lim\limits_{x \to \pi/2} \dfrac{-\sin x}{2\cos(2x)} = \dfrac{1}{2}.$

39. $\lim\limits_{x \to \infty} e^{-x}(x^3 - x^2 + 9)$

SOLUTION

$$\lim\limits_{x \to \infty} e^{-x}(x^3 - x^2 + 9) = \lim\limits_{x \to \infty} \frac{x^3 - x^2 + 9}{e^x} = \lim\limits_{x \to \infty} \frac{3x^2 - 2x}{e^x} = \lim\limits_{x \to \infty} \frac{6x - 2}{e^x} = \lim\limits_{x \to \infty} \frac{6}{e^x} = 0.$$

41. $\lim\limits_{x \to 1} \dfrac{x^{1/3} - 1}{x^2 + 3x - 4}$

SOLUTION $\lim\limits_{x \to 1} \dfrac{x^{1/3} - 1}{x^2 + 3x - 4} = \lim\limits_{x \to 1} \dfrac{\frac{1}{3}x^{-\frac{2}{3}}}{2x + 3} = \dfrac{\frac{1}{3}}{5} = \dfrac{1}{15}.$

43. $\lim\limits_{x \to \infty} x^{1/x}$

SOLUTION $\lim\limits_{x \to \infty} \ln x^{1/x} = \lim\limits_{x \to \infty} \dfrac{1}{x} \ln x = \lim\limits_{x \to \infty} \dfrac{\ln x}{x} = \lim\limits_{x \to \infty} \dfrac{\frac{1}{x}}{1} = 0.$ Hence,

$$\lim\limits_{x \to \infty} x^{1/x} = \lim\limits_{x \to \infty} e^{\ln(x^{(1/x)})} = e^0 = 1.$$

45. $\lim\limits_{t \to 0+} (\sin t)(\ln t)$

SOLUTION

$$\lim\limits_{t \to 0+} (\sin t)(\ln t) = \lim\limits_{t \to 0+} \frac{\ln t}{\csc t} = \lim\limits_{t \to 0+} \frac{\frac{1}{t}}{-\csc t \cot t} = \lim\limits_{t \to 0+} \frac{-\sin^2 t}{t \cos t} = \lim\limits_{t \to 0+} \frac{-2\sin t \cos t}{\cos t - t \sin t} = 0.$$

47. $\lim\limits_{x \to 1} \dfrac{(1 + 3x)^{1/2} - 2}{(1 + 7x)^{1/3} - 2}$

SOLUTION   Apply L'Hôpital's Rule once:

$$\lim_{x \to 1} \frac{(1+3x)^{1/2} - 2}{(1+7x)^{1/3} - 2} = \lim_{x \to 1} \frac{\frac{3}{2}(1+3x)^{-1/2}}{\frac{7}{3}(1+7x)^{-2/3}}$$

$$= \frac{(\frac{3}{2})\frac{1}{2}}{(\frac{7}{3})(\frac{1}{4})} = \frac{9}{7}$$

49. Evaluate $\lim\limits_{x \to \pi/2} \dfrac{\cos mx}{\cos nx}$, where m and n are nonzero whole numbers.

SOLUTION   Suppose m and n are even. Then there exist integers k and l such that $m = 2k$ and $n = 2l$ and

$$\lim_{x \to \pi/2} \frac{\cos mx}{\cos nx} = \frac{\cos k\pi}{\cos l\pi} = (-1)^{k-l}.$$

Now, suppose m is even and n is odd. Then

$$\lim_{x \to \pi/2} \frac{\cos mx}{\cos nx}$$

does not exist (from one side the limit tends toward $-\infty$, while from the other side the limit tends toward $+\infty$). Third, suppose m is odd and n is even. Then

$$\lim_{x \to \pi/2} \frac{\cos mx}{\cos nx} = 0.$$

Finally, suppose m and n are odd. This is the only case when the limit is indeterminate. Then there exist integers k and l such that $m = 2k + 1$, $n = 2l + 1$ and, by L'Hôpital's Rule,

$$\lim_{x \to \pi/2} \frac{\cos mx}{\cos nx} = \lim_{x \to \pi/2} \frac{-m \sin mx}{-n \sin nx} = (-1)^{k-l} \frac{m}{n}.$$

To summarize,

$$\lim_{x \to \pi/2} \frac{\cos mx}{\cos nx} = \begin{cases} (-1)^{(m-n)/2}, & m, n \text{ even} \\ \text{does not exist}, & m \text{ even}, n \text{ odd} \\ 0 & m \text{ odd}, n \text{ even} \\ (-1)^{(m-n)/2}\frac{m}{n}, & m, n \text{ odd} \end{cases}$$

51. $\boxed{\text{GU}}$ Can L'Hôpital's Rule be applied to $\lim\limits_{x \to 0+} x^{\sin(1/x)}$? Does a graphical or numerical investigation suggest that the limit exists?

SOLUTION   Since $\sin(1/x)$ oscillates as $x \to 0+$, L'Hôpital's Rule cannot be applied. Both numerical and graphical investigations suggest that the limit does not exist due to the oscillation.

x	1	0.1	0.01	0.001	0.0001	0.00001
$x^{\sin(1/x)}$	1	3.4996	10.2975	0.003316	16.6900	0.6626

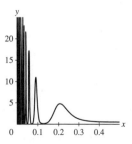

53. Let $f(x) = x^{1/x}$ in the domain $\{x : x > 0\}$.

(a) Calculate $\lim\limits_{x \to 0+} f(x)$ and $\lim\limits_{x \to \infty} f(x)$.

(b) Find the maximum value of $f(x)$ and determine the intervals on which $f(x)$ is increasing or decreasing.

(c) Use (a) and (b) to prove that $x^{1/x} = c$ has a unique solution if $0 < c \le 1$ or $c = e^{1/e}$, two solutions if $1 < c < e^{1/e}$, and no solutions if $c > e^{1/e}$.

(d) $\boxed{\text{GU}}$ Plot the graph of $f(x)$. Explain how the graph confirms the conclusions in (c).

SOLUTION

(a) Let $f(x) = x^{1/x}$. Note that $\lim_{x \to 0+} x^{1/x}$ is not indeterminate. As $x \to 0+$, the base of the function tends toward 0 and the exponent tends toward $+\infty$. Both of these factors force $x^{1/x}$ toward 0. Thus, $\lim_{x \to 0+} f(x) = 0$. On the other hand, $\lim_{x \to \infty} f(x)$ is indeterminate. We calculate this limit as follows:

$$\lim_{x \to \infty} \ln f(x) = \lim_{x \to \infty} \frac{\ln x}{x} = \lim_{x \to \infty} \frac{1}{x} = 0,$$

so $\lim_{x \to \infty} f(x) = e^0 = 1$.

(b) Again, let $f(x) = x^{1/x}$, so that $\ln f(x) = \frac{1}{x} \ln x$. To find the derivative f', we apply the derivative to both sides:

$$\frac{d}{dx} \ln f(x) = \frac{d}{dx} \left(\frac{1}{x} \ln x \right)$$

$$\frac{1}{f(x)} f'(x) = -\frac{\ln x}{x^2} + \frac{1}{x^2}$$

$$f'(x) = f(x) \left(-\frac{\ln x}{x^2} + \frac{1}{x^2} \right) = \frac{x^{1/x}}{x^2} (1 - \ln x)$$

Thus, f is increasing for $0 < x < e$, is decreasing for $x > e$ and has a maximum at $x = e$. The maximum value is $f(e) = e^{1/e} \approx 1.444668$.

(c) Because $(e, e^{1/e})$ is the only maximum, no solution exists for $c > e^{1/e}$ and only one solution exists for $c = e^{1/e}$. Moreover, because $f(x)$ increases from 0 to $e^{1/e}$ as x goes from 0 to e and then decreases from $e^{1/e}$ to 1 as x goes from e to $+\infty$, it follows that there are two solutions for $1 < c < e^{1/e}$, but only one solution for $0 < c \le 1$.

(d) Observe that if we sketch the horizontal line $y = c$, this line will intersect the graph of $y = f(x)$ only once for $0 < c \le 1$ and $c = e^{1/e}$ and will intersect the graph of $y = f(x)$ twice for $1 < c < e^{1/e}$. There are no points of intersection for $c > e^{1/e}$.

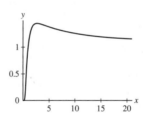

55. Show that $(\ln x)^2 \ll \sqrt{x}$ and $(\ln x)^4 \ll x^{1/10}$.

SOLUTION

- $\sqrt{x} \gg (\ln x)^2$:

$$\lim_{x \to \infty} \frac{\sqrt{x}}{(\ln x)^2} = \lim_{x \to \infty} \frac{\frac{1}{2\sqrt{x}}}{\frac{2}{x} \ln x} = \lim_{x \to \infty} \frac{\sqrt{x}}{4 \ln x} = \lim_{x \to \infty} \frac{\frac{1}{2\sqrt{x}}}{\frac{4}{x}} = \lim_{x \to \infty} \frac{\sqrt{x}}{8} = \infty.$$

- $x^{1/10} \gg (\ln x)^4$:

$$\lim_{x \to \infty} \frac{x^{1/10}}{(\ln x)^4} = \lim_{x \to \infty} \frac{\frac{1}{10 x^{9/10}}}{\frac{4}{x} (\ln x)^3} = \lim_{x \to \infty} \frac{x^{1/10}}{40 (\ln x)^3} = \lim_{x \to \infty} \frac{\frac{1}{10 x^{9/10}}}{\frac{120}{x} (\ln x)^2} = \lim_{x \to \infty} \frac{x^{1/10}}{1200 (\ln x)^2}$$

$$= \lim_{x \to \infty} \frac{\frac{1}{10 x^{9/10}}}{\frac{2400}{x} (\ln x)} = \lim_{x \to \infty} \frac{x^{1/10}}{24000 \ln x} = \lim_{x \to \infty} \frac{\frac{1}{10 x^{9/10}}}{\frac{24000}{x}} = \lim_{x \to \infty} \frac{x^{1/10}}{240000} = \infty.$$

57. Show that $(\ln x)^N \ll x^a$ for all N and all $a > 0$.

SOLUTION

$$\lim_{x \to \infty} \frac{x^a}{(\ln x)^N} = \lim_{x \to \infty} \frac{a x^{a-1}}{\frac{N}{x} (\ln x)^{N-1}} = \lim_{x \to \infty} \frac{a x^a}{N (\ln x)^{N-1}} = \cdots$$

If we continue in this manner, L'Hôpital's Rule will give a factor of x^a in the numerator, but the power on $\ln x$ in the denominator will eventually be zero. Thus,

$$\lim_{x \to \infty} \frac{x^a}{(\ln x)^N} = \infty,$$

so $x^a \gg (\ln x)^N$ for all N and for all $a > 0$.

59. Show that $\lim_{x \to \infty} x^n e^{-x} = 0$ for all whole numbers $n > 0$.

SOLUTION

$$\lim_{x \to \infty} x^n e^{-x} = \lim_{x \to \infty} \frac{x^n}{e^x} = \lim_{x \to \infty} \frac{nx^{n-1}}{e^x}$$

$$= \lim_{x \to \infty} \frac{n(n-1)x^{n-2}}{e^x}$$

$$\vdots$$

$$= \lim_{x \to \infty} \frac{n!}{e^x} = 0.$$

61. Use Eq. (2) of Section 7.5 to show that the PV of an investment which pays out income continuously at a constant rate of R dollars/year for T years is $PV = R\dfrac{1 - e^{-rT}}{r}$, where r is the interest rate. Use L'Hôpital's Rule to prove that the PV approaches RT as $r \to 0$.

SOLUTION By Eq. (2) of Section 7.5,

$$PV = \int_0^T Re^{-rt}\, dt = \frac{R}{-r}e^{-rt}\Big|_0^T = \frac{R}{r}(1 - e^{-rT}).$$

Using L'Hôpital's Rule,

$$\lim_{r \to 0} \frac{R(1 - e^{-rT})}{r} = \lim_{r \to 0} \frac{RTe^{-rT}}{1} = RT.$$

63. Show that $\lim_{t \to \infty} t^k e^{-t^2} = 0$ for all k. *Hint:* Compare with $\lim_{t \to \infty} t^k e^{-t} = 0$.

SOLUTION Because we are interested in the limit as $t \to +\infty$, we will restrict attention to $t > 1$. Then, for all k,

$$0 \le t^k e^{-t^2} \le t^k e^{-t}.$$

As $\lim_{t \to \infty} t^k e^{-t} = 0$, it follows from the Squeeze Theorem that

$$\lim_{t \to \infty} t^k e^{-t^2} = 0.$$

In Exercises 64–66, let

$$f(x) = \begin{cases} e^{-1/x^2} & \text{for } x \ne 0 \\ 0 & \text{for } x = 0 \end{cases}$$

These exercises show that $f(x)$ has an unusual property: All of its higher derivatives at $x = 0$ exist and are equal to zero.

65. Show that $f'(0)$ exists and is equal to zero. Also, verify that $f''(0)$ exists and is equal to zero.

SOLUTION Working from the definition,

$$f'(0) = \lim_{x \to 0} \frac{f(x) - f(0)}{x - 0} = \lim_{x \to 0} \frac{f(x)}{x} = 0$$

by the previous exercise. Thus, $f'(0)$ exists and is equal to 0. Moreover,

$$f'(x) = \begin{cases} e^{-1/x^2}\left(\frac{2}{x^3}\right) & \text{for } x \ne 0 \\ 0 & \text{for } x = 0 \end{cases}$$

Now,

$$f''(0) = \lim_{x \to 0} \frac{f'(x) - f'(0)}{x - 0} = \lim_{x \to 0} e^{-1/x^2} \left(\frac{2}{x^4}\right) = 2 \lim_{x \to 0} \frac{f(x)}{x^4} = 0$$

by the previous exercise. Thus, $f''(0)$ exists and is equal to 0.

Further Insights and Challenges

67. Show that L'Hôpital's Rule applies to $\lim\limits_{x \to \infty} \dfrac{x}{\sqrt{x^2 + 1}}$, but that it is of no help. Then evaluate the limit directly.

SOLUTION Both the numerator $f(x) = x$ and the denominator $g(x) = \sqrt{x^2 + 1}$ tend to infinity as $x \to \infty$, and $g'(x) = x/\sqrt{x^2 + 1}$ is nonzero for $x > 0$. Therefore, L'Hôpital's Rule applies:

$$\lim_{x \to \infty} \frac{x}{\sqrt{x^2 + 1}} = \lim_{x \to \infty} \frac{1}{x(x^2 + 1)^{-1/2}} = \lim_{x \to \infty} \frac{(x^2 + 1)^{1/2}}{x}$$

We may apply L'Hôpital's Rule again: $\lim\limits_{x \to \infty} \dfrac{(x^2 + 1)^{1/2}}{x} = \lim\limits_{x \to \infty} \dfrac{x(x^2 + 1)^{-1/2}}{1} = \lim\limits_{x \to \infty} \dfrac{x}{\sqrt{x^2 + 1}}$. This takes us back to the original limit, so L'Hôpital's Rule is ineffective. However, we can evaluate the limit directly by observing that

$$\frac{x}{\sqrt{x^2 + 1}} = \frac{x^{-1}(x)}{x^{-1}\sqrt{x^2 + 1}} = \frac{1}{\sqrt{1 + x^{-2}}} \quad \text{and hence} \quad \lim_{x \to \infty} \frac{x}{\sqrt{x^2 + 1}} = \lim_{x \to \infty} \frac{1}{\sqrt{1 + x^{-2}}} = 1.$$

69. Resonance A spring oscillates with a natural frequency $\lambda/2\pi$. If we drive the spring with a sinusoidal force $C \sin(\omega t)$, where $\omega \neq \lambda$, then the spring oscillates according to

$$y(t) = \frac{C}{\lambda^2 - \omega^2}\left(\lambda \sin(\omega t) - \omega \sin(\lambda t)\right)$$

(a) Use L'Hôpital's Rule to determine $y(t)$ in the limit as $\omega \to \lambda$.
(b) Define $y_0(t) = \lim\limits_{\omega \to \lambda} y(t)$. Show that $y_0(t)$ ceases to be periodic and that its amplitude $|y_0(t)|$ tends to infinity as $t \to \infty$ (the system is said to be in resonance; eventually, the spring is stretched beyond its limits).
(c) $\boxed{CAS}$ Plot $y(t)$ for $\lambda = 1$ and $\omega = 0.5, 0.8, 0.9, 0.99,$ and 0.999. How do the graphs change? Do the graphs confirm your conclusion in (b)?

SOLUTION

(a) $\lim\limits_{\omega \to \lambda} y(t) = \lim\limits_{\omega \to \lambda} C \dfrac{\lambda \sin(\omega t) - \omega \sin(\lambda t)}{\lambda^2 - \omega^2} = C \lim\limits_{\omega \to \lambda} \dfrac{\frac{d}{d\omega}(\lambda \sin(\omega t) - \omega \sin(\lambda t))}{\frac{d}{d\omega}(\lambda^2 - \omega^2)}$

$$= C \lim_{\omega \to \lambda} \frac{\lambda t \cos(\omega t) - \sin(\lambda t)}{-2\omega} = C \frac{\lambda t \cos(\lambda t) - \sin(\lambda t)}{-2\lambda}$$

(b) From part (a)

$$y_0(t) = \lim_{\omega \to \lambda} y(t) = C \frac{\lambda t \cos(\lambda t) - \sin(\lambda t)}{-2\lambda}.$$

This may be rewritten as

$$y_0(t) = C \frac{\sqrt{\lambda^2 t^2 + 1}}{-2\lambda} \cos(\lambda t + \phi),$$

where $\cos \phi = \dfrac{\lambda t}{\sqrt{\lambda^2 t^2 + 1}}$ and $\sin \phi = \dfrac{1}{\sqrt{\lambda^2 t^2 + 1}}$. Since the amplitude varies with t, $y_0(t)$ is not periodic. Also note that

$$C \frac{\sqrt{\lambda^2 t^2 + 1}}{-2\lambda} \to \infty \quad \text{as} \quad t \to \infty.$$

(c) The graphs below were produced with $C = 1$ and $\lambda = 1$. Moving from left to right and from top to bottom, $\omega = 0.5, 0.8, 0.9, 0.99, 0.999, 1$.

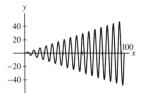
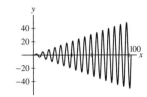

71. Suppose that f and g are polynomials such that $f(a) = g(a) = 0$. In this case, it is a fact from algebra that $f(x) = (x - a)f_1(x)$ and $g(x) = (x - a)g_1(x)$ for some polynomials f_1 and g_1. Use this to verify L'Hôpital's Rule directly for f and g.

SOLUTION As in the problem statement, let $f(x)$ and $g(x)$ be two polynomials such that $f(a) = g(a) = 0$, and let $f_1(x)$ and $g_1(x)$ be the polynomials such that $f(x) = (x - a)f_1(x)$ and $g(x) = (x - a)g_1(x)$. By the product rule, we have the following facts,

$$f'(x) = (x - a)f_1'(x) + f_1(x)$$

$$g'(x) = (x - a)g_1'(x) + g_1(x)$$

so

$$\lim_{x \to a} f'(x) = f_1(a) \quad \text{and} \quad \lim_{x \to a} g'(x) = g_1(a).$$

L'Hôpital's Rule stated for f and g is: if $\lim_{x \to a} g'(x) \neq 0$, so that $g_1(a) \neq 0$,

$$\lim_{x \to a} \frac{f(x)}{g(x)} = \lim_{x \to a} \frac{f'(x)}{g'(x)} = \frac{f_1(a)}{g_1(a)}.$$

Suppose $g_1(a) \neq 0$. Then, by direct computation,

$$\lim_{x \to a} \frac{f(x)}{g(x)} = \lim_{x \to a} \frac{(x - a)f_1(x)}{(x - a)g_1(x)} = \lim_{x \to a} \frac{f_1(x)}{g_1(x)} = \frac{f_1(a)}{g_1(a)},$$

exactly as predicted by L'Hôpital's Rule.

73. CAS The integral on the left in Exercise 72 is equal to $f_n(x) = \dfrac{x^{n+1} - 1}{n + 1}$. Investigate the limit graphically by plotting $f_n(x)$ for $n = 0, -0.3, -0.6$, and -0.9 together with $\ln x$ on a single plot.

SOLUTION

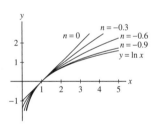

7.8 Inverse Trigonometric Functions

Preliminary Questions

1. Which of the following quantities is undefined?

(a) $\sin^{-1}\left(-\frac{1}{2}\right)$

(b) $\cos^{-1}(2)$

(c) $\csc^{-1}\left(\frac{1}{2}\right)$

(d) $\csc^{-1}(2)$

SOLUTION (b) and (c) are undefined. $\sin^{-1}\left(-\frac{1}{2}\right) = -\frac{\pi}{6}$ and $\csc^{-1}(2) = \frac{\pi}{6}$.

2. Give an example of an angle θ such that $\cos^{-1}(\cos \theta) \neq \theta$. Does this contradict the definition of inverse function?

SOLUTION Any angle $\theta < 0$ or $\theta > \pi$ will work. No, this does not contradict the definition of inverse function.

3. What is the geometric interpretation of the identity $\sin^{-1} x + \cos^{-1} x = \pi/2$?

SOLUTION Angles whose sine and cosine are x are complementary.

Exercises

In Exercises 1–6, evaluate without using a calculator.

1. $\cos^{-1} 1$

SOLUTION $\cos^{-1} 1 = 0$.

3. $\cot^{-1} 1$

SOLUTION $\cot^{-1} 1 = \frac{\pi}{4}$.

5. $\tan^{-1} \sqrt{3}$

SOLUTION $\tan^{-1} \sqrt{3} = \tan^{-1}\left(\frac{\sqrt{3}/2}{1/2}\right) = \frac{\pi}{3}$.

In Exercises 7–16, compute without using a calculator.

7. $\sin^{-1}\left(\sin \dfrac{\pi}{3}\right)$

SOLUTION $\sin^{-1}(\sin \frac{\pi}{3}) = \frac{\pi}{3}$.

9. $\cos^{-1}\left(\cos \dfrac{3\pi}{2}\right)$

SOLUTION $\cos^{-1}(\cos \frac{3\pi}{2}) = \cos^{-1}(0) = \frac{\pi}{2}$. The answer is not $\frac{3\pi}{2}$ because $\frac{3\pi}{2}$ is not in the range of the inverse cosine function.

11. $\tan^{-1}\left(\tan \dfrac{3\pi}{4}\right)$

SOLUTION $\tan^{-1}(\tan \frac{3\pi}{4}) = \tan^{-1}(-1) = -\frac{\pi}{4}$. The answer is not $\frac{3\pi}{4}$ because $\frac{3\pi}{4}$ is not in the range of the inverse tangent function.

13. $\sec^{-1}(\sec 3\pi)$

SOLUTION $\sec^{-1}(\sec 3\pi) = \sec^{-1}(-1) = \pi$. The answer is not 3π because 3π is not in the range of the inverse secant function.

15. $\csc^{-1}\big(\csc(-\pi)\big)$

SOLUTION   No inverse since $\csc(-\pi) = \frac{1}{\sin(-\pi)} = \frac{1}{0} \longrightarrow \infty$.

In Exercises 17–20, simplify by referring to the appropriate triangle or trigonometric identity.

17. $\tan(\cos^{-1} x)$

SOLUTION   Let $\theta = \cos^{-1} x$. Then $\cos \theta = x$ and we generate the triangle shown below. From the triangle,

$$\tan(\cos^{-1} x) = \tan \theta = \frac{\sqrt{1 - x^2}}{x}.$$

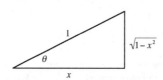

19. $\cot(\sec^{-1} x)$

SOLUTION   Let $\theta = \sec^{-1} x$. Then $\sec \theta = x$ and we generate the triangle shown below. From the triangle,

$$\cot(\sec^{-1} x) = \cot \theta = \frac{1}{\sqrt{x^2 - 1}}.$$

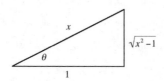

In Exercises 21–28, refer to the appropriate triangle or trigonometric identity to compute the given value.

21. $\cos\left(\sin^{-1}\frac{2}{3}\right)$

SOLUTION Let $\theta = \sin^{-1}\frac{2}{3}$. Then $\sin\theta = \frac{2}{3}$ and we generate the triangle shown below. From the triangle,

$$\cos\left(\sin^{-1}\frac{2}{3}\right) = \cos\theta = \frac{\sqrt{5}}{3}.$$

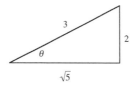

23. $\tan\left(\sin^{-1}0.8\right)$

SOLUTION Let $\theta = \sin^{-1}0.8$. Then $\sin\theta = 0.8 = \frac{4}{5}$ and we generate the triangle shown below. From the triangle,

$$\tan(\sin^{-1}0.8) = \tan\theta = \frac{4}{3}.$$

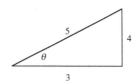

25. $\cot\left(\csc^{-1}2\right)$

SOLUTION $\csc^{-1}2 = \frac{\pi}{6}$. Hence, $\cot(\csc^{-1}2) = \cot\frac{\pi}{6} = \sqrt{3}$.

27. $\cot\left(\tan^{-1}20\right)$

SOLUTION Let $\theta = \tan^{-1}20$. Then $\tan\theta = 20$, so $\cot(\tan^{-1}20) = \cot\theta = \frac{1}{\tan\theta} = \frac{1}{20}$.

In Exercises 29–32, compute the derivative at the point indicated without using a calculator.

29. $y = \sin^{-1}x$, $x = \frac{3}{5}$

SOLUTION Let $y = \sin^{-1}x$. Then $y' = \frac{1}{\sqrt{1-x^2}}$ and

$$y'\left(\frac{3}{5}\right) = \frac{1}{\sqrt{1-9/25}} = \frac{1}{4/5} = \frac{5}{4}.$$

31. $y = \sec^{-1}x$, $x = 4$

SOLUTION Let $y = \sec^{-1}x$. Then $y' = \frac{1}{|x|\sqrt{x^2-1}}$ and

$$y'(4) = \frac{1}{4\sqrt{15}}.$$

In Exercises 33–48, find the derivative.

33. $y = \sin^{-1}(7x)$

SOLUTION $\frac{d}{dx}\sin^{-1}(7x) = \frac{1}{\sqrt{1-(7x)^2}} \cdot \frac{d}{dx}7x = \frac{7}{\sqrt{1-(7x)^2}}.$

35. $y = \cos^{-1}(x^2)$

SOLUTION $\frac{d}{dx}\cos^{-1}(x^2) = \frac{-1}{\sqrt{1-x^4}} \cdot \frac{d}{dx}x^2 = \frac{-2x}{\sqrt{1-x^4}}.$

37. $y = x\tan^{-1}x$

SOLUTION $\frac{d}{dx} x \tan^{-1} x = x \left(\frac{1}{1+x^2} \right) + \tan^{-1} x.$

39. $y = \arcsin(e^x)$

SOLUTION $\frac{d}{dx} \sin^{-1}(e^x) = \frac{1}{\sqrt{1-e^{2x}}} \cdot \frac{d}{dx} e^x = \frac{e^x}{\sqrt{1-e^{2x}}}.$

41. $y = \tan^{-1} \left(\frac{z}{1-z^2} \right)$

SOLUTION $\frac{d}{dx} \tan^{-1} \left(\frac{z}{1-z^2} \right) = \frac{1}{\left(\frac{z}{1-z^2} \right)^2 + 1} \cdot \left(\frac{(1-z^2) - z(-2z)}{(1-z^2)^2} \right) = \frac{1+z^2}{z^2 + (1-z^2)^2}.$

43. $y = \cos^{-1} t^{-1} - \sec^{-1} t$

SOLUTION $\frac{d}{dx}(\cos^{-1} t^{-1} - \sec^{-1} t) = \frac{-1}{\sqrt{1 - (1/t)^2}} \left(\frac{-1}{t^2} \right) - \frac{1}{|t|\sqrt{t^2-1}}$

$$= \frac{1}{\sqrt{t^4 - t^2}} - \frac{1}{|t|\sqrt{t^2-1}} = \frac{1}{|t|\sqrt{t^2-1}} - \frac{1}{|t|\sqrt{t^2-1}} = 0.$$

Alternately, let $t = \sec \theta$. Then $t^{-1} = \cos \theta$ and $\cos^{-1} t^{-1} - \sec^{-1} t = \theta - \theta = 0$. Consequently,

$$\frac{d}{dx}(\cos^{-1} t^{-1} - \sec^{-1} t) = 0.$$

45. $y = \cos^{-1}(x + \sin^{-1} x)$

SOLUTION $\frac{d}{dx} \cos^{-1}(x + \sin^{-1} x) = \frac{-1}{\sqrt{1 - (x + \sin^{-1} x)^2}} \left(1 + \frac{1}{\sqrt{1-x^2}} \right).$

47. $y = \sqrt{1 - t^2} + \sin^{-1} t$

SOLUTION $\frac{d}{dx} \left(\sqrt{1 - t^2} + \sin^{-1} t \right) = \frac{1}{2}(1 - t^2)^{-1/2}(-2t) + \frac{1}{\sqrt{1-t^2}} = \frac{-t}{\sqrt{1-t^2}} + \frac{1}{\sqrt{1-t^2}} = \frac{1-t}{\sqrt{1-t^2}}.$

49. Use Figure 9 to prove that $(\cos^{-1} x)' = -\frac{1}{\sqrt{1-x^2}}.$

FIGURE 9 Right triangle with $\theta = \cos^{-1} x$.

SOLUTION Let $\theta = \cos^{-1} x$. Then $\cos \theta = x$ and

$$-\sin \theta \frac{d\theta}{dx} = 1 \quad \text{or} \quad \frac{d\theta}{dx} = -\frac{1}{\sin \theta} = -\frac{1}{\sin(\cos^{-1} x)}.$$

From Figure 9, we see that $\sin(\cos^{-1} x) = \sin \theta = \sqrt{1-x^2}$; hence,

$$\frac{d}{dx} \cos^{-1} x = \frac{1}{-\sin(\cos^{-1} x)} = -\frac{1}{\sqrt{1-x^2}}.$$

51. Let $\theta = \sec^{-1} x$. Show that $\tan \theta = \sqrt{x^2 - 1}$ if $x \geq 1$ and $\tan \theta = -\sqrt{x^2 - 1}$ if $x \leq -1$. *Hint:* $\tan \theta \geq 0$ on $[0, \pi/2)$ and $\tan \theta \leq 0$ on $(\pi/2, \pi]$.

SOLUTION In general, $1 + \tan^2 \theta = \sec^2 \theta$, so $\tan \theta = \pm \sqrt{\sec^2 \theta - 1}$. With $\theta = \sec^{-1} x$, it follows that $\sec \theta = x$, so $\tan \theta = \pm \sqrt{x^2 - 1}$. Finally, if $x \geq 1$ then $\theta = \sec^{-1} x \in [0, \pi/2)$ so $\tan \theta$ is positive; on the other hand, if $x \leq 1$ then $\theta = \sec^{-1} x \in (-\pi/2, 0]$ so $\tan \theta$ is negative.

53. Let $f(x) = \tan^{-1} x$. Compute $f'(x)$ and $f''(x)$, and determine the increasing/decreasing and concavity behavior of $f(x)$.

SOLUTION $\frac{d}{dx} \tan^{-1} x = \frac{1}{1+x^2} > 0$ for all x, so $\tan^{-1} x$ is always increasing. Moreover, $\frac{d^2}{dx^2} \tan^{-1} x = \frac{-2x}{(1+x^2)^2}$, so $\tan^{-1} x$ is concave down for $x > 0$ and is concave up for $x < 0$.

55. Find the minimum value of $y = \tan^{-1}(x^2 - x)$.

SOLUTION Let $f(x) = \tan^{-1}(x^2 - x)$. Then

$$f'(x) = \frac{2x - 1}{(x^2 - x)^2 + 1}.$$

Hence, $f(x)$ is decreasing for all $x < \frac{1}{2}$, is increasing for all $x > \frac{1}{2}$ and has an absolute minimum at $x = \frac{1}{2}$. The minimum value is

$$f\left(\frac{1}{2}\right) = \tan^{-1}\left(-\frac{1}{4}\right) \approx -0.244979.$$

57. Use the Linear Approximation to estimate

$$\tan^{-1}(1.05) - \tan^{-1} 1$$

SOLUTION $\tan^{-1}(1.05) - \tan^{-1} 1 \approx \left(\frac{d}{dx} \tan^{-1} x\right)\Big|_{x=1} (0.05) \approx \frac{1}{2}(0.05) = 0.025.$

In Exercises 59–62, evaluate the limit using L'Hôpital's Rule if necessary.

59. $\lim\limits_{x \to 0} \dfrac{\sin^{-1} x}{x}$

SOLUTION $\lim\limits_{x \to 0} \dfrac{\sin^{-1} x}{x} = \lim\limits_{x \to 0} \dfrac{\frac{1}{\sqrt{1-x^2}}}{1} = 1.$

61. $\lim\limits_{x \to 1} \dfrac{\tan^{-1} x - \frac{\pi}{4}}{\tan \frac{\pi}{4} x - 1}$

SOLUTION $\lim\limits_{x \to 1} \dfrac{\tan^{-1} x - \frac{\pi}{4}}{\tan(\pi x/4) - 1} = \lim\limits_{x \to 1} \dfrac{\frac{1}{1+x^2}}{\frac{\pi}{4}\sec^2(\pi x/4)} = \dfrac{\frac{1}{2}}{\frac{\pi}{2}} = \dfrac{1}{\pi}.$

In Exercises 63–66, calculate the definite integral.

63. $\displaystyle\int_0^{1/2} \dfrac{1}{\sqrt{1-x^2}} \, dx$

SOLUTION $\displaystyle\int_0^{1/2} \dfrac{1}{\sqrt{1-x^2}} \, dx = \sin^{-1} x \Big|_0^{1/2} = \sin^{-1}\dfrac{1}{2} - \sin^{-1} 0 = \dfrac{\pi}{6}.$

65. $\displaystyle\int_{1/2}^{\sqrt{3}/2} \dfrac{1}{\sqrt{1-x^2}} \, dx$

SOLUTION $\displaystyle\int_{1/2}^{\sqrt{3}/2} \dfrac{1}{\sqrt{1-x^2}} \, dx = \sin^{-1} x \Big|_{1/2}^{\sqrt{3}/2} = \sin^{-1}\dfrac{\sqrt{3}}{2} - \sin^{-1}\dfrac{1}{2} = \dfrac{\pi}{3} - \dfrac{\pi}{6} = \dfrac{\pi}{6}.$

67. Use the substitution $u = x/3$ to prove

$$\int \dfrac{dx}{9 + x^2} = \dfrac{1}{3}\tan^{-1}\dfrac{x}{3} + C$$

SOLUTION Let $u = x/3$. Then, $x = 3u$, $dx = 3\,du$, $9 + x^2 = 9(1 + u^2)$, and

$$\int \dfrac{dx}{9 + x^2} = \int \dfrac{3\,du}{9(1 + u^2)} = \dfrac{1}{3}\int \dfrac{du}{1 + u^2} = \dfrac{1}{3}\tan^{-1} u + C = \dfrac{1}{3}\tan^{-1}\dfrac{x}{3} + C.$$

In Exercises 69–80, calculate the indefinite integral.

69. $\displaystyle\int \dfrac{dt}{\sqrt{16 - t^2}}$

SOLUTION Let $t = 4u$. Then $dt = 4\,du$, and

$$\int \dfrac{dt}{\sqrt{16 - t^2}} = \int \dfrac{4\,du}{\sqrt{16 - (4u)^2}} = \int \dfrac{4\,du}{4\sqrt{1 - u^2}} = \int \dfrac{du}{\sqrt{1 - u^2}} = \sin^{-1} u + C = \sin^{-1}\left(\dfrac{t}{4}\right) + C.$$

71. $\displaystyle\int \frac{dt}{\sqrt{25 - 4t^2}}$

SOLUTION   Let $t = (5/2)u$. Then $dt = (5/2)\,du$, and

$$\int \frac{dt}{\sqrt{25 - 4t^2}} = \int \frac{(5/2)du}{\sqrt{25 - 4(\frac{5}{2}u)^2}} = \int \frac{5/2}{\sqrt{25 - 25u^2}}\,du = \int \frac{du}{2\sqrt{1 - u^2}}$$

$$= \frac{1}{2}\sin^{-1} u + C = \frac{1}{2}\sin^{-1}\left(\frac{2t}{5}\right) + C.$$

73. $\displaystyle\int \frac{dx}{\sqrt{1 - 4x^2}}$

SOLUTION   Let $u = 2x$. Then $du = 2\,dx$, and

$$\int \frac{dx}{\sqrt{1 - 4x^2}} = \int \frac{du}{2\sqrt{1 - u^2}} = \frac{1}{2}\sin^{-1} u + C = \frac{1}{2}\sin^{-1}(2x) + C.$$

75. $\displaystyle\int \frac{(x+1)dx}{\sqrt{1 - x^2}}$

SOLUTION   Observe that

$$\int \frac{(x+1)\,dx}{\sqrt{1 - x^2}} = \int \frac{x\,dx}{\sqrt{1 - x^2}} + \int \frac{dx}{\sqrt{1 - x^2}}.$$

In the first integral on the right, we let $u = 1 - x^2$, $du = -2x\,dx$. Thus

$$\int \frac{(x+1)\,dx}{\sqrt{1 - x^2}} = -\frac{1}{2}\int \frac{du}{u^{1/2}} + \int \frac{1\,dx}{\sqrt{1 - x^2}} = -\sqrt{1 - x^2} + \sin^{-1} x + C.$$

77. $\displaystyle\int \frac{e^x\,dx}{1 + e^{2x}}$

SOLUTION   Let $u = e^x$. Then $du = e^x dx$, and

$$\int \frac{e^x}{1 + e^{2x}} = \int \frac{du}{1 + u^2} = \tan^{-1} u + C = \tan^{-1} e^x + C.$$

79. $\displaystyle\int \frac{\tan^{-1} x\,dx}{1 + x^2}$

SOLUTION   Let $u = \tan^{-1} x$. Then $du = \dfrac{dx}{1 + x^2}$, and

$$\int \frac{\tan^{-1} x\,dx}{1 + x^2} = \int u\,du = \frac{1}{2}u^2 + C = \frac{(\tan^{-1} x)^2}{2} + C.$$

81. Use Figure 11 to prove the formula

$$\int_0^x \sqrt{1 - t^2}\,dt = \frac{1}{2}x\sqrt{1 - x^2} + \frac{1}{2}\sin^{-1} x$$

Hint: The area represented by the integral is the sum of a triangle and a sector.

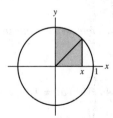

FIGURE 11

SOLUTION The definite integral $\int_0^x \sqrt{1-t^2}\, dt$ represents the area of the region under the upper half of the unit circle from 0 to x. The region consists of a sector of the circle and a right triangle. The sector has a central angle of $\frac{\pi}{2} - \theta$, where $\cos\theta = x$. Hence, the sector has an area of

$$\frac{1}{2}(1)^2\left(\frac{\pi}{2} - \cos^{-1}x\right) = \frac{1}{2}\sin^{-1}x.$$

The right triangle has a base of length x, a height of $\sqrt{1-x^2}$, and hence an area of $\frac{1}{2}x\sqrt{1-x^2}$. Thus,

$$\int_0^x \sqrt{1-t^2}\, dt = \frac{1}{2}x\sqrt{1-x^2} + \frac{1}{2}\sin^{-1}x.$$

83. A painting of length b is located at a height h above eye level (Figure 12). Find the distance x at which the viewing angle θ is maximized (this coincides with Exercise 47 in Section 4.6; solve it this time using inverse trigonometric functions).

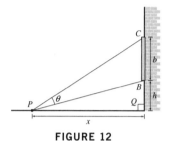

FIGURE 12

SOLUTION From the figure, we see that

$$\theta(x) = \tan^{-1}\frac{b+h}{x} - \tan^{-1}\frac{h}{x}.$$

Thus,

$$\theta'(x) = \frac{h}{x^2 + h^2} - \frac{b+h}{x^2 + (b+h)^2}.$$

Setting $\theta'(x) = 0$ yields

$$\frac{h}{x^2 + h^2} = \frac{b+h}{x^2 + (b+h)^2},$$

which simplifies to $x = \sqrt{bh + h^2}$.

Further Insights and Challenges

85. A cylindrical tank of radius R and length L lying horizontally as in Figure 13 is filled with oil to height h.

(a) Show that the volume $V(h)$ of oil in the tank as a function of height h is

$$V(h) = L\left(R^2\cos^{-1}\left(1 - \frac{h}{R}\right) - (R-h)\sqrt{2hR - h^2}\right)$$

(b) Show that $\dfrac{dV}{dh} = 2L\sqrt{h(2R - h)}$.

(c) Suppose that $R = 4$ ft and $L = 30$ ft, and that the tank is filled at a constant rate of 10 ft³/min. How fast is the height h increasing when $h = 5$?

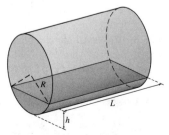

FIGURE 13 Oil in the tank has level h.

SOLUTION

(a) From Figure 13, we see that the volume of oil in the tank, $V(h)$, is equal to L times $A(h)$, the area of that portion of the circular cross section occupied by the oil. Now,

$$A(h) = \text{area of sector} - \text{area of triangle} = \frac{R^2\theta}{2} - \frac{R^2\sin\theta}{2},$$

where θ is the central angle of the sector. Referring to the diagram below,

$$\cos\frac{\theta}{2} = \frac{R-h}{R} \quad \text{and} \quad \sin\frac{\theta}{2} = \frac{\sqrt{2hR-h^2}}{R}.$$

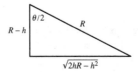

Thus,

$$\theta = 2\cos^{-1}\left(1 - \frac{h}{R}\right),$$

$$\sin\theta = 2\sin\frac{\theta}{2}\cos\frac{\theta}{2} = 2\frac{(R-h)\sqrt{2hR-h^2}}{R^2},$$

and

$$V(h) = L\left(R^2\cos^{-1}\left(1 - \frac{h}{R}\right) - (R-h)\sqrt{2hR-h^2}\right).$$

(b) Recalling that $\frac{d}{dx}\cos^{-1}u = -\frac{1}{\sqrt{1-x^2}}\frac{du}{dx}$,

$$\frac{dV}{dh} = L\left(\frac{d}{dh}\left(R^2\cos^{-1}\left(1 - \frac{h}{R}\right)\right) - \frac{d}{dh}\left((R-h)\sqrt{2hR-h^2}\right)\right)$$

$$= L\left(-R\frac{-1}{\sqrt{1-(1-(h/R))^2}} + \sqrt{2hR-h^2} - \frac{(R-h)^2}{\sqrt{2hR-h^2}}\right)$$

$$= L\left(\frac{R^2}{\sqrt{2hR-h^2}} + \sqrt{2hR-h^2} - \frac{R^2-2Rh+h^2}{\sqrt{2hR-h^2}}\right)$$

$$= L\left(\frac{R^2 + (2hR-h^2) - (R^2-2Rh+h^2)}{\sqrt{2hR-h^2}}\right)$$

$$= L\left(\frac{4hR-2h^2}{\sqrt{2hR-h^2}}\right) = L\left(\frac{2(2hR-h^2)}{\sqrt{2hR-h^2}}\right) = 2L\sqrt{2hR-h^2}.$$

(c) $\frac{dV}{dt} = \frac{dV}{dh}\frac{dh}{dt}$, so $\frac{dh}{dt} = \frac{1}{dV/dh}\frac{dV}{dt}$. From part (b) with $R = 4$, $L = 30$ and $h = 5$,

$$\frac{dV}{dh} = 2(30)\sqrt{2(5)(4)-5^2} = 60\sqrt{15}\ \text{ft}^2.$$

Thus,

$$\frac{dh}{dt} = \frac{1}{60\sqrt{15}}(10) = \frac{\sqrt{15}}{90} \approx 0.043\ \text{ft/min}.$$

87. Tom drives with his friend Ali along a highway represented by the graph of a differentiable function $y = f(x)$ as in Figure 14. During the trip, Ali views a billboard represented by the segment $\overline{BC}$ along the y-axis. Let Q be the y-intercept of the tangent line to $y = f(x)$. Show that θ is maximized at the value of x for which the angles $\angle QPB$ and $\angle QCP$ are equal. This is a generalization of Exercise 84 [which corresponds to the case $f(x) = 0$]. *Hints:*

(a) Compute $d\theta/dx$ and check that it equals

$$(b-c)\cdot\frac{(x^2+(xf'(x))^2)-(b-(f(x)-xf'(x)))(c-(f(x)-xf'(x)))}{(x^2+(b-f(x))^2)(x^2+(c-f(x))^2)}$$

(b) Show that the y-coordinate of Q is $f(x) - xf'(x)$.

(c) Show that the condition $\dfrac{d\theta}{dx} = 0$ is equivalent to

$$PQ^2 = BQ \cdot CQ$$

(d) Use (c) to conclude that triangles $\triangle QPB$ and $\triangle QCP$ are similar.

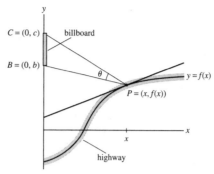

FIGURE 14

SOLUTION

(a) From the figure, we see that

$$\theta(x) = \tan^{-1} \frac{c - f(x)}{x} - \tan^{-1} \frac{b - f(x)}{x}.$$

Then

$$\theta'(x) = \frac{b - (f(x) - xf'(x))}{x^2 + (b - f(x))^2} - \frac{c - (f(x) - xf'(x))}{x^2 + (c - f(x))^2}$$

$$= (b - c)\frac{x^2 - bc + (b + c)(f(x) - xf'(x)) - (f(x))^2 + 2xf(x)f'(x)}{(x^2 + (b - f(x))^2)(x^2 + (c - f(x))^2)}$$

$$= (b - c)\frac{(x^2 + (xf'(x))^2) - (bc - (b + c)(f(x) - xf'(x)) + (f(x) - xf'(x))^2)}{(x^2 + (b - f(x))^2)(x^2 + (c - f(x))^2)}$$

$$= (b - c)\frac{(x^2 + (xf'(x))^2) - (b - (f(x) - xf'(x)))(c - (f(x) - xf'(x)))}{(x^2 + (b - f(x))^2)(x^2 + (c - f(x))^2)}.$$

(b) The point Q is the y-intercept of the line tangent to the graph of $f(x)$ at point P. The equation of this tangent line is

$$Y - f(x) = f'(x)(X - x).$$

The y-coordinate of Q is then $f(x) - xf'(x)$.

(c) From the figure, we see that

$$BQ = b - (f(x) - xf'(x)),$$

$$CQ = c - (f(x) - xf'(x))$$

and

$$PQ = \sqrt{x^2 + (f(x) - (f(x) - xf'(x)))^2} = \sqrt{x^2 + (xf'(x))^2}.$$

Comparing these expressions with the numerator of $d\theta/dx$, it follows that $\dfrac{d\theta}{dx} = 0$ is equivalent to

$$PQ^2 = BQ \cdot CQ.$$

(d) The equation $PQ^2 = BQ \cdot CQ$ is equivalent to

$$\frac{PQ}{BQ} = \frac{CQ}{PQ}.$$

In other words, the sides CQ and PQ from the triangle $\triangle QCP$ are proportional in length to the sides PQ and BQ from the triangle $\triangle QPB$. As $\angle PQB = \angle CQP$, it follows that triangles $\triangle QCP$ and $\triangle QPB$ are similar.

7.9 Hyperbolic Functions

Preliminary Questions

1. Which hyperbolic functions take on only positive values?

SOLUTION $\cosh x$ and $\operatorname{sech} x$ take on only positive values.

2. Which hyperbolic functions are increasing on their domains?

SOLUTION $\sinh x$ and $\tanh x$ are increasing on their domains.

3. Describe three properties of hyperbolic functions that have trigonometric analogs.

SOLUTION Hyperbolic functions have the following analogs with trigonometric functions: parity, identities and derivative formulas.

4. Which hyperbolic derivative formulas differ from their trigonometric counterparts by a minus sign?

SOLUTION The derivatives of $\cosh x$ and $\operatorname{sech} x$ differ from their trigonometric counterparts by a minus sign.

Exercises

1. Use a calculator to compute $\sinh x$ and $\cosh x$ for $x = -3, 0, 5$.

SOLUTION

x		-3	0	5
$\sinh x =$	$\dfrac{e^x - e^{-x}}{2}$	$\dfrac{e^{-3} - e^3}{2} = -10.0179$	$\dfrac{e^0 - e^0}{2} = 0$	$\dfrac{e^5 - e^{-5}}{2} = 74.203$
$\cosh x =$	$\dfrac{e^x + e^{-x}}{2}$	$\dfrac{e^{-3} + e^3}{2} = 10.0677$	$\dfrac{e^0 + e^0}{2} = 1$	$\dfrac{e^5 + e^{-5}}{2} = 74.210$

3. For which values of x are $y = \sinh x$ and $y = \cosh x$ increasing and decreasing?

SOLUTION $\dfrac{d}{dx}\sinh x = \cosh x$. Since e^x and e^{-x} are positive for all x, it follows that $\cosh x = \frac{1}{2}(e^x + e^{-x}) > 0$ for all x. Thus, $\sinh x$ is increasing for all x. On the other hand, $\dfrac{d}{dx}\cosh x = \sinh x$. Since $e^x > e^{-x}$ for $x > 0$ while $e^x < e^{-x}$ for $x < 0$, it follows that $\sinh x > 0$ for $x > 0$ but $\sinh x < 0$ for $x < 0$. Thus, $\cosh x$ is decreasing for $x < 0$ and is increasing for $x > 0$.

In Exercises 5–30, calculate the derivative.

5. $y = \sinh(3x)$

SOLUTION $\dfrac{d}{dx}\sinh(3x) = 3\cosh(3x)$.

7. $y = \cosh(1 - 4t)$

SOLUTION $\dfrac{d}{dt}\cosh(1 - 4t) = -4\sinh(1 - 4t)$.

9. $y = \sqrt{\cosh x + 1}$

SOLUTION $\dfrac{d}{dx}\sqrt{\cosh x + 1} = \dfrac{1}{2}(\cosh x + 1)^{-1/2}\sinh x$.

11. $y = \dfrac{\sinh t}{1 + \cosh t}$

SOLUTION $\dfrac{d}{dt}\dfrac{\sinh t}{1 + \cosh t} = \dfrac{\cosh t(1 + \cosh t) - \sinh t(\sinh t)}{(1 + \cosh t)^2} = \dfrac{1 + \cosh t}{(1 + \cosh t)^2} = \dfrac{1}{1 + \cosh t}$.

13. $y = \tanh(2x)$

SOLUTION $\dfrac{d}{dx}\tanh(2x) = 2\operatorname{sech}^2(2x)$.

15. $y = e^{\tanh x}$

SOLUTION $\dfrac{d}{dx}e^{\tanh x} = \operatorname{sech}^2 x \cdot e^{\tanh x}$.

17. $y = \sinh(xe^x)$

SOLUTION $\dfrac{d}{dx}\sinh(xe^x) = (e^x + xe^x)\cosh(xe^x).$

19. $y = \operatorname{sech}(\sqrt{x})$

SOLUTION $\dfrac{d}{dx}\operatorname{sech}(\sqrt{x}) = -\dfrac{1}{2}x^{-1/2}\operatorname{sech}\sqrt{x}\tanh\sqrt{x}.$

21. $y = \ln(\coth x)$

SOLUTION $\dfrac{d}{dx}\ln(\coth x) = \dfrac{-\operatorname{csch}^2 x}{\coth x} = \dfrac{-1}{\sinh^2 x(\frac{\cosh x}{\sinh x})} = \dfrac{-1}{\sinh x \cosh x}.$

23. $y = \tanh(3x^2 - 9)$

SOLUTION $\dfrac{d}{dx}\tanh(3x^2 - 9) = 6x\operatorname{sech}^2(3x^2 - 9).$

25. $y = \tanh^{-1} 3x$

SOLUTION $\dfrac{d}{dx}\tanh^{-1}(3x) = \dfrac{3}{1 - 9x^2}.$

27. $y = (\operatorname{csch}^{-1} 3x)^4$

SOLUTION $\dfrac{d}{dx}(\operatorname{csch}^{-1} 3x)^4 = 4(\operatorname{csch}^{-1} 3x)^3 \left(\dfrac{-1}{|3x|\sqrt{1+9x^2}}\right)(3) = \dfrac{-4(\operatorname{csch}^{-1} 3x)^3}{|x|\sqrt{1+9x^2}}.$

29. $y = \sinh^{-1}(\sqrt{x^2 + 1})$

SOLUTION $\dfrac{d}{dx}\sinh^{-1}(\sqrt{x^2 + 1}) = \dfrac{1}{\sqrt{x^2 + 1 + 1}}\left(\dfrac{1}{2\sqrt{x^2 + 1}}\right)(2x) = \dfrac{x}{\sqrt{x^2 + 2}\cdot\sqrt{x^2 + 1}}.$

In Exercises 31–42, calculate the integral.

31. $\displaystyle\int \cosh(3x)\,dx$

SOLUTION $\displaystyle\int \cosh(3x)\,dx = \dfrac{1}{3}\sinh 3x + C.$

33. $\displaystyle\int x\sinh(x^2 + 1)\,dx$

SOLUTION $\displaystyle\int x\sinh(x^2 + 1)\,dx = \dfrac{1}{2}\cosh(x^2 + 1) + C.$

35. $\displaystyle\int \operatorname{sech}^2(1 - 2x)\,dx$

SOLUTION $\displaystyle\int \operatorname{sech}^2(1 - 2x)\,dx = -\dfrac{1}{2}\tanh(1 - 2x) + C.$

37. $\displaystyle\int \tanh x \operatorname{sech}^2 x\,dx$

SOLUTION Let $u = \tanh x$. Then $du = \operatorname{sech}^2 x\,dx$ nd

$$\int \tanh x \operatorname{sech}^2 x\,dx = \int u\,du = \dfrac{1}{2}u^2 + C = \dfrac{\tanh^2 x}{2} + C.$$

39. $\displaystyle\int \tanh x\,dx$

SOLUTION $\displaystyle\int \tanh x\,dx = \ln\cosh x + C.$

41. $\displaystyle\int e^{-x}\sinh x\,dx$

SOLUTION Since $\sinh x = \dfrac{e^x - e^{-x}}{2}$ we can combine the two functions to get

$$\int e^{-x}\sinh x\,dx = \dfrac{1}{2}\int e^{-x}(e^x - e^{-x})\,dx = \dfrac{1}{2}\int \left(1 - e^{-2x}\right)dx = \dfrac{1}{2}x + \dfrac{1}{4}e^{-2x} + C.$$

43. Verify the formula $\dfrac{d}{dx}(\coth x) = -\operatorname{csch}^2 x.$

SOLUTION $\dfrac{d}{dx}\coth x = \dfrac{d}{dx}\dfrac{\cosh x}{\sinh x} = \dfrac{\sinh^2 x - \cosh^2 x}{\sinh^2 x} = \dfrac{-1}{\sinh^2 x} = -\operatorname{csch}^2 x.$

45. Refer to the graphs to explain why the equation $\sinh x = t$ has a unique solution for every t and $\cosh x = t$ has two solutions for every $t > 1$.

SOLUTION From its graph we see that $\sinh x$ is a one-to-one function with $\lim\limits_{x \to -\infty} \sinh x = -\infty$ and $\lim\limits_{x \to \infty} \sinh x = \infty$. Thus, for every real number t, the equation $\sinh x = t$ has a unique solution. On the other hand, from its graph, we see that $\cosh x$ is not one-to-one. Rather, it is an even function with a minimum value of $\cosh 0 = 1$. Thus, for every $t > 1$, the equation $\cosh x = t$ has two solutions: one positive, the other negative.

47. Prove the addition formula for $\cosh x$.

SOLUTION

$$
\begin{aligned}
\cosh(x+y) &= \frac{e^{x+y} + e^{-(x+y)}}{2} = \frac{2e^{x+y} + 2e^{-(x+y)}}{4} \\
&= \frac{e^{x+y} + e^{-x+y} + e^{x-y} + e^{-(x+y)}}{4} + \frac{e^{x+y} - e^{-x+y} - e^{x-y} + e^{-(x+y)}}{4} \\
&= \left(\frac{e^x + e^{-x}}{2}\right)\left(\frac{e^y + e^{-y}}{2}\right) + \left(\frac{e^x - e^{-x}}{2}\right)\left(\frac{e^y - e^{-y}}{2}\right) \\
&= \cosh x \cosh y + \sinh x \sinh y.
\end{aligned}
$$

In Exercises 49–52, prove the formula.

49. $\cosh(\sinh^{-1} t) = \sqrt{t^2 + 1}$

SOLUTION Note

$$
\frac{d}{dt}\left[\cosh(\sinh^{-1} t)\right] = \sinh(\sinh^{-1} t)\frac{1}{\sqrt{1+t^2}} = \frac{t}{\sqrt{1+t^2}} = \frac{d}{dt}\sqrt{1+t^2},
$$

so the functions $\cosh(\sinh^{-1} t)$ and $\sqrt{1+t^2}$ differ by a constant; substituting $t = 0$ we find that the constant is 0. Therefore,

$$
\cosh(\sinh^{-1} t) = \sqrt{t^2 + 1}.
$$

51. $\dfrac{d}{dt}\sinh^{-1} t = \dfrac{1}{\sqrt{t^2 + 1}}$

SOLUTION Let $x = \sinh^{-1} t$. Then $t = \sinh x$ and

$$
1 = \cosh x \frac{dx}{dt} \quad \text{or} \quad \frac{dx}{dt} = \frac{1}{\cosh x}.
$$

Thus,

$$
\frac{d}{dt}\sinh^{-1} t = \frac{1}{\cosh(\sinh^{-1} t)} = \frac{1}{\sqrt{t^2 + 1}}
$$

by Exercise 49.

In Exercises 53–60, calculate the integral in terms of inverse hyperbolic functions.

53. $\displaystyle\int_2^4 \frac{dx}{\sqrt{x^2 - 1}}$

SOLUTION $\displaystyle\int_2^4 \frac{dx}{\sqrt{x^2 - 1}} = \cosh^{-1} x \Big|_2^4 = \cosh^{-1} 4 - \cosh^{-1} 2.$

55. $\displaystyle\int \frac{dx}{\sqrt{9 + x^2}}$

SOLUTION $\displaystyle\int \frac{dt}{\sqrt{9 + x^2}} = \int \frac{dx}{3\sqrt{1 + (x/3)^2}} = \sinh^{-1}\frac{x}{3} + C.$

57. $\displaystyle\int_{1/3}^{1/2} \frac{dx}{1 - x^2}$

SOLUTION $\displaystyle\int_{1/3}^{1/2}\frac{dx}{1-x^2} = \tanh^{-1}x\Big|_{1/3}^{1/2} = \tanh^{-1}\frac{1}{2} - \tanh^{-1}\frac{1}{3}.$

59. $\displaystyle\int_{2}^{10}\frac{dx}{4x^2-1}$

SOLUTION $\displaystyle\int_{2}^{10}\frac{dx}{4x^2-1} = -\frac{1}{2}\coth^{-1}(2x)\Big|_{2}^{10} = \frac{1}{2}(\coth^{-1}4 - \coth^{-1}20).$

61. Prove that $\sinh^{-1}t = \ln(t + \sqrt{t^2+1})$. *Hint:* Let $t = \sinh x$. Prove that $\cosh x = \sqrt{t^2+1}$ and use the relation

$$\sinh x + \cosh x = e^x.$$

SOLUTION Let $t = \sinh x$. Then

$$\cosh x = \sqrt{1+\sinh^2 x} = \sqrt{1+t^2}.$$

Moreover, because

$$\sinh x + \cosh x = \frac{e^x - e^{-x}}{2} + \frac{e^x + e^{-x}}{2} = e^x,$$

it follows that

$$\sinh^{-1}t = x = \ln(\sinh x + \cosh x) = \ln(t + \sqrt{t^2+1}).$$

63. Prove that $\tanh^{-1}t = \frac{1}{2}\ln\left(\frac{1+t}{1-t}\right)$ for $|t| < 1$.

SOLUTION Let $A = \tanh^{-1}t$. Then

$$t = \tanh A = \frac{\sinh A}{\cosh A} = \frac{e^A - e^{-A}}{e^A + e^{-A}}.$$

Solving for A yields

$$A = \frac{1}{2}\ln\frac{t+1}{1-t};$$

hence,

$$\tanh^{-1}t = \frac{1}{2}\ln\frac{t+1}{1-t}.$$

65. An (imaginary) train moves along a track at velocity u, and a woman walks down the aisle of the train with velocity v in the direction of the train's motion. Compute the velocity w of the woman relative to the ground using the laws of both Galileo and Einstein in the following cases.
(a) $u = 1,000$ mph and $v = 50$ mph. Is your calculator accurate enough to detect the difference between the two laws?
(b) $u = 100,000$ miles/sec and $v = 50$ miles/sec.

SOLUTION
(a) By Galileo's law, $1,000 + 50 = 1,050$ mph. The speed of light is $c = 186,000$ miles/sec $= 669,600,000$ mph. Using Einstein's law and a calculator,

$$\tanh^{-1}\frac{w}{c} = \tanh^{-1}\frac{1,000}{669,600,000} + \tanh^{-1}\frac{50}{669,600,000} = 1.568100358 \times 10^{-6};$$

so $w = 1,050$ mph. No, the calculator was not accurate enough to detect the difference between the two laws.
(b) By Galileo's law, $100,000 + 50 = 100,050$ miles/sec. By Einstein's law,

$$\tanh^{-1}\frac{w}{c} = \tanh^{-1}\frac{100,000}{186,000} + \tanh^{-1}\frac{50}{186,000} = 0.601091074,$$

so $w \approx 100,035.5$ miles/sec.

Further Insights and Challenges

67. (a) Use the addition formulas for $\sinh x$ and $\cosh x$ to prove

$$\tanh(u + v) = \frac{\tanh u + \tanh v}{1 + \tanh u \tanh v}$$

(b) Use (a) to show that Einstein's Law of Velocity Addition [Eq. (2)] is equivalent to

$$w = \frac{u + v}{1 + \dfrac{uv}{c^2}}$$

SOLUTION

(a)

$$\tanh(u + v) = \frac{\sinh(u + v)}{\cosh(u + v)} = \frac{\sinh u \cosh v + \cosh u \sinh v}{\cosh u \cosh v + \sinh u \sinh v}$$

$$= \frac{\sinh u \cosh v + \cosh u \sinh v}{\cosh u \cosh v + \sinh u \sinh v} \cdot \frac{1/(\cosh u \cosh v)}{1/(\cosh u \cosh v)} = \frac{\tanh u + \tanh v}{1 + \tanh u \tanh v}$$

(b) Einstein's law states: $\tanh^{-1}(w/c) = \tanh^{-1}(u/c) + \tanh^{-1}(v/c)$. Thus

$$\frac{w}{c} = \tanh\left(\tanh^{-1}(u/c) + \tanh^{-1}(v/c)\right) = \frac{\tanh(\tanh^{-1}(v/c)) + \tanh(\tanh^{-1}(u/c))}{1 + \tanh(\tanh^{-1}(v/c))\tanh(\tanh^{-1}(u/c))}$$

$$= \frac{\frac{v}{c} + \frac{u}{c}}{1 + \frac{v}{c}\frac{u}{c}} = \frac{(1/c)(u + v)}{1 + \frac{uv}{c^2}}.$$

Hence,

$$w = \frac{u + v}{1 + \dfrac{uv}{c^2}}.$$

69. (a) Show that $y = \tanh t$ satisfies the differential equation $\dfrac{dy}{dt} = 1 - y^2$ with initial condition $y(0) = 0$.

(b) Show that for arbitrary constants A, B, the function

$$y = A \tanh(Bt)$$

satisfies

$$\frac{dy}{dt} = AB - \frac{B}{A}y^2, \qquad y(0) = 0$$

(c) Let $v(t)$ be the velocity of a falling object of mass m. For large velocities, air resistance is proportional to the square of velocity $v(t)^2$. If we choose coordinates so that $v(t) > 0$ for a falling object, then by Newton's Law of Motion, there is a constant $k > 0$ such that

$$\frac{dv}{dt} = g - \frac{k}{m}v^2$$

Solve for $v(t)$ by applying the result of (b) with $A = \sqrt{gm/k}$ and $B = \sqrt{gk/m}$.

(d) Calculate the terminal velocity $\lim_{t \to \infty} v(t)$.

(e) Find k if $m = 150$ lb and the terminal velocity is 100 mph.

SOLUTION

(a) First, note that if we divide the identity $\cosh^2 t - \sinh^2 t = 1$ by $\cosh^2 t$, we obtain the identity $1 - \tanh^2 t = \operatorname{sech}^2 t$. Now, let $y = \tanh t$. Then

$$\frac{dy}{dt} = \operatorname{sech}^2 t = 1 - \tanh^2 t = 1 - y^2.$$

Furthermore, $y(0) = \tanh 0 = 0$.

(b) Let $y = A \tanh(Bt)$. Then

$$\frac{dy}{dt} = AB \operatorname{sech}^2(Bt) = AB(1 - \tanh^2(Bt)) = AB\left(1 - \frac{y^2}{A^2}\right) = AB - \frac{By^2}{A}.$$

Furthermore, $y(0) = A \tanh(0) = 0$.

(c) Matching the differential equation

$$\frac{dv}{dt} = g - \frac{k}{m}v^2$$

with the template

$$\frac{dv}{dt} = AB - \frac{B}{A}v^2$$

from part (b) yields

$$AB = g \quad \text{and} \quad \frac{B}{A} = \frac{k}{m}.$$

Solving for A and B gives

$$A = \sqrt{\frac{mg}{k}} \quad \text{and} \quad B = \sqrt{\frac{kg}{m}}.$$

Thus

$$v(t) = A \tanh(Bt) = \sqrt{\frac{mg}{k}} \tanh\left(\sqrt{\frac{kg}{m}}t\right).$$

(d) $\lim_{t \to \infty} v(t) = \sqrt{\frac{mg}{k}} \lim_{t \to \infty} \tanh\left(\sqrt{\frac{kg}{m}}t\right) = \sqrt{\frac{mg}{k}}$

(e) Substitute $m = 150$ lb and $g = 32$ ft/sec^2 = 78545.5 miles/hr^2 into the equation for the terminal velocity obtained in part (d) and then solve for k. This gives

$$k = \frac{150(78545.5)}{100^2} = 1178.18 \text{ lb/mile.}$$

*In Exercises 70–72, a flexible chain of length L is suspended between two poles of equal height separated by a distance $2M$ (Figure 9). By Newton's laws, the chain describes a curve (called a **catenary**) with equation $y = a \cosh\left(\dfrac{x}{a}\right) + C$.*

The constant C is arbitrary and a is the number such that $L = 2a \sinh\left(\dfrac{M}{a}\right)$. The sag s is the vertical distance from the highest to the lowest point on the chain.

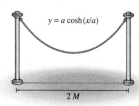

$y = a \cosh(x/a)$

$2M$

FIGURE 9 Chain hanging between two poles describes the curve $y = a \cosh(x/a)$.

71. Let M be a fixed constant. Show that the sag is given by $s = a \cosh\left(\dfrac{M}{a}\right) - a$.

(a) Calculate $\dfrac{ds}{da}$.

(b) Calculate $\dfrac{da}{dL}$ by implicit differentiation using the relation $L = 2a \sinh\left(\dfrac{M}{a}\right)$.

(c) Use (a) and (b) and the Chain Rule to show that

$$\frac{ds}{dL} = \frac{ds}{da}\frac{da}{dL} = \frac{\cosh(M/a) - (M/a)\sinh(M/a) - 1}{2\sinh(M/a) - (2M/a)\cosh(M/a)} \qquad \boxed{6}$$

SOLUTION The sag in the curve is

$$s = y(M) - y(0) = a \cosh\left(\frac{M}{a}\right) + C - (a \cosh 0 + C) = a \cosh\left(\frac{M}{a}\right) - a.$$

(a) $\dfrac{ds}{da} = \cosh\left(\dfrac{M}{a}\right) - \dfrac{M}{a}\sinh\left(\dfrac{M}{a}\right) - 1$

(b) If we differentiate the relation $L = 2a \sinh\left(\dfrac{M}{a}\right)$ with respect to a, we find

$$0 = 2\frac{da}{dL}\sinh\left(\frac{M}{a}\right) - \frac{2M}{a}\frac{da}{dL}\cosh\left(\frac{M}{a}\right).$$

Solving for da/dL yields

$$\frac{da}{dL} = \left(2\sinh\left(\frac{M}{a}\right) - \frac{2M}{a}\cosh\left(\frac{M}{a}\right)\right)^{-1}.$$

(c) By the Chain Rule,

$$\frac{ds}{dL} = \frac{ds}{da}\cdot\frac{da}{dL}.$$

The formula for ds/dL follows upon substituting the results from parts (a) and (b).

73. Prove that every function $f(x)$ is the sum of an even function $f_+(x)$ and an odd function $f_-(x)$. [*Hint:* $f_\pm(x) = \frac{1}{2}(f(x)\pm f(-x))$.] Express $f(x) = 5e^x + 8e^{-x}$ in terms of $\cosh x$ and $\sinh x$.

SOLUTION   Let $f_+(x) = \frac{f(x)+f(-x)}{2}$ and $f_-(x) = \frac{f(x)-f(-x)}{2}$. Then $f_+ + f_- = \frac{2f(x)}{2} = f(x)$. Moreover,

$$f_+(-x) = \frac{f(-x)+f(-(-x))}{2} = \frac{f(-x)+f(x)}{2} = f_+(x),$$

so $f_+(x)$ is an even function, while

$$f_-(-x) = \frac{f(-x)-f(-(-x))}{2} = \frac{f(-x)-f(x)}{2} = -\frac{(f(x)-f(-x))}{2} = -f_-(x),$$

so $f_-(x)$ is an odd function.
   For $f(x) = 5e^x + 8e^{-x}$, we have

$$f_+(x) = \frac{5e^x + 8e^{-x} + 5e^{-x} + 8e^x}{2} = 8\cosh x + 5\cosh x = 13\cosh x$$

and

$$f_-(x) = \frac{5e^x + 8e^{-x} - 5e^{-x} - 8e^x}{2} = 5\sinh x - 8\sinh x = -3\sinh x.$$

Therefore, $f(x) = f_+(x) + f_-(x) = 13\cosh x - 3\sinh x$.

75. In the Excursion, we discussed the relations

$$\cosh(it) = \cos t \quad \text{and} \quad \sinh(it) = i\sin t$$

Use these relations to show that the identity $\cos^2 t + \sin^2 t = 1$ results from the identity $\cosh^2 x - \sinh^2 x = 1$ by setting $x = it$.

SOLUTION   Substituting $x = it$ into $\cosh^2 x - \sinh^2 x = 1$ yields $\cosh^2(it) - \sinh^2(it) = 1$. Since $\cosh^2(it) = \cos^2 t$ and $\sinh^2(it) = (i\sin t)^2 = -\sin^2 t$, it follows that $\cos^2 t + \sin^2 t = 1$.

CHAPTER REVIEW EXERCISES

1. Match each quantity (a)–(d) with (i), (ii), or (iii) if possible, or state that no match exists.

(a) $2^a 3^b$

(b) $\dfrac{2^a}{3^b}$

(c) $(2^a)^b$

(d) $2^{a-b}3^{b-a}$

(i) 2^{ab} **(ii)** 6^{a+b} **(iii)** $\left(\frac{2}{3}\right)^{a-b}$

SOLUTION
(a) No match.
(b) No match.
(c) (i): $(2^a)^b = 2^{ab}$.
(d) (iii): $2^{a-b}3^{b-a} = 2^{a-b}\left(\frac{1}{3}\right)^{a-b} = \left(\frac{2}{3}\right)^{a-b}$.

3. Which of the following is equal to $\dfrac{d}{dx}2^x$?

(a) 2^x

(b) $(\ln 2)2^x$

(c) $x2^{x-1}$

(d) $\dfrac{1}{\ln 2}2^x$

SOLUTION   The derivative of $f(x) = 2^x$ is

$$\frac{d}{dx}2^x = 2^x \ln 2.$$

Hence, the correct answer is **(b)**.

5. Find the inverse of $f(x) = \dfrac{x-2}{x-1}$ and determine its domain and range.

SOLUTION   To find the inverse of $f(x) = \frac{x-2}{x-1}$, we solve $y = \frac{x-2}{x-1}$ for x as follows:

$$x - 2 = y(x - 1) = yx - y$$
$$x - yx = 2 - y$$
$$x = \frac{2 - y}{1 - y}.$$

Therefore,

$$f^{-1}(x) = \frac{2 - x}{1 - x} = \frac{x - 2}{x - 1}.$$

The domain of f^{-1} is the range of f, namely $\{x : x \neq 1\}$; the range of f^{-1} is the domain of f, namely $\{y : y \neq 1\}$.

7. Show that $g(x) = \dfrac{x}{x-1}$ is equal to its inverse on the domain $\{x : x \neq -1\}$.

SOLUTION   To show that $g(x) = \frac{x}{x-1}$ is equal to its inverse, we need to show that for $x \neq 1$,

$$g(g(x)) = x.$$

First, we notice that for $x \neq 1$, $g(x) \neq 1$. Therefore,

$$g(g(x)) = g\left(\frac{x}{x-1}\right) = \frac{\frac{x}{x-1}}{\frac{x}{x-1} - 1} = \frac{x}{x - (x-1)} = \frac{x}{1} = x.$$

9. Suppose that $g(x)$ is the inverse of $f(x)$. Match the functions (a)–(d) with their inverses (i)–(iv).

(a) $f(x) + 1$
(b) $f(x + 1)$
(c) $4f(x)$
(d) $f(4x)$

(i) $g(x)/4$
(ii) $g(x/4)$
(iii) $g(x - 1)$
(iv) $g(x) - 1$

SOLUTION
(a) (iii): $f(x) + 1$ and $g(x - 1)$ are inverse functions:

$$f(g(x - 1)) + 1 = (x - 1) + 1 = x;$$
$$g(f(x) + 1 - 1) = g(f(x)) = x.$$

(b) (iv): $f(x + 1)$ and $g(x) - 1$ are inverse functions:

$$f(g(x) - 1 + 1) = f(g(x)) = x;$$
$$g(f(x + 1)) - 1 = (x + 1) - 1 = x.$$

(c) (ii): $4f(x)$ and $g(x/4)$ are inverse functions:

$$4f(g(x/4)) = 4(x/4) = x;$$
$$g(4f(x)/4) = g(f(x)) = x.$$

(d) (i): $f(4x)$ and $g(x)/4$ are inverse functions:

$$f(4 \cdot g(x)/4) = f(g(x)) = x;$$

$$\frac{1}{4}g(f(4x)) = \frac{1}{4}(4x) = x.$$

In Exercises 11–32, find the derivative.

11. $f(x) = 9e^{-4x}$

SOLUTION $\dfrac{d}{dx}9e^{-4x} = -36e^{-4x}.$

13. $f(x) = \dfrac{e^{-x}}{x}$

SOLUTION $\dfrac{d}{dx}\left(\dfrac{e^{-x}}{x}\right) = \dfrac{-xe^{-x} - e^{-x}}{x^2} = -\dfrac{e^{-x}(x+1)}{x^2}.$

15. $G(s) = (\ln(s))^2$

SOLUTION $\dfrac{d}{ds}(\ln s)^2 = \dfrac{2\ln s}{s}.$

17. $g(t) = e^{4t - t^2}$

SOLUTION $\dfrac{d}{dt}e^{4t - t^2} = (4 - 2t)e^{4t - t^2}.$

19. $f(\theta) = \ln(\sin \theta)$

SOLUTION $\dfrac{d}{d\theta}\ln(\sin\theta) = \dfrac{\cos\theta}{\sin\theta} = \cot\theta.$

21. $f(x) = e^{x + \ln x}$

SOLUTION $\dfrac{d}{dx}e^{x + \ln x} = \left(1 + \dfrac{1}{x}\right)e^{x + \ln x}.$

23. $h(y) = 2^{1-y}$

SOLUTION $\dfrac{d}{dy}2^{1-y} = -2^{1-y}\ln 2.$

25. $G(s) = \cos^{-1}(s^{-1})$

SOLUTION $\dfrac{d}{ds}\cos^{-1}(s^{-1}) = \dfrac{-1}{\sqrt{1 - \left(\frac{1}{s}\right)^2}}\left(-\dfrac{1}{s^2}\right) = \dfrac{1}{\sqrt{s^4 - s^2}}.$

27. $f(x) = \ln(\csc^{-1} x)$

SOLUTION $\dfrac{d}{dx}\ln(\csc^{-1} x) = -\dfrac{1}{|x|\sqrt{x^2 - 1}\,\csc^{-1} x}.$

29. $g(t) = \sinh(t^2)$

SOLUTION $\dfrac{d}{dt}\sinh(t^2) = 2t\cosh(t^2).$

31. $g(x) = \tanh^{-1}(e^x)$

SOLUTION $\dfrac{d}{dx}\tanh^{-1}(e^x) = \dfrac{1}{1 - (e^x)^2}e^x = \dfrac{e^x}{1 - e^{2x}}.$

33. Suppose that $f(g(x)) = e^{x^2}$, where $g(1) = 2$ and $g'(1) = 4$. Find $f'(2)$.

SOLUTION We differentiate both sides of the equation $f(g(x)) = e^{x^2}$ to obtain,

$$f'(g(x))\,g'(x) = 2xe^{x^2}.$$

Setting $x = 1$ yields

$$f'(g(1))\,g'(1) = 2e.$$

Since $g(1) = 2$ and $g'(1) = 4$, we find

$$f'(2) \cdot 4 = 2e,$$

or

$$f'(2) = \frac{e}{2}.$$

35. Find the points of inflection of $f(x) = \ln(x^2 + 1)$ and determine whether the concavity changes from up to down or vice versa.

SOLUTION   With $f(x) = \ln(x^2 + 1)$, we find

$$f'(x) = \frac{2x}{x^2 + 1}; \quad \text{and}$$

$$f''(x) = \frac{2\left(x^2 + 1\right) - 2x \cdot 2x}{\left(x^2 + 1\right)^2} = \frac{2(1 - x^2)}{\left(x^2 + 1\right)^2}$$

Thus, $f''(x) > 0$ for $-1 < x < 1$, whereas $f''(x) < 0$ for $x < -1$ and for $x > 1$. It follows that there are points of inflection at $x = \pm 1$, and that the concavity of f changes from down to up at $x = -1$ and from up to down at $x = 1$.

In Exercises 36–38, let $f(x) = xe^{-x}$.

37. Show that $f(x)$ has an inverse on $[1, \infty)$. Let $g(x)$ be this inverse. Find the domain and range of $g(x)$ and compute $g'(2e^{-2})$.

SOLUTION   Let $f(x) = xe^{-x}$. Then $f'(x) = e^{-x}(1 - x)$. On $[1, \infty)$, $f'(x) < 0$, so $f(x)$ is decreasing and therefore one-to-one. It follows that $f(x)$ has an inverse on $[1, \infty)$. Let $g(x)$ denote this inverse. Because $f(1) = e^{-1}$ and $f(x) \to 0$ as $x \to \infty$, the domain of $g(x)$ is $(0, e^{-1}]$, and the range is $[1, \infty)$.

To determine $g'(2e^{-2})$, we use the formula $g'(x) = 1/f'(g(x))$. Because $f(2) = 2e^{-2}$, it follows that $g(2e^{-2}) = 2$. Then,

$$g'(2e^{-2}) = \frac{1}{f'(g(2e^{-2}))} = \frac{1}{f'(2)} = \frac{1}{-e^{-2}} = -e^2.$$

In Exercises 39–42, find the local extrema and points of inflection, and sketch the graph over the interval specified. Use L'Hôpital's Rule to determine the limits as $x \to 0+$ or $x \to \pm\infty$ if necessary.

39. $y = x \ln x, \quad x > 0$

SOLUTION   Let $y = x \ln x$. Then

$$y' = \ln x + x\left(\frac{1}{x}\right) = 1 + \ln x,$$

and $y'' = \frac{1}{x}$. Solving $y' = 0$ yields the critical point $x = e^{-1}$. Since $y''(e^{-1}) = e > 0$, the function has a local minimum at $x = e^{-1}$. y'' is positive for $x > 0$, hence the function is concave up for $x > 0$ and there are no points of inflection. As $x \to 0+$ and as $x \to \infty$, we find

$$\lim_{x \to 0+} x \ln x = \lim_{x \to 0+} \frac{\ln x}{x^{-1}} = \lim_{x \to 0+} \frac{x^{-1}}{-x^{-2}} = \lim_{x \to 0+} (-x) = 0;$$

$$\lim_{x \to \infty} x \ln x = \infty.$$

The graph is shown below:

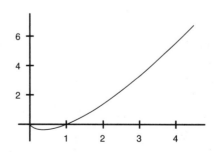

41. $y = x(\log x)^2, \quad x > 0$

SOLUTION Let $y = x(\log x)^2$. Then

$$y' = (\log x)^2 + x \cdot \frac{2 \log x}{x \ln 10} = (\log x)\left(\frac{2}{\ln 10} + \log x\right),$$

and

$$y'' = \frac{2 \log x}{x \ln 10} + \frac{2}{\ln 10} \cdot \frac{1}{x \ln 10} = \frac{2}{x \ln 10}\left(\log x + \frac{1}{\ln 10}\right).$$

Solving $y' = 0$ yields the critical points $x = 1$ and $x = e^{-2}$. Because

$$y''(1) = \frac{2}{(\ln 10)^2} > 0 \quad \text{and} \quad y''(e^{-2}) = -\frac{2(\log e)^2}{e^{-2}} < 0,$$

we conclude that the function has a local minimum at $x = 1$ and a local maximum at $x = e^{-2}$. We see that $y'' > 0$ for $x > e^{-1}$ and $y'' < 0$ for $0 < x < e^{-1}$. Therefore, there is a point of inflection at $x = e^{-1}$. As $x \to 0+$ and as $x \to \infty$, we find

$$\lim_{x \to 0+} x(\log x)^2 = \lim_{x \to 0+} \frac{(\log x)^2}{1/x} = \lim_{x \to 0+} \frac{2 \log x \frac{1}{\ln 10} \cdot \frac{1}{x}}{-1/x^2}$$

$$= -\frac{2}{\ln 10} \lim_{x \to 0+} \frac{\log x}{1/x} = -\frac{2}{\ln 10} \lim_{x \to 0+} \frac{\frac{1}{\ln 10} \cdot \frac{1}{x}}{-1/x^2}$$

$$= \frac{2}{(\ln 10)^2} \lim_{x \to 0+} x = 0; \quad \text{and}$$

$$\lim_{x \to \infty} x(\log x)^2 = \infty.$$

The graph is shown below:

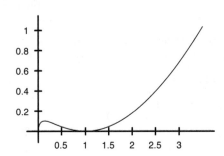

In Exercises 43–48, use logarithmic differentiation to find the derivative.

43. $y = \dfrac{(x + 1)^3}{(4x - 2)^2}$

SOLUTION Let $y = \dfrac{(x + 1)^3}{(4x - 2)^2}$. Then

$$\ln y = \ln \left(\frac{(x + 1)^3}{(4x - 2)^2}\right) = \ln (x + 1)^3 - \ln (4x - 2)^2 = 3 \ln(x + 1) - 2 \ln(4x - 2).$$

By logarithmic differentiation,

$$\frac{y'}{y} = \frac{3}{x + 1} - \frac{2}{4x - 2} \cdot 4 = \frac{3}{x + 1} - \frac{4}{2x - 1},$$

so

$$y' = \frac{(x + 1)^3}{(4x - 2)^2}\left(\frac{3}{x + 1} - \frac{4}{2x - 1}\right).$$

45. $y = e^{(x-1)^2} e^{(x-3)^2}$

SOLUTION   Let $y = e^{(x-1)^2} e^{(x-3)^2}$. Then

$$\ln y = \ln\left(e^{(x-1)^2} e^{(x-3)^2}\right) = \ln\left(e^{(x-1)^2 + (x-3)^2}\right) = (x-1)^2 + (x-3)^2.$$

By logarithmic differentiation,

$$\frac{y'}{y} = 2(x-1) + 2(x-3) = 4x - 8,$$

so

$$y' = 4e^{(x-1)^2} e^{(x-3)^2} (x-2).$$

47. $y = \dfrac{e^{3x}(x-2)^2}{(x+1)^2}$

SOLUTION   Let $y = \dfrac{e^{3x}(x-2)^2}{(x+1)^2}$. Then

$$\ln y = \ln\left(\frac{e^{3x}(x-2)^2}{(x+1)^2}\right) = \ln e^{3x} + \ln(x-2)^2 - \ln(x+1)^2$$

$$= 3x + 2\ln(x-2) - 2\ln(x+1).$$

By logarithmic differentiation,

$$\frac{y'}{y} = 3 + \frac{2}{x-2} - \frac{2}{x+1},$$

so

$$y = \frac{e^{3x}(x-2)^2}{(x+1)^2}\left(3 + \frac{2}{x-2} - \frac{2}{x+1}\right).$$

In Exercises 49–54, use the given substitution to evaluate the integral.

49. $\displaystyle\int \frac{(\ln x)^2 dx}{x}, \quad u = \ln x$

SOLUTION   Let $u = \ln x$. Then $du = \frac{dx}{x}$, and

$$\int \frac{(\ln x)^2\, dx}{x} = \int u^2\, du = \frac{u^3}{3} + C = \frac{(\ln x)^3}{3} + C.$$

51. $\displaystyle\int \frac{dx}{\sqrt{e^{2x} - 1}}, \quad u = e^{-x}$

SOLUTION   We first rewrite the integrand in terms of e^{-x}. That is,

$$\int \frac{1}{\sqrt{e^{2x} - 1}}\, dx = \int \frac{1}{\sqrt{e^{2x}\left(1 - e^{-2x}\right)}}\, dx = \int \frac{1}{e^x\sqrt{1 - e^{-2x}}}\, dx = \int \frac{e^{-x}\, dx}{\sqrt{1 - e^{-2x}}}$$

Now, let $u = e^{-x}$. Then $du = -e^{-x}\, dx$, and

$$\int \frac{1}{\sqrt{e^{2x} - 1}}\, dx = -\int \frac{du}{\sqrt{1 - u^2}} = -\sin^{-1} u + C = -\sin^{-1}(e^{-x}) + C.$$

53. $\displaystyle\int \frac{dt}{t(1 + (\ln t)^2)}, \quad u = \ln t$

SOLUTION   Let $u = \ln t$. Then, $du = \frac{1}{t}\, dt$ and

$$\int \frac{dt}{t(1 + (\ln t)^2)} = \int \frac{du}{1 + u^2} = \tan^{-1} u + C = \tan^{-1}(\ln t) + C.$$

In Exercises 55–74, calculate the integral.

55. $\int e^{9-2x}\, dx$

SOLUTION Let $u = 9 - 2x$. Then $du = -2\, dx$, and

$$\int e^{9-2x}\, dx = -\frac{1}{2}\int e^u\, du = -\frac{1}{2}e^u + C = -\frac{1}{2}e^{9-2x} + C.$$

57. $\int e^{-2x}\sin(e^{-2x})\, dx$

SOLUTION Let $u = e^{-2x}$. Then $du = -2e^{-2x}\, dx$, and

$$\int e^{-2x}\sin\left(e^{-2x}\right)\, dx = -\frac{1}{2}\int \sin u\, du = \frac{\cos u}{2} + C = \frac{1}{2}\cos\left(e^{-2x}\right) + C.$$

59. $\int_1^e \frac{\ln x\, dx}{x}$

SOLUTION Let $u = \ln x$. Then $du = \frac{dx}{x}$ and the new limits of integration are $u = \ln 1 = 0$ and $u = \ln e = 1$. Thus,

$$\int_1^e \frac{\ln x\, dx}{x} = \int_0^1 u\, du = \frac{1}{2}u^2\Big|_0^1 = \frac{1}{2}.$$

61. $\int_{1/3}^{2/3} \frac{dx}{\sqrt{1-x^2}}$

SOLUTION $\displaystyle\int_{1/3}^{2/3} \frac{dx}{\sqrt{1-x^2}} = \sin^{-1}x\Big|_{1/3}^{2/3} = \sin^{-1}\frac{2}{3} - \sin^{-1}\frac{1}{3}.$

63. $\int_0^1 \cosh(2t)\, dt$

SOLUTION Let $u = 2t$. Then $t = \frac{u}{2}$ and $dt = \frac{du}{2}$. The new limits of integration are $u = 0$ and $u = 2$. Thus,

$$\int_0^1 \cosh(2t)\, dt = \frac{1}{2}\int_0^2 \cosh u\, du = \frac{1}{2}\sinh u\Big|_0^2 = \frac{1}{2}\left(\sinh 2 - \sinh 0\right) = \frac{1}{2}\sinh 2.$$

65. $\int_0^3 \frac{x\, dx}{x^2+9}$

SOLUTION Let $u = x^2 + 9$. Then $du = 2x\, dx$, and the new limits of integration are $u = 9$ and $u = 18$. Thus,

$$\int_0^3 \frac{x\, dx}{x^2+9} = \frac{1}{2}\int_9^{18} \frac{du}{u} = \frac{1}{2}\ln u\Big|_9^{18} = \frac{1}{2}(\ln 18 - \ln 9) = \frac{1}{2}\ln\frac{18}{9} = \frac{1}{2}\ln 2.$$

67. $\int \frac{x\, dx}{\sqrt{1-x^4}}$

SOLUTION Let $u = x^2$. Then $du = 2x\, dx$, and $\sqrt{1-x^4} = \sqrt{1-u^2}$. Thus,

$$\int \frac{x\, dx}{\sqrt{1-x^4}} = \frac{1}{2}\int \frac{du}{\sqrt{1-u^2}} = \frac{1}{2}\sin^{-1}u + C = \frac{1}{2}\sin^{-1}(x^2) + C.$$

69. $\int \frac{\sin^{-1}x\, dx}{\sqrt{1-x^2}}$

SOLUTION Let $u = \sin^{-1}x$. Then $du = \frac{1}{\sqrt{1-x^2}}\, dx$ and

$$\int \frac{\sin^{-1}x\, dx}{\sqrt{1-x^2}} = \int u\, du = \frac{1}{2}u^2 + C = \frac{1}{2}(\sin^{-1}x)^2 + C.$$

71. $\int \sinh^3 x \cosh x\, dx$

SOLUTION   Let $u = \sinh x$. Then $du = \cosh x \, dx$ and

$$\int \sinh^3 x \cosh x \, dx = \int u^3 \, du = \frac{u^4}{4} + C = \frac{\sinh^4 x}{4} + C.$$

73. $\displaystyle\int_0^4 \frac{dx}{2x^2 + 1}$

SOLUTION   Let $u = \sqrt{2}x$. Then $du = \sqrt{2}\,dx$, and the new limits of integration are $u = 0$ and $u = 4\sqrt{2}$. Thus,

$$\int_0^4 \frac{dx}{2x^2 + 1} = \int_0^{4\sqrt{2}} \frac{\frac{1}{\sqrt{2}}\,du}{u^2 + 1} = \frac{1}{\sqrt{2}}\int_0^{4\sqrt{2}} \frac{du}{u^2 + 1}$$

$$= \frac{1}{\sqrt{2}}\tan^{-1}u\Big|_0^{4\sqrt{2}} = \frac{1}{\sqrt{2}}\left(\tan^{-1}(4\sqrt{2}) - \tan^{-1}0\right) = \frac{1}{\sqrt{2}}\tan^{-1}(4\sqrt{2}).$$

75. The isotope Thorium-234 has a half-life of 24.5 days.

(a) Find the differential equation satisfied by the amount $y(t)$ of Thorium-234 in a sample at time t.

(b) At $t = 0$, a sample contains 2 kg of Thorium-234. How much remains after 1 year?

SOLUTION

(a) By the equation for half-life,

$$24.5 = \frac{\ln 2}{k}, \quad \text{so} \quad k = \frac{\ln 2}{24.5} \approx 0.028 \text{ days}^{-1}.$$

Therefore, the differential equation for $y(t)$ is

$$y' = -0.028y.$$

(b) If there are 2 kg of Thorium-234 at $t = 0$, then $y(t) = 2e^{-0.028t}$. After one year (365 days), the amount of Thorium-234 is

$$y(365) = 2e^{-0.028(365)} = 7.29 \times 10^{-5} \text{ kg} = 0.0729 \text{ grams}.$$

77. The C^{14} to C^{12} ratio of a sample is proportional to the disintegration rate (number of beta particles emitted per minute) that is measured directly with a Geiger counter. The disintegration rate of carbon in a living organism is 15.3 beta particles/min per gram. Find the age of a sample that emits 9.5 beta particles/min per gram.

SOLUTION   Let t be the age of the sample in years. Because the disintegration rate for the sample has dropped from 15.3 beta particles/min per gram to 9.5 beta particles/min per gram and the C^{14} to C^{12} ratio is proportional to the disintegration rate, it follows that

$$e^{-0.000121t} = \frac{9.5}{15.3},$$

so

$$t = -\frac{1}{0.000121}\ln\frac{9.5}{15.3} \approx 3938.5.$$

We conclude that the sample is approximately 3938.5 years old.

79. In a first-order chemical reaction, the quantity $y(t)$ of reactant at time t satisfies $y' = -ky$, where $k > 0$. The dependence of k on temperature T (in kelvins) is given by the **Arrhenius equation** $k = Ae^{-E_a/(RT)}$, where E_a is the activation energy (J-mol^{-1}), $R = 8.314$ J-mol^{-1}-K^{-1}, and A is a constant. Assume that $A = 72 \times 10^{12}$ hour^{-1} and $E_a = 1.1 \times 10^5$. Calculate $\dfrac{dk}{dT}$ for $T = 500$ and use the Linear Approximation to estimate the change in k if T is raised from 500 to 510 K.

SOLUTION   Let

$$k = Ae^{-E_a/(RT)}.$$

Then

$$\frac{dk}{dT} = \frac{AE_a}{RT^2}e^{-E_a/(RT)}.$$

For $A = 72 \times 10^{12}$, $R = 8.314$ and $E_a = 1.1 \times 10^5$ we have

$$\frac{dk}{dT} = \frac{72 \times 10^{12} \cdot 1.1 \times 10^5}{8.314} \frac{e^{-\frac{1.1 \times 10^5}{8.314T}}}{T^2} = \frac{9.53 \times 10^{17} e^{-\frac{1.32 \times 10^4}{T}}}{T^2}.$$

The derivative for $T = 500$ is thus

$$\left. \frac{dk}{dT} \right|_{T=500} = \frac{9.53 \times 10^{17} e^{-\frac{1.32 \times 10^4}{500}}}{500^2} \approx 12.27 \text{ hours}^{-1}\text{K}^{-1}.$$

Using the linear approximation we find

$$\Delta k \approx \left. \frac{dk}{dT} \right|_{T=500} \cdot (510 - 500) = 12.27 \cdot 10 = 122.7 \text{ hours}^{-1}.$$

81. Find the solutions to $y' = -2y + 8$ satisfying $y(0) = 3$ and $y(0) = 4$, and sketch their graphs.

SOLUTION First, rewrite the differential equation as $y' = -2(y - 4)$; from here we see that the general solution is

$$y(t) = 4 + Ce^{-2t},$$

for some constant C. If $y(0) = 3$, then

$$3 = 4 + Ce^0 \quad \text{and} \quad C = -1.$$

Thus, $y(t) = 4 - e^{-2t}$. If $y(0) = 4$, then

$$4 = 4 + Ce^0 \quad \text{and} \quad C = 0;$$

hence, $y(t) = 4$. The graphs of the two solutions are shown below.

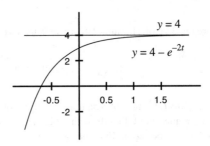

In Exercises 83–86, let $P(t)$ denote the balance at time t (years) of an annuity that earns 5% interest continuously compounded and pays out \$2000/year continuously.

83. Find the differential equation satisfied by $P(t)$.

SOLUTION Since money is withdrawn continuously at a rate of \$2000 a year and the growth due to interest is $0.05P$, the rate of change of the balance is

$$P'(t) = 0.05P - 2000.$$

Thus, the differential equation satisfied by $P(t)$ is

$$P'(t) = 0.05(P - 40,000).$$

85. When does the annuity run out of money if $P(0) = \$2,000$?

SOLUTION In the previous exercise, we found that

$$P(t) = 40,000 + Ce^{0.05t}.$$

If $P(0) = 2000$, then

$$2000 = 40,000 + Ce^{0.05 \cdot 0} = 40,000 + C$$

or

$$C = -38,000.$$

Thus,

$$P(t) = 40,000 - 38,000e^{0.05t}.$$

The annuity runs out of money when $P(t) = 0$; that is, when

$$40,000 - 38,000e^{0.05t} = 0.$$

Solving for t yields

$$t = \frac{1}{0.05} \ln\left(\frac{40,000}{38,000}\right) \approx 1.03.$$

The money runs out after roughly 1.03 years.

In Exercises 87–98, verify that L'Hôpital's Rule applies and evaluate the limit.

87. $\displaystyle\lim_{x \to 3} \frac{4x - 12}{x^2 - 5x + 6}$

SOLUTION   The given expression is an indeterminate form of type $\frac{0}{0}$, therefore L'Hôpital's Rule applies. We find

$$\lim_{x \to 3} \frac{4x - 12}{x^2 - 5x + 6} = \lim_{x \to 3} \frac{4}{2x - 5} = \frac{4}{2 \cdot 3 - 5} = 4.$$

89. $\displaystyle\lim_{x \to 0+} x^{1/2} \ln x$

SOLUTION   First rewrite the function as $\frac{\ln x}{x^{-1/2}}$. The limit is now an indeterminate form of type $\frac{\infty}{\infty}$, hence we may apply L'Hôpital's Rule. We find

$$\lim_{x \to 0+} x^{1/2} \ln x = \lim_{x \to 0+} \frac{\ln x}{x^{-1/2}} = \lim_{x \to 0+} \frac{x^{-1}}{-\frac{1}{2}x^{-3/2}} = \lim_{x \to 0+} -2x^{1/2} = 0.$$

91. $\displaystyle\lim_{\theta \to 0} \frac{2\sin\theta - \sin 2\theta}{\sin\theta - \theta\cos\theta}$

SOLUTION   The given expression is an indeterminate form of type $\frac{0}{0}$; hence, we may apply L'Hôpital's Rule. We find

$$\lim_{\theta \to 0} \frac{2\sin\theta - \sin 2\theta}{\sin\theta - \theta\cos\theta} = \lim_{\theta \to 0} \frac{2\cos\theta - 2\cos 2\theta}{\cos\theta - (\cos\theta - \theta\sin\theta)} = \lim_{\theta \to 0} \frac{2\cos\theta - 2\cos 2\theta}{\theta\sin\theta}$$

$$= \lim_{\theta \to 0} \frac{-2\sin\theta + 4\sin 2\theta}{\sin\theta + \theta\cos\theta} = \lim_{\theta \to 0} \frac{-2\cos\theta + 8\cos 2\theta}{\cos\theta + \cos\theta - \theta\sin\theta} = \frac{-2 + 8}{1 + 1 - 0} = 3.$$

93. $\displaystyle\lim_{t \to \infty} \frac{\ln(t + 2)}{\log_2 t}$

SOLUTION   The limit is an indeterminate form of type $\frac{\infty}{\infty}$; hence, we may apply L'Hôpital's Rule. We find

$$\lim_{t \to \infty} \frac{\ln(t + 2)}{\log_2 t} = \lim_{t \to \infty} \frac{\frac{1}{t+2}}{\frac{1}{t\ln 2}} = \lim_{t \to \infty} \frac{t\ln 2}{t + 2} = \lim_{t \to \infty} \frac{\ln 2}{1} = \ln 2.$$

95. $\displaystyle\lim_{y \to 0} \frac{\sin^{-1} y - y}{y^3}$

SOLUTION   The limit is an indeterminate form of type $\frac{0}{0}$; hence, we may apply L'Hôpital's Rule. We find

$$\lim_{y \to 0} \frac{\sin^{-1} y - y}{y^3} = \lim_{y \to 0} \frac{\frac{1}{\sqrt{1-y^2}} - 1}{3y^2} = \lim_{y \to 0} \frac{y(1 - y^2)^{-3/2}}{6y} = \lim_{y \to 0} \frac{(1 - y^2)^{-3/2}}{6} = \frac{1}{6}.$$

97. $\displaystyle\lim_{x \to 0} \frac{\sinh(x^2)}{\cosh x - 1}$

SOLUTION   The limit is an indeterminate form of type $\frac{0}{0}$; hence, we may apply L'Hôpital's Rule. We find

$$\lim_{x \to 0} \frac{\sinh(x^2)}{\cosh x - 1} = \lim_{x \to 0} \frac{2x\cosh(x^2)}{\sinh x} = \lim_{x \to 0} \frac{2\cosh(x^2) + 4x^2\sinh(x^2)}{\cosh x} = \frac{2 + 0}{1} = 2.$$

99. Explain why L'Hôpital's Rule gives no information about $\lim\limits_{x\to\infty} \dfrac{2x - \sin x}{3x + \cos 2x}$. Evaluate the limit by another method.

SOLUTION As $x \to \infty$, both $2x - \sin x$ and $3x + \cos 2x$ tend toward infinity, so L'Hôpital's Rule applies to $\lim\limits_{x\to\infty} \dfrac{2x - \sin x}{3x + \cos 2x}$; however, the resulting limit, $\lim\limits_{x\to\infty} \dfrac{2 - \cos x}{3 - 2\sin 2x}$, does not exist due to the oscillation of $\sin x$ and $\cos x$.

To evaluate the limit, we note

$$\lim_{x\to\infty} \frac{2x - \sin x}{3x + \cos 2x} = \lim_{x\to\infty} \frac{2 - \frac{\sin x}{x}}{3 + \frac{\cos 2x}{x}} = \frac{2}{3}.$$

101. In this exercise, we prove that for all $x > 0$,

$$x - \frac{x^2}{2} \le \ln(1 + x) \le x \qquad \boxed{1}$$

(a) Show that $\ln(1 + x) = \displaystyle\int_0^x \frac{dt}{1 + t}$ for $x > 0$.

(b) Verify that $1 - t \le \dfrac{1}{1 + t} \le 1$ for all $t > 0$.

(c) Use (b) to prove Eq. (1).

(d) Verify Eq. (1) for $x = 0.5, 0.1$, and 0.01.

SOLUTION

(a) Let $x > 0$. Then

$$\int_0^x \frac{dt}{1 + t} = \ln(1 + t)\Big|_0^x = \ln(1 + x) - \ln 1 = \ln(1 + x).$$

(b) For $t > 0$, $1 + t > 1$, so $\frac{1}{1+t} < 1$. Moreover, $(1 - t)(1 + t) = 1 - t^2 < 1$. Because $1 + t > 0$, it follows that $1 - t < \frac{1}{1+t}$. Hence,

$$1 - t \le \frac{1}{1 + t} \le 1.$$

(c) Integrating each expression in the result from part (b) from $t = 0$ to $t = x$ yields

$$x - \frac{x^2}{2} \le \ln(1 + x) \le x.$$

(d) For $x = 0.5$, $x = 0.1$ and $x = 0.01$, we obtain the string of inequalities

$$0.375 \le 0.405465 \le 0.5$$
$$0.095 \le 0.095310 \le 0.1$$
$$0.00995 \le 0.00995033 \le 0.01,$$

respectively.

*In Exercises 103–106, let $gd(y) = \tan^{-1}(\sinh y)$ be the so-called **gudermannian**, which arises in cartography. In a map of the earth constructed by Mercator projection, points located y radial units from the equator correspond to points on the globe of latitude $gd(y)$.*

103. Prove that $\dfrac{d}{dy} gd(y) = \operatorname{sech} y$.

SOLUTION Let $gd(y) = \tan^{-1}(\sinh y)$. Then

$$\frac{d}{dy} gd(y) = \frac{1}{1 + \sinh^2 y} \cosh y = \frac{1}{\cosh y} = \operatorname{sech} y,$$

where we have used the identity $1 + \sinh^2 y = \cosh^2 y$.

105. Show that $t(y) = \sinh^{-1}(\tan y)$ is the inverse of $gd(y)$ for $0 \le y < \pi/2$.

SOLUTION Let $x = gd(y) = \tan^{-1}(\sinh y)$. Solving for y yields $y = \sinh^{-1}(\tan x)$. Therefore,

$$gd^{-1}(y) = \sinh^{-1}(\tan y).$$

107. Let

$$F(x) = \int_2^x \frac{dt}{\ln t} \quad \text{and} \quad G(x) = \frac{x}{\ln x}$$

Verify that L'Hôpital's Rule may be applied to the limit $L = \lim_{x \to \infty} \frac{F(x)}{G(x)}$ and evaluate L.

SOLUTION   Because $t > \ln t$ for $t > 2$,

$$F(x) = \int_2^x \frac{dt}{\ln t} > \int_2^x \frac{dt}{t} > \ln x.$$

Thus, $F(x) \to \infty$ as $x \to \infty$. Moreover,

$$\lim_{x \to \infty} G(x) = \lim_{x \to \infty} \frac{1}{1/x} = \lim_{x \to \infty} x = \infty.$$

Thus, $\lim_{x \to \infty} \frac{F(x)}{G(x)}$ is of the form ∞/∞, and L'Hôpital's Rule applies. Finally,

$$L = \lim_{x \to \infty} \frac{F(x)}{G(x)} = \lim_{x \to \infty} \frac{\frac{1}{\ln x}}{\frac{\ln x - 1}{(\ln x)^2}} = \lim_{x \to \infty} \frac{\ln x}{\ln x - 1} = 1.$$

8 | TECHNIQUES OF INTEGRATION

8.1 Numerical Integration

Preliminary Questions

1. What are T_1 and T_2 for a function on $[0, 2]$ such that $f(0) = 3$, $f(1) = 4$, and $f(2) = 3$?

SOLUTION Using the given function values

$$T_1 = \frac{1}{2}(2)(3 + 3) = 6 \quad \text{and} \quad T_2 = \frac{1}{2}(1)(3 + 8 + 3) = 7.$$

2. For which graph in Figure 16 will T_N *overestimate* the integral? What about M_N?

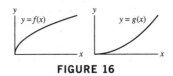

FIGURE 16

SOLUTION T_N overestimates the value of the integral when the integrand is concave up; thus, T_N will overestimate the integral of $y = g(x)$. On the other hand, M_N overestimates the value of the integral when the integrand is concave down; thus, M_N will overestimate the integral of $y = f(x)$.

3. How large is the error when the Trapezoidal Rule is applied to a linear function? Explain graphically.

SOLUTION The Trapezoidal Rule integrates linear functions exactly, so the error will be zero.

4. Suppose T_4 is used to approximate $\int_0^3 f(x)\,dx$, where $|f''(x)| \leq 2$ for all x. What is the maximum possible error?

SOLUTION The maximum possible error in T_4 is

$$\max |f''(x)| \frac{(b-a)^3}{12n^2} = \frac{2(3-0)^3}{12(4)^2} = \frac{9}{32}.$$

5. What are the two graphical interpretations of the Midpoint Rule?

SOLUTION The two graphical interpretations of the Midpoint Rule are the sum of the areas of the midpoint rectangles and the sum of the areas of the tangential trapezoids.

Exercises

In Exercises 1–12, calculate T_N and M_N for the value of N indicated.

1. $\int_0^2 x^2\,dx, \quad N = 4$

SOLUTION Let $f(x) = x^2$. We divide $[0, 2]$ into 4 subintervals of width

$$\Delta x = \frac{2 - 0}{4} = \frac{1}{2}$$

with endpoints $0, 0.5, 1, 1.5, 2$, and midpoints $0.25, 0.75, 1.25, 1.75$. With this data, we get

$$T_4 = \frac{1}{2} \cdot \frac{1}{2}\left(0^2 + 2(0.5)^2 + 2(1)^2 + 2(1.5)^2 + 2^2\right) = 2.75; \text{ and}$$

$$M_4 = \frac{1}{2}\left(0.25^2 + 0.75^2 + 1.25^2 + 1.75^2\right) = 2.625.$$

3. $\int_1^4 x^3\,dx, \quad N = 6$

SOLUTION Let $f(x) = x^3$. We divide $[1, 4]$ into 6 subintervals of width

$$\Delta x = \frac{4 - 1}{6} = \frac{1}{2}$$

with endpoints 1, 1.5, 2, 2.5, 3, 3.5, 4, and midpoints 1.25, 1.75, 2.25, 2.75, 3.25, 3.75. With this data, we get

$$T_6 = \frac{1}{2}\left(\frac{1}{2}\right)\left(1^3 + 2(1.5)^3 + 2(2)^3 + 2(2.5)^3 + 2(3)^3 + 2(3.5)^3 + 4^3\right) = 64.6875; \text{ and}$$

$$M_6 = \frac{1}{2}\left(1.25^3 + 1.75^3 + 2.25^3 + 2.75^3 + 3.25^3 + 3.75^3\right) = 63.28125.$$

5. $\int_1^4 \frac{dx}{x}, \quad N = 6$

SOLUTION Let $f(x) = 1/x$. We divide $[1, 4]$ into 6 subintervals of width

$$\Delta x = \frac{4 - 1}{6} = \frac{1}{2}$$

with endpoints 1, 1.5, 2, 2.5, 3, 3.5, 4, and midpoints 1.25, 1.75, 2.25, 2.75, 3.25, 3.75. With this data, we get

$$T_6 = \frac{1}{2}\left(\frac{1}{2}\right)\left(\frac{1}{1} + \frac{2}{1.5} + \frac{2}{2} + \frac{2}{2.5} + \frac{2}{3} + \frac{2}{3.5} + \frac{1}{4}\right) \approx 1.40536; \text{ and}$$

$$M_6 = \frac{1}{2}\left(\frac{1}{1.25} + \frac{1}{1.75} + \frac{1}{2.25} + \frac{1}{2.75} + \frac{1}{3.25} + \frac{1}{3.75}\right) \approx 1.37693.$$

7. $\int_0^{\pi/4} \sec x \, dx, \quad N = 6$

SOLUTION Let $f(x) = \sec x$. We divide $[0, \pi/4]$ into 6 subintervals of width

$$\Delta x = \frac{\frac{\pi}{4} - 0}{6} = \frac{\pi}{24}$$

with endpoints

$$0, \frac{\pi}{24}, \frac{2\pi}{24}, \dots, \frac{6\pi}{24} = \frac{\pi}{4},$$

and midpoints

$$\frac{\pi}{48}, \frac{3\pi}{48}, \dots, \frac{11\pi}{48}.$$

With this data, we get

$$T_6 = \frac{1}{2}\left(\frac{\pi}{24}\right)\left(\sec(0) + 2\sec(\pi/24) + 2\sec(2\pi/24) + \dots + \sec(6\pi/24)\right) \approx 0.883387; \text{ and}$$

$$M_6 = \frac{\pi}{24}\left(\sec(\pi/48) + \sec(3\pi/48) + \sec(5\pi/48) + \dots + \sec(11\pi/48)\right) \approx 0.880369.$$

9. $\int_2^3 \frac{dx}{\ln x}, \quad N = 5$

SOLUTION Let $f(x) = 1/\ln x$. We divide $[2, 3]$ into 5 subintervals of width

$$\Delta x = \frac{3 - 2}{5} = \frac{1}{5} = 0.2$$

with endpoints 2, 2.2, 2.4, 2.6, 2.8, 3, and midpoints 2.1, 2.3, 2.5, 2.7, 2.9. With this data, we get

$$T_5 = \frac{1}{2}\left(\frac{1}{5}\right)\left(\frac{1}{\ln 2} + \frac{2}{\ln 2.2} + \frac{2}{\ln 2.4} + \frac{2}{\ln 2.6} + \frac{2}{\ln 2.8} + \frac{1}{\ln 3}\right) \approx 1.12096; \text{ and}$$

$$M_5 = \frac{1}{5}\left(\frac{1}{\ln 2.1} + \frac{1}{\ln 2.3} + \frac{1}{\ln 2.5} + \frac{1}{\ln 2.7} + \frac{1}{\ln 2.9}\right) \approx 1.11716.$$

11. $\int_0^2 \frac{e^x}{x + 1} \, dx, \quad N = 8$

SOLUTION Let $f(x) = e^x/(x+1)$. We divide $[0, 2]$ into 8 subintervals of width

$$\Delta x = \frac{2-0}{8} = \frac{1}{4} = 0.25$$

with endpoints 0, 0.25, 0.5, 0.75, 1, 1.25, 1.5, 1.75, 2, and midpoints 0.125, 0.375, 0.625, 0.875, 1.125, 1.375, 1.625, 1.875. With this data, we get

$$T_8 = \frac{1}{2}\left(\frac{1}{4}\right)\left(\frac{e^0}{0+1} + \frac{2e^{0.25}}{0.25+1} + \frac{2e^{0.5}}{0.5+1} + \cdots + \frac{2e^{1.75}}{1.75+1} + \frac{e^2}{2+1}\right) \approx 2.96581; \text{ and}$$

$$M_8 = \frac{1}{4}\left(\frac{e^{0.125}}{0.125+1} + \frac{e^{0.375}}{0.375+1} + \frac{e^{0.625}}{0.625+1} + \cdots + \frac{e^{1.875}}{1.875+1}\right) \approx 2.95302.$$

In Exercises 13–22, calculate S_N given by Simpson's Rule for the value of N indicated.

13. $\displaystyle\int_0^4 \sqrt{x}\, dx, \quad N = 4$

SOLUTION Let $f(x) = \sqrt{x}$. We divide $[0, 4]$ into 4 subintervals of width

$$\Delta x = \frac{4-0}{4} = 1$$

with endpoints 0, 1, 2, 3, 4. With this data, we get

$$S_4 = \frac{1}{3}(1)\left(\sqrt{0} + 4\sqrt{1} + 2\sqrt{2} + 4\sqrt{3} + \sqrt{4}\right) \approx 5.25221.$$

15. $\displaystyle\int_0^2 \frac{dx}{x^4+1}, \quad N = 4$

SOLUTION Let $f(x) = 1/(x^4+1)$. We divide $[0, 2]$ into 4 subintervals of length

$$\Delta x = \frac{2-0}{4} = \frac{1}{2} = 0.5$$

with endpoints 0, 0.5, 1, 1.5, 2. With this data, we get

$$S_4 = \frac{1}{3}\left(\frac{1}{2}\right)\left[\frac{1}{0^4+1} + \frac{4}{0.5^4+1} + \frac{2}{1^4+1} + \frac{4}{1.5^4+1} + \frac{1}{2^4+1}\right] \approx 1.08055.$$

17. $\displaystyle\int_0^1 e^{-x^2}\, dx, \quad N = 6$

SOLUTION Let $f(x) = e^{-x^2}$. We divide $[0, 1]$ into 6 subintervals of length

$$\Delta x = \frac{1-0}{6} = \frac{1}{6}$$

with endpoints $0, \frac{1}{6}, \frac{2}{6}, \dots, \frac{6}{6} = 1$. With this data, we get

$$S_6 = \frac{1}{3}\left(\frac{1}{6}\right)\left[e^{-0^2} + 4e^{-(1/6)^2} + 2e^{-(2/6)^2} + 4e^{-(3/6)^2} + 2e^{-(4/6)^2} + 4e^{-(5/6)^2} + e^{-(1)^2}\right] \approx 0.746830.$$

19. $\displaystyle\int_1^4 \ln x\, dx, \quad N = 8$

SOLUTION Let $f(x) = \ln x$. We divide $[1, 4]$ into 8 subintervals of length

$$\Delta x = \frac{4-1}{8} = \frac{3}{8} = 0.375$$

with endpoints 1, 1.375, 1.75, 2.125, 2.5, 2.875, 3.25, 3.625, 4. With this data, we get

$$S_8 = \frac{1}{3}\left(\frac{3}{8}\right)\left[\ln 1 + 4\ln(1.375) + 2\ln(1.75) + \cdots + 4\ln(3.625) + \ln 4\right] \approx 2.54499.$$

21. $\displaystyle\int_0^{\pi/4} \sec x \, dx, \quad N = 10$

SOLUTION Let $f(x) = \sec x$. We divide $[0, \frac{\pi}{4}]$ into 10 subintervals of width

$$\Delta x = \frac{\frac{\pi}{4} - 0}{10} = \frac{\pi}{40}$$

with endpoints $0, \frac{\pi}{40}, \frac{2\pi}{40}, \frac{3\pi}{40}, \ldots, \frac{10\pi}{40} = \frac{\pi}{4}$. With this data, we get

$$S_{10} = \frac{1}{3}\left(\frac{\pi}{40}\right)\left[\sec(0) + 4\sec\left(\frac{\pi}{40}\right) + 2\sec\left(\frac{2\pi}{40}\right) + \cdots + 4\sec\left(\frac{9\pi}{40}\right) + \sec\left(\frac{10\pi}{40}\right)\right] \approx 0.881377.$$

In Exercises 23–26, calculate the approximation to the volume of the solid obtained by rotating the graph around the given axis.

23. $y = \cos x$; $\left[0, \frac{\pi}{2}\right]$; x-axis; M_8

SOLUTION Using the disk method, the volume is given by

$$V = \int_0^{\pi/2} \pi r^2 \, dx = \pi \int_0^{\pi/2} (\cos x)^2 \, dx$$

which can be estimated as

$$\pi \int_0^{\pi/2} (\cos x)^2 \, dx \approx \pi[M_8].$$

Let $f(x) = \cos^2 x$. We divide $[0, \pi/2]$ into 8 subintervals of length

$$\Delta x = \frac{\frac{\pi}{2} - 0}{8} = \frac{\pi}{16}$$

with midpoints

$$\frac{\pi}{32}, \frac{3\pi}{32}, \frac{5\pi}{32}, \ldots, \frac{15\pi}{32}.$$

With this data, we get

$$V \approx \pi[M_8] = \pi\left[\Delta x(y_1 + y_2 + \cdots + y_8)\right] = \frac{\pi^2}{16}\left[\cos^2\left(\frac{\pi}{32}\right) + \cos^2\left(\frac{3\pi}{32}\right) + \cdots + \cos^2\left(\frac{15\pi}{32}\right)\right] \approx 2.46740.$$

25. $y = e^{-x^2}$; $[0, 1]$; x-axis; T_8

SOLUTION Using the disk method, the volume is given by

$$V = \int_0^1 \pi r^2 \, dx = \pi \int_0^1 \left(e^{-x^2}\right)^2 dx = \pi \int_0^1 e^{-2x^2} \, dx.$$

We can use the approximation

$$V = \pi \int_0^1 e^{-2x^2} \, dx \approx \pi[T_8],$$

where $f(x) = e^{-2x^2}$. Divide $[0, 1]$ into 8 subintervals of length

$$\Delta x = \frac{1 - 0}{8} = \frac{1}{8},$$

with endpoints

$$0, \frac{1}{8}, \frac{2}{8}, \ldots, 1.$$

With this data, we get

$$V \approx \pi[T_8] = \pi\left[\frac{1}{2} \cdot \frac{1}{8}\left(e^{-2(0^2)} + 2e^{-2(1/8)^2} + \cdots + 2e^{-2(7/8)^2} + e^{-2(1)^2}\right)\right] \approx 1.87691.$$

27. Use S_8 to estimate $\int_0^{\pi/2} \dfrac{\sin x}{x}\, dx$, taking the value of $(\sin x)/x$ at $x = 0$ to be 1.

SOLUTION Divide $[0, \pi/2]$ into 8 subintervals of length

$$\Delta x = \frac{\frac{\pi}{2} - 0}{8} = \frac{\pi}{16}$$

with endpoints

$$0,\ \frac{\pi}{16},\ \frac{2\pi}{16}, \ldots, \frac{8\pi}{16}.$$

Taking the value of $(\sin x)/x$ at $x = 0$ to be 1, we get

$$S_8 = \frac{1}{3}\left(\frac{\pi}{16}\right)\left[1 + 4\frac{\sin(\pi/16)}{\pi/16} + 2\frac{\sin(2\pi/16)}{2\pi/16} + \cdots + \frac{\sin(\pi/2)}{\pi/2}\right] \approx 1.37076.$$

29. Calculate M_6 for the integral $I = \int_0^{\pi/2} \cos x\, dx$.

(a) Is M_6 too large or too small? Explain graphically.
(b) Show that $K_2 = 1$ may be used in the Error Bound and find a bound for the error.
(c) Evaluate I and check that the actual error is less than the bound computed in (b).

SOLUTION Let $f(x) = \cos x$. Divide $[0, \pi/2]$ into 6 subintervals of length

$$\Delta x = \frac{\frac{\pi}{2} - 0}{6} = \frac{\pi}{12}$$

with midpoints

$$\frac{\pi}{24},\ \frac{3\pi}{24}, \ldots, \frac{11\pi}{24}.$$

With this data, we get

$$M_6 = \frac{\pi}{12}\left[\cos\left(\frac{\pi}{24}\right) + \cos\left(\frac{3\pi}{24}\right) + \cdots + \cos\left(\frac{11\pi}{24}\right)\right] \approx 1.0028615.$$

(a) Since $f(x) = \cos x$ is concave down on $[0, \pi/2]$, M_6 is too large.
(b) We have $f'(x) = -\sin x$ and $f''(x) = -\cos x$. Since $|f''(x)| = |-\cos x| \leq 1$ on $[0, \pi/2]$, we may take $K_2 = 1$. Then

$$\text{Error}(M_6) \leq \frac{K_2(b - a)^3}{24N^2} = \frac{(1)(\pi/2 - 0)^3}{24(6)^2} = \frac{\pi^3}{6912} \approx 0.00448586.$$

(c) The exact value is

$$\int_0^{\pi/2} \cos x\, dx = \sin x \Big|_0^{\pi/2} = \sin\left(\frac{\pi}{2}\right) - \sin 0 = 1 - 0 = 1.$$

We can use this to compute the actual error:

$$\text{Error}(M_6) = |M_6 - 1| \approx |1.0028615 - 1| \approx 0.0028615.$$

Since $0.0028615 < 0.00448586$, the actual error is indeed less than the maximum possible error.

In Exercises 30–33, state whether T_N or M_N underestimates or overestimates the integral and find a bound for the error (but do not calculate T_N or M_N).

31. $\int_0^2 e^{-x/4}\, dx,\quad T_{20}$

SOLUTION Let $f(x) = e^{-x/4}$. Then $f'(x) = -(1/4)e^{-x/4}$ and

$$f''(x) = \frac{1}{16}e^{-x/4} > 0$$

on $[0, 2]$, so $f(x)$ is concave up, and T_{20} overestimates the integral. Since $|f''(x)| = |(1/16)e^{-x/4}|$ has its maximum value on $[0, 2]$ at $x = 0$, we can take $K_2 = |(1/16)e^0| = 1/16$, and

$$\text{Error}(T_{20}) \leq \frac{K_2(2 - 0)^3}{12N^2} = \frac{\frac{1}{16}(2)^3}{12(20)^2} = 1.04167 \times 10^{-4}.$$

33. $\int_0^{\pi/4} \cos x, \quad M_{20}$

SOLUTION Let $f(x) = \cos x$. Then $f'(x) = -\sin x$ and $f''(x) = -\cos x < 0$ on $[0, \pi/4]$, so $f(x)$ is concave down, and M_{20} overestimates the integral. Since $|f''(x)| = |-\cos x|$ has its maximum value on $[0, \pi/4]$ at $x = 0$, we can take $K_2 = |-\cos(0)| = 1$, and

$$\text{Error}(M_{20}) \leq \frac{K_2(\pi/4 - 0)^3}{24N^2} = \frac{(1)(\pi/4)^3}{24(20)^2} = 5.04659 \times 10^{-5}.$$

CAS *In Exercises 34–37, use the Error Bound to find a value of N for which $Error(T_N) \leq 10^{-6}$. If you have a computer algebra system, calculate the corresponding approximation and confirm that the error satisfies the required bound.*

35. $\int_0^{\pi/6} \cos x \, dx$

SOLUTION Let $f(x) = \cos x$. Then $f'(x) = -\sin x$ and $|f''(x)| = |-\cos x|$, which has its maximum value on $[0, \pi/6]$ at $x = 0$, so we can take $K_2 = |-\cos(0)| = 1$. Then we have

$$\text{Error}(T_N) \leq \frac{K_2(\pi/6 - 0)^3}{12N^2} = \frac{\pi^3}{12 \cdot 6^3 N^2}.$$

To ensure that the error is at most 10^{-6}, we must choose N such that

$$\frac{\pi^3}{12 \cdot 6^3 N^2} \leq \frac{1}{10^6}.$$

This gives us

$$N^2 \geq \frac{\pi^3 10^6}{12 \cdot 6^3} \Rightarrow N \geq \sqrt{\frac{\pi^3 10^6}{12 \cdot 6^3}} \approx 109.37.$$

Thus let $N = 110$. The exact value of the integral is

$$\int_0^{\pi/6} \cos x \, dx = \sin x \Big|_0^{\pi/6} = 0.5.$$

Using a CAS, we find that

$$T_{110} \approx 0.4999990559.$$

The error is approximately $|0.5 - 0.4999990559| \approx 9.441 \times 10^{-7}$ and is indeed less than 10^{-6}.

37. $\int_0^3 e^{-x} \, dx$

SOLUTION Let $f(x) = e^{-x}$. Then $f'(x) = -e^{-x}$ and $|f''(x)| = |e^{-x}| = e^{-x}$, which has its maximum value on $[0, 3]$ at $x = 0$, so we can take $K_2 = e^0 = 1$. Then we have

$$\text{Error}(T_N) \leq \frac{K_2(3 - 0)^3}{12N^2} = \frac{(1)3^3}{12N^2} = \frac{9}{4N^2}.$$

To ensure that the error is at most 10^{-6}, we must choose N such that

$$\frac{9}{4N^2} \leq \frac{1}{10^6}.$$

This gives us

$$N^2 \geq \frac{9 \cdot 10^6}{4} \Rightarrow N \geq \sqrt{\frac{9 \cdot 10^6}{4}} = 1500.$$

Thus let $N = 1500$. The exact value of the integral is

$$\int_0^3 e^{-x} \, dx = \left(-e^{-3}\right) - \left(-e^{-0}\right) = 1 - e^{-3} \approx 0.9502129316.$$

Using a CAS, we find that

$$T_{1500} \approx 0.9502132468.$$

The error is approximately

$$|0.9502129316 - 0.9502132468| \approx 3.152 \times 10^{-7}$$

and is indeed less than 10^{-6}.

39. (a) Compute S_6 for the integral $I = \int_0^1 e^{-2x}\, dx$.

(b) Show that $K_4 = 16$ may be used in the Error Bound and find a bound for the error.

(c) Evaluate I and check that the actual error is less than the bound for the error computed in (b).

SOLUTION

(a) Let $f(x) = e^{-2x}$. We divide $[0, 1]$ into six subintervals of length $\Delta x = (1 - 0)/6 = 1/6$, with endpoints $0, 1/6, \ldots, 5/6, 1$. With this data, we get

$$S_6 = \frac{1}{3} \cdot \frac{1}{6}\left[e^{-2(0)} + 4e^{-2(1/6)} + 2e^{-2(2/6)} + \cdots + e^{-2(1)}\right] \approx 0.432361.$$

(b) Taking derivatives, we get

$$f'(x) = -2e^{-2x}, \quad f''(x) = 4e^{-2x}, \quad f^{(3)}(x) = -8e^{-2x}, \quad f^{(4)}(x) = 16e^{-2x}.$$

Since $|f^{(4)}(x)| = |16e^{-2x}|$ assumes its maximum value on $[0, 1]$ at $x = 0$, we can set $K_4 = |16e^0| = 16$. Then we have

$$\text{Error}(S_6) \leq \frac{K_4(1 - 0)^5}{180N^4} = \frac{16}{180 \cdot 6^4} \approx 6.86 \times 10^{-5}.$$

(c) The exact value of the integral is

$$\int_0^1 e^{-2x}\, dx = \frac{e^{-2x}}{-2}\bigg|_0^1 = \frac{1 - e^{-2}}{2} \approx 0.432332.$$

The actual error is

$$\text{Error}(S_6) \approx |0.432361 - 0.432332| \approx 2.9 \times 10^{-5}.$$

The error is indeed less than the maximum possible error.

41. Calculate S_8 for $\int_1^5 \ln x\, dx$ and find a bound for the error. Then find a value of N such that S_N has an error of at most 10^{-6}.

SOLUTION Let $f(x) = \ln x$. We divide $[1, 5]$ into eight subintervals of length $\Delta x = (5 - 1)/8 = 0.5$, with endpoints $1, 1.5, 2, \ldots, 5$. With this data, we get

$$S_8 = \frac{1}{3} \cdot \frac{1}{2}\left[\ln 1 + 4\ln 1.5 + 2\ln 2 + \cdots + 4\ln 4.5 + \ln 5\right] \approx 4.046655.$$

To find the maximum possible error, we first take derivatives:

$$f'(x) = \frac{1}{x}, \quad f''(x) = -\frac{1}{x^2}, \quad f^{(3)}(x) = \frac{2}{x^3}, \quad f^{(4)}(x) = -\frac{6}{x^4}.$$

Since $|f^{(4)}(x)| = |-6x^{-4}| = 6x^{-4}$, assumes its maximum value on $[1, 5]$ at $x = 1$, we can set $K_4 = 6(1)^{-4} = 6$. Then we have

$$\text{Error}(S_8) \leq \frac{K_4(5 - 1)^5}{180N^4} = \frac{6 \cdot 4^5}{180 \cdot 8^4} \approx 0.0083333.$$

To ensure that S_N has error at most 10^{-6}, we must find N such that

$$\frac{6 \cdot 4^5}{180N^4} \leq \frac{1}{10^6}.$$

This gives us

$$N^4 \geq \frac{6 \cdot 4^5 \cdot 10^6}{180} \Rightarrow N \geq \left(\frac{6 \cdot 4^5 \cdot 10^6}{180}\right)^{1/4} \approx 76.435.$$

Thus let $N = 78$ (remember that N must be even when using Simpson's Rule).

43. CAS Use a computer algebra system to compute and graph $f^{(4)}(x)$ for $f(x) = \sqrt{1+x^4}$ and find a bound for the error in the approximation S_{40} to $\int_0^5 f(x)\,dx$.

SOLUTION From the graph of $f^{(4)}(x)$ shown below, we see that $|f^{(4)}(x)| \le 15$ on $[0,5]$. Therefore we set $K_4 = 15$. Now we have

$$\text{Error}(S_{40}) \le \frac{15(5-0)^5}{180(40)^4} = \frac{5}{49152} \approx 1.017 \times 10^{-4}.$$

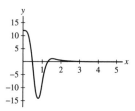

In Exercises 45–48, use the Error Bound to find a value of N for which $Error(S_N) \le 10^{-9}$.

45. $\displaystyle\int_1^7 x^{3/2}\,dx$

SOLUTION Let $f(x) = x^{3/2}$. To find K_4, we first take derivatives:

$$f'(x) = \frac{3}{2}x^{1/2}, \quad f''(x) = \frac{3}{4}x^{-1/2}, \quad f^{(3)}(x) = -\frac{3}{8}x^{-3/2}, \quad f^{(4)}(x) = \frac{9}{16}x^{-5/2}.$$

Since $f^{(5)}(x) = -(45/32)x^{-7/2} < 0$ on $[1,7]$, $f^{(4)}$ is decreasing on $[1,7]$ and therefore assumes its maximum value at $x = 1$. Thus we can set

$$K_4 = |f^{(4)}(1)| = \left|\frac{9}{16}(1)^{-5/2}\right| = \frac{9}{16}.$$

Then we have

$$\text{Error}(S_N) \le \frac{K_4(7-1)^5}{180N^4} = \frac{9}{16} \cdot \frac{6^5}{180N^4}.$$

To ensure that S_N has error at most 10^{-9}, we must find N such that

$$\frac{9 \cdot 6^5}{16 \cdot 180N^4} \le \frac{1}{10^9}.$$

This gives us

$$N^4 \ge \frac{9 \cdot 6^5 \cdot 10^9}{16 \cdot 180} \Rightarrow N \ge \left(\frac{9 \cdot 6^5 \cdot 10^9}{16 \cdot 180}\right)^{1/4} \approx 394.82.$$

Thus let $N = 396$ (remember that N must be even when using Simpson's Rule).

47. $\displaystyle\int_0^1 e^{x^2}\,dx$

SOLUTION Let $f(x) = e^{x^2}$. To find K_4, we first take derivatives:

$$f'(x) = 2xe^{x^2}$$

$$f''(x) = 4x^2e^{x^2} + 2e^{x^2}$$

$$f^{(3)}(x) = 8x^3e^{x^2} + 12xe^{x^2}$$

$$f^{(4)}(x) = 16x^4e^{x^2} + 48x^2e^{x^2} + 12e^{x^2}.$$

On the interval $[0,1]$, $|f^{(4)}(x)|$ assumes its maximum value at $x = 1$. Therefore we set

$$K_4 = |f^{(4)}(1)| = 16e + 48e + 12e = 76e.$$

Now we have

$$\text{Error}(S_N) \le \frac{K_4(1-0)^5}{180N^4} = \frac{76e}{180N^4}.$$

To ensure that S_N has error at most 10^{-9}, we must find N such that

$$\frac{76e}{180N^4} \le \frac{1}{10^9}.$$

This gives us

$$N^4 \ge \frac{76e \cdot 10^9}{180} \Rightarrow N \ge \left(\frac{76e \cdot 10^9}{180}\right)^{1/4} \approx 184.06.$$

Thus we let $N = 186$ (remember that N must be even when using Simpson's Rule).

49. CAS Show that $\int_0^1 \frac{dx}{1+x^2} = \frac{\pi}{4}$ [use Eq. (5) in Section 7.8].

(a) Use a computer algebra system to graph $f^{(4)}(x)$ for $f(x) = (1+x^2)^{-1}$ and find its maximum on $[0, 1]$.

(b) Find a value of N such that S_N approximates the integral with an error of at most 10^{-6}. Calculate the corresponding approximation and confirm that you have computed $\frac{\pi}{4}$ to at least four places.

SOLUTION Recall from Section 7.8 that

$$\frac{d}{dx}\tan^{-1}(x) = \frac{1}{1+x^2}.$$

So then

$$\int_0^1 \frac{dx}{1+x^2} = \tan^{-1}x\Big|_0^1 = \tan^{-1}(1) - \tan^{-1}(0) = \frac{\pi}{4}.$$

(a) From the graph of $f^{(4)}(x)$ shown below, we can see that the maximum value of $|f^{(4)}(x)|$ on the interval $[0, 1]$ is 24.

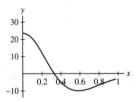

(b) From part (a), we set $K_4 = 24$. Then we have

$$\text{Error}(S_N) \le \frac{24(1-0)^5}{180N^4} = \frac{2}{15N^4}.$$

To ensure that S_N has error at most 10^{-6}, we must find N such that

$$\frac{2}{15N^4} \le \frac{1}{10^6}.$$

This gives us

$$N^4 \ge \frac{2 \cdot 10^6}{15} \Rightarrow N \ge \left(\frac{2 \cdot 10^6}{15}\right)^{1/4} \approx 19.1.$$

Thus let $N = 20$. To compute S_{20}, let $\Delta x = (1-0)/20 = 0.05$. The endpoints of $[0, 1]$ are 0, 0.05, $\ldots$, 1. With this data, we get

$$S_{20} = \frac{1}{3}\left(\frac{1}{20}\right)\left[\frac{1}{1+0^2} + \frac{4}{1+(0.05)^2} + \frac{2}{1+(0.1)^2} + \cdots + \frac{1}{1+1^2}\right] \approx 0.785398163242.$$

The actual error is

$$|0.785398163242 - \pi/4| = |0.785398163242 - 0.785398163397| = 1.55 \times 10^{-10}.$$

51. **CAS** The Error Bound for M_N is proportional to $1/N^2$, so the Error Bound decreases by $\frac{1}{4}$ if N is increased to $2N$. Compute the actual error in M_N for $\int_0^\pi \sin x \, dx$ for $N = 4, 8, 16, 32,$ and 64. Does the actual error seem to decrease by $\frac{1}{4}$ as N is doubled?

SOLUTION The exact value of the integral is

$$\int_0^\pi \sin x \, dx = -\cos x \Big|_0^\pi = -(-1) - (1) = 2.$$

To compute M_4, we have $\Delta x = (\pi - 0)/4 = \pi/4$, and midpoints $\pi/8, \ 3\pi/8, \ 5\pi/8, \ 7\pi/8$. With this data, we get

$$M_4 = \frac{\pi}{4}\left[\sin\left(\frac{\pi}{8}\right) + \sin\left(\frac{3\pi}{8}\right) + \sin\left(\frac{5\pi}{8}\right) + \sin\left(\frac{7\pi}{8}\right)\right] \approx 2.052344.$$

The values for $M_8, M_{16}, M_{32},$ and M_{64} are computed similarly:

$$M_8 = \frac{\pi}{8}\left[\sin\left(\frac{\pi}{16}\right) + \sin\left(\frac{3\pi}{16}\right) + \cdots + \sin\left(\frac{15\pi}{16}\right)\right] \approx 2.012909;$$

$$M_{16} = \frac{\pi}{16}\left[\sin\left(\frac{\pi}{32}\right) + \sin\left(\frac{3\pi}{32}\right) + \cdots + \sin\left(\frac{31\pi}{32}\right)\right] \approx 2.0032164;$$

$$M_{32} = \frac{\pi}{32}\left[\sin\left(\frac{\pi}{64}\right) + \sin\left(\frac{3\pi}{64}\right) + \cdots + \sin\left(\frac{63\pi}{64}\right)\right] \approx 2.00080342;$$

$$M_{64} = \frac{\pi}{64}\left[\sin\left(\frac{\pi}{128}\right) + \sin\left(\frac{3\pi}{128}\right) + \cdots + \sin\left(\frac{127\pi}{128}\right)\right] \approx 2.00020081.$$

Now we can compute the actual errors for each N:

$$\text{Error}(M_4) = |2 - 2.052344| = 0.052344$$

$$\text{Error}(M_8) = |2 - 2.012909| = 0.012909$$

$$\text{Error}(M_{16}) = |2 - 2.0032164| = 0.0032164$$

$$\text{Error}(M_{32}) = |2 - 2.00080342| = 0.00080342$$

$$\text{Error}(M_{64}) = |2 - 2.00020081| = 0.00020081$$

The actual error does in fact decrease by about $1/4$ each time N is doubled.

53. **CAS** Explain why the Error Bound for S_N decreases by $\frac{1}{16}$ if N is increased to $2N$. Compute the actual error in S_N for $\int_0^\pi \sin x \, dx$ for $N = 4, 8, 16, 32,$ and 64. Does the actual error seem to decrease by $\frac{1}{16}$ as N is doubled?

SOLUTION If we plug in $2N$ for N in the formula for the error bound for S_N, we get

$$\frac{K_4(b - a)^5}{180(2N)^4} = \frac{K_4(b - a)^5}{180 \cdot 2^4 \cdot N^4} = \frac{1}{16}\left(\frac{K_4(b - a)^5}{180N^4}\right).$$

Thus we see that, since N is raised to the fourth power in the denominator, the Error Bound for S_N decreases by $1/16$ if N is increased to $2N$. The exact value of the integral is

$$\int_0^\pi \sin x \, dx = -\cos x \Big|_0^\pi = -(-1) - (1) = 2.$$

To compute S_4, we have $\Delta x = (\pi - 0)/4 = \pi/4$, and endpoints $0, \ \pi/4, \ 2\pi/4, \ 3\pi/4, \ \pi$. With this data, we get

$$S_4 = \frac{1}{3} \cdot \frac{\pi}{4}\left[\sin(0) + 4\sin\left(\frac{\pi}{4}\right) + 2\sin\left(\frac{2\pi}{4}\right) + 4\sin\left(\frac{3\pi}{4}\right) + \sin(\pi)\right] \approx 2.004560.$$

The values for $S_8, S_{16}, S_{32},$ and S_{64} are computed similarly:

$$S_8 = \frac{1}{3} \cdot \frac{\pi}{8}\left[\sin(0) + 4\sin\left(\frac{\pi}{8}\right) + 2\sin\left(\frac{2\pi}{8}\right) + \cdots + 4\sin\left(\frac{7\pi}{8}\right) + \sin(\pi)\right] \approx 2.0002692;$$

$$S_{16} = \frac{1}{3} \cdot \frac{\pi}{16}\left[\sin(0) + 4\sin\left(\frac{\pi}{16}\right) + 2\sin\left(\frac{2\pi}{16}\right) + \cdots + 4\sin\left(\frac{15\pi}{16}\right) + \sin(\pi)\right] \approx 2.00001659;$$

$$S_{32} = \frac{1}{3} \cdot \frac{\pi}{32}\left[\sin(0) + 4\sin\left(\frac{\pi}{32}\right) + 2\sin\left(\frac{2\pi}{32}\right) + \cdots + 4\sin\left(\frac{31\pi}{32}\right) + \sin(\pi)\right] \approx 2.000001033;$$

$$S_{64} = \frac{1}{3} \cdot \frac{\pi}{64} \left[\sin(0) + 4\sin\left(\frac{\pi}{64}\right) + 2\sin\left(\frac{2\pi}{64}\right) + \cdots + 4\sin\left(\frac{63\pi}{64}\right) + \sin(\pi) \right] \approx 2.00000006453.$$

Now we can compute the actual errors for each N:

$$\text{Error}(S_4) = |2 - 2.004560| = 0.004560$$

$$\text{Error}(S_8) = |2 - 2.0002692| = 2.692 \times 10^{-4}$$

$$\text{Error}(S_{16}) = |2 - 2.00001659| = 1.659 \times 10^{-5}$$

$$\text{Error}(S_{32}) = |2 - 2.000001033| = 1.033 \times 10^{-6}$$

$$\text{Error}(S_{64}) = |2 - 2.00000006453| = 6.453 \times 10^{-8}$$

The actual error does in fact decrease by about $1/16$ each time N is doubled. For example, $0.004560/16 = 2.85 \times 10^{-4}$, which is roughly the same as 2.692×10^{-4}.

55. Use Simpson's Rule to determine the average temperature in a museum over a 3-hour period, if the temperatures (in degrees Celsius), recorded at 15-min intervals, are

$$21, 21.3, 21.5, 21.8, 21.6, 21.2, 20.8, 20.6, 20.9, 21.2, 21.1, 21.3, 21.2$$

SOLUTION If $T(t)$ represents the temperature at time t, then the average temperature T_{ave} from $t = 0$ to $t = 3$ hours is given by

$$T_{\text{ave}} = \frac{1}{3-0} \int_0^3 T(t)\,dt.$$

To use Simpson's Rule to approximate this, let $\Delta t = 1/4$ (15 minute intervals). Then we have

$$T_{\text{ave}} = \frac{1}{3}[S_{12}] = \frac{1}{3} \cdot \frac{1}{3} \cdot \frac{1}{4}\left[21 + 4 \cdot 21.3 + 2 \cdot 21.5 + \cdots + 4 \cdot 21.3 + 21.2\right] \approx 21.2111.$$

The average temperature is approximately $21.2°$ C.

Further Insights and Challenges

57. Show that $T_N = \int_a^b f(x)\,dx$ for all N and all endpoints a, b if $f(x) = rx + s$ is a linear function (r, s constants).

SOLUTION First, note that

$$\int_a^b (rx + s)\,dx = \frac{r(b^2 - a^2)}{2} + s(b - a).$$

Now,

$$T_N(rx + s) = \frac{b-a}{2N}\left[f(a) + 2\sum_{i=1}^{N-1} f(x_i) + f(b)\right] = \frac{r(b-a)}{2N}\left[a + 2\sum_{i=1}^{N-1} a + 2\frac{b-a}{N}\sum_{i=1}^{N-1} i + b\right] + s\frac{b-a}{2N}(2N)$$

$$= \frac{r(b-a)}{2N}\left[(2N-1)a + 2\frac{b-a}{N}\frac{(N-1)N}{2} + b\right] + s(b-a) = \frac{r(b^2 - a^2)}{2} + s(b - a).$$

59. For N even, divide $[a, b]$ into N subintervals of width $\Delta x = \frac{b-a}{N}$. Set $x_j = a + j\,\Delta x$, $y_j = f(x_j)$, and

$$S_2^{2j} = \frac{b-a}{3N}\left(y_{2j} + 4y_{2j+1} + y_{2j+2}\right)$$

(a) Show that S_N is the sum of the approximations on the intervals $[x_{2j}, x_{2j+2}]$, that is, $S_N = S_2^0 + S_2^2 + \cdots + S_2^{N-2}$.

(b) By Exercise 58, $S_2^{2j} = \int_{x_{2j}}^{x_{2j+2}} f(x)\,dx$ if $f(x)$ is a quadratic polynomial. Use (a) to show that S_N is exact *for all* N if $f(x)$ is a quadratic polynomial.

SOLUTION

(a) This result follows because the even-numbered interior endpoints overlap:

$$\sum_{i=0}^{(N-2)/2} S_2^{2j} = \frac{b-a}{6} \left[(y_0 + 4y_1 + y_2) + (y_2 + 4y_3 + y_4) + \cdots \right]$$

$$= \frac{b-a}{6} \left[y_0 + 4y_1 + 2y_2 + 4y_3 + 2y_4 + \cdots + 4y_{N-1} + y_N \right] = S_N.$$

(b) If $f(x)$ is a quadratic polynomial, then by part (a) we have

$$S_N = S_2^0 + S_2^2 + \cdots + S_2^{N-2} = \int_{x_0}^{x_2} f(x)\, dx + \int_{x_2}^{x_4} f(x)\, dx + \cdots + \int_{x_{N-2}}^{x_N} f(x)\, dx = \int_a^b f(x)\, dx.$$

61. Use the Error Bound for S_N to obtain another proof that Simpson's Rule is exact for all cubic polynomials.

SOLUTION Let $f(x) = ax^3 + bx^2 + cx + d$, with $a \neq 0$, be any cubic polynomial. Then, $f^{(4)}(x) = 0$, so we can take $K_4 = 0$. This yields

$$\text{Error}(S_N) \leq \frac{0}{180N^4} = 0.$$

In other words, S_N is exact for all cubic polynomials for all N.

8.2 Integration by Parts

Preliminary Questions

1. Which derivative rule is used to derive the Integration by Parts formula?

SOLUTION The Integration by Parts formula is derived from the Product Rule.

2. For each of the following integrals, state whether substitution or Integration by Parts should be used:

$$\int x \cos(x^2)\, dx, \qquad \int x \cos x\, dx, \qquad \int x^2 e^x\, dx, \qquad \int x e^{x^2}\, dx$$

SOLUTION

(a) $\int x \cos(x^2)\, dx$: use the substitution $u = x^2$.

(b) $\int x \cos x\, dx$: use Integration by Parts.

(c) $\int x^2 e^x\, dx$; use Integration by Parts.

(d) $\int x e^{x^2}\, dx$; use the substitution $u = x^2$.

3. Why is $u = \cos x$, $v' = x$ a poor choice for evaluating $\int x \cos x\, dx$?

SOLUTION Transforming $v' = x$ into $v = \frac{1}{2}x^2$ increases the power of x and makes the new integral harder than the original.

Exercises

In Exercises 1–6, evaluate the integral using the Integration by Parts formula with the given choice of u and v'.

1. $\int x \sin x\, dx$; $\quad u = x, v' = \sin x$

SOLUTION Using the given choice of u and v' results in

$$u = x \qquad v = -\cos x$$
$$u' = 1 \qquad v' = \sin x$$

Using Integration by Parts,

$$\int x \sin x\, dx = x(-\cos x) - \int (1)(-\cos x)\, dx = -x \cos x + \int \cos x\, dx = -x \cos x + \sin x + C.$$

3. $\int (2x + 9)e^x \, dx; \quad u = 2x + 9, v' = e^x$

SOLUTION Using $u = 2x + 9$ and $v' = e^x$ gives us

$$u = 2x + 9 \quad v = e^x$$
$$u' = 2 \qquad v' = e^x$$

Integration by Parts gives us

$$\int (2x + 9)e^x \, dx = (2x + 9)e^x - \int 2e^x \, dx = (2x + 9)e^x - 2e^x + C = e^x(2x + 7) + C.$$

5. $\int x^3 \ln x \, dx; \quad u = \ln x, v' = x^3$

SOLUTION Using $u = \ln x$ and $v' = x^3$ gives us

$$u = \ln x \quad v = \frac{1}{4}x^4$$
$$u' = \frac{1}{x} \quad v' = x^3$$

Integration by Parts gives us

$$\int x^3 \ln x \, dx = (\ln x)\left(\frac{1}{4}x^4\right) - \int \left(\frac{1}{x}\right)\left(\frac{1}{4}x^4\right) dx$$

$$= \frac{1}{4}x^4 \ln x - \frac{1}{4}\int x^3 \, dx = \frac{1}{4}x^4 \ln x - \frac{1}{16}x^4 + C = \frac{x^4}{16}(4\ln x - 1) + C.$$

In Exercises 7–32, use Integration by Parts to evaluate the integral.

7. $\int (3x - 1)e^{-x} \, dx$

SOLUTION Let $u = 3x - 1$ and $v' = e^{-x}$. Then we have

$$u = 3x - 1 \quad v = -e^{-x}$$
$$u' = 3 \qquad v' = e^{-x}$$

Using Integration by Parts, we get

$$\int (3x - 1)e^{-x} \, dx = (3x - 1)(-e^{-x}) - \int (3)(-e^{-x}) \, dx$$

$$= -e^{-x}(3x - 1) + 3\int e^{-x} \, dx = -e^{-x}(3x - 1) - 3e^{-x} + C = -e^{-x}(3x + 2) + C.$$

9. $\int x^2 e^x \, dx$

SOLUTION Let $u = x^2$ and $v' = e^x$. Then we have

$$u = x^2 \quad v = e^x$$
$$u' = 2x \quad v' = e^x$$

Using Integration by Parts, we get

$$\int x^2 e^x \, dx = x^2 e^x - 2\int x e^x \, dx.$$

We must apply Integration by Parts again to evaluate $\int x e^x \, dx$. Taking $u = x$ and $v' = e^x$, we get

$$\int x e^x \, dx = x e^x - \int (1)e^x \, dx = x e^x - e^x + C.$$

Plugging this into the original equation gives us

$$\int x^2 e^x \, dx = x^2 e^x - 2\left(x e^x - e^x\right) + C = e^x(x^2 - 2x + 2) + C.$$

11. $\displaystyle\int x \cos 2x \, dx$

SOLUTION   Let $u = x$ and $v' = \cos 2x$. Then we have

$$u = x \quad v = \tfrac{1}{2}\sin 2x$$
$$u' = 1 \quad v' = \cos 2x$$

Using Integration by Parts, we get

$$\int x \, \cos 2x \, dx = x \left(\frac{1}{2}\sin 2x\right) - \int (1)\left(\frac{1}{2}\sin 2x\right) dx$$
$$= \frac{1}{2}x \sin 2x - \frac{1}{2}\int \sin 2x \, dx = \frac{1}{2}x \sin 2x + \frac{1}{4}\cos 2x + C.$$

13. $\displaystyle\int e^{-x} \sin x \, dx$

SOLUTION   Let $u = e^{-x}$ and $v' = \sin x$. Then we have

$$u = e^{-x} \qquad v = -\cos x$$
$$u' = -e^{-x} \quad v' = \sin x$$

Using Integration by Parts, we get

$$\int e^{-x} \sin x \, dx = -e^{-x} \cos x - \int (-e^{-x})(-\cos x) \, dx = -e^{-x} \cos x - \int e^{-x} \cos x \, dx.$$

We must apply Integration by Parts again to evaluate $\displaystyle\int e^{-x} \cos x \, dx$. Using $u = e^{-x}$ and $v' = \cos x$, we get

$$\int e^{-x} \cos x \, dx = e^{-x} \sin x - \int (-e^{-x})(\sin x) \, dx = e^{-x} \sin x + \int e^{-x} \sin x \, dx.$$

Plugging this into the original equation, we get

$$\int e^{-x} \sin x \, dx = -e^{-x} \cos x - \left[e^{-x} \sin x + \int e^{-x} \sin x \, dx \right].$$

Solving this equation for $\displaystyle\int e^{-x} \sin x \, dx$ gives us

$$\int e^{-x} \sin x \, dx = -\frac{1}{2}e^{-x}(\sin x + \cos x) + C.$$

15. $\displaystyle\int x \ln x \, dx$

SOLUTION   Let $u = \ln x$ and $v' = x$. Then we have

$$u = \ln x \quad v = \tfrac{1}{2}x^2$$
$$u' = \tfrac{1}{x} \quad v' = x$$

Using Integration by Parts, we get

$$\int x \ln x \, dx = \frac{1}{2}x^2 \ln x - \int \left(\frac{1}{x}\right)\left(\frac{1}{2}x^2\right) dx$$
$$= \frac{1}{2}x^2 \ln x - \frac{1}{2}\int x \, dx = \frac{1}{2}x^2 \ln x - \frac{1}{2}\left(\frac{x^2}{2}\right) + C = \frac{1}{4}x^2(2\ln x - 1) + C.$$

17. $\displaystyle\int x^{-9} \ln x \, dx$

SOLUTION Let $u = \ln x$ and $v' = x^{-9}$. Then we have

$$u = \ln x \qquad v = -\tfrac{1}{8}x^{-8}$$
$$u' = \tfrac{1}{x} \qquad v' = x^{-9}$$

Using Integration by Parts, we get

$$\int x^{-9} \ln x \, dx = -\frac{1}{8}x^{-8} \ln x - \int \frac{1}{x}\left(-\frac{1}{8}x^{-8}\right) dx = -\frac{\ln x}{8x^8} + \frac{1}{8}\int x^{-9}\, dx$$

$$= -\frac{\ln x}{8x^8} + \frac{1}{8}\left(\frac{x^{-8}}{-8}\right) + C = -\frac{1}{8x^8}\left(\ln x + \frac{1}{8}\right) + C.$$

19. $\displaystyle\int x \cos(2 - x)\, dx$

SOLUTION Let $u = x$ and $v' = \cos(2 - x)$. Then we have

$$u = x \qquad v = -\sin(2 - x)$$
$$u' = 1 \qquad v' = \cos(2 - x)$$

Using Integration by Parts, we get

$$\int x \cos(2 - x)\, dx = -x \sin(2 - x) - \int (1)(-\sin(2 - x))\, dx$$

$$= -x \sin(2 - x) + \int \sin(2 - x)\, dx = -x \sin(2 - x) + \cos(2 - x) + C.$$

21. $\displaystyle\int x\, 2^x\, dx$

SOLUTION Let $u = x$ and $v' = 2^x$. Then we have

$$u = x \qquad v = \frac{2^x}{\ln 2}$$

$$u' = 1 \qquad v' = 2^x$$

Using Integration by Parts, we get

$$\int x\, 2^x\, dx = x\left(\frac{2^x}{\ln 2}\right) - \int (1)\frac{2^x}{\ln 2}\, dx = \frac{x\, 2^x}{\ln 2} - \frac{1}{\ln 2}\int 2^x\, dx = \frac{x\, 2^x}{\ln 2} - \frac{1}{\ln 2}\left(\frac{2^x}{\ln 2}\right) + C = \frac{2^x}{\ln 2}\left(x - \frac{1}{\ln 2}\right) + C.$$

23. $\displaystyle\int (\ln x)^2\, dx$

SOLUTION Let $u = (\ln x)^2$ and $v' = 1$. Then we have

$$u = (\ln x)^2 \qquad v = x$$
$$u' = \frac{2}{x}\ln x \qquad v' = 1$$

Using Integration by Parts, we get

$$\int (\ln x)^2\, dx = (\ln x)^2 (x) - \int \left(\frac{2}{x}\ln x\right) x\, dx = x(\ln x)^2 - 2\int \ln x\, dx.$$

We must apply Integration by Parts again to evaluate $\displaystyle\int \ln x\, dx$. Using $u = \ln x$ and $v' = 1$, we have

$$\int \ln x\, dx = x \ln x - \int \frac{1}{x}\cdot x\, dx = x \ln x - \int dx = x \ln x - x + C.$$

Plugging this into the original equation, we get

$$\int (\ln x)^2\, dx = x(\ln x)^2 - 2\,(x \ln x - x) + C = x\left[(\ln x)^2 - 2 \ln x + 2\right] + C.$$

25. $\int \sin^{-1} x \, dx$

SOLUTION Let $u = \sin^{-1} x$ and $v' = 1$. Then we have

$$u = \sin^{-1} x \qquad v = x$$
$$u' = \frac{1}{\sqrt{1-x^2}} \qquad v' = 1$$

Using Integration by Parts, we get

$$\int \sin^{-1} x \, dx = x \sin^{-1} x - \int \frac{x}{\sqrt{1-x^2}} \, dx.$$

We can evaluate $\int \frac{x}{\sqrt{1-x^2}} \, dx$ by making the substitution $w = 1 - x^2$. Then $dw = -2x \, dx$, and we have

$$\int \sin^{-1} x \, dx = x \sin^{-1} x + \frac{1}{2} \int \frac{-2x \, dx}{\sqrt{1-x^2}} = x \sin^{-1} x + \frac{1}{2} \int w^{-1/2} \, dw$$

$$= x \sin^{-1} x + \frac{1}{2}(2w^{1/2}) + C = x \sin^{-1} x + \sqrt{1-x^2} + C.$$

27. $\int x \, 5^x \, dx$

SOLUTION Let $u = x$ and $v' = 5^x$. Then we have

$$u = x \qquad v = \frac{5^x}{\ln 5}$$
$$u' = 1 \qquad v' = 5^x$$

Using Integration by Parts, we get

$$\int x \, 5^x \, dx = x \left(\frac{5^x}{\ln 5} \right) - \int (1) \frac{5^x}{\ln 5} \, dx = \frac{x \, 5^x}{\ln 5} - \frac{1}{\ln 5} \int 5^x \, dx$$

$$= \frac{x \, 5^x}{\ln 5} - \frac{1}{\ln 5} \left(\frac{5^x}{\ln 5} \right) + C = \frac{5^x}{\ln 5} \left(x - \frac{1}{\ln 5} \right) + C.$$

29. $\int x \cosh 2x \, dx$

SOLUTION Let $u = x$ and $v' = \cosh 2x$. Then we have

$$u = x \qquad v = \tfrac{1}{2} \sinh 2x$$
$$u' = 1 \qquad v' = \cosh 2x$$

Using Integration by Parts, we get

$$\int x \cosh 2x \, dx = x \left(\frac{1}{2} \sinh 2x \right) - \int (1) \left(\frac{1}{2} \sinh 2x \right) dx$$

$$= \frac{1}{2} x \sinh 2x - \frac{1}{2} \int \sinh 2x \, dx = \frac{1}{2} x \sinh 2x - \frac{1}{4} \cosh 2x + C.$$

31. $\int \sinh^{-1} x \, dx$

SOLUTION Using $u = \sinh^{-1} x$ and $v' = 1$ gives us

$$u = \sinh^{-1} x \qquad v = x$$
$$u' = \frac{1}{\sqrt{1+x^2}} \qquad v' = 1$$

Integration by Parts gives us

$$\int \sinh^{-1} x \, dx = x \sinh^{-1} x - \int \left(\frac{1}{\sqrt{1+x^2}} \right) x \, dx.$$

For the integral on the right we'll use the substitution $w = 1 + x^2$, $dw = 2x\,dx$. Then we have

$$\int \sinh^{-1} x\,dx = x\sinh^{-1} x - \frac{1}{2}\int \frac{dw}{\sqrt{w}} = x\sinh^{-1} x - \sqrt{w} + C$$

$$= x\sinh^{-1} x - \sqrt{1 + x^2} + C.$$

33. Use the substitution $u = x^{1/2}$ and then Integration by Parts to evaluate $\int e^{\sqrt{x}}\,dx$.

SOLUTION Let $w = x^{1/2}$. Then $dw = \frac{1}{2}x^{-1/2}dx$, or $dx = 2x^{1/2}\,dw = 2w\,dw$. Now,

$$\int e^{\sqrt{x}}\,dx = 2\int we^{w}\,dw.$$

Using Integration by Parts with $u = w$ and $v' = e^w$, we get

$$2\int we^{w}\,dw = 2(we^{w} - e^{w}) + C.$$

Substituting back, we find

$$\int e^{\sqrt{x}}\,dx = 2e^{\sqrt{x}}(\sqrt{x} - 1) + C.$$

In Exercises 35–44, evaluate using Integration by Parts, substitution, or both if necessary.

35. $\int x\cos 4x\,dx$

SOLUTION Let $u = x$ and $v' = \cos 4x$. Then we have

$$u = x \quad v = \tfrac{1}{4}\sin 4x$$
$$u' = 1 \quad v' = \cos 4x$$

Using Integration by Parts, we get

$$\int x\cos 4x\,dx = \frac{1}{4}x\sin 4x - \int (1)\frac{1}{4}\sin 4x\,dx = \frac{1}{4}x\sin 4x - \frac{1}{4}\left(-\frac{1}{4}\cos 4x\right) + C$$

$$= \frac{1}{4}x\sin 4x + \frac{1}{16}\cos 4x + C.$$

37. $\int \frac{x\,dx}{\sqrt{x+1}}$

SOLUTION Let $u = x + 1$. Then $du = dx$, $x = u - 1$, and

$$\int \frac{x\,dx}{\sqrt{x+1}} = \int \frac{(u-1)\,du}{\sqrt{u}} = \int \left(\frac{u}{\sqrt{u}} - \frac{1}{\sqrt{u}}\right)du = \int (u^{1/2} - u^{-1/2})\,du$$

$$= \frac{2}{3}u^{3/2} - 2u^{1/2} + C = \frac{2}{3}(x+1)^{3/2} - 2(x+1)^{1/2} + C.$$

39. $\int \cos x\,\ln(\sin x)\,dx$

SOLUTION Let $w = \sin x$. Then $dw = \cos x\,dx$, and

$$\int \cos x\,\ln(\sin x)\,dx = \int \ln w\,dw.$$

Now use Integration by Parts with $u = \ln w$ and $v' = 1$. Then $u' = 1/w$ and $v = w$, which gives us

$$\int \cos x\,\ln(\sin x)\,dx = \int \ln w\,dw = w\ln w - w + C = \sin x\,\ln(\sin x) - \sin x + C.$$

41. $\int \sin\sqrt{x}\,dx$

SOLUTION First use substitution, with $w = \sqrt{x}$ and $dw = dx/(2\sqrt{x})$. This gives us

$$\int \sin \sqrt{x}\, dx = \int \frac{(2\sqrt{x}) \sin \sqrt{x}\, dx}{(2\sqrt{x})} = 2 \int w \sin w\, dw.$$

Now use Integration by Parts, with $u = w$ and $v' = \sin w$. Then we have

$$\int \sin \sqrt{x}\, dx = 2 \int w \sin w\, dw = 2\left(-w \cos w - \int -\cos w\, dw \right)$$

$$= 2(-w \cos w + \sin w) + C = 2 \sin \sqrt{x} - 2\sqrt{x} \cos \sqrt{x} + C.$$

43. $\displaystyle \int \frac{\ln(\ln x)\, \ln x\, dx}{x}$

SOLUTION Let $w = \ln x$. Then $dw = dx/x$, and

$$\int \frac{\ln(\ln x)\, \ln x\, dx}{x} = \int w \ln w\, dw.$$

Now use Integration by Parts, with $u = \ln w$ and $v' = w$. Then,

$$u = \ln w \qquad v = \frac{1}{2}w^2$$

$$u' = w^{-1} \qquad v' = w$$

and

$$\int \frac{\ln(\ln x)\, \ln x\, dx}{x} = \frac{1}{2}w^2 \ln w - \frac{1}{2}\int w\, dw = \frac{1}{2}w^2 \ln w - \frac{1}{2}\left(\frac{w^2}{2}\right) + C$$

$$= \frac{1}{2}(\ln x)^2 \ln(\ln x) - \frac{1}{4}(\ln x)^2 + C = \frac{1}{4}(\ln x)^2[2 \ln(\ln x) - 1] + C.$$

In Exercises 45–50, compute the definite integral.

45. $\displaystyle \int_0^2 xe^{9x}\, dx$

SOLUTION Let $u = x$ and $v' = e^{9x}$. Then $u' = 1$ and $v = \frac{1}{9}e^{9x}$. Using Integration by Parts,

$$\int_0^2 xe^9 x\, dx = \frac{1}{9}xe^{9x}\Big|_0^2 - \int_0^2 (1)\frac{1}{9}e^{9x}\, dx = \left(\frac{1}{9}xe^{9x} - \frac{1}{81}e^{9x}\right)\Big|_0^2$$

$$= \left(\frac{2}{9}e^{18} - \frac{1}{81}e^{18}\right) - \left(0 - \frac{1}{81}(1)\right) = \frac{1}{81}(18e^{18} - e^{18} + 1) = \frac{1}{18}(17e^{18} + 1).$$

47. $\displaystyle \int_0^4 x\sqrt{4 - x}\, dx$

SOLUTION Let $u = 4 - x$. Then $x = 4 - u$, $du = -dx$ and

$$\int_0^4 x\sqrt{4 - x}\, dx = \int_{u=4}^{u=0} (4 - u)u^{1/2}\, (-du) = -\int_{u=4}^{u=0} (4u^{1/2} - u^{3/2})\, du$$

$$= \int_{u=0}^{u=4} (4u^{1/2} - u^{3/2})\, du = \left((4)\frac{2u^{3/2}}{3} - \frac{2u^{5/2}}{5}\right)\Big|_0^4$$

$$= \frac{8}{3}(4^{3/2}) - \frac{2}{5}(4^{5/2}) = \frac{64}{3} - \frac{64}{5} = \frac{128}{15}.$$

49. $\displaystyle \int_1^4 \sqrt{x}\, \ln x\, dx$

SOLUTION Let $u = \ln x$ and $v' = \sqrt{x}$. Then $u' = 1/x$ and $v = \frac{2}{3}x^{3/2}$. Using Integration by Parts,

$$\int_1^4 \sqrt{x}\, \ln x\, dx = \frac{2}{3}x^{3/2} \ln x\Big|_1^4 - \int_1^4 \frac{2}{3}x^{1/2}\, dx = \left(\frac{2}{3}x^{3/2} \ln x - \frac{2}{3}\cdot\frac{2}{3}x^{3/2}\right)\Big|_1^4$$

$$= \frac{2}{3}x^{3/2}\left(\ln x - \frac{2}{3}\right)\Big|_1^4 = \frac{2}{3}4^{3/2}\left(\ln 4 - \frac{2}{3}\right) - \frac{2}{3}(1)\left(0 - \frac{2}{3}\right) = \frac{16}{3}\ln 4 - \frac{28}{9}.$$

51. Use Eq. (5) to evaluate $\int x^4 e^x\, dx$.

SOLUTION

$$\int x^4 e^x\, dx = x^4 e^x - 4\int x^3 e^x\, dx = x^4 e^x - 4\left[x^3 e^x - 3\int x^2 e^x\, dx\right]$$

$$= x^4 e^x - 4x^3 e^x + 12\int x^2 e^x\, dx = x^4 e^x - 4x^3 e^x + 12\left[x^2 e^x - 2\int x e^x\, dx\right]$$

$$= x^4 e^x - 4x^3 e^x + 12x^2 e^x - 24\int x e^x\, dx = x^4 e^x - 4x^3 e^x + 12x^2 e^x - 24\left[x e^x - \int e^x\, dx\right]$$

$$= x^4 e^x - 4x^3 e^x + 12x^2 e^x - 24\left[x e^x - e^x\right] + C.$$

Thus,

$$\int x^4 e^x\, dx = e^x(x^4 - 4x^3 + 12x^2 - 24x + 24) + C.$$

53. Find a reduction formula for $\int x^n e^{-x}\, dx$ similar to Eq. (5).

SOLUTION Let $u = x^n$ and $v' = e^{-x}$. Then

$$u = x^n \qquad v = -e^{-x}$$
$$u' = nx^{n-1} \quad v' = e^{-x}$$

Using Integration by Parts, we get

$$\int x^n e^{-x}\, dx = -x^n e^{-x} - \int nx^{n-1}(-e^{-x})\, dx = -x^n e^{-x} + n\int x^{n-1} e^{-x}\, dx.$$

In Exercises 55–62, indicate a good method for evaluating the integral (but do not evaluate). Your choices are algebraic manipulation, substitution (specify u and du), and Integration by Parts (specify u and v′). If it appears that the techniques you have learned thus far are not sufficient, state this.

55. $\int \sqrt{x}\ln x\, dx$

SOLUTION Use Integration by Parts, with $u = \ln x$ and $v' = \sqrt{x}$.

57. $\int \dfrac{x^3}{\sqrt{4 - x^2}}\, dx$

SOLUTION Use substitution, followed by algebraic manipulation: Let $u = 4 - x^2$. Then $du = -2x\, dx$, $x^2 = 4 - u$, and

$$\int \frac{x^3}{\sqrt{4 - x^2}}\, dx = -\frac{1}{2}\int \frac{(x^2)(-2x\, dx)}{\sqrt{u}} = -\frac{1}{2}\int \frac{(4 - u)(du)}{\sqrt{u}} = -\frac{1}{2}\int \left(\frac{4}{\sqrt{u}} - \frac{u}{\sqrt{u}}\right) du.$$

59. $\int \dfrac{2x + 3}{x^2 + 3x + 6}\, dx$

SOLUTION Use substitution, with $u = x^2 + 3x + 6$ and $du = (2x + 3)\, dx$.

61. $\int x\sin(3x + 4)\, dx$

SOLUTION Use Integration by Parts, with $u = x$ and $v' = \sin(3x + 4)$.

63. Evaluate $\int (\sin^{-1} x)^2\, dx$. *Hint:* First use Integration by Parts and then substitution.

SOLUTION First use integration by parts with $v' = 1$ to get

$$\int (\sin^{-1} x)^2 \, dx = x(\sin^{-1} x)^2 - 2 \int \frac{x \sin^{-1} x \, dx}{\sqrt{1 - x^2}}.$$

Now use substitution on the integral on the right, with $u = \sin^{-1} x$. Then $du = dx/\sqrt{1 - x^2}$ and $x = \sin u$, and we get (using Integration by Parts again)

$$\int \frac{x \sin^{-1} x \, dx}{\sqrt{1 - x^2}} = \int u \sin u \, du = -u \cos u + \sin u + C = -\sqrt{1 - x^2} \, \sin^{-1} x + x + C.$$

where $\cos u = \sqrt{1 - \sin^2 u} = \sqrt{1 - x^2}$. So the final answer is

$$\int (\sin^{-1} x)^2 \, dx = x(\sin^{-1} x)^2 + 2\sqrt{1 - x^2} \, \sin^{-1} x - 2x + C.$$

65. Evaluate $\int x^7 \cos(x^4) \, dx$.

SOLUTION First, let $w = x^4$. Then $dw = 4x^3 \, dx$ and

$$\int x^7 \cos(x^4) \, dx = \frac{1}{4} \int w \cos x \, dw.$$

Now, use Integration by Parts with $u = w$ and $v' = \cos w$. Then

$$\int x^7 \cos(x^4) \, dx = \frac{1}{4} \left(w \sin w - \int \sin w \, dw \right) = \frac{1}{4} w \sin w + \frac{1}{4} \cos w + C = \frac{1}{4} x^4 \sin(x^4) + \frac{1}{4} \cos(x^4) + C.$$

67. Find the volume of the solid obtained by revolving $y = e^x$ for $0 \le x \le 2$ about the y-axis.

SOLUTION By the Method of Cylindrical Shells, the volume V of the solid is

$$V = \int_a^b (2\pi r)h \, dx = 2\pi \int_0^2 xe^x \, dx.$$

Using Integration by Parts with $u = x$ and $v' = e^x$, we find

$$V = 2\pi \left. (xe^x - e^x) \right|_0^2 = 2\pi [(2e^2 - e^2) - (0 - 1)] = 2\pi(e^2 + 1).$$

69. Recall that the *present value* of an investment which pays out income continuously at a rate $R(t)$ for T years is $\int_0^T R(t)e^{-rt} \, dt$, where r is the interest rate. Find the present value if income is produced at a rate $R(t) = 5,000 + 100t$ dollars/year for 10 years.

SOLUTION The present value is given by

$$PV = \int_0^T R(t)e^{-rt} \, dt = \int_0^{10} (5000 + 100t)e^{-rt} \, dt = 5000 \int_0^{10} e^{-rt} \, dt + 100 \int_0^{10} te^{-rt} \, dt.$$

Using Integration by Parts for the integral on the right, with $u = t$ and $v' = e^{-rt}$, we find

$$PV = 5000 \left. \left(-\frac{1}{r} e^{-rt} \right) \right|_0^{10} + 100 \left[\left. \left(-\frac{t}{r} e^{-rt} \right) \right|_0^{10} - \int_0^{10} \frac{-1}{r} e^{-rt} \, dt \right]$$

$$= -\frac{5000}{r} e^{-rt} \Big|_0^{10} - \frac{100}{r} \left. \left(te^{-rt} + \frac{1}{r} e^{-rt} \right) \right|_0^{10}$$

$$= -\frac{5000}{r} (e^{-10r} - 1) - \frac{100}{r} \left[\left(10e^{-10r} + \frac{1}{r} e^{-10r} \right) - \left(0 + \frac{1}{r} \right) \right]$$

$$= e^{-10r} \left[-\frac{5000}{r} - \frac{1000}{r} - \frac{100}{r^2} \right] + \frac{5000}{r} + \frac{100}{r^2}$$

$$= \frac{5000r + 100 - e^{-10r}(6000r + 100)}{r^2}.$$

71. Use Eq. (6) to calculate $\int (\ln x)^k \, dx$ for $k = 2, 3$.

SOLUTION

$$\int (\ln x)^2 \, dx = x(\ln x)^2 - 2\int \ln x \, dx = x(\ln x)^2 - 2(x \ln x - x) + C = x(\ln x)^2 - 2x \ln x + 2x + C;$$

$$\int (\ln x)^3 \, dx = x(\ln x)^3 - 3\int (\ln x)^2 \, dx = x(\ln x)^3 - 3\left[x(\ln x)^2 - 2x \ln x + 2x\right] + C$$

$$= x(\ln x)^3 - 3x(\ln x)^2 + 6x \ln x - 6x + C.$$

73. Prove $\int xb^x \, dx = b^x \left(\dfrac{x}{\ln b} - \dfrac{1}{\ln^2 b} \right) + C.$

SOLUTION Let $u = x$ and $v' = b^x$. Then $u' = 1$ and $v = b^x / \ln b$. Using Integration by Parts, we get

$$\int x \, b^x \, dx = \frac{xb^x}{\ln b} - \frac{1}{\ln b}\int b^x \, dx = \frac{xb^x}{\ln b} - \frac{1}{\ln b} \cdot \frac{b^x}{\ln b} + C = b^x \left(\frac{x}{\ln b} - \frac{1}{(\ln b)^2} \right) + C.$$

Further Insights and Challenges

75. Prove in two ways

$$\int_0^a f(x) \, dx = af(a) - \int_0^a xf'(x) \, dx \qquad \boxed{8}$$

First use Integration by Parts. Then assume $f(x)$ is increasing. Use the substitution $u = f(x)$ to prove that $\int_0^a xf'(x) \, dx$ is equal to the area of the shaded region in Figure 1 and derive Eq. (8) a second time.

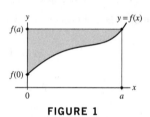

FIGURE 1

SOLUTION Let $u = f(x)$ and $v' = 1$. Then Integration by Parts gives

$$\int_0^a f(x) \, dx = xf(x) \Big|_0^a - \int_0^a xf'(x) \, dx = af(a) - \int_0^a xf'(x) \, dx.$$

Alternately, let $u = f(x)$. Then $du = f'(x) \, dx$, and if $f(x)$ is either increasing or decreasing, it has an inverse function, and $x = f^{-1}(u)$. Thus,

$$\int_{x=0}^{x=a} xf'(x) \, dx = \int_{f(0)}^{f(a)} f^{-1}(u) \, du$$

which is precisely the area of the shaded region in Figure 1 (integrating along the vertical axis). Since the area of the entire rectangle is $af(a)$, the difference between the areas of the two regions is $\int_0^a f(x) \, dx$.

77. Set $I(a, b) = \displaystyle\int_0^1 x^a (1 - x)^b \, dx$, where a, b are whole numbers.

(a) Use substitution to show that $I(a, b) = I(b, a)$.

(b) Show that $I(a, 0) = I(0, a) = \dfrac{1}{a + 1}$.

(c) Prove that for $a \geq 1$ and $b \geq 0$,

$$I(a, b) = \frac{a}{b + 1} I(a - 1, b + 1)$$

(d) Use (b) and (c) to calculate $I(1, 1)$ and $I(3, 2)$.

(e) Show that $I(a, b) = \dfrac{a! \, b!}{(a + b + 1)!}.$

SOLUTION

(a) Let $u = 1 - x$. Then $du = -dx$ and

$$I(a, b) = \int_{u=1}^{u=0} (1 - u)^a u^b (-du) = \int_0^1 u^b (1 - u)^a \, du = I(b, a).$$

(b) $I(a, 0) = I(0, a)$ by part (a). Further,

$$I(a, 0) = \int_0^1 x^a (1 - x)^0 \, dx = \int_0^1 x^a \, dx = \frac{1}{a + 1}.$$

(c) Using Integration by Parts with $u = (1 - x)^b$ and $v' = x^a$ gives

$$I(a, b) = (1 - x)^b \frac{x^{a+1}}{a + 1} \Big|_0^1 + \frac{b}{a + 1} \int_0^1 x^{a+1} (1 - x)^{b-1} \, dx = \frac{b}{a + 1} I(a + 1, b - 1).$$

The other equality arises from Integration by Parts with $u = x^a$ and $v' = (1 - x)^b$.

(d)

$$I(1, 1) = \frac{1}{1 + 1} I(1 - 1, 1 + 1) = \frac{1}{2} I(0, 2) = \frac{1}{2} \cdot \frac{1}{3} = \frac{1}{6}$$

$$I(3, 2) = \frac{1}{2} I(4, 2) = \frac{1}{2} \cdot \frac{1}{5} I(5, 0) = \frac{1}{10} \cdot \frac{1}{6} = \frac{1}{60}.$$

(e) We proceed as follows:

$$I(a, b) = \frac{a}{b + 1} I(a - 1, b + 1) = \frac{a}{b + 1} \cdot \frac{a - 1}{b + 2} I(a - 2, b + 2)$$

$$\vdots$$

$$= \frac{a}{b + 1} \cdot \frac{a - 1}{b + 2} \cdots \frac{1}{b + a} I(0, b + a)$$

$$= \frac{a(a - 1) \cdots (1)}{(b + 1)(b + 2) \cdots (b + a)} \cdot \frac{1}{b + a + 1}$$

$$= \frac{b! \, a!}{b! \, (b + 1)(b + 2) \cdots (b + a)(b + a + 1)} = \frac{a! \, b!}{(a + b + 1)!}.$$

79. Let $I_n = \int x^n \cos(x^2) \, dx$ and $J_n = \int x^n \sin(x^2) \, dx$.

(a) Find a reduction formula that expresses I_n in terms of J_{n-2}. *Hint:* Write $x^n \cos(x^2)$ as $x^{n-1}(x \cos(x^2))$.

(b) 🖼 Use the result of (a) to show that I_n can be evaluated explicitly if n is odd.

(c) Evaluate I_3.

SOLUTION

(a) Integration by Parts with $u = x^{n-1}$ and $v' = x \cos(x^2) \, dx$ yields

$$I_n = \frac{1}{2} x^{n-1} \sin(x^2) - \frac{n - 1}{2} \int x^{n-2} \sin(x^2) \, dx = \frac{1}{2} x^{n-1} \sin(x^2) - \frac{n - 1}{2} J_{n-2}.$$

(b) If n is odd, the reduction process will eventually lead to either

$$\int x \cos(x^2) \, dx \quad \text{or} \quad \int x \sin(x^2) \, dx,$$

both of which can be evaluated using the substitution $u = x^2$.

(c) Starting with the reduction formula from part (a), we find

$$I_3 = \frac{1}{2} x^2 \sin(x^2) - \frac{2}{2} \int x \sin(x^2) \, dx = \frac{1}{2} x^2 \sin(x^2) + \frac{1}{2} \cos(x^2) + C.$$

8.3 Trigonometric Integrals

Preliminary Questions

1. Describe the technique used to evaluate $\int \sin^5 x \, dx$.

SOLUTION Because the sine function is raised to an odd power, rewrite $\sin^5 x = \sin x \sin^4 x = \sin x (1 - \cos^2 x)^2$ and then substitute $u = \cos x$.

2. Describe a way of evaluating $\int \sin^6 x \, dx$.

SOLUTION Repeatedly use the reduction formula for powers of $\sin x$.

3. Are reduction formulas needed to evaluate $\int \sin^7 x \cos^2 x \, dx$? Why or why not?

SOLUTION No, a reduction formula is not needed because the sine function is raised to an odd power.

4. Describe a way of evaluating $\int \sin^6 x \cos^2 x \, dx$.

SOLUTION Because both trigonometric functions are raised to even powers, write $\cos^2 x = 1 - \sin^2 x$ and then apply the reduction formula for powers of the sine function.

5. Which integral requires more work to evaluate?

$$\int \sin^{798} x \cos x \, dx \qquad \text{or} \qquad \int \sin^4 x \cos^4 x \, dx$$

Explain your answer.

SOLUTION The first integral can be evaluated using the substitution $u = \sin x$, whereas the second integral requires the use of reduction formulas. The second integral therefore requires more work to evaluate.

Exercises

In Exercises 1–6, use the method for odd powers to evaluate the integral.

1. $\int \cos^3 x \, dx$.

SOLUTION Use the identity $\cos^2 x = 1 - \sin^2 x$ to rewrite the integrand:

$$\int \cos^3 x \, dx = \int \left(1 - \sin^2 x\right) \cos x \, dx.$$

Now use the substitution $u = \sin x$, $du = \cos x \, dx$:

$$\int \cos^3 x \, dx = \int \left(1 - u^2\right) du = u - \frac{1}{3} u^3 + C = \sin x - \frac{1}{3} \sin^3 x + C.$$

3. $\int \sin^3 \theta \cos^2 \theta \, d\theta$

SOLUTION Write $\sin^3 \theta = \sin^2 \theta \sin \theta = (1 - \cos^2 \theta) \sin \theta$. Then

$$\int \sin^3 \theta \cos^2 \theta \, d\theta = \int \left(1 - \cos^2 \theta\right) \cos^2 \theta \sin \theta \, d\theta.$$

Now use the substitution $u = \cos \theta$, $du = -\sin \theta \, d\theta$:

$$\int \sin^3 \theta \cos^2 \theta \, d\theta = -\int \left(1 - u^2\right) u^2 \, du = -\int \left(u^2 - u^4\right) du$$

$$= -\frac{1}{3} u^3 + \frac{1}{5} u^5 + C = -\frac{1}{3} \cos^3 \theta + \frac{1}{5} \cos^5 \theta + C.$$

5. $\int \sin^3 t \cos^3 t \, dt$

SOLUTION Write $\sin^3 t = (1 - \cos^2 t)\sin t \, dt$. Then

$$\int \sin^3 t \cos^3 t \, dt = \int (1 - \cos^2 t)\cos^3 t \sin t \, dt = \int \left(\cos^3 t - \cos^5 t\right)\sin t \, dt.$$

Now use the substitution $u = \cos t$, $du = -\sin t \, dt$:

$$\int \sin^3 t \cos^3 t \, dt = -\int \left(u^3 - u^5\right) du = -\frac{1}{4}u^4 + \frac{1}{6}u^6 + C = -\frac{1}{4}\cos^4 t + \frac{1}{6}\cos^6 t + C.$$

7. Find the area of the shaded region in Figure 1.

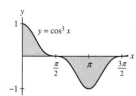

FIGURE 1 Graph of $y = \cos^3 x$.

SOLUTION First evaluate the indefinite integral by writing $\cos^3 x = (1 - \sin^2 x)\cos x$, and using the substitution $u = \sin x$, $du = \cos x \, dx$:

$$\int \cos^3 x \, dx = \int \left(1 - \sin^2 x\right)\cos x \, dx = \int \left(1 - u^2\right) du = u - \frac{1}{3}u^3 + C = \sin x - \frac{1}{3}\sin^3 x + C.$$

The area is given by

$$A = \int_0^{\pi/2} \cos^3 x \, dx - \int_{\pi/2}^{3\pi/2} \cos^3 x \, dx = \left(\sin x - \frac{1}{3}\sin^3 x\right)\Big|_0^{\pi/2} - \left(\sin x - \frac{1}{3}\sin^3 x\right)\Big|_{\pi/2}^{3\pi/2}$$

$$= \left[\left(\sin \frac{\pi}{2} - \frac{1}{3}\sin^3 \frac{\pi}{2}\right) - 0\right] - \left[\left(\sin \frac{3\pi}{2} - \frac{1}{3}\sin^3 \frac{3\pi}{2}\right) - \left(\sin \frac{\pi}{2} - \frac{1}{3}\sin^3 \frac{\pi}{2}\right)\right]$$

$$= 1 - \frac{1}{3}(1)^3 - (-1) + \frac{1}{3}(-1)^3 + 1 - \frac{1}{3}(1)^3 = 2.$$

In Exercises 9–12, evaluate the integral using the methods employed in Examples 3 and 4.

9. $\displaystyle\int \cos^4 y \, dy$

SOLUTION Using the reduction formula for $\cos^m y$, we get

$$\int \cos^4 y \, dy = \frac{1}{4}\cos^3 y \sin y + \frac{3}{4}\int \cos^2 y \, dy = \frac{1}{4}\cos^3 y \sin y + \frac{3}{4}\left(\frac{1}{2}\cos y \sin y + \frac{1}{2}\int dy\right)$$

$$= \frac{1}{4}\cos^3 y \sin y + \frac{3}{8}\cos y \sin y + \frac{3}{8}y + C.$$

11. $\displaystyle\int \sin^4 x \cos^2 x \, dx$

SOLUTION Use the identity $\cos^2 x = 1 - \sin^2 x$ to write:

$$\int \sin^4 x \cos^2 x \, dx = \int \sin^4 x \left(1 - \sin^2 x\right) dx = \int \sin^4 x \, dx - \int \sin^6 x \, dx.$$

Using the reduction formula for $\sin^m x$:

$$\int \sin^4 x \cos^2 x \, dx = \int \sin^4 x \, dx - \left[-\frac{1}{6}\sin^5 x \cos x + \frac{5}{6}\int \sin^4 x \, dx\right]$$

$$= \frac{1}{6}\sin^5 x \cos x + \frac{1}{6}\int \sin^4 x \, dx = \frac{1}{6}\sin^5 x \cos x + \frac{1}{6}\left(-\frac{1}{4}\sin^3 x \cos x + \frac{3}{4}\int \sin^2 x \, dx\right)$$

$$= \frac{1}{6}\sin^5 x \cos x - \frac{1}{24}\sin^3 x \cos x + \frac{1}{8}\int \sin^2 x \, dx$$

$$= \frac{1}{6}\sin^5 x \cos x - \frac{1}{24}\sin^3 x \cos x + \frac{1}{8}\left(-\frac{1}{2}\sin x \cos x + \frac{1}{2}\int dx\right)$$

$$= \frac{1}{6}\sin^5 x \cos x - \frac{1}{24}\sin^3 x \cos x - \frac{1}{16}\sin x \cos x + \frac{1}{16}x + C.$$

In Exercises 13–20, evaluate the integral using the reduction formulas on p. 438 as necessary.

13. $\int \tan^2 x \sec^2 x \, dx$

SOLUTION Use the substitution $u = \tan x$, $du = \sec^2 x \, dx$. Then

$$\int \tan^2 x \sec^2 x \, dx = \int u^2 \, du = \frac{1}{3}u^3 + C = \frac{1}{3}\tan^3 x + C.$$

15. $\int \tan^4 \theta \sec^2 \theta \, d\theta$

SOLUTION Use the substitution $u = \tan \theta$, $du = \sec^2 \theta \, d\theta$. Then

$$\int \tan^4 \theta \sec^2 \theta \, d\theta = \int u^4 \, du = \frac{1}{5}u^5 + C = \frac{1}{5}\tan^5 \theta + C.$$

17. $\int \tan^2 t \, dt$

SOLUTION Using the reduction formula for $\tan^m t$, we get

$$\int \tan^2 t \, dt = \tan t - \int dt = \tan t - t + C.$$

19. $\int \sec^3 x \, dx$

SOLUTION Using the reduction formula for $\sec^m x$, we get

$$\int \sec^3 x \, dx = \frac{1}{2}\tan x \sec x + \frac{1}{2}\int \sec x \, dx = \frac{1}{2}\tan x \sec x + \frac{1}{2}\ln|\sec x + \tan x| + C.$$

In Exercises 21–54, use the techniques and reduction formulas necessary to evaluate the integral.

21. $\int \cos^5 x \sin x \, dx$

SOLUTION Use the substitution $u = \cos x$, $du = -\sin x \, dx$. Then

$$\int \cos^5 x \sin x \, dx = -\int u^5 \, du = -\frac{1}{6}u^6 + C = -\frac{1}{6}\cos^6 x + C.$$

23. $\int \cos^4(3x) \, dx$

SOLUTION Use the substitution $u = 3x$, $du = 3\,dx$, followed by the reduction formula for $\cos^m x$:

$$\int \cos^4(3x)\,dx = \frac{1}{3}\int \cos^4 u\,du = \frac{1}{12}\cos^3 u \sin u + \frac{1}{4}\int \cos^2 u\,du = \frac{1}{12}\cos^3 u \sin u + \frac{1}{4}\left(\frac{1}{2}\cos u \sin u + \frac{1}{2}\int du\right)$$

$$= \frac{1}{12}\cos^3 u \sin u + \frac{1}{8}\cos u \sin u + \frac{1}{8}u + C = \frac{1}{12}\cos^3(3x)\sin(3x) + \frac{1}{8}\cos(3x)\sin(3x) + \frac{3}{8}x + C.$$

25. $\int \cos^3(\pi\theta)\sin^4(\pi\theta)\,d\theta$

SOLUTION Use the substitution $u = \pi\theta$, $du = \pi\,d\theta$, and the identity $\cos^2 u = 1 - \sin^2 u$ to write

$$\int \cos^3(\pi\theta)\sin^4(\pi\theta)\,d\theta = \frac{1}{\pi}\int \cos^3 u \sin^4 u\,du = \frac{1}{\pi}\int \left(1 - \sin^2 u\right)\sin^4 u \cos u\,du.$$

Now use the substitution $w = \sin u$, $dw = \cos u\,du$:

$$\int \cos^3(\pi\theta)\sin^4(\pi\theta)\,d\theta = \frac{1}{\pi}\int \left(1 - w^2\right)w^4\,dw = \frac{1}{\pi}\int \left(w^4 - w^6\right)dw = \frac{1}{5\pi}w^5 - \frac{1}{7\pi}w^7 + C$$

$$= \frac{1}{5\pi}\sin^5(\pi\theta) - \frac{1}{7\pi}\sin^7(\pi\theta) + C.$$

27. $\int \sin^4(3x)\,dx$

SOLUTION Use the substitution $u = 3x$, $du = 3\,dx$ and the reduction formula for $\sin^m x$:

$$\int \sin^4(3x)\,dx = \frac{1}{3}\int \sin^4 u\,du = -\frac{1}{12}\sin^3 u \cos u + \frac{1}{4}\int \sin^2 u\,du$$

$$= -\frac{1}{12}\sin^3 u \cos u + \frac{1}{4}\left(-\frac{1}{2}\sin u \cos u + \frac{1}{2}\int du\right)$$

$$= -\frac{1}{12}\sin^3 u \cos u - \frac{1}{8}\sin u \cos u + \frac{1}{8}u + C$$

$$= -\frac{1}{12}\sin^3(3x)\cos(3x) - \frac{1}{8}\sin(3x)\cos(3x) + \frac{3}{8}x + C.$$

29. $\int \sec 7t\,dt$

SOLUTION Use the substitution $u = 7t$, $du = 7\,dt$. Then

$$\int \sec 7t\,dt = \frac{1}{7}\int \sec u\,du = \frac{1}{7}\ln|\sec u + \tan u| + C = \frac{1}{7}\ln|\sec 7t + \tan 7t| + C.$$

31. $\int \tan x \sec^2 x\,dx$

SOLUTION Use the substitution $u = \tan x$, $du = \sec^2 x\,dx$. Then

$$\int \tan x \sec^2 x\,dx = \int u\,du = \frac{1}{2}u^2 + C = \frac{1}{2}\tan^2 x + C.$$

33. $\int \tan^5 x \sec^4 x\,dx$

SOLUTION Use the identity $\tan^2 x = \sec^2 x - 1$ to write

$$\int \tan^5 x \sec^4 x\,dx = \int \left(\sec^2 x - 1\right)^2 \sec^3 x(\sec x \tan x\,dx).$$

Now use the substitution $u = \sec x$, $du = \sec x \tan x\,dx$:

$$\int \tan^5 x \sec^4 x\,dx = \int \left(u^2 - 1\right)^2 u^3\,du = \int \left(u^7 - 2u^5 + u^3\right)du$$

$$= \frac{1}{8}u^8 - \frac{1}{3}u^6 + \frac{1}{4}u^4 + C = \frac{1}{8}\sec^8 x - \frac{1}{3}\sec^6 x + \frac{1}{4}\sec^4 x + C.$$

35. $\int \tan^4 x \sec x\,dx$

SOLUTION Use the identity $\tan^2 x = \sec^2 x - 1$ to write

$$\int \tan^4 x \sec x\,dx = \int \left(\sec^2 x - 1\right)^2 \sec x\,dx = \int \sec^5 x\,dx - 2\int \sec^3 x\,dx + \int \sec x\,dx.$$

Now use the reduction formula for $\sec^m x$:

$$\int \tan^4 x \sec x\,dx = \left(\frac{1}{4}\tan x \sec^3 x + \frac{3}{4}\int \sec^3 x\,dx\right) - 2\int \sec^3 x\,dx + \int \sec x\,dx$$

$$= \frac{1}{4}\tan x \sec^3 x - \frac{5}{4}\int \sec^3 x\,dx + \int \sec x\,dx$$

$$= \frac{1}{4}\tan x \sec^3 x - \frac{5}{4}\left(\frac{1}{2}\tan x \sec x + \frac{1}{2}\int \sec x\,dx\right) + \int \sec x\,dx$$

$$= \frac{1}{4}\tan x \sec^3 x - \frac{5}{8}\tan x \sec x + \frac{3}{8}\int \sec x\,dx$$

$$= \frac{1}{4}\tan x \sec^3 x - \frac{5}{8}\tan x \sec x + \frac{3}{8}\ln|\sec x + \tan x| + C.$$

37. $\displaystyle\int \tan^4 x \sec^3 x \, dx$

SOLUTION Use the identity $\tan^2 x = \sec^2 x - 1$ to write

$$\int \tan^4 x \sec^3 x \, dx = \int \left(\sec^2 x - 1\right)^2 \sec^3 x \, dx = \int \sec^7 x \, dx - 2 \int \sec^5 x \, dx + \int \sec^3 x \, dx.$$

Now use the reduction formula for $\sec^m x$:

$$\int \tan^4 x \sec^3 x \, dx = \left(\frac{1}{6} \tan x \sec^5 x + \frac{5}{6} \int \sec^5 x \, dx\right) - 2 \int \sec^5 x \, dx + \int \sec^3 x \, dx$$

$$= \frac{1}{6} \tan x \sec^5 x - \frac{7}{6} \int \sec^5 x \, dx + \int \sec^3 x \, dx$$

$$= \frac{1}{6} \tan x \sec^5 x - \frac{7}{6} \left(\frac{1}{4} \tan x \sec^3 x + \frac{3}{4} \int \sec^3 x \, dx\right) + \int \sec^3 x \, dx$$

$$= \frac{1}{6} \tan x \sec^5 x - \frac{7}{24} \tan x \sec^3 x + \frac{1}{8} \int \sec^3 x \, dx$$

$$= \frac{1}{6} \tan x \sec^5 x - \frac{7}{24} \tan x \sec^3 x + \frac{1}{8} \left(\frac{1}{2} \tan x \sec x + \frac{1}{2} \int \sec x \, dx\right)$$

$$= \frac{1}{6} \tan x \sec^5 x - \frac{7}{24} \tan x \sec^3 x + \frac{1}{16} \tan x \sec x + \frac{1}{16} \ln|\sec x + \tan x| + C.$$

39. $\displaystyle\int \sin 2x \cos 2x \, dx$

SOLUTION Use the substitution $u = \sin 2x$, $du = 2\cos 2x \, dx$:

$$\int \sin 2x \cos 2x \, dx = \frac{1}{2} \int \sin 2x (2\cos 2x \, dx) = \frac{1}{2} \int u \, du = \frac{1}{4} u^2 + C = \frac{1}{4} \sin^2 2x + C.$$

41. $\displaystyle\int \sin 2x \cos 4x \, dx$

SOLUTION Use the formula for $\int \sin mx \cos nx \, dx$:

$$\int \sin 2x \cos 4x \, dx = -\frac{\cos(2-4)x}{2(2-4)} - \frac{\cos(2+4)x}{2(2+4)} + C = -\frac{\cos(-2x)}{-4} - \frac{\cos 6x}{12} + C = \frac{1}{4} \cos 2x - \frac{1}{12} \cos 6x + C.$$

Here we've used the fact that $\cos x$ is an even function: $\cos(-x) = \cos x$.

43. $\displaystyle\int \frac{\tan^3(\ln t)}{t} \, dt$

SOLUTION Use the substitution $u = \ln t$, $du = \frac{1}{t} dt$, followed by the reduction formula for $\tan^n x$:

$$\int \frac{\tan^3(\ln t)}{t} \, dt = \int \tan^3 u \, du = \frac{1}{2} \tan^2 u - \int \tan u \, du$$

$$= \frac{1}{2} \tan^2 u - \ln|\sec u| + C = \frac{1}{2} \tan^2(\ln t) - \ln|\sec(\ln t)| + C.$$

45. $\displaystyle\int_0^{2\pi} \sin^2 x \, dx$

SOLUTION Use the formula for $\int \sin^2 x \, dx$:

$$\int_0^{2\pi} \sin^2 x \, dx = \left(\frac{x}{2} - \frac{\sin 2x}{4}\right)\Big|_0^{2\pi} = \left(\frac{2\pi}{2} - \frac{\sin 4\pi}{4}\right) - \left(\frac{0}{2} - \frac{\sin 0}{4}\right) = \pi.$$

47. $\displaystyle\int_0^{\pi/3} \sin^3 x \, dx$

SOLUTION Use the reduction formula for $\sin^m x$:

$$\int_0^{\pi/3} \sin^3 x \, dx = -\frac{1}{3} \sin^2 x \cos x \Big|_0^{\pi/3} + \frac{2}{3} \int_0^{\pi/3} \sin x \, dx$$

$$= \left[-\frac{1}{3} \left(\frac{\sqrt{3}}{2} \right)^2 \left(\frac{1}{2} \right) - 0 \right] - \frac{2}{3} \cos x \Big|_0^{\pi/3} = -\frac{1}{8} - \frac{2}{3} \left(\frac{1}{2} - 1 \right) = \frac{5}{24}.$$

49. $\displaystyle\int_{\pi/4}^{\pi/2} \frac{dx}{\sin x}$

SOLUTION Use the definition of $\csc x$ to simplify the integral:

$$\int_{\pi/4}^{\pi/2} \frac{dx}{\sin x} = \int_{\pi/4}^{\pi/2} \csc x \, dx = \ln |\csc x - \cot x| \Big|_{\pi/4}^{\pi/2} = \ln |1 - 0| - \ln \left| \sqrt{2} - 1 \right| = -\ln \left| \sqrt{2} - 1 \right|$$

$$= \ln \left(\frac{1}{\sqrt{2} - 1} \right) = \ln \left(\frac{(\sqrt{2} + 1)}{(\sqrt{2} - 1)(\sqrt{2} + 1)} \right) = \ln(\sqrt{2} + 1).$$

51. $\displaystyle\int_0^{\pi/4} \tan^5 x \, dx$

SOLUTION First use the reduction formula for $\tan^m x$ to evaluate the indefinite integral:

$$\int \tan^5 x \, dx = \frac{1}{4} \tan^4 x - \int \tan^3 x \, dx = \frac{1}{4} \tan^4 x - \left(\frac{1}{2} \tan^2 x - \int \tan x \, dx \right)$$

$$= \frac{1}{4} \tan^4 x - \frac{1}{2} \tan^2 x + \ln |\sec x| + C.$$

Now compute the definite integral:

$$\int_0^{\pi/4} \tan^5 x \, dx = \left(\frac{1}{4} \tan^4 x - \frac{1}{2} \tan^2 x + \ln |\sec x| \right) \Big|_0^{\pi/4}$$

$$= \left(\frac{1}{4} \left(1^4 \right) - \frac{1}{2} \left(1^2 \right) + \ln \sqrt{2} \right) - (0 - 0 + \ln 1)$$

$$= \frac{1}{4} - \frac{1}{2} + \ln \sqrt{2} - 0 = \frac{1}{2} \ln 2 - \frac{1}{4}.$$

53. $\displaystyle\int_0^{\pi} \sin 3x \cos 4x \, dx$

SOLUTION Use the formula for $\int \sin mx \cos nx \, dx$:

$$\int_0^{\pi} \sin 3x \cos 4x \, dx = \left(-\frac{\cos(3 - 4)x}{2(3 - 4)} - \frac{\cos(3 + 4)x}{2(3 + 4)} \right) \Big|_0^{\pi} = \left(-\frac{\cos(-x)}{-2} - \frac{\cos 7x}{14} \right) \Big|_0^{\pi}$$

$$= \left(\frac{1}{2} \cos x - \frac{1}{14} \cos 7x \right) \Big|_0^{\pi} = \left[\frac{1}{2}(-1) - \frac{1}{14}(-1) \right] - \left[\frac{1}{2}(1) - \frac{1}{14}(1) \right] = -\frac{6}{7}.$$

55. Use the identities for $\sin 2x$ and $\cos 2x$ listed on page 434 to verify that the first of the following formulas is equivalent to the second:

$$\int \sin^4 x \, dx = \frac{1}{32} (12x - 8 \sin 2x + \sin 4x) + C$$

$$\int \sin^4 x \, dx = -\frac{1}{4} \sin^3 x \cos x - \frac{3}{8} \sin x \cos x + \frac{3}{8} x + C$$

SOLUTION First, observe

$$\sin 4x = 2 \sin 2x \cos 2x = 2 \sin 2x (1 - 2 \sin^2 x)$$

$$= 2 \sin 2x - 4 \sin 2x \sin^2 x = 2 \sin 2x - 8 \sin^3 x \cos x.$$

Then

$$\frac{1}{32}(12x - 8\sin 2x + \sin 4x) + C = \frac{3}{8}x - \frac{3}{16}\sin 2x - \frac{1}{4}\sin^3 x \cos x + C$$

$$= \frac{3}{8}x - \frac{3}{8}\sin x \cos x - \frac{1}{4}\sin^3 x \cos x + C.$$

In Exercises 57–60, evaluate using the identity $\cot^2 x + 1 = \csc^2 x$ *and methods similar to those for integrating* $\tan^m x \sec^n x$.

57. $\int \cot^3 x \csc x \, dx$

SOLUTION Use the identity $\cot^2 x = \csc^2 x - 1$ to write

$$\int \cot^3 x \csc x \, dx = \int \left(\csc^2 x - 1\right)\csc x \cot x \, dx.$$

Now use the substitution $u = \csc x$, $du = -\csc x \cot x \, dx$:

$$\int \cot^3 x \csc x \, dx = -\int \left(u^2 - 1\right) du = \int \left(1 - u^2\right) du = u - \frac{1}{3}u^3 + C = \csc x - \frac{1}{3}\csc^3 x + C.$$

59. $\int \cot^2 x \csc^2 x \, dx$

SOLUTION Use the substitution $u = \cot x$, $du = -\csc^2 x \, dx$:

$$\int \cot^2 x \csc^2 x \, dx = -\int \cot^2 x \left(-\csc^2 x \, dx\right) = -\int u^2 \, du = -\frac{1}{3}u^3 + C = -\frac{1}{3}\cot^3 x + C.$$

61. Find the volume of the solid obtained by revolving $y = \sin x$ for $0 \le x \le \pi$ about the x-axis.

SOLUTION Using the disk method, the volume is given by

$$V = \int_0^\pi \pi(\sin x)^2 \, dx = \pi \int_0^\pi \sin^2 x \, dx = \pi\left(\frac{x}{2} - \frac{\sin 2x}{4}\right)\Big|_0^\pi = \pi\left[\left(\frac{\pi}{2} - 0\right) - (0)\right] = \frac{\pi^2}{2}.$$

63. Here is another reduction method for evaluating the integral $J = \int \sin^m x \cos^n x \, dx$ when m and n are even. Use the identities

$$\sin^2 x = \frac{1}{2}(1 - \cos 2x), \qquad \cos^2 x = \frac{1}{2}(1 + \cos 2x)$$

to write $J = \frac{1}{4}\int (1 - \cos 2x)^{m/2}(1 + \cos 2x)^{n/2} \, dx$. Then expand the right-hand side as a sum of integrals involving smaller powers of sine and cosine in the variable $2x$. Use this method to evaluate $J = \int \sin^2 x \cos^2 x \, dx$.

SOLUTION Using the identities $\sin^2 x = \frac{1}{2}(1 - \cos 2x)$ and $\cos^2 x = \frac{1}{2}(1 + \cos 2x)$, we have

$$J = \int \sin^2 x \cos^2 x \, dx = \frac{1}{4}\int (1 - \cos 2x)(1 + \cos 2x) \, dx$$

$$= \frac{1}{4}\int \left(1 - \cos^2 2x\right) dx = \frac{1}{4}\int dx - \frac{1}{4}\int \cos^2 2x \, dx.$$

Now use the substitution $u = 2x$, $du = 2 \, dx$, and the formula for $\int \cos^2 u \, du$:

$$J = \frac{1}{4}x - \frac{1}{8}\int \cos^2 u \, du = \frac{1}{4}x - \frac{1}{8}\left(\frac{u}{2} + \frac{1}{2}\sin u \cos u\right) + C$$

$$= \frac{1}{4}x - \frac{1}{16}(2x) - \frac{1}{16}\sin 2x \cos 2x + C = \frac{1}{8}x - \frac{1}{16}\sin 2x \cos 2x + C.$$

65. Use the method of Exercise 63 to evaluate $\int \sin^4 x \cos^2 x \, dx$.

SOLUTION Using the identities $\sin^2 x = \frac{1}{2}(1 - \cos 2x)$ and $\cos^2 x = \frac{1}{2}(1 + \cos 2x)$, we have

$$J = \int \sin^4 x \cos^2 x \, dx = \frac{1}{8} \int (1 - \cos 2x)^2 (1 + \cos 2x) \, dx$$

$$= \frac{1}{8} \int \left(1 - 2\cos 2x + \cos^2 2x\right)(1 + \cos 2x) \, dx$$

$$= \frac{1}{8} \int \left(1 - \cos 2x - \cos^2 2x + \cos^3 2x\right) dx.$$

Now use the substitution $u = 2x$, $du = 2\,dx$, together with the reduction formula for $\cos^m x$:

$$J = \frac{1}{8}x - \frac{1}{16} \int \cos u \, du - \frac{1}{16} \int \cos^2 u \, du + \frac{1}{16} \int \cos^3 u \, du$$

$$= \frac{1}{8}x - \frac{1}{16}\sin u - \frac{1}{16}\left(\frac{u}{2} + \frac{1}{2}\sin u \cos u\right) + \frac{1}{16}\left(\frac{1}{3}\cos^2 u \sin u + \frac{2}{3}\int \cos u \, du\right)$$

$$= \frac{1}{8}x - \frac{1}{16}\sin 2x - \frac{1}{32}(2x) - \frac{1}{32}\sin 2x \cos 2x + \frac{1}{48}\cos^2 2x \sin 2x + \frac{1}{24}\sin 2x + C$$

$$= \frac{1}{16}x - \frac{1}{48}\sin 2x - \frac{1}{32}\sin 2x \cos 2x + \frac{1}{48}\cos^2 2x \sin 2x + C.$$

67. Prove the reduction formula

$$\int \tan^k x \, dx = \frac{\tan^{k-1} x}{k - 1} - \int \tan^{k-2} x \, dx$$

Hint: Use the identity $\tan^2 x = (\sec^2 x - 1)$ to write

$$\tan^k x = (\sec^2 x - 1) \tan^{k-2} x.$$

SOLUTION Use the identity $\tan^2 x = \sec^2 x - 1$ to write

$$\int \tan^k x \, dx = \int \tan^{k-2} x \left(\sec^2 x - 1\right) dx = \int \tan^{k-2} x \sec^2 x \, dx - \int \tan^{k-2} x \, dx.$$

Now use the substitution $u = \tan x$, $du = \sec^2 x \, dx$:

$$\int \tan^k x \, dx = \int u^{k-2} \, du - \int \tan^{k-2} x \, dx = \frac{1}{k-1}u^{k-1} - \int \tan^{k-2} x \, dx = \frac{\tan^{k-1} x}{k-1} - \int \tan^{k-2} x \, dx.$$

69. Use the substitution $u = \csc x - \cot x$ to evaluate $\int \csc x \, dx$ (see Example 5).

SOLUTION Using the substitution $u = \csc x - \cot x$,

$$du = \left(-\csc x \cot x + \csc^2 x\right)dx = \csc x(\csc x - \cot x)\,dx,$$

we have

$$\int \csc x \, dx = \int \frac{\csc x(\csc x - \cot x)\,dx}{\csc x - \cot x} = \int \frac{du}{u} = \ln|u| + C = \ln|\csc x - \cot x| + C.$$

71. Let m, n be integers with $m \neq \pm n$. Use formulas (25)–(27) in the table of trigonometric integrals to prove that

$$\int_0^\pi \sin mx \sin nx \, dx = 0, \qquad \int_0^\pi \cos mx \cos nx \, dx = 0$$

$$\int_0^{2\pi} \sin mx \cos nx \, dx = 0$$

These formulas, known as the **orthogonality relations**, play a basic role in the theory of Fourier Series (Figure 2).

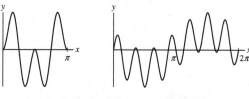

$y = \sin 2x \sin 4x$ $y = \sin 3x \cos 4x$

FIGURE 2 By the orthogonality relations, the signed area under these graphs is zero.

SOLUTION If m, n are integers, then $m - n$ and $m + n$ are integers, and therefore $\sin(m - n)\pi = \sin(m + n)\pi = 0$, since $\sin k\pi = 0$ if k is an integer. Thus we have

$$\int_0^\pi \sin mx \sin nx \, dx = \left(\frac{\sin(m - n)x}{2(m - n)} - \frac{\sin(m + n)x}{2(m + n)} \right)\Big|_0^\pi = \left(\frac{\sin(m - n)\pi}{2(m - n)} - \frac{\sin(m + n)\pi}{2(m + n)} \right) - 0 = 0;$$

$$\int_0^\pi \cos mx \cos nx \, dx = \left(\frac{\sin(m - n)x}{2(m - n)} + \frac{\sin(m + n)x}{2(m + n)} \right)\Big|_0^\pi = \left(\frac{\sin(m - n)\pi}{2(m - n)} + \frac{\sin(m + n)\pi}{2(m + n)} \right) - 0 = 0.$$

If k is an integer, then $\cos 2k\pi = 1$. Using this fact, we have

$$\int_0^{2\pi} \sin mx \cos nx \, dx = \left(-\frac{\cos(m - n)x}{2(m - n)} - \frac{\cos(m + n)x}{2(m + n)} \right)\Big|_0^{2\pi}$$

$$= \left(-\frac{\cos(m - n)2\pi}{2(m - n)} - \frac{\cos(m + n)2\pi}{2(m + n)} \right) - \left(-\frac{1}{2(m - n)} - \frac{1}{2(m + n)} \right)$$

$$= \left(-\frac{1}{2(m - n)} - \frac{1}{2(m + n)} \right) - \left(-\frac{1}{2(m - n)} - \frac{1}{2(m + n)} \right) = 0.$$

Further Insights and Challenges

73. Evaluate $\int_0^\pi \sin^2 mx \, dx$ for m an arbitrary integer.

SOLUTION Use the substitution $u = mx$, $du = m \, dx$. Then

$$\int_0^\pi \sin^2 mx \, dx = \frac{1}{m} \int_{x=0}^{x=\pi} \sin^2 u \, du = \frac{1}{m} \left(\frac{u}{2} - \frac{\sin 2u}{4} \right)\Big|_{x=0}^{x=\pi} = \frac{1}{m} \left(\frac{mx}{2} - \frac{\sin 2mx}{4} \right)\Big|_0^\pi$$

$$= \left(\frac{x}{2} - \frac{\sin 2mx}{4m} \right)\Big|_0^\pi = \left(\frac{\pi}{2} - \frac{\sin 2\pi m}{4} \right) - (0).$$

If m is an arbitrary integer, then $\sin 2m\pi = 0$. Thus

$$\int_0^\pi \sin^2 mx \, dx = \frac{\pi}{2}.$$

75. Set $I_m = \int_0^{\pi/2} \sin^m x \, dx$.

(a) Show that $I_1 = 1$, $I_2 = \left(\frac{1}{2} \right)\left(\frac{\pi}{2} \right)$ and use Eq. (28) to prove that for $m > 1$, $I_m = \left(\frac{m - 1}{m} \right) I_{m-2}$.

(b) Show that $I_3 = \frac{2}{3}$ and $I_4 = \left(\frac{3}{4} \right)\left(\frac{1}{2} \right)\left(\frac{\pi}{2} \right)$.

(c) Show more generally:

$$I_{2m} = \frac{2m - 1}{2m} \frac{2m - 3}{2m - 2} \cdots \frac{1}{2} \cdot \frac{\pi}{2}$$

$$I_{2m+1} = \frac{2m}{2m + 1} \frac{2m - 2}{2m - 1} \cdots \frac{2}{3}$$

(d) Conclude that

$$\frac{\pi}{2} = \frac{2 \cdot 2}{1 \cdot 3} \cdot \frac{4 \cdot 4}{3 \cdot 5} \cdots \frac{2m \cdot 2m}{(2m - 1)(2m + 1)} \frac{I_{2m}}{I_{2m+1}}$$

SOLUTION

(a)

$$I_1 = \int_0^{\pi/2} \sin x \, dx = -\cos x \Big|_0^{\pi/2} = 0 - (-1) = 1;$$

$$I_2 = \int_0^{\pi/2} \sin^2 x \, dx = \left(\frac{x}{2} - \frac{1}{2} \sin x \cos x \right)\Big|_0^{\pi/2} = \left(\frac{\pi}{4} - \frac{1}{2}(1)(0) \right) - (0 - 0) = \frac{\pi}{4} = \left(\frac{1}{2} \right)\left(\frac{\pi}{2} \right).$$

Using Eq. (28), we have

$$I_m = \int_0^{\pi/2} \sin^m x \, dx = \frac{m - 1}{m} \int_0^{\pi/2} \sin^{m-2} x \, dx = \left(\frac{m - 1}{m} \right) I_{m-2}.$$

(b) Using the result from (a), we get

$$I_3 = \frac{3-1}{3} I_1 = \frac{2}{3}(1) = \frac{2}{3};$$

$$I_4 = \frac{4-1}{4} I_2 = \frac{3}{4}\left(\frac{1}{2}\right)\left(\frac{\pi}{2}\right).$$

(c) We'll use induction to show these results. For I_{2m}, the result is true for $m = 1$ and $m = 2$. Now assume the result is true for $m = k - 1$:

$$I_{2(k-1)} = I_{2k-2} = \frac{2k-3}{2k-2} \cdot \frac{2k-5}{2k-4} \cdots \frac{1}{2} \cdot \frac{\pi}{2}$$

Using the relation $I_m = ((m-1)/m)I_{m-2}$, we have

$$I_{2k} = \frac{2k-1}{2k} I_{2k-2} = \frac{2k-1}{2k} \cdot \left(\frac{2k-3}{2k-2} \cdot \frac{2k-5}{2k-4} \cdots \frac{1}{2} \cdot \frac{\pi}{2}\right).$$

For I_{2m+1}, the result is true for $m = 1$. Now assume the result is true for $m = k - 1$:

$$I_{2(k-1)+1} = I_{2k-1} = \frac{2k-2}{2k-1} \cdot \frac{2k-4}{2k-3} \cdots \frac{2}{3}$$

Again using the relation $I_m = ((m-1)/m)I_{m-2}$, we have

$$I_{2k+1} = \left(\frac{2k+1-1}{2k+1}\right) I_{2k-1} = \frac{2k}{2k+1}\left(\frac{2k-2}{2k-1} \cdot \frac{2k-4}{2k-3} \cdots \frac{2}{3}\right).$$

(d) First divide the two results from part (c) to obtain:

$$\frac{I_{2m}}{I_{2m+1}} = \frac{(2m-1)(2m+1)}{2m \cdot 2m} \cdot \frac{(2m-3)(2m-1)}{(2m-2)(2m-2)} \cdots \frac{1 \cdot 3}{2 \cdot 2} \cdot \frac{\pi}{2}.$$

Solving for $\pi/2$, we get the desired result:

$$\frac{\pi}{2} = \frac{2 \cdot 2}{1 \cdot 3} \cdot \frac{4 \cdot 4}{3 \cdot 5} \cdots \frac{2m \cdot 2m}{(2m-1)(2m+1)} \cdot \frac{I_{2m}}{I_{2m+1}}.$$

8.4 Trigonometric Substitution

Preliminary Questions

1. Explain why trigonometric substitution is *not needed* to evaluate $\int x\sqrt{9-x^2}\,dx$.

SOLUTION Because there is a factor of x in the integrand outside the radical and the derivative of $9 - x^2$ is $-2x$, we may use the substitution $u = 9 - x^2$, $du = -2x\,dx$ to evaluate this integral.

2. State the trigonometric substitution appropriate to the given integral:

(a) $\int \sqrt{9-x^2}\,dx$

(b) $\int x^2(x^2 - 16)^{3/2}\,dx$

(c) $\int x^2(x^2 + 16)^{3/2}\,dx$

(d) $\int (x^2 - 5)^{-2}\,dx$

SOLUTION

(a) $x = 3\sin\theta$

(b) $x = 4\sec\theta$

(c) $x = 4\tan\theta$

(d) $x = \sqrt{5}\sec\theta$

3. Which of the triangles in Figure 6 would be used together with the substitution $x = 3\sin\theta$?

SOLUTION The substitution $x = 3\sin\theta$ implies $\sin\theta = \frac{x}{3}$. We therefore need a triangle whose opposite side has length x and whose hypotenuse has length 3. This describes the triangle in Figure 6(A).

4. Express $\tan \theta$ in terms of x for the angle in Figure 6(A).

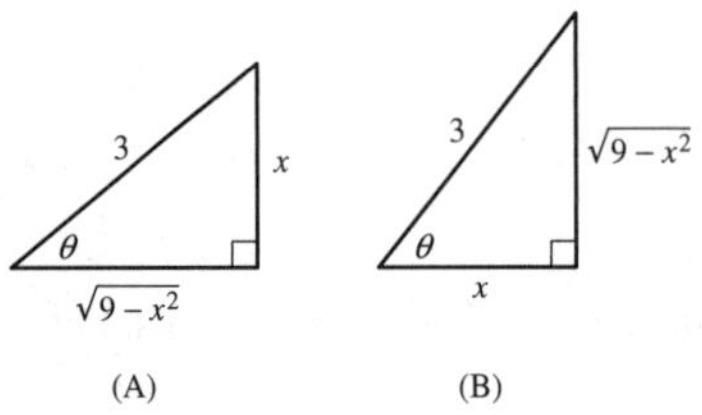

(A)   (B)

FIGURE 6

SOLUTION Tangent is the ratio of the length of the opposite side to the length of the adjacent side. For the triangle in Figure 6(A), it then follows that

$$\tan \theta = \frac{x}{\sqrt{9 - x^2}}.$$

5. Express $\sec \theta$ in terms of x for the angle in Figure 6(B).

SOLUTION Secant is the ratio of the length of the hypotenuse to the length of the adjacent side. For the triangle in Figure 6(B), it then follows that

$$\sec \theta = \frac{3}{x}.$$

6. Express $\sin 2\theta$ in terms of x, where $x = \sin \theta$.

SOLUTION First note that if $\sin \theta = x$, then $\cos \theta = \sqrt{1 - \sin^2 \theta} = \sqrt{1 - x^2}$. Thus,

$$\sin 2\theta = 2 \sin \theta \cos \theta = 2x\sqrt{1 - x^2}.$$

Exercises

In Exercises 1–4, evaluate the integral by following the steps.

1. $I = \displaystyle\int \frac{dx}{\sqrt{9 - x^2}}$

(a) Show that the substitution $x = 3 \sin \theta$ transforms I into $\displaystyle\int d\theta$ and evaluate I in terms of θ.

(b) Evaluate I in terms of x.

SOLUTION

(a) Let $x = 3 \sin \theta$. Then $dx = 3 \cos \theta \, d\theta$, and

$$\sqrt{9 - x^2} = \sqrt{9 - 9 \sin^2 \theta} = 3\sqrt{1 - \sin^2 \theta} = 3\sqrt{\cos^2 \theta} = 3 \cos \theta.$$

Thus,

$$I = \int \frac{dx}{\sqrt{9 - x^2}} = \int \frac{3 \cos \theta \, d\theta}{3 \cos \theta} = \int d\theta = \theta + C.$$

(b) If $x = 3 \sin \theta$, then $\theta = \sin^{-1}(\frac{x}{3})$. Thus,

$$I = \theta + C = \sin^{-1}\left(\frac{x}{3}\right) + C.$$

3. $I = \displaystyle\int \frac{dx}{\sqrt{x^2 + 9}}$

(a) Show that the substitution $x = 3 \tan \theta$ transforms I into $\displaystyle\int \sec \theta \, d\theta$ and evaluate I in terms of θ (refer to the table of integrals in Section 8.3 if necessary).

(b) Show that if $x = 3 \tan \theta$, then $\sec \theta = \frac{1}{3}\sqrt{x^2 + 9}$.

(c) Express I in terms of x.

SOLUTION

(a) If $x = 3 \tan \theta$, then $dx = 3 \sec^2 \theta \, d\theta$, and

$$\sqrt{x^2 + 9} = \sqrt{9 \tan^2 \theta + 9} = \sqrt{9(\tan^2 \theta + 1)} = \sqrt{9 \sec^2 \theta} = 3 \sec \theta.$$

Thus,

$$I = \int \frac{dx}{\sqrt{x^2 + 9}} = \int \frac{3 \sec^2 \theta \, d\theta}{3 \sec \theta} = \int \sec \theta \, d\theta = \ln |\sec \theta + \tan \theta| + C.$$

(b) Since $x = 3 \tan \theta$, we construct a right triangle with $\tan \theta = \frac{x}{3}$:

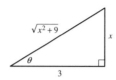

From this triangle we see that $\sec \theta = \frac{1}{3}\sqrt{x^2 + 9}$.

(c) Combining the results from parts (a) and (b),

$$I = \ln \left| \frac{1}{3}\sqrt{x^2 + 9} + \frac{x}{3} \right| + C = \ln \left| \frac{\sqrt{x^2 + 9} + x}{3} \right| + C$$

$$= \ln |\sqrt{x^2 + 9} + x| - \ln 3 + C = \ln |\sqrt{x^2 + 9} + x| + C.$$

In Exercises 5–10, use the indicated substitution to evaluate the integral.

5. $\displaystyle \int \sqrt{4 - x^2} \, dx, \quad x = 2 \sin \theta$

SOLUTION   Let $x = 2 \sin \theta$. Then $dx = 2 \cos \theta \, d\theta$, and

$$\int \sqrt{4 - x^2} \, dx = \int \sqrt{4 - 4 \sin^2 \theta}(2 \cos \theta \, d\theta) = 4 \int \sqrt{1 - \sin^2 \theta} \cos \theta \, d\theta$$

$$= 4 \int \sqrt{\cos^2 \theta} \cos \theta \, d\theta = 4 \int \cos^2 \theta \, d\theta$$

$$= 4 \left[\frac{1}{2}\theta + \frac{1}{2} \sin \theta \cos \theta \right] + C = 2\theta + 2 \sin \theta \cos \theta + C.$$

Since $x = 2 \sin \theta$, we construct a right triangle with $\sin \theta = \frac{x}{2}$:

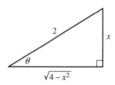

From this triangle we see that $\cos \theta = \frac{1}{2}\sqrt{4 - x^2}$, so we have

$$I = 2\theta + 2 \sin \theta \cos \theta + C = 2 \sin^{-1} \left(\frac{x}{2} \right) + 2 \left(\frac{x}{2} \right) \left(\frac{\sqrt{4 - x^2}}{2} \right) + C$$

$$= 2 \sin^{-1} \left(\frac{x}{2} \right) + \frac{1}{2}x\sqrt{4 - x^2} + C.$$

7. $\displaystyle \int \frac{dx}{x\sqrt{x^2 - 9}}, \quad x = 3 \sec \theta$

SOLUTION   Let $x = 3 \sec \theta$. Then $dx = 3 \sec \theta \tan \theta \, d\theta$, and

$$\sqrt{x^2 - 9} = \sqrt{9 \sec^2 \theta - 9} = 3\sqrt{\sec^2 \theta - 1} = 3\sqrt{\tan^2 \theta} = 3 \tan \theta.$$

Thus,

$$\int \frac{dx}{x\sqrt{x^2 - 9}} = \int \frac{(3 \sec \theta \tan \theta \, d\theta)}{(3 \sec \theta)(3 \tan \theta)} = \frac{1}{3} \int d\theta = \frac{1}{3}\theta + C.$$

Since $x = 3 \sec \theta$, $\theta = \sec^{-1}\left(\frac{x}{3}\right)$, and

$$\int \frac{dx}{x\sqrt{x^2 - 9}} = \frac{1}{3} \sec^{-1}\left(\frac{x}{3}\right) + C.$$

9. $\displaystyle\int \frac{dx}{(x^2 - 4)^{3/2}}, \quad x = 2 \sec \theta$

SOLUTION Let $x = 2 \sec \theta$. Then $dx = 2 \sec \theta \tan \theta \, d\theta$, and

$$x^2 - 4 = 4 \sec^2 \theta - 4 = 4(\sec^2 \theta - 1) = 4 \tan^2 \theta.$$

This gives

$$I = \int \frac{dx}{(x^2 - 4)^{3/2}} = \int \frac{2 \sec \theta \tan \theta \, d\theta}{(4 \tan^2 \theta)^{3/2}} = \int \frac{2 \sec \theta \tan \theta \, d\theta}{8 \tan^3 \theta} = \frac{1}{4} \int \frac{\sec \theta \, d\theta}{\tan^2 \theta} = \frac{1}{4} \int \frac{\cos \theta}{\sin^2 \theta} \, d\theta.$$

Now use substitution with $u = \sin \theta$ and $du = \cos \theta \, d\theta$. Then

$$I = \frac{1}{4} \int u^{-2} \, du = -\frac{1}{4} u^{-1} + C = \frac{-1}{4 \sin \theta} + C.$$

Since $x = 2 \sec \theta$, we construct a right triangle with $\sec \theta = \frac{x}{2}$:

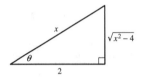

From this triangle we see that $\sin \theta = \sqrt{x^2 - 4}/x$, so therefore

$$I = \frac{-1}{4(\sqrt{x^2 - 4}/x)} + C = \frac{-x}{4\sqrt{x^2 - 4}} + C.$$

11. Is the substitution $u = x^2 - 4$ effective for evaluating the integral $\displaystyle\int \frac{x^2 \, dx}{\sqrt{x^2 - 4}}$? If not, evaluate using trigonometric substitution.

SOLUTION If $u = x^2 - 4$, then $du = 2x \, dx$, $x^2 = u + 4$, $dx = du/2x = du/2\sqrt{u + 4}$, and

$$I = \int \frac{x^2 \, dx}{\sqrt{x^2 - 4}} = \int \frac{(u + 4)}{\sqrt{u}} \left(\frac{du}{2\sqrt{u + 4}}\right) = \frac{1}{2} \int \frac{u + 4}{\sqrt{u^2 + 4u}} \, du$$

This substitution is clearly not effective for evaluating this integral.

Instead, use the trigonometric substitution $x = 2 \sec \theta$. Then $dx = 2 \sec \theta \tan \theta$,

$$\sqrt{x^2 - 4} = \sqrt{4 \sec^2 \theta - 4} = 2 \tan \theta,$$

and we have

$$I = \int \frac{x^2 \, dx}{\sqrt{x^2 - 4}} = \int \frac{4 \sec^2 \theta (2 \sec \theta \tan \theta \, d\theta)}{2 \tan \theta} = 4 \int \sec^3 \theta \, d\theta.$$

Now use the reduction formula for $\int \sec^m x \, dx$ from Section 8.8.3:

$$4 \int \sec^3 \theta \, d\theta = 4 \left[\frac{\tan \theta \sec \theta}{2} + \frac{1}{2} \int \sec \theta \, d\theta\right] = 2 \tan \theta \sec \theta + 2\left[\ln|\sec \theta + \tan \theta|\right] + C.$$

Since $x = 2 \sec \theta$, we construct a right triangle with $\sec \theta = \frac{x}{2}$:

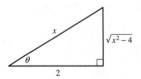

From this triangle we see that $\tan\theta = \frac{1}{2}\sqrt{x^2 - 4}$. Therefore

$$I = 2\left(\frac{1}{2}\sqrt{x^2 - 4}\right)\left(\frac{x}{2}\right) + 2\ln\left|\frac{x}{2} + \frac{1}{2}\sqrt{x^2 - 4}\right| + C = \frac{1}{2}x\sqrt{x^2 - 4} + 2\ln\left|\frac{1}{2}\left(x + \sqrt{x^2 - 4}\right)\right| + C.$$

Finally, since

$$\ln\left|\frac{1}{2}(x + \sqrt{x^2 - 4})\right| = \ln\left(\frac{1}{2}\right) + \ln|x + \sqrt{x^2 - 4}|,$$

and $\ln(\frac{1}{2})$ is a constant, we can "absorb" this constant into the constant of integration, so that

$$I = \frac{1}{2}x\sqrt{x^2 - 4} + 2\ln|x + \sqrt{x^2 - 4}| + C.$$

In Exercises 13–30, evaluate the integral using trigonometric substitution. Refer to the table of trigonometric integrals as necessary.

13. $\displaystyle\int \frac{x^2\,dx}{\sqrt{9 - x^2}}$

SOLUTION Let $x = 3\sin\theta$. Then $dx = 3\cos\theta\,d\theta$,

$$9 - x^2 = 9 - 9\sin^2\theta = 9(1 - \sin^2\theta) = 9\cos^2\theta,$$

and

$$I = \int \frac{x^2\,dx}{\sqrt{9 - x^2}} = \int \frac{9\sin^2\theta(3\cos\theta\,d\theta)}{3\cos\theta} = 9\int \sin^2\theta\,d\theta = 9\left[\frac{1}{2}\theta - \frac{1}{2}\sin\theta\cos\theta\right] + C.$$

Since $x = 3\sin\theta$, we construct a right triangle with $\sin\theta = \frac{x}{3}$:

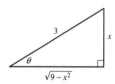

From this we see that $\cos\theta = \sqrt{9 - x^2}/3$, and so

$$I = \frac{9}{2}\sin^{-1}\left(\frac{x}{3}\right) - \frac{9}{2}\left(\frac{x}{3}\right)\left(\frac{\sqrt{9 - x^2}}{3}\right) + C = \frac{9}{2}\sin^{-1}\left(\frac{x}{3}\right) - \frac{1}{2}x\sqrt{9 - x^2} + C.$$

15. $\displaystyle\int \sqrt{12 + 4x^2}\,dx$

SOLUTION First simplify the integral:

$$I = \int \sqrt{12 + 4x^2}\,dx = 2\int \sqrt{3 + x^2}\,dx$$

Now let $x = \sqrt{3}\tan\theta$. Then $dx = \sqrt{3}\sec^2\theta\,d\theta$,

$$3 + x^2 = 3 + 3\tan^2\theta = 3(1 + \tan^2\theta) = 3\sec^2\theta,$$

and

$$I = 2\int \sqrt{3\sec^2\theta}\left(\sqrt{3}\sec^2\theta\,d\theta\right) = 6\int \sec^3\theta\,d\theta = 6\left[\frac{\tan\theta\sec\theta}{2} + \frac{1}{2}\int \sec\theta\,d\theta\right]$$

$$= 3\tan\theta\sec\theta + 3\ln|\sec\theta + \tan\theta| + C.$$

Since $x = \sqrt{3}\tan\theta$, we construct a right triangle with $\tan\theta = \frac{x}{\sqrt{3}}$:

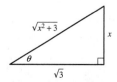

From this we see that $\sec\theta = \sqrt{x^2+3}/\sqrt{3}$. Therefore,

$$I = 3\left(\frac{x}{\sqrt{3}}\right)\left(\frac{\sqrt{x^2+3}}{\sqrt{3}}\right) + 3\ln\left|\frac{\sqrt{x^2+3}}{\sqrt{3}} + \frac{x}{\sqrt{3}}\right| + C_1 = x\sqrt{x^2+3} + 3\ln\left|\sqrt{x^2+3}+x\right| + 3\ln\left(\frac{1}{\sqrt{3}}\right) + C_1$$

$$= x\sqrt{x^2+3} + 3\ln\left|\sqrt{x^2+3}+x\right| + C,$$

where $C = 3\ln(\frac{1}{\sqrt{3}}) + C_1$.

17. $\displaystyle\int \frac{dt}{(4-t^2)^{3/2}}$

SOLUTION   Let $t = 2\sin\theta$. Then $dt = 2\cos\theta\,d\theta$,

$$4 - t^2 = 4 - 4\sin^2\theta = 4(1-\sin^2\theta) = 4\cos^2\theta,$$

and

$$I = \int \frac{dt}{(4-t^2)^{3/2}} = \int \frac{2\cos\theta\,d\theta}{(4\cos^2\theta)^{3/2}} = \int \frac{2\cos\theta\,d\theta}{8\cos^3\theta} = \frac{1}{4}\int \frac{d\theta}{\cos^2\theta} = \frac{1}{4}\int \sec^2\theta\,d\theta = \frac{1}{4}\tan\theta + C.$$

Since $t = 2\sin\theta$, we construct a right triangle with $\sin\theta = \frac{t}{2}$:

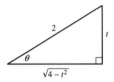

From this we see that $\tan\theta = t/\sqrt{4-t^2}$, which gives us

$$I = \frac{1}{4}\left(\frac{t}{\sqrt{4-t^2}}\right) + C = \frac{t}{4\sqrt{4-t^2}} + C.$$

19. $\displaystyle\int \frac{dy}{y^2\sqrt{5-y^2}}$

SOLUTION   Let $y = \sqrt{5}\sin\theta$. Then $dy = \sqrt{5}\cos\theta\,d\theta$,

$$5 - y^2 = 5 - 5\sin^2\theta = 5(1-\sin^2\theta) = 5\cos^2\theta,$$

and

$$I = \int \frac{dy}{y^2\sqrt{5-y^2}} = \int \frac{\sqrt{5}\cos\theta\,d\theta}{(5\sin^2\theta)(\sqrt{5}\cos\theta)} = \frac{1}{5}\int \frac{d\theta}{\sin^2\theta} = \frac{1}{5}\int \csc^2\theta\,d\theta = \frac{1}{5}(-\cot\theta) + C.$$

Since $y = \sqrt{5}\sin\theta$, we construct a right triangle with $\sin\theta = \frac{y}{\sqrt{5}}$:

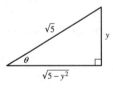

From this we see that $\cot\theta = \sqrt{5-y^2}/y$, which gives us

$$I = \frac{1}{5}\left(\frac{-\sqrt{5-y^2}}{y}\right) + C = -\frac{\sqrt{5-y^2}}{5y} + C.$$

21. $\displaystyle\int \frac{dz}{z^3\sqrt{z^2-4}}$

SOLUTION   Let $z = 2\sec\theta$. Then $dz = 2\sec\theta\tan\theta\,d\theta$,

$$z^2 - 4 = 4\sec^2\theta - 4 = 4(\sec^2\theta - 1) = 4\tan^2\theta,$$

and

$$I = \int \frac{dz}{z^3\sqrt{z^2 - 4}} = \int \frac{2\sec\theta\tan\theta\,d\theta}{(8\sec^3\theta)(2\tan\theta)} = \frac{1}{8}\int \frac{d\theta}{\sec^2\theta} = \frac{1}{8}\int \cos^2\theta\,d\theta$$

$$= \frac{1}{8}\left[\frac{1}{2}\theta + \frac{1}{2}\sin\theta\cos\theta\right] + C = \frac{1}{16}\theta + \frac{1}{16}\sin\theta\cos\theta + C.$$

Since $z = 2\sec\theta$, we construct a right triangle with $\sec\theta = \frac{z}{2}$:

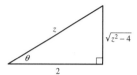

From this we see that $\sin\theta = \sqrt{z^2 - 4}/z$ and $\cos\theta = 2/z$. Then

$$I = \frac{1}{16}\sec^{-1}\left(\frac{z}{2}\right) + \frac{1}{16}\left(\frac{\sqrt{z^2 - 4}}{z}\right)\left(\frac{2}{z}\right) + C = \frac{1}{16}\sec^{-1}\left(\frac{z}{2}\right) + \frac{\sqrt{z^2 - 4}}{8z^2} + C.$$

23. $\int \dfrac{x^2\,dx}{\sqrt{x^2 + 1}}$

SOLUTION Let $x = \tan\theta$. Then $dx = \sec^2\theta\,d\theta$, $x^2 + 1 = \tan^2\theta + 1 = \sec^2\theta$, and

$$I = \int \frac{x^2\,dx}{\sqrt{x^2 + 1}} = \int \frac{\tan^2\theta(\sec^2\theta\,d\theta)}{\sec\theta} = \int \tan^2\theta\sec\theta\,d\theta.$$

Now use the identity $\tan^2\theta = \sec^2\theta - 1$ along with the reduction formula for $\int \sec^3\theta\,d\theta$:

$$I = \int \left(\sec^2\theta - 1\right)\sec\theta\,d\theta = \int \sec^3\theta\,d\theta - \int \sec\theta\,d\theta = \left(\frac{\tan\theta\sec\theta}{2} + \frac{1}{2}\int \sec\theta\,d\theta\right) - \int \sec\theta\,d\theta$$

$$= \frac{1}{2}\tan\theta\sec\theta - \frac{1}{2}\int \sec\theta\,d\theta = \frac{1}{2}\tan\theta\sec\theta - \frac{1}{2}\ln|\sec\theta + \tan\theta| + C.$$

Since $x = \tan\theta$, we construct a right triangle with $\tan\theta = \frac{x}{1}$:

From this we see that $\sec\theta = \sqrt{x^2 + 1}$. Therefore,

$$I = \frac{1}{2}x\sqrt{x^2 + 1} - \frac{1}{2}\ln\left|\sqrt{x^2 + 1} + x\right| + C.$$

25. $\int \dfrac{dx}{(x^2 + 9)^2}$

SOLUTION Let $x = 3\tan\theta$. Then $dx = 3\sec^2\theta\,d\theta$,

$$x^2 + 9 = 9\tan^2\theta + 9 = 9(\tan^2\theta + 1) = 9\sec^2\theta,$$

and

$$I = \int \frac{dx}{(x^2 + 9)^2} = \int \frac{3\sec^2\theta\,d\theta}{81\sec^4\theta} = \frac{1}{27}\int \cos^2\theta\,d\theta = \frac{1}{27}\left[\frac{1}{2}\theta + \frac{1}{2}\sin\theta\cos\theta\right] + C.$$

Since $x = 3\tan\theta$, we construct a right triangle with $\tan\theta = \frac{x}{3}$:

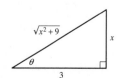

From this we see that $\sin\theta = x/\sqrt{x^2+9}$ and $\cos\theta = 3/\sqrt{x^2+9}$. Thus

$$I = \frac{1}{54}\tan^{-1}\left(\frac{x}{3}\right) + \frac{1}{54}\left(\frac{x}{\sqrt{x^2+9}}\right)\left(\frac{3}{\sqrt{x^2+9}}\right) + C = \frac{1}{54}\tan^{-1}\left(\frac{x}{3}\right) + \frac{x}{18(x^2+9)} + C.$$

27. $\displaystyle\int \frac{dx}{(x^2-4)^2}$

SOLUTION Let $x = 2\sec\theta$. Then $dx = 2\sec\theta\tan\theta\,d\theta$,

$$x^2 - 4 = 4\sec^2\theta - 4 = 4(\sec^2\theta - 1) = 4\tan^2\theta,$$

and

$$I = \int \frac{dx}{(x^2-4)^2} = \int \frac{2\sec\theta\tan\theta\,d\theta}{16\tan^4\theta} = \frac{1}{8}\int \frac{\sec\theta\,d\theta}{\tan^3\theta}$$

$$= \frac{1}{8}\int \frac{\cos^2\theta}{\sin^3\theta}\,d\theta = \frac{1}{8}\int \frac{1-\sin^2\theta}{\sin^3\theta}\,d\theta = \frac{1}{8}\int \csc^3\theta\,d\theta - \frac{1}{8}\int \csc\theta\,d\theta.$$

Now use the reduction formula for $\displaystyle\int \csc^3\theta\,d\theta$:

$$I = \frac{1}{8}\left[-\frac{\cot\theta\csc\theta}{2} + \frac{1}{2}\int \csc\theta\,d\theta\right] - \frac{1}{8}\int \csc\theta\,d\theta = -\frac{1}{16}\cot\theta\csc\theta - \frac{1}{16}\int \csc\theta\,d\theta$$

$$= -\frac{1}{16}\cot\theta\csc\theta - \frac{1}{16}\ln|\csc\theta - \cot\theta| + C.$$

Since $x = 2\sec\theta$, we construct a right triangle with $\sec\theta = \frac{x}{2}$:

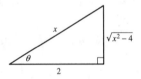

From this we see that $\cot\theta = 2/\sqrt{x^2-4}$ and $\csc\theta = x/\sqrt{x^2-4}$. Thus

$$I = -\frac{1}{16}\left(\frac{2}{\sqrt{x^2-4}}\right)\left(\frac{x}{\sqrt{x^2-4}}\right) - \frac{1}{16}\ln\left|\frac{x}{\sqrt{x^2-4}} - \frac{2}{\sqrt{x^2-4}}\right| + C$$

$$= \frac{-x}{8(x^2-4)} - \frac{1}{16}\ln\left|\frac{x-2}{\sqrt{x^2-4}}\right| + C.$$

29. $\displaystyle\int x^3\sqrt{9-x^2}\,dx$

SOLUTION Let $x = 3\sin\theta$. Then $dx = 3\cos\theta\,d\theta$,

$$9 - x^2 = 9 - 9\sin^2\theta = 9(1 - \sin^2\theta) = 9\cos^2\theta,$$

and

$$I = \int x^3\sqrt{9-x^2}\,dx = \int (27\sin^3\theta)(3\cos\theta)(3\cos\theta\,d\theta)$$

$$= 243\int \sin^3\theta\cos^2\theta\,d\theta = 243\int (1-\cos^2\theta)\cos^2\theta\sin\theta\,d\theta$$

$$= 243\left[\int \cos^2\theta\sin\theta\,d\theta - \int \cos^4\theta\sin\theta\,d\theta\right].$$

Now use substitution, with $u = \cos\theta$ and $du = -\sin\theta\,d\theta$ for both integrals:

$$I = 243\left[-\frac{1}{3}\cos^3\theta + \frac{1}{5}\cos^5\theta\right] + C.$$

Since $x = 3\sin\theta$, we construct a right triangle with $\sin\theta = \frac{x}{3}$:

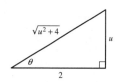

From this we see that $\cos\theta = \sqrt{9-x^2}/3$. Thus

$$I = 243\left[-\frac{1}{3}\left(\frac{\sqrt{9-x^2}}{3}\right)^3 + \frac{1}{5}\left(\frac{\sqrt{9-x^2}}{3}\right)^5\right] + C = -3(9-x^2)^{3/2} + \frac{1}{5}(9-x^2)^{5/2} + C.$$

Alternately, let $u = 9 - x^2$. Then

$$I = \int x^3\sqrt{9-x^2}\,dx = -\frac{1}{2}\int (9-u)\sqrt{u}\,du = -\frac{1}{2}\left(6u^{3/2} - \frac{2}{5}u^{5/2}\right) + C$$

$$= \frac{1}{5}u^{5/2} - 3u^{3/2} + C = \frac{1}{5}(9-x^2)^{5/2} - 3(9-x^2)^{3/2} + C.$$

31. Prove the following for $a > 0$:

$$\int \frac{dx}{x^2 + a} = \frac{1}{\sqrt{a}}\tan^{-1}\frac{x}{\sqrt{a}} + C$$

SOLUTION Let $x = \sqrt{a}\,u$. Then, $x^2 = au^2$, $dx = \sqrt{a}\,du$, and

$$\int \frac{dx}{x^2 + a} = \frac{1}{\sqrt{a}}\int \frac{du}{u^2 + 1} = \frac{1}{\sqrt{a}}\tan^{-1}u + C = \frac{1}{\sqrt{a}}\tan^{-1}\left(\frac{x}{\sqrt{a}}\right) + C.$$

33. Let $I = \displaystyle\int \frac{dx}{\sqrt{x^2 - 4x + 8}}$.

(a) Complete the square to show that $x^2 - 4x + 8 = (x-2)^2 + 4$.

(b) Use the substitution $u = x - 2$ to show that $I = \displaystyle\int \frac{du}{\sqrt{u^2 + 2^2}}$. Evaluate I.

SOLUTION

(a) Completing the square, we get

$$x^2 - 4x + 8 = x^2 - 4x + 4 + 4 = (x-2)^2 + 4.$$

(b) Let $u = x - 2$. Then $du = dx$, and

$$I = \int \frac{dx}{\sqrt{x^2 - 4x + 8}} = \int \frac{dx}{\sqrt{(x-2)^2 + 4}} = \int \frac{du}{\sqrt{u^2 + 4}}.$$

Now let $u = 2\tan\theta$. Then $du = 2\sec^2\theta\,d\theta$,

$$u^2 + 4 = 4\tan^2\theta + 4 = 4(\tan^2\theta + 1) = 4\sec^2\theta,$$

and

$$I = \int \frac{2\sec^2\theta\,d\theta}{2\sec\theta} = \int \sec\theta\,d\theta = \ln|\sec\theta + \tan\theta| + C.$$

Since $u = 2\tan\theta$, we construct a right triangle with $\tan\theta = \frac{u}{2}$:

From this we see that $\sec\theta = \sqrt{u^2 + 4}/2$. Thus

$$I = \ln\left|\frac{\sqrt{u^2 + 4}}{2} + \frac{u}{2}\right| + C_1 = \ln\left|\sqrt{u^2 + 4} + u\right| + \left(\ln\frac{1}{2} + C_1\right) = \ln\left|\sqrt{u^2 + 4} + u\right| + C.$$

Finally, we substitute back for x:

$$I = \ln\left|\sqrt{(x-2)^2 + 4} + x - 2\right| + C.$$

In Exercises 35–40, evaluate the integral by completing the square and using trigonometric substitution.

35. $\displaystyle\int \frac{dx}{\sqrt{x^2 + 4x + 13}}$

SOLUTION   First complete the square:

$$x^2 + 4x + 13 = x^2 + 4x + 4 + 9 = (x+2)^2 + 9.$$

Let $u = x + 2$. Then $du = dx$, and

$$I = \int \frac{dx}{\sqrt{x^2 + 4x + 13}} = \int \frac{dx}{\sqrt{(x+2)^2 + 9}} = \int \frac{du}{\sqrt{u^2 + 9}}.$$

Now let $u = 3\tan\theta$. Then $du = 3\sec^2\theta\, d\theta$,

$$u^2 + 9 = 9\tan^2\theta + 9 = 9(\tan^2\theta + 1) = 9\sec^2\theta,$$

and

$$I = \int \frac{3\sec^2\theta\, d\theta}{3\sec\theta} = \int \sec\theta\, d\theta = \ln|\sec\theta + \tan\theta| + C.$$

Since $u = 3\tan\theta$, we construct the following right triangle:

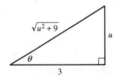

From this we see that $\sec\theta = \sqrt{u^2 + 9}/3$. Thus

$$I = \ln\left|\frac{\sqrt{u^2 + 9}}{3} + \frac{u}{3}\right| + C_1 = \ln\left|\sqrt{u^2 + 9} + u\right| + \left(\ln\frac{1}{3} + C_1\right)$$

$$= \ln\left|\sqrt{(x+2)^2 + 9} + x + 2\right| + C = \ln\left|\sqrt{x^2 + 4x + 13} + x + 2\right| + C.$$

37. $\displaystyle\int \frac{dx}{\sqrt{x + x^2}}$

SOLUTION   First complete the square:

$$x^2 + x = x^2 + x + \frac{1}{4} - \frac{1}{4} = \left(x + \frac{1}{2}\right)^2 - \frac{1}{4}.$$

Let $u = x + \frac{1}{2}$. Then $du = dx$, and

$$I = \int \frac{dx}{\sqrt{x + x^2}} = \int \frac{dx}{\sqrt{(x + \frac{1}{2})^2 - \frac{1}{4}}} = \int \frac{du}{\sqrt{u^2 - \frac{1}{4}}}.$$

Now let $u = \frac{1}{2}\sec\theta$. Then $du = \frac{1}{2}\sec\theta\tan\theta\, d\theta$,

$$u^2 - \frac{1}{4} = \frac{1}{4}\sec^2\theta - \frac{1}{4} = \frac{1}{4}(\sec^2\theta - 1) = \frac{1}{4}\tan^2\theta,$$

and

$$I = \int \frac{\frac{1}{2}\sec\theta\tan\theta\, d\theta}{\frac{1}{2}\tan\theta} = \int \sec\theta\, d\theta = \ln|\sec\theta + \tan\theta| + C.$$

Since $u = \frac{1}{2}\sec\theta$, we construct the following right triangle:

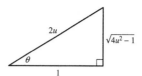

From this we see that $\tan\theta = \sqrt{4u^2 - 1}$. Then

$$I = \ln\left|2u + \sqrt{4u^2 - 1}\right| + C = \ln\left|2\left(x + \frac{1}{2}\right) + \sqrt{4\left(x + \frac{1}{2}\right)^2 - 1}\right| + C$$

$$= \ln\left|2x + 1 + \sqrt{4\left(x^2 + x + \frac{1}{4}\right) - 1}\right| + C = \ln\left|2x + 1 + 2\sqrt{x^2 + x}\right| + C.$$

39. $\int \sqrt{x^2 - 4x + 3}\,dx$

SOLUTION First complete the square:

$$x^2 - 4x + 3 = x^2 - 4x + 4 - 1 = (x - 2)^2 - 1.$$

Let $u = x - 2$. Then $du = dx$, and

$$I = \int \sqrt{x^2 - 4x + 3}\,dx = \int \sqrt{(x - 2)^2 - 1}\,dx = \int \sqrt{u^2 - 1}\,du.$$

Now let $u = \sec\theta$. Then $du = \sec\theta\tan\theta\,d\theta$, $u^2 - 1 = \sec^2\theta - 1 = \tan^2\theta$, and

$$I = \int \sqrt{\tan^2\theta}(\sec\theta\tan\theta\,d\theta) = \int \tan^2\theta\sec\theta\,d\theta = \int\left(\sec^2\theta - 1\right)\sec\theta\,d\theta$$

$$= \int \sec^3\theta\,d\theta - \int \sec\theta\,d\theta = \left(\frac{\tan\theta\sec\theta}{2} + \frac{1}{2}\int \sec\theta\,d\theta\right) - \int \sec\theta\,d\theta$$

$$= \frac{1}{2}\tan\theta\sec\theta - \frac{1}{2}\int \sec\theta\,d\theta = \frac{1}{2}\tan\theta\sec\theta - \frac{1}{2}\ln|\sec\theta + \tan\theta| + C.$$

Since $u = \sec\theta$, we construct the following right triangle:

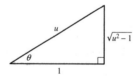

From this we see that $\tan\theta = \sqrt{u^2 - 1}$. Thus

$$I = \frac{1}{2}u\sqrt{u^2 - 1} - \frac{1}{2}\ln\left|u + \sqrt{u^2 - 1}\right| + C = \frac{1}{2}(x - 2)\sqrt{(x - 2)^2 - 1} - \frac{1}{2}\ln\left|x - 2 + \sqrt{(x - 2)^2 - 1}\right| + C$$

$$= \frac{1}{2}(x - 2)\sqrt{x^2 - 4x + 3} - \frac{1}{2}\ln\left|x - 2 + \sqrt{x^2 - 4x + 3}\right| + C.$$

41. Evaluate $\int \sec^{-1}x\,dx$. *Hint:* First use Integration by Parts.

SOLUTION Let $u = \sec^{-1}x$ and $v' = 1$. Then $v = x$, $u' = 1/x\sqrt{x^2 - 1}$, and

$$I = \int \sec^{-1}x\,dx = x\sec^{-1}x - \int \frac{x}{x\sqrt{x^2 - 1}}\,dx = x\sec^{-1}x - \int \frac{dx}{\sqrt{x^2 - 1}}.$$

To evaluate the integral on the right, let $x = \sec\theta$. Then $dx = \sec\theta\tan\theta\,d\theta$, $x^2 - 1 = \sec^2\theta - 1 = \tan^2\theta$, and

$$\int \frac{dx}{\sqrt{x^2 - 1}} = \int \frac{\sec\theta\tan\theta\,d\theta}{\tan\theta} = \int \sec\theta\,d\theta = \ln|\sec\theta + \tan\theta| + C = \ln\left|x + \sqrt{x^2 - 1}\right| + C.$$

Thus, the final answer is

$$I = x\sec^{-1}x - \ln\left|x + \sqrt{x^2 - 1}\right| + C.$$

In Exercises 43–52, indicate a good method for evaluating the integral (but do not evaluate). Your choices are recognizing a basic integration formula, algebraic manipulation, substitution (specify u and du), Integration by Parts (specify u and v′), a trigonometric method, or trigonometric substitution (specify). If it appears that the techniques you have learned thus far are not sufficient, state this.

43. $\displaystyle\int \frac{dx}{\sqrt{12 - 6x - x^2}}$

SOLUTION For this integral use a combination of three methods: Algebraic manipulation (complete the square), substitution (let $u = x + 3$), then trigonometric substitution (let $u = \sqrt{21} \sin \theta$).

45. $\displaystyle\int x \sec^2 x \, dx$

SOLUTION Use Integration by Parts, with $u = x$ and $v' = \sec^2 x$.

47. $\displaystyle\int \frac{e^{2x}}{e^{4x} + 1} \, dx$

SOLUTION First use substitution, with $u = e^{2x}$. Then either use trigonometric substitution (with $u = \tan \theta$) or recognize the formula for the inverse tangent:

$$\int \frac{du}{1 + u^2} = \tan^{-1} u + C.$$

49. $\displaystyle\int \cot x \csc x \, dx$

SOLUTION Recognize the formula

$$\int \cot x \csc x \, dx = -\csc x + C.$$

51. $\displaystyle\int \frac{dx}{(x + 2)^3}$

SOLUTION Use the substitution $u = x + 2$, and then recognize the formula

$$\int u^{-3} \, du = -\frac{1}{2u^2} + C.$$

53. Which of the following integrals can be evaluated using the substitution $u = 1 - x^2$ and which require trigonometric substitution? Determine the integral obtained after substitution in each case.

(a) $\displaystyle\int x^3 \sqrt{1 - x^2} \, dx$ **(b)** $\displaystyle\int x^2 \sqrt{1 - x^2} \, dx$

(c) $\displaystyle\int \frac{x^4}{\sqrt{1 - x^2}} \, dx$ **(d)** $\displaystyle\int \frac{x}{\sqrt{1 - x^2}} \, dx$

SOLUTION

(a) Use the substitution $u = 1 - x^2$. Then $du = -2x \, dx$, $x^2 = 1 - u$, and so

$$\int x^3 \sqrt{1 - x^2} \, dx = -\frac{1}{2} \int x^2 \sqrt{1 - x^2} (-2x \, dx) = -\frac{1}{2} \int (1 - u) u^{1/2} \, du.$$

(b) Let $x = \sin \theta$. Then $dx = \cos \theta \, d\theta$, $1 - x^2 = \cos^2 \theta$, and so

$$\int x^2 \sqrt{1 - x^2} \, dx = \int \sin^2 \theta (\cos \theta) \cos \theta \, d\theta = \int \sin^2 \theta \cos^2 \theta \, d\theta.$$

(c) Let $x = \sin \theta$. Then $dx = \cos \theta \, d\theta$, $1 - x^2 = \cos^2 \theta$, and so

$$\int \frac{x^4}{\sqrt{1 - x^2}} \, dx = \int \frac{\sin^4 \theta}{\cos \theta} \cos \theta \, d\theta = \int \sin^4 \theta \, d\theta.$$

(d) Let $u = 1 - x^2$. Then $du = -2x \, dx$, and we have

$$\int \frac{x}{\sqrt{1 - x^2}} \, dx = -\frac{1}{2} \int \frac{-2x \, dx}{\sqrt{1 - x^2}} = -\frac{1}{2} \int \frac{du}{u^{1/2}}.$$

55. Find the volume of the solid obtained by revolving the graph of $y = x\sqrt{1 - x^2}$ over $[0, 1]$ about the y-axis.

SOLUTION Using the method of cylindrical shells, the volume is given by

$$V = 2\pi \int_0^1 x \left(x\sqrt{1 - x^2}\right) dx = 2\pi \int_0^1 x^2\sqrt{1 - x^2}\, dx.$$

To evaluate this integral, let $x = \sin\theta$. Then $dx = \cos\theta\, d\theta$,

$$1 - x^2 = 1 - \sin^2\theta = \cos^2\theta,$$

and

$$I = \int x^2\sqrt{1 - x^2}\, dx = \int \sin^2\theta \cos^2\theta\, d\theta = \int \left(1 - \cos^2\theta\right)\cos^2\theta\, d\theta = \int \cos^2\theta\, d\theta - \int \cos^4\theta\, d\theta.$$

Now use the reduction formula for $\int \cos^4\theta\, d\theta$:

$$I = \int \cos^2\theta\, d\theta - \left[\frac{\cos^3\theta \sin\theta}{4} + \frac{3}{4}\int \cos^2\theta\, d\theta\right] = -\frac{1}{4}\cos^3\theta \sin\theta + \frac{1}{4}\int \cos^2\theta\, d\theta$$

$$= -\frac{1}{4}\cos^3\theta \sin\theta + \frac{1}{4}\left[\frac{1}{2}\theta + \frac{1}{2}\sin\theta \cos\theta\right] + C = -\frac{1}{4}\cos^3\theta \sin\theta + \frac{1}{8}\theta + \frac{1}{8}\sin\theta \cos\theta + C.$$

Since $\sin\theta = x$, we know that $\cos\theta = \sqrt{1 - x^2}$. Then we have

$$I = -\frac{1}{4}(1 - x^2)^{3/2}x + \frac{1}{8}\sin^{-1}x + \frac{1}{8}x\sqrt{1 - x^2} + C.$$

Now we can complete the volume:

$$V = 2\pi \left(-\frac{1}{4}x(1 - x^2)^{3/2} + \frac{1}{8}\sin^{-1}x + \frac{1}{8}x\sqrt{1 - x^2}\right)\Big|_0^1 = 2\pi\left[\left(0 + \frac{\pi}{16} + 0\right) - (0)\right] = \frac{\pi^2}{8}.$$

57. Find the volume of revolution for the region in Exercise 56, but revolve around $y = 3$.

SOLUTION Using the washer method, the volume is given by

$$V = \int_{-\sqrt{3}}^{\sqrt{3}} \pi\left(R^2 - r^2\right) dx = 2\pi \int_0^{\sqrt{3}} \left[\left(3 - \sqrt{x^2 + 1}\right)^2 - 1^2\right] dx$$

$$= 2\pi \int_0^{\sqrt{3}} \left(9 - 6\sqrt{x^2 + 1} + \left(x^2 + 1\right) - 1\right) dx = 2\pi \int_0^{\sqrt{3}} \left(9 - 6\sqrt{x^2 + 1} + x^2\right) dx$$

$$= 2\pi\left[9x - 6\left(\frac{1}{2}x\sqrt{x^2 + 1} + \frac{1}{2}\ln\left|\sqrt{x^2 + 1} + x\right|\right) + \frac{1}{3}x^3\right]\Big|_0^{\sqrt{3}}$$

$$= 2\pi\left[\left(9\sqrt{3} - 3\sqrt{3}(2) - 3\ln\left|2 + \sqrt{3}\right| + \sqrt{3}\right) - (0)\right] = 8\pi\sqrt{3} - 6\pi\ln\left|2 + \sqrt{3}\right|.$$

59. ⌐⌐⌐⌐ Having ordered an 18-in. pizza for yourself and two friends, you want to divide it up using vertical slices as in Figure 8. Use Eq. (7) in Exercise 63 below and a computer algebra system to find the value of x that divides the pizza into equal parts.

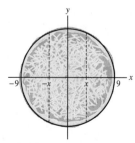

FIGURE 8 Dividing a pizza into three equal parts.

SOLUTION   First find the value of x which divides evenly a pizza with a 1-inch radius. By proportionality, we can then take this answer and multiply by 9 to get the answer for the 18-inch pizza. The total area of a 1-inch radius pizza is $\pi \cdot 1^2 = \pi$ (in square inches). The three equal pieces will have an area of $\pi/3$. The center piece is further divided into 4 equal pieces, each of area $\pi/12$. From Example 1, we know that

$$\int_0^x \sqrt{1 - x^2}\, dx = \frac{1}{2} \sin^{-1} x + \frac{1}{2} x \sqrt{1 - x^2}.$$

Setting this expression equal to $\pi/12$ and solving for x using a computer algebra system, we find $x = 0.265$. For the 18-inch pizza, the value of x should be

$$x = 9(0.265) = 2.385 \text{ inches.}$$

Further Insights and Challenges

61. Hyperbolic Substitution   Hyperbolic functions can be used instead of trigonometric substitution to treat integrals involving $\sqrt{x^2 \pm a^2}$. Let $I = \displaystyle\int \frac{dx}{\sqrt{x^2 - 1}}$.

(a) Show that the substitution $x = \cosh t$ transforms I into the integral $\displaystyle\int dt = t + C$.

(b) Show that $I = \cosh^{-1} x + C$.

(c) Trigonometric substitution with $x = \sec \theta$ leads to

$$I = \ln |x + \sqrt{x^2 - 1}| + C$$

Show that the two answers coincide.

SOLUTION

(a) Let $x = \cosh t$. Then, $dx = \sinh t\, dt$, and $x^2 - 1 = \cosh^2 t - 1 = \sinh^2 t$. Thus,

$$\int \frac{dx}{\sqrt{x^2 - 1}} = \int \frac{\sinh t}{\sinh t}\, dt = \int dt = t + C.$$

(b) Since $x = \cosh t$, by definition $t = \cosh^{-1} x$. Thus $I = \cosh^{-1} x + C$.

(c) To establish that $\cosh^{-1} x = \ln |x + \sqrt{x^2 - 1}|$, first note that

$$\frac{d}{dx} \cosh^{-1} x = \frac{1}{\sqrt{x^2 - 1}},$$

and

$$\frac{d}{dx} \ln |x + \sqrt{x^2 - 1}| = \frac{1}{x + \sqrt{x^2 - 1}} \left(1 + \frac{x}{\sqrt{x^2 - 1}} \right) = \frac{1}{x + \sqrt{x^2 - 1}} \frac{x + \sqrt{x^2 - 1}}{\sqrt{x^2 - 1}} = \frac{1}{\sqrt{x^2 - 1}};$$

in other words, $\cosh^{-1} x$ and $\ln |x + \sqrt{x^2 - 1}|$ have the same derivative. The two functions can therefore differ by at most an additive constant; however, $\cosh^{-1} 1 = 0 = \ln |1 + \sqrt{1 - 1}|$, so that constant must be zero and the two functions must be equal.

63. In Example 1, we proved the formula

$$\int \sqrt{1 - x^2}\, dx = \frac{1}{2} \sin^{-1} x + \frac{1}{2} x \sqrt{1 - x^2} + C \qquad \boxed{7}$$

Derive this formula using geometry rather than calculus by interpreting the integral as the area of part of the unit circle.

SOLUTION   The integral $\displaystyle\int_0^a \sqrt{1 - x^2}\, dx$ is the area bounded by the unit circle, the x-axis, the y-axis, and the line $x = a$. This area can be divided into two regions as follows:

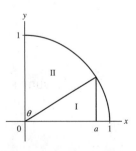

Region I is a triangle with base a and height $\sqrt{1 - a^2}$. Region II is a sector of the unit circle with central angle $\theta = \frac{\pi}{2} - \cos^{-1} a = \sin^{-1} a$. Thus,

$$\int_0^a \sqrt{1 - x^2}\, dx = \frac{1}{2} a\sqrt{1 - a^2} + \frac{1}{2}\sin^{-1} a = \left(\frac{1}{2}x\sqrt{1 - x^2} + \frac{1}{2}\sin^{-1} x\right)\Bigg|_0^a.$$

8.5 The Method of Partial Fractions

Preliminary Questions

1. Suppose that $\int f(x)\, dx = \ln x + \sqrt{x + 1} + C$. Can $f(x)$ be a rational function? Explain.

SOLUTION No, $f(x)$ cannot be a rational function because the integral of a rational function cannot contain a term with a non-integer exponent such as $\sqrt{x + 1}$.

2. Which of the following are *proper* rational functions?

(a) $\dfrac{x}{x - 3}$

(b) $\dfrac{4}{9 - x}$

(c) $\dfrac{x^2 + 12}{(x + 2)(x + 1)(x - 3)}$

(d) $\dfrac{4x^3 - 7x}{(x - 3)(2x + 5)(9 - x)}$

SOLUTION

(a) No, this is not a proper rational function because the degree of the numerator is not less than the degree of the denominator.

(b) Yes, this is a proper rational function.

(c) Yes, this is a proper rational function.

(d) No, this is not a proper rational function because the degree of the numerator is not less than the degree of the denominator.

3. Which of the following quadratic polynomials are irreducible? To check, complete the square if necessary.

(a) $x^2 + 5$

(b) $x^2 - 5$

(c) $x^2 + 4x + 6$

(d) $x^2 + 4x + 2$

SOLUTION

(a) Square is already completed; irreducible.

(b) Square is already completed; factors as $(x - \sqrt{5})(x + \sqrt{5})$.

(c) $x^2 + 4x + 6 = (x + 2)^2 + 2$; irreducible.

(d) $x^2 + 4x + 2 = (x + 2)^2 - 2$; factors as $(x + 2 - \sqrt{2})(x + 2 + \sqrt{2})$.

4. Let $P(x)/Q(x)$ be a proper rational function where $Q(x)$ factors as a product of distinct linear factors $(x - a_i)$. Then

$$\int \frac{P(x)\, dx}{Q(x)}$$

(choose correct answer):

(a) is a sum of logarithmic terms $A_i \ln(x - a_i)$ for some constants A_i.

(b) may contain a term involving the arctangent.

SOLUTION The correct answer is **(a)**: the integral is a sum of logarithmic terms $A_i \ln(x - a_i)$ for some constants A_i.

Exercises

1. Match the rational function (a)–(d) with the corresponding partial fraction decomposition (i)–(iv).

(a) $\dfrac{x^2 + 4x + 12}{(x + 2)(x^2 + 4)}$

(b) $\dfrac{2x^2 + 8x + 24}{(x + 2)^2(x^2 + 4)}$

(c) $\dfrac{x^2 - 4x + 8}{(x - 1)^2(x - 2)^2}$

(d) $\dfrac{x^4 - 4x + 8}{(x + 2)(x^2 + 4)}$

(i) $x - 2 + \dfrac{4}{x + 2} - \dfrac{4x - 4}{x^2 + 4}$

(ii) $\dfrac{-8}{x - 2} + \dfrac{4}{(x - 2)^2} + \dfrac{8}{x - 1} + \dfrac{5}{(x - 1)^2}$

(iii) $\dfrac{1}{x + 2} + \dfrac{2}{(x + 2)^2} + \dfrac{-x + 2}{x^2 + 4}$

(iv) $\dfrac{1}{x + 2} + \dfrac{4}{x^2 + 4}$

SOLUTION

(a) $\dfrac{x^2 + 4x + 12}{(x + 2)(x^2 + 4)} = \dfrac{1}{x + 2} + \dfrac{4}{x^2 + 4}$.

(b) $\dfrac{2x^2 + 8x + 24}{(x + 2)^2(x^2 + 4)} = \dfrac{1}{x + 2} + \dfrac{2}{(x + 2)^2} + \dfrac{-x + 2}{x^2 + 4}$.

(c) $\dfrac{x^2 - 4x + 8}{(x - 1)^2(x - 2)^2} = \dfrac{-8}{x - 2} + \dfrac{4}{(x - 2)^2} + \dfrac{8}{x - 1} + \dfrac{5}{(x - 1)^2}$.

(d) $\dfrac{x^4 - 4x + 8}{(x + 2)(x^2 + 4)} = x - 2 + \dfrac{4}{x + 2} - \dfrac{4x - 4}{x^2 + 4}$.

3. Clear denominators in the following partial fraction decomposition and determine the constant B (substitute a value of x or use the method of undetermined coefficients):

$$\frac{3x^2 + 11x + 12}{(x + 1)(x + 3)^2} = \frac{1}{x + 1} - \frac{B}{x + 3} - \frac{3}{(x + 3)^2}$$

SOLUTION   Clearing denominators gives

$$3x^2 + 11x + 12 = (x + 3)^2 - B(x + 1)(x + 3) - 3(x + 1).$$

Setting $x = 0$ then yields

$$12 = 9 - B(1)(3) - 3(1) \quad \text{or} \quad B = -2.$$

To use the method of undetermined coefficients, expand the right-hand side and gather like terms:

$$3x^2 + 11x + 12 = (1 - B)x^2 + (3 - 4B)x + (6 - 3B).$$

Equating x^2-coefficients on both sides, we find

$$3 = 1 - B \quad \text{or} \quad B = -2.$$

In Exercises 5–8, use long division to write $f(x)$ as the sum of a polynomial and a proper rational function. Then calculate $\int f(x)\, dx$.

5. $f(x) = \dfrac{x}{3x - 9}$

SOLUTION   Long division gives us

$$\frac{x}{3x - 9} = \frac{1}{3} + \frac{1}{x - 3}.$$

Therefore the integral is

$$\int \frac{x}{3x - 9}\, dx = \frac{1}{3} \int dx + \int \frac{dx}{x - 3} = \frac{1}{3}x + \ln|x - 3| + C.$$

7. $f(x) = \dfrac{x^3 + x + 1}{x - 2}$

SOLUTION   Long division gives us

$$\frac{x^3 + x + 1}{x - 2} = x^2 + 2x + 5 + \frac{11}{x - 2}.$$

Therefore the integral is

$$\int \frac{x^3 + x + 1}{x - 2}\, dx = \int (x^2 + 2x + 5)\, dx + 11 \int \frac{dx}{x - 2} = \frac{1}{3}x^3 + x^2 + 5x + 11 \ln|x - 2| + C.$$

In Exercises 9–46, evaluate the integral.

9. $\displaystyle\int \frac{dx}{(x-2)(x-4)}$

SOLUTION   The partial fraction decomposition has the form:

$$\frac{1}{(x-2)(x-4)} = \frac{A}{x-2} + \frac{B}{x-4}.$$

Clearing denominators gives us

$$1 = A(x-4) + B(x-2).$$

Setting $x=2$ then yields

$$1 = A(2-4) + 0 \quad \text{or} \quad A = -\frac{1}{2},$$

while setting $x=4$ yields

$$1 = 0 + B(4-2) \quad \text{or} \quad B = \frac{1}{2}.$$

The result is:

$$\frac{1}{(x-2)(x-4)} = \frac{-\frac{1}{2}}{x-2} + \frac{\frac{1}{2}}{x-4}.$$

Thus,

$$\int \frac{dx}{(x-2)(x-4)} = -\frac{1}{2}\int \frac{dx}{x-2} + \frac{1}{2}\int \frac{dx}{x-4} = -\frac{1}{2}\ln|x-2| + \frac{1}{2}\ln|x-4| + C.$$

11. $\displaystyle\int \frac{dx}{x(2x+1)}$

SOLUTION   The partial fraction decomposition has the form:

$$\frac{1}{x(2x+1)} = \frac{A}{x} + \frac{B}{2x+1}.$$

Clearing denominators gives us

$$1 = A(2x+1) + Bx.$$

Setting $x=0$ then yields

$$1 = A(1) + 0 \quad \text{or} \quad A = 1,$$

while setting $x = -\frac{1}{2}$ yields

$$1 = 0 + B\left(-\frac{1}{2}\right) \quad \text{or} \quad B = -2.$$

The result is:

$$\frac{1}{x(2x+1)} = \frac{1}{x} + \frac{-2}{2x+1}.$$

Thus,

$$\int \frac{dx}{x(2x+1)} = \int \frac{dx}{x} - \int \frac{2\,dx}{2x+1} = \ln|x| - \ln|2x+1| + C.$$

For the integral on the right, we have used the substitution $u = 2x+1$, $du = 2\,dx$.

13. $\displaystyle\int \frac{(2x-1)\,dx}{x^2 - 5x + 6}$

SOLUTION The partial fraction decomposition has the form:

$$\frac{2x-1}{x^2-5x+6} = \frac{2x-1}{(x-2)(x-3)} = \frac{A}{x-2} + \frac{B}{x-3}.$$

Clearing denominators gives us

$$2x-1 = A(x-3) + B(x-2).$$

Setting $x = 2$ then yields

$$3 = A(-1) + 0 \quad \text{or} \quad A = -3,$$

while setting $x = 3$ yields

$$5 = 0 + B(1) \quad \text{or} \quad B = 5.$$

The result is:

$$\frac{2x-1}{x^2-5x+6} = \frac{-3}{x-2} + \frac{5}{x-3}.$$

Thus,

$$\int \frac{(2x-1)\,dx}{x^2-5x+6} = -3\int \frac{dx}{x-2} + 5\int \frac{dx}{x-3} = -3\ln|x-2| + 5\ln|x-3| + C.$$

15. $\displaystyle \int \frac{(x^2+3x-44)\,dx}{(x+3)(x+5)(3x-2)}$

SOLUTION The partial fraction decomposition has the form:

$$\frac{x^2+3x-44}{(x+3)(x+5)(3x-2)} = \frac{A}{x+3} + \frac{B}{x+5} + \frac{C}{3x-2}.$$

Clearing denominators gives us

$$x^2+3x-44 = A(x+5)(3x-2) + B(x+3)(3x-2) + C(x+3)(x+5).$$

Setting $x = -3$ then yields

$$9 - 9 - 44 = A(2)(-11) + 0 + 0 \quad \text{or} \quad A = 2,$$

while setting $x = -5$ yields

$$25 - 15 - 44 = 0 + B(-2)(-17) + 0 \quad \text{or} \quad B = -1,$$

and setting $x = \frac{2}{3}$ yields

$$\frac{4}{9} + 2 - 44 = 0 + 0 + C\left(\frac{11}{3}\right)\left(\frac{17}{3}\right) \quad \text{or} \quad C = -2.$$

The result is:

$$\frac{x^2+3x-44}{(x+3)(x+5)(3x-2)} = \frac{2}{x+3} + \frac{-1}{x+5} + \frac{-2}{3x-2}.$$

Thus,

$$\int \frac{(x^2+3x-44)\,dx}{(x+3)(x+5)(3x-2)} = 2\int \frac{dx}{x+3} - \int \frac{dx}{x+5} - 2\int \frac{dx}{3x-2} = 2\ln|x+3| - \ln|x+5| - \frac{2}{3}\ln|3x-2| + C.$$

To evaluate the last integral, we have made the substitution $u = 3x - 2$, $du = 3\,dx$.

17. $\displaystyle \int \frac{(x^2+11x)\,dx}{(x-1)(x+1)^2}$

SOLUTION The partial fraction decomposition has the form:

$$\frac{x^2+11x}{(x-1)(x+1)^2} = \frac{A}{x-1} + \frac{B}{x+1} + \frac{C}{(x+1)^2}.$$

Clearing denominators gives us

$$x^2 + 11x = A(x+1)^2 + B(x-1)(x+1) + C(x-1).$$

Setting $x = 1$ then yields

$$12 = A(4) + 0 + 0 \qquad \text{or} \qquad A = 3,$$

while setting $x = -1$ yields

$$-10 = 0 + 0 + C(-2) \qquad \text{or} \qquad C = 5.$$

Plugging in these values results in

$$x^2 + 11x = 3(x+1)^2 + B(x-1)(x+1) + 5(x-1).$$

The constant B can be determined by plugging in for x any value other than 1 or -1. If we plug in $x = 0$, we get

$$0 = 3 + B(-1)(1) + 5(-1) \qquad \text{or} \qquad B = -2.$$

The result is

$$\frac{x^2 + 11x}{(x-1)(x+1)^2} = \frac{3}{x-1} + \frac{-2}{x+1} + \frac{5}{(x+1)^2}.$$

Thus,

$$\int \frac{(x^2 + 11x)\, dx}{(x-1)(x+1)^2} = 3\int \frac{dx}{x-1} - 2\int \frac{dx}{x+1} + 5\int \frac{dx}{(x+1)^2} = 3\ln|x-1| - 2\ln|x+1| - \frac{5}{x+1} + C.$$

19. $\displaystyle \int \frac{dx}{(x-1)^2(x-2)^2}$

SOLUTION   The partial fraction decomposition has the form:

$$\frac{1}{(x-1)^2(x-2)^2} = \frac{A}{x-1} + \frac{B}{(x-1)^2} + \frac{C}{x-2} + \frac{D}{(x-2)^2}.$$

Clearing denominators gives us

$$1 = A(x-1)(x-2)^2 + B(x-2)^2 + C(x-2)(x-1)^2 + D(x-1)^2.$$

Setting $x = 1$ then yields

$$1 = B(1) \qquad \text{or} \qquad B = 1,$$

while setting $x = 2$ yields

$$1 = D(1) \qquad \text{or} \qquad D = 1.$$

Plugging in these values gives us

$$1 = A(x-1)(x-2)^2 + (x-2)^2 + C(x-2)(x-1)^2 + (x-1)^2.$$

Setting $x = 0$ now yields

$$1 = A(-1)(4) + 4 + C(-2)(1) + 1 \qquad \text{or} \qquad -4 = -4A - 2C,$$

while setting $x = 3$ yields

$$1 = A(2)(1) + 1 + C(1)(4) + 4 \qquad \text{or} \qquad -4 = 2A + 4C.$$

Solving this system of two equations in two unknowns gives $A = 2$ and $C = -2$. The result is

$$\frac{1}{(x-1)^2(x-2)^2} = \frac{2}{x-1} + \frac{1}{(x-1)^2} + \frac{-2}{x-2} + \frac{1}{(x-2)^2}.$$

Thus,

$$\int \frac{dx}{(x-1)^2(x-2)^2} = 2\int \frac{dx}{x-1} + \int \frac{dx}{(x-1)^2} - 2\int \frac{dx}{x-2} + \int \frac{dx}{(x-2)^2}$$

$$= 2\ln|x-1| - \frac{1}{x-1} - 2\ln|x-2| - \frac{1}{x-2} + C.$$

21. $\displaystyle \int \frac{48\,dx}{x(x+4)^2}$

SOLUTION   The partial fraction decomposition has the form:

$$\frac{48}{x(x+4)^2} = \frac{A}{x} + \frac{B}{x+4} + \frac{C}{(x+4)^2}.$$

Clearing denominators gives us

$$48 = A(x+4)^2 + Bx(x+4) + Cx.$$

Setting $x = -4$ then yields

$$48 = 0 + 0 - 4C \qquad \text{or} \qquad C = -12,$$

while setting $x = 0$ yields

$$48 = A(16) + 0 + 0 \qquad \text{or} \qquad A = 3.$$

Plugging in $A = 3$ and $C = -12$ gives us

$$48 = 3(x+4)^2 + Bx(x+4) - 12x.$$

Setting $x = 1$ now yields

$$48 = 3(25) + B(1)(5) - 12 \qquad \text{or} \qquad B = -3.$$

The result is

$$\frac{48}{x(x+4)^2} = \frac{3}{x} + \frac{-3}{x+4} + \frac{-12}{(x+4)^2}.$$

Thus,

$$\int \frac{48\,dx}{x(x+4)^2} = 3\int \frac{dx}{x} - 3\int \frac{dx}{x+4} - 12\int \frac{dx}{(x+4)^2} = 3\ln|x| - 3\ln|x+4| + \frac{12}{x+4} + C.$$

23. $\displaystyle \int \frac{dx}{(x-4)^2(x-1)}$

SOLUTION   The partial fraction decomposition has the form:

$$\frac{1}{(x-4)^2(x-1)} = \frac{A}{x-4} + \frac{B}{(x-4)^2} + \frac{C}{(x-1)}.$$

Clearing denominators, we get

$$1 = A(x-4)(x-1) + B(x-1) + C(x-4)^2.$$

Setting $x = 1$ then yields

$$1 = 0 + 0 + C(9) \qquad \text{or} \qquad C = \frac{1}{9},$$

while setting $x = 4$ yields

$$1 = 0 + B(3) + 0 \qquad \text{or} \qquad B = \frac{1}{3}.$$

Plugging in $B = \frac{1}{3}$ and $C = \frac{1}{9}$, and setting $x = 5$, we find

$$1 = A(1)(4) + \frac{1}{3}(4) + \frac{1}{9}(1) \qquad \text{or} \qquad A = -\frac{1}{9}.$$

The result is

$$\frac{1}{(x-4)^2(x-1)} = \frac{-\frac{1}{9}}{x-4} + \frac{\frac{1}{3}}{(x-4)^2} + \frac{\frac{1}{9}}{x-1}.$$

Thus,

$$\int \frac{dx}{(x-4)^2(x-1)} = -\frac{1}{9}\int \frac{dx}{x-4} + \frac{1}{3}\int \frac{dx}{(x-4)^2} + \frac{1}{9}\int \frac{dx}{x-1} = -\frac{1}{9}\ln|x-4| - \frac{1}{3(x-4)} + \frac{1}{9}\ln|x-1| + C.$$

25. $\int \dfrac{3x+6}{x^2(x-1)(x-3)} \, dx$

SOLUTION The partial fraction decomposition has the form:

$$\frac{3x+6}{x^2(x-1)(x-3)} = \frac{A}{x} + \frac{B}{x^2} + \frac{C}{x-1} + \frac{D}{x-3}.$$

Clearing denominators gives us

$$3x+6 = Ax(x-1)(x-3) + B(x-1)(x-3) + Cx^2(x-3) + Dx^2(x-1).$$

Setting $x = 0$, then yields

$$6 = 0 + B(-1)(-3) + 0 + 0 \qquad \text{or} \qquad B = 2,$$

while setting $x = 1$ yields

$$9 = 0 + 0 + C(1)(-2) + 0 \qquad \text{or} \qquad C = -\frac{9}{2},$$

and setting $x = 3$ yields

$$15 = 0 + 0 + 0 + D(9)(2) \qquad \text{or} \qquad D = \frac{5}{6}.$$

In order to find A, let's look at the x^3-coefficient on the right-hand side (which must equal 0, since there's no x^3 term on the left):

$$0 = A + C + D = A - \frac{9}{2} + \frac{5}{6}, \qquad \text{so} \qquad A = \frac{11}{3}.$$

The result is

$$\frac{3x+6}{x^2(x-1)(x-3)} = \frac{\frac{11}{3}}{x} + \frac{2}{x^2} + \frac{-\frac{9}{2}}{x-1} + \frac{\frac{5}{6}}{x-3}.$$

Thus,

$$\int \frac{(3x+6)\,dx}{x^2(x-1)(x-3)} = \frac{11}{3}\int \frac{dx}{x} + 2\int \frac{dx}{x^2} - \frac{9}{2}\int \frac{dx}{x-1} + \frac{5}{6}\int \frac{dx}{x-3}$$

$$= \frac{11}{3}\ln|x| - \frac{2}{x} - \frac{9}{2}\ln|x-1| + \frac{5}{6}\ln|x-3| + C.$$

27. $\int \dfrac{(3x^2-2)\,dx}{x-4}$

SOLUTION First we use long division to write

$$\frac{3x^2-2}{x-4} = 3x + 12 + \frac{46}{x-4}.$$

Then the integral becomes

$$\int \frac{(3x^2-2)\,dx}{x-4} = \int (3x+12)\,dx + 46\int \frac{dx}{x-4} = \frac{3}{2}x^2 + 12x + 46\ln|x-4| + C.$$

29. $\int \dfrac{dx}{x(x^2+1)}$

SOLUTION The partial fraction decomposition has the form:

$$\frac{1}{x(x^2+1)} = \frac{A}{x} + \frac{Bx+C}{x^2+1}.$$

Clearing denominators, we get

$$1 = A(x^2+1) + (Bx+C)x.$$

Setting $x = 0$ then yields

$$1 = A(1) + 0 \qquad \text{or} \qquad A = 1.$$

This gives us

$$1 = x^2 + 1 + Bx^2 + Cx = (B+1)x^2 + Cx + 1.$$

Equating x^2-coefficients, we find

$$B + 1 = 0 \qquad \text{or} \qquad B = -1;$$

while equating x-coefficients yields $C = 0$. The result is

$$\frac{1}{x(x^2+1)} = \frac{1}{x} + \frac{-x}{x^2+1}.$$

Thus,

$$\int \frac{dx}{x(x^2+1)} = \int \frac{dx}{x} - \int \frac{x\,dx}{x^2+1}.$$

For the integral on the right, use the substitution $u = x^2 + 1$, $du = 2x\,dx$. Then we have

$$\int \frac{dx}{x(x^2+1)} = \int \frac{dx}{x} - \frac{1}{2}\int \frac{du}{u} = \ln|x| - \frac{1}{2}\ln|x^2+1| + C.$$

31. $\displaystyle\int \frac{x^2\,dx}{x^2+3}$

SOLUTION   First use long division to obtain

$$\frac{x^2}{x^2+3} = \frac{x^2+3-3}{x^2+3} = \frac{x^2+3}{x^2+3} + \frac{-3}{x^2+3} = 1 - \frac{3}{x^2+3}.$$

The integral becomes

$$\int \frac{x^2\,dx}{x^2+3} = \int dx - 3\int \frac{dx}{x^2+3} = x - 3\left(\frac{1}{\sqrt{3}}\right)\tan^{-1}\left(\frac{x}{\sqrt{3}}\right) + C = x - \sqrt{3}\tan^{-1}\left(\frac{x}{\sqrt{3}}\right) + C.$$

33. $\displaystyle\int \frac{x^2}{(x+1)(x^2+1)}\,dx$

SOLUTION   The partial fraction decomposition has the form

$$\frac{x^2}{(x+1)(x^2+1)} = \frac{A}{x+1} + \frac{Bx+C}{x^2+1}.$$

Clearing denominators, we get

$$x^2 = A(x^2+1) + (Bx+C)(x+1).$$

Setting $x = -1$ then yields

$$1 = A(2) + 0 \qquad \text{or} \qquad A = \frac{1}{2}.$$

This gives us

$$x^2 = \frac{1}{2}x^2 + \frac{1}{2} + Bx^2 + Bx + Cx + C = \left(B + \frac{1}{2}\right)x^2 + (B+C)x + \left(C + \frac{1}{2}\right).$$

Equating x^2-coefficients, we find

$$1 = B + \frac{1}{2} \qquad \text{or} \qquad B = \frac{1}{2},$$

while equating constant coefficients yields

$$0 = C + \frac{1}{2} \qquad \text{or} \qquad C = -\frac{1}{2}.$$

The result is

$$\frac{x^2}{(x+1)(x^2+1)} = \frac{\frac{1}{2}}{x+1} + \frac{\frac{1}{2}x - \frac{1}{2}}{x^2+1}.$$

Thus,

$$\int \frac{x^2 \, dx}{(x+1)(x^2+1)} = \frac{1}{2} \int \frac{dx}{x+1} + \frac{1}{2} \int \frac{(x-1) \, dx}{x^2+1} = \frac{1}{2} \int \frac{dx}{x+1} + \frac{1}{2} \int \frac{x \, dx}{x^2+1} - \frac{1}{2} \int \frac{dx}{x^2+1}$$

$$= \frac{1}{2} \ln|x+1| + \frac{1}{4} \ln|x^2+1| - \frac{1}{2} \tan^{-1} x + C.$$

Here we used $u = x^2 + 1$, $du = 2x \, dx$ for the second integral.

35. $\int \dfrac{x^2 \, dx}{(3x+7)^3}$

SOLUTION   This problem can be done without partial fraction decomposition. Let $u = 3x + 7$, so that $x = \frac{1}{3}(u - 7)$, $x^2 = \frac{1}{9}(u - 7)^2$ and $du = 3 \, dx$. Then

$$\int \frac{x^2 \, dx}{(3x+7)^3} = \frac{1}{27} \int \frac{(u-7)^2}{u^3} \, du = \frac{1}{27} \int \frac{u^2 - 14u + 49}{u^3} \, du = \frac{1}{27} \int \left(\frac{1}{u} - 14u^{-2} + 49u^{-3} \right) du$$

$$= \frac{1}{27} \left(\ln|u| + 14u^{-1} - \frac{49}{2}u^{-2} + C \right) = \frac{1}{27} \ln|3x+7| + \frac{14}{27}(3x+7)^{-1} - \frac{49}{54}(3x+7)^{-2} + C.$$

37. $\int \dfrac{dx}{x^2(x^2+25)}$

SOLUTION   The partial fraction decomposition has the form:

$$\frac{1}{x^2(x^2+25)} = \frac{A}{x} + \frac{B}{x^2} + \frac{Cx+D}{x^2+25}.$$

Clearing denominators, we get

$$1 = Ax(x^2+25) + B(x^2+25) + (Cx+D)x^2.$$

Setting $x = 0$ then yields

$$1 = 0 + B(25) + 0 \qquad \text{or} \qquad B = \frac{1}{25}.$$

This gives us

$$1 = Ax^3 + 25Ax + \frac{1}{25}x^2 + 1 + Cx^3 + Dx^2 = (A+C)x^3 + \left(D + \frac{1}{25} \right)x^2 + 25Ax + 1.$$

Equating x-coefficients yields

$$0 = 25A \qquad \text{or} \qquad A = 0,$$

while equating x^3-coefficients yields

$$0 = A + C = 0 + C \qquad \text{or} \qquad C = 0,$$

and equating x^2-coefficients yields

$$0 = D + \frac{1}{25} \qquad \text{or} \qquad D = \frac{-1}{25}.$$

The result is

$$\frac{1}{x^2(x^2+25)} = \frac{\frac{1}{25}}{x^2} + \frac{\frac{-1}{25}}{x^2+25}.$$

Thus,

$$\int \frac{dx}{x^2(x^2+25)} = \frac{1}{25} \int \frac{dx}{x^2} - \frac{1}{25} \int \frac{dx}{x^2+25} = -\frac{1}{25x} - \frac{1}{125} \tan^{-1} \left(\frac{x}{5} \right) + C.$$

39. $\int \dfrac{10 \, dx}{(x+1)(x^2+9)^2}$

SOLUTION   The partial fraction decomposition has the form:

$$\frac{10}{(x+1)(x^2+9)^2} = \frac{A}{x+1} + \frac{Bx+C}{x^2+9} + \frac{Dx+E}{(x^2+9)^2}.$$

Clearing denominators gives us

$$10 = A(x^2+9)^2 + (Bx+C)(x+1)(x^2+9) + (Dx+E)(x+1).$$

Setting $x = -1$ then yields

$$10 = A(100) + 0 + 0 \qquad \text{or} \qquad A = \frac{1}{10}.$$

Expanding the right-hand side, we find

$$10 = \left(B + \frac{1}{10}\right)x^4 + (B+C)x^3 + \left(9B + C + D + \frac{18}{10}\right)x^2(9B + 9C + D + E)x + \left(9C + E + \frac{81}{10}\right).$$

Equating x^4-coefficients yields

$$B + \frac{1}{10} = 0 \qquad \text{or} \qquad B = -\frac{1}{10},$$

while equating x^3-coefficients yields

$$-\frac{1}{10} + C = 0 \qquad \text{or} \qquad C = \frac{1}{10},$$

and equating x^2-coefficients yields

$$-\frac{9}{10} + \frac{1}{10} + D + \frac{18}{10} = 0 \qquad \text{or} \qquad D = -1.$$

Finally, equating constant coefficients, we find

$$10 = \frac{9}{10} + E + \frac{81}{10} \qquad \text{or} \qquad E = 1.$$

The result is

$$\frac{10}{(x+1)(x^2+9)^2} = \frac{\frac{1}{10}}{x+1} + \frac{-\frac{1}{10}x + \frac{1}{10}}{x^2+9} + \frac{-x+1}{(x^2+9)^2}.$$

Thus,

$$\int \frac{10\,dx}{(x+1)(x^2+9)^2} = \frac{1}{10}\int \frac{dx}{x+1} - \frac{1}{10}\int \frac{x\,dx}{x^2+9} + \frac{1}{10}\int \frac{dx}{x^2+9} - \int \frac{x\,dx}{(x^2+9)^2} + \int \frac{dx}{(x^2+9)^2}.$$

For the second and fourth integrals, use the substitution $u = x^2 + 9$, $du = 2x\,dx$. Then we have

$$\int \frac{10\,dx}{(x+1)(x^2+9)^2} = \frac{1}{10}\ln|x+1| - \frac{1}{20}\ln|x^2+9| + \frac{1}{30}\tan^{-1}\left(\frac{x}{3}\right) + \frac{1}{2(x^2+9)} + \int \frac{dx}{(x^2+9)^2}.$$

For the last integral, use the trigonometric substitution

$$x = 3\tan\theta, \qquad dx = 3\sec^2\theta\,d\theta, \qquad x^2 + 9 = \tan^2\theta + 9 = 9\sec^2\theta.$$

Then,

$$\int \frac{dx}{(x^2+9)^2} = \int \frac{3\sec^2\theta\,d\theta}{(9\sec^2\theta)^2} = \frac{1}{27}\int \frac{d\theta}{\sec^2\theta} = \frac{1}{27}\int \cos^2\theta\,d\theta = \frac{1}{27}\left[\frac{1}{2}\theta + \frac{1}{2}\sin\theta\cos\theta\right] + C.$$

Now we construct a right triangle with $\tan\theta = \frac{x}{3}$:

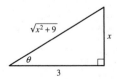

From this we see that $\sin\theta = x/\sqrt{x^2+9}$ and $\cos\theta = 3/\sqrt{x^2+9}$. Thus

$$\int \frac{dx}{(x^2+9)^2} = \frac{1}{54}\tan^{-1}\left(\frac{x}{3}\right) + \frac{1}{54}\left(\frac{x}{\sqrt{x^2+9}}\right)\left(\frac{3}{\sqrt{x^2+9}}\right) + C = \frac{1}{54}\tan^{-1}\left(\frac{x}{3}\right) + \frac{x}{18(x^2+9)} + C.$$

Collecting all the terms, we obtain

$$\int \frac{10\,dx}{(x+1)(x^2+9)^2} = \frac{1}{10}\ln|x+1| - \frac{1}{20}\ln|x^2+9| + \frac{1}{30}\tan^{-1}\left(\frac{x}{3}\right) + \frac{1}{2(x^2+9)}$$

$$+ \frac{1}{54}\tan^{-1}\left(\frac{x}{3}\right) + \frac{x}{18(x^2+9)} + C$$

$$= \frac{1}{10}\ln|x+1| - \frac{1}{20}\ln|x^2+9| + \frac{7}{135}\tan^{-1}\left(\frac{x}{3}\right) + \frac{x+9}{18(x^2+9)} + C.$$

41. $\displaystyle\int \frac{100x\,dx}{(x-3)(x^2+1)^2}$

SOLUTION   The partial fraction decomposition has the form:

$$\frac{100x}{(x-3)(x^2+1)^2} = \frac{A}{x-3} + \frac{Bx+C}{x^2+1} + \frac{Dx+E}{(x^2+1)^2}.$$

Clearing denominators, we get

$$100x = A(x^2+1)^2 + (Bx+C)(x-3)(x^2+1) + (Dx+E)(x-3).$$

Setting $x=3$ then yields

$$300 = A(100) + 0 + 0 \qquad \text{or} \qquad A = 3.$$

Expanding the right-hand side, we find

$$100x = (B+3)x^4 + (C-3B)x^3 + (B-3C+D+6)x^2 + (C-3B-3D+E)x + (3-3C-3E).$$

Equating coefficients of like powers of x then yields

$$B + 3 = 0$$

$$C - 3B = 0$$

$$B - 3C + D + 6 = 0$$

$$C - 3B - 3D + E = 100$$

$$3 - 3C - 3E = 0$$

The solution to this system of equations is

$$B = -3, \qquad C = -9, \qquad D = -30, \qquad E = 10.$$

Therefore

$$\frac{100x}{(x-3)(x^2+1)^2} = \frac{3}{x-3} + \frac{-3x-9}{x^2+1} + \frac{-30x+10}{(x^2+1)^2},$$

and

$$\int \frac{100x\,dx}{(x-3)(x^2+1)^2} = 3\int \frac{dx}{x-3} + \int \frac{(-3x-9)\,dx}{x^2+1} + \int \frac{(-30x+10)\,dx}{(x^2+1)^2}$$

$$= 3\int \frac{dx}{x-3} - 3\int \frac{x\,dx}{x^2+1} - 9\int \frac{dx}{x^2+1} - 30\int \frac{x\,dx}{(x^2+1)^2} + 10\int \frac{dx}{(x^2+1)^2}.$$

For the second and fourth integrals, use the substitution $u = x^2 + 1$, $du = 2x\,dx$. Then we have

$$\int \frac{100x\,dx}{(x-3)(x^2+1)^2} = 3\ln|x-3| - \frac{3}{2}\ln|x^2+1| - 9\tan^{-1}x + \frac{15}{x^2+1} + 10\int \frac{dx}{(x^2+1)^2}.$$

For the last integral, use the trigonometric substitution $x = \tan\theta$, $dx = \sec^2\theta\,d\theta$. Then $x^2 + 1 = \tan^2\theta + 1 = \sec^2\theta$, and

$$\int \frac{dx}{(x^2+1)^2} = \int \frac{\sec^2\theta\,d\theta}{\sec^4\theta} = \int \cos^2\theta = \frac{1}{2}\theta + \frac{1}{2}\sin\theta\cos\theta + C.$$

We construct the following right triangle with $\tan\theta = x$:

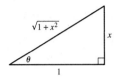

From this we see that $\sin\theta = x/\sqrt{1+x^2}$ and $\cos\theta = 1/\sqrt{1+x^2}$. Thus

$$\int \frac{dx}{(x^2+1)^2} = \frac{1}{2}\tan^{-1}x + \frac{1}{2}\left(\frac{x}{\sqrt{1+x^2}}\right)\left(\frac{1}{\sqrt{1+x^2}}\right) + C = \frac{1}{2}\tan^{-1}x + \frac{x}{2(x^2+1)} + C.$$

Collecting all the terms, we obtain

$$\int \frac{100x\,dx}{(x-3)(x^2+1)^2} = 3\ln|x-3| - \frac{3}{2}\ln|x^2+1| - 9\tan^{-1}x + \frac{15}{x^2+1} + 10\left(\frac{1}{2}\tan^{-1}x + \frac{x}{2(x^2+1)}\right) + C$$

$$= 3\ln|x-3| - \frac{3}{2}\ln|x^2+1| - 4\tan^{-1}x + \frac{5x+15}{x^2+1} + C.$$

43. $\displaystyle\int \frac{9\,dx}{(x+1)(x^2-2x+6)}$

SOLUTION The partial fraction decomposition has the form:

$$\frac{9}{(x+1)(x^2-2x+6)} = \frac{A}{x+1} + \frac{Bx+C}{x^2-2x+6}.$$

Clearing denominators gives us

$$9 = A(x^2-2x+6) + (Bx+C)(x+1).$$

Setting $x=-1$ then yields

$$9 = A(9) + 0 \qquad \text{or} \qquad A = 1.$$

Expanding the right-hand side gives us

$$9 = (1+B)x^2 + (-2+B+C)x + (6+C).$$

Equating x^2-coefficients yields

$$0 = 1+B \qquad \text{or} \qquad B = -1,$$

while equating constant coefficients yields

$$9 = 6+C \qquad \text{or} \qquad C = 3.$$

The result is

$$\frac{9}{(x+1)(x^2-2x+6)} = \frac{1}{x+1} + \frac{-x+3}{x^2-2x+6}.$$

Thus,

$$\int \frac{9\,dx}{(x+1)(x^2-2x+6)} = \int \frac{dx}{x+1} + \int \frac{(-x+3)\,dx}{x^2-2x+6}.$$

To evaluate the integral on the right, we first write

$$\int \frac{(-x+3)\,dx}{x^2-2x+6} = -\int \frac{(x-1-2)\,dx}{x^2-2x+6} = -\int \frac{(x-1)\,dx}{x^2-2x+6} + 2\int \frac{dx}{x^2-2x+6}.$$

For the first integral, use the substitution $u = x^2-2x+6$, $du = (2x-2)\,dx$. Then

$$-\int \frac{(x-1)\,dx}{x^2-2x+6} = -\frac{1}{2}\int \frac{(2x-2)\,dx}{x^2-2x+6} = -\frac{1}{2}\ln|x^2-2x+6| + C.$$

For the second integral, we first complete the square:

$$2\int \frac{dx}{x^2-2x+6} = 2\int \frac{dx}{(x^2-2x+1)+5} = 2\int \frac{dx}{(x-1)^2+5}.$$

Now let $u = x - 1$, $du = dx$. Then

$$2 \int \frac{dx}{(x-1)^2 + 5} = 2 \int \frac{du}{u^2 + 5} = 2 \left(\frac{1}{\sqrt{5}} \right) \tan^{-1} \left(\frac{u}{\sqrt{5}} \right) + C = \frac{2}{\sqrt{5}} \tan^{-1} \left(\frac{x-1}{\sqrt{5}} \right) + C.$$

Collecting all the terms, we have

$$\int \frac{9 \, dx}{(x+1)(x^2 - 2x + 6)} = \ln|x+1| - \frac{1}{2} \ln|x^2 - 2x + 6| + \frac{2}{\sqrt{5}} \tan^{-1} \left(\frac{x-1}{\sqrt{5}} \right) + C.$$

45. $\displaystyle \int \frac{(x^2 + 3) \, dx}{(x^2 + 2x + 3)^2}$

SOLUTION   The partial fraction decomposition has the form:

$$\frac{x^2 + 3}{(x^2 + 2x + 3)^2} = \frac{Ax + B}{x^2 + 2x + 3} + \frac{Cx + D}{(x^2 + 2x + 3)^2}.$$

Clearing denominators gives us

$$x^2 + 3 = (Ax + B)(x^2 + 2x + 3) + Cx + D.$$

Expanding the right-hand side, we get

$$x^2 + 3 = Ax^3 + (2A + B)x^2 + (3A + 2B + C)x + (3B + D).$$

Equating coefficients of like powers of x then yields

$$A = 0$$
$$2A + B = 1$$
$$3A + 2B + C = 0$$
$$3B + D = 3$$

The solution to this system of equations is

$$A = 0, \qquad B = 1, \qquad C = -2, \qquad D = 0.$$

Therefore

$$\frac{x^2 + 3}{(x^2 + 2x + 3)^2} = \frac{1}{x^2 + 2x + 3} + \frac{-2x}{(x^2 + 2x + 3)^2},$$

and

$$\int \frac{(x^2 + 3) \, dx}{(x^2 + 2x + 3)^2} = \int \frac{dx}{x^2 + 2x + 3} - \int \frac{2x \, dx}{(x^2 + 2x + 3)^2}.$$

The first integral can be evaluated by completing the square:

$$\int \frac{dx}{x^2 + 2x + 3} = \int \frac{dx}{x^2 + 2x + 1 + 2} = \int \frac{dx}{(x+1)^2 + 2}.$$

Now use the substitution $u = x + 1$, $du = dx$. Then

$$\int \frac{dx}{x^2 + 2x + 3} = \int \frac{du}{u^2 + 2} = \frac{1}{\sqrt{2}} \tan^{-1} \left(\frac{x+1}{\sqrt{2}} \right) + C.$$

For the second integral, let $u = x^2 + 2x + 3$. We want $du = (2x + 2) \, dx$ to appear in the numerator, so we write

$$\int \frac{2x \, dx}{(x^2 + 2x + 3)^2} = \int \frac{(2x + 2 - 2) \, dx}{(x^2 + 2x + 3)^2} = \int \frac{(2x + 2) \, dx}{(x^2 + 2x + 3)^2} - 2 \int \frac{dx}{(x^2 + 2x + 3)^2}$$

$$= \int \frac{du}{u^2} - 2 \int \frac{dx}{(x^2 + 2x + 3)^2} = -\frac{1}{u} - 2 \int \frac{dx}{(x^2 + 2x + 3)^2}$$

$$= \frac{-1}{x^2 + 2x + 3} - 2 \int \frac{dx}{(x^2 + 3x + 3)^2}.$$

Finally, for this last integral, complete the square, then substitute $u = x + 1$, $du = dx$:

$$\int \frac{dx}{(x^2 + 2x + 3)^2} = \int \frac{dx}{((x+1)^2 + 2)^2} = \int \frac{du}{(u^2 + 2)^2}.$$

Now use the trigonometric substitution $u = \sqrt{2}\tan\theta$. Then $du = \sqrt{2}\sec^2\theta\,d\theta$, and $u^2 + 2 = 2\tan^2\theta + 2 = 2\sec^2\theta$. Thus

$$\int \frac{du}{(u^2+2)^2} = \int \frac{\sqrt{2}\sec^2\theta\,d\theta}{4\sec^4\theta} = \frac{\sqrt{2}}{4}\int\cos^2\theta\,d\theta = \frac{\sqrt{2}}{4}\left[\frac{1}{2}\theta + \frac{1}{2}\sin\theta\cos\theta\right] = \frac{\sqrt{2}}{8}\theta + \frac{\sqrt{2}}{8}\sin\theta\cos\theta + C.$$

We construct a right triangle with $\tan\theta = u/\sqrt{2}$:

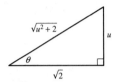

From this we see that $\sin\theta = u/\sqrt{u^2+2}$ and $\cos\theta = \sqrt{2}/\sqrt{u^2+2}$. Therefore

$$\int \frac{du}{(u^2+2)^2} = \frac{\sqrt{2}}{8}\tan^{-1}\left(\frac{u}{\sqrt{2}}\right) + \frac{\sqrt{2}}{8}\left(\frac{u}{\sqrt{u^2+2}}\right)\left(\frac{\sqrt{2}}{\sqrt{u^2+2}}\right) + C$$

$$= \frac{\sqrt{2}}{8}\tan^{-1}\left(\frac{u}{\sqrt{2}}\right) + \frac{u}{4(u^2+2)} + C = \frac{\sqrt{2}}{8}\tan^{-1}\left(\frac{x+1}{\sqrt{2}}\right) + \frac{x+1}{4(x^2+2x+3)} + C.$$

Collecting all the terms, we have

$$\int \frac{(x^2+3)\,dx}{(x^2+2x+3)^2} = \frac{1}{\sqrt{2}}\tan^{-1}\left(\frac{x+1}{\sqrt{2}}\right) - \left[\frac{-1}{x^2+2x+3} - 2\left(\frac{\sqrt{2}}{8}\tan^{-1}\left(\frac{x+1}{\sqrt{2}}\right) + \frac{x+1}{4(x^2+2x+3)}\right)\right] + C$$

$$= \left(\frac{1}{\sqrt{2}} + \frac{\sqrt{2}}{4}\right)\tan^{-1}\left(\frac{x+1}{\sqrt{2}}\right) + \frac{2+(x+1)}{2(x^2+2x+3)} + C$$

$$= \frac{3\sqrt{2}}{4}\tan^{-1}\left(\frac{x+1}{\sqrt{2}}\right) + \frac{x+3}{2(x^2+2x+3)} + C.$$

47. Evaluate $\displaystyle\int \frac{dx}{x^2-1}$ in two ways: using partial fractions and trigonometric substitution. Verify that the two answers agree.

SOLUTION The partial fraction decomposition has the form:

$$\frac{1}{x^2-1} = \frac{1}{(x-1)(x+1)} = \frac{A}{x-1} + \frac{B}{x+1}.$$

Clearing denominators gives us

$$1 = A(x+1) + B(x-1).$$

Setting $x = 1$, we get $1 = A(2)$ or $A = \frac{1}{2}$; while setting $x = -1$, we get $1 = B(-2)$ or $B = -\frac{1}{2}$. The result is

$$\frac{1}{x^2-1} = \frac{\frac{1}{2}}{x-1} + \frac{-\frac{1}{2}}{x+1}.$$

Thus,

$$\int \frac{dx}{x^2-1} = \frac{1}{2}\int \frac{dx}{x-1} - \frac{1}{2}\int \frac{dx}{x+1} = \frac{1}{2}\ln|x-1| - \frac{1}{2}\ln|x+1| + C.$$

Using trigonometric substitution, let $x = \sec\theta$. Then $dx = \tan\theta\sec\theta\,d\theta$, and $x^2 - 1 = \sec^2\theta - 1 = \tan^2\theta$. Thus

$$\int \frac{dx}{x^2-1} = \int \frac{\tan\theta\sec\theta\,d\theta}{\tan^2\theta} = \int \frac{\sec\theta\,d\theta}{\tan\theta} = \int \frac{\cos\theta\,d\theta}{\sin\theta\cos\theta}$$

$$= \int \csc\theta\,d\theta = \ln|\csc\theta - \cot\theta| + C.$$

Now we construct a right triangle with $\sec\theta = x$:

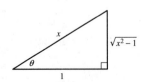

From this we see that $\csc\theta = x/\sqrt{x^2-1}$ and $\cot\theta = 1/\sqrt{x^2-1}$. Thus

$$\int \frac{dx}{x^2-1} = \ln\left|\frac{x}{\sqrt{x^2-1}} - \frac{1}{\sqrt{x^2-1}}\right| + C = \ln\left|\frac{x-1}{\sqrt{x^2-1}}\right| + C.$$

To check that these two answers agree, we write

$$\frac{1}{2}\ln|x-1| - \frac{1}{2}\ln|x+1| = \frac{1}{2}\left|\frac{x-1}{x+1}\right| = \ln\left|\sqrt{\frac{x-1}{x+1}}\right| = \ln\left|\frac{\sqrt{x-1}}{\sqrt{x+1}} \cdot \frac{\sqrt{x-1}}{\sqrt{x-1}}\right| = \ln\left|\frac{x-1}{\sqrt{x^2-1}}\right|.$$

49. Evaluate $\int \dfrac{\sqrt{x}\,dx}{x-1}$. *Hint:* Use the substitution $u = \sqrt{x}$ (sometimes called a **rationalizing substitution**).

SOLUTION Let $u = \sqrt{x}$. Then $du = (1/2\sqrt{x})\,dx = (1/2u)\,dx$. Thus

$$\int \frac{\sqrt{x}\,dx}{x-1} = \int \frac{u(2u\,du)}{u^2-1} = 2\int \frac{u^2\,du}{u^2-1} = 2\int \frac{(u^2-1+1)\,du}{u^2-1}$$

$$= 2\int \left(\frac{u^2-1}{u^2-1} + \frac{1}{u^2-1}\right)du = 2\int du + \int \frac{2\,du}{u^2-1} = 2u + \int \frac{2\,du}{u^2-1}.$$

The partial fraction decomposition of the remaining integral has the form:

$$\frac{2}{u^2-1} = \frac{2}{(u-1)(u+1)} = \frac{A}{u-1} + \frac{B}{u+1}.$$

Clearing denominators gives us

$$2 = A(u+1) + B(u-1).$$

Setting $u = 1$ yields $2 = A(2) + 0$ or $A = 1$, while setting $u = -1$ yields $2 = 0 + B(-2)$ or $B = -1$. The result is

$$\frac{2}{u^2-1} = \frac{1}{u-1} + \frac{-1}{u+1}.$$

Thus,

$$\int \frac{2\,du}{u^2-1} = \int \frac{du}{u-1} - \int \frac{du}{u+1} = \ln|u-1| - \ln|u+1| + C.$$

The final answer is

$$\int \frac{\sqrt{x}\,dx}{x-1} = 2u + \ln|u-1| - \ln|u+1| + C = 2\sqrt{x} + \ln|\sqrt{x}-1| - \ln|\sqrt{x}+1| + C.$$

In Exercises 51–66, evaluate the integral using the appropriate method or combination of methods covered thus far in the text.

51. $\int \dfrac{dx}{x^2\sqrt{4-x^2}}$

SOLUTION Use the trigonometric substitution $x = 2\sin\theta$. Then $dx = 2\cos\theta\,d\theta$,

$$4 - x^2 = 4 - 4\sin^2\theta = 4(1 - \sin^2\theta) = 4\cos^2\theta,$$

and

$$\int \frac{dx}{x^2\sqrt{4-x^2}} = \int \frac{2\cos\theta\,d\theta}{(4\sin^2\theta)(2\cos\theta)} = \frac{1}{4}\int \csc^2\theta\,d\theta = -\frac{1}{4}\cot\theta + C.$$

Now construct a right triangle with $\sin\theta = x/2$:

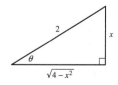

From this we see that $\cot\theta = \sqrt{4-x^2}/x$. Thus

$$\int \frac{dx}{x^2\sqrt{4-x^2}} = -\frac{1}{4}\left(\frac{\sqrt{4-x^2}}{x}\right) + C = -\frac{\sqrt{4-x^2}}{4x} + C.$$

53. $\displaystyle\int \frac{dx}{x(x-1)^2}$

SOLUTION Using partial fractions, we first write

$$\frac{1}{x(x-1)^2} = \frac{A}{x} + \frac{B}{x-1} + \frac{C}{(x-1)^2}.$$

Clearing denominators gives us

$$1 = A(x-1)^2 + Bx(x-1) + Cx.$$

Setting $x=0$ yields

$$1 = A(1) + 0 + 0 \quad\text{or}\quad A = 1,$$

while setting $x=1$ yields

$$1 = 0 + 0 + C \quad\text{or}\quad C = 1,$$

and setting $x=2$ yields

$$1 = 1 + 2B + 2 \quad\text{or}\quad B = -1.$$

The result is

$$\frac{1}{x(x-1)^2} = \frac{1}{x} + \frac{-1}{x-1} + \frac{1}{(x-1)^2}.$$

Thus,

$$\int \frac{dx}{x(x-1)^2} = \int \frac{dx}{x} - \int \frac{dx}{x-1} + \int \frac{dx}{(x-1)^2} = \ln|x| - \ln|x-1| - \frac{1}{x-1} + C.$$

55. $\displaystyle\int \frac{dx}{(x^2+9)^2}$

SOLUTION Use the trigonometric substitution $x = 3\tan\theta$. Then $dx = 3\sec^2\theta\,d\theta$,

$$x^2 + 9 = 9\tan^2\theta + 9 = 9(\tan^2\theta + 1) = 9\sec^2\theta,$$

and

$$\int \frac{dx}{(x^2+9)^2} = \int \frac{3\sec^2\theta\,d\theta}{(9\sec^2\theta)^2} = \frac{3}{81}\int \frac{\sec^2\theta\,d\theta}{\sec^4\theta} = \frac{1}{27}\int \cos^2\theta\,d\theta = \frac{1}{27}\left(\frac{1}{2}\theta + \frac{1}{2}\sin\theta\cos\theta\right) + C.$$

Now construct a right triangle with $\tan\theta = x/3$:

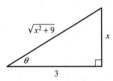

From this we see that $\sin\theta = x/\sqrt{x^2+9}$ and $\cos\theta = 3/\sqrt{x^2+9}$. Thus

$$\int \frac{dx}{\sqrt{x^2+9}^2} = \frac{1}{54}\tan^{-1}\left(\frac{x}{3}\right) + \frac{1}{54}\left(\frac{x}{\sqrt{x^2+9}}\right)\left(\frac{3}{\sqrt{x^2+9}}\right) + C = \frac{1}{54}\tan^{-1}\left(\frac{x}{3}\right) + \frac{x}{18(x^2+9)} + C.$$

57. $\displaystyle\int \tan^5 x\sec x\,dx$

SOLUTION Use the trigonometric identity $\tan^2 x = \sec^2 x - 1$ to write

$$\int \tan^5 x \sec x \, dx = \int \left(\sec^2 x - 1\right)^2 \tan x \sec x \, dx.$$

Now use the substitution $u = \sec x$, $du = \sec x \tan x \, dx$:

$$\int \tan^5 x \sec x \, dx = \int (u^2 - 1)^2 \, du = \int \left(u^4 - 2u^2 + 1\right) du$$

$$= \frac{1}{5}u^5 - \frac{2}{3}u^3 + u + C = \frac{1}{5}\sec^5 x - \frac{2}{3}\sec^3 x + \sec x + C.$$

59. $\displaystyle\int \frac{x^2 \, dx}{(x^2 - 1)^{3/2}}$

SOLUTION Use the trigonometric substitution $x = \sec\theta$. Then $dx = \sec\theta \tan\theta \, d\theta$,

$$x^2 - 1 = \sec^2\theta - 1 = \tan^2\theta,$$

and

$$\int \frac{x^2 \, dx}{(x^2 - 1)^{3/2}} = \int \frac{(\sec^2\theta)\sec\theta\tan\theta \, d\theta}{(\tan^2\theta)^{3/2}} = \int \frac{\sec^3\theta \, d\theta}{\tan^2\theta} = \int \frac{(\tan^2\theta + 1)\sec\theta \, d\theta}{\tan^2\theta}$$

$$= \int \frac{\tan^2\theta \sec\theta \, d\theta}{\tan^2\theta} + \int \frac{\sec\theta \, d\theta}{\tan^2\theta} = \int \sec\theta \, d\theta + \int \csc\theta \cot\theta \, d\theta$$

$$= \ln|\sec\theta + \tan\theta| - \csc\theta + C.$$

Now construct a right triangle with $\sec\theta = x$:

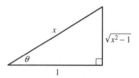

From this we see that $\tan\theta = \sqrt{x^2 - 1}$ and $\csc\theta = x/\sqrt{x^2 - 1}$. So the final answer is

$$\int \frac{x^2 \, dx}{(x^2 - 1)^{3/2}} = \ln\left|x + \sqrt{x^2 - 1}\right| - \frac{x}{\sqrt{x^2 - 1}} + C.$$

61. $\displaystyle\int \frac{dx}{x(x^2 - 1)}$

SOLUTION Using partial fractions, we first write

$$\frac{1}{x(x^2 - 1)} = \frac{1}{x(x - 1)(x + 1)} = \frac{A}{x} + \frac{B}{x - 1} + \frac{C}{x + 1}.$$

Clearing denominators gives us

$$1 = A(x - 1)(x + 1) + Bx(x + 1) + Cx(x - 1).$$

Setting $x = 0$ then yields

$$1 = A(-1) + 0 + 0 \quad\text{or}\quad A = -1,$$

while setting $x = 1$ yields

$$1 = 0 + B(2) + 0 \quad\text{or}\quad B = \frac{1}{2},$$

and setting $x = -1$ yields

$$1 = 0 + 0 + C(2) \quad\text{or}\quad C = \frac{1}{2}.$$

The result is

$$\frac{1}{x(x^2 - 1)} = \frac{-1}{x} + \frac{\frac{1}{2}}{x - 1} + \frac{\frac{1}{2}}{x + 1}.$$

Thus,

$$\int \frac{dx}{x(x^2-1)} = -\int \frac{dx}{x} + \frac{1}{2}\int \frac{dx}{x-1} + \frac{1}{2}\int \frac{dx}{x+1} = -\ln|x| + \frac{1}{2}\ln|x-1| + \frac{1}{2}\ln|x+1| + C.$$

63. $\displaystyle\int \frac{\sqrt{x}\,dx}{x^3+1}$

SOLUTION Use the substitution $u = x^{3/2}$, $du = \frac{3}{2}x^{1/2}\,dx$. Then $x^3 = (x^{3/2})^2 = u^2$, so we have

$$\int \frac{\sqrt{x}\,dx}{x^3+1} = \frac{2}{3}\int \frac{du}{u^2+1} = \frac{2}{3}\tan^{-1}u + C = \frac{2}{3}\tan^{-1}(x^{3/2}) + C.$$

65. $\displaystyle\int \frac{dx}{x^4(x^3+1)}$

SOLUTION Use the substitution $u = x^3$, $du = 3x^2\,dx$. Then $dx = du/3x^2$, and we have

$$\int \frac{dx}{x^4(x^3+1)} = \int \frac{du}{x^4(3x^2)(u+1)} = \frac{1}{3}\int \frac{du}{x^6(u+1)} = \frac{1}{3}\int \frac{du}{u^2(u+1)}.$$

Now we can use partial fractions:

$$\frac{1}{u^2(u+1)} = \frac{A}{u} + \frac{B}{u^2} + \frac{C}{u+1}.$$

Clearing denominators gives us

$$1 = Au(u+1) + B(u+1) + Cu^2.$$

Setting $u = 0$ then yields

$$1 = 0 + B(1) + 0 \qquad \text{or} \qquad B = 1,$$

while setting $u = -1$ yields

$$1 = 0 + 0 + C(1) \qquad \text{or} \qquad C = 1,$$

and setting $u = 1$ yields

$$1 = 2A + 2 + 1 \qquad \text{or} \qquad A = -1.$$

The result is

$$\frac{1}{u^2(u+1)} = \frac{-1}{u} + \frac{1}{u^2} + \frac{1}{u+1}.$$

Thus,

$$\frac{1}{3}\int \frac{du}{u^2(u+1)} = \frac{1}{3}\left[-\int \frac{du}{u} + \int \frac{du}{u^2} + \int \frac{du}{u+1} \right] = \frac{1}{3}\left[-\ln|u| - \frac{1}{u} + \ln|u+1| \right] + C.$$

The final answer is

$$\int \frac{dx}{x^4(x^3+1)} = -\frac{1}{3}\ln|x^3| - \frac{1}{3x^3} + \frac{1}{3}\ln|x^3+1| + C.$$

67. Show that the substitution $\theta = 2\tan^{-1}t$ (Figure 2) yields the formulas

$$\cos\theta = \frac{1-t^2}{1+t^2}, \qquad \sin\theta = \frac{2t}{1+t^2}, \qquad d\theta = \frac{2\,dt}{1+t^2} \qquad \boxed{10}$$

This substitution transforms the integral of any rational function of $\cos\theta$ and $\sin\theta$ into an integral of a rational function of t (which can then be evaluated using partial fractions). Use it to evaluate $\displaystyle\int \frac{d\theta}{\cos\theta + (3/4)\sin\theta}$.

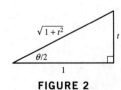

FIGURE 2

SOLUTION   If $\theta = 2 \tan^{-1} t$, then $d\theta = 2\,dt/(1 + t^2)$. We also have that $\cos(\frac{\theta}{2}) = 1/\sqrt{1+t^2}$ and $\sin(\frac{\theta}{2}) = t/\sqrt{1+t^2}$. To find $\cos\theta$, we use the double angle identity $\cos\theta = 1 - 2\sin^2(\frac{\theta}{2})$. This gives us

$$\cos\theta = 1 - 2\left(\frac{t}{\sqrt{1+t^2}}\right)^2 = 1 - \frac{2t^2}{1+t^2} = \frac{1+t^2-2t^2}{1+t^2} = \frac{1-t^2}{1+t^2}.$$

To find $\sin\theta$, we use the double angle identity $\sin\theta = 2\sin(\frac{\theta}{2})\cos(\frac{\theta}{2})$. This gives us

$$\sin\theta = 2\left(\frac{t}{\sqrt{1+t^2}}\right)\left(\frac{1}{\sqrt{1+t^2}}\right) = \frac{2t}{1+t^2}.$$

With these formulas, we have

$$\int \frac{d\theta}{\cos\theta + (3/4)\sin\theta} = \int \frac{\frac{2\,dt}{1+t^2}}{\left(\frac{1-t^2}{1+t^2}\right) + \frac{3}{4}\left(\frac{2t}{1+t^2}\right)} = \int \frac{8\,dt}{4(1-t^2)+3(2t)} = \int \frac{8\,dt}{4+6t-4t^2} = \int \frac{4\,dt}{2+3t-2t^2}.$$

The partial fraction decomposition has the form

$$\frac{4}{2+3t-2t^2} = \frac{A}{2-t} + \frac{B}{1+2t}.$$

Clearing denominators gives us

$$4 = A(1+2t) + B(2-t).$$

Setting $t = 2$ then yields

$$4 = A(5) + 0 \qquad \text{or} \qquad A = \frac{4}{5},$$

while setting $t = -\frac{1}{2}$ yields

$$4 = 0 + B\left(\frac{5}{2}\right) \qquad \text{or} \qquad B = \frac{8}{5}.$$

The result is

$$\frac{4}{2+3t-2t^2} = \frac{\frac{4}{5}}{2-t} + \frac{\frac{8}{5}}{1+2t}.$$

Thus,

$$\int \frac{4}{2+3t-2t^2}\,dt = \frac{4}{5}\int\frac{dt}{2-t} + \frac{8}{5}\int\frac{dt}{1+2t} = -\frac{4}{5}\ln|2-t| + \frac{4}{5}\ln|1+2t| + C.$$

The original substitution was $\theta = 2\tan^{-1} t$, which means that $t = \tan(\frac{\theta}{2})$. The final answer is then

$$\int \frac{d\theta}{\cos\theta + \frac{3}{4}\sin\theta} = -\frac{4}{5}\ln\left|2-\tan\left(\frac{\theta}{2}\right)\right| + \frac{4}{5}\ln\left|1+2\tan\left(\frac{\theta}{2}\right)\right| + C.$$

Further Insights and Challenges

69. Prove the general formula

$$\int \frac{dx}{(x-a)(x-b)} = \frac{1}{a-b}\ln\frac{x-a}{x-b} + C$$

where a, b are constants such that $a \neq b$.

SOLUTION   The partial fraction decomposition has the form:

$$\frac{1}{(x-a)(x-b)} = \frac{A}{x-a} + \frac{B}{x-b}.$$

Clearing denominators, we get

$$1 = A(x-b) + B(x-a).$$

Setting $x = a$ then yields

$$1 = A(a - b) + 0 \qquad \text{or} \qquad A = \frac{1}{a - b},$$

while setting $x = b$ yields

$$1 = 0 + B(b - a) \qquad \text{or} \qquad B = \frac{1}{b - a}.$$

The result is

$$\frac{1}{(x - a)(x - b)} = \frac{\frac{1}{a-b}}{x - a} + \frac{\frac{1}{b-a}}{x - b}.$$

Thus,

$$\int \frac{dx}{(x - a)(x - b)} = \frac{1}{a - b} \int \frac{dx}{x - a} + \frac{1}{b - a} \int \frac{dx}{x - b} = \frac{1}{a - b} \ln|x - a| + \frac{1}{b - a} \ln|x - b| + C$$

$$= \frac{1}{a - b} \ln|x - a| - \frac{1}{a - b} \ln|x - b| + C = \frac{1}{a - b} \ln\left|\frac{x - a}{x - b}\right| + C.$$

71. Suppose that $Q(x) = (x - a)(x - b)$, where $a \neq b$, and let $\dfrac{P(x)}{Q(x)}$ be a proper rational function so that

$$\frac{P(x)}{Q(x)} = \frac{A}{(x - a)} + \frac{B}{(x - b)}$$

(a) Show that $A = \dfrac{P(a)}{Q'(a)}$ and $B = \dfrac{P(b)}{Q'(b)}$.

(b) Use this result to find the partial fraction decomposition for $P(x) = 3x - 2$ and $Q(x) = x^2 - 4x - 12$.

SOLUTION

(a) Clearing denominators gives us

$$P(x) = A(x - b) + B(x - a).$$

Setting $x = a$ then yields

$$P(a) = A(a - b) + 0 \qquad \text{or} \qquad A = \frac{P(a)}{a - b},$$

while setting $x = b$ yields

$$P(b) = 0 + B(b - a) \qquad \text{or} \qquad B = \frac{P(b)}{b - a}.$$

Now use the product rule to differentiate $Q(x)$:

$$Q'(x) = (x - a)(1) + (1)(x - b) = x - a + x - b = 2x - a - b;$$

therefore,

$$Q'(a) = 2a - a - b = a - b$$
$$Q'(b) = 2b - a - b = b - a$$

Substituting these into the above results, we find

$$A = \frac{P(a)}{Q'(a)} \quad \text{and} \quad B = \frac{P(b)}{Q'(b)}.$$

(b) The partial fraction decomposition has the form:

$$\frac{P(x)}{Q(x)} = \frac{3x - 2}{x^2 - 4x - 12} = \frac{3x - 2}{(x - 6)(x + 2)} = \frac{A}{x - 6} + \frac{B}{x + 2};$$

$$A = \frac{P(6)}{Q'(6)} = \frac{3(6) - 2}{2(6) - 4} = \frac{16}{8} = 2;$$

$$B = \frac{P(-2)}{Q'(-2)} = \frac{3(-2) - 2}{2(-2) - 4} = \frac{-8}{-8} = 1.$$

The result is

$$\frac{3x - 2}{x^2 - 4x - 12} = \frac{2}{x - 6} + \frac{1}{x + 2}.$$

8.6 Improper Integrals

Preliminary Questions

1. State whether the integral converges or diverges:

(a) $\int_1^\infty x^{-3}\, dx$

(b) $\int_0^1 x^{-3}\, dx$

(c) $\int_1^\infty x^{-2/3}\, dx$

(d) $\int_0^1 x^{-2/3}\, dx$

SOLUTION

(a) The integral is improper because one of the limits of integration is infinite. Because the power of x in the integrand is less than -1, this integral converges.

(b) The integral is improper because the integrand is undefined at $x = 0$. Because the power of x in the integrand is less than -1, this integral diverges.

(c) The integral is improper because one of the limits of integration is infinite. Because the power of x in the integrand is greater than -1, this integral diverges.

(d) The integral is improper because the integrand is undefined at $x = 0$. Because the power of x in the integrand is greater than -1, this integral converges.

2. Is $\int_0^{\pi/2} \cot x\, dx$ an improper integral? Explain.

SOLUTION Because the integrand $\cot x$ is undefined at $x = 0$, this is an improper integral.

3. Find a value of $b > 0$ that makes $\int_0^b \dfrac{1}{x^2 - 4}\, dx$ an improper integral.

SOLUTION Any value of b satisfying $|b| \geq 2$ will make this an improper integral.

4. Which comparison would show that $\int_0^\infty \dfrac{dx}{x + e^x}$ converges?

SOLUTION Note that, for $x > 0$,

$$\frac{1}{x + e^x} < \frac{1}{e^x} = e^{-x}.$$

Moreover

$$\int_0^\infty e^{-x}\, dx$$

converges. Therefore,

$$\int_0^\infty \frac{1}{x + e^x}\, dx$$

converges by the comparison test.

5. Explain why it is not possible to draw any conclusions about the convergence of $\int_1^\infty \dfrac{e^{-x}}{x}\, dx$ by comparing with the integral $\int_1^\infty \dfrac{dx}{x}$.

SOLUTION For $1 \leq x < \infty$,

$$\frac{e^{-x}}{x} < \frac{1}{x},$$

but

$$\int_1^\infty \frac{dx}{x}$$

diverges. Knowing that an integral is smaller than a divergent integral does not allow us to draw any conclusions using the comparison test.

Exercises

1. Which of the following integrals is improper? Explain your answer, but do not evaluate the integral.

(a) $\displaystyle\int_0^2 \frac{dx}{x^{1/3}}\,dx$

(b) $\displaystyle\int_1^\infty \frac{dx}{x^{0.2}}$

(c) $\displaystyle\int_{-1}^\infty e^{-x}\,dx$

(d) $\displaystyle\int_0^1 e^{-x}$

(e) $\displaystyle\int_0^{\pi/2} \sec x\,dx$

(f) $\displaystyle\int_0^\infty \sin x\,dx$

(g) $\displaystyle\int_0^1 \sin x\,dx$

(h) $\displaystyle\int_0^1 \frac{dx}{\sqrt{3-x^2}}$

(i) $\displaystyle\int_1^\infty \ln x\,dx$

(j) $\displaystyle\int_0^3 \ln x\,dx$

SOLUTION

(a) Improper. The function $x^{-1/3}$ is infinite at 0.

(b) Improper. Infinite interval of integration.

(c) Improper. Infinite interval of integration.

(d) Proper. The function e^{-x} is continuous on the finite interval $[0, 1]$.

(e) Improper. The function $\sec x$ is infinite at $\frac{\pi}{2}$.

(f) Improper. Infinite interval of integration.

(g) Proper. The function $\sin x$ is continuous on the finite interval $[0, 1]$.

(h) Proper. The function $1/\sqrt{3-x^2}$ is continuous on the finite interval $[0, 1]$.

(i) Improper. Infinite interval of integration.

(j) Improper. The function $\ln x$ is infinite at 0.

3. Prove that $\displaystyle\int_1^\infty x^{-2/3}\,dx$ diverges by showing that

$$\lim_{R\to\infty}\int_1^R x^{-2/3}\,dx = \infty$$

SOLUTION   First compute the proper integral:

$$\int_1^R x^{-2/3}\,dx = 3x^{1/3}\Big|_1^R = 3R^{1/3} - 3 = 3\left(R^{1/3} - 1\right).$$

Then show divergence:

$$\int_1^\infty x^{-2/3}\,dx = \lim_{R\to\infty}\int_1^R x^{-2/3}\,dx = \lim_{R\to\infty}3\left(R^{1/3} - 1\right) = \infty.$$

In Exercises 5–46, determine whether the improper integral converges and, if so, evaluate it.

5. $\displaystyle\int_1^\infty \frac{dx}{x^{19/20}}$

SOLUTION   First evaluate the integral over the finite interval $[1, R]$ for $R > 1$:

$$\int_1^R \frac{dx}{x^{19/20}} = 20x^{1/20}\Big|_1^R = 20R^{1/20} - 20.$$

Now compute the limit as $R \to \infty$:

$$\int_1^\infty \frac{dx}{x^{19/20}} = \lim_{R\to\infty}\int_1^R \frac{dx}{x^{19/20}} = \lim_{R\to\infty}\left(20R^{1/20} - 20\right) = \infty.$$

The integral does not converge.

7. $\displaystyle\int_{-\infty}^4 e^{0.0001t}\,dt$

SOLUTION First evaluate the integral over the finite interval $[R, 4]$ for $R < 4$:

$$\int_R^4 e^{(0.0001)t}\, dt = \frac{e^{(0.0001)t}}{0.0001}\bigg|_R^4 = 10{,}000\left(e^{0.0004} - e^{(0.0001)R}\right).$$

Now compute the limit as $R \to -\infty$:

$$\int_{-\infty}^4 e^{(0.0001)t}\, dt = \lim_{R\to-\infty}\int_R^4 e^{(0.0001)t}\, dt = \lim_{R\to-\infty} 10{,}000\left(e^{0.0004} - e^{(0.0001)R}\right)$$

$$= 10{,}000\left(e^{0.0004} - 0\right) = 10{,}000 e^{0.0004}.$$

9. $\displaystyle\int_0^5 \frac{dx}{x^{20/19}}$

SOLUTION The function $x^{-20/19}$ is infinite at the endpoint 0, so we'll first evaluate the integral on the finite interval $[R, 5]$ for $0 < R < 5$:

$$\int_R^5 \frac{dx}{x^{20/19}} = -19 x^{-1/19}\bigg|_R^5 = -19\left(5^{-1/19} - R^{-1/19}\right) = 19\left(\frac{1}{R^{1/19}} - \frac{1}{5^{1/19}}\right).$$

Now compute the limit as $R \to 0^+$:

$$\int_0^5 \frac{dx}{x^{20/19}} = \lim_{R\to 0^+}\int_R^5 \frac{dx}{x^{20/19}} = \lim_{R\to 0^+} 19\left(\frac{1}{R^{1/19}} - \frac{1}{5^{1/19}}\right) = \infty;$$

thus, the integral does not converge.

11. $\displaystyle\int_0^4 \frac{dx}{\sqrt{4-x}}$

SOLUTION The function $1/\sqrt{4-x}$ is infinite at $x = 4$, but is left-continuous at $x = 4$, so we'll first evaluate the integral on the interval $[0, R]$ for $0 < R < 4$:

$$\int_0^R \frac{dx}{\sqrt{4-x}} = -2\sqrt{4-x}\bigg|_0^R = -2\sqrt{4-R} - (-2)\sqrt{4} = 4 - 2\sqrt{4-R}.$$

Now compute the limit as $R \to 4^-$:

$$\int_0^4 \frac{dx}{\sqrt{4-x}} = \lim_{R\to 4^-}\int_0^R \frac{dx}{\sqrt{4-x}} = \lim_{R\to 4^-}\left(4 - 2\sqrt{4-R}\right) = 4 - 0 = 4.$$

13. $\displaystyle\int_2^\infty x^{-3}\, dx$

SOLUTION First evaluate the integral on the finite interval $[2, R]$ for $2 < R$:

$$\int_2^R x^{-3}\, dx = \frac{x^{-2}}{-2}\bigg|_2^R = \frac{-1}{2R^2} - \frac{-1}{2(2^2)} = \frac{1}{8} - \frac{1}{2R^2}.$$

Now compute the limit as $R \to \infty$:

$$\int_2^\infty x^{-3}\, dx = \lim_{R\to\infty}\int_2^R x^{-3}\, dx = \lim_{R\to\infty}\left(\frac{1}{8} - \frac{1}{2R^2}\right) = \frac{1}{8}.$$

15. $\displaystyle\int_{-3}^\infty \frac{dx}{(x+4)^{3/2}}$

SOLUTION First evaluate the integral on the finite interval $[-3, R]$ for $R > -3$:

$$\int_{-3}^R \frac{dx}{(x+4)^{3/2}} = -2(x+4)^{-1/2}\bigg|_{-3}^R = \frac{-2}{\sqrt{R+4}} - \frac{-2}{\sqrt{1}} = 2 - \frac{2}{\sqrt{R+4}}.$$

Now compute the limit as $R \to \infty$:

$$\int_{-3}^\infty \frac{dx}{(x+4)^{3/2}} = \lim_{R\to\infty}\int_{-3}^R \frac{dx}{(x+4)^{3/2}} = \lim_{R\to\infty}\left(2 - \frac{2}{\sqrt{R+4}}\right) = 2 - 0 = 2.$$

17. $\displaystyle\int_0^1 \frac{dx}{x^{0.2}}$

SOLUTION The function $x^{-0.2}$ is infinite at $x = 0$ and right-continuous at $x = 0$, so we'll first evaluate the integral on the interval $[R, 1]$ for $0 < R < 1$:

$$\int_R^1 \frac{dx}{x^{0.2}} = \frac{x^{0.8}}{0.8}\bigg|_R^1 = 1.25\left(1 - R^{0.8}\right).$$

Now compute the limit as $R \to 0^+$:

$$\int_0^1 \frac{dx}{x^{0.2}} = \lim_{R \to 0^+} \int_R^1 \frac{dx}{x^{0.2}} = \lim_{R \to 0^+} 1.25\left(1 - R^{0.8}\right) = 1.25(1 - 0) = 1.25.$$

19. $\displaystyle\int_4^\infty e^{-3x}\, dx$

SOLUTION First evaluate the integral on the finite interval $[4, R]$ for $R > 4$:

$$\int_4^R e^{-3x}\, dx = \frac{e^{-3x}}{-3}\bigg|_4^R = -\frac{1}{3}\left(e^{-3R} - e^{-12}\right) = \frac{1}{3}\left(e^{-12} - e^{-3R}\right).$$

Now compute the limit as $R \to \infty$:

$$\int_4^\infty e^{-3x}\, dx = \lim_{R \to \infty} \int_4^R e^{-3x}\, dx = \lim_{R \to \infty} \frac{1}{3}\left(e^{-12} - e^{-3R}\right) = \frac{1}{3}\left(e^{-12} - 0\right) = \frac{1}{3e^{12}}.$$

21. $\displaystyle\int_{-\infty}^0 e^{3x}\, dx$

SOLUTION First evaluate the integral on the finite interval $[R, 0]$ for $R < 0$:

$$\int_R^0 e^{3x}\, dx = \frac{e^{3x}}{3}\bigg|_R^0 = \frac{1}{3} - \frac{e^{3R}}{3}.$$

Now compute the limit as $R \to -\infty$:

$$\int_{-\infty}^0 e^{3x}\, dx = \lim_{R \to -\infty} \int_R^0 e^{3x}\, dx = \lim_{R \to -\infty} \left(\frac{1}{3} - \frac{e^{3R}}{3}\right) = \frac{1}{3} - 0 = \frac{1}{3}.$$

23. $\displaystyle\int_2^\infty \frac{dx}{(x+3)^4}$

SOLUTION First evaluate the integral on the finite interval $[2, R]$ for $R > 2$:

$$\int_2^R \frac{dx}{(x+3)^4} = \frac{(x+3)^{-3}}{-3}\bigg|_2^R = \frac{-1}{3}\left[\frac{1}{(R+3)^3} - \frac{1}{5^3}\right].$$

Now compute the limit as $R \to \infty$:

$$\int_2^\infty \frac{dx}{(x+3)^4} = \lim_{R \to \infty} \int_2^R \frac{dx}{(x+3)^4} = \lim_{R \to -\infty} \frac{1}{3}\left[\frac{1}{5^3} - \frac{1}{(R+3)^3}\right] = \frac{1}{3}\left[\frac{1}{5^3} - 0\right] = \frac{1}{375}.$$

25. $\displaystyle\int_1^3 \frac{dx}{\sqrt{3-x}}$

SOLUTION The function $f(x) = 1/\sqrt{3-x}$ is infinite at $x = 3$ and is left continuous at $x = 3$, so we first evaluate the integral on the interval $[1, R]$ for $1 < R < 3$:

$$\int_1^R \frac{dx}{\sqrt{3-x}} = -2\sqrt{3-x}\bigg|_1^R = -2\sqrt{3-R} + 2\sqrt{2}.$$

Now compute the limit as $R \to 3^-$:

$$\int_1^3 \frac{dx}{\sqrt{3-x}} = \lim_{R \to 3^-} \int_1^R \frac{dx}{\sqrt{3-x}} = 0 + 2\sqrt{2} = 2\sqrt{2}.$$

27. $\displaystyle\int_0^\infty \frac{dx}{1+x}$

SOLUTION   First evaluate the integral on the finite interval $[0, R]$ for $R > 0$:

$$\int_0^R \frac{dx}{1+x} = \ln|1+x|\Big|_0^R = \ln|1+R| - \ln 1 = \ln|1+R|.$$

Now compute the limit as $R \to \infty$:

$$\int_0^\infty \frac{dx}{1+x} = \lim_{R\to\infty} \int_0^R \frac{dx}{1+x} = \lim_{R\to\infty} \ln|1+R| = \infty;$$

thus, the integral does not converge.

29. $\displaystyle\int_0^\infty \frac{dx}{(1+x^2)^2}$

SOLUTION   First evaluate the indefinite integral using the trigonometric substitution $x = \tan\theta$. Then $dx = \sec^2\theta\,d\theta$, $1 + x^2 = 1 + \tan^2\theta = \sec^2\theta$, and we have

$$\int \frac{dx}{(1+x^2)^2} = \int \frac{\sec^2\theta\,d\theta}{\sec^4\theta} = \int \cos^2\theta\,d\theta = \frac{1}{2}\theta + \frac{1}{2}\sin\theta\cos\theta + C.$$

Now construct a right triangle with $\tan\theta = x$:

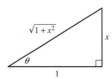

From this we see that $\sin\theta = x/\sqrt{1+x^2}$ and $\cos\theta = 1/\sqrt{1+x^2}$. Thus

$$\int \frac{dx}{(1+x^2)^2} = \frac{1}{2}\tan^{-1}x + \frac{1}{2}\left(\frac{x}{\sqrt{1+x^2}}\right)\left(\frac{1}{\sqrt{1+x^2}}\right) + C = \frac{1}{2}\tan^{-1}x + \frac{x}{2(1+x^2)} + C.$$

Finally,

$$\int_0^\infty \frac{dx}{(1+x^2)^2} = \lim_{R\to\infty}\int_0^R \frac{dx}{(1+x^2)^2} = \lim_{R\to\infty}\left[\left(\frac{1}{2}\tan^{-1}R + \frac{R}{2(1+R^2)}\right) - 0\right]$$

$$= \lim_{R\to\infty}\left(\frac{1}{2}\tan^{-1}R + \frac{1/R}{2/R^2+2}\right) = \frac{1}{2}\left(\frac{\pi}{2}\right) + \frac{0}{0+2} = \frac{\pi}{4} + 0 = \frac{\pi}{4}.$$

31. $\displaystyle\int_0^\infty e^{-x}\cos x\,dx$

SOLUTION   First evaluate the indefinite integral using Integration by Parts, with $u = e^{-x}$, $v' = \cos x$. Then $u' = -e^{-x}$, $v = \sin x$, and

$$\int e^{-x}\cos x\,dx = e^{-x}\sin x - \int \sin x(-e^{-x})\,dx = e^{-x}\sin x + \int e^{-x}\sin x\,dx.$$

Now use Integration by Parts again, with $u = e^{-x}$, $v' = \sin x$. Then $u' = -e^{-x}$, $v = -\cos x$, and

$$\int e^{-x}\cos x\,dx = e^{-x}\sin x + \left[-e^{-x}\cos x - \int e^{-x}\cos x\,dx\right].$$

Solving this equation for $\int e^{-x}\cos x\,dx$, we find

$$\int e^{-x}\cos x\,dx = \frac{1}{2}e^{-x}(\sin x - \cos x) + C.$$

Thus,

$$\int_0^R e^{-x}\cos x\,dx = \frac{1}{2}e^{-x}(\sin x - \cos x)\Big|_0^R = \frac{\sin R - \cos R}{2e^R} - \frac{\sin 0 - \cos 0}{2} = \frac{\sin R - \cos R}{2e^R} + \frac{1}{2},$$

and

$$\int_0^\infty e^{-x}\cos x\,dx = \lim_{R\to\infty}\left(\frac{\sin R - \cos R}{2e^R} + \frac{1}{2}\right) = 0 + \frac{1}{2} = \frac{1}{2}.$$

33. $\displaystyle\int_0^1 \frac{dx}{\sqrt{1-x^2}}$

SOLUTION   The function $(1-x^2)^{-1/2}$ is infinite at $x=1$, and is left-continuous at $x=1$, so we'll first evaluate the integral on the interval $[0,R]$ for $0 < R < 1$:

$$\int_0^R \frac{dx}{\sqrt{1-x^2}} = \sin^{-1} x \Big|_0^R = \sin^{-1} R - \sin^{-1} 0 = \sin^{-1} R.$$

Thus,

$$\int_0^1 \frac{dx}{\sqrt{1-x^2}} = \lim_{R\to 1^-} \sin^{-1} R = \sin^{-1} 1 = \frac{\pi}{2}.$$

35. $\displaystyle\int_1^\infty \frac{e^{\sqrt{x}}\, dx}{\sqrt{x}}$

SOLUTION   Let $u = \sqrt{x}, du = \frac{1}{2}x^{-1/2}\, dx$. Then

$$\int \frac{e^{\sqrt{x}}\, dx}{\sqrt{x}} = 2 \int e^{\sqrt{x}} \left(\frac{dx}{2\sqrt{x}}\right) = 2 \int e^u\, du = 2e^u + C = 2e^{\sqrt{x}} + C,$$

and

$$\int_0^\infty \frac{e^{\sqrt{x}}\, dx}{\sqrt{x}} = \lim_{R\to\infty} \int_0^R \frac{e^{\sqrt{x}}\, dx}{\sqrt{x}} = \lim_{R\to\infty} 2e^{\sqrt{x}} \Big|_0^R = \lim_{R\to\infty} \left(2e^{\sqrt{R}} - 2\right) = \infty.$$

The integral does not converge.

37. $\displaystyle\int_0^\infty \frac{x\, dx}{x^2+1}$

SOLUTION   First, evaluate the integral on the finite interval $[0,R]$ for $R > 0$:

$$\int_0^R \frac{x}{x^2+1}\, dx = \frac{1}{2}\ln(1+x^2) \Big|_0^R = \frac{1}{2}\ln(1+R^2).$$

Now compute the limit as $R \to \infty$:

$$\int_0^\infty \frac{x}{x^2+1}\, dx = \lim_{R\to\infty} \int_0^R \frac{x}{x^2+1}\, dx = \lim_{R\to\infty} \frac{1}{2}\ln(1+R^2) = \infty.$$

The integral does not converge.

39. $\displaystyle\int_0^\infty \sin x\, dx$

SOLUTION   First evaluate the integral on the finite interval $[0,R]$ for $R > 0$:

$$\int_0^R \sin x\, dx = -\cos x \Big|_0^R = -\cos R + \cos 0 = 1 - \cos R.$$

Thus,

$$\int_0^R \sin x\, dx = \lim_{R\to\infty} (1-\cos R) = 1 - \lim_{R\to\infty} \cos R.$$

This limit does not exist, since the value of $\cos R$ oscillates between 1 and -1 as R approaches infinity. Hence the integral does not converge.

41. $\displaystyle\int_0^{\pi/2} \tan x \sec x\, dx$

SOLUTION   The function $\tan x \sec x$ is infinite and left-continuous at $x = \frac{\pi}{2}$ so we'll first evaluate the integral on $[0,R]$ for $0 < R < \frac{\pi}{2}$:

$$\int_0^R \tan x \sec x\, dx = \sec x \Big|_0^R = \sec R - 1.$$

Thus,

$$\int_0^{\pi/2} \tan x \sec x \, dx = \lim_{R \to \frac{\pi}{2}^-} (\sec R - 1) = \infty.$$

The integral does not converge.

43. $\int_0^1 x \ln x \, dx$

SOLUTION The function $x \ln x$ is undefined at $x = 0$, so we'll first evaluate the integral on $[R, 1]$ for $0 < R < 1$. Use Integration by Parts with $u = \ln x$ and $v' = x$. Then $u' = 1/x$, $v = x^2/2$, and

$$\int_R^1 x \ln x \, dx = \frac{1}{2} x^2 \ln x \Big|_R^1 - \frac{1}{2} \int_R^1 x \, dx = \left(\frac{1}{2} x^2 \ln x - \frac{1}{4} x^2 \right) \Big|_R^1$$

$$= \left(\frac{1}{2} \ln 1 - \frac{1}{4} \right) - \left(\frac{1}{2} R^2 \ln R - \frac{1}{4} R^2 \right) = \frac{1}{4} R^2 - \frac{1}{2} R^2 \ln R - \frac{1}{4}.$$

Thus,

$$\int_0^1 \ln x \, dx = \lim_{R \to 0^+} \left(\frac{1}{4} R^2 - \frac{1}{2} R^2 \ln R - \frac{1}{4} \right) = -\frac{1}{4} - \lim_{R \to 0^+} \frac{1}{2} R^2 \ln R.$$

To evaluate the limit, rewrite the function as a quotient and apply L'Hôpital's Rule:

$$\int_0^1 x \ln x \, dx = -\frac{1}{4} - \lim_{R \to 0^+} \frac{\ln R}{2 \frac{1}{R^2}} = -\frac{1}{4} - \lim_{R \to 0^+} \frac{\frac{1}{R}}{-4 \frac{1}{R^3}} = -\frac{1}{4} - \lim_{R \to 0^+} \frac{R^2}{-4} = -\frac{1}{4} - 0 = -\frac{1}{4}.$$

45. $\int_0^1 \frac{\ln x}{x^2} \, dx$

SOLUTION Use Integration by Parts, with $u = \ln x$ and $v' = x^{-2}$. Then $u' = 1/x$, $v = -x^{-1}$, and

$$\int \frac{\ln x}{x^2} \, dx = -\frac{1}{x} \ln x + \int \frac{dx}{x^2} = -\frac{1}{x} \ln x - \frac{1}{x} + C.$$

The function is infinite and right-continuous at $x = 0$, so we'll first evaluate the integral on $[R, 1]$ for $0 < R < 1$:

$$\int_a^1 \frac{\ln x}{x^2} \, dx = \left(-\frac{1}{x} \ln x - \frac{1}{x} \right) \Big|_R^1 = \left(-\frac{1}{1} \ln 1 - \frac{1}{1} \right) - \left(-\frac{1}{R} \ln R - \frac{1}{R} \right) = \frac{1}{R} \ln R + \frac{1}{R} - 1.$$

Thus,

$$\int_0^1 \frac{\ln x}{x^2} \, dx = \lim_{R \to 0^+} \frac{1}{R} \ln R + \frac{1}{R} - 1 = -1 + \lim_{R \to 0^+} \frac{\ln R + 1}{R} = -\infty.$$

The integral does not converge.

47. Let $I = \int_4^\infty \frac{dx}{(x-2)(x-3)}$.

(a) Show that for $R > 4$,

$$\int_4^R \frac{dx}{(x-2)(x-3)} = \ln \left| \frac{R-3}{R-2} \right| - \ln \frac{1}{2}$$

(b) Then show that $I = \ln 2$.

SOLUTION

(a) The partial fraction decomposition takes the form

$$\frac{1}{(x-2)(x-3)} = \frac{A}{x-2} + \frac{B}{x-3}.$$

Clearing denominators gives us

$$1 = A(x-3) + B(x-2).$$

Setting $x = 2$ then yields $A = -1$, while setting $x = 3$ yields $B = 1$. Thus,

$$\int \frac{dx}{(x-2)(x-3)} = \int \frac{dx}{x-3} - \int \frac{dx}{x-2} = \ln|x-3| - \ln|x-2| + C = \ln \left| \frac{x-3}{x-2} \right| + C,$$

and, for $R > 4$,

$$\int_4^R \frac{dx}{(x-2)(x-3)} = \ln\left|\frac{x-3}{x-2}\right|\Big|_4^R = \ln\left|\frac{R-3}{R-2}\right| - \ln\frac{1}{2}.$$

(b) Using the result from part (a),

$$I = \lim_{R\to\infty}\left(\ln\left|\frac{R-3}{R-2}\right| - \ln\frac{1}{2}\right) = \ln 1 - \ln\frac{1}{2} = \ln 2.$$

49. Determine if $I = \displaystyle\int_0^1 \frac{dx}{x(2x+5)}$ converges and, if so, evaluate.

SOLUTION   The partial fraction decomposition takes the form

$$\frac{1}{x(2x+5)} = \frac{A}{x} + \frac{B}{2x+5}.$$

Clearing denominators gives us

$$1 = A(2x+5) + Bx.$$

Setting $x = 0$ then yields $A = \frac{1}{5}$, while setting $x = -\frac{5}{2}$ yields $B = -\frac{2}{5}$. Thus,

$$\int \frac{dx}{x(2x+5)} = \frac{1}{5}\int\frac{dx}{x} - \frac{2}{5}\int\frac{dx}{2x+5} = \frac{1}{5}\ln|x| - \frac{1}{5}\ln|2x+5| + C = \frac{1}{5}\ln\left|\frac{x}{2x+5}\right| + C,$$

and, for $0 < R < 1$,

$$\int_R^1 \frac{dx}{x(2x+5)} = \frac{1}{5}\ln\left|\frac{x}{2x+5}\right|\Big|_R^1 = \frac{1}{5}\ln\frac{1}{7} - \frac{1}{5}\ln\left|\frac{R}{2R+5}\right|.$$

Thus,

$$I = \lim_{R\to 0+}\left(\frac{1}{5}\ln\frac{1}{7} - \frac{1}{5}\ln\left|\frac{R}{2R+5}\right|\right) = \infty.$$

The integral does not converge.

In Exercises 51–54, determine if the doubly infinite improper integral converges and, if so, evaluate it. Use definition (2).

51. $\displaystyle\int_{-\infty}^{\infty} \frac{dx}{1+x^2}$

SOLUTION   First note that

$$\int \frac{dx}{1+x^2} = \tan^{-1}x + C.$$

Thus,

$$\int_0^\infty \frac{dx}{1+x^2} = \lim_{R\to\infty}\int_0^R \frac{dx}{1+x^2} = \lim_{R\to\infty}\left(\tan^{-1}R - \tan^{-1}0\right) = \frac{\pi}{2};$$

$$\int_{-\infty}^0 \frac{dx}{1+x^2} = \lim_{R\to-\infty}\int_R^0 \frac{dx}{1+x^2} = \lim_{R\to-\infty}\left(\tan^{-1}0 - \tan^{-1}R\right) = \frac{\pi}{2};$$

and

$$\int_\infty^\infty \frac{dx}{1+x^2} = \frac{\pi}{2} + \frac{\pi}{2} = \pi.$$

53. $\displaystyle\int_{-\infty}^{\infty} xe^{-x^2}\,dx$

SOLUTION   First note that

$$\int xe^{-x^2}\,dx = -\frac{1}{2}e^{-x^2} + C.$$

Thus,

$$\int_0^\infty xe^{-x^2}\,dx = \lim_{R\to\infty}\int_0^R xe^{-x^2}\,dx = \lim_{R\to\infty}\left(\frac{1}{2}-\frac{1}{2}e^{-R^2}\right) = \frac{1}{2};$$

$$\int_{-\infty}^0 xe^{-x^2}\,dx = \lim_{R\to-\infty}\int_R^0 xe^{-x^2}\,dx = \lim_{R\to-\infty}\left(-\frac{1}{2}+\frac{1}{2}e^{-R^2}\right) = -\frac{1}{2};$$

and

$$\int_{-\infty}^\infty xe^{-x^2}\,dx = \frac{1}{2}-\frac{1}{2} = 0.$$

55. For which values of a does $\displaystyle\int_0^\infty e^{ax}\,dx$ converge?

SOLUTION   First evaluate the integral on the finite interval $[0, R]$ for $R > 0$:

$$\int_0^R e^{ax}\,dx = \frac{1}{a}e^{ax}\Big|_0^R = \frac{1}{a}\left(e^{aR}-1\right).$$

Thus,

$$\int_0^\infty e^{ax}\,dx = \lim_{R\to\infty}\frac{1}{a}\left(e^{aR}-1\right).$$

If $a > 0$, then $e^{aR}\to\infty$ as $R\to\infty$. If $a < 0$, then $e^{aR}\to 0$ as $R\to\infty$, and

$$\int_0^\infty e^{ax}\,dx = \lim_{R\to\infty}\frac{1}{a}\left(e^{aR}-1\right) = -\frac{1}{a}.$$

The integral converges for $a < 0$.

57. Sketch the region under the graph of $f(x) = \dfrac{1}{1+x^2}$ for $-\infty < x < \infty$ and show that its area is π.

SOLUTION   The graph is shown below.

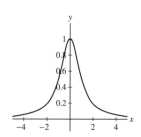

Since $(1+x^2)^{-1}$ is an even function, we can first compute the area under the graph for $x > 0$:

$$\int_0^R \frac{dx}{1+x^2} = \tan^{-1}x\Big|_0^R = \tan^{-1}R - \tan^{-1}0 = \tan^{-1}R.$$

Thus,

$$\int_0^\infty \frac{dx}{1+x^2} = \lim_{R\to\infty}\tan^{-1}R = \frac{\pi}{2}.$$

By symmetry, we have

$$\int_{-\infty}^\infty \frac{dx}{1+x^2} = \int_{-\infty}^0 \frac{dx}{1+x^2} + \int_0^\infty \frac{dx}{1+x^2} = \frac{\pi}{2}+\frac{\pi}{2} = \pi.$$

59. Show that $\displaystyle\int_1^\infty \frac{dx}{x^3+4}$ converges by comparing with $\displaystyle\int_1^\infty x^{-3}\,dx$.

SOLUTION   The integral $\displaystyle\int_1^\infty x^{-3}\,dx$ converges because $3 > 1$. Since $x^3+4\geq x^3$, it follows that

$$\frac{1}{x^3+4} \leq \frac{1}{x^3}.$$

Therefore, by the comparison test,

$$\int_1^\infty \frac{dx}{x^3 + 4} \text{ converges.}$$

61. ✎ Show that $0 \le e^{-x^2} \le e^{-x}$ for $x \ge 1$ (Figure 11). Then use the Comparison Test and (3) to show that $\int_0^\infty e^{-x^2}\,dx$ converges.

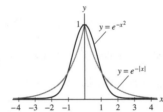

FIGURE 11 Comparison of $y = e^{-|x|}$ and $y = e^{-x^2}$.

SOLUTION For $x \ge 1$, $x^2 \ge x$, so $-x^2 \le -x$ and $e^{-x^2} \le e^{-x}$. Now

$$\int_1^\infty e^{-x}\,dx \text{ converges,} \quad \text{so} \quad \int_1^\infty e^{-x^2}\,dx \text{ converges}$$

by the comparison test. Finally, because e^{-x^2} is continuous on $[0, 1]$,

$$\int_0^\infty e^{-x^2}\,dx \text{ converges.}$$

63. Show that $\int_1^\infty \frac{1 - \sin x}{x^2}\,dx$ converges.

SOLUTION Let $f(x) = \dfrac{1 - \sin x}{x^2}$. Since $f(x) \le \dfrac{2}{x^2}$ and $\int_1^\infty 2x^{-2}\,dx = 2$, it follows that

$$\int_1^\infty \frac{1 - \sin x}{x^2}\,dx \text{ converges}$$

by the comparison test.

In Exercises 65–74, use the Comparison Test to determine whether or not the integral converges.

65. $\int_1^\infty \dfrac{1}{\sqrt{x^5 + 2}}\,dx$

SOLUTION Since $\sqrt{x^5 + 2} \ge \sqrt{x^5} = x^{5/2}$, it follows that

$$\frac{1}{\sqrt{x^5 + 2}} \le \frac{1}{x^{5/2}}.$$

The integral $\int_1^\infty dx/x^{5/2}$ converges because $\frac{5}{2} > 1$. Therefore, by the comparison test:

$$\int_1^\infty \frac{dx}{\sqrt{x^5 + 2}} \text{ also converges.}$$

67. $\int_3^\infty \dfrac{dx}{\sqrt{x} - 1}$

SOLUTION Since $\sqrt{x} \ge \sqrt{x} - 1$, we have (for $x > 1$)

$$\frac{1}{\sqrt{x}} \le \frac{1}{\sqrt{x} - 1}.$$

The integral $\int_1^\infty dx/\sqrt{x} = \int_1^\infty dx/x^{1/2}$ diverges because $\frac{1}{2} < 1$. Since the function $x^{-1/2}$ is continuous (and therefore finite) on $[1, 3]$, we also know that $\int_3^\infty dx/x^{1/2}$ diverges. Therefore, by the comparison test,

$$\int_3^\infty \frac{dx}{\sqrt{x} - 1} \text{ also diverges.}$$

69. $\int_1^\infty e^{-(x+x^{-1})} dx$

SOLUTION For all $x \geq 1$, $\frac{1}{x} > 0$ so $x + \frac{1}{x} \geq x$. Then

$$-\left(x + x^{-1}\right) \leq -x \quad \text{and} \quad e^{-(x+x^{-1})} \leq e^{-x}.$$

The integral $\int_1^\infty e^{-x} dx$ converges by direct computation:

$$\int_1^\infty e^{-x} dx = \lim_{R \to \infty} \int_1^R e^{-x} dx = \lim_{R \to \infty} -e^{-x}\Big|_1^R = \lim_{R \to \infty} -e^{-R} + e^{-1} = 0 + e^{-1} = e^{-1}.$$

Therefore, by the comparison test,

$$\int_1^\infty e^{-(x+x^{-1})} \text{ also converges.}$$

71. $\int_0^1 \frac{e^x}{x^2} dx$

SOLUTION For $0 < x < 1$, $e^x > 1$, and therefore

$$\frac{1}{x^2} < \frac{e^x}{x^2}.$$

The integral $\int_0^1 dx/x^2$ diverges since $2 > 1$. Therefore, by the comparison test,

$$\int_0^1 \frac{e^x}{x^2} \text{ also diverges.}$$

73. $\int_0^1 \frac{1}{x^4 + \sqrt{x}} dx$

SOLUTION For $0 < x < 1$, $x^4 + \sqrt{x} \geq \sqrt{x}$, and

$$\frac{1}{x^4 + \sqrt{x}} \leq \frac{1}{\sqrt{x}}.$$

The integral $\int_0^1 (1/\sqrt{x}) dx$ converges, since $p = \frac{1}{2} < 1$. Therefore, by the comparison test,

$$\int_0^1 \frac{dx}{x^4 + \sqrt{x}} \text{ also converges.}$$

75. An investment pays a dividend of \$250/year continuously forever. If the interest rate is 7%, what is the present value of the entire income stream generated by the investment?

SOLUTION The present value of the income stream after T years is

$$\int_0^T 250e^{-0.07t} dt = \frac{250e^{-0.07t}}{-0.07}\Big|_0^T = \frac{-250}{0.07}\left(e^{-0.07T} - 1\right) = \frac{250}{0.07}\left(1 - e^{-0.07T}\right).$$

Therefore the present value of the entire income stream is

$$\int_0^\infty 250e^{-0.07t} = \lim_{T \to \infty} \int_0^T 250e^{-0.07t} = \lim_{T \to \infty} \frac{250}{0.07}\left(1 - e^{-0.07T}\right) = \frac{250}{0.07}(1 - 0) = \frac{250}{0.07} = \$3{,}571.43.$$

77. Compute the present value of an investment that generates income at a rate of $5{,}000te^{0.01t}$ dollars/year forever, assuming an interest rate of 6%.

SOLUTION The present value of the income stream after T years is

$$\int_0^T \left(5000te^{0.01t}\right) e^{-0.06t}\, dt = 5000 \int_0^T te^{-0.05t}\, dt$$

Compute the indefinite integral using Integration by Parts, with $u = t$ and $v' = e^{-0.05t}$. Then $u' = 1$, $v = (-1/0.05)e^{-0.05t}$, and

$$\int te^{-0.05t}\, dt = \frac{-t}{0.05}e^{-0.05t} + \frac{1}{0.05}\int e^{-0.05t}\, dt = -20te^{-0.05t} + \frac{20}{-0.05}e^{-0.05t} + C$$

$$= e^{-0.05t}(-20t - 400) + C.$$

Thus,

$$5000 \int_0^T te^{-0.05t}\, dt = 5000e^{-0.05t}(-20t - 400)\Big|_0^T = 5000e^{-0.05T}(-20T - 400) - 5000(-400)$$

$$= 2{,}000{,}000 - 5000e^{-0.05T}(20T + 400).$$

Use L'Hôpital's Rule to compute the limit:

$$\lim_{T\to\infty} \left(2{,}000{,}000 - \frac{5000(20T + 400)}{e^{0.05T}}\right) = 2{,}000{,}000 - \lim_{T\to\infty} \frac{5000(20)}{0.05e^{0.05T}} = 2{,}000{,}000 - 0 = \$2{,}000{,}000.$$

79. Let S be the solid obtained by rotating the region below the graph of $y = x^{-1}$ about the x-axis for $1 \le x < \infty$ (Figure 12).
(a) Use the Disk Method (Section 6.3) to compute the volume of S. Note that the volume is finite even though S is an infinite region.
(b) It can be shown that the surface area of S is

$$A = 2\pi \int_1^\infty x^{-1}\sqrt{1 + x^{-4}}\, dx.$$

Show that A is infinite.

If S were a container, you could fill its interior with a finite amount of paint, but you could not paint its surface with a finite amount of paint.

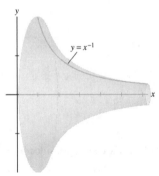

$y = x^{-1}$

FIGURE 12

SOLUTION
(a) The volume is given by

$$V = \int_1^\infty \pi \left(\frac{1}{x}\right)^2 dx.$$

First compute the volume over a finite interval:

$$\int_1^R \pi \left(\frac{1}{x}\right)^2 dx = \pi \int_1^R x^{-2}\, dx = \pi \frac{x^{-1}}{-1}\Big|_1^R = \pi\left(\frac{-1}{R} - \frac{-1}{1}\right) = \pi\left(1 - \frac{1}{R}\right).$$

Thus,

$$V = \lim_{R\to\infty} \int_1^\infty \pi x^{-2}\, dx = \lim_{R\to\infty} \pi\left(1 - \frac{1}{R}\right) = \pi.$$

(b) For $x > 1$, we have

$$\frac{1}{x}\sqrt{1 + \frac{1}{x^4}} = \frac{1}{x}\sqrt{\frac{x^4 + 1}{x^4}} = \frac{\sqrt{x^4 + 1}}{x^3} \geq \frac{\sqrt{x^4}}{x^3} = \frac{x^2}{x^3} = \frac{1}{x}.$$

The integral $\int_1^\infty \frac{1}{x}\,dx$ diverges, since $p = 1 \geq 1$. Therefore, by the comparison test,

$$\int_1^\infty \frac{1}{x}\sqrt{1 + \frac{1}{x^4}}\,dx \text{ also diverges.}$$

Finally,

$$A = 2\pi \int_1^\infty \frac{1}{x}\sqrt{1 + \frac{1}{x^4}}\,dx$$

diverges.

81. When a capacitor of capacitance C is charged by a source of voltage V, the power expended at time t is

$$P(t) = \frac{V^2}{R}(e^{-t/RC} - e^{-2t/RC}),$$

where R is the resistance in the circuit. The total energy stored in the capacitor is

$$W = \int_0^\infty P(t)\,dt$$

Show that $W = \frac{1}{2}CV^2$.

SOLUTION The total energy contained after the capacitor is fully charged is

$$W = \frac{V^2}{R}\int_0^\infty \left(e^{-t/RC} - e^{-2t/RC}\right)dt.$$

The energy after a finite amount of time $(t = T)$ is

$$\frac{V^2}{R}\int_0^T \left(e^{-t/RC} - e^{-2t/RC}\right)dt = \frac{V^2}{R}\left(-RCe^{-t/RC} + \frac{RC}{2}e^{-2t/RC}\right)\Bigg|_0^T$$

$$= V^2 C\left[\left(-e^{-T/RC} + \frac{1}{2}e^{-2T/RC}\right) - \left(-1 + \frac{1}{2}\right)\right]$$

$$= CV^2\left(\frac{1}{2} - e^{-T/RC} + \frac{1}{2}e^{-2T/RC}\right).$$

Thus,

$$W = \lim_{T\to\infty} CV^2\left(\frac{1}{2} - e^{-T/RC} + \frac{1}{2}e^{-2T/RC}\right) = CV^2\left(\frac{1}{2} - 0 + 0\right) = \frac{1}{2}CV^2.$$

83. When a radioactive substance decays, the fraction of atoms present at time t is $f(t) = e^{-kt}$, where $k > 0$ is the decay constant. It can be shown that the *average* life of an atom (until it decays) is $A = -\int_0^\infty tf'(t)\,dt$. Use Integration by Parts to show that $A = \int_0^\infty f(t)\,dt$ and compute A. What is the average decay time of Radon-222, whose half-life is 3.825 days?

SOLUTION Let $u = t$, $v' = f'(t)$. Then $u' = 1$, $v = f(t)$, and

$$A = -\int_0^\infty tf'(t)\,dt = -tf(t)\Bigg|_0^\infty + \int_0^\infty f(t)\,dt.$$

Since $f(t) = e^{-kt}$, we have

$$-tf(t)\Big|_0^\infty = \lim_{R\to\infty} -te^{-kt}\Bigg|_0^R = \lim_{R\to\infty} -Re^{-Rt} + 0 = \lim_{R\to\infty} \frac{-R}{e^{Rt}} = \lim_{R\to\infty} \frac{-1}{Re^{Rt}} = 0.$$

Here we used L'Hôpital's Rule to compute the limit. Thus

$$A = \int_0^\infty f(t)\,dt = \int_0^\infty e^{-kt}\,dt.$$

Now,

$$\int_0^R e^{-kt}\,dt = -\frac{1}{k}e^{-kt}\bigg|_0^R = -\frac{1}{k}\left(e^{-kR} - 1\right) = \frac{1}{k}\left(1 - e^{-kR}\right),$$

so

$$A = \lim_{R\to\infty} \frac{1}{k}\left(1 - e^{-kR}\right) = \frac{1}{k}(1 - 0) = \frac{1}{k}.$$

Because k has units of $(\text{time})^{-1}$, A does in fact have the appropriate units of time. To find the average decay time of Radon-222, we need to determine the decay constant k, given the half-life of 3.825 days. Recall that

$$k = \frac{\ln 2}{t_n}$$

where t_n is the half-life. Thus,

$$A = \frac{1}{k} = \frac{t_n}{\ln 2} = \frac{3.825}{\ln 2} \approx 5.518 \text{ days.}$$

85. Let $J_n = \int_0^\infty x^n e^{-\alpha x}\,dx$, where $n \geq 1$ is an integer and $\alpha > 0$. Prove that $J_n = (n/\alpha)J_{n-1}$ and $J_0 = 1/\alpha$. Use this to compute J_4. Show that $J_n = \dfrac{n!}{\alpha^{n+1}}$.

SOLUTION Using Integration by Parts, with $u = x^n$ and $v' = e^{-\alpha x}$, we get $u' = nx^{n-1}$, $v = -\frac{1}{\alpha}e^{-\alpha x}$, and

$$\int x^n e^{-\alpha x}\,dx = -\frac{1}{\alpha}x^n e^{-\alpha x} + \frac{n}{\alpha}\int x^{n-1}e^{-\alpha x}\,dx.$$

Thus,

$$J_n = \int_0^\infty x^n e^{-\alpha x}\,dx = \lim_{R\to\infty}\left(-\frac{1}{\alpha}x^n e^{-\alpha x}\right)\bigg|_0^R + \frac{n}{\alpha}\int_0^\infty x^{n-1}e^{-\alpha x}\,dx = \lim_{R\to\infty}\frac{-R^n}{\alpha e^{\alpha R}} + 0 + \frac{n}{\alpha}J_{n-1}.$$

Use L'Hôpital's Rule repeatedly to compute the limit:

$$\lim_{R\to\infty}\frac{-R^n}{\alpha e^{\alpha R}} = \lim_{R\to\infty}\frac{-nR^{n-1}}{\alpha^2 e^{\alpha R}} = \lim_{R\to\infty}\frac{-n(n-1)R^{n-2}}{\alpha^3 e^{\alpha R}} = \cdots = \lim_{R\to\infty}\frac{-n(n-1)(n-2)\cdots(3)(2)(1)}{\alpha^{n+1}e^{\alpha R}} = 0.$$

Finally,

$$J_n = 0 + \frac{n}{\alpha}J_{n-1} = \frac{n}{\alpha}J_{n-1}.$$

J_0 can be computed directly:

$$J_0 = \int_0^\infty e^{-\alpha x}\,dx = \lim_{R\to\infty}\int_0^R e^{-\alpha x}\,dx = \lim_{R\to\infty} -\frac{1}{\alpha}e^{-\alpha x}\bigg|_0^R = \lim_{R\to\infty} -\frac{1}{\alpha}\left(e^{-\alpha R} - 1\right) = -\frac{1}{\alpha}(0 - 1) = \frac{1}{\alpha}.$$

With this starting point, we can work up to J_4:

$$J_1 = \frac{1}{\alpha}J_0 = \frac{1}{\alpha}\left(\frac{1}{\alpha}\right) = \frac{1}{\alpha^2};$$

$$J_2 = \frac{2}{\alpha}J_1 = \frac{2}{\alpha}\left(\frac{1}{\alpha^2}\right) = \frac{2}{\alpha^3} = \frac{2!}{\alpha^{2+1}};$$

$$J_3 = \frac{3}{\alpha}J_2 = \frac{3}{\alpha}\left(\frac{2}{\alpha^3}\right) = \frac{6}{\alpha^4} = \frac{3!}{\alpha^{3+1}};$$

$$J_4 = \frac{4}{\alpha}J_3 = \frac{4}{\alpha}\left(\frac{6}{\alpha^4}\right) = \frac{24}{\alpha^5} = \frac{4!}{\alpha^{4+1}}.$$

We can use induction to prove the formula for J_n. If

$$J_{n-1} = \frac{(n-1)!}{\alpha^n},$$

then we have

$$J_n = \frac{n}{\alpha} J_{n-1} = \frac{n}{\alpha} \cdot \frac{(n-1)!}{\alpha^n} = \frac{n!}{\alpha^{n+1}}.$$

87. There is a function $F(v)$ such that the amount of electromagnetic energy with frequency between v and $v + \Delta v$ radiated by a so-called black body at temperature T is proportional to $F(v) \Delta v$. The total radiated energy is $E = \int_0^\infty F(v) \, dv$. According to **Planck's Radiation Law**,

$$F(v) = \left(\frac{8\pi h}{c^3}\right) \frac{v^3}{e^{hv/kT} - 1}$$

where c, h, k are physical constants. To derive this law, Planck introduced the quantum hypothesis in 1900, which thus marked the birth of quantum mechanics. Show that E is finite (use Exercise 86).

SOLUTION The total radiated energy E is given by

$$E = \int_0^\infty F(v) \, dv = \frac{8\pi h}{c^3} \int_0^\infty \frac{v^3}{e^{hv/kT} - 1} \, dv.$$

Let $\alpha = h/kT$. Then

$$E = \frac{8\pi h}{c^3} \int_0^\infty \frac{v^3}{e^{\alpha v} - 1} \, dv.$$

Because $\alpha > 0$ and $8\pi h/c^3$ is a constant, we know E is finite by Exercise 86.

89. A **probability density** function on $[0, \infty)$ is a function $f(x)$ defined for $x \geq 0$ such that $f(x) \geq 0$ and $\int_0^\infty f(x) \, dx = 1$. The *mean value* of $f(x)$ is the quantity $\mu = \int_0^\infty x f(x) \, dx$. For $k > 0$, find a constant C such that Ce^{-kx} is a probability density and compute μ.

SOLUTION For Ce^{-kx} to be a probability distribution, it is required that

$$\int_0^\infty Ce^{-kx} \, dx = \lim_{R \to \infty} \int_0^R Ce^{-kx} \, dx = \lim_{R \to \infty} \left(-\frac{C}{k} e^{-kx} \Big|_0^R \right) = \frac{C}{k} = 1.$$

This requires $C = k$, so that $f(x) = ke^{-kx}$. To compute

$$\mu = \int_0^\infty x(ke^{-kx}) \, dx = k \int_0^\infty xe^{-kx} \, dx$$

we must use integration by parts with $u = x$, $v' = e^{-kx}$, hence $du = dx$ and $v = -\frac{1}{k}e^{-kx}$:

$$\mu = \int_0^\infty x f(x) \, dx = \int_0^\infty kxe^{-kx} \, dx = \lim_{R \to \infty} k \int_0^R xe^{-kx} \, dx = \lim_{R \to \infty} \left(-xe^{-kx} \Big|_0^R + \int_0^R e^{-kx} dx \right)$$

$$= \lim_{R \to \infty} \left(-Re^{-Rk} - \frac{1}{k} \left(e^{-kx} \Big|_0^R \right) \right) = \lim_{R \to \infty} \left(-Re^{-Rk} - \frac{1}{k} \left(e^{-Rk} - 1 \right) \right) = \left(0 - \frac{1}{k}(-1) \right) = \frac{1}{k}.$$

*In Exercises 90–93, the **Laplace transform** of a function $f(x)$ is the function $Lf(s)$ of the variable s defined by the improper integral (if it converges):*

$$Lf(s) = \int_0^\infty f(x)e^{-sx} \, dx$$

Laplace transforms are widely used in physics and engineering.

91. Show that if $f(x) = \sin \alpha x$, then $Lf(s) = \dfrac{\alpha}{s^2 + \alpha^2}$.

SOLUTION If $f(x) = \sin \alpha x$, then the Laplace transform of $f(x)$ is

$$Lf(s) = \int_0^\infty e^{-sx} \sin \alpha x \, dx$$

First evaluate the indefinite integral using Integration by Parts, with $u = \sin \alpha x$ and $v' = e^{-sx}$. Then $u' = \alpha \cos \alpha x$, $v = -\frac{1}{s} e^{-sx}$, and

$$\int e^{-sx} \sin \alpha x \, dx = -\frac{1}{s} e^{-sx} \sin \alpha x + \frac{\alpha}{s} \int e^{-sx} \cos \alpha x \, dx.$$

Use Integration by Parts again, with $u = \cos \alpha x$, $v' = e^{-sx}$. Then $u' = -\alpha \sin \alpha x$, $v = -\frac{1}{s} e^{-sx}$, and

$$\int e^{-sx} \cos \alpha x \, dx = -\frac{1}{s} e^{-sx} \cos \alpha x - \frac{\alpha}{s} \int e^{-sx} \sin \alpha x \, dx.$$

Substituting this into the first equation and solving for $\int e^{-sx} \sin \alpha x \, dx$, we get

$$\int e^{-sx} \sin \alpha x \, dx = -\frac{1}{s} e^{-sx} \sin \alpha x - \frac{\alpha}{s^2} e^{-sx} \cos \alpha x - \frac{\alpha^2}{s^2} \int e^{-sx} \sin \alpha x \, dx$$

$$\int e^{-sx} \sin \alpha x \, dx = \frac{-e^{-sx} \left(\frac{1}{s} \sin \alpha x + \frac{\alpha}{s^2} \cos \alpha x \right)}{\left(1 + \frac{\alpha^2}{s^2} \right)} = \frac{-e^{-sx} (s \sin \alpha x + \alpha \cos \alpha x)}{s^2 + \alpha^2}$$

Thus,

$$\int_0^R e^{-sx} \sin \alpha x \, dx = \frac{1}{s^2 + \alpha^2} \left[\frac{s \sin \alpha R + \alpha \cos \alpha R}{-e^{sR}} - \frac{0 + \alpha}{-1} \right] = \frac{1}{s^2 + \alpha^2} \left[\alpha - \frac{s \sin \alpha R + \alpha \cos \alpha R}{e^{sR}} \right].$$

Finally we take the limit, noting the fact that, for all values of R, $|s \sin \alpha R + \alpha \cos \alpha R| \leq s + |\alpha|$

$$Lf(s) = \lim_{R \to \infty} \frac{1}{s^2 + \alpha^2} \left[\alpha - \frac{s \sin \alpha R + \alpha \cos \alpha R}{e^{sR}} \right] = \frac{1}{s^2 + \alpha^2} (\alpha - 0) = \frac{\alpha}{s^2 + \alpha^2}.$$

93. Compute $Lf(s)$, where $f(x) = \cos \alpha x$ and $s > 0$.

SOLUTION   If $f(x) = \cos \alpha x$, then the Laplace transform of $f(x)$ is

$$Lf(x) = \int_0^\infty e^{-sx} \cos \alpha x \, dx$$

First evaluate the indefinite integral using Integration by Parts, with $u = \cos \alpha x$ and $v' - e^{-sx}$. Then $u' = -\alpha \sin \alpha x$, $v = -\frac{1}{s} e^{-sx}$, and

$$\int e^{-sx} \cos \alpha x \, dx = -\frac{1}{s} e^{-sx} \cos \alpha x - \frac{\alpha}{s} \int e^{-sx} \sin \alpha x \, dx.$$

Use Integration by Parts again, with $u = \sin \alpha x \, dx$ and $v' = -e^{-sx}$. Then $u' = \alpha \cos \alpha x$, $v = -\frac{1}{s} e^{-sx}$, and

$$\int e^{-sx} \sin \alpha x \, dx = -\frac{1}{s} e^{-sx} \sin \alpha x + \frac{\alpha}{s} \int e^{-sx} \cos \alpha x \, dx.$$

Substituting this into the first equation and solving for $\int e^{-sx} \cos \alpha x \, dx$, we get

$$\int e^{-sx} \cos \alpha x \, dx = -\frac{1}{s} e^{-sx} \cos \alpha x - \frac{\alpha}{s} \left[-\frac{1}{s} e^{-sx} \sin \alpha x + \frac{\alpha}{s} \int e^{-sx} \cos \alpha \, dx \right]$$

$$= -\frac{1}{s} e^{-sx} \cos \alpha x + \frac{\alpha}{s^2} e^{-sx} \sin \alpha x - \frac{\alpha^2}{s^2} \int e^{-sx} \cos \alpha x \, dx$$

$$\int e^{-sx} \cos \alpha x \, dx = \frac{e^{-sx} \left(\frac{\alpha}{s^2} \sin \alpha x - \frac{1}{s} \cos \alpha x \right)}{1 + \frac{\alpha^2}{s^2}} = \frac{e^{-sx} (\alpha \sin \alpha x - s \cos \alpha x)}{s^2 + \alpha^2}$$

Thus,

$$\int_0^R e^{-sx} \cos \alpha x \, dx = \frac{1}{s^2 + \alpha^2} \left[\frac{\alpha \sin \alpha R - s \cos \alpha R}{e^{sR}} - \frac{0 - s}{1} \right].$$

Finally we take the limit, noting the fact that, for all values of R, $|\alpha \sin \alpha R - s \cos \alpha R| \leq |\alpha| + s$

$$Lf(s) = \lim_{R \to \infty} \frac{1}{s^2 + \alpha^2} \left[s + \frac{\alpha \sin \alpha R - s \cos \alpha R}{e^{sR}} \right] = \frac{1}{s^2 + \alpha^2} (s + 0) = \frac{s}{s^2 + \alpha^2}.$$

Further Insights and Challenges

95. Let $I = \int_0^1 x^p \ln x \, dx$.

(a) Show that I diverges for $p = -1$.

(b) Show that if $p \neq -1$, then

$$\int x^p \ln x \, dx = \frac{x^{p+1}}{p+1}\left(\ln x - \frac{1}{p+1}\right) + C$$

(c) Use L'Hôpital's Rule to show that I converges if $p > -1$ and diverges if $p < -1$.

SOLUTION

(a) If $p = -1$, then

$$I = \int_0^1 x^{-1} \ln x \, dx = \int_0^1 \frac{\ln x}{x} \, dx.$$

Let $u = \ln x$, $du = (1/x) \, dx$. Then

$$\int \frac{\ln x}{x} \, dx = \int u \, du = \frac{u^2}{2} + C = \frac{1}{2}(\ln x)^2 + C.$$

Thus,

$$\int_R^1 \frac{\ln x}{x} \, dx = \frac{1}{2}(\ln 1)^2 - \frac{1}{2}(\ln R)^2 = -\frac{1}{2}(\ln R)^2,$$

and

$$I = \lim_{R \to 0^+} -\frac{1}{2}(\ln R)^2 = \infty.$$

The integral diverges for $p = -1$.

(b) If $p \neq 1$, then use Integration by Parts, with $u = \ln x$ and $v' = x^p$. Then $u' = 1/x$, $v = x^{p+1}/p + 1$, and

$$\int x^p \ln x \, dx = \frac{x^{p+1}}{p+1} \ln x - \frac{1}{p+1} \int \left(x^{p+1}\right)\left(\frac{1}{x}\right) dx = \frac{x^{p+1}}{p+1} \ln x - \frac{1}{p+1} \int x^p \, dx$$

$$= \frac{x^{p+1}}{p+1} \ln x - \frac{1}{p+1}\left(\frac{x^{p+1}}{p+1}\right) + C = \frac{x^{p+1}}{p+1}\left(\ln x - \frac{1}{p+1}\right) + C.$$

(c) Let $p < -1$. Then

$$I = \lim_{R \to 0^+} \int_R^1 x^p \ln x = \lim_{R \to 0^+}\left[\frac{1}{p+1}\left(\ln 1 - \frac{1}{p+1}\right) - \frac{R^{p+1}}{p+1}\left(\ln R - \frac{1}{p+1}\right)\right]$$

$$= \lim_{R \to 0^+}\left(\frac{-1}{(p+1)^2} - \frac{R^{p+1}}{p+1}\ln R + \frac{R^{p+1}}{(p+1)^2}\right).$$

Since $p < -1$, $p + 1 < 0$, and we have

$$I = \lim_{R \to 0^+}\left(\frac{-1}{(p+1)^2} - \frac{\ln R}{(p+1)R^{-p-1}} + \frac{1}{(p+1)^2 R^{-p-1}}\right) = \infty.$$

The integral diverges for $p < -1$. On the other hand, if $p > -1$, then $p + 1 > 0$, and

$$I = \frac{-1}{(p+1)^2} + \frac{1}{p+1}\lim_{R \to 0+} R^{p+1}\ln R + \frac{1}{(p+1)^2}\lim_{R \to 0^+} R^{p+1} = \frac{-1}{(p+1)^2} + 0 = \frac{-1}{(p+1)^2}.$$

*In Exercises 96–98, an improper integral $I = \int_a^\infty f(x) \, dx$ is called **absolutely convergent** if $\int_a^\infty |f(x)| \, dx$ converges. It can be shown that if I is absolutely convergent, then it is convergent.*

97. Show that $\int_1^\infty e^{-x^2} \cos x \, dx$ is absolutely convergent.

SOLUTION By the result of Exercise 61, we know that $\int_0^\infty e^{-x^2} dx$ is convergent. Then $\int_1^\infty e^{-x^2} dx$ is also convergent. Because $|\cos x| \leq 1$ for all x, we have

$$\left| e^{-x^2} \cos x \right| = |\cos x| \left| e^{-x^2} \right| \leq \left| e^{-x^2} \right| = e^{-x^2}.$$

Therefore, by the comparison test, we have

$$\int_1^\infty \left| e^{-x^2} \cos x \right| dx \text{ also converges.}$$

Since $\int_1^\infty e^{-x^2} \cos x \, dx$ converges absolutely, it itself converges.

99. The **gamma function**, which plays an important role in advanced applications, is defined for $n \geq 1$ by

$$\Gamma(n) = \int_0^\infty t^{n-1} e^{-t} \, dt$$

(a) Show that the integral defining $\Gamma(n)$ converges for $n \geq 1$ (it actually converges for all $n > 0$). *Hint:* Show that $t^{n-1} e^{-t} < t^{-2}$ for t sufficiently large.
(b) Show that $\Gamma(n+1) = n\Gamma(n)$ using Integration by Parts.
(c) Show that $\Gamma(n+1) = n!$ if $n \geq 1$ is an integer. *Hint:* Use (a) repeatedly. Thus, $\Gamma(n)$ provides a way of defining n-factorial when n is not an integer.

SOLUTION
(a) By repeated use of L'Hôpital's Rule, we can compute the following limit:

$$\lim_{t \to \infty} \frac{e^t}{t^{n+1}} = \lim_{t \to \infty} \frac{e^t}{(n+1)t^n} = \cdots = \lim_{t \to \infty} \frac{e^t}{(n+1)!} = \infty.$$

This implies that, for t sufficiently large, we have

$$e^t \geq t^{n+1};$$

therefore

$$\frac{e^t}{t^{n-1}} \geq \frac{t^{n+1}}{t^{n-1}} = t^2 \quad \text{or} \quad t^{n-1} e^{-t} \leq t^{-2}.$$

The integral $\int_1^\infty t^{-2} \, dt$ converges because $p = 2 > 1$. Therefore, by the comparison test,

$$\int_M^\infty t^{n-1} e^{-t} \, dt \text{ also converges,}$$

where M is the value above which the above comparisons hold. Finally, because the function $t^{n-1} e^{-t}$ is continuous for all t, we know that

$$\Gamma(n) = \int_0^\infty t^{n-1} e^{-t} \, dt \text{ converges} \quad \text{for all } n \geq 1.$$

(b) Using Integration by Parts, with $u = t^n$ and $v' - e^{-t}$, we have $u' = nt^{n-1}$, $v = -e^{-t}$, and

$$\Gamma(n+1) = \int_0^\infty t^n e^{-t} \, dt = -t^n e^{-t} \Big|_0^\infty + n \int_0^\infty t^{n-1} e^{-t} \, dt$$

$$= \lim_{R \to \infty} \left(\frac{-R^n}{e^R} - 0 \right) + n\Gamma(n) = 0 + n\Gamma(n) = n\Gamma(n).$$

Here, we've computed the limit as in part (a) with repeated use of L'Hôpital's Rule.
(c) By the result of part (b), we have

$$\Gamma(n+1) = n\Gamma(n) = n(n-1)\Gamma(n-1) = n(n-1)(n-2)\Gamma(n-2) = \cdots = n! \, \Gamma(1).$$

If $n = 1$, then

$$\Gamma(1) = \int_0^\infty e^{-t} \, dt = \lim_{R \to \infty} -e^{-t} \Big|_0^R = \lim_{R \to \infty} \left(1 - e^{-R} \right) = 1.$$

Thus

$$\Gamma(n+1) = n! \, (1) = n!$$

CHAPTER REVIEW EXERCISES

1. Estimate $\int_2^5 f(x)\,dx$ by computing T_2, M_3, T_6, and S_6 for a function $f(x)$ taking on the values in the table below:

x	2	2.5	3	3.5	4	4.5	5
$f(x)$	$\frac{1}{2}$	2	1	0	$-\frac{3}{2}$	-4	-2

SOLUTION   To calculate T_2, divide $[2, 5]$ into two subintervals of length $\Delta x = \frac{3}{2}$ with endpoints $x_0 = 2$, $x_1 = 3.5$, $x_2 = 5$. Then

$$T_2 = \frac{1}{2} \cdot \frac{3}{2}\,(f(2) + 2f(3.5) + f(5)) = 0.75\left(\frac{1}{2} + 2 \cdot 0 + (-2)\right) = -\frac{9}{8}.$$

To calculate M_3, divide $[2, 5]$ into three subintervals of length $\Delta x = 1$ with midpoints $c_1 = 2.5$, $c_2 = 3.5$, $c_3 = 4.5$. Then

$$M_3 = 1 \cdot (f(2.5) + f(3.5) + f(4.5)) = 2 + 0 - 4 = -2.$$

To calculate T_6, divide $[2, 5]$ into 6 subintervals of length $\frac{5-2}{6} = \frac{1}{2}$ with endpoints $x_0 = 2$, $x_1 = 2.5$, $x_2 = 3$, $x_3 = 3.5$, $x_4 = 4$, $x_5 = 4.5$, $x_6 = 5$. Then

$$T_6 = \frac{1}{2} \cdot \frac{1}{2}\,(f(2) + 2f(2.5) + 2f(3) + 2f(3.5) + 2f(4) + 2f(4.5) + f(5))$$

$$= \frac{1}{4}\left(\frac{1}{2} + 2 \cdot 2 + 2 \cdot 1 + 2 \cdot 0 + 2 \cdot \left(-\frac{3}{2}\right) + 2(-4) + (-2)\right) = -\frac{13}{8}.$$

Finally, to calculate S_6, divide $[2, 5]$ into 6 subintervals of length $\Delta x = \frac{5-2}{6} = \frac{1}{2}$ with endpoints $x_0 = 2$, $x_1 = 2.5$, $x_2 = 3$, $x_3 = 3.5$, $x_4 = 4$, $x_5 = 4.5$, $x_6 = 5$. Then

$$S_6 = \frac{1}{3} \cdot \frac{1}{2}\,(f(2) + 4f(2.5) + 2f(3) + 4f(3.5) + 2f(4) + 4f(4.5) + f(5))$$

$$= \frac{1}{6}\left(\frac{1}{2} + 4 \cdot 2 + 2 \cdot 1 + 4 \cdot 0 + 2 \cdot \left(-\frac{3}{2}\right) + 4(-4) + (-2)\right) = -\frac{7}{4}.$$

3. The rainfall rate (in inches per hour) was measured hourly during a 10-hour thunderstorm with the following results:

$$0,\ 0.41,\ 0.49,\ 0.32,\ 0.3,\ 0.23,\ 0.09,\ 0.08,\ 0.05,\ 0.11,\ 0.12$$

Use Simpson's Rule to estimate the total rainfall during the 10-hour period.

SOLUTION   We have 10 subintervals of length $\Delta x = 1$. Thus, the total rainfall during the 10-hour period is approximately

$$S_{10} = \frac{1}{3} \cdot 1\,[0 + 4 \cdot 0.41 + 2 \cdot 0.49 + 4 \cdot 0.32 + 2 \cdot 0.3 + 4 \cdot 0.23 + +2 \cdot 0.09 + 4 \cdot 0.08 + 2 \cdot 0.05 + 4 \cdot 0.11 + 0.12]$$

$$= 2.19 \text{ inches.}$$

In Exercises 4–9, compute the given approximation to the integral.

5. $\int_2^4 \sqrt{6t^3 + 1}\,dt$, T_3

SOLUTION   Divide the interval $[2, 4]$ into 3 subintervals of length $\Delta x = \frac{4-2}{3} = \frac{2}{3}$, with endpoints $2, \frac{8}{3}, \frac{10}{3}, 4$. Then,

$$T_3 = \frac{1}{2}\Delta x\left(f(2) + 2f\left(\frac{8}{3}\right) + 2f\left(\frac{10}{3}\right) + f(4)\right)$$

$$= \frac{1}{2} \cdot \frac{2}{3}\left(\sqrt{6 \cdot 2^3 + 1} + 2\sqrt{6 \cdot \left(\frac{8}{3}\right)^3 + 1} + 2\sqrt{6 \cdot \left(\frac{10}{3}\right)^3 + 1} + \sqrt{6 \cdot 4^3 + 1}\right) = 25.976514.$$

7. $\int_1^4 \frac{dx}{x^3 + 1}$, T_6

SOLUTION Divide the interval $[1, 4]$ into 6 subintervals of length $\Delta x = \frac{4-1}{6} = \frac{1}{2}$ with endpoints $1, \frac{3}{2}, 2, \frac{5}{2}, 3, \frac{7}{2}, 4$. Then

$$T_6 = \frac{1}{2}\Delta x \left(f(1) + 2f\left(\frac{3}{2}\right) + 2f(2) + 2f\left(\frac{5}{2}\right) + 2f(3) + 2f\left(\frac{7}{2}\right) + f(4) \right)$$

$$= \frac{1}{2}\cdot\frac{1}{2}\left(\frac{1}{1^3+1} + 2\frac{1}{\left(\frac{3}{2}\right)^3+1} + 2\frac{1}{2^3+1} + 2\frac{1}{\left(\frac{5}{2}\right)^3+1} + 2\frac{1}{3^3+1} + 2\frac{1}{\left(\frac{7}{2}\right)^2+1} + \frac{1}{4^3+1} \right) = 0.358016.$$

9. $\displaystyle\int_5^9 \cos(x^2)\,dx, \quad S_8$

SOLUTION Divide the interval $[5, 9]$ into 8 subintervals of length $\Delta x = \frac{9-5}{8} = \frac{1}{2}$ with endpoints $5, \frac{11}{2}, 6, \frac{13}{2}, 7, \frac{15}{2}, 8, \frac{17}{2}, 9$. Then

$$S_8 = \frac{1}{3}\Delta x \left(f(5) + 4f\left(\frac{11}{2}\right) + 2f(6) + 4f\left(\frac{13}{2}\right) + 2f(7) + 4f\left(\frac{15}{2}\right) + 2f(8) + 4f\left(\frac{17}{2}\right) + f(9) \right)$$

$$= \frac{1}{3}\cdot\frac{1}{2}\Big(\cos(5^2) + 4\cos(5.5^2) + 2\cos(6^2) + 4\cos(6.5^2)$$

$$+ 2\cos(7^2) + 4\cos(7.5^2) + 2\cos(8^2) + 4\cos(8.5^2) + \cos(9^2) \Big)$$

$$= 0.608711.$$

11. Suppose that the second derivative of the function $A(h)$ in Exercise 10 satisfies $|A''(h)| \le 1.5$. Use the Error Bound to find the maximum possible error in your estimate of the volume V of the pond.

SOLUTION The Error Bound for the Trapezoidal Rule states that

$$\text{Error}(T_N) \le \frac{K_2(b-a)^3}{12N^2},$$

where K_2 is a number such that $|f''(x)| \le K_2$ for all $x \in [a, b]$. We estimated the volume of the pond by T_9; hence $N = 9$. The interval of depth is $[0, 18]$ hence $b - a = 18 - 0 = 18$. Since $|A''(h)| \le 1.5$ acres/ft^2 we may take $K_2 = 1.5$, to find that the error cannot exceed

$$\frac{K_2(b-a)^3}{12N^2} = \frac{1.5 \cdot 18^3}{12 \cdot 9^2} = 9 \text{ acre} \cdot \text{ft} = 392{,}040 \text{ ft}^3,$$

where we have used the fact that 1 acre $= 43560$ ft^2.

13. [GU] Let $f(x) = \sin(x^3)$. Find a bound for the error

$$\left| T_{24} - \int_0^{\pi/2} f(x)\,dx \right|.$$

Hint: Find a bound K_2 for $|f''(x)|$ by plotting $f''(x)$ with a graphing utility.

SOLUTION Using the error bound for T_{24} we obtain:

$$\left| T_{24} - \int_0^{\pi/2} f(x)\,dx \right| \le \frac{K_2\left(\frac{\pi}{2} - 0\right)^3}{12 \cdot 24^2} = \frac{K_2\pi^3}{55{,}296},$$

where K_2 is a number such that $|f''(x)| < k_2$ for all $x \in \left[0, \frac{\pi}{2}\right]$. We compute the first and second derivative of $f(x) = \sin(x^3)$:

$$f'(x) = 3x^2\cos(x^3)$$

$$f''(x) = 6x\cos(x^3) + 3x^2 \cdot 3x^2\left(-\sin(x^3)\right) = 6x\cos(x^3) - 9x^4\sin(x^3)$$

The graph of $f''(x) = 6x\cos(x^3) - 9x^4\sin(x^3)$ on the interval $\left[0, \frac{\pi}{2}\right]$ shows that $|f''(x)| \le 30$ on this interval. We may choose $K_2 = 30$ and find

$$\left| T_{24} - \int_0^{\pi/2} f(x)\,dx \right| \le \frac{30\pi^3}{55{,}296} = \frac{5\pi^3}{9216} = 0.0168220.$$

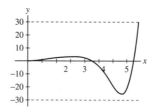

15. Find a value of N such that S_N approximates $\displaystyle\int_2^5 x^{-1/4}\,dx$ with an error of at most 10^{-2} (but do not calculate S_N).

SOLUTION To use the error bound we must find the fourth derivative $f^{(4)}(x)$. We differentiate $f(x) = x^{-1/4}$ four times to obtain:

$$f'(x) = -\frac{1}{4}x^{-5/4}, \quad f''(x) = \frac{5}{16}x^{-9/4}, \quad f'''(x) = -\frac{45}{64}x^{-13/4}, \quad f^{(4)}(x) = \frac{585}{256}x^{-17/4}.$$

For $2 \le x \le 5$ we have:

$$\left|f^{(4)}(x)\right| = \frac{585}{256x^{17/4}} \le \frac{585}{256 \cdot 2^{17/4}} = 0.120099.$$

Using the error bound with $b = 5$, $a = 2$ and $K_4 = 0.120099$ we have:

$$\text{Error}\,(S_N) \le \frac{0.120099(5-2)^5}{180N^4} = \frac{0.162134}{N^4}.$$

We must choose a value of N such that:

$$\frac{0.162134}{N^4} \le 10^{-2}$$

$$N^4 \ge 16.2134$$

$$N \ge 2.00664$$

The smallest even value of N that is needed to obtain the required precision is $N = 4$.

In Exercises 17–25, compute the integral using the suggested method.

17. $\displaystyle\int \cos^3\theta \sin^8\theta\,d\theta$ [write $\cos^3\theta$ as $\cos\theta(1 - \sin^2\theta)$]

SOLUTION We use the identity $\cos^2\theta = 1 - \sin^2\theta$ to rewrite the integral:

$$\int \cos^3\theta\sin^8\theta\,d\theta = \int \cos^2\theta\sin^8\theta\cos\theta\,d\theta = \int \left(1 - \sin^2\theta\right)\sin^8\theta\cos\theta\,d\theta.$$

Now, we use the substitution $u = \sin\theta$, $du = \cos\theta\,d\theta$:

$$\int \cos^3\theta\sin^8\theta\,d\theta = \int \left(1 - u^2\right)u^8\,du = \int \left(u^8 - u^{10}\right)\,du = \frac{u^9}{9} - \frac{u^{11}}{11} + C = \frac{\sin^9\theta}{9} - \frac{\sin^{11}\theta}{11} + C.$$

19. $\displaystyle\int \sec^3\theta\tan^4\theta\,d\theta$ (trigonometric identity, reduction formula)

SOLUTION We use the identity $1 + \tan^2\theta = \sec^2\theta$ to write $\tan^4\theta = \left(\sec^2\theta - 1\right)^2$ and to rewrite the integral as

$$\int \sec^3\theta\tan^4\theta\,d\theta \int \sec^3\theta\left(1 - \sec^2\theta\right)^2 d\theta = \int \sec^3\theta\left(1 - 2\sec^2\theta + \sec^4\theta\right)\,d\theta$$

$$= \int \sec^7\theta\,d\theta - 2\int \sec^5\theta\,d\theta + \int \sec^3\theta\,d\theta.$$

Now we use the reduction formula

$$\int \sec^m\theta\,d\theta = \frac{\tan\theta\sec^{m-2}\theta}{m-1} + \frac{m-2}{m-1}\int \sec^{m-2}\theta\,d\theta.$$

We have

$$\int \sec^5\theta\,d\theta = \frac{\tan\theta\sec^3\theta}{4} + \frac{3}{4}\int \sec^3\theta\,d\theta + C,$$

and

$$\int \sec^7\theta\,d\theta = \frac{\tan\theta\sec^5\theta}{6} + \frac{5}{6}\int \sec^5\theta\,d\theta = \frac{\tan\theta\sec^5\theta}{6} + \frac{5}{6}\left(\frac{\tan\theta\sec^3\theta}{4} + \frac{3}{4}\int \sec^3\theta\,d\theta\right) + C$$

$$= \frac{\tan\theta\sec^5\theta}{6} + \frac{5}{24}\tan\theta\sec^3\theta + \frac{5}{8}\int \sec^3\theta\,d\theta + C.$$

Therefore,

$$\int \sec^3\theta\tan^4\theta\,d\theta = \left(\frac{\tan\theta\sec^5\theta}{6} + \frac{5}{24}\tan\theta\sec^3\theta + \frac{5}{8}\int \sec^3\theta\,d\theta\right)$$

$$- 2\left(\frac{\tan\theta\sec^3\theta}{4} + \frac{3}{4}\int \sec^3\theta\,d\theta\right) + \int \sec^3\theta\,d\theta$$

$$= \frac{\tan\theta\sec^5\theta}{6} - \frac{7\tan\theta\sec^3\theta}{24} + \frac{1}{8}\int \sec^3\theta\,d\theta.$$

We again use the reduction formula to compute

$$\int \sec^3\theta\,d\theta = \frac{\tan\theta\sec\theta}{2} + \frac{1}{2}\int \sec\theta\,d\theta = \frac{\tan\theta\sec\theta}{2} + \frac{1}{2}\ln|\sec\theta + \tan\theta| + C.$$

Finally,

$$\int \sec^3\theta\tan^4\theta\,d\theta = \frac{\tan\theta\sec^5\theta}{6} - \frac{7\tan\theta\sec^3\theta}{24} + \frac{\tan\theta\sec\theta}{16} + \frac{1}{16}\ln|\sec\theta + \tan\theta| + C.$$

21. $\displaystyle\int \frac{dx}{x(x^2-1)^{3/2}}\,dx$ (trigonometric substitution)

SOLUTION   Substitute $x = \sec\theta$, $dx = \sec\theta\tan\theta\,d\theta$. Then,

$$\left(x^2-1\right)^{3/2} = \left(\sec^2\theta - 1\right)^{3/2} = \left(\tan^2\theta\right)^{3/2} = \tan^3\theta,$$

and

$$\int \frac{dx}{x\left(x^2-1\right)^{3/2}} = \int \frac{\sec\theta\tan\theta\,d\theta}{\sec\theta\tan^3\theta} = \int \frac{d\theta}{\tan^2\theta} = \int \cot^2\theta\,d\theta.$$

Using a reduction formula we find that:

$$\int \cot^2\theta\,d\theta = -\cot\theta - \theta + C$$

so

$$\int \frac{dx}{x\left(x^2-1\right)^{3/2}} = -\cot\theta - \theta + C.$$

We now must return to the original variable x. We use the relation $x = \sec\theta$ and the figure to obtain:

$$\int \frac{dx}{x\left(x^2-1\right)^{3/2}} = -\frac{1}{\sqrt{x^2-1}} - \sec^{-1}x + C.$$

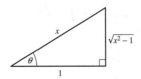

23. $\displaystyle\int \frac{dx}{x+x^{-1}}$ (rewrite integrand)

SOLUTION   We rewrite the integrand as follows:

$$\int \frac{dx}{x+x^{-1}} = \int \frac{x\,dx}{x^2+1}.$$

Now, we substitute $u = x^2 + 1$. Then $du = 2x\,dx$ and

$$\int \frac{dx}{x + x^{-1}} = \int \frac{\frac{1}{2}\,du}{u} = \frac{1}{2}\int \frac{du}{u} = \frac{1}{2}\ln|u| + C = \frac{1}{2}\ln\left(1 + x^2\right) + C.$$

25. $\displaystyle\int \frac{dx}{x^2 + 4x - 5}$ (complete square, substitution, partial fractions)

SOLUTION   The partial fraction decomposition takes the form

$$\frac{1}{x^2 + 4x - 5} = \frac{A}{x - 1} + \frac{B}{x + 5}.$$

Clearing denominators gives us

$$1 = A(x + 5) + B(x - 1).$$

Setting $x = 1$ then yields $A = \frac{1}{6}$, while setting $x = -5$ yields $B = -\frac{1}{6}$. Therefore,

$$\int \frac{dx}{x^2 + 4x - 5} = \frac{1}{6}\int \frac{dx}{x - 1} - \frac{1}{6}\int \frac{dx}{x + 5} = \frac{1}{6}\ln|x - 1| - \frac{1}{6}\ln|x + 5| + C = \frac{1}{6}\ln\left|\frac{x - 1}{x + 5}\right| + C.$$

In Exercises 26–69, compute the integral using the appropriate method or combination of methods.

27. $\displaystyle\int \frac{x^2}{\sqrt{9 - x^2}}\,dx$

SOLUTION   Substitute $x = 3\sin\theta$, $dx = 3\cos\theta\,d\theta$. Then

$$\sqrt{9 - x^2} = \sqrt{9 - 9\sin^2\theta} = \sqrt{9\left(1 - \sin^2\theta\right)} = \sqrt{9\cos^2\theta} = 3\cos\theta,$$

and

$$\int \frac{x^2}{\sqrt{9 - x^2}}\,dx = \int \frac{9\sin^2\theta \cdot 3\cos\theta\,d\theta}{3\cos\theta} = 9\int \sin^2\theta\,d\theta$$

$$= 9\left(\frac{\theta}{2} - \frac{\sin 2\theta}{4}\right) + C = \frac{9\theta}{2} - \frac{9\sin\theta\cos\theta}{2} + C.$$

We now must return to the original variable x. Since $x = 3\sin\theta$, we have $t = \sin^{-1}\frac{x}{3}$. Using the figure we obtain

$$\int \frac{x^2}{\sqrt{9 - x^2}}\,dx = \frac{9}{2}\sin^{-1}\left(\frac{x}{3}\right) - \frac{9}{2}\cdot\frac{x}{3}\cdot\frac{\sqrt{9 - x^2}}{3} + C = \frac{9}{2}\sin^{-1}\left(\frac{x}{3}\right) - \frac{x\sqrt{9 - x^2}}{2} + C.$$

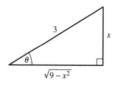

29. $\displaystyle\int \sec^2\theta\,\tan^4\theta\,d\theta$

SOLUTION   We substitute $u = \tan\theta$, $du = \sec^2\theta\,d\theta$ to obtain

$$\int \sec^2\theta\tan^4\theta\,d\theta = \int u^4\,du = \frac{u^5}{5} + C = \frac{\tan^5\theta}{5} + C.$$

31. $\displaystyle\int \frac{dt}{(t^2 - 1)^2}$

SOLUTION   Substitute $t = \sin\theta$, $dt = \cos\theta\,d\theta$. Then

$$\left(t^2 - 1\right)^2 = \left(1 - t^2\right)^2 = \left(1 - \sin^2\theta\right)^2 = \left(\cos^2\theta\right)^2 = \cos^4\theta,$$

and

$$\int \frac{dt}{(t^2-1)^2} = \int \frac{\cos\theta\,d\theta}{\cos^4\theta} = \int \frac{d\theta}{\cos^3\theta} = \int \sec^3\theta\,d\theta.$$

We use a reduction formula to compute the resulting integral:

$$\int \frac{dt}{(t^2-1)^2} = \int \sec^3\theta\,d\theta = \frac{\tan\theta\sec\theta}{2} + \frac{1}{2}\int \sec\theta\,d\theta = \frac{\tan\theta\sec\theta}{2} + \frac{1}{2}\ln|\sec\theta + \tan\theta| + C.$$

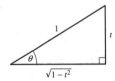

We now must return to the original variable t. Using the relation $t = \sin\theta$ and the accompanying figure,

$$\int \frac{dt}{(t^2-1)^2} = \frac{1}{2}\cdot\frac{t}{\sqrt{1-t^2}}\cdot\frac{1}{\sqrt{1-t^2}} + \frac{1}{2}\ln\left|\frac{1}{\sqrt{1-t^2}} + \frac{t}{\sqrt{1-t^2}}\right| + C = \frac{1}{2}\frac{t}{1-t^2} + \frac{1}{2}\ln\frac{|1+t|}{\sqrt{1-t^2}} + C.$$

33. $\displaystyle\int \sin 2\theta \sin^2\theta\,d\theta$

SOLUTION   We use the trigonometric identity $\sin 2\theta = 2\sin\theta\cos\theta$ to rewrite the integral:

$$\int \sin 2\theta\sin^2\theta\,d\theta = \int 2\sin\theta\cos\theta\sin^2\theta\,d\theta = \int 2\sin^3\theta\cos\theta\,d\theta.$$

Now, we substitute $u = \sin\theta$. Then $du = \cos\theta\,d\theta$ and

$$\int \sin 2\theta\sin^2\theta\,d\theta = 2\int u^3\,du = \frac{u^4}{2} + C = \frac{\sin^4\theta}{2} + C.$$

35. $\displaystyle\int (\ln(x+1))^2\,dx$

SOLUTION   First, substitute $w = x+1, dw = dx$. Then

$$\int (\ln(x+1))^2\,dx = \int (\ln w)^2\,dw.$$

Now, we use Integration by Parts with $u = (\ln w)^2$ and $v' = 1$. We find $u' = 2\frac{\ln w}{w}$, $v = w$, and

$$\int (\ln w)^2\,dw = w(\ln w)^2 - 2\int \ln w\,dw.$$

We use Integration by Parts again, this time with $u = \ln w$ and $v' = 1$. We find $u' = \frac{1}{w}$, $v = w$, and

$$\int \ln w\,dx = w\ln w - \int dw = w\ln w - w + C.$$

Thus,

$$\int (\ln w)^2\,dw = w(\ln w)^2 - 2w\ln w + 2w + C,$$

and

$$\int (\ln(x+1))^2\,dx = (x+1)\left[\ln(x+1)\right]^2 - 2(x+1)\ln(x+1) + 2(x+1) + C.$$

37. $\displaystyle\int \cos^4(9x-2)\,dx$

SOLUTION   We substitute $u = 9x-2, du = 9\,dx$ and then use a reduction formula to evaluate the resulting integral. We obtain:

$$\int \cos^4(9x-2)\,dx = \frac{1}{9}\int \cos^4 u\,du = \frac{1}{9}\left(\frac{\cos^3 u\sin u}{4} + \frac{3}{4}\int \cos^2 u\,du\right)$$

$$= \frac{\cos^3 u \sin u}{36} + \frac{1}{12} \int \cos^2 u \, du = \frac{\cos^3 u \sin u}{36} + \frac{1}{12} \left(\frac{u}{2} + \frac{\sin 2u}{4} \right) + C$$

$$= \frac{\cos^3(9x - 2) \sin(9x - 2)}{36} + \frac{9x - 2}{24} + \frac{\sin(18x - 4)}{48} + C.$$

39. $\int \sin 2x \sec^2 x \, dx$

SOLUTION   We use the trigonometric identity $\sin 2x = 2 \cos x \sin x$ to rewrite the integrand:

$$\sin 2x \sec^2 x = 2 \sin x \cos x \sec^2 x = \frac{2 \sin x \cos x}{\cos^2 x} = \frac{2 \sin x}{\cos x} = 2 \tan x.$$

Hence,

$$\int \sin 2x \sec^2 x \, dx = \int 2 \tan x \, dx = 2 \ln | \sec x | + C.$$

41. $\int (\sec x + \tan x)^2 \, dx$

SOLUTION   We rewrite the integrand as

$$(\sec x + \tan x)^2 = \sec^2 x + 2 \sec x \tan x + \tan^2 x = 2 \sec x \tan x + 2 \sec^2 x - 1.$$

Therefore,

$$\int (\sec x + \tan x)^2 \, dx = 2 \int \sec x \tan x \, dx + 2 \int \sec^2 x \, dx - \int dx = 2 \sec x + 2 \tan x - x + C.$$

43. $\int \cot^2 \frac{\theta}{2} \, d\theta$

SOLUTION   We substitute $u = \frac{\theta}{2}$. Then $du = \frac{1}{2} d\theta$ and

$$\int \cot^2 \frac{\theta}{2} \, d\theta = 2 \int \cot^2 u \, du.$$

Now, we use a reduction formula to compute

$$\int \cot^2 \frac{\theta}{2} \, d\theta = 2 \int \cot^2 u \, du = 2(-\cot u - u) + C = -2 \cot \frac{\theta}{2} - \theta + C.$$

45. $\int \frac{dt}{(t - 3)^2(t + 4)}$

SOLUTION   The partial fraction decomposition has the form

$$\frac{1}{(t - 3)^2(t + 4)} = \frac{A}{t + 4} + \frac{B}{t - 3} + \frac{C}{(t - 3)^2}.$$

Clearing denominators gives us

$$1 = A(t - 3)^2 + B(t - 3)(t + 4) + C(t + 4).$$

Setting $t = 3$ then yields $C = \frac{1}{7}$, while setting $t = -4$ yields $A = \frac{1}{49}$. Lastly, setting $t = 0$ yields

$$1 = 9A - 12B + 4C \qquad \text{or} \qquad B = -\frac{1}{49}.$$

Hence,

$$\int \frac{dt}{(t - 3)^2(t + 4)} = \frac{1}{49} \int \frac{dt}{t + 4} - \frac{1}{49} \int \frac{dt}{t - 3} + \frac{1}{7} \int \frac{dt}{(t - 3)^2}$$

$$= \frac{1}{49} \ln |t + 4| - \frac{1}{49} \ln |t - 3| + \frac{1}{7} \cdot \frac{-1}{t - 3} + C = \frac{1}{49} \ln \left| \frac{t + 4}{t - 3} \right| - \frac{1}{7} \cdot \frac{1}{t - 3} + C.$$

47. $\int \frac{dx}{x \sqrt{x^2 - 4}}$

SOLUTION Substitute $x = 2 \sec \theta$, $dx = 2 \sec \theta \tan \theta \, d\theta$. Then

$$\sqrt{x^2 - 4} = \sqrt{4\sec^2\theta - 4} = \sqrt{4\left(\sec^2\theta - 1\right)} = \sqrt{4\tan^2\theta} = 2 \tan \theta,$$

and

$$\int \frac{dx}{x\sqrt{x^2 - 4}} = \int \frac{2 \sec \theta \tan \theta \, d\theta}{2 \sec \theta \cdot 2 \tan \theta} = \frac{1}{2} \int d\theta = \frac{1}{2}\theta + C.$$

Now, return to the original variable x. Since $x = 2 \sec \theta$, we have $\sec \theta = \frac{x}{2}$ or $\theta = \sec^{-1}\frac{x}{2}$. Thus,

$$\int \frac{dx}{x\sqrt{x^2 - 4}} = \frac{1}{2}\sec^{-1}\frac{x}{2} + C.$$

49. $\displaystyle\int \frac{dx}{x^{3/2} + ax^{1/2}}$

SOLUTION Let $u = x^{1/2}$ or $x = u^2$. Then $dx = 2u \, du$ and

$$\int \frac{dx}{x^{3/2} + ax^{1/2}} = \int \frac{2u \, du}{u^3 + au} = 2 \int \frac{du}{u^2 + a}.$$

If $a > 0$, then

$$\int \frac{dx}{x^{3/2} + ax^{1/2}} = 2 \int \frac{du}{u^2 + a} = \frac{2}{\sqrt{a}}\tan^{-1}\left(\frac{u}{\sqrt{a}}\right) + C = \frac{2}{\sqrt{a}}\tan^{-1}\sqrt{\frac{x}{a}} + C.$$

If $a = 0$, then

$$\int \frac{dx}{x^{3/2}} = -\frac{2}{\sqrt{x}} + C.$$

Finally, if $a < 0$, then

$$\int \frac{du}{u^2 + a} = \int \frac{du}{u^2 - \left(\sqrt{-a}\right)^2},$$

and the partial fraction decomposition takes the form

$$\frac{1}{u^2 - \left(\sqrt{-a}\right)^2} = \frac{A}{u - \sqrt{-a}} + \frac{B}{u + \sqrt{-a}}.$$

Clearing denominators gives us

$$1 = A(u + \sqrt{-a}) + B(u - \sqrt{-a}).$$

Setting $u = \sqrt{-a}$ then yields $A = \frac{1}{2\sqrt{-a}}$, while setting $u = -\sqrt{-a}$ yields $B = -\frac{1}{2\sqrt{-a}}$. Hence,

$$\int \frac{dx}{x^{3/2} + ax^{1/2}} = 2 \int \frac{du}{u^2 + a} = \frac{1}{\sqrt{-a}} \int \frac{du}{u - \sqrt{-a}} - \frac{1}{\sqrt{-a}} \int \frac{du}{u + \sqrt{-a}}$$

$$= \frac{1}{\sqrt{-a}} \ln|u - \sqrt{-a}| - \frac{1}{\sqrt{-a}} \ln|u + \sqrt{-a}| + C$$

$$= \frac{1}{\sqrt{-a}} \ln\left|\frac{u - \sqrt{-a}}{u + \sqrt{-a}}\right| + C = \frac{1}{\sqrt{-a}} \ln\left|\frac{\sqrt{x} - \sqrt{-a}}{\sqrt{x} + \sqrt{-a}}\right| + C.$$

In summary,

$$\int \frac{dx}{x^{3/2} + ax^{1/2}} = \begin{cases} \frac{2}{\sqrt{a}}\tan^{-1}\sqrt{\frac{x}{a}} + C & a > 0 \\ \frac{1}{\sqrt{-a}} \ln\left|\frac{\sqrt{x}-\sqrt{-a}}{\sqrt{x}+\sqrt{-a}}\right| + C & a < 0 \\ -\frac{2}{\sqrt{x}} + C & a = 0 \end{cases}$$

51. $\displaystyle\int \frac{(x^2 - x) \, dx}{(x + 2)^3}$

SOLUTION The partial fraction decomposition has the form

$$\frac{x^2 - x}{(x+2)^3} = \frac{A}{x+2} + \frac{B}{(x+2)^2} + \frac{C}{(x+2)^3}.$$

Clearing denominators gives us

$$x^2 - x = A(x+2)^2 + B(x+2) + C.$$

Setting $x = -2$ then yields $C = 6$. Equating x^2-coefficients gives us $A = 1$, and equating x-coefficients yields $4A + B = -1$, or $B = -5$. Thus,

$$\int \frac{x^2 - x}{(x+2)^3}\, dx = \int \frac{dx}{x+2} + \int \frac{-5\, dx}{(x+2)^2} + \int \frac{6\, dx}{(x+2)^3} = \ln|x+2| + \frac{5}{x+2} - \frac{3}{(x+2)^2} + C.$$

53. $\displaystyle \int \frac{16\, dx}{(x-2)^2(x^2+4)}$

SOLUTION The partial fraction decomposition has the form

$$\frac{16}{(x-2)^2\left(x^2+4\right)} = \frac{A}{x-2} + \frac{B}{(x-2)^2} + \frac{Cx+D}{x^2+4}.$$

Clearing denominators gives us

$$16 = A(x-2)\left(x^2+4\right) + B\left(x^2+4\right) + (Cx+D)(x-2)^2.$$

Setting $x = 2$ then yields $B = 2$. With $B = 2$,

$$16 = A\left(x^3 - 2x^2 + 4x - 8\right) + 2\left(x^2+4\right) + Cx^3 + (D-4C)x^2 + (4C-4D)x + 4D$$

$$16 = (A+C)x^3 + (-2A + 2 + D - 4C)\, x^2 + (4A + 4C - 4D)x + (-8A + 8 + 4D)$$

Equating coefficients of like powers of x now gives us the system of equations

$$A + C = 0$$
$$-2A - 4C + D + 2 = 0$$
$$4A + 4C - 4D = 0$$
$$-8A + 4D + 8 = 1$$

whose solution is

$$A = -1,\ C = 1,\ D = 0.$$

Thus,

$$\int \frac{dx}{(x-2)^2\left(x^2+4\right)} = -\int \frac{dx}{x-2} + 2\int \frac{dx}{(x-2)^2} + \int \frac{x}{x^2+4}\, dx$$

$$= -\ln|x-2| - 2\frac{1}{x-2} + \frac{1}{2}\ln\left(x^2+4\right) + C.$$

55. $\displaystyle \int \frac{dx}{x^2 + 8x + 25}$

SOLUTION Complete the square to rewrite the denominator as

$$x^2 + 8x + 25 = (x+4)^2 + 9.$$

Now, let $u = x + 4$, $du = dx$. Then,

$$\int \frac{dx}{x^2 + 8x + 25} = \int \frac{du}{u^2 + 9} = \frac{1}{3}\tan^{-1}\frac{u}{3} + C = \frac{1}{3}\tan^{-1}\left(\frac{x+4}{3}\right) + C.$$

57. $\displaystyle \int \frac{(x^2 - x)\, dx}{(x+2)^3}$

SOLUTION   The partial fraction decomposition has the form

$$\frac{x^2 - x}{(x+2)^3} = \frac{A}{x+2} + \frac{B}{(x+2)^2} + \frac{C}{(x+2)^3}.$$

Clearing denominators gives us

$$x^2 - x = A(x+2)^2 + B(x+2) + C.$$

Setting $x = -2$ then yields $C = 6$. Equating x^2-coefficients gives us $A = 1$, and equating x-coefficients yields $4A + B = -1$, or $B = -5$. Thus,

$$\int \frac{x^2 - x}{(x+2)^3}\, dx = \int \frac{dx}{x+2} + \int \frac{-5\,dx}{(x+2)^2} + \int \frac{6\,dx}{(x+2)^3} = \ln|x+2| + \frac{5}{x+2} - \frac{3}{(x+2)^2} + C.$$

59. $\displaystyle\int \frac{dx}{x^4\sqrt{x^2+4}}$

SOLUTION   Substitute $x = 2\tan\theta$, $dx = 2\sec^2\theta\,d\theta$. Then

$$\sqrt{x^2+4} = \sqrt{4\tan^2\theta + 4} = \sqrt{4\left(\tan^2\theta + 1\right)} = 2\sqrt{\sec^2\theta} = 2\sec\theta,$$

and

$$\int \frac{dx}{x^4\sqrt{x^2+4}} = \int \frac{2\sec^2\theta\,d\theta}{16\tan^4\theta \cdot 2\sec\theta} = \int \frac{\sec\theta\,d\theta}{16\tan^4\theta}.$$

We have

$$\frac{\sec\theta}{\tan^4\theta} = \frac{\cos^3\theta}{\sin^4\theta}.$$

Hence,

$$\int \frac{dx}{x^4\sqrt{x^2+4}} = \frac{1}{16}\int \frac{\cos^3\theta\,d\theta}{\sin^4\theta} = \frac{1}{16}\int \frac{\cos^2\theta\cos\theta\,d\theta}{\sin^4\theta} = \frac{1}{16}\int \frac{\left(1 - \sin^2\theta\right)\cos\theta\,d\theta}{\sin^4\theta}.$$

Now substitute $u = \sin\theta$ and $du = \cos\theta\,d\theta$ to obtain

$$\int \frac{dx}{x^4\sqrt{x^2+4}} = \frac{1}{16}\int \frac{1 - u^2}{u^4}\,du = \frac{1}{16}\int \left(u^{-4} - u^{-2}\right)du = -\frac{1}{48u^3} + \frac{1}{16}\frac{1}{u} + C$$

$$= -\frac{1}{48}\cdot\frac{1}{\sin^3\theta} + \frac{1}{16}\frac{1}{\sin\theta} + C = -\frac{1}{48}\csc^3\theta + \frac{1}{16}\csc\theta + C.$$

Finally, return to the original to the original variable x using the relation $x = 2\tan\theta$ and the figure below.

$$\int \frac{dx}{x^4\sqrt{x^2+4}} = -\frac{1}{48}\left(\frac{\sqrt{x^2+4}}{x}\right)^3 + \frac{1}{16}\frac{\sqrt{x^2+4}}{x} + C = -\frac{\left(x^2+4\right)^{3/2}}{48x^3} + \frac{\sqrt{x^2+4}}{16x} + C.$$

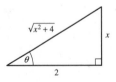

61. $\displaystyle\int (x+1)e^{4-3x}\, dx$

SOLUTION   We compute the integral using Integration by Parts with $u = x+1$ and $v' = e^{4-3x}$. Then $u' = 1$, $v = -\frac{1}{3}e^{4-3x}$ and

$$\int (x+1)e^{4-3x}\, dx = -\frac{1}{3}(x+1)e^{4-3x} + \frac{1}{3}\int e^{4-3x}\, dx = -\frac{1}{3}(x+1)e^{4-3x} + \frac{1}{3}\cdot\left(-\frac{1}{3}\right)e^{4-3x} + C$$

$$= -\frac{1}{9}e^{4-3x}(3x+4) + C.$$

63. $\displaystyle\int x^{-2}\tan^{-1}x\,dx$

SOLUTION We use Integration by Parts with $u = \tan^{-1}x$ and $v' = x^{-2}$. Then $u' = \frac{1}{1+x^2}$, $v = -x^{-1}$ and

$$\int x^{-2}\tan^{-1}x\,dx = -\frac{\tan^{-1}x}{x} + \int \frac{dx}{x\left(1+x^2\right)}.$$

For the remaining integral, the partial fraction decomposition takes the form

$$\frac{1}{x(1+x^2)} = \frac{A}{x} + \frac{Bx+C}{1+x^2}.$$

Clearing denominators gives us

$$1 = A(1+x^2) + (Bx+C)x.$$

Setting $x = 0$ then yields $A = 1$. Next, equating the x^2-coefficients gives

$$0 = A + B \qquad \text{so} \qquad B = -1,$$

while equating x-coefficients gives $C = 0$. Hence,

$$\frac{1}{x\left(1+x^2\right)} = \frac{1}{x} - \frac{x}{1+x^2},$$

and

$$\int \frac{dx}{x\left(1+x^2\right)} = \int \frac{1}{x}\,dx - \int \frac{x\,dx}{1+x^2} = \ln|x| - \frac{1}{2}\ln\left(1+x^2\right) + C.$$

Therefore,

$$\int x^{-2}\tan^{-1}x\,dx = -\frac{\tan^{-1}x}{x} + \ln|x| - \frac{1}{2}\ln\left(1+x^2\right) + C.$$

65. $\displaystyle\int x^2(\ln x)^2\,dx$

SOLUTION We use Integration by Parts with $u = (\ln x)^2$ and $v' = x^2$. Then $u' = \frac{2\ln x}{x}$, $v = \frac{x^3}{3}$ and

$$\int x^2(\ln x)^2\,dx = \frac{x^3}{3}(\ln x)^2 - \frac{2}{3}\int x^2\ln x\,dx.$$

To calculate the resulting integral, we again use Integration by Parts, this time with $u = \ln x$ and $v' = x^2$. Then, $u' = \frac{1}{x}$, $v = \frac{x^3}{3}$, and

$$\int x^2\ln x\,dx = \frac{x^3}{3}\ln x - \frac{1}{3}\int x^2\,dx = \frac{x^3}{3}\ln x - \frac{x^3}{9} + C.$$

Finally,

$$\int x^2(\ln x)^2\,dx = \frac{x^3}{3}(\ln x)^2 - \frac{2}{3}\left(\frac{x^3}{3}\ln x - \frac{x^3}{9}\right) + C = \frac{x^3}{3}\left((\ln x)^2 - \frac{2}{3}\ln x + \frac{2}{9}\right) + C.$$

67. $\displaystyle\int \frac{\tan^{-1}t\,dt}{1+t^2}$

SOLUTION Substitute $u = \tan^{-1}t$. Then, $du = \frac{dt}{1+t^2}$ and

$$\int \frac{\tan^{-1}t\,dt}{1+t^2} = \int u\,du = \frac{1}{2}u^2 + C = \frac{1}{2}\left(\tan^{-1}t\right)^2 + C.$$

69. $\displaystyle\int (\sin x)(\cosh x)\,dx$

SOLUTION We compute the integral using Integration by Parts with $u = \sin x$ and $v' = \cosh x$. Then $u' = \cos x$, $v = \sinh x$ and

$$\int \sin x \cosh x \, dx = \sin x \sinh x - \int \cos x \sinh x \, dx.$$

We compute the resulting integral using Integration by Parts, this time with $u = \cos x$ and $v' = \sinh x$. Then $u' = -\sin x$, $v = \cosh x$ and

$$\int \cos x \sinh x \, dx = \cos x \cosh x + \int \sin x \cosh x \, dx.$$

Therefore,

$$\int \sin x \cosh x \, dx = \sin x \sinh x - \cos x \cosh x - \int \sin x \cosh x \, dx.$$

Solving for $\int (\sin x)(\cosh x) \, dx$, we find

$$2 \int \sin x \cosh x \, dx = \sin x \sinh x - \cos x \cosh x + C$$

$$\int \sin x \cosh x \, dx = \frac{1}{2} \sin x \sinh x - \frac{1}{2} \cos x \cosh x + C$$

In Exercises 70–79, determine whether the improper integral converges and, if so, evaluate it.

71. $\int_4^\infty \dfrac{dx}{x^{2/3}}$

SOLUTION The integral $\int_a^\infty \dfrac{dx}{x^p} \ (a > 0)$ converges if $p > 1$ and diverges if $p \le 1$. Here, $p = \frac{2}{3} < 1$, hence the integral diverges.

73. $\int_9^\infty \dfrac{dx}{x^{12/5}}$

SOLUTION

$$\int_9^\infty \frac{dx}{x^{12/5}} = \lim_{R \to \infty} \int_9^R \frac{dx}{x^{12/5}} = \lim_{R \to \infty} -\frac{5}{7} x^{-7/5} \Big|_9^R = \lim_{R \to \infty} \left(-\frac{5}{7} R^{-7/5} + \frac{5}{7} \cdot 9^{-7/5} \right)$$

$$= 0 + \frac{5}{7} \cdot 9^{-7/5} = \frac{5}{7 \cdot 9 \cdot 9^{2/5}} = \frac{5}{63 \cdot 9^{2/5}}.$$

75. $\int_{-\infty}^9 e^{4x} \, dx$

SOLUTION

$$\int_{-\infty}^9 e^{4x} \, dx = \lim_{R \to -\infty} \int_R^9 e^{4x} \, dx = \lim_{R \to -\infty} \frac{1}{4} e^{4x} \Big|_R^9 = \lim_{R \to -\infty} \frac{1}{4} e^{36} - \frac{1}{4} e^{4R} = \frac{e^{36}}{4}.$$

77. $\int_1^\infty \dfrac{dx}{(x + 2)(2x + 3)}$

SOLUTION First, evaluate the indefinite integral. The following partial fraction decomposition has the form

$$\frac{1}{(x + 2)(2x + 3)} = -\frac{1}{x + 2} + \frac{2}{2x + 3}.$$

Clearing denominators gives us

$$1 = A(2x + 3) + B(x + 2).$$

Setting $x = -2$ then yields $A = -1$, while setting $x = -\frac{3}{2}$ yields $B = 2$. Hence,

$$\int \frac{dx}{(x + 2)(2x + 3)} = -\int \frac{dx}{x + 2} + 2 \int \frac{dx}{2x + 3} = -\ln|x + 2| + \ln|2x + 3| + C = \ln\left|\frac{2x + 3}{x + 2}\right| + C.$$

Now, for $R > 1$,

$$\int_1^R \frac{dx}{(x+2)(2x+3)} = \ln\left|\frac{2x+3}{x+2}\right|\Big|_1^R = \ln\frac{2R+3}{R+2} - \ln\frac{5}{3},$$

and

$$\int_1^\infty \frac{dx}{(x+2)(2x+3)} = \lim_{R\to\infty}\left(\ln\frac{2R+3}{R+2}\right) - \ln\frac{5}{3} = \ln 2 + \ln\frac{3}{5} = \ln\frac{6}{5}.$$

79. $\displaystyle\int_2^5 (5-x)^{-1/3}\,dx$

SOLUTION

$$\int_2^5 (5-x)^{-1/3}\,dx = \lim_{R\to 5-}\int_2^R (5-x)^{-1/3}\,dx = \lim_{R\to 5-} -\frac{3}{2}(5-x)^{2/3}\Big|_2^R$$

$$= \lim_{R\to 5-} -\frac{3}{2}\left((5-R)^{2/3} - 3^{2/3}\right) = -\frac{3}{2}\left(0 - 3^{2/3}\right) = \frac{3^{5/3}}{2}.$$

In Exercises 80–85, use the Comparison Test to determine if the improper integral converges or diverges.

81. $\displaystyle\int_8^\infty (\sin^2 x)e^{-x}\,dx$

SOLUTION   The following inequality holds for all x,

$$0 \le \left(\sin^2 x\right)e^{-x} \le e^{-x}.$$

We use direct computation to show that the improper integral of e^{-x} over the interval $[8, \infty)$ converges:

$$\int_8^\infty e^{-x}\,dx = \lim_{R\to\infty}\int_8^R e^{-x}\,dx = \lim_{R\to\infty} -e^{-x}\Big|_8^R = \lim_{R\to\infty}\left(-e^{-R} + e^{-8}\right) = 0 + e^{-8} = e^{-8}.$$

Therefore, by the Comparison Test, the improper integral $\displaystyle\int_8^\infty (\sin^2 x)e^{-x}\,dx$ also converges.

83. $\displaystyle\int_1^\infty \frac{dx}{x^{1/3} + x^{2/3}}$

SOLUTION   If $x \ge 1$, then $x^{1/3} \ge 1$; therefore,

$$x^{1/3} + x^{2/3} = x^{1/3}\left(1 + x^{1/3}\right) \le x^{1/3}\left(x^{1/3} + x^{1/3}\right) = x^{1/3}\cdot 2x^{1/3} = 2x^{2/3}.$$

Hence,

$$\frac{1}{x^{1/3} + x^{2/3}} \ge \frac{1}{2x^{2/3}}.$$

The integral $\displaystyle\int_1^\infty \frac{dx}{x^{2/3}}$ diverges; hence $\displaystyle\int_1^\infty \frac{dx}{2x^{2/3}}$ also diverges. Therefore, by the Comparison Test, the improper integral $\displaystyle\int_1^\infty \frac{dx}{x^{1/3} + x^{2/3}}$ also diverges.

85. $\displaystyle\int_0^\infty e^{-x^3}\,dx$

SOLUTION   For $x > 1$, $e^x \ge x$; hence $e^{x^3} \ge x^3$, therefore $0 \le e^{-x^3} \le x^{-3}$. Since $\displaystyle\int_1^\infty \frac{dx}{x^3}$ converges, the integral $\displaystyle\int_1^\infty e^{-x^3}\,dx$ also converges by the Comparison Test. We write

$$\int_0^\infty e^{-x^3}\,dx = \int_0^1 e^{-x^3}\,dx + \int_1^\infty e^{-x^3}\,dx.$$

The first integral on the right hand side has a finite value and the second integral converges. We conclude that the integral $\displaystyle\int_0^\infty e^{-x^3}\,dx$ converges.

87. Let R be the region under the graph of $y = (x + 1)^{-1}$ for $0 \le x < \infty$. Which of the following quantities is finite?

(a) The area of R

(b) The volume of the solid obtained by rotating R about the x-axis

(c) The volume of the solid obtained by rotating R about the y-axis

SOLUTION

(a) The area of R is

$$\int_0^\infty \frac{dx}{x+1} = \lim_{R \to \infty} \int_0^R \frac{dx}{x+1} = \lim_{R \to \infty} \ln|x+1|\Big|_0^R = \lim_{R \to \infty} \left(\ln(R+1) - \ln 1\right) = \infty.$$

Hence, the area of R is not finite.

(b) Using the Disk Method, the volume of the solid obtained by rotating R about the x-axis is

$$\pi \int_0^\infty \frac{dx}{(x+1)^2} = \pi \lim_{R \to \infty} \int_0^R \frac{dx}{(x+1)^2} = \pi \lim_{R \to \infty} -\frac{1}{x+1}\Big|_0^R = \pi \lim_{R \to \infty} \left(-\frac{1}{R+1} + 1\right) = \pi.$$

Hence, the volume of the solid obtained by rotating R about the x-axis is finite.

(c) Using the Shell Method, the volume of the solid obtained by rotating R about the y-axis is

$$2\pi \int_0^\infty \frac{x}{x+1}\, dx = 2\pi \lim_{R \to \infty} \int_0^R \frac{x\, dx}{x+1}.$$

Now,

$$\int_0^R \frac{x\, dx}{x+1} = \int_0^R \frac{(x+1) - 1}{x+1}\, dx = \int_0^R \left(1 - \frac{1}{x+1}\right) dx = (x - \ln(x+1))\Big|_0^R$$
$$= R - (\ln(R+1) - \ln 1) = R - \ln(R+1).$$

Thus,

$$2\pi \lim_{R \to \infty} \int_0^R \frac{x\, dx}{x+1} = 2\pi \lim_{R \to \infty} (R - \ln(R+1)) = 2\pi \lim_{R \to \infty} R \left(1 - \frac{\ln(R+1)}{R}\right) = \infty.$$

Hence, the volume of the solid obtained by rotating R about the y-axis is not finite.

89. According to kinetic theory, the molecules of ordinary matter are in constant random motion. The probability that a molecule has kinetic energy in a small interval $[E, E + \Delta E]$ is approximately $\frac{1}{kT} e^{-E/(kT)}$, where T is the temperature (in kelvins) and k Boltzmann's constant. Compute the *average* kinetic energy $\overline{E}$ in terms of k and T, where

$$\overline{E} = \frac{1}{kT} \int_0^\infty E e^{-E/kT}\, dE$$

SOLUTION By definition,

$$\int_0^\infty E e^{-E/kT}\, dE = \lim_{R \to \infty} \int_0^R E e^{-E/kT}\, dE.$$

We compute the definite integral using Integration by Parts with $u = E$, $v' = e^{-E/kT}$. Then $u' = 1$, $v = -kT e^{-E/kT}$ and

$$\int_0^R E e^{-E/kT}\, dE = -kT e^{-E/kT} E\Big|_{E=0}^R + \int_0^R kT e^{-E/kT}\, dE = -kT e^{-R/kT} R - (kT)^2 e^{-E/kT}\Big|_{E=0}^R$$
$$= -kT R e^{-R/kT} - \left(k^2 T^2 e^{-R/kT} - k^2 T^2 e^0\right) = k^2 T^2 - kT R e^{-R/kT} - k^2 T^2 e^{-R/kT}.$$

We now let $R \to \infty$, obtaining:

$$\int_0^\infty E e^{-E/RT}\, dE = \lim_{R \to \infty} \int_0^R E e^{-E/RT}\, dE = \lim_{R \to \infty} \left(k^2 T^2 - kT R e^{-R/kT} - k^2 T^2 e^{-R/kT}\right)$$
$$= k^2 T^2 - kT \lim_{R \to \infty} R e^{-R/kT} - 0 = k^2 T^2 - kT \lim_{R \to \infty} R e^{-R/kT}.$$

We compute the remaining limit using L'Hôpital's Rule:

$$\lim_{R \to \infty} R e^{-R/kT} = \lim_{R \to \infty} \frac{R}{e^{R/kT}} = \lim_{R \to \infty} \frac{\frac{dR}{dR}}{\frac{d}{dR}\left(e^{R/kT}\right)} = \lim_{R \to \infty} \frac{1}{\frac{1}{kT} e^{R/kT}} = 0.$$

Thus,

$$\int_0^\infty E e^{-E/RT}\, dE = k^2 T^2,$$

and

$$\overline{E} = \frac{1}{kT} \int_0^\infty E e^{-E/kT}\, dE = \frac{1}{kT} \cdot k^2 T^2 = kT.$$

91. Compute the Laplace transform $Lf(s)$ of the function $f(x) = x^2 e^{\alpha x}$ for $s > \alpha$.

SOLUTION   The Laplace transform is the following integral:

$$L\left(x^2 e^{\alpha x}\right)(s) = \int_0^\infty x^2 e^{\alpha x} e^{-sx}\, dx = \int_0^\infty x^2 e^{(\alpha - s)x}\, dx = \lim_{R \to \infty} \int_0^R x^2 e^{(\alpha - s)x}\, dx.$$

We compute the definite integral using Integration by Parts with $u = x^2$, $v' = e^{(\alpha - s)x}$. Then $u' = 2x$, $v = \frac{1}{\alpha - s} e^{(\alpha - s)x}$ and

$$\int_0^R x^2 e^{(\alpha - s)x}\, dx = \frac{1}{\alpha - s} x^2 e^{(\alpha - s)x}\Big|_{x=0}^R - \int_0^R 2x \cdot \frac{1}{\alpha - s} e^{(\alpha - s)x}\, dx$$

$$= \frac{1}{\alpha - s} R^2 e^{(\alpha - s)R} - \frac{2}{\alpha - s} \int_0^R x e^{(\alpha - s)x}\, dx.$$

We compute the resulting integral using Integration by Parts again, this time with $u = x$ and $v' = e^{(\alpha - s)x}$. Then $u' = 1$, $v = \frac{1}{\alpha - s} e^{(\alpha - s)x}$ and

$$\int_0^R x e^{(\alpha - s)x}\, dx = x \cdot \frac{1}{\alpha - s} e^{(\alpha - s)x}\Big|_{x=0}^R - \frac{1}{\alpha - s} \int_0^R e^{(\alpha - s)x}\, dx = \left(\frac{x}{\alpha - s} e^{(\alpha - s)x} - \frac{1}{(\alpha - s)^2} e^{(\alpha - s)x} \right)\Big|_{x=0}^R$$

$$= \frac{R}{\alpha - s} e^{(\alpha - s)R} - \frac{1}{(\alpha - s)^2} \left(e^{(\alpha - s)R} - e^0 \right) = \frac{1}{(\alpha - s)^2} - \frac{1}{(\alpha - s)^2} e^{(\alpha - s)R} + \frac{R}{\alpha - s} e^{(\alpha - s)R}.$$

Thus,

$$\int_0^R x^2 e^{(\alpha - s)x}\, dx = \frac{1}{\alpha - s} R^2 e^{(\alpha - s)R} - \frac{2}{\alpha - s} \left(\frac{1}{(\alpha - s)^2} - \frac{1}{(\alpha - s)^2} e^{(\alpha - s)R} + \frac{R}{\alpha - s} e^{(\alpha - s)R} \right)$$

$$= \frac{1}{\alpha - s} R^2 e^{(\alpha - s)R} - \frac{2}{(\alpha - s)^3} + \frac{2}{(\alpha - s)^3} e^{(\alpha - s)R} - \frac{2R}{(\alpha - s)^2} e^{(\alpha - s)R},$$

and

$$L\left(x^2 e^{\alpha x}\right)(s) = \frac{2}{(s - \alpha)^3} - \frac{1}{s - \alpha} \lim_{R \to \infty} R^2 e^{-(s - \alpha)R} - \frac{2}{(s - \alpha)^3} \lim_{R \to \infty} e^{-(s - \alpha)R} - \frac{2}{(s - \alpha)^2} \lim_{R \to \infty} R e^{-(s - \alpha)R}.$$

Now, since $s > \alpha$, $\displaystyle\lim_{R \to \infty} e^{-(s - \alpha)R} = 0$. We use L'Hôpital's Rule to compute the other two limits:

$$\lim_{R \to \infty} R e^{-(s - \alpha)R} = \lim_{R \to \infty} \frac{R}{e^{(s - \alpha)R}} = \lim_{R \to \infty} \frac{1}{(s - \alpha) e^{(s - \alpha)R}} = 0;$$

$$\lim_{R \to \infty} R^2 e^{-(s - \alpha)R} = \lim_{R \to \infty} \frac{R^2}{e^{(s - \alpha)R}} = \lim_{R \to \infty} \frac{2R}{(s - \alpha) e^{(s - \alpha)R}} = \lim_{R \to \infty} \frac{2}{(s - \alpha)^2 e^{(s - \alpha)R}} = 0.$$

Finally,

$$L\left(x^2 e^{\alpha x}\right)(s) = \frac{2}{(s - \alpha)^3} - 0 - 0 - 0 = \frac{2}{(s - \alpha)^3}.$$

93. Let $J_n = \displaystyle\int x^n e^{-x^2/2}\, dx$.

(a) Show that $J_1 = -e^{-x^2/2} + C$.

(b) Prove $J_n = -x^{n-1} e^{-x^2/2} + (n - 1) J_{n-2}$.

(c) Use (a) and (b) to compute J_3 and J_5.

SOLUTION

(a) Let $u = -\frac{x^2}{2}$. Then $du = -x\,dx$ and

$$J_1 = \int xe^{-x^2/2}\,dx = -\int e^u\,du = -e^u + C = -e^{-x^2/2} + C.$$

(b) Using Integration by Parts with $u = x^{n-1}$ and $v' = xe^{-x^2/2}$, we find

$$J_n = -x^{n-1}e^{-x^2/2} + (n-1)\int x^{n-2}e^{-x^2/2}\,dx = -x^{n-1}e^{-x^2/2} + (n-1)J_{n-2}.$$

(c) Using the results from parts (a) and (b),

$$J_3 = -x^{3-1}e^{-x^2/2} + (3-1)J_{3-2} = -x^2 e^{-x^2/2} + 2J_1$$

$$= -x^2 e^{-x^2/2} - 2e^{-x^2/2} + C = -e^{-x^2/2}(x^2 + 2) + C$$

and then

$$J_5 = -x^{5-1}e^{-x^2/2} + (5-1)J_{5-2} = -x^4 e^{-x^2/2} + 4J_3$$

$$= -x^4 e^{-x^2/2} - 4e^{-x^2/2}(x^2 + 2) + C = -e^{-x^2/2}(x^4 + 4x^2 + 8) + C.$$

9 | FURTHER APPLICATIONS OF THE INTEGRAL AND TAYLOR POLYNOMIALS

9.1 Arc Length and Surface Area

Preliminary Questions

1. Which integral represents the length of the curve $y = \cos x$ between 0 and π?

$$\int_0^\pi \sqrt{1 + \cos^2 x}\, dx, \qquad \int_0^\pi \sqrt{1 + \sin^2 x}\, dx$$

SOLUTION Let $y = \cos x$. Then $y' = -\sin x$, and $1 + (y')^2 = 1 + \sin^2 x$. Thus, the length of the curve $y = \cos x$ between 0 and π is

$$\int_0^\pi \sqrt{1 + \sin^2 x}\, dx.$$

2. How do the arc lengths of the curves $y = f(x)$ and $y = f(x) + C$ over an interval $[a, b]$ differ (C is a constant)? Explain geometrically and then justify using the arc length formula.

SOLUTION The graph of $y = f(x) + C$ is a vertical translation of the graph of $y = f(x)$; hence, the two graphs should have the same arc length. We can explicitly establish this as follows:

$$\text{length of } y = f(x) + C = \int_a^b \sqrt{1 + \left[\frac{d}{dx}(f(x) + C)\right]^2}\, dx = \int_a^b \sqrt{1 + [f'(x)]^2}\, dx = \text{length of } y = f(x).$$

Exercises

1. Express the arc length of the curve $y = x^4$ between $x = 2$ and $x = 6$ as an integral (but do not evaluate).

SOLUTION Let $y = x^4$. Then $y' = 4x^3$ and

$$s = \int_2^6 \sqrt{1 + (4x^3)^2}\, dx = \int_2^6 \sqrt{1 + 16x^6}\, dx.$$

3. Find the arc length of $y = \frac{1}{12}x^3 + x^{-1}$ for $1 \le x \le 2$. *Hint:* Show that $1 + (y')^2 = \left(\frac{1}{4}x^2 + x^{-2}\right)^2$.

SOLUTION Let $y = \frac{1}{12}x^3 + x^{-1}$. Then $y' = \frac{x^2}{4} - x^{-2}$, and

$$(y')^2 + 1 = \left(\frac{x^2}{4} - x^{-2}\right)^2 + 1 = \frac{x^4}{16} - \frac{1}{2} + x^{-4} + 1 = \frac{x^4}{16} + \frac{1}{2} + x^{-4} = \left(\frac{x^2}{4} + x^{-2}\right)^2.$$

Thus,

$$s = \int_1^2 \sqrt{1 + (y')^2}\, dx = \int_1^2 \sqrt{\left(\frac{x^2}{4} + \frac{1}{x^2}\right)^2}\, dx = \int_1^2 \left|\frac{x^2}{4} + \frac{1}{x^2}\right| dx$$

$$= \int_1^2 \left(\frac{x^2}{4} + \frac{1}{x^2}\right) dx \quad \text{since} \quad \frac{x^2}{4} + \frac{1}{x^2} > 0$$

$$= \left(\frac{x^3}{12} - \frac{1}{x}\right)\Bigg|_1^2 = \frac{13}{12}.$$

In Exercises 5–10, calculate the arc length over the given interval.

5. $y = 3x + 1$, $[0, 3]$

SOLUTION   Let $y = 3x + 1$. Then $y' = 3$, and $s = \int_0^3 \sqrt{1 + 9} \, dx = 3\sqrt{10}$.

7. $y = x^{3/2}$, $[1, 2]$

SOLUTION   Let $y = x^{3/2}$. Then $y' = \frac{3}{2} x^{1/2}$, and

$$s = \int_1^2 \sqrt{1 + \frac{9}{4} x} \, dx = \frac{8}{27} \left(1 + \frac{9}{4} x\right)^{3/2} \Big|_1^2 = \frac{8}{27} \left(\left(\frac{11}{2}\right)^{3/2} - \left(\frac{13}{4}\right)^{3/2}\right) = \frac{1}{27} \left(22\sqrt{22} - 13\sqrt{13}\right).$$

9. $y = \frac{1}{4} x^2 - \frac{1}{2} \ln x$, $[1, 2e]$

SOLUTION   Let $y = \frac{1}{4} x^2 - \frac{1}{2} \ln x$. Then

$$y' = \frac{x}{2} - \frac{1}{2x},$$

and

$$1 + (y')^2 = 1 + \left(\frac{x}{2} - \frac{1}{2x}\right)^2 = \frac{x^2}{4} + \frac{1}{2} + \frac{1}{4x^2} = \left(\frac{x}{2} + \frac{1}{2x}\right)^2.$$

Hence,

$$s = \int_1^{2e} \sqrt{1 + (y')^2} \, dx = \int_1^{2e} \sqrt{\left(\frac{x}{2} + \frac{1}{2x}\right)^2} \, dx = \int_1^{2e} \left|\frac{x}{2} + \frac{1}{2x}\right| dx$$

$$= \int_1^{2e} \left(\frac{x}{2} + \frac{1}{2x}\right) dx \quad \text{since} \quad \frac{x}{2} + \frac{1}{2x} > 0 \text{ on } [1, 2e]$$

$$= \left(\frac{x^2}{4} + \frac{1}{2} \ln x\right) \Big|_1^{2e} = e^2 + \frac{\ln 2}{2} + \frac{1}{4}.$$

In Exercises 11–14, approximate the arc length of the curve over the interval using the Trapezoidal Rule T_N, the Midpoint Rule M_N, or Simpson's Rule S_N as indicated.

11. $y = \frac{1}{4} x^4$, $[1, 2]$, T_5

SOLUTION   Let $y = \frac{1}{4} x^4$. Then

$$1 + (y')^2 = 1 + (x^3)^2 = 1 + x^6.$$

Therefore, the arc length over $[1, 2]$ is

$$\int_1^2 \sqrt{1 + x^6} \, dx.$$

Now, let $f(x) = \sqrt{1 + x^6}$. With $n = 5$,

$$\Delta x = \frac{2 - 1}{5} = \frac{1}{5} \quad \text{and} \quad \{x_i\}_{i=0}^5 = \left\{1, \frac{6}{5}, \frac{7}{5}, \frac{8}{5}, \frac{9}{5}, 2\right\}.$$

Using the Trapezoidal Rule,

$$\int_1^2 \sqrt{1 + x^6} \, dx \approx \frac{\Delta x}{2} \left[f(x_0) + 2 \sum_{i=1}^4 f(x_i) + f(x_5)\right] = 3.957736.$$

The arc length is approximately 3.957736 units.

13. $y = x^{-1}$, $[1, 2]$, S_8

SOLUTION   Let $y = x^{-1}$. Then $y' = -x^{-2}$ and

$$1 + (y')^2 = 1 + \frac{1}{x^4}.$$

Therefore, the arc length over [1, 2] is

$$\int_1^2 \sqrt{1 + \frac{1}{x^4}}\, dx.$$

Now, let $f(x) = \sqrt{1 + \frac{1}{x^4}}$. With $n = 8$,

$$\Delta x = \frac{2 - 1}{8} = \frac{1}{8} \quad \text{and} \quad \{x_i\}_{i=0}^8 = \left\{1, \frac{9}{8}, \frac{5}{4}, \frac{11}{8}, \frac{3}{2}, \frac{13}{8}, \frac{7}{4}, \frac{15}{8}, 2\right\}.$$

Using Simpson's Rule,

$$\int_1^2 \sqrt{1 + \frac{1}{x^4}}\, dx \approx \frac{\Delta x}{3}\left[f(x_0) + 4\sum_{i=1}^4 f(x_{2i-1}) + 2\sum_{i=1}^3 f(x_{2i}) + f(x_8)\right] = 1.132123.$$

The arc length is approximately 1.132123 units.

15. Calculate the length of the astroid $x^{2/3} + y^{2/3} = 1$ (Figure 11).

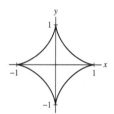

FIGURE 11 Graph of $x^{2/3} + y^{2/3} = 1$.

SOLUTION We will calculate the arc length of the portion of the asteroid in the first quadrant and then multiply by 4. By implicit differentiation

$$\frac{2}{3}x^{-1/3} + \frac{2}{3}y^{-1/3}y' = 0,$$

so

$$y' = -\frac{x^{-1/3}}{y^{-1/3}} = -\frac{y^{1/3}}{x^{1/3}}.$$

Thus

$$1 + (y')^2 = 1 + \frac{y^{2/3}}{x^{2/3}} = \frac{x^{2/3} + y^{2/3}}{x^{2/3}} = \frac{1}{x^{2/3}},$$

and

$$s = \int_0^1 \frac{1}{x^{1/3}}\, dx = \frac{3}{2}.$$

The total arc length is therefore $4 \cdot \frac{3}{2} = 6$.

17. Find the arc length of the curve shown in Figure 12.

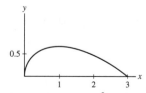

FIGURE 12 Graph of $9y^2 = x(x - 3)^2$.

SOLUTION Using implicit differentiation,

$$18yy' = x(2)(x - 3) + (x - 3)^2 = 3(x - 3)(x - 1).$$

Hence,

$$(y')^2 = \frac{(x - 3)^2(x - 1)^2}{36y^2} = \frac{(x - 3)^2(x - 1)^2}{4(9y^2)} = \frac{(x - 3)^2(x - 1)^2}{4x(x - 3)^2} = \frac{(x - 1)^2}{4x}$$

and

$$1 + (y')^2 = \frac{(x-1)^2 + 4x}{4x} = \frac{(x+1)^2}{4x}.$$

Finally,

$$s = \int_0^3 \sqrt{\frac{(x+1)^2}{4x}}\, dx = \int_0^3 \frac{|x+1|}{2\sqrt{x}}\, dx$$

$$= \int_0^3 \frac{x+1}{2\sqrt{x}}\, dx \quad \text{since} \quad x+1 > 0 \text{ on } [0,3]$$

$$= \int_0^3 \left(\frac{1}{2}x^{1/2} + \frac{1}{2}x^{-1/2}\right) dx = \left(\frac{1}{3}x^{3/2} + x^{1/2}\right)\Big|_0^3 = 2\sqrt{3}.$$

19. Let $f(x) = mx + r$ be a linear function (Figure 13). Use the Pythagorean Theorem to verify that the arc length over $[a, b]$ is equal to

$$\int_a^b \sqrt{1 + f'(x)^2}\, dx$$

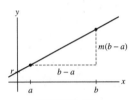

FIGURE 13

SOLUTION Let h denote the length of the hypotenuse. Then, by Pythagoras' Theorem,

$$h^2 = (b-a)^2 + m^2(b-a)^2 = (b-a)^2(1+m^2),$$

or

$$h = (b-a)\sqrt{1+m^2}$$

since $b > a$. Moreover, $(f'(x))^2 = m^2$, so

$$s = \int_a^b \sqrt{1+m^2}\, dx = (b-a)\sqrt{1+m^2} = h.$$

21. Show that the circumference C of the circle of radius r is $C = 2\int_{-r}^r \frac{dx}{\sqrt{1 - x^2/r^2}}$. Use a substitution to show that C is proportional to r.

SOLUTION Let $y = \sqrt{r^2 - x^2}$ denote the upper half of a circle of radius r centered at the origin. Then

$$1 + (y')^2 = 1 + \frac{x^2}{r^2 - x^2} = \frac{r^2}{r^2 - x^2} = \frac{1}{1 - \frac{x^2}{r^2}},$$

and the circumference of the circle is given by

$$C = 2\int_{-r}^r \frac{dx}{\sqrt{1 - x^2/r^2}}.$$

Using the substitution $u = x/r$, $du = dx/r$, we find

$$C = 2r\int_{-1}^1 \frac{du}{\sqrt{1 - u^2}} = kr,$$

where

$$k = 2\int_{-1}^1 \frac{du}{\sqrt{1 - u^2}}$$

is the circumference of the unit circle. Thus, the circumference is proportional to r.

23. Express the arc length of $g(x) = \sqrt{x}$ over $[0, 1]$ as a definite integral. Then use the substitution $u = \sqrt{x}$ to show that this arc length is equal to the arc length of x^2 over $[0, 1]$ (but do not evaluate the integrals). Explain this result graphically.

SOLUTION Let $g(x) = \sqrt{x}$. Then

$$1 + g'(x)^2 = \frac{1 + 4x}{4x} \quad \text{and} \quad s = \int_0^1 \sqrt{\frac{1 + 4x}{4x}}\, dx = \int_0^1 \frac{\sqrt{1 + 4x}}{2\sqrt{x}}\, dx.$$

With the substitution $u = \sqrt{x}$, $du = \dfrac{1}{2\sqrt{x}}\, dx$, this becomes

$$s = \int_0^1 \sqrt{1 + 4u^2}\, du.$$

Now, let $f(x) = x^2$. Then $1 + f'(x)^2 = 1 + 4x^2$, and

$$s = \int_0^1 \sqrt{1 + 4x^2}\, dx.$$

Thus, the two arc lengths are equal. This is explained graphically by the fact that for $x \geq 0$, x^2 and $\sqrt{x}$ are inverses of each other. This means that the two graphs are symmetric with respect to the line $y = x$. Moreover, the graphs of x^2 and $\sqrt{x}$ intersect at $x = 0$ and at $x = 1$. Thus, it is clear that the arc length of the two graphs on $[0, 1]$ are equal.

25. Find the arc length of $y = e^x$ over $[0, a]$. *Hint:* Try the substitution $u = \sqrt{1 + e^{2x}}$ followed by partial fractions.

SOLUTION Let $y = e^x$. Then $1 + (y')^2 = 1 + e^{2x}$, and the arc length over $[0, a]$ is

$$\int_0^a \sqrt{1 + e^{2x}}\, dx.$$

Now, let $u = \sqrt{1 + e^{2x}}$. Then

$$du = \frac{1}{2} \cdot \frac{2e^{2x}}{\sqrt{1 + e^{2x}}}\, dx = \frac{u^2 - 1}{u}\, dx$$

and the arc length is

$$\int_0^a \sqrt{1 + e^{2x}}\, dx = \int_{x=0}^{x=a} u \cdot \frac{u}{u^2 - 1}\, du = \int_{x=0}^{x=a} \frac{u^2}{u^2 - 1}\, du = \int_{x=0}^{x=a} \left(1 + \frac{1}{u^2 - 1}\right) du$$

$$= \int_{x=0}^{x=a} \left(1 + \frac{1}{2}\frac{1}{u - 1} - \frac{1}{2}\frac{1}{u + 1}\right) du = \left(u + \frac{1}{2}\ln(u - 1) - \frac{1}{2}\ln(u + 1)\right)\Bigg|_{x=0}^{x=a}$$

$$= \left[\sqrt{1 + e^{2x}} + \frac{1}{2}\ln\left(\frac{\sqrt{1 + e^{2x}} - 1}{\sqrt{1 + e^{2x}} + 1}\right)\right]\Bigg|_0^a$$

$$= \sqrt{1 + e^{2a}} + \frac{1}{2}\ln\frac{\sqrt{1 + e^{2a}} - 1}{\sqrt{1 + e^{2a}} + 1} - \sqrt{2} + \frac{1}{2}\ln\frac{1 + \sqrt{2}}{\sqrt{2} - 1}$$

$$= \sqrt{1 + e^{2a}} + \frac{1}{2}\ln\frac{\sqrt{1 + e^{2a}} - 1}{\sqrt{1 + e^{2a}} + 1} - \sqrt{2} + \ln(1 + \sqrt{2}).$$

27. Use Eq. (4) to compute the arc length of $y = \ln(\sin x)$ for $\frac{\pi}{4} \leq x \leq \frac{\pi}{2}$.

SOLUTION With $f(x) = \sin x$, Eq. (4) yields

$$s = \int_{\pi/4}^{\pi/2} \frac{\sqrt{\sin^2 x + \cos^2 x}}{\sin x}\, dx = \int_{\pi/4}^{\pi/2} \csc x\, dx = \ln(\csc x - \cot x)\Bigg|_{\pi/4}^{\pi/2}$$

$$= \ln 1 - \ln(\sqrt{2} - 1) = \ln\frac{1}{\sqrt{2} - 1} = \ln(\sqrt{2} + 1).$$

29. Show that if $0 \leq f'(x) \leq 1$ for all x, then the arc length of $y = f(x)$ over $[a, b]$ is at most $\sqrt{2}(b - a)$. Show that for $f(x) = x$, the arc length equals $\sqrt{2}(b - a)$.

SOLUTION If $0 \le f'(x) \le 1$ for all x, then

$$s = \int_a^b \sqrt{1 + f'(x)^2}\, dx \le \int_a^b \sqrt{1+1}\, dx = \sqrt{2}(b-a).$$

If $f(x) = x$, then $f'(x) = 1$ and

$$s = \int_a^b \sqrt{1+1}\, dx = \sqrt{2}(b-a).$$

31. Approximate the arc length of one-quarter of the unit circle (which we know is $\frac{\pi}{2}$) by computing the length of the polygonal approximation with $N = 4$ segments (Figure 14).

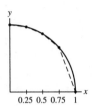

FIGURE 14 One-quarter of the unit circle

SOLUTION With $y = \sqrt{1 - x^2}$, the five points along the curve are

$$P_0(0, 1),\ P_1(1/4, \sqrt{15}/4),\ P_2(1/2, \sqrt{3}/2),\ P_3(3/4, \sqrt{7}/4),\ P_4(1, 0)$$

Then

$$\overline{P_0 P_1} = \sqrt{\frac{1}{16} + \left(\frac{4 - \sqrt{15}}{4}\right)^2} \qquad \approx .252009$$

$$\overline{P_1 P_2} = \sqrt{\frac{1}{16} + \left(\frac{2\sqrt{3} - \sqrt{15}}{4}\right)^2} \qquad \approx .270091$$

$$\overline{P_2 P_3} = \sqrt{\frac{1}{16} + \left(\frac{2\sqrt{3} - \sqrt{7}}{4}\right)^2} \qquad \approx .323042$$

$$\overline{P_3 P_4} = \sqrt{\frac{1}{16} + \frac{7}{16}} \qquad \approx .707108$$

and the total approximate distance is 1.552250 whereas $\pi/2 \approx 1.570796$.

In Exercises 32–39, compute the surface area of revolution about the x-axis over the interval.

33. $y = 4x + 3,\quad [2, 4]$

SOLUTION Let $y = 4x + 3$. Then $1 + (y')^2 = 17$ and

$$SA = 2\pi \int_2^4 (4x + 3)\sqrt{17}\, dx = 2\pi\sqrt{17}\left(2x^2 + 3x\right)\Big|_2^4 = 60\pi\sqrt{17}.$$

35. $y = (4 - x^{2/3})^{3/2},\quad [0, 8]$

SOLUTION Let $y = (4 - x^{2/3})^{3/2}$. Then

$$y' = -x^{-1/3}(4 - x^{2/3})^{1/2},$$

and

$$1 + (y')^2 = 1 + \frac{4 - x^{2/3}}{x^{2/3}} = \frac{4}{x^{2/3}}.$$

Therefore,

$$SA = 2\pi \int_0^8 (4 - x^{2/3})^{3/2}\left(\frac{2}{x^{1/3}}\right) dx.$$

Using the substitution $u = 4 - x^{2/3}$, $du = -\frac{2}{3}x^{-1/3}\,dx$, we find

$$SA = 2\pi \int_4^0 u^{3/2}(-3)\,du = 6\pi \int_0^4 u^{3/2}\,du = \frac{12}{5}\pi u^{5/2}\Big|_0^4 = \frac{384\pi}{5}.$$

37. $y = e^x$, $[0, 1]$

SOLUTION Let $y = e^x$. Then $y' = e^x$ and

$$SA = 2\pi \int_0^1 e^x \sqrt{1 + e^{2x}}\,dx.$$

Using the substitution $e^x = \tan\theta$, $e^x\,dx = \sec^2\theta\,d\theta$, we find that

$$\int e^x \sqrt{1 + e^{2x}}\,dx = \int \sec^3\theta\,d\theta = \frac{1}{2}\sec\theta\tan\theta + \frac{1}{2}\ln|\sec\theta + \tan\theta| + C$$

$$= \frac{1}{2}e^x\sqrt{1 + e^{2x}} + \frac{1}{2}\ln|\sqrt{1 + e^{2x}} + e^x| + C.$$

Finally,

$$SA = \left(\pi e^x\sqrt{1 + e^{2x}} + \pi\ln|\sqrt{1 + e^{2x}} + e^x|\right)\Big|_0^1$$

$$= \pi e\sqrt{1 + e^2} + \pi\ln(\sqrt{1 + e^2} + e) - \pi\sqrt{2} - \pi\ln(\sqrt{2} + 1)$$

$$= \pi e\sqrt{1 + e^2} - \pi\sqrt{2} + \pi\ln\left(\frac{\sqrt{1 + e^2} + e}{\sqrt{2} + 1}\right).$$

39. $y = \sin x$, $[0, \pi]$

SOLUTION Let $y = \sin x$. Then $y' = \cos x$, and

$$SA = 2\pi \int_0^\pi \sin x \sqrt{1 + \cos^2 x}\,dx.$$

Using the substitution $\cos x = \tan\theta$, $-\sin x\,dx = \sec^2\theta\,d\theta$, we find that

$$\int \sin x\sqrt{1 + \cos^2 x}\,dx = -\int \sec^3\theta\,d\theta = -\frac{1}{2}\sec\theta\tan\theta - \frac{1}{2}\ln|\sec\theta + \tan\theta| + C$$

$$= -\frac{1}{2}\cos x\sqrt{1 + \cos^2 x} - \frac{1}{2}\ln|\sqrt{1 + \cos^2 x} + \cos x| + C.$$

Finally,

$$SA = 2\pi\left(-\frac{1}{2}\cos x\sqrt{1 + \cos^2 x} - \frac{1}{2}\ln|\sqrt{1 + \cos^2 x} + \cos x|\right)\Big|_0^\pi$$

$$= 2\pi\left(\frac{1}{2}\sqrt{2} - \frac{1}{2}\ln(\sqrt{2} - 1) + \frac{1}{2}\sqrt{2} + \frac{1}{2}\ln(\sqrt{2} + 1)\right) = 2\pi\left(\sqrt{2} + \ln(\sqrt{2} + 1)\right).$$

41. Prove that the surface area of a sphere of radius r is $4\pi r^2$ by rotating the top half of the unit circle $x^2 + y^2 = r^2$ about the x-axis.

SOLUTION Let $y = \sqrt{r^2 - x^2}$. Then $y' = -\dfrac{x}{\sqrt{r^2 - x^2}}$, and

$$1 + (y')^2 = 1 + \frac{x^2}{r^2 - x^2} = \frac{r^2}{r^2 - x^2}.$$

Finally,

$$SA = 2\pi \int_{-r}^r \sqrt{r^2 - x^2}\,\frac{r}{\sqrt{r^2 - x^2}}\,dx = 2\pi r \int_{-r}^r dx = 2\pi r(2r) = 4\pi r^2.$$

CAS In Exercises 42–45, use a computer algebra system to find the exact or approximate surface area of the solid generated by rotating the curve about the x-axis.

43. $y = x^2$ for $0 \le x \le 4$, exact area

SOLUTION Let $y = x^2$. Then $y' = 2x$, $1 + (y')^2 = 1 + 4x^2$, and

$$SA = 2\pi \int_0^4 x^2 \sqrt{1 + 4x^2}\, dx.$$

Using a computer algebra system to evaluate the definite integral, we find

$$SA = \frac{129\sqrt{65}}{4}\pi - \frac{\pi}{32} \ln(8 + \sqrt{65}).$$

45. $y = \tan x$ for $0 \le x \le \frac{\pi}{4}$, approximate area

SOLUTION Let $y = \tan x$. Then $y' = \sec^2 x$, $1 + (y')^2 = 1 + \sec^4 x$, and

$$SA = 2\pi \int_0^{\pi/4} \tan x \sqrt{1 + \sec^4 x}\, dx.$$

Using a computer algebra system to approximate the value of the definite integral, we find

$$SA \approx 3.83908.$$

47. *CAS* A merchant intends to produce specialty carpets in the shape of the region in Figure 15, bounded by the axes and graph of $y = 1 - x^n$ (units in yards). Assume that material costs \$50/yd^2 and that it costs $50L$ dollars to cut the carpet, where L is the length of the curved side of the carpet. The carpet can be sold for $150A$ dollars, where A is the carpet's area. Using numerical integration with a computer algebra system, find the whole number n for which the merchant's profits are maximal.

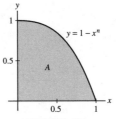

FIGURE 15

SOLUTION The area of the carpet is

$$A = \int_0^1 (1 - x^n)\, dx = \left(x - \frac{x^{n+1}}{n+1} \right)\Big|_0^1 = 1 - \frac{1}{n+1} = \frac{n}{n+1},$$

while the length of the curved side of the carpet is

$$L = \int_0^1 \sqrt{1 + (nx^{n-1})^2}\, dx = \int_0^1 \sqrt{1 + n^2 x^{2n-2}}\, dx.$$

Using these formulas, we find that the merchant's profit is given by

$$150A - (50A + 50L) = 100A - 50L = \frac{100n}{n+1} - 50 \int_0^1 \sqrt{1 + n^2 x^{2n-2}}\, dx.$$

Using a CAS, we find that the merchant's profit is maximized (approximately \$3.31 per carpet) when $n = 13$. The table below lists the profit for $1 \le n \le 15$.

n	Profit	n	Profit
1	−20.71067810	9	3.06855532
2	−7.28047621	10	3.18862208
3	−2.39328273	11	3.25953632
4	−0.01147138	12	3.29668137
5	1.30534545	13	3.31024566
6	2.08684099	14	3.30715476
7	2.57017349	15	3.29222024
8	2.87535925		

49. [icon] Suppose that the observer in Exercise 48 moves off to infinity, that is, $d \to \infty$. What do you expect the limiting value of the observed area to be? Check your guess by calculating the limit using the formula for the area in the previous exercise.

SOLUTION We would assume the observed surface area would approach $2\pi R^2$ which is the surface area of a hemisphere of radius R. To verify this, observe:

$$\lim_{d \to \infty} SA = \lim_{d \to \infty} \frac{2\pi R^2 d}{R + d} = \lim_{d \to \infty} \frac{2\pi R^2}{1} = 2\pi R^2.$$

Further Insights and Challenges

51. Find the surface area of the ellipsoid obtained by rotating the ellipse $\left(\frac{x}{a}\right)^2 + \left(\frac{y}{b}\right)^2 = 1$ about the x-axis.

SOLUTION Taking advantage of symmetry, we can find the surface area of the ellipsoid by doubling the surface area obtained by rotating the portion of the ellipse in the first quadrant about the x-axis. The equation for the portion of the ellipse in the first quadrant is

$$y = \frac{b}{a}\sqrt{a^2 - x^2}.$$

Thus,

$$1 + (y')^2 = 1 + \frac{b^2 x^2}{a^2(a^2 - x^2)} = \frac{a^4 + (b^2 - a^2)x^2}{a^2(a^2 - x^2)},$$

and

$$SA = 4\pi \int_0^a \frac{b}{a}\sqrt{a^2 - x^2}\frac{\sqrt{a^4 + (b^2 - a^2)x^2}}{a\sqrt{a^2 - x^2}}\,dx = 4\pi b \int_0^a \sqrt{1 + \left(\frac{b^2 - a^2}{a^4}\right)x^2}\,dx.$$

We now consider two cases. If $b^2 > a^2$, then we make the substitution

$$\frac{\sqrt{b^2 - a^2}}{a^2}x = \tan\theta, \quad dx = \frac{a^2}{\sqrt{b^2 - a^2}}\sec^2\theta\,d\theta,$$

and find that

$$SA = 4\pi b \frac{a^2}{\sqrt{b^2 - a^2}} \int_{x=0}^{x=a} \sec^3\theta\,d\theta = 2\pi b\frac{a^2}{\sqrt{b^2 - a^2}}\left(\sec\theta\tan\theta + \ln|\sec\theta + \tan\theta|\right)\Big|_{x=0}^{x=a}$$

$$= \left(2\pi bx\sqrt{1 + \left(\frac{b^2 - a^2}{a^4}\right)x^2} + 2\pi b\frac{a^2}{\sqrt{b^2 - a^2}}\ln\left|\sqrt{1 + \left(\frac{b^2 - a^2}{a^4}\right)x^2} + \frac{\sqrt{b^2 - a^2}}{a^2}x}\right|\right)\Big|_0^a$$

$$= 2\pi b^2 + 2\pi b\frac{a^2}{\sqrt{b^2 - a^2}}\ln\left(\frac{b}{a} + \frac{\sqrt{b^2 - a^2}}{a}\right).$$

On the other hand, if $a^2 > b^2$, then we make the substitution

$$\frac{\sqrt{a^2 - b^2}}{a^2}x = \sin\theta, \quad dx = \frac{a^2}{\sqrt{a^2 - b^2}}\cos\theta\,d\theta,$$

and find that

$$SA = 4\pi b \frac{a^2}{\sqrt{a^2 - b^2}} \int_{x=0}^{x=a} \cos^2\theta\,d\theta = 2\pi b\frac{a^2}{\sqrt{a^2 - b^2}}(\theta + \sin\theta\cos\theta)\Big|_{x=0}^{x=a}$$

$$= \left[2\pi bx\sqrt{1 - \left(\frac{a^2 - b^2}{a^4}\right)x^2} + 2\pi b\frac{a^2}{\sqrt{a^2 - b^2}}\sin^{-1}\left(\frac{\sqrt{a^2 - b^2}}{a^2}x\right)\right]\Big|_0^a$$

$$= 2\pi b^2 + 2\pi b\frac{a^2}{\sqrt{a^2 - b^2}}\sin^{-1}\left(\frac{\sqrt{a^2 - b^2}}{a}\right).$$

Observe that in both cases, as a approaches b, the value of the surface area of the ellipsoid approaches $4\pi b^2$, the surface area of a sphere of radius b.

53. **CAS**   Let L be the arc length of the upper half of the ellipse with equation $y = \left(\dfrac{b}{a}\right)\sqrt{a^2 - x^2}$ (Figure 17) and

let $\eta = \sqrt{1 - \dfrac{b^2}{a^2}}$. Use substitution to show that

$$L = a \int_{-\pi/2}^{\pi/2} \sqrt{1 - \eta^2 \sin^2 \theta}\, d\theta$$

Use a computer algebra system to approximate L for $a = 2$, $b = 1$.

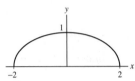

FIGURE 17 Graph of the ellipse $y = \frac{1}{2}\sqrt{4 - x^2}$.

SOLUTION   Let $y = \dfrac{b}{a}\sqrt{a^2 - x^2}$. Then

$$1 + (y')^2 = \frac{b^2 x^2 + a^2(a^2 - x^2)}{a^2(a^2 - x^2)}$$

and

$$s = \int_{-a}^{a} \sqrt{\frac{b^2 x^2 + a^2(a^2 - x^2)}{a^2(a^2 - x^2)}}\, dx.$$

With the substitution $x = a \sin t$, $dx = a \cos t\, dt$, $a^2 - x^2 = a^2 \cos^2 t$ and

$$s = a \int_{-\pi/2}^{\pi/2} \cos t \sqrt{\frac{a^2 b^2 \sin^2 t + a^2 a^2 \cos^2 t}{a^2(a^2 \cos^2 t)}}\, dt = a \int_{\pi/2}^{\pi/2} \sqrt{\frac{b^2 \sin^2 t}{a^2} + \cos^2 t}\, dt$$

Because

$$\eta = \sqrt{1 - \frac{b^2}{a^2}}, \quad \eta^2 = 1 - \frac{b^2}{a^2}$$

we then have

$$1 - \eta^2 \sin^2 t = 1 - \left(1 - \frac{b^2}{a^2}\right)\sin^2 t = 1 - \sin^2 t + \frac{b^2}{a^2}\sin^2 t = \cos^2 t + \frac{b^2}{a^2}\sin^2 t$$

which is the same as the expression under the square root above. Substituting, we get

$$s = a \int_{-\pi/2}^{\pi/2} \sqrt{1 - \eta^2 \sin^2 t}\, dt$$

When $a = 2$ and $b = 1$, $\eta^2 = \frac{3}{4}$. Using a computer algebra system to approximate the value of the definite integral, we find $s \approx 4.84422$.

55.   Let $f(x)$ be an increasing function on $[a, b]$ and let $g(x)$ be its inverse. Argue on the basis of arc length that the following equality holds:

$$\int_{a}^{b} \sqrt{1 + f'(x)^2}\, dx = \int_{f(a)}^{f(b)} \sqrt{1 + g'(y)^2}\, dy \qquad \boxed{5}$$

Then use the substitution $u = f(x)$ to prove Eq. (5).

SOLUTION   Since the graphs of $f(x)$ and $g(x)$ are symmetric with respect to the line $y = x$, the arc length of the curves will be equal on the respective domains. Since the domain of g is the range of f, on $f(a)$ to $f(b)$, $g(x)$ will have the same arc length as $f(x)$ on a to b. If $g(x) = f^{-1}(x)$ and $u = f(x)$, then $x = g(u)$ and $du = f'(x)\, dx$. But

$$g'(u) = \frac{1}{f'(g(u))} = \frac{1}{f'(x)} \Rightarrow f'(x) = \frac{1}{g'(u)}$$

Now substituting $u = f(x)$,

$$s = \int_{a}^{b} \sqrt{1 + f'(x)^2}\, dx = \int_{f(a)}^{f(b)} \sqrt{1 + \left(\frac{1}{g'(u)}\right)^2}\, g'(u)\, du = \int_{f(a)}^{f(b)} \sqrt{g'(u)^2 + 1}\, du$$

9.2 Fluid Pressure and Force

Preliminary Questions

1. How is pressure defined?

SOLUTION Pressure is defined as force per unit area.

2. Fluid pressure is proportional to depth. What is the factor of proportionality?

SOLUTION The factor of proportionality is the weight density of the fluid, $w = \rho g$, where ρ is the mass density of the fluid.

3. When fluid force acts on the side of a submerged object, in which direction does it act?

SOLUTION Fluid force acts in the direction perpendicular to the side of the submerged object.

4. Why is fluid pressure on a surface calculated using thin horizontal strips rather than thin vertical strips?

SOLUTION Pressure depends only on depth and does not change horizontally at a given depth.

5. If a thin plate is submerged horizontally, then the fluid force on one side of the plate is equal to pressure times area. Is this true if the plate is submerged vertically?

SOLUTION When a plate is submerged vertically, the pressure is not constant along the plate, so the fluid force is not equal to the pressure times the area.

Exercises

1. A box of height 6 ft and square base of side 3 ft is submerged in a pool of water. The top of the box is 2 ft below the surface of the water.

(a) Calculate the fluid force on the top and bottom of the box.

(b) Write a Riemann sum that approximates the fluid force on a side of the box by dividing the side into N horizontal strips of thickness $\Delta y = 6/N$.

(c) To which integral does the Riemann sum converge?

(d) Compute the fluid force on a side of the box.

SOLUTION

(a) At a depth of 2 feet, the pressure on the top of the box is $(62.5)(2) = 125$ lb/ft^2; therefore, the force on the top of the box is $(125)(3^2) = 1125$ lb. At a depth of 8 feet, the pressure on the bottom of the box is $(62.5)(8) = 500$ lb/ft^2; therefore, the force on the bottom of the box is $(500)(3^2) = 4500$ lb.

(b) Let y_j denote the depth of the jth strip, for $j = 1, 2, 3, \ldots, N$; the pressure at this depth is $62.5y_j$. Because the strip has thickness Δy feet and width 3 feet, the area of the strip is $3\Delta y$, and the force exerted by the fluid on the strip is $187.5y_j\Delta y$. Summing over the strips, we find that the fluid force on the side of the box is approximately

$$F \approx \sum_{j=1}^{N} 187.5 y_j \Delta y.$$

(c) As $N \to \infty$, the Riemann sum $\displaystyle\sum_{j=1}^{N} 187.5 y_j \Delta y$ converges to the definite integral $187.5 \displaystyle\int_{2}^{8} y \, dy$.

(d) Using the result from part (c), the fluid force on the side of the box is

$$F = 187.5 \int_{2}^{8} y \, dy = 187.5 \left. \frac{y^2}{2} \right|_{2}^{8} = 187.5(32 - 2) = 5625 \text{ lb}.$$

3. Repeat Exercise 2, but assume that the top of the triangle is located 3 ft below the surface of the water.

SOLUTION

(a) Examine the figure below. By similar triangles, $\dfrac{y-3}{2} = \dfrac{f(y)}{1}$ so $f(y) = \dfrac{y-3}{2}$.

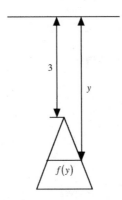

(b) The pressure at a depth of y feet is wy lb/ft^2, and the area of the strip is approximately $f(y) \Delta y = \frac{1}{2}(y - 3) \Delta y$ ft^2. Therefore, the fluid force on this strip is approximately

$$wy \left(\frac{1}{2}(y - 3) \Delta y \right) = \frac{1}{2} wy(y - 3) \Delta y.$$

(c) $F \approx \sum_{j=1}^{N} w \frac{y_j^2 - 3y_j}{2} \Delta y$. As $N \to \infty$, the Riemann sum converges to the definite integral

$$\frac{w}{2} \int_3^5 (y^2 - 3y) \, dy.$$

(d) Using the result of part (c),

$$F = \frac{w}{2} \int_3^5 (y^2 - 3y) \, dy = \frac{w}{2} \left(\frac{y^3}{3} - \frac{3y^2}{2} \right) \Big|_3^5 = \frac{62.5}{2} \left[\left(\frac{125}{3} - \frac{75}{2} \right) - \left(9 - \frac{27}{2} \right) \right] = \frac{1625}{6} \text{ lb.}$$

5. Let F be the fluid force (in Newtons) on a side of a semicircular plate of radius r meters, submerged vertically in water so that its diameter is level with the water's surface (Figure 9).

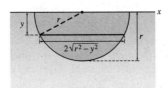

FIGURE 9

(a) Show that the width of the plate at depth y is $2\sqrt{r^2 - y^2}$.

(b) Calculate F using Eq. (2).

SOLUTION

(a) Place the origin at the center of the semicircle and point the positive y-axis downward. The equation for the edge of the semicircular plate is then $x^2 + y^2 = r^2$. At a depth of y, the plate extends from the point $(-\sqrt{r^2 - y^2}, y)$ on the left to the point $(\sqrt{r^2 - y^2}, y)$ on the right. The width of the plate at depth y is then

$$\sqrt{r^2 - y^2} - \left(-\sqrt{r^2 - y^2} \right) = 2\sqrt{r^2 - y^2}.$$

(b) With $w = 9800$ N/m^3,

$$F = 2w \int_0^r y\sqrt{r^2 - y^2} \, dy = -\frac{19600}{3} (r^2 - y^2)^{3/2} \Big|_0^r = \frac{19600 r^3}{3} \text{ N.}$$

7. A semicircular plate of radius r, oriented as in Figure 9, is submerged in water so that its diameter is located at a depth of m feet. Calculate the force on one side of the plate in terms of m and r.

SOLUTION   Place the origin at the center of the semicircular plate with the positive y-axis pointing downward. The water surface is then at $y = -m$. Moreover, at location y, the width of the plate is $2\sqrt{r^2 - y^2}$ and the depth is $y + m$. Thus,

$$F = 2w \int_0^r (y + m)\sqrt{r^2 - y^2} \, dy.$$

Now,

$$\int_0^r y\sqrt{r^2 - y^2}\, dy = -\frac{1}{3}(r^2 - y^2)^{3/2}\Big|_0^r = \frac{1}{3}r^3.$$

Geometrically,

$$\int_0^r \sqrt{r^2 - y^2}\, dy$$

represents the area of one quarter of a circle of radius r, and thus has the value $\frac{\pi r^2}{4}$. Bringing these results together, we find that

$$F = \frac{2}{3}wr^3 + \frac{1}{2}m\pi wr^2 = \frac{125}{3}r^3 + \frac{125}{4}m\pi r^2 \text{ lb.}$$

9. Calculate the total force (in Newtons) on a side of the plate in Figure 11(A), submerged in water.

SOLUTION The width of the plate varies linearly from 4 meters at a depth of 3 meters to 7 meters at a depth of 5 meters. Thus, at depth y, the width of the plate is

$$4 + \frac{3}{2}(y - 3) = \frac{3}{2}y - \frac{1}{2}.$$

Finally, the force on a side of the plate is

$$F = w\int_3^5 y\left(\frac{3}{2}y - \frac{1}{2}\right) dy = w\left(\frac{1}{2}y^3 - \frac{1}{4}y^2\right)\Big|_3^5 = 45w = 441000 \text{ N.}$$

11. The plate in Figure 12 is submerged in water with its top level with the surface of the water. The left and right edges of the plate are the curves $y = x^{1/3}$ and $y = -x^{1/3}$. Find the fluid force on a side of the plate.

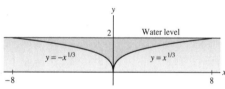

FIGURE 12

SOLUTION At height y, the plate extends from the point $(-y^3, y)$ on the left to the point (y^3, y) on the right; thus, the width of the plate is $f(y) = y^3 - (-y^3) = 2y^3$. Because the water surface is at height $y = 2$, the horizontal strip at height y is at a depth of $2 - y$. Consequently,

$$F = w\int_0^2 (2 - y)(2y^3)\, dy = 2w\left(\frac{1}{2}y^4 - \frac{1}{5}y^5\right)\Big|_0^2 = \frac{16w}{5}.$$

If distances are in feet, then $w = 62.5$ lb/ft^3 and $F = 200$ lb; if distances are in meters, then $w = 9800$ N/m^3 and $F = 31360$ N.

13. In the notation of Exercise 12, calculate the fluid force on a side of the plate R if it is oriented as in Figure 13(A). You may need to use Integration by Parts and trigonometric substitution.

SOLUTION Place the origin at the lower left corner of the plate. Because the fluid surface is at height $y = 1$, the horizontal strip at height y is at a depth of $1 - y$. Moreover, this strip has a width of

$$\frac{\pi}{2} - \sin^{-1} y = \cos^{-1} y.$$

Thus,

$$F = w\int_0^1 (1 - y)\cos^{-1} y\, dy.$$

Starting with integration by parts, we find

$$\int_0^1 (1 - y)\cos^{-1} y\, dy = \left(y - \frac{1}{2}y^2\right)\cos^{-1} y\Big|_0^1 + \int_0^1 \frac{y - \frac{1}{2}y^2}{\sqrt{1 - y^2}}\, dy$$

$$= \frac{1}{2}\cos^{-1} 1 + \int_0^1 \frac{y - \frac{1}{2}y^2}{\sqrt{1 - y^2}}\, dy = \int_0^1 \frac{y}{\sqrt{1 - y^2}}\, dy - \frac{1}{2}\int_0^1 \frac{y^2}{\sqrt{1 - y^2}}\, dy.$$

Now,

$$\int_0^1 \frac{y}{\sqrt{1-y^2}}\,dy = \left. -\sqrt{1-y^2}\,\right|_0^1 = 1.$$

For the remaining integral, we use the trigonometric substitution $y = \sin\theta$, $dy = \cos\theta\,d\theta$ and find

$$\frac{1}{2}\int_0^1 \frac{y^2}{\sqrt{1-y^2}}\,dy = \frac{1}{2}\int_{y=0}^{y=1}\sin^2\theta\,d\theta = \frac{1}{4}(\theta - \sin\theta\cos\theta)\Big|_{y=0}^{y=1}$$

$$= \frac{1}{4}\left(\sin^{-1}y - y\sqrt{1-y^2}\right)\Big|_0^1 = \frac{\pi}{8}.$$

Finally,

$$F = \left(1 - \frac{\pi}{8}\right)w \approx 85.0\text{ lb}.$$

15. Calculate the fluid force on one side of the "infinite" plate B in Figure 14.

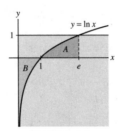

FIGURE 14

SOLUTION   Because the fluid surface is at height $y = 1$, the horizontal strip at height y is at a depth of $1 - y$. Moreover, this strip has a width of e^y. Thus,

$$F = w\int_{-\infty}^0 (1-y)e^y\,dy.$$

Using integration by parts, we find

$$\int_{-\infty}^0 (1-y)e^y\,dy = \left[(1-y)e^y + e^y\right]\Big|_{-\infty}^0 = 2.$$

Thus, $F = 2w$. If distances are in feet, then $w = 62.5\text{ lb/ft}^3$ and $F = 125$ lb; if distances are in meters, then $w = 9800$ N/m^3 and $F = 19600$ N.

17. Repeat Exercise 16, but assume that the top edge of the plate lies at a depth of 6 m.

SOLUTION   Because the plate is 3 meters on a side, is submerged at a horizontal angle of $30°$, and has its top edge located at a depth of 6 meters, the bottom edge of the plate is located at a depth of $6 + 3\sin 30° = \frac{15}{2}$ meters. Let y denote the depth at any point of the plate. The width of each horizontal strip of the plate is then

$$\frac{\Delta y}{\sin 30°} = 2\Delta y,$$

and

$$F = 2w\int_6^{15/2} 3y\,dy = \frac{243}{4}w = 595350\text{ N}.$$

19. Calculate the fluid force on one side of the plate (an isosceles triangle) shown in Figure 15(B).

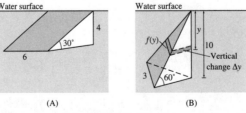

(A)   (B)

FIGURE 15

SOLUTION   A horizontal strip at depth y has length $f(y) = \frac{3}{10}y$ and width

$$\frac{\Delta y}{\sin 60°} = \frac{2}{\sqrt{3}}\Delta y.$$

Thus,

$$F = \frac{\sqrt{3}}{5}w\int_0^{10} y^2\, dy = \frac{200\sqrt{3}}{3}w.$$

If distances are in feet, then $w = 62.5$ lb/ft^3 and $F \approx 7216.88$ lb; if distances are in meters, then $w = 9800$ N/m^3 and $F \approx 1{,}131{,}606.5$ N.

21. Calculate the fluid pressure on one of the slanted sides of the trough in Figure 16, filled with corn syrup as in Exercise 20.

FIGURE 16

SOLUTION

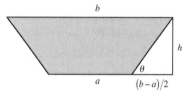

The diagram above displays a side view of the trough. From this diagram, we see that

$$\sin\theta = \frac{h}{\sqrt{\left(\frac{b-a}{2}\right)^2 + h^2}}.$$

Thus,

$$F = \frac{w}{\sin\theta}\int_0^h d\cdot y\, dy = \frac{90\sqrt{\left(\frac{b-a}{2}\right)^2 + h^2}}{h}\frac{dh^2}{2} = 45dh\sqrt{\left(\frac{b-a}{2}\right)^2 + h^2}.$$

Further Insights and Challenges

23. The end of the trough in Figure 18 is an equilateral triangle of side 3. Assume that the trough is filled with water to height y. Calculate the fluid force on each side of the trough as a function of the level y and the length l of the trough.

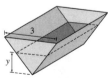

FIGURE 18

SOLUTION   Because we want the answer to be a function of y, to avoid confusion, we will use h to denote the variable in the vertical direction. Now, place the origin at the lower vertex of the trough and orient the positive h-axis pointing upward. First, consider the faces at the front and back ends of the trough. A horizontal strip at height h has a length of $\frac{2h}{\sqrt{3}}$ and is at a depth of $y - h$. Thus,

$$F = w\int_0^y (y-h)\frac{2h}{\sqrt{3}}\, dy = w\left(\frac{y}{\sqrt{3}}h^2 - \frac{2}{3\sqrt{3}}h^3\right)\Big|_0^y = \frac{\sqrt{3}}{9}wy^3.$$

For the slanted sides, we note that each side makes an angle of $60°$ with the horizontal. If we let ℓ denote the length of the trough, then

$$F = \frac{2w\ell}{\sqrt{3}} \int_0^y (y - h) \, dy = \frac{\sqrt{3}}{3} \ell w y^2.$$

25. Prove that the force on the side of a rectangular plate of area A submerged vertically in a fluid is equal to $p_0 A$, where p_0 is the fluid pressure at the center point of the rectangle.

SOLUTION Let ℓ denote the length of the vertical side of the rectangle, x denote the length of the horizontal side of the rectangle, and suppose the top edge of the rectangle is at depth $y = m$. The pressure at the center of the rectangle is then

$$p_0 = w\left(m + \frac{\ell}{2}\right),$$

and the force on the side of the rectangular plate is

$$F = \int_m^{\ell+m} wxy \, dy = \frac{wx}{2}\left[(\ell+m)^2 - m^2\right] = \frac{wx\ell}{2}(\ell + 2m) = Aw\left(\frac{\ell}{2} + m\right) = Ap_0.$$

9.3 Center of Mass

Preliminary Questions

1. What are the x- and y-moments of a lamina whose center of mass is located at the origin?

SOLUTION Because the center of mass is located at the origin, it follows that $M_x = M_y = 0$.

2. A thin plate has mass 3. What is the x-moment of the plate if its center of mass has coordinates $(2, 7)$?

SOLUTION The x-moment of the plate is the product of the mass of the plate and the y-coordinate of the center of mass. Thus, $M_x = 3(7) = 21$.

3. The center of mass of a lamina of total mass 5 has coordinates $(2, 1)$. What are the lamina's x- and y-moments?

SOLUTION The x-moment of the plate is the product of the mass of the plate and the y-coordinate of the center of mass, whereas the y-moment is the product of the mass of the plate and the x-coordinate of the center of mass. Thus, $M_x = 5(1) = 5$, and $M_y = 5(2) = 10$.

4. Explain how the Symmetry Principle is used to conclude that the centroid of a rectangle is the center of the rectangle.

SOLUTION Because a rectangle is symmetric with respect to both the vertical line and the horizontal line through the center of the rectangle, the Symmetry Principle guarantees that the centroid of the rectangle must lie along both of these lines. The only point in common to both lines of symmetry is the center of the rectangle, so the centroid of the rectangle must be the center of the rectangle.

Exercises

1. Four particles are located at points

$$(1, 1) \quad (1, 2) \quad (4, 0) \quad (3, 1)$$

(a) Find the moments M_x and M_y and the center of mass of the system, assuming that the particles have equal mass m.
(b) Find the center of mass of the system, assuming the particles have mass 3, 2, 5, and 7, respectively.

SOLUTION
(a) Because each particle has mass m,

$$M_x = m(1) + m(2) + m(0) + m(1) = 4m;$$
$$M_y = m(1) + m(1) + m(4) + m(3) = 9m;$$

and the total mass of the system is $4m$. Thus, the coordinates of the center of mass are

$$\left(\frac{M_y}{M}, \frac{M_x}{M}\right) = \left(\frac{9m}{4m}, \frac{4m}{4m}\right) = \left(\frac{9}{4}, 1\right).$$

(b) With the indicated masses of the particles,

$$M_x = 3(1) + 2(2) + 5(0) + 7(1) = 14;$$

$$M_y = 3(1) + 2(1) + 5(4) + 7(3) = 46;$$

and the total mass of the system is 17. Thus, the coordinates of the center of mass are

$$\left(\frac{M_y}{M}, \frac{M_x}{M}\right) = \left(\frac{46}{17}, \frac{14}{17}\right).$$

3. Point masses of equal size are placed at the vertices of the triangle with coordinates $(a, 0)$, $(b, 0)$, and $(0, c)$. Show that the center of mass of the system of masses has coordinates $(\frac{1}{3}(a + b), \frac{1}{3}c)$.

SOLUTION Let each particle have mass m. The total mass of the system is then $3m$. and the moments are

$$M_x = 0(m) + 0(m) + c(m) = cm; \text{ and}$$

$$M_y = a(m) + b(m) + 0(m) = (a + b)m.$$

Thus, the coordinates of the center of mass are

$$\left(\frac{M_y}{M}, \frac{M_x}{M}\right) = \left(\frac{(a + b)m}{3m}, \frac{cm}{3m}\right) = \left(\frac{a + b}{3}, \frac{c}{3}\right).$$

5. Sketch the lamina S of constant density $\rho = 3$ g/cm^2 occupying the region beneath the graph of $y = x^2$ for $0 \le x \le 3$.
(a) Use formulas (1) and (2) to compute M_x and M_y.
(b) Find the area and the center of mass of S.

SOLUTION A sketch of the lamina is shown below

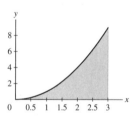

(a) Using (2),

$$M_x = 3 \int_0^9 y(3 - \sqrt{y}) \, dy = \left(\frac{9y^2}{2} - \frac{6}{5} y^{5/2}\right)\Big|_0^9 = \frac{729}{10}.$$

Using (1),

$$M_y = 3 \int_0^3 x(x^2) \, dx = \frac{3x^4}{4}\Big|_0^3 = \frac{243}{4}.$$

(b) The area of the lamina is

$$A = \int_0^3 x^2 \, dx = \frac{x^3}{3}\Big|_0^3 = 9 \text{ cm}^2.$$

With a constant density of $\rho = 3$ g/cm^2, the mass of the lamina is $M = 27$ grams, and the coordinates of the center of mass are

$$\left(\frac{M_y}{M}, \frac{M_x}{M}\right) = \left(\frac{243/4}{27}, \frac{729/10}{27}\right) = \left(\frac{9}{4}, \frac{27}{10}\right).$$

7. Find the moments and center of mass of the lamina of uniform density ρ occupying the region underneath $y = x^3$ for $0 \le x \le 2$.

SOLUTION With uniform density ρ,

$$M_x = \frac{1}{2}\rho \int_0^2 (x^3)^2 \, dx = \frac{64\rho}{7} \quad \text{and} \quad M_y = \rho \int_0^2 x(x^3) \, dx = \frac{32\rho}{5}.$$

The mass of the lamina is

$$M = \rho \int_0^2 x^3 \, dx = 4\rho,$$

so the coordinates of the center of mass are

$$\left(\frac{M_y}{M}, \frac{M_x}{M} \right) = \left(\frac{8}{5}, \frac{16}{7} \right).$$

9. Let T be the triangular lamina in Figure 17.

(a) Show that the horizontal cut at height y has length $4 - \frac{2}{3}y$ and use Eq. (2) to compute M_x (with $\rho = 1$).

(b) Use the Symmetry Principle to show that $M_y = 0$ and find the center of mass.

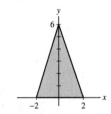

FIGURE 17 Isosceles triangle.

SOLUTION

(a) The equation of the line from $(2, 0)$ to $(0, 6)$ is $y = -3x + 6$, so

$$x = 2 - \frac{1}{3}y.$$

The length of the horizontal cut at height y is then

$$2 \left(2 - \frac{1}{3}y \right) = 4 - \frac{2}{3}y,$$

and

$$M_x = \int_0^6 y \left(4 - \frac{2}{3}y \right) dy = 24.$$

(b) Because the triangular lamina is symmetric with respect to the y-axis, $x_{cm} = 0$, which implies that $M_y = 0$. The total mass of the lamina is

$$M = 2 \int_0^2 (-3x + 6) \, dx = 12,$$

so $y_{cm} = 24/12$. Finally, the coordinates of the center of mass are $(0, 2)$.

In Exercises 10–17, find the centroid of the region lying underneath the graph of the function over the given interval.

11. $f(x) = \sqrt{x}$, $[4, 9]$

SOLUTION The moments of the region are

$$M_x = \frac{1}{2} \int_4^9 x \, dx = \frac{65}{4} \quad \text{and} \quad M_y = \int_4^9 x\sqrt{x} \, dx = \frac{422}{5}.$$

The area of the region is

$$A = \int_4^9 \sqrt{x} \, dx = \frac{38}{3},$$

so the coordinates of the centroid are

$$\left(\frac{M_y}{A}, \frac{M_x}{A} \right) = \left(\frac{633}{95}, \frac{195}{152} \right).$$

13. $f(x) = 9 - x^2$, $[0, 3]$

SOLUTION   The moments of the region are

$$M_x = \frac{1}{2}\int_0^3 (9 - x^2)^2\,dx = \frac{324}{5} \quad \text{and} \quad M_y = \int_0^3 x(9 - x^2)\,dx = \frac{81}{4}.$$

The area of the region is

$$A = \int_0^3 (9 - x^2)\,dx = 18,$$

so the coordinates of the centroid are

$$\left(\frac{M_y}{A}, \frac{M_x}{A}\right) = \left(\frac{9}{8}, \frac{18}{5}\right).$$

15. $f(x) = e^{-x}, \quad [0, 4]$

SOLUTION   The moments of the region are

$$M_x = \frac{1}{2}\int_0^4 e^{-2x}\,dx = \frac{1}{4}\left(1 - e^{-8}\right) \quad \text{and} \quad M_y = \int_0^4 xe^{-x}\,dx = -e^{-x}(x + 1)\Big|_0^4 = 1 - 5e^{-4}.$$

The area of the region is

$$A = \int_0^4 e^{-x}\,dx = 1 - e^{-4},$$

so the coordinates of the centroid are

$$\left(\frac{M_y}{A}, \frac{M_x}{A}\right) = \left(\frac{1 - 5e^{-4}}{1 - e^{-4}}, \frac{1 - e^{-8}}{4(1 - e^{-4})}\right).$$

17. $f(x) = \sin x, \quad [0, \pi]$

SOLUTION   The moments of the region are

$$M_x = \frac{1}{2}\int_0^\pi \sin^2 x\,dx = \frac{1}{4}(x - \sin x \cos x)\Big|_0^\pi = \frac{\pi}{4}; \quad \text{and}$$

$$M_y = \int_0^\pi x \sin x\,dx = (-x \cos x + \sin x)\Big|_0^\pi = \pi.$$

The area of the region is

$$A = \int_0^\pi \sin x\,dx = 2,$$

so the coordinates of the centroid are

$$\left(\frac{M_y}{A}, \frac{M_x}{A}\right) = \left(\frac{\pi}{2}, \frac{\pi}{8}\right).$$

19. Sketch the region between $y = x + 4$ and $y = 2 - x$ for $0 \le x \le 2$. Using symmetry, explain why the centroid of the region lies on the line $y = 3$. Verify this by computing the moments and the centroid.

SOLUTION   A sketch of the region is shown below.

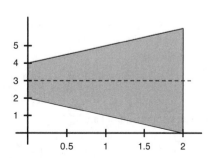

The region is clearly symmetric about the line $y = 3$, so we expect the centroid of the region to lie along this line. We find

$$M_x = \frac{1}{2} \int_0^2 \left((x+4)^2 - (2-x)^2 \right) dx = 24;$$

$$M_y = \int_0^2 x \left((x+4) - (2-x) \right) dx = \frac{28}{3}; \text{ and}$$

$$A = \int_0^2 \left((x+4) - (2-x) \right) dx = 8.$$

Thus, the coordinates of the centroid are $\left(\frac{7}{6}, 3 \right)$.

In Exercises 20–25, find the centroid of the region lying between the graphs of the functions over the given interval.

21. $y = x^2$, $y = \sqrt{x}$, $[0, 1]$

SOLUTION The moments of the region are

$$M_x = \frac{1}{2} \int_0^1 (x - x^4) \, dx = \frac{3}{20} \quad \text{and} \quad M_y = \int_0^1 x(\sqrt{x} - x^2) \, dx = \frac{3}{20}.$$

The area of the region is

$$A = \int_0^1 (\sqrt{x} - x^2) \, dx = \frac{1}{3},$$

so the coordinates of the centroid are

$$\left(\frac{9}{20}, \frac{9}{20} \right).$$

Note: This makes sense, since the functions are inverses of each other. This makes the region symmetric with respect to the line $y = x$. Thus, by the symmetry principle, the center of mass must lie on that line.

23. $y = e^x$, $y = 1$, $[0, 1]$

SOLUTION The moments of the region are

$$M_x = \frac{1}{2} \int_0^1 (e^{2x} - 1) \, dx = \frac{e^2 - 3}{4} \quad \text{and} \quad M_y = \int_0^1 x(e^x - 1) \, dx = \left(xe^x - e^x - \frac{1}{2}x^2 \right)\Big|_0^1 = \frac{1}{2}.$$

The area of the region is

$$A = \int_0^1 (e^x - 1) \, dx = e - 2,$$

so the coordinates of the centroid are

$$\left(\frac{1}{2(e-2)}, \frac{e^2 - 3}{4(e-2)} \right).$$

25. $y = \sin x$, $y = \cos x$, $[0, \pi/4]$

SOLUTION The moments of the region are

$$M_x = \frac{1}{2} \int_0^{\pi/4} (\cos^2 x - \sin^2 x) \, dx = \frac{1}{2} \int_0^{\pi/4} \cos 2x \, dx = \frac{1}{4}; \text{ and}$$

$$M_y = \int_0^{\pi/4} x(\cos x - \sin x) \, dx = [(x-1)\sin x + (x+1)\cos x]\Big|_0^{\pi/4} = \frac{\pi\sqrt{2}}{4} - 1.$$

The area of the region is

$$A = \int_0^{\pi/4} (\cos x - \sin x) \, dx = \sqrt{2} - 1,$$

so the coordinates of the centroid are

$$\left(\frac{\pi\sqrt{2} - 4}{4(\sqrt{2} - 1)}, \frac{1}{4(\sqrt{2} - 1)} \right).$$

27. Sketch the region enclosed by $y = 0$, $y = (x + 1)^3$, and $y = (1 - x)^3$ and find its centroid.

SOLUTION   A sketch of the region is shown below.

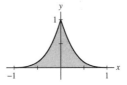

The moments of the region are

$$M_x = \frac{1}{2}\left(\int_{-1}^{0} (x + 1)^6 \, dx + \int_{0}^{1} (1 - x)^6 \, dx \right) = \frac{1}{7}; \text{ and}$$

$$M_y = 0 \text{ by the Symmetry Principle.}$$

The area of the region is

$$A = \int_{-1}^{0} (x + 1)^3 \, dx + \int_{0}^{1} (1 - x)^3 \, dx = \frac{1}{2},$$

so the coordinates of the centroid are $\left(0, \frac{2}{7}\right)$.

In Exercises 28–32, find the centroid of the region.

29. Top half of the ellipse $\left(\dfrac{x}{a}\right)^2 + \left(\dfrac{y}{b}\right)^2 = 1$ for arbitrary $a, b > 0$

SOLUTION   The equation of the top half of the ellipse is

$$y = \sqrt{b^2 - \frac{b^2 x^2}{a^2}}$$

Thus,

$$M_x = \frac{1}{2}\int_{-a}^{a} \left(\sqrt{b^2 - \frac{b^2 x^2}{a^2}} \right)^2 dx = \frac{2ab^2}{3}.$$

By the Symmetry Principle, $M_y = 0$. The area of the region is one-half the area of an ellipse with axes of length a and b; i.e., $\frac{1}{2}\pi ab$. Finally, the coordinates of the centroid are

$$\left(0, \frac{4b}{3\pi}\right).$$

31. Quarter of the unit circle lying in the first quadrant

SOLUTION   By the Symmetry Principle, the center of mass must lie on the line $y = x$ in the first quadrant. Therefore, we need only calculate one of the moments of the region. With $y = \sqrt{1 - x^2}$, we find

$$M_y = \int_{0}^{1} x\sqrt{1 - x^2} \, dx = \frac{1}{3}.$$

The area of the region is one-quarter of the area of a unit circle; i.e., $\frac{1}{4}\pi$. Thus, the coordinates of the centroid are

$$\left(\frac{4}{3\pi}, \frac{4}{3\pi}\right).$$

33. Find the centroid for the shaded region of the semicircle of radius r in Figure 18. What is the centroid when $r = 1$ and $h = \frac{1}{2}$? *Hint:* Use geometry rather than integration to show that the *area* of the region is $r^2 \sin^{-1}(\sqrt{1 - h^2/r^2}) - h\sqrt{r^2 - h^2}$.

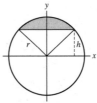

FIGURE 18

SOLUTION From the symmetry of the region, it is obvious that the centroid lies along the y-axis. To determine the y-coordinate of the centroid, we must calculate the moment about the x-axis and the area of the region. Now, the length of the horizontal cut of the semicircle at height y is

$$\sqrt{r^2 - y^2} - \left(-\sqrt{r^2 - y^2}\right) = 2\sqrt{r^2 - y^2}.$$

Therefore, taking $\rho = 1$, we find

$$M_x = 2\int_h^r y\sqrt{r^2 - y^2}\,dy = \frac{2}{3}(r^2 - h^2)^{3/2}.$$

Observe that the region is comprised of a sector of the circle with the triangle between the two radii removed. The angle of the sector is 2θ, where $\theta = \sin^{-1}\sqrt{1 - h^2/r^2}$, so the area of the sector is $\frac{1}{2}r^2(2\theta) = r^2\sin^{-1}\sqrt{1 - h^2/r^2}$. The triangle has base $2\sqrt{r^2 - h^2}$ and height h, so the area is $h\sqrt{r^2 - h^2}$. Therefore,

$$Y_{CM} = \frac{M_x}{A} = \frac{\frac{2}{3}(r^2 - h^2)^{3/2}}{r^2\sin^{-1}\sqrt{1 - h^2/r^2} - h\sqrt{r^2 - h^2}}.$$

When $r = 1$ and $h = 1/2$, we find

$$Y_{CM} = \frac{\frac{2}{3}(3/4)^{3/2}}{\sin^{-1}\frac{\sqrt{3}}{2} - \frac{\sqrt{3}}{4}} = \frac{3\sqrt{3}}{4\pi - 3\sqrt{3}}.$$

In Exercises 34–36, use the additivity of moments to find the COM of the region.

35. An ice cream cone consisting of a semicircle on top of an equilateral triangle of side 6 (Figure 20)

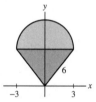

FIGURE 20

SOLUTION The region is symmetric with respect to the y-axis, so $M_y = 0$ by the Symmetry Principle. The moment about the x-axis for the triangle is

$$M_x^{\text{triangle}} = \frac{2}{\sqrt{3}}\int_0^{3\sqrt{3}} y^2\,dy = 54.$$

For the semicircle, first note that the center is $(0, 3\sqrt{3})$, so the equation is $x^2 + (y - 3\sqrt{3})^2 = 9$, and

$$M_x^{\text{semi}} = 2\int_{3\sqrt{3}}^{3+3\sqrt{3}} y\sqrt{9 - (y - 3\sqrt{3})^2}\,dy.$$

Using the substitution $w = y - 3\sqrt{3}, dw = dy$, we find

$$M_x^{\text{semi}} = 2\int_0^3 (w + 3\sqrt{3})\sqrt{9 - w^2}\,dw$$

$$= 2\int_0^3 w\sqrt{9 - w^2}\,dw + 6\sqrt{3}\int_0^3 \sqrt{9 - w^2}\,dw = 18 + \frac{27\pi\sqrt{3}}{2},$$

where we have used the fact that $\int_0^3 \sqrt{9 - w^2}\,dw$ represents the area of one-quarter of a circle of radius 3. The total moment about the x-axis is then

$$M_x = M_x^{\text{triangle}} + M_x^{\text{semi}} = 72 + \frac{27\pi\sqrt{3}}{2}.$$

Because the area of the region is $9\sqrt{3} + \frac{9\pi}{2}$, the coordinates of the center of mass are

$$\left(0, \frac{16 + 3\pi\sqrt{3}}{\pi + 2\sqrt{3}}\right).$$

37. Let S be the lamina of mass density $\rho = 1$ obtained by removing a circle of radius r from the circle of radius $2r$ shown in Figure 21. Let M_x^S and M_y^S denote the moments of S. Similarly, let M_y^{big} and M_y^{small} be the y-moments of the larger and smaller circles.

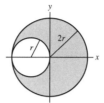

FIGURE 21

(a) Use the Symmetry Principle to show that $M_x^S = 0$.

(b) Show that $M_y^S = M_y^{\text{big}} - M_y^{\text{small}}$ using the additivity of moments.

(c) Find M_y^{big} and M_y^{small} using the fact that the COM of a circle is its center. Then compute M_y^S using (b).

(d) Determine the COM of S.

SOLUTION

(a) Because S is symmetric with respect to the x-axis, $M_x^S = 0$.

(b) Because the small circle together with the region S comprise the big circle, by the additivity of moments,

$$M_y^S + M_y^{\text{small}} = M_y^{\text{big}}.$$

Thus $M_y^S = M_y^{\text{big}} - M_y^{\text{small}}$.

(c) The center of the big circle is the origin, so $x_{\text{cm}}^{\text{big}} = 0$; consequently, $M_y^{\text{big}} = 0$. On the other hand, the center of the small circle is $(-r, 0)$, so $x_{\text{cm}}^{\text{small}} = -r$; consequently

$$M_y^{\text{small}} = x_{\text{cm}}^{\text{small}} \cdot A^{\text{small}} = -r \cdot \pi r^2 = -\pi r^3.$$

By the result of part (b), it follows that $M_y^S = 0 - (-\pi r^3) = \pi r^3$.

(d) The area of the region S is $4\pi r^2 - \pi r^2 = 3\pi r^2$. The coordinates of the center of mass of the region S are then

$$\left(\frac{\pi r^3}{3\pi r^2}, 0 \right) = \left(\frac{r}{3}, 0 \right).$$

Further Insights and Challenges

39. A **median** of a triangle is a segment joining a vertex to the midpoint of the opposite side. Show that the centroid of a triangle lies on each of its medians, at a distance two-thirds down from the vertex. Then use this fact to prove that the three medians intersect at a single point. *Hint:* Simplify the calculation by assuming that one vertex lies at the origin and another on the x-axis.

SOLUTION Orient the triangle by placing one vertex at $(0, 0)$ and the long side of the triangle along the x-axis. Label the vertices $(0, 0)$, $(a, 0)$, (b, c). Thus, the equations of the short sides are $y = \frac{cx}{b}$ and $y = \frac{cx}{b-a} - \frac{ac}{b-a}$. Now,

$$M_x = \frac{1}{2} \int_0^b (cx/b)^2 \, dx + \frac{1}{2} \int_b^a \left(\frac{cx - ac}{b - a} \right)^2 dx = \frac{ac^2}{6};$$

$$M_y = \int_0^b x(cx/b) \, dx + \int_b^a x \left(\frac{cx - ac}{b - a} \right) dx = \frac{ac(a + b)}{6}; \quad \text{and}$$

$$M = \frac{ac}{2}.$$

so the center of mass is $\left(\dfrac{a+b}{3}, \dfrac{c}{3} \right)$. To show that the centroid lies on each median, let y_1 be the median from (b, c), y_2 the median from $(0, 0)$ and y_3 the median from $(a, 0)$. We find

$$y_1(x) = \frac{2c}{2b - a}(x - a/2), \qquad \text{so} \qquad y_1\left(\frac{a+b}{3} \right) = \frac{c}{3};$$

$$y_2(x) = \frac{c}{a + b}x, \qquad \text{so} \qquad y_2\left(\frac{a+b}{3} \right) = \frac{c}{3};$$

$$y_3(x) = \frac{c}{b - 2a}(x - a), \qquad \text{so} \qquad y_3\left(\frac{a+b}{3}\right) = \frac{c}{3}.$$

This shows that the center of mass lies on each median. We now show that the center of mass is $\frac{2}{3}$ of the way from each vertex. For y_1, note that $x = b$ gives the vertex and $x = \frac{a}{2}$ gives the midpoint of the opposite side, so two-thirds of this distance is

$$x = b + \frac{2}{3}\left(\frac{a}{2} - b\right) = \frac{a+b}{3},$$

the x-coordinate of the center of mass. Likewise, for y_2, two-thirds of the distance from $x = 0$ to $x = \frac{a+b}{2}$ is $\frac{a+b}{3}$, and for y_3, the two-thirds point is

$$x = a + \frac{2}{3}\left(\frac{b}{2} - a\right) = \frac{a+b}{3}.$$

A similar method shows that the y-coordinate is also two-thirds of the way along the median. Thus, since the centroid lies on all three medians, we can conclude that all three medians meet at a single point, namely the centroid.

41. Find the COM of a system of two weights of masses m_1 and m_2 connected by a lever of length d whose mass density ρ is uniform. *Hint:* The moment of the system is the sum of the moments of the weights and the lever.

SOLUTION   Let A be the cross-sectional area of the rod. Place the rod with m_1 at the origin and rod lying on the positive x-axis. The y-moment of the rod is $M_y = \frac{1}{2}\rho A d^2$, the y-moment of the mass m_2 is $M_y = m_2 d$, and the total mass of the system is $M = m_1 + m_2 + \rho A d$. Therefore, the x-coordinate of the center of mass is

$$\frac{m_2 d + \frac{1}{2}\rho A d^2}{m_1 + m_2 + \rho A d}.$$

43. Prove directly that Eqs. (2) and (3) are equivalent in the following situation. Let $f(x)$ be a positive decreasing function on $[0, b]$ such that $f(b) = 0$. Set $d = f(0)$ and $g(y) = f^{-1}(y)$. Show that

$$\frac{1}{2}\int_0^b f(x)^2\, dx = \int_0^d y g(y)\, dy$$

Hint: First apply the substitution $y = f(x)$ to the integral on the left and observe that $dx = g'(y)\, dy$. Then apply Integration by Parts.

SOLUTION $f(x) \geq 0$ and $f'(x) < 0$ shows that f has an inverse g on $[a, b]$. Because $f(b) = 0$, $f(0) = d$, and $f^{-1}(x) = g(x)$, it follows that $g(d) = 0$ and $g(0) = b$. If we let $x = g(y)$, then $dx = g'(y)\, dy$. Thus, with $y = f(x)$,

$$\frac{1}{2}\int_0^b f(x)^2\, dx = \frac{1}{2}\int_0^b y^2\, dx = \frac{1}{2}\int_d^0 y^2 g'(y)\, dy.$$

Using Integration by Parts with $u = y^2$ and $v' = g'(y)\, dy$, we find

$$\frac{1}{2}\int_d^0 y^2 g'(y)\, dy = \frac{1}{2}\left[y^2 g(y)\Big|_d^0 - 2\int_d^0 y g(y)\, dy\right] = \frac{1}{2}\left[0 - d^2 g(d)\right] - \int_d^0 y g(y)\, dy = \int_0^d y g(y)\, dy.$$

9.4 Taylor Polynomials

Preliminary Questions

1. What is $T_3(x)$ centered at $a = 3$ for a function $f(x)$ such that $f(3) = 9$, $f'(3) = 8$, $f''(3) = 4$, and $f'''(3) = 12$.

SOLUTION   In general, with $a = 3$,

$$T_3(x) = f(3) + f'(3)(x - 3) + \frac{f''(3)}{2}(x - 3)^2 + \frac{f'''(3)}{6}(x - 3)^3.$$

Using the information provided, we find

$$T_3(x) = 9 + 8(x - 3) + 2(x - 3)^2 + 2(x - 3)^3.$$

2. The dashed graphs in Figure 3 are Taylor polynomials for a function $f(x)$. Which of the two is a MacLaurin polynomial?

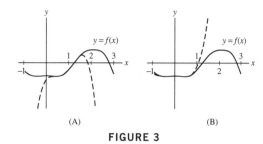

(A)   (B)

FIGURE 3

SOLUTION   A MacLaurin polynomial always gives the value of $f(0)$ exactly. This is true for the Taylor polynomial sketched in (B); thus, this is the MacLaurin polynomial.

3. For which value of x does the MacLaurin polynomial $T_n(x)$ satisfy $T_n(x) = f(x)$, no matter what $f(x)$ is?

SOLUTION   A MacLaurin polynomial always gives the value of $f(0)$ exactly.

4. Let $T_n(x)$ be the MacLaurin polynomial of a function $f(x)$ satisfying $|f^{(4)}(x)| \leq 1$ for all x. Which of the following statements follow from the Error Bound?

(a) $|T_4(2) - f(2)| \leq 2^4/24$

(b) $|T_3(2) - f(2)| \leq 2^3/6$

(c) $|T_3(2) - f(2)| \leq 1/3$

SOLUTION   For a function $f(x)$ satisfying $|f^{(4)}(x)| \leq 1$ for all x,

$$|T_3(2) - f(2)| \leq \frac{1}{24}|f^{(4)}(x)|2^4 \leq \frac{2^3}{12} < \frac{2^3}{6}.$$

Thus, **(b)** is the correct answer.

Exercises

In Exercises 1–14, calculate the Taylor polynomials $T_2(x)$ and $T_3(x)$ centered at $x = a$ for the given function and value of a.

1. $f(x) = \sin x, \quad a = 0$

SOLUTION   First, we calculate and evaluate the needed derivatives:

$$f(x) = \sin x \qquad f(a) = 0$$
$$f'(x) = \cos x \qquad f'(a) = 1$$
$$f''(x) = -\sin x \qquad f''(a) = 0$$
$$f'''(x) = -\cos x \qquad f'''(a) = -1$$

Now,

$$T_2(x) = f(a) + f'(a)(x - a) + \frac{f''(a)}{2}(x - a)^2 = 0 + 1(x - 0) + \frac{0}{2}(x - 0)^2 = x; \text{ and}$$

$$T_3(x) = f(a) + f'(a)(x - a) + \frac{f''(a)}{2}(x - a)^2 + \frac{f'''(a)}{6}(x - a)^3$$

$$= 0 + 1(x - 0) + \frac{0}{2}(x - 0)^2 + \frac{-1}{6}(x - 0)^3 = x - \frac{1}{6}x^3.$$

3. $f(x) = \dfrac{1}{1 + x}, \quad a = 0$

SOLUTION   First, we calculate and evaluate the needed derivatives:

$$f(x) = \frac{1}{1 + x} \qquad f(a) = 1$$

$$f'(x) = \frac{-1}{(1 + x)^2} \qquad f'(a) = -1$$

$$f''(x) = \frac{2}{(1+x)^3} \qquad f''(a) = 2$$

$$f'''(x) = \frac{-6}{(1+x)^4} \qquad f'''(a) = -6$$

Now,

$$T_2(x) = f(a) + f'(a)(x-a) + \frac{f''(a)}{2}(x-a)^2$$

$$= 1 + (-1)(x-0) + \frac{2}{2}(x-0)^2 = 1 - x + x^2; \text{ and}$$

$$T_3(x) = f(a) + f'(a)(x-a) + \frac{f''(a)}{2}(x-a)^2 + \frac{f'''(a)}{6}(x-a)^3$$

$$= 1 + (-1)(x-0) + \frac{2}{2}(x-0)^2 + \frac{-6}{6}(x-0)^3 = 1 - x + x^2 - x^3.$$

5. $f(x) = \tan x, \quad a = 0$

SOLUTION First, we calculate and evaluate the needed derivatives:

$$f(x) = \tan x \qquad\qquad\qquad f(a) = 0$$

$$f'(x) = \sec^2 x \qquad\qquad\qquad f'(a) = 1$$

$$f''(x) = 2\sec^2 x \tan x \qquad\qquad f''(a) = 0$$

$$f'''(x) = 2\sec^4 x + 4\sec^2 x \tan^2 x \qquad f'''(a) = 2$$

Now,

$$T_2(x) = f(a) + f'(a)(x-a) + \frac{f''(a)}{2}(x-a)^2 = 0 + 1(x-0) + \frac{0}{2}(x-0)^2 = x; \text{ and}$$

$$T_3(x) = f(a) + f'(a)(x-a) + \frac{f''(a)}{2}(x-a)^2 + \frac{f'''(a)}{6}(x-a)^3$$

$$= 0 + 1(x-0) + \frac{0}{2}(x-0)^2 + \frac{2}{6}(x-0)^3 = x + \frac{1}{3}x^3.$$

7. $f(x) = \dfrac{1}{1+x^2}, \quad a = 0$

SOLUTION First, we calculate and evaluate the needed derivatives:

$$f(x) = \frac{1}{1+x^2} \qquad\qquad f(a) = 1$$

$$f'(x) = \frac{-2x}{(x^2+1)^2} \qquad\qquad f'(a) = 0$$

$$f''(x) = \frac{2(3x^2-1)}{x^2+1} \qquad\qquad f''(a) = -2$$

$$f'''(x) = \frac{-24x(x^2-1)}{(x^2+1)^4} \qquad f'''(a) = 0$$

Now,

$$T_2(x) = f(a) + f'(a)(x-a) + \frac{f''(a)}{2}(x-a)^2 = 1 + 0(x-0) + \frac{-2}{2}(x-0)^2 = 1 - x^2; \text{ and}$$

$$T_3(x) = f(a) + f'(a)(x-a) + \frac{f''(a)}{2}(x-a)^2 + \frac{f'''(a)}{6}(x-a)^3$$

$$= 1 + 0(x-0) + \frac{-2}{2}(x-0)^2 + \frac{0}{6}(x-0)^3 = 1 - x^2.$$

9. $f(x) = e^x, \quad a = 0$

SOLUTION First, we calculate and evaluate the needed derivatives:

$$f(x) = e^x \qquad f(a) = 1$$

$$f'(x) = e^x \qquad f'(a) = 1$$
$$f''(x) = e^x \qquad f''(a) = 1$$
$$f'''(x) = e^x \qquad f'''(a) = 1$$

Now,

$$T_2(x) = f(a) + f'(a)(x-a) + \frac{f''(a)}{2}(x-a)^2$$

$$= 1 + 1(x-0) + \frac{1}{2}(x-0)^2 = 1 + x + \frac{1}{2}x^2; \text{ and}$$

$$T_3(x) = f(a) + f'(a)(x-a) + \frac{f''(a)}{2}(x-a)^2 + \frac{f'''(a)}{6}(x-a)^3$$

$$= 1 + 1(x-0) + \frac{1}{2}(x-0)^2 + \frac{1}{6}(x-0)^3 = 1 + x + \frac{1}{2}x^2 + \frac{1}{6}x^3.$$

11. $f(x) = e^{-x} + e^{-2x}, \quad a = 0$

SOLUTION First, we calculate and evaluate the needed derivatives:

$$f(x) = e^{-x} + e^{-2x} \qquad f(a) = 2$$
$$f'(x) = -e^{-x} - 2e^{-2x} \qquad f'(a) = -3$$
$$f''(x) = e^{-x} + 4e^{-2x} \qquad f''(a) = 5$$
$$f'''(x) = -e^{-x} - 8e^{-2x} \qquad f'''(a) = -9$$

Now,

$$T_2(x) = f(a) + f'(a)(x-a) + \frac{f''(a)}{2}(x-a)^2$$

$$= 2 + (-3)(x-0) + \frac{5}{2}(x-0)^2 = 2 - 3x + \frac{5}{2}x^2; \text{ and}$$

$$T_3(x) = f(a) + f'(a)(x-a) + \frac{f''(a)}{2}(x-a)^2 + \frac{f'''(a)}{6}(x-a)^3$$

$$= 2 + (-3)(x-0) + \frac{5}{2}(x-0)^2 + \frac{-9}{6}(x-0)^3 = 2 - 3x + \frac{5}{2}x^2 - \frac{3}{2}x^3.$$

13. $f(x) = \ln(x+1), \quad a = 0$

SOLUTION First, we calculate and evaluate the needed derivatives:

$$f(x) = \ln(x+1) \qquad f(a) = 0$$
$$f'(x) = \frac{1}{x+1} \qquad f'(a) = 1$$
$$f''(x) = \frac{-1}{(x+1)^2} \qquad f''(a) = -1$$
$$f'''(x) = \frac{2}{(x+1)^3} \qquad f'''(a) = 2$$

Now,

$$T_2(x) = f(a) + f'(a)(x-a) + \frac{f''(a)}{2}(x-a)^2 = 0 + 1(x-0) + \frac{-1}{2}(x-0)^2 = x - \frac{1}{2}x^2; \text{ and}$$

$$T_3(x) = f(a) + f'(a)(x-a) + \frac{f''(a)}{2}(x-a)^2 + \frac{f'''(a)}{6}(x-a)^3$$

$$= 0 + 1(x-0) + \frac{-1}{2}(x-0)^2 + \frac{2}{6}(x-0)^3 = x - \frac{1}{2}x^2 + \frac{1}{3}x^3.$$

In Exercises 15–18, compute $T_2(x)$ at $x = a$ and use a calculator to compute the error $|f(x) - T_2(x)|$ at the given value of x.

15. $y = e^x, \quad x = -0.5, \quad a = 0$

SOLUTION Let $f(x) = e^x$. Then $f'(x) = e^x$, $f''(x) = e^x$, $f(a) = 1$, $f'(a) = 1$ and $f''(a) = 1$. Therefore

$$T_2(x) = 1 + 1(x - 0) + \frac{1}{2}(x - 0)^2 = 1 + x + \frac{1}{2}x^2,$$

and

$$T_2(-0.5) = 1 + (-0.5) + \frac{1}{2}(-0.5)^2 = 0.625.$$

Using a calculator, we find

$$f(-0.5) = \frac{1}{\sqrt{e}} = 0.606531,$$

so

$$|T_2(-0.5) - f(-0.5)| = 0.0185.$$

17. $y = x^{-3/2}$, $x = 0.3$, $a = 1$

SOLUTION Let $f(x) = x^{-3/2}$. Then $f'(x) = -\frac{3}{2}x^{-5/2}$, $f''(x) = \frac{15}{4}x^{-7/2}$, $f(a) = 1$, $f'(a) = -\frac{3}{2}$ and $f''(a) = \frac{15}{4}$. Therefore

$$T_2(x) = 1 + \left(-\frac{3}{2}\right)(x - 1) + \frac{15/4}{2}(x - 1)^2 = 1 - \frac{3}{2}(x - 1) + \frac{15}{8}(x - 1)^2,$$

and

$$T_2(0.3) = 1 - \frac{3}{2}(-0.7) + \frac{15}{8}(-0.7)^2 = 2.96875.$$

Using a calculator, $f(0.3) = 6.08581$, so

$$|T_2(0.3) - f(0.3)| = 3.11706.$$

19. Show that the nth MacLaurin polynomial for $f(x) = e^x$ is

$$T_n(x) = 1 + \frac{x}{1!} + \frac{x^2}{2!} + \cdots + \frac{x^n}{n!}$$

SOLUTION With $f(x) = e^x$, it follows that $f^{(n)}(x) = e^x$ and $f^{(n)}(0) = 1$ for all n. Thus,

$$T_n(x) = 1 + 1(x - 0) + \frac{1}{2}(x - 0)^2 + \cdots + \frac{1}{n!}(x - 0)^n = 1 + x + \frac{x^2}{2} + \cdots + \frac{x^n}{n!}.$$

In Exercises 21–26, find $T_n(x)$ at $x = a$ for all n.

21. $f(x) = \frac{1}{1 - x}$, $a = 0$

SOLUTION Let $f(x) = \frac{1}{1-x}$. Then

$$f(x) = \frac{1}{1 - x} \qquad\qquad f(0) = 1 = 0!$$

$$f'(x) = \frac{1}{(1 - x)^2} \qquad\qquad f'(0) = 1 = 1!$$

$$f''(x) = \frac{2}{(1 - x)^3} \qquad\qquad f''(0) = 2 = 2!$$

$$f'''(x) = \frac{6}{(1 - x)^4} \qquad\qquad f'''(0) = 6 = 3!$$

$$\vdots \qquad\qquad\qquad \vdots$$

$$f^{(n)}(x) = \frac{n!}{(1 - x)^{n+1}} \qquad\qquad f^{(n)}(0) = n!$$

Therefore,

$$T_n(x) = 1 + 1(x - 0) + \frac{2}{2}(x - 0)^2 + \frac{6}{6}(x - 0)^3 + \cdots + \frac{n!}{n!}(x - 0)^n$$

$$= 1 + x + x^2 + x^3 + \cdots + x^n.$$

23. $f(x) = e^x$, $a = 1$

SOLUTION Let $f(x) = e^x$. Then $f^{(n)}(x) = e^x$ and $f^{(n)}(1) = e$ for all n. Therefore,

$$T_n(x) = e + e(x - 1) + \frac{e}{2!}(x - 1)^2 + \cdots + \frac{e}{n!}(x - 1)^n.$$

25. $f(x) = x^{5/2}$, $a = 2$

SOLUTION Let $f(x) = x^{5/2}$. Then

$$f(x) = x^{5/2} \qquad\qquad f(2) = 2^{5/2}$$

$$f'(x) = \frac{5}{2}x^{3/2} \qquad\qquad f'(2) = \frac{5}{2}2^{3/2}$$

$$f''(x) = \frac{15}{4}x^{1/2} \qquad\qquad f''(2) = \frac{15}{4}2^{1/2}$$

$$f'''(x) = \frac{15}{8}x^{-1/2} \qquad\qquad f'''(x) = \frac{15}{8}2^{-1/2}$$

As we continue taking derivatives, a pattern emerges:

$$f^{(4)}(x) = \frac{15}{8}\left(-\frac{1}{2}\right)x^{-3/2} \qquad\qquad f^{(4)}(2) = \frac{15}{8}\left(-\frac{1}{2}\right)2^{-3/2}$$

$$f^{(5)}(x) = \frac{15}{8}\left(-\frac{1}{2}\right)\left(-\frac{3}{2}\right)x^{-5/2} \qquad\qquad f^{(5)}(2) = \frac{15}{8}\left(-\frac{1}{2}\right)\left(-\frac{3}{2}\right)2^{-5/2}$$

$$f^{(6)}(x) = \frac{15}{8}\left(-\frac{1}{2}\right)\left(-\frac{3}{2}\right)\left(-\frac{5}{2}\right)x^{-7/2} \qquad\qquad f^{(6)}(2) = \frac{15}{8}\left(-\frac{1}{2}\right)\left(-\frac{3}{2}\right)\left(-\frac{5}{2}\right)2^{-7/2}$$

$$\vdots \qquad\qquad\qquad\qquad \vdots$$

Thus, for $n \geq 4$,

$$f^{(n)}(2) = \frac{15}{8}(-1)^{n-3}\frac{1}{2^{n-3}}\left(1 \cdot 3 \cdot 5 \cdot \cdots \cdot (2n - 7)\right)2^{-(2n-5)/2}.$$

To express $1 \cdot 3 \cdot \cdots \cdot (2n - 7)$ in closed form, we note that the product of the even numbers $2 \cdot 4 \cdot 6 \cdot 2n - 8$ is $2^{n-4}(n - 4)!$ so that the product of the odd numbers is

$$1 \cdot 3 \cdot \cdots \cdot (2n - 7) = \frac{(2n - 7)!}{2^{n-4}(n - 4)!}$$

Therefore, for $n \geq 4$,

$$f''(2) = 15(-1)^{n-3}\frac{(2n - 7)!}{(n - 4)!}2^{(13-6n)/2}.$$

Finally,

$$T_n(x) = 4\sqrt{2} + 5\sqrt{2}(x - 2) + \frac{15(x - 2)^2}{4\sqrt{2}} + \frac{5(x - 2)^3}{16\sqrt{2}} + \cdots + 15(-1)^{n-3}\frac{(2n - 7)!}{n!(n - 4)!}2^{(13-6n)/2}(x - 2)^n.$$

27. CAS Plot $y = e^x$ together with the MacLaurin polynomials $T_n(x)$ for $n = 1, 3, 5$ and then for $n = 2, 4, 6$ on the interval $[-3, 3]$. What difference do you notice between the even and odd MacLaurin polynomials?

SOLUTION The odd polynomials are concave down on the left and lie beneath the graph of $y = e^x$; the even polynomials are concave up on the left and lie above the graph of $y = e^x$.

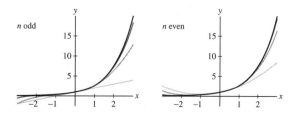

29. Use the Error Bound to find the maximum possible size of $|\cos 0.3 - T_5(0.3)|$, where $T_5(x)$ is the MacLaurin polynomial. Verify your result with a calculator.

SOLUTION Using the Error Bound, we have

$$|\cos 0.3 - T_5(0.3)| \le \frac{K|0.3|^6}{6!},$$

where K is a number such that $|f^{(6)}(x)| \le K$ for x between 0 and 0.3. With $f(x) = \cos x$, $f^{(6)}(x) = -\cos x$, so $|f^{(6)}(x)| \le 1$ for all x. We can therefore take $K = 1$ and then

$$|\cos 0.3 - T_5(0.3)| \le \frac{|0.3|^6}{6!} = 1.0125 \times 10^{-6}.$$

Now, with $f(x) = \cos x$,

$$T_5(x) = 1 - \frac{1}{2}x^2 + \frac{1}{24}x^4,$$

and

$$T_5(0.3) = 1 - \frac{1}{2}(0.3)^2 + \frac{1}{24}(0.3)^4 = 0.9553375.$$

Using a calculator, we find $\cos 0.3 \approx 0.955336489$, so

$$|\cos 0.3 - T_5(0.3)| = 1.010874 \times 10^{-6}.$$

31. Let T_n be the MacLaurin polynomial of $f(x) = e^x$ and let $c > 0$. Show that we may take $K = e^c$ in the Error Bound for $|T_n(c) - f(c)|$. Then use the Error Bound to determine a number n such that $|T_n(0.1) - e^{0.1}| \le 10^{-5}$.

SOLUTION Let $f(x) = e^x$. We know that $f^{(n+1)}(x) = e^x$ is always increasing, and, because $c > 0$, we can take $K = e^c$ in the error bound. Thus,

$$\left|T_n(x) - e^x\right| \le \frac{e^c \cdot x^{n+1}}{(n+1)!} \quad \text{so} \quad \left|T_n(0.1) - e^{0.1}\right| \le \frac{e^{0.1} \cdot (0.1)^{n+1}}{(n+1)!}.$$

By trial and error, we find

$$\left|T_2(0.1) - e^{0.1}\right| \le \frac{e^{0.1} \cdot (0.1)^3}{3!} \approx 1.842 \times 10^{-4} > 10^{-5}$$

but

$$\left|T_3(0.1) - e^{0.1}\right| \le \frac{e^{0.1} \cdot (0.1)^4}{4!} \approx 4.605 \times 10^{-6} < 10^{-5}.$$

Thus, $n = 3$.

33. Calculate $T_3(x)$ at $a = 0$ for $f(x) = \tan^{-1} x$. Then compute $T_3(\frac{1}{2})$ and use the Error Bound to find a bound for $|T_3(\frac{1}{2}) - \tan^{-1}(\frac{1}{2})|$. Refer to the graph in Figure 4 to find an acceptable value of K.

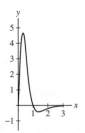

FIGURE 4 Graph of $f^{(4)}(x) = -24x(x^2 - 1)/(x^2 + 1)^4$, where $f(x) = \tan^{-1} x$.

SOLUTION Let $f(x) = \tan^{-1} x$. Then

$$f(x) = \tan^{-1} x \qquad\qquad f(0) = 0$$

$$f'(x) = \frac{1}{1 + x^2} \qquad\qquad f'(0) = 1$$

$$f''(x) = \frac{-2x}{(1 + x^2)^2} \qquad\qquad f''(0) = 0$$

$$f'''(x) = \frac{(1 + x^2)^2(-2) - (-2x)(2)(1 + x^2)(2x)}{(1 + x^2)^4} \qquad f'''(0) = -2$$

and

$$T_3(x) = 0 + 1(x - 0) + \frac{0}{2}(x - 0)^2 + \frac{-2}{6}(x - 0)^3 = x - \frac{x^3}{3}.$$

Since $f^{(4)}(x) \le 5$ for $x \ge 0$, we may take $K = 5$ in the error bound; then,

$$\left| T_3\left(\frac{1}{2}\right) - \tan^{-1}\left(\frac{1}{2}\right) \right| \le \frac{5(1/2)^4}{4!} = \frac{5}{384}.$$

35. Show that the MacLaurin polynomials for $f(x) = \sin x$ are

$$T_{2n-1}(x) = T_{2n} = x - \frac{x^3}{3!} + \frac{x^5}{5!} - \cdots + (-1)^{n-1}\frac{x^{2n-1}}{(2n-1)!}$$

Use the Error Bound with $n = 4$ to show that

$$\left| \sin x - \left(x - \frac{x^3}{6}\right) \right| \le \frac{|x|^5}{120} \quad \text{(for all } x\text{)}$$

SOLUTION Let $f(x) = \sin x$. Then

$$\begin{aligned}
f(x) &= \sin x & f(0) &= 0 \\
f'(x) &= \cos x & f'(0) &= 1 \\
f''(x) &= -\sin x & f''(0) &= 0 \\
f'''(x) &= -\cos x & f'''(0) &= -1 \\
f^{(4)}(x) &= \sin x & f^{(4)}(0) &= 0 \\
f^{(5)}(x) &= \cos x & f^{(5)}(0) &= 1
\end{aligned}$$

$$\vdots \qquad\qquad \vdots$$

Consequently,

$$T_{2n-1}(x) = x - \frac{x^3}{3!} + \frac{x^5}{5!} + \cdots + (-1)^{n+1}\frac{x^{2n-1}}{(2n-1)!}$$

and

$$T_{2n}(x) = x - \frac{x^3}{3!} + \frac{x^5}{5!} + \cdots + (-1)^{n+1}\frac{x^{2n-1}}{(2n-1)!} + 0 = T_{2n-1}(x).$$

With $n = 4$ and $K = 1$,

$$\left| \sin x - \left(x - \frac{x^3}{6}\right) \right| \le K\frac{|x|^5}{5!} \le \frac{|x|^5}{5!}.$$

37. Find n such that $|T_n(1.3) - \ln(1.3)| \le 10^{-4}$, where T_n is the Taylor polynomial for $f(x) = \ln x$ at $a = 1$.

SOLUTION Let $f(x) = \ln x$. Then $f'(x) = x^{-1}$, $f''(x) = -x^{-2}$, $f'''(x) = 2x^{-3}$, $f^{(4)}(x) = -6x^{-4}$, etc. In general,

$$f^{(n)}(x) = (-1)^{n+1}(n-1)!x^{-n}.$$

Now, $|f^{(n+1)}(x)|$ is decreasing on the interval $[1, 1.3]$, so $|f^{(n+1)}(x)| \le |f^{(n+1)}(1)| = n!$ for all $x \in [1, 1.3]$. We can therefore take $K = n!$ in the error bound, and

$$|T_n(1.3) - \ln(1.3)| \le n!\frac{|1.3 - 1|^{n+1}}{(n+1)!} = \frac{(0.3)^{n+1}}{n+1}.$$

With $n = 5$,

$$\frac{(0.3)^6}{6} = 1.215 \times 10^{-4} > 10^{-4},$$

but with $n = 6$,

$$\frac{(0.3)^7}{7} = 3.124 \times 10^{-5} < 10^{-4}.$$

Therefore, the error is guaranteed to be below 10^{-4} for $n = 6$.

39. Find n such that $|T_n(1.3) - \sqrt{1.3}| \leq 10^{-6}$, where $T_n(x)$ is the Taylor polynomial for $f(x) = \sqrt{x}$ at $a = 1$.

SOLUTION   Using the Error Bound, we have

$$|T_n(1.3) - \sqrt{1.3}| \leq K \frac{|1.3 - 1|^{n+1}}{(n+1)!} = K \frac{|0.3|^{n+1}}{(n+1)!},$$

where K is a number such that $|f^{(n+1)}(x)| \leq K$ for x between 1 and 1.3. For $f(x) = \sqrt{x}$, $|f^{(n)}(x)|$ is decreasing for $x > 1$, hence the maximum value of $|f^{(n+1)}(x)|$ occurs at $x = 1$. We may therefore take

$$K = |f^{(n+1)}(1)| = \frac{1 \cdot 3 \cdot 5 \cdots (2n+1)}{2^{n+1}}$$

$$= \frac{1 \cdot 3 \cdot 5 \cdots (2n+1)}{2^{n+1}} \cdot \frac{2 \cdot 4 \cdot 6 \cdots (2n+2)}{2 \cdot 4 \cdot 6 \cdots (2n+2)} = \frac{(2n+2)!}{(n+1)!2^{2n+2}}.$$

Then

$$|T_n(1.3) - \sqrt{1.3}| \leq \frac{(2n+2)!}{(n+1)!2^{2n+2}} \cdot \frac{|.3|^{n+1}}{(n+1)!} = \frac{(2n+2)!}{[(n+1)!]^2}(0.075)^{n+1}.$$

With $n = 9$

$$\frac{(20)!}{[(10)!]^2}(0.075)^{10} = 1.040 \times 10^{-6} > 10^{-6},$$

but with $n = 10$

$$\frac{(22)!}{[(11)!]^2}(0.075)^{11} = 2.979 \times 10^{-7} < 10^{-6}.$$

Hence, $n = 10$ will guarantee the desired accuracy.

41. Let $n \geq 1$. Show that if $|x|$ is small, then

$$(x+1)^{1/n} \approx 1 + \frac{x}{n} + \frac{1-n}{2n^2}x^2$$

Use this approximation with $n = 6$ to estimate $1.5^{1/6}$.

SOLUTION   Let $f(x) = (x+1)^{1/n}$. Then

$$f(x) = (x+1)^{1/n} \qquad\qquad f(0) = 1$$

$$f'(x) = \frac{1}{n}(x+1)^{1/n-1} \qquad\qquad f'(0) = \frac{1}{n}$$

$$f''(x) = \frac{1}{n}\left(\frac{1}{n} - 1\right)(x+1)^{1/n-2} \qquad f''(0) = \frac{1}{n}\left(\frac{1}{n} - 1\right)$$

and

$$T_2(x) = 1 + \frac{1}{n}(x) + \left(\frac{1}{n^2} - \frac{1}{n}\right)\frac{x^2}{2} = 1 + \frac{x}{n} + \left(\frac{1-n}{2n^2}\right)x^2.$$

With $n = 6$ and $x = 0.5$,

$$1.5^{1/6} \approx T_2(0.5) = \frac{307}{288} \approx 1.065972.$$

43. Find the fourth MacLaurin polynomial for $f(x) = \sin x \cos x$ by multiplying the fourth MacLaurin polynomials for $f(x) = \sin x$ and $f(x) = \cos x$.

SOLUTION   The fourth MacLaurin polynomial for $\sin x$ is $x - \frac{x^3}{6}$, and the fourth MacLaurin polynomial for $\cos x$ is $1 - \frac{x^2}{2} + \frac{x^4}{24}$. Multiplying these two polynomials, and then discarding terms of degree greater than 4, we find that the fourth MacLaurin polynomial for $f(x) = \sin x \cos x$ is

$$T_4(x) = x - \frac{2x^3}{3}.$$

45. Find the MacLaurin polynomials of $\dfrac{1}{1+x^2}$ by substituting $-x^2$ for x in the MacLaurin polynomials of $\dfrac{1}{1-x}$ (see Exercise 21).

SOLUTION The MacLaurin polynomials for $\frac{1}{1-x}$ are of the form

$$T_n(x) = 1 + x + x^2 + \cdots + x^n.$$

Accordingly, the MacLaurin polynomials for $\frac{1}{1+x^2}$ are of the form

$$T_{2n}(x) = 1 - x^2 + x^4 - x^6 + \cdots + (-x^2)^n.$$

47. Let $f(x) = 3x^3 + 2x^2 - x - 4$. Calculate $T_j(x)$ for $j = 1, 2, 3, 4, 5$ at both $a = 0$ and $a = 1$. Show that $T_3(x) = f(x)$ in both cases.

SOLUTION Let $f(x) = 3x^3 + 2x^2 - x - 4$. Then

$$
\begin{array}{lll}
f(x) = 3x^3 + 2x^2 - x - 4 & f(0) = -4 & f(1) = 0 \\
f'(x) = 9x^2 + 4x - 1 & f'(0) = -1 & f'(1) = 12 \\
f''(x) = 18x + 4 & f''(0) = 4 & f''(1) = 22 \\
f'''(x) = 18 & f'''(0) = 18 & f'''(1) = 18 \\
f^{(4)}(x) = 0 & f^{(4)}(0) = 0 & f^{(4)}(1) = 0 \\
f^{(5)}(x) = 0 & f^{(5)}(0) = 0 & f^{(5)}(1) = 0
\end{array}
$$

At $a = 0$,

$$T_1(x) = -4 - x;$$
$$T_2(x) = -4 - x + 2x^2;$$
$$T_3(x) = -4 - x + 2x^2 + 3x^3 = f(x);$$
$$T_4(x) = T_3(x); \text{ and}$$
$$T_5(x) = T_3(x).$$

At $a = 1$,

$$T_1(x) = 12(x - 1);$$
$$T_2(x) = 12(x - 1) + 11(x - 1)^2;$$
$$T_3(x) = 12(x - 1) + 11(x - 1)^2 + 3(x - 1)^3 = -4 - x + 2x^2 + 3x^3 = f(x);$$
$$T_4(x) = T_3(x); \text{ and}$$
$$T_5(x) = T_3(x).$$

49. Taylor polynomials can be used instead of L'Hôpital's Rule to evaluate limits. Consider $L = \lim\limits_{x \to 0} \dfrac{e^x + e^{-x} - 2}{1 - \cos x}$.

(a) Show that the second MacLaurin polynomial for

$$f(x) = e^x + e^{-x} - 2$$

is $T_2(x) = x^2$. Use the Error Bound with $n = 2$ to show that

$$e^x + e^{-x} - 2 = x^2 + g_1(x)$$

where $\lim\limits_{x \to 0} \dfrac{g_1(x)}{x^2} = 0$. Similarly, prove that

$$1 - \cos x = \frac{1}{2}x^2 + g_2(x)$$

where $\lim\limits_{x \to 0} \dfrac{g_2(x)}{x^2} = 0$.

(b) Evaluate L by using (a) to show that $L = \lim\limits_{x \to 0} \dfrac{1 + \dfrac{g_1(x)}{x^2}}{\dfrac{1}{2} + \dfrac{g_2(x)}{x^2}}$.

(c) Evaluate L again using L'Hôpital's Rule.

SOLUTION

(a) Let $f(x) = e^x + e^{-x} - 2$. Then $f'(x) = e^x - e^{-x}$, $f''(x) = e^x + e^{-x}$, $f(0) = 0$, $f'(x) = 0$ and $f''(0) = 2$. Thus, $T_2(x) = x^2$. Let $g_1(x) = f(x) - T_2(x) = e^x + e^{-x} - 2 - x^2$. By the Error Bound, for $|x| < 1$,

$$|g_1(x)| \le \frac{K_1 |x|^3}{3!},$$

where $K_1 = \max\{f'''(x) : |x| \le 1\}$, so

$$\frac{|g_1(x)|}{x^2} \le \frac{K_1 |x|}{3!} \to 0$$

as $x \to 0$. Likewise, $1 - \cos x$ has MacLaurin polynomial

$$T_2(x) = 1 - \left(1 - \frac{x^2}{2}\right) = \frac{x^2}{2},$$

so, for $g_2(x) = 1 - \cos x - \frac{1}{2}x^2$,

$$\frac{|g_2(x)|}{x^2} \le \frac{K_2 |x|}{3!} \to 0$$

as $x \to 0$. Here, we may take $K_2 = 1$.

(b) By substitution

$$L = \lim_{x \to 0} \frac{x^2 + g_1(x)}{x^2/2 + g_2(x)} = \lim_{x \to 0} \frac{1 + g_1(x)/x^2}{1/2 + g_2(x)/x^2} = 2.$$

(c) Using L'Hôpital's Rule

$$L = \lim_{x \to 0} \frac{e^x + e^{-x} - 2}{1 - \cos x} = \lim_{x \to 0} \frac{e^x - e^{-x}}{\sin x} = \lim_{x \to 0} \frac{e^x + e^{-x}}{\cos x} = 2.$$

51. A light wave of wavelength λ travels from A to B by passing through an aperture (see Figure 6 for the notation). The aperture (circular region) is located in a plane that is perpendicular to $\overline{AB}$. Let $f(r) = d' + h'$, that is, $f(r)$ is the distance $AC + CB$ as a function of r. The **Fresnel zones**, used to determine the optical disturbance at B, are the concentric bands bounded by the circles of radius R_n such that $f(R_n) = AB + n\lambda/2 = d + h + n\lambda/2$.

(a) Show that $f(r) = \sqrt{d^2 + r^2} + \sqrt{h^2 + r^2}$, and use the MacLaurin polynomial of order 2 to show

$$f(r) \approx d + h + \frac{1}{2}\left(\frac{1}{d} + \frac{1}{h}\right)r^2$$

(b) Deduce that $R_n \approx \sqrt{n\lambda L}$, where $L = (d^{-1} + h^{-1})^{-1}$.
(c) Estimate the radii R_1 and R_{100} for blue light ($\lambda = 475 \times 10^{-7}$ cm) if $d = h = 100$ cm.

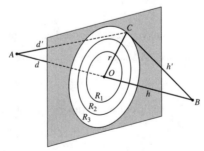

FIGURE 6 The Fresnel zones are the regions between the circles of radius R_n.

SOLUTION

(a) From the diagram, we see that $\overline{AC} = \sqrt{d^2 + r^2}$ and $\overline{CB} = \sqrt{h^2 + r^2}$. Therefore, $f(r) = \sqrt{d^2 + r^2} + \sqrt{h^2 + r^2}$. Moreover,

$$f'(r) = \frac{r}{\sqrt{d^2 + r^2}} + \frac{r}{\sqrt{h^2 + r^2}}, \qquad f''(r) = \frac{d^2}{(d^2 + r^2)^{3/2}} + \frac{h^2}{(h^2 + r^2)^{3/2}},$$

$f(0) = d + h$, $f'(0) = 0$ and $f''(0) = d^{-1} + h^{-1}$. Thus,

$$f(r) \approx T_2(r) = d + h + \frac{1}{2}\left(\frac{1}{d} + \frac{1}{h}\right)r^2.$$

(b) Solving

$$f(R_n) \approx d + h + \frac{1}{2}\left(\frac{1}{d} + \frac{1}{h}\right)R_n^2 = d + h + \frac{n\lambda}{2}$$

yields

$$R_n = \sqrt{n\lambda(d^{-1} + h^{-1})^{-1}} = \sqrt{n\lambda L},$$

where $L = (d^{-1} + h^{-1})^{-1}$.

(c) With $d = h = 100$ cm, $L = 50$ cm. Taking $\lambda = 475 \times 10^{-7}$ cm, it follows that

$$R_1 \approx \sqrt{\lambda L} = 0.04873 \text{ cm; and}$$
$$R_{100} \approx \sqrt{100\lambda L} = 0.4873 \text{ cm.}$$

Further Insights and Challenges

53. Use Taylor's Theorem to show that if $f^{(n+1)}(t) \geq 0$ for all t, then the nth MacLaurin polynomial $T_n(x)$ satisfies $T_n(x) \leq f(x)$ for all $x \geq 0$.

SOLUTION From Taylor's Theorem,

$$R_n(x) = f(x) - T_n(x) = \frac{1}{n!}\int_0^x (x - u)^n f^{(n+1)}(u)\, du.$$

If $f^{(n+1)}(t) \geq 0$ for all t then

$$\frac{1}{n!}\int_0^x (x - u)^n f^{(n+1)}(u)\, du \geq 0$$

since $(x - u)^n \geq 0$ for $0 \leq u \leq x$. Thus, $f(x) - T_n(x) \geq 0$, or $f(x) \geq T_n(x)$.

55. This exercise is intended to reinforce the proof of Taylor's Theorem.

(a) Show that $f(x) = T_0(x) + \displaystyle\int_a^x f'(u)\, du$.

(b) Use Integration by Parts to prove the formula

$$\int_a^x (x - u)f^{(2)}(u)\, du = -f'(a)(x - a) + \int_a^x f'(u)\, du$$

(c) Prove the case $n = 2$ of Taylor's Theorem:

$$f(x) = T_1(x) + \int_a^x (x - u)f^{(2)}(u)\, du.$$

SOLUTION

(a)

$$T_0(x) + \int_a^x f'(u)\, du = T_0(x) + f(x) - f(a) \quad \text{(from FTC2)}$$
$$= f(a) + f(x) - f(a) = f(x).$$

(b) Using Integration by Parts with $w = x - u$ and $v' = f''(u)\, du$,

$$\int_a^x (x - u)f''(u)\, du = f'(u)(x - u)\Big|_a^x + \int_a^x f'(u)\, du$$
$$= f'(x)(x - x) - f'(a)(x - a) + \int_a^x f'(u)\, du$$
$$= -f'(a)(x - a) + \int_a^x f'(u)\, du.$$

(c)

$$T_1(x) + \int_a^x (x - u)f''(u)\, du = f(a) + f'(a)(x - a) + \left(-f'(a)(x - a)\right) + \int_a^x f'(u)\, du$$
$$= f(a) + f(x) - f(a) = f(x).$$

57. Use the fourth MacLaurin polynomial and the method of Exercise 56 to approximate $\int_0^{1/2} \sin(x^2)\,dx$. Find a bound for the error.

SOLUTION   The fourth MacLaurin polynomial for $\sin(x^2)$ is x^2. Thus, by Exercise 56,

$$\int_0^{1/2} \sin(x^2)\,dx \approx \int_0^{1/2} x^2\,dx = \frac{1}{24}.$$

On $[0, 1/2]$,

$$\left|\sin(x^2) - T_2(x)\right| \le \frac{(1/2)^6}{6},$$

so the error in the approximation is bounded by

$$\frac{(1/2)^6}{6}\left(\frac{1}{2} - 0\right) \approx 1.302 \times 10^{-3}.$$

59. Prove by induction that for all k,

$$\frac{d^j}{dx^j}\left(\frac{(x-a)^k}{k!}\right) = \frac{k(k-1)\cdots(k-j+1)(x-a)^{k-j}}{k!}$$

$$\frac{d^j}{dx^j}\left(\frac{(x-a)^k}{k!}\right)\Bigg|_{x=a} = \begin{cases} 1 & \text{for } k = j \\ 0 & \text{for } k \ne j \end{cases}$$

Use this to prove that $T_n(x)$ agrees with $f(x)$ at $x = a$ to order n.

SOLUTION   The first formula is clearly true for $j = 0$. Suppose the formula is true for an arbitrary j. Then

$$\frac{d^{j+1}}{dx^{j+1}}\left(\frac{(x-a)^k}{k!}\right) = \frac{d}{dx}\frac{d^j}{dx^j}\left(\frac{(x-a)^k}{k!}\right) = \frac{d}{dx}\left(\frac{k(k-1)\cdots(k-j+1)(x-a)^{k-j}}{k!}\right)$$

$$= \frac{k(k-1)\cdots(k-j+1)(k-(j+1)+1)(x-a)^{k-(j+1)}}{k!}$$

as desired. Note that if $k = j$, then the numerator is $k!$, the denominator is $k!$ and the value of the derivative is 1; otherwise, the value of the derivative is 0 at $x = a$. In other words,

$$\frac{d^j}{dx^j}\left(\frac{(x-a)^k}{k!}\right)\Bigg|_{x=a} = \begin{cases} 1 & \text{for } k = j \\ 0 & \text{for } k \ne j \end{cases}$$

Applying this latter formula, it follows that

$$\frac{d^j}{dx^j}T_n(a)\Bigg|_{x=a} = \sum_{k=0}^{n}\frac{d^j}{dx^j}\left(\frac{f^{(k)}(a)}{k!}(x-a)^k\right)\Bigg|_{x=a} = f^{(j)}(a)$$

as required.

CHAPTER REVIEW EXERCISES

In Exercises 1–4, calculate the arc length over the given interval.

1. $y = \dfrac{x^5}{10} + \dfrac{x^{-3}}{6}$, $[3, 5]$

SOLUTION   Let $y = \dfrac{x^5}{10} + \dfrac{x^{-3}}{6}$. Then

$$1 + (y')^2 = 1 + \left(\frac{x^4}{2} - \frac{x^{-4}}{2}\right)^2 = 1 + \frac{x^8}{4} - \frac{1}{2} + \frac{x^{-8}}{4}$$

$$= \frac{x^8}{4} + \frac{1}{2} + \frac{x^{-8}}{4} = \left(\frac{x^4}{2} + \frac{x^{-4}}{2}\right)^2.$$

Because $\frac{1}{2}(x^4 + x^{-4}) > 0$ on $[3, 5]$, the arc length is

$$s = \int_3^5 \sqrt{1 + (y')^2}\, dx = \int_3^5 \left(\frac{x^4}{2} + \frac{x^{-4}}{2} \right) dx = \left(\frac{x^5}{10} - \frac{x^{-3}}{6} \right) \Big|_3^5 = \frac{2918074}{10125}.$$

3. $y = 4x - 2$, $[-2, 2]$

SOLUTION Let $y = 4x - 2$. Then

$$\sqrt{1 + (y')^2} = \sqrt{1 + 4^2} = \sqrt{17}.$$

Hence,

$$s = \int_{-2}^2 \sqrt{17}\, dx = 4\sqrt{17}.$$

5. Show that the arc length of $y = 2\sqrt{x}$ over $[0, a]$ is equal to $\sqrt{a(a+1)} + \ln(\sqrt{a} + \sqrt{a+1})$. *Hint:* Apply the substitution $x = \tan^2 \theta$ to the arc length integral.

SOLUTION Let $y = 2\sqrt{x}$. Then $y' = \frac{1}{\sqrt{x}}$, and

$$\sqrt{1 + (y')^2} = \sqrt{1 + \frac{1}{x}} = \sqrt{\frac{x+1}{x}} = \frac{1}{\sqrt{x}}\sqrt{x+1}.$$

Thus,

$$s = \int_0^a \frac{1}{\sqrt{x}}\sqrt{1+x}\, dx.$$

We make the substitution $x = \tan^2 \theta$, $dx = 2\tan\theta \sec^2\theta\, d\theta$. Then

$$s = \int_{x=0}^{x=a} \frac{1}{\tan\theta} \sec\theta \cdot 2\tan\theta \sec^2\theta\, d\theta = 2\int_{x=0}^{x=a} \sec^3\theta\, d\theta.$$

We use a reduction formula to obtain

$$s = 2\left(\frac{\tan\theta \sec\theta}{2} + \frac{1}{2}\ln|\sec\theta + \tan\theta| \right) \Big|_{x=0}^{|x=a} = (\sqrt{x}\sqrt{1+x} + \ln|\sqrt{1+x} + \sqrt{x}|)\Big|_0^a$$

$$= \sqrt{a}\sqrt{1+a} + \ln|\sqrt{1+a} + \sqrt{a}| = \sqrt{a(a+1)} + \ln\left(\sqrt{a} + \sqrt{a+1} \right).$$

In Exercises 7–10, calculate the surface area of the solid obtained by rotating the curve over the given interval about the x-axis.

7. $y = x + 1$, $[0, 4]$

SOLUTION Let $y = x + 1$. Then $y' = 1$, and

$$y\sqrt{1 + y'^2} = (x+1)\sqrt{1+1} = \sqrt{2}(x+1).$$

Thus,

$$SA = 2\pi \int_0^4 \sqrt{2}(x+1)\, dx = 2\sqrt{2}\pi \left(\frac{x^2}{2} + x \right) \Big|_0^4 = 24\sqrt{2}\pi.$$

9. $y = \frac{2}{3}x^{3/2} - \frac{1}{2}x^{1/2}$, $[1, 2]$

SOLUTION Let $y = \frac{2}{3}x^{3/2} - \frac{1}{2}x^{1/2}$. Then

$$y' = \sqrt{x} - \frac{1}{4\sqrt{x}},$$

and

$$1 + (y')^2 = 1 + \left(\sqrt{x} - \frac{1}{4\sqrt{x}} \right)^2 = 1 + \left(x - \frac{1}{2} + \frac{1}{16x} \right) = x + \frac{1}{2} + \frac{1}{16x} = \left(\sqrt{x} + \frac{1}{4\sqrt{x}} \right)^2.$$

Because $\sqrt{x} + \frac{1}{\sqrt{x}} \geq 0$, the surface area is

$$2\pi \int_a^b y\sqrt{1+(y')^2}\,dx = 2\pi \int_1^2 \left(\frac{2}{3}x^{3/2} - \frac{\sqrt{x}}{2}\right)\left(\sqrt{x} + \frac{1}{4\sqrt{x}}\right)dx$$

$$= 2\pi \int_1^2 \left(\frac{2}{3}x^2 + \frac{1}{6}x - \frac{1}{2}x - \frac{1}{8}\right)dx = 2\pi\left(\frac{2x^3}{9} - \frac{x^2}{6} - \frac{1}{8}x\right)\Big|_1^2 = \frac{67}{36}\pi.$$

11. Compute the total surface area of the coin obtained by rotating the region in Figure 1 about the x-axis. The top and bottom parts of the region are semicircles with a radius of 1 mm.

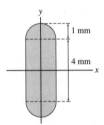

1 mm

4 mm

FIGURE 1

SOLUTION The generating half circle of the edge is $y = 2 + \sqrt{1 - x^2}$. Then,

$$y' = \frac{-2x}{2\sqrt{1-x^2}} = \frac{-x}{\sqrt{1-x^2}},$$

and

$$1 + (y')^2 = 1 + \frac{x^2}{1-x^2} = \frac{1}{1-x^2}.$$

The surface area of the edge of the coin is

$$2\pi \int_{-1}^1 y\sqrt{1+(y')^2}\,dx = 2\pi \int_{-1}^1 \left(2 + \sqrt{1-x^2}\right)\frac{1}{\sqrt{1-x^2}}\,dx$$

$$= 2\pi\left(2\int_{-1}^1 \frac{dx}{\sqrt{1-x^2}} + \int_{-1}^1 \frac{\sqrt{1-x^2}}{\sqrt{1-x^2}}\,dx\right)$$

$$= 2\pi\left(2\arcsin x\big|_{-1}^1 + \int_{-1}^1 dx\right)$$

$$= 2\pi(2\pi + 2) = 4\pi^2 + 4\pi.$$

We now add the surface area of the two sides of the disk, which are circles of radius 2. Hence the surface area of the coin is:

$$\left(4\pi^2 + 4\pi\right) + 2\pi \cdot 2^2 = 4\pi^2 + 12\pi.$$

13. Calculate the fluid force on the side of a right triangle of height 3 ft and base 2 ft submerged in water vertically, with its upper vertex located at a depth of 4 ft.

SOLUTION We need to find an expression for the horizontal width $f(y)$ at depth y.

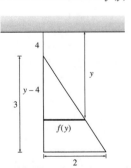

4

y

$y - 4$

3

$f(y)$

2

By similar triangles we have:

$$\frac{f(y)}{y-4} = \frac{2}{3} \qquad \text{so} \qquad f(y) = \frac{2(y-4)}{3}.$$

Hence, the force on the side of the triangle is

$$F = w \int_4^7 y f(y)\,dy = \frac{2w}{3} \int_4^7 \left(y^2 - 4y\right) dy = \frac{2w}{3} \left(\frac{y^3}{3} - 2y^2\right)\Bigg|_4^7 = 18w.$$

The density of water is $w = 62.5\,\text{lb/ft}^2$, hence,

$$F = 18 \cdot 62.5 = 1125\,\text{lb}.$$

15. Calculate the moments and COM of the lamina occupying the region under $y = x(4-x)$ for $0 \le x \le 4$, assuming a density of $\rho = 5\,\text{lb/ft}^2$.

SOLUTION   Because the lamina is symmetric with respect to the vertical line $x = 2$, by the symmetry principle, we know that $x_{\text{cm}} = 2$. Now,

$$M_x = \frac{\rho}{2} \int_0^4 f(x)^2\,dx = \frac{5}{2} \int_0^4 x^2(4-x)^2\,dx = \frac{5}{2} \left(\frac{16}{3}x^3 - 2x^4 + \frac{1}{5}x^5\right)\Bigg|_0^4 = \frac{256}{3}.$$

Moreover, the mass of the lamina is

$$M = \rho \int_0^4 f(x)\,dx = 5 \int_0^4 x(4-x)\,dx = 5 \left(2x^2 - \frac{1}{3}x^3\right)\Bigg|_0^4 = \frac{160}{3}.$$

Thus, the coordinates of the center of mass are

$$\left(2, \frac{256/3}{160/3}\right) = \left(2, \frac{8}{5}\right).$$

17. Find the centroid of the region between the semicircle $y = \sqrt{1-x^2}$ and the top half of the ellipse $y = \frac{1}{2}\sqrt{1-x^2}$ (Figure 3).

SOLUTION   Since the region is symmetric with respect to the y-axis, the centroid lies on the y-axis. To find y_{cm} we calculate

$$M_x = \frac{1}{2} \int_{-1}^1 \left[\left(\sqrt{1-x^2}\right)^2 - \left(\frac{\sqrt{1-x^2}}{2}\right)^2\right] dx$$

$$= \frac{1}{2} \int_{-1}^1 \frac{3}{4}\left(1-x^2\right) dx = \frac{3}{8}\left(x - \frac{1}{3}x^3\right)\Bigg|_{-1}^1 = \frac{1}{2}.$$

The area of the lamina is $\frac{\pi}{2} - \frac{\pi}{4} = \frac{\pi}{4}$, so the coordinates of the centroid are

$$\left(0, \frac{1/2}{\pi/4}\right) = \left(0, \frac{2}{\pi}\right).$$

19. Find the centroid of the shaded region in Figure 4 bounded on the left by $x = 2y^2 - 2$ and on the right by a semicircle of radius 1. *Hint:* Use symmetry and additivity of moments.

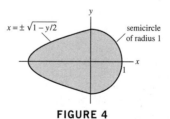

FIGURE 4

SOLUTION   The region is symmetric with respect to the x-axis, hence the centroid lies on the x-axis; that is, $y_{\text{cm}} = 0$. To compute the area and the moment with respect to the y-axis, we treat the left side and the right side of the region separately. Starting with the left side, we find

$$M_y^{\text{left}} = 2 \int_{-2}^0 x\sqrt{\frac{x}{2} + 1}\,dx \qquad \text{and} \qquad A^{\text{left}} = 2 \int_{-2}^0 \sqrt{\frac{x}{2} + 1}\,dx.$$

In each integral we make the substitution $u = \frac{x}{2} + 1$, $du = \frac{1}{2} dx$, and find

$$M_y^{\text{left}} = 8 \int_0^1 (u - 1)u^{1/2} \, du = 8 \int_0^1 \left(u^{3/2} - u^{1/2} \right) du = 8 \left(\frac{2}{5} u^{5/2} - \frac{2}{3} u^{3/2} \right) \Big|_0^1 = -\frac{32}{15}$$

and

$$A^{\text{left}} = 4 \int_0^1 u^{1/2} \, du = \frac{8}{3} u^{3/2} \Big|_0^1 = \frac{8}{3}.$$

On the right side of the region

$$M_y^{\text{right}} = 2 \int_0^1 x\sqrt{1 - x^2} \, dx = -\frac{2}{3}(1 - x^2)^{3/2} \Big|_0^1 = \frac{2}{3},$$

and $A^{\text{right}} = \frac{\pi}{2}$ (because the right side of the region is one-half of a circle of radius 1). Thus,

$$M_y = M_y^{\text{left}} + M_y^{\text{right}} = -\frac{32}{15} + \frac{2}{3} = -\frac{22}{15};$$

$$A = A^{\text{left}} + A^{\text{right}} = \frac{8}{3} + \frac{\pi}{2} = \frac{16 + 3\pi}{6};$$

and the coordinates of the centroid are

$$\left(\frac{-22/15}{(16 + 3\pi)/6}, 0 \right) = \left(-\frac{44}{80 + 15\pi}, 0 \right).$$

In Exercises 20–25, find the Taylor polynomial at $x = a$ for the given function.

21. $f(x) = 3(x + 2)^3 - 5(x + 2)$, $T_3(x)$, $a = -2$

SOLUTION $T_3(x)$ is the Taylor polynomial of f consisting of powers of $(x + 2)$ up to three. Since $f(x)$ is already in this form we conclude that $T_3(x) = f(x)$.

23. $f(x) = (3x + 2)^{1/3}$, $T_3(x)$, $a = 2$

SOLUTION   We start by computing the first three derivatives of $f(x) = (3x + 2)^{1/3}$:

$$f'(x) = \frac{1}{3}(3x + 2)^{-2/3} \cdot 3 = (3x + 2)^{-2/3}$$

$$f''(x) = -\frac{2}{3}(3x + 2)^{-5/3} \cdot 3 = -2(3x + 2)^{-5/3}$$

$$f'''(x) = \frac{10}{3}(3x + 2)^{-8/3} \cdot 3 = 10(3x + 2)^{-8/3}$$

Evaluating the function and its derivatives at $x = 2$, we find

$$f(2) = 2, \quad f'(2) = \frac{1}{4}, \quad f''(2) = -\frac{1}{16}, \quad f'''(2) = \frac{5}{128}.$$

Therefore,

$$T_3(x) = f(2) + f'(2)(x - 2) + \frac{f''(2)}{2!}(x - 2)^2 + \frac{f'''(2)}{3!}(x - 2)^3$$

$$= 2 + \frac{1}{4}(x - 2) + \frac{-1/16}{2!}(x - 2)^2 + \frac{5/128}{3!}(x - 2)^3$$

$$= 2 + \frac{1}{4}(x - 2) - \frac{1}{32}(x - 2)^2 - \frac{5}{768}(x - 2)^3.$$

25. $f(x) = \ln(\cos x)$, $T_3(x)$, $a = 0$

SOLUTION   We start by computing the first three derivatives of $f(x) = \ln(\cos x)$:

$$f'(x) = -\frac{\sin x}{\cos x} = -\tan x$$

$$f''(x) = -\sec^2 x$$

$$f'''(x) = -2 \sec^2 x \tan x$$

Evaluating the function and its derivatives at $x = 0$, we find

$$f(0) = 0, \ f'(0) = 0, \ f''(0) = -1, \ f'''(0) = 0.$$

Therefore,

$$T_3(x) = f(0) + f'(0)x + \frac{f''(0)}{2!}x^2 + \frac{f'''(0)}{3!}x^3 = 0 + \frac{0}{1!}x - \frac{1}{2!}x^2 + \frac{0}{3!}x^3 = -\frac{x^2}{2}.$$

27. Use the fifth MacLaurin polynomial of $f(x) = e^x$ to approximate $\sqrt{e}$. Use a calculator to determine the error.

SOLUTION Let $f(x) = e^x$. Then $f^{(n)}(x) = e^x$ and $f^{(n)}(0) = 1$ for all n. Hence,

$$T_5(x) = f(0) + f'(0)x + \frac{f''(0)}{2!}x^2 + \frac{f'''(0)}{3!}x^3 + \frac{f^{(4)}(0)}{4!}x^4 + \frac{f^{(5)}(0)}{5!}x^5$$

$$= 1 + x + \frac{x^2}{2!} + \frac{x^3}{3!} + \frac{x^4}{4!} + \frac{x^5}{5!}.$$

For $x = \frac{1}{2}$ we have

$$T_5\left(\frac{1}{2}\right) = 1 + \frac{1}{2} + \frac{\left(\frac{1}{2}\right)^2}{2!} + \frac{\left(\frac{1}{2}\right)^3}{3!} + \frac{\left(\frac{1}{2}\right)^4}{4!} + \frac{\left(\frac{1}{2}\right)^5}{5!}$$

$$= 1 + \frac{1}{2} + \frac{1}{8} + \frac{1}{48} + \frac{1}{384} + \frac{1}{3840} = 1.648697917$$

Using a calculator, we find that $\sqrt{e} = 1.648721271$. The error in the Taylor polynomial approximation is

$$|1.648697917 - 1.648721271| = 2.335 \times 10^{-5}.$$

29. Let $T_4(x)$ be the Taylor polynomial for $f(x) = \sqrt{x}$ at $a = 16$. Use the Error Bound to find the maximum possible size of $|f(17) - T_4(17)|$.

SOLUTION Using the Error Bound, we have

$$|f(17) - T_4(17)| \le K\frac{(17 - 16)^5}{5!} = \frac{K}{5!},$$

where K is a number such that $\left|f^{(5)}(x)\right| \le K$ for all $16 \le x \le 17$. Starting from $f(x) = \sqrt{x}$ we find

$$f'(x) = \frac{1}{2}x^{-1/2}, \ f''(x) = -\frac{1}{4}x^{-3/2}, \ f'''(x) = \frac{3}{8}x^{-5/2}, \ f^{(4)}(x) = -\frac{15}{16}x^{-7/2},$$

and

$$f^{(5)}(x) = \frac{105}{32}x^{-9/2}.$$

For $16 \le x \le 17$,

$$\left|f^{(5)}(x)\right| = \frac{105}{32x^{9/2}} \le \frac{105}{32 \cdot 16^{9/2}} = \frac{105}{8388608}.$$

Therefore, we may take

$$K = \frac{105}{8388608}.$$

Finally,

$$|f(17) - T_4(17)| \le \frac{105}{8388608} \cdot \frac{1}{5!} \approx 1.044 \cdot 10^{-7}.$$

31. Let $T_4(x)$ be the Taylor polynomial for $f(x) = x \ln x$ at $x = 1$ computed in Exercise 22. Use the Error Bound to find a bound for $|f(1.2) - T_4(1.2)|$.

SOLUTION   Using the Error Bound, we have

$$|f(1.2) - T_4(1.2)| \leq K \frac{(1.2 - 1)^5}{5!} = \frac{(0.2)^5}{120} K,$$

where K is a number such that $\left| f^{(5)} x \right| \leq K$ for all $1 \leq x \leq 1.2$. Starting from $f(x) = x \ln x$, we find

$$f'(x) = \ln x + x \frac{1}{x} = \ln x + 1, \; f''(x) = \frac{1}{x}, \; f'''(x) = -\frac{1}{x^2}, \; f^{(4)}(x) = \frac{2}{x^3},$$

and

$$f^{(5)}(x) = \frac{-6}{x^4}.$$

For $1 \leq x \leq 1.2$,

$$\left| f^{(5)}(x) \right| = \frac{6}{x^4} \leq \frac{6}{1^4} = 6.$$

Hence we may take $K = 6$ to obtain:

$$|f(1.2) - T_4(1.2)| \leq \frac{(0.2)^5}{120} 6 = 1.6 \times 10^{-5}.$$

33. Show that the nth MacLaurin polynomial for $f(x) = \dfrac{1}{1-x}$ is $T_n(x) = 1 + x + x^2 + \cdots + x^n$. Conclude by substituting $x/4$ for x so that the nth MacLaurin polynomial for $f(x) = \dfrac{1}{1 + x/4}$ is

$$T_n(x) = 1 + \frac{1}{4}x + \frac{1}{4^2}x^2 + \cdots + \frac{1}{4^n}x^n$$

What is the nth MacLaurin polynomial for $g(x) = \dfrac{1}{1+x}$? *Hint:* $g(x) = f(-x)$.

SOLUTION   Let $f(x) = (1 - x)^{-1}$. Then, $f'(x) = (1 - x)^{-2}$, $f''(x) = 2(1 - x)^{-3}$, $f'''(x) = 3!(1 - x)^{-4}$, and, in general, $f^{(n)}(x) = n!(1 - x)^{-(n+1)}$. Therefore, $f^{(n)}(0) = n!$ and

$$T_n(x) = 1 + \frac{1!}{1!}x + \frac{2!}{2!}x^2 + \cdots + \frac{n!}{n!}x^n = 1 + x + x^2 + \cdots + x^n.$$

Upon substituting $x/4$ for x, we find that the nth MacLaurin polynomial for $f(x) = \dfrac{1}{1 - x/4}$ is

$$T_n(x) = 1 + \frac{1}{4}x + \frac{1}{4^2}x^2 + \cdots + \frac{1}{4^n}x^n.$$

Substituting $-x$ for x, the nth MacLaurin polynomial for $g(x) = \dfrac{1}{1+x}$ is

$$T_n(x) = 1 - x + x^2 - x^3 + - \cdots + (-x)^n.$$

10 INTRODUCTION TO DIFFERENTIAL EQUATIONS

10.1 Solving Differential Equations

Preliminary Questions

1. Determine the order of the following differential equations:

(a) $x^5 y' = 1$

(b) $(y')^3 + x = 1$

(c) $y''' + x^4 y' = 2$

SOLUTION

(a) The highest order derivative that appears in this equation is a first derivative, so this is a first order equation.

(b) The highest order derivative that appears in this equation is a first derivative, so this is a first order equation.

(c) The highest order derivative that appears in this equation is a third derivative, so this is a third order equation.

2. Is $y' = \sin x$ a linear differential equation?

SOLUTION Yes.

3. Give an example of a nonlinear differential equation of the form $y' = f(y)$.

SOLUTION One possibility is $y' = y^2$.

4. Can a nonlinear differential equation be separable? If so, give an example.

SOLUTION Yes. An example is $y' = y^2$.

5. Give an example of a linear, nonseparable differential equation.

SOLUTION One example is $y' + y = x$.

Exercises

1. Which of the following differential equations are first-order?

(a) $y' = x^2$

(b) $y'' = y^2$

(c) $(y')^3 + yy' = \sin x$

(d) $x^2 y' - e^x y = \sin y$

(e) $y'' + 3y' = \dfrac{y}{x}$

(f) $yy' + x + y = 0$

SOLUTION

(a) The highest order derivative that appears in this equation is a first derivative, so this is a first order equation.

(b) The highest order derivative that appears in this equation is a second derivative, so this is not a first order equation.

(c) The highest order derivative that appears in this equation is a first derivative, so this is a first order equation.

(d) The highest order derivative that appears in this equation is a first derivative, so this is a first order equation.

(e) The highest order derivative that appears in this equation is a second derivative, so this is not a first order equation.

(f) The highest order derivative that appears in this equation is a first derivative, so this is a first order equation.

In Exercises 3–9, verify that the given function is a solution of the differential equation.

3. $y' - 8x = 0, \quad y = 4x^2$

SOLUTION Let $y = 4x^2$. Then $y' = 8x$ and

$$y' - 8x = 8x - 8x = 0.$$

5. $yy' + 4x = 0, \quad y = \sqrt{12 - 4x^2}$

SOLUTION   Let $y = \sqrt{12 - 4x^2}$. Then

$$y' = \frac{-4x}{\sqrt{12 - 4x^2}},$$

and

$$yy' + 4x = \sqrt{12 - 4x^2}\frac{-4x}{\sqrt{12 - 4x^2}} + 4x = -4x + 4x = 0.$$

7. $(x^2 - 1)y' + xy = 0, \quad y = 4(x^2 - 1)^{-1/2}$

SOLUTION   Let $y = 4(x^2 - 1)^{-1/2}$. Then $y' = -4x(x^2 - 1)^{-3/2}$, and

$$(x^2 - 1)y' + xy = (x^2 - 1)(-4x)(x^2 - 1)^{-3/2} + 4x(x^2 - 1)^{-1/2}$$
$$= -4x(x^2 - 1)^{-1/2} + 4x(x^2 - 1)^{-1/2} = 0.$$

9. $y'' - 2y' + 5y = 0, \quad y = e^x \sin 2x$

SOLUTION   Let $y = e^x \sin 2x$. Then

$$y' = 2e^x \cos 2x + e^x \sin 2x,$$
$$y'' = -4e^x \sin 2x + 2e^x \cos 2x + 2e^x \cos 2x + e^x \sin 2x = -3e^x \sin 2x + 4e^x \cos 2x,$$

and

$$y'' - 2y' + 5y = -3e^x \sin 2x + 4e^x \cos 2x - 4e^x \cos 2x - 2e^x \sin 2x + 5e^x \sin 2x$$
$$= (-3e^x - 2e^x + 5e^x) \sin 2x + (4e^x - 4e^x) \cos 2x = 0.$$

11. Consider the differential equation $y' = e^y \cos x$.
(a) Write it as $e^{-y} \, dy = \cos x \, dx$.
(b) Integrate both sides of $\int e^{-y} \, dy = \int \cos x \, dx$.
(c) Show that $y = -\ln|C - \sin x|$ is the general solution.
(d) Find the particular solution satisfying $y(0) = 0$.

SOLUTION

(a) As $y' = \frac{dy}{dx}$, we regroup:

$$\frac{dy}{dx} = e^y \cos x$$
$$e^{-y} \, dy = \cos x \, dx$$

(b) Integrating both sides of the resulting equation yields

$$-e^{-y} + C_1 = \sin x + C_2$$
$$e^{-y} = -\sin x + C$$

(Note that a linear sum of arbitrary constants is another arbitrary constant.)
(c) Solving the last equation for y yields

$$-y = \ln|-\sin x + C|$$
$$y = -\ln|C - \sin x|$$

(d) Setting $y(0) = 0$ in the last equation yields

$$0 = -\ln(C)$$

so that $C = 1$. Therefore, $y(x) = -\ln|1 - \sin x|$.

In Exercises 13–28, solve using separation of variables.

13. $y' = xy^2$

SOLUTION Rewrite

$$\frac{dy}{dx} = xy^2 \qquad \text{as} \qquad \frac{dy}{y^2} = x\,dx.$$

Integrating both sides of this equation yields

$$\int \frac{dy}{y^2} = \int x\,dx$$

$$-\frac{1}{y} = \frac{1}{2}x^2 + C.$$

Thus,

$$y = -\frac{1}{\frac{1}{2}x^2 + C},$$

where C is an arbitrary constant.

15. $y' = 9y$

SOLUTION Rewrite

$$\frac{dy}{dx} = 9y \qquad \text{as} \qquad \frac{dy}{y} = 9\,dx.$$

Integrating both sides of this equation yields

$$\int \frac{dy}{y} = \int 9\,dx$$

$$\ln|y| = 9x + C.$$

Solving for y, we find

$$|y| = e^{9x+C} = e^C e^{9x}$$

$$y = \pm e^C e^{9x} = Ae^{9x},$$

where $A = \pm e^C$ is an arbitrary constant.

17. $2\dfrac{dy}{dx} + 6y + 4 = 0$

SOLUTION Rewrite

$$2\frac{dy}{dx} + (6y + 4) = 0$$

as

$$\frac{dy}{dx} = -(3y + 2) \qquad \text{and then} \qquad \frac{dy}{3y + 2} = -dx.$$

Integrating both sides of this equation yields

$$\int \frac{dy}{3y + 2} = -\int dx$$

$$\ln|3y + 2| = -x + C.$$

Solving for y, we find

$$|3y + 2| = e^{-x+C} = e^C e^{-x}$$

$$3y + 2 = \pm e^C e^{-x}$$

$$y = \pm\frac{1}{3}e^C e^{-x} - \frac{2}{3} = Ae^{-x} - \frac{2}{3},$$

where $A = \pm\frac{1}{3}e^C$ is an arbitrary constant.

19. $\dfrac{dy}{dt} - te^y = 0$

SOLUTION   Rewrite

$$\frac{dy}{dt} = te^y \quad \text{as} \quad e^{-y}\,dy = t\,dt.$$

Integrating both sides of this equation yields

$$\int e^{-y}\,dy = \int t\,dt$$

$$-e^{-y} = \frac{1}{2}t^2 + C.$$

Solving for y, we find

$$-y = \ln\left|-\frac{1}{2}t^2 + C\right|$$

$$y = -\ln\left|C - \frac{1}{2}t^2\right|,$$

where C is an arbitrary constant.

21. $y' = y^2(1 - x^2)$

SOLUTION   Rewrite

$$\frac{dy}{dx} = y^2(1 - x^2) \quad \text{as} \quad \frac{dy}{y^2} = (1 - x^2)\,dx.$$

Integrating both sides of this equation yields

$$\int \frac{dy}{y^2} = \int (1 - x^2)\,dx$$

$$-y^{-1} = x - \frac{1}{3}x^3 + C.$$

Solving for y, we find

$$y^{-1} = \frac{1}{3}x^3 - x + C$$

$$y = \frac{1}{\frac{1}{3}x^3 - x + C},$$

where C is an arbitrary constant.

23. $(t^2 + 1)\dfrac{dx}{dt} = x^2 + 1$

SOLUTION   Rewrite

$$(t^2 + 1)\frac{dx}{dt} = (x^2 + 1) \quad \text{as} \quad \frac{1}{x^2 + 1}\,dx = \frac{1}{t^2 + 1}\,dt.$$

Integrating both sides of this equation yields

$$\int \frac{1}{x^2 + 1}\,dx = \int \frac{1}{t^2 + 1}\,dt$$

$$\tan^{-1}x = \tan^{-1}t + C.$$

Solving for x, we find

$$x = \tan\left(\tan^{-1}t + C\right).$$

We can simplify this expression by applying the sum formula for the tangent function:

$$x = \frac{\tan(\tan^{-1}t) + \tan C}{1 - \tan(\tan^{-1}t)\tan C} = \frac{t + \tan C}{1 - t\tan C} = \frac{t + A}{1 - At},$$

where $A = \tan C$ is an arbitrary constant.

25. $y' = x\sec y$

SOLUTION Rewrite

$$\frac{dy}{dx} = x \sec y \qquad \text{as} \qquad \cos y \, dy = x \, dx.$$

Integrating both sides of this equation yields

$$\int \cos y \, dy = \int x \, dx$$

$$\sin y = \frac{1}{2}x^2 + C.$$

Solving for y, we find

$$y = \sin^{-1}\left(\frac{1}{2}x^2 + C\right),$$

where C is an arbitrary constant.

27. $\dfrac{dy}{dt} = y \tan t$

SOLUTION Rewrite

$$\frac{dy}{dt} = y \tan t \qquad \text{as} \qquad \frac{1}{y} \, dy = \tan t \, dt.$$

Integrating both sides of this equation yields

$$\int \frac{1}{y} \, dy = \int \tan t \, dt$$

$$\ln |y| = \ln |\sec t| + C.$$

Solving for y, we find

$$|y| = e^{\ln |\sec t| + C} = e^C \, |\sec t|$$

$$y = \pm e^C \sec t = A \sec t,$$

where $A = \pm e^C$ is an arbitrary constant.

In Exercises 29–41, solve the initial value problem.

29. $y' + 2y = 0, \quad y(\ln 2) = 3$

SOLUTION First, we find the general solution of the differential equation. Rewrite

$$\frac{dy}{dx} + 2y = 0 \qquad \text{as} \qquad \frac{1}{y} \, dy = -2 \, dx,$$

and then integrate to obtain

$$\ln |y| = -2x + C.$$

Thus,

$$y = Ae^{-2x},$$

where $A = \pm e^C$ is an arbitrary constant. The initial condition $y(\ln 2) = 3$ allows us to determine the value of A.

$$3 = Ae^{-2(\ln 2)}; \quad 3 = A\frac{1}{4}; \quad \text{so} \quad 12 = A.$$

Finally,

$$y = 12e^{-2x}.$$

31. $yy' = xe^{-y^2}, \quad y(0) = -1$

SOLUTION   First, we find the general solution of the differential equation. Rewrite

$$y \frac{dy}{dx} = xe^{-y^2} \quad \text{as} \quad ye^{y^2} \, dy = x \, dx,$$

and then integrate to obtain

$$\frac{1}{2} e^{y^2} = \frac{1}{2} x^2 + C.$$

Thus,

$$y = \pm \sqrt{\ln(x^2 + A)},$$

where $A = 2C$ is an arbitrary constant. The initial condition $y(0) = -1$ allows us to determine the value of A. Since $y(0) < 0$, we have $y = -\sqrt{\ln(x^2 + A)}$, and

$$-1 = -\sqrt{\ln(A)}; \quad 1 = \ln(A); \quad \text{so} \quad e = A.$$

Finally,

$$y = -\sqrt{\ln(x^2 + e)}.$$

33. $y' = (x - 1)(y - 2), \quad y(0) = 3$

SOLUTION   First, we find the general solution of the differential equation. Rewrite

$$\frac{dy}{dx} = (x - 1)(y - 2) \quad \text{as} \quad \frac{1}{y - 2} \, dy = (x - 1) \, dx,$$

and then integrate to obtain

$$\ln|y - 2| = \frac{1}{2} x^2 - x + C.$$

Thus,

$$y = Ae^{(1/2)x^2 - x} + 2,$$

where $A = \pm e^C$ is an arbitrary constant. The initial condition $y(0) = 3$ allows us to determine the value of A.

$$3 = Ae^0 + 2 = A + 2 \quad \text{so} \quad A = 1.$$

Finally,

$$y = e^{(1/2)x^2 - x} + 2.$$

35. $\dfrac{dy}{dt} = ye^{-t}, \quad y(0) = 1$

SOLUTION   First, we find the general solution of the differential equation. Rewrite

$$\frac{dy}{dt} = ye^{-t} \quad \text{as} \quad \frac{1}{y} \, dy = e^{-t} \, dt,$$

and then integrate to obtain

$$\ln|y| = -e^{-t} + C.$$

Thus,

$$y = Ae^{-e^{-t}},$$

where $A = \pm e^C$ is an arbitrary constant. The initial condition $y(0) = 1$ allows us to determine the value of A.

$$1 = Ae^{-1} \quad \text{so} \quad A = e.$$

Finally,

$$y = (e)e^{-e^{-t}} = e^{1 - e^{-t}}.$$

37. $t^2 \dfrac{dy}{dt} - t = 1 + y + ty, \quad y(1) = 0$

SOLUTION First, we find the general solution of the differential equation. Rewrite

$$t^2 \frac{dy}{dt} = 1 + t + y + ty = (1 + t)(1 + y)$$

as

$$\frac{1}{1 + y} \, dy = \frac{1 + t}{t^2} \, dt,$$

and then integrate to obtain

$$\ln|1 + y| = -t^{-1} + \ln|t| + C.$$

Thus,

$$y = A \frac{t}{e^{1/t}} - 1,$$

where $A = \pm e^C$ is an arbitrary constant. The initial condition $y(1) = 0$ allows us to determine the value of A.

$$0 = A\left(\frac{1}{e}\right) - 1 \quad \text{so} \quad A = e.$$

Finally,

$$y = \frac{et}{e^{1/t}} - 1.$$

39. $\sqrt{1 - x^2} \, y' = y^2, \quad y(0) = 1$

SOLUTION First, we find the general solution of the differential equation. Rewrite

$$\sqrt{1 - x^2} \frac{dy}{dx} = y^2 \quad \text{as} \quad \frac{dy}{y^2} = \frac{dx}{\sqrt{1 - x^2}},$$

and then integrate to obtain

$$-\frac{1}{y} = \sin^{-1} x + C.$$

Thus,

$$y = -\frac{1}{\sin^{-1} x + C},$$

where C is an arbitrary constant. The initial condition $y(0) = 1$ allows us to determine the value of C.

$$1 = -\frac{1}{\sin^{-1} 0 + C} = -\frac{1}{C} \quad \text{so} \quad C = -1.$$

Finally,

$$y = -\frac{1}{\sin^{-1} x - 1} = \frac{1}{1 - \sin^{-1} x}.$$

41. $y' = y^2 \sin x, \quad y(0) = 3$

SOLUTION First, we find the general solution of the differential equation. Rewrite

$$\frac{dy}{dx} = y^2 \sin x \quad \text{as} \quad y^{-2} \, dy = \sin x \, dx,$$

and then integrate to obtain

$$-y^{-1} = -\cos x + C.$$

Thus,

$$y = \frac{1}{A + \cos x},$$

where $A = -C$ is an arbitrary constant. The initial condition $y(0) = 3$ allows us to determine the value of A.

$$3 = \frac{1}{A+1}; \quad A + 1 = \frac{1}{3} \quad \text{so} \quad A = \frac{1}{3} - 1 = -\frac{2}{3}.$$

Finally,

$$y = \frac{1}{\cos x - (2/3)}.$$

43. Find all values of a such that $y = e^{ax}$ is a solution of

$$y'' + 2y' - 8y = 0$$

SOLUTION   Let $y = e^{ax}$. Then

$$y' = ae^{ax} \quad \text{and} \quad y'' = a^2 e^{ax}.$$

Substituting into the differential equation, we find

$$y'' + 2y' - 8y = e^{ax}(a^2 + 2a - 8).$$

Because e^{ax} is never zero, $y'' + 2y' - 8y = 0$ if only if $a^2 + 2a - 8 = (a + 4)(a - 2) = 0$. Hence, $y = e^{ax}$ is a solution of the differential equation $y'' + 2y' - 8y = 0$ provided $a = -4$ or $a = 2$.

In Exercises 45–48, use Eq. (4) and Torricelli's Law [Eq. (5)].

45. A cylindrical tank filled with water has height 10 ft and a base of area 30 ft^2. Water leaks through a hole in the bottom of area $\frac{1}{3}$ ft^2. How long does it take (a) for half of the water to leak out and (b) for the tank to empty?

SOLUTION   Because the tank has a constant cross-sectional area of 30 ft^2 and the hole has an area of $\frac{1}{3}$ ft^2, the differential equation for the height of the water in the tank is

$$\frac{dy}{dt} = \frac{\frac{1}{3}v}{30} = \frac{v}{90}.$$

By Torricelli's Law,

$$v = -\sqrt{2gy} = -8\sqrt{y},$$

using $g = 32$ ft/s^2. Thus,

$$\frac{dy}{dt} = -\frac{4}{45}\sqrt{y}.$$

Separating variables and then integrating yields

$$y^{-1/2}\, dy = -\frac{4}{45}\, dt$$

$$2y^{1/2} = -\frac{4}{45}t + C$$

Solving for y, we find

$$y(t) = \left(C - \frac{2}{45}t\right)^2.$$

Since the tank is originally full, we have the initial condition $y(0) = 10$, whence $\sqrt{10} = C$. Therefore,

$$y(t) = \left(\sqrt{10} - \frac{2}{45}t\right)^2.$$

When half of the water is out of the tank, $y = 5$, so we solve:

$$5 = \left(\sqrt{10} - \frac{2}{45}t\right)^2$$

for t, finding

$$t = \frac{45}{2}(\sqrt{10} - \sqrt{5}) \approx 20.84 \text{ sec.}$$

When all of the water is out of the tank, $y = 0$, so

$$\sqrt{10} - \frac{2}{45}t = 0 \quad \text{and} \quad t = \frac{45}{2}\sqrt{10} \approx 71.15 \text{ sec.}$$

47. The tank in Figure 7(B) is a cylinder of radius 10 ft and length 40 ft. Assume that the tank is half-filled with water and that water leaks through a hole in the bottom of area $B = 3$ in.^{2}. Determine the water level $y(t)$ and the time t_e when the tank is empty.

SOLUTION When the water is at height y over the bottom, the top cross section is a rectangle with length 40 ft, and with width x satisfying the equation:

$$(x/2)^2 + (y - 10)^2 = 100.$$

Thus, $x = 2\sqrt{20y - y^2}$, and

$$A(y) = 40x = 80\sqrt{20y - y^2}.$$

With $B = 3$ in$^2 = \frac{1}{48}$ ft^2 and $v = -\sqrt{2gy} = -8\sqrt{y}$, it follows that

$$\frac{dy}{dt} = -\frac{\frac{1}{48}8\sqrt{y}}{80\sqrt{20y - y^2}} = -\frac{1}{480\sqrt{20 - y}}.$$

Separating variables and integrating then yields:

$$\sqrt{20 - y}\, dy = -\frac{1}{480}\, dt$$

$$-\frac{2}{3}(20 - y)^{3/2} = -\frac{t}{480} + C$$

When $t = 0$, $y = 10$, so $C = -\frac{2}{3}10^{3/2}$, and

$$-\frac{2}{3}(20 - y)^{3/2} = -\frac{t}{480} - \frac{2}{3}10^{3/2}$$

$$y(t) = 20 - \left(\frac{t}{320} + 10^{3/2}\right)^{2/3}.$$

The tank is empty when $y = 0$. Thus, t_e satisfies the equation

$$20 - \left(\frac{t_e}{320} + 10^{3/2}\right)^{2/3} = 0.$$

It follows that

$$t_e = 320(20^{3/2} - 10^{3/2}) \approx 18502.4 \text{ seconds.}$$

49. Figure 8 shows a circuit consisting of a resistor of R ohms, a capacitor of C farads, and a battery of voltage V. When the circuit is completed, the amount of charge $q(t)$ (in coulombs) on the plates of the capacitor varies according to the differential equation (t in seconds)

$$R\frac{dq}{dt} + \frac{1}{C}q = V$$

(a) Solve for $q(t)$.

(b) Show that $\lim_{t \to \infty} q(t) = CV$.

(c) Find $q(t)$, assuming that $q(0) = 0$. Show that the capacitor charges to approximately 63% of its final value CV after a time period of length $\tau = RC$ (τ is called the time constant of the capacitor).

FIGURE 8 An RC circuit.

SOLUTION

(a) Upon rearranging the terms of the differential equation, we have

$$\frac{dq}{dt} = -\frac{q - CV}{RC}.$$

Separating the variables and integrating both sides, we obtain

$$\frac{dq}{q - CV} = -\frac{dt}{RC}$$

$$\int \frac{dq}{q - CV} = -\int \frac{dt}{RC}$$

and

$$\ln|q - CV| = -\frac{t}{RC} + k,$$

where k is an arbitrary constant. Solving for $q(t)$ yields

$$q(t) = CV + Ke^{-\frac{1}{RC}t},$$

where $K = \pm e^k$.

(b) Using the result from part (a), we calculate

$$\lim_{t \to \infty} q(t) = \lim_{t \to \infty}\left(CV + Ke^{-\frac{1}{RC}t}\right) = CV + K\lim_{t \to \infty} e^{-\frac{1}{RC}t} = CV + K \cdot 0 = CV.$$

(c) Using the result from part (a), the condition $q(0) = 0$ determines $K = -CV$. Thus,

$$q(t) = CV\left(1 - e^{-\frac{1}{RC}t}\right),$$

and

$$q(\tau) = q(RC) = CV(1 - e^{-1}) \approx 0.632CV.$$

51. One hypothesis for the growth rate of the volume V of a cell is that $\dfrac{dV}{dt}$ is proportional to the cell's surface area A. Since V has cubic units such as cm^3 and A has square units such as cm^2, we may assume roughly that $A \propto V^{2/3}$, and hence $\dfrac{dV}{dt} = kV^{2/3}$ for some constant k. If this hypothesis is correct, which dependence of volume on time would we expect to see (again, roughly speaking) in the laboratory?

(a) Linear   **(b)** Quadratic   **(c)** Cubic

SOLUTION Rewrite

$$\frac{dV}{dt} = kV^{2/3} \qquad \text{as} \qquad V^{-2/3}\, dv = k\, dt,$$

and then integrate both sides to obtain

$$3V^{1/3} = kt + C$$

$$V = (kt/3 + C)^3.$$

Thus, we expect to see V increasing roughly like the cube of time.

53. In general, $(fg)'$ is not equal to $f'g'$, but let $f(x) = e^{2x}$ and find a function $g(x)$ such that $(fg)' = f'g'$. Do the same for $f(x) = x$.

SOLUTION If $(fg)' = f'g'$, we have

$$f'(x)g(x) + g'(x)f(x) = f'(x)g'(x)$$

$$g'(x)(f(x) - f'(x)) = -g(x)f'(x)$$

$$\frac{g'(x)}{g(x)} = \frac{f'(x)}{f'(x) - f(x)}$$

Now, let $f(x) = e^{2x}$. Then $f'(x) = 2e^{2x}$ and

$$\frac{g'(x)}{g(x)} = \frac{2e^{2x}}{2e^{2x} - e^{2x}} = 2.$$

Integrating and solving for $g(x)$, we find

$$\frac{dg}{dg} = 2\,dx$$

$$\ln|g| = 2x + C$$

$$g(x) = Ae^{2x},$$

where $A = \pm e^C$ is an arbitrary constant.

If $f(x) = x$, then $f'(x) = 1$, and

$$\frac{g'(x)}{g(x)} = \frac{1}{1-x}.$$

Thus,

$$\frac{dg}{g} = \frac{1}{1-x}\,dx$$

$$\ln|g| = -\ln|1-x| + C$$

$$g(x) = \frac{A}{1-x},$$

where $A = \pm e^C$ is an arbitrary constant.

55. If a bucket of water spins about a vertical axis with constant angular velocity ω (in radians per second), the water climbs up the side of the bucket until it reaches an equilibrium position (Figure 10). Two forces act on a particle located at a distance x from the vertical axis: the gravitational force $-mg$ acting downward and the force of the bucket on the

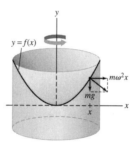

FIGURE 10

particle (transmitted indirectly through the liquid) in the direction perpendicular to the surface of the water. These two forces must combine to supply a centripetal force $m\omega^2 x$, and this occurs if the diagonal of the rectangle in Figure 10 is normal to the water's surface (that is, perpendicular to the tangent line). Prove that if $y = f(x)$ is the equation of the curve obtained by taking a vertical cross section through the axis, then $-1/y' = -g/(\omega^2 x)$. Show that $y = f(x)$ is a parabola.

SOLUTION At any point along the surface of the water, the slope of the tangent line is given by the value of y' at that point; hence, the slope of the line perpendicular to the surface of the water is given by $-1/y'$. The slope of the resultant force generated by the gravitational force and the centrifugal force is

$$\frac{-mg}{m\omega^2 x} = -\frac{g}{\omega^2 x}.$$

Therefore, the curve obtained by taking a vertical cross-section of the water surface is determined by the equation

$$-\frac{1}{y'} = -\frac{g}{\omega^2 x} \qquad \text{or} \qquad y' = \frac{\omega^2}{g}x.$$

Performing one integration yields

$$y = f(x) = \frac{\omega^2}{g}x^2 + C,$$

where C is a constant of integration. Thus, $y = f(x)$ is a parabola.

57. Find the family of curves satisfying $y' = x/y$ and sketch several members of the family. Then find the differential equation for the orthogonal family (see Exercise 56), find its general solution, and add some members of this orthogonal family to your plot.

SOLUTION Separation of variables and integration applied to $y' = x/y$ gives

$$y\,dy = x\,dx$$

$$\frac{1}{2}y^2 = \frac{1}{2}x^2 + C$$

$$y = \pm\sqrt{x^2 + C}$$

If $y(x)$ is a curve of the family orthogonal to these, it must have tangent lines of slope $-y/x$ at every point (x, y). This gives

$$y' = -y/x$$

Separation of variables and integration give

$$\frac{dy}{y} = -\frac{dx}{x}$$

$$\ln|y| = -\ln|x| + C$$

$$y = \frac{A}{x}$$

Several solution curves of both differential equations appear below:

59. Let $v(t)$ be the velocity of an object of mass m in free fall near the earth's surface. If we assume that air resistance is proportional to v^2, then v satisfies the differential equation $m\dfrac{dv}{dt} = -g + kv^2$ for some constant $k > 0$.

(a) Set $\alpha = (g/k)^{1/2}$ and rewrite the differential equation as

$$\frac{dv}{dt} = -\frac{k}{m}(\alpha^2 - v^2)$$

Then solve using separation of variables with initial condition $v(0) = 0$.

(b) Show that the terminal velocity $\lim\limits_{t\to\infty} v(t)$ is equal to $-\alpha$.

SOLUTION

(a) Let $\alpha = (g/k)^{1/2}$. Then

$$\frac{dv}{dt} = -\frac{g}{m} + \frac{k}{m}v^2 = -\frac{k}{m}\left(\frac{g}{k} - v^2\right) = -\frac{k}{m}\left(\alpha^2 - v^2\right)$$

Separating variables and integrating yields

$$\int \frac{dv}{\alpha^2 - v^2} = -\frac{k}{m}\int dt = -\frac{k}{m}t + C$$

We now use partial fraction decomposition for the remaining integral to obtain

$$\int \frac{dv}{\alpha^2 - v^2} = \frac{1}{2\alpha}\int\left(\frac{1}{\alpha + v} + \frac{1}{\alpha - v}\right)dv = \frac{1}{2\alpha}\ln\left|\frac{\alpha + v}{\alpha - v}\right|$$

Therefore,

$$\frac{1}{2\alpha}\ln\left|\frac{\alpha + v}{\alpha - v}\right| = -\frac{k}{m}t + C.$$

The initial condition $v(0) = 0$ allows us to determine the value of C:

$$\frac{1}{2\alpha}\ln\left|\frac{\alpha + 0}{\alpha - 0}\right| = -\frac{k}{m}(0) + C$$

$$C = \frac{1}{2\alpha} \ln 1 = 0.$$

Finally, solving for v, we find

$$v(t) = -\alpha \left(\frac{1 - e^{-2(\sqrt{gk}/m)t}}{1 + e^{-2(\sqrt{gk}/m)t}} \right).$$

(b) As $t \to \infty$, $e^{-2(\sqrt{gk}/m)t} \to 0$, so

$$v(t) \to -\alpha \left(\frac{1-0}{1+0} \right) = -\alpha.$$

Further Insights and Challenges

61. In most cases of interest, the general solution of a differential equation of order n depends on n arbitrary constants. This exercise shows there are exceptions.

(a) Show that $(y')^2 + y^2 = 0$ is a first-order equation with only one solution $y = 0$.

(b) Show that $(y')^2 + y^2 + 1 = 0$ is a first-order equation with no solutions.

SOLUTION

(a) $(y')^2 + y^2 \geq 0$ and equals zero if and only if $y' = 0$ and $y = 0$

(b) $(y')^2 + y^2 + 1 \geq 1 > 0$ for all y' and y, so $(y')^2 + y^2 + 1 = 0$ has no solution

63. A spherical tank of radius R is half-filled with water. Suppose that water leaks through a hole in the bottom of area B ft^2. Let $y(t)$ be the water level at time t (seconds).

(a) Show that $\dfrac{dy}{dt} = \dfrac{-8B\sqrt{y}}{\pi(2Ry - y^2)}$.

(b) Show that for some constant C,

$$\frac{\pi}{4B} \left(\frac{2}{3} Ry^{3/2} - \frac{1}{5} y^{5/2} \right) = C - t$$

(c) Use the initial condition $y(0) = R$ to compute C and show that $C = t_e$, the time at which the tank is empty.

(d) Show that t_e is proportional to $R^{5/2}$ and inversely proportional to B.

SOLUTION

(a) At height y above the bottom of the tank, the cross section is a circle of radius

$$r = \sqrt{R^2 - (R - y)^2} = \sqrt{2Ry - y^2}.$$

The cross-sectional area function is then $A(y) = \pi(2Ry - y^2)$. The differential equation for the height of the water in the tank is then

$$\frac{dy}{dt} = -\frac{8B\sqrt{y}}{\pi(2Ry - y^2)}.$$

(b) Rewrite the differential equation as

$$\frac{\pi}{8B} \left(2Ry^{1/2} - y^{3/2} \right) dy = -dt,$$

and then integrate both sides to obtain

$$\frac{\pi}{4B} \left(\frac{2}{3} Ry^{3/2} - \frac{1}{5} y^{5/2} \right) = C - t,$$

where C is an arbitrary constant.

(c) The initial condition $y(0) = R$ allows us to determine the value of C:

$$C = \frac{\pi}{4B} \left(\frac{2}{3} R^{5/2} - \frac{1}{5} R^{5/2} \right) = \frac{7\pi}{60B} R^{5/2}.$$

Moreover, note that $y = 0$ when $t = C$, $C = t_e$, the time at which the tank is empty.

(d) From part (c),

$$t_e = \frac{7\pi}{60B} R^{5/2},$$

from which it is clear that t_e is proportional to $R^{5/2}$ and inversely proportional to B.

10.2 Graphical and Numerical Methods

Preliminary Questions

1. What is the slope of the segment in the slope field for $\dot{y} = ty + 1$ at the point $(2, 3)$?

SOLUTION The slope of the segment in the slope field for $\dot{y} = ty + 1$ at the point $(2, 3)$ is $(2)(3) + 1 = 7$.

2. What is the equation of the isocline of slope $c = 1$ for $\dot{y} = y^2 - t$?

SOLUTION The isocline of slope $c = 1$ has equation $y^2 - t = 1$, or $y = \pm\sqrt{1 + t}$.

3. True or false? In the slope field for $\dot{y} = \ln y$, the slopes at points on a vertical line $t = C$ are all equal.

SOLUTION This statement is true because the right-hand side of the differential equation does not explicitly depend on the variable t.

4. What about the slope field for $\dot{y} = \ln t$? Are the slopes at points on a vertical line $t = C$ all equal?

SOLUTION No, because the right-hand side of this differential equation does explicitly depend on the variable t.

5. Let $y(t)$ be the solution to $\dot{y} = F(t, y)$ with $y(1) = 3$. How many iterations of Euler's Method are required to approximate $y(3)$ if the time step is $h = 0.1$?

SOLUTION The initial condition is specified at $t = 1$ and we want to obtain an approximation to the value of the solution at t = 3. With a time step of $h = 0.1$,

$$\frac{3 - 1}{0.1} = 20$$

iterations of Euler's method are required.

Exercises

1. Figure 8 shows the slope field for $\dot{y} = \sin y \sin t$. Sketch the graphs of the solutions with initial conditions $y(0) = 1$ and $y(0) = -1$. Show that $y(t) = 0$ is a solution and add its graph to the plot.

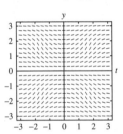

FIGURE 8 Slope field for $\dot{y} = \sin y \sin t$.

SOLUTION The sketches of the solutions appear below.

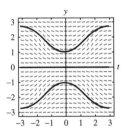

If $y(t) = 0$, then $y' = 0$; moreover, $\sin 0 \sin t = 0$. Thus, $y(t) = 0$ is a solution of $\dot{y} = \sin y \sin t$.

3. Show that $f(t) = \frac{1}{2}(t - \frac{1}{2})$ is a solution to $\dot{y} = t - 2y$. Sketch the four solutions with $y(0) = \pm 0.5, \pm 1$ on the slope field in Figure 10. The slope field suggests that every solution approaches $f(t)$ as $t \to \infty$. Confirm this by showing that $y = f(t) + Ce^{-2t}$ is the general solution.

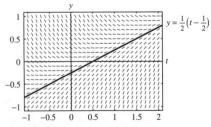

FIGURE 10 Slope field for $\dot{y} = t - 2y$.

SOLUTION Let $y = f(t) = \frac{1}{2}(t - \frac{1}{2})$. Then $\dot{y} = \frac{1}{2}$ and

$$\dot{y} + 2y = \frac{1}{2} + t - \frac{1}{2} = t,$$

so $f(t) = \frac{1}{2}(t - \frac{1}{2})$ is a solution to $\dot{y} = t - 2y$. The slope field with the four required solutions is shown below.

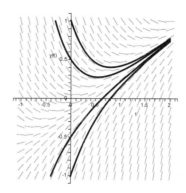

Now, let $y = f(t) + Ce^{-2t} = \frac{1}{2}(t - \frac{1}{2}) + Ce^{-2t}$. Then

$$\dot{y} = \frac{1}{2} - 2Ce^{-2t},$$

and

$$\dot{y} + 2y = \frac{1}{2} - 2Ce^{-2t} + \left(t - \frac{1}{2}\right) + 2Ce^{-2t} = t.$$

Thus, $y = f(t) + Ce^{-2t}$ is the general solution to the equation $\dot{y} = t - 2y$.

5. Show that the isoclines of $\dot{y} = 1/y$ are horizontal lines. Sketch the slope field for $-2 \le t, y \le 2$ and plot the solutions with initial conditions $y(0) = 0$ and $y(0) = 1$.

SOLUTION The isocline of slope c is defined by $\frac{1}{y} = c$. This is equivalent to $y = \frac{1}{c}$, which is a horizontal line. The slope field and the solutions are shown below.

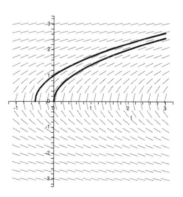

7. Sketch the slope field of $\dot{y} = ty$ for $-2 \le t, y \le 2$. Based on the sketch, determine $\lim_{t \to \infty} y(t)$, where $y(t)$ is a solution with $y(0) > 0$. What is $\lim_{t \to \infty} y(t)$ if $y(0) < 0$?

SOLUTION The slope field for $\dot{y} = ty$ is shown below.

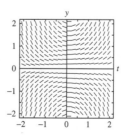

With $y(0) > 0$, the slope field indicates that y is an always increasing, always concave up function; consequently, $\lim_{t \to \infty} y = \infty$. On the other hand, when $y(0) < 0$, the slope field indicates that y is an always decreasing, always concave down function; consequently, $\lim_{t \to \infty} y = -\infty$.

9. One of the slope fields in Figures 12(A) and (B) is the slope field for $\dot{y} = t^2$. The other is for $\dot{y} = y^2$. Identify which is which. In each case, sketch the solutions with initial conditions $y(0) = 1$, $y(0) = 0$, and $y(0) = -1$.

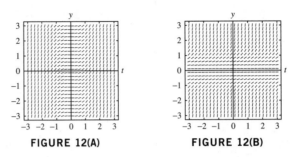

FIGURE 12(A)   **FIGURE 12(B)**

SOLUTION   For $y' = t^2$, y' only depends on t. The isoclines of any slope c will be the two vertical lines $t = \pm\sqrt{c}$. This indicates that the slope field will be the one given in Figure 12(A). The solutions are sketched below:

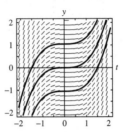

For $y' = y^2$, y' only depends on y. The isoclines of any slope c will be the two *horizontal* lines $y = \pm\sqrt{c}$. This indicates that the slope field will be the one given in Figure 12(B). The solutions are sketched below:

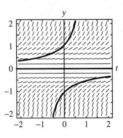

11. Sketch the slope field of $\dot{y} = t^2 - y$ in the region $-3 \leq t \leq 3$, $-3 \leq y \leq 3$ and sketch the solutions satisfying $y(1) = 0$, $y(1) = 1$, and $y(1) = -1$.

SOLUTION   The slope field for $\dot{y} = t^2 - y$, together with the required solution curves, is shown below.

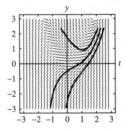

13. Let $y(t)$ be the solution to $\dot{y} = te^{-y}$ satisfying $y(0) = 0$.

(a) Use Euler's Method with time step $h = 0.1$ to approximate $y(0.1)$, $y(0.2)$, ... , $y(1)$.

(b) Use separation of variables to find $y(t)$ exactly.

(c) Compute the error in the approximations to $y(0.1)$, $y(0.5)$, and $y(1)$.

SOLUTION

(a) With $y_0 = 0$, $t_0 = 0$, $h = 0.1$, and $F(t, y) = te^{-y}$, we compute

n	t_n	y_n
0	0	0
1	0.1	$y_0 + hF(t_0, y_0) = 0$
2	0.2	$y_1 + hF(t_1, y_1) = 0.01$
3	0.3	$y_2 + hF(t_2, y_2) = 0.029801$
4	0.4	$y_3 + hF(t_3, y_3) = 0.058920$
5	0.5	$y_4 + hF(t_4, y_4) = 0.096631$
6	0.6	$y_5 + hF(t_5, y_5) = 0.142026$
7	0.7	$y_6 + hF(t_6, y_6) = 0.194082$
8	0.8	$y_7 + hF(t_7, y_7) = 0.251733$
9	0.9	$y_8 + hF(t_8, y_8) = 0.313929$
10	1.0	$y_9 + hF(t_9, y_9) = 0.379681$

(b) Rewrite

$$\frac{dy}{dt} = te^{-y} \quad \text{as} \quad e^y \, dy = t \, dt,$$

and then integrate both sides to obtain

$$e^y = \frac{1}{2}t^2 + C.$$

Thus,

$$y = \ln \left| \frac{1}{2}t^2 + C \right|.$$

Applying the initial condition $y(0) = 0$ yields $0 = \ln |C|$, so $C = 1$. The exact solution to the initial value problem is then $y = \ln \left(\frac{1}{2}t^2 + 1 \right)$.

(c) The three errors requested are computed here:

$$|y(0.1) - y_1| = |0.00498754 - 0| = 0.00498754;$$

$$|y(0.5) - y_5| = |0.117783 - 0.0966314| = 0.021152;$$

$$|y(1) - y_{10}| = |0.405465 - 0.379681| = 0.025784.$$

In Exercises 14–19, use Euler's Method with $h = 0.1$ to approximate the given value of $y(t)$.

15. $y(1)$; $\dot{y} = y$, $y(0) = 0$

SOLUTION   Let $F(t, y) = y$. With $t_0 = 0$, $y_0 = 0$, and $h = 0.1$, we compute

$$y(0.0) = y_0 \qquad\qquad\qquad = 0$$
$$y(0.1) \approx y_1 = y_0 + 0.1(y_0) = 0$$
$$y(0.2) \approx y_2 = y_1 + 0.1(y_1) = 0$$
$$y(0.3) \approx y_3 = y_2 + 0.1(y_2) = 0$$
$$y(0.4) \approx y_4 = y_3 + 0.1(y_3) = 0$$
$$y(0.5) \approx y_5 = y_4 + 0.1(y_4) = 0$$
$$y(0.6) \approx y_6 = y_5 + 0.1(y_5) = 0$$
$$y(0.7) \approx y_7 = y_6 + 0.1(y_6) = 0$$
$$y(0.8) \approx y_8 = y_7 + 0.1(y_7) = 0$$
$$y(0.9) \approx y_9 = y_8 + 0.1(y_8) = 0$$
$$y(1.0) \approx y_{10} = y_9 + 0.1(y_9) = 0$$

17. $y(1.5)$; $\dot{y} = t \sin y$, $y(1) = 2$

SOLUTION Let $F(t, y) = t \sin y$. With $t_0 = 1$, $y_0 = 2$ and $h = 0.1$, we compute

$$y(1.0) = y_0 \qquad\qquad\qquad = 2$$
$$y(1.1) \approx y_0 + 0.1(t_0 \sin(y_0)) = 2.09093$$
$$y(1.2) \approx y_1 + 0.1(t_1 \sin(y_1)) = 2.18638$$
$$y(1.3) \approx y_2 + 0.1(t_2 \sin(y_2)) = 2.28435$$
$$y(1.4) \approx y_3 + 0.1(t_3 \sin(y_3)) = 2.38264$$
$$y(1.5) \approx y_4 + 0.1(t_4 \sin(y_4)) = 2.47898$$

19. $y(0.5);$ $\quad \dot{y} = t - y, \quad y(0) = 1$

SOLUTION Let $F(t, y) = t - y$. With $t_0 = 0$, $y_0 = 1$, and $h = 0.1$, we compute

$$y(0.0) = y_0 \qquad\qquad\qquad = 1$$
$$y(0.1) \approx y_1 = y_0 + 0.1(t_0 - y_0) = 0.9$$
$$y(0.2) \approx y_2 = y_1 + 0.1(t_1 - y_1) = 0.82$$
$$y(0.3) \approx y_3 = y_2 + 0.1(t_2 - y_2) = 0.758$$
$$y(0.4) \approx y_4 = y_3 + 0.1(t_3 - y_3) = 0.7122$$
$$y(0.5) \approx y_5 = y_4 + 0.1(t_4 - y_4) = 0.68098$$

Further Insights and Challenges

In Exercises 21–22, use a modification of Euler's Method, Euler's Midpoint Method, that gives a significant improvement in accuracy. With time step h and initial value $y_0 = y(t_0)$, the values y_k are defined successively by the equation

$$y_k = y_{k-1} + h m_{k-1}$$

where $m_{k-1} = F\left(t_{k-1} + \dfrac{h}{2}, y_{k-1} + \dfrac{h}{2} F(t_{k-1}, y_{k-1})\right).$

21. Apply both Euler's Method and the Euler Midpoint Method with $h = 0.1$ to estimate $y(1.5)$, where $y(t)$ satisfies $\dot{y} = y$ with $y(0) = 1$. Find $y(t)$ exactly and compute the errors in these two approximations.

SOLUTION Let $F(t, y) = y$. With $t_0 = 0$, $y_0 = 1$, and $h = 0.1$, fifteen iterations of Euler's method yield

$$y(1.5) \approx y_{15} = 4.177248.$$

The Euler midpoint approximation with $F(t, y) = y$ is

$$m_{k-1} = F\left(t_{k-1} + \frac{h}{2}, y_{k-1} + \frac{h}{2} F(t_{k-1}, y_{k-1})\right) = y_{k-1} + \frac{h}{2} y_{k-1}$$

$$y_k = y_{k-1} + h\left(y_{k-1} + \frac{h}{2} y_{k-1}\right) = y_{k-1} + h y_{k-1} + \frac{h^2}{2} y_{k-1}$$

Fifteen iterations of Euler's midpoint method yield:

$$y(1.5) \approx y_{15} = 4.471304.$$

The exact solution to $y' = y$, $y(0) = 1$ is $y(t) = e^t$; therefore $y(1.5) = 4.481689$. The error from Euler's method is $|4.177248 - 4.481689| = 0.304441$, while the error from Euler's midpoint method is $|4.471304 - 4.481689| = 0.010385$.

In Exercises 23–26, use Euler's Midpoint Method with $h = 0.1$ to approximate the given value of $y(t)$.

23. $y(0.5);$ $\quad \dot{y} = y^2, \quad y(0) = 0$

SOLUTION Let $F(t, y) = y^2$. With $t_0 = 0$, $y_0 = 0$, and $h = 0.1$, we compute

k	t_k	m_k	y_k
0	0.0	0	0
1	0.1	0	0
2	0.2	0	0
3	0.3	0	0
4	0.4	0	0
5	0.5	0	0

Thus, $y(0.5) \approx y_5 = 0$.

25. $y(1)$; $\dot{y} = t + y$, $y(0) = 0$

SOLUTION Let $F(t, y) = t + y$. With $t_0 = 0$, $y_0 = 0$, and $h = 0.1$, we compute

k	t_k	m_k	y_k
0	0.0	0	0
1	0.1	0.05	0.005
2	0.2	0.16025	0.021025
3	0.3	0.282076	0.0492326
4	0.4	0.416694	0.0909021
5	0.5	0.565447	0.147447
6	0.6	0.729819	0.220429
7	0.7	0.911450	0.311574
8	0.8	1.112152	0.422789
9	0.9	1.333928	0.556182
10	1.0	1.578991	0.714081

Hence $y(1) \approx y_{10} = 0.714081$.

10.3 The Logistic Equation

Preliminary Questions

1. Which of the following is a logistic differential equation?

(a) $\dot{y} = 2y(1 - y^2)$

(b) $\dot{y} = 2y\left(1 - \dfrac{y}{3}\right)$

(c) $\dot{y} = 2y\left(1 - \dfrac{x}{4}\right)$

(d) $\dot{y} = 2y(1 - 3y)$

SOLUTION The differential equations in **(b)** and **(d)** are logistic equations. The equation in **(a)** is not a logistic equation because of the y^2 term inside the parentheses on the right-hand side; the equation in **(c)** is not a logistic equation because of the presence of the independent variable on the right-hand side.

2. True or false? The logistic equation is linear.

SOLUTION False, the logistic equation is not a linear differential equation.

3. True or false? The logistic equation is separable.

SOLUTION True, the logistic equation is a separable differential equation.

4. Let $y(t)$ be a solution to $\dot{y} = 4y(3 - y)$. What is $\lim_{t \to \infty} y(t)$ in the following three cases:

(a) $y(0) = 3$

(b) $y(0) = 4$

(c) $y(0) = -2$

SOLUTION

(a) If $y(0) = 3$, then $\dot{y} = 0$, and $y(t) = 3$ for all t. Thus, $\lim_{t \to \infty} y(t) = 3$.

(b) If $y(0) = 4$, then $\dot{y} < 0$, and $\lim_{t \to \infty} y(t) = 3$.

(c) If $y(0) = -2$, then $\dot{y} < 0$, and $\lim_{t \to \infty} y(t) \to -\infty$.

Exercises

1. Find the general solution of the logistic equation

$$\dot{y} = 3y(1 - y/5)$$

Then find the particular solution satisfying $y(0) = 2$.

SOLUTION $\dot{y} = 3y(1 - y/5)$ is a logistic equation with $k = 3$ and $A = 5$; therefore, the general solution is

$$y = \frac{5}{1 - e^{-3t}/C}.$$

The initial condition $y(0) = 2$ allows us to determine the value of C:

$$2 = \frac{5}{1 - 1/C}; \quad 1 - \frac{1}{C} = \frac{5}{2}; \quad \text{so} \quad C = -\frac{2}{3}.$$

The particular solution is then

$$y = \frac{5}{1 + \frac{3}{2}e^{-3t}} = \frac{10}{2 + 3e^{-3t}}.$$

3. Let $y(t)$ be a solution of $\dot{y} = 0.5y(1 - 0.5y)$ such that $y(0) = 4$. Determine $\lim_{t \to \infty} y(t)$ without finding $y(t)$ explicitly.

SOLUTION The equation is a logistic equation with $k = 0.5$ and $A = 2$. Let $F(y) = 0.5y(1 - 0.5y)$, and take $y_0 = 4$. Then, $F(y_0) = 0.5(4)(1 - 0.5(4)) < 0$, so, $y(t)$ is strictly decreasing and

$$\lim_{t \to \infty} y(t) = A = 2.$$

5. A population of squirrels lives in a forest with a carrying capacity of 2,000. Assume logistic growth with growth constant $k = 0.6 \text{ yr}^{-1}$.

(a) Find a formula for the squirrel population $P(t)$, assuming an initial population of 500 squirrels.

(b) How long will it take for the squirrel population to double?

SOLUTION

(a) Since $k = 0.6$ and the carrying capacity is $A = 2000$, the population $P(t)$ of the squirrels satisfies the differential equation

$$P'(t) = 0.6P(t)(1 - P(t)/2000),$$

with general solution

$$P(t) = \frac{2000}{1 - e^{-0.6t}/C}.$$

The initial condition $P(0) = 500$ allows us to determine the value of C:

$$500 = \frac{2000}{1 - 1/C}; \quad 1 - \frac{1}{C} = 4; \quad \text{so} \quad C = -\frac{1}{3}.$$

The formula for the population is then

$$P(t) = \frac{2000}{1 + 3e^{-0.6t}}.$$

(b) The squirrel population will have doubled at the time t where $P(t) = 1000$. This gives

$$1000 = \frac{2000}{1 + 3e^{-0.6t}}; \quad 1 + 3e^{-0.6t} = 2; \quad \text{so} \quad t = \frac{5}{3} \ln 3 \approx 1.83.$$

It therefore takes approximately 1.83 years for the squirrel population to double.

7. Sunset Lake is stocked with 2,000 rainbow trout and after 1 year the population has grown to 4,500. Assuming logistic growth with a carrying capacity of 20,000, find the growth constant k (specify the units) and determine when the population will increase to 10,000.

SOLUTION Since $A = 20000$, the trout population $P(t)$ satisfies the logistic equation

$$P'(t) = kP(t)(1 - P(t)/20000),$$

with general solution

$$P(t) = \frac{20000}{1 - e^{-kt}/C}.$$

The initial condition $P(0) = 2000$ allows us to determine the value of C:

$$2000 = \frac{20000}{1 - 1/C}; \quad 1 - \frac{1}{C} = 10; \quad \text{so} \quad C = -\frac{1}{9}.$$

After one year, we know the population has grown to 4500. Let's measure time in years. Then

$$4500 = \frac{20000}{1 + 9e^{-k}}$$

$$1 + 9e^{-k} = \frac{40}{9}$$

$$e^{-k} = \frac{31}{81}$$

$$k = \ln \frac{81}{31} \approx 0.9605 \text{ years}^{-1}.$$

The population will increase to 10000 at time t where $P(t) = 10000$. This gives

$$10000 = \frac{20000}{1 + 9e^{-0.9605t}}$$

$$1 + 9e^{-0.9605t} = 2$$

$$e^{-0.9605t} = \frac{1}{9}$$

$$t = \frac{1}{0.9605} \ln 9 \approx 2.29 \text{ years.}$$

9. A rumor spreads through a school with 1,000 students. At 8 AM, 80 students have heard the rumor, and by noon, half the school has heard it. Using the logistic model of Exercise 8, determine when 90% of the students will have heard the rumor.

SOLUTION Let $y(t)$ be the proportion of students that have heard the rumor at a time t hours after 8 AM. In the logistic model of Exercise 8, we have a capacity of $A = 1$ (100% of students) and an unknown growth factor of k. Hence,

$$y(t) = \frac{1}{1 - e^{-kt}/C}.$$

The initial condition $y(0) = 0.08$ allows us to determine the value of C:

$$\frac{2}{25} = \frac{1}{1 - 1/C}; \quad 1 - \frac{1}{C} = \frac{25}{2}; \quad \text{so} \quad C = -\frac{2}{23}.$$

so that

$$y(t) = \frac{2}{2 + 23e^{-kt}}.$$

The condition $y(4) = .5$ now allows us to determine the value of k:

$$\frac{1}{2} = \frac{2}{2 + 23e^{-4k}}; \quad 2 + 23e^{-4k} = 4; \quad \text{so} \quad k = \frac{1}{4} \ln \frac{23}{2} \approx 0.6106 \text{ hours}^{-1}.$$

90% of the students have heard the rumor when $y(t) = .9$. Thus

$$\frac{9}{10} = \frac{2}{2 + 23e^{-0.6106t}}$$

$$2 + 23e^{-0.6106t} = \frac{20}{9}$$

$$t = \frac{1}{0.6106} \ln \frac{207}{2} \approx 7.6 \text{ hours.}$$

Thus, 90% of the students have heard the rumor after 7.6 hours, or at 3:36 PM.

11. Let $k = 1$ and $A = 1$ in the logistic equation.
(a) Find the solutions satisfying $y_1(0) = 10$ and $y_2(0) = -1$.
(b) Find the time t when $y_1(t) = 5$.
(c) When does $y_2(t)$ become infinite?

SOLUTION The general solution of the logistic equation with $k = 1$ and $A = 1$ is

$$y(t) = \frac{1}{1 - e^{-t}/C}.$$

(a) Given $y_1(0) = 10$, we find $C = \frac{10}{9}$, and

$$y_1(t) = \frac{1}{1 - \frac{10}{9}e^{-t}} = \frac{10}{10 - 9e^{-t}}.$$

On the other hand, given $y_2(0) = -1$, we find $C = \frac{1}{2}$, and

$$y_2(t) = \frac{1}{1 - 2e^{-t}}.$$

(b) From part (a), we have

$$y_1(t) = \frac{10}{10 - 9e^{-t}}.$$

Thus, $y_1(t) = 5$ when

$$5 = \frac{10}{10 - 9e^{-t}}; \quad 10 - 9e^{-t} = 2; \quad \text{so} \quad t = \ln\frac{9}{8}.$$

(c) From part (a), we have

$$y_2(t) = \frac{1}{1 - 2e^{-t}}.$$

Thus, $y_2(t)$ becomes infinite when

$$1 - 2e^{-t} = 0 \quad \text{or} \quad t = \ln 2.$$

13. A tissue culture grows until it has a maximum area of M cm^2. The area $A(t)$ of the culture at time t may be modeled by the differential equation

$$\dot{A} = k\sqrt{A}\left(1 - \frac{A}{M}\right) \qquad \boxed{8}$$

where k is a growth constant.

(a) By setting $A = u^2$, show that the equation can be rewritten

$$\dot{u} = \frac{1}{2}k\left(1 - \frac{u^2}{M}\right)$$

Then find the general solution using separation of variables.

(b) Show that the general solution to Eq. (8) is

$$A(t) = M\left(\frac{Ce^{(k/\sqrt{M})t} - 1}{Ce^{(k/\sqrt{M})t} + 1}\right)^2 \qquad \boxed{9}$$

SOLUTION

(a) Let $A = u^2$. This gives $\dot{A} = 2u\dot{u}$, so that Eq. (8) becomes:

$$2u\dot{u} = ku\left(1 - \frac{u^2}{M}\right)$$

$$\dot{u} = \frac{k}{2}\left(1 - \frac{u^2}{M}\right)$$

Now, rewrite

$$\frac{du}{dt} = \frac{k}{2}\left(1 - \frac{u^2}{M}\right) \quad \text{as} \quad \frac{du}{1 - u^2/M} = \frac{1}{2}k\,dt.$$

The partial fraction decomposition for the term on the left-hand side is

$$\frac{1}{1 - u^2/M} = \frac{\sqrt{M}}{2}\left(\frac{1}{\sqrt{M} + u} + \frac{1}{\sqrt{M} - u}\right),$$

so after integrating both sides, we obtain

$$\frac{\sqrt{M}}{2} \ln \left| \frac{\sqrt{M} + u}{\sqrt{M} - u} \right| = \frac{1}{2}kt + C.$$

Thus,

$$\frac{\sqrt{M} + u}{\sqrt{M} - u} = Ce^{(k/\sqrt{M})t}$$

$$u(Ce^{(k/\sqrt{M})t} + 1) = \sqrt{M}(Ce^{(k/\sqrt{M})t} - 1)$$

and

$$u = \sqrt{M} \frac{Ce^{(k/\sqrt{M})t} - 1}{Ce^{(k/\sqrt{M})t} + 1}.$$

(b) Recall $A = u^2$. Therefore,

$$A(t) = M \left(\frac{Ce^{(k/\sqrt{M})t} - 1}{Ce^{(k/\sqrt{M})t} + 1} \right)^2.$$

15. Show that if a tissue culture grows according to Eq. (8), then the growth rate reaches a maximum when $A = M/3$.

SOLUTION   According to Equation (8), the growth rate of the tissue culture is $k\sqrt{A}(1 - \frac{A}{M})$. Therefore

$$\frac{d}{dA}\left(k\sqrt{A}\left(1 - \frac{A}{M}\right)\right) = \frac{1}{2}kA^{-1/2} - \frac{3}{2}kA^{1/2}/M = \frac{1}{2}kA^{-1/2}\left(1 - \frac{3A}{M}\right) = 0$$

when $A = M/3$. Because the growth rate is zero for $A = 0$ and for $A = M$ and is positive for $0 < A < M$, it follows that the maximum growth rate occurs when $A = M/3$.

Further Insights and Challenges

17. Let $y(t)$ be a solution of the logistic equation

$$\frac{dy}{dt} = ky\left(1 - \frac{y}{A}\right) \qquad \boxed{10}$$

(a) Differentiate Eq. (10) with respect to t and use the Chain Rule to show that

$$\frac{d^2y}{dt^2} = k^2y\left(1 - \frac{y}{A}\right)\left(1 - \frac{2y}{A}\right)$$

(b) Show that $y(t)$ is concave up if $0 < y < A/2$ and concave down if $A/2 < y < A$.

(c) Show that if $0 < y(0) < A/2$, then $y(t)$ has a point of inflection at $y = A/2$ (Figure 5).

(d) Assume that $0 < y(0) < A/2$. Find the time t when $y(t)$ reaches the inflection point.

SOLUTION

(a) The derivative of Eq. (10) with respect to t is

$$y'' = ky' - \frac{2kyy'}{A} = ky'\left(1 - \frac{2y}{A}\right) = k\left(1 - \frac{y}{A}\right)ky\left(1 - \frac{2y}{A}\right) = k^2y\left(1 - \frac{y}{A}\right)\left(1 - \frac{2y}{A}\right).$$

(b) If $0 < y < A/2$, $1 - \frac{y}{A}$ and $1 - \frac{2y}{A}$ are both positive, so $y'' > 0$. Therefore, y is concave up. If $A/2 < y < A$, $1 - \frac{y}{A} > 0$, but $1 - \frac{2y}{A} < 0$, so $y'' < 0$, so y is concave down.

(c) If $y_0 < A$, y grows and $\lim_{t\to\infty} y(t) = A$. If $0 < y < A/2$, y is concave up at first. Once y passes $A/2$, y becomes concave down, so y has an inflection point at $y = A/2$.

(d) The general solution to Eq. (10) is

$$y = \frac{A}{1 - e^{-kt}/C};$$

thus, $y = A/2$ when

$$\frac{A}{2} = \frac{A}{1 - e^{-kt}/C}$$

$$1 - e^{-kt}/C = 2$$

$$t = -\frac{1}{k}\ln(-C)$$

Now, $C = y_0/(y_0 - A)$, so

$$t = -\frac{1}{k}\ln\frac{y_0}{A - y_0} = \frac{1}{k}\ln\frac{A - y_0}{y_0}.$$

10.4 First-Order Linear Equations

Preliminary Questions

1. Which of the following are first-order linear equations?

(a) $y' + x^2 y = 1$ 　　　　　　　　　　　　　　**(b)** $y' + xy^2 = 1$

(c) $x^5 y' + y = e^x$ 　　　　　　　　　　　　　　**(d)** $x^5 y' + y = e^y$

SOLUTION　The equations in **(a)** and **(c)** are first-order linear differential equations. The equation in **(b)** is not linear because of the y^2 factor in the second term on the left-hand side of the equation; the equation in **(d)** is not linear because of the e^y term on the right-hand side of the equation.

2. If $\alpha(x)$ is an integrating factor for $y' + A(x)y = B(x)$, then $\alpha'(x)$ is equal to (choose the correct answer):

(a) $B(x)$ 　　　　　　　　　　　　　　　　　　**(b)** $\alpha(x)A(x)$

(c) $\alpha(x)A'(x)$ 　　　　　　　　　　　　　　**(d)** $\alpha(x)B(x)$

SOLUTION　The correct answer is **(b)**: $\alpha(x)A(x)$.

Exercises

1. Consider $y' + x^{-1}y = x^3$.

(a) Verify that $\alpha(x) = x$ is an integrating factor.

(b) Show that when multiplied by $\alpha(x)$, the differential equation can be written $(xy)' = x^4$.

(c) Conclude that xy is an antiderivative of x^4 and use this information to find the general solution.

(d) Find the particular solution satisfying $y(1) = 0$.

SOLUTION

(a) The equation is of the form

$$y' + A(x)y = B(x)$$

for $A(x) = x^{-1}$ and $B(x) = x^3$. By Theorem 1, $\alpha(x)$ is defined by

$$\alpha(x) = e^{\int A(x)\,dx} = e^{\ln x} = x.$$

(b) When multiplied by $\alpha(x)$, the equation becomes:

$$xy' + y = x^4.$$

Now, $xy' + y = xy' + (x)'y = (xy)'$, so

$$(xy)' = x^4.$$

(c) Since $(xy)' = x^4$, $(xy) = \frac{x^5}{5} + C$ and

$$y = \frac{x^4}{5} + \frac{C}{x}$$

(d) If $y(1) = 0$, we find

$$0 = \frac{1}{5} + C \quad \text{so} \quad -\frac{1}{5} = C.$$

The solution, therefore, is

$$y = \frac{x^4}{5} - \frac{1}{5x}.$$

3. Let $\alpha(x) = e^{x^2}$. Verify the identity

$$(\alpha(x)y)' = \alpha(x)(y' + 2xy)$$

and explain how it is used to find the general solution of

$$y' + 2xy = x.$$

SOLUTION Let $\alpha(x) = e^{x^2}$. Then

$$(\alpha(x)y)' = (e^{x^2}y)' = 2xe^{x^2}y + e^{x^2}y' = e^{x^2}(2xy + y') = \alpha(x)(y' + 2xy).$$

If we now multiply both sides of the differential equation $y' + 2xy = x$ by $\alpha(x)$, we obtain

$$\alpha(x)(y' + 2xy) = x\alpha(x) = xe^{x^2}.$$

But $\alpha(x)(y' + 2xy) = (\alpha(x)y)'$, so by integration we find

$$\alpha(x)y = \int xe^{x^2}\,dx = \frac{1}{2}e^{x^2} + C.$$

Finally,

$$y(x) = \frac{1}{2} + Ce^{-x^2}.$$

In Exercises 5–18, find the general solution of the first-order linear differential equation.

5. $xy' + y = x$

SOLUTION Rewrite the equation as

$$y' + \frac{1}{x}y = 1,$$

which is in standard linear form with $A(x) = \frac{1}{x}$ and $B(x) = 1$. By Theorem 1, the integrating factor is

$$\alpha(x) = e^{\int A(x)\,dx} = e^{\ln x} = x.$$

When multiplied by the integrating factor, the rewritten differential equation becomes

$$xy' + y = x \qquad \text{or} \qquad (xy)' = x.$$

Integration of both sides now yields

$$xy = \frac{1}{2}x^2 + C.$$

Finally,

$$y(x) = \frac{1}{2}x + \frac{C}{x}.$$

7. $3xy' - y = x^{-1}$

SOLUTION Rewrite the equation as

$$y' - \frac{1}{3x}y = \frac{1}{3x^2},$$

which is in standard form with $A(x) = -\frac{1}{3}x^{-1}$ and $B(x) = \frac{1}{3}x^{-2}$. By Theorem 1, the integrating factor is

$$\alpha(x) = e^{\int A(x)\,dx} = e^{-(1/3)\ln x} = x^{-1/3}.$$

When multiplied by the integrating factor, the rewritten differential equation becomes

$$x^{-1/3}y' - \frac{1}{3}x^{-4/3} = \frac{1}{3}x^{-7/3} \qquad \text{or} \qquad (x^{-1/3}y)' = \frac{1}{3}x^{-7/3}.$$

Integration of both sides now yields

$$x^{-1/3}y = -\frac{1}{4}x^{-4/3} + C.$$

Finally,

$$y(x) = -\frac{1}{4}x^{-1} + Cx^{1/3}.$$

9. $y' + 3x^{-1}y = x + x^{-1}$

SOLUTION   This equation is in standard form with $A(x) = 3x^{-1}$ and $B(x) = x + x^{-1}$. By Theorem 1, the integrating factor is

$$\alpha(x) = e^{\int 3x^{-1}} = e^{3\ln x} = x^3.$$

When multiplied by the integrating factor, the original differential equation becomes

$$x^3 y' + 3x^2 y = x^4 + x^2 \qquad \text{or} \qquad (x^3 y)' = x^4 + x^3.$$

Integration of both sides now yields

$$x^3 y = \frac{1}{5}x^5 + \frac{1}{3}x^3 + C.$$

Finally,

$$y(x) = \frac{1}{5}x^2 + \frac{1}{3} + Cx^{-3}.$$

11. $xy' = y - x$

SOLUTION   Rewrite the equation as

$$y' - \frac{1}{x}y = -1,$$

which is in standard form with $A(x) = -\frac{1}{x}$ and $B(x) = -1$. By Theorem 1, the integrating factor is

$$\alpha(x) = e^{\int -(1/x)\,dx} = e^{-\ln x} = x^{-1}.$$

When multiplied by the integrating factor, the rewritten differential equation becomes

$$\frac{1}{x}y' - \frac{1}{x^2}y = -\frac{1}{x} \qquad \text{or} \qquad \left(\frac{1}{x}y\right)' = -\frac{1}{x}.$$

Integration on both sides now yields

$$\frac{1}{x}y = -\ln x + C.$$

Finally,

$$y(x) = -x \ln x + Cx.$$

13. $y' + y = e^x$

SOLUTION   This equation is in standard form with $A(x) = 1$ and $B(x) = e^x$. By Theorem 1, the integrating factor is

$$\alpha(x) = e^{\int 1\,dx} = e^x.$$

When multiplied by the integrating factor, the original differential equation becomes

$$e^x y' + e^x y = e^{2x} \qquad \text{or} \qquad (e^x y)' = e^{2x}.$$

Integration on both sides now yields

$$e^x y = \frac{1}{2}e^{2x} + C.$$

Finally,

$$y(x) = \frac{1}{2}e^x + Ce^{-x}.$$

15. $y' + (\tan x)y = \cos x$

SOLUTION This equation is in standard form with $A(x) = \tan x$ and $B(x) = \cos x$. By Theorem 1, the integrating factor is

$$\alpha(x) = e^{\int \tan x \, dx} = e^{\ln \sec x} = \sec x.$$

When multiplied by the integrating factor, the original differential equation becomes

$$\sec x y' + \sec x \tan x y = 1 \qquad \text{or} \qquad (y \sec x)' = 1.$$

Integration on both sides now yields

$$y \sec x = x + C.$$

Finally,

$$y(x) = x \cos x + C \cos x.$$

17. $y' - (\ln x)y = x^x$

SOLUTION This equation is in standard form with $A(x) = -\ln x$ and $B(x) = x^x$. By Theorem 1, the integrating factor is

$$\alpha(x) = e^{\int -\ln x \, dx} = e^{x - x \ln x} = \frac{e^x}{x^x}.$$

When multiplied by the integrating factor, the original differential equation becomes

$$x^{-x}e^x y' - (\ln x)x^{-x}e^x y = e^x \qquad \text{or} \qquad (x^{-x}e^x y)' = e^x.$$

Integration on both sides now yields

$$x^{-x}e^x y = e^x + C.$$

Finally,

$$y(x) = x^x + Cx^x e^{-x}.$$

In Exercises 19–26, solve the initial value problem.

19. $y' + 3y = e^{2x}, \quad y(0) = -1$

SOLUTION First, we find the general solution of the differential equation. This linear equation is in standard form with $A(x) = 3$ and $B(x) = e^{2x}$. By Theorem 1, the integrating factor is

$$\alpha(x) = e^{3x}.$$

When multiplied by the integrating factor, the original differential equation becomes

$$(e^{3x}y)' = e^{5x}.$$

Integration on both sides now yields

$$(e^{3x}y) = \frac{1}{5}e^{5x} + C;$$

hence,

$$y(x) = \frac{1}{5}e^{2x} + Ce^{-3x}.$$

The initial condition $y(0) = -1$ allows us to determine the value of C:

$$-1 = \frac{1}{5} + C \qquad \text{so} \qquad C = -\frac{6}{5}.$$

The solution to the initial value problem is therefore

$$y(x) = \frac{1}{5}e^{2x} - \frac{6}{5}e^{-3x}.$$

21. $y' + \dfrac{1}{x+1}y = x^{-2}, \quad y(1) = 2$

SOLUTION First, we find the general solution of the differential equation. This linear equation is in standard form with $A(x) = \frac{1}{x+1}$ and $B(x) = x^{-2}$. By Theorem 1, the integrating factor is

$$\alpha(x) = e^{\int 1/(x+1)\,dx} = e^{\ln(x+1)} = x + 1.$$

When multiplied by the integrating factor, the original differential equation becomes

$$((x+1)y)' = x^{-1} + x^{-2}.$$

Integration on both sides now yields

$$(x+1)y = \ln x - x^{-1} + C;$$

hence,

$$y(x) = \frac{1}{x+1}\left(C + \ln x - \frac{1}{x}\right).$$

The initial condition $y(1) = 2$ allows us to determine the value of C:

$$2 = \frac{1}{2}(C - 1) \qquad \text{so} \qquad C = 5.$$

The solution to the initial value problem is therefore

$$y(x) = \frac{1}{x+1}\left(5 + \ln x - \frac{1}{x}\right).$$

23. $(\sin x)y' = (\cos x)y + 1, \quad y(\frac{\pi}{4}) = 0$

SOLUTION First, we find the general solution of the differential equation. Rewrite the equation as

$$y' - (\cot x)y = \csc x,$$

which is in standard form with $A(x) = -\cot x$ and $B(x) = \csc x$. By Theorem 1, the integrating factor is

$$\alpha(x) = e^{\int -\cot x\,dx} = e^{-\ln \sin x} = \csc x.$$

When multiplied by the integrating factor, the rewritten differential equation becomes

$$(\csc x\, y)' = \csc^2 x.$$

Integration on both sides now yields

$$(\csc x)y = -\cot x + C;$$

hence,

$$y(x) = -\cos x + C \sin x.$$

The initial condition $y(\pi/4) = 0$ allows us to determine the value of C:

$$0 = -\frac{\sqrt{2}}{2} + C\frac{\sqrt{2}}{2} \qquad \text{so} \qquad C = 1.$$

The solution to the initial value problem is therefore

$$y(x) = -\cos x + \sin x.$$

25. $y' + (\tanh x)y = 1, \quad y(0) = 3$

SOLUTION First, we find the general solution of the differential equation. This equation is in standard form with $A(x) = \tanh x$ and $B(x) = 1$. By Theorem 1, the integrating factor is

$$\alpha(x) = e^{\int \tanh x\,dx} = e^{\ln \cosh x} = \cosh x.$$

When multiplied by the integrating factor, the original differential equation becomes

$$(\cosh x\, y)' = \cosh x.$$

Integration on both sides now yields

$$(\cosh x\, y) = \sinh x + C;$$

hence,

$$y(x) = \tanh x + C \operatorname{sech} x.$$

The initial condition $y(0) = 3$ allows us to determine the value of C:

$$3 = C.$$

The solution to the initial value problem is therefore

$$y(x) = \tanh x + 3 \operatorname{sech} x.$$

27. Find the general solution of $y' + ny = e^{mx}$ for all m, n. *Note:* The case $m = -n$ must be treated separately.

SOLUTION For any m, n, Theorem 1 gives us the formula for $\alpha(x)$:

$$\alpha(x) = e^{\int n\, dx} = e^{nx}.$$

When multiplied by the integrating factor, the original differential equation becomes

$$(e^{nx} y)' = e^{(m+n)x}.$$

If $m \neq -n$, integration on both sides yields

$$e^{nx} y = \frac{1}{m+n} e^{(m+n)x} + C,$$

so

$$y(x) = \frac{1}{m+n} e^{mx} + Ce^{-nx}.$$

However, if $m = -n$, then $m + n = 0$ and the equation reduces to

$$(e^{nx} y)' = 1,$$

so integration yields

$$e^{nx} y = x + C \qquad \text{or} \qquad y(x) = (x + C)e^{-nx}.$$

29. Repeat Exercise 28(a), assuming that water is pumped out at a rate of 20 gal/min. What is the limiting salt concentration for large t?

SOLUTION Because water flows into the tank at the same rate as water flows out of the tank, the amount of water in the tank remains a constant 100 gallons. Now, let $y(t)$ be the amount of salt in the tank (in pounds) at any time t (in minutes). The net flow of salt into the tank t is

$$\frac{dy}{dt} = \text{salt rate in} - \text{salt rate out} = \left(20\frac{\text{gal}}{\text{min}}\right)\left(0.4\frac{\text{lb}}{\text{gal}}\right) - \left(20\frac{\text{gal}}{\text{min}}\right)\left(\frac{y\,\text{lb}}{100\,\text{gal}}\right) = 8 - \frac{1}{5}y.$$

Rewriting this linear equation in standard form, we have

$$\frac{dy}{dt} + \frac{1}{5}y = 8,$$

so $A(t) = \frac{1}{5}$ and $B(t) = 8$. By Theorem 1, the integrating factor is

$$\alpha(t) = e^{\int (1/5)\, dt} = e^{t/5}.$$

When multiplied by the integrating factor, the rewritten differential equation becomes

$$(e^{t/5} y)' = 8e^{t/5}.$$

Integration on both sides now yields

$$e^{t/5} y = 40e^{t/5} + C;$$

hence,

$$y(t) = 40 + Ce^{-t/5}.$$

The initial condition $y(0) = 10$ allows us to determine the value of C:

$$10 = 40 + C \qquad \text{so} \qquad C = -30.$$

The solution to the initial value problem is therefore

$$y(t) = 40 - 30e^{-t/5}.$$

From here, we see that $y(t) \to 40$ as $t \to \infty$, so that the limiting concentration of salt in the tank is $\frac{40}{100} = 0.4$ pounds per gallon. This is completely intuitive; as time goes on, the concentration in the tank should resemble the concentration pouring into it.

31. Water flows into a tank at the variable rate $R_{\text{in}} = \dfrac{20}{1+t}$ gal/min and out at the constant rate $R_{\text{out}} = 5$ gal/min. Let $V(t)$ be the volume of water in the tank at time t.

(a) Set up a differential equation for $V(t)$ and solve it with the initial condition $V(0) = 100$.

(b) Find the maximum value of V.

(c) $\mathcal{CAS}$ Plot $V(t)$ and estimate the time t when the tank is empty.

SOLUTION

(a) The rate of change of the volume of water in the tank is given by

$$\frac{dV}{dt} = R_{\text{in}} - R_{\text{out}} = \frac{20}{1+t} - 5.$$

Because the right-hand side depends only on the independent variable t, we integrate to obtain

$$V(t) = 20\ln(1+t) - 5t + C.$$

The initial condition $V(0) = 100$ allows us to determine the value of C:

$$100 = 20\ln 1 - 0 + C \qquad \text{so} \qquad C = 100.$$

Therefore

$$V(t) = 20\ln(1+t) - 5t + 100.$$

(b) Using the result from part (a),

$$\frac{dV}{dt} = \frac{20}{1+t} - 5 = 0$$

when $t = 3$. Because $\frac{dV}{dt} > 0$ for $t < 3$ and $\frac{dV}{dt} < 0$ for $t > 3$, it follows that

$$V(3) = 20\ln 4 - 15 + 100 \approx 112.726 \text{ gal}$$

is the maximum volume.

(c) $V(t)$ is plotted in the figure below at the left. On the right, we zoom in near the location where the curve crosses the t-axis. From this graph, we estimate that the tank is empty after roughly 34.25 minutes.

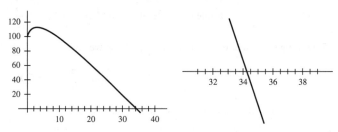

In Exercises 33–35, consider a series circuit (Figure 4) consisting of a resistor of R ohms, an inductor of L henries, and variable voltage source of V(t) volts (time t in seconds). The current through the circuit I(t) (in amperes) satisfies the differential equation:

$$\frac{dI}{dt} + \frac{R}{L}I = \frac{1}{L}V(t) \qquad \boxed{12}$$

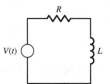

FIGURE 4 *RL* circuit.

33. Find the solution to Eq. (12) with initial condition $I(0) = 0$, assuming that $R = 100 \, \Omega$, $L = 5$ H, and $V(t)$ is constant with $V(t) = 10$ V.

SOLUTION If $R = 100$, $V(t) = 10$, and $L = 5$, the differential equation becomes

$$\frac{dI}{dt} + 20I = 2,$$

which is a linear equation in standard form with $A(t) = 20$ and $B(t) = 2$. The integrating factor is $\alpha(t) = e^{20t}$, and when multiplied by the integrating factor, the differential equation becomes

$$(e^{20t} I)' = 2e^{20t}.$$

Integration of both sides now yields

$$e^{20t} I = \frac{1}{10} e^{20t} + C;$$

hence,

$$I(t) = \frac{1}{10} + Ce^{-20t}.$$

The initial condition $I(0) = 0$ allows us to determine the value of C:

$$0 = \frac{1}{10} + C \qquad \text{so} \qquad C = -\frac{1}{10}.$$

Finally,

$$I(t) = \frac{1}{10}\left(1 - e^{-20t}\right).$$

35. Assume that $V(t) = V$ is constant and $I(0) = 0$.
(a) Solve for $I(t)$.
(b) Show that $\lim_{t \to \infty} I(t) = V/R$ and that $I(t)$ reaches approximately 63% of its limiting value after L/R seconds.
(c) How long does it take for $I(t)$ to reach 90% of its limiting value if $R = 500 \, \Omega$, $L = 4$ H, and $V = 20$ V?

SOLUTION
(a) The equation

$$\frac{dI}{dt} + \frac{R}{L}I = \frac{1}{L}V$$

is a linear equation in standard form with $A(t) = \frac{R}{L}$ and $B(t) = \frac{1}{L}V(t)$. By Theorem 1, the integrating factor is

$$\alpha(t) = e^{\int (R/L)\,dt} = e^{(R/L)t}.$$

When multiplied by the integrating factor, the original differential equation becomes

$$(e^{(R/L)t} I)' = e^{(R/L)t}\frac{V}{L}.$$

Integration on both sides now yields

$$(e^{(R/L)t} I) = \frac{V}{R}e^{(R/L)t} + C;$$

hence,

$$I(t) = \frac{V}{R} + Ce^{-(R/L)t}.$$

The initial condition $I(0) = 0$ allows us to determine the value of C:

$$0 = \frac{V}{R} + C \qquad \text{so} \qquad C = -\frac{V}{R}.$$

Therefore the current is given by

$$I(t) = \frac{V}{R}\left(1 - e^{-(R/L)t}\right).$$

(b) As $t \to \infty$, $e^{-(R/L)t} \to 0$, so $I(t) \to \frac{V}{R}$. Moreover, when $t = (L/R)$ seconds, we have

$$I\left(\frac{L}{R}\right) = \frac{V}{R}\left(1 - e^{-(R/L)\,(L/R)}\right) = \frac{V}{R}\left(1 - e^{-1}\right) \approx 0.632\frac{V}{R}.$$

(c) Using the results from part (a) and part (b), $I(t)$ reaches 90% of its limiting value when

$$\frac{9}{10} = 1 - e^{-(R/L)t},$$

or when

$$t = \frac{L}{R}\ln 10.$$

With $L = 4$ and $R = 500$, this takes approximately 0.0184 seconds.

37. Continuing with the previous exercise, let Tank 2 be another tank filled with V_2 gallons of water. Assume that the inky water from Tank 1 empties into Tank 2 as in Figure 5, mixes instantaneously, and leaves Tank 2 at the same rate R. Let $y_2(t)$ be the amount of ink in Tank 2 at time t.

R(gal/min)

Tank 1

R(gal/min)

Tank 2

R(gal/min)

FIGURE 5

(a) Explain why y_2 satisfies the differential equation

$$\frac{dy_2}{dt} = R\left(\frac{y_1}{V_1} - \frac{y_2}{V_2}\right)$$

(b) Use the solution to Exercise 36 to solve for $y_2(t)$ if $V_1 = 100$, $V_2 = 200$, $R = 10$, $I = 2$, and $y_2(0) = 0$.
(c) GU Plot the solution for $0 \le t \le 120$.
(d) Find the maximum ink concentration in Tank 2.

SOLUTION

(a) The water flowing into Tank 2 has an ink concentration of $y_1(t)/V_1$ and the water flowing out from the tank has an ink concentration of $y_2(t)/V_2$. Thus,

$$\frac{dy_2}{dt} = R\left(\frac{y_1}{V_1} - \frac{y_2}{V_2}\right).$$

(b) With $V_1 = 100$, $R = 10$ and $I = 2$, we know from the previous exercise that $y_1(t) = 2e^{-t/10}$. Substituting this expression and the given parameter values into the differential equation obtained in part (a), we have

$$\frac{dy_2}{dt} = 10\left(\frac{2}{100}e^{-t/10} - \frac{1}{200}y_2\right) = \frac{1}{5}e^{-t/10} - \frac{1}{20}y_2.$$

Hence,

$$\frac{dy_2}{dt} + \frac{1}{20}y_2 = \frac{1}{5}e^{-t/10}.$$

The integrating factor for this linear equation is $\alpha(t) = e^{t/20}$ so we get

$$\left(e^{t/20}y_2\right)' = \frac{1}{5}e^{-t/20},$$

and

$$e^{t/20}y_2 = -4e^{-t/20} + C.$$

Thus,

$$y_2(t) = -4e^{-t/10} + Ce^{-t/20}.$$

The initial condition $y_2(0) = 0$ allows us to determine $C = 4$; consequently,

$$y_2(t) = 4\left(e^{-t/20} - e^{-t/10}\right).$$

(c) A plot of $y_2(t)$ is shown below.

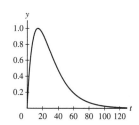

(d) Using the result of part (b), we find that

$$\frac{dy_2}{dt} = 4\left(\frac{1}{10}e^{-t/10} - \frac{1}{20}e^{-t/20}\right) = 0$$

when

$$t = 20\ln 2.$$

Thus, the maximum ink concentration in Tank 2 is

$$y_2(20\ln 2) = 4\left(\frac{1}{2} - \frac{1}{4}\right) = 1.$$

Further Insights and Challenges

39. Use the Fundamental Theorem of Calculus and the Product Rule to verify directly that for any x_0, the function

$$f(x) = \alpha(x)^{-1} \int_{x_0}^{x} \alpha(t)B(t)\,dt$$

is a solution of the initial value problem

$$y' + A(x)y = B(x), \qquad y(x_0) = 0,$$

where $\alpha(x)$ is an integrating factor [a solution to Eq. (3)].

SOLUTION   Remember that $\alpha'(x) = A(x)\alpha(x)$. Now, let

$$y(x) = \frac{1}{\alpha(x)} \int_{x_0}^{x} \alpha(t)B(t)\,dt.$$

Then,

$$y(x_0) = \frac{1}{\alpha(x)} \int_{x_0}^{x_0} \alpha(t)B(t)\,dt = 0,$$

and

$$y' + A(x)y = -\frac{\alpha'(x)}{(\alpha(x))^2} \int_{x_0}^{x} \alpha(t)B(t)\,dt + B(x) + \frac{A(x)}{\alpha(x)} \int_{x_0}^{x} \alpha(t)B(t)\,dt$$

$$= B(x) + \left(-\frac{A(x)}{\alpha(x)} + \frac{A(x)}{\alpha(x)}\right) \int_{x_0}^{x} \alpha(t)B(t)\,dt = B(x).$$

CHAPTER REVIEW EXERCISES

1. Which of the following differential equations are linear? Determine the order of each equation.

(a) $y' = y^5 - 3x^4 y$

(b) $y' = x^5 - 3x^4 y$

(c) $y = y''' - 3x\sqrt{y}$

(d) $\sin x \cdot y'' = y - 1$

SOLUTION

(a) y^5 is a nonlinear term involving the dependent variable, so this is not a linear equation; the highest order derivative that appears in the equation is a first derivative, so this is a first-order equation.

(b) This is linear equation; the highest order derivative that appears in the equation is a first derivative, so this is a first-order equation.

(c) $\sqrt{y}$ is a nonlinear term involving the dependent variable, so this is not a linear equation; the highest order derivative that appears in the equation is a third derivative, so this is a third-order equation.

(d) This is linear equation; the highest order derivative that appears in the equation is a second derivative, so this is a second-order equation.

In Exercises 3–6, solve using separation of variables.

3. $\dfrac{dy}{dt} = t^2 y^{-3}$

SOLUTION Rewrite the equation as

$$y^3\, dy = t^2\, dt.$$

Upon integrating both sides of this equation, we obtain:

$$\int y^3\, dy = \int t^2\, dt$$

$$\frac{y^4}{4} = \frac{t^3}{3} + C.$$

Thus,

$$y = \pm \left(\frac{4}{3} t^3 + C \right)^{1/4},$$

where C is an arbitrary constant.

5. $x\dfrac{dy}{dx} - y = 1$

SOLUTION Rewrite the equation as

$$\frac{dy}{1+y} = \frac{dx}{x}.$$

upon integrating both sides of this equation, we obtain

$$\int \frac{dy}{1+y} = \int \frac{dx}{x}$$

$$\ln|1 + y| = \ln|x| + C.$$

Thus,

$$y = -1 + Ax,$$

where $A = \pm e^C$ is an arbitrary constant.

In Exercises 7–10, solve the initial value problem using separation of variables.

7. $y' = \cos^2 x, \quad y(0) = \dfrac{\pi}{4}$

SOLUTION First, we find the general solution of the differential equation. Because the variables are already separated, we integrate both sides to obtain

$$y = \int \cos^2 x\, dx = \int \left(\frac{1}{2} + \frac{1}{2} \cos 2x \right) dx = \frac{x}{2} + \frac{\sin 2x}{4} + C.$$

The initial condition $y(0) = \frac{\pi}{4}$ allows us to determine $C = \frac{\pi}{4}$. Thus, the solution is:

$$y(x) = \frac{x}{2} + \frac{\sin 2x}{4} + \frac{\pi}{4}.$$

9. $y' = xy^2, \quad y(1) = 2$

SOLUTION   First, we find the general solution of the differential equation. Rewrite

$$\frac{dy}{dx} = xy^2 \qquad \text{as} \qquad \frac{dy}{y^2} = x \, dx.$$

Upon integrating both sides of this equation, we find

$$\int \frac{dy}{y^2} = \int x \, dx$$

$$-\frac{1}{y} = \frac{1}{2}x^2 + C.$$

Thus,

$$y = -\frac{1}{\frac{1}{2}x^2 + C}.$$

The initial condition $y(1) = 2$ allows us to determine the value of C:

$$2 = -\frac{1}{\frac{1}{2} \cdot 1^2 + C} = -\frac{2}{1 + 2C}$$

$$1 + 2C = -1$$

$$C = -1$$

Hence, the solution to the initial value problem is

$$y = -\frac{1}{\frac{1}{2}x^2 - 1} = -\frac{2}{x^2 - 2}.$$

11. Figure 1 shows the slope field for $\dot{y} = \sin y + ty$. Sketch the graphs of the solutions with the initial conditions $y(0) = 1$, $y(0) = 0$, and $y(0) = -1$.

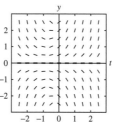

FIGURE 1

SOLUTION

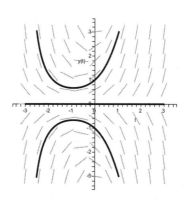

13. Let $y(t)$ be the solution to the differential equation with slope field as shown in Figure 2, satisfying $y(0) = 0$. Sketch the graph of $y(t)$. Then use your answer to Exercise 12 to solve for $y(t)$.

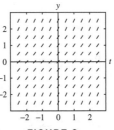

FIGURE 2

SOLUTION As explained in the previous exercise, the slope field in Figure 2 corresponds to the equation $\dot{y} = 1 + y^2$. The graph of the solution satisfying $y(0) = 0$ is:

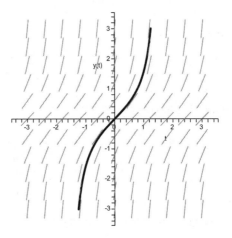

To solve the initial value problem $\dot{y} = 1 + y^2$, $y(0) = 0$, we first find the general solution of the differential equation. Separating variables yields:

$$\frac{dy}{1 + y^2} = dt.$$

Upon integrating both sides of this equation, we find

$$\tan^{-1} y = t + C \qquad \text{or} \qquad y = \tan(t + C).$$

The initial condition gives $C = 0$, so the solution is $y = \tan x$.

15. Let $y(t)$ be the solution of $(x^3 + 1)\dot{y} = y$ satisfying $y(0) = 1$. Compute approximations to $y(0.1)$, $y(0.2)$, and $y(0.3)$ using Euler's Method with time step $h = 0.1$.

SOLUTION Rewriting the equation as $\dot{y} = \frac{y}{x^3+1}$ we have $F(x, y) = \frac{y}{x^3+1}$. Using Euler's Method with $x_0 = 0$, $y_0 = 1$ and $h = 0.1$, we calculate

$$y(0.1) \approx y_1 = y_0 + hF(x_0, y_0) = 1 + 0.1 \cdot \frac{1}{0^3 + 1} = 1.1$$

$$y(0.2) \approx y_2 = y_1 + hF(x_1, y_1) = 1.209890$$

$$y(0.3) \approx y_3 = y_2 + hF(x_2, y_2) = 1.329919$$

In Exercises 16–19, solve using the method of integrating factors.

17. $\dfrac{dy}{dx} = \dfrac{y}{x} + x$, $\quad y(1) = 3$

SOLUTION First, we find the general solution of the differential equation. Rewrite the equation as

$$y' - \frac{1}{x}y = x,$$

which is in standard form with $A(x) = -\frac{1}{x}$ and $B(x) = x$. The integrating factor is

$$\alpha(x) = e^{\int -\frac{1}{x}\,dx} = e^{-\ln x} = \frac{1}{x}.$$

When multiplied by the integrating factor, the rewritten differential equation becomes

$$\left(\frac{1}{x}y\right)' = 1.$$

Integration on both sides now yields

$$\frac{1}{x}y = x + C;$$

hence,

$$y(x) = x^2 + Cx.$$

The initial condition $y(1) = 3$ allows us to determine the value of C:

$$3 = 1 + C \qquad \text{so} \qquad C = 2.$$

The solution to the initial value problem is then

$$y = x^2 + 2x.$$

19. $y' + 2y = 1 + e^{-x}, \quad y(0) = -4$

SOLUTION   The equation is already in standard form with $A(x) = 2$ and $B(x) = 1 + e^{-x}$. The integrating factor is

$$\alpha(x) = e^{\int 2\,dx} = e^{2x}.$$

When multiplied by the integrating factor, the original differential equation becomes

$$(e^{2x}y)' = e^{2x} + e^x.$$

Integration on both sides now yields

$$e^{2x}y = \frac{1}{2}e^{2x} + e^x + C;$$

hence,

$$y(x) = \frac{1}{2} + e^{-x} + Ce^{-2x}.$$

The initial condition $y(0) = -4$ allows us to determine the value of C:

$$-4 = \frac{1}{2} + 1 + C \qquad \text{so} \qquad C = -\frac{11}{2}.$$

The solution to the initial value problem is then

$$y(x) = \frac{1}{2} + e^{-x} - \frac{11}{2}e^{-2x}.$$

In Exercises 20–27, solve using the appropriate method.

21. $y' + (\tan x)y = \cos^2 x, \quad y(\pi) = 2$

SOLUTION   First, we find the general solution of the differential equation. As this is a first order linear equation with $A(x) = \tan x$ and $B(x) = \cos^2 x$, we compute the integrating factor

$$\alpha(x) = e^{\int A(x)\,dx} = e^{\int \tan x\,dx} = e^{-\ln \cos x} = \frac{1}{\cos x}.$$

When multiplied by the integrating factor, the original differential equation becomes

$$\left(\frac{1}{\cos x}y\right)' = \cos x.$$

Integration on both sides now yields

$$\frac{1}{\cos x} y = \sin x + C;$$

hence,

$$y(x) = \sin x \cos x + C \cos x = \frac{1}{2} \sin 2x + C \cos x.$$

The initial condition $y(\pi) = 2$ allows us to determine the value of C:

$$2 = 0 + C(-1) \qquad \text{so} \qquad C = -2.$$

The solution to the initial value problem is then

$$y = \frac{1}{2} \sin 2x - 2 \cos x.$$

23. $(y - 1)y' = t, \quad y(1) = -3$

SOLUTION First, we find the general solution of the differential equation. This is a separable equation that we rewrite as

$$(y - 1) \, dy = t \, dt.$$

Upon integrating both sides of this equation, we find

$$\int (y - 1) \, dy = \int t \, dt$$

$$\frac{y^2}{2} - y = \frac{1}{2}t^2 + C$$

$$y^2 - 2y + 1 = t^2 + C$$

$$(y - 1)^2 = t^2 + C$$

$$y(t) = \pm\sqrt{t^2 + C} + 1$$

To satisfy the initial condition $y(1) = -3$ we must choose the negative square root; moreover,

$$-3 = -\sqrt{1 + C} + 1 \qquad \text{so} \qquad C = 15.$$

The solution to the initial value problem is then

$$y(t) = -\sqrt{t^2 + 15} + 1$$

25. $\dfrac{dw}{dx} = k \dfrac{1 + w^2}{x}, \quad w(1) = 1$

SOLUTION First, we find the general solution of the differential equation. This is a separable equation that we rewrite as

$$\frac{dw}{1 + w^2} = \frac{k}{x} \, dx.$$

Upon integrating both sides of this equation, we find

$$\int \frac{dw}{1 + w^2} = \int \frac{k}{x} \, dx$$

$$\tan^{-1} w = k \ln x + C$$

$$w(x) = \tan(k \ln x + C).$$

Because the initial condition is specified at $x = 1$, we are interested in the solution for $x > 0$; we can therefore omit the absolute value within the natural logarithm function. The initial condition $w(1) = 1$ allows us to determine the value of C:

$$1 = \tan(k \ln 1 + C) \qquad \text{so} \qquad C = \tan^{-1} 1 = \frac{\pi}{4}.$$

The solution to the initial value problem is then

$$w = \tan\left(k \ln x + \frac{\pi}{4}\right).$$

27. $y' + \dfrac{y}{x} = \sin x$

SOLUTION This is a first order linear equation with $A(x) = \frac{1}{x}$ and $B(x) = \sin x$. The integrating factor is

$$\alpha(x) = e^{\int A(x)\,dx} = e^{\ln x} = x.$$

When multiplied by the integrating factor, the original differential equation becomes

$$(xy)' = x \sin x.$$

Integration on both sides (integration by parts is needed for the integral on the right-hand side) now yields

$$xy = -x \cos x + \sin x + C;$$

hence,

$$y(x) = -\cos x + \frac{\sin x}{x} + \frac{C}{x}.$$

29. Let A and B be constants. Prove that if $A > 0$, then all solution of $\dfrac{dy}{dt} + Ay = B$ approach the same limit as $t \to \infty$.

SOLUTION This is a linear first-order equation in standard form with integrating factor

$$\alpha(t) = e^{\int A\,dt} = e^{At}.$$

When multiplied by the integrating factor, the original differential equation becomes

$$(e^{At} y)' = Be^{At}.$$

Integration on both sides now yields

$$e^{At} y = \frac{B}{A} e^{At} + C;$$

hence,

$$y(t) = \frac{B}{A} + Ce^{-At}.$$

Because $A > 0$,

$$\lim_{t\to\infty} y(t) = \lim_{t\to\infty}\left(\frac{B}{A} + Ce^{-At}\right) = \frac{B}{A}.$$

We conclude that if $A > 0$, all solutions approach the limit $\frac{B}{A}$ as $t \to \infty$.

31. Find the solution of the logistic equation $\dot{y} = 0.4y(4 - y)$ satisfying $y(0) = 8$.

SOLUTION We can write the given equation as

$$\dot{y} = 1.6y\left(1 - \frac{y}{4}\right).$$

This is a logistic equation with $k = 1.6$ and $A = 4$. Therefore,

$$y(t) = \frac{A}{1 - e^{-kt}/C} = \frac{4}{1 - e^{-1.6t}/C}.$$

The initial condition $y(0) = 8$ allows us to determine the value of C:

$$8 = \frac{4}{1 - \frac{1}{C}}; \quad 1 - \frac{1}{C} = \frac{1}{2}; \quad \text{so} \quad C = 2.$$

Thus,

$$y(t) = \frac{4}{1 - e^{-1.6t}/2} = \frac{8}{2 - e^{-1.6t}}.$$

33. Suppose that $y' = ky(1 - y/8)$ has a solution satisfying $y(0) = 12$ and $y(10) = 24$. Find k.

SOLUTION   The given differential equation is a logistic equation with $A = 8$. Thus,

$$y(t) = \frac{8}{1 - e^{-kt}/C}.$$

The initial condition $y(0) = 12$ allows us to determine the value of C:

$$12 = \frac{8}{1 - \frac{1}{C}}; \quad 1 - \frac{1}{C} = \frac{2}{3}; \quad \text{so} \quad C = 3.$$

Hence,

$$y(t) = \frac{8}{1 - e^{-kt}/3} = \frac{24}{3 - e^{-kt}}.$$

Now, the condition $y(10) = 24$ allows us to determine the value of k:

$$24 = \frac{24}{3 - e^{-10k}}$$

$$3 - e^{-10k} = 1$$

$$k = -\frac{\ln 2}{10} \approx -0.0693.$$

35. 🖊️   A rabbit population on a deserted island increases exponentially with growth rate $k = 0.12$ months^{-1}. When the population reaches 150 rabbits (say, at time $t = 0$), hunters begin killing the rabbits at a rate of r rabbits per month.
(a) Find a differential equation satisfied by the rabbit population $P(t)$.
(b) How large can r be without the rabbit population becoming extinct?

SOLUTION
(a) The rabbit population $P(t)$ obeys the differential equation

$$\frac{dP}{dt} = 0.12P - r,$$

where the term $0.12P$ accounts for the exponential growth of the population and the term $-r$ accounts for the rate of decline in the rabbit population due to hunting.
(b) Rewrite the linear differential equation from part (a) as

$$\frac{dP}{dt} - 0.12P = -r,$$

which is in standard form with $A = -0.12$ and $B = -r$. The integrating factor is

$$\alpha(t) = e^{\int A\, dt} = e^{\int -0.12\, dt} = e^{-0.12t}.$$

When multiplied by the integrating factor, the rewritten differential equation becomes

$$(e^{-0.12t} P)' = -re^{-0.12t}.$$

Integration on both sides now yields

$$e^{-0.12t} P = \frac{r}{0.12}e^{-0.12t} + C;$$

hence,

$$P(t) = \frac{r}{0.12} + Ce^{0.12t}.$$

The initial condition $P(0) = 150$ allows us to determine the value of C:

$$150 = \frac{r}{0.12} + C \quad \text{so} \quad C = 150 - \frac{r}{0.12}.$$

The solution to the initial value problem is then

$$P(t) = \left(150 - \frac{r}{0.12}\right)e^{0.12t} + \frac{r}{0.12}.$$

Now, if $150 - \frac{r}{0.12} < 0$, then $\lim_{t \to \infty} P(t) = -\infty$, and the population becomes extinct. Therefore, in order for the population to survive, we must have

$$150 - \frac{r}{0.12} \geq 0 \quad \text{or} \quad r \leq 18.$$

We conclude that the maximum rate at which the hunters can kill the rabbits without driving the rabbits to extinction is $r = 18$ rabbits per month.

37. 🖊️ A tank contains 100 gal of pure water. Water is pumped out at a rate of 20 gal/min, and polluted water with a toxin concentration of 0.1 lb/gal is pumped in at a rate of 15 gal/min. Let $y(t)$ be the amount of toxin present in the tank at time t.

(a) Find a differential equation satisfied by $y(t)$.

(b) Solve for $y(t)$.

(c) CAS Plot $y(t)$ and find the time t at which the amount of toxin is maximal.

SOLUTION

(a) Because water flows into the tank at a rate of 15 gallons per minute but flows out at a rate of 20 gallons per minute, there is a new outflow of 5 gallons per minute from the tank. Therefore, at any time t, there are $100 - 5t$ gallons of water in the tank. Now, to determine the differential equation satisfied by $y(t)$, we employ the basic rate balance

$$\frac{dy}{dt} = \text{toxin rate in} \ - \ \text{toxin rate out.}$$

The amount of toxin per minute coming into the tank is

$$\text{toxin rate in} = \text{concentration} \ \cdot \ \text{water rate in} = 0.1 \cdot 15 = 1.5 \ \frac{\text{lb}}{\text{min}}.$$

To determine the toxin rate out we first compute the toxin concentration at time t:

$$\text{toxin concentration} = \frac{\text{lbs of toxin}}{\text{gallons of water}} = \frac{y(t)}{100 - 5t}.$$

Since water flows out at a rate of 20 gallons per minute,

$$\text{toxin rate out} = 20 \frac{y}{100 - 5t} = \frac{4y}{20 - t}.$$

Thus, the differential equation is

$$\frac{dy}{dt} = 1.5 - \frac{4y}{20 - t}.$$

(b) Rewrite the linear first-order differential equation found in part (a) as

$$\frac{dy}{dt} + \frac{4}{20 - t} y = \frac{3}{2}.$$

The integrating factor for this equation is

$$\alpha(t) = e^{4 \int (20-t)^{-1} \, dt} = \frac{1}{(20 - t)^4}.$$

When multiplied by the integrating factor, the rewritten differential equation becomes

$$\left(\frac{1}{(20 - t)^4} y \right)' = \frac{3}{2(20 - t)^4}.$$

Integration on both sides now yields

$$\frac{1}{(20 - t)^4} y = \frac{1}{2(20 - t)^3} + C;$$

hence,

$$y(t) = \frac{20 - t}{2} + C(20 - t)^4.$$

The initial condition $y(0) = 0$ allows us to determine the value of C:

$$0 = 10 + 20^4 C \quad \text{so} \quad C = -\frac{1}{16000}.$$

Therefore,

$$y(t) = \frac{20 - t}{2} - \frac{(20 - t)^4}{16000}.$$

(c) A plot of $y(t)$ is shown below.

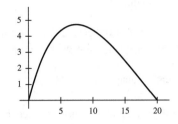

To determine the time at which the amount of toxin is maximal, we note that

$$\frac{dy}{dt} = \frac{d}{dt}\left(\frac{20-t}{2} - \frac{(20-t)^4}{16000}\right) = -\frac{1}{2} + \frac{(20-t)^3}{4000} = 0$$

when

$$t = 20 - 10\sqrt[3]{2} \approx 7.4 \text{ minutes.}$$

11 | INFINITE SERIES

11.1 Sequences

Preliminary Questions

1. What is a_4 for the sequence $a_n = n^2 - n$?

SOLUTION Substituting $n = 4$ in the expression for a_n gives

$$a_4 = 4^2 - 4 = 12.$$

2. Which of the following sequences converge to zero?

(a) $\dfrac{n^2}{n^2 + 1}$ 　　　　　　 **(b)** 2^n 　　　　　　 **(c)** $\left(\dfrac{-1}{2}\right)^n$

SOLUTION

(a) This sequence does not converge to zero:

$$\lim_{n \to \infty} \frac{n^2}{n^2 + 1} = \lim_{x \to \infty} \frac{x^2}{x^2 + 1} = \lim_{x \to \infty} \frac{1}{1 + \frac{1}{x^2}} = \frac{1}{1 + 0} = 1.$$

(b) This sequence does not converge to zero: this is a geometric sequence with $r = 2 > 1$; hence, the sequence diverges to ∞.

(c) Recall that if $|a_n|$ converges to 0, then a_n must also converge to zero. Here,

$$\left|\left(-\frac{1}{2}\right)^n\right| = \left(\frac{1}{2}\right)^n,$$

which is a geometric sequence with $0 < r < 1$; hence, $(\frac{1}{2})^n$ converges to zero. It therefore follows that $(-\frac{1}{2})^n$ converges to zero.

3. Let a_n be the nth decimal approximation to $\sqrt{2}$. That is, $a_1 = 1$, $a_2 = 1.4$, $a_3 = 1.41$, etc. What is $\lim_{n \to \infty} a_n$?

SOLUTION $\lim_{n \to \infty} a_n = \sqrt{2}$.

4. Which sequence is defined recursively?

(a) $a_n = \sqrt{2 + n^{-1}}$ 　　　　　　 **(b)** $b_n = \sqrt{4 + b_{n-1}}$

SOLUTION

(a) a_n can be computed directly, since it depends on n only and not on preceding terms. Therefore a_n is defined explicitly and not recursively.

(b) b_n is computed in terms of the preceding term b_{n-1}, hence the sequence $\{b_n\}$ is defined recursively.

5. Theorem 5 says that every convergent sequence is bounded. Which of the following statements follow from Theorem 5 and which are false? If false, give a counterexample.

(a) If $\{a_n\}$ is bounded, then it converges.

(b) If $\{a_n\}$ is not bounded, then it diverges.

(c) If $\{a_n\}$ diverges, then it is not bounded.

SOLUTION

(a) This statement is false. The sequence $a_n = \cos \pi n$ is bounded since $-1 \leq \cos \pi n \leq 1$ for all n, but it does not converge: since $a_n = \cos n\pi = (-1)^n$, the terms assume the two values 1 and -1 alternately, hence they do not approach one value.

(b) By Theorem 5, a converging sequence must be bounded. Therefore, if a sequence is not bounded, it certainly does not converge.

(c) The statement is false. The sequence $a_n = (-1)^n$ is bounded, but it does not approach one limit.

Exercises

1. Match the sequence with the general term:

$a_1, a_2, a_3, a_4, \ldots$	General term
(a) $\frac{1}{2}, \frac{2}{3}, \frac{3}{4}, \frac{4}{5}, \ldots$	(i) $\cos \pi n$
(b) $-1, 1, -1, 1, \ldots$	(ii) $\dfrac{n!}{2^n}$
(c) $1, -1, 1, -1, \ldots$	(iii) $(-1)^{n+1}$
(d) $\frac{1}{2}, \frac{2}{4}, \frac{6}{8}, \frac{24}{16} \ldots$	(iv) $\dfrac{n}{n+1}$

SOLUTION

(a) The numerator of each term is the same as the index of the term, and the denominator is one more than the numerator; hence $a_n = \frac{n}{n+1}, n = 1, 2, 3, \ldots$.

(b) The terms of this sequence are alternating between -1 and 1 so that the positive terms are in the even places. Since $\cos \pi n = 1$ for even n and $\cos \pi n = -1$ for odd n, we have $a_n = \cos \pi n, n = 1, 2, \ldots$.

(c) The terms a_n are 1 for odd n and -1 for even n. Hence, $a_n = (-1)^{n+1}, n = 1, 2, \ldots$

(d) The numerator of each term is $n!$, and the denominator is 2^n; hence, $a_n = \frac{n!}{2^n}, n = 1, 2, 3, \ldots$.

In Exercises 3–10, calculate the first four terms of the following sequences, starting with $n = 1$.

3. $c_n = \dfrac{2^n}{n!}$

SOLUTION Setting $n = 1, 2, 3, 4$ in the formula for c_n gives

$$c_1 = \frac{2^1}{1!} = \frac{2}{1} = 2, \qquad c_2 = \frac{2^2}{2!} = \frac{4}{2} = 2,$$

$$c_3 = \frac{2^3}{3!} = \frac{8}{6} = \frac{4}{3}, \qquad c_4 = \frac{2^4}{4!} = \frac{16}{24} = \frac{2}{3}.$$

5. $a_1 = 3, \quad a_{n+1} = 1 + a_n^2$

SOLUTION For $n = 1, 2, 3$ we have:

$$a_2 = a_{1+1} = 1 + a_1^2 = 1 + 3^2 = 10;$$

$$a_3 = a_{2+1} = 1 + a_2^2 = 1 + 10^2 = 101;$$

$$a_4 = a_{3+1} = 1 + a_3^2 = 1 + 101^2 = 10{,}202.$$

The first four terms of $\{a_n\}$ are $3, 10, 101, 10{,}202$.

7. $c_n = 1 + \dfrac{1}{2} + \dfrac{1}{3} + \cdots + \dfrac{1}{n}$

SOLUTION

$$c_1 = 1;$$

$$c_2 = 1 + \frac{1}{2} = \frac{3}{2};$$

$$c_3 = 1 + \frac{1}{2} + \frac{1}{3} = \frac{3}{2} + \frac{1}{3} = \frac{11}{6};$$

$$c_4 = 1 + \frac{1}{2} + \frac{1}{3} + \frac{1}{4} = \frac{11}{6} + \frac{1}{4} = \frac{25}{12}.$$

9. $b_1 = 2, \quad b_2 = 5, \quad b_n = b_{n-1} + 2b_{n-2}$

SOLUTION We need to find b_3 and b_4. Setting $n = 3$ and $n = 4$ and using the given values for b_1 and b_2 we obtain:

$$b_3 = b_{3-1} + 2b_{3-2} = b_2 + 2b_1 = 5 + 2 \cdot 2 = 9;$$

$$b_4 = b_{4-1} + 2b_{4-2} = b_3 + 2b_2 = 9 + 2 \cdot 5 = 19.$$

The first four terms of the sequence $\{b_n\}$ are $2, 5, 9, 19$.

11. Find a formula for the nth term of the following sequence:

(a) $\dfrac{1}{1}, \dfrac{-1}{8}, \dfrac{1}{27}, \ldots$

(b) $\dfrac{2}{6}, \dfrac{3}{7}, \dfrac{4}{8}, \ldots$

SOLUTION

(a) The denominators are the third powers of the positive integers starting with $n = 1$. Also, the sign of the terms is alternating with the sign of the first term being positive. Thus,

$$a_1 = \frac{1}{1^3} = \frac{(-1)^{1+1}}{1^3}; \quad a_2 = -\frac{1}{2^3} = \frac{(-1)^{2+1}}{2^3}; \quad a_3 = \frac{1}{3^3} = \frac{(-1)^{3+1}}{3^3}.$$

This rule leads to the following formula for the nth term:

$$a_n = \frac{(-1)^{n+1}}{n^3}.$$

(b) Assuming a starting index of $n = 1$, we see that each numerator is one more than the index and the denominator is four more than the numerator. Thus, the general term a_n is

$$a_n = \frac{n+1}{n+5}.$$

In Exercises 13–26, use Theorem 1 to determine the limit of the sequence or state that the sequence diverges.

13. $a_n = 4$

SOLUTION We have $a_n = f(n)$ where $f(x) = 4$; thus,

$$\lim_{n \to \infty} a_n = \lim_{x \to \infty} f(x) = \lim_{x \to \infty} 4 = 4.$$

15. $a_n = 5 - \dfrac{9}{n^2}$

SOLUTION We have $a_n = f(n)$ where $f(x) = 5 - \dfrac{9}{x^2}$; thus,

$$\lim_{n \to \infty} \left(5 - \frac{9}{n^2}\right) = \lim_{x \to \infty} \left(5 - \frac{9}{x^2}\right) = 5 - 0 = 5.$$

17. $c_n = -2^{-n}$

SOLUTION We have $c_n = f(n)$ where $f(x) = -2^{-x}$; thus,

$$\lim_{n \to \infty} \left(-2^{-n}\right) = \lim_{x \to \infty} -2^{-x} = \lim_{x \to \infty} -\frac{1}{2^x} = 0.$$

19. $z_n = \left(\dfrac{1}{3}\right)^n$

SOLUTION We have $z_n = f(n)$ where $f(x) = \left(\dfrac{1}{3}\right)^x$; thus,

$$\lim_{n \to \infty} \left(\frac{1}{3}\right)^n = \lim_{x \to \infty} \left(\frac{1}{3}\right)^x = 0.$$

21. $a_n = \dfrac{(-1)^n n^2 + n}{4n^2 + 1}$

SOLUTION We examine the terms with n odd and n even separately. With n odd, the terms take the form

$$\frac{-n^2 + n}{4n^2 + 1} \quad \text{and} \quad \lim_{n \to \infty} \frac{-n^2 + n}{4n^2 + 1} = \lim_{x \to \infty} \frac{-x^2 + x}{4x^2 + 1} = -\frac{1}{4}.$$

On the other hand, with n even, the terms take the form

$$\frac{n^2 + n}{4n^2 + 1} \quad \text{and} \quad \lim_{n \to \infty} \frac{n^2 + n}{4n^2 + 1} = \lim_{x \to \infty} \frac{x^2 + x}{4x^2 + 1} = \frac{1}{4}.$$

Because these two limits are different, the general term of the given sequence does not approach one value. Hence, the sequence diverges.

23. $a_n = \dfrac{n}{\sqrt{n^3 + 1}}$

SOLUTION We have $a_n = f(n)$ where $f(x) = \dfrac{x}{\sqrt{x^3 + 1}}$; thus,

$$\lim_{n\to\infty} \frac{n}{\sqrt{n^3 + 1}} = \lim_{x\to\infty} \frac{x}{\sqrt{x^3 + 1}} = \lim_{x\to\infty} \frac{\frac{x}{x^{3/2}}}{\frac{\sqrt{x^3+1}}{x^{3/2}}} = \lim_{x\to\infty} \frac{\frac{1}{\sqrt{x}}}{\sqrt{1 + \frac{1}{x^3}}} = \frac{0}{\sqrt{1+0}} = \frac{0}{1} = 0.$$

25. $a_n = \cos \pi n$

SOLUTION The terms in the odd places are -1 and the terms in the even places are 1. Therefore, the general term of the sequence does not approach one limit. Hence, the sequence diverges.

27. Let $a_n = \dfrac{n}{n+1}$. Find a number M such that:

(a) $|a_n - 1| \le 0.001$ for $n \ge M$.

(b) $|a_n - 1| \le 0.00001$ for $n \ge M$.

Then use the limit definition to prove that $\lim_{n\to\infty} a_n = 1$.

SOLUTION

(a) We have

$$|a_n - 1| = \left| \frac{n}{n+1} - 1 \right| = \left| \frac{n - (n+1)}{n+1} \right| = \left| \frac{-1}{n+1} \right| = \frac{1}{n+1}.$$

Therefore $|a_n - 1| \le 0.001$ provided $\frac{1}{n+1} \le 0.001$, that is, $n \ge 999$. It follows that we can take $M = 999$.

(b) By part (a), $|a_n - 1| \le 0.00001$ provided $\frac{1}{n+1} \le 0.00001$, that is, $n \ge 99999$. It follows that we can take $M = 99999$.

We now prove formally that $\lim_{n\to\infty} a_n = 1$. Using part (a), we know that

$$|a_n - 1| = \frac{1}{n+1} < \epsilon,$$

provided $n > \frac{1}{\epsilon} - 1$. Thus, Let $\epsilon > 0$ and take $M = \frac{1}{\epsilon} - 1$. Then, for $n > M$, we have

$$|a_n - 1| = \frac{1}{n+1} < \frac{1}{M+1} = \epsilon.$$

29. Use the limit definition to prove that $\lim_{n\to\infty} n^{-2} = 0$.

SOLUTION We see that

$$|n^{-2} - 0| = \left| \frac{1}{n^2} \right| = \frac{1}{n^2} < \epsilon$$

provided

$$n > \frac{1}{\sqrt{\epsilon}}.$$

Thus, let $\epsilon > 0$ and take $M = \frac{1}{\sqrt{\epsilon}}$. Then, for $n > M$, we have

$$|n^{-2} - 0| = \left| \frac{1}{n^2} \right| = \frac{1}{n^2} < \frac{1}{M^2} = \epsilon.$$

31. Find $\lim_{n\to\infty} 2^{1/n}$.

SOLUTION Because 2^x is a continuous function,

$$\lim_{n\to\infty} 2^{1/n} = \lim_{x\to\infty} 2^{1/x} = 2^{\lim_{x\to\infty}(1/x)} = 2^0 = 1.$$

33. Find $\lim_{n\to\infty} n^{1/n}$.

SOLUTION Let $a_n = n^{1/n}$. Take the natural logarithm of both sides of this expression to obtain

$$\ln a_n = \ln n^{1/n} = \frac{\ln n}{n}.$$

Thus,

$$\lim_{n \to \infty} (\ln a_n) = \lim_{n \to \infty} \frac{\ln n}{n} = \lim_{x \to \infty} \frac{\ln x}{x} = \lim_{x \to \infty} \frac{1}{x} = 0.$$

Because $f(x) = e^x$ is a continuous function, it follows that

$$\lim_{n \to \infty} a_n = \lim_{n \to \infty} e^{\ln a_n} = e^{\lim_{n \to \infty}(\ln a_n)} = e^0 = 1.$$

That is,

$$\lim_{n \to \infty} n^{1/n} = 1.$$

35. Find $\displaystyle\lim_{n \to \infty} \left(1 + \frac{1}{n}\right)^n$.

SOLUTION Let $a_n = \left(1 + \frac{1}{n}\right)^n$. Taking the natural logarithm of both sides of this expression yields

$$\ln a_n = \ln \left(1 + \frac{1}{n}\right)^n = n \ln \left(1 + \frac{1}{n}\right) = \frac{\ln \left(1 + \frac{1}{n}\right)}{\frac{1}{n}}.$$

Thus,

$$\lim_{n \to \infty} (\ln a_n) = \lim_{x \to \infty} \frac{\ln \left(1 + \frac{1}{x}\right)}{\frac{1}{x}} = \lim_{x \to \infty} \frac{\frac{d}{dx}\left(\ln\left(1 + \frac{1}{x}\right)\right)}{\frac{d}{dx}\left(\frac{1}{x}\right)} = \lim_{x \to \infty} \frac{\frac{1}{1 + \frac{1}{x}} \cdot \left(-\frac{1}{x^2}\right)}{-\frac{1}{x^2}} = \lim_{x \to \infty} \frac{1}{1 + \frac{1}{x}} = \frac{1}{1 + 0} = 1.$$

Because $f(x) = e^x$ is a continuous function, it follows that

$$\lim_{n \to \infty} a_n = \lim_{n \to \infty} e^{\ln a_n} = e^{\lim_{n \to \infty}(\ln a_n)} = e^1 = e.$$

37. Use the Squeeze Theorem to find $\displaystyle\lim_{n \to \infty} a_n$, where

$$a_n = \frac{1}{\sqrt{n^4 + n^8}} \text{ by proving that}$$

$$\frac{1}{\sqrt{2}n^4} \le a_n \le \frac{1}{\sqrt{2}n^2}$$

SOLUTION For all $n > 1$ we have $n^4 < n^8$, so the quotient $\frac{1}{\sqrt{n^4 + n^8}}$ is smaller than $\frac{1}{\sqrt{n^4 + n^4}}$ and larger than $\frac{1}{\sqrt{n^8 + n^8}}$. That is,

$$a_n < \frac{1}{\sqrt{n^4 + n^4}} = \frac{1}{\sqrt{n^4 \cdot 2}} = \frac{1}{\sqrt{2}n^2}; \text{ and}$$

$$a_n > \frac{1}{\sqrt{n^8 + n^8}} = \frac{1}{\sqrt{2n^8}} = \frac{1}{\sqrt{2}n^4}.$$

Now, since $\displaystyle\lim_{n \to \infty} \frac{1}{\sqrt{2}n^4} = \lim_{n \to \infty} \frac{1}{\sqrt{2}n^2} = 0$, the Squeeze Theorem for Sequences implies that $\displaystyle\lim_{n \to \infty} a_n = 0$.

39. Evaluate $\displaystyle\lim_{n \to \infty} n \sin \frac{1}{n}$.

SOLUTION We have $a_n = f(n)$ where $f(x) = x \sin \frac{1}{x}$. Thus,

$$\lim_{n \to \infty} n \sin \frac{1}{n} = \lim_{x \to \infty} x \sin \frac{1}{x} = \lim_{x \to \infty} \frac{\sin \frac{1}{x}}{\frac{1}{x}}.$$

The limit $\lim_{x \to 0} \frac{\sin x}{x} = 1$ implies that $\lim_{x \to \infty} \frac{\sin \frac{1}{x}}{\frac{1}{x}} = 1$. Hence,

$$\lim_{n \to \infty} n \sin \frac{1}{n} = 1.$$

41. [icon] Which statement is equivalent to the assertion $\lim_{n\to\infty} a_n = L$? Explain.

(a) For every $\epsilon > 0$, the interval $(L - \epsilon, L + \epsilon)$ contains at least one element of the sequence $\{a_n\}$.

(b) For every $\epsilon > 0$, the interval $(L - \epsilon, L + \epsilon)$ contains all but at most finitely many elements of the sequence $\{a_n\}$.

SOLUTION Statement (b) is equivalent to Definition 1 of the limit, since the assertion "$|a_n - L| < \epsilon$ for all $n > M$" means that $L - \epsilon < a_n < L + \epsilon$ for all $n > M$; that is, the interval $(L - \epsilon, L + \epsilon)$ contains all the elements a_n except (maybe) the finite number of elements $a_1, a_2, \ldots, a_M$.

Statement (a) is not equivalent to the assertion $\lim_{n\to\infty} a_n = L$. We show this, by considering the following sequence:

$$a_n = \begin{cases} \dfrac{1}{n} & \text{for odd } n \\[2mm] 1 + \dfrac{1}{n} & \text{for even } n \end{cases}$$

Clearly for every $\epsilon > 0$, the interval $(-\epsilon, \epsilon) = (L - \epsilon, L + \epsilon)$ for $L = 0$ contains at least one element of $\{a_n\}$, but the sequence diverges (rather than converges to $L = 0$). Since the terms in the odd places converge to 0 and the terms in the even places converge to 1. Hence, a_n does not approach one limit.

In Exercises 43–63, determine the limit of the sequence or show that the sequence diverges by using the appropriate Limit Laws or theorems.

43. $a_n = \dfrac{3n^2 + n + 2}{2n^2 - 3}$

SOLUTION

$$\lim_{n\to\infty} \frac{3n^2 + n + 2}{2n^2 - 3} = \lim_{x\to\infty} \frac{3x^2 + x + 2}{2x^2 - 3} = \frac{3}{2}.$$

45. $a_n = 3 + \left(-\dfrac{1}{2}\right)^n$

SOLUTION By the Limit Laws for Sequences we have:

$$\lim_{n\to\infty} \left(3 + \left(-\frac{1}{2}\right)^n\right) = \lim_{n\to\infty} 3 + \lim_{n\to\infty} \left(-\frac{1}{2}\right)^n = 3 + \lim_{n\to\infty} \left(-\frac{1}{2}\right)^n.$$

Now,

$$-\left(\frac{1}{2}\right)^n \le \left(-\frac{1}{2}\right)^n \le \left(\frac{1}{2}\right)^n.$$

Because

$$\lim_{n\to\infty} \left(\frac{1}{2}\right)^n = 0,$$

by the Limit Laws for Sequences,

$$\lim_{n\to\infty} -\left(\frac{1}{2}\right)^n = -\lim_{n\to\infty} \left(\frac{1}{2}\right)^n = 0.$$

Thus, we have

$$\lim_{n\to\infty} \left(-\frac{1}{2}\right)^n = 0,$$

and

$$\lim_{n\to\infty} \left(3 + \left(-\frac{1}{2}\right)^n\right) = 3 + 0 = 3.$$

47. $b_n = \tan^{-1}\left(1 - \dfrac{2}{n}\right)$

SOLUTION Because $f(x) = \tan^{-1} x$ is a continuous function, it follows that

$$\lim_{n\to\infty} a_n = \lim_{x\to\infty} \tan^{-1}\left(1 - \frac{2}{x}\right) = \tan^{-1}\left(\lim_{x\to\infty} \left(1 - \frac{2}{x}\right)\right) = \tan^{-1} 1 = \frac{\pi}{4}.$$

49. $c_n = \ln\left(\dfrac{2n+1}{3n+4}\right)$

SOLUTION Because $f(x) = \ln x$ is a continuous function, it follows that

$$\lim_{n\to\infty} c_n = \lim_{x\to\infty} \ln\left(\frac{2x+1}{3x+4}\right) = \ln\left(\lim_{x\to\infty}\frac{2x+1}{3x+4}\right) = \ln\frac{2}{3}.$$

51. $y_n = \dfrac{e^n + 3^n}{5^n}$

SOLUTION We rewrite the general term of the sequence as follows:

$$\frac{e^n + 3^n}{5^n} = \frac{e^n}{5^n} + \frac{3^n}{5^n} = \left(\frac{e}{5}\right)^n + \left(\frac{3}{5}\right)^n.$$

Because $0 < \frac{e}{5} < 1$ and $0 < \frac{3}{5} < 1$, the geometric sequences $(\frac{e}{5})^n$ and $(\frac{3}{5})^n$ converge to 0; hence,

$$\lim_{n\to\infty}\frac{e^n+3^n}{5^n} = \lim_{n\to\infty}\left(\left(\frac{e}{5}\right)^n + \left(\frac{3}{5}\right)^n\right) = \lim_{n\to\infty}\left(\frac{e}{5}\right)^n + \lim_{n\to\infty}\left(\frac{3}{5}\right)^n = 0 + 0 = 0.$$

53. $a_n = \dfrac{e^n}{2^{n^2}}$

SOLUTION Using the natural logarithm, we find

$$\ln a_n = \ln\frac{e^n}{2^{n^2}} = \ln e^n - \ln 2^{n^2} = n - n^2 \ln 2.$$

Thus,

$$\lim_{n\to\infty} \ln a_n = \lim_{n\to\infty} n\,(1 - n\ln 2) = \lim_{x\to\infty} x\,(1 - x\ln 2) = -\infty.$$

Because $f(x) = e^x$ is a continuous function, it follows that

$$\lim_{n\to\infty} a_n = \lim_{n\to\infty} e^{\ln a_n} = e^{\lim_{n\to\infty}(\ln a_n)} = 0.$$

55. $b_n = \dfrac{n^3 + 2e^{-n}}{3n^3 + 4e^{-n}}$

SOLUTION We rewrite the general term of the sequence as follows:

$$a_n = \frac{n^3 + 2e^{-n}}{3n^3 + 4e^{-n}} = \frac{\frac{n^3}{n^3} + \frac{2e^{-n}}{n^3}}{\frac{3n^3}{n^3} + \frac{4e^{-n}}{n^3}} = \frac{1 + \frac{2}{n^3 e^n}}{3 + \frac{4}{n^3 e^n}}.$$

Thus,

$$\lim_{n\to\infty} a_n = \lim_{x\to\infty}\frac{1 + \frac{2}{x^3 e^x}}{3 + \frac{4}{x^3 e^x}} = \frac{\lim_{x\to\infty}\left(1 + \frac{2}{x^3 e^x}\right)}{\lim_{x\to\infty}\left(3 + \frac{4}{x^3 e^x}\right)} = \frac{\lim_{x\to\infty} 1 + 2\lim_{x\to\infty}\frac{1}{x^3 e^x}}{\lim_{x\to\infty} 3 + 4\lim_{x\to\infty}\frac{1}{x^3 e^x}} = \frac{1 + 2\cdot 0}{3 + 4\cdot 0} = \frac{1}{3}.$$

57. $A_n = \dfrac{3 - 4^n}{2 + 7\cdot 3^n}$

SOLUTION Divide the numerator and denominator by 3^n to obtain

$$a_n = \frac{3 - 4^n}{2 + 7\cdot 3^n} = \frac{\frac{3}{3^n} - \frac{4^n}{3^n}}{\frac{2}{3^n} + \frac{7\cdot 3^n}{3^n}} = \frac{\frac{3}{3^n} - \left(\frac{4}{3}\right)^n}{\frac{2}{3^n} + 7}.$$

We examine the limits of the numerator and the denominator:

$$\lim_{n\to\infty}\left(\frac{3}{3^n} - \left(\frac{4}{3}\right)^n\right) = 3\lim_{n\to\infty}\left(\frac{1}{3}\right)^n - 3\lim_{n\to\infty}\left(\frac{4}{3}\right)^n = 3\cdot 0 - \infty = -\infty,$$

whereas

$$\lim_{n\to\infty}\left(\frac{2}{3^n} + 7\right) = \lim_{n\to\infty}\frac{2}{3^n} + \lim_{n\to\infty} 7 = 2\lim_{n\to\infty}\left(\frac{1}{3}\right)^n + \lim_{n\to\infty} 7 = 2\cdot 0 + 7 = 7.$$

Thus, $\lim_{n\to\infty} a_n = -\infty$; that is, the sequence diverges.

59. $A_n = \dfrac{(-4,000)^n}{n!}$

SOLUTION   Let $a_n = \dfrac{(-4000)^n}{n!}$ and note

$$-\left(\frac{4000^n}{n!}\right) \le a_n \le \frac{4000^n}{n!}.$$

For $n > 4001$,

$$\left(\frac{4000}{1} \cdot \frac{4000}{2} \cdots \frac{4000}{4000}\right)\left(\frac{4000}{4001} \cdot \frac{4000}{4002} \cdots \frac{4000}{n-1}\right) \cdot \frac{4000}{n} = \frac{4000^{4000}}{4000!}\left(\frac{4000}{4001} \cdot \frac{4000}{4002} \cdots \frac{4000}{n-1}\right) \cdot \frac{4000}{n}.$$

Each one of the factors in the brackets is less than 1, so we have

$$0 < \frac{4000^n}{n!} < \frac{4000^{4000}}{4000!} \cdot \frac{4000}{n} = \frac{4000^{4001}}{4000!} \cdot \frac{1}{n}.$$

Since $\frac{4000^{4001}}{4000!} \cdot \frac{1}{n}$ tends to zero, the Squeeze Theorem guarantees that

$$\lim_{n\to\infty} A_n = \lim_{n\to\infty} \frac{(-4000)^n}{n!} = 0.$$

61. $a_n = \cos\dfrac{\pi}{n}$

SOLUTION   By the Theorem on Sequences Defined by a Function, we have:

$$\lim_{n\to\infty} \cos\frac{\pi}{n} = \lim_{x\to\infty} \cos\frac{\pi}{x}.$$

The limit $\lim\limits_{x\to\infty} \frac{\pi}{x} = 0$ and the continuity of $\cos u$ at $u = 0$ imply:

$$\lim_{x\to\infty} \cos\frac{\pi}{x} = \cos\left(\lim_{x\to\infty} \frac{\pi}{x}\right) = \cos 0 = 1.$$

Thus,

$$\lim_{n\to\infty} \cos\frac{\pi}{n} = 1.$$

63. $a_n = \sqrt[n]{n}$

SOLUTION   Let $a_n = n^{1/n}$. Taking the natural logarithm of both sides of this expression yields

$$\ln a_n = \ln n^{1/n} = \frac{1}{n} \ln n = \frac{\ln n}{n}.$$

Thus,

$$\lim_{n\to\infty} (\ln a_n) = \lim_{n\to\infty} \frac{\ln n}{n} = \lim_{x\to\infty} \frac{\ln x}{x} = \lim_{x\to\infty} \frac{1/x}{1} = 0.$$

Because $f(x) = e^x$ is a continuous function, it follows that

$$\lim_{n\to\infty} a_n = \lim_{n\to\infty} e^{\ln a_n} = e^{\lim_{n\to\infty} (\ln a_n)} = e^0 = 1.$$

65. Show that $a_n = \dfrac{3n^2}{n^2 + 2}$ is strictly increasing. Find an upper bound.

SOLUTION   Let $f(x) = \frac{3x^2}{x^2+2}$. Then

$$f'(x) = \frac{6x(x^2 + 2) - 3x^2 \cdot 2x}{(x^2 + 2)^2} = \frac{12x}{(x^2 + 2)^2}.$$

$f'(x) > 0$ for $x > 0$, hence f is strictly increasing on this interval. It follows that $a_n = f(n)$ is also strictly increasing. We now show that $M = 3$ is an upper bound for a_n, by writing:

$$a_n = \frac{3n^2}{n^2 + 2} \le \frac{3n^2 + 6}{n^2 + 2} = \frac{3(n^2 + 2)}{n^2 + 2} = 3.$$

That is, $a_n \le 3$ for all n.

67. Use the limit definition to prove that the limit does not change if a finite number of terms are added or removed from a convergent sequence.

SOLUTION Suppose that $\{a_n\}$ is a sequence such that $\lim_{n\to\infty} a_n = L$. For every $\epsilon > 0$, there is a number M such that $|a_n - L| < \epsilon$ for all $n > M$. That is, the inequality $|a_n - L| < \epsilon$ holds for all the terms of $\{a_n\}$ except possibly a finite number of terms. If we add a finite number of terms, these terms may not satisfy the inequality $|a_n - L| < \epsilon$, but there are still only a finite number of terms that do not satisfy this inequality. By removing terms from the sequence, the number of terms in the new sequence that do not satisfy $|a_n - L| < \epsilon$ are no more than in the original sequence. Hence the new sequence also converges to L.

69. Let $\{a_n\}$ be a sequence such that $\lim_{n\to\infty} |a_n|$ exists and is nonzero. Show that $\lim_{n\to\infty} a_n$ exists if and only if there exists an integer M such that the sign of a_n does not change for $n > M$.

SOLUTION Let $\{a_n\}$ be a sequence such that $\lim_{n\to\infty} |a_n|$ exists and is nonzero. Suppose $\lim_{n\to\infty} a_n$ exists and let $L = \lim_{n\to\infty} a_n$. Note that L cannot be zero for then $\lim_{n\to\infty} |a_n|$ would also be zero. Now, choose $\epsilon < |L|$. Then there exists an integer M such that $|a_n - L| < \epsilon$, or $L - \epsilon < a_n < L + \epsilon$, for all $n > M$. If $L < 0$, then $-2L < a_n < 0$, whereas if $L > 0$, then $0 < a_n < 2L$; that is, a_n does not change for $n > M$.

Now suppose that there exists an integer M such that a_n does not change for $n > M$. If $a_n > 0$ for $n > M$, then $a_n = |a_n|$ for $n > M$ and

$$\lim_{n\to\infty} a_n = \lim_{n\infty} |a_n|.$$

On the other hand, if $a_n < 0$ for $n > M$, then $a_n = -|a_n|$ for $n > M$ and

$$\lim_{n\to\infty} a_n = \lim_{n\infty} -|a_n| = -\lim_{n\to\infty} |a_n|.$$

In either case, $\lim_{n\to\infty} a_n$ exists. Thus, $\lim_{n\to\infty} a_n$ exists if and only if there exists an integer M such that the sign of a_n does not change for $n > M$.

71. Show, by giving an example, that there exist *divergent* sequences $\{a_n\}$ and $\{b_n\}$ such that $\{a_n + b_n\}$ converges.

SOLUTION Let $a_n = 2^n$ and $b_n = -2^n$. Then $\{a_n\}$ and $\{b_n\}$ are divergent geometric sequences. However, since $a_n + b_n = 2^n - 2^n = 0$, the sequence $\{a_n + b_n\}$ is the constant sequence with all the terms equal zero, so it converges to zero.

73. Use the limit definition to prove that if $\{a_n\}$ is a convergent sequence of integers with limit L, then there exists a number M such that $a_n = L$ for all $n \geq M$.

SOLUTION Suppose $\{a_n\}$ converges to L, and let $\epsilon = \frac{1}{2}$. Then, there exists a number M such that

$$|a_n - L| < \frac{1}{2}$$

for all $n \geq M$. In other words, for all $n \geq M$,

$$L - \frac{1}{2} < a_n < L + \frac{1}{2}.$$

However, we are given that $\{a_n\}$ is a sequence of integers. Thus, it must be that $a_n = L$ for all $n \geq M$.

75. Prove that the following sequence is bounded and increasing. Then find its limit:

$$a_1 = \sqrt{5}, \quad a_2 = \sqrt{5 + \sqrt{5}}, \quad a_3 = \sqrt{5 + \sqrt{5 + \sqrt{5}}}, \ldots$$

SOLUTION Notice that this sequence is defined recursively by the formula:

$$a_{n+1} = \sqrt{5 + a_n}.$$

First, let's show that the sequence is bounded. All the terms in the sequence are positive, so 0 is a lower bound. Now, $a_1 = \sqrt{5} < 3$. If we suppose that $a_n < 3$ for some n, it then follows that

$$a_{n+1} = \sqrt{5 + a_n} < \sqrt{5 + 3} = \sqrt{8} < \sqrt{9} = 3.$$

Thus, by mathematical induction, $a_n < 3$ for all n, and 3 is an upper bound. Next, let's show that the sequence is increasing. Observe that $a_1 = \sqrt{5} \approx 2.236$ and $a_2 = \sqrt{5 + \sqrt{5}} \approx 2.690$. Thus, $a_2 > a_1$. If we suppose that $a_n > a_{n-1}$ for some n, it then follows that

$$a_{n+1} = \sqrt{5 + a_n} > \sqrt{5 + a_{n-1}} = a_n;$$

hence, by mathematical induction, $a_{n+1} > a_n$ for all n.

Now, since $\{a_n\}$ is increasing with upper bound, this sequence is convergent. Let $\lim_{n\to\infty} a_n = L$. Then, by Exercise 68, $\lim_{n\to\infty} a_{n+1} = L$ as well. It follows that

$$L = \lim_{n\to\infty} a_{n+1} = \lim_{n\to\infty} \sqrt{5+a_n} = \sqrt{\lim_{n\to\infty}(5+a_n)} = \sqrt{5+\lim_{n\to\infty} a_n} = \sqrt{5+L}.$$

That is,

$$L^2 = 5+L$$

$$L^2 - L - 5 = 0 \Rightarrow L_{1,2} = \frac{1\pm\sqrt{1+20}}{2}.$$

Since $a_n \geq 0$ for all n, the appropriate solution is:

$$\lim_{n\to\infty} a_n = \frac{1+\sqrt{21}}{2}.$$

77. Find the limit of the sequence

$$c_n = \frac{1}{\sqrt{n^2+1}} + \frac{1}{\sqrt{n^2+2}} + \cdots + \frac{1}{\sqrt{n^2+n}}$$

Hint: Show that

$$\frac{n}{\sqrt{n^2+n}} \leq c_n \leq \frac{n}{\sqrt{n^2+1}}$$

SOLUTION Since each of the n terms in the sum defining c_n is not smaller than $\frac{1}{\sqrt{n^2+n}}$ and not larger than $\frac{1}{\sqrt{n^2+1}}$ we obtain the following inequalities:

$$c_n \geq \underbrace{\frac{1}{\sqrt{n^2+n}} + \cdots + \frac{1}{\sqrt{n^2+n}}}_{n \text{ terms}} = n\cdot\frac{1}{\sqrt{n^2+n}} = \frac{n}{\sqrt{n^2+n}};$$

$$c_n \leq \underbrace{\frac{1}{\sqrt{n^2+1}} + \cdots + \frac{1}{\sqrt{n^2+1}}}_{n \text{ terms}} = n\cdot\frac{1}{\sqrt{n^2+1}} = \frac{n}{\sqrt{n^2+1}}.$$

Thus,

$$\frac{n}{\sqrt{n^2+n}} \leq c_n \leq \frac{n}{\sqrt{n^2+1}}.$$

We now compute the limits of the two sequences:

$$\lim_{n\to\infty}\frac{n}{\sqrt{n^2+1}} = \lim_{n\to\infty}\frac{\frac{n}{n}}{\frac{\sqrt{n^2+1}}{n}} = \lim_{n\to\infty}\frac{1}{\frac{\sqrt{n^2+1}}{\sqrt{n^2}}} = \lim_{n\to\infty}\frac{1}{\sqrt{1+\frac{1}{n^2}}} = 1;$$

$$\lim_{n\to\infty}\frac{n}{\sqrt{n^2+n}} = \lim_{n\to\infty}\frac{\frac{n}{n}}{\frac{\sqrt{n^2+n}}{n}} = \lim_{n\to\infty}\frac{1}{\frac{\sqrt{n^2+n}}{\sqrt{n^2}}} = \lim_{n\to\infty}\frac{1}{\sqrt{1+\frac{1}{n}}} = 1.$$

By the Squeeze Theorem we conclude that:

$$\lim_{n\to\infty} c_n = 1.$$

Further Insights and Challenges

79. Let $b_n = \dfrac{\sqrt[n]{n!}}{n}$

(a) Show that $\ln b_n = \dfrac{\ln(n!) - n\ln n}{n} = \dfrac{1}{n}\sum_{k=1}^{n}\ln\dfrac{k}{n}$.

(b) Show that $\ln b_n$ converges to $\displaystyle\int_0^1 \ln x\, dx$ and conclude that $b_n \to e^{-1}$.

SOLUTION

(a) Let $b_n = \frac{(n!)^{1/n}}{n}$. Then

$$\ln b_n = \ln (n!)^{1/n} - \ln n = \frac{1}{n} \ln (n!) - \ln n = \frac{\ln (n!) - n \ln n}{n} = \frac{1}{n}\left[\ln (n!) - \ln n^n\right] = \frac{1}{n} \ln \frac{n!}{n^n}$$

$$= \frac{1}{n} \ln \left(\frac{1}{n} \cdot \frac{2}{n} \cdot \frac{3}{n} \cdots \cdots \frac{n}{n}\right) = \frac{1}{n}\left(\ln \frac{1}{n} + \ln \frac{2}{n} + \ln \frac{3}{n} + \cdots + \ln \frac{n}{n}\right) = \frac{1}{n}\sum_{k=1}^{n} \ln \frac{k}{n}.$$

(b) By part (a) we have,

$$\lim_{n\to\infty} (\ln b_n) = \lim_{n\to\infty} \frac{1}{n}\sum_{k=1}^{n} \ln \frac{k}{n}.$$

Notice that $\frac{1}{n}\sum_{k=1}^{n} \ln \frac{k}{n}$ is the nth right-endpoint approximation of the integral of $\ln x$ over the interval $[0, 1]$. Hence,

$$\lim_{n\to\infty} \frac{1}{n}\sum_{k=1}^{n} \ln \frac{k}{n} = \int_0^1 \ln x \, dx.$$

We compute the improper integral using integration by parts, with $u = \ln x$ and $v' = 1$. Then $u' = \frac{1}{x}$, $v = x$ and

$$\int_0^1 \ln x \, dx = x \ln x \Big|_0^1 - \int_0^1 \frac{1}{x} x \, dx = 1 \cdot \ln 1 - \lim_{x\to 0+} (x \ln x) - \int_0^1 dx$$

$$= 0 - \lim_{x\to 0+} (x \ln x) - x \Big|_0^1 = -1 - \lim_{x\to 0+} (x \ln x).$$

We compute the remaining limit using L'Hôpital's Rule. This gives:

$$\lim_{x\to 0+} (x \cdot \ln x) = \lim_{x\to 0+} \frac{\ln x}{\frac{1}{x}} = \lim_{x\to 0+} \frac{\frac{1}{x}}{-\frac{1}{x^2}} = \lim_{x\to 0+} (-x) = 0.$$

Thus,

$$\lim_{n\to\infty} \ln b_n = \int_0^1 \ln x \, dx = -1,$$

and

$$\lim_{n\to\infty} b_n = e^{-1}.$$

81. Let $c_n = \frac{1}{n} + \frac{1}{n+1} + \frac{1}{n+2} + \cdots + \frac{1}{2n}$.

(a) Calculate c_1, c_2, c_3, c_4.

(b) Use a comparison of rectangles with the area under $y = x^{-1}$ over the interval $[n, 2n]$ to prove that

$$\int_n^{2n} \frac{dx}{x} + \frac{1}{2n} \le c_n \le \int_n^{2n} \frac{dx}{x} + \frac{1}{n}$$

(c) Use the Squeeze Theorem to determine $\lim_{n\to\infty} c_n$.

SOLUTION

(a)

$$c_1 = 1 + \frac{1}{2} = \frac{3}{2};$$

$$c_2 = \frac{1}{2} + \frac{1}{3} + \frac{1}{4} = \frac{13}{12};$$

$$c_3 = \frac{1}{3} + \frac{1}{4} + \frac{1}{5} + \frac{1}{6} = \frac{19}{20};$$

$$c_4 = \frac{1}{4} + \frac{1}{5} + \frac{1}{6} + \frac{1}{7} + \frac{1}{8} = \frac{743}{840};$$

(b) We consider the left endpoint approximation to the integral of $y = \frac{1}{x}$ over the interval $[n, 2n]$. Since the function $y = \frac{1}{x}$ is decreasing, the left endpoint approximation is greater than $\int_n^{2n} \frac{dx}{x}$; that is,

$$\int_n^{2n} \frac{dx}{x} \le \frac{1}{n} \cdot 1 + \frac{1}{n+1} \cdot 1 + \frac{1}{n+2} \cdot 1 + \cdots + \frac{1}{2n-1} \cdot 1.$$

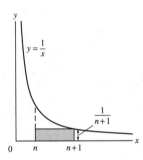

We express the right hand-side of the inequality in terms of c_n, obtaining:

$$\int_n^{2n} \frac{dx}{x} \le c_n - \frac{1}{2n}.$$

We now consider the right endpoint approximation to the integral $\int_n^{2n} \frac{dx}{x}$; that is,

$$\frac{1}{n+1} \cdot 1 + \frac{1}{n+2} \cdot 1 + \cdots + \frac{1}{2n} \cdot 1 \le \int_n^{2n} \frac{dx}{x}.$$

We express the left hand-side of the inequality in terms of c_n, obtaining:

$$c_n - \frac{1}{n} \le \int_n^{2n} \frac{dx}{x}.$$

Thus,

$$\int_n^{2n} \frac{dx}{x} + \frac{1}{2n} \le c_n \le \int_n^{2n} \frac{dx}{x} + \frac{1}{n}.$$

(c) With

$$\int_n^{2n} \frac{dx}{x} = \ln x \big|_n^{2n} = \ln 2n - \ln n = \ln \frac{2n}{n} = \ln 2,$$

the result from part (b) becomes

$$\ln 2 + \frac{1}{2n} \le c_n \le \ln 2 + \frac{1}{n}.$$

Because

$$\lim_{n \to \infty} \frac{1}{2n} = \lim_{n \to \infty} \frac{1}{n} = 0,$$

it follows from the Squeeze Theorem that

$$\lim_{n \to \infty} c_n = \ln 2.$$

11.2 Summing an Infinite Series

Preliminary Questions

1. What role do partial sums play in defining the sum of an infinite series?

SOLUTION   The sum of an infinite series is defined as the limit of the sequence of partial sums. If the limit of this sequence does not exist, the series is said to diverge.

2. What is the sum of the following infinite series?

$$\frac{1}{4} + \frac{1}{8} + \frac{1}{16} + \frac{1}{32} + \frac{1}{64} + \cdots$$

SOLUTION   This is a geometric series with $c = \frac{1}{4}$ and $r = \frac{1}{2}$. The sum of the series is therefore

$$\frac{\frac{1}{4}}{1 - \frac{1}{2}} = \frac{\frac{1}{4}}{\frac{1}{2}} = \frac{1}{2}.$$

3. What happens if you apply the formula for the sum of a geometric series to the following series? Is the formula valid?

$$1 + 3 + 3^2 + 3^3 + 3^4 + \cdots$$

SOLUTION   This is a geometric series with $c = 1$ and $r = 3$. Applying the formula for the sum of a geometric series then gives

$$\sum_{n=0}^{\infty} 3^n = \frac{1}{1 - 3} = -\frac{1}{2}.$$

Clearly, this is not valid: a series with all positive terms cannot have a negative sum. The formula is not valid in this case because a geometric series with $r = 3$ diverges.

4. Arvind asserts that $\sum_{n=1}^{\infty} \frac{1}{n^2} = 0$ because $\frac{1}{n^2}$ tends to zero. Is this valid reasoning?

SOLUTION   Arvind's reasoning is not valid. Though the terms in the series do tend to zero, the general term in the sequence of partial sums,

$$S_n = 1 + \frac{1}{2^2} + \frac{1}{3^2} + \cdots + \frac{1}{n^2},$$

is clearly larger than 1. The sum of the series therefore cannot be zero.

5. Colleen claims that $\sum_{n=1}^{\infty} \frac{1}{\sqrt{n}}$ converges because $\lim_{n \to \infty} \frac{1}{\sqrt{n}} = 0$. Is this valid reasoning?

SOLUTION   Colleen's reasoning is not valid. Although the general term of a convergent series must tend to zero, a series whose general term tends to zero need not converge. In the case of $\sum_{n=1}^{\infty} \frac{1}{\sqrt{n}}$, the series diverges even though its general term tends to zero.

6. Find an N such that $S_N > 25$ for the series $\sum_{n=1}^{\infty} 2$.

SOLUTION   The Nth partial sum of the series is:

$$S_N = \sum_{n=1}^{N} 2 = \underbrace{2 + \cdots + 2}_{N} = 2N.$$

7. Does there exist an N such that $S_N > 25$ for the series $\sum_{n=1}^{\infty} 2^{-n}$? Explain.

SOLUTION   The series $\sum_{n=1}^{\infty} 2^{-n}$ is a convergent geometric series with the common ratio $r = \frac{1}{2}$. The sum of the series is:

$$S = \frac{\frac{1}{2}}{1 - \frac{1}{2}} = 1.$$

Notice that the sequence of partial sums $\{S_N\}$ is increasing and converges to 1; therefore $S_N \leq 1$ for all N. Thus, there does not exist an N such that $S_N > 25$.

8. Give an example of a divergent infinite series whose general term tends to zero.

SOLUTION Consider the series $\sum_{n=1}^{\infty} \frac{1}{n^{\frac{9}{10}}}$. The general term tends to zero, since $\lim_{n \to \infty} \frac{1}{n^{\frac{9}{10}}} = 0$. However, the Nth partial sum satisfies the following inequality:

$$S_N = \frac{1}{1^{\frac{9}{10}}} + \frac{1}{2^{\frac{9}{10}}} + \cdots + \frac{1}{N^{\frac{9}{10}}} \geq \frac{N}{N^{\frac{9}{10}}} = N^{1-\frac{9}{10}} = N^{\frac{1}{10}}.$$

That is, $S_N \geq N^{\frac{1}{10}}$ for all N. Since $\lim_{N \to \infty} N^{\frac{1}{10}} = \infty$, the sequence of partial sums S_n diverges; hence, the series $\sum_{n=1}^{\infty} \frac{1}{n^{\frac{9}{10}}}$ diverges.

Exercises

1. Find a formula for the general term a_n (not the partial sum) of the infinite series.

(a) $\frac{1}{3} + \frac{1}{9} + \frac{1}{27} + \frac{1}{81} + \cdots$

(b) $\frac{1}{1} + \frac{5}{2} + \frac{25}{4} + \frac{125}{8} + \cdots$

(c) $\frac{1}{1} - \frac{2^2}{2 \cdot 1} + \frac{3^3}{3 \cdot 2 \cdot 1} - \frac{4^4}{4 \cdot 3 \cdot 2 \cdot 1} + \cdots$

(d) $\frac{2}{1^2 + 1} + \frac{1}{2^2 + 1} + \frac{2}{3^2 + 1} + \frac{1}{4^2 + 1} + \cdots$

SOLUTION

(a) The denominators of the terms are powers of 3, starting with the first power. Hence, the general term is:

$$a_n = \frac{1}{3^n}.$$

(b) The numerators are powers of 5, and the denominators are the same powers of 2. The first term is $a_1 = 1$ so,

$$a_n = \left(\frac{5}{2}\right)^{n-1}.$$

(c) The general term of this series is,

$$a_n = (-1)^{n+1} \frac{n^n}{n!}.$$

(d) Notice that the numerators of a_n equal 2 for odd values of n and 1 for even values of n. Thus,

$$a_n = \begin{cases} \dfrac{2}{n^2 + 1} & \text{odd } n \\ \dfrac{1}{n^2 + 1} & \text{even } n \end{cases}$$

The formula can also be rewritten as follows:

$$a_n = \frac{1 + \frac{(-1)^{n+1}+1}{2}}{n^2 + 1}.$$

In Exercises 3–6, compute the partial sums S_2, S_4, and S_6.

3. $1 + \frac{1}{2^2} + \frac{1}{3^2} + \frac{1}{4^2} + \cdots$

SOLUTION

$$S_2 = 1 + \frac{1}{2^2} = \frac{5}{4};$$

$$S_4 = 1 + \frac{1}{2^2} + \frac{1}{3^2} + \frac{1}{4^2} = \frac{205}{144};$$

$$S_6 = 1 + \frac{1}{2^2} + \frac{1}{3^2} + \frac{1}{4^2} + \frac{1}{5^2} + \frac{1}{6^2} = \frac{5369}{3600}.$$

5. $\dfrac{1}{1\cdot 2}+\dfrac{1}{2\cdot 3}+\dfrac{1}{3\cdot 4}+\cdots$

SOLUTION

$$S_2 = \frac{1}{1\cdot 2}+\frac{1}{2\cdot 3}=\frac{1}{2}+\frac{1}{6}=\frac{4}{6}=\frac{2}{3};$$

$$S_4 = S_2 + a_3 + a_4 = \frac{2}{3}+\frac{1}{3\cdot 4}+\frac{1}{4\cdot 5}=\frac{2}{3}+\frac{1}{12}+\frac{1}{20}=\frac{4}{5};$$

$$S_6 = S_4 + a_5 + a_6 = \frac{4}{5}+\frac{1}{5\cdot 6}+\frac{1}{6\cdot 7}=\frac{4}{5}+\frac{1}{30}+\frac{1}{42}=\frac{6}{7}.$$

7. Compute S_5, S_{10}, and S_{15} for the series

$$S = \frac{1}{2\cdot 3\cdot 4}-\frac{1}{4\cdot 5\cdot 6}+\frac{1}{6\cdot 7\cdot 8}-\frac{1}{8\cdot 9\cdot 10}+\cdots$$

This series S is known to converge to $\dfrac{\pi-3}{4}$. Do your calculations support this conclusion?

SOLUTION The formula for the general term in the series is

$$a_n = \frac{(-1)^{n+1}}{2n(2n+1)(2n+2)}.$$

Thus,

$$S_5 = \frac{1}{2(3)(4)}-\frac{1}{4(5)(6)}+\cdots-\frac{1}{8(9)(10)}+\frac{1}{10(11)(12)}=0.035678;$$

$$S_{10} = \frac{1}{2(3)(4)}-\frac{1}{4(5)(6)}+\cdots+\frac{1}{18(19)(20)}-\frac{1}{20(21)(22)}=0.035352;$$

$$S_{15} = \frac{1}{2(3)(4)}-\frac{1}{4(5)(6)}+\cdots-\frac{1}{28(29)(30)}+\frac{1}{30(31)(32)}=0.035413.$$

Using a calculator we find

$$\frac{\pi-3}{4}=0.035398,$$

which is consistent with our partial sum calculations.

9. Calculate S_3, S_4, and S_5 and then find the sum of the telescoping series

$$S = \sum_{n=1}^{\infty}\left(\frac{1}{n+1}-\frac{1}{n+2}\right)$$

SOLUTION

$$S_3 = \left(\frac{1}{2}-\frac{1}{3}\right)+\left(\frac{1}{3}-\frac{1}{4}\right)+\left(\frac{1}{4}-\frac{1}{5}\right)=\frac{1}{2}-\frac{1}{5}=\frac{3}{10};$$

$$S_4 = S_3 + \left(\frac{1}{5}-\frac{1}{6}\right)=\frac{1}{2}-\frac{1}{6}=\frac{1}{3};$$

$$S_5 = S_4 + \left(\frac{1}{6}-\frac{1}{7}\right)=\frac{1}{2}-\frac{1}{7}=\frac{5}{14}.$$

The general term in the sequence of partial sums is

$$S_N = \left(\frac{1}{2}-\frac{1}{3}\right)+\left(\frac{1}{3}-\frac{1}{4}\right)+\left(\frac{1}{4}-\frac{1}{5}\right)+\cdots+\left(\frac{1}{N+1}-\frac{1}{N+2}\right)=\frac{1}{2}-\frac{1}{N+2};$$

thus,

$$S = \lim_{N\to\infty} S_N = \lim_{N\to\infty}\left(\frac{1}{2}-\frac{1}{N+2}\right)=\frac{1}{2}.$$

The sum of the telescoping series is therefore $\frac{1}{2}$.

11. Write $\displaystyle\sum_{n=3}^{\infty}\frac{1}{n(n-1)}$ as a telescoping series and find its sum.

SOLUTION By partial fraction decomposition

$$\frac{1}{n(n-1)} = \frac{1}{n-1} - \frac{1}{n},$$

so

$$\sum_{n=3}^{\infty} \frac{1}{n(n-1)} = \sum_{n=3}^{\infty} \left(\frac{1}{n-1} - \frac{1}{n} \right).$$

The general term in the sequence of partial sums for this series is

$$S_N = \left(\frac{1}{2} - \frac{1}{3} \right) + \left(\frac{1}{3} - \frac{1}{4} \right) + \left(\frac{1}{4} - \frac{1}{5} \right) + \cdots + \left(\frac{1}{N-1} - \frac{1}{N} \right) = \frac{1}{2} - \frac{1}{N};$$

thus,

$$S = \lim_{N \to \infty} S_N = \lim_{N \to \infty} \left(\frac{1}{2} - \frac{1}{N} \right) = \frac{1}{2}.$$

In Exercises 13–16, use Theorem 2 to prove that the following series diverge.

13. $\displaystyle\sum_{n=1}^{\infty} (-1)^n n^2$

SOLUTION The general term $a_n = (-1)^n n^2$ does not tend to zero. In fact, because $\lim_{n \to \infty} n^2 = \infty$, $\lim_{n \to \infty} a_n$ does not exist. By Theorem 2, we conclude that the given series diverges.

15. $\displaystyle\sum_{n=0}^{\infty} \left(\sqrt{4n^2 + 1} - n \right)$

SOLUTION The general term of the series satisfies

$$\sqrt{4n^2 + 1} - n > \sqrt{4n^2} - n = n$$

Thus the general term tends to infinity. The series diverges by Theorem 2.

17. Which of these series converge?

(a) $\displaystyle\sum_{n=1}^{\infty} \left(\frac{1}{\sqrt{n}} - \frac{1}{\sqrt{n+1}} \right)$

(b) $\displaystyle\sum_{n=1}^{\infty} (\ln n - \ln(n+1))$

SOLUTION

(a) This series converges. The general term in the sequence of partial sums is

$$S_N = \left(1 - \frac{1}{\sqrt{2}} \right) + \left(\frac{1}{\sqrt{2}} - \frac{1}{\sqrt{3}} \right) + \left(\frac{1}{\sqrt{3}} - \frac{1}{\sqrt{4}} \right) + \cdots + \left(\frac{1}{\sqrt{N}} - \frac{1}{\sqrt{N+1}} \right) = 1 - \frac{1}{\sqrt{N+1}}.$$

Because

$$\lim_{N \to \infty} S_N = \lim_{N \to \infty} \left(1 - \frac{1}{\sqrt{N+1}} \right) = 1,$$

the series converges to 1.

(b) This series diverges. The general term in the sequence of partial sums is

$$S_N = (\ln 1 - \ln 2) + (\ln 2 - \ln 3) + (\ln 3 - \ln 4) + \cdots + (\ln N - \ln(N+1)) = -\ln(N+1).$$

Because

$$\lim_{N \to \infty} S_N = \lim_{N \to \infty} -\ln(N+1) = -\infty,$$

we conclude the given series diverges.

In Exercises 18–31, use the formula for the sum of a geometric series to find the sum or state that the series diverges.

19. $\dfrac{1}{3^3} + \dfrac{1}{3^4} + \dfrac{1}{3^5} + \cdots$

SOLUTION This is a geometric series with $c = \frac{1}{27}$ and $r = \frac{1}{3}$. Thus,

$$\sum_{n=3}^{\infty} \left(\frac{1}{3}\right)^n = \frac{\frac{1}{27}}{1 - \frac{1}{3}} = \frac{\frac{1}{27}}{\frac{2}{3}} = \frac{1}{18}.$$

21. $\displaystyle\sum_{n=3}^{\infty} \frac{3^n}{11^n}$

SOLUTION This is a geometric series with $c = \left(\frac{3}{11}\right)^3 = \frac{27}{1331}$ and $r = \frac{3}{11}$. Thus,

$$\sum_{n=3}^{\infty} \frac{3^n}{11^n} = \frac{\frac{27}{1331}}{1 - \frac{3}{11}} = \frac{\frac{27}{1331}}{\frac{8}{11}} = \frac{27}{968}.$$

23. $1 + \dfrac{2}{7} + \dfrac{2^2}{7^2} + \dfrac{2^3}{7^3} + \cdots$

SOLUTION This is a geometric series with $c = 1$ and $r = \frac{2}{7}$. Thus,

$$\sum_{n=0}^{\infty} \left(\frac{2}{7}\right)^n = \frac{1}{1 - \frac{2}{7}} = \frac{1}{\frac{5}{7}} = \frac{7}{5}.$$

25. $\displaystyle\sum_{n=2}^{\infty} e^{3-2n}$

SOLUTION Rewrite the series as

$$\sum_{n=2}^{\infty} e^3 e^{-2n} = \sum_{n=2}^{\infty} e^3 \left(\frac{1}{e^2}\right)^n$$

to recognize it as a geometric series with $c = e^3 \left(\frac{1}{e^2}\right)^2 = \frac{1}{e}$ and $r = \frac{1}{e^2}$. Thus,

$$\sum_{n=2}^{\infty} e^{3-2n} = \frac{\frac{1}{e}}{1 - \frac{1}{e^2}} = \frac{e}{e^2 - 1}.$$

27. $\displaystyle\sum_{n=0}^{\infty} \frac{93^n + 4^{n-2}}{5^n}$

SOLUTION Rewrite the series as

$$\sum_{n=0}^{\infty} \frac{93^n + 4^{n-2}}{5^n} = \sum_{n=0}^{\infty} \left(\frac{93^n}{5^n} + \frac{4^{n-2}}{5^n}\right) = \sum_{n=0}^{\infty} \left(\frac{93}{5}\right)^n + \sum_{n=0}^{\infty} \frac{1}{16} \cdot \left(\frac{4}{5}\right)^n,$$

which is a sum of two geometric series. The first series has $c = \left(\frac{93}{5}\right)^0 = 1$ and $r = \frac{93}{5}$; the second has $c = \frac{1}{16} \left(\frac{4}{5}\right)^0 = \frac{1}{16}$ and $r = \frac{4}{5}$. Because the first series has $r > 1$, that series diverges. Consequently, the original series also diverges.

29. $\dfrac{2^3}{7} + \dfrac{2^4}{7^2} + \dfrac{2^5}{7^3} + \dfrac{2^6}{7^4} + \cdots$

SOLUTION This is a geometric series with $c = \frac{8}{7}$ and $r = \frac{2}{7}$. Thus,

$$\sum_{n=0}^{\infty} \frac{8}{7} \cdot \left(\frac{2}{7}\right)^n = \frac{\frac{8}{7}}{1 - \frac{2}{7}} = \frac{\frac{8}{7}}{\frac{5}{7}} = \frac{8}{5}.$$

31. $\dfrac{64}{49} + \dfrac{8}{7} + 1 + \dfrac{7}{8} + \dfrac{49}{64} + \dfrac{343}{512} + \cdots$

SOLUTION This is a geometric series with $c = \frac{64}{49}$ and $r = \frac{7}{8}$. Thus,

$$\sum_{n=0}^{\infty} \frac{64}{49} \cdot \left(\frac{7}{8}\right)^n = \frac{\frac{64}{49}}{1 - \frac{7}{8}} = \frac{\frac{64}{49}}{\frac{1}{8}} = \frac{512}{49}.$$

33. Which of the following series are divergent?

(a) $\displaystyle\sum_{n=0}^{\infty} \frac{2^n}{5^n}$

(b) $\displaystyle\sum_{n=3}^{\infty} 1.5^n$

(c) $\displaystyle\sum_{n=0}^{\infty} \frac{5^n}{2^n}$

(d) $\displaystyle\sum_{n=0}^{\infty} (0.4)^n$

SOLUTION

(a) The series $\displaystyle\sum_{n=0}^{\infty} \frac{2^n}{5^n} = \sum_{n=0}^{\infty} \left(\frac{2}{5}\right)^n$ is a geometric series with common ratio $r = \frac{2}{5}$. Since $|r| < 1$, the series converges.

(b) The series $\displaystyle\sum_{n=3}^{\infty} 1.5^n$ is a geometric series with common ratio $r = 1.5$. Since $r > 1$, the series diverges.

(c) The series $\displaystyle\sum_{n=0}^{\infty} \frac{5^n}{2^n} = \sum_{n=0}^{\infty} \left(\frac{5}{2}\right)^n$ is a geometric series with common ratio $r = \frac{5}{2}$. Since $r > 1$, the series diverges.

(d) The series $\displaystyle\sum_{n=0}^{\infty} 0.4^n$ is a geometric series with common ratio $r = 0.4$. Since $|r| < 1$, the series converges.

35. Let $S = \displaystyle\sum_{n=1}^{\infty} a_n$ be an infinite series such that $S_N = 5 - \dfrac{2}{N^2}$.

(a) What are the values of $\displaystyle\sum_{n=1}^{10} a_n$ and $\displaystyle\sum_{n=4}^{16} a_n$?

(b) What is the value of a_3?

(c) Find a general formula for a_n.

(d) Find the sum $\displaystyle\sum_{n=1}^{\infty} a_n$.

SOLUTION

(a)

$$\sum_{n=1}^{10} a_n = S_{10} = 5 - \frac{2}{10^2} = \frac{249}{50};$$

$$\sum_{n=4}^{16} a_n = (a_1 + \cdots + a_{16}) - (a_1 + a_2 + a_3) = S_{16} - S_3 = \left(5 - \frac{2}{16^2}\right) - \left(5 - \frac{2}{3^2}\right) = \frac{2}{9} - \frac{2}{256} = \frac{494}{2304}.$$

(b)

$$a_3 = (a_1 + a_2 + a_3) - (a_1 + a_2) = S_3 - S_2 = \left(5 - \frac{2}{3^2}\right) - \left(5 - \frac{2}{2^2}\right) = \frac{1}{2} - \frac{2}{9} = \frac{5}{18}.$$

(c) Since $a_n = S_n - S_{n-1}$, we have:

$$a_n = S_n - S_{n-1} = \left(5 - \frac{2}{n^2}\right) - \left(5 - \frac{2}{(n-1)^2}\right) = \frac{2}{(n-1)^2} - \frac{2}{n^2}$$

$$= \frac{2\left(n^2 - (n-1)^2\right)}{(n(n-1))^2} = \frac{2\left(n^2 - n^2 + 2n - 1\right)}{(n(n-1))^2} = \frac{2(2n-1)}{n^2(n-1)^2}.$$

(d) The sum $\displaystyle\sum_{n=1}^{\infty} a_n$ is the limit of the sequence of partial sums $\{S_N\}$. Hence:

$$\sum_{n=1}^{\infty} a_n = \lim_{N\to\infty} S_N = \lim_{N\to\infty} \left(5 - \frac{2}{N^2}\right) = 5.$$

37. Use the method of Example 5 to show that $\displaystyle\sum_{k=1}^{\infty} \frac{1}{k^{1/3}}$ diverges.

SOLUTION Each term in the Nth partial sum is greater than or equal to $\dfrac{1}{N^{\frac{1}{3}}}$, hence:

$$S_N = \frac{1}{1^{1/3}} + \frac{1}{2^{1/3}} + \frac{3}{3^{1/3}} + \cdots + \frac{1}{N^{1/3}} \geq \frac{1}{N^{1/3}} + \frac{1}{N^{1/3}} + \frac{1}{N^{1/3}} + \cdots + \frac{1}{N^{1/3}} = N \cdot \frac{1}{N^{1/3}} = N^{2/3}.$$

Since $\displaystyle\lim_{N\to\infty} N^{2/3} = \infty$, it follows that

$$\lim_{N\to\infty} S_N = \infty.$$

Thus, the series $\displaystyle\sum_{k=1}^{\infty} \frac{1}{k^{1/3}}$ diverges.

39. A ball dropped from a height of 10 ft begins to bounce. Each time it strikes the ground, it returns to two-thirds of its previous height. What is the total distance traveled by the ball if it bounces infinitely many times?

SOLUTION The distance traveled by the ball is shown in the accompanying figure:

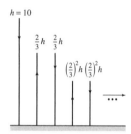

The total distance d traveled by the ball is given by the following infinite sum:

$$d = h + 2 \cdot \frac{2}{3}h + 2 \cdot \left(\frac{2}{3}\right)^2 h + 2 \cdot \left(\frac{2}{3}\right)^3 h + \cdots = h + 2h\left(\frac{2}{3} + \left(\frac{2}{3}\right)^2 + \left(\frac{2}{3}\right)^3 + \cdots\right) = h + 2h\sum_{n=1}^{\infty}\left(\frac{2}{3}\right)^n.$$

We use the formula for the sum of a geometric series to compute the sum of the resulting series:

$$d = h + 2h \cdot \frac{\left(\frac{2}{3}\right)^1}{1 - \frac{2}{3}} = h + 2h(2) = 5h.$$

With $h = 10$ feet, it follows that the total distance traveled by the ball is 50 feet.

41. Find the sum of $\dfrac{1}{1 \cdot 3} + \dfrac{1}{3 \cdot 5} + \dfrac{1}{5 \cdot 7} + \cdots$.

SOLUTION We may write this sum as

$$\sum_{n=1}^{\infty} \frac{1}{(2n-1)(2n+1)} = \sum_{n=1}^{\infty} \frac{1}{2}\left(\frac{1}{2n-1} - \frac{1}{2n+1}\right).$$

The general term in the sequence of partial sums is

$$S_N = \frac{1}{2}\left(\frac{1}{1} - \frac{1}{3}\right) + \frac{1}{2}\left(\frac{1}{3} - \frac{1}{5}\right) + \frac{1}{2}\left(\frac{1}{5} - \frac{1}{7}\right) + \cdots + \frac{1}{2}\left(\frac{1}{2N-1} - \frac{1}{2N+1}\right) = \frac{1}{2}\left(1 - \frac{1}{2N+1}\right);$$

thus,

$$\lim_{N\to\infty} S_N = \lim_{N\to\infty} \frac{1}{2}\left(1 - \frac{1}{2N+1}\right) = \frac{1}{2},$$

and

$$\sum_{n=1}^{\infty} \frac{1}{(2n-1)(2n+1)} = \frac{1}{2}.$$

43. Let $\{b_n\}$ be a sequence and let $a_n = b_n - b_{n-1}$. Show that $\displaystyle\sum_{n=1}^{\infty} a_n$ converges if and only if $\displaystyle\lim_{n\to\infty} b_n$ exists.

SOLUTION   Let $a_n = b_n - b_{n-1}$. The general term in the sequence of partial sums for the series $\displaystyle\sum_{n=1}^{\infty} a_n$ is then

$$S_N = (b_1 - b_0) + (b_2 - b_1) + (b_3 - b_2) + \cdots + (b_N - b_{N-1}) = b_N - b_0.$$

Now, if $\displaystyle\lim_{N\to\infty} b_N$ exists, then so does $\displaystyle\lim_{N\to\infty} S_N$ and $\displaystyle\sum_{n=1}^{\infty} a_n$ converges. On the other hand, if $\displaystyle\sum_{n=1}^{\infty} a_n$ converges, then $\displaystyle\lim_{N\to\infty} S_N$ exists, which implies that $\displaystyle\lim_{N\to\infty} b_N$ also exists. Thus, $\displaystyle\sum_{n=1}^{\infty} a_n$ converges if and only if $\displaystyle\lim_{n\to\infty} b_n$ exists.

45. Find the total length of the infinite zigzag path in Figure 4 (each zag occurs at an angle of $\frac{\pi}{4}$).

FIGURE 4

SOLUTION   Because the angle at the lower left in Figure 4 has measure $\frac{\pi}{4}$ and each zag in the path occurs at an angle of $\frac{\pi}{4}$, every triangle in the figure is an isosceles right triangle. Accordingly, the length of each new segment in the path is $\frac{1}{\sqrt{2}}$ times the length of the previous segment. Since the first segment has length 1, the total length of the path is

$$\sum_{n=0}^{\infty} \left(\frac{1}{\sqrt{2}}\right)^n = \frac{1}{1 - \frac{1}{\sqrt{2}}} = \frac{\sqrt{2}}{\sqrt{2} - 1} = 2 + \sqrt{2}.$$

47. Show that if a is a positive integer, then

$$\sum_{n=1}^{\infty} \frac{1}{n(n+a)} = \frac{1}{a}\left(1 + \frac{1}{2} + \cdots + \frac{1}{a}\right)$$

SOLUTION   By partial fraction decomposition

$$\frac{1}{n(n+a)} = \frac{A}{n} + \frac{B}{n+a};$$

clearing the denominators gives

$$1 = A(n+a) + Bn.$$

Setting $n = 0$ then yields $A = \frac{1}{a}$, while setting $n = -a$ yields $B = -\frac{1}{a}$. Thus,

$$\frac{1}{n(n+a)} = \frac{\frac{1}{a}}{n} - \frac{\frac{1}{a}}{n+a} = \frac{1}{a}\left(\frac{1}{n} - \frac{1}{n+a}\right),$$

and

$$\sum_{n=1}^{\infty} \frac{1}{n(n+a)} = \sum_{n=1}^{\infty} \frac{1}{a}\left(\frac{1}{n} - \frac{1}{n+a}\right).$$

For $N > a$, the Nth partial sum is

$$S_N = \frac{1}{a}\left(1 + \frac{1}{2} + \frac{1}{3} + \cdots + \frac{1}{a}\right) - \frac{1}{a}\left(\frac{1}{N+1} + \frac{1}{N+2} + \frac{1}{N+3} + \cdots + \frac{1}{N+a}\right).$$

Thus,

$$\sum_{n=1}^{\infty} \frac{1}{n(n+a)} = \lim_{N\to\infty} S_N = \frac{1}{a}\left(1 + \frac{1}{2} + \frac{1}{3} + \cdots + \frac{1}{a}\right).$$

Further Insights and Challenges

49. Professor George Andrews of Pennsylvania State University observed that geometric sums can be used to calculate the derivative of $f(x) = x^N$ in a new way. By Eq. (2),

$$1 + r + r^2 + \cdots + r^{N-1} = \frac{1 - r^N}{1 - r}$$

7

Assume that $a \neq 0$ and let $x = ra$. Show that

$$f'(a) = \lim_{x \to a} \frac{x^N - a^N}{x - a} = a^{N-1} \lim_{r \to 1} \frac{r^N - 1}{r - 1}$$

Then use Eq. (7) to evaluate the limit on the right.

SOLUTION   According to the definition of derivative of $f(x)$ at $x = a$

$$f'(a) = \lim_{x \to a} \frac{x^N - a^N}{x - a}.$$

Now, let $x = ra$. Then $x \to a$ if and only if $r \to 1$, and

$$f'(a) = \lim_{x \to a} \frac{x^N - a^N}{x - a} = \lim_{r \to 1} \frac{(ra)^N - a^N}{ra - a} = \lim_{r \to 1} \frac{a^N \left(r^N - 1 \right)}{a \left(r - 1 \right)} = a^{N-1} \lim_{r \to 1} \frac{r^N - 1}{r - 1}.$$

By Eq. (7),

$$\frac{1 - r^N}{1 - r} = \frac{r^N - 1}{r - 1} = 1 + r + r^2 + \cdots + r^{N-1},$$

so

$$\lim_{r \to 1} \frac{r^N - 1}{r - 1} = \lim_{r \to 1} \left(1 + r + r^2 + \cdots + r^{N-1} \right) = 1 + 1 + 1^2 + \cdots + 1^{N-1} = N.$$

Therefore, $f'(a) = a^{N-1} \cdot N = N a^{N-1}$

51. Cantor's Disappearing Table (following Larry Knop of Hamilton College)   Take a table of length L (Figure 6). At stage 1, remove the section of length $L/4$ centered at the midpoint. Two sections remain, each with a length less than $L/2$. At stage 2, remove sections of length $L/4^2$ from each of these two sections (this stage removes $L/8$ of the table). Now four sections remain, each of length less than $L/4$. At stage 3, remove the four central sections of length $L/4^3$, etc.

(a) Show that at the Nth stage, each remaining section has length less than $L/2^N$ and that the total amount of table removed is

$$L \left(\frac{1}{4} + \frac{1}{8} + \frac{1}{16} + \cdots + \frac{1}{2^{N+1}} \right)$$

(b) Show that in the limit as $N \to \infty$, precisely one-half of the table remains.

This result is curious, because there are no nonzero intervals of table left (at each stage, the remaining sections have a length less than $L/2^N$). So the table has "disappeared." However, we can place any object longer than $L/4$ on the table and it will not fall through since it will not fit through any of the removed sections.

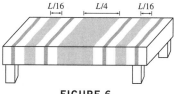

FIGURE 6

SOLUTION

(a) After the Nth stage, the total amount of table that has been removed is

$$\frac{L}{4} + \frac{2L}{4^2} + \frac{4L}{4^3} + \cdots + \frac{2^{N-1}L}{4^N} = L \left(\frac{1}{4} + \frac{1}{8} + \frac{1}{16} + \cdots + \frac{2^{N-1}}{2^{2N}} \right) = L \left(\frac{1}{4} + \frac{1}{8} + \frac{1}{16} + \cdots + \frac{1}{2^{N+1}} \right)$$

At the first stage ($N = 1$), there are two remaining sections each of length

$$\frac{L - \frac{L}{4}}{2} = \frac{3L}{8} < \frac{L}{2}.$$

Suppose that at the Kth stage, each of the 2^K remaining sections has length less than $\dfrac{L}{2^K}$. The $(K+1)$st stage is obtained by removing the section of length $\dfrac{L}{4^{K+1}}$ centered at the midpoint of each segment in the Kth stage. Let a_k and a_{K+1}, respectively, denote the length of each segment in the Kth and $(K+1)$st stage. Then,

$$a_{K+1} = \frac{a_K - \frac{L}{4^{K+1}}}{2} < \frac{\frac{L}{2^K} - \frac{L}{4^{K+1}}}{2} = \frac{L}{2^K}\left(\frac{1 - \frac{1}{2^{K+2}}}{2}\right) < \frac{L}{2^K} \cdot \frac{1}{2} = \frac{L}{2^{K+1}}.$$

Thus, by mathematical induction, each remaining section at the Nth stage has length less than $\dfrac{L}{2^N}$.

(b) From part (a), we know that after N stages, the amount of the table that has been removed is

$$L\left(\frac{1}{4} + \frac{1}{8} + \frac{1}{16} + \cdots + \frac{1}{2^{N+1}}\right) = \sum_{n=1}^{N} \frac{1}{2^{n+1}}.$$

As $N \to \infty$, the amount of the table that has been removed becomes a geometric series whose sum is

$$L\sum_{n=1}^{\infty} \frac{1}{2}\left(\frac{1}{2}\right)^n = L\frac{\frac{1}{4}}{1 - \frac{1}{2}} = \frac{1}{2}L.$$

Thus, the amount of table that remains is $L - \frac{1}{2}L = \frac{1}{2}L$.

11.3 Convergence of Series with Positive Terms

Preliminary Questions

1. Let $S = \displaystyle\sum_{n=1}^{\infty} a_n$. If the partial sums S_N are increasing, then (choose correct conclusion)

(a) $\{a_n\}$ is an increasing sequence.

(b) $\{a_n\}$ is a positive sequence.

SOLUTION The correct response is **(b)**. Recall that $S_N = a_1 + a_2 + a_3 + \cdots + a_N$; thus, $S_N - S_{N-1} = a_N$. If S_N is increasing, then $S_N - S_{N-1} \geq 0$. It then follows that $a_N \geq 0$; that is, $\{a_n\}$ is a positive sequence.

2. What are the hypotheses of the Integral Test?

SOLUTION The hypotheses for the Integral Test are: A function $f(x)$ such that $a_n = f(n)$ must be positive, decreasing, and continuous for $x \geq 1$.

3. Which test would you use to determine whether $\displaystyle\sum_{n=1}^{\infty} n^{-3.2}$ converges?

SOLUTION Because $n^{-3.2} = \frac{1}{n^{3.2}}$, we see that the indicated series is a p-series with $p = 3.2 > 1$. Therefore, the series converges.

4. Which test would you use to determine whether $\displaystyle\sum_{n=1}^{\infty} \frac{1}{2^n + \sqrt{n}}$ converges?

SOLUTION Because

$$\frac{1}{2^n + \sqrt{n}} < \frac{1}{2^n} = \left(\frac{1}{2}\right)^n,$$

and

$$\sum_{n=1}^{\infty} \left(\frac{1}{2}\right)^n$$

is a convergent geometric series, the comparison test would be an appropriate choice to establish that the given series converges.

5. Ralph hopes to investigate the convergence of $\displaystyle\sum_{n=1}^{\infty} \frac{e^{-n}}{n}$ by comparing it with $\displaystyle\sum_{n=1}^{\infty} \frac{1}{n}$. Is Ralph on the right track?

SOLUTION No, Ralph is not on the right track. For $n \geq 1$,

$$\frac{e^{-n}}{n} < \frac{1}{n};$$

however, $\sum_{n=1}^{\infty} \frac{1}{n}$ is a divergent series. The Comparison Test therefore does not allow us to draw a conclusion about the

convergence or divergence of the series $\sum_{n=1}^{\infty} \frac{e^{-n}}{n}$.

Exercises

In Exercises 1–14, use the Integral Test to determine whether the infinite series is convergent.

1. $\sum_{n=1}^{\infty} \frac{1}{n^4}$

SOLUTION Let $f(x) = \frac{1}{x^4}$. This function is continuous, positive and decreasing on the interval $x \geq 1$, so the Integral Test applies. Moreover,

$$\int_1^{\infty} \frac{dx}{x^4} = \lim_{R \to \infty} \int_1^R x^{-4}\, dx = -\frac{1}{3} \lim_{R \to \infty} \left(\frac{1}{R^3} - 1 \right) = \frac{1}{3}.$$

The integral converges; hence, the series $\sum_{n=1}^{\infty} \frac{1}{n^4}$ also converges.

3. $\sum_{n=1}^{\infty} n^{-1/3}$

SOLUTION Let $f(x) = x^{-\frac{1}{3}} = \frac{1}{\sqrt[3]{x}}$. This function is continuous, positive and decreasing on the interval $x \geq 1$, so the Integral Test applies. Moreover,

$$\int_1^{\infty} x^{-1/3}\, dx = \lim_{R \to \infty} \int_1^R x^{-1/3}\, dx = \frac{3}{2} \lim_{R \to \infty} \left(R^{2/3} - 1 \right) = \infty.$$

The integral diverges; hence, the series $\sum_{n=1}^{\infty} n^{-1/3}$ also diverges.

5. $\sum_{n=25}^{\infty} \frac{n^2}{(n^3 + 9)^{5/2}}$

SOLUTION Let $f(x) = \frac{x^2}{(x^3 + 9)^{5/2}}$. This function is positive and continuous for $x \geq 25$. Moreover, because

$$f'(x) = \frac{2x(x^3 + 9)^{5/2} - x^2 \cdot \frac{5}{2}(x^3 + 9)^{3/2} \cdot 3x^2}{(x^3 + 9)^5} = \frac{x(36 - 11x^3)}{2(x^3 + 9)^{7/2}},$$

we see that $f'(x) < 0$ for $x \geq 25$, so f is decreasing on the interval $x \geq 25$. The Integral Test therefore applies. To evaluate the improper integral, we use the substitution $u = x^3 + 9$, $du = 3x^2 dx$. We then find

$$\int_{25}^{\infty} \frac{x^2}{(x^3 + 9)^{5/2}}\, dx = \lim_{R \to \infty} \int_{25}^R \frac{x^2}{(x^3 + 9)^{5/2}}\, dx = \frac{1}{3} \lim_{R \to \infty} \int_{15634}^{R^3 + 9} \frac{du}{u^{5/2}}$$

$$= -\frac{2}{9} \lim_{R \to \infty} \left(\frac{1}{(R^3 + 9)^{3/2}} - \frac{1}{15634^{3/2}} \right) = \frac{2}{9 \cdot 15634^{3/2}}.$$

The integral converges; hence, the series $\sum_{n=25}^{\infty} \frac{n^2}{(n^3 + 9)^{5/2}}$ also converges.

7. $\sum_{n=1}^{\infty} \frac{1}{n^2 + 1}$

SOLUTION Let $f(x) = \dfrac{1}{x^2 + 1}$. This function is positive, decreasing and continuous on the interval $x \geq 1$, hence the Integral Test applies. Moreover,

$$\int_1^\infty \frac{dx}{x^2 + 1} = \lim_{R \to \infty} \int_1^R \frac{dx}{x^2 + 1} = \lim_{R \to \infty} \left(\tan^{-1} R - \frac{\pi}{4} \right) = \frac{\pi}{2} - \frac{\pi}{4} = \frac{\pi}{4}.$$

The integral converges; hence, the series $\displaystyle\sum_{n=1}^\infty \frac{1}{n^2 + 1}$ also converges.

9. $\displaystyle\sum_{n=1}^\infty n e^{-n^2}$

SOLUTION Let $f(x) = xe^{-x^2}$. This function is continuous and positive on the interval $x \geq 1$. Moreover, because

$$f'(x) = 1 \cdot e^{-x^2} + x \cdot e^{-x^2} \cdot (-2x) = e^{-x^2} \left(1 - 2x^2 \right),$$

we see that $f'(x) < 0$ for $x \geq 1$, so f is decreasing on this interval. To compute the improper integral we make the substitution $u = x^2$, $du = 2x \, dx$. Then, we find

$$\int_1^\infty x e^{-x^2} \, dx = \lim_{R \to \infty} \int_1^R x e^{-x^2} \, dx = \frac{1}{2} \int_1^{R^2} e^{-u} \, du = -\frac{1}{2} \lim_{R \to \infty} \left(e^{-R^2} - e^{-1} \right) = \frac{1}{2e}.$$

The integral converges; hence, the series $\displaystyle\sum_{n=1}^\infty n e^{-n^2}$ also converges.

11. $\displaystyle\sum_{n=1}^\infty \frac{1}{2^{\ln n}}$

SOLUTION Note that

$$2^{\ln n} = (e^{\ln 2})^{\ln n} = (e^{\ln n})^{\ln 2} = n^{\ln 2}.$$

Thus,

$$\sum_{n=1}^\infty \frac{1}{2^{\ln n}} = \sum_{n=1}^\infty \frac{1}{n^{\ln 2}}.$$

Now, let $f(x) = \dfrac{1}{x^{\ln 2}}$. This function is positive, continuous and decreasing on the interval $x \geq 1$; therefore, the Integral Test applies. Moreover,

$$\int_1^\infty \frac{dx}{x^{\ln 2}} = \lim_{R \to \infty} \int_1^R \frac{dx}{x^{\ln 2}} = \frac{1}{1 - \ln 2} \lim_{R \to \infty} (R^{1 - \ln 2} - 1) = \infty,$$

because $1 - \ln 2 > 0$. The integral diverges; hence, the series $\displaystyle\sum_{n=1}^\infty \frac{1}{2^{\ln n}}$ also diverges.

13. $\displaystyle\sum_{n=1}^\infty \frac{\ln n}{n^2}$

SOLUTION Let $f(x) = \dfrac{\ln x}{x^2}$. Because

$$f'(x) = \frac{\frac{1}{x} \cdot x^2 - 2x \ln x}{x^4} = \frac{x (1 - 2 \ln x)}{x^4} = \frac{1 - 2 \ln x}{x^3},$$

we see that $f'(x) < 0$ for $x > \sqrt{e} \approx 1.65$. We conclude that f is decreasing on the interval $x \geq 2$. Since f is also positive and continuous on this interval, the Integral Test can be applied. By Integration by Parts, we find

$$\int \frac{\ln x}{x^2} \, dx = -\frac{\ln x}{x} + \int x^{-2} \, dx = -\frac{\ln x}{x} - \frac{1}{x} + C;$$

therefore,

$$\int_2^\infty \frac{\ln x}{x^2} \, dx = \lim_{R \to \infty} \int_2^R \frac{\ln x}{x^2} \, dx = \lim_{R \to \infty} \left(\frac{1}{2} + \frac{\ln 2}{2} - \frac{1}{R} - \frac{\ln R}{R} \right) = \frac{1 + \ln 2}{2} - \lim_{R \to \infty} \frac{\ln R}{R}.$$

We compute the resulting limit using L'Hôpital's Rule:

$$\lim_{R\to\infty} \frac{\ln R}{R} = \lim_{R\to\infty} \frac{1/R}{1} = 0.$$

Hence,

$$\int_2^\infty \frac{\ln x}{x^2}\, dx = \frac{1 + \ln 2}{2}.$$

The integral converges; therefore, the series $\sum_{n=2}^\infty \frac{\ln n}{n^2}$ also converges. Since the convergence of the series is not affected by adding the finite sum $\sum_{n=1}^1 \frac{\ln n}{n^2}$, the series $\sum_{n=1}^\infty \frac{\ln n}{n^2}$ also converges.

15. Use the Comparison Test to show that $\sum_{n=1}^\infty \frac{1}{n^3 + 8n}$ converges. *Hint:* Compare with $\sum_{n=1}^\infty n^{-3}$.

SOLUTION We compare the series with the *p*-series $\sum_{n=1}^\infty n^{-3}$. For $n \geq 1$,

$$\frac{1}{n^3 + 8n} \leq \frac{1}{n^3}.$$

Since $\sum_{n=1}^\infty \frac{1}{n^3}$ converges (it is a *p*-series with $p = 3 > 1$), the series $\sum_{n=1}^\infty \frac{1}{n^3 + 8n}$ also converges by the Comparison Test.

17. Let $S = \sum_{n=1}^\infty \frac{1}{n + \sqrt{n}}$. Verify that for $n \geq 1$

$$\frac{1}{n + \sqrt{n}} \leq \frac{1}{n}, \qquad \frac{1}{n + \sqrt{n}} \leq \frac{1}{\sqrt{n}}$$

Can either inequality be used to show that S diverges? Show that $\frac{1}{n + \sqrt{n}} \geq \frac{1}{2n}$ and conclude that S diverges.

SOLUTION For $n \geq 1$, $n + \sqrt{n} \geq n$ and $n + \sqrt{n} \geq \sqrt{n}$. Taking the reciprocal of each of these inequalities yields

$$\frac{1}{n + \sqrt{n}} \leq \frac{1}{n} \quad \text{and} \quad \frac{1}{n + \sqrt{n}} \leq \frac{1}{\sqrt{n}}.$$

These inequalities indicate that the series $\sum_{n=1}^\infty \frac{1}{n + \sqrt{n}}$ is smaller than both $\sum_{n=1}^\infty \frac{1}{n}$ and $\sum_{n=1}^\infty \frac{1}{\sqrt{n}}$; however, $\sum_{n=1}^\infty \frac{1}{n}$ and $\sum_{n=1}^\infty \frac{1}{\sqrt{n}}$ both diverge so neither inequality allows us to show that S diverges.

On the other hand, for $n \geq 1$, $n \geq \sqrt{n}$, so $2n \geq n + \sqrt{n}$ and

$$\frac{1}{n + \sqrt{n}} \geq \frac{1}{2n}.$$

The series $\sum_{n=1}^\infty \frac{1}{2n} = 2 \sum_{n=1}^\infty \frac{1}{n}$ diverges, since the harmonic series diverges. The Comparison Test then lets us conclude that the larger series $\sum_{n=1}^\infty \frac{1}{n + \sqrt{n}}$ also diverges.

In Exercises 19–31, use the Comparison Test to determine whether the infinite series is convergent.

19. $\sum_{n=1}^\infty \frac{1}{n2^n}$

SOLUTION We compare with the geometric series $\sum_{n=1}^\infty \left(\frac{1}{2}\right)^n$. For $n \geq 1$,

$$\frac{1}{n2^n} \leq \frac{1}{2^n} = \left(\frac{1}{2}\right)^n.$$

Since $\displaystyle\sum_{n=1}^{\infty}\left(\frac{1}{2}\right)^n$ converges (it is a geometric series with $r = \frac{1}{2}$), we conclude by the Comparison Test that $\displaystyle\sum_{n=1}^{\infty}\frac{1}{n2^n}$ also converges.

21. $\displaystyle\sum_{k=1}^{\infty}\frac{k^{1/3}}{k^2 + k}$

SOLUTION For $k \geq 1$,

$$\frac{k^{1/3}}{k^2 + k} \leq \frac{k^{1/3}}{k^2} = \frac{1}{k^{5/3}}.$$

The series $\displaystyle\sum_{k=1}^{\infty}\frac{1}{k^{5/3}}$ is a p-series with $p = \frac{5}{3} > 1$, so it converges. By the Comparison Test we can therefore conclude that the series $\displaystyle\sum_{k=1}^{\infty}\frac{k^{1/3}}{k^2 + k}$ also converges.

23. $\displaystyle\sum_{n=1}^{\infty}\frac{1}{\sqrt{n^3 + 1}}$

SOLUTION For $n \geq 1$,

$$\frac{1}{\sqrt{n^3 + 1}} \leq \frac{1}{\sqrt{n^3}} = \frac{1}{n^{3/2}}.$$

The series $\displaystyle\sum_{n=1}^{\infty}\frac{1}{n^{\frac{3}{2}}}$ converges, since it is a p-series with $p = \frac{3}{2} > 1$. By the Comparison Test we can therefore conclude that the series $\displaystyle\sum_{n=1}^{\infty}\frac{1}{\sqrt{n^3 + 1}}$ also converges.

25. $\displaystyle\sum_{k=1}^{\infty}\frac{\sin^2 k}{k^2}$

SOLUTION For $k \geq 1, 0 \leq \sin^2 k \leq 1$, so

$$0 \leq \frac{\sin^2 k}{k^2} \leq \frac{1}{k^2}.$$

The series $\displaystyle\sum_{k=1}^{\infty}\frac{1}{k^2}$ is a p-series with $p = 2 > 1$, so it converges. By the Comparison Test we can therefore conclude that the series $\displaystyle\sum_{k=1}^{\infty}\frac{\sin^2 k}{k^2}$ also converges.

27. $\displaystyle\sum_{m=1}^{\infty}\frac{4}{m! + 4^m}$

SOLUTION For $m \geq 1$,

$$\frac{4}{m! + 4^m} \leq \frac{4}{4^m} = \left(\frac{1}{4}\right)^{m-1}.$$

The series $\displaystyle\sum_{m=1}^{\infty}\left(\frac{1}{4}\right)^{m-1}$ is a geometric series with $r = \frac{1}{4}$, so it converges. By the Comparison Test we can therefore conclude that the series $\displaystyle\sum_{m=1}^{\infty}\frac{4}{m! + 4^m}$ also converges.

29. $\displaystyle\sum_{k=1}^{\infty}2^{-k^2}$

SOLUTION For $k \geq 1, k^2 \geq k$ and

$$\frac{1}{2^{k^2}} \leq \frac{1}{2^k} = \left(\frac{1}{2}\right)^k.$$

The series $\sum_{k=1}^{\infty} \left(\frac{1}{2}\right)^k$ is a geometric series with $r = \frac{1}{2}$, so it converges. By the Comparison Test we can therefore conclude that the series $\sum_{k=1}^{\infty} \frac{1}{2^{k^2}} = \sum_{k=1}^{\infty} 2^{-k^2}$ also converges.

31. $\sum_{n=1}^{\infty} \frac{\ln n}{n^3 + 3\ln n}$

SOLUTION For $n \geq 1, 0 \leq \ln n \leq n$ and

$$0 \leq \frac{\ln n}{n^3 + 3\ln n} \leq \frac{n}{n^3} = \frac{1}{n^2}.$$

The series $\sum_{n=1}^{\infty} \frac{1}{n^2}$ is a p-series with $p = 2 > 1$, so it converges. By the Comparison Test we can therefore conclude that the series $\sum_{n=1}^{\infty} \frac{\ln n}{n^3 + 3\ln n}$ also converges.

33. Does $\sum_{n=1}^{\infty} \frac{n}{\sqrt{n^2 + c}}$ converge for any c?

SOLUTION Because

$$\lim_{n \to \infty} \frac{n}{\sqrt{n^2 + c}} = \lim_{n \to \infty} \frac{n}{n} = 1 \neq 0,$$

it follows from the Divergence Test that the series $\sum_{n=1}^{\infty} \frac{n}{\sqrt{n^2 + c}}$ diverges for all values of c.

In Exercises 34–42, use the Limit Comparison Test to prove convergence or divergence of the infinite series.

35. $\sum_{n=2}^{\infty} \frac{1}{n^2 - \sqrt{n}}$

SOLUTION Let $a_n = \frac{1}{n^2 - \sqrt{n}}$. For large n, $\frac{1}{n^2 - \sqrt{n}} \approx \frac{1}{n^2}$, so we apply the Limit Comparison Test with $b_n = \frac{1}{n^2}$. We find

$$L = \lim_{n \to \infty} \frac{a_n}{b_n} = \lim_{n \to \infty} \frac{\frac{1}{n^2 - \sqrt{n}}}{\frac{1}{n^2}} = \lim_{n \to \infty} \frac{n^2}{n^2 - \sqrt{n}} = 1.$$

The series $\sum_{n=1}^{\infty} \frac{1}{n^2}$ is a p-series with $p = 2 > 1$, so it converges; hence, the series $\sum_{n=2}^{\infty} \frac{1}{n^2}$ also converges. Because L exists, by the Limit Comparison Test we can conclude that the series $\sum_{n=2}^{\infty} \frac{1}{n^2 - \sqrt{n}}$ converges.

37. $\sum_{n=2}^{\infty} \frac{n^3}{\sqrt{n^7 - 2n^2 + 1}}$

SOLUTION Let a_n be the general term of our series. Observe that

$$a_n = \frac{n^3}{\sqrt{n^7 - 2n^2 + 1}} = \frac{n^{-3} \cdot n^3}{n^{-3} \cdot \sqrt{n^7 - 2n^2 + 1}} = \frac{1}{\sqrt{n - 2n^{-4} + n^{-6}}}$$

This suggests that apply the Limit Comparison Test, comparing our series with

$$\sum_{n=2}^{\infty} b_n = \sum_{n=2}^{\infty} \frac{1}{n^{1/2}}$$

The ratio of the terms is

$$\frac{a_n}{b_n} = \frac{1}{\sqrt{n - 2n^{-4} + n^{-6}}} \cdot \frac{\sqrt{n}}{1} = \frac{1}{\sqrt{1 - 2n^{-5} + n^{-7}}}$$

Hence

$$\lim_{n \to \infty} \frac{a_n}{b_n} = \lim_{n \to \infty} \frac{1}{\sqrt{1 - 2n^{-5} + n^{-7}}} = 1$$

The p-series $\displaystyle\sum_{n=2}^{\infty} \frac{1}{n^{1/2}}$ diverges since $p = 1/2 < 1$. Therefore, our original series diverges.

39. $\displaystyle\sum_{n=1}^{\infty} \frac{e^n + n}{e^{2n} - n^2}$

SOLUTION Let

$$a_n = \frac{e^n + n}{e^{2n} - n^2} = \frac{e^n + n}{(e^n - n)(e^n + n)} = \frac{1}{e^n - n}.$$

For large n,

$$\frac{1}{e^n - n} \approx \frac{1}{e^n} = e^{-n},$$

so we apply the Limit Comparison Test with $b_n = e^{-n}$. We find

$$L = \lim_{n \to \infty} \frac{a_n}{b_n} = \lim_{n \to \infty} \frac{\frac{1}{e^n - n}}{e^{-n}} = \lim_{n \to \infty} \frac{e^n}{e^n - n} = 1.$$

The series $\displaystyle\sum_{n=1}^{\infty} e^{-n} = \sum_{n=1}^{\infty} \left(\frac{1}{e}\right)^n$ is a geometric series with $r = \frac{1}{e} < 1$, so it converges. Because L exists, by the Limit

Comparison Test we can conclude that the series $\displaystyle\sum_{n=1}^{\infty} \frac{e^n + n}{e^{2n} - n^2}$ also converges.

41. $\displaystyle\sum_{n=1}^{\infty} \left(1 - \cos \frac{1}{n}\right)$ *Hint:* Compare with $\displaystyle\sum_{n=1}^{\infty} n^{-2}$.

SOLUTION Let $a_n = 1 - \cos \dfrac{1}{n}$, and apply the Limit Comparison Test with $b_n = \dfrac{1}{n^2}$. We find

$$L = \lim_{n \to \infty} \frac{a_n}{b_n} = \lim_{n \to \infty} \frac{1 - \cos \frac{1}{n}}{\frac{1}{n^2}} = \lim_{x \to \infty} \frac{1 - \cos \frac{1}{x}}{\frac{1}{x^2}} = \lim_{x \to \infty} \frac{-\frac{1}{x^2} \sin \frac{1}{x}}{-\frac{2}{x^3}} = \frac{1}{2} \lim_{x \to \infty} \frac{\sin \frac{1}{x}}{\frac{1}{x}}.$$

As $x \to \infty$, $u = \frac{1}{x} \to 0$, so

$$L = \frac{1}{2} \lim_{x \to \infty} \frac{\sin \frac{1}{x}}{\frac{1}{x}} = \frac{1}{2} \lim_{u \to 0} \frac{\sin u}{u} = \frac{1}{2}.$$

The series $\displaystyle\sum_{n=1}^{\infty} \frac{1}{n^2}$ is a p-series with $p = 2 > 1$, so it converges. Because L exists, by the Limit Comparison Test we can

conclude that the series $\displaystyle\sum_{n=1}^{\infty} \left(1 - \cos \frac{1}{n}\right)$ also converges.

43. Show that if $a_n \geq 0$ and $\displaystyle\lim_{n \to \infty} n^2 a_n$ exists, then $\displaystyle\sum_{n=1}^{\infty} a_n$ converges. *Hint:* Show that if M is larger than $\displaystyle\lim_{n \to \infty} n^2 a_n$,

then $a_n \leq M/n^2$ for n sufficiently large.

SOLUTION Let $\displaystyle\lim_{n \to \infty} n^2 a_n = L$. Then there exists an integer N such that, for all $n \geq N$,

$$|n^2 a_n - L| < \frac{1}{2}.$$

Thus,

$$0 \leq n^2 a_n < L + \frac{1}{2} \quad \text{or} \quad 0 \leq a_n < \frac{L + \frac{1}{2}}{n^2}.$$

The series $\sum_{n=1}^{\infty} \frac{1}{n^2}$ is a p-series with $p = 2 > 1$, so it converges; hence, the series $\sum_{n=N}^{\infty} \frac{L + \frac{1}{2}}{n^2}$ also converges. By the Comparison Test we can therefore conclude that the series $\sum_{n=N}^{\infty} a_n$ converges. This series continues to converge if we add the finite sum $\sum_{n=1}^{N-1} a_n$; hence, the series $\sum_{n=1}^{\infty} a_n$ converges.

45. Show that $\sum_{n=2}^{\infty} (\ln n)^{-2}$ diverges. *Hint:* Show that for x sufficiently large, $\ln x < x^{1/2}$.

SOLUTION Using L'Hôpital's Rule,

$$\lim_{x \to \infty} \frac{x^{1/2}}{\ln x} = \lim_{x \to \infty} \frac{\frac{1}{2}x^{-1/2}}{\frac{1}{x}} = \lim_{x \to \infty} \frac{1}{2}x^{1/2} = \infty;$$

thus, there exists an integer N such that

$$\frac{n^{1/2}}{\ln n} > 1 \quad \text{or} \quad n^{1/2} > \ln n$$

for all $n \geq N$. Hence,

$$\frac{1}{\ln n} > \frac{1}{n^{1/2}} \quad \text{and} \quad \frac{1}{(\ln n)^2} > \frac{1}{n}$$

for all $n \geq N$. The harmonic series diverges, so the series $\sum_{n=N}^{\infty} \frac{1}{n}$ also diverges. By the Comparison Test we can therefore conclude that the series $\sum_{n=N}^{\infty} \frac{1}{(\ln n)^2}$ diverges. It follows that the series $\sum_{n=2}^{\infty} (\ln n)^{-2}$ also diverges.

47. For which a does $\sum_{n=2}^{\infty} \frac{1}{n^a \ln n}$ converge?

SOLUTION First consider the case $a > 1$. For $n \geq 3$, $\ln n > 1$ and

$$\frac{1}{n^a \ln n} < \frac{1}{n^a}.$$

The series $\sum_{n=1}^{\infty} \frac{1}{n^a}$ is a p-series with $p = a > 1$, so it converges; hence, $\sum_{n=3}^{\infty} \frac{1}{n^a}$ also converges. By the Comparison Test we can therefore conclude that the series $\sum_{n=3}^{\infty} \frac{1}{n^a \ln n}$ converges, which implies the series $\sum_{n=2}^{\infty} \frac{1}{n^a \ln n}$ also converges.

For $a \leq 1$, $n^a \leq n$ so

$$\frac{1}{n^a \ln n} \geq \frac{1}{n \ln n}$$

for $n \geq 2$. Let $f(x) = \frac{1}{x \ln x}$. For $x \geq 2$, this function is continuous, positive and decreasing, so the Integral Test applies. Using the substitution $u = \ln x$, $du = \frac{1}{x} dx$, we find

$$\int_2^{\infty} \frac{dx}{x \ln x} = \lim_{R \to \infty} \int_2^R \frac{dx}{x \ln x} = \lim_{R \to \infty} \int_{\ln 2}^{\ln R} \frac{du}{u} = \lim_{R \to \infty} (\ln(\ln R) - \ln(\ln 2)) = \infty.$$

The integral diverges; hence, the series $\sum_{n=2}^{\infty} \frac{1}{n \ln n}$ also diverges. By the Comparison Test we can therefore conclude that the series $\sum_{n=2}^{\infty} \frac{1}{n^a \ln n}$ diverges.

To summarize,

$$\sum_{n=2}^{\infty} \frac{1}{n^a \ln n} \text{ converges for } a > 1 \text{ and diverges for } a \leq 1.$$

In Exercises 49–74, determine convergence or divergence using any method covered so far.

49. $\displaystyle\sum_{n=4}^{\infty} \frac{1}{n^2 - 9}$

SOLUTION Apply the Limit Comparison Test with $a_n = \dfrac{1}{n^2 - 9}$ and $b_n = \dfrac{1}{n^2}$:

$$L = \lim_{n\to\infty} \frac{a_n}{b_n} = \lim_{n\to\infty} \frac{\frac{1}{n^2-9}}{\frac{1}{n^2}} = \lim_{n\to\infty} \frac{n^2}{n^2 - 9} = 1.$$

Since the p-series $\displaystyle\sum_{n=1}^{\infty} \frac{1}{n^2}$ converges, the series $\displaystyle\sum_{n=4}^{\infty} \frac{1}{n^2}$ also converges. Because L exists, by the Limit Comparison Test we can conclude that the series $\displaystyle\sum_{n=4}^{\infty} \frac{1}{n^2 - 9}$ converges.

51. $\displaystyle\sum_{n=1}^{\infty} \frac{\sqrt{n}}{4n + 9}$

SOLUTION Apply the Limit Comparison Test with $a_n = \dfrac{\sqrt{n}}{4n + 9}$ and $b_n = \dfrac{1}{\sqrt{n}}$:

$$L = \lim_{n\to\infty} \frac{a_n}{b_n} = \lim_{n\to\infty} \frac{\frac{\sqrt{n}}{4n+9}}{\frac{1}{\sqrt{n}}} = \lim_{n\to\infty} \frac{n}{4n + 9} = \frac{1}{4}.$$

The series $\displaystyle\sum_{n=1}^{\infty} \frac{1}{\sqrt{n}}$ is a divergent p-series. Because $L > 0$, by the Limit Comparison Test we can conclude that the series $\displaystyle\sum_{n=1}^{\infty} \frac{\sqrt{n}}{4n + 9}$ also diverges.

53. $\displaystyle\sum_{n=1}^{\infty} \frac{1}{3^{n^2}}$

SOLUTION Because $n^2 \geq n$ for $n \geq 1$, $3^{n^2} \geq 3^n$ and

$$\frac{1}{3^{n^2}} \leq \frac{1}{3^n} = \left(\frac{1}{3}\right)^n.$$

The series $\displaystyle\sum_{n=1}^{\infty} \left(\frac{1}{3}\right)^n$ is a geometric series with $r = \dfrac{1}{3}$, so it converges. By the Comparison Test we can therefore conclude that the series $\displaystyle\sum_{n=1}^{\infty} \frac{1}{3^{n^2}}$ also converges.

55. $\displaystyle\sum_{n=2}^{\infty} \frac{1}{n^{3/2} \ln n}$

SOLUTION For $n \geq 3$, $\ln n > 1$, so $n^{3/2} \ln n > n^{3/2}$ and

$$\frac{1}{n^{3/2} \ln n} < \frac{1}{n^{3/2}}.$$

The series $\displaystyle\sum_{n=1}^{\infty} \frac{1}{n^{3/2}}$ is a convergent p-series, so the series $\displaystyle\sum_{n=3}^{\infty} \frac{1}{n^{3/2}}$ also converges. By the Comparison Test we can therefore conclude that the series $\displaystyle\sum_{n=3}^{\infty} \frac{1}{n^{3/2} \ln n}$ converges. Hence, the series $\displaystyle\sum_{n=2}^{\infty} \frac{1}{n^{3/2} \ln n}$ also converges.

57. $\displaystyle\sum_{n=2}^{\infty} \frac{1}{n^{1/2} \ln n}$

SOLUTION By L'Hôpital's Rule

$$\lim_{x\to\infty} \frac{x^{1/4}}{\ln x} = \lim_{x\to\infty} \frac{\frac{1}{4}x^{-3/4}}{\frac{1}{x}} = \lim_{x\to\infty} \frac{1}{4}x^{1/4} = \infty;$$

so there exists an integer N such that for all $n \geq N$

$$\frac{n^{1/4}}{\ln n} \geq 1 \quad \text{or} \quad n^{1/4} \geq \ln n.$$

Therefore, for $n \geq N$,

$$\frac{1}{n^{1/2} \ln n} \geq \frac{1}{n^{3/4}}.$$

The series $\sum_{n=1}^{\infty} \frac{1}{n^{3/4}}$ is a divergent p-series, so $\sum_{n=N}^{\infty} \frac{1}{n^{3/4}}$ also diverges. By the Comparison Test we can therefore conclude that $\sum_{n=N}^{\infty} \frac{1}{n^{1/2} \ln n}$ diverges. Hence, the series $\sum_{n=2}^{\infty} \frac{1}{n^{1/2} \ln n}$ also diverges.

59. $\sum_{n=2}^{\infty} \frac{n}{e^{n^2}}$

SOLUTION Let $f(x) = xe^{-x^2}$. This function is continuous and positive for $x \geq 2$. Moreover, as

$$f'(x) = x(-2xe^{-x^2}) + e^{-x^2} = (1 - 2x^2)e^{-x^2},$$

we see that $f'(x) < 0$ for $x \geq 2$, so f is decreasing on this interval. The Integral Test therefore applies. Now,

$$\int_{2}^{\infty} xe^{-x^2} \, dx = \lim_{R \to \infty} \int_{2}^{R} xe^{-x^2} \, dx = -\frac{1}{2} \lim_{R \to \infty} \left(e^{-R^2} - e^{-4} \right) = \frac{1}{2} e^{-4}.$$

The integral converges; hence, the series $\sum_{n=2}^{\infty} \frac{n}{e^{n^2}}$ also converges.

61. $\sum_{n=1}^{\infty} \frac{2^n}{3^n - n}$

SOLUTION Apply the Limit Comparison Test with $a_n = \frac{2^n}{3^n - n}$ and $b_n = \frac{2^n}{3^n}$:

$$L = \lim_{n \to \infty} \frac{a_n}{b_n} = \lim_{n \to \infty} \frac{\frac{2^n}{3^n - n}}{\frac{2^n}{3^n}} = \lim_{n \to \infty} \frac{1}{1 - \frac{n}{3^n}}.$$

Now,

$$\lim_{n \to \infty} \frac{n}{3^n} = \lim_{x \to \infty} \frac{x}{3^x} = \lim_{x \to \infty} \frac{1}{3^x \ln 3} = 0,$$

so

$$L = \lim_{n \to \infty} \frac{a_n}{b_n} = \frac{1}{1 - 0} = 1.$$

The series $\sum_{n=1}^{\infty} \left(\frac{2}{3} \right)^n$ is a convergent geometric series. Because L exists, by the Limit Comparison Test we can conclude that the series $\sum_{n=1}^{\infty} \frac{2^n}{3^n - n}$ also converges.

63. $\sum_{n=1}^{\infty} \frac{\tan^{-1} n}{n^2}$

SOLUTION Apply the Limit Comparison Test with $a_n = \frac{\tan^{-1} n}{n^2}$ and $b_n = \frac{1}{n^2}$:

$$L = \lim_{n \to \infty} \frac{a_n}{b_n} = \lim_{n \to \infty} \frac{\frac{\tan^{-1} n}{n^2}}{\frac{1}{n^2}} = \lim_{n \to \infty} \tan^{-1} n = \frac{\pi}{4}.$$

The series $\sum_{n=1}^{\infty} \frac{1}{n^2}$ is a convergent p-series. Because L exists, by the Limit Comparison Test we can conclude that the series $\sum_{n=1}^{\infty} \frac{\tan^{-1} n}{n^2}$ also converges.

65. $\displaystyle\sum_{n=1}^{\infty} \frac{\ln n}{n^3}$

SOLUTION Apply the Limit Comparison Test with $a_n = \dfrac{\ln n}{n^3}$ and $b_n = \dfrac{1}{n^2}$:

$$L = \lim_{n\to\infty} \frac{a_n}{b_n} = \lim_{n\to\infty} \frac{\frac{\ln n}{n^3}}{\frac{1}{n^2}} = \lim_{n\to\infty} \frac{\ln n}{n} = 0.$$

The series $\displaystyle\sum_{n=1}^{\infty} \frac{1}{n^2}$ is a convergent p-series. Because L exists, by the Limit Comparison Test we can conclude that the

series $\displaystyle\sum_{n=1}^{\infty} \frac{\ln n}{n^3}$ also converges.

67. $\displaystyle\sum_{n=1}^{\infty} \frac{2 + (-1)^n}{n^{3/2}}$

SOLUTION For $n \geq 1$

$$0 < \frac{2 + (-1)^n}{n^{3/2}} \leq \frac{2+1}{n^{3/2}} = \frac{3}{n^{3/2}}.$$

The series $\displaystyle\sum_{n=1}^{\infty} \frac{1}{n^{3/2}}$ is a convergent p-series; hence, the series $\displaystyle\sum_{n=1}^{\infty} \frac{3}{n^{3/2}}$ also converges. By the Comparison Test we can

therefore conclude that the series $\displaystyle\sum_{n=1}^{\infty} \frac{2 + (-1)^n}{n^{3/2}}$ converges.

69. $\displaystyle\sum_{n=1}^{\infty} \frac{2n + 1}{4^n}$

SOLUTION For $n \geq 3$, $2n + 1 < 2^n$, so

$$\frac{2n + 1}{4^n} < \frac{2^n}{4^n} = \left(\frac{1}{2}\right)^n.$$

The series $\displaystyle\sum_{n=1}^{\infty} \left(\frac{1}{2}\right)^n$ is a convergent geometric series, so $\displaystyle\sum_{n=3}^{\infty} \left(\frac{1}{2}\right)^n$ also converges. By the Comparison Test we can

therefore conclude that the series $\displaystyle\sum_{n=3}^{\infty} \frac{2n + 1}{4^n}$ converges. Finally, the series $\displaystyle\sum_{n=1}^{\infty} \frac{2n + 1}{4^n}$ converges.

71. $\displaystyle\sum_{n=1}^{\infty} \frac{n^2 - n}{n^5 + n}$

SOLUTION First rewrite $a_n = \dfrac{n^2 - n}{n^5 + n} = \dfrac{n(n-1)}{n(n^4 + 1)} = \dfrac{n-1}{n^4 + 1}$ and observe

$$\frac{n-1}{n^4 + 1} < \frac{n}{n^4} = \frac{1}{n^3}$$

for $n \geq 1$. The series $\displaystyle\sum_{n=1}^{\infty} \frac{1}{n^3}$ is a convergent p-series, so by the Comparison Test we can conclude that the series

$\displaystyle\sum_{n=1}^{\infty} \frac{n^2 - n}{n^5 + n}$ also converges.

73. $\displaystyle\sum_{n=2}^{\infty} \frac{1}{n^{1.2} \ln n}$

SOLUTION For $n \geq 3$, $\ln n > 1$, so

$$\frac{1}{n^{1.2} \ln n} < \frac{1}{n^{1.2}}.$$

The series $\sum\limits_{n=1}^{\infty} \dfrac{1}{n^{1.2}}$ is a convergent p-series, so the series $\sum\limits_{n=3}^{\infty} \dfrac{1}{n^{1.2}}$ also converges. By the Comparison Test we can therefore conclude that $\sum\limits_{n=3}^{\infty} \dfrac{1}{n^{1.2}\ln n}$ converges. Finally, $\sum\limits_{n=2}^{\infty} \dfrac{1}{n^{1.2}\ln n}$ also converges.

Approximating Infinite Sums In Exercises 75–77, let $a_n = f(n)$, where $f(x)$ is a continuous, decreasing function such that

$$\int_1^{\infty} f(x)\,dx$$

converges.

75. Show that

$$\int_1^{\infty} f(x)\,dx \leq \sum_{n=1}^{\infty} a_n \leq a_1 + \int_1^{\infty} f(x)\,dx \qquad \boxed{4}$$

SOLUTION From the proof of the Integral Test, we know that

$$a_2 + a_3 + a_4 + \cdots + a_N \leq \int_1^N f(x)\,dx \leq \int_1^{\infty} f(x)\,dx;$$

that is,

$$S_N - a_1 \leq \int_1^{\infty} f(x)\,dx \quad \text{or} \quad S_N \leq a_1 + \int_1^{\infty} f(x)\,dx.$$

Also from the proof of the Integral test, we know that

$$\int_1^N f(x)\,dx \leq a_1 + a_2 + a_3 + \cdots + a_{N-1} = S_N - a_N \leq S_N.$$

Thus,

$$\int_1^N f(x)\,dx \leq S_N \leq a_1 + \int_1^{\infty} f(x)\,dx.$$

Taking the limit as $N \to \infty$ yields Eq. (4), as desired.

77. Let $S = \sum\limits_{n=1}^{\infty} a_n$. Arguing as in Exercise 75, show that

$$\sum_{n=1}^{M} a_n + \int_{M+1}^{\infty} f(x)\,dx \leq S \leq \sum_{n=1}^{M+1} a_n + \int_{M+1}^{\infty} f(x)\,dx \qquad \boxed{5}$$

Conclude that

$$0 \leq S - \left(\sum_{n=1}^{M} a_n + \int_{M+1}^{\infty} f(x)\,dx \right) \leq a_{M+1} \qquad \boxed{6}$$

This yields a method for approximating S with an error of at most a_{M+1}.

SOLUTION Following the proof of the Integral Test and the argument in Exercise 75, but starting with $n = M + 1$ rather than $n = 1$, we obtain

$$\int_{M+1}^{\infty} f(x)\,dx \leq \sum_{n=M+1}^{\infty} a_n \leq a_{M+1} + \int_{M+1}^{\infty} f(x)\,dx.$$

Adding $\sum\limits_{n=1}^{M} a_n$ to each part of this inequality yields

$$\sum_{n=1}^{M} a_n + \int_{M+1}^{\infty} f(x)\,dx \leq \sum_{n=1}^{\infty} a_n = S \leq \sum_{n=1}^{M+1} a_n + \int_{M+1}^{\infty} f(x)\,dx.$$

Subtracting $\displaystyle\sum_{n=1}^{M} a_n + \int_{M+1}^{\infty} f(x)\,dx$ from each part of this last inequality then gives us

$$0 \le S - \left(\sum_{n=1}^{M} a_n + \int_{M+1}^{\infty} f(x)\,dx\right) \le a_{M+1}.$$

79. CAS Apply Eq. (5) with $M = 40{,}000$ to show that

$$1.644934066 \le \sum_{n=1}^{\infty} \frac{1}{n^2} \le 1.644934068$$

Is this consistent with Euler's result, according to which this infinite series has sum $\pi^2/6$?

SOLUTION Using Eq. (5) with $f(x) = \dfrac{1}{x^2}$, $a_n = \dfrac{1}{n^2}$ and $M = 40000$, we find

$$S_{40000} + \int_{40001}^{\infty} \frac{dx}{x^2} \le \sum_{n=1}^{\infty} \frac{1}{n^2} \le S_{40001} + \int_{40001}^{\infty} \frac{dx}{x^2}.$$

Now,

$$S_{40000} = 1.6449090672;$$

$$S_{40001} = S_{40000} + \frac{1}{40001} = 1.6449090678;$$

and

$$\int_{40001}^{\infty} \frac{dx}{x^2} = \lim_{R \to \infty} \int_{40001}^{R} \frac{dx}{x^2} = -\lim_{R \to \infty} \left(\frac{1}{R} - \frac{1}{40001}\right) = \frac{1}{40001} = 0.0000249994.$$

Thus,

$$1.6449090672 + 0.0000249994 \le \sum_{n=1}^{\infty} \frac{1}{n^2} \le 1.6449090678 + 0.0000249994,$$

or

$$1.6449340665 \le \sum_{n=1}^{\infty} \frac{1}{n^2} \le 1.6449340672.$$

Since $\dfrac{\pi^2}{6} \approx 1.6449340668$, our approximation is consistent with Euler's result.

81. CAS Using a CAS and Eq. (6), determine the value of $\displaystyle\sum_{n=1}^{\infty} n^{-5}$ to within an error less than 10^{-4}.

SOLUTION Using Eq. (6) with $f(x) = x^{-5}$ and $a_n = n^{-5}$, we have

$$0 \le \sum_{n=1}^{\infty} n^{-5} - \left(\sum_{n=1}^{M+1} n^{-5} + \int_{M+1}^{\infty} x^{-5}\,dx\right) \le (M+1)^{-5}.$$

To guarantee an error less than 10^{-4}, we need $(M+1)^{-5} \le 10^{-4}$. This yields $M \ge 10^{4/5} - 1 \approx 5.3$, so we choose $M = 6$. Now,

$$\sum_{n=1}^{7} n^{-5} = 1.0368498887,$$

and

$$\int_{7}^{\infty} x^{-5}\,dx = \lim_{R \to \infty} \int_{7}^{R} x^{-5}\,dx = -\frac{1}{4} \lim_{R \to \infty} \left(R^{-4} - 7^{-4}\right) = \frac{1}{4 \cdot 7^4} = 0.0001041233.$$

Thus,

$$\sum_{n=1}^{\infty} n^{-5} \approx \sum_{n=1}^{7} n^{-5} + \int_{7}^{\infty} x^{-5}\,dx = 1.0368498887 + 0.0001041233 = 1.0369540120.$$

83. Let p_n denote the nth prime number ($p_1 = 2$, $p_2 = 3$, etc.). It is known that there is a constant C such that $p_n \leq Cn \ln n$. Prove the divergence of

$$\sum_{n=1}^{\infty} \frac{1}{p_n} = \frac{1}{2} + \frac{1}{3} + \frac{1}{5} + \frac{1}{7} + \frac{1}{11} + \cdots$$

SOLUTION Since $p_n \leq Cn \ln n$ for $n \geq 2$,

$$\frac{1}{p_n} \geq \frac{1}{C} \cdot \frac{1}{n \ln n}.$$

Now, let $f(x) = \dfrac{1}{x \ln x}$. For $x \geq 2$, this function is continuous, positive and decreasing, so the Integral Test applies. Using the substitution $u = \ln x$, $du = \frac{1}{x} dx$, we find

$$\int_2^{\infty} \frac{dx}{x \ln x} = \lim_{R \to \infty} \int_2^R \frac{dx}{x \ln x} = \lim_{R \to \infty} \int_{\ln 2}^{\ln R} \frac{du}{u} = \lim_{R \to \infty} (\ln(\ln R) - \ln(\ln 2)) = \infty.$$

The integral diverges; hence, the series $\displaystyle\sum_{n=2}^{\infty} \frac{1}{n \ln n}$ also diverges. By the Comparison Test we can therefore conclude that the series $\displaystyle\sum_{n=2}^{\infty} \frac{1}{p_n}$ diverges; hence, $\displaystyle\sum_{n=1}^{\infty} \frac{1}{p_n}$ also diverges.

Further Insights and Challenges

85. Use the Integral Test to prove again that the geometric series $\displaystyle\sum_{n=1}^{\infty} r^n$ converges if $0 < r < 1$ and diverges if $r > 1$.

SOLUTION Let $f(x) = r^x$ on the interval $x \geq 1$. For $0 < r < 1$ this function is decreasing, continuous and positive, so the Integral Test applies.

$$\int_1^{\infty} r^x \, dx = \lim_{R \to \infty} \int_1^R r^x \, dx = \frac{1}{\ln r} \lim_{R \to \infty} (r^R - r) = -\frac{r}{\ln r}.$$

The integral converges; hence, the series $\displaystyle\sum_{n=1}^{\infty} r^n$ also converges. Now, if $r > 1$, $\lim_{n \to \infty} r^n = \infty$, and the series $\displaystyle\sum_{n=1}^{\infty} r^n$ diverges by the Divergence Test.

87. Kummer's Acceleration Method Suppose we wish to approximate $S = \displaystyle\sum_{n=1}^{\infty} \frac{1}{n^2}$. There is a similar telescoping series whose value can be computed exactly (see Example 1 in Section 11.2):

$$\sum_{n=1}^{\infty} \frac{1}{n(n+1)} = 1$$

(a) Verify that

$$S = \sum_{n=1}^{\infty} \frac{1}{n(n+1)} + \sum_{n=1}^{\infty} \left(\frac{1}{n^2} - \frac{1}{n(n+1)} \right)$$

Thus for M large,

$$S \approx 1 + \sum_{n=1}^{M} \frac{1}{n^2(n+1)} \qquad \boxed{7}$$

(b) Explain what has been gained. Why is (7) a better approximation to S than $\displaystyle\sum_{n=1}^{M} \frac{1}{n^2}$?

(c) ⊂⊃⊃ Compute

$$\sum_{n=1}^{1,000} \frac{1}{n^2}, \qquad 1 + \sum_{n=1}^{100} \frac{1}{n^2(n+1)}$$

Which is a better approximation to S, whose exact value is $\pi^2/6$?

SOLUTION

(a) Because the series $\sum_{n=1}^{\infty} \frac{1}{n^2}$ and $\sum_{n=1}^{\infty} \frac{1}{n(n+1)}$ both converge,

$$\sum_{n=1}^{\infty} \frac{1}{n(n+1)} + \sum_{n=1}^{\infty} \left(\frac{1}{n^2} - \frac{1}{n(n+1)} \right) = \sum_{n=1}^{\infty} \frac{1}{n(n+1)} + \sum_{n=1}^{\infty} \frac{1}{n^2} - \sum_{n=1}^{\infty} \frac{1}{n(n+1)} = \sum_{n=1}^{\infty} \frac{1}{n^2} = S.$$

Now,

$$\frac{1}{n^2} - \frac{1}{n(n+1)} = \frac{n+1}{n^2(n+1)} - \frac{n}{n^2(n+1)} = \frac{1}{n^2(n+1)},$$

so, for M large,

$$S \approx 1 + \sum_{n=1}^{M} \frac{1}{n^2(n+1)}.$$

(b) The series $\sum_{n=1}^{\infty} \frac{1}{n^2(n+1)}$ converges more rapidly than $\sum_{n=1}^{\infty} \frac{1}{n^2}$ since the degree of n in the denominator is larger.

(c) Using a computer algebra system, we find

$$\sum_{n=1}^{1000} \frac{1}{n^2} = 1.6439345667 \quad \text{and} \quad 1 + \sum_{n=1}^{100} \frac{1}{n^2(n+1)} = 1.6448848903.$$

The second sum is more accurate because it is closer to the exact solution $\frac{\pi^2}{6} \approx 1.6449340668$.

11.4 Absolute and Conditional Convergence

Preliminary Questions

1. Suppose that $S = \sum_{n=0}^{\infty} a_n$ is conditionally convergent. Which of the following statements are correct?

(a) $\sum_{n=0}^{\infty} |a_n|$ may or may not converge.

(b) S may or may not converge.

(c) $\sum_{n=0}^{\infty} |a_n|$ diverges.

SOLUTION By definition, because $\sum_{n=1}^{\infty} a_n$ is conditionally convergent, we know that $\sum_{n=1}^{\infty} a_n$ converges but $\sum_{n=1}^{\infty} |a_n|$ diverges. Thus:

(a) This statement is incorrect: $\sum_{n=0}^{\infty} |a_n|$ must diverge.

(b) This statement is incorrect: S must converge.

(c) This statement is correct.

2. Which of the following statements is equivalent to Theorem 1?

(a) If $\sum_{n=0}^{\infty} |a_n|$ diverges, then $\sum_{n=0}^{\infty} a_n$ also diverges.

(b) If $\sum_{n=0}^{\infty} a_n$ diverges, then $\sum_{n=0}^{\infty} |a_n|$ also diverges.

(c) If $\sum_{n=0}^{\infty} a_n$ converges, then $\sum_{n=0}^{\infty} |a_n|$ also converges.

SOLUTION The correct answer is **(b)**: If $\sum_{n=0}^{\infty} a_n$ diverges, then $\sum_{n=0}^{\infty} |a_n|$ also diverges. Take $a_n = (-1)^n \frac{1}{n}$ to see that statements **(a)** and **(c)** are not true in general.

3. Lathika argues that $\sum_{n=1}^{\infty} (-1)^n \sqrt{n}$ is an alternating series and therefore converges. Is Lathika right?

SOLUTION No. Although $\sum_{n=1}^{\infty} (-1)^n \sqrt{n}$ is an alternating series, the terms $a_n = \sqrt{n}$ do not form a decreasing sequence that tends to zero. In fact, $a_n = \sqrt{n}$ is an increasing sequence that tends to ∞, so $\sum_{n=1}^{\infty} (-1)^n \sqrt{n}$ diverges by the Divergence Test.

4. Give an example of a series such that $\sum a_n$ converges but $\sum |a_n|$ diverges.

SOLUTION The series $\sum \dfrac{(-1)^n}{\sqrt[3]{n}}$ converges by the Leibniz Test, but the positive series $\sum \dfrac{1}{\sqrt[3]{n}}$ is a divergent p-series.

Exercises

1. Show that $\sum_{n=0}^{\infty} \dfrac{(-1)^n}{2^n}$ converges absolutely.

SOLUTION The positive series $\sum_{n=0}^{\infty} \dfrac{1}{2^n}$ is a geometric series with $r = \dfrac{1}{2}$. Thus, the positive series converges, and the given series converges absolutely.

In Exercises 3–12, determine whether the series converges absolutely, conditionally, or not at all.

3. $\sum_{n=1}^{\infty} \dfrac{(-1)^n}{\sqrt{n}}$

SOLUTION Let $a_n = \dfrac{1}{\sqrt{n}}$. Then a_n forms a decreasing sequence that tends to zero; hence, the series $\sum_{n=1}^{\infty} \dfrac{(-1)^n}{\sqrt{n}}$ converges by the Leibniz Test. However, the positive series $\sum_{n=1}^{\infty} \dfrac{1}{\sqrt{n}}$ is a divergent p-series, so the original series converges conditionally.

5. $\sum_{n=1}^{\infty} \dfrac{(-1)^{n-1}}{(1.1)^n}$

SOLUTION The positive series $\sum_{n=1}^{\infty} \left(\dfrac{1}{1.1} \right)^n$ is a convergent geometric series; thus, the original series converges absolutely.

7. $\sum_{n=2}^{\infty} \dfrac{(-1)^{n+1}}{n \ln n}$

SOLUTION Let $a_n = \dfrac{1}{n \ln n}$. Then a_n forms a decreasing sequence (note that n and $\ln n$ are both increasing functions of n) that tends to zero; hence, the series $\sum_{n=2}^{\infty} \dfrac{(-1)^{n+1}}{n \ln n}$ converges by the Leibniz Test. However, the positive series $\sum_{n=2}^{\infty} \dfrac{1}{n \ln n}$ diverges, so the original series converges conditionally.

9. $\sum_{n=1}^{\infty} \dfrac{\sin n\pi}{\sqrt{n}}$

SOLUTION $\sin n\pi = 0$ for all n, so the general term of the series is zero. Consequently, the series, as well as the positive series, converge to zero; that is, the series converges absolutely.

11. $\sum_{n=1}^{\infty} \dfrac{\cos \frac{1}{n}}{n^2}$

SOLUTION The positive series is $\sum_{n=1}^{\infty} \dfrac{|\cos \frac{1}{n}|}{n^2}$. Because

$$\frac{|\cos \frac{1}{n}|}{n^2} \le \frac{1}{n^2}$$

for all n, the Comparison Test and the convergence of the p-series $\sum_{n=1}^{\infty} \frac{1}{n^2}$ imply that the series $\sum_{n=1}^{\infty} \frac{|\cos \frac{1}{n}|}{n^2}$ converges. Hence, the original series converges absolutely.

13. Let $S = \sum_{n=1}^{\infty} (-1)^{n+1} \frac{1}{n^3}$.

(a) Calculate S_n for $1 \le n \le 10$.

(b) Use Eq. (2) to show that $0.9 \le S \le 0.902$.

SOLUTION

(a)

$$S_1 = 1 \qquad\qquad S_6 = S_5 - \frac{1}{6^3} = 0.899782407$$

$$S_2 = 1 - \frac{1}{2^3} = \frac{7}{8} = 0.875 \qquad\qquad S_7 = S_6 + \frac{1}{7^3} = 0.902697859$$

$$S_3 = S_2 + \frac{1}{3^3} = 0.912037037 \qquad\qquad S_8 = S_7 - \frac{1}{8^3} = 0.900744734$$

$$S_4 = S_3 - \frac{1}{4^3} = 0.896412037 \qquad\qquad S_9 = S_8 + \frac{1}{9^3} = 0.902116476$$

$$S_5 = S_4 + \frac{1}{5^3} = 0.904412037 \qquad\qquad S_{10} = S_9 - \frac{1}{10^3} = 0.901116476$$

(b) By Eq. (2),

$$|S_{10} - S| \le a_{11} = \frac{1}{11^3},$$

so

$$S_{10} - \frac{1}{11^3} \le S \le S_{10} + \frac{1}{11^3},$$

or

$$0.900365161 \le S \le 0.901867791.$$

15. Approximate $\sum_{n=1}^{\infty} \frac{(-1)^{n+1}}{n^4}$ to three decimal places.

SOLUTION Let $S = \sum_{n=1}^{\infty} \frac{(-1)^{n+1}}{n^4}$, so that $a_n = \frac{1}{n^4}$. By Eq. (2),

$$|S_N - S| \le a_{N+1} = \frac{1}{(N+1)^4}.$$

To guarantee accuracy to three decimal places, we must choose N so that

$$\frac{1}{(N+1)^4} < 5 \times 10^{-4} \quad \text{or} \quad N > \sqrt[4]{2000} - 1 \approx 5.7.$$

The smallest value that satisfies the required inequality is then $N = 6$. Thus,

$$S \approx S_6 = 1 - \frac{1}{2^4} + \frac{1}{3^4} - \frac{1}{4^4} + \frac{1}{5^4} - \frac{1}{6^4} = 0.946767824.$$

In Exercises 17–18, use Eq. (2) to approximate the value of the series to within an error of at most 10^{-5}.

17. $\sum_{n=1}^{\infty} \frac{(-1)^{n+1}}{n(n+2)(n+3)}$

SOLUTION Let $S = \sum_{n=1}^{\infty} \frac{(-1)^{n+1}}{n(n+2)(n+3)}$, so that $a_n = \frac{1}{n(n+2)(n+3)}$. By Eq. (2),

$$|S_N - S| \le a_{N+1} = \frac{1}{(N+1)(N+3)(N+4)}.$$

We must choose N so that

$$\frac{1}{(N+1)(N+3)(N+4)} \leq 10^{-5} \quad \text{or} \quad (N+1)(N+3)(N+4) \geq 10^5.$$

For $N = 43$, the product on the left hand side is 95,128, while for $N = 44$ the product is 101,520; hence, the smallest value of N which satisfies the required inequality is $N = 44$. Thus,

$$S \approx S_{44} = \sum_{n=1}^{44} \frac{(-1)^{n+1}}{n(n+2)(n+3)} = 0.0656746.$$

In Exercises 19–26, determine convergence or divergence by any method.

19. $\displaystyle\sum_{n=1}^{\infty} \frac{1}{3^n + 5^n}$

SOLUTION For $n \geq 1$

$$\frac{1}{3^n + 5^n} \leq \frac{1}{3^n} = \left(\frac{1}{3}\right)^n.$$

The series $\displaystyle\sum_{n=1}^{\infty} \left(\frac{1}{3}\right)^n$ is a convergent geometric series, so the Comparison Test implies that the series $\displaystyle\sum_{n=1}^{\infty} \frac{1}{3^n + 5^n}$ also converges.

21. $\displaystyle\sum_{n=1}^{\infty} \frac{(-1)^n}{\sqrt{n^2 + 1}}$

SOLUTION This is an alternating series with $a_n = \dfrac{1}{\sqrt{n^2 + 1}}$. Because a_n is a decreasing sequence that converges to zero, the series $\displaystyle\sum_{n=1}^{\infty} \frac{(-1)^n}{\sqrt{n^2 + 1}}$ converges by the Leibniz Test.

23. $\displaystyle\sum_{n=1}^{\infty} \frac{3^n + (-1)^n 2^n}{5^n}$

SOLUTION The series $\displaystyle\sum_{n=1}^{\infty} \frac{3^n}{5^n} = \sum_{n=1}^{\infty} \left(\frac{3}{5}\right)^n$ is a convergent geometric series, as is the series $\displaystyle\sum_{n=1}^{\infty} \frac{(-1)^n 2^n}{5^n} = \sum_{n=1}^{\infty} \left(-\frac{2}{5}\right)^n$. Hence,

$$\sum_{n=1}^{\infty} \frac{3^n + (-1)^n 2^n}{5^n} = \sum_{n=1}^{\infty} \left(\frac{3}{5}\right)^n + \sum_{n=1}^{\infty} \left(-\frac{2}{5}\right)^n$$

also converges.

25. $\displaystyle\sum_{n=1}^{\infty} (-1)^n n e^{-n}$

SOLUTION This is an alternating series with $a_n = n e^{-n}$. Consider the function $f(x) = x e^{-x}$. Using L'Hôpital's Rule,

$$\lim_{x \to \infty} \frac{x}{e^x} = \lim_{x \to \infty} \frac{1}{e^x} = 0.$$

Moreover,

$$f'(x) = \frac{e^x - x e^x}{e^{2x}} = \frac{1 - x}{e^x},$$

so $f'(x) < 0$ and f is decreasing for $x > 1$. Therefore, $\{a_n\}$ is a decreasing sequence which converges to zero, and $\displaystyle\sum_{n=1}^{\infty} (-1)^n n e^{-n}$ converges by the Leibniz Test.

27. Show that

$$S = \frac{1}{2} - \frac{1}{2} + \frac{1}{3} - \frac{1}{3} + \frac{1}{4} - \frac{1}{4}$$

converges by computing the partial sums. Does it converge absolutely?

SOLUTION The sequence of partial sums is

$$S_1 = \frac{1}{2}$$

$$S_2 = S_1 - \frac{1}{2} = 0$$

$$S_3 = S_2 + \frac{1}{3} = \frac{1}{3}$$

$$S_4 = S_3 - \frac{1}{3} = 0$$

and, in general,

$$S_N = \begin{cases} \dfrac{1}{N}, & \text{for odd } N \\[2mm] 0, & \text{for even } N \end{cases}$$

Thus, $\lim\limits_{N \to \infty} S_N = 0$, and the series converges to 0. The positive series is

$$\frac{1}{2} + \frac{1}{2} + \frac{1}{3} + \frac{1}{3} + \frac{1}{4} + \frac{1}{4} + \cdots = 2 \sum_{n=2}^{\infty} \frac{1}{n};$$

which diverges. Therefore, the original series converges conditionally, not absolutely.

29. Determine whether the following series converges conditionally:

$$1 - \frac{1}{3} + \frac{1}{2} - \frac{1}{5} + \frac{1}{3} - \frac{1}{7} + \frac{1}{4} - \frac{1}{9} + \frac{1}{5} - \frac{1}{11} + \cdots$$

SOLUTION Although this is an alternating series, the sequence of terms $\{a_n\}$ is not decreasing, so we cannot apply the Leibniz Test. However, we may express the series as

$$\sum_{n=1}^{\infty} \left(\frac{1}{n} - \frac{1}{2n+1} \right) = \sum_{n=1}^{\infty} \frac{n+1}{n(2n+1)}.$$

Using the Limit Comparison Test and comparing with the harmonic series, we find

$$L = \lim_{n \to \infty} \frac{\frac{n+1}{n(2n+1)}}{\frac{1}{n}} = \lim_{n \to \infty} \frac{n+1}{2n+1} = \frac{1}{2}.$$

Because $L > 0$, we conclude that the series

$$1 - \frac{1}{3} + \frac{1}{2} - \frac{1}{5} + \frac{1}{3} - \frac{1}{7} + \frac{1}{4} - \frac{1}{9} + \frac{1}{5} - \frac{1}{11} + \cdots$$

diverges.

Further Insights and Challenges

31. Prove the following variant of the Leibniz Test: If $\{a_n\}$ is a positive, decreasing sequence with $\lim\limits_{n \to \infty} a_n = 0$, then the series

$$a_1 + a_2 - 2a_3 + a_4 + a_5 - 2a_6 + \cdots$$

converges. *Hint:* Show that S_{3N} is increasing and bounded by $a_1 + a_2$, and continue as in the proof of the Leibniz Test.

SOLUTION Following the hint, we first examine the sequence $\{S_{3N}\}$. Now,

$$S_{3N+3} = S_{3(N+1)} = S_{3N} + a_{3N+1} + a_{3N+2} - 2a_{3N+3} = S_{3N} + \left(a_{3N+1} - a_{3N+3} \right) + \left(a_{3N+2} - a_{3N+3} \right) \geq S_{3N}$$

because $\{a_n\}$ is a decreasing sequence. Moreover,

$$S_{3N} = a_1 + a_2 - \sum_{k=1}^{N-1} \left(2a_{3k} - a_{3k+1} - a_{3k+2} \right) - 2a_{3N}$$

$$= a_1 + a_2 - \sum_{k=1}^{N-1} \left[\left(a_{3k} - a_{3k+1} \right) + \left(a_{3k} - a_{3k+2} \right) - 2a_{3N} \right] \leq a_1 + a_2$$

again because $\{a_n\}$ is a decreasing sequence. Thus, $\{S_{3N}\}$ is an increasing sequence with an upper bound; hence, $\{S_{3N}\}$ converges. Next,

$$S_{3N+1} = S_{3N} + a_{3N+1} \quad \text{and} \quad S_{3N+2} = S_{3N} + a_{3N+1} + a_{3N+2}.$$

Given that $\lim_{n\to\infty} a_n = 0$, it follows that

$$\lim_{N\to\infty} S_{3N+1} = \lim_{N\to\infty} S_{3N+2} = \lim_{N\to\infty} S_{3N}.$$

Having just established that $\lim_{N\to\infty} S_{3N}$ exists, it follows that the sequences $\{S_{3N+1}\}$ and $\{S_{3N+2}\}$ converge to the same limit. Finally, we can conclude that the sequence of partial sums $\{S_N\}$ converges, so the given series converges.

33. Prove the conditional convergence of

$$R = 1 + \frac{1}{2} + \frac{1}{3} - \frac{3}{4} + \frac{1}{5} + \frac{1}{6} + \frac{1}{7} - \frac{3}{8} + \cdots$$

SOLUTION Using Exercise 31 as a template, we first examine the sequence $\{R_{4N}\}$. Now,

$$R_{4N+4} = R_{4(N+1)} = R_{4N} + \frac{1}{4N+1} + \frac{1}{4N+2} + \frac{1}{4N+3} - \frac{3}{4N+4}$$

$$= R_N + \left(\frac{1}{4N+1} - \frac{1}{4N+4}\right) + \left(\frac{1}{4N+2} - \frac{1}{4N+4}\right) + \left(\frac{1}{4N+3} - \frac{1}{4N+4}\right) \geq R_{4N}.$$

Moreover,

$$R_{4N} = 1 + \frac{1}{2} + \frac{1}{3} - \sum_{k=1}^{N-1}\left(\frac{3}{4k} - \frac{1}{4k+1} - \frac{1}{4k+2} - \frac{1}{4k+3}\right) - \frac{3}{4N} \leq 1 + \frac{1}{2} + \frac{1}{3}.$$

Thus, $\{R_{4N}\}$ is an increasing sequence with an upper bound; hence, $\{R_{4N}\}$ converges. Next,

$$R_{4N+1} = R_{4N} + \frac{1}{4N+1};$$

$$R_{4N+2} = R_{4N} + \frac{1}{4N+1} + \frac{1}{4N+2}; \quad \text{and}$$

$$R_{4N+3} = R_{4N} + \frac{1}{4N+1} + \frac{1}{4N+2} + \frac{1}{4N+3},$$

so

$$\lim_{n\to\infty} R_{4N+1} = \lim_{N\to\infty} R_{4N+2} = \lim_{N\to\infty} R_{4N+3} = \lim_{N\to\infty} R_{4N}.$$

Having just established that $\lim_{N\to\infty} R_{4N}$ exists, it follows that the sequences $\{R_{4N+1}\}$, $\{R_{4N+2}\}$ and $\{R_{4N+3}\}$ converge to the same limit. Finally, we can conclude that the sequence of partial sums $\{R_N\}$ converges, so the series R converges. Now, consider the positive series

$$R^+ = 1 + \frac{1}{2} + \frac{1}{3} + \frac{3}{4} + \frac{1}{5} + \frac{1}{6} + \frac{1}{7} + \frac{3}{8} + \cdots$$

Because the terms in this series are greater than or equal to the corresponding terms in the divergent harmonic series, it follows from the Comparison Test that R^+ diverges. Thus, by definition, R converges conditionally.

35. **Assumptions Matter** Show by counterexample that the Leibniz Test does not remain true if $\{a_n\}$ tends to zero but we drop the assumption that the sequence a_n is decreasing. *Hint:* Consider

$$R = \frac{1}{2} - \frac{1}{4} + \frac{1}{3} - \frac{1}{8} + \frac{1}{4} - \frac{1}{16} + \cdots + \left(\frac{1}{n} - \frac{1}{2^n}\right) + \cdots$$

SOLUTION Let

$$R = \frac{1}{2} - \frac{1}{4} + \frac{1}{3} - \frac{1}{8} + \frac{1}{4} - \frac{1}{16} + \cdots + \left(\frac{1}{n+1} - \frac{1}{2^{n+1}}\right) + \cdots$$

This is an alternating series with

$$a_n = \begin{cases} \dfrac{1}{k+1}, & n = 2k-1 \\[2mm] \dfrac{1}{2^{k+1}}, & n = 2k \end{cases}$$

Note that $a_n \to 0$ as $n \to \infty$, but the sequence $\{a_n\}$ is not decreasing. We will now establish that R diverges.

For sake of contradiction, suppose that R converges. The geometric series

$$\sum_{n=1}^{\infty} \frac{1}{2^{n+1}}$$

converges, so the sum of R and this geometric series must also converge; however,

$$R + \sum_{n=1}^{\infty} \frac{1}{2^{n+1}} = \sum_{n=2}^{\infty} \frac{1}{n},$$

which diverges because the harmonic series diverges. Thus, the series R must diverge.

37. We say that $\{b_n\}$ is a rearrangement of $\{a_n\}$ if $\{b_n\}$ has the same terms as $\{a_n\}$ but occurring in a different order. Show that if $\{b_n\}$ is a rearrangement of $\{a_n\}$ and $S = \sum_{n=1}^{\infty} a_n$ converges absolutely, then $T = \sum_{n=1}^{\infty} b_n$ also converges absolutely.

(This result does not hold if S is only conditionally convergent.) *Hint:* Prove that the partial sums $\sum_{n=1}^{N} |b_n|$ are bounded. It can be shown further that $S = T$.

SOLUTION Suppose the series $S = \sum_{n=1}^{\infty} a_n$ converges absolutely and denote the corresponding positive series by

$$S^+ = \sum_{n=1}^{\infty} |a_n|.$$

Further, let $T_N = \sum_{n=1}^{N} |b_n|$ denote the Nth partial sum of the series $\sum_{n=1}^{\infty} |b_n|$. Because $\{b_n\}$ is a rearrangement of $\{a_n\}$, we know that

$$0 \le T_N \le \sum_{n=1}^{\infty} |a_n| = S^+;$$

that is, the sequence $\{T_N\}$ is bounded. Moreover,

$$T_{N+1} = \sum_{n=1}^{N+1} |b_n| = T_N + |b_{N+1}| \ge T_N;$$

that is, $\{T_N\}$ is increasing. It follows that $\{T_N\}$ converges, so the series $\sum_{n=1}^{\infty} |b_n|$ converges, which means the series $\sum_{n=1}^{\infty} b_n$ converges absolutely.

11.5 The Ratio and Root Tests

Preliminary Questions

1. In the Ratio Test, is ρ equal to $\lim_{n\to\infty} \left| \frac{a_{n+1}}{a_n} \right|$ or $\lim_{n\to\infty} \left| \frac{a_n}{a_{n+1}} \right|$?

SOLUTION In the Ratio Test ρ is the limit $\lim_{n\to\infty} \left| \frac{a_{n+1}}{a_n} \right|$.

2. Is the Ratio Test conclusive for $\sum_{n=1}^{\infty} \frac{1}{2^n}$? Is it conclusive for $\sum_{n=1}^{\infty} \frac{1}{n}$?

SOLUTION The general term of $\sum_{n=1}^{\infty} \frac{1}{2^n}$ is $a_n = \frac{1}{2^n}$; thus,

$$\left| \frac{a_{n+1}}{a_n} \right| = \frac{1}{2^{n+1}} \cdot \frac{2^n}{1} = \frac{1}{2},$$

and

$$\rho = \lim_{n\to\infty} \left| \frac{a_{n+1}}{a_n} \right| = \frac{1}{2} < 1.$$

Consequently, the Ratio Test guarantees that the series $\sum_{n=1}^{\infty} \frac{1}{2^n}$ converges.

The general term of $\sum_{n=1}^{\infty} \frac{1}{n}$ is $a_n = \frac{1}{n}$; thus,

$$\left| \frac{a_{n+1}}{a_n} \right| = \frac{1}{n+1} \cdot \frac{n}{1} = \frac{n}{n+1},$$

and

$$\rho = \lim_{n \to \infty} \left| \frac{a_{n+1}}{a_n} \right| = \lim_{n \to \infty} \frac{n}{n+1} = 1.$$

The Ratio Test is therefore inconclusive for the series $\sum_{n=1}^{\infty} \frac{1}{n}$.

3. Can the Ratio Test be used to show convergence if the series is only conditionally convergent?

SOLUTION No. The Ratio Test can only establish absolute convergence and divergence, not conditional convergence.

Exercises

In Exercises 1–18, apply the Ratio Test to determine convergence or divergence, or state that the Ratio Test is inconclusive.

1. $\sum_{n=1}^{\infty} \frac{1}{5^n}$

SOLUTION With $a_n = \frac{1}{5^n}$,

$$\left| \frac{a_{n+1}}{a_n} \right| = \frac{1}{5^{n+1}} \cdot \frac{5^n}{1} = \frac{1}{5} \quad \text{and} \quad \rho = \lim_{n \to \infty} \left| \frac{a_{n+1}}{a_n} \right| = \frac{1}{5} < 1.$$

Therefore, the series $\sum_{n=1}^{\infty} \frac{1}{5^n}$ converges by the Ratio Test.

3. $\sum_{n=1}^{\infty} \frac{(-1)^{n-1}}{n^n}$

SOLUTION With $a_n = \frac{(-1)^{n-1}}{n^n}$,

$$\left| \frac{a_{n+1}}{a_n} \right| = \frac{1}{(n+1)^{n+1}} \cdot \frac{n^n}{1} = \frac{1}{n+1} \left(\frac{n}{n+1} \right)^n = \frac{1}{n+1} \left(1 + \frac{1}{n} \right)^{-n},$$

and

$$\rho = \lim_{n \to \infty} \left| \frac{a_{n+1}}{a_n} \right| = 0 \cdot \frac{1}{e} = 0 < 1.$$

Therefore, the series $\sum_{n=1}^{\infty} \frac{(-1)^{n-1}}{n^n}$ converges by the Ratio Test.

5. $\sum_{n=1}^{\infty} \frac{n}{n^2 + 1}$

SOLUTION With $a_n = \frac{n}{n^2+1}$,

$$\left| \frac{a_{n+1}}{a_n} \right| = \frac{n+1}{(n+1)^2 + 1} \cdot \frac{n^2 + 1}{n} = \frac{n+1}{n} \cdot \frac{n^2 + 1}{n^2 + 2n + 2},$$

and

$$\rho = \lim_{n \to \infty} \left| \frac{a_{n+1}}{a_n} \right| = 1 \cdot 1 = 1.$$

Therefore, for the series $\sum_{n=1}^{\infty} \frac{n}{n^2 + 1}$, the Ratio Test is inconclusive.

We can show that this series diverges by using the Limit Comparison Test and comparing with the divergent harmonic series.

7. $\displaystyle\sum_{n=1}^{\infty} \frac{2^n}{n^{100}}$

SOLUTION With $a_n = \frac{2^n}{n^{100}}$,

$$\left|\frac{a_{n+1}}{a_n}\right| = \frac{2^{n+1}}{(n+1)^{100}} \cdot \frac{n^{100}}{2^n} = 2\left(\frac{n}{n+1}\right)^{100} \quad \text{and} \quad \rho = \lim_{n\to\infty} \left|\frac{a_{n+1}}{a_n}\right| = 2 \cdot 1^{100} = 2 > 1.$$

Therefore, the series $\displaystyle\sum_{n=1}^{\infty} \frac{2^n}{n^{100}}$ diverges by the Ratio Test.

9. $\displaystyle\sum_{n=1}^{\infty} \frac{10^n}{2^{n^2}}$

SOLUTION With $a_n = \frac{10^n}{2^{n^2}}$,

$$\left|\frac{a_{n+1}}{a_n}\right| = \frac{10^{n+1}}{2^{(n+1)^2}} \cdot \frac{2^{n^2}}{10^n} = 10 \cdot \frac{1}{2^{2n+1}} \quad \text{and} \quad \rho = \lim_{n\to\infty} \left|\frac{a_{n+1}}{a_n}\right| = 10 \cdot 0 = 0 < 1.$$

Therefore, the series $\displaystyle\sum_{n=1}^{\infty} \frac{10^n}{2^{n^2}}$ converges by the Ratio Test.

11. $\displaystyle\sum_{n=1}^{\infty} \frac{e^n}{n^n}$

SOLUTION With $a_n = \frac{e^n}{n^n}$,

$$\left|\frac{a_{n+1}}{a_n}\right| = \frac{e^{n+1}}{(n+1)^{n+1}} \cdot \frac{n^n}{e^n} = \frac{e}{n+1}\left(\frac{n}{n+1}\right)^n = \frac{e}{n+1}\left(1 + \frac{1}{n}\right)^{-n},$$

and

$$\rho = \lim_{n\to\infty} \left|\frac{a_{n+1}}{a_n}\right| = 0 \cdot \frac{1}{e} = 0 < 1.$$

Therefore, the series $\displaystyle\sum_{n=1}^{\infty} \frac{e^n}{n^n}$ converges by the Ratio Test.

13. $\displaystyle\sum_{n=0}^{\infty} (-1)^n \frac{n!}{4^n}$

SOLUTION With $a_n = (-1)^n \frac{n!}{4^n}$,

$$\left|\frac{a_{n+1}}{a_n}\right| = \frac{(n+1)!}{4^{n+1}} \cdot \frac{4^n}{n!} = \frac{n+1}{4} \quad \text{and} \quad \rho = \lim_{n\to\infty} \left|\frac{a_{n+1}}{a_n}\right| = \infty > 1.$$

Therefore, the series $\displaystyle\sum_{n=0}^{\infty} (-1)^n \frac{n!}{4^n}$ diverges by the Ratio Test.

15. $\displaystyle\sum_{n=2}^{\infty} \frac{1}{n \ln n}$

SOLUTION With $a_n = \frac{1}{n \ln n}$,

$$\left|\frac{a_{n+1}}{a_n}\right| = \frac{1}{(n+1) \ln(n+1)} \cdot \frac{n \ln n}{1} = \frac{n}{n+1} \frac{\ln n}{\ln(n+1)},$$

and

$$\rho = \lim_{n\to\infty} \left|\frac{a_{n+1}}{a_n}\right| = 1 \cdot \lim_{n\to\infty} \frac{\ln n}{\ln(n+1)}.$$

Now,

$$\lim_{n\to\infty} \frac{\ln n}{\ln(n+1)} = \lim_{x\to\infty} \frac{\ln x}{\ln(x+1)} = \lim_{x\to\infty} \frac{1/(x+1)}{1/x} = \lim_{x\to\infty} \frac{x}{x+1} = 1.$$

Thus, $\rho = 1$, and the Ratio Test is inconclusive for the series $\displaystyle\sum_{n=2}^{\infty} \frac{1}{n \ln n}$.

Using the Integral Test, we can show that the series $\displaystyle\sum_{n=2}^{\infty} \frac{1}{n \ln n}$ diverges.

17. $\displaystyle\sum_{n=1}^{\infty} \frac{n^2}{(2n+1)!}$

SOLUTION With $a_n = \frac{n^2}{(2n+1)!}$,

$$\left| \frac{a_{n+1}}{a_n} \right| = \frac{(n+1)^2}{(2n+3)!} \cdot \frac{(2n+1)!}{n^2} = \left(\frac{n+1}{n} \right)^2 \frac{1}{(2n+3)(2n+2)},$$

and

$$\rho = \lim_{n\to\infty} \left| \frac{a_{n+1}}{a_n} \right| = 1^2 \cdot 0 = 0 < 1.$$

Therefore, the series $\displaystyle\sum_{n=1}^{\infty} \frac{n^2}{(2n+1)!}$ converges by the Ratio Test.

19. Show that $\displaystyle\sum_{n=1}^{\infty} n^k \, 3^{-n}$ converges for all exponents k.

SOLUTION With $a_n = n^k 3^{-n}$,

$$\left| \frac{a_{n+1}}{a_n} \right| = \frac{(n+1)^k 3^{-(n+1)}}{n^k 3^{-n}} = \frac{1}{3} \left(1 + \frac{1}{n} \right)^k,$$

and, for all k,

$$\rho = \lim_{n\to\infty} \left| \frac{a_{n+1}}{a_n} \right| = \frac{1}{3} \cdot 1 = \frac{1}{3} < 1.$$

Therefore, the series $\displaystyle\sum_{n=1}^{\infty} n^k \, 3^{-n}$ converges for all exponents k by the Ratio Test.

21. Show that $\displaystyle\sum_{n=1}^{\infty} 2^n x^n$ converges if $|x| < \frac{1}{2}$.

SOLUTION With $a_n = 2^n x^n$,

$$\left| \frac{a_{n+1}}{a_n} \right| = \frac{2^{n+1}|x|^{n+1}}{2^n |x|^n} = 2|x| \quad \text{and} \quad \rho = \lim_{n\to\infty} \left| \frac{a_{n+1}}{a_n} \right| = 2|x|.$$

Therefore, $\rho < 1$ and the series $\displaystyle\sum_{n=1}^{\infty} 2^n x^n$ converges by the Ratio Test provided $|x| < \frac{1}{2}$.

23. Show that $\displaystyle\sum_{n=1}^{\infty} \frac{r^n}{n}$ converges if $|r| < 1$.

SOLUTION With $a_n = \frac{r^n}{n}$,

$$\left| \frac{a_{n+1}}{a_n} \right| = \frac{|r|^{n+1}}{n+1} \cdot \frac{n}{|r|^n} = |r| \frac{n}{n+1} \quad \text{and} \quad \rho = \lim_{n\to\infty} \left| \frac{a_{n+1}}{a_n} \right| = 1 \cdot |r| = |r|.$$

Therefore, by the Ratio Test, the series $\displaystyle\sum_{n=1}^{\infty} \frac{r^n}{n}$ converges provided $|r| < 1$.

25. Show that $\displaystyle\sum_{n=1}^{\infty} \frac{n!}{n^n}$ converges. *Hint:* Use $\displaystyle\lim_{n\to\infty} \left(1 + \frac{1}{n} \right)^n = e$.

SOLUTION With $a_n = \frac{n!}{n^n}$,

$$\left| \frac{a_{n+1}}{a_n} \right| = \frac{(n+1)!}{(n+1)^{n+1}} \cdot \frac{n^n}{n!} = \left(\frac{n}{n+1} \right)^n = \left(1 + \frac{1}{n} \right)^{-n},$$

and

$$\rho = \lim_{n \to \infty} \left| \frac{a_{n+1}}{a_n} \right| = \frac{1}{e} < 1.$$

Therefore, the series $\sum_{n=1}^{\infty} \frac{n!}{n^n}$ converges by the Ratio Test.

In Exercises 26–31, assume that $|a_{n+1}/a_n|$ converges to $\rho = \frac{1}{3}$. What can you say about the convergence of the given series?

27. $\sum_{n=1}^{\infty} n^3 a_n$

SOLUTION Let $b_n = n^3 a_n$. Then

$$\rho = \lim_{n \to \infty} \left| \frac{b_{n+1}}{b_n} \right| = \lim_{n \to \infty} \left(\frac{n+1}{n} \right)^3 \left| \frac{a_{n+1}}{a_n} \right| = 1^3 \cdot \frac{1}{3} = \frac{1}{3} < 1.$$

Therefore, the series $\sum_{n=1}^{\infty} n^3 a_n$ converges by the Ratio Test.

29. $\sum_{n=1}^{\infty} 3^n a_n$

SOLUTION Let $b_n = 3^n a_n$. Then

$$\rho = \lim_{n \to \infty} \left| \frac{b_{n+1}}{b_n} \right| = \lim_{n \to \infty} \frac{3^{n+1}}{3^n} \left| \frac{a_{n+1}}{a_n} \right| = 3 \cdot \frac{1}{3} = 1.$$

Therefore, the Ratio Test is inconclusive for the series $\sum_{n=1}^{\infty} 3^n a_n$.

31. $\sum_{n=1}^{\infty} a_n^2$

SOLUTION Let $b_n = a_n^2$. Then

$$\rho = \lim_{n \to \infty} \left| \frac{b_{n+1}}{b_n} \right| = \lim_{n \to \infty} \left| \frac{a_{n+1}}{a_n} \right|^2 = \left(\frac{1}{3} \right)^2 = \frac{1}{9} < 1.$$

Therefore, the series $\sum_{n=1}^{\infty} a_n^2$ converges by the Ratio Test.

33. Is the Ratio Test conclusive for the *p*-series $\sum_{n=1}^{\infty} \frac{1}{n^p}$?

SOLUTION With $a_n = \frac{1}{n^p}$,

$$\left| \frac{a_{n+1}}{a_n} \right| = \frac{1}{(n+1)^p} \cdot \frac{n^p}{1} = \left(\frac{n}{n+1} \right)^p \quad \text{and} \quad \rho = \lim_{n \to \infty} \left| \frac{a_{n+1}}{a_n} \right| = 1^p = 1.$$

Therefore, the Ratio Test is inconclusive for the *p*-series $\sum_{n=1}^{\infty} \frac{1}{n^p}$.

In Exercises 34–39, use the Root Test to determine convergence or divergence (or state that the test is inconclusive).

35. $\sum_{n=1}^{\infty} \frac{1}{n^n}$

SOLUTION With $a_n = \frac{1}{n^n}$,

$$\sqrt[n]{a_n} = \sqrt[n]{\frac{1}{n^n}} = \frac{1}{n} \quad \text{and} \quad \lim_{n\to\infty} \sqrt[n]{a_n} = 0 < 1.$$

Therefore, the series $\sum_{n=0}^{\infty} \frac{1}{n^n}$ converges by the Root Test.

37. $\sum_{k=0}^{\infty} \left(\frac{k}{3k+1} \right)^k$

SOLUTION With $a_k = \left(\frac{k}{3k+1} \right)^k$,

$$\sqrt[k]{a_k} = \sqrt[k]{\left(\frac{k}{3k+1} \right)^k} = \frac{k}{3k+1} \quad \text{and} \quad \lim_{k\to\infty} \sqrt[k]{a_k} = \frac{1}{3} < 1.$$

Therefore, the series $\sum_{k=0}^{\infty} \left(\frac{k}{3k+1} \right)^k$ converges by the Root Test.

39. $\sum_{n=4}^{\infty} \left(1 + \frac{1}{n} \right)^{-n^2}$

SOLUTION With $a_k = \left(1 + \frac{1}{n} \right)^{-n^2}$,

$$\sqrt[n]{a_n} = \sqrt[n]{\left(1 + \frac{1}{n} \right)^{-n^2}} = \left(1 + \frac{1}{n} \right)^{-n} \quad \text{and} \quad \lim_{n\to\infty} \sqrt[n]{a_n} = e^{-1} < 1.$$

Therefore, the series $\sum_{k=0}^{\infty} \left(1 + \frac{1}{n} \right)^{-n^2}$ converges by the Root Test.

In Exercises 41–52, determine convergence or divergence using any method covered in the text so far.

41. $\sum_{n=1}^{\infty} \frac{2^n + 4^n}{7^n}$

SOLUTION Because the series

$$\sum_{n=1}^{\infty} \frac{2^n}{7^n} = \sum_{n=1}^{\infty} \left(\frac{2}{7} \right)^n \quad \text{and} \quad \sum_{n=1}^{\infty} \frac{4^n}{7^n} = \sum_{n=1}^{\infty} \left(\frac{4}{7} \right)^n$$

are both convergent geometric series, it follows that

$$\sum_{n=1}^{\infty} \frac{2^n + 4^n}{7^n} = \sum_{n=1}^{\infty} \left(\frac{2}{7} \right)^n + \sum_{n=1}^{\infty} \left(\frac{4}{7} \right)^n$$

also converges.

43. $\sum_{n=1}^{\infty} \frac{n^3}{5^n}$

SOLUTION The presence of the exponential term suggests applying the Ratio Test. With $a_n = \frac{n^3}{5^n}$,

$$\left| \frac{a_{n+1}}{a_n} \right| = \frac{(n+1)^3}{5^{n+1}} \cdot \frac{5^n}{n^3} = \frac{1}{5} \left(1 + \frac{1}{n} \right)^3 \quad \text{and} \quad \rho = \lim_{n\to\infty} \left| \frac{a_{n+1}}{a_n} \right| = \frac{1}{5} \cdot 1^3 = \frac{1}{5} < 1.$$

Therefore, the series $\sum_{n=1}^{\infty} \frac{n^3}{5^n}$ converges by the Ratio Test.

45. $\sum_{n=2}^{\infty} \frac{1}{\sqrt{n^3 - n^2}}$

SOLUTION This series is similar to a p-series; because

$$\frac{1}{\sqrt{n^3 - n^2}} \approx \frac{1}{\sqrt{n^3}} = \frac{1}{n^{3/2}}$$

for large n, we will apply the Limit Comparison Test comparing with the p-series with $p = \frac{3}{2}$. Now,

$$L = \lim_{n \to \infty} \frac{\frac{1}{\sqrt{n^3 - n^2}}}{\frac{1}{n^{3/2}}} = \lim_{n \to \infty} \sqrt{\frac{n^3}{n^3 - n^2}} = 1.$$

The p-series with $p = \frac{3}{2}$ converges and L exists; therefore, the series $\sum_{n=2}^{\infty} \frac{1}{\sqrt{n^3 - n^2}}$ also converges.

47. $\sum_{n=1}^{\infty} \frac{n^2 + 4n}{3n^4 + 9}$

SOLUTION This series is similar to a p-series; because

$$\frac{n^2 + 4n}{3n^4 + 9} \approx \frac{n^2}{\sqrt{3n^4}} = \frac{1}{3n^2}$$

for large n, we will apply the Limit Comparison Test comparing with the p-series with $p = 2$. Now,

$$L = \lim_{n \to \infty} \frac{\frac{n^2 + 4n}{3n^4 + 9}}{\frac{1}{n^2}} = \lim_{n \to \infty} \frac{n^4 + 4n^3}{3n^4 + 9} = \frac{1}{3}.$$

The p-series with $p = 2$ converges and L exists; therefore, the series $\sum_{n=1}^{\infty} \frac{n^2 + 4n}{3n^4 + 9}$ also converges.

49. $\sum_{n=1}^{\infty} \sin \frac{1}{n^2}$

SOLUTION Here, we will apply the Limit Comparison Test, comparing with the p-series with $p = 2$. Now,

$$L = \lim_{n \to \infty} \frac{\sin \frac{1}{n^2}}{\frac{1}{n^2}} = \lim_{u \to 0} \frac{\sin u}{u} = 1,$$

where $u = \frac{1}{n^2}$. The p-series with $p = 2$ converges and L exists; therefore, the series $\sum_{n=1}^{\infty} \sin \frac{1}{n^2}$ also converges.

51. $\sum_{n=1}^{\infty} \left(\frac{n}{n + 12} \right)^n$

SOLUTION Because the general term has the form of a function of n raised to the nth power, we might be tempted to use the Root Test; however, the Root Test is inconclusive for this series. Instead, note

$$\lim_{n \to \infty} a_n = \lim_{n \to \infty} \left(1 + \frac{12}{n} \right)^{-n} = \lim_{n \to \infty} \left[\left(1 + \frac{12}{n} \right)^{n/12} \right]^{-12} = e^{-12} \neq 0.$$

Therefore, the series diverges by the Divergence Test.

Further Insights and Challenges

53. 🖊️ **Proof of the Root Test** Let $S = \sum_{n=0}^{\infty} a_n$ be a positive series and assume that $L = \lim_{n \to \infty} \sqrt[n]{a_n}$ exists.

(a) Show that S converges if $L < 1$. *Hint:* Choose R with $\rho < R < 1$ and show that $a_n \leq R^n$ for n sufficiently large. Then compare with the geometric series $\sum R^n$.

(b) Show that S diverges if $L > 1$.

SOLUTION Suppose $\lim_{n \to \infty} \sqrt[n]{a_n} = L$ exists.

(a) If $L < 1$, let $\epsilon = \dfrac{1-L}{2}$. By the definition of a limit, there is a positive integer N such that

$$-\epsilon \le \sqrt[n]{a_n} - L \le \epsilon$$

for $n \ge N$. From this, we conclude that

$$0 \le \sqrt[n]{a_n} \le L + \epsilon$$

for $n \ge N$. Now, let $R = L + \epsilon$. Then

$$R = L + \frac{1-L}{2} = \frac{L+1}{2} < \frac{1+1}{2} = 1,$$

and

$$0 \le \sqrt[n]{a_n} \le R \quad \text{or} \quad 0 \le a_n \le R^n$$

for $n \ge N$. Because $0 \le R < 1$, the series $\displaystyle\sum_{n=N}^{\infty} R^n$ is a convergent geometric series, so the series $\displaystyle\sum_{n=N}^{\infty} a_n$ converges by the Comparison Test. Therefore, the series $\displaystyle\sum_{n=0}^{\infty} a_n$ also converges.

(b) If $L > 1$, let $\epsilon = \dfrac{L-1}{2}$. By the definition of a limit, there is a positive integer N such that

$$-\epsilon \le \sqrt[n]{a_n} - L \le \epsilon$$

for $n \ge N$. From this, we conclude that

$$L - \epsilon \le \sqrt[n]{a_n}$$

for $n \ge N$. Now, let $R = L - \epsilon$. Then

$$R = L - \frac{L-1}{2} = \frac{L+1}{2} > \frac{1+1}{2} = 1,$$

and

$$R \le \sqrt[n]{a_n} \quad \text{or} \quad R^n \le a_n$$

for $n \ge N$. Because $R > 1$, the series $\displaystyle\sum_{n=N}^{\infty} R^n$ is a divergent geometric series, so the series $\displaystyle\sum_{n=N}^{\infty} a_n$ diverges by the Comparison Test. Therefore, the series $\displaystyle\sum_{n=0}^{\infty} a_n$ also diverges.

55. Let $S = \displaystyle\sum_{n=1}^{\infty} \frac{c^n n!}{n^n}$, where c is a constant.

(a) Prove that S converges absolutely if $|c| < e$ and diverges if $|c| > e$.

(b) It is known that $\displaystyle\lim_{n \to \infty} \frac{e^n n!}{n^{n+1/2}} = \sqrt{2\pi}$. Verify this numerically.

(c) Use the Limit Comparison Test to prove that S diverges for $c = e$.

SOLUTION

(a) With $a_n = \frac{c^n n!}{n^n}$,

$$\left| \frac{a_{n+1}}{a_n} \right| = \frac{|c|^{n+1}(n+1)!}{(n+1)^{n+1}} \cdot \frac{n^n}{|c|^n n!} = |c| \left(\frac{n}{n+1} \right)^n = |c| \left(1 + \frac{1}{n} \right)^{-n},$$

and

$$\rho = \lim_{n \to \infty} \left| \frac{a_{n+1}}{a_n} \right| = |c| e^{-1}.$$

Thus, by the Ratio Test, the series $\displaystyle\sum_{n=1}^{\infty} \frac{c^n n!}{n^n}$ converges when $|c|e^{-1} < 1$, or when $|c| < e$. The series diverges when $|c| > e$.

(b) The table below lists the value of $\frac{e^n n!}{n^{n+1/2}}$ for several increasing values of n. Since $\sqrt{2\pi} = 2.506628275$, the numerical evidence verifies that

$$\lim_{n \to \infty} \frac{e^n n!}{n^{n+1/2}} = \sqrt{2\pi}.$$

n	100	1000	10000	100000
$\frac{e^n n!}{n^{n+1/2}}$	2.508717995	2.506837169	2.506649163	2.506630363

(c) With $c = e$, the series S becomes $\displaystyle\sum_{n=1}^{\infty} \frac{e^n n!}{n^n}$. Using the result from part (b),

$$L = \lim_{n \to \infty} \frac{\frac{e^n n!}{n^n}}{\sqrt{n}} = \lim_{n \to \infty} \frac{e^n n!}{n^{n+1/2}} = \sqrt{2\pi}.$$

Because the series $\displaystyle\sum_{n=1}^{\infty} \sqrt{n}$ diverges by the Divergence Test and $L > 0$, we conclude that $\displaystyle\sum_{n=1}^{\infty} \frac{e^n n!}{n^n}$ diverges by the Limit Comparison Test.

11.6 Power Series

Preliminary Questions

1. Suppose that $\sum a_n x^n$ converges for $x = 5$. Must it also converge for $x = 4$? What about $x = -3$?

SOLUTION The power series $\sum a_n x^n$ is centered at $x = 0$. Because the series converges for $x = 5$, the radius of convergence must be at least 5 and the series converges absolutely at least for the interval $|x| < 5$. Both $x = 4$ and $x = -3$ are inside this interval, so the series converges for $x = 4$ and for $x = -3$.

2. Suppose that $\sum a_n (x - 6)^n$ converges for $x = 10$. At which of the points (a)–(d) must it also converge?

(a) $x = 8$ **(b)** $x = 12$ **(c)** $x = 2$ **(d)** $x = 0$

SOLUTION The given power series is centered at $x = 6$. Because the series converges for $x = 10$, the radius of convergence must be at least $|10 - 6| = 4$ and the series converges absolutely at least for the interval $|x - 6| < 4$, or $2 < x < 10$.

(a) $x = 8$ is inside the interval $2 < x < 10$, so the series converges for $x = 8$.

(b) $x = 12$ is not inside the interval $2 < x < 10$, so the series may or may not converge for $x = 12$.

(c) $x = 2$ is an endpoint of the interval $2 < x < 10$, so the series may or may not converge for $x = 2$.

(d) $x = 0$ is not inside the interval $2 < x < 10$, so the series may or may not converge for $x = 0$.

3. Suppose that $F(x)$ is a power series with radius of convergence $R = 12$. What is the radius of convergence of $F(3x)$?

SOLUTION If the power series $F(x)$ has radius of convergence $R = 12$, then the power series $F(3x)$ has radius of convergence $R = \frac{12}{3} = 4$.

4. The power series $F(x) = \displaystyle\sum_{n=1}^{\infty} n x^n$ has radius of convergence $R = 1$. What is the power series expansion of $F'(x)$ and what is its radius of convergence?

SOLUTION We obtain the power series expansion for $F'(x)$ by differentiating the power series expansion for $F(x)$ term-by-term. Thus,

$$F'(x) = \sum_{n=1}^{\infty} n^2 x^{n-1}.$$

The radius of convergence for this series is $R = 1$, the same as the radius of convergence for the series expansion for $F(x)$.

Exercises

1. Use the Ratio Test to determine the radius of convergence of $\displaystyle\sum_{n=0}^{\infty} \frac{x^n}{2^n}$.

SOLUTION With $a_n = \frac{x^n}{2^n}$,

$$\left| \frac{a_{n+1}}{a_n} \right| = \frac{|x|^{n+1}}{2^{n+1}} \cdot \frac{2^n}{|x|^n} = \frac{|x|}{2} \quad \text{and} \quad \rho = \lim_{n\to\infty} \left| \frac{a_{n+1}}{a_n} \right| = \frac{|x|}{2}.$$

By the Ratio Test, the series converges when $\rho = \frac{|x|}{2} < 1$, or $|x| < 2$, and diverges when $\rho = \frac{|x|}{2} > 1$, or $|x| > 2$. The radius of convergence is therefore $R = 2$.

3. Show that the following three power series have the same radius of convergence. Then show that (a) diverges at both endpoints, (b) converges at one endpoint but diverges at the other, and (c) converges at both endpoints.

(a) $\displaystyle\sum_{n=1}^{\infty} \frac{x^n}{3^n}$ **(b)** $\displaystyle\sum_{n=1}^{\infty} \frac{x^n}{n3^n}$ **(c)** $\displaystyle\sum_{n=1}^{\infty} \frac{x^n}{n^2 3^n}$

SOLUTION

(a) With $a_n = \frac{1}{3^n}$,

$$\left| \frac{a_{n+1}}{a_n} \right| = \frac{1}{3^{n+1}} \cdot \frac{3^n}{1} = \frac{1}{3} \quad \text{and} \quad r = \lim_{n\to\infty} \left| \frac{a_{n+1}}{a_n} \right| = \frac{1}{3}.$$

The radius of convergence is therefore $R = r^{-1} = 3$. For the endpoint $x = 3$, the series becomes

$$\sum_{n=1}^{\infty} \frac{3^n}{3^n} = \sum_{n=1}^{\infty} 1,$$

which diverges by the Divergence Test. For the endpoint $x = -3$, the series becomes

$$\sum_{n=1}^{\infty} \frac{(-3)^n}{3^n} = \sum_{n=1}^{\infty} (-1)^n,$$

which also diverges by the Divergence Test.

(b) With $a_n = \frac{1}{n3^n}$,

$$\left| \frac{a_{n+1}}{a_n} \right| = \frac{1}{(n+1)3^{n+1}} \cdot \frac{n3^n}{1} = \frac{1}{3}\left(\frac{n}{n+1} \right) \quad \text{and} \quad r = \lim_{n\to\infty} \left| \frac{a_{n+1}}{a_n} \right| = \frac{1}{3} \cdot 1 = \frac{1}{3}.$$

The radius of convergence is therefore $R = r^{-1} = 3$. For the endpoint $x = 3$, the series becomes

$$\sum_{n=1}^{\infty} \frac{3^n}{n3^n} = \sum_{n=1}^{\infty} \frac{1}{n},$$

which is the divergent harmonic series. For the endpoint $x = -3$, the series becomes

$$\sum_{n=1}^{\infty} \frac{(-3)^n}{n3^n} = \sum_{n=1}^{\infty} \frac{(-1)^n}{n},$$

which converges by the Leibniz Test.

(c) With $a_n = \frac{1}{n^2 3^n}$,

$$\left| \frac{a_{n+1}}{a_n} \right| = \frac{1}{(n+1)^2 3^{n+1}} \cdot \frac{n^2 3^n}{1} = \frac{1}{3}\left(\frac{n}{n+1} \right)^2 \quad \text{and} \quad r = \lim_{n\to\infty} \left| \frac{a_{n+1}}{a_n} \right| = \frac{1}{3} \cdot 1^2 = \frac{1}{3}.$$

The radius of convergence is therefore $R = r^{-1} = 3$. For the endpoint $x = 3$, the series becomes

$$\sum_{n=1}^{\infty} \frac{3^n}{n^2 3^n} = \sum_{n=1}^{\infty} \frac{1}{n^2},$$

which is a convergent p-series. For the endpoint $x = -3$, the series becomes

$$\sum_{n=1}^{\infty} \frac{(-3)^n}{n^2 3^n} = \sum_{n=1}^{\infty} \frac{(-1)^n}{n^2},$$

which converges by the Leibniz Test.

5. Show that $\sum_{n=0}^{\infty} n^n x^n$ diverges for all $x \neq 0$.

SOLUTION With $a_n = n^n$,

$$\left| \frac{a_{n+1}}{a_n} \right| = \frac{(n+1)^{n+1}}{n^n} = \left(1 + \frac{1}{n}\right)^n (n+1) \quad \text{and} \quad r = \lim_{n \to \infty} \left| \frac{a_{n+1}}{a_n} \right| = \infty.$$

The radius of convergence is therefore $R = r^{-1} = 0$. In other words, the power series converges only for $x = 0$.

In Exercises 7–26, find the values of x for which the following power series converge.

7. $\sum_{n=1}^{\infty} n x^n$

SOLUTION With $a_n = n$,

$$\left| \frac{a_{n+1}}{a_n} \right| = \frac{n+1}{n} \quad \text{and} \quad r = \lim_{n \to \infty} \left| \frac{a_{n+1}}{a_n} \right| = 1.$$

The radius of convergence is therefore $R = r^{-1} = 1$, and the series converges absolutely on the interval $|x| < 1$, or $-1 < x < 1$. For the endpoint $x = 1$, the series becomes $\sum_{n=1}^{\infty} n$, which diverges by the Divergence Test. For the endpoint $x = -1$, the series becomes $\sum_{n=1}^{\infty} (-1)^n n$, which also diverges by the Divergence Test. Thus, the series $\sum_{n=1}^{\infty} n x^n$ converges for $-1 < x < 1$ and diverges elsewhere.

9. $\sum_{n=1}^{\infty} \frac{2^n x^n}{n}$

SOLUTION With $a_n = \frac{2^n}{n}$,

$$\left| \frac{a_{n+1}}{a_n} \right| = \frac{2^{n+1}}{n+1} \cdot \frac{n}{2^n} = 2 \frac{n}{n+1} \quad \text{and} \quad r = \lim_{n \to \infty} \left| \frac{a_{n+1}}{a_n} \right| = 2 \cdot 1 = 2.$$

The radius of convergence is therefore $R = r^{-1} = \frac{1}{2}$, and the series converges absolutely on the interval $|x| < \frac{1}{2}$, or $-\frac{1}{2} < x < \frac{1}{2}$. For the endpoint $x = \frac{1}{2}$, the series becomes $\sum_{n=1}^{\infty} \frac{1}{n}$, which is the divergent harmonic series. For the endpoint $x = -\frac{1}{2}$, the series becomes $\sum_{n=1}^{\infty} \frac{(-1)^n}{n}$, which converges by the Leibniz Test. Thus, the series $\sum_{n=1}^{\infty} \frac{2^n x^n}{n}$ converges for $-\frac{1}{2} \leq x < \frac{1}{2}$ and diverges elsewhere.

11. $\sum_{n=2}^{\infty} \frac{x^n}{\ln n}$

SOLUTION With $a_n = \frac{1}{\ln n}$,

$$\left| \frac{a_{n+1}}{a_n} \right| = \frac{1}{\ln(n+1)} \cdot \frac{\ln n}{1} = \frac{\ln n}{\ln(n+1)} \quad \text{and} \quad r = \lim_{n \to \infty} \left| \frac{a_{n+1}}{a_n} \right| = 1.$$

The radius of convergence is therefore $R = r^{-1} = 1$, and the series converges absolutely on the interval $|x| < 1$, or $-1 < x < 1$. For the endpoint $x = 1$, the series becomes $\sum_{n=1}^{\infty} \frac{1}{\ln n}$. Because $\frac{1}{\ln n} > \frac{1}{n}$ and $\sum_{n=1}^{\infty} \frac{1}{n}$ is the divergent harmonic series, the endpoint series diverges by the Comparison Test. For the endpoint $x = -1$, the series becomes $\sum_{n=1}^{\infty} \frac{(-1)^n}{\ln n}$, which converges by the Leibniz Test. Thus, the series $\sum_{n=2}^{\infty} \frac{x^n}{\ln n}$ converges for $-1 \leq x < 1$ and diverges elsewhere.

13. $\sum_{n=1}^{\infty} \frac{x^n}{(n!)^2}$

SOLUTION With $a_n = \frac{1}{(n!)^2}$,

$$\left| \frac{a_{n+1}}{a_n} \right| = \frac{1}{((n+1)!)^2} \cdot \frac{(n!)^2}{1} = \left(\frac{1}{n+1} \right)^2 \quad \text{and} \quad r = \lim_{n \to \infty} \left| \frac{a_{n+1}}{a_n} \right| = 0.$$

The radius of convergence is therefore $R = r^{-1} = \infty$, and the series converges absolutely for all x.

15. $\displaystyle\sum_{n=1}^{\infty} (-1)^n n^4 (x+4)^n$

SOLUTION With $a_n = (-1)^n n^4$,

$$\left|\frac{a_{n+1}}{a_n}\right| = \frac{(n+1)^4}{n^4} = \left(1 + \frac{1}{n}\right)^4 \quad \text{and} \quad r = \lim_{n\to\infty} \left|\frac{a_{n+1}}{a_n}\right| = 1^4 = 1.$$

The radius of convergence is therefore $R = r^{-1} = 1$, and the series converges absolutely on the interval $|x+4| < 1$, or $-5 < x < -3$. For the endpoint $x = -3$, the series becomes $\displaystyle\sum_{n=1}^{\infty} (-1)^n n^4$, which diverges by the Divergence Test.

For the endpoint $x = -5$, the series becomes $\displaystyle\sum_{n=1}^{\infty} n^4$, which also diverges by the Divergence Test. Thus, the series $\displaystyle\sum_{n=1}^{\infty} (-1)^n n^4 (x+4)^n$ converges for $-5 < x < -3$ and diverges elsewhere.

17. $\displaystyle\sum_{n=0}^{\infty} \frac{n}{2^n} x^n$

SOLUTION With $a_n = \frac{n}{2^n}$,

$$\left|\frac{a_{n+1}}{a_n}\right| = \frac{n+1}{2^{n+1}} \cdot \frac{2^n}{n} = \frac{1}{2}\left(1 + \frac{1}{n}\right) \quad \text{and} \quad r = \lim_{n\to\infty} \left|\frac{a_{n+1}}{a_n}\right| = \frac{1}{2} \cdot 1 = \frac{1}{2}.$$

The radius of convergence is therefore $R = r^{-1} = 2$, and the series converges absolutely on the interval $|x| < 2$, or $-2 < x < 2$. For the endpoint $x = 2$, the series becomes $\displaystyle\sum_{n=1}^{\infty} n$, which diverges by the Divergence Test. For the endpoint $x = -2$, the series becomes $\displaystyle\sum_{n=1}^{\infty} (-1)^n n$, which also diverges by the Divergence Test. Thus, the series $\displaystyle\sum_{n=0}^{\infty} \frac{n}{2^n} x^n$ converges for $-2 < x < 2$ and diverges elsewhere.

19. $\displaystyle\sum_{n=1}^{\infty} \frac{(x-4)^n}{n^4}$

SOLUTION With $a_n = \frac{1}{n^4}$,

$$\left|\frac{a_{n+1}}{a_n}\right| = \frac{1}{(n+1)^4} \cdot \frac{n^4}{1} = \left(\frac{n}{n+1}\right)^4 \quad \text{and} \quad r = \lim_{n\to\infty} \left|\frac{a_{n+1}}{a_n}\right| = 1^4 = 1.$$

The radius of convergence is therefore $R = r^{-1} = 1$, and the series converges absolutely on the interval $|x-4| < 1$, or $3 < x < 5$. For the endpoint $x = 5$, the series becomes $\displaystyle\sum_{n=1}^{\infty} \frac{1}{n^4}$, which is a convergent p-series. For the endpoint $x = 3$, the series becomes $\displaystyle\sum_{n=1}^{\infty} \frac{(-1)^n}{n^4}$, which converges by the Leibniz Test. Thus, the series $\displaystyle\sum_{n=1}^{\infty} \frac{(x-4)^n}{n^4}$ converges for $3 \leq x \leq 5$ and diverges elsewhere.

21. $\displaystyle\sum_{n=10}^{\infty} n!\,(x+5)^n$

SOLUTION With $a_n = n!$,

$$\left|\frac{a_{n+1}}{a_n}\right| = \frac{(n+1)!}{n!} = n+1 \quad \text{and} \quad r = \lim_{n\to\infty} \left|\frac{a_{n+1}}{a_n}\right| = \infty.$$

The radius of convergence is therefore $R = r^{-1} = 0$, and the series converges absolutely only for $x = -5$.

23. $\displaystyle\sum_{n=12}^{\infty} e^n (x-2)^n$

SOLUTION With $a_n = e^n$,

$$\left|\frac{a_{n+1}}{a_n}\right| = \frac{e^{n+1}}{e^n} = e \quad \text{and} \quad r = \lim_{n \to \infty} \left|\frac{a_{n+1}}{a_n}\right| = e.$$

The radius of convergence is therefore $R = r^{-1} = e^{-1}$, and the series converges absolutely on the interval $|x - 2| < e^{-1}$, or $2 - e^{-1} < x < 2 + e^{-1}$. For the endpoint $x = 2 + e^{-1}$, the series becomes $\sum_{n=1}^{\infty} 1$, which diverges by the Divergence Test. For the endpoint $x = 2 - e^{-1}$, the series becomes $\sum_{n=1}^{\infty} (-1)^n$, which also diverges by the Divergence Test. Thus, the series $\sum_{n=12}^{\infty} e^n (x - 2)^n$ converges for $2 - e^{-1} < x < 2 + e^{-1}$ and diverges elsewhere.

25. $\displaystyle\sum_{n=1}^{\infty} \frac{x^n}{n - 4\ln n}$

SOLUTION With $a_n = \frac{1}{n - 4\ln n}$,

$$\left|\frac{a_{n+1}}{a_n}\right| = \frac{1}{n + 1 - 4\ln(n+1)} \cdot \frac{n - 4\ln n}{1} = \frac{1 - 4\frac{\ln n}{n}}{1 + \frac{1}{n} - 4\frac{\ln(n+1)}{n}},$$

and

$$r = \lim_{n \to \infty} \left|\frac{a_{n+1}}{a_n}\right| = 1.$$

The radius of convergence is therefore $R = r^{-1} = 1$, and the series converges absolutely on the interval $|x| < 1$, or $-1 < x < 1$. For the endpoint $x = 1$, the series becomes $\sum_{n=1}^{\infty} \frac{1}{n - 4\ln n}$. Because $\frac{1}{n - 4\ln n} > \frac{1}{n}$ and $\sum_{n=1}^{\infty} \frac{1}{n}$ is the divergent harmonic series, the endpoint series diverges by the Comparison Test. For the endpoint $x = -1$, the series becomes $\sum_{n=1}^{\infty} \frac{(-1)^n}{n - 4\ln n}$, which converges by the Leibniz Test. Thus, the series $\sum_{n=1}^{\infty} \frac{x^n}{n - 4\ln n}$ converges for $-1 \leq x < 1$ and diverges elsewhere.

In Exercises 27–34, use Eq. (1) to expand the function in a power series with center $c = 0$ and determine the set of x for which the expansion is valid.

27. $f(x) = \dfrac{1}{1 - 3x}$

SOLUTION Substituting $3x$ for x in Eq. (1), we obtain

$$\frac{1}{1 - 3x} = \sum_{n=0}^{\infty} (3x)^n = \sum_{n=0}^{\infty} 3^n x^n.$$

This series is valid for $|3x| < 1$, or $|x| < \frac{1}{3}$.

29. $f(x) = \dfrac{1}{3 - x}$

SOLUTION First write

$$\frac{1}{3 - x} = \frac{1}{3} \cdot \frac{1}{1 - \frac{x}{3}}.$$

Substituting $\frac{x}{3}$ for x in Eq. (1), we obtain

$$\frac{1}{1 - \frac{x}{3}} = \sum_{n=0}^{\infty} \left(\frac{x}{3}\right)^n = \sum_{n=0}^{\infty} \frac{x^n}{3^n};$$

Thus,

$$\frac{1}{3 - x} = \frac{1}{3} \sum_{n=0}^{\infty} \frac{x^n}{3^n} = \sum_{n=0}^{\infty} \frac{x^n}{3^{n+1}}.$$

This series is valid for $|x/3| < 1$, or $|x| < 3$.

31. $f(x) = \dfrac{1}{1 + x^9}$

SOLUTION Substituting $-x^9$ for x in Eq. (1), we obtain

$$\frac{1}{1 + x^9} = \sum_{n=0}^{\infty} (-x^9)^n = \sum_{n=0}^{\infty} (-1)^n x^{9n}.$$

This series is valid for $|-x^9| < 1$, or $|x| < 1$.

33. $f(x) = \dfrac{1}{1 + 3x^7}$

SOLUTION Substituting $-3x^7$ for x in Eq. (1), we obtain

$$\frac{1}{1 + 3x^7} = \sum_{n=0}^{\infty} (-3x^7)^n = \sum_{n=0}^{\infty} (-3)^n x^{7n}.$$

This series is valid for $|-3x^7| < 1$, or $|x| < \frac{1}{\sqrt[7]{3}}$.

35. Use the equalities

$$\frac{1}{1 - x} = \frac{1}{-3 - (x - 4)} = \frac{-\frac{1}{3}}{1 + (\frac{x-4}{3})}$$

to show that for $|x - 4| < 3$

$$\frac{1}{1 - x} = \sum_{n=0}^{\infty} (-1)^{n+1} \frac{(x - 4)^n}{3^{n+1}}$$

SOLUTION Substituting $-\frac{x-4}{3}$ for x in Eq. (1), we obtain

$$\frac{1}{1 + \left(\frac{x-4}{3}\right)} = \sum_{n=0}^{\infty} \left(-\frac{x - 4}{3}\right)^n = \sum_{n=0}^{\infty} (-1)^n \frac{(x - 4)^n}{3^n}.$$

Thus,

$$\frac{1}{1 - x} = -\frac{1}{3} \sum_{n=0}^{\infty} (-1)^n \frac{(x - 4)^n}{3^n} = \sum_{n=0}^{\infty} (-1)^{n+1} \frac{(x - 4)^n}{3^{n+1}}.$$

This series is valid for $|-\frac{x-4}{3}| < 1$, or $|x - 4| < 3$.

37. Use the method of Exercise 35 to expand $\dfrac{1}{4 - x}$ in a power series with center $c = 5$. Determine the set of x for which the expansion is valid.

SOLUTION First write

$$\frac{1}{4 - x} = \frac{1}{-1 - (x - 5)} = -\frac{1}{1 + (x - 5)}.$$

Substituting $-(x - 5)$ for x in Eq. (1), we obtain

$$\frac{1}{1 + (x - 5)} = \sum_{n=0}^{\infty} (-(x - 5))^n = \sum_{n=0}^{\infty} (-1)^n (x - 5)^n.$$

Thus,

$$\frac{1}{4 - x} = -\sum_{n=0}^{\infty} (-1)^n (x - 5)^n = \sum_{n=0}^{\infty} (-1)^{n+1} (x - 5)^n.$$

This series is valid for $|-(x - 5)| < 1$, or $|x - 5| < 1$.

39. Give an example of a power series that converges for x in $[2, 6)$.

SOLUTION   The power series must be centered at $c = \dfrac{6 + 2}{2} = 4$, with radius of convergence $R = 2$. Consider the following series:

$$\sum_{n=1}^{\infty} \frac{(x-4)^n}{n2^n}.$$

With $a_n = \dfrac{1}{n2^n}$,

$$r = \lim_{n \to \infty} \frac{n2^n}{(n+1)2^{n+1}} = \frac{1}{2} \lim_{n \to \infty} \frac{n}{n+1} = \frac{1}{2}.$$

The radius of convergence is therefore $R = r^{-1} = 2$, and the series converges absolutely for $|x - 4| < 2$, or $2 < x < 6$. For the endpoint $x = 6$, the series becomes $\sum_{n=1}^{\infty} \dfrac{(6-4)^n}{n \cdot 2^n} = \sum_{n=1}^{\infty} \dfrac{1}{n}$, which is the divergent harmonic series. For the endpoint $x = 2$, the series becomes $\sum_{n=1}^{\infty} \dfrac{(2-4)^n}{n \cdot 2^n} = \sum_{n=1}^{\infty} \dfrac{(-1)^n}{n}$, which converges by the Leibniz Test. Therefore, the series converges for $2 \le x < 6$, as desired.

41. Use Exercise 40 to prove that

$$\ln \frac{3}{2} = \frac{1}{2} - \frac{1}{2 \cdot 2^2} + \frac{1}{3 \cdot 2^3} - \frac{1}{4 \cdot 2^4} + \cdots$$

Use your knowledge of alternating series to find an N such that the partial sum S_N approximates $\ln \frac{3}{2}$ to within an error of at most 10^{-3}. Confirm this using a calculator to compute both S_N and $\ln \frac{3}{2}$.

SOLUTION   In the previous exercise we found that

$$\ln(1 + x) = \sum_{n=0}^{\infty} (-1)^n \frac{x^{n+1}}{n+1}.$$

Setting $x = \frac{1}{2}$ yields:

$$\ln \frac{3}{2} = \sum_{n=1}^{\infty} (-1)^{n-1} \frac{\left(\frac{1}{2}\right)^n}{n} = \sum_{n=1}^{\infty} \frac{(-1)^{n-1}}{n2^n} = \frac{1}{2} - \frac{1}{2 \cdot 2^2} + \frac{1}{3 \cdot 2^3} - \frac{1}{4 \cdot 2^4} + \cdots$$

Note that the series for $\ln \frac{3}{2}$ is an alternating series with $a_n = \dfrac{1}{n2^n}$. The error in approximating $\ln \frac{3}{2}$ by the partial sum S_N is therefore bounded by

$$\left| \ln \frac{3}{2} - S_N \right| < a_{N+1} = \frac{1}{(N+1)2^{N+1}}.$$

To obtain an error of at most 10^{-3}, we must find an N such that

$$\frac{1}{(N+1)2^{N+1}} < 10^{-3} \quad \text{or} \quad (N+1)2^{N+1} > 1000.$$

For $N = 6$, $(N+1)2^{N+1} = 7 \cdot 2^7 = 896 < 1000$, but for $N = 7$, $(N+1)2^{N+1} = 8 \cdot 2^8 = 2048 > 1000$; hence, the smallest value for N is $N = 7$. The corresponding approximation is

$$S_7 = \frac{1}{2} - \frac{1}{2 \cdot 2^2} + \frac{1}{3 \cdot 2^3} - \frac{1}{4 \cdot 2^4} + \frac{1}{5 \cdot 2^5} - \frac{1}{6 \cdot 2^6} + \frac{1}{7 \cdot 2^7} = 0.405803571.$$

Now, $\ln \frac{3}{2} = 0.405465108$, so

$$\left| \ln \frac{3}{2} - S_7 \right| = 3.385 \times 10^{-4} < 10^{-3}.$$

43. Show that for $|x| < 1$

$$\frac{1 + 2x}{1 + x + x^2} = 1 + x - 2x^2 + x^3 + x^4 - 2x^5 + x^6 + x^7 - 2x^8 + \cdots$$

Hint: Use the hint from Exercise 42.

SOLUTION The terms in the series on the right-hand side are either of the form x^n or $-2x^n$ for some n. Because

$$\lim_{n\to\infty} \sqrt[n]{2} = \lim_{n\to\infty} \sqrt[n]{1} = 1,$$

it follows that

$$\lim_{n\to\infty} \sqrt[n]{|a_n|} = |x|.$$

Hence, by the Root Test, the series converges absolutely for $|x| < 1$.

By Exercise 37 of Section 11.4, any rearrangement of the terms of an absolutely convergent series yields another absolutely convergent series with the same sum as the original series. If we let S denote the sum of the series, then

$$S = \left(1 + x^3 + x^6 + \cdots\right) + \left(x + x^4 + x^7 + \cdots\right) - 2\left(x^2 + x^5 + x^8 + \cdots\right)$$

$$= \frac{1}{1-x^3} + \frac{x}{1-x^3} - \frac{2x^2}{1-x^3} = \frac{1+x-2x^2}{1-x^3} = \frac{(1-x)(2x+1)}{(1-x)(1+x+x^2)} = \frac{2x+1}{1+x+x^2}.$$

45. Use the power series for $y = e^x$ to show that

$$\frac{1}{e} = \frac{1}{2!} - \frac{1}{3!} + \frac{1}{4!} - \cdots$$

Use your knowledge of alternating series to find an N such that the partial sum S_N approximates e^{-1} to within an error of at most 10^{-3}. Confirm this using a calculator to compute both S_N and e^{-1}.

SOLUTION Recall that the series for e^x is

$$\sum_{n=0}^{\infty} \frac{x^n}{n!} = 1 + x + \frac{x^2}{2!} + \frac{x^3}{3!} + \frac{x^4}{4!} + \cdots.$$

Setting $x = -1$ yields

$$e^{-1} = 1 - 1 + \frac{1}{2!} - \frac{1}{3!} + \frac{1}{4!} - + \cdots = \frac{1}{2!} - \frac{1}{3!} + \frac{1}{4!} - + \cdots.$$

This is an alternating series with $a_n = \frac{1}{(n+1)!}$. The error in approximating e^{-1} with the partial sum S_N is therefore bounded by

$$|S_N - e^{-1}| \le a_{N+1} = \frac{1}{(N+2)!}.$$

To make the error at most 10^{-3}, we must choose N such that

$$\frac{1}{(N+2)!} \le 10^{-3} \quad \text{or} \quad (N+2)! \ge 1000.$$

For $N = 4$, $(N+2)! = 6! = 720 < 1000$, but for $N = 5$, $(N+2)! = 7! = 5040$; hence, $N = 5$ is the smallest value that satisfies the error bound. The corresponding approximation is

$$S_5 = \frac{1}{2!} - \frac{1}{3!} + \frac{1}{4!} - \frac{1}{5!} + \frac{1}{6!} = 0.368055555$$

Now, $e^{-1} = 0.367879441$, so

$$|S_5 - e^{-1}| = 1.761 \times 10^{-4} < 10^{-3}.$$

47. Find a power series $P(x)$ satisfying the differential equation:

$$y'' - xy' + y = 0 \qquad \boxed{10}$$

with initial condition $y(0) = 1$, $y'(0) = 0$. What is the radius of convergence of the power series?

SOLUTION Let $P(x) = \displaystyle\sum_{n=0}^{\infty} a_n x^n$. Then

$$P'(x) = \sum_{n=1}^{\infty} n a_n x^{n-1} \quad \text{and} \quad P''(x) = \sum_{n=2}^{\infty} n(n-1) a_n x^{n-2}.$$

Note that $P(0) = a_0$ and $P'(0) = a_1$; in order to satisfy the initial conditions $P(0) = 1$, $P'(0) = 0$, we must have $a_0 = 1$ and $a_1 = 0$. Now,

$$P''(x) - xP'(x) + P(x) = \sum_{n=2}^{\infty} n(n-1)a_n x^{n-2} - \sum_{n=1}^{\infty} na_n x^n + \sum_{n=0}^{\infty} a_n x^n$$

$$= \sum_{n=0}^{\infty} (n+2)(n+1)a_{n+2} x^n - \sum_{n=1}^{\infty} na_n x^n + \sum_{n=0}^{\infty} a_n x^n$$

$$= 2a_2 + a_0 + \sum_{n=1}^{\infty} \left[(n+2)(n+1)a_{n+2} - na_n + a_n \right] x^n.$$

In order for this series to be equal to zero, the coefficient of x^n must be equal to zero for each n; thus, $2a_2 + a_0 = 0$ and $(n+2)(n+1)a_{n+2} - (n-1)a_n = 0$, or

$$a_2 = -\frac{1}{2}a_0 \quad \text{and} \quad a_{n+2} = \frac{n-1}{(n+2)(n+1)}a_n.$$

Starting from $a_1 = 0$, we calculate

$$a_3 = \frac{1-1}{(3)(2)}a_1 = 0;$$

$$a_5 = \frac{2}{(5)(4)}a_3 = 0;$$

$$a_7 = \frac{4}{(7)(6)}a_5 = 0;$$

and, in general, all of the odd coefficients are zero. As for the even coefficients, we have $a_0 = 1$, $a_2 = -\frac{1}{2}$,

$$a_4 = \frac{1}{(4)(3)}a_2 = -\frac{1}{4!};$$

$$a_6 = \frac{3}{(6)(5)}a_4 = -\frac{3}{6!};$$

$$a_8 = \frac{5}{(8)(7)}a_6 = -\frac{15}{8!}$$

and so on. Thus,

$$P(x) = 1 - \frac{1}{2}x^2 - \frac{1}{4!}x^4 - \frac{3}{6!}x^6 - \frac{15}{8!}x^8 - \cdots$$

To determine the radius of convergence, treat this as a series in the variable x^2, and observe that

$$r = \lim_{k \to \infty} \left| \frac{a_{2k+2}}{a_{2k}} \right| = \lim_{k \to \infty} \frac{2k-1}{(2k+2)(2k+1)} = 0.$$

Thus, the radius of convergence is $R = r^{-1} = \infty$.

49. Prove that $J_2(x) = \displaystyle\sum_{k=0}^{\infty} \frac{(-1)^k}{2^{2k+2} \, k! \, (k+3)!} x^{2k+2}$ is a solution of the Bessel differential equation of order two:

$$x^2 y'' + xy' + (x^2 - 4)y = 0$$

SOLUTION Let $J_2(x) = \displaystyle\sum_{k=0}^{\infty} \frac{(-1)^k}{2^{2k+2} \, k! \, (k+2)!} x^{2k+2}$. Then

$$J_2'(x) = \sum_{k=0}^{\infty} \frac{(-1)^k (k+1)}{2^{2k+1} \, k! \, (k+2)!} x^{2k+1}$$

$$J_2''(x) = \sum_{k=0}^{\infty} \frac{(-1)^k (k+1)(2k+1)}{2^{2k+1} \, k! \, (k+2)!} x^{2k}$$

and

$$x^2 J_2''(x) + x J_2'(x) + (x^2 - 4)J_2(x) = \sum_{k=0}^{\infty} \frac{(-1)^k (k+1)(2k+1)}{2^{2k+1} \, k! \, (k+2)!} x^{2k+2} + \sum_{k=0}^{\infty} \frac{(-1)^k (k+1)}{2^{2k+1} \, k! \, (k+2)!} x^{2k+2}$$

$$-\sum_{k=0}^{\infty}\frac{(-1)^k}{2^{2k+2}\,k!\,(k+2)!}x^{2k+4}-\sum_{k=0}^{\infty}\frac{(-1)^k}{2^{2k}\,k!\,(k+2)!}x^{2k+2}$$

$$=\sum_{k=0}^{\infty}\frac{(-1)^k k(k+2)}{2^{2k}k!(k+2)!}x^{2k+2}+\sum_{k=1}^{\infty}\frac{(-1)^{k-1}}{2^{2k}(k-1)!(k+1)!}x^{2k+2}$$

$$=\sum_{k=1}^{\infty}\frac{(-1)^k}{2^{2k}(k-1)!(k+1)!}x^{2k+2}-\sum_{k=1}^{\infty}\frac{(-1)^k}{2^{2k}(k-1)!(k+1)!}x^{2k+2}=0.$$

51. Let $C(x)=1-\dfrac{x^2}{2!}+\dfrac{x^4}{4!}-\dfrac{x^6}{6!}+\cdots$.

(a) Show that $C(x)$ has an infinite radius of convergence.

(b) Prove that $C(x)$ and $f(x)=\cos x$ are both solutions of $y''=-y$ with initial conditions $y(0)=1$, $y'(0)=0$. This initial value problem has a unique solution, so it follows that $C(x)=\cos x$ for all x.

SOLUTION

(a) Consider the series

$$C(x)=1-\frac{x^2}{2!}+\frac{x^4}{4!}-\frac{x^6}{6!}+\cdots=\sum_{n=0}^{\infty}(-1)^n\frac{x^{2n}}{(2n)!}.$$

With $a_n=(-1)^n\dfrac{x^{2n}}{(2n)!}$,

$$\left|\frac{a_{n+1}}{a_n}\right|=\frac{|x|^{2n+2}}{(2n+2)!}\cdot\frac{(2n)!}{|x|^{2n}}=\frac{|x|^2}{(2n+2)(2n+1)},$$

and

$$r=\lim_{n\to\infty}\left|\frac{a_{n+1}}{a_n}\right|=0.$$

The radius of convergence for $C(x)$ is therefore $R=r^{-1}=\infty$.

(b) Differentiating the series defining $C(x)$ term-by-term, we find

$$C'(x)=\sum_{n=1}^{\infty}(-1)^n(2n)\frac{x^{2n-1}}{(2n)!}=\sum_{n=1}^{\infty}(-1)^n\frac{x^{2n-1}}{(2n-1)!}$$

and

$$C''(x)=\sum_{n=1}^{\infty}(-1)^n(2n-1)\frac{x^{2n-2}}{(2n-1)!}=\sum_{n=1}^{\infty}(-1)^n\frac{x^{2n-2}}{(2n-2)!}$$

$$=\sum_{n=0}^{\infty}(-1)^{n+1}\frac{x^{2n}}{(2n)!}=-\sum_{n=0}^{\infty}(-1)^n\frac{x^{2n}}{(2n)!}=-C(x).$$

Moreover, $C(0)=1$ and $C'(0)=0$.

53. Find all values of x such that the following series converges:

$$F(x)=1+3x+x^2+27x^3+x^4+243x^5+\cdots$$

SOLUTION Observe that $F(x)$ can be written as the sum of two geometric series:

$$F(x)=\left(1+x^2+x^4+\cdots\right)+\left(3x+27x^3+243x^5+\cdots\right)=\sum_{n=0}^{\infty}(x^2)^n+\sum_{n=0}^{\infty}3x(9x^2)^n$$

The first geometric series converges for $|x^2|<1$, or $|x|<1$; the second geometric series converges for $|9x^2|<1$, or $|x|<\frac{1}{3}$. Since both geometric series must converge for $F(x)$ to converge, we find that $F(x)$ converges for $|x|<\frac{1}{3}$, the intersection of the intervals of convergence for the two geometric series.

55. Why is it impossible to expand $f(x)=|x|$ as a power series that converges in an interval around $x=0$? Explain this using Theorem 3.

SOLUTION Suppose that there exists a $c > 0$ such that f can be represented by a power series on the interval $(-c, c)$; that is,

$$|x| = \sum_{n=0}^{\infty} a_n x^n$$

for $|x| < c$. Then it follows by Theorem 3 that $|x|$ is differentiable on $(-c, c)$. This contradicts the well known property that $f(x) = |x|$ is not differentiable at the point $x = 0$.

Further Insights and Challenges

57. Suppose that the coefficients of $F(x) = \displaystyle\sum_{n=0}^{\infty} a_n x^n$ are *periodic*, that is, for some whole number $M > 0$, we have $a_{M+n} = a_n$. Prove that $F(x)$ converges absolutely for $|x| < 1$ and that

$$F(x) = \frac{a_0 + a_1 x + \cdots + a_{M-1} x^{M-1}}{1 - x^M}$$

Hint: Use the hint for Exercise 42.

SOLUTION Suppose the coefficients of $F(x)$ are periodic, with $a_{M+n} = a_n$ for some whole number M and all n. The $F(x)$ can be written as the sum of M geometric series:

$$F(x) = a_0 \left(1 + x^M + x^{2M} + \cdots\right) + a_1 \left(x + x^{M+1} + x^{2M+1} + \cdots\right) +$$
$$= a_2 \left(x^2 + x^{M+2} + x^{2M+2} + \cdots\right) + \cdots + a_{M-1}\left(x^{M-1} + x^{2M-1} + x^{3M-1} + \cdots\right)$$
$$= \frac{a_0}{1 - x^M} + \frac{a_1 x}{1 - x^M} + \frac{a_2 x^2}{1 - x^M} + \cdots + \frac{a_{M-1} x^{M-1}}{1 - x^M} = \frac{a_0 + a_1 x + a_2 x^2 + \cdots + a_{M-1} x^{M-1}}{1 - x^M}.$$

As each geometric series converges absolutely for $|x| < 1$, it follows that $F(x)$ also converges absolutely for $|x| < 1$.

11.7 Taylor Series

Preliminary Questions

1. Determine $f(0)$ and $f'''(0)$ for a function $f(x)$ with Maclaurin series

$$T(x) = 3 + 2x + 12x^2 + 5x^3 + \cdots$$

SOLUTION The Maclaurin series for a function f has the form

$$f(0) + \frac{f'(0)}{1!}x + \frac{f''(0)}{2!}x^2 + \frac{f'''(0)}{3!}x^3 + \cdots$$

Matching this general expression with the given series, we find $f(0) = 3$ and $\dfrac{f'''(0)}{3!} = 5$. From this latter equation, it follows that $f'''(0) = 30$.

2. Determine $f(-2)$ and $f^{(4)}(-2)$ for a function with Taylor series

$$T(x) = 3(x + 2) + (x + 2)^2 - 4(x + 2)^3 + 2(x + 2)^4 + \cdots$$

SOLUTION The Taylor series for a function f centered at $x = -2$ has the form

$$f(-2) + \frac{f'(-2)}{1!}(x + 2) + \frac{f''(-2)}{2!}(x + 2)^2 + \frac{f'''(-2)}{3!}(x + 2)^3 + \frac{f^{(4)}(-2)}{4!}(x + 2)^4 + \cdots$$

Matching this general expression with the given series, we find $f(-2) = 0$ and $\dfrac{f^{(4)}(-2)}{4!} = 2$. From this latter equation, it follows that $f^{(4)}(-2) = 48$.

3. What is the easiest way to find the Maclaurin series for the function $f(x) = \sin(x^2)$?

SOLUTION The easiest way to find the Maclaurin series for $\sin\left(x^2\right)$ is to substitute x^2 for x in the Maclaurin series for $\sin x$.

4. What is the Taylor series for $f(x)$ centered at $c = 3$ if $f(3) = 4$ and $f'(x)$ has a Taylor expansion

$$f'(x) = \sum_{n=1}^{\infty} \frac{(x-3)^n}{n}$$

SOLUTION Integrating the series for $f'(x)$ term-by-term gives

$$f(x) = C + \sum_{n=1}^{\infty} \frac{(x-3)^{n+1}}{n(n+1)}.$$

Substituting $x = 3$ then yields

$$f(3) = C = 4;$$

so

$$f(x) = 4 + \sum_{n=1}^{\infty} \frac{(x-3)^{n+1}}{n(n+1)}.$$

5. Let $T(x)$ be the Maclaurin series of $f(x)$. Which of the following guarantees that $f(2) = T(2)$?
(a) $T(x)$ converges for $x = 2$.
(b) The remainder $R_k(2)$ approaches a limit as $k \to \infty$.
(c) The remainder $R_k(2)$ approaches zero as $k \to \infty$.

SOLUTION The correct response is **(c)**: $f(2) = T(2)$ if and only if the remainder $R_k(2)$ approaches zero as $k \to \infty$.

Exercises

1. Write out the first four terms of the Maclaurin series of $f(x)$ if

$$f(0) = 2, \quad f'(0) = 3, \quad f''(0) = 4, \quad f'''(0) = 12$$

SOLUTION The first four terms of the Maclaurin series of $f(x)$ are

$$f(0) + f'(0)x + \frac{f''(0)}{2!}x^2 + \frac{f'''(0)}{3!}x^3 = 2 + 3x + \frac{4}{2}x^2 + \frac{12}{6}x^3 = 2 + 3x + 2x^2 + 2x^3.$$

In Exercises 3–20, find the Maclaurin series.

3. $f(x) = \dfrac{1}{1-2x}$

SOLUTION Substituting $2x$ for x in the Maclaurin series for $\frac{1}{1-x}$ gives

$$\frac{1}{1-2x} = \sum_{n=0}^{\infty} (2x)^n = \sum_{n=0}^{\infty} 2^n x^n.$$

This series is valid for $|2x| < 1$, or $|x| < \frac{1}{2}$.

5. $f(x) = \cos 3x$

SOLUTION Substituting $3x$ for x in the Maclaurin series for $\cos x$ gives

$$\cos 3x = \sum_{n=0}^{\infty} (-1)^n \frac{(3x)^{2n}}{(2n)!} = \sum_{n=0}^{\infty} (-1)^n \frac{9^n x^{2n}}{(2n)!}.$$

This series is valid for all x.

7. $f(x) = \sin(x^2)$

SOLUTION Substituting x^2 for x in the Maclaurin series for $\sin x$ gives

$$\sin x^2 = \sum_{n=0}^{\infty} (-1)^n \frac{(x^2)^{2n+1}}{(2n+1)!} = \sum_{n=0}^{\infty} (-1)^n \frac{x^{4n+2}}{(2n+1)!}.$$

This series is valid for all x.

9. $f(x) = \ln(1 - x^2)$

SOLUTION Substituting $-x^2$ for x in the Maclaurin series for $\ln(1+x)$ gives

$$\ln(1-x^2) = \sum_{n=1}^{\infty} \frac{(-1)^{n-1}(-x^2)^n}{n} = \sum_{n=1}^{\infty} \frac{(-1)^{2n-1}x^{2n}}{n} = -\sum_{n=1}^{\infty} \frac{x^{2n}}{n}.$$

This series is valid for $|x| < 1$.

11. $f(x) = \tan^{-1}(x^2)$

SOLUTION Substituting x^2 for x in the Maclaurin series for $\tan^{-1}x$ gives

$$\tan^{-1}(x^2) = \sum_{n=0}^{\infty} (-1)^n \frac{(x^2)^{2n+1}}{2n+1} = \sum_{n=0}^{\infty} (-1)^n \frac{x^{4n+2}}{2n+1}.$$

This series is valid for $|x| \le 1$.

13. $f(x) = e^{x-2}$

SOLUTION $e^{x-2} = e^{-2}e^x$; thus,

$$e^{x-2} = e^{-2} \sum_{n=0}^{\infty} \frac{x^n}{n!} = \sum_{n=0}^{\infty} \frac{x^n}{e^2 n!}.$$

This series is valid for all x.

15. $f(x) = \ln(1-5x)$

SOLUTION Substituting $-5x$ for x in the Maclaurin series for $\ln(1+x)$ gives

$$\ln(1-5x) = \sum_{n=1}^{\infty} \frac{(-1)^{n-1}(-5x)^n}{n} = \sum_{n=1}^{\infty} \frac{(-1)^{2n-1}5^n x^n}{n} = -\sum_{n=1}^{\infty} \frac{5^n x^n}{n}.$$

This series is valid for $|5x| < 1$, or $|x| < \frac{1}{5}$, and for $x = -\frac{1}{5}$.

17. $f(x) = \sinh x$

SOLUTION Recall that

$$\sinh x = \frac{1}{2}(e^x - e^{-x}).$$

Therefore,

$$\sinh x = \frac{1}{2}\left(\sum_{n=0}^{\infty} \frac{x^n}{n!} - \sum_{n=0}^{\infty} \frac{(-x)^n}{n!} \right) = \sum_{n=0}^{\infty} \frac{x^n}{2(n!)} \left(1 - (-1)^n\right).$$

Now,

$$1 - (-1)^n = \begin{cases} 0, & n \text{ even} \\ 2, & n \text{ odd} \end{cases}$$

so

$$\sinh x = \sum_{k=0}^{\infty} 2\frac{x^{2k+1}}{2(2k+1)!} = \sum_{k=0}^{\infty} \frac{x^{2k+1}}{(2k+1)!}.$$

This series is valid for all x.

19. $f(x) = \dfrac{1 - \cos(x^2)}{x}$

SOLUTION Substituting x^2 for x in the Maclaurin series for $\cos x$ gives

$$\cos x^2 = \sum_{n=0}^{\infty} (-1)^n \frac{(x^2)^{2n}}{(2n)!} = \sum_{n=0}^{\infty} (-1)^n \frac{x^{4n}}{(2n)!} = 1 + \sum_{n=1}^{\infty} (-1)^n \frac{x^{4n}}{(2n)!}.$$

Thus,

$$1 - \cos x^2 = 1 - \left(1 + \sum_{n=1}^{\infty} (-1)^n \frac{x^{4n}}{(2n)!} \right) = \sum_{n=1}^{\infty} (-1)^{n+1} \frac{x^{4n}}{(2n)!},$$

and

$$\frac{1 - \cos(x^2)}{x} = \frac{1}{x}\sum_{n=1}^{\infty}(-1)^{n+1}\frac{x^{4n}}{(2n)!} = \sum_{n=1}^{\infty}(-1)^{n+1}\frac{x^{4n-1}}{(2n)!}.$$

21. Use multiplication to find the first four terms in the Maclaurin series for $f(x) = e^x \sin x$.

SOLUTION Multiply the fifth-order Taylor Polynomials for e^x and $\sin x$:

$$\left(1 + x + \frac{x^2}{2} + \frac{x^3}{6} + \frac{x^4}{24} + \frac{x^5}{120}\right)\left(x - \frac{x^3}{6} + \frac{x^5}{120}\right)$$

$$= x + x^2 - \frac{x^3}{6} + \frac{x^3}{2} - \frac{x^4}{6} + \frac{x^4}{6} + \frac{x^5}{120} - \frac{x^5}{12} + \frac{x^5}{24} + \text{higher-order terms}$$

$$= x + x^2 + \frac{x^3}{3} - \frac{x^5}{30} + \text{higher-order terms}.$$

The first four terms in the Maclaurin series for $f(x) = e^x \sin x$ are therefore

$$x + x^2 + \frac{x^3}{3} - \frac{x^5}{30}.$$

23. Find the first four terms of the Maclaurin series for $f(x) = e^x \ln(1 - x)$.

SOLUTION Multiply the fourth order Taylor Polynomials for e^x and $\ln(1 - x)$:

$$\left(1 + x + \frac{x^2}{2} + \frac{x^3}{6} + \frac{x^4}{24}\right)\left(-x - \frac{x^2}{2} - \frac{x^3}{3} - \frac{x^4}{4}\right)$$

$$= -x - \frac{x^2}{2} - x^2 - \frac{x^3}{3} - \frac{x^3}{2} - \frac{x^3}{2} - \frac{x^4}{4} - \frac{x^4}{3} - \frac{x^4}{4} - \frac{x^4}{6} + \text{higher-order terms}$$

$$= -x - \frac{3x^2}{2} - \frac{4x^3}{3} - x^4 + \text{higher-order terms}.$$

The first four terms of the Maclaurin series for $f(x) = e^x \ln(1 - x)$ are therefore

$$-x - \frac{3x^2}{2} - \frac{4x^3}{3} - x^4.$$

25. Write out the first five terms of the binomial series for $f(x) = (1 + x)^{-3/2}$.

SOLUTION The first five generalized binomial coefficients for $a = -\frac{3}{2}$ are

$$1, \quad -\frac{3}{2}, \quad \frac{-\frac{3}{2}(-\frac{5}{2})}{2!} = \frac{15}{8}, \quad \frac{-\frac{3}{2}(-\frac{5}{2})(-\frac{7}{2})}{3!} = -\frac{35}{16}, \quad \frac{-\frac{3}{2}(-\frac{5}{2})(-\frac{7}{2})(-\frac{9}{2})}{4!} = \frac{315}{128}.$$

Therefore, the first five terms in the binomial series for $f(x) = (1 + x)^{-3/2}$ are

$$1 - \frac{3}{2}x + \frac{15}{8}x^2 - \frac{35}{16}x^3 + \frac{315}{128}x^4.$$

27. Find the first four terms of the Maclaurin for $f(x) = e^{(e^x)}$.

SOLUTION With $f(x) = e^{(e^x)}$, we find

$$f'(x) = e^{(e^x)} \cdot e^x$$

$$f''(x) = e^{(e^x)} \cdot e^x + e^{(e^x)} \cdot e^{2x} = e^{(e^x)}\left(e^{2x} + e^x\right)$$

$$f'''(x) = e^{(e^x)}\left(2e^{2x} + e^x\right) + e^{(e^x)}\left(e^{2x} + e^x\right)e^x$$

$$= e^{(e^x)}\left(e^{3x} + 3e^{2x} + e^x\right)$$

and

$$f(0) = e, \quad f'(0) = e, \quad f''(0) = 2e, \quad f'''(0) = 5e.$$

Therefore, the first four terms of the Maclaurin for $f(x) = e^{(e^x)}$ are

$$e + ex + ex^2 + \frac{5e}{6}x^3.$$

29. Find the Taylor series for $\sin x$ at $c = \dfrac{\pi}{2}$.

SOLUTION Because

$$\sin x = \cos\left(x - \frac{\pi}{2}\right),$$

we obtain

$$\sin x = \sum_{n=0}^{\infty} \frac{(-1)^n}{(2n)!}\left(x - \frac{\pi}{2}\right)^{2n},$$

by substituting $x - \frac{\pi}{2}$ for x in the Maclaurin series for $\cos x$.

In Exercises 31–40, find the Taylor series centered at c.

31. $f(x) = \dfrac{1}{x}, \quad c = 1$

SOLUTION Write

$$\frac{1}{x} = \frac{1}{1 + (x - 1)},$$

and then substitute $-(x - 1)$ for x in the Maclaurin series for $\frac{1}{1-x}$ to obtain

$$\frac{1}{x} = \sum_{n=0}^{\infty} [-(x - 1)]^n = \sum_{n=0}^{\infty} (-1)^n (x - 1)^n.$$

This series is valid for $|x - 1| < 1$.

33. $f(x) = \dfrac{1}{1 - x}, \quad c = 5$

SOLUTION Write

$$\frac{1}{1 - x} = \frac{1}{-4 - (x - 5)} = -\frac{1}{4} \cdot \frac{1}{1 + \frac{x-5}{4}}.$$

Substituting $-\frac{x-5}{4}$ for x in the Maclaurin series for $\frac{1}{1-x}$ yields

$$\frac{1}{1 + \frac{x-5}{4}} = \sum_{n=0}^{\infty} \left(-\frac{x - 5}{4}\right)^n = \sum_{n=0}^{\infty} (-1)^n \frac{(x - 5)^n}{4^n}.$$

Thus,

$$\frac{1}{1 - x} = -\frac{1}{4} \sum_{n=0}^{\infty} (-1)^n \frac{(x - 5)^n}{4^n} = \sum_{n=0}^{\infty} (-1)^{n+1} \frac{(x - 5)^n}{4^{n+1}}.$$

This series is valid for $\left|\frac{x-5}{4}\right| < 1$, or $|x - 5| < 4$.

35. $f(x) = x^4 + 3x - 1, \quad c = 2$

SOLUTION To determine the Taylor series with center $c = 2$, we compute

$$f'(x) = 4x^3 + 3, \quad f''(x) = 12x^2, \quad f'''(x) = 24x,$$

and $f^{(4)}(x) = 24$. All derivatives of order five and higher are zero. Now,

$$f(2) = 21, \quad f'(2) = 35, \quad f''(2) = 48, \quad f'''(2) = 48,$$

and $f^{(4)}(2) = 24$. Therefore, the Taylor series is

$$21 + 35(x - 2) + \frac{48}{2}(x - 2)^2 + \frac{48}{6}(x - 2)^3 + \frac{24}{24}(x - 2)^4,$$

or

$$21 + 35(x - 2) + 24(x - 2)^2 + 8(x - 2)^3 + (x - 2)^4.$$

37. $f(x) = e^{3x}, \quad c = -1$

SOLUTION Write

$$e^{3x} = e^{3(x+1)-3} = e^{-3}e^{3(x+1)}.$$

Now, substitute $3(x + 1)$ for x in the Maclaurin series for e^x to obtain

$$e^{3(x+1)} = \sum_{n=0}^{\infty} \frac{(3(x + 1))^n}{n!} = \sum_{n=0}^{\infty} \frac{3^n}{n!} (x + 1)^n.$$

Thus,

$$e^{3x} = e^{-3} \sum_{n=0}^{\infty} \frac{3^n}{n!} (x + 1)^n = \sum_{n=0}^{\infty} \frac{3^n e^{-3}}{n!} (x + 1)^n,$$

This series is valid for all x.

39. $f(x) = \dfrac{1}{1 - x^2}, \quad c = 3$

SOLUTION By partial fraction decomposition

$$\frac{1}{1 - x^2} = \frac{\frac{1}{2}}{1 - x} + \frac{\frac{1}{2}}{1 + x},$$

so

$$\frac{1}{1 - x^2} = \frac{\frac{1}{2}}{-2 - (x - 3)} + \frac{\frac{1}{2}}{4 + (x - 3)} = -\frac{1}{4} \cdot \frac{1}{1 + \frac{x-3}{2}} + \frac{1}{8} \cdot \frac{1}{1 + \frac{x-3}{4}}.$$

Substituting $-\frac{x-3}{2}$ for x in the Maclaurin series for $\frac{1}{1-x}$ gives

$$\frac{1}{1 + \frac{x-3}{2}} = \sum_{n=0}^{\infty} \left(-\frac{x - 3}{2}\right)^n = \sum_{n=0}^{\infty} \frac{(-1)^n}{2^n} (x - 3)^n,$$

while substituting $-\frac{x-3}{4}$ for x in the same series gives

$$\frac{1}{1 + \frac{x-3}{4}} = \sum_{n=0}^{\infty} \left(-\frac{x - 3}{4}\right)^n = \sum_{n=0}^{\infty} \frac{(-1)^n}{4^n} (x - 3)^n.$$

Thus,

$$\frac{1}{1 - x^2} = -\frac{1}{4} \sum_{n=0}^{\infty} \frac{(-1)^n}{2^n} (x - 3)^n + \frac{1}{8} \sum_{n=0}^{\infty} \frac{(-1)^n}{4^n} (x - 3)^n = \sum_{n=0}^{\infty} \frac{(-1)^{n+1}}{2^{n+2}} (x - 3)^n + \sum_{n=0}^{\infty} \frac{(-1)^n}{2^{2n+3}} (x - 3)^n$$

$$= \sum_{n=0}^{\infty} \left(\frac{(-1)^{n+1}}{2^{n+2}} + \frac{(-1)^n}{2^{2n+3}}\right) (x - 3)^n = \sum_{n=0}^{\infty} \frac{(-1)^{n+1}(2^{n+1} - 1)}{2^{2n+3}} (x - 3)^n.$$

This series is valid for $|x - 3| < 2$.

41. Find the Maclaurin series for $f(x) = \dfrac{1}{\sqrt{1 - 9x^2}}$ (see Example 10).

SOLUTION From Example 10, we know that for $|x| < 1$,

$$\frac{1}{\sqrt{1 - x^2}} = \sum_{n=0}^{\infty} \frac{1 \cdot 3 \cdot 5 \cdots (2n - 1)}{2 \cdot 4 \cdot 6 \cdots (2n)} x^{2n},$$

so

$$\frac{1}{\sqrt{1 - 9x^2}} = \frac{1}{\sqrt{1 - (3x)^2}} = \sum_{n=0}^{\infty} \frac{1 \cdot 3 \cdot 5 \cdots (2n - 1)}{2 \cdot 4 \cdot 6 \cdots (2n)} (3x)^{2n} = \sum_{n=0}^{\infty} \frac{1 \cdot 3 \cdot 5 \cdots (2n - 1)}{2 \cdot 4 \cdot 6 \cdots (2n)} 9^n x^{2n}.$$

This series is valid for $9x^2 < 1$, or $|x| < \frac{1}{3}$.

43. Use the first five terms of the Maclaurin series in Exercise 42 to approximate $\sin^{-1} \frac{1}{2}$. Compare the result with the calculator value.

SOLUTION   From Exercise 42 we know that for $|x| < 1$,

$$\sin^{-1} x = x + \sum_{n=1}^{\infty} \frac{1 \cdot 3 \cdot 5 \cdots (2n-1)}{2 \cdot 4 \cdot 6 \cdots (2n)} \frac{x^{2n+1}}{2n+1}.$$

The first five terms of the series are:

$$x + \frac{1}{2} \frac{x^3}{3} + \frac{1 \cdot 3}{2 \cdot 4} \frac{x^5}{5} + \frac{1 \cdot 3 \cdot 5}{2 \cdot 4 \cdot 6} \frac{x^7}{7} + \frac{1 \cdot 3 \cdot 5 \cdot 7}{2 \cdot 4 \cdot 6 \cdot 8} \frac{x^9}{9} = x + \frac{x^3}{6} + \frac{3x^5}{40} + \frac{5x^7}{112} + \frac{35x^9}{1152}$$

Setting $x = \frac{1}{2}$, we obtain the following approximation:

$$\sin^{-1} \frac{1}{2} \approx \frac{1}{2} + \frac{\left(\frac{1}{2}\right)^3}{6} + \frac{3 \cdot \left(\frac{1}{2}\right)^5}{40} + \frac{5 \cdot \left(\frac{1}{2}\right)^7}{112} + \frac{35 \cdot \left(\frac{1}{2}\right)^9}{1152} \approx 0.52358519539.$$

The calculator value is $\sin^{-1} \frac{1}{2} \approx 0.5235988775$.

45. Use the Maclaurin series for $\ln(1+x)$ and $\ln(1-x)$ to show that

$$\frac{1}{2} \ln\left(\frac{1+x}{1-x}\right) = x + \frac{x^3}{3} + \frac{x^5}{5} + \cdots$$

What can you conclude by comparing this result with that of Exercise 44?

SOLUTION   Using the Maclaurin series for $\ln(1+x)$ and $\ln(1-x)$, we have for $|x| < 1$

$$\ln(1+x) - \ln(1-x) = \sum_{n=1}^{\infty} \frac{(-1)^{n-1}}{n} x^n - \sum_{n=1}^{\infty} \frac{(-1)^{n-1}}{n} (-x)^n$$

$$= \sum_{n=1}^{\infty} \frac{(-1)^{n-1}}{n} x^n + \sum_{n=1}^{\infty} \frac{x^n}{n} = \sum_{n=1}^{\infty} \frac{1 + (-1)^{n-1}}{n} x^n.$$

Since $1 + (-1)^{n-1} = 0$ for even n and $1 + (-1)^{n-1} = 2$ for odd n,

$$\ln(1+x) - \ln(1-x) = \sum_{k=0}^{\infty} \frac{2}{2k+1} x^{2k+1}.$$

Thus,

$$\frac{1}{2} \ln\left(\frac{1+x}{1-x}\right) = \frac{1}{2} (\ln(1+x) - \ln(1-x)) = \frac{1}{2} \sum_{k=0}^{\infty} \frac{2}{2k+1} x^{2k+1} = \sum_{k=0}^{\infty} \frac{x^{2k+1}}{2k+1}.$$

Observe that this is the same series we found in Exercise 44; therefore,

$$\frac{1}{2} \ln\left(\frac{1+x}{1-x}\right) = \tanh^{-1} x.$$

47. Use the Maclaurin expansion for e^{-t^2} to express $\int_0^x e^{-t^2} \, dt$ as an alternating power series in t.

(a) How many terms of the infinite series are needed to approximate the integral for $x = 1$ to within an error of at most 0.001?

(b) *CAS*   Carry out the computation and check your answer using a computer algebra system.

SOLUTION   Substituting $-t^2$ for t in the Maclaurin series for e^t yields

$$e^{-t^2} = \sum_{n=0}^{\infty} \frac{(-t^2)^n}{n!} = \sum_{n=0}^{\infty} (-1)^n \frac{t^{2n}}{n!};$$

thus,

$$\int_0^x e^{-t^2} \, dt = \sum_{n=0}^{\infty} (-1)^n \frac{x^{2n+1}}{n!(2n+1)}.$$

(a) For $x = 1$,

$$\int_0^1 e^{-t^2}\, dt = \sum_{n=0}^{\infty} (-1)^n \frac{1}{n!(2n+1)}.$$

This is an alternating series with $a_n = \frac{1}{n!(2n+1)}$; therefore, the error incurred by using S_N to approximate the value of the definite integral is bounded by

$$\left| \int_0^1 e^{-t^2}\, dt - S_N \right| \leq a_{N+1} = \frac{1}{(N+1)!(2N+3)}.$$

To guarantee the error is at most 0.001, we must choose N so that

$$\frac{1}{(N+1)!(2N+3)} < 0.001 \quad \text{or} \quad (N+1)!(2N+3) > 1000.$$

For $N = 3$, $(N+1)!(2N+3) = 4! \cdot 9 = 216 < 1000$ and for $N = 4$, $(N+1)!(2N+3) = 5! \cdot 11 = 1320 > 1000$; thus, the smallest acceptable value for N is $N = 4$. The corresponding approximation is

$$S_4 = \sum_{n=0}^{4} \frac{(-1)^n}{n!(2n+1)} = 1 - \frac{1}{3} + \frac{1}{2! \cdot 5} - \frac{1}{3! \cdot 7} + \frac{1}{4! \cdot 9} = 0.747486772.$$

(b) Using a computer algebra system, we find

$$\int_0^1 e^{-t^2}\, dt = 0.746824133;$$

therefore

$$\left| \int_0^1 e^{-t^2}\, dt - S_4 \right| = 6.626 \times 10^{-4} < 10^{-3}.$$

In Exercises 49–52, express the definite integral as an infinite series and find its value to within an error of at most 10^{-4}.

49. $\displaystyle \int_0^1 \cos(x^2)\, dx$

SOLUTION Substituting x^2 for x in the Maclaurin series for $\cos x$ yields

$$\cos(x^2) = \sum_{n=0}^{\infty} (-1)^n \frac{(x^2)^{2n}}{(2n)!} = \sum_{n=0}^{\infty} (-1)^n \frac{x^{4n}}{(2n)!};$$

therefore,

$$\int_0^1 \cos(x^2)\, dx = \sum_{n=0}^{\infty} (-1)^n \left. \frac{x^{4n+1}}{(2n)!(4n+1)} \right|_0^1 = \sum_{n=0}^{\infty} \frac{(-1)^n}{(2n)!(4n+1)}.$$

This is an alternating series with $a_n = \frac{1}{(2n)!(4n+1)}$; therefore, the error incurred by using S_N to approximate the value of the definite integral is bounded by

$$\left| \int_0^1 \cos(x^2)\, dx - S_N \right| \leq a_{N+1} = \frac{1}{(2N+2)!(4N+5)}.$$

To guarantee the error is at most 0.0001, we must choose N so that

$$\frac{1}{(2N+2)!(4N+5)} < 0.0001 \quad \text{or} \quad (2N+2)!(4N+5) > 10000.$$

For $N = 2$, $(2N+2)!(4N+5) = 6! \cdot 13 = 9360 < 10000$ and for $N = 3$, $(2N+2)!(4N+5) = 8! \cdot 17 = 685440 > 10000$; thus, the smallest acceptable value for N is $N = 3$. The corresponding approximation is

$$S_3 = \sum_{n=0}^{3} \frac{(-1)^n}{(2n)!(4n+1)} = 1 - \frac{1}{5 \cdot 2!} + \frac{1}{9 \cdot 4!} - \frac{1}{13 \cdot 6!} = 0.904522792.$$

51. $\displaystyle \int_0^2 e^{-x^3}\, dx$

SOLUTION Substituting $-x^3$ for x in the Maclaurin series for e^x yields

$$e^{-x^3} = \sum_{n=0}^{\infty} \frac{(-x^3)^n}{n!} = \sum_{n=0}^{\infty} (-1)^n \frac{x^{3n}}{n!};$$

therefore,

$$\int_0^1 e^{-x^3}\, dx = \sum_{n=0}^{\infty} (-1)^n \left. \frac{x^{3n+1}}{n!(3n+1)} \right|_0^1 = \sum_{n=0}^{\infty} \frac{(-1)^n}{n!(3n+1)}.$$

This is an alternating series with $a_n = \frac{1}{n!(3n+1)}$; therefore, the error incurred by using S_N to approximate the value of the definite integral is bounded by

$$\left| \int_0^1 e^{-x^3}\, dx - S_N \right| \le a_{N+1} = \frac{1}{(N+1)!(3N+4)}.$$

To guarantee the error is at most 0.0001, we must choose N so that

$$\frac{1}{(N+1)!(3N+4)} < 0.0001 \quad \text{or} \quad (N+1)!(3N+4) > 10000.$$

For $N = 4$, $(N+1)!(3N+4) = 5! \cdot 16 = 1920 < 10000$ and for $N = 5$, $(N+1)!(3N+4) = 6! \cdot 19 = 13680 > 10000$; thus, the smallest acceptable value for N is $N = 5$. The corresponding approximation is

$$S_5 = \sum_{n=0}^{5} \frac{(-1)^n}{n!(3n+1)} = 0.807446200.$$

In Exercises 53–56, express the integral as an infinite series.

53. $\displaystyle \int_0^x \frac{1 - \cos(t)}{t}\, dt$, for all x

SOLUTION The Maclaurin series for $\cos t$ is

$$\cos t = \sum_{n=0}^{\infty} (-1)^n \frac{t^{2n}}{(2n)!} = 1 + \sum_{n=1}^{\infty} (-1)^n \frac{t^{2n}}{(2n)!},$$

so

$$1 - \cos t = -\sum_{n=1}^{\infty} (-1)^n \frac{t^{2n}}{(2n)!} = \sum_{n=1}^{\infty} (-1)^{n+1} \frac{t^{2n}}{(2n)!},$$

and

$$\frac{1 - \cos t}{t} = \frac{1}{t} \sum_{n=1}^{\infty} (-1)^{n+1} \frac{t^{2n}}{(2n)!} = \sum_{n=1}^{\infty} (-1)^{n+1} \frac{t^{2n-1}}{(2n)!}.$$

Thus,

$$\int_0^x \frac{1 - \cos(t)}{t}\, dt = \sum_{n=1}^{\infty} (-1)^{n+1} \left. \frac{t^{2n}}{(2n)!\, 2n} \right|_0^x = \sum_{n=1}^{\infty} (-1)^{n+1} \frac{x^{2n}}{(2n)!\, 2n}.$$

55. $\displaystyle \int_0^x \ln(1 + t^2)\, dt$, for $|x| < 1$

SOLUTION Substituting t^2 for t in the Maclaurin series for $\ln(1 + t)$ yields

$$\ln(1 + t^2) = \sum_{n=1}^{\infty} (-1)^{n-1} \frac{(t^2)^n}{n} = \sum_{n=1}^{\infty} (-1)^n \frac{t^{2n}}{n}.$$

Thus,

$$\int_0^x \ln(1 + t^2)\, dt = \sum_{n=1}^{\infty} (-1)^n \left. \frac{t^{2n+1}}{n(2n+1)} \right|_0^x = \sum_{n=1}^{\infty} (-1)^n \frac{x^{2n+1}}{n(2n+1)}.$$

57. Which function has Maclaurin series $\displaystyle\sum_{n=0}^{\infty}(-1)^n 2^n x^n$?

SOLUTION We recognize that

$$\sum_{n=0}^{\infty}(-1)^n 2^n x^n = \sum_{n=0}^{\infty}(-2x)^n$$

is the Maclaurin series for $\frac{1}{1-x}$ with x replaced by $-2x$. Therefore,

$$\sum_{n=0}^{\infty}(-1)^n 2^n x^n = \frac{1}{1-(-2x)} = \frac{1}{1+2x}.$$

In Exercises 59–62, find the first four terms of the Taylor series.

59. $f(x) = \sin(x^2)\cos(x^2)$

SOLUTION Substituting x^2 for x in the Maclaurin series for $\sin x$ and $\cos x$, we find

$$\sin(x^2)\cos(x^2) = \left(x^2 - \frac{x^6}{6} + \frac{x^{10}}{120} - \frac{x^{14}}{5040} + \cdots\right)\left(1 - \frac{x^4}{2} + \frac{x^8}{24} - \frac{x^{12}}{720} + \cdots\right)$$

$$= x^2 - \frac{x^6}{6} - \frac{x^6}{2} + \frac{x^{10}}{24} + \frac{x^{10}}{12} + \frac{x^{10}}{120} - \frac{x^{14}}{720} - \frac{x^{14}}{144} - \frac{x^{14}}{240} - \frac{x^{14}}{5040} + \cdots$$

$$= x^2 - \frac{2}{3}x^6 + \frac{2}{15}x^{10} - \frac{4}{315}x^{14} + \cdots.$$

61. $f(x) = e^{\sin x}$

SOLUTION Substituting $\sin x$ for x in the Maclaurin series for e^x and then using the Maclaurin series for $\sin x$, we find

$$e^{\sin x} = 1 + \sin x + \frac{\sin^2 x}{2} + \frac{\sin^3 x}{6} + \frac{\sin^4 x}{24} + \cdots$$

$$= 1 + \left(x - \frac{x^3}{6} + \cdots\right) + \frac{1}{2}\left(x - \frac{x^3}{6} + \cdots\right)^2 + \frac{1}{6}(x - \cdots)^3 + \frac{1}{24}(x - \cdots)^4$$

$$= 1 + x + \frac{1}{2}x^2 - \frac{1}{6}x^3 + \frac{1}{6}x^3 - \frac{1}{6}x^4 + \frac{1}{24}x^4 + \cdots$$

$$= 1 + x + \frac{1}{2}x^2 - \frac{1}{8}x^4 + \cdots.$$

In Exercises 63–66, find the functions with the following Maclaurin series (refer to Table 1).

63. $1 + x^3 + \dfrac{x^6}{2!} + \dfrac{x^9}{3!} + \dfrac{x^{12}}{4!} + \cdots$

SOLUTION We recognize

$$1 + x^3 + \frac{x^6}{2!} + \frac{x^9}{3!} + \frac{x^{12}}{4!} + \cdots = \sum_{n=0}^{\infty} \frac{x^{3n}}{n!} = \sum_{n=0}^{\infty} \frac{(x^3)^n}{n!}$$

as the Maclaurin series for e^x with x replaced by x^3. Therefore,

$$1 + x^3 + \frac{x^6}{2!} + \frac{x^9}{3!} + \frac{x^{12}}{4!} + \cdots = e^{x^3}.$$

65. $1 - \dfrac{5^3 x^3}{3!} + \dfrac{5^5 x^5}{5!} - \dfrac{5^7 x^7}{7!} + \cdots$

SOLUTION Note

$$1 - \frac{5^3 x^3}{3!} + \frac{5^5 x^5}{5!} - \frac{5^7 x^7}{7!} + \cdots = 1 - 5x + \left(5x - \frac{5^3 x^3}{3!} + \frac{5^5 x^5}{5!} - \frac{5^7 x^7}{7!} + \cdots\right)$$

$$= 1 - 5x + \sum_{n=0}^{\infty} (-1)^n \frac{(5x)^{2n+1}}{(2n+1)!}.$$

The series is the Maclaurin series for $\sin x$ with x replaced by $5x$, so

$$1 - \frac{5^3 x^3}{3!} + \frac{5^5 x^5}{5!} - \frac{5^7 x^7}{7!} + \cdots = 1 - 5x + \sin(5x).$$

67. When a voltage V is applied to a series circuit consisting of a resistor R and an inductor L, the current at time t is

$$I(t) = \left(\frac{V}{R}\right)\left(1 - e^{-Rt/L}\right)$$

Expand $I(t)$ in a Maclaurin series. Show that $I(t) \approx Vt/L$ if R is small.

SOLUTION Substituting $-\frac{Rt}{L}$ for t in the Maclaurin series for e^t gives

$$e^{-Rt/L} = \sum_{n=0}^{\infty} \frac{\left(-\frac{Rt}{L}\right)^n}{n!} = \sum_{n=0}^{\infty} \frac{(-1)^n}{n!}\left(\frac{R}{L}\right)^n t^n = 1 + \sum_{n=1}^{\infty} \frac{(-1)^n}{n!}\left(\frac{R}{L}\right)^n t^n$$

Thus,

$$1 - e^{-Rt/L} = 1 - \left(1 + \sum_{n=1}^{\infty} \frac{(-1)^n}{n!}\left(\frac{R}{L}\right)^n t^n\right) = \sum_{n=1}^{\infty} \frac{(-1)^{n+1}}{n!}\left(\frac{Rt}{L}\right)^n,$$

and

$$I(t) = \frac{V}{R}\sum_{n=1}^{\infty} \frac{(-1)^{n+1}}{n!}\left(\frac{Rt}{L}\right)^n = \frac{Vt}{L} + \frac{V}{R}\sum_{n=2}^{\infty} \frac{(-1)^{n+1}}{n!}\left(\frac{Rt}{L}\right)^n.$$

If Rt/L is small, then the terms in the series are even smaller, and we find

$$V(t) \approx \frac{Vt}{L}.$$

69. Find the Maclaurin series for $f(x) = \cos(\sqrt{x})$ and use it to determine $f^{(5)}(0)$.

SOLUTION Substituting $\sqrt{x}$ for x in the Maclaurin series for $\cos x$

$$\cos(\sqrt{x}) = \sum_{n=0}^{\infty} (-1)^n \frac{(\sqrt{x})^{2n}}{(2n)!} = \sum_{n=0}^{\infty} (-1)^n \frac{x^n}{(2n)!}.$$

The coefficient of x^5 in this series is

$$\frac{(-1)^5}{10!} = -\frac{1}{10!} = \frac{f^{(5)}(0)}{5!},$$

so

$$f^{(5)}(0) = -\frac{5!}{10!} = -\frac{1}{6 \cdot 7 \cdot 8 \cdot 9 \cdot 10} = -\frac{1}{30240}.$$

71. Use the binomial series to find $f^{(8)}(0)$ for $f(x) = \sqrt{1 - x^2}$.

SOLUTION We obtain the Maclaurin series for $f(x) = \sqrt{1 - x^2}$ by substituting $-x^2$ for x in the binomial series with $a = \frac{1}{2}$. This gives

$$\sqrt{1 - x^2} = \sum_{n=0}^{\infty} \binom{\frac{1}{2}}{n}\left(-x^2\right)^n = \sum_{n=0}^{\infty} (-1)^n \binom{\frac{1}{2}}{n} x^{2n}.$$

The coefficient of x^8 is

$$(-1)^4 \binom{\frac{1}{2}}{4} = \frac{\frac{1}{2}\left(\frac{1}{2}-1\right)\left(\frac{1}{2}-2\right)\left(\frac{1}{2}-3\right)}{4!} = -\frac{15}{16 \cdot 4!} = \frac{f^{(8)}(0)}{8!},$$

so

$$f^{(8)}(0) = \frac{-15 \cdot 8!}{16 \cdot 4!} = -1575.$$

73. Does the Taylor series for $f(x) = (1 + x)^{3/4}$ converge to $f(x)$ at $x = 2$? Give numerical evidence to support your answer.

SOLUTION The Taylor series for $f(x) = (1 + x)^{3/4}$ converges to $f(x)$ for $|x| < 1$; because $x = 2$ is not contained on this interval, the series does not converge to $f(x)$ at $x = 2$. The graph below displays

$$S_N = \sum_{n=0}^{N} \binom{\frac{3}{4}}{n} 2^n$$

for $0 \le N \le 14$. The divergent nature of the sequence of partial sums is clear.

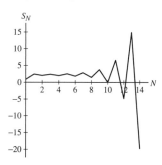

75. Explain the steps required to verify that the Maclaurin series for $f(x) = \tan^{-1} x$ converges to $f(x)$ at $x = 0.5$.

SOLUTION Recall that the Maclaurin series for $\tan^{-1} x$ can be obtained by term-by-term integration of the Maclaurin series for $\frac{1}{1+x^2}$. Now, we know that the geometric series

$$\sum_{n=0}^{\infty} (-x^2)^n$$

converges to $\frac{1}{1+x^2}$ for $|x| < 1$. It then follows from Theorem 3 of Section 11.6 that the Maclaurin series for $f(x) = \tan^{-1} x$ converges to $f(x)$ for $|x| < 1$. Because $x = 0.5$ is inside this interval, the Maclaurin series for $f(x) = \tan^{-1} x$ converges to $f(x)$ at $x = 0.5$.

77. How many terms of the Maclaurin series of $f(x) = \ln(1 + x)$ are needed to compute $\ln 1.2$ to within an error of at most 0.0001? Make the computation and compare the result with the calculator value.

SOLUTION Substitute $x = 0.2$ into the Maclaurin series for $\ln(1 + x)$ to obtain:

$$\ln 1.2 = \sum_{n=1}^{\infty} (-1)^{n-1} \frac{(0.2)^n}{n} = \sum_{n=1}^{\infty} (-1)^{n-1} \frac{1}{5^n n}.$$

This is an alternating series with $a_n = \dfrac{1}{n \cdot 5^n}$. Using the error bound for alternating series

$$|\ln 1.2 - S_N| \le a_{N+1} = \frac{1}{(N+1)5^{N+1}},$$

so we must choose N so that

$$\frac{1}{(N+1)5^{N+1}} < 0.0001 \quad \text{or} \quad (N+1)5^{N+1} > 10000.$$

For $N = 3$, $(N + 1)5^{N+1} = 4 \cdot 5^4 = 2500 < 10,000$, and for $N = 4$, $(N + 1)5^{N+1} = 5 \cdot 5^5 = 15,625 > 10,000$; thus, the smallest acceptable value for N is $N = 4$. The corresponding approximation is:

$$S_4 = \sum_{n=1}^{4} \frac{(-1)^{n-1}}{5^n \cdot n} = \frac{1}{5} - \frac{1}{5^2 \cdot 2} + \frac{1}{5^3 \cdot 3} - \frac{1}{5^4 \cdot 4} = 0.182266666.$$

Now, $\ln 1.2 = 0.182321556$, so

$$|\ln 1.2 - S_4| = 5.489 \times 10^{-5} < 0.0001.$$

In Exercises 78–79, let

$$f(x) = \frac{1}{(1 - x)(1 - 2x)}$$

79. Find the Taylor series for $f(x)$ at $c = 2$. *Hint:* Rewrite the identity of Exercise 78 as

$$f(x) = \frac{2}{-3 - 2(x-2)} - \frac{1}{-1 - (x-2)}$$

SOLUTION Using the given identity,

$$f(x) = \frac{2}{-3 - 2(x-2)} - \frac{1}{-1 - (x-2)} = -\frac{2}{3}\frac{1}{1 + \frac{2}{3}(x-2)} + \frac{1}{1 + (x-2)}.$$

Substituting $-\frac{2}{3}(x-2)$ for x in the Maclaurin series for $\frac{1}{1-x}$ yields

$$\frac{1}{1 + \frac{2}{3}(x-2)} = \sum_{n=0}^{\infty}(-1)^n\left(\frac{2}{3}\right)^n(x-2)^n,$$

and substituting $-(x-2)$ for x in the same Maclaurin series yields

$$\frac{1}{1 + (x-2)} = \sum_{n=0}^{\infty}(-1)^n(x-2)^n.$$

The first series is valid for $\left|-\frac{2}{3}(x-2)\right| < 1$, or $|x-2| < \frac{3}{2}$, and the second series is valid for $|x-2| < 1$; therefore, the two series together are valid for $|x-2| < 1$. Finally, for $|x-2| < 1$,

$$f(x) = -\frac{2}{3}\sum_{n=0}^{\infty}(-1)^n\left(\frac{2}{3}\right)^n(x-2)^n + \sum_{n=0}^{\infty}(-1)^n(x-2)^n = \sum_{n=0}^{\infty}(-1)^n\left[1 - \left(\frac{2}{3}\right)^{n+1}\right](x-2)^n.$$

81. Use Example 11 and the approximation $\sin x \approx x$ to show that the period T of a pendulum released at an angle θ has the following second-order approximation:

$$T \approx 2\pi\sqrt{\frac{L}{g}}\left(1 + \frac{\theta^2}{16}\right)$$

SOLUTION The period T of a pendulum of length L released from an angle θ is

$$T = 4\sqrt{\frac{L}{g}}E(k),$$

where $g \approx 9.8$ m/s^2 is the acceleration due to gravity, $E(k)$ is the elliptic function of the first kind and $k = \sin\frac{\theta}{2}$. From Example 11, we know that

$$E(k) = \frac{\pi}{2}\sum_{n=0}^{\infty}\left(\frac{1 \cdot 3 \cdot 5 \cdots (2n-1)}{2 \cdot 4 \cdot 6 \cdots (2n)}\right)^2 k^{2n}.$$

Using the approximation $\sin x \approx x$, we have

$$k = \sin\frac{\theta}{2} \approx \frac{\theta}{2};$$

moreover, using the first two terms of the series for $E(k)$, we find

$$E(k) \approx \frac{\pi}{2}\left[1 + \left(\frac{1}{2}\right)^2\left(\frac{\theta}{2}\right)^2\right] = \frac{\pi}{2}\left(1 + \frac{\theta^2}{16}\right).$$

Therefore,

$$T = 4\sqrt{\frac{L}{g}}E(k) \approx 2\pi\sqrt{\frac{L}{g}}\left(1 + \frac{\theta^2}{16}\right).$$

Further Insights and Challenges

In Exercises 83–84, we investigate the convergence of the binomial series

$$T_a(x) = \sum_{n=0}^{\infty}\binom{a}{n}x^n$$

83. Prove that $T_a(x)$ has radius of convergence $R = 1$ if a is not a whole number. What is the radius of convergence if a is a whole number?

SOLUTION Suppose that a is not a whole number. Then

$$\binom{a}{n} = \frac{a(a-1)\cdots(a-n+1)}{n!}$$

is never zero. Moreover,

$$\left| \frac{\binom{a}{n+1}}{\binom{a}{n}} \right| = \left| \frac{a(a-1)\cdots(a-n+1)(a-n)}{(n+1)!} \cdot \frac{n!}{a(a-1)\cdots(a-n+1)} \right| = \left| \frac{a-n}{n+1} \right|,$$

so, by the formula for the radius of convergence

$$r = \lim_{n \to \infty} \left| \frac{a-n}{n+1} \right| = 1.$$

The radius of convergence of $T_a(x)$ is therefore $R = r^{-1} = 1$.

If a is a whole number, then $\binom{a}{n} = 0$ for all $n > a$. The infinite series then reduces to a polynomial of degree a, so it converges for all x (i.e. $R = \infty$).

85. The function $G(k) = \int_0^{\pi/2} \sqrt{1 - k^2 \sin^2 t}\, dt$ is called an **elliptic function of the second kind**. Prove that for $|k| < 1$

$$G(k) = \frac{\pi}{2} - \frac{\pi}{2} \sum_{n=1}^{\infty} \left(\frac{1 \cdot 3 \cdots (2n-1)}{2 \cdots 4 \cdot (2n)} \right)^2 \frac{k^{2n}}{2n-1}$$

SOLUTION For $|k| < 1$, $|k^2 \sin^2 t| < 1$ for all t. Substituting $-k^2 \sin^2 t$ for t in the binomial series for $a = \frac{1}{2}$, we find

$$\sqrt{1 - k^2 \sin^2 t} = 1 + \sum_{n=1}^{\infty} \binom{\frac{1}{2}}{n} \left(-k^2 \sin^2 t \right)^n$$

$$= 1 + \sum_{n=1}^{\infty} (-1)^n \frac{\frac{1}{2}\left(\frac{1}{2}-1\right)\left(\frac{1}{2}-2\right)\cdots\left(\frac{1}{2}-n+1\right)}{n!} k^{2n} \sin^{2n} t$$

$$= 1 + \sum_{n=1}^{\infty} (-1)^n \frac{1(1-2)(1-4)\cdots(1-2(n-1))}{2^n n!} k^{2n} \sin^{2n} t$$

$$= 1 + \sum_{n=1}^{\infty} (-1)^n (-1)^{n-1} \frac{(2-1)(4-1)\cdots(2n-3)}{2^n n!} k^{2n} \sin^{2n} t$$

$$= 1 - \sum_{n=1}^{\infty} \frac{1 \cdot 3 \cdot 5 \cdots (2n-3)}{2 \cdot 4 \cdot 6 \cdots (2n)} k^{2n} \sin^{2n} t.$$

Integrating from 0 to $\frac{\pi}{2}$ term-by-term, we obtain

$$G(k) = \frac{\pi}{2} - \sum_{n=1}^{\infty} \frac{1 \cdot 3 \cdot 5 \cdots (2n-3)}{2 \cdot 4 \cdot 6 \cdots (2n)} k^{2n} \int_0^{\pi/2} \sin^{2n} t\, dt.$$

Finally, using the formula

$$\int_0^{\pi/2} \sin^{2n} t\, dt = \frac{1 \cdot 3 \cdot 5 \cdots (2n-1)}{2 \cdot 4 \cdot 6 \cdots (2n)} \frac{\pi}{2},$$

we arrive at

$$G(k) = \frac{\pi}{2} - \frac{\pi}{2} \sum_{n=1}^{\infty} \left(\frac{1 \cdot 3 \cdot 5 \cdots (2n-3)}{2 \cdot 4 \cdot 6 \cdots (2n)} \right)^2 (2n-1) k^{2n} = \frac{\pi}{2} - \frac{\pi}{2} \sum_{n=1}^{\infty} \left(\frac{1 \cdot 3 \cdot 5 \cdots (2n-1)}{2 \cdot 4 \cdot 6 \cdots (2n)} \right)^2 \frac{k^{2n}}{2n-1}.$$

87. Use Exercise 85 to prove that if $a < b$ and a/b is near 1 (a nearly circular ellipse), then

$$L \approx \frac{\pi}{2}\left(3b + \frac{a^2}{b}\right)$$

Hint: Use the first two terms of the series for $G(k)$.

SOLUTION From the previous exercise, we know that

$$L = 4bG(k), \quad \text{where} \quad k = \sqrt{1 - \frac{a^2}{b^2}}.$$

Following the hint and using only the first two terms of the series expansion for $G(k)$ from Exercise 85, we find

$$L \approx 4b\left(\frac{\pi}{2} - \frac{\pi}{2}\left(\frac{1}{2}\right)^2 k^2\right) = \frac{\pi}{2}\left(4b - b\left(1 - \frac{a^2}{b^2}\right)\right) = \frac{\pi}{2}\left(3b + \frac{a^2}{b}\right).$$

CHAPTER REVIEW EXERCISES

1. Let $a_n = \dfrac{n-3}{n!}$ and $b_n = a_{n+3}$. Calculate the first three terms in the sequence:

(a) a_n^2

(b) b_n

(c) $a_n b_n$

(d) $2a_{n+1} - 3a_n$

SOLUTION

(a)

$$a_1^2 = \left(\frac{1-3}{1!}\right)^2 = (-2)^2 = 4;$$

$$a_2^2 = \left(\frac{2-3}{2!}\right)^2 = \left(-\frac{1}{2}\right)^2 = \frac{1}{4};$$

$$a_3^2 = \left(\frac{3-3}{3!}\right)^2 = 0.$$

(b)

$$b_1 = a_4 = \frac{4-3}{4!} = \frac{1}{24};$$

$$b_2 = a_5 = \frac{5-3}{5!} = \frac{1}{60};$$

$$b_3 = a_6 = \frac{6-3}{6!} = \frac{1}{240}.$$

(c) Using the formula for a_n and the values in (b) we obtain:

$$a_1 b_1 = \frac{1-3}{1!} \cdot \frac{1}{24} = -\frac{1}{12};$$

$$a_2 b_2 = \frac{2-3}{2!} \cdot \frac{1}{60} = -\frac{1}{120};$$

$$a_3 b_3 = \frac{3-3}{3!} \cdot \frac{1}{240} = 0.$$

(d)

$$2a_2 - 3a_1 = 2\left(-\frac{1}{2}\right) - 3(-2) = 5;$$

$$2a_3 - 3a_2 = 2 \cdot 0 - 3\left(-\frac{1}{2}\right) = \frac{3}{2};$$

$$2a_4 - 3a_3 = 2 \cdot \frac{1}{24} - 3 \cdot 0 = \frac{1}{12}.$$

In Exercises 3–8, compute the limit (or state that it does not exist) assuming that $\lim_{n\to\infty} a_n = 2$.

3. $\lim_{n\to\infty} (5a_n - 2a_n^2)$

SOLUTION

$$\lim_{n\to\infty} \left(5a_n - 2a_n^2\right) = 5 \lim_{n\to\infty} a_n - 2 \lim_{n\to\infty} a_n^2 = 5 \lim_{n\to\infty} a_n - 2\left(\lim_{n\to\infty} a_n\right)^2 = 5 \cdot 2 - 2 \cdot 2^2 = 2.$$

5. $\lim_{n\to\infty} e^{a_n}$

SOLUTION   The function $f(x) = e^x$ is continuous, hence:

$$\lim_{n\to\infty} e^{a_n} = e^{\lim_{n\to\infty} a_n} = e^2.$$

7. $\lim_{n\to\infty} (-1)^n a_n$

SOLUTION   Because $\lim_{n\to\infty} a_n \neq 0$, it follows that $\lim_{n\to\infty} (-1)^n a_n$ does not exist.

In Exercises 9–22, determine the limit of the sequence or show that the sequence diverges.

9. $a_n = \sqrt{n+5} - \sqrt{n+2}$

SOLUTION   First rewrite a_n as follows:

$$a_n = \frac{\left(\sqrt{n+5} - \sqrt{n+2}\right)\left(\sqrt{n+5} + \sqrt{n+2}\right)}{\sqrt{n+5} + \sqrt{n+2}} = \frac{(n+5) - (n+2)}{\sqrt{n+5} + \sqrt{n+2}} = \frac{3}{\sqrt{n+5} + \sqrt{n+2}}.$$

Thus,

$$\lim_{n\to\infty} a_n = \lim_{n\to\infty} \frac{3}{\sqrt{n+5} + \sqrt{n+2}} = 0.$$

11. $a_n = 2^{1/n^2}$

SOLUTION   The function $f(x) = 2^x$ is continuous, so

$$\lim_{n\to\infty} a_n = \lim_{n\to\infty} 2^{1/n^2} = 2^{\lim_{n\to\infty}(1/n^2)} = 2^0 = 1.$$

13. $b_m = 1 + (-1)^m$

SOLUTION   Because $1 + (-1)^m$ is equal to 0 for m odd and is equal to 2 for m even, the sequence $\{b_m\}$ does not approach one limit; hence this sequence diverges.

15. $b_n = \tan^{-1}\left(\dfrac{n+2}{n+5}\right)$

SOLUTION   The function $\tan^{-1} x$ is continuous, so

$$\lim_{n\to\infty} b_n = \lim_{n\to\infty} \tan^{-1}\left(\frac{n+2}{n+5}\right) = \tan^{-1}\left(\lim_{n\to\infty} \frac{n+2}{n+5}\right) = \tan^{-1} 1 = \frac{\pi}{4}.$$

17. $b_n = \sqrt{n^2 + n} - \sqrt{n^2 + 1}$

SOLUTION   Rewrite b_n as

$$b_n = \frac{\left(\sqrt{n^2+n} - \sqrt{n^2+1}\right)\left(\sqrt{n^2+n} + \sqrt{n^2+1}\right)}{\sqrt{n^2+n} + \sqrt{n^2+1}} = \frac{\left(n^2+n\right) - \left(n^2+1\right)}{\sqrt{n^2+n} + \sqrt{n^2+1}} = \frac{n-1}{\sqrt{n^2+n} + \sqrt{n^2+1}}.$$

Then

$$\lim_{n\to\infty} b_n = \lim_{n\to\infty} \frac{\frac{n}{n} - \frac{1}{n}}{\sqrt{\frac{n^2}{n^2} + \frac{n}{n^2}} + \sqrt{\frac{n^2}{n^2} + \frac{1}{n^2}}} = \lim_{n\to\infty} \frac{1 - \frac{1}{n}}{\sqrt{1 + \frac{1}{n}} + \sqrt{1 + \frac{1}{n^2}}} = \frac{1 - 0}{\sqrt{1+0} + \sqrt{1+0}} = \frac{1}{2}.$$

19. $a_n = \dfrac{100^n}{n!} - \dfrac{3 + \pi^n}{5^n}$

SOLUTION For $n > 100$,

$$0 \leq \frac{100^n}{n!} = \left(\frac{100}{1} \cdot \frac{100}{2} \cdots \frac{100}{100} \right) \frac{100}{101} \cdot \frac{100}{102} \cdot \frac{100}{n} < \frac{100^{100}}{99!n};$$

therefore,

$$\lim_{n \to \infty} \frac{100^n}{n!} = 0$$

by the Squeeze Theorem. Moreover,

$$\lim_{n \to \infty} \left(\frac{3 + \pi^n}{5^n} \right) = \lim_{n \to \infty} \frac{3}{5^n} + \lim_{n \to \infty} \left(\frac{\pi}{5} \right)^n = 0 + 0 = 0.$$

Thus,

$$\lim_{n \to \infty} a_n = 0 + 0 = 0.$$

21. $c_n = \left(1 + \dfrac{3}{n} \right)^n$

SOLUTION Write

$$c_n = \left(1 + \frac{1}{n/3} \right)^n = \left[\left(1 + \frac{1}{n/3} \right)^{n/3} \right]^3.$$

Then, because x^3 is a continuous function,

$$\lim_{n \to \infty} c_n = \left[\lim_{n \to \infty} \left(1 + \frac{1}{n/3} \right)^{n/3} \right]^3 = e^3.$$

23. Use the Squeeze Theorem to show that $\displaystyle\lim_{n \to \infty} \frac{\arctan(n^2)}{\sqrt{n}} = 0$.

SOLUTION For all x,

$$-\frac{\pi}{2} < \arctan x < \frac{\pi}{2},$$

so

$$-\frac{\pi/2}{\sqrt{n}} < \frac{\arctan(n^2)}{\sqrt{n}} < \frac{\pi/2}{\sqrt{n}},$$

for all n. Because

$$\lim_{n \to \infty} \left(-\frac{\pi/2}{\sqrt{n}} \right) = \lim_{n \to \infty} \frac{\pi/2}{\sqrt{n}} = 0,$$

it follows by the Squeeze Theorem that

$$\lim_{n \to \infty} \frac{\arctan(n^2)}{\sqrt{n}} = 0.$$

25. Given $a_n = \dfrac{1}{2} 3^n - \dfrac{1}{3} 2^n$,

(a) Calculate $\displaystyle\lim_{n \to \infty} a_n$.

(b) Calculate $\displaystyle\lim_{n \to \infty} \frac{a_{n+1}}{a_n}$.

SOLUTION

(a) Because

$$\frac{1}{2} 3^n - \frac{1}{3} 2^n \geq \frac{1}{2} 3^n - \frac{1}{3} 3^n = \frac{3^n}{6}$$

and

$$\lim_{n \to \infty} \frac{3^n}{6} = \infty,$$

we conclude that

$$\lim_{n \to \infty} a_n = \infty.$$

(b) $\lim\limits_{n \to \infty} \dfrac{a_{n+1}}{a_n} = \lim\limits_{n \to \infty} \dfrac{\frac{1}{2}3^{n+1} - \frac{1}{3}2^{n+1}}{\frac{1}{2}3^n - \frac{1}{3}2^n} = \lim\limits_{n \to \infty} \dfrac{3^{n+2} - 2^{n+2}}{3^{n+1} - 2^{n+1}} = \lim\limits_{n \to \infty} \dfrac{3 - 2\left(\frac{2}{3}\right)^{n+1}}{1 - \left(\frac{2}{3}\right)^{n+1}} = \dfrac{3 - 0}{1 - 0} = 3.$

27. Calculate the partial sums S_4 and S_7 of the series $\displaystyle\sum_{n=1}^{\infty} \dfrac{n-2}{n^2 + 2n}$.

SOLUTION

$$S_4 = -\frac{1}{3} + 0 + \frac{1}{15} + \frac{2}{24} = -\frac{11}{60} = -0.183333;$$

$$S_7 = -\frac{1}{3} + 0 + \frac{1}{15} + \frac{2}{24} + \frac{3}{35} + \frac{4}{48} + \frac{5}{63} = \frac{287}{4410} = 0.065079.$$

29. Find the sum $\dfrac{4}{9} + \dfrac{8}{27} + \dfrac{16}{81} + \dfrac{32}{243} + \cdots$.

SOLUTION This is a geometric series with common ratio $r = \frac{2}{3}$. Therefore,

$$\frac{4}{9} + \frac{8}{27} + \frac{16}{81} + \frac{32}{243} + \cdots = \frac{\frac{4}{9}}{1 - \frac{2}{3}} = \frac{4}{3}.$$

31. Find the sum $\displaystyle\sum_{n=0}^{\infty} \dfrac{2^{n+1}}{3^n}$.

SOLUTION Note

$$\sum_{n=0}^{\infty} \frac{2^{n+1}}{3^n} = 2 \sum_{n=0}^{\infty} \frac{2^n}{3^n} = 2 \sum_{n=0}^{\infty} \left(\frac{2}{3}\right)^n;$$

therefore,

$$\sum_{n=0}^{\infty} \frac{2^{n+1}}{3^n} = 2 \frac{1}{1 - \frac{2}{3}} = 2 \cdot 3 = 6.$$

33. Give an example of divergent series $\displaystyle\sum_{n=1}^{\infty} a_n$, $\displaystyle\sum_{n=1}^{\infty} b_n$ such that $\displaystyle\sum_{n=1}^{\infty}(a_n + b_n) = 1$.

SOLUTION Let $a_n = \left(\frac{1}{2}\right)^n + 1$, $b_n = -1$. The corresponding series diverge by the Divergence Test; however,

$$\sum_{n=1}^{\infty}(a_n + b_n) = \sum_{n=1}^{\infty} \left(\frac{1}{2}\right)^n = \frac{\frac{1}{2}}{1 - \frac{1}{2}} = 1.$$

In Exercises 35–38, use the Integral Test to determine if the infinite series converges.

35. $\displaystyle\sum_{n=1}^{\infty} \dfrac{n^2}{n^3 + 1}$

SOLUTION Let $f(x) = \frac{x^2}{x^3+1}$. This function is continuous and positive for $x \geq 1$. Because

$$f'(x) = \frac{(x^3 + 1)(2x) - x^2(3x^2)}{(x^3 + 1)^2} = \frac{x(2 - x^3)}{(x^3 + 1)^2},$$

we see that $f'(x) < 0$ and f is decreasing on the interval $x \geq 2$. Therefore, the Integral Test applies on the interval $x \geq 2$. Now,

$$\int_2^{\infty} \frac{x^2}{x^3 + 1}\,dx = \lim_{R \to \infty} \int_2^R \frac{x^2}{x^3 + 1}\,dx = \frac{1}{3}\lim_{R \to \infty}\left(\ln(R^3 + 1) - \ln 9\right) = \infty.$$

The integral diverges; hence, the series $\displaystyle\sum_{n=2}^{\infty} \dfrac{n^2}{n^3 + 1}$ diverges, as does the series $\displaystyle\sum_{n=1}^{\infty} \dfrac{n^2}{n^3 + 1}$.

37. $\displaystyle\sum_{n=1}^{\infty} \frac{n^3}{e^{n^4}}$

SOLUTION Let $f(x) = x^3 e^{-x^4}$. This function is continuous and positive for $x \geq 1$. Because

$$f'(x) = x^3 \left(-4x^3 e^{-x^4}\right) + 3x^2 e^{-x^4} = x^2 e^{-x^4} \left(3 - 4x^4\right),$$

we see that $f'(x) < 0$ and f is decreasing on the interval $x \geq 1$. Therefore, the Integral Test applies on the interval $x \geq 1$. Now,

$$\int_1^{\infty} x^3 e^{-x^4}\, dx = \lim_{R \to \infty} \int_1^R x^3 e^{-x^4}\, dx = -\frac{1}{4} \lim_{R \to \infty} \left(e^{-R^4} - e^{-1}\right) = \frac{1}{4e}.$$

The integral converges; hence, the series $\displaystyle\sum_{n=1}^{\infty} \frac{n^3}{e^{n^4}}$ also converges.

In Exercises 39–46, use the Comparison or Limit Comparison Test to determine whether the infinite series converges.

39. $\displaystyle\sum_{n=1}^{\infty} \frac{1}{(n+1)^2}$

SOLUTION For all $n \geq 1$,

$$0 < \frac{1}{n+1} < \frac{1}{n} \quad \text{so} \quad \frac{1}{(n+1)^2} < \frac{1}{n^2}.$$

The series $\displaystyle\sum_{n=1}^{\infty} \frac{1}{n^2}$ is a convergent p-series, so the series $\displaystyle\sum_{n=1}^{\infty} \frac{1}{(n+1)^2}$ converges by the Comparison Test.

41. $\displaystyle\sum_{n=2}^{\infty} \frac{n^2+1}{n^{3.5}-2}$

SOLUTION Apply the Limit Comparison Test with $a_n = \frac{n^2+1}{n^{3.5}-2}$ and $b_n = \frac{1}{n^{1.5}}$. Now,

$$L = \lim_{n \to \infty} \frac{\frac{n^2+1}{n^{3.5}-2}}{\frac{1}{n^{1.5}}} = \lim_{n \to \infty} \frac{n^{3.5} + n^{1.5}}{n^{3.5} - 2} = 1.$$

Because L exists and $\displaystyle\sum_{n=1}^{\infty} \frac{1}{n^{1.5}}$ is a convergent p-series, we conclude by the Limit Comparison Test that the series $\displaystyle\sum_{n=2}^{\infty} \frac{n^2+1}{n^{3.5}-2}$ also converges.

43. $\displaystyle\sum_{n=2}^{\infty} \frac{\ln n}{1.5^n}$

SOLUTION Apply the Limit Comparison Test with $a_n = \frac{\ln n}{1.5^n}$ and $b_n = \frac{1}{1.25^n}$. Then,

$$L = \lim_{n \to \infty} \frac{a_n}{b_n} = \lim_{n \to \infty} \frac{\ln n}{1.5^n} \cdot \frac{1.25^n}{1} = \lim_{n \to \infty} \frac{\ln n}{(6/5)^n} = 0,$$

because the logarithm grows more slowly than the exponential function. The series $\displaystyle\sum_{n=2}^{\infty} \frac{1}{1.25^n}$ is a convergent geometric series; because L exists, we may therefore conclude by the Limit Comparison Test that the series $\displaystyle\sum_{n=2}^{\infty} \frac{\ln n}{1.5^n}$ also converges.

45. $\displaystyle\sum_{n=1}^{\infty} \frac{1}{3^n - 2^n}$

SOLUTION Apply the Limit Comparison Test with $a_n = \frac{1}{3^n - 2^n}$ and $b_n = \frac{1}{3^n}$. Then,

$$L = \lim_{n \to \infty} \frac{a_n}{b_n} = \lim_{n \to \infty} \frac{3^n}{3^n - 2^n} = \lim_{n \to \infty} \frac{1}{1 - \left(\frac{2}{3}\right)^n} = 1.$$

The series $\displaystyle\sum_{n=1}^{\infty} \frac{1}{3^n}$ is a convergent geometric series; because L exists, we may therefore conclude by the Limit Comparison Test that the series $\displaystyle\sum_{n=1}^{\infty} \frac{1}{3^n - 2^n}$ also converges.

47. Show that $\displaystyle\sum_{n=2}^{\infty} \left(1 - \sqrt{1 - \frac{1}{n}}\right)$ diverges. *Hint:* Show that

$$1 - \sqrt{1 - \frac{1}{n}} \geq \frac{1}{2n}$$

SOLUTION

$$1 - \sqrt{1 - \frac{1}{n}} = 1 - \sqrt{\frac{n-1}{n}} = \frac{\sqrt{n} - \sqrt{n-1}}{\sqrt{n}} = \frac{n - (n-1)}{\sqrt{n}(\sqrt{n} + \sqrt{n-1})} = \frac{1}{n + \sqrt{n^2 - n}}$$

$$\geq \frac{1}{n + \sqrt{n^2}} = \frac{1}{2n}.$$

The series $\displaystyle\sum_{n=2}^{\infty} \frac{1}{2n}$ diverges, so the series $\displaystyle\sum_{n=2}^{\infty} \left(1 - \sqrt{1 - \frac{1}{n}}\right)$ also diverges by the Comparison Test.

49. Let $\displaystyle S = \sum_{n=1}^{\infty} \frac{n}{(n^2 + 1)^2}$.

(a) Show that S converges.

(b) *CAS* Use Eq. (5) in Exercise 77 of Section 11.3 with $M = 99$ to approximate S. What is the maximum size of the error?

SOLUTION

(a) For $n \geq 1$,

$$\frac{n}{(n^2 + 1)^2} < \frac{n}{(n^2)^2} = \frac{1}{n^3}.$$

The series $\displaystyle\sum_{n=1}^{\infty} \frac{1}{n^3}$ is a convergent p-series, so the series $\displaystyle\sum_{n=1}^{\infty} \frac{n}{(n^2 + 1)^2}$ also converges by the Comparison Test.

(b) With $a_n = \frac{n}{(n^2+1)^2}$, $f(x) = \frac{x}{(x^2+1)^2}$ and $M = 99$, Eq. (5) in Exercise 77 of Section 11.3 becomes

$$\sum_{n=1}^{99} \frac{n}{(n^2 + 1)^2} + \int_{100}^{\infty} \frac{x}{(x^2 + 1)^2}\,dx \leq S \leq \sum_{n=1}^{100} \frac{n}{(n^2 + 1)^2} + \int_{100}^{\infty} \frac{x}{(x^2 + 1)^2}\,dx,$$

or

$$0 \leq S - \left(\sum_{n=1}^{99} \frac{n}{(n^2 + 1)^2} + \int_{100}^{\infty} \frac{x}{(x^2 + 1)^2}\,dx\right) \leq \frac{100}{(100^2 + 1)^2}.$$

Now,

$$\sum_{n=1}^{99} \frac{n}{(n^2 + 1)^2} = 0.397066274; \text{ and}$$

$$\int_{100}^{\infty} \frac{x}{(x^2 + 1)^2}\,dx = \lim_{R \to \infty} \int_{100}^{R} \frac{x}{(x^2 + 1)^2}\,dx = \frac{1}{2} \lim_{R \to \infty} \left(-\frac{1}{R^2 + 1} + \frac{1}{100^2 + 1}\right)$$

$$= \frac{1}{20002} = 0.000049995;$$

thus,

$$S \approx 0.397066274 + 0.000049995 = 0.397116269.$$

The bound on the error in this approximation is

$$\frac{100}{(100^2 + 1)^2} = 9.998 \times 10^{-7}.$$

In Exercises 50–53, determine whether the series converges absolutely. If not, determine whether it converges conditionally.

51. $\displaystyle\sum_{n=1}^{\infty} \frac{(-1)^n}{n^{1.1} \ln(n+1)}$

SOLUTION Consider the corresponding positive series $\displaystyle\sum_{n=1}^{\infty} \frac{1}{n^{1.1} \ln(n+1)}$. Because

$$\frac{1}{n^{1.1} \ln(n+1)} < \frac{1}{n^{1.1}}$$

and $\displaystyle\sum_{n=1}^{\infty} \frac{1}{n^{1.1}}$ is a convergent p-series, we can conclude by the Comparison Test that $\displaystyle\sum_{n=1}^{\infty} \frac{(-1)^n}{n^{1.1} \ln(n+1)}$ also converges. Thus, $\displaystyle\sum_{n=1}^{\infty} \frac{(-1)^n}{n^{1.1} \ln(n+1)}$ converges absolutely.

53. $\displaystyle\sum_{n=1}^{\infty} \frac{\cos\left(\frac{\pi}{4} + 2\pi n\right)}{\sqrt{n}}$

SOLUTION $\cos\left(\frac{\pi}{4} + 2\pi n\right) = \cos\frac{\pi}{4} = \frac{\sqrt{2}}{2}$, so

$$\sum_{n=1}^{\infty} \frac{\cos\left(\frac{\pi}{4} + 2\pi n\right)}{\sqrt{n}} = \frac{\sqrt{2}}{2} \sum_{n=1}^{\infty} \frac{1}{\sqrt{n}}.$$

This is a divergent p-series, so the series $\displaystyle\sum_{n=1}^{\infty} \frac{\cos\left(\frac{\pi}{4} + 2\pi n\right)}{\sqrt{n}}$ diverges.

55. How many terms of the series are needed to calculate Catalan's constant $K = \displaystyle\sum_{k=0}^{\infty} \frac{(-1)^k}{(2k+1)^2}$ to three decimal places? Carry out the calculation.

SOLUTION Using the error bound for an alternating series, we have

$$|S_N - K| \le \frac{1}{(2(N+1)+1)^2} = \frac{1}{(2N+3)^2}.$$

For accuracy to three decimal places, we must choose N so that

$$\frac{1}{(2N+3)^2} < 5 \times 10^{-3} \quad \text{or} \quad (2N+3)^2 > 2000.$$

Solving for N yields

$$N > \frac{1}{2}\left(\sqrt{2000} - 3\right) \approx 20.9.$$

Thus,

$$K \approx \sum_{k=0}^{21} \frac{(-1)^k}{(2k+1)^2} = 0.915707728.$$

57. Let $\displaystyle\sum_{n=1}^{\infty} a_n$ be an absolutely convergent series. Determine whether the following series are convergent or divergent:

(a) $\displaystyle\sum_{n=1}^{\infty}\left(a_n + \frac{1}{n^2}\right)$ **(b)** $\displaystyle\sum_{n=1}^{\infty}(-1)^n a_n$

(c) $\displaystyle\sum_{n=1}^{\infty} \frac{1}{1+a_n^2}$ **(d)** $\displaystyle\sum_{n=1}^{\infty} \frac{|a_n|}{n}$

SOLUTION Because $\displaystyle\sum_{n=1}^{\infty} a_n$ converges absolutely, we know that $\displaystyle\sum_{n=1}^{\infty} a_n$ converges and that $\displaystyle\sum_{n=1}^{\infty} |a_n|$ converges.

(a) Because we know that $\sum\limits_{n=1}^{\infty} a_n$ converges and the series $\sum\limits_{n=1}^{\infty} \dfrac{1}{n^2}$ is a convergent p-series, the sum of these two series,

$\sum\limits_{n=1}^{\infty} \left(a_n + \dfrac{1}{n^2} \right)$ also converges.

(b) We have,

$$\sum_{n=1}^{\infty} \left| (-1)^n a_n \right| = \sum_{n=1}^{\infty} |a_n|$$

Because $\sum\limits_{n=1}^{\infty} |a_n|$ converges, it follows that $\sum\limits_{n=1}^{\infty} (-1)^n a_n$ converges absolutely, which implies that $\sum\limits_{n=1}^{\infty} (-1)^n a_n$ converges.

(c) Because $\sum\limits_{n=1}^{\infty} a_n$ converges, $\lim_{n\to\infty} a_n = 0$. Therefore,

$$\lim_{n\to\infty} \frac{1}{1+a_n^2} = \frac{1}{1+0^2} = 1 \neq 0,$$

and the series $\sum\limits_{n=1}^{\infty} \dfrac{1}{1+a_n^2}$ diverges by the Divergence Test.

(d) $\dfrac{|a_n|}{n} \leq |a_n|$ and the series $\sum\limits_{n=1}^{\infty} |a_n|$ converges, so the series $\sum\limits_{n=1}^{\infty} \dfrac{|a_n|}{n}$ also converges by the Comparison Test.

In Exercises 58–65, apply the Ratio Test to determine convergence or divergence, or state that the Ratio Test is inconclusive.

59. $\sum\limits_{n=1}^{\infty} \dfrac{\sqrt{n+1}}{n^8}$

SOLUTION With $a_n = \dfrac{\sqrt{n+1}}{n^8}$,

$$\left| \frac{a_{n+1}}{a_n} \right| = \frac{\sqrt{n+2}}{(n+1)^8} \cdot \frac{n^8}{\sqrt{n+1}} = \sqrt{\frac{n+2}{n+1}} \left(\frac{n}{n+1} \right)^8,$$

and

$$\rho = \lim_{n\to\infty} \left| \frac{a_{n+1}}{a_n} \right| = 1 \cdot 1^8 = 1.$$

Because $\rho = 1$, the Ratio Test is inconclusive.

61. $\sum\limits_{n=1}^{\infty} \dfrac{n^4}{n!}$

SOLUTION With $a_n = \dfrac{n^4}{n!}$,

$$\left| \frac{a_{n+1}}{a_n} \right| = \frac{(n+1)^4}{(n+1)!} \cdot \frac{n!}{n^4} = \frac{(n+1)^3}{n^4} \quad \text{and} \quad \rho = \lim_{n\to\infty} \frac{a_{n+1}}{a_n} = 0.$$

Because $\rho < 1$, the series converges by the Ratio Test.

63. $\sum\limits_{n=4}^{\infty} \dfrac{\ln n}{n^{3/2}}$

SOLUTION With $a_n = \dfrac{\ln n}{n^{3/2}}$,

$$\left| \frac{a_{n+1}}{a_n} \right| = \frac{\ln(n+1)}{(n+1)^{3/2}} \cdot \frac{n^{3/2}}{\ln n} = \left(\frac{n}{n+1} \right)^{3/2} \frac{\ln(n+1)}{\ln n},$$

and

$$\rho = \lim_{n\to\infty} \left| \frac{a_{n+1}}{a_n} \right| = 1^{3/2} \cdot 1 = 1.$$

Because $\rho = 1$, the Ratio Test is inconclusive.

65. $\displaystyle\sum_{n=1}^{\infty} \left(\frac{n}{4}\right)^n \frac{1}{n!}$

SOLUTION With $a_n = \left(\frac{n}{4}\right)^n \frac{1}{n!}$,

$$\left|\frac{a_{n+1}}{a_n}\right| = \left(\frac{n+1}{4}\right)^{n+1} \frac{1}{(n+1)!} \cdot \left(\frac{4}{n}\right)^n n! = \frac{1}{4}\left(\frac{n+1}{n}\right)^n = \frac{1}{4}\left(1 + \frac{1}{n}\right)^n,$$

and

$$\rho = \lim_{n\to\infty}\left|\frac{a_{n+1}}{a_n}\right| = \frac{1}{4}e.$$

Because $\rho = \frac{e}{4} < 1$, the series converges by the Ratio Test.

In Exercises 66–69, apply the Root Test to determine convergence or divergence, or state that the Root Test is inconclusive.

67. $\displaystyle\sum_{n=1}^{\infty} \left(\frac{2}{n}\right)^n$

SOLUTION With $a_n = \left(\frac{2}{n}\right)^n$,

$$L = \lim_{n\to\infty} \sqrt[n]{\left(\frac{2}{n}\right)^n} = \lim_{n\to\infty} \frac{2}{n} = 0.$$

Because $L < 1$, the series converges by the Root Test.

69. $\displaystyle\sum_{n=1}^{\infty} \left(\cos\frac{1}{n}\right)^{n^3}$

SOLUTION With $a_n = \left(\cos\frac{1}{n}\right)^{n^3}$,

$$L = \lim_{n\to\infty} \sqrt[n]{a_n} = \lim_{n\to\infty} \sqrt[n]{\cos\left(\frac{1}{n}\right)^{n^3}} = \lim_{n\to\infty} \cos\left(\frac{1}{n}\right)^{n^2} = \lim_{x\to\infty} \cos\left(\frac{1}{x}\right)^{x^2}.$$

Now,

$$\ln L = \lim_{x\to\infty} x^2 \ln\cos\left(\frac{1}{x}\right) = \lim_{x\to\infty} \frac{\ln\cos\left(\frac{1}{x}\right)}{\frac{1}{x^2}} = \lim_{x\to\infty} \frac{\frac{1}{\cos\left(\frac{1}{x}\right)}\left(-\sin\left(\frac{1}{x}\right)\right)\left(-\frac{1}{x^2}\right)}{-\frac{2}{x^3}}$$

$$= -\frac{1}{2}\lim_{x\to\infty}\frac{1}{\cos\left(\frac{1}{x}\right)} \cdot \lim_{x\to\infty}\frac{\sin\left(\frac{1}{x}\right)}{\frac{1}{x}} = -\frac{1}{2}\cdot 1 \cdot 1 = -\frac{1}{2}.$$

Therefore, $L = e^{-1/2}$. Because $L < 1$, the series converges by the Root Test.

In Exercises 71–84, determine convergence or divergence using any method covered in the text.

71. $\displaystyle\sum_{n=1}^{\infty} \left(\frac{2}{3}\right)^n$

SOLUTION This is a geometric series with ratio $r = \frac{2}{3} < 1$; hence, the series converges.

73. $\displaystyle\sum_{n=1}^{\infty} e^{-0.02n}$

SOLUTION This is a geometric series with common ratio $r = \frac{1}{e^{0.02}} \approx 0.98 < 1$; hence, the series converges.

75. $\displaystyle\sum_{n=1}^{\infty} \frac{(-1)^{n-1}}{\sqrt{n} + \sqrt{n+1}}$

SOLUTION In this alternating series, $a_n = \frac{1}{\sqrt{n}+\sqrt{n+1}}$. The sequence $\{a_n\}$ is decreasing, and

$$\lim_{n\to\infty} a_n = 0;$$

therefore the series converges by the Leibniz Test.

77. $\displaystyle\sum_{n=10}^{\infty} \frac{(-1)^n}{\log n}$

SOLUTION The sequence $a_n = \frac{1}{\log n}$ is decreasing for $n \geq 10$ and

$$\lim_{n \to \infty} a_n = 0;$$

therefore, the series converges by the Leibniz Test.

79. $\displaystyle\sum_{n=1}^{\infty} \frac{e^n}{n!}$

SOLUTION With $a_n = \frac{e^n}{n!}$,

$$\left| \frac{a_{n+1}}{a_n} \right| = \frac{e^{n+1}}{(n+1)!} \cdot \frac{n!}{e^n} = \frac{e}{n+1} \quad \text{and} \quad \rho = \lim_{n \to \infty} \left| \frac{a_{n+1}}{a_n} \right| = 0.$$

Because $\rho < 1$, the series converges by the Ratio Test.

81. $\displaystyle\sum_{n=1}^{\infty} \frac{1}{n - 100.1}$

SOLUTION For $n \geq 101$, the sequence $\frac{1}{n-100.1}$ is positive and

$$\frac{1}{n - 100.1} > \frac{1}{n}.$$

Now, $\displaystyle\sum_{n=101}^{\infty} \frac{1}{n}$ diverges, so $\displaystyle\sum_{n=101}^{\infty} \frac{1}{n - 100.1}$ diverges by the Comparison Test. Since adding the finite sum $\displaystyle\sum_{n=1}^{100} \frac{1}{n - 100.1}$

does not affect the convergence of the series, it follows that $\displaystyle\sum_{n=1}^{\infty} \frac{1}{n - 100.1}$ diverges as well.

83. $\displaystyle\sum_{n=1}^{\infty} \sin^2 \frac{\pi}{n}$

SOLUTION For all $x > 0$, $\sin x < x$. Therefore, $\sin^2 x < x^2$, and for $x = \frac{\pi}{n}$,

$$\sin^2 \frac{\pi}{n} < \frac{\pi^2}{n^2} = \pi^2 \cdot \frac{1}{n^2}.$$

The series $\displaystyle\sum_{n=1}^{\infty} \frac{1}{n^2}$ is a convergent p-series, so the series $\displaystyle\sum_{n=1}^{\infty} \sin^2 \frac{\pi}{n}$ also converges by the Comparison Test.

In Exercises 85–90, find the values of x for which the power series converges.

85. $\displaystyle\sum_{n=0}^{\infty} \frac{2^n x^n}{n!}$

SOLUTION With $a_n = \frac{2^n}{n!}$,

$$\left| \frac{a_{n+1}}{a_n} \right| = \frac{2^{n+1}}{(n+1)!} \cdot \frac{n!}{2^n} = \frac{2}{n+1} \quad \text{and} \quad r = \lim_{n \to \infty} \left| \frac{a_{n+1}}{a_n} \right| = 0.$$

The radius of convergence is therefore $R = r^{-1} = \infty$, and the series converges for all x.

87. $\displaystyle\sum_{n=0}^{\infty} \frac{n^6(x - 3)^n}{n^8 + 1}$

SOLUTION With $a_n = \frac{n^6}{n^8+1}$,

$$r = \lim_{n \to \infty} \left| \frac{a_{n+1}}{a_n} \right| = \lim_{n \to \infty} \left(\frac{n+1}{n} \right)^6 \frac{n^8 + 1}{(n+1)^8 + 1} = 1.$$

The radius of convergence is therefore $R = r^{-1} = 1$, and the series converges absolutely for $|x - 3| < 1$, or $2 < x < 4$.

For the endpoint $x = 4$, the series becomes $\displaystyle\sum_{n=0}^{\infty} \frac{n^6}{n^8 + 1}$, which converges by the Comparison Test comparing with the convergent p-series $\displaystyle\sum_{n=1}^{\infty} \frac{1}{n^2}$. For the endpoint $x = 2$, the series becomes $\displaystyle\sum_{n=0}^{\infty} \frac{n^6(-1)^n}{n^8 + 1}$, which converges by the Leibniz Test. The series $\displaystyle\sum_{n=0}^{\infty} \frac{n^6(x-3)^n}{n^8 + 1}$ therefore converges for $2 \le x \le 4$.

89. $\displaystyle\sum_{n=0}^{\infty} (nx)^n$

SOLUTION With $a_n = n^n$,

$$\left| \frac{a_{n+1}}{a_n} \right| = \frac{(n+1)^{n+1}}{n^n} = (n+1)\left(\frac{n+1}{n}\right)^n = (n+1)\left(1 + \frac{1}{n}\right)^n,$$

and

$$r = \lim_{n \to \infty} \left| \frac{a_{n+1}}{a_n} \right| = \infty.$$

The radius of convergence is therefore $R = r^{-1} = 0$, and the series converges only for $x = 0$.

91. Expand the function $f(x) = \dfrac{2}{4 - 3x}$ as a power series centered at $c = 0$. Determine the values of x for which the series converges.

SOLUTION Write

$$\frac{2}{4 - 3x} = \frac{1}{2}\frac{1}{1 - \frac{3}{4}x}.$$

Substituting $\frac{3}{4}x$ for x in the Maclaurin series for $\frac{1}{1-x}$, we obtain

$$\frac{1}{1 - \frac{3}{4}x} = \sum_{n=0}^{\infty} \left(\frac{3}{4}\right)^n x^n.$$

This series converges for $\left|\frac{3}{4}x\right| < 1$, or $|x| < \frac{4}{3}$. Hence, for $|x| < \frac{4}{3}$,

$$\frac{2}{4 - 3x} = \frac{1}{2}\sum_{n=0}^{\infty} \left(\frac{3}{4}\right)^n x^n.$$

93. Let $F(x) = \displaystyle\sum_{k=0}^{\infty} \frac{x^{2k}}{2^k \cdot k!}$.

(a) Show that $F(x)$ has infinite radius of convergence.

(b) Show that $y = F(x)$ is a solution to the differential equation

$$y'' = xy' + y$$

satisfying $y(0) = 1$, $y'(0) = 0$.

(c) **CAS** Plot the partial sums S_N for $N = 1, 3, 5, 7$ on the same set of axes.

SOLUTION

(a) With $a_k = \frac{x^{2k}}{2^k \cdot k!}$,

$$\left| \frac{a_{k+1}}{a_k} \right| = \frac{|x|^{2k+2}}{2^{k+1} \cdot (k+1)!} \cdot \frac{2^k \cdot k!}{|x|^{2k}} = \frac{x^2}{2(k+1)},$$

and

$$\rho = \lim_{n \to \infty} \left| \frac{a_{k+1}}{a_k} \right| = x^2 \cdot 0 = 0.$$

Because $\rho < 1$ for all x, we conclude that the series converges for all x; that is, $R = \infty$.

(b) Let

$$y = F(x) = \sum_{k=0}^{\infty} \frac{x^{2k}}{2^k \cdot k!}.$$

Then

$$y' = \sum_{k=1}^{\infty} \frac{2kx^{2k-1}}{2^k k!} = \sum_{k=1}^{\infty} \frac{x^{2k-1}}{2^{k-1}(k-1)!},$$

$$y'' = \sum_{k=1}^{\infty} \frac{(2k-1)x^{2k-2}}{2^{k-1}(k-1)!},$$

and

$$xy' + y = x\sum_{k=1}^{\infty} \frac{x^{2k-1}}{2^{k-1}(k-1)!} + \sum_{k=0}^{\infty} \frac{x^{2k}}{2^k k!} = \sum_{k=1}^{\infty} \frac{x^{2k}}{2^{k-1}(k-1)!} + 1 + \sum_{k=1}^{\infty} \frac{x^{2k}}{2^k k!}$$

$$= 1 + \sum_{k=1}^{\infty} \frac{(2k+1)x^{2k}}{2^k k!} = \sum_{k=0}^{\infty} \frac{(2k+1)x^{2k}}{2^k k!} = \sum_{k=1}^{\infty} \frac{(2k-1)x^{2k-2}}{2^{k-1}(k-1)!} = y''.$$

Moreover,

$$y(0) = 1 + \sum_{k=1}^{\infty} \frac{0^{2k}}{2^k k!} = 1 \quad \text{and} \quad y'(0) = \sum_{k=1}^{\infty} \frac{0^{2k-1}}{2^{k-1}(k-1)!} = 0.$$

Thus, $\sum_{k=0}^{\infty} \frac{x^{2k}}{2^k k!}$ is the solution to the equation $y'' = xy' + y$ satisfying $y(0) = 1$, $y'(0) = 0$.

(c) The partial sums S_1, S_3, S_5 and S_7 are plotted in the figure below.

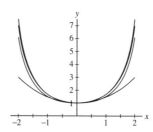

95. Use the Maclaurin series for $f(x) = \cos x$ to calculate the limit $\lim_{x \to 0} \frac{1 - \frac{x^2}{2} - \cos x}{x^4}$.

SOLUTION Using the Maclaurin series for $\cos x$, we find

$$\cos x = \sum_{n=0}^{\infty} (-1)^n \frac{x^{2n}}{(2n)!} = 1 - \frac{x^2}{2} + \frac{x^4}{24} + \sum_{n=3}^{\infty} (-1)^n \frac{x^{2n}}{(2n)!}.$$

Thus,

$$1 - \frac{x^2}{x} - \cos x = -\frac{x^4}{24} - \sum_{n=3}^{\infty} (-1)^n \frac{x^{2n}}{(2n)!},$$

and

$$\frac{1 - \frac{x^2}{2} - \cos x}{x^4} = -\frac{1}{24} - \sum_{n=3}^{\infty} (-1)^n \frac{x^{2n-4}}{(2n)!}.$$

Finally,

$$\lim_{x \to 0} \frac{1 - \frac{x^2}{2} - \cos x}{x^4} = \lim_{x \to 0} \left(-\frac{1}{24} - \sum_{n=3}^{\infty} (-1)^n \frac{x^{2n-4}}{(2n)!} \right) = -\frac{1}{24} + 0 = -\frac{1}{24}.$$

In Exercises 96–103, find the Taylor series centered at c.

97. $f(x) = e^{2x}, \quad c = -1$

SOLUTION Write:

$$e^{2x} = e^{2(x+1)-2} = e^{-2}e^{2(x+1)}.$$

Substituting $2(x + 1)$ for x in the Maclaurin series for e^x yields

$$e^{2(x+1)} = \sum_{n=0}^{\infty} \frac{(2(x + 1))^n}{n!} = \sum_{n=0}^{\infty} \frac{2^n}{n!}(x + 1)^n;$$

hence,

$$e^{2x} = e^{-2} \sum_{n=0}^{\infty} \frac{2^n (x + 1)^n}{n!}.$$

99. $f(x) = \ln \dfrac{x}{2}, \quad c = 2$

SOLUTION   Write

$$\ln \frac{x}{2} = \ln \left(\frac{(x - 2) + 2}{2} \right) = \ln \left(1 + \frac{x - 2}{2} \right).$$

Substituting $\frac{x-2}{2}$ for x in the Maclaurin series for $\ln(1 + x)$ yields

$$\ln \frac{x}{2} = \sum_{n=1}^{\infty} \frac{(-1)^{n+1} \left(\frac{x-2}{2} \right)^n}{n} = \sum_{n=1}^{\infty} \frac{(-1)^{n+1} (x - 2)^n}{n \cdot 2^n}.$$

This series is valid for $|x - 2| < 2$.

101. $f(x) = \sqrt{x} \arctan \sqrt{x}, \quad c = 0$

SOLUTION   Substituting $\sqrt{x}$ for x in the Maclaurin series for $\arctan x$ yields

$$\arctan \sqrt{x} = \sum_{n=0}^{\infty} \frac{(-1)^n \sqrt{x}^{2n+1}}{2n + 1} = \sum_{n=0}^{\infty} \frac{(-1)^n x^{n+\frac{1}{2}}}{2n + 1}.$$

Thus,

$$\sqrt{x} \arctan \sqrt{x} = \sqrt{x} \sum_{n=0}^{\infty} \frac{(-1)^n x^{n+\frac{1}{2}}}{2n + 1} = \sum_{n=0}^{\infty} \frac{(-1)^n x^{n+1}}{2n + 1}.$$

103. $f(x) = e^{x-1}, \quad c = -1$

SOLUTION   Write

$$e^{x-1} = e^{x+1-1-1} = e^{-2} e^{x+1}.$$

Substituting $x + 1$ for x in the Maclaurin series for e^x yields

$$e^{x+1} = \sum_{n=0}^{\infty} \frac{(x + 1)^n}{n!};$$

hence,

$$e^{x-1} = e^{-2} \sum_{n=0}^{\infty} \frac{(x + 1)^n}{n!} = \sum_{n=0}^{\infty} \frac{(x + 1)^n}{n! e^2}.$$

105. Use the Maclaurin series of $\sin x$ and $\sqrt{1 + x}$ to calculate $f^{(4)}(0)$, where $f(x) = (\sin x)\sqrt{1 + x}$.

SOLUTION   Recall that the coefficient of x^4 in the Maclaurin series for $f(x) = (\sin x)\sqrt{1 + x}$ is $\frac{f^{(4)}(0)}{4!}$. Now, we can obtain the Maclaurin series for $f(x)$ by multiplying the Maclaurin series for $\sqrt{1 + x}$:

$$\sqrt{1 + x} = 1 + \frac{1}{2}x - \frac{1}{8}x^2 + \frac{1}{16}x^3 - \frac{5}{128}x^4 + \cdots$$

by the Maclaurin series for $\sin x$:

$$\sin x = x - \frac{x^3}{3!} + \cdots$$

The term involving x^4 in this product is

$$x\left(\frac{1}{16}x^3\right) - \frac{x^3}{3!}\left(\frac{1}{2}x\right) = x^4\left(\frac{1}{16} - \frac{1}{12}\right) = -\frac{1}{48}x^4.$$

Therefore,

$$\frac{f^{(4)}(0)}{4!} = -\frac{1}{48} \quad \text{and} \quad f^{(4)}(0) = -\frac{1}{48} \cdot 24 = -\frac{1}{2}.$$

107. Find the Maclaurin series of the function $F(x) = \int_0^x \frac{e^t - 1}{t}\,dt$.

SOLUTION   Subtracting 1 from the Maclaurin series for e^t yields

$$e^t - 1 = \sum_{n=0}^{\infty} \frac{t^n}{n!} - 1 = 1 + \sum_{n=1}^{\infty} \frac{t^n}{n!} - 1 = \sum_{n=1}^{\infty} \frac{t^n}{n!}.$$

Thus,

$$\frac{e^t - 1}{t} = \frac{1}{t}\sum_{n=1}^{\infty} \frac{t^n}{n!} = \sum_{n=1}^{\infty} \frac{t^{n-1}}{n!}.$$

Finally, integrating term-by-term yields

$$\int_0^x \frac{e^t - 1}{t}\,dt = \int_0^x \sum_{n=1}^{\infty} \frac{t^{n-1}}{n!}\,dt = \sum_{n=1}^{\infty} \int_0^x \frac{t^{n-1}}{n!}\,dt = \sum_{n=1}^{\infty} \frac{x^n}{n!\,n}.$$

12 PARAMETRIC EQUATIONS, POLAR COORDINATES, AND CONIC SECTIONS

12.1 Parametric Equations

Preliminary Questions

1. Describe the shape of the curve $x = 3\cos t$, $y = 3\sin t$.

SOLUTION For all t,

$$x^2 + y^2 = (3\cos t)^2 + (3\sin t)^2 = 9(\cos^2 t + \sin^2 t) = 9 \cdot 1 = 9,$$

therefore the curve is on the circle $x^2 + y^2 = 9$. Also, each point on the circle $x^2 + y^2 = 9$ can be represented in the form $(3\cos t, 3\sin t)$ for some value of t. We conclude that the curve $x = 3\cos t$, $y = 3\sin t$ is the circle of radius 3 centered at the origin.

2. How does $x = 4 + 3\cos t$, $y = 5 + 3\sin t$ differ from the curve in the previous question?

SOLUTION In this case we have

$$(x - 4)^2 + (y - 5)^2 = (3\cos t)^2 + (3\sin t)^2 = 9(\cos^2 t + \sin^2 t) = 9 \cdot 1 = 9$$

Therefore, the given equations parametrize the circle of radius 3 centered at the point $(4, 5)$.

3. What is the maximum height of a particle whose path has parametric equations $x = t^9$, $y = 4 - t^2$?

SOLUTION The particle's height is $y = 4 - t^2$. To find the maximum height we set the derivative equal to zero and solve:

$$\frac{dy}{dt} = \frac{d}{dt}(4 - t^2) = -2t = 0 \quad \text{or} \quad t = 0$$

The maximum height is $y(0) = 4 - 0^2 = 4$.

4. Can the parametric curve $(t, \sin t)$ be represented as a graph $y = f(x)$? What about $(\sin t, t)$?

SOLUTION In the parametric curve $(t, \sin t)$ we have $x = t$ and $y = \sin t$, therefore, $y = \sin x$. That is, the curve can be represented as a graph of a function. In the parametric curve $(\sin t, t)$ we have $x = \sin t$, $y = t$, therefore $x = \sin y$. This equation does not define y as a function of x, therefore the parametric curve $(\sin t, t)$ cannot be represented as a graph of a function $y = f(x)$.

5. Match the derivatives with a verbal description:

(a) $\dfrac{dx}{dt}$ **(b)** $\dfrac{dy}{dt}$ **(c)** $\dfrac{dy}{dx}$

(i) Slope of the tangent line to the curve

(ii) Vertical rate of change with respect to time

(iii) Horizontal rate of change with respect to time

SOLUTION

(a) The derivative $\dfrac{dx}{dt}$ is the horizontal rate of change with respect to time.

(b) The derivative $\dfrac{dy}{dt}$ is the vertical rate of change with respect to time.

(c) The derivative $\dfrac{dy}{dx}$ is the slope of the tangent line to the curve.

Hence, (a) $\leftrightarrow$ (iii), (b) $\leftrightarrow$ (ii), (c) $\leftrightarrow$ (i)

Exercises

1. Find the coordinates at times $t = 0, 2, 4$ of a particle following the path $x = 1 + t^3$, $y = 9 - 3t^2$.

SOLUTION Substituting $t = 0$, $t = 2$, and $t = 4$ into $x = 1 + t^3$, $y = 9 - 3t^2$ gives the coordinates of the particle at these times respectively. That is,

$$
\begin{aligned}
(t = 0) \quad & x = 1 + 0^3 = 1, \ y = 9 - 3 \cdot 0^2 = 9 && \Rightarrow (1, 9) \\
(t = 2) \quad & x = 1 + 2^3 = 9, \ y = 9 - 3 \cdot 2^2 = -3 && \Rightarrow (9, -3) \\
(t = 4) \quad & x = 1 + 4^3 = 65, \ y = 9 - 3 \cdot 4^2 = -39 && \Rightarrow (65, -39).
\end{aligned}
$$

3. Show that the path traced by the bullet in Example 2 is a parabola by eliminating the parameter.

SOLUTION The path traced by the bullet is given by the following parametric equations:

$$ x = 200t, \ y = 400t - 16t^2 $$

We eliminate the parameter. Since $x = 200t$, we have $t = \dfrac{x}{200}$. Substituting into the equation for y we obtain:

$$ y = 400t - 16t^2 = 400 \cdot \frac{x}{200} - 16 \left(\frac{x}{200} \right)^2 = 2x - \frac{x^2}{2500} $$

The equation $y = -\dfrac{x^2}{2500} + 2x$ is the equation of a parabola.

5. Graph the parametric curves. Include arrows indicating the direction of motion.

(a) (t, t), $\quad -\infty < t < \infty$

(b) $(\sin t, \sin t)$, $\quad 0 \le t \le 2\pi$

(c) (e^t, e^t), $\quad -\infty < t < \infty$

(d) (t^3, t^3), $\quad -1 \le t \le 1$

SOLUTION

(a) For the trajectory $c(t) = (t, t)$, $-\infty < t < \infty$ we have $y = x$. Also the two coordinates tend to ∞ and $-\infty$ as $t \to \infty$ and $t \to -\infty$ respectively. The graph is shown next:

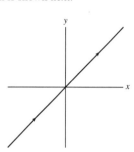

(b) For the curve $c(t) = (\sin t, \sin t)$, $0 \le t \le 2\pi$, we have $y = x$. $\sin t$ is increasing for $0 < t \le \frac{\pi}{2}$, decreasing for $\frac{\pi}{2} \le t \le \frac{3\pi}{2}$ and increasing again for $\frac{3\pi}{2} \le t < 2\pi$. Hence the particle moves from $c(0) = (0, 0)$ to $c(\frac{\pi}{2}) = (1, 1)$, then moves back to $c(\frac{3\pi}{2}) = (-1, -1)$ and then returns to $c(2\pi) = (0, 0)$. We obtain the following trajectory:

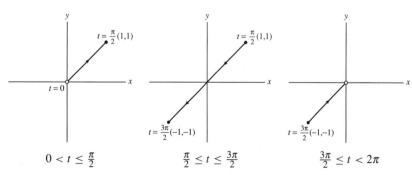

$$ 0 < t \le \frac{\pi}{2} \qquad\qquad \frac{\pi}{2} \le t \le \frac{3\pi}{2} \qquad\qquad \frac{3\pi}{2} \le t < 2\pi $$

These three parts of the trajectory are shown together in the next figure:

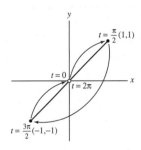

(c) For the trajectory $c(t) = (e^t, e^t)$, $-\infty < t < \infty$, we have $y = x$. However since $\lim\limits_{t \to -\infty} e^t = 0$ and $\lim\limits_{t \to \infty} e^t = \infty$, the trajectory is the part of the line $y = x$, $0 < x$.

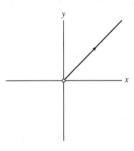

(d) For the trajectory $c(t) = (t^3, t^3)$, $-1 \le t \le 1$, we have again $y = x$. Since the function t^3 is increasing the particle moves in one direction starting at $((-1)^3, (-1)^3) = (-1, -1)$ and ending at $(1^3, 1^3) = (1, 1)$. The trajectory is shown next:

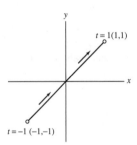

In Exercises 7–14, express in the form $y = f(x)$ by eliminating the parameter.

7. $x = t + 3$, $\quad y = 4t$

SOLUTION We eliminate the parameter. Since $x = t + 3$, we have $t = x - 3$. Substituting into $y = 4t$ we obtain

$$y = 4t = 4(x - 3) \Rightarrow y = 4x - 12$$

9. $x = t$, $\quad y = \tan^{-1}(t^3 + e^t)$

SOLUTION Replacing t by x in the equation for y we obtain $y = \tan^{-1}(x^3 + e^x)$.

11. $x = e^{-2t}$, $\quad y = 6e^{4t}$

SOLUTION We eliminate the parameter. Since $x = e^{-2t}$, we have $-2t = \ln x$ or $t = -\frac{1}{2} \ln x$. Substituting in $y = 6e^{4t}$ we get

$$y = 6e^{4t} = 6e^{4 \cdot (-\frac{1}{2} \ln x)} = 6e^{-2 \ln x} = 6e^{\ln x^{-2}} = 6x^{-2} \Rightarrow y = \frac{6}{x^2}, \quad x > 0.$$

13. $x = \ln t$, $\quad y = 2 - t$

SOLUTION Since $x = \ln t$ we have $t = e^x$. Substituting in $y = 2 - t$ we obtain $y = 2 - e^x$.

In Exercises 15–18, graph the curve and draw an arrow specifying the direction corresponding to motion.

15. $x = \frac{1}{2}t$, $\quad y = 2t^2$

SOLUTION Let $c(t) = (x(t), y(t)) = (\frac{1}{2}t, 2t^2)$. Then $c(-t) = (-x(t), y(t))$ so the curve is symmetric with respect to the y-axis. Also, the function $\frac{1}{2}t$ is increasing. Hence there is only one direction of motion on the curve. The corresponding function is the parabola $y = 2 \cdot (2x)^2 = 8x^2$. We obtain the following trajectory:

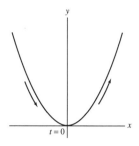

17. $x = \pi t, \quad y = \sin t$

SOLUTION We find the function by eliminating t. Since $x = \pi t$, we have $t = \frac{x}{\pi}$. Substituting $t = \frac{x}{\pi}$ into $y = \sin t$ we get $y = \sin \frac{x}{\pi}$. We obtain the following curve:

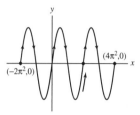

19. Match the parametrizations (a)–(d) below with their plots in Figure 14 and draw an arrow indicating the direction of motion.

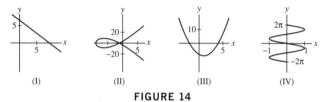

FIGURE 14

(a) $c(t) = (\sin t, -t)$

(b) $c(t) = (t^2 - 9, -t^3 - 8)$

(c) $c(t) = (1 - t, t^2 - 9)$

(d) $c(t) = (4t + 2, 5 - 3t)$

SOLUTION

(a) In the curve $c(t) = (\sin t, -t)$ the x-coordinate is varying between -1 and 1 so this curve corresponds to plot IV. As t increases, the y-coordinate $y = -t$ is decreasing so the direction of motion is downward.

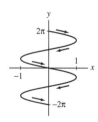

$$(IV) \; c(t) = (\sin t, -t)$$

(b) The curve $c(t) = (t^2 - 9, -t^3 - 8)$ intersects the x-axis where $y = -t^3 - 8 = 0$, or $t = -2$. The x-intercept is $(-5, 0)$. The y-intercepts are obtained where $x = t^2 - 9 = 0$, or $t = \pm 3$. The y-intercepts are $(0, -35)$ and $(0, 19)$. As t increases from $-\infty$ to 0, x and y decrease, and as t increases from 0 to ∞, x increases and y decreases. We obtain the following trajectory:

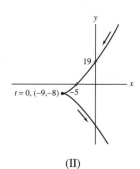

$$(II)$$

(c) The curve $c(t) = (1 - t, t^2 - 9)$ intersects the y-axis where $x = 1 - t = 0$, or $t = 1$. The y-intercept is $(0, -8)$. The x-intercepts are obtained where $t^2 - 9 = 0$ or $t = \pm 3$. These are the points $(-2, 0)$ and $(4, 0)$. Setting $t = 1 - x$ we get

$$y = t^2 - 9 = (1 - x)^2 - 9 = x^2 - 2x - 8.$$

As t increases the x coordinate decreases and we obtain the following trajectory:

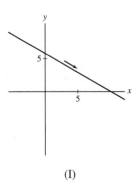

(III)

(d) The curve $c(t) = (4t + 2, 5 - 3t)$ is a straight line, since eliminating t in $x = 4t + 2$ and substituting in $y = 5 - 3t$ gives $y = 5 - 3 \cdot \frac{x-2}{4} = -\frac{3}{4}x + \frac{13}{2}$ which is the equation of a line. As t increases, the x coordinate $x = 4t + 2$ increases and the y-coordinate $y = 5 - 3t$ decreases. We obtain the following trajectory:

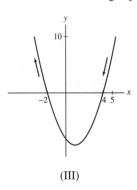

(I)

21. Find an interval of t-values such that $c(t) = (\cos t, \sin t)$ traces the lower half of the unit circle.

SOLUTION For $t = \pi$, we have $c(\pi) = (-1, 0)$. As t increases from π to 2π, the x-coordinate of $c(t)$ increases from -1 to 1, and the y-coordinate decreases from 0 to -1 (at $t = 3\pi/2$) and then returns to 0. Thus, for t in $[\pi, 2\pi]$, the equation traces the lower part of the circle.

In Exercises 23–34, find parametric equations for the given curve.

23. $y = 9 - 4x$

SOLUTION This is a line through $P = (0, 9)$ with slope $m = -4$. Using the parametric representation of a line, as given in Example 3, we obtain $c(t) = (t, 9 - 4t)$.

25. $4x - y^2 = 5$

SOLUTION We define the parameter $t = y$. Then, $x = \dfrac{5 + y^2}{4} = \dfrac{5 + t^2}{4}$, giving us the parametrization $c(t) = \left(\dfrac{5 + t^2}{4}, t \right)$.

27. $(x + 9)^2 + (y - 4)^2 = 49$

SOLUTION This is a circle of radius 7 centered at $(-9, 4)$. Using the parametric representation of a circle we get $c(t) = (-9 + 7\cos t, 4 + 7\sin t)$.

29. Line through $(2, 5)$ perpendicular to $y = 3x$

SOLUTION The line perpendicular to $y = 3x$ has slope $m = -\frac{1}{3}$. We use the parametric representation of a line given in Example 3 to obtain the parametrization $c(t) = (2 + t, 5 - \frac{1}{3}t)$.

31. $\left(\dfrac{x}{4} \right)^2 + \left(\dfrac{y}{9} \right)^2 = 1$

SOLUTION   This is an ellipse with $a = 4$ and $b = 9$. Hence using the parametric representation given in Example 4 we get $c(t) = (4 \cos t, 9 \sin t)$.

33. The parabola $y = x^2$ translated so that its minimum occurs at $(2, 3)$

SOLUTION   The equation of the translated parabola is

$$y - 3 = (x - 2)^2 \Rightarrow y = 3 + (x - 2)^2.$$

We let $t = x - 2$, hence $x = t + 2$ and $y = 3 + t^2$. We obtain the representation $c(t) = (t + 2, t^2 + 3)$.

35. Describe the parametrized curve $c(t) = (\sin^2 t, \cos^2 t)$ for $0 \le t \le \pi$.

SOLUTION   The graphs of $x = \sin^2 t$ and $y = \cos^2 t$ for $0 \le t \le \pi$ are shown in the following figures:

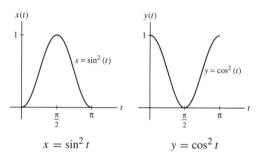

$$x = \sin^2 t \qquad\qquad y = \cos^2 t$$

Now $\sin^2 t + \cos^2 t = 1$, hence $x + y = 1$ and so $y = 1 - x$ (for $0 \le x \le 1$). The path starts at the point $c(0) = (0, 1)$. As t increases from $t = 0$ to $t = \frac{\pi}{2}$, the x coordinate increases and the y coordinate decreases. As t increases from $t = \frac{\pi}{2}$ to $t = \pi$, the x coordinate decreases and the y coordinate increases, while returning from $c(\frac{\pi}{2}) = (1, 0)$ back to $c(\pi) = (0, 1)$. We obtain the following curve:

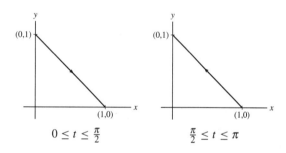

$$0 \le t \le \frac{\pi}{2} \qquad\qquad \frac{\pi}{2} \le t \le \pi$$

37. Find a parametrization $c(t)$ of the line $y = 3x - 4$ such that $c(0) = (2, 2)$.

SOLUTION   Let $x(t) = t + a$ and $y(t) = 3x - 4 = 3(t + a) - 4$. We want $x(0) = 2$, thus we must use $a = 2$. Our line is $c(t) = (x(t), y(t)) = (t + 2, 3(t + 2) - 4) = (t + 2, 3t + 2)$.

39. Show that $x = \cosh t$, $y = \sinh t$ parametrizes the hyperbola $x^2 - y^2 = 1$. Calculate $\dfrac{dy}{dx}$ as a function of t. Generalize to obtain a parametrization of $\left(\dfrac{x}{a}\right)^2 - \left(\dfrac{y}{b}\right)^2 = 1$.

SOLUTION   We check that $x = \cosh t$, $y = \sinh t$ satisfy the equation of the hyperbola

$$x^2 - y^2 = (\cosh t)^2 - (\sinh t)^2 = 1.$$

We now find $\frac{dy}{dx}$ using the formula for the slope of the tangent line:

$$\frac{dy}{dx} = \frac{\frac{dy}{dt}}{\frac{dx}{dt}} = \frac{\frac{d}{dt}(\sinh t)}{\frac{d}{dt}(\cosh t)} = \frac{\cosh t}{\sinh t} = \coth t.$$

We can generalize to obtain the following parametrization of $(\frac{x}{a})^2 - (\frac{y}{b})^2 = 1$:

$$x = a \cosh t, \quad y = b \sinh t.$$

Next, we check that $x = a \cosh t$ and $y = b \sin t$ satisfy the equation $(\frac{x}{a})^2 - (\frac{y}{b})^2 = 1$:

$$\left(\frac{x}{a}\right)^2 - \left(\frac{y}{b}\right)^2 = \left(\frac{a \cosh t}{a}\right)^2 - \left(\frac{b \sinh t}{b}\right)^2 = (\cosh t)^2 - (\sinh t)^2 = 1.$$

41. Which of (I) or (II) is the graph of $x(t)$ for the parametric curve in Figure 16(A)? Which represents $y(t)$?

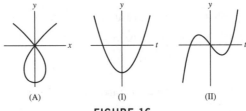

(A)   (I)   (II)

FIGURE 16

SOLUTION As indicated by Figure 16(A), the y-coordinate is decreasing and then increasing, so plot I is the graph of y. Figure 16(A) also shows that the x-coordinate is increasing, decreasing and then increasing, so plot II is the graph for x.

43. Sketch $c(t) = (t^2, \sin t)$ for $-2\pi \le t \le 2\pi$.

SOLUTION The graphs of $x(t) = t^2$ and $y = \sin t$ on the interval $-2\pi \le t \le 2\pi$ are shown in the following figures:

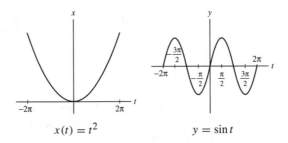

$$x(t) = t^2 \qquad\qquad y = \sin t$$

Since $c(-t) = (x(-t), y(-t)) = (x(t), -y(t))$, $c(t)$ and $c(-t)$ are symmetric with respect to the x-axis, it suffices to graph the curve for $t \ge 0$. We have $c(0) = (0, 0)$. The x-coordinate increases as t increases so the curve is directed to the right. The y-coordinate is positive and increasing for $0 < t < \frac{\pi}{2}$, positive and decreasing for $\frac{\pi}{2} < t < \pi$, negative and decreasing for $\pi < t < \frac{3\pi}{2}$, and negative and increasing for $\frac{3\pi}{2} < t < 2\pi$.

Therefore, starting at the origin, the curve first goes up then at $c(\frac{\pi}{2}) = (\frac{\pi^2}{4}, 1)$ it turns down, crosses the x-axis at $c(\pi) = (\pi^2, 0)$, turns up again at $c(\frac{3\pi}{2}) = (\frac{9\pi^2}{4}, -1)$ ending at $c(2\pi) = (4\pi^2, 0)$. The part of the path for $t \le 0$ is obtained by reflecting across the x-axis. We obtain the following path:

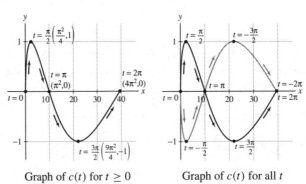

Graph of $c(t)$ for $t \ge 0$ Graph of $c(t)$ for all t

In Exercises 45–48, use Eq. (7) to find $\dfrac{dy}{dx}$ at the given point.

45. $(t^3, t^2 - 1), \quad t = -4$

SOLUTION By Eq. (7) we have

$$\frac{dy}{dx} = \frac{y'(t)}{x'(t)} = \frac{(t^2 - 1)'}{(t^3)'} = \frac{2t}{3t^2} = \frac{2}{3t}$$

Substituting $t = -4$ we get

$$\frac{dy}{dx} = \frac{2}{3t}\bigg|_{t=-4} = \frac{2}{3 \cdot (-4)} = -\frac{1}{6}.$$

47. $(s^{-1} - 3s, s^3), \quad s = -1$

SOLUTION Using Eq. (7) we get

$$\frac{dy}{dx} = \frac{y'(s)}{x'(s)} = \frac{(s^3)'}{(s^{-1} - 3s)'} = \frac{3s^2}{-s^{-2} - 3} = \frac{3s^4}{-1 - 3s^2}$$

Substituting $s = -1$ we obtain

$$\frac{dy}{dx} = \frac{3s^4}{-1 - 3s^2}\bigg|_{s=-1} = \frac{3 \cdot (-1)^4}{-1 - 3 \cdot (-1)^2} = -\frac{3}{4}.$$

In Exercises 49–52, find an equation $y = f(x)$ for the parametric curve and compute $\dfrac{dy}{dx}$ in two ways: using Eq. (7) and by differentiating $f(x)$.

49. $c(t) = (2t + 1, 1 - 9t)$

SOLUTION Since $x = 2t + 1$, we have $t = \dfrac{x - 1}{2}$. Substituting in $y = 1 - 9t$ we have

$$y = 1 - 9\left(\frac{x - 1}{2}\right) = -\frac{9}{2}x + \frac{11}{2}$$

Differentiating $y = -\dfrac{9}{2}x + \dfrac{11}{2}$ gives $\dfrac{dy}{dx} = -\dfrac{9}{2}$. We now find $\dfrac{dy}{dx}$ using Eq. (7):

$$\frac{dy}{dx} = \frac{y'(t)}{x'(t)} = \frac{(1 - 9t)'}{(2t + 1)'} = -\frac{9}{2}$$

51. $x = s^3, \quad y = s^6 + s^{-3}$

SOLUTION We find y as a function of x:

$$y = s^6 + s^{-3} = \left(s^3\right)^2 + \left(s^3\right)^{-1} = x^2 + x^{-1}.$$

We now differentiate $y = x^2 + x^{-1}$. This gives

$$\frac{dy}{dx} = 2x - x^{-2}.$$

Alternatively, we can use Eq. (7) to obtain the following derivative:

$$\frac{dy}{dx} = \frac{y'(s)}{x'(s)} = \frac{\left(s^6 + s^{-3}\right)'}{\left(s^3\right)'} = \frac{6s^5 - 3s^{-4}}{3s^2} = 2s^3 - s^{-6}.$$

Hence, since $x = s^3$,

$$\frac{dy}{dx} = 2x - x^{-2}.$$

In Exercises 53–56, let $c(t) = (t^2 - 9, t^2 - 8t)$ (see Figure 17).

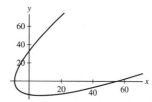

FIGURE 17 Plot of $c(t) = (t^2 - 9, t^2 - 8t)$.

53. Draw an arrow indicating the direction of motion and determine the interval of t-values corresponding to the portion of the curve in each of the four quadrants.

SOLUTION We plot the functions $x(t) = t^2 - 9$ and $y(t) = t^2 - 8t$:

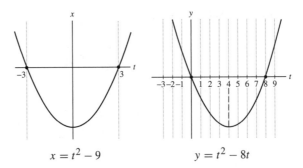

$$x = t^2 - 9 \qquad\qquad y = t^2 - 8t$$

We note carefully where each of these graphs are positive or negative, increasing or decreasing. In particular, $x(t)$ is decreasing for $t < 0$, increasing for $t > 0$, positive for $|t| > 3$, and negative for $|t| < 3$. Likewise, $y(t)$ is decreasing for $t < 4$, increasing for $t > 4$, positive for $t > 8$ or $t < 0$, and negative for $0 < t < 8$. We now draw arrows on the path following the decreasing/increasing behavior of the coordinates as indicated above. We obtain:

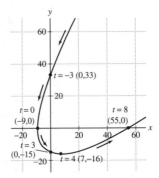

This plot also shows that:

- The graph is in the first quadrant for $t < -3$ or $t > 8$.
- The graph is in the second quadrant for $-3 < t < 0$.
- The graph is in the third quadrant for $0 < t < 3$.
- The graph is in the fourth quadrant for $3 < t < 8$.

55. Find the points where the tangent has slope $\frac{1}{2}$.

SOLUTION   The slope of the tangent at t is

$$\frac{dy}{dx} = \frac{\left(t^2 - 8t\right)'}{\left(t^2 - 9\right)'} = \frac{2t - 8}{2t} = 1 - \frac{4}{t}$$

The point where the tangent has slope $\frac{1}{2}$ corresponds to the value of t that satisfies

$$\frac{dy}{dx} = 1 - \frac{4}{t} = \frac{1}{2} \Rightarrow \frac{4}{t} = \frac{1}{2} \Rightarrow t = 8.$$

We substitute $t = 8$ in $x(t) = t^2 - 9$ and $y(t) = t^2 - 8t$ to obtain the following point:

$$\begin{array}{l} x(8) = 8^2 - 9 = 55 \\ y(8) = 8^2 - 8 \cdot 8 = 0 \end{array} \quad \Rightarrow \quad (55, 0)$$

57. Find the equation of the ellipse represented parametrically by $x = 4\cos t$, $y = 7\sin t$. Calculate the slope of the tangent line at the point $(2\sqrt{2}, 7\sqrt{2}/2)$.

SOLUTION   In Example 4 it is shown that the ellipse

$$\left(\frac{x}{a}\right)^2 + \left(\frac{y}{b}\right)^2 = 1$$

has the parametrization $x = a\cos t$, $y = b\sin t$. Therefore for the ellipse represented by the parametric equations $x = 4\cos t$, $y = 7\sin t$, we have $a = 4$, $b = 7$. The equation of the ellipse is therefore

$$\left(\frac{x}{4}\right)^2 + \left(\frac{y}{7}\right)^2 = 1.$$

The slope of the tangent line at t is:

$$m = \frac{dy}{dx} = \frac{y'(t)}{x'(t)} = \frac{(7\sin t)'}{(4\cos t)'} = \frac{7\cos t}{-4\sin t} \qquad (1)$$

we compute the value of t corresponding to the point $(2\sqrt{2}, \frac{7\sqrt{2}}{2})$, by solving the following equations:

$$x(t) = 2\sqrt{2} = 4\cos t$$

$$y(t) = \frac{7\sqrt{2}}{2} = 7\sin t$$

The first equation implies that $\cos t = \frac{\sqrt{2}}{2}$ and the second equation implies that $\sin t = \frac{\sqrt{2}}{2}$. Setting these values in (1) gives the following slope:

$$m = \frac{7 \cdot \frac{\sqrt{2}}{2}}{-4 \cdot \frac{\sqrt{2}}{2}} = -\frac{7}{4}$$

In Exercises 58–60, refer to the Bézier curve defined by Eqs. (8) and (9).

59. Find and plot the Bézier curve $c(t)$ passing through the control points

$$P_0 = (3, 2), \quad P_1 = (0, 2), \quad P_2 = (5, 4), \quad P_3 = (2, 4)$$

SOLUTION Setting $a_0 = 3$, $a_1 = 0$, $a_2 = 5$, $a_3 = 2$, and $b_0 = 2$, $b_1 = 2$, $b_2 = 4$, $b_3 = 4$ into Eq. (8)–(9) and simplifying gives

$$
\begin{aligned}
x(t) &= 3(1-t)^3 + 0 + 15t^2(1-t) + 2t^3 \\
&= 3(1 - 3t + 3t^2 - t^3) + 15t^2 - 15t^3 + 2t^3 = 3 - 9t + 24t^2 - 16t^3 \\
y(t) &= 2(1-t)^3 + 6t(1-t)^2 + 12t^2(1-t) + 4t^3 \\
&= 2(1 - 3t + 3t^2 - t^3) + 6t(1 - 2t + t^2) + 12t^2 - 12t^3 + 4t^3 \\
&= 2 - 6t + 6t^2 - 2t^3 + 6t - 12t^2 + 6t^3 + 12t^2 - 12t^3 + 4t^3 = 2 + 6t^2 - 4t^3
\end{aligned}
$$

We obtain the following equation

$$c(t) = (3 - 9t + 24t^2 - 16t^3, 2 + 6t^2 - 4t^3), \quad 0 \le t \le 1.$$

The graph of the Bézier curve is shown in the following figure:

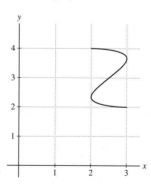

61. A bullet fired from a gun follows the trajectory

$$x = at, \quad y = bt - 16t^2 \quad (a, b > 0)$$

Show that the bullet leaves the gun at an angle $\theta = \tan^{-1}\left(\frac{b}{a}\right)$ and lands at a distance $\frac{ab}{16}$ from the origin.

SOLUTION The height of the bullet equals the value of the y-coordinate. When the bullet leaves the gun, $y(t) = t(b - 16t) = 0$. The solutions to this equation are $t = 0$ and $t = \frac{b}{16}$, with $t = 0$ corresponding to the moment the bullet leaves the gun. We find the slope m of the tangent line at $t = 0$:

$$\frac{dy}{dx} = \frac{y'(t)}{x'(t)} = \frac{b - 32t}{a} \Rightarrow m = \left.\frac{b - 32t}{a}\right|_{t=0} = \frac{b}{a}$$

It follows that $\tan \theta = \frac{b}{a}$ or $\theta = \tan^{-1}\left(\frac{b}{a}\right)$. The bullet lands at $t = \frac{b}{16}$. We find the distance of the bullet from the origin at this time, by substituting $t = \frac{b}{16}$ in $x(t) = at$. This gives

$$x\left(\frac{b}{16}\right) = \frac{ab}{16}$$

63. **CAS** Plot the astroid $x = \cos^3 \theta$, $y = \sin^3 \theta$ and find the equation of the tangent line at $\theta = \frac{\pi}{3}$.

SOLUTION The graph of the astroid $x = \cos^3 \theta$, $y = \sin^3 \theta$ is shown in the following figure:

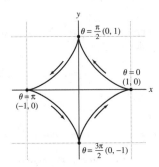

The slope of the tangent line at $\theta = \frac{\pi}{3}$ is

$$m = \left.\frac{dy}{dx}\right|_{\theta=\pi/3} = \left.\frac{(\sin^3 \theta)'}{(\cos^3 \theta)'}\right|_{\theta=\pi/3} = \left.\frac{3\sin^2 \theta \cos \theta}{3\cos^2 \theta(-\sin \theta)}\right|_{\theta=\pi/3} = \left.-\frac{\sin \theta}{\cos \theta}\right|_{\theta=\pi/3} = \left.-\tan \theta\right|_{\pi/3} = -\sqrt{3}$$

We find the point of tangency:

$$\left(x\left(\frac{\pi}{3}\right), y\left(\frac{\pi}{3}\right)\right) = \left(\cos^3 \frac{\pi}{3}, \sin^3 \frac{\pi}{3}\right) = \left(\frac{1}{8}, \frac{3\sqrt{3}}{8}\right)$$

The equation of the tangent line at $\theta = \frac{\pi}{3}$ is, thus,

$$y - \frac{3\sqrt{3}}{8} = -\sqrt{3}\left(x - \frac{1}{8}\right) \Rightarrow y = -\sqrt{3}x + \frac{\sqrt{3}}{2}$$

65. Find the points with horizontal tangent line on the cycloid with parametric equation (5).

SOLUTION The parametric equations of the cycloid are

$$x = t - \sin t, \quad y = 1 - \cos t$$

We find the slope of the tangent line at t:

$$\frac{dy}{dx} = \frac{(1 - \cos t)'}{(t - \sin t)'} = \frac{\sin t}{1 - \cos t}$$

The tangent line is horizontal where it has slope zero. That is,

$$\frac{dy}{dx} = \frac{\sin t}{1 - \cos t} = 0 \quad \Rightarrow \quad \begin{array}{l} \sin t = 0 \\ \cos t \neq 1 \end{array} \quad \Rightarrow \quad t = (2k - 1)\pi, \quad k = 0, \pm 1, \pm 2, \ldots$$

We find the coordinates of the points with horizontal tangent line, by substituting $t = (2k - 1)\pi$ in $x(t)$ and $y(t)$. This gives

$$x = (2k - 1)\pi - \sin(2k - 1)\pi = (2k - 1)\pi$$

$$y = 1 - \cos((2k - 1)\pi) = 1 - (-1) = 2$$

The required points are

$$((2k - 1)\pi, 2), \quad k = 0, \pm 1, \pm 2, \ldots$$

67. A *curtate cycloid* (Figure 19) is the curve traced by a point at a distance h from the center of a circle of radius R rolling along the x-axis where $h < R$. Show that this curve has parametric equations $x = Rt - h \sin t$, $y = R - h \cos t$.

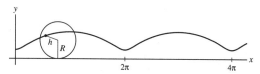

FIGURE 19 Curtate cycloid.

SOLUTION Let P be a point at a distance h from the center C of the circle. Assume that at $t = 0$, the line of CP is passing through the origin. When the circle rolls a distance Rt along the x-axis, the length of the arc $\overset{\frown}{SQ}$ (see figure) is also Rt and the angle $\angle SCQ$ has radian measure t. We compute the coordinates x and y of P.

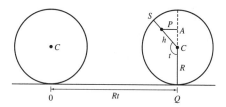

$$x = Rt - \overline{PA} = Rt - h \sin(\pi - t) = Rt - h \sin t$$
$$y = R + \overline{AC} = R + h \cos(\pi - t) = R - h \cos t$$

We obtain the following parametrization:

$$x = Rt - h \sin t, \quad y = R - h \cos t.$$

69. Show that the line of slope t through $(-1, 0)$ intersects the unit circle in the point with coordinates

$$x = \frac{1 - t^2}{t^2 + 1}, \quad y = \frac{2t}{t^2 + 1}$$

$\boxed{10}$

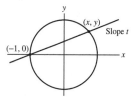

FIGURE 20 Unit circle.

Conclude that these equations parametrize the unit circle with the point $(-1, 0)$ excluded (Figure 20). Show further that $t = \dfrac{y}{x + 1}$.

SOLUTION The equation of the line of slope t through $(-1, 0)$ is $y = t(x + 1)$. The equation of the unit circle is $x^2 + y^2 = 1$. Hence, the line intersects the unit circle at the points (x, y) that satisfy the equations:

$$y = t(x + 1) \tag{1}$$
$$x^2 + y^2 = 1 \tag{2}$$

Substituting y from equation (1) into equation (2) and solving for x we obtain

$$x^2 + t^2(x + 1)^2 = 1$$
$$x^2 + t^2 x^2 + 2t^2 x + t^2 = 1$$
$$(1 + t^2)x^2 + 2t^2 x + (t^2 - 1) = 0$$

This gives

$$x_{1,2} = \frac{-2t^2 \pm \sqrt{4t^4 - 4(t^2 + 1)(t^2 - 1)}}{2(1 + t^2)} = \frac{-2t^2 \pm 2}{2(1 + t^2)} = \frac{\pm 1 - t^2}{1 + t^2}$$

So $x_1 = -1$ and $x_2 = \dfrac{1 - t^2}{t^2 + 1}$. The solution $x = -1$ corresponds to the point $(-1, 0)$. We are interested in the second point of intersection that is varying as t varies. Hence the appropriate solution is

$$x = \frac{1 - t^2}{t^2 + 1}$$

We find the y-coordinate by substituting x in equation (1). This gives

$$y = t(x + 1) = t\left(\frac{1 - t^2}{t^2 + 1} + 1\right) = t \cdot \frac{1 - t^2 + t^2 + 1}{t^2 + 1} = \frac{2t}{t^2 + 1}$$

We conclude that the line and the unit circle intersect, besides at $(-1, 0)$, at the point with the following coordinates:

$$x = \frac{1 - t^2}{t^2 + 1}, \quad y = \frac{2t}{t^2 + 1} \tag{3}$$

Since these points determine all the points on the unit circle except for $(-1, 0)$ and no other points, the equations in (3) parametrize the unit circle with the point $(-1, 0)$ excluded.

We show that $t = \dfrac{y}{x + 1}$. Using (3) we have

$$\frac{y}{x + 1} = \frac{\frac{2t}{t^2+1}}{\frac{1-t^2}{t^2+1} + 1} = \frac{\frac{2t}{t^2+1}}{\frac{1-t^2+t^2+1}{t^2+1}} = \frac{\frac{2t}{t^2+1}}{\frac{2}{t^2+1}} = \frac{2t}{2} = t.$$

71. Use the results of Exercise 70 to show that the asymptote of the folium is the line $x + y = -a$. Hint: show that $\lim\limits_{t \to -1} (x + y) = -a$.

SOLUTION   We must show that as $x \to \infty$ or $x \to -\infty$ the graph of the folium is getting arbitrarily close to the line $x + y = -a$, and the derivative $\dfrac{dy}{dx}$ is approaching the slope -1 of the line.

In Exercise 70 we showed that $x \to \infty$ when $t \to (-1^-)$ and $x \to -\infty$ when $t \to (-1^+)$. We first show that the graph is approaching the line $x + y = -a$ as $x \to \infty$ or $x \to -\infty$, by showing that $\lim\limits_{t \to -1-} x + y = \lim\limits_{t \to -1+} x + y = -a$.

For $x(t) = \dfrac{3at}{1 + t^3}, y(t) = \dfrac{3at^2}{1 + t^3}, a > 0$, calculated in Exercise 70, we obtain using L'Hôpital's Rule:

$$\lim_{t \to -1-} (x + y) = \lim_{t \to -1-} \frac{3at + 3at^2}{1 + t^3} = \lim_{t \to -1-} \frac{3a + 6at}{3t^2} = \frac{3a - 6a}{3} = -a$$

$$\lim_{t \to -1+} (x + y) = \lim_{t \to -1+} \frac{3at + 3at^2}{1 + t^3} = \lim_{t \to -1+} \frac{3a + 6at}{3t^2} = \frac{3a - 6a}{3} = -a$$

We now show that $\dfrac{dy}{dx}$ is approaching -1 as $t \to -1-$ and as $t \to -1+$. We use $\dfrac{dy}{dx} = \dfrac{6at - 3at^4}{3a - 6at^3}$ computed in Exercise 70 to obtain

$$\lim_{t \to -1-} \frac{dy}{dx} = \lim_{t \to -1-} \frac{6at - 3at^4}{3a - 6at^3} = \frac{-9a}{9a} = -1$$

$$\lim_{t \to -1+} \frac{dy}{dx} = \lim_{t \to -1+} \frac{6at - 3at^4}{3a - 6at^3} = \frac{-9a}{9a} = -1$$

We conclude that the line $x + y = -a$ is an asymptote of the folium as $x \to \infty$ and as $x \to -\infty$.

73. Verify that the **tractrix** curve ($\ell > 0$)

$$c(t) = \left(t - \ell \tanh\frac{t}{\ell}, \ell \operatorname{sech}\frac{t}{\ell}\right)$$

has the following property: For all t, the segment from $c(t)$ to $(0, t)$ is tangent to the curve and has length ℓ (Figure 22).

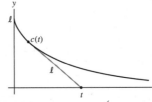

FIGURE 22 The tractrix $c(t) = \left(t - \ell \tanh\dfrac{t}{\ell}, \ell \operatorname{sech}\dfrac{t}{\ell}\right)$.

SOLUTION Let $P = c(t)$ and $Q = (t, 0)$.

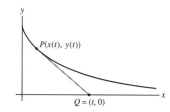

The slope of the segment $\overline{PQ}$ is

$$m_1 = \frac{y(t) - 0}{x(t) - t} = \frac{\ell \operatorname{sech}\left(\frac{t}{\ell}\right)}{-\ell \tanh\left(\frac{t}{\ell}\right)} = -\frac{1}{\sinh\left(\frac{t}{\ell}\right)}$$

We compute the slope of the tangent line at P:

$$m_2 = \frac{dy}{dx} = \frac{y'(t)}{x'(t)} = \frac{\left(\ell \operatorname{sech}\left(\frac{t}{\ell}\right)\right)'}{\left(t - \ell \tanh\left(\frac{t}{\ell}\right)\right)'} = \frac{\ell \cdot \frac{1}{\ell}\left(-\operatorname{sech}\left(\frac{t}{\ell}\right)\tanh\left(\frac{t}{\ell}\right)\right)}{1 - \ell \cdot \frac{1}{\ell}\operatorname{sech}^2\left(\frac{t}{\ell}\right)}$$

$$= -\frac{-\operatorname{sech}\left(\frac{t}{\ell}\right)\tanh\left(\frac{t}{\ell}\right)}{1 - \operatorname{sech}^2\left(\frac{t}{\ell}\right)} = \frac{-\operatorname{sech}\left(\frac{t}{\ell}\right)\tanh\left(\frac{t}{\ell}\right)}{-\tanh^2\left(\frac{t}{\ell}\right)} = \frac{-\operatorname{sech}\left(\frac{t}{\ell}\right)}{\tanh\left(\frac{t}{\ell}\right)} = -\frac{1}{\sinh\left(\frac{t}{\ell}\right)}$$

Since $m_1 = m_2$, we conclude that the segment from $c(t)$ to $(t, 0)$ is tangent to the curve.
We now show that $|\overline{PQ}| = \ell$:

$$|\overline{PQ}| = \sqrt{(x(t) - t)^2 + (y(t) - 0)^2} = \sqrt{\left(-\ell \tanh\frac{t}{\ell}\right)^2 + \left(\ell \operatorname{sech}\left(\frac{t}{\ell}\right)\right)^2}$$

$$= \sqrt{\ell^2\left(\tanh^2\left(\frac{t}{\ell}\right) + \operatorname{sech}^2\left(\frac{t}{\ell}\right)\right)} = \ell\sqrt{\operatorname{sech}^2\left(\frac{t}{\ell}\right)\sinh^2\left(\frac{t}{\ell}\right) + \operatorname{sech}^2\left(\frac{t}{\ell}\right)}$$

$$= \ell \operatorname{sech}\left(\frac{t}{\ell}\right)\sqrt{\sin h^2\left(\frac{t}{\ell}\right) + 1} = \ell \operatorname{sech}\left(\frac{t}{\ell}\right)\cosh\left(\frac{t}{\ell}\right) = \ell \cdot 1 = \ell$$

75. Let A and B be the points where the ray of angle θ intersects the two concentric circles of radii $r < R$ centered at the origin (Figure 23). Let P be the point of intersection of the horizontal line through A and the vertical line through B. Express the coordinates of P as a function of θ and describe the curve traced by P for $0 \le \theta \le 2\pi$.

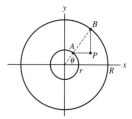

FIGURE 23

SOLUTION We use the parametric representation of a circle to determine the coordinates of the points A and B. That is,

$$A = (r\cos\theta, r\sin\theta), \quad B = (R\cos\theta, R\sin\theta)$$

The coordinates of P are therefore

$$P = (R\cos\theta, r\sin\theta)$$

In order to identify the curve traced by P, we notice that the x and y coordinates of P satisfy $\frac{x}{R} = \cos\theta$ and $\frac{y}{r} = \sin\theta$. Hence

$$\left(\frac{x}{R}\right)^2 + \left(\frac{y}{r}\right)^2 = \cos^2\theta + \sin^2\theta = 1.$$

The equation

$$\left(\frac{x}{R}\right)^2 + \left(\frac{y}{r}\right)^2 = 1$$

is the equation of ellipse. Hence, the coordinates of P, $(R\cos\theta, r\sin\theta)$ describe an ellipse for $0 \le \theta \le 2\pi$.

Further Insights and Challenges

77. Derive the formula for the slope of the tangent line to a parametric curve $c(t) = (x(t), y(t))$ using a different method than that presented in the text. Assume that $x'(t_0)$ and $y'(t_0)$ exist and that $x'(t_0) \neq 0$. Show that

$$\lim_{h \to 0} \frac{y(t_0 + h) - y(t_0)}{x(t_0 + h) - x(t_0)} = \frac{y'(t_0)}{x'(t_0)}$$

Then explain why this limit is equal to the slope dy/dx. Draw a diagram showing that the ratio in the limit is the slope of a secant line.

SOLUTION Since $y'(t_0)$ and $x'(t_0)$ exist, we have the following limits:

$$\lim_{h \to 0} \frac{y(t_0 + h) - y(t_0)}{h} = y'(t_0), \quad \lim_{h \to 0} \frac{x(t_0 + h) - x(t_0)}{h} = x'(t_0) \tag{1}$$

We use Basic Limit Laws, the limits in (1) and the given data $x'(t_0) \neq 0$, to write

$$\lim_{h \to 0} \frac{y(t_0 + h) - y(t_0)}{x(t_0 + h) - x(t_0)} = \lim_{h \to 0} \frac{\frac{y(t_0+h)-y(t_0)}{h}}{\frac{x(t_0+h)-x(t_0)}{h}} = \frac{\lim_{h \to 0} \frac{y(t_0+h)-y(t_0)}{h}}{\lim_{h \to 0} \frac{x(t_0+h)-x(t_0)}{h}} = \frac{y'(t_0)}{x'(t_0)}$$

Notice that the quotient $\dfrac{y(t_0 + h) - y(t_0)}{x(t_0 + h) - x(t_0)}$ is the slope of the secant line determined by the points $P = (x(t_0), y(t_0))$ and $Q = (x(t_0 + h), y(t_0 + h))$. Hence, the limit of the quotient as $h \to 0$ is the slope of the tangent line at P, that is the derivative $\dfrac{dy}{dx}$.

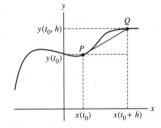

In Exercises 79–82, use Eq. (11) to find $\dfrac{d^2 y}{dx^2}$.

79. $x = t^3 + t^2$, $y = 7t^2 - 4$, $t = 2$

SOLUTION We find the first and second derivatives of $x(t)$ and $y(t)$:

$$x'(t) = 3t^2 + 2t \Rightarrow x'(2) = 3 \cdot 2^2 + 2 \cdot 2 = 16$$
$$x''(t) = 6t + 2 \quad \Rightarrow x''(2) = 6 \cdot 2 + 2 = 14$$
$$y'(t) = 14t \qquad \Rightarrow y'(2) = 14 \cdot 2 = 28$$
$$y''(t) = 14 \qquad \Rightarrow y''(2) = 14$$

Using Eq. (11) we get

$$\left.\frac{d^2 y}{dx^2}\right|_{t=2} = \left.\frac{x'(t)y''(t) - y'(t)x''(t)}{x'(t)^3}\right|_{t=2} = \frac{16 \cdot 14 - 28 \cdot 14}{16^3} = \frac{-21}{512}$$

81. $x = 8t + 9$, $y = 1 - 4t$, $t = -3$

SOLUTION We compute the first and second derivatives of $x(t)$ and $y(t)$:

$$x'(t) = 8 \quad \Rightarrow x'(-3) = 8$$
$$x''(t) = 0 \quad \Rightarrow x''(-3) = 0$$
$$y'(t) = -4 \Rightarrow y'(-3) = -4$$
$$y''(t) = 0 \quad \Rightarrow y''(-3) = 0$$

Using Eq. (11) we get

$$\left.\frac{d^2 y}{dx^2}\right|_{t=-3} = \frac{x'(-3)y''(-3) - y'(-3)x''(-3)}{x'(-3)^3} = \frac{8 \cdot 0 - (-4) \cdot 0}{8^3} = 0$$

83. Use Eq. (11) to find the t-intervals on which $c(t) = (t^2, t^3 - 4t)$ is concave up.

SOLUTION The curve is concave up where $\dfrac{d^2y}{dx^2} > 0$. Thus,

$$\frac{x'(t)y''(t) - y'(t)x''(t)}{x'(t)^3} > 0 \tag{1}$$

We compute the first and second derivatives:

$$x'(t) = 2t, \qquad x''(t) = 2$$
$$y'(t) = 3t^2 - 4, \quad y''(t) = 6t$$

Substituting in (1) and solving for t gives

$$\frac{12t^2 - (6t^2 - 8)}{8t^3} = \frac{6t^2 + 8}{8t^3}$$

Since $6t^2 + 8 > 0$ for all t, the quotient is positive if $8t^3 > 0$. We conclude that the curve is concave up for $t > 0$.

85. Area under a Parametrized Curve Let $c(t) = (x(t), y(t))$ be a parametrized curve such that $x'(t) > 0$ and $y(t) > 0$ (Figure 25). Show that the area A under $c(t)$ for $t_0 \le t \le t_1$ is

$$A = \int_{t_0}^{t_1} y(t)x'(t)\,dt \tag{12}$$

Hint: $x(t)$ is increasing and therefore has an inverse, say, $t = g(x)$. Observe that $c(t)$ is the graph of the function $y(g(x))$ and apply the Change of Variables formula to $A = \displaystyle\int_{x(t_0)}^{x(t_1)} y(g(x))\,dx$.

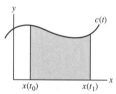

FIGURE 25

SOLUTION Let $x_0 = x(t_0)$ and $x_1 = x(t_1)$. We are given that $x'(t) > 0$, hence $x = x(t)$ is an increasing function of t, so it has an inverse function $t = g(x)$. The area A is given by $\int_{x_0}^{x_1} y(g(x))\,dx$. Recall that y is a function of t and $t = g(x)$, so the height y at any point x is given by $y = y(g(x))$. We find the new limits of integration. Since $x_0 = x(t_0)$ and $x_1 = x(t_1)$, the limits for t are t_0 and t_1, respectively. Also since $x'(t) = \frac{dx}{dt}$, we have $dx = x'(t)dt$. Performing this substitution gives

$$A = \int_{x_0}^{x_1} y(g(x))\,dx = \int_{t_0}^{t_1} y(g(x))x'(t)\,dt.$$

Since $g(x) = t$, we have $A = \displaystyle\int_{t_0}^{t_1} y(t)x'(t)\,dt$.

87. What does Eq. (12) say if $c(t) = (t, f(t))$?

SOLUTION In the parametrization $x(t) = t$, $y(t) = f(t)$ we have $x'(t) = 1$, $t_0 = x(t_0)$, $t_1 = x(t_1)$. Hence Eq. (12) becomes

$$A = \int_{t_0}^{t_1} y(t)x'(t)\,dt = \int_{x(t_0)}^{x(t_1)} f(t)\,dt$$

We see that in this parametrization Eq. (12) is the familiar formula for the area under the graph of a positive function.

89. Use Eq. (12) to show that the area under one arch of the cycloid $c(t)$ (Figure 26) generated by a circle of radius R is equal to three times the area of the circle. Recall that

$$c(t) = (Rt - R\sin t, R - R\cos t)$$

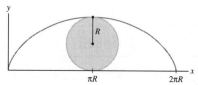

FIGURE 26 The area of the generating circle is one-third the area of one arch of the cycloid.

SOLUTION This reduces to

$$\int_0^{2\pi} (R - R\cos t)(Rt - R\sin t)' \, dt = \int_0^{2\pi} R^2(1 - \cos t)^2 \, dt = 3\pi R^2.$$

In Exercises 90–91, refer to Figure 27.

91. Show that the parametrization of the ellipse by the angle θ is

$$x = \frac{ab\cos\theta}{\sqrt{a^2\sin^2\theta + b^2\cos^2\theta}}$$

$$y = \frac{ab\sin\theta}{\sqrt{a^2\sin^2\theta + b^2\cos^2\theta}}$$

SOLUTION We consider the ellipse

$$\frac{x^2}{a^2} + \frac{y^2}{b^2} = 1.$$

For the angle θ we have $\tan\theta = \frac{y}{x}$, hence,

$$y = x\tan\theta \tag{1}$$

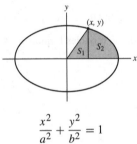

$$\frac{x^2}{a^2} + \frac{y^2}{b^2} = 1$$

Substituting in the equation of the ellipse and solving for x we obtain

$$\frac{x^2}{a^2} + \frac{x^2\tan^2\theta}{b^2} = 1$$

$$b^2 x^2 + a^2 x^2 \tan^2\theta = a^2 b^2$$

$$(a^2\tan^2\theta + b^2)x^2 = a^2 b^2$$

$$x^2 = \frac{a^2 b^2}{a^2\tan^2\theta + b^2} = \frac{a^2 b^2\cos^2\theta}{a^2\sin^2\theta + b^2\cos^2\theta}$$

We now take the square root. Since the sign of the x-coordinate is the same as the sign of $\cos\theta$, we take the positive root, obtaining

$$x = \frac{ab\cos\theta}{\sqrt{a^2\sin^2\theta + b^2\cos^2\theta}} \tag{2}$$

Hence by (1), the y-coordinate is

$$y = x\tan\theta = \frac{ab\cos\theta\tan\theta}{\sqrt{a^2\sin^2\theta + b^2\cos^2\theta}} = \frac{ab\sin\theta}{\sqrt{a^2\sin^2\theta + b^2\cos^2\theta}} \tag{3}$$

Equalities (2) and (3) give the following parametrization for the ellipse:

$$c_1(\theta) = \left(\frac{ab\cos\theta}{\sqrt{a^2\sin^2\theta + b^2\cos^2\theta}}, \frac{ab\sin\theta}{\sqrt{a^2\sin^2\theta + b^2\cos^2\theta}} \right)$$

12.2 Arc Length and Speed

Preliminary Questions

1. What is the definition of arc length?

SOLUTION A curve can be approximated by a polygonal path obtained by connecting points

$$p_0 = c(t_0), \; p_1 = c(t_1), \ldots, \; p_N = c(t_N)$$

on the path with segments. One gets an approximation by summing the lengths of the segments. The definition of arc length is the limit of that approximation when increasing the number of points so that the lengths of the segments approach zero. In doing so, we obtain the following theorem for the arc length:

$$S = \int_a^b \sqrt{x'(t)^2 + y'(t)^2}\, dt,$$

which is the length of the curve $c(t) = (x(t), y(t))$ for $a \le t \le b$.

2. What is the interpretation of $\sqrt{x'(t)^2 + y'(t)^2}$ for a particle following the trajectory $(x(t), y(t))$?

SOLUTION The expression $\sqrt{x'(t)^2 + y'(t)^2}$ denotes the speed at time t of a particle following the trajectory $(x(t), y(t))$.

3. A particle travels along a path from $(0, 0)$ to $(3, 4)$. What is the displacement? Can the distance traveled be determined from the information given?

SOLUTION The net displacement is the distance between the initial point $(0, 0)$ and the endpoint $(3, 4)$. That is

$$\sqrt{(3 - 0)^2 + (4 - 0)^2} = \sqrt{25} = 5.$$

The distance traveled can be determined only if the trajectory $c(t) = (x(t), y(t))$ of the particle is known.

4. A particle traverses the parabola $y = x^2$ with constant speed 3 cm/s. What is the distance traveled during the first minute? *Hint:* No computation is necessary.

SOLUTION Since the speed is constant, the distance traveled is the following product: $L = st = 3 \cdot 60 = 180$ cm.

Exercises

1. Use Eq. (4) to calculate the length of the semicircle

$$x = 3\sin t, \quad y = 3\cos t, \qquad 0 \le t \le \pi$$

SOLUTION We substitute $x' = 3\cos t$, $y' = -3\sin t$, $a = 0$, and $b = \pi$ in the equation and compute the resulting integral. We obtain the following length:

$$S = \int_0^\pi \sqrt{x'(t)^2 + y'(t)^2}\, dt = \int_0^\pi \sqrt{9\cos^2 t + 9\sin^2 t}\, dt = \int_0^\pi \sqrt{9}\, dt = 3\pi.$$

In Exercises 3–12, use Eq. (4) to find the length of the path over the given interval.

3. $(3t + 1, 9 - 4t), \quad 0 \le t \le 2$

SOLUTION Since $x = 3t + 1$ and $y = 9 - 4t$ we have $x' = 3$ and $y' = -4$. Hence, the length of the path is

$$S = \int_0^2 \sqrt{3^2 + (-4)^2}\, dt = 5\int_0^2 dt = 10.$$

5. $(2t^2, 3t^2 - 1), \quad 0 \le t \le 4$

SOLUTION Since $x = 2t^2$ and $y = 3t^2 - 1$, we have $x' = 4t$ and $y' = 6t$. By the formula for the arc length we get

$$S = \int_0^4 \sqrt{x'(t)^2 + y'(t)^2}\, dt = \int_0^4 \sqrt{16t^2 + 36t^2}\, dt = \sqrt{52}\int_0^4 t\, dt = \sqrt{52} \cdot \left.\frac{t^2}{2}\right|_0^4 = 16\sqrt{13}$$

7. $(3t^2, 4t^3), \quad 1 \le t \le 4$

SOLUTION We have $x = 3t^2$ and $y = 4t^3$. Hence $x' = 6t$ and $y' = 12t^2$. By the formula for the arc length we get

$$S = \int_1^4 \sqrt{x'(t)^2 + y'(t)^2}\, dt = \int_1^4 \sqrt{36t^2 + 144t^4}\, dt = 6\int_1^4 \sqrt{1 + 4t^2}\, t\, dt.$$

Using the substitution $u = 1 + 4t^2$, $du = 8t\, dt$ we obtain

$$S = \frac{6}{8}\int_5^{65} \sqrt{u}\, du = \frac{3}{4} \cdot \frac{2}{3}u^{3/2}\Big|_5^{65} = \frac{1}{2}(65^{3/2} - 5^{3/2}) \approx 256.43$$

9. $(\sin 3t, \cos 3t), \quad 0 \leq t \leq \pi$

SOLUTION We have $x = \sin 3t$, $y = \cos 3t$, hence $x' = 3\cos 3t$ and $y' = -3\sin 3t$. By the formula for the arc length we obtain:

$$S = \int_0^\pi \sqrt{x'(t)^2 + y'(t)^2}\, dt = \int_0^\pi \sqrt{9\cos^2 3t + 9\sin^2 3t}\, dt = \int_0^\pi \sqrt{9}\, dt = 3\pi$$

11. $(\sin\theta - \theta\cos\theta, \cos\theta + \theta\sin\theta), \quad 0 \leq \theta \leq 2$

SOLUTION We have $x = \sin\theta - \theta\cos\theta$ and $y = \cos\theta + \theta\sin\theta$. Hence, $x' = \cos\theta - (\cos\theta - \theta\sin\theta) = \theta\sin\theta$ and $y' = -\sin\theta + \sin\theta + \theta\cos\theta = \theta\cos\theta$. Using the formula for the arc length we obtain:

$$S = \int_0^2 \sqrt{x'(\theta)^2 + y'(\theta)^2}\, d\theta = \int_0^2 \sqrt{(\theta\sin\theta)^2 + (\theta\cos\theta)^2}\, d\theta$$

$$= \int_0^2 \sqrt{\theta^2(\sin^2\theta + \cos^2\theta)}\, d\theta = \int_0^2 \theta\, d\theta = \left.\frac{\theta^2}{2}\right|_0^2 = 2$$

13. Show that one arch of a cycloid generated by a circle of radius R has length $8R$.

SOLUTION Recall from earlier that the cycloid generated by a circle of radius R has parametric equations $x = Rt - R\sin t$, $y = R - R\cos t$. Hence, $x' = R - R\cos t$, $y' = R\sin t$. Using the identity $\sin^2 \dfrac{t}{2} = \dfrac{1 - \cos t}{2}$, we get

$$x'(t)^2 + y'(t)^2 = R^2(1 - \cos t)^2 + R^2 \sin^2 t = R^2(1 - 2\cos t + \cos^2 t + \sin^2 t)$$

$$= R^2(1 - 2\cos t + 1) = 2R^2(1 - \cos t) = 4R^2 \sin^2 \frac{t}{2}$$

One arch of the cycloid is traced as t varies from 0 to 2π. Hence, using the formula for the arc length we obtain:

$$S = \int_0^{2\pi} \sqrt{x'(t)^2 + y'(t)^2}\, dt = \int_0^{2\pi} \sqrt{4R^2 \sin^2 \frac{t}{2}}\, dt = 2R\int_0^{2\pi} \sin\frac{t}{2}\, dt = 4R\int_0^\pi \sin u\, du$$

$$= -4R\cos u \Big|_0^\pi = -4R(\cos\pi - \cos 0) = 8R$$

15. Find the length of the spiral $c(t) = (t\cos t, t\sin t)$ for $0 \leq t \leq 2\pi$ to three decimal places (Figure 8). *Hint:* Use the formula

$$\int \sqrt{1 + t^2}\, dt = \frac{1}{2}t\sqrt{1 + t^2} + \frac{1}{2}\ln\left(t + \sqrt{1 + t^2}\right)$$

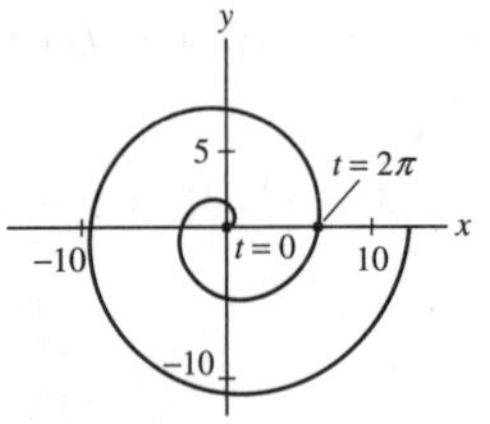

FIGURE 8 The spiral $c(t) = (t\cos t, t\sin t)$.

SOLUTION We use the formula for the arc length:

$$S = \int_0^{2\pi} \sqrt{x'(t)^2 + y'(t)^2}\, dt \tag{1}$$

Differentiating $x = t\cos t$ and $y = t\sin t$ yields

$$x'(t) = \frac{d}{dt}(t\cos t) = \cos t - t\sin t$$

$$y'(t) = \frac{d}{dt}(t\sin t) = \sin t + t\cos t$$

Thus,

$$\sqrt{x'(t)^2 + y'(t)^2} = \sqrt{(\cos t - t\sin t)^2 + (\sin t + t\cos t)^2}$$

$$= \sqrt{\cos^2 t - 2t\cos t\sin t + t^2\sin^2 t + \sin^2 t + 2t\sin t\cos t + t^2\cos^2 t}$$

$$= \sqrt{(\cos^2 t + \sin^2 t)(1 + t^2)} = \sqrt{1 + t^2}$$

We substitute into (1) and use the integral given in the hint to obtain the following arc length:

$$S = \int_0^{2\pi} \sqrt{1 + t^2}\, dt = \frac{1}{2} t\sqrt{1 + t^2} + \frac{1}{2}\ln\left(t + \sqrt{1 + t^2}\right)\Big|_0^{2\pi}$$

$$= \frac{1}{2} \cdot 2\pi\sqrt{1 + (2\pi)^2} + \frac{1}{2}\ln\left(2\pi + \sqrt{1 + (2\pi)^2}\right) - \left(0 + \frac{1}{2}\ln 1\right)$$

$$= \pi\sqrt{1 + 4\pi^2} + \frac{1}{2}\ln\left(2\pi + \sqrt{1 + 4\pi^2}\right) \approx 21.256$$

In Exercises 16–19, determine the speed $s(t)$ of a particle with a given trajectory at time t_0 (in units of meters and seconds).

17. $(3\sin 5t, 8\cos 5t)$, $t = \frac{\pi}{4}$

SOLUTION   We have $x = 3\sin 5t$, $y = 8\cos 5t$, hence $x' = 15\cos 5t$, $y' = -40\sin 5t$. Thus, the speed of the particle at time t is

$$\frac{ds}{dt} = \sqrt{x'(t)^2 + y'(t)^2} = \sqrt{225\cos^2 5t + 1600\sin^2 5t}$$

$$= \sqrt{225(\cos^2 5t + \sin^2 5t) + 1375\sin^2 5t} = 5\sqrt{9 + 55\sin^2 5t}$$

Thus,

$$\frac{ds}{dt} = 5\sqrt{9 + 55\sin^2 5t}.$$

The speed at time $t = \frac{\pi}{4}$ is thus

$$\frac{ds}{dt}\Big|_{t=\pi/4} = 5\sqrt{9 + 55\sin^2\left(5 \cdot \frac{\pi}{4}\right)} \cong 30.21 \text{ m/s}$$

19. $(\ln(t^2 + 1), t^3)$, $t = 1$

SOLUTION   We have $x = \ln(t^2 + 1)$, $y = t^3$, so $x' = \dfrac{2t}{t^2 + 1}$ and $y' = 3t^2$. The speed of the particle at time t is thus

$$\frac{ds}{dt} = \sqrt{x'(t)^2 + y'(t)^2} = \sqrt{\frac{4t^2}{(t^2 + 1)^2} + 9t^4} = t\sqrt{\frac{4}{(t^2 + 1)^2} + 9t^2}.$$

The speed at time $t = 1$ is

$$\frac{ds}{dt}\Big|_{t=1} = \sqrt{\frac{4}{2^2} + 9} = \sqrt{10} \approx 3.16 \text{ m/s}.$$

21. Find the minimum speed of a particle with trajectory $c(t) = (t^3 - 4t, t^2 + 1)$ for $t \geq 0$. *Hint:* It is easier to find the minimum of the square of the speed.

SOLUTION   We first find the speed of the particle. We have $x(t) = t^3 - 4t$, $y(t) = t^2 + 1$, hence $x'(t) = 3t^2 - 4$ and $y'(t) = 2t$. The speed is thus

$$\frac{ds}{dt} = \sqrt{(3t^2 - 4)^2 + (2t)^2} = \sqrt{9t^4 - 24t^2 + 16 + 4t^2} = \sqrt{9t^4 - 20t^2 + 16}.$$

The square root function is an increasing function, hence the minimum speed occurs at the value of t where the function $f(t) = 9t^4 - 20t^2 + 16$ has minimum value. Since $\lim_{t\to\infty} f(t) = \infty$, f has a minimum value on the interval $0 \leq t < \infty$, and it occurs at a critical point or at the endpoint $t = 0$. We find the critical point of f on $t \geq 0$:

$$f'(t) = 36t^3 - 40t = 4t(9t^2 - 10) = 0 \Rightarrow t = 0, t = \sqrt{\frac{10}{9}}.$$

We compute the values of f at these points:

$$f(0) = 9 \cdot 0^4 - 20 \cdot 0^2 + 16 = 16$$

$$f\left(\sqrt{\frac{10}{9}}\right) = 9\left(\sqrt{\frac{10}{9}}\right)^4 - 20\left(\sqrt{\frac{10}{9}}\right)^2 + 16 = \frac{44}{9} \approx 4.89$$

We conclude that the minimum value of f on $t \geq 0$ is 4.89. The minimum speed is therefore

$$\left(\frac{ds}{dt}\right)_{min} \approx \sqrt{4.89} \approx 2.21.$$

23. Find the speed of the cycloid $c(t) = (4t - 4\sin t, 4 - 4\cos t)$ at points where the tangent line is horizontal.

SOLUTION We first find the points where the tangent line is horizontal. The slope of the tangent line is the following quotient:

$$\frac{dy}{dx} = \frac{dy/dt}{dx/dt} = \frac{4\sin t}{4 - 4\cos t} = \frac{\sin t}{1 - \cos t}.$$

To find the points where the tangent line is horizontal we solve the following equation for $t \geq 0$:

$$\frac{dy}{dx} = 0, \quad \frac{\sin t}{1 - \cos t} = 0 \Rightarrow \sin t = 0 \quad \text{and} \quad \cos t \neq 1.$$

Now, $\sin t = 0$ and $t \geq 0$ at the points $t = \pi k$, $k = 0, 1, 2, \ldots$. Since $\cos \pi k = (-1)^k$, the points where $\cos t \neq 1$ are $t = \pi k$ for k odd. The points where the tangent line is horizontal are, therefore:

$$t = \pi(2k - 1), \quad k = 1, 2, 3, \ldots$$

The speed at time t is given by the following expression:

$$\frac{ds}{dt} = \sqrt{x'(t)^2 + y'(t)^2} = \sqrt{(4 - 4\cos t)^2 + (4\sin t)^2}$$

$$= \sqrt{16 - 32\cos t + 16\cos^2 t + 16\sin^2 t} = \sqrt{16 - 32\cos t + 16}$$

$$= \sqrt{32(1 - \cos t)} = \sqrt{32 \cdot 2\sin^2 \frac{t}{2}} = 8\left|\sin \frac{t}{2}\right|$$

That is, the speed of the cycloid at time t is

$$\frac{ds}{dt} = 8\left|\sin \frac{t}{2}\right|.$$

We now substitute

$$t = \pi(2k - 1), \quad k = 1, 2, 3, \ldots$$

to obtain

$$\frac{ds}{dt} = 8\left|\sin \frac{\pi(2k - 1)}{2}\right| = 8|(-1)^{k+1}| = 8$$

CAS In Exercises 25–28, plot the curve and use the Midpoint Rule with $N = 10, 20, 30,$ and 50 to approximate its length.

25. $c(t) = (\cos t, e^{\sin t})$ for $0 \leq t \leq 2\pi$

SOLUTION The curve of $c(t) = (\cos t, e^{\sin t})$ for $0 \leq t \leq 2\pi$ is shown in the figure below:

$c(t) = (\cos t, e^{\sin t}), 0 \leq t \leq 2\pi.$

The length of the curve is given by the following integral:

$$S = \int_0^{2\pi} \sqrt{x'(t)^2 + y'(t)^2} \, dt = \int_0^{2\pi} \sqrt{(-\sin t)^2 + (\cos t \, e^{\sin t})^2} \, dt.$$

That is, $S = \int_0^{2\pi} \sqrt{\sin^2 t + \cos^2 t \, e^{2\sin t}} \, dt$. We approximate the integral using the Mid-Point Rule with $N = 10, 20, 30, 50$. For $f(t) = \sqrt{\sin^2 t + \cos^2 t \, e^{2\sin t}}$ we obtain

$$(N = 10): \quad \Delta x = \frac{2\pi}{10} = \frac{\pi}{5}, \, c_i = \left(i - \frac{1}{2}\right) \cdot \frac{\pi}{5}$$

$$M_{10} = \frac{\pi}{5} \sum_{i=1}^{10} f(c_i) = 6.903734$$

$$(N = 20): \quad \Delta x = \frac{2\pi}{20} = \frac{\pi}{10}, \, c_i = \left(i - \frac{1}{2}\right) \cdot \frac{\pi}{10}$$

$$M_{20} = \frac{\pi}{10} \sum_{i=1}^{20} f(c_i) = 6.915035$$

$$(N = 30): \quad \Delta x = \frac{2\pi}{30} = \frac{\pi}{15}, \, c_i = \left(i - \frac{1}{2}\right) \cdot \frac{\pi}{15}$$

$$M_{30} = \frac{\pi}{15} \sum_{i=1}^{30} f(c_i) = 6.914949$$

$$(N = 50): \quad \Delta x = \frac{2\pi}{50} = \frac{\pi}{25}, \, c_i = \left(i - \frac{1}{2}\right) \cdot \frac{\pi}{25}$$

$$M_{50} = \frac{\pi}{25} \sum_{i=1}^{50} f(c_i) = 6.914951$$

27. The ellipse $\left(\frac{x}{5}\right)^2 + \left(\frac{y}{3}\right)^2 = 1$

SOLUTION We use the parametrization given in Example 4, section 12.1, that is, $c(t) = (5\cos t, 3\sin t), 0 \le t \le 2\pi$. The curve is shown in the figure below:

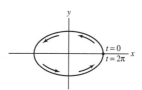

$$c(t) = (5\cos t, 3\sin t), 0 \le t \le 2\pi.$$

The length of the curve is given by the following integral:

$$S = \int_0^{2\pi} \sqrt{x'(t)^2 + y'(t)^2} \, dt = \int_0^{2\pi} \sqrt{(-5\sin t)^2 + (3\cos t)^2} \, dt$$

$$= \int_0^{2\pi} \sqrt{25\sin^2 t + 9\cos^2 t} \, dt = \int_0^{2\pi} \sqrt{9(\sin^2 t + \cos^2 t) + 16\sin^2 t} \, dt = \int_0^{2\pi} \sqrt{9 + 16\sin^2 t} \, dt.$$

That is,

$$S = \int_0^{2\pi} \sqrt{9 + 16\sin^2 t} \, dt.$$

We approximate the integral using the Mid-Point Rule with $N = 10, 20, 30, 50$, for $f(t) = \sqrt{9 + 16\sin^2 t}$. We obtain

$$(N = 10): \quad \Delta x = \frac{2\pi}{10} = \frac{\pi}{5}, \, c_i = \left(i - \frac{1}{2}\right) \cdot \frac{\pi}{5}$$

$$M_{10} = \frac{\pi}{5} \sum_{i=1}^{10} f(c_i) = 25.528309$$

$$(N = 20): \quad \Delta x = \frac{2\pi}{20} = \frac{\pi}{10}, \, c_i = \left(i - \frac{1}{2}\right) \cdot \frac{\pi}{10}$$

$$M_{20} = \frac{\pi}{10} \sum_{i=1}^{20} f(c_i) = 25.526999$$

$$(N = 30): \quad \Delta x = \frac{2\pi}{30} = \frac{\pi}{15}, c_i = \left(i - \frac{1}{2}\right) \cdot \frac{\pi}{15}$$

$$M_{30} = \frac{\pi}{15} \sum_{i=1}^{30} f(c_i) = 25.526999$$

$$(N = 50): \quad \Delta x = \frac{2\pi}{50} = \frac{\pi}{25}, c_i = \left(i - \frac{1}{2}\right) \cdot \frac{\pi}{25}$$

$$M_{50} = \frac{\pi}{25} \sum_{i=1}^{50} f(c_i) = 25.526999$$

29. Let $a > b$ and set $k = \sqrt{1 - \frac{b^2}{a^2}}$. Use a parametric representation to show that the ellipse $\left(\frac{x}{a}\right)^2 + \left(\frac{y}{b}\right)^2 = 1$ has length $L = 4aG\left(\frac{\pi}{2}, k\right)$, where

$$G(\theta, k) = \int_0^\theta \sqrt{1 - k^2 \sin^2 t} \, dt$$

is the *elliptic integral of the second kind.*

SOLUTION Since the ellipse is symmetric with respect to the x and y axis, its length L is four times the length of the part of the ellipse which is in the first quadrant. This part is represented by the following parametrization: $x(t) = a \sin t$, $y(t) = b \cos t$, $0 \le t \le \frac{\pi}{2}$. Using the formula for the arc length we get:

$$L = 4 \int_0^{\pi/2} \sqrt{x'(t)^2 + y'(t)^2} \, dt = 4 \int_0^{\pi/2} \sqrt{(a \cos t)^2 + (-b \sin t)^2} \, dt$$

$$= 4 \int_0^{\pi/2} \sqrt{a^2 \cos^2 t + b^2 \sin^2 t} \, dt$$

We rewrite the integrand as follows:

$$L = 4 \int_0^{\pi/2} \sqrt{a^2 \cos^2 t + a^2 \sin^2 t + (b^2 - a^2) \sin^2 t} \, dt$$

$$= 4 \int_0^{\pi/2} \sqrt{a^2(\cos^2 t + \sin^2 t) + (b^2 - a^2) \sin^2 t} \, dt$$

$$= 4 \int_0^{\pi/2} \sqrt{a^2 + (b^2 - a^2) \sin^2 t} \, dt = 4a \int_0^{\pi/2} \sqrt{\frac{a^2}{a^2} + \frac{b^2 - a^2}{a^2} \sin^2 t} \, dt$$

$$= 4a \int_0^{\pi/2} \sqrt{1 - \left(1 - \frac{b^2}{a^2}\right) \sin^2 t} \, dt = 4a \int_0^{\pi/2} \sqrt{1 - k^2 \sin^2 t} \, dt = 4aG\left(\frac{\pi}{2}, k\right)$$

where $k = \sqrt{1 - \frac{b^2}{a^2}}$.

In Exercises 30–33, use Eq. (5) to compute the surface area of the given surface.

31. The surface generated by revolving the astroid with parametrization $c(t) = (\cos^3 t, \sin^3 t)$ about the x-axis for $0 \le t \le \frac{\pi}{2}$

SOLUTION We have $x(t) = \cos^3 t$, $y(t) = \sin^3 t$, $x'(t) = -3 \cos^2 t \sin t$, $y'(t) = 3 \sin^2 t \cos t$. Hence,

$$x'(t)^2 + y'(t)^2 = 9 \cos^4 t \sin^2 t + 9 \sin^4 t \cos^2 t = 9 \cos^2 t \sin^2 t (\cos^2 t + \sin^2 t) = 9 \cos^2 t \sin^2 t$$

Using the formula for the surface area we get

$$S = 2\pi \int_0^{\pi/2} y(t) \sqrt{x'(t)^2 + y'(t)^2} \, dt = 2\pi \int_0^{\pi/2} \sin^3 t \cdot 3 \cos t \sin t \, dt = 6\pi \int_0^{\pi/2} \sin^4 t \cos t \, dt$$

We compute the integral using the substitution $u = \sin t$ $du = \cos t \, dt$. We obtain

$$S = 6\pi \int_0^1 u^4 \, du = 6\pi \frac{u^5}{5} \Big|_0^1 = \frac{6\pi}{5}.$$

33. The surface generated by revolving one arch of the cycloid $c(t) = (t - \sin t, 1 - \cos t)$ about the x-axis

SOLUTION One arch of the cycloid is traced as t varies from 0 to 2π. Since $x(t) = t - \sin t$ and $y(t) = 1 - \cos t$, we have $x'(t) = 1 - \cos t$ and $y'(t) = \sin t$. Hence, using the identity $1 - \cos t = 2\sin^2 \frac{t}{2}$, we get

$$x'(t)^2 + y'(t)^2 = (1 - \cos t)^2 + \sin^2 t = 1 - 2\cos t + \cos^2 t + \sin^2 t = 2 - 2\cos t = 4\sin^2 \frac{t}{2}$$

By the formula for the surface area we obtain:

$$S = 2\pi \int_0^{2\pi} y(t)\sqrt{x'(t)^2 + y'(t)^2}\, dt = 2\pi \int_0^{2\pi} (1 - \cos t) \cdot 2\sin \frac{t}{2}\, dt$$

$$= 2\pi \int_0^{2\pi} 2\sin^2 \frac{t}{2} \cdot 2\sin \frac{t}{2}\, dt = 8\pi \int_0^{2\pi} \sin^3 \frac{t}{2}\, dt = 16\pi \int_0^{\pi} \sin^3 u\, du$$

We use a reduction formula to compute this integral, obtaining

$$S = 16\pi \left[\frac{1}{3}\cos^3 u - \cos u \right]\Big|_0^{\pi} = 16\pi \left[\frac{4}{3} \right] = \frac{64\pi}{3}$$

Further Insights and Challenges

35. $\boxed{\text{CAS}}$ Let $a \geq b > 0$ and set $k = \dfrac{2\sqrt{ab}}{a - b}$. Show that the **trochoid**

$$x = at - b\sin t, \qquad y = a - b\cos t, \qquad 0 \leq t \leq T$$

has length $2(a - b)G\left(\dfrac{T}{2}, k\right)$ with $G(\theta, k)$ as in Exercise 29.

SOLUTION We have $x'(t) = a - b\cos t$, $y'(t) = b\sin t$. Hence,

$$x'(t)^2 + y'(t)^2 = (a - b\cos t)^2 + (b\sin t)^2 = a^2 - 2ab\cos t + b^2\cos^2 t + b^2\sin^2 t$$

$$= a^2 + b^2 - 2ab\cos t$$

The length of the trochoid for $0 \leq t \leq T$ is

$$L = \int_0^T \sqrt{a^2 + b^2 - 2ab\cos t}\, dt$$

We rewrite the integrand as follows to bring it to the required form. We use the identity $1 - \cos t = 2\sin^2 \frac{t}{2}$ to obtain

$$L = \int_0^T \sqrt{(a - b)^2 + 2ab - 2ab\cos t}\, dt = \int_0^T \sqrt{(a - b)^2 + 2ab(1 - \cos t)}\, dt$$

$$= \int_0^T \sqrt{(a - b)^2 + 4ab\sin^2 \frac{t}{2}}\, dt = \int_0^T \sqrt{(a - b)^2 \left(1 + \frac{4ab}{(a - b)^2}\sin^2 \frac{t}{2}\right)}\, dt$$

$$= (a - b)\int_0^T \sqrt{1 + k^2\sin^2 \frac{t}{2}}\, dt$$

(where $k = \frac{2\sqrt{ab}}{a - b}$).

Substituting $u = \frac{t}{2}$, $du = \frac{1}{2}\, dt$, we get

$$L = 2(a - b)\int_0^{T/2} \sqrt{1 + k^2\sin^2 u}\, du = 2(a - b)E(T/2, k)$$

---

37. The acceleration due to gravity on the surface of the earth is $g = \dfrac{Gm_e}{R_e^2} = 9.8 \text{ m/s}^2$, where $R_e = 6{,}378$ km. Use Exercise 36(b) to show that a satellite orbiting at the earth's surface would have period $T_e = 2\pi\sqrt{R_e/g} \approx 84.5$ min. Then estimate the distance R_m from the moon to the center of the earth. Assume that the period of the moon (sidereal month) is $T_m \approx 27.43$ days.

SOLUTION By part (b) of Exercise 36, it follows that

$$\frac{R_e^3}{T_e^2} = \frac{Gm_e}{4\pi^2} \Rightarrow T_e^2 = \frac{4\pi^2 R_e^3}{Gm_e} = \frac{4\pi^2 R_e}{\frac{Gm_e}{R_e^2}} = \frac{4\pi^2 R_e}{g}$$

Hence,

$$T_e = 2\pi \sqrt{\frac{R_e}{g}} = 2\pi \sqrt{\frac{6378 \cdot 10^3}{9.8}} \approx 5068.8 \text{ s} \approx 84.5 \text{ min.}$$

In part (b) of Exercise 36 we showed that $\dfrac{R^3}{T^2}$ is the same for all orbits. It follows that this quotient is the same for the satellite orbiting at the earth's surface and for the moon orbiting around the earth. Thus,

$$\frac{R_m^3}{T_m^2} = \frac{R_e^3}{T_e^2} \Rightarrow R_m = R_e \left(\frac{T_m}{T_e}\right)^{2/3}.$$

Setting $T_m = 27.43 \cdot 1440 = 39499.2$ minutes, $T_e = 84.5$ minutes, and $R_e = 6378$ km we get

$$R_m = 6378 \left(\frac{39499.2}{84.5}\right)^{2/3} \approx 384154 \text{ km.}$$

12.3 Polar Coordinates

Preliminary Questions

1. If P and Q have the same radial coordinate, then (choose the correct answer):

(a) P and Q lie on the same circle with the center at the origin.

(b) P and Q lie on the same ray based at the origin.

SOLUTION Two points with the same radial coordinate are equidistant from the origin, therefore they lie on the same circle centered at the origin. The angular coordinate defines a ray based at the origin. Therefore, if the two points have the same angular coordinate, they lie on the same ray based at the origin.

2. Give two polar coordinate representations for the point $(x, y) = (0, 1)$, one with negative r and one with positive r.

SOLUTION The point $(0, 1)$ is on the y-axis, distant one unit from the origin, hence the polar representation with positive r is $(r, \theta) = \left(1, \frac{\pi}{2}\right)$. The point $(r, \theta) = \left(-1, \frac{\pi}{2}\right)$ is the reflection of $(r, \theta) = \left(1, \frac{\pi}{2}\right)$ through the origin, hence we must add π to return to the original point.

We obtain the following polar representation of $(0, 1)$ with negative r:

$$(r, \theta) = \left(-1, \frac{\pi}{2} + \pi\right) = \left(-1, \frac{3\pi}{2}\right).$$

3. Does a point (r, θ) have more than one representation in rectangular coordinates?

SOLUTION The rectangular coordinates are determined uniquely by the relations $x = r \cos \theta$, $y = r \sin \theta$. Therefore a point (r, θ) has exactly one representation in rectangular coordinates.

4. Describe the curves with polar equations

(a) $r = 2$ **(b)** $r^2 = 2$ **(c)** $r \cos \theta = 2$

SOLUTION

(a) Converting to rectangular coordinates we get

$$\sqrt{x^2 + y^2} = 2 \quad \text{or} \quad x^2 + y^2 = 2^2.$$

This is the equation of the circle of radius 2 centered at the origin.

(b) We convert to rectangular coordinates, obtaining $x^2 + y^2 = 2$. This is the equation of the circle of radius $\sqrt{2}$, centered at the origin.

(c) We convert to rectangular coordinates. Since $x = r \cos \theta$ we obtain the following equation: $x = 2$. This is the equation of the vertical line through the point $(2, 0)$.

5. If $f(-\theta) = f(\theta)$, then the curve $r = f(\theta)$ is symmetric with respect to the (choose the correct answer):

(a) x-axis **(b)** y-axis **(c)** origin

SOLUTION The equality $f(-\theta) = f(\theta)$ for all θ implies that whenever a point (r, θ) is on the curve, also the point $(r, -\theta)$ is on the curve. Since the point $(r, -\theta)$ is the reflection of (r, θ) with respect to the x-axis, we conclude that the curve is symmetric with respect to the x-axis.

Exercises

1. Find polar coordinates for each of the seven points plotted in Figure 17.

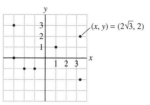

FIGURE 17

SOLUTION   We mark the points as shown in the figure.

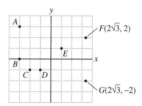

Using the data given in the figure for the x and y coordinates and the quadrants in which the point are located, we obtain:

(A): $\begin{array}{l} r = \sqrt{(-3)^2 + 3^2} = \sqrt{18} \\ \theta = \pi - \frac{\pi}{4} = \frac{3\pi}{4} \end{array} \Rightarrow (r, \theta) = \left(3\sqrt{2}, \frac{3\pi}{4}\right)$

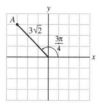

(B): $\begin{array}{l} r = 3 \\ \theta = \pi \end{array} \Rightarrow (r, \theta) = (3, \pi)$

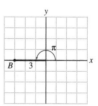

(C): $\begin{array}{l} r = \sqrt{2^2 + 1^2} = \sqrt{5} \approx 2.2 \\ \theta = \tan^{-1}\left(\frac{-1}{-2}\right) = \tan^{-1}\left(\frac{1}{2}\right) = \pi + 0.46 \approx 3.6 \end{array} \Rightarrow (r, \theta) \approx \left(\sqrt{5}, 3.6\right)$

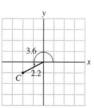

(D): $\begin{array}{l} r = \sqrt{1^2 + 1^2} = \sqrt{2} \approx 1.4 \\ \theta = \pi + \frac{\pi}{4} = \frac{5\pi}{4} \end{array} \Rightarrow (r, \theta) \approx \left(\sqrt{2}, \frac{5\pi}{4}\right)$

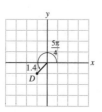

(E): $r = \sqrt{1^2 + 1^2} = \sqrt{2} \approx 1.4$
$\theta = \tan^{-1}\left(\frac{1}{1}\right) = \frac{\pi}{4}$ $\Rightarrow (r, \theta) \approx \left(\sqrt{2}, \frac{\pi}{4}\right)$

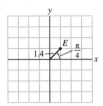

(F): $r = \sqrt{\left(2\sqrt{3}\right)^2 + 2^2} = \sqrt{16} = 4$
$\theta = \tan^{-1}\left(\frac{2}{2\sqrt{3}}\right) = \tan^{-1}\left(\frac{1}{\sqrt{3}}\right) = \frac{\pi}{6}$ $\Rightarrow (r, \theta) = \left(4, \frac{\pi}{6}\right)$

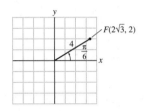

(G): G is the reflection of F about the x axis, hence the two points have equal radial coordinates, and the angular coordinate of G is obtained from the angular coordinate of F: $\theta = 2\pi - \frac{\pi}{6} = \frac{11\pi}{6}$. Hence, the polar coordinates of G are $\left(4, \frac{11\pi}{6}\right)$.

3. Convert from rectangular to polar coordinates:

(a) $(1, 0)$ (b) $(3, \sqrt{3})$

(c) $(-2, 2)$ (d) $(-1, \sqrt{3})$

SOLUTION

(a) The point $(1, 0)$ is on the positive x axis distanced one unit from the origin. Hence, $r = 1$ and $\theta = 0$. Thus, $(r, \theta) = (1, 0)$.

(b) The point $\left(3, \sqrt{3}\right)$ is in the first quadrant so $\theta = \tan^{-1}\left(\frac{\sqrt{3}}{3}\right) = \frac{\pi}{6}$. Also, $r = \sqrt{3^2 + \left(\sqrt{3}\right)^2} = \sqrt{12}$. Hence, $(r, \theta) = \left(\sqrt{12}, \frac{\pi}{6}\right)$.

(c) The point $(-2, 2)$ is in the second quadrant. Hence,

$$\theta = \tan^{-1}\left(\frac{2}{-2}\right) = \tan^{-1}(-1) = \pi - \frac{\pi}{4} = \frac{3\pi}{4}.$$

Also, $r = \sqrt{(-2)^2 + 2^2} = \sqrt{8}$. Hence, $(r, \theta) = \left(\sqrt{8}, \frac{3\pi}{4}\right)$.

(d) The point $\left(-1, \sqrt{3}\right)$ is in the second quadrant, hence,

$$\theta = \tan^{-1}\left(\frac{\sqrt{3}}{-1}\right) = \tan^{-1}\left(-\sqrt{3}\right) = \pi - \frac{\pi}{3} = \frac{2\pi}{3}.$$

Also, $r = \sqrt{(-1)^2 + \left(\sqrt{3}\right)^2} = \sqrt{4} = 2$. Hence, $(r, \theta) = \left(2, \frac{2\pi}{3}\right)$.

5. Convert from polar to rectangular coordinates:

(a) $\left(3, \frac{\pi}{6}\right)$ (b) $\left(6, \frac{3\pi}{4}\right)$ (c) $\left(5, -\frac{\pi}{2}\right)$

SOLUTION

(a) Since $r = 3$ and $\theta = \frac{\pi}{6}$, we have:

$$x = r\cos\theta = 3\cos\frac{\pi}{6} = 3 \cdot \frac{\sqrt{3}}{2} \approx 2.6$$
$$\Rightarrow \quad (x, y) \approx (2.6, 1.5).$$
$$y = r\sin\theta = 3\sin\frac{\pi}{6} = 3 \cdot \frac{1}{2} = 1.5$$

(b) For $\left(6, \frac{3\pi}{4}\right)$ we have $r = 6$ and $\theta = \frac{3\pi}{4}$. Hence,

$$x = r\cos\theta = 6\cos\frac{3\pi}{4} \approx -4.24$$

$$y = r\sin\theta = 6\sin\frac{3\pi}{4} \approx 4.24 \qquad \Rightarrow \qquad (x, y) \approx (-4.24, 4.24).$$

(c) Since $r = 5$ and $\theta = -\frac{\pi}{2}$ we have

$$x = r\cos\theta = 5\cos\left(-\frac{\pi}{2}\right) = 5 \cdot 0 = 0$$

$$y = r\sin\theta = 5\sin\left(-\frac{\pi}{2}\right) = 5 \cdot (-1) = -5 \qquad \Rightarrow \qquad (x, y) = (0, -5)$$

7. Which of the following are possible polar coordinates for the point P with rectangular coordinates $(0, -2)$?

(a) $\left(2, \frac{\pi}{2}\right)$ **(b)** $\left(2, \frac{7\pi}{2}\right)$

(c) $\left(-2, -\frac{3\pi}{2}\right)$ **(d)** $\left(-2, \frac{7\pi}{2}\right)$

(e) $\left(-2, -\frac{\pi}{2}\right)$ **(f)** $\left(2, -\frac{7\pi}{2}\right)$

SOLUTION The point P has distance 2 from the origin and the angle between $\overline{OP}$ and the positive x-axis in the positive direction is $\frac{3\pi}{2}$. Hence, $(r, \theta) = \left(2, \frac{3\pi}{2}\right)$ is one choice for the polar coordinates for P.

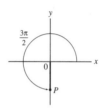

The polar coordinates $(2, \theta)$ are possible for P if $\theta - \frac{3\pi}{2}$ is a multiple of 2π. The polar coordinate $(-2, \theta)$ are possible for P if $\theta - \frac{3\pi}{2}$ is an odd multiple of π. These considerations lead to the following conclusions:

(a) $\left(2, \frac{\pi}{2}\right)$ $\frac{\pi}{2} - \frac{3\pi}{2} = -\pi \Rightarrow \left(2, \frac{\pi}{2}\right)$ does not represent P.

(b) $\left(2, \frac{7\pi}{2}\right)$ $\frac{7\pi}{2} - \frac{3\pi}{2} = 2\pi \Rightarrow \left(2, \frac{7\pi}{2}\right)$ represents P.

(c) $\left(-2, -\frac{3\pi}{2}\right)$ $-\frac{3\pi}{2} - \frac{3\pi}{2} = -3\pi \Rightarrow \left(-2, -\frac{3\pi}{2}\right)$ represents P.

(d) $\left(-2, \frac{7\pi}{2}\right)$ $\frac{7\pi}{2} - \frac{3\pi}{2} = 2\pi \Rightarrow \left(-2, \frac{7\pi}{2}\right)$ does not represent P.

(e) $\left(-2, -\frac{\pi}{2}\right)$ $-\frac{\pi}{2} - \frac{3\pi}{2} = -2\pi \Rightarrow \left(-2, -\frac{\pi}{2}\right)$ does not represent P.

(f) $\left(2, -\frac{7\pi}{2}\right)$ $-\frac{7\pi}{2} - \frac{3\pi}{2} = -5\pi \Rightarrow \left(2, -\frac{7\pi}{2}\right)$ does not represent P.

9. Find the equation in polar coordinates of the line through the origin with slope $\frac{1}{2}$.

SOLUTION A line of slope $m = \frac{1}{2}$ makes an angle $\theta_0 = \tan^{-1}\frac{1}{2} \approx 0.46$ with the positive x-axis. The equation of the line is $\theta \approx 0.46$, while r is arbitrary.

11. Which of the two equations, $r = 2\sec\theta$ and $r = 2\csc\theta$, defines a horizontal line?

SOLUTION The equation $r = 2\csc\theta$ is the polar equation of a horizontal line, as it can be written as $r = 2/\sin\theta$, so $r\sin\theta = 2$, which becomes $y = 2$. On the other hand, the equation $r = 2\sec\theta$ is the polar equation of a vertical line, as it can be written as $r = 2/\cos\theta$, so $r\cos\theta = 2$, which becomes $x = 2$.

In Exercises 12–17, convert to an equation in rectangular coordinates.

13. $r = \sin\theta$

SOLUTION Multiplying by r and substituting $y = r\sin\theta$ and $r^2 = x^2 + y^2$ gives

$$r^2 = r\sin\theta$$

$$x^2 + y^2 = y$$

We move the y and then complete the square to obtain

$$x^2 + y^2 - y = 0$$

$$x^2 + \left(y - \frac{1}{2}\right)^2 = \left(\frac{1}{2}\right)^2$$

Thus, $r = \sin\theta$ is the equation of a circle of radius $\frac{1}{2}$ and center $\left(0, \frac{1}{2}\right)$.

15. $r = 2\csc\theta$

SOLUTION We multiply the equation by $\sin\theta$ and substitute $y = r\sin\theta$. We get

$$r\sin\theta = 2$$

$$y = 2$$

Thus, $r = 2\csc\theta$ is the equation of the line $y = 2$.

17. $r = \dfrac{1}{2 - \cos\theta}$

SOLUTION We multiply the equation by $2 - \cos\theta$. Then we substitute $x = r\cos\theta$ and $r = \sqrt{x^2 + y^2}$, to obtain

$$r\left(2 - \cos\theta\right) = 1$$

$$2r - r\cos\theta = 1$$

$$2\sqrt{x^2 + y^2} - x = 1$$

Moving the x, then squaring and simplifying, we obtain

$$2\sqrt{x^2 + y^2} = x + 1$$

$$4\left(x^2 + y^2\right) = x^2 + 2x + 1$$

$$3x^2 - 2x + 4y^2 = 1$$

We complete the square:

$$3\left(x^2 - \frac{2}{3}x\right) + 4y^2 = 1$$

$$3\left(x - \frac{1}{3}\right)^2 + 4y^2 = \frac{4}{3}$$

$$\frac{\left(x - \frac{1}{3}\right)^2}{\frac{4}{9}} + \frac{y^2}{\frac{1}{3}} = 1$$

This is the equation of the ellipse shown in the figure:

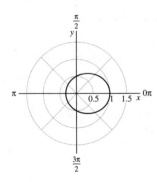

In Exercises 18–21, convert to an equation in polar coordinates.

19. $x = 5$

SOLUTION Substituting $x = r\cos\theta$ gives the polar equation $r\cos\theta = 5$ or $r = 5\sec\theta$.

21. $xy = 1$

SOLUTION We substitute $x = r \cos \theta$, $y = r \sin \theta$ to obtain

$$(r \cos \theta)(r \sin \theta) = 1$$

$$r^2 \cos \theta \sin \theta = 1$$

Using the identity $\cos \theta \sin \theta = \frac{1}{2} \sin 2\theta$ yields

$$r^2 \cdot \frac{\sin 2\theta}{2} = 1 \Rightarrow r^2 = 2 \csc 2\theta.$$

23. Find the values of θ in the plot of $r = 4 \cos \theta$ corresponding to points A, B, C, D in Figure 19. Then indicate the portion of the graph traced out as θ varies in the following intervals:

(a) $0 \le \theta \le \frac{\pi}{2}$ **(b)** $\frac{\pi}{2} \le \theta \le \pi$ **(c)** $\pi \le \theta \le \frac{3\pi}{2}$

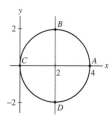

FIGURE 19 Plot of $r = 4 \cos \theta$.

SOLUTION The point A is on the x-axis hence $\theta = 0$. The point B is in the first quadrant with $x = y = 2$ hence $\theta = \tan^{-1}\left(\frac{2}{2}\right) = \tan^{-1}(1) = \frac{\pi}{4}$. The point C is at the origin. Thus,

$$r = 0 \Rightarrow 4 \cos \theta = 0 \Rightarrow \theta = \frac{\pi}{2}, \frac{3\pi}{2}.$$

The point D is in the fourth quadrant with $x = 2$, $y = -2$, hence

$$\theta = \tan^{-1}\left(\frac{-2}{2}\right) = \tan^{-1}(-1) = 2\pi - \frac{\pi}{4} = \frac{7\pi}{4}.$$

$0 \le \theta \le \frac{\pi}{2}$ represents the first quadrant, hence the points (r, θ) where $r = 4 \cos \theta$ and $0 \le \theta \le \frac{\pi}{2}$ are the points on the circle which are in the first quadrant, as shown below:

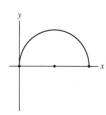

If we insist that $r \ge 0$, then since $\frac{\pi}{2} \le \theta \le \pi$ represents the second quadrant and $\pi \le \theta \le \frac{3\pi}{2}$ represents the third quadrant, and since the circle $r = 4 \cos \theta$ has no points in the left xy-plane, then there are no points for (b) and (c). However, if we allow $r < 0$ then (b) represents the semi-circle

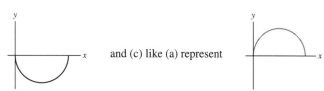

and (c) like (a) represent

25. Match each equation in rectangular coordinates with its equation in polar coordinates.

(a) $x^2 + y^2 = 2$ **(i)** $r^2(1 + 2 \sin^2 \theta) = 4$

(b) $x^2 + (y - 1)^2 = 1$ **(ii)** $r(\cos \theta + \sin \theta) = 4$

(c) $x^2 - y^2 = 4$ **(iii)** $r = 2 \sin \theta$

(d) $x + y = 4$ **(iv)** $r = 2$

SOLUTION

(a) Since $x^2 + y^2 = r^2$, we have $r^2 = 4$ or $r = 2$.

(b) Using Example 7, the equation of the circle $x^2 + (y-1)^2 = 1$ has polar equation $r = 2\sin\theta$.

(c) Setting $x = r\cos\theta$, $y = r\sin\theta$ in $x^2 - y^2 = 4$ gives

$$x^2 - y^2 = r^2\cos^2\theta - r^2\sin^2\theta = r^2\left(\cos^2\theta - \sin^2\theta\right) = 4.$$

We now use the identity $\cos^2\theta - \sin^2\theta = 1 - 2\sin^2\theta$ to obtain the following equation:

$$r^2\left(1 - 2\sin^2\theta\right) = 4.$$

(d) Setting $x = r\cos\theta$ and $y = r\sin\theta$ in $x + y = 4$ we get:

$$x + y = 4$$
$$r\cos\theta + r\sin\theta = 4$$

so

$$r\left(\cos\theta + \sin\theta\right) = 4$$

27. Show that $r = \sin\theta + \cos\theta$ is the equation of the circle of radius $1/\sqrt{2}$ whose center in rectangular coordinates is $(\frac{1}{2}, \frac{1}{2})$. Then find the values of θ between 0 and π such that $(\theta, r(\theta))$ yields the points A, B, C, and D in Figure 20.

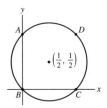

FIGURE 20 Plot of $r = \sin\theta + \cos\theta$.

SOLUTION We show that the rectangular equation of $r = \sin\theta + \cos\theta$ is

$$\left(x - \frac{1}{2}\right)^2 + \left(y - \frac{1}{2}\right)^2 = \frac{1}{2}.$$

We multiply the polar equation by r and substitute $r^2 = x^2 + y^2$, $r\sin\theta = y$, $r\cos\theta = x$. This gives

$$r = \sin\theta + \cos\theta$$
$$r^2 = r\sin\theta + r\cos\theta$$
$$x^2 + y^2 = y + x$$

Transferring sides and completing the square yields

$$x^2 - x + y^2 - y = 0$$
$$\left(x - \frac{1}{2}\right)^2 + \left(y - \frac{1}{2}\right)^2 = \frac{1}{4} + \frac{1}{4} = \frac{1}{2}$$

The point A corresponds to $\theta = \frac{\pi}{2}$. Hence, $r = \sin\frac{\pi}{2} + \cos\frac{\pi}{2} = 1 + 0 = 1$. That is, for the point A we have $(\theta, r) = \left(\frac{\pi}{2}, 1\right)$. The point B corresponds to $r = 0$, that is, $\sin\theta + \cos\theta = 0$. Solving for $0 \le \theta \le \pi$ we get $\sin\theta = -\cos\theta$ or $\tan\theta = -1$, hence $\theta = \frac{3\pi}{4}$. That is, $(\theta, r) = \left(\frac{3\pi}{4}, 0\right)$. The point C corresponds to $\theta = 0$. Hence, $r = \sin 0 + \cos 0 = 1$. That is, $(\theta, r) = (0, 1)$. The point D is on the line $y = x$, hence $\theta = \frac{\pi}{4}$. The corresponding value of r is

$$r = \sin\frac{\pi}{4} + \cos\frac{\pi}{4} = 2\frac{\sqrt{2}}{2} = \sqrt{2}.$$

Thus, $(\theta, r) = \left(\frac{\pi}{4}, \sqrt{2}\right)$.

29. Sketch the graph of $r = 3\cos\theta - 1$ (see Example 8).

SOLUTION We first choose some values of θ between 0 and π and mark the corresponding points on the graph. Then we use symmetry (due to $\cos(2\pi - \theta) = \cos\theta$) to plot the other half of the graph by reflecting the first half through the x-axis. Since $r = 3\cos\theta - 1$ is periodic, the entire curve is obtained as θ varies from 0 to 2π. We start with the values $\theta = 0, \frac{\pi}{6}, \frac{\pi}{3}, \frac{\pi}{2}, \frac{2\pi}{3}, \frac{5\pi}{6}, \pi$, and compute the corresponding values of r:

$$r = 3\cos 0 - 1 = 3 - 1 = 2 \Rightarrow A = (2, 0)$$

$$r = 3\cos\frac{\pi}{6} - 1 = \frac{3\sqrt{3}}{2} - 1 \approx 1.6 \Rightarrow B = \left(1.6, \frac{\pi}{6}\right)$$

$$r = 3\cos\frac{\pi}{3} - 1 = \frac{3}{2} - 1 = 0.5 \Rightarrow C = \left(0.5, \frac{\pi}{3}\right)$$

$$r = 3\cos\frac{\pi}{2} - 1 = 3 \cdot 0 - 1 = -1 \Rightarrow D = \left(-1, \frac{\pi}{2}\right)$$

$$r = 3\cos\frac{2\pi}{3} - 1 = -2.5 \Rightarrow E = \left(-2.5, \frac{2\pi}{3}\right)$$

$$r = 3\cos\frac{5\pi}{6} - 1 = -3.6 \Rightarrow F = \left(-3.6, \frac{5\pi}{6}\right)$$

$$r = 3\cos\pi - 1 = -4 \Rightarrow G = (-4, \pi)$$

The graph begins at the point $(r, \theta) = (2, 0)$ and moves toward the other points in this order, as θ varies from 0 to π. Since r is negative for $\frac{\pi}{2} \leq \theta \leq \pi$, the curve continues into the fourth quadrant, rather than into the second quadrant. We obtain the following graph:

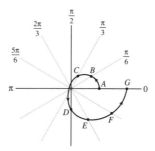

Now we have half the curve and we use symmetry to plot the rest. Reflecting the first half through the x axis we obtain the whole curve:

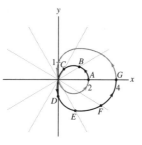

31. Figure 21 displays the graphs of $r = \sin 2\theta$ in rectangular coordinates and in polar coordinates, where it is a "rose with four petals." Identify (a) the points in (B) corresponding to the points labeled A–I in (A), and (b) the parts of the curve in (B) corresponding to the angle intervals $\left[0, \frac{\pi}{2}\right]$, $\left[\frac{\pi}{2}, \pi\right]$, $\left[\pi, \frac{3\pi}{2}\right]$, and $\left[\frac{3\pi}{2}, 2\pi\right]$.

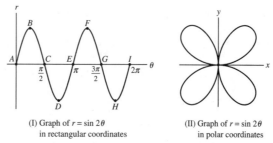

(I) Graph of $r = \sin 2\theta$
in rectangular coordinates

(II) Graph of $r = \sin 2\theta$
in polar coordinates

FIGURE 21 Rose with four petals.

SOLUTION

(a) The graph (I) gives the following polar coordinates of the labeled points:

$$A: \quad \theta = 0, \quad r = 0$$

$$B: \quad \theta = \frac{\pi}{4}, \quad r = \sin\frac{2\pi}{4} = 1$$

$$C: \quad \theta = \frac{\pi}{2}, \quad r = 0$$

$$D: \quad \theta = \frac{3\pi}{4}, \quad r = \sin\frac{2 \cdot 3\pi}{4} = -1$$

$$E: \quad \theta = \pi, \quad r = 0$$

$$F: \quad \theta = \frac{5\pi}{4}, \quad r = 1$$

$$G: \quad \theta = \frac{3\pi}{2}, \quad r = 0$$

$$H: \quad \theta = \frac{7\pi}{4}, \quad r = -1$$

$$I: \quad \theta = 2\pi, \quad r = 0.$$

Since the maximal value of $|r|$ is 1, the points with $r = 1$ or $r = -1$ are the furthest points from the origin. The corresponding quadrant is determined by the value of θ and the sign of r. If $r_0 < 0$, the point (r_0, θ_0) is on the ray $\theta = -\theta_0$. These considerations lead to the following identification of the points in the xy plane. Notice that A, C, G, E, and I are the same point.

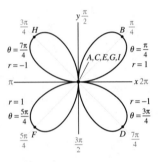

(b) We use the graph (I) to find the sign of $r = \sin 2\theta$: $0 \leq \theta \leq \frac{\pi}{2} \Rightarrow r \geq 0 \Rightarrow (r, \theta)$ is in the first quadrant. $\frac{\pi}{2} \leq \theta \leq \pi \Rightarrow r \leq 0 \Rightarrow (r, \theta)$ is in the fourth quadrant. $\pi \leq \theta \leq \frac{3\pi}{2} \Rightarrow r \geq 0 \Rightarrow (r, \theta)$ is in the third quadrant. $\frac{3\pi}{2} \leq \theta \leq 2\pi \Rightarrow r \leq 0 \Rightarrow (r, \theta)$ is in the second quadrant. That is,

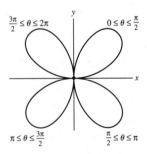

33. $\boxed{CAS}$ Plot the **cissoid** $r = 2\sin\theta\tan\theta$ and show that its equation in rectangular coordinates is $y^2 = \dfrac{x^3}{2 - x}$.

SOLUTION Using a CAS we obtain the following curve of the cissoid:

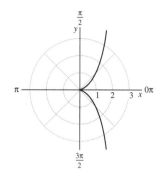

We substitute $\sin \theta = \frac{y}{r}$ and $\tan \theta = \frac{y}{x}$ in $r = 2 \sin \theta \tan \theta$ to obtain

$$r = 2 \frac{y}{r} \cdot \frac{y}{x}.$$

Multiplying by rx, setting $r^2 = x^2 + y^2$ and simplifying, yields

$$r^2 x = 2y^2$$
$$(x^2 + y^2)x = 2y^2$$
$$x^3 + y^2 x = 2y^2$$
$$y^2 (2 - x) = x^3$$

so

$$y^2 = \frac{x^3}{2 - x}$$

35. Show that $r = a \cos \theta + b \sin \theta$ is the equation of a circle passing through the origin. Express the radius and center (in rectangular coordinates) in terms of a and b.

SOLUTION   We multiply the equation by r and then make the substitution $x = r \cos \theta$, $y = r \sin \theta$, and $r^2 = x^2 + y^2$. This gives

$$r^2 = ar \cos \theta + br \sin \theta$$
$$x^2 + y^2 = ax + by$$

Transferring sides and completing the square yields

$$x^2 - ax + y^2 - by = 0$$
$$\left(x^2 - 2 \cdot \frac{a}{2}x + \left(\frac{a}{2}\right)^2\right) + \left(y^2 - 2 \cdot \frac{b}{2}y + \left(\frac{b}{2}\right)^2\right) = \left(\frac{a}{2}\right)^2 + \left(\frac{b}{2}\right)^2$$
$$\left(x - \frac{a}{2}\right)^2 + \left(y - \frac{b}{2}\right)^2 = \frac{a^2 + b^2}{4}$$

This is the equation of the circle with radius $\frac{\sqrt{a^2 + b^2}}{2}$ centered at the point $\left(\frac{a}{2}, \frac{b}{2}\right)$. By plugging in $x = 0$ and $y = 0$ it is clear that the circle passes through the origin.

37. Use the identity $\cos 2\theta = \cos^2 \theta - \sin^2 \theta$ to find a polar equation of the hyperbola $x^2 - y^2 = 1$.

SOLUTION   We substitute $x = r \cos \theta$, $y = r \sin \theta$ in $x^2 - y^2 = 1$ to obtain

$$r^2 \cos^2 \theta - r^2 \sin^2 \theta = 1$$
$$r^2 (\cos^2 \theta - \sin^2 \theta) = 1$$

Using the identity $\cos 2\theta = \cos^2 \theta - \sin^2 \theta$ we obtain the following equation of the hyperbola:

$$r^2 \cos 2\theta = 1 \quad \text{or} \quad r^2 = \sec 2\theta.$$

39. Show that $\cos 3\theta = \cos^3 \theta - 3 \cos \theta \sin^2 \theta$ and use this identity to find an equation in rectangular coordinates for the curve $r = \cos 3\theta$.

SOLUTION We use the identities $\cos(\alpha + \beta) = \cos \alpha \cos \beta - \sin \alpha \sin \beta$, $\cos 2\alpha = \cos^2 \alpha - \sin^2 \alpha$, and $\sin 2\alpha = 2 \sin \alpha \cos \alpha$ to write

$$\cos 3\theta = \cos(2\theta + \theta) = \cos 2\theta \cos \theta - \sin 2\theta \sin \theta$$
$$= (\cos^2 \theta - \sin^2 \theta) \cos \theta - 2 \sin \theta \cos \theta \sin \theta$$
$$= \cos^3 \theta - \sin^2 \theta \cos \theta - 2 \sin^2 \theta \cos \theta$$
$$= \cos^3 \theta - 3 \sin^2 \theta \cos \theta$$

Using this identity we may rewrite the equation $r = \cos 3\theta$ as follows:

$$r = \cos^3 \theta - 3 \sin^2 \theta \cos \theta \qquad (1)$$

Since $x = r \cos \theta$ and $y = r \sin \theta$, we have $\cos \theta = \frac{x}{r}$ and $\sin \theta = \frac{y}{r}$. Substituting into (1) gives:

$$r = \left(\frac{x}{r}\right)^3 - 3\left(\frac{y}{r}\right)^2 \left(\frac{x}{r}\right)$$
$$r = \frac{x^3}{r^3} - \frac{3y^2 x}{r^3}$$

We now multiply by r^3 and make the substitution $r^2 = x^2 + y^2$ to obtain the following equation for the curve:

$$r^4 = x^3 - 3y^2 x$$
$$(x^2 + y^2)^2 = x^3 - 3y^2 x$$

In Exercises 41–45, find an equation in polar coordinates of the line $\mathcal{L}$ with given description.

41. The point on $\mathcal{L}$ closest to the origin has polar coordinates $\left(2, \frac{\pi}{9}\right)$.

SOLUTION In Example 5, it is shown that the polar equation of the line where (r, α) is the point on the line closest to the origin is $r = d \sec(\theta - \alpha)$. Setting $(d, \alpha) = \left(2, \frac{\pi}{9}\right)$ we obtain the following equation of the line:

$$r = 2 \sec\left(\theta - \frac{\pi}{9}\right).$$

43. $\mathcal{L}$ is tangent to the circle $r = 2\sqrt{10}$ at the point with rectangular coordinates $(-2, -6)$.

SOLUTION

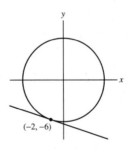

Since $\mathcal{L}$ is tangent to the circle at the point $(-2, -6)$, this is the point on $\mathcal{L}$ closest to the center of the circle which is at the origin. Therefore, we may use the polar coordinates (d, α) of this point in the equation of the line:

$$r = d \sec(\theta - \alpha) \qquad (1)$$

We thus must convert the coordinates $(-2, -6)$ to polar coordinates. This point is in the third quadrant so $\pi < \alpha < \frac{3\pi}{2}$. We get

$$d = \sqrt{(-2)^2 + (-6)^2} = \sqrt{40} = 2\sqrt{10}$$

$$\alpha = \tan^{-1}\left(\frac{-6}{-2}\right) = \tan^{-1} 3 \approx \pi + 1.25 \approx 4.39$$

Substituting in (1) yields the following equation of the line:

$$r = 2\sqrt{10} \sec(\theta - 4.39).$$

45. $y = 4x - 9$.

SOLUTION Substituting $y = r \sin \theta$ and $x = r \cos \theta$ in $y = 4x - 9$, gives

$$r \sin \theta = 4r \cos \theta - 9$$

$$4r \cos \theta - r \sin \theta = 9$$

$$r (4 \cos \theta - \sin \theta) = 9$$

so

$$r = \frac{9}{4 \cos \theta - \sin \theta}$$

47. Distance Formula Use the Law of Cosines (Figure 23) to show that the distance d between two points with polar coordinates (r, θ) and (r_0, θ_0) is

$$d^2 = r^2 + r_0^2 - 2rr_0 \cos(\theta - \theta_0)$$ ⟦2⟧

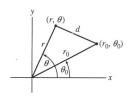

FIGURE 23

SOLUTION Note that the angle between the line segments r and r_0 has measurement $\theta - \theta_0$. Thus, by the Law of Cosines,

$$d^2 = r^2 + r_0^2 - 2rr_0 \cos(\theta - \theta_0)$$

49. Show that the cardiod $r = 1 + \sin \theta$ has equation

$$x^2 + y^2 = (x^2 + y^2 - x)^2$$

SOLUTION We write the equation of the cardioid in the form $1 = r \sin \theta$ and multiply by r. This gives $r = r^2 - r \sin \theta$. We now make the substitution $r = \sqrt{x^2 + y^2}$ and $r \sin \theta = y$ to obtain

$$\sqrt{x^2 + y^2} = x^2 + y^2 - y.$$

Finally, we square both sides to obtain

$$x^2 + y^2 = \left(x^2 + y^2 - y\right)^2.$$

51. The Derivative in Polar Coordinates A polar curve $r = f(\theta)$ has parametric equations (since $x = r \cos \theta$ and $y = r \sin \theta$):

$$x = f(\theta) \cos \theta, \quad y = f(\theta) \sin \theta$$

Apply Theorem 1 of Section 12.1 to prove the formula

$$\frac{dy}{dx} = \frac{f(\theta) \cos \theta + f'(\theta) \sin \theta}{-f(\theta) \sin \theta + f'(\theta) \cos \theta}$$ ⟦3⟧

where $f'(\theta) = df/d\theta$.

SOLUTION By the formula for the derivative we have

$$\frac{dy}{dx} = \frac{y'(\theta)}{x'(\theta)}$$ (1)

We differentiate the functions $x = f(\theta) \cos \theta$ and $y = f(\theta) \sin \theta$ using the Product Rule for differentiation. This gives

$$y'(\theta) = f'(\theta) \sin \theta + f(\theta) \cos \theta$$

$$x'(\theta) = f'(\theta) \cos \theta - f(\theta) \sin \theta$$

Substituting in (1) gives

$$\frac{dy}{dx} = \frac{f'(\theta)\sin\theta + f(\theta)\cos\theta}{f'(\theta)\cos\theta - f(\theta)\sin\theta} = \frac{f(\theta)\cos\theta + f'(\theta)\sin\theta}{-f(\theta)\sin\theta + f'(\theta)\cos\theta}.$$

53. Find the equation in rectangular coordinates of the tangent line to $r = 4\cos 3\theta$ at $\theta = \frac{\pi}{6}$.

SOLUTION   We have $f(\theta) = 4\cos 3\theta$. By Eq. (3),

$$m = \frac{4\cos 3\theta\cos\theta - 12\sin 3\theta\sin\theta}{-4\cos 3\theta\sin\theta - 12\sin 3\theta\cos\theta}.$$

Setting $\theta = \frac{\pi}{6}$ yields

$$m = \frac{4\cos\left(\frac{\pi}{2}\right)\cos\left(\frac{\pi}{6}\right) - 12\sin\left(\frac{\pi}{2}\right)\sin\left(\frac{\pi}{6}\right)}{-4\cos\left(\frac{\pi}{2}\right)\sin\left(\frac{\pi}{6}\right) - 12\sin\left(\frac{\pi}{2}\right)\cos\left(\frac{\pi}{6}\right)} = \frac{-12\sin\frac{\pi}{6}}{-12\cos\frac{\pi}{6}} = \tan\frac{\pi}{6} = \frac{1}{\sqrt{3}}.$$

We identify the point of tangency. For $\theta = \frac{\pi}{6}$ we have $r = 4\cos\frac{3\pi}{6} = 4\cos\frac{\pi}{2} = 0$. The point of tangency is the origin. The tangent line is the line through the origin with slope $\frac{1}{\sqrt{3}}$. This is the line $y = \frac{x}{\sqrt{3}}$.

Further Insights and Challenges

55. ✎ Let c be a fixed constant. Explain the relationship between the graphs of:

(a) $y = f(x + c)$ and $y = f(x)$ (rectangular)
(b) $r = f(\theta + c)$ and $r = f(\theta)$ (polar)
(c) $y = f(x) + c$ and $y = f(x)$ (rectangular)
(d) $r = f(\theta) + c$ and $r = f(\theta)$ (polar)

SOLUTION

(a) For $c > 0$, $y = f(x + c)$ shifts the graph of $y = f(x)$ by c units to the left. If $c < 0$, the result is a shift to the right. It is a horizontal translation.

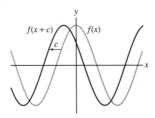

(b) As in part (a), the graph of $r = f(\theta + c)$ is a shift of the graph of $r = f(\theta)$ by c units in θ. Thus, the graph in polar coordinates is rotated by angle c as shown in the following figure:

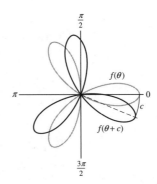

(c) $y = f(x) + c$ shifts the graph vertically upward by c units if $c > 0$, and downward by $(-c)$ units if $c < 0$. It is a vertical translation.

(d) The graph of $r = f(\theta) + c$ is a shift of the graph of $r = f(\theta)$ by c units in r. In the corresponding graph, in polar coordinates, each point with $f(\theta) > 0$ moves on the ray connecting it to the origin c units away from the origin if $c > 0$ and $(-c)$ units toward the origin if $c < 0$, and vice-versa for $f(\theta) < 0$.

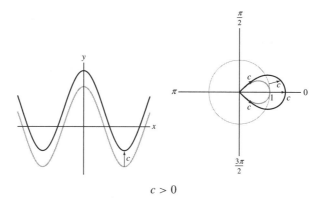

$$c > 0$$

57. GU   Use a graphing utility to convince yourself that graphs of the polar equations $r = f_1(\theta) = 2\cos\theta - 1$ and $r = f_2(\theta) = 2\cos\theta + 1$ have the same graph. Then explain why. *Hint:* Show that the points $(f_1(\theta + \pi), \theta + \pi)$ and $(f_2(\theta), \theta)$ coincide.

SOLUTION   The graphs of $r = 2\cos\theta - 1$ and $r = 2\cos\theta + 1$ in the xy-plane coincide as shown in the graph obtained using a CAS.

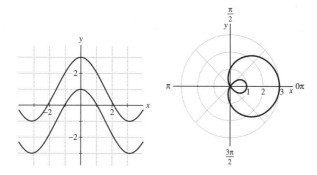

Recall that (r, θ) and $(-r, \theta + \pi)$ represent the same point. Replacing θ by $\theta + \pi$ and r by $(-r)$ in $r = 2\cos\theta - 1$ we obtain

$$-r = 2\cos(\theta + \pi) - 1$$
$$-r = -2\cos\theta - 1$$
$$r = 2\cos\theta + 1$$

Thus, the two equations define the same graph. (One could also convert both equations to rectangular coordinates and note that they come out identical.)

12.4 Area and Arc Length in Polar Coordinates

Preliminary Questions

1. True or False: The area under the curve with polar equation $r = f(\theta)$ is equal to the integral of $f(\theta)$.

SOLUTION   The statement is false. Consider the circle $r = 1$. Its area is π, yet the integral of 1 from 0 to 2π is 2π. Thus, we see that the integral does not give the area under the curve.

2. Polar coordinates are best suited to finding the area (choose one):
(a) Under a curve between $x = a$ and $x = b$.
(b) Bounded by a curve and two rays through the origin.

SOLUTION   Polar coordinates are best suited to finding the area bounded by a curve and two rays through the origin. The formula for the area in polar coordinates gives the area of this region.

3. True or False: The formula for area in polar coordinates is valid only if $f(\theta) \geq 0$.

SOLUTION   The statement is false. The formula for the area

$$\frac{1}{2}\int_\alpha^\beta f(\theta)^2\,d\theta$$

always gives the actual (positive) area, even if $f(\theta)$ takes on negative values.

4. The horizontal line $y = 1$ has polar equation $r = \csc \theta$. Which area is represented by the integral $\dfrac{1}{2} \displaystyle\int_{\pi/6}^{\pi/2} \csc^2 \theta \, d\theta$ (Figure 13)?

(a) $\square ABCD$ **(b)** $\triangle ABC$ **(c)** $\triangle ACD$

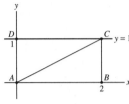

FIGURE 13

SOLUTION This integral represents an area taken from $\theta = \pi/6$ to $\theta = \pi/2$, which can only be the triangle $\triangle ACD$, as seen in part (c).

Exercises

1. Sketch the area bounded by the circle $r = 5$ and the rays $\theta = \frac{\pi}{2}$ and $\theta = \pi$, and compute its area as an integral in polar coordinates.

SOLUTION The region bounded by the circle $r = 5$ and the rays $\theta = \frac{\pi}{2}$ and $\theta = \pi$ is the shaded region in the figure. The area of the region is given by the following integral:

$$\frac{1}{2} \int_{\pi/2}^{\pi} r^2 \, d\theta = \frac{1}{2} \int_{\pi/2}^{\pi} 5^2 \, d\theta = \frac{25}{2} \left(\pi - \frac{\pi}{2} \right) = \frac{25\pi}{4}$$

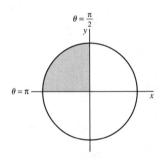

3. Calculate the area of the circle $r = 4 \sin \theta$ as an integral in polar coordinates (see Figure 4). Be careful to choose the correct limits of integration.

SOLUTION The equation $r = 4 \sin \theta$ defines a circle of radius 2 tangent to the x-axis at the origin as shown in the figure:

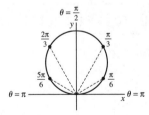

The circle is traced as θ varies from 0 to π. We use the area in polar coordinates and the identity

$$\sin^2 \theta = \frac{1}{2} (1 - \cos 2\theta)$$

to obtain the following area:

$$A = \frac{1}{2} \int_0^{\pi} r^2 \, d\theta = \frac{1}{2} \int_0^{\pi} (4 \sin \theta)^2 \, d\theta = 8 \int_0^{\pi} \sin^2 \theta \, d\theta = 4 \int_0^{\pi} (1 - \cos 2\theta) \, d\theta = 4 \left[\theta - \frac{\sin 2\theta}{2} \right]_0^{\pi}$$

$$= 4 \left(\left(\pi - \frac{\sin 2\pi}{2} \right) - 0 \right) = 4\pi.$$

5. Find the total area enclosed by the cardioid $r = 1 - \cos\theta$ (Figure 15).

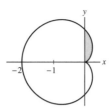

FIGURE 15 The cardioid $r = 1 - \cos\theta$.

SOLUTION We graph $r = 1 - \cos\theta$ in r and θ (cartesian, not polar, this time):

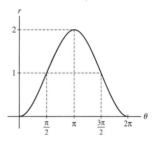

We see that as θ varies from 0 to π, the radius r increases from 0 to 2, so we get the upper half of the cardioid (the lower half is obtained as θ varies from π to 2π and consequently r decreases from 2 to 0). Since the cardioid is symmetric with respect to the x-axis we may compute the upper area and double the result. Using

$$\cos^2\theta = \frac{\cos 2\theta + 1}{2}$$

we get

$$A = 2 \cdot \frac{1}{2} \int_0^\pi r^2\, d\theta = \int_0^\pi (1 - \cos\theta)^2\, d\theta = \int_0^\pi \left(1 - 2\cos\theta + \cos^2\theta\right) d\theta$$

$$= \int_0^\pi \left(1 - 2\cos\theta + \frac{\cos 2\theta + 1}{2}\right) d\theta = \int_0^\pi \left(\frac{3}{2} - 2\cos\theta + \frac{1}{2}\cos 2\theta\right) d\theta$$

$$= \frac{3}{2}\theta - 2\sin\theta + \frac{1}{4}\sin 2\theta \Big|_0^\pi = \frac{3\pi}{2}$$

The total area enclosed by the cardioid is $A = \frac{3\pi}{2}$.

7. Find the area of one leaf of the "four-petaled rose" $r = \sin 2\theta$ (Figure 16).

SOLUTION We consider the graph of $r = \sin 2\theta$ in cartesian and in polar coordinates:

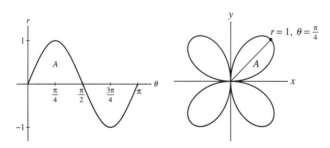

We see that as θ varies from 0 to $\frac{\pi}{4}$ the radius r is increasing from 0 to 1, and when θ varies from $\frac{\pi}{4}$ to $\frac{\pi}{2}$, r is decreasing back to zero. Hence, the leaf in the first quadrant is traced as θ varies from 0 to $\frac{\pi}{2}$. The area of the leaf (the four leaves have equal areas) is thus

$$A = \frac{1}{2}\int_0^{\pi/2} r^2\, d\theta = \frac{1}{2}\int_0^{\pi/2} \sin^2 2\theta\, d\theta.$$

Using the identity

$$\sin^2 2\theta = \frac{1 - \cos 4\theta}{2}$$

we get

$$A = \frac{1}{2} \int_0^{\pi/2} \left(\frac{1}{2} - \frac{\cos 4\theta}{2} \right) d\theta = \frac{1}{2} \left(\frac{\theta}{2} - \frac{\sin 4\theta}{8} \right) \Big|_0^{\pi/2} = \frac{1}{2} \left(\left(\frac{\pi}{4} - \frac{\sin 2\pi}{8} \right) - 0 \right) = \frac{\pi}{8}$$

The area of one leaf is $A = \frac{\pi}{8} \approx 0.39$.

9. Find the area enclosed by one loop of the lemniscate with equation $r^2 = \cos 2\theta$ (Figure 17). Choose your limits of integration carefully.

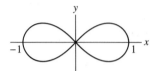

FIGURE 17 The lemniscate $r^2 = \cos 2\theta$.

SOLUTION   We sketch the graph of $r^2 = \cos 2\theta$ in the $\left(r^2, \theta \right)$ plane; for $-\frac{\pi}{4} \le \theta \le \frac{\pi}{4}$:

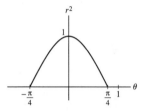

We see that as θ varies from $-\frac{\pi}{4}$ to 0, r^2 increases from 0 to 1, hence r also increases from 0 to 1. Then, as θ varies from 0 to $\frac{\pi}{4}$, r^2, so r decreases from 1 to 0. This gives the right-hand loop of the lemniscate.

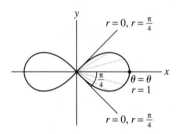

We compute the area enclosed by the right-hand loop, by the following integral:

$$\frac{1}{2} \int_{-\pi/4}^{\pi/4} r^2 \, d\theta = \frac{1}{2} \int_{-\pi/4}^{\pi/4} \cos 2\theta \, d\theta = \frac{1}{2} \frac{\sin 2\theta}{2} \Big|_{-\pi/4}^{\pi/4} = \frac{1}{4} \left(\sin \frac{\pi}{2} - \sin \left(-\frac{\pi}{2} \right) \right) = \frac{1}{2}$$

Since the lemniscate is symmetric with respect to the y-axis, the total area A enclosed by the lemniscate is twice the area enclosed by the right-hand loop. Thus,

$$A = 2 \cdot \frac{1}{2} = 1.$$

11. Find the area enclosed by the cardioid $r = a(1 + \cos \theta)$, where $a > 0$.

SOLUTION   The graph of $r = a(1 + \cos \theta)$ in the $r\theta$-plane for $0 \le \theta \le 2\pi$ and the cardioid in the xy-plane are shown in the following figures:

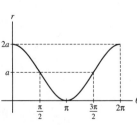

$r = a(1 + \cos \theta)$ The cardioid $r = a(1 + \cos \theta)$, $a > 0$

As θ varies from 0 to π the radius r decreases from $2a$ to 0, and this gives the upper part of the cardioid.

The lower part is traced as θ varies from π to 2π and consequently r increases from 0 back to $2a$. We compute the area enclosed by the upper part of the cardioid and the x-axis, using the following integral (we use the identity $\cos^2\theta = \frac{1}{2} + \frac{1}{2}\cos 2\theta$):

$$\frac{1}{2}\int_0^\pi r^2\,d\theta = \frac{1}{2}\int_0^\pi a^2(1+\cos\theta)^2\,d\theta = \frac{a^2}{2}\int_0^\pi \left(1+2\cos\theta+\cos^2\theta\right)\,d\theta$$

$$= \frac{a^2}{2}\int_0^\pi \left(1+2\cos\theta+\frac{1}{2}+\frac{1}{2}\cos 2\theta\right)\,d\theta = \frac{a^2}{2}\int_0^\pi \left(\frac{3}{2}+2\cos\theta+\frac{1}{2}\cos 2\theta\right)\,d\theta$$

$$= \frac{a^2}{2}\left[\frac{3\theta}{2}+2\sin\theta+\frac{1}{4}\sin 2\theta\right]\Bigg|_0^\pi = \frac{a^2}{2}\left[\frac{3\pi}{2}+2\sin\pi+\frac{1}{4}\sin 2\pi-0\right] = \frac{3\pi a^2}{4}$$

Using symmetry, the total area A enclosed by the cardioid is

$$A = 2\cdot\frac{3\pi a^2}{4} = \frac{3\pi a^2}{2}$$

13. Find the area of region A in Figure 18.

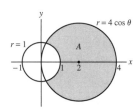

FIGURE 18

SOLUTION We first find the values of θ at the points of intersection of the two circles, by solving the following equation for $-\frac{\pi}{2}\le x\le\frac{\pi}{2}$:

$$4\cos\theta = 1 \Rightarrow \cos\theta = \frac{1}{4} \Rightarrow \theta_1 = \cos^{-1}\left(\frac{1}{4}\right)$$

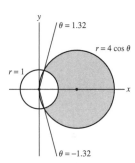

We now compute the area using the formula for the area between two curves:

$$A = \frac{1}{2}\int_{-\theta_1}^{\theta_1}\left((4\cos\theta)^2-1^2\right)\,d\theta = \frac{1}{2}\int_{-\theta_1}^{\theta_1}\left(16\cos^2\theta-1\right)\,d\theta$$

Using the identity $\cos^2\theta = \frac{\cos 2\theta+1}{2}$ we get

$$A = \frac{1}{2}\int_{-\theta_1}^{\theta_1}\left(\frac{16(\cos 2\theta+1)}{2}-1\right)\,d\theta = \frac{1}{2}\int_{-\theta_1}^{\theta_1}(8\cos 2\theta+7)\,d\theta = \frac{1}{2}(4\sin 2\theta+7\theta)\Bigg|_{-\theta_1}^{\theta_1}$$

$$= 4\sin 2\theta_1+7\theta_1 = 8\sin\theta_1\cos\theta_1+7\theta_1 = 8\sqrt{1-\cos^2\theta_1}\cos\theta_1+7\theta_1$$

Using the fact that $\cos\theta_1 = \frac{1}{4}$ we get

$$A = \frac{\sqrt{15}}{2}+7\cos^{-1}\left(\frac{1}{4}\right)\approx 11.163$$

15. Find the area of the inner loop of the limaçon with polar equation $r = 2\cos\theta - 1$ (Figure 20).

SOLUTION We consider the graph of $r = 2\cos\theta - 1$ in cartesian and in polar, for $-\frac{\pi}{2}\le x\le\frac{\pi}{2}$:

$$r = 2\cos\theta - 1$$

As θ varies from $-\frac{\pi}{3}$ to 0, r increases from 0 to 1. As θ varies from 0 to $\frac{\pi}{3}$, r decreases from 1 back to 0. Hence, the inner loop of the limaçon is traced as θ varies from $-\frac{\pi}{3}$ to $\frac{\pi}{3}$. The area of the shaded region is thus

$$A = \frac{1}{2}\int_{-\pi/3}^{\pi/3} r^2\, d\theta = \frac{1}{2}\int_{-\pi/3}^{\pi/3}(2\cos\theta - 1)^2\, d\theta = \frac{1}{2}\int_{-\pi/3}^{\pi/3}\left(4\cos^2\theta - 4\cos\theta + 1\right) d\theta$$

$$= \frac{1}{2}\int_{-\pi/3}^{\pi/3}(2(\cos 2\theta + 1) - 4\cos\theta + 1)\, d\theta = \frac{1}{2}\int_{-\pi/3}^{\pi/3}(2\cos 2\theta - 4\cos\theta + 3)\, d\theta$$

$$= \frac{1}{2}(\sin 2\theta - 4\sin\theta + 3\theta)\Big|_{-\pi/3}^{\pi/3} = \frac{1}{2}\left(\left(\sin\frac{2\pi}{3} - 4\sin\frac{\pi}{3} + \pi\right) - \left(\sin\left(-\frac{2\pi}{3}\right) - 4\sin\left(-\frac{\pi}{3}\right) - \pi\right)\right)$$

$$= \frac{\sqrt{3}}{2} - \frac{4\sqrt{3}}{2} + \pi = \pi - \frac{3\sqrt{3}}{2} \approx 0.54$$

17. Find the area of the part of the circle $r = \sin\theta + \cos\theta$ in the fourth quadrant (see Exercise 27 in Section 12.3).

SOLUTION   The value of θ corresponding to the point B is the solution of $r = \sin\theta + \cos\theta = 0$ for $-\pi \le \theta \le \pi$.

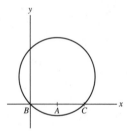

That is,

$$\sin\theta + \cos\theta = 0 \Rightarrow \sin\theta = -\cos\theta \Rightarrow \tan\theta = -1 \Rightarrow \theta = -\frac{\pi}{4}$$

At the point C, we have $\theta = 0$. The part of the circle in the fourth quadrant is traced if θ varies between $-\frac{\pi}{4}$ and 0. This leads to the following area:

$$A = \frac{1}{2}\int_{-\pi/4}^{0} r^2\, d\theta = \frac{1}{2}\int_{-\pi/4}^{0}(\sin\theta + \cos\theta)^2\, d\theta = \frac{1}{2}\int_{-\pi/4}^{0}\left(\sin^2\theta + 2\sin\theta\cos\theta + \cos^2\theta\right) d\theta$$

Using the identities $\sin^2\theta + \cos^2\theta = 1$ and $2\sin\theta\cos\theta = \sin 2\theta$ we get:

$$A = \frac{1}{2}\int_{-\pi/4}^{0}(1 + \sin 2\theta)\, d\theta = \frac{1}{2}\left(\theta - \frac{\cos 2\theta}{2}\right)\Big|_{-\pi/4}^{0}$$

$$= \frac{1}{2}\left(\left(0 - \frac{1}{2}\right) - \left(-\frac{\pi}{4} - \frac{\cos\left(\frac{-\pi}{2}\right)}{2}\right)\right) = \frac{1}{2}\left(\frac{\pi}{4} - \frac{1}{2}\right) = \frac{\pi}{8} - \frac{1}{4} \approx 0.14.$$

19. Find the area between the two curves in Figure 22(A).

SOLUTION   We compute the area A between the two curves as the difference between the area A_1 of the region enclosed in the outer curve $r = 2 + \cos 2\theta$ and the area A_2 of the region enclosed in the inner curve $r = \sin 2\theta$. That is,

$$A = A_1 - A_2.$$

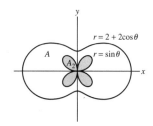

In Exercise 8 we showed that $A_2 = \frac{\pi}{2}$, hence,

$$A = A_1 - \frac{\pi}{2} \tag{1}$$

We compute the area A_1.

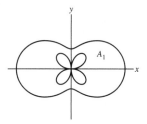

Using symmetry, the area is four times the area enclosed in the first quadrant. That is,

$$A_1 = 4 \cdot \frac{1}{2} \int_0^{\pi/2} r^2 \, d\theta = 2 \int_0^{\pi/2} (2 + \cos 2\theta)^2 \, d\theta = 2 \int_0^{\pi/2} \left(4 + 4\cos 2\theta + \cos^2 2\theta\right) d\theta$$

Using the identity $\cos^2 2\theta = \frac{1}{2} \cos 4\theta + \frac{1}{2}$ we get

$$A_1 = 2 \int_0^{\pi/2} \left(4 + 4\cos 2\theta + \frac{1}{2}\cos 4\theta + \frac{1}{2}\right) d\theta = 2 \int_0^{\pi/2} \left(\frac{9}{2} + \frac{1}{2}\cos 4\theta + 4\cos 2\theta\right) d\theta$$

$$= 2\left(\frac{9\theta}{2} + \frac{\sin 4\theta}{8} + 2\sin 2\theta\right)\Bigg|_0^{\pi/2} = 2\left(\left(\frac{9\pi}{4} + \frac{\sin 2\pi}{8} + 2\sin \pi\right) - 0\right) = \frac{9\pi}{2} \tag{2}$$

Combining (1) and (2) we obtain

$$A = \frac{9\pi}{2} - \frac{\pi}{2} = 4\pi.$$

21. Find the area inside both curves in Figure 23.

SOLUTION The area we need to find is the area of the shaded region in the figure.

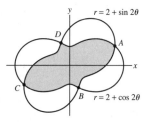

We first find the values of θ at the points of intersection A, B, C, and D of the two curves, by solving the following equation for $-\pi \leq \theta \leq \pi$:

$$2 + \cos 2\theta = 2 + \sin 2\theta$$

$$\cos 2\theta = \sin 2\theta$$

$$\tan 2\theta = 1 \Rightarrow 2\theta = \frac{\pi}{4} + \pi k \Rightarrow \theta = \frac{\pi}{8} + \frac{\pi k}{2}$$

The solutions for $-\pi \leq \theta \leq \pi$ are

$$A: \quad \theta = \frac{\pi}{8}.$$

$$B: \quad \theta = -\frac{3\pi}{8}.$$

$$C: \quad \theta = -\frac{7\pi}{8}.$$

$$D: \quad \theta = \frac{5\pi}{8}.$$

Using symmetry, we compute the shaded area in the figure below and multiply it by 4:

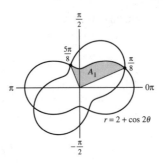

$$A = 4 \cdot A_1 = 4 \cdot \frac{1}{2} \cdot \int_{\pi/8}^{5\pi/8} (2 + \cos 2\theta)^2 \, d\theta = 2 \int_{\pi/8}^{5\pi/8} \left(4 + 4\cos 2\theta + \cos^2 2\theta \right) d\theta$$

$$= 2 \int_{\pi/8}^{5\pi/8} \left(4 + 4\cos 2\theta + \frac{1 + \cos 4\theta}{2} \right) d\theta = \int_{\pi/8}^{5\pi/8} (9 + 8\cos 2\theta + \cos 4\theta) \, d\theta$$

$$= 9\theta + 4\sin 2\theta + \frac{\sin 4\theta}{4} \Big|_{\pi/8}^{5\pi/8} = 9\left(\frac{5\pi}{8} - \frac{\pi}{8} \right) + 4\left(\sin \frac{5\pi}{4} - \sin \frac{\pi}{4} \right) + \frac{1}{4}\left(\sin \frac{5\pi}{2} - \sin \frac{\pi}{2} \right) = \frac{9\pi}{2} - 4\sqrt{2}$$

23. Figure 24 suggests that the circle $r = \sin\theta$ lies inside the spiral $r = \theta$. Which inequality from Chapter 2 assures us that this is the case? Find the area between the curves $r = \theta$ and $r = \sin\theta$ in the first quadrant.

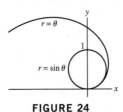

FIGURE 24

SOLUTION   The inequality $|\sin\theta| \le |\theta|$ assures us that for all θ, the corresponding point on the spiral has radial co-ordinate with absolute value greater than that of the corresponding point on the circle. Hence, the circle lies inside the spiral.

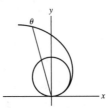

The area between the circle and the spiral in the first quadrant is traced as θ varies from 0 to $\frac{\pi}{2}$.

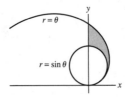

Using the formula for the area between two curves and the identity $\sin^2\theta = \frac{1}{2} - \frac{1}{2}\cos 2\theta$ we obtain

$$A = \frac{1}{2} \int_0^{\pi/2} \left(\theta^2 - \sin^2\theta \right) d\theta = \frac{1}{2} \int_0^{\pi/2} \left(\theta^2 - \frac{1}{2} + \frac{1}{2}\cos 2\theta \right) d\theta$$

$$= \frac{1}{2} \left(\frac{\theta^3}{3} - \frac{\theta}{2} + \frac{1}{4} \sin 2\theta \right) \Big|_0^{\pi/2} = \frac{1}{2} \left(\frac{\pi^3}{24} - \frac{\pi}{4} + \frac{\sin \pi}{4} - 0 \right) = \frac{\pi^3}{48} - \frac{\pi}{8} \approx 0.25$$

25. Find the length of the spiral $r = \theta$ for $0 \le \theta \le A$.

SOLUTION We use the formula for the arc length. In this case $f(\theta) = \theta$, $f'(\theta) = 1$. Using integration formulas we get:

$$S = \int_0^A \sqrt{\theta^2 + 1^2} \, d\theta = \int_0^A \sqrt{\theta^2 + 1} \, d\theta = \frac{\theta}{2} \sqrt{\theta^2 + 1} + \frac{1}{2} \ln |\theta + \sqrt{\theta^2 + 1}| \, \Big|_0^A$$

$$= \frac{A}{2} \sqrt{A^2 + 1} + \frac{1}{2} \ln |A + \sqrt{A^2 + 1}|$$

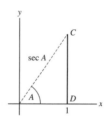

The spiral $r = \theta$

27. Sketch the segment $r = \sec \theta$ for $0 \le \theta \le A$. Then compute its length in two ways: as an integral in polar coordinates and using trigonometry.

SOLUTION The line $r = \sec \theta$ has the rectangular equation $x = 1$. The segment AB for $0 \le \theta \le A$ is shown in the figure.

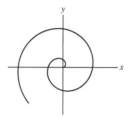

Using trigonometry, the length of the segment $\overline{AB}$ is

$$L = \overline{AB} = \overline{OB} \tan A = 1 \cdot \tan A = \tan A$$

Alternatively, we use the integral in polar coordinates with $f(\theta) = \sec(\theta)$ and $f'(\theta) = \tan \theta \sec \theta$. This gives

$$L = \int_0^A \sqrt{(\sec \theta)^2 + (\tan \theta \sec \theta)^2} \, d\theta = \int_0^A \sqrt{1 + \tan^2 \theta} \, \sec \theta \, d\theta = \int_0^A \sec^2 \theta \, d\theta = \tan \theta \Big|_0^A = \tan A.$$

The two answers agree, as expected.

29. Find the length of the cardioid $r = 1 + \cos \theta$. *Hint:* Use the identity $1 + \cos \theta = 2 \cos^2 \left(\frac{\theta}{2} \right)$ to evaluate the arc length integral.

SOLUTION In the equation of the cardioid, $f(\theta) = 1 + \cos \theta$. Using the formula for arc length in polar coordinates we have:

$$L = \int_\alpha^\beta \sqrt{f(\theta)^2 + f'(\theta)^2} \, d\theta \tag{1}$$

We compute the integrand:

$$\sqrt{f(\theta)^2 + f'(\theta)^2} = \sqrt{(1 + \cos \theta)^2 + (-\sin \theta)^2} = \sqrt{1 + 2 \cos \theta + \cos^2 \theta + \sin^2 \theta} = \sqrt{2(1 + \cos \theta)}$$

We identify the interval of θ. Since $-1 \le \cos \theta \le 1$, every $0 \le \theta \le 2\pi$ corresponds to nonnegative value of r. Hence, θ varies from 0 to 2π. By (1) we obtain

$$L = \int_0^{2\pi} \sqrt{2(1 + \cos \theta)} \, d\theta$$

Since $1 + \cos\theta = 2\cos^2(\theta/2)$, and by the symmetry of the graph,

$$L = 2\int_0^\pi \sqrt{2(1 + \cos\theta)}\, d\theta = 2\int_0^\pi 2\cos(\theta/2)\, d\theta = 8\sin\frac{\theta}{2}\bigg|_0^\pi = 8$$

31. Find the length of the *equiangular spiral* $r = e^\theta$ for $0 \le \theta \le 2\pi$.

SOLUTION   Since $f(\theta) = e^\theta$, by the formula for the arc length we have:

$$L = \int_0^{2\pi}\sqrt{f'(\theta)^2 + f(\theta)^2}\, d\theta + \int_0^{2\pi}\sqrt{\left(e^\theta\right)^2 + \left(e^\theta\right)^2}\, d\theta = \int_0^{2\pi}\sqrt{2e^{2\theta}}\, d\theta$$

$$= \sqrt{2}\int_0^{2\pi} e^\theta\, d\theta = \sqrt{2}e^\theta\bigg|_0^{2\pi} = \sqrt{2}\left(e^{2\pi} - e^0\right) = \sqrt{2}\left(e^{2\pi} - 1\right) \approx 755.9$$

In Exercises 33–36, express the length of the curve as an integral but do not evaluate it.

33. $r = e^{a\theta}, \quad 0 \le \theta \le \pi$

SOLUTION   We use the formula for the arc length in polar coordinates. For this curve $f(\theta) = e^{a\theta}$, $f'(\theta) = ae^{a\theta}$, hence,

$$L = \int_0^\pi \sqrt{f(\theta)^2 + f'(\theta)^2}\, d\theta = \int_0^\pi \sqrt{e^{2a\theta} + a^2 e^{2a\theta}}\, d\theta = \int_0^\pi \sqrt{1 + a^2}\, e^{a\theta}\, d\theta$$

35. $r = (2 - \cos\theta)^{-1}, \quad 0 \le \theta \le 2\pi$

SOLUTION   We have $f(\theta) = (2 - \cos\theta)^{-1}$, $f'(\theta) = -(2 - \cos\theta)^{-2}\sin\theta$, hence,

$$\sqrt{f^2(\theta) + f'(\theta)^2} = \sqrt{(2 - \cos\theta)^{-2} + (2 - \cos\theta)^{-4}\sin^2\theta} = \sqrt{(2 - \cos\theta)^{-4}\left((2 - \cos\theta)^2 + \sin^2\theta\right)}$$

$$= (2 - \cos\theta)^{-2}\sqrt{4 - 4\cos\theta + \cos^2\theta + \sin^2\theta} = (2 - \cos\theta)^{-2}\sqrt{5 - 4\cos\theta}$$

Using the integral for the arc length we get

$$L = \int_0^{2\pi} \sqrt{5 - 4\cos\theta}(2 - \cos\theta)^{-2}\, d\theta.$$

Further Insights and Challenges

37. Suppose that the polar coordinates of a moving particle at time t are $(r(t), \theta(t))$. Prove that the particle's speed is equal to $\sqrt{(dr/dt)^2 + r^2(d\theta/dt)^2}$.

SOLUTION   The speed of the particle in rectangular coordinates is:

$$\frac{ds}{dt} = \sqrt{x'(t)^2 + y'(t)^2} \tag{1}$$

We need to express the speed in polar coordinates. The x and y coordinates of the moving particles as functions of t are

$$x(t) = r(t)\cos\theta(t), \quad y(t) = r(t)\sin\theta(t)$$

We differentiate $x(t)$ and $y(t)$, using the Product Rule for differentiation. We obtain (omitting the independent variable t)

$$x' = r'\cos\theta - r(\sin\theta)\theta'$$
$$y' = r'\sin\theta - r(\cos\theta)\theta'$$

Hence,

$$x'^2 + y'^2 = \left(r'\cos\theta - r\theta'\sin\theta\right)^2 + \left(r'\sin\theta + r\theta'\cos\theta\right)^2$$

$$= r'^2\cos^2\theta - 2r'r\theta'\cos\theta\sin\theta + r^2\theta'^2\sin^2\theta + r'^2\sin^2\theta + 2r'r\theta'\sin^2\theta\cos\theta + r^2\theta'^2\cos^2\theta$$

$$= r'^2\left(\cos^2\theta + \sin^2\theta\right) + r^2\theta'^2\left(\sin^2\theta + \cos^2\theta\right) = r'^2 + r^2\theta'^2 \tag{2}$$

Substituting (2) into (1) we get

$$\frac{ds}{dt} = \sqrt{r'^2 + r^2\theta'^2} = \sqrt{\left(\frac{dr}{dt}\right)^2 + r^2\left(\frac{d\theta}{dt}\right)^2}$$

12.5 Conic Sections

Preliminary Questions

1. Which of the following equations defines an ellipse? Which does not define a conic section?

(a) $4x^2 - 9y^2 = 12$

(b) $-4x + 9y^2 = 0$

(c) $4y^2 + 9x^2 = 12$

(d) $4x^3 + 9y^3 = 12$

SOLUTION

(a) This is the equation of the hyperbola $\left(\frac{x}{\sqrt{3}}\right)^2 - \left(\frac{y}{\frac{2}{\sqrt{3}}}\right)^2 = 1$, which is a conic section.

(b) The equation $-4x + 9y^2 = 0$ can be rewritten as $x = \frac{9}{4}y^2$, which defines a parabola. This is a conic section.

(c) The equation $4y^2 + 9x^2 = 12$ can be rewritten in the form $\left(\frac{y}{\sqrt{3}}\right)^2 + \left(\frac{x}{\frac{2}{\sqrt{3}}}\right)^2 = 1$, hence it is the equation of an ellipse, which is a conic section.

(d) This is not the equation of a conic section, since it is not an equation of degree two in x and y.

2. For which conic sections do the vertices lie between the foci?

SOLUTION If the vertices lie between the foci, the conic section is a hyperbola.

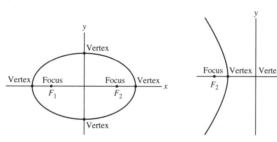

ellipse: foci between vertices   hyperbola: vertices between foci

3. What are the foci of $\left(\frac{x}{a}\right)^2 + \left(\frac{y}{b}\right)^2 = 1$ if $a < b$?

SOLUTION If $a < b$ the foci of the ellipse $\left(\frac{x}{a}\right)^2 + \left(\frac{y}{b}\right)^2 = 1$ are at the points $(0, c)$ and $(0, -c)$ on the y-axis, where $c = \sqrt{b^2 - a^2}$.

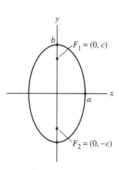

$$\left(\tfrac{x}{a}\right)^2 + \left(\tfrac{y}{b}\right)^2 = 1; \, a < b$$

4. For a hyperbola in standard position, the set of points equidistant from the foci is the y-axis. Use the definition $PF_1 - PF_2 = \pm K$ to explain why the hyperbola does not intersect the y-axis.

SOLUTION The points on the hyperbola are the point such that the difference of the distances to the two foci is $\pm k$, for some constant $k > 0$. For the points on the y-axis, this difference is zero, hence they are not on the hyperbola. Therefore, the hyperbola does not intersect the y-axis.

5. What is the geometric interpretation of the quantity $\frac{b}{a}$ in the equation of a hyperbola in standard position?

SOLUTION The vertices, i.e., the points where the focal axis intersects the hyperbola, are at the points $(a, 0)$ and $(-a, 0)$. The values $\pm\frac{b}{a}$ are the slopes of the two asymptotes of the hyperbola.

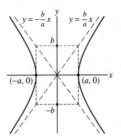

Hyperbola in standard position

Exercises

In Exercises 1–8, find the vertices and foci of the conic section.

1. $\left(\dfrac{x}{9}\right)^2 + \left(\dfrac{y}{4}\right)^2 = 1$

SOLUTION   This is an ellipse in standard position with $a = 9$ and $b = 4$. Hence, $c = \sqrt{9^2 - 4^2} = \sqrt{65} \approx 8.06$. The foci are at $F_1 = (-8.06, 0)$ and $F_2 = (8.06, 0)$, and the vertices are $(9, 0)$, $(-9, 0)$, $(0, 4)$, $(0, -4)$.

3. $\dfrac{x^2}{9} + \dfrac{y^2}{4} = 1$

SOLUTION   Writing the equation in the from $\left(\dfrac{x}{3}\right)^2 + \left(\dfrac{y}{2}\right)^2 = 1$ we get an ellipse with $a = 3$ and $b = 2$. Hence $c = \sqrt{3^2 - 2^2} = \sqrt{5} \approx 2.24$. The foci are at $F_1 = (-2.24, 0)$ and $F_2 = (2.24, 0)$ and the vertices are $(3, 0)$, $(-3, 0)$, $(0, 2)$, $(0, -2)$.

5. $\left(\dfrac{x}{4}\right)^2 - \left(\dfrac{y}{9}\right)^2 = 1$

SOLUTION   For this hyperbola $a = 4$, $b = 9$ and $c = \sqrt{4^2 + 9^2} \approx 9.85$. The foci are at $F_1 = (9.85, 0)$ and $F_2 = (-9.85, 0)$ and the vertices are $A = (4, 0)$ and $A' = (-4, 0)$.

7. $\left(\dfrac{x - 3}{7}\right)^2 - \left(\dfrac{y + 1}{4}\right)^2 = 1$

SOLUTION   We first consider the hyperbola $\left(\dfrac{x}{7}\right)^2 - \left(\dfrac{y}{4}\right)^2 = 1$. For this hyperbola, $a = 7$, $b = 4$ and $c = \sqrt{7^2 + 4^2} \approx 8.06$. Hence, the foci are at $(8.06, 0)$ and $(-8.06, 0)$ and the vertices are at $(7, 0)$ and $(-7, 0)$. Since the given hyperbola is obtained by translating the center of the hyperbola $\left(\dfrac{x}{7}\right)^2 - \left(\dfrac{y}{4}\right)^2 = 1$ to the point $(3, -1)$, the foci are at $F_1 = (8.06 + 3, 0 - 1) = (11.06, -1)$ and $F_2 = (-8.06 + 3, 0 - 1) = (-5.06, -1)$ and the vertices are $A = (7 + 3, 0 - 1) = (10, -1)$ and $A' = (-7 + 3, 0 - 1) = (-4, -1)$.

In Exercises 9–12, consider the ellipse

$$\left(\dfrac{x - 12}{5}\right)^2 + \left(\dfrac{y - 9}{7}\right)^2 = 1$$

Find the equation of the translated ellipse.

9. Translated so that its center is at the origin

SOLUTION   Recall that the equation

$$\dfrac{(x - h)^2}{a^2} + \dfrac{(y - k)^2}{b^2} = 1$$

describes an ellipse with center (h, k). Thus, for our ellipse to be located at the origin, it must have equation

$$\dfrac{x^2}{5^2} + \dfrac{y^2}{7^2} = 1$$

11. Translated three units down

SOLUTION   Recall that the equation

$$\frac{(x-h)^2}{a^2} + \frac{(y-k)^2}{b^2} = 1$$

describes an ellipse with center (h, k). Thus, for our ellipse to be moved 3 units down, it must have equation

$$\frac{(x-12)^2}{5^2} + \frac{(y-6)^2}{7^2} = 1$$

In Exercises 13–16, find the equation of the ellipse with the given properties.

13. Vertices at $(\pm 9, 0)$ and $(0, \pm 16)$

SOLUTION   The equation is $\left(\frac{x}{a}\right)^2 + \left(\frac{y}{b}\right)^2 = 1$ with $a = 9$ and $b = 16$. That is,

$$\left(\frac{x}{9}\right)^2 + \left(\frac{y}{16}\right)^2 = 1$$

15. Foci $(0, \pm 6)$ and two vertices at $(\pm 4, 0)$

SOLUTION   The equation of the ellipse is $\left(\frac{x}{a}\right)^2 + \left(\frac{y}{b}\right)^2 = 1$. The foci are $(0, \pm c)$ on the y-axis with $c = 6$, and two vertices are at $(\pm a, 0)$ with $a = 4$. We use the relation $c = \sqrt{b^2 - a^2}$ to find b:

$$b = \sqrt{a^2 + c^2} = \sqrt{4^2 + 6^2} \approx 7.2$$

Therefore the equation is

$$\left(\frac{x}{4}\right)^2 + \left(\frac{y}{7.2}\right)^2 = 1.$$

In Exercises 17–22, find the equation of the hyperbola with the given properties.

17. Vertices $(\pm 3, 0)$ and foci at $(\pm 5, 0)$

SOLUTION   The equation is $\left(\frac{x}{a}\right)^2 - \left(\frac{y}{b}\right)^2 = 1$. The vertices are $(\pm a, 0)$ with $a = 3$ and the foci $(\pm c, 0)$ with $c = 5$. We use the relation $c = \sqrt{a^2 + b^2}$ to find b:

$$b = \sqrt{c^2 - a^2} = \sqrt{5^2 - 3^2} = \sqrt{16} = 4$$

Therefore, the equation of the hyperbola is

$$\left(\frac{x}{3}\right)^2 - \left(\frac{y}{4}\right)^2 = 1.$$

19. Vertices $(\pm 4, 0)$ and asymptotes $y = \pm 3x$

SOLUTION   The equation is $\left(\frac{x}{a}\right)^2 - \left(\frac{y}{b}\right)^2 = 1$. The vertices are at $(\pm a, 0)$ with $a = 4$ and the asymptotes are $y = \pm \frac{b}{a}x$ with $\frac{b}{a} = 3$. Hence $b = 3a = 3 \cdot 4 = 12$, and the equation of the hyperbola is

$$\left(\frac{x}{4}\right)^2 - \left(\frac{y}{12}\right)^2 = 1$$

21. Vertices $(0, -5)$, $(0, 4)$ and foci $(0, -8)$, $(0, 7)$

SOLUTION   The center of the parabola is at $\frac{-5+4}{2} = -0.5$. Thus, the equation has the form $\left(\frac{y+0.5}{b}\right)^2 - \left(\frac{x}{a}\right)^2 = 1$. Since $b = 4.5$ and $c = 7.5$, we quickly find that $a = 6$, giving us the equation

$$\left(\frac{y+0.5}{4.5}\right)^2 - \left(\frac{x}{6}\right)^2 = 1.$$

In Exercises 23–30, find the equation of the parabola with the given properties.

23. Vertex $(0, 0)$, focus $(2, 0)$

SOLUTION   Since the focus is on the x-axis rather than on the y-axis, the equation is $x = \frac{y^2}{4c}$. Since $c = 2$ we get $x = \frac{y^2}{8}$.

25. Vertex $(0, 0)$, directrix $y = -5$

SOLUTION   The equation is $y = \frac{1}{4c}x^2$. The directrix is $y = -c$ with $c = 5$, hence $y = \frac{1}{20}x^2$.

27. Focus $(0, 4)$, directrix $y = -4$

SOLUTION   The focus is $(0, c)$ with $c = 4$ and the directrix is $y = -c$ with $c = 4$, hence the equation of the parabola is

$$y = \frac{1}{4c}x^2 = \frac{x^2}{16}.$$

29. Focus $(2, 0)$, directrix $x = -2$

SOLUTION   The focus is on the x-axis rather than on the y-axis and the directrix is a vertical line rather than horizontal as in the parabola in standard position. Therefore, the equation of the parabola is obtained by interchanging x and y in $y = \frac{1}{4c}x^2$. Also, by the given information $c = 2$. Hence, $x = \frac{1}{4c}y^2 = \frac{1}{4 \cdot 2}y^2$ or $x = \frac{y^2}{8}$.

In Exercises 31–40, find the vertices, foci, axes, center (if an ellipse or a hyperbola) and asymptotes (if a hyperbola) of the conic section.

31. $x^2 + 4y^2 = 16$

SOLUTION   We first divide the equation by 16 to convert it to the equation in standard form:

$$\frac{x^2}{16} + \frac{4y^2}{16} = 1 \Rightarrow \frac{x^2}{16} + \frac{y^2}{4} = 1 \Rightarrow \left(\frac{x}{4}\right)^2 + \left(\frac{y}{2}\right)^2 = 1$$

For this ellipse, $a = 4$ and $b = 2$ hence $c = \sqrt{4^2 - 2^2} = \sqrt{12} \approx 3.5$. Since $a > b$ we have:

- The vertices are at $(\pm 4, 0)$, $(0, \pm 2)$.
- The foci are $F_1 = (-3.5, 0)$ and $F_2 = (3.5, 0)$.
- The focal axis is the x-axis and the conjugate axis is the y-axis.
- The ellipse is centered at the origin.

33. $4x^2 + y^2 = 16$

SOLUTION   We divide the equation by 16 to rewrite it in the standard form:

$$\frac{4x^2}{16} + \frac{y^2}{16} = 1 \Rightarrow \frac{x^2}{4} + \frac{y^2}{16} = 1 \Rightarrow \left(\frac{x}{2}\right)^2 + \left(\frac{y}{4}\right)^2 = 1$$

This is the equation of an ellipse with $a = 2$, $b = 4$. Since $a < b$ the focal axis is the y-axis. Also, $c = \sqrt{4^2 - 2^2} = \sqrt{12} \approx 3.5$. We get:

- The vertices are at $(\pm 2, 0)$, $(0, \pm 4)$.
- The foci are $(0, \pm 3.5)$.
- The focal axis is the y-axis and the conjugate axis is the x-axis.
- The center is at the origin.

35. $4x^2 - 3y^2 + 8x + 30y = 215$

SOLUTION   Since there is no cross term, we complete the square of the terms involving x and y separately:

$$4x^2 - 3y^2 + 8x + 30y = 4\left(x^2 + 2x\right) - 3\left(y^2 - 10y\right) = 4(x + 1)^2 - 4 - 3(y - 5)^2 + 75 = 215$$

Hence,

$$4(x + 1)^2 - 3(y - 5)^2 = 144$$

$$\frac{4(x + 1)^2}{144} - \frac{3(y - 5)^2}{144} = 1$$

$$\left(\frac{x + 1}{6}\right)^2 - \left(\frac{y - 5}{\sqrt{48}}\right)^2 = 1$$

This is the equation of the hyperbola obtained by translating the hyperbola $\left(\frac{x}{6}\right)^2 - \left(\frac{y}{\sqrt{48}}\right)^2 = 1$ one unit to the left and five units upwards. Since $a = 6$, $b = \sqrt{48}$, we have $c = \sqrt{36 + 48} = \sqrt{84} \sim 9.2$. We obtain the following table:

	Standard position	Translated hyperbola
vertices	$(6, 0)$, $(-6, 0)$	$(5, 5)$, $(-7, 5)$
foci	$(\pm 9.2, 0)$	$(8.2, 5)$, $(-10.2, 5)$
focal axis	The x-axis	$y = 5$
conjugate axis	The y-axis	$x = -1$
center	The origin	$(-1, 5)$
asymptotes	$y = \pm 1.15x$	$y = -1.15x + 3.85$ $y = 1.15x + 6.15$

37. $y = 4(x - 4)^2$

SOLUTION By Exercise 36, the parabola $y = 4x^2$ has the vertex at the origin, the focus at $\left(0, \frac{1}{16}\right)$ and its axis is the y-axis. Our parabola is a translation of the standard parabola four units to the right. Hence its vertex is at $(4, 0)$, the focus is at $\left(4, \frac{1}{16}\right)$ and its axis is the vertical line $x = 4$.

39. $4x^2 + 25y^2 - 8x - 10y = 20$

SOLUTION Since there are no cross terms this conic section is obtained by translating a conic section in standard position. To identify the conic section we complete the square of the terms involving x and y separately:

$$4x^2 + 25y^2 - 8x - 10y = 4\left(x^2 - 2x\right) + 25\left(y^2 - \frac{2}{5}y\right)$$

$$= 4(x - 1)^2 - 4 + 25\left(y - \frac{1}{5}\right)^2 - 1$$

$$= 4(x - 1)^2 + 25\left(y - \frac{1}{5}\right)^2 - 5 = 20$$

Hence,

$$I4(x - 1)^2 + 25\left(y - \frac{1}{5}\right)^2 = 25$$

$$\frac{4}{25}(x - 1)^2 + \left(y - \frac{1}{5}\right)^2 = 1$$

$$\left(\frac{x - 1}{\frac{5}{2}}\right)^2 + \left(y - \frac{1}{5}\right)^2 = 1$$

This is the equation of the ellipse obtained by translating the ellipse in standard position $\left(\frac{x}{\frac{5}{2}}\right)^2 + y^2 = 1$ one unit to the right and $\frac{1}{5}$ unit upward. Since $a = \frac{5}{2}$, $b = 1$ we have $c = \sqrt{\left(\frac{5}{2}\right)^2 - 1} \approx 2.3$, so we obtain the following table:

	Standard position	Translated ellipse
Vertices	$\left(\pm \frac{5}{2}, 0\right)$, $(0, \pm 1)$	$\left(1 \pm \frac{5}{2}, \frac{1}{5}\right)$, $\left(1, \frac{1}{5} \pm 1\right)$
Foci	$(-2.3, 0)$, $(2.3, 0)$	$\left(-1.3, \frac{1}{5}\right)$, $\left(3.3, \frac{1}{5}\right)$
Focal axis	The x-axis	$y = \frac{1}{5}$
Conjugate axis	The y-axis	$x = 1$
Center	The origin	$\left(1, \frac{1}{5}\right)$

In Exercises 41–44, use the Discriminant Test to determine the type of the conic section defined by the equation. You may assume that the equation is nondegenerate. Plot the curve if you have a computer algebra system.

41. $4x^2 + 5xy + 7y^2 = 24$

SOLUTION Here, $D = 25 - 4 \cdot 4 \cdot 7 = -87$, so the conic section is an ellipse.

43. $2x^2 - 8xy - 3y^2 - 4 = 0$

SOLUTION   Here, $D = 64 - 4 \cdot 2 \cdot (-3) = 88$, giving us a hyperbola.

45. Show that $\dfrac{b}{a} = \sqrt{1 - e^2}$ for a standard ellipse of eccentricity e.

SOLUTION   By the definition of eccentricity:

$$e = \frac{c}{a} \tag{1}$$

For the ellipse in standard position, $c = \sqrt{a^2 - b^2}$. Substituting into (1) and simplifying yields

$$e = \frac{\sqrt{a^2 - b^2}}{a} = \sqrt{\frac{a^2 - b^2}{a^2}} = \sqrt{1 - \left(\frac{b}{a}\right)^2}$$

We square the two sides and solve for $\dfrac{b}{a}$:

$$e^2 = 1 - \left(\frac{b}{a}\right)^2 \;\Rightarrow\; \left(\frac{b}{a}\right)^2 = 1 - e^2 \;\Rightarrow\; \frac{b}{a} = \sqrt{1 - e^2}$$

47. Explain why the dots in Figure 22 lie on a parabola. Where are the focus and directrix located?

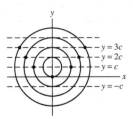

FIGURE 22

SOLUTION   All the circles are centered at $(0, c)$ and the kth circle has radius kc. Hence the indicated point P_k on the kth circle has a distance kc from the point $F = (0, c)$. The point P_k also has distance kc from the line $y = -c$. That is, the indicated point on each circle is equidistant from the point $F = (0, c)$ and the line $y = -c$, hence it lies on the parabola with focus at $F = (0, c)$ and directrix $y = -c$.

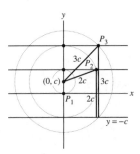

49. Kepler's First Law states that the orbits of the planets around the sun are ellipses with the sun at one focus. The orbit of Pluto has an eccentricity of approximately $e = 0.25$ and the **perihelion** (closest distance to the sun) of Pluto's orbit is approximately 2.7 billion miles. Find the **aphelion** (farthest distance from the sun).

SOLUTION   We define an xy-coordinate system so that the orbit is an ellipse in standard position, as shown in the figure.

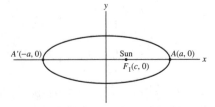

The aphelion is the length of $\overline{A'F_1}$, that is $a + c$. By the given data, we have

$$0.25 = e = \frac{c}{a} \;\Rightarrow\; c = 0.25a$$

$$a - c = 2.7 \;\Rightarrow\; c = a - 2.7$$

Equating the two expressions for c we get

$$0.25a = a - 2.7$$

$$0.75a = 2.7 \Rightarrow a = \frac{2.7}{0.75} = 3.6, \ c = 3.6 - 2.7 = 0.9$$

The aphelion is thus

$$\overline{A'F_0} = a + c = 3.6 + 0.9 = 4.5 \text{ billion miles.}$$

In Exercises 51–54, find the polar equation of the conic with given eccentricity and directrix.

51. $e = \frac{1}{2}, \ x = 3$

SOLUTION Substituting $e = \frac{1}{2}$ and $d = 3$ in the polar equation of a conic section we obtain

$$r = \frac{ed}{1 + e\cos\theta} = \frac{\frac{1}{2} \cdot 3}{1 + \frac{1}{2}\cos\theta} = \frac{3}{2 + \cos\theta} \Rightarrow r = \frac{3}{2 + \cos\theta}$$

53. $e = 1, \ x = 4$

SOLUTION We substitute $e = 1$ and $d = 4$ in the polar equation of a conic section to obtain

$$r = \frac{ed}{1 + e\cos\theta} = \frac{1 \cdot 4}{1 + 1 \cdot \cos\theta} = \frac{4}{1 + \cos\theta} \Rightarrow r = \frac{4}{1 + \cos\theta}$$

In Exercises 55–58, identify the type of conic, the eccentricity, and the equation of the directrix.

55. $r = \dfrac{8}{1 + 4\cos\theta}$

SOLUTION Matching with the polar equation $r = \frac{ed}{1 + e\cos\theta}$ we get $ed = 8$ and $e = 4$ yielding $d = 2$. Since $e > 1$, the conic section is a hyperbola, having eccentricity $e = 4$ and directrix $x = 2$ (referring to the focus-directrix definition (11)).

57. $r = \dfrac{8}{4 + 3\cos\theta}$

SOLUTION We first rewrite the equation in the form $r = \frac{ed}{1 + e\cos\theta}$, obtaining

$$r = \frac{2}{1 + \frac{3}{4}\cos\theta}$$

Hence, $ed = 2$ and $e = \frac{3}{4}$ yielding $d = \frac{8}{3}$. Since $e < 1$, the conic section is an ellipse, having eccentricity $e = \frac{3}{4}$ and directrix $x = \frac{8}{3}$.

59. ✏️ Show that $r = f_1(\theta)$ and $r = f_2(\theta)$ define the same curves in polar coordinates if $f_1(\theta) = -f_2(\theta + \pi)$, and use this to show that the following define the same conic section:

$$r = \frac{de}{1 - e\cos\theta}, \qquad r = \frac{-de}{1 + e\cos\theta}$$

SOLUTION The curve $r = f_2(\theta)$ can be parametrized using θ as a parameter:

$$x = f_2(\theta)\cos\theta \tag{1}$$

$$y = f_2(\theta)\sin\theta$$

Using the identities $\cos(\theta + \pi) = -\cos\theta$ and $\sin(\theta + \pi) = -\sin\theta$, we obtain the following parametrization for $r = f_1(\theta)$:

$$x = f_1(\theta)\cos\theta = -f_2(\theta + \pi)\cos\theta$$

$$= f_2(\theta + \pi)(-\cos\theta) = f_2(\theta + \pi)\cos(\theta + \pi)$$

$$y = f_1(\theta)\sin\theta = -f_2(\theta + \pi)\sin\theta$$

$$= f_2(\theta + \pi)(-\sin\theta) = f_2(\theta + \pi)\sin(\theta + \pi)$$

Using $t = \theta + \pi$ as the parameter we get

$$x = f_2(t)\cos t \tag{2}$$

$$y = f_2(t) \sin t$$

The parametrizations (1) and (2) define the same curve. We now consider the polar equations:

$$r = \frac{de}{1 - e \cos \theta} = f_1(\theta), \quad r = \frac{-de}{1 + e \cos \theta} = f_2(\theta) \qquad (3)$$

The following equality holds:

$$-f_2(\theta + \pi) = -\frac{-de}{1 + e \cos(\theta + \pi)} = \frac{de}{1 - e \cos \theta} = f_1(\theta)$$

We conclude that the polar equations in (3) define the same conic section.

61. Find the equation of the ellipse $r = \dfrac{4}{2 + \cos \theta}$ in rectangular coordinates.

SOLUTION We use the polar equation of the ellipse with focus at the origin and directrix at $x = d$:

$$r = \frac{ed}{1 + e \cos \theta}$$

Dividing the numerator and denominator of the given equation by 2 yields

$$r = \frac{2}{1 + \frac{1}{2} \cos \theta}.$$

We identify the values $e = \frac{1}{2}$ and $ed = 2$. Hence, $d = \frac{2}{e} = 4$. The focus-directrix of the ellipse is $\overline{PO} = eD$. That is, $\sqrt{x^2 + y^2} = \frac{1}{2}|4 - x|$.

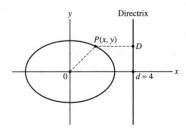

We square the two sides, simplify, and complete the square to obtain

$$4\left(x^2 + y^2\right) = 16 - 8x + x^2$$

$$3x^2 + 8x + 4y^2 = 16$$

$$3\left(x^2 + \frac{8}{3}x\right) + 4y^2 = 16$$

$$3\left(x + \frac{4}{3}\right)^2 + 4y^2 = \frac{64}{3}$$

$$\frac{\left(x + \frac{4}{3}\right)^2}{\frac{64}{9}} + \frac{y^2}{\frac{16}{3}} = 1 \Rightarrow \left(\frac{x + \frac{4}{3}}{\frac{8}{3}}\right)^2 + \left(\frac{y}{\frac{4}{\sqrt{3}}}\right)^2 = 1.$$

63. Let $e > 1$. Show that the vertices of the hyperbola $r = \dfrac{de}{1 + e \cos \theta}$ have x-coordinates $\dfrac{ed}{e + 1}$ and $\dfrac{ed}{e - 1}$.

SOLUTION Since the focus is at the origin and the hyperbola is to the right (see figure), the two vertices have positive x coordinates. The corresponding values of θ at the vertices are $\theta = 0$ and $\theta = \pi$. Hence, since $e > 1$ we obtain

$$x_A = |r(0)| = \left| \frac{de}{1 + e \cos 0} \right| = \frac{de}{1 + e}$$

$$x_{A'} = |r(\pi)| = \left| \frac{de}{1 + e \cos \pi} \right| = \left| \frac{de}{1 - e} \right| = \frac{de}{e - 1}$$

Further Insights and Challenges

65. Verify Theorem 4 in the case $0 < e < 1$. *Hint:* Repeat the proof of Theorem 4, but set $c = d/(e^{-2} - 1)$.

SOLUTION   We follow closely the proof of Theorem 4 in the book, which covered the case $e > 1$. This time, for $0 < e < 1$, we prove that $PF = ePD$ defines an ellipse. We choose our coordinate axes so that the focus F lies on the x-axis with coordinates $F = (c, 0)$ and so that the directrix is vertical, lying to the right of F at a distance d from F. As suggested by the hint, we set $c = \frac{d}{e^{-2}-1}$, but since we are working towards an ellipse, we will also need to let $b = \sqrt{a^2 - c^2}$ as opposed to the $\sqrt{c^2 - a^2}$ from the original proof of Theorem 4. Here's the complete list of definitions:

$$c = \frac{d}{e^{-2} - 1}, \quad a = \frac{c}{e}, \quad b = \sqrt{a^2 - c^2}$$

The directrix is the line

$$x = c + d = c + c(e^{-2} - 1) = ce^{-2} = \frac{a}{e}$$

Now, the equation

$$PF = e \cdot PD$$

for the points $P = (x, y)$, $F = (c, 0)$, and $D = (a/e, y)$ becomes

$$\sqrt{(x - c)^2 + y^2} = e \cdot \sqrt{(x - (a/e))^2}$$

Returning to the proof of Theorem 4, we see that this is the same equation that appears in the middle of the proof of the Theorem. As seen there, this equation can be transformed into

$$\frac{x^2}{a^2} - \frac{y^2}{a^2(e^2 - 1)} = 1$$

and this is equivalent to

$$\frac{x^2}{a^2} + \frac{y^2}{a^2(1 - e^2)} = 1$$

Since $a^2(1 - e^2) = a^2 - a^2 e^2 = a^2 - c^2 = b^2$, then we obtain the equation of the ellipse

$$\frac{x^2}{a^2} + \frac{y^2}{b^2} = 1$$

Reflective Property of the Ellipse   *In Exercises 67–69, we prove that the focal radii at a point on an ellipse make equal angles with the tangent line. Let $P = (x_0, y_0)$ be a point on the ellipse $\left(\frac{x}{a}\right)^2 + \left(\frac{y}{b}\right)^2 = 1$ $(a > b)$ with foci $F_1 = (-c, 0)$ and $F_2 = (c, 0)$, and eccentricity e (Figure 24).*

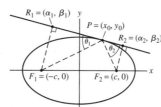

FIGURE 24 The ellipse $\left(\frac{x}{a}\right)^2 + \left(\frac{y}{b}\right)^2 = 1$.

67. Show that $PF_1 = a + x_0 e$ and $PF_2 = a - x_0 e$. *Hints:*
(a) Show that $PF_1{}^2 - PF_2{}^2 = 4x_0 c$.
(b) Divide the previous relation by $PF_1 + PF_2 = 2a$, and conclude that $PF_1 - PF_2 = 2x_0 e$.

SOLUTION   Using the distance formula we have

$$\overline{PF_1}^2 = (x_0 + c)^2 + y^2; \quad \overline{PF_2}^2 = (x_0 - c)^2 + y^2$$

Hence,

$$\overline{PF_1}^2 - \overline{PF_2}^2 = (x_0 + c)^2 + y^2 - (x_0 - c)^2 - y^2$$

$$= (x_0 + c)^2 - (x_0 - c)^2$$

$$= x_0{}^2 + 2x_0 c + c^2 - x_0{}^2 + 2x_0 c - c^2 = 4x_0 c$$

That is, $\overline{PF_1} + \overline{PF_2} = 4x_0 c$. We use the identity $u^2 - v^2 = (u - v)(u + v)$ to write this as

$$\left(\overline{PF_1} - \overline{PF_2}\right)\left(\overline{PF_1} + \overline{PF_2}\right) = 4x_0 c \tag{1}$$

Since P lies on the ellipse $\left(\frac{x}{a}\right)^2 + \left(\frac{y}{b}\right)^2 = 1$ we have

$$\overline{PF_1} + \overline{PF_2} = 2a \tag{2}$$

Substituting in (1) gives

$$\left(\overline{PF_1} - \overline{PF_2}\right) \cdot 2a = 4x_0 c$$

We divide by a and use the eccentricity $e = \frac{c}{a}$ to obtain

$$\overline{PF_1} - \overline{PF_2} = 2x_0 e$$

69. Define R_1 and R_2 as in the figure, so that $\overline{F_1 R_1}$ and $\overline{F_2 R_2}$ are perpendicular to the tangent line.

(a) Show that $\dfrac{\alpha_1 + c}{\beta_1} = \dfrac{\alpha_2 - c}{\beta_2} = \dfrac{A}{B}$.

(b) Use (a) and the distance formula to show that

$$\frac{F_1 R_1}{F_2 R_2} = \frac{\beta_1}{\beta_2}$$

(c) Solve for β_1 and β_2:

$$\beta_1 = \frac{B(1 + Ac)}{A^2 + B^2}, \qquad \beta_2 = \frac{B(1 - Ac)}{A^2 + B^2}$$

(d) Show that $\dfrac{F_1 R_1}{F_2 R_2} = \dfrac{PF_1}{PF_2}$. Conclude that $\theta_1 = \theta_2$.

SOLUTION

(a) Since $R_1 = (\alpha_1, \beta_1)$ and $R_2 = (\alpha_2, \beta_2)$ lie on the tangent line at P, that is on the line $Ax + By = 1$, we have

$$A\alpha_1 + B\beta_1 = 1 \quad \text{and} \quad A\alpha_2 + B\beta_2 = 1$$

The slope of the line $R_1 F_1$ is $\frac{\beta_1}{\alpha_1 + c}$ and it is perpendicular to the tangent line having slope $-\frac{A}{B}$. Similarly, the slope of the line $R_2 F_2$ is $\frac{\beta_2}{\alpha_2 - c}$ and it is also perpendicular to the tangent line. Hence,

$$\frac{\alpha_1 + c}{\beta_1} = \frac{A}{B} \quad \text{and} \quad \frac{\alpha_2 - c}{\beta_2} = \frac{A}{B}.$$

(b) Using the distance formula, we have

$$\overline{R_1 F_1}^2 = (\alpha_1 + c)^2 + \beta_1^2$$

Thus,

$$\overline{R_1 F_1}^2 = \beta_1^2 \left(\left(\frac{\alpha_1 + c}{\beta_1} \right)^2 + 1 \right) \tag{1}$$

By part (a), $\frac{\alpha_1 + c}{\beta_1} = \frac{A}{B}$. Substituting in (1) gives

$$\overline{R_1 F_1}^2 = \beta_1^2 \left(\frac{A^2}{B^2} + 1 \right) = \beta_1^2 \left(1 + B^{-2} A^2 \right) \tag{2}$$

Likewise,

$$\overline{R_2 F_2}^2 = (\alpha_2 - c)^2 + \beta_2{}^2 = \beta_2{}^2 \left(\left(\frac{\alpha_2 - c}{\beta_2} \right)^2 + 1 \right)$$

but since $\frac{\alpha_2 - c}{\beta_2} = \frac{A}{B}$, we get that

$$\overline{R_2 F_2}^2 = \beta_2^2 \left(\frac{A^2}{B^2} + 1 \right). \tag{3}$$

Dividing, we find that

$$\frac{\overline{R_1 F_1}^2}{\overline{R_2 F_2}^2} = \frac{\beta_1^2}{\beta_2^2} \quad \text{so} \quad \frac{\overline{R_1 F_1}}{\overline{R_2 F_2}} = \frac{\beta_1}{\beta_2},$$

as desired.

(c), (d) In part (a) we show that

$$\begin{cases} A\alpha_1 + B\beta_1 = 1 \\ \dfrac{\beta_1}{\alpha_1 + c} = \dfrac{B}{A} \end{cases}$$

Solving for β_1 gives $\beta_1 = \frac{B(1+Ac)}{A^2+B^2}$. Substituting in (2) we obtain

$$\overline{R_1 F_1}^2 = \frac{B^2(1 + Ac)^2}{(A^2 + B^2)^2} \left(1 + \frac{A^2}{B^2} \right) = \frac{B^2(1 + Ac)^2(A^2 + B^2)}{(A^2 + B^2)^2 B^2} = \frac{(1 + Ac)^2}{A^2 + B^2}$$

Similarly solving the equations in part (a) for β_2 and using equation (3) yields

$$\begin{cases} A\alpha_2 + B\beta_2 = 1 \\ \dfrac{\beta_2}{\alpha_2 - c} = \dfrac{B}{A} \end{cases} \Rightarrow \quad \beta_2 = \frac{B(1 - Ac)}{A^2 + B^2} \tag{4}$$

Using the distance formula for $\overline{R_2 F_2}$ we have

$$\overline{R_2 F_2}^2 = (\alpha_2 - c)^2 + \beta_2^2 = \beta_2^2 \left(\left(\frac{A_2 - c}{\beta_2} \right)^2 + 1 \right)$$

Substituting $\frac{A_2 - c}{\beta_2} = \frac{A}{B}$ from part (a) and β_2 in (3) we get

$$\overline{R_2 F_2}^2 = \frac{B^2(1 - Ac)^2}{\left(A^2 + B^2 \right)^2} \left(\frac{A^2}{B^2} + 1 \right) = \frac{B^2(1 - Ac)^2 \left(A^2 + B^2 \right)}{\left(A^2 + B^2 \right)^2 B^2} = \frac{(1 - Ac)^2}{A^2 + B^2}$$

Using the expression for $\overline{R_1 F_1}$ and $\overline{R_2 F_2}$ obtained above, we get

$$\frac{\overline{R_1 F_1}}{\overline{R_2 F_2}} = \frac{\frac{1+Ac}{\sqrt{A^2+B^2}}}{\frac{1-Ac}{\sqrt{A^2+B^2}}} = \frac{1 + Ac}{1 - Ac}$$

Substituting $c = ea$ and $A = \frac{x_0}{a^2}$ we obtain

$$\frac{\overline{R_1 F_1}}{\overline{R_2 F_2}} = \frac{1 + \frac{x_0 ea}{a^2}}{1 - \frac{x_0 ea}{a^2}} = \frac{1 + \frac{x_0 e}{a}}{1 - \frac{x_0 e}{a}} = \frac{a + x_0 e}{a - x_0 e}$$

Now by Exercise 67, we have $\overline{P F_1} = a + x_0 e$ and $\overline{P F_2} = a - x_0 e$, where $P = (x_0, y_0)$. By Exercise 69, $\frac{\overline{R_1 F_1}}{\overline{R_2 F_2}} = \frac{a + ex_0}{a - ex_0}$. Using these results we derive the following equality:

$$\frac{\overline{P F_1}}{\overline{P F_2}} = \frac{\overline{R_1 F_1}}{\overline{R_2 F_2}} \Rightarrow \frac{\overline{R_1 F_1}}{\overline{P F_1}} = \frac{\overline{R_2 F_2}}{\overline{P F_2}}$$

By $\frac{\overline{R_1 F_1}}{\overline{P F_1}} = \sin \theta_1$ and $\frac{\overline{R_2 F_2}}{\overline{P F_2}} = \sin \theta_2$ we get $\sin \theta_1 = \sin \theta_2$, which implies that $\theta_1 = \theta_2$ since the two angles are acute.

71. Show that $y = \dfrac{x^2}{4c}$ is the equation of a parabola with directrix $y = -c$, focus $(0, c)$, and the vertex at the origin, as stated in Theorem 3.

SOLUTION The points $P = (x, y)$ on the parabola are equidistant from $F = (0, c)$ and the line $y = -c$.

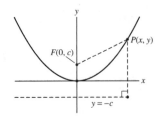

That is, by the distance formula, we have

$$\overline{PF} = \overline{PD}$$

$$\sqrt{x^2 + (y - c)^2} = |y + c|$$

Squaring and simplifying yields

$$x^2 + (y - c)^2 = (y + c)^2$$

$$x^2 + y^2 - 2yc + c^2 = y^2 + 2yc + c^2$$

$$x^2 - 2yc = 2yc$$

$$x^2 = 4yc \Rightarrow y = \frac{x^2}{4c}$$

Thus, we showed that the points that are equidistant from the focus $F = (0, c)$ and the directrix $y = -c$ satisfy the equation $y = \frac{x^2}{4c}$.

73. If we rewrite the general equation of degree two (13) in terms of variables x' and y' that are related to x and y by equations (14) and (15), we obtain a new equation of degree two in x' and y' of the same form but with different coefficients:

$$A'x^2 + B'xy + C'y^2 + D'x + E'y + F' = 0$$

(a) Show that $B' = B \cos 2\theta + (C - A) \sin 2\theta$.

(b) Show that if $B \neq 0$, then we obtain $B' = 0$ for

$$\theta = \frac{1}{2} \cot^{-1} \frac{A - C}{B}$$

This proves that it is always possible to eliminate the cross term Bxy by rotating the axes through a suitable angle.

SOLUTION

(a) If we plug in $x = x' \cos \theta - y' \sin \theta$ and $y = x' \sin \theta + y' \cos \theta$ into the equation $Ax^2 + Bxy + Cy^2 + Dx + Ey + F = 0$, we will get a very ugly mess. Fortunately, we only care about the $x'y'$ term, so we really only need to look at the $Ax^2 + Bxy + Cy^2$ part of the formula. In fact, we only need to pull out those terms which have an $x'y'$ in them. From the Ax^2 term, after replacing x with $x = x' \cos \theta - y' \sin \theta$, we will get an $x'y'$ term of $-2Ax'y' \cos \theta \sin \theta$. Likewise, from the Bxy term we will get $Bx'y' \cos^2 \theta - \sin^2 \theta$, and from the Cy^2 term we get $2Cx'y' \cos \theta \sin \theta$. Adding these together, we see that the (new) B', the coefficient of the (new) $x'y'$ term, will be

$$-2A \cos \theta \sin \theta + B \cos^2 \theta - \sin^2 \theta + 2Cx'y' \cos \theta \sin \theta$$

which simplifies to $B \cos 2\theta + (C - A) \sin 2\theta$, as desired.

(b) Setting $B' = 0$, we get $0 = B \cos 2\theta + (C - A) \sin 2\theta$, so $B \cos 2\theta = (A - C) \sin 2\theta$, so $\cot 2\theta = \frac{A-C}{B}$, giving us $2\theta = \cot^{-1} \frac{A-C}{B}$, and thus $\theta = \frac{1}{2} \cot^{-1} \frac{A-C}{B}$.

CHAPTER REVIEW EXERCISES

1. Which of the following curves pass through the point $(1, 4)$?

(a) $c(t) = (t^2, t + 3)$

(b) $c(t) = (t^2, t - 3)$

(c) $c(t) = (t^2, 3 - t)$

(d) $c(t) = (t - 3, t^2)$

SOLUTION To check whether it passes through the point $(1, 4)$, we solve the equations $c(t) = (1, 4)$ for the given curves.

(a) Comparing the second coordinate of the curve and the point yields:

$$t + 3 = 4$$

$$t = 1$$

We substitute $t = 1$ in the first coordinate, to obtain

$$t^2 = 1^2 = 1$$

Hence the curve passes through $(1, 4)$.

(b) Comparing the second coordinate of the curve and the point yields:

$$t - 3 = 4$$

$$t = 7$$

We substitute $t = 7$ in the first coordinate to obtain

$$t^2 = 7^2 = 49 \neq 1$$

Hence the curve does not pass through $(1, 4)$.

(c) Comparing the second coordinate of the curve and the point yields

$$3 - t = 4$$

$$t = -1$$

We substitute $t = -1$ in the first coordinate, to obtain

$$t^2 = (-1)^2 = 1$$

Hence the curve passes through $(1, 4)$.

(d) Comparing the first coordinate of the curve and the point yields

$$t - 3 = 1$$

$$t = 4$$

We substitute $t = 4$ in the second coordinate, to obtain:

$$t^2 = 4^2 = 16 \neq 4$$

Hence the curve does not pass through $(1, 4)$.

3. Find parametric equations for the circle of radius 2 with center $(1, 1)$. Use the equations to find the points of intersection of the circle with the x- and the y-axes.

SOLUTION Using the standard technique for parametric equations of curves, we obtain

$$c(t) = (1 + 2\cos t, 1 + 2\sin t)$$

We compare the x coordinate of $c(t)$ to 0:

$$1 + 2\cos t = 0$$

$$\cos t = -\frac{1}{2}$$

$$t = \pm\frac{2\pi}{3}$$

Substituting in the y coordinate yields

$$1 + 2\sin\left(\pm\frac{2\pi}{3}\right) = 1 \pm 2\frac{\sqrt{3}}{2} = 1 \pm \sqrt{3}$$

Hence, the intersection points with the y-axis are $(0, 1 \pm \sqrt{3})$. We compare the y coordinate of $c(t)$ to 0:

$$1 + 2\sin t = 0$$

$$\sin t = -\frac{1}{2}$$

$$t = -\frac{\pi}{6} \quad \text{or} \quad \frac{7}{6}\pi$$

Substituting in the x coordinates yields

$$1 + 2\cos\left(-\frac{\pi}{6}\right) = 1 + 2\frac{\sqrt{3}}{2} = 1 + \sqrt{3}$$

$$1 + 2\cos\left(\frac{7}{6}\pi\right) = 1 - 2\cos\left(\frac{\pi}{6}\right) = 1 - 2\frac{\sqrt{3}}{2} = 1 - \sqrt{3}$$

Hence, the intersection points with the x-axis are $(1 \pm \sqrt{3}, 0)$.

5. Find a parametrization $c(\theta)$ of the unit circle such that $c(0) = (-1, 0)$.

SOLUTION The unit circle has the parametrization

$$c(t) = (\cos t, \sin t)$$

This parametrization does not satisfy $c(0) = (-1, 0)$. We replace the parameter t by a parameter θ so that $t = \theta + \alpha$, to obtain another parametrization for the circle:

$$c^*(\theta) = (\cos(\theta + \alpha), \sin(\theta + \alpha)) \tag{1}$$

We need that $c^*(0) = (1, 0)$, that is,

$$c^*(0) = (\cos \alpha, \sin \alpha) = (-1, 0)$$

Hence

$$\begin{matrix} \cos \alpha = -1 \\ \sin \alpha = 0 \end{matrix} \quad \Rightarrow \quad \alpha = \pi$$

Substituting in (1) we obtain the following parametrization:

$$c^*(\theta) = (\cos(\theta + \pi), \sin(\theta + \pi))$$

7. Find a path $c(t)$ that traces the line $y = 2x + 1$ from $(1, 3)$ to $(3, 7)$ for $0 \le t \le 1$.

SOLUTION Solution 1: By one of the examples in section 12.1, the line through $P = (1, 3)$ with slope 2 has the parametrization

$$c(t) = (1 + t, 3 + 2t)$$

But this parametrization does not satisfy $c(1) = (3, 7)$. We replace the parameter t by a parameter s so that $t = \alpha s + \beta$. We get

$$c^*(s) = \left(1 + \alpha s + \beta, 3 + 2(\alpha s + \beta)\right) = (\alpha s + \beta + 1, 2\alpha s + 2\beta + 3)$$

We need that $c^*(0) = (1, 3)$ and $c^*(1) = (3, 7)$. Hence,

$$c^*(0) = (1 + \beta, 3 + 2\beta) = (1, 3)$$
$$c^*(1) = (\alpha + \beta + 1, 2\alpha + 2\beta + 3) = (3, 7)$$

We obtain the equations

$$\begin{matrix} 1 + \beta = 1 \\ 3 + 2\beta = 3 \\ \alpha + \beta + 1 = 3 \\ 2\alpha + 2\beta + 3 = 7 \end{matrix} \quad \Rightarrow \quad \beta = 0, \alpha = 2$$

Substituting in (1) gives

$$c^*(s) = (2s + 1, 4s + 3)$$

Solution 2: The segment from $(1, 3)$ to $(3, 7)$ has the following vector parametrization:

$$(1 - t)\langle 1, 3 \rangle + t\langle 3, 7 \rangle = \langle 1 - t + 3t, 3(1 - t) + 7t \rangle = \langle 1 + 2t, 3 + 4t \rangle$$

The parametrization is thus

$$c(t) = (1 + 2t, 3 + 4t)$$

In Exercises 9–12, express the parametric curve in the form $y = f(x)$.

9. $c(t) = (4t - 3, 10 - t)$

SOLUTION   We use the given equation to express t in terms of x.

$$x = 4t - 3$$

$$4t = x + 3$$

$$t = \frac{x + 3}{4}$$

Substituting in the equation of y yields

$$y = 10 - t = 10 - \frac{x + 3}{4} = -\frac{x}{4} + \frac{37}{4}$$

That is,

$$y = -\frac{x}{4} + \frac{37}{4}$$

11. $c(t) = \left(3 - \frac{2}{t}, t^3 + \frac{1}{t}\right)$

SOLUTION   We use the given equation to express t in terms of x:

$$x = 3 - \frac{2}{t}$$

$$\frac{2}{t} = 3 - x$$

$$t = \frac{2}{3 - x}$$

Substituting in the equation of y yields

$$y = \left(\frac{2}{3 - x}\right)^3 + \frac{1}{2/(3 - x)} = \frac{8}{(3 - x)^3} + \frac{3 - x}{2}$$

13. Find all points visited twice by the path $c(t) = (t^2, \sin t)$. Plot $c(t)$ with a graphing utility.

SOLUTION   For every point, if the curve passes through it twice, then its x coordinate and y coordinate are obtained twice by the functions $x(t)$, $y(t)$. We first calculate the x coordinate of these points. Since $x(t) = t^2$, every x coordinate is obtained twice—once for t and once for $-t$. We now check for which of the above x coordinates, the y coordinates are equal as well. We substitute the above condition in the formula for the y coordinates. We have $\sin t = \sin(-t)$. Since $\sin t = -\sin(-t)$ for all $t \in \mathbb{R}$, we can add it to the former equation. We obtain $2 \sin t = 0$ or $t = \pi k$ where $k \in \mathbb{Z}$. We substitute the t values in the parametric equation to find the desired points. We obtain

$$(\pi^2 k^2, 0) \quad \text{where } k \in \mathbb{Z}$$

The path $c(t)$ is shown in the following figure.

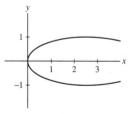

In Exercises 14–17, calculate $\dfrac{dy}{dx}$ at the point indicated.

15. $c(\theta) = (\tan^2 \theta, \cos \theta), \quad \theta = \frac{\pi}{4}$

SOLUTION The parametric equations are $x = \tan^2 \theta$, $y = \cos \theta$. We use the theorem on the slope of the tangent line to find $\frac{dy}{dx}$:

$$\frac{dy}{dx} = \frac{\frac{dy}{d\theta}}{\frac{dx}{d\theta}} = \frac{-\sin \theta}{2 \tan \theta \sec^2 \theta} = -\frac{\cos^3 \theta}{2}$$

We now substitute $\theta = \frac{\pi}{4}$ to obtain

$$\frac{dy}{dx}\bigg|_{\theta = \pi/4} = -\frac{\cos^3 \frac{\pi}{4}}{2} = -\frac{1}{4\sqrt{2}}$$

17. $c(t) = (\ln t, 3t^2 - t)$, $\quad P = (0, 2)$

SOLUTION The parametric equations are $x = \ln t$, $y = 3t^2 - t$. We use the theorem for the slope of the tangent line to find $\frac{dy}{dx}$:

$$\frac{dy}{dx} = \frac{\frac{dy}{dt}}{\frac{dx}{dt}} = \frac{6t - 1}{\frac{1}{t}} = 6t^2 - t \tag{1}$$

We now must identify the value of t corresponding to the point $P = (0, 2)$ on the curve. We solve the following equations:

$$\begin{aligned} \ln t &= 0 \\ 3t^2 - t &= 2 \end{aligned} \quad \Rightarrow \quad t = 1$$

Substituting $t = 1$ in (1) we obtain

$$\frac{dy}{dx}\bigg|_P = 6 \cdot 1^2 - 1 = 5$$

19. Find the points on $(t + \sin t, t - 2 \sin t)$ where the tangent is vertical or horizontal.

SOLUTION We use the theorem for the slope of the tangent line to find $\frac{dy}{dx}$:

$$\frac{dy}{dx} = \frac{\frac{dy}{dt}}{\frac{dx}{dt}} = \frac{1 - 2 \cos t}{1 + \cos t}$$

We find the values of t for which the denominator is zero. We ignore the numerator, since when $1 + \cos t = 0$, $1 - 2 \cos t = 3 \neq 0$.

$$1 + \cos t = 0$$
$$\cos t = -1$$
$$t = \pi + 2\pi k \quad \text{where } k \in \mathbb{Z}$$

We now find the values of t for which the numerator is 0:

$$1 - 2 \cos t = 0$$
$$1 = 2 \cos t$$
$$\frac{1}{2} = \cos t$$
$$t = \pm \frac{\pi}{3} + 2\pi k \quad \text{where } k \in \mathbb{Z}$$

Note that the denominator is not zero at these points. Thus, we have vertical tangents at $t = \pi + 2\pi k$ and horizontal tangents at $t = \pm \pi/3 + 2\pi k$.

21. Find the speed at $t = \frac{\pi}{4}$ of a particle whose position at time t seconds is $c(t) = (\sin 4t, \cos 3t)$.

SOLUTION We use the parametric definition to find the speed. We obtain

$$\frac{ds}{dt} = \sqrt{((\sin 4t)')^2 + ((\cos 3t)')^2} = \sqrt{(4 \cos 4t)^2 + (-3 \sin 3t)^2} = \sqrt{16 \cos^2 4t + 9 \sin^2 3t}$$

At time $t = \frac{\pi}{4}$ the speed is

$$\frac{ds}{dt}\bigg|_{t=\pi/4} = \sqrt{16 \cos^2 \pi + 9 \sin^2 \frac{3\pi}{4}} = \sqrt{16 + 9 \cdot \frac{1}{2}} = \sqrt{20.5} \approx 4.53$$

23. Find the length of $(3e^t - 3, 4e^t + 7)$ for $0 \leq t \leq 1$.

SOLUTION   We use the formula for arc length, to obtain

$$s = \int_0^1 \sqrt{((3e^t - 3)')^2 + ((4e^t + 7)')^2}\, dt = \int_0^1 \sqrt{(3e^t)^2 + (4e^t)^2}\, dt$$

$$= \int_0^1 \sqrt{9e^{2t} + 16e^{2t}}\, dt = \int_0^1 \sqrt{25e^{2t}}\, dt = \int_0^1 5e^t\, dt = 5e^t \Big|_0^1 = 5(e - 1)$$

In Exercises 24–25, let $c(t) = (e^{-t}\cos t, e^{-t}\sin t)$.

25. Find the first positive value of t_0 such that the tangent line to $c(t_0)$ is vertical and calculate the speed at $t = t_0$.

SOLUTION   The curve has a vertical tangent where $\lim\limits_{t \to t_0} \left|\frac{dy}{dx}\right| = \infty$. We first find $\frac{dy}{dx}$ using the theorem for the slope of a tangent line:

$$\frac{dy}{dx} = \frac{\frac{dy}{dt}}{\frac{dx}{dt}} = \frac{(e^{-t}\sin t)'}{(e^{-t}\cos t)'} = \frac{-e^{-t}\sin t + e^{-t}\cos t}{-e^{-t}\cos t - e^{-t}\sin t}$$

$$= -\frac{\cos t - \sin t}{\cos t + \sin t} = \frac{\sin t - \cos t}{\sin t + \cos t}$$

We now search for t_0 such that $\lim\limits_{t \to t_0}\left|\frac{dy}{dx}\right| = \infty$. In our case, this happens when the denominator is 0, but the numerator is not, thus:

$$\sin t_0 + \cos t_0 = 0$$

$$\cos t_0 = -\sin t_0$$

$$\cos -t_0 = \sin -t_0$$

$$-t_0 = \frac{\pi}{4} - \pi$$

$$t_0 = \frac{3}{4}\pi$$

We now use the formula for the speed, to find the speed at t_0.

$$\frac{ds}{dt} = \sqrt{((e^{-t}\sin t)')^2 + ((e^{-t}\cos t)')^2}$$

$$= \sqrt{(-e^{-t}\cos t - e^{-t}\sin t)^2 + (-e^{-t}\sin t + e^{-t}\cos t)^2}$$

$$= \sqrt{e^{-2t}(\cos t + \sin t)^2 + e^{-2t}(\cos t - \sin t)^2}$$

$$= e^{-t}\sqrt{\cos^2 t + 2\sin t\cos t + \sin^2 t + \cos^2 t - 2\sin t\cos t + \sin^2 t} = e^{-t}\sqrt{2}$$

Next we substitute $t = \frac{3}{4}\pi$, to obtain

$$e^{-t_0}\sqrt{2} = e^{-3\pi/4}\sqrt{2}$$

27. Convert the points $(x, y) = (1, -3)$, $(3, -1)$ from rectangular to polar coordinates.

SOLUTION   We convert the given points from cartesian coordinates to polar coordinates. For the first point we have

$$r = \sqrt{x^2 + y^2} = \sqrt{1^2 + (-3)^2} = \sqrt{10}$$

$$\theta = \arctan\frac{y}{x} = \arctan -3 = 5.034$$

For the second point we have

$$r = \sqrt{x^2 + y^2} = \sqrt{3^2 + (-1)^2}$$

$$= \sqrt{10} \qquad \theta = \arctan\frac{y}{x} = \arctan\frac{-1}{3} = -0.321,\ 5.961$$

29. Write $(x + y)^2 = xy + 6$ as an equation in polar coordinates.

SOLUTION   We use the formula for converting from cartesian coordinates to polar coordinates to substitute r and θ for x and y:

$$(x + y)^2 = xy + 6$$
$$x^2 + 2xy + y^2 = xy + 6$$
$$x^2 + y^2 = -xy + 6$$
$$r^2 = -(r \cos \theta)(r \sin \theta) + 6$$
$$r^2 = -r^2 \cos \theta \sin \theta + 6$$
$$r^2(1 + \sin \theta \cos \theta) = 6$$
$$r^2 = \frac{6}{1 + \sin \theta \cos \theta}$$
$$r^2 = \frac{6}{1 + \frac{\sin 2\theta}{2}}$$
$$r^2 = \frac{12}{2 + \sin 2\theta}$$

31. Show that $r = \dfrac{4}{7 \cos \theta - \sin \theta}$ is the polar equation of a line.

SOLUTION   We use the formula for converting from polar coordinates to cartesian coordinates to substitute x and y for r and θ:

$$r = \frac{4}{7 \cos \theta - \sin \theta}$$
$$1 = \frac{4}{7r \cos \theta - r \sin \theta}$$
$$1 = \frac{4}{7x - y}$$
$$7x - y = 4$$
$$y = 7x - 4$$

We obtained a linear function. Since the original equation in polar coordinates represents the same curve, it represents a straight line as well.

33. Calculate the area of the circle $r = 3 \sin \theta$ bounded by the rays $\theta = \frac{\pi}{3}$ and $\theta = \frac{2\pi}{3}$.

SOLUTION   We use the formula for area in polar coordinates to obtain

$$A = \frac{1}{2} \int_{\pi/3}^{2\pi/3} (3 \sin \theta)^2 \, d\theta = \frac{9}{2} \int_{\pi/3}^{2\pi/3} \sin^2 \theta \, d\theta = \frac{9}{4} \int_{\pi/3}^{2\pi/3} (1 - \cos 2\theta) \, d\theta = \frac{9}{4} \left(\theta - \frac{\sin 2\theta}{2} \Big|_{\pi/3}^{2\pi/3} \right)$$

$$= \frac{9}{4} \left(\frac{\pi}{3} - \frac{1}{2} \left(\sin \frac{4\pi}{3} - \sin \frac{2\pi}{3} \right) \right) = \frac{9}{4} \left(\frac{\pi}{3} - \frac{1}{2} \left(-\frac{\sqrt{3}}{2} - \frac{\sqrt{3}}{2} \right) \right) = \frac{9}{4} \left(\frac{\pi}{3} + \frac{\sqrt{3}}{2} \right)$$

35. The equation $r = \sin(n\theta)$, where $n \geq 2$ is even, is a "rose" of $2n$ petals (Figure 1). Compute the total area of the flower and show that it does not depend on n.

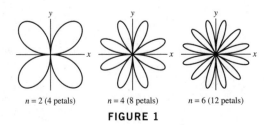

$n = 2$ (4 petals) $n = 4$ (8 petals) $n = 6$ (12 petals)

FIGURE 1

SOLUTION   We calculate the total area of the flower, that is, the area between the rays $\theta = 0$ and $\theta = 2\pi$, using the formula for area in polar coordinates:

$$A = \frac{1}{2} \int_0^{2\pi} \sin^2 2n\theta \, d\theta = \frac{1}{4} \int_0^{2\pi} (1 - \cos 4n\theta) \, d\theta = \frac{1}{4} \left(\theta - \frac{\sin 4n\theta}{4n} \Big|_0^{2\pi} \right)$$

$$= \frac{\pi}{2} - \frac{1}{16n} (\sin 8n\pi - \sin 0) = \frac{\pi}{2}$$

Since the area is $\dfrac{\pi}{2}$ for every $n \in \mathbb{Z}$, the area is independent of n.

37. Find the shaded area in Figure 3.

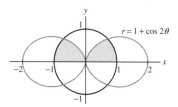

$r = 1 + \cos 2\theta$

FIGURE 3

SOLUTION   We first find the points of intersection between the unit circle and the function.

$$1 = 1 + \cos 2\theta$$

$$\cos 2\theta = 0$$

$$2\theta = \frac{\pi}{2} + \pi n$$

$$\theta = \frac{\pi}{4} + \frac{\pi}{2} n$$

We now find the area of the shaded figure in the first quadrant. This has two parts. The first, from 0 to $\pi/4$, is just an octant of the unit circle, and thus has area $\pi/8$. The second, from $\pi/4$ to $\pi/2$, is found as follows:

$$A = \frac{1}{2} \int_{\pi/4}^{\pi/2} (1 + \cos 2\theta)^2 \, d\theta = \frac{1}{2} \int_{\pi/4}^{\pi/2} 1 + 2\cos 2\theta + \cos^2 2\theta \, d\theta = \frac{1}{2} \int_{\pi/4}^{\pi/2} \frac{3}{2} + 2\cos 2\theta + \frac{1}{2} \cos 4\theta \, d\theta$$

$$= \frac{1}{2} \left(\frac{3\theta}{2} + \sin 2\theta + \frac{1}{8} \sin 4\theta \right) \Big|_{\pi/4}^{\pi/2} = \frac{1}{2} \left(\frac{3\pi}{8} - 1 \right)$$

The total area in the first quadrant is thus $\frac{5\pi}{16} - \frac{1}{2}$; multiply by 2 to get the total area of $\frac{5\pi}{8} - 1$.

39. **CAS**   Figure 5 shows the graph of $r = e^{0.5\theta} \sin \theta$ for $0 \le \theta \le 2\pi$. Use a CAS to approximate the difference in length between the outer and inner loops.

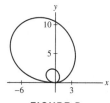

FIGURE 5

SOLUTION   We note that the inner loop is the curve for $\theta \in [0, \pi]$, and the outer loop is the curve for $\theta \in [\pi, 2\pi]$. We express the length of these loops using the formula for the arc length. The length of the inner loop is

$$s_1 = \int_0^\pi \sqrt{(e^{0.5\theta} \sin \theta)^2 + ((e^{0.5\theta} \sin \theta)')^2} \, d\theta = \int_0^\pi \sqrt{e^\theta \sin^2 \theta + \left(\frac{e^{0.5\theta} \sin \theta}{2} + e^{0.5\theta} \cos \theta \right)^2} \, d\theta$$

and the length of the outer loop is

$$s_2 = \int_\pi^{2\pi} \sqrt{e^\theta \sin^2 \theta + \left(\frac{e^{0.5\theta} \sin \theta}{2} + e^{0.5\theta} \cos \theta \right)^2} \, d\theta$$

We now use the CAS to calculate the arc length of each of the loops. We obtain that the length of the inner loop is 7.5087 and the length of the outer loop is 36.121, hence the outer one is 4.81 times longer than the inner one.

In Exercises 40–43, identify the conic section. Find the vertices and foci.

41. $x^2 - 2y^2 = 4$

SOLUTION   We divide the equation by 4 to obtain

$$\left(\frac{x}{2}\right)^2 - \left(\frac{y}{\sqrt{2}}\right)^2 = 1$$

This is a hyperbola in standard position, its foci are $\left(\pm\sqrt{2^2 + \sqrt{2}^2}, 0\right) = (\pm\sqrt{6}, 0)$, and its vertices are $(\pm 2, 0)$.

43. $(y - 3)^2 = 2x^2 - 1$

SOLUTION   We simplify the equation:

$$(y - 3)^2 = 2x^2 - 1$$
$$2x^2 - (y - 3)^2 = 1$$
$$\left(\frac{x}{\frac{1}{\sqrt{2}}}\right)^2 - (y - 3)^2 = 1$$

This is a hyperbola shifted 3 units on the y-axis. Therefore, its foci are $\left(\pm\sqrt{\left(\frac{1}{\sqrt{2}}\right)^2 + 1}, 3\right) = \left(\pm\sqrt{\frac{3}{2}}, 3\right)$ and its

vertices are $\left(\pm\frac{1}{\sqrt{2}}, 3\right)$.

45. Find the equation of a standard hyperbola with vertices at $(\pm 8, 0)$ and asymptotes $y = \pm\frac{3}{4}x$.

SOLUTION   Since the asymptotes of the hyperbola are $y = \pm\frac{3}{4}x$, and the equation of the asymptotes for a general hyperbola in standard position is $y = \pm\frac{b}{a}x$, we conclude that $\frac{b}{a} = \frac{3}{4}$. We are given that the vertices are $(\pm 8, 0)$, thus $a = 8$. We substitute and solve for b:

$$\frac{b}{a} = \frac{3}{4}$$
$$\frac{b}{8} = \frac{3}{4}$$
$$b = 6$$

Next we use a and b to construct the equation of the hyperbola:

$$\left(\frac{x}{8}\right)^2 - \left(\frac{y}{6}\right)^2 = 1.$$

47. Find the equation of a standard ellipse with foci at $(\pm 8, 0)$ and eccentricity $\frac{1}{8}$.

SOLUTION   If the foci are on the x-axis, then $a > b$, and $c = \sqrt{a^2 - b^2}$. We are given that $e = \frac{1}{8}$, and $c = 8$. Substituting and solving for a and b yields

$$e = \frac{c}{a}$$
$$c = \sqrt{a^2 - b^2}$$
$$\frac{1}{8} = \frac{8}{a}$$
$$64 = a$$
$$8 = \sqrt{64^2 - b^2}$$
$$64 = 64^2 - b^2$$
$$b^2 = 64 \cdot 63$$
$$b = 8\sqrt{63}$$

We use a and b to construct the equation of the ellipse:

$$\left(\frac{x}{64}\right)^2 + \left(\frac{y}{8\sqrt{63}}\right)^2 = 1.$$

49. Show that the "conic section" with equation $x^2 - 4x + y^2 + 5 = 0$ has no points.

SOLUTION   We complete the squares in the given equation:

$$x^2 - 4x + 4y^2 + 5 = 0$$
$$x^2 - 4x + 4 - 4 + 4y^2 + 5 = 0$$
$$(x - 2)^2 + 4y^2 = -1$$

Since $(x - 2)^2 \geq 0$ and $y^2 \geq 0$, there is no point satisfying the equation, hence it cannot represent a conic section.

51. The orbit of Jupiter is an ellipse with the sun at a focus. Find the eccentricity of the orbit if the perihelion (closest distance to the sun) equals 740×10^6 km and the aphelion (farthest distance to the sun) equals 816×10^6 km.

SOLUTION   For the sake of simplicity, we treat all numbers in units of 10^6 km. By Kepler's First Law we conclude that the sun is at one of the foci of the ellipse. Therefore, the closest and farthest points to the sun are vertices. Moreover, they are the vertices on the x-axis, hence we conclude that the distance between the two vertices is

$$2a = 740 + 816 = 1556$$

Since the distance between each focus and the vertex that is closest to it is the same distance, and since $a = 778$, we conclude that the distance between the foci is

$$c = a - 740 = 38$$

We substitute this in the formula for the eccentricity to obtain:

$$e = \frac{c}{a} = 0.0488.$$